Studies in Economic Theory

Editors

Charalambos D. Aliprantis
Purdue University
Department of Economics
West Lafayette, IN 47907-2076
USA

Nicholas C. Yannelis
University of Illinois
Department of Economics
Champaign, IL 61820
USA

Springer

Berlin
Heidelberg
New York
Hong Kong
London
Milan
Paris
Tokyo

Titles in the Series

Charalambos D. Aliprantis
Kenneth J. Arrow · Peter Hammond
Felix Kubler · Ho-Mou Wu
Nicholas C. Yannelis
Editors

Assets, Beliefs, and Equilibria in Economic Dynamics

Essays in Honor
of Mordecai Kurz

 Springer

Professor Dr. Charalambos D. Aliprantis
Purdue University, Krannert School of Management
Department of Economics
Rawls Hall
West Lafayette, IN 47907-2076, USA et@mgmt.purdue.edu

Professor Kenneth J. Arrow   arrow@leland.stanford.edu
Professor Peter Hammond   peter.hammond@stanford.edu
Professor Felix Kubler   fkubler@stanford.edu

Stanford University
Department of Economics
Stanford, CA 94305-6072, USA

Professor Ho-Mou Wu   homowu@ccms.ntu.edu.tw
National Taiwan University
Department of International Business
Taipei, Taiwan R.O.C.

Professor Nicholas C. Yannelis   nyanneli@uiuc.edu
University of Illinois
Department of Economics
330 Commerce West Building
Champaign, IL 61820, USA

Parts of the papers of this volume have been published in the journal *Economic Theory*.

ISBN 3-540-00911-6 Springer-Verlag Berlin Heidelberg New York

Cataloging-in-Publication Data applied for
A catalog record for this book is available from the Library of Congress.

Bibliographic information published by Die Deutsche Bibliothek
Die Deutsche Bibliothek lists this publication in the Deutsche Nationalbibliografie;
detailed bibliographic data available in the internet at *http.//dnb.ddb.de*

Springer-Verlag Berlin Heidelberg New York
a member of BertelsmannSpringer Science + Business Media GmbH

http://www.springer.de

© Springer-Verlag Berlin Heidelberg 2004
Printed in Germany

Cover design: Erich Kirchner, Heidelberg

SPIN 10922361   42/3130 – 5 4 3 2 1 0 – Printed on acid-free paper

Table of Contents

Foreword to the Symposium in Honor of Mordecai Kurz

Charalambos D. Aliprantis[1], Kenneth J. Arrow[2], Peter J. Hammond[2], Felix Kubler[2], Ho-Mou Wu[3], and Nicholas C. Yannelis[4]

[1] Department of Economics, Purdue University, West Lafayette, IN 47907-2076, USA
   (e-mail: aliprantis@mgmt.purdue.edu)
[2] Department of Economics, Stanford University, Stanford, CA 94305–6072, USA
   (email: arrow@leland.stanford.edu, {peter.hammond;fkubler}@stanford.edu)
[3] Department of International Business, National Taiwan University, Taipei, TAIWAN
   (e-mail: homouwu@ccms.ntu.edu.tw)
[4] Department of Economics, University of Illinois at Urbana-Champaign, IL 61820, USA
   (e-mail: nyanneli@uiuc.edu)

This collection of papers is dedicated to Mordecai Kurz – our friend, mentor, and colleague. They are to mark our appreciation of his contributions to economic science, from which we have all benefited enormously. A conference in his honor took place on August 2–3, 2002, co-hosted by the Stanford University Department of Economics, the Stanford Institute of Economic Policy Research, and Springer-Verlag. At this conference a selection of these papers was presented, along with some more recent work by several of the same authors and by other invitees.

Mordecai has enjoyed a long and outstandingly productive career. His favorable influence on the economics profession has already lasted almost forty years, and shows no sign of diminishing any time soon. He has made significant and valuable contributions to many areas in economics, including growth and capital theory, general equilibrium, game theory, and public economics. He has also formalized the important idea of endogenous economic uncertainty in his theory of rational belief equilibria. This has been very fruitfully applied in the development of novel macroeconomic models of empirical financial phenomena.

Apart from his research, Mordecai has provided intellectual leadership in many other areas. In 1969 he became the director of the Economics Section of the Institute for Mathematical Studies in the Social Sciences (IMSSS) at Stanford University, and instigated annual summer workshops which he organized and presided over for the next twenty successive years. These extremely lively workshops were the focal point for rigorous reasoning of the highest standard covering many fields in economics, including applications as well as theory. The workshops continue to this day, though the organization has since changed its name to the Stanford Institute for Theoretical Economics (SITE).

In addition, a large majority of the graduate students trained in the economics Ph.D. program at Stanford University have benefited from Mordecai's enthusiastic and exemplary teaching, especially of his first-year core course. Many students proceeded to his more advanced courses, and some of these then went on to write dissertations under Mordecai's supervision. His support and dedication to his many dissertation students are legendary. To this day, he continues to attract numerous students to the regular group meeting he holds with his advisees.

On behalf of everybody who has agreed to contribute to this volume, we wish him all the best for a long and healthy life, as well as for the continuation of his professional activities at the extraordinarily high level we have learned to admire so much.

Mordecai Kurz, Joan Kenney Professor of Economics at Stanford University, USA

The genuine savings criterion
and the value of population*

Kenneth J. Arrow[1], Partha Dasgupta[2], and Karl-Göran Mäler[3]

[1] Department of Economics, Stanford University, Stanford, CA 94305-6072, USA
  (e-mail: arrow@stanford.edu)
[2] Faculty of Economics and Politics, University of Cambridge, Cambridge CB3 9DE, UK
  (e-mail: partha.dasgupta@econ.cam.ac.uk)
[3] Stockholm School of Economics and Beijer International Institute for Ecological Economics,
  Royal Academy of Sciences, Box 50005, 10691 Stockholm, SWEDEN
  (e-mai: Karl@Beijer.kva.se)

Received: June 1, 2002; revised version: September 27, 2002

Summary. In any dynamic model of the economy with changing population, the latter should properly be one of the state variables of the system. It enters both in the maximand, at least under total utilitarianism, and into the production function in one way or another. If population growth is exponential and constant returns prevails, then a simple transformation to *per capita* variables can be used to eliminate one state variable, but this ceases to be true if growth is not exponential, as it obviously is not and cannot be. If the growth of population is exogenous, then introducing it into the system does not affect the optimal policy. However, if one asks whether the system is sustainable, in the sense of at least maintaining total welfare (integral of discounted utilities), then the criterion is that that the value of the rates of change of the state variables is non-negative, so that the shadow price of population becomes relevant. In this paper, we derive explicit formulas in a simple model, showing that the rate of growth of *per capita* capital is not the correct formula but must have another terms added to it. We also study the question under an alternative criterion of long-run average utilitarianism.

Keywords and Phrases: Optimal control, Population, Genuine savings, Accounting prices.

JEL Classification Numbers: D90, H43.

---

  * Research support was provided by the William and Flora Hewlett Foundation. An earlier version of this paper was presented at a celebration of Mordecai Kurz's 66th birthday at Stanford University, 1–3 August 2002.
Correspondence to: K.J. Arrow

The idea of systematic planning for the future (whether by individual economic agents or by the collectivity) was implicit in economic theory since the latter nineteenth century. It had been given more explict though still not very usable form in the 1930s with the work of Lindahl (1929, 1939) and Hicks (1939). But dynamic planning became a practical possibility with the nearly contemporaneous work of two mathematicians, Bellman on dynamic programming (1957) and Pontryagin and associates on optimal control theory (1962). The two approaches are equivalent; each has technical advantages and disadvantages of its own. However, in many ways, the formulations of optimal control theory are closer to standard economic thinking, and it has been the preferred approach, particularly in theoretical work.

Optimal control theory started being applied by economists fairly soon after being published in book form. One of the earliest applications was the work of Kenneth J. Arrow and Mordecai Kurz (1970).[1] It discussed the criteria for optimal public investment policy using the tools of optimal control theory to clarify much of the existing literature and to introduce new concepts. The present paper continues the intellectual impetus of the Arrow–Kurz book and brings some new considerations to bear.

1 Introduction

We deal with a set of ideas with regard to control of the economy. Much of the stimulus has come from an increasing attention to the role of the environment and ecological factors in general. The importance of these factors is given recognition by arguing that there are many forms of capital supplied by nature beyond the reproducible capital usually emphasized in growth models.

Two traditions are drawn on, though our conclusions go beyond those in the literature. One is the study of the criteria for evaluating policies when population is varying. The other is the question, whether and what extent a given policy is causing a gain in aggregate welfare, what has come to be called the measurement of "genuine savings." The term, "sustainability," has been been much used, especially since its adoption in the Brundtland Commission report of 1987. One interpretation of sustainability is a positive value of genuine savings. The stimulus for measuring this new concept has come from an increasing attention to the role of the environment and ecological factors in general. The importance of these factors is given recognition by arguing that there are many forms of capital supplied by nature beyond the reproducible capital usually emphasized in growth models.

The aim of this paper to give a rigorous analysis of the role of varying population in measuring genuine savings, i.e., in giving a criterion for improvement in welfare. We argue that the only consistent approach is to recognize population as another form of capital (state variable); this does not exclude *a priori* its having a negative value, as many have argued. This will hold even if we do not consider population policy and regard the evolution of population as exogenous to the economic and policy variables. The main aim of the paper is to derive the accounting price for population (costate variable).

---

[1] It was a pleasure and an educational experience to have the opportunity for this collaboration. (KJA)

It should be emphasized that we make no claim that our approach is relevant to choice of population policy. There are deep ethical problems in comparing alternative sizes and compositions of population, and we make no claim to having addressed them. As already stated, we restrict ourselves to cases in which population growth is not affected by any control variables.

In the existing literature, varying population is usually modelled by a constant rate of growth of population. In this case, it is easy to measure the various kinds of capital on a *per capita* basis. The implications of constant exponential growth are obviously absurd, and certainly the dramatic reductions in birth rates throughout the world make such an assumption a poor guide to analysis.

As an introduction to the subject, we confine ourselves to the case where an economy is following an optimal course according to the fairly conventional criterion of maximizing the sum of discounted utilities of consumption. Further, we confine ourselves to the case of one form of capital made of a good which can be used indifferently for consumption and capital formation. The extension to many forms of capital does not offer any essential difficulties. The extension to accounting for growth in non-optimal policies is probably more difficult in practice if not in theory, but we think the present work provides a beginning.

In the next section, we review a broad model and the concept of genuine savings as expressed in it. Then we consider in particular the role of population, which enters both into the maximand and into the production function as labor. The evolution of accounting prices is then found.

The basic analysis is in the tradition of "total utilitarianism," i.e., the criterion is the sum of the utilities. In the last section, we see how the sustainability criterion is modified when a dynamic form of "average utilitarianism" is considered.

This work is part of a broader research program of the authors (see Arrow, Dasgupta, and Mäler (forthcoming) on the measurement of genuine savings as a criterion for sustainability. The full program includes measurement in non-optimal paths and under non-convex environments.

2 Genuine savings

A general class of models of the economy over time takes as its criterion for choice among alternative policies,

$$\text{Max!} \int_0^\infty e^{-\delta t} U(c_t) dt, \tag{1}$$

subject to various constraints. Among these constraints is a conservation law for produced goods, output equals consumption plus capital formation,

$$\dot{K} = F(\cdot) - c, \tag{2}$$

where K is produced capital, c is consumption; for the moment, we leave the arguments of the production function F unspecified, but they are all forms of capital, including produced capital. Let $V(t)$ be the integral of utility from t on discounted to t.

$$V(t) = \int_t^\infty e^{-\delta(u-t)} U(c_u) du. \tag{3}$$

In an autonomous system (where all the capital variables completely determine the future for any given policy, including the optimal policy), $V(t)$ is completely determined by the values of the capital variables at time t. Then the costate variable (accounting price) for state variable K is,

$$p_K(t) = \partial V / \partial K,$$

and, for other capital variables (forms of natural capital, including mineral resources, human capital, knowledge, etc.), similarly, the shadow prices, p_i are the partial derivatives of V with respect to the corresponding capital variables, K_i. Then,

$$\dot{V} = p_K \dot{K} + \sum p_i \dot{K}_i. \tag{4}$$

In this context, Pezzey (1992) proposed a reasonable definition of "sustainability."

Definition 1. *The path is sustainable at time t if $\dot{V} > 0$ at that time.*

From (4) and Definition 1, evaluation requires determining the net formation of each kind of capital and the corresponding accounting prices.

Theorem 1. *A path is sustainable at time t if and only if,*

$$p_K \dot{K} + \sum p_i \dot{K}_i > 0$$

at time t.

Since this expression is a weighted sum of the net formations of all the different kinds of capital, it has come to be called, "genuine savings." It has been developed with varying degrees of formality by Hamilton (1994), Pearce, Hamilton, and Atkinson (1996), and Dasgupta and Mäler (2000). Empirical estimates based on this concept have appeared in Hamilton and Clemens (1999).

Note that V is measured in utility units. A variation of (4), with the same sign, is obtained by dividing through by p_K. If we let $q_i = p_i/p_K$, then sustainability is defined by the condition that,

$$\dot{V}/p_K = \dot{K} + \sum q_i \dot{K}_i > 0, \tag{5}$$

which is expressed in commodity terms. Also note that the current-value Hamiltonian is given (in part) by,

$$H = U(c) + p_K[F(\cdot) - c] + \dots,$$

where the omitted terms are based on the equations of motion of the other types of capital. If c does not occur in the equations of motion for the other capital stocks, we have, by the Maximum principles, that,

$$U'(c) = p_K, \tag{6}$$

an occasionally useful relation.

3 Population as a variable

In our model, population is assumed to be independent of economic conditions but evolving according to some laws. The analysis will treat it as another form of capital. Let $N(t)$ be population at time t, and also labor force (they can be distinguished in a more sophisticated model). Then, N enters both the maximand (though this has sometimes been disputed) and the production function.

As usual, we assume that individual consumption, $c(t)$, is the same for all. There is a single good, the production function for which is $F(K, N)$, concave with constant returns to scale.

The objective (felicity) for a single period will be taken to be $NU(c)$. The literature on this subject is vast and will not be reviewed here; the dispute goes back to the pioneers of utilitarianism, Henry Sidgwick and Francis Y. Edgeworth. In determining the optimal accumulation policy, it seems hard to deny something like this. Any idea of treating people more or less equally implies that if tomorrow's population is bigger, it should get proportionately more weight. We are still weighing people according to their futurity (discounting) but not according to the numbers of their contemporaries. Then (3) becomes,

$$V(t) = \int_t^\infty e^{-\delta(u-t)} N(u) U(c_u) du, \tag{7}$$

and the criterion of optimality is,

$$\text{Max! } V(0). \tag{8}$$

The equation of motion for produced capital is the obvious modification of (2),

$$\dot{K} = F(K, N) - Nc, \tag{9}$$

We need to make an assumption about the evolution of population. Since we want to exclude the dependence of population on economic conditions, clearly population growth must be a function of N only.

$$\dot{N} = \phi(N) \tag{10}$$

Some but not all formulas will simplify if we write,

$$\phi(N) = \nu(N)N, \tag{11}$$

where $\nu(N)$ is the (relative) rate of growth. As far as we know, virtually all models which have introduced changing population have assumed $\nu(N)$ constant. A somewhat more acceptable formulation is that giving rise to the logistic curve,

$$\phi(N) = AN(N^* - N). \tag{12}$$

Then, the genuine increase in wealth in commodity units, (5), is,

$$\dot{K} + q \dot{N}, \tag{13}$$

where q is the ratio of the costate variable for N to that for K.

Expression (15) takes a slightly simpler form when divided by N (*per capita* genuine savings); the sign is unaltered. Let $k = K/N$.

$$(1/N)(\dot{K} + q\dot{N}) = \dot{k} + (q + k)\nu(t). \tag{14}$$

Before giving an explicit expression for $q(t) + k(t)$, it is worth thinking about (14). It would strike most people as reasonable to look at the increase in *per capita* capital as a measure of sustainability. This would hold if $q + k = 0$. This has an intuitive basis; indeed, the demographic literature (see Leibenstein, 1971) has noted that faster population growth is costly (apart from Malthusian effects) because it requires higher capital accumulation (and therefore lower consumption) to maintain the same capital–labor ratio. This is precisely and more accurately captured by the second term in (14). A careful analysis shows that we cannot take $q(t) + k(t)$ to be zero. In fact, the appendix shows that,

$$q(t) + k(t) = \int_t^\infty \frac{R(u)}{R(t)} \frac{\phi[N(u)]}{\phi[N(t)]} \{L(c_u) - \nu'[N(u)]K(u)\}du, \tag{15}$$

where,

$$R(t) = \exp\left\{-\int_0^t F_K[K(u), N(u)]du\right\}, \tag{16}$$

$$L(c) = U(c)/U'(c), \tag{17}$$

Thus the "benefit term" is discounted at the marginal productivity of capital. Note that if we assume, as is natural, that $U(c)$ is positive in the relevant range, and that the rate of growth of population, ν, decreases as population grows (at least for large populations), then $q(t) + k(t) > 0$, so that genuine savings exceed increases in *per capita* capital. This does not mean that population itself is a good; that depends on the sign of q, which may itself easily be negative.

We have not succeeded in making the terms and factors in (15) entirely intuitive. However, the important term defined in (17) does have an interpretation as being, in a sense, the "value of life." As is commonly done, this is interpreted to mean the (compensated) willingness to pay for a marginal increase in the probability of survival. The indifferences curves between consumption and probability, p, of survival are the curves on which $p\,U(c)$ is constant, so that, $p\,dc/dp = -U(c)/U'(c)$. If we start from a situation where the probability of survival is 1, then indeed $L(c)$ is the value of life, and its presence in the accounting price for population is natural.

Theorem 2. *The optimal path for the model with varying population is sustainable if and only if,*

$$\dot{k} + (q + k)\nu(t) > 0,$$

where $q(t) + k(t)$ is defined by (15–17). If the rate of growth of population decreases as population increases, then $q(t) + k(t) > 0$, so that sustainability is possible even if per capita capital is decreasing.

The only example simple enough to be illustrative is the case of constant population growth. Then, $\nu'(N) = 0$, and also,

$$\phi[N(u)]/\phi[N(t) = N(u)/N(t),$$

so that, $q(t) + k(t)$ is the discounted value of the total value of life for the entire population.

Note that the value of life is evidenced by willingness to spend on avoiding death (e.g., medical expenditures) and on raising children. Thus, (i) it reflects a revealed preference, and (ii) it is capable of measurement from observed quantities.

4 An alternative criterion

Dasgupta (2001, pp. 258–259) presents an alternative criterion for measuring welfare and consequently genuine savings in a world of changing population. Rather than the total, he takes the expected utility of a random individual out of the present and future chosen with probabilities weighted by futurity, i.e., the probability density of an individual t years hence is proportional to $e^{-\delta t}$. This may be regarded as a dynamic version of average utilitarianism. Define, then,

$$V^*(t) = V(t)/N^*(t), \tag{18}$$

where

$$N^*(t) = \int_t^\infty e^{-\delta(u-t)} N(u) du, \tag{19}$$

and $V(t)$ is defined by (7). He shows that with this criterion, production under constant returns to scale, and exponential rate of growth of population, then genuine savings are measured by *per capita* wealth.

Here, we reexamine the issue for our more general assumptions about population growth. Since we are taking the time path $N(t)$ to be determined exogenously, the optimal policy is unaltered. However, the accounting prices for K and N become,

$$p_K^* = \partial V^*/\partial K, \quad p_N^* = \partial V^*/\partial N.$$

If we define,

$$q^* = p_N^*/p_K^*,$$

the sustainability criterion becomes,

$$\dot{K} + q^* \dot{N} > 0,$$

or, equivalently, as in (14),

$$\dot{k} + (q^* + k)\nu(t) > 0. \tag{20}$$

It will be shown in the Appendix that,

$$q^* = q - (V^*/p_K)[\delta N^*(t) - N(t)]/\phi[N(t)]. \tag{21}$$

Definition 2. *The Dynamic Average Utilitarian criterion, at any time t is,*

$$V^*(t) = V(t)/N^*(t),$$

where $V(t)$ and $N^*(t)$ are defined in (7) and (18), respectively,

Theorem 3. *Under the Dynamic Average Utilitarian criterion, sustainability is defined by (20) and (21).*

It is important to note that the sustainability criterion of Theorem 2 is not invariant under an additive shift in the utility function, even though the optimal path is. If one adds a constant h to the utility function, then $V(t)$ is increased by $hN^*(t)$, which depends on $N(t)$, so that the accounting price of N is altered. However, $V^*(t)$ is increased by the constant h, so that the accounting prices p_N^*, p_K^* and q^* are unaltered.

Appendix

As promised, we here sketch the derivation of equations (15) and (21).

For equation (15), the Hamiltonian for the total utilitarian criterion (7) with the dynamic equations (8) and (9) is,

$$H = NU(c) + p_K[F(K, N) - Nc] + p_N\phi(N).$$

Hence, the equations of motion for the accounting prices are,

$$\dot{p}_K = p_K(\delta - F_K), \tag{22}$$
$$\dot{p}_N = p_N[\delta - \phi'(N)] - U(c) - p_K(F_N - c). \tag{23}$$

Since $q = p_N/p_K$,

$$\dot{q}/q = \dot{p}_N/p_N - \dot{p}_K/p_K. \tag{24}$$

Divide through in (22) and (23) by $p_K = U'(c)$ and by p_N, respectively, then substitute into (24), use the definition of q, and multiply both sides by q. Then,

$$\dot{q} = (F_K - \phi')q - [U(c)/U'(c)] + c - F_N.$$

In the accumulation equation for capital, (9), use Euler's theorem to replace $F(K, N)$ by $F_K K + F_N N$; then we can deduce,

$$\dot{k} = F_K k + F_N - c - \nu(N)k.$$

Adding the last two equations yields, with the help of (11),

$$d(q + k)/dt = (F_K - \phi')(q + k) + \nu'K - [U(c)/U'(c)].$$

Replace t by u, integrate from t to infinity, and use the transversality conditions. Then (15–17) follow.

To deduce (21), first take the partial derivatives of (18) with respect to K and N.

$$p_K^* = p_K/N^*,$$
$$p_N^* = (p_N/N^*) - (V^*/N^*)(dN^*/dN).$$

Then,

$$q^* = q - (V^*/p_K)(dN^*/dN).$$

It remains to compute the last factor. Since $N(u)$ is completely determined by $N(t)$ for all $u \geq t$, $N^*(t)$ is determined by $N(t)$. It follows that,

$$dN^*/dN = \dot{N}^*/\dot{N}. \qquad (25)$$

But, from the definition (19), it follows immediately that,

$$\dot{N}^* = \delta N^* - N,$$

while,

$$\dot{N} = \phi(N),$$

by (10), so that (21) follows from (25).

References

Arrow, K.J., Dasgupta, P., Mäler, K.G.: Evaluating projects and assessing sustainable development in imperfect economies. Forthcoming

Arrow, K.J., Kurz, M.: Public investment, the rate of return, and optimal fiscal policy. Baltimore London: Johns Hopkins Press 1970

Bellman, R.: Dynamic programming. Princeton, NJ: Princeton University Press 1957

Dasgupta, P.: Human well-being and the natural environment. Oxford New York: Oxford Unversity Press 2001

Dasgupta, P., Mäler, K.-G.: Net national product, wealth, and social well-being. Environment and Development Economics **5**, 69–93 (2000)

Hamilton, K.: Green adjustments to GDP. Resources Policy **20**, 155–168 (1994)

Hamilton, K., Clemens, M.: Genuine savings rates in developing countries. World Bank Economic Review **13**, 333–356 (1999)

Hicks, J.R.: Value and capital. Oxford: Clarendon Press 1939

Leibenstein, H.: The impact of population growth on economic welfare — nontraditional elements. In: Rapid population growth: consequences and policy implications, Vol. 2, pp. 175–198. Baltimore: Published for the National Academy of Sciences by the Johns Hopkins Press 1971

Lindahl, E.: Prisbildungsproblemets uppläggning från kapitalteorisk synpunkt. Ekonomisk Tidskrift **31**, 31–81 (1939). Translated as Part III in: Studies in the theory of money and capital. London: George Allen & Unwin 1939

Pearce, D.W., Hamilton, K., Atkinson, G.: Measuring sustainable development: progress on indicators. Environment and Development Economics **1**, 85–101 (1996)

Pezzey, J.: Sustainable development concepts: an economic analysis. World Bank Environment Paper No. 2, World Bank (1992)

Pontryagin, L.S., Boltyanskii, V.G., Gamkrelidze, R.V., Mischenko, E.F.: The mathematical theory of optimal processes. New York London: Interscience 1962

Macro foundations of micro-economics[*]

Frank Hahn

16 Adams Road, Cambridge, CB3 9AD, UK (e-mail: dorothyhahn@compuserve.com)

Received: September 19, 2001; revised version: July 24, 2002

Summary. This paper attempts to circumvent the nonsense of the representative agent which arises in macroeconomics. It recognises that macro data are relevant to agents' decisions, and so excess demands should contain macro variables as arguments. The macro variables I consider are the price index, unemployment and GNP. This paper should be regarded as a tentative beginning to make macroeconomic theory literate.

Keywords and Phrases: Representative agents, Price index, Labour market search.

JEL Classification Numbers: E13, E24, E31.

1 Introduction

If one asks "what does micro economics contribute to our understanding and study of an economy?" the answer is that it furnishes us with a theory of the actions, and inter-actions, of agents. Moreover, it seeks to account for any regularities in the aggregated behaviour of agents. At present the simple aggregation into a representative agent who behaves just like any other agent leaves one bereft of understanding why micro-economics should not suffice. And indeed it seems to have done so, (for instance, real trade cycle theory). Moreover the representative agent allows meaning to be given to perfect foresight or rational expectations. Also one can legitimately apply rigorous theory to this fictional character and, I suppose, hope for the best.

All these difficulties of aggregation are of course well known, but it is surprising that so little attention has been paid to them by the proponents of classical macro-

---

[*] I am delighted to contribute to this Festschrift for Mordecai now that he has reached the appropriate age.

economics. The aggregation seems to be of the essence of producing a macro-economics of the kind we have become used to, and yet there is no justification for endowing the abstract aggregate with the behaviour satisfying the rigorously modelled individual. It occurs to one that this may be due to a quite unclear notion of what macro-economic theory is to be about. On the one hand we could think of it as an attempt to simplify sufficiently for civil servants and politicians to understand the fundamentals of the economic world they observe. On the other there may be the much grander project of explaining why certain macro-variables appear to be subject to a certain regularity. The first project is so loose that one really should not object to anything as long as one doesn't practise it oneself. It requires intuition, wisdom, experience and judgment, and not the skills of technical economics.

My purpose is limited: it is to show that in general macro variables need to enter the relations explaining the actions of agents. Moreover there is very little theory to tell us exactly how they enter. So that it will be difficult to deliver a 'new' theory with any of the certainty of classical macro-economics.

2 Macro variables as signals

The idea that I want to put forward is this: an agent cannot observe every single price nor form an expectation of its future value. This it would have do in a canonical sequence economy version of GE theory. Not only is it implausible that any agent has enough information to learn the history of all prices but also to form a view of whether any one sequence is stationary. [Kurz (1994) has noted the difficulties in learning the true character of any non-stationary process from empirical evidence.]

In the spirit in which macro-economists proceed, price expectations are always expectations concerning a price index. But if this is to be recognised as useful one needs to be given a good deal of argument. For instance why should a perfectly competitive motor car manufacturer base his supply decision on the value he expects a price index, which includes prices of all sorts of goods not related to cars, to have? It may be so, but it surely requires a good deal of justification, especially if one is searching for "micro-foundations".

In any case I here reject such short cuts and take the "foundation" question seriously, indeed so seriously that I want the answer to relate to GE. On the other hand I shall not be searching for conditions of 'perfect' aggregation since Gorman (1953) and others have taught us that they are far too demanding for a practising macro-economist.

Surprisingly there is a great deal of theory already where macro-variables enter into the functions describing the decisions of agents. I shall briefly discuss two examples.

(a) A labour search model

Consider the simplest labour search model. Workers know the distribution of job offers but not where any particular offer is located. They engage in optimum search of offers, which given that every search has a cost is found by the following rule "set a minimum acceptance wage by the rule that if one turned down an offer of that wage and searched again, the expected return net of search costs would leave one indifferent between accepting the critical wage and making one more search." (As I have noted this is the simplest search theory which no-one takes to be descriptive, for one thing offers need to be taken for the infinite future). But there are many more satisfactory theories and as long as they postulate knowledge of the distribution of offers all is OK for this part of my argument. A worker's action will be affected by the distribution of offers, in particular the calculated acceptance rule. This in turn will affect the amount of search unemployment. It seems reasonable to suppose that the distribution is itself affected by the level of unemployment. Let u' be the level of unemployment and $F(w, u')$ be the wage-offer distribution. Both sides of the market take u as given by the actions of other agents. These decisions will include the optimum search rule and so total unemployment thus, say, $U(F(., u')) = u$ where the function $U(.)$ is derived from the optimum search rule when $F(., u')$ is given. It now follows that for long run equilibrium one has u^*, the equilibrium level of unemployment given by:

$$U(F(., u^*)) = u^* .$$

It is seen that we have arrived at an outcome which is not the usual one.

The macro-variable is regarded as beyond the control of the agent...it acts as an externality which is due to all choosing their optimum search strategy independently. Hence unemployment is a macro-variable on which choice depends and of course it is the outcome of choice – so one is not surprised by the fixed point. On the other hand I have proceeded rather informally; and I ought to cross some of the t's and put dots on some of the i's.

Let us first consider a notorious difficulty with search: where does the distribution come from and how can it be sustained in a competitive economy? I do not believe that credible answers are available for a market with homogeneous labour. A cowardly way out which I shall take is to measure labour in efficiency units and take w to be the wage of a unit of labour in these units. In practice differences in efficiency will be due to training and natural endowment. But I do not wish to build a full-scale model of the labour market and rest satisfied with differences due to inborn endowment differences. Some producers are looking for high quality labour, say, because it will have to use expensive high quality machines, while others having less sophisticated machines are more willing to take the risk on less efficient labour (at a lower wage).

The second modification we must make is this: a firm has a certain number of jobs on offer at a certain efficiency wage but the number applying is not the same as the number of jobs offered. Indeed an offer is made because the marginal product of efficiency labour exceeds the ruling wage. My hypothesis is that a firm in that position will raise the wage and one in the reverse position will lower the offer. With the further assumption that at any one time most firms are in the same position, one obtains the process which I have sketched.

In fact the whole matter is made easier if the uncertainty of being able to make the desired hires at a given wage, as well as the uncertainty of being hired, is explicitly modelled. The critical wage must then be calculated in an obvious but different way than has been indicated above. The searching worker must now take into consideration that being turned down is costly in new search costs.

I do not want to do all the calculations involved, partly because they will be tedious and partly because this is not a paper on the labour market. I do not want to lose sight of the main message. That is that macro-variables act as measures of particular externalities and so must be considered by micro theory. In a sense I am arguing the "Lucas critique" by noting that changes in the values of macro variables may lead to structural changes in the GE description of an economy. But my grounds are not the same as his: Lucas assumes that agents know everything they need to know for correct market prediction for the long run. I do not know what the world is like in that situation (when the market has no more to teach), much less whether actual dynamics lead to it, so that I do not put my eggs in the Lucas basket. That is why I rest satisfied with a sketch of dynamics and make no claims regarding its asymptotic behaviour. It may take a very long time before some workers reduce their reservation wage, for instance. Policy may be able to accelerate the process.

(b) Price indices as signals

It has been increasingly realised in the literature that agents learn more about their immediate 'neighbours' in the appropriate space than they do about the economy as a whole. So that a firm learns something about prices which are pay-off relevant to it and not much about others. The same will be true of consumers, and so it happens that if there is a publicly available price index, it becomes relevant to the formation of price expectations. The price index will reveal some information about unobserved prices. For instance some health conscious consumer will at any date observe the prices of different health supplements, but his/her expectation of these prices will in part depend on the prices not observed and so on the price index.

The story I am about to tell is pretty stylised. Let m_t stand for a vector of macro variables at time t which affect agents' expectations when they are each only partially informed about current variables like prices or employment. I postulate that m_t is known with a lag of one period. Let p_t be the vector of current prices which agent h observes – just those that are relevant for him. It is assumed that every price is relevant to someone and I write X_t as the aggregate demand vector at t.

Assume that $X_t = H(p_t, m_{t-1})$, so that it is taken to be true that macro variables are treated as exogenously given – they are announced after the event. Write p_t^* as the (assumed to be unique) solution to

$$H(p_t, m_{t-1}) = 0.$$

Hence $p^* = G(m_{t-1})$ since, under standard regularity assumptions, the above allows us to use an implicit function. Then $m_t = J(G(m_{t-1}), m_{t-1})$, that is the micro theory will give the demand and supply of every good at the values of these variables and so all the components of the macro vector to be announced in the next period. It is clear how the dynamics proceeds but it will not be easy to implement.

Moreover it will be clear that this example is not easily reconciled with the previous one since there labour markets do not necessarily clear.

As I have indicated a really satisfactory example will need to model all markets together, when some are in equilibrium and others are not. That I fear is beyond me and so I shall stick to what is essentially a sequence of equilibria indexed by predetermined values of macro variables, but I should wish to exclude the labour market for reasons of relevance and realism. Hence unemployment will have to be treated by agents as exogenous. What I mean by this is the aggregate level of unemployment U, so that U becomes a variable in the excess demands. The excess supply of labour is in equilibrium when it is equal to U. In general the excess demands must in equilibrium be consistent with the macro variables – for instance the prices together with the macro-variables such as the price level which give zero excess demands for goods must be consistent with the vector of equilibrium prices. That is the case for all excess demands except that for labour, which has already been explained.

The notion of equilibrium is somewhat more sophisticated than usual, because it is claimed that aggregates like the price level and unemployment affect the choices of agents, but must be treated parametrically by that individual agent. On the other hand they are plainly endogenous but not under the control of the agent.

3 A formal definition of equilibrium

Let the price index be the only macro variable agents take into account when making their plans. Write it as P. Then since it consists of a linear combination of prices with weights of consumption, the conditions of market clearing, that is prices given the previous price index will provide the required information.

Hence the price vector p^* and price index P^* together constitute an equilibrium if

$$P_t = P^* \text{ all } t, \text{ and } H(p^*, P^*) = 0.$$

It will be seen that this definition is not entirely satisfactory since it demands stationary conditions and we are implicitly assuming a clearing labour market. Some of these shortcomings are rectifiable, in particular the labour market is simply dealt with by entering both P and U into the excess demand functions and asking for consistency in the function representing the excess demand for labour. Assuming that labour efficiency is constant as well as production conditions one can let w stand for a distribution of efficiency labour. But if agents are always in the process of learning then the stationary solution does not make much sense. I believe that it is possible to proceed in a more satisfactory way but this does not seem the occasion for yet more speculation.

The simplest explanation is to write a future expected macro-variable as conditioned on information, the ω of the new classicals. I have taken the values of the macro variable that agents learn. It is natural to assume a lag between the occurrence and publication, and dissemination to agents. This explains the manner in which I take macro variables to enter the excess demand functions. The manner in which they enter is important but not easily specified. At t the history of a macro variable is known up to date $t - 1$ and the announcement at that date is the new macro information. This is used to condition prices and other variables expected at $t+1$ and so consumption and production planned for that date. But this will induce the appropriate saving and investment at t.

4 Conclusions

The present procedure to study macro is based on the representative agent. For instance in growth theory he knows his production function which is the same for him as for every other agent. It will be argued that on average it all comes out in the wash. I have never seen this demonstrated and I doubt that there are general results to this effect.

In the mean time we are all endowed with rational expectations, which seems only credible with Gorman utility functions. It seems to me the stock market alone makes this implausible. There one attempts to make sense of what is observed by referring to the division of the market into bulls and bears. But in general the hypothesis is absurd. (I realise how provocative this must sound, but it is time for a justification other than that the assumption makes econometrics possible without ad hockery.) We get into ad hockery as I do in this paper because we have too few facts and are pretty ignorant. I am prepared to admit that rather than calling on identical rational expectations. In this paper I have taken the first tentative steps.

They are not satisfactory for many reasons. The most important is the postulate of perfect competition. Macro-economics does not comfortably coexist with that. I have proceeded in this way for the same reason that the American current vintage of macroeconomists also does so: it is a model that is well understood. It is now time to go beyond it.

While there cannot be any doubt that some macro variables affect agents' decisions, one needs to decide in what manner they do so. I am convinced that making agents into Ramsey maximisers is not to be recommended. It is an obvious fact that most agents are myopic in most of their decisions. However this remark does not get us anywhere and one needs to experiment with various ideas. Theorising in economics is one manner of experimenting.

Lastly I want the reader to know that this paper has been written for purely theoretical reasons. Policy is to be undertaken in one way or another, and it may well be that following the new macro-economics is second best. But unless the subject of economics can digest some of the points made here then it is in a poor state.

References

Gorman, W.M.: Community preference fields. Econometrica **21**, 63–80 (1953)
Kurz, M.: On the structure and diversity of rational beliefs. Economic Theory **4**, 877–900 (1994)

Risk aversion in the Talmud*

Robert J. Aumann

Center for Rationality and Interactive Decision Theory, The Hebrew University of Jerusalem,
91904 Jerusalem, ISRAEL (e-mail: raumann@math.huji.ac.il)

Received: April 10, 2002; revised version: May 7, 2002

Summary. Evidence is adduced that the sages of the ancient Babylonian Talmud, as well as some of the medieval commentators thereon, were well aware of sophisticated concepts of modern theories of risk-bearing.

Keywords and Phrases: Talmud, Risk aversion, Subjective value.
JEL Classification Numbers: B11, D80.

"If an evil witness arise against a man, to testify against him ... and the judges investigate the matter, and find that ... his testimony ... is a lie, then shall ye do unto him as he plotted to do unto his brother, and eradicate evil from your midst" (Deutoronomy 19, 16–19).

In Jewish law, a perjurer is treated exactly like the person against whom he testified would have been treated, had the perjury not been discovered. This rule is universal; it applies to the most trivial civil claim as well as to capital cases. Thus if Adams falsely testifies that Brown committed a capital crime, then Adams is executed; and if Adams falsely testifies that Brown owes Cox $100, then Adams must pay Cox $100.

---

* Presented at the Institute for Mathematical Studies in the Social Sciences-Economics, Stanford University, August 4, 1981. Subsequent to that presentation, the author's attention was drawn to an article by Zvi Ilani, "Models in the Economics of Uncertainty: The Cost of Concluding a Conditional Contract, according to the Talmud and the Halachic Literature," *Iyunim Bekalkala (Investigations in Economics)*, The Israel Association for Economics, Jerusalem, Nissan 5740 (April 1980), 246–261 (in Hebrew). Inter alia, Ilani treats the Talmudic passage that forms the subject of this paper, and provides a fairly comprehensive review of the medieval commentaries thereon; undoubtedly, he was the first to recognize in print the relevance of this passage to modern economic theories of uncertainty. It is not clear, though, whether or not his understanding of the passage agrees with ours. The current paper appeared in January 2002 in the Research Bulletin Series of the Research Center on Jewish Law and Economics, Department of Economics, Bar Ilan University.

There are, however, instances in which it is not a priori clear how the rule should be applied. Consider the following passage from the Babylonian Talmud[1]:

"'We testify[2] that John Doe divorced his wife and did not pay [the amount stipulated in] her marriage contract'[3] [when in fact, he did not divorce her]. But [in punishing the perjurers, one should take into account that] he may eventually have to pay anyway!" (Makkoth 3a, in the Mishna[4] in the middle of the page; the material in square brackets is not explicit in the original).

The point raised here is that on the face of it, the extent of the putative damage to the husband is unclear. If the perjury had not been discovered, the husband would have had to pay the stipulated amount immediately. But the full amount cannot be considered damage that the witnesses plotted to cause, since eventually the husband may have to pay it anyway – even now, after the perjury was discovered. What, then, *should* the perjurers pay? Clearly, there *is* damage, of two kinds: First, in that he would have had to pay immediately, rather than after some (indefinite) time; and second, in that perhaps in the end he might really not have to pay anything – if the wife predeceases him. Thus the damage has two components: *impatience* and *uncertainty* (or *risk*). To be sure, modern economics recognizes such questions and is able to deal with them, at least in theory. But in the 2000-year-old Talmud, it is remarkable that the question is at all recognized – let alone answered. Indeed, the passage continues as follows:

"One estimates how much a person would pay for [her rights under] her marriage contract, [taking into account that she will only be paid] if she is widowed or divorced, but that if she dies [before the husband], her husband is her heir [and the perjurers pay that estimate]" (op. cit.).

This appears to solve both components of the difficulty with aplomb. An estimate is made of the value – presumably market value – of this asset, in the absence of an immediate divorce. The amount that the perjurers must pay is the damage they plotted to cause the husband: i.e., the face value of the contract (the amount

---

[1] An ancient document that forms the basis for Jewish religious, criminal, and civil law. It consists of the *Mishna*, put into definitive form about 1,800 years ago, which sets forth the basic rules; and the *Gemara*, put into definitive form some two or three hundred years later, which discusses the Mishna and expands on it. These two parts form the nucleus of an enormous literature that has been evolving ever since, and continues to evolve to this day. In most editions of the Talmud, passages of the Mishna are intertwined with the corresponding passages of the Gemara; selected medieval (and later) commentaries appear as extensive marginal notes; and selected additional commentaries appear as extensive endnotes. There are altogether sixty "tractates" or books, usually bound in twenty separate folio volumes, taking up a linear meter of bookcase space. Whole libraries are filled with thousands of additional volumes of commentaries.

[2] The passage sets forth what the law is if such testimony should be rendered in court.

[3] The marriage contract, concluded at the time of the wedding, provides a fixed sum to the wife, payable by the husband or his estate, if the marriage should be terminated during her life, either by divorce or by his death.

[4] The word "Mishna" is used both for the entire text on which the Talmud is based, and for specific passages dealing with particular issues. Similar ambiguities occur in many languages; one may say "my son studied law" as well as "yesterday Congress passed a law."

to be paid in case of an immediate divorce), less the amount of this estimate (the current value of his obligation in the absence of an immediate divorce)[5].

So far, so good. But in interpreting the Mishna a century later, the Gemara complicates matters considerably: "How does one estimate? Rabbi Khisda says, in accordance with the husband; Rabbi Nathan ben Oshaya says, in accordance with the wife" (op. cit., in the Gemara).

Most commentators on the Talmud had considerable difficulty with this passage (see below). Just what does "in accordance with the husband (or wife)" mean? But in terms of modern theories of risk-bearing, the passage becomes beautifully clear. Evidently, we are estimating not market value, but subjective value; an amount of money such that a person would as soon have the risky asset as that amount.[6] In Rabbi Nathan's opinion, the calculation is made as above, but substituting the wife's subjective evaluation of the contract for the market value. In Rabbi Khisda's opinion, one uses the husband's subjective evaluation; i.e., the maximum amount of money that the husband would pay to get rid of his obligations under the contract. Under risk aversion, the two amounts are very different, even if husband and wife have identical utility functions. The wife will be willing to accept less than the actuarial value in exchange for the contract; the husband will be willing to pay more than the actuarial value to get rid of his obligations.

In symbols[7], denote the random variable[8] consisting of payoffs to her by x, let the face value of the contract be 200 zuz[9], and let f be the function that associates with each risk its certainty equivalent[10]. It is convenient to normalize by supposing that the husband has set aside 200 zuz in a separate account, for the specific purpose of paying her the amount specified in the marriage contract, if and when it should become necessary[11]; thus if she dies before him, he "inherits" the 200 zuz. If the witnesses had been successful in their plot, he would have been left with nothing; as it is, he is left with $200 - x$. It is this amount of $200 - x$ of which the witnesses wanted to deprive him. The certainty equivalent of this amount – what it is worth to *him* – is $f(200 - x)$, and that is what Rabbi Khisda says that the witnesses should pay.

---

[5] The impatience component is dealt with also in the following passage: "'We testify against John Doe that he owes his friend 1000 zuz, with payment due in 30 days;' but he [Doe] asserts that the payment is due in 10 years. Then [if the testimony is false] one estimates how much more a person would pay for a 1000-zuz note due in 30 days than for one due in 10 years [and this is the amount to be paid by the perjurers]" (Makkoth 3a, in the Mishna near the bottom of the page). From this it is clear that the Talmud was well aware of impatience, in spite of an absolute prohibition against payment of interest for loans.

[6] Subjective value is usually less than market value. Indeed, if the price of an asset were the same as its subjective value to the buyer, the buyer would be indifferent between buying it and not buying it. But clearly, in most transactions, both seller and buyer distinctly *prefer* that the transaction be consummated.

[7] Readers unfamiliar with mathematical notation and jargon may skip the next two paragraphs, continuing with "A numerical example...".

[8] In this paper we are interested mainly in uncertainty, and so will henceforth ignore the time factor.

[9] The zuz was a unit of currency at the time the Talmud evolved. Two hundred zuz is, in most cases, the statutory minimum for a marriage contract.

[10] If $u(t)$ is the von Neumann-Morgenstern utility of t zuz, and z is a random variable whose values are expressed in zuz units, then $f(\mathbf{z}) = u^{-1}(Eu(\mathbf{z}))$.

[11] Indeed, it was customary to set aside a piece of real estate – usually a field – for this purpose.

On the other hand, the certainty equivalent of what she will receive – now that the perjury has been discovered – is $f(\mathbf{x})$. If the witnesses had been successful, she would have gotten 200; now that they are not, what she gets is worth $f(\mathbf{x})$ to her. Thus what *she* stood to gain from the perjury is $200 - f(\mathbf{x})$, and that is what Rabbi Nathan says that the witnesses should pay. Note that if the utility function u that underlies the definition of f is strictly concave – i.e., if the protagonists are risk averse – then

(1) $$f(200 - \mathbf{x}) < 200 - f(\mathbf{x});$$

that is, Rabbi Nathan requires a larger payment from the witnesses than does Rabbi Khisda.

A numerical example is perhaps in place. Suppose that both husband and wife are indifferent between a sure payoff of 80 and a lottery whose outcome is 200 or 0 with $1/2 - 1/2$ probabilities. Thus both are risk averse, to the same degree: each is willing to pay a "premium" of 20 zuz to avoid the uncertainty of the lottery. As above, assume that the husband has set aside a piece of property worth 200 to pay the wife's contract in case that that should become necessary. Let the probability be $1/2$ that the marriage will be terminated before the wife dies. Before the false testimony, the positions of the husband and wife are identical: both will receive 200 with probability $1/2$, and 0 with probability $1/2$. So the husband's position is worth 80 to him, and the wife's position is worth 80 to her. If the false testimony had been accepted, the property set aside by the husband would have been given to the wife, making his position worthless. The wife, on the other hand, would have gotten the full value of the property, namely 200. In terms of certainty equivalents, the husband would have lost 80, while the wife would have gained $200 - 80 = 120$. Rabbi Khisda says that the perjurers must pay "according to the husband," namely 80; Rabbi Nathan, that they must pay "according to the wife," namely 120.

More precisely, one should perhaps take cognizance of the fact that husband and wife may have different utility functions u_h and u_w, with correspondingly different certainty-equivalent functions f_h and f_w. Rabbi Khisda and Rabbi Nathan then say that the witnesses must pay $f_h(200 - \mathbf{x})$ and $200 - f_w(\mathbf{x})$ respectively; and if both u_h and u_w are strictly concave, then the latter is always larger than the former. Note, however, that the disagreement between Rabbi Khisda and Rabbi Nathan does not hinge on a possible difference between their utility functions; as the above example shows, the two Rabbis differ even when husband and wife have the same utility function.

Two difficulties, one textual and one conceptual, remain.

The textual difficulty is that the Mishna specifically refers to an estimate of how much "a person" would be willing to pay for the asset. This implies that the discussion is about a neutral third party, rather than the specific husband and wife before us. The conceptual difficulty is that while the opinion of Rabbi Khisda seems natural enough, it is difficult to understand why Rabbi Nathan relies on the subjective evaluation of the wife, who is really not a party to the litigation between the witnesses and the husband. "Ye shall do unto him as he plotted to do unto his brother." This plot was directed against the husband; *he* would have had to pay if it had been successful, not she. Why, then, should *her* subjective evaluation be relevant?

The two difficulties can be cleared up simultaneously. Neither Rabbi Khisda nor Rabbi Nathan refer to the specific flesh-and-blood husband and wife before us. The estimate appertains to how an outside, neutral party who has a "typical" degree of risk aversion would evaluate the risky assets $200 - x$ and x respectively. Thus we must indeed use a single function f as in (1), rather than separate functions f_h and f_w. The words "according to the husband" and "according to the wife" refer only to the formal places of the husband and wife in the scheme, rather than to their individual utility functions (which it would, indeed, be almost impossible to estimate). They are, in fact, merely a convenient way of referring to the two sides of inequality (1).

In these terms, the conceptual difficulty also disappears. Rabbi Khisda hews more closely to the letter and spirit of the Mosaic rule, since the husband's loss would have been $200 - x$, and it is the certainty equivalent of this that the witnesses must pay. But Rabbi Nathan hews more closely to the text of the Mishna, which states specifically that the amount to be estimated is x, not $200 - x$.

While we cannot be sure why the Mishna was in fact phrased in this way, one possible reason has to do with moral hazard. People might be loath to purchase the husband's position, since doing so would decrease his disincentives to divorce her. Even the gedanken-experiment in which the court estimates how the husband himself evaluates his position is beclouded by moral hazard. Since there are conceptual difficulties in evaluating the husband's position, Rabbi Nathan "rolls over" the estimate to the other side, a procedure not uncommon in Talmudic law[12].

More convincing, perhaps, is a suggestion of Professor Michael Keren[13], based on fundamental principles of criminal law. Rabbi Khisda and Rabbi Nathan agree that "the punishment should fit the crime," but differ on whether to evaluate the crime by its effect (or putative effect) on the victim or on the perpetrator. If punishment is "retribution," then the yardstick should be the effect on the victim; if "deterrent," then it should be the effect on the perpetrator. The "retribution" philosophy would dictate that it is the putative damage to the husband that counts, in accordance with Rabbi Khisda; the "deterrent" philosophy, that it is the putative benefit to the false witnesses. And while we do not know the motives of the witnesses, it is the wife who stood to gain from their testimony; in a sense, they were acting on her behalf, even if it was without her consent. Therefore it is *her* putative gain that should determine the punishment, which is in accordance with Rabbi Nathan.

A final comment has to do with why the Gemara was not satisfied with the obvious interpretation of the Mishna, in terms of market value, described above. Again, there are two reasons, one textual, one conceptual.

---

[12] For example: No allegation can be established in court without the testimony of at least two witnesses. In criminal cases, one witness is as good as none. But in civil cases, if a single witness testifies against, say, the defendant, then the defendant must pay up, or swear a solemn oath that the witness is lying. Suppose, now, that the defendant has a criminal record that prevents his oath from being acceptable in court. Then the oath is "rolled over" onto the plaintiff; if he swears that the witness is right, then the defendant must pay up.

[13] Privately communicated.

Textually, the Mishna uses the word "omdin," which generally means a rough, subjective estimate. For the more accurate objective assessments that are possible in cases of market value, the Mishna generally uses "shummin[14]." Conceptually, the Mosaic injunction quoted above is best implemented by a subjective certainty equivalent; market value seems pretty irrelevant to the spirit and letter of this injunction.

In conclusion, there seems to be little doubt that fairly sophisticated concepts of the modern theory of risk-bearing underlie this Talmudic passage, and were well understood by its authors.

A glimpse at the medieval commentary

Most commentators, unfamiliar with modern theories of risk-bearing, had considerable difficulty with this passage. Rabbi Isaac Alfasi, living in Spain about a millenium ago, wrote, "Rabbi Khisda says, 'in accordance with the husband': Is he old, and so more likely soon to die than a young man? Is he ill or well? Does he have property, which would enable him to divorce her and give her the amount stipulated in the marriage contract[15]? Are they on poor terms, so that he has an incentive to divorce her? Whereas Rabbi Nathan says the opposite: Is *she* ill or well? Is *she* old or young? ..." (Alfasi Makkoth 1b).

Commenting[16] on this passage from Alfasi, Nachmanides (Spain, about 800 years ago) objects that it makes no sense to look only at the husband's condition, or only at the wife's; that a reasonable estimate of the situation must take into account the conditions of both husband *and* wife. Though the remainder of his comment is not entirely transparent, it does include the assertion that "a person does not want to endanger his money by buying risks, unless he can do so for very little money[17]" – a succinct statement of the principle of risk aversion.

The greatest of all commentators – Rashi (Rabbi Shlomo Yitzchaki) – living in France about 900 years ago, wrote as follows:

"... It is possible to understand our Mishna in two ways. How?

Well, both the man and the woman have a doubtful privilege[18] in this contract. She expects that if he dies or divorces her she will get the entire amount; and he

---

[14] The difference between "omdin" (one estimates) and "shummin" (one assesses) is beautifully illustrated in the following Mishna: "He who causes bodily injury is liable on five counts: for permanent disability, pain, medical expenses, temporary disability, and humiliation. Permanent Disability: suppose he took out his [the victim's] eye, cut off his hand, or broke his leg; one considers him [the victim] as if he were a slave to be sold in the market place, and assesses ("shummin") how much he was worth [before the incident] and how much he is worth [now]. Pain: suppose he burned him with a spit or a nail, even on his fingernail, where no wound is caused; one estimates ("omdin") how much a man like the victim would take in return for suffering such pain [voluntarily]. Medical expenses: . . . "(Op. Cit., Baba Kama 83b, in the Mishna).

[15] The husband's obligations under the marriage contract often constituted a significant barrier to divorce.

[16] In Milchamoth Hashem ("The Wars of the Lord"), Alfasi Makkoth 1a.

[17] "Ein adam rotzeh lessaken bemamono likach sfekot ela bedamim kalim" (op. cit.).

[18] "Z'chut safek;" literally, the privilege of the risk.

expects that if she dies during his lifetime, he will inherit[19] her rights.... . So this is the meaning of it: Does one estimate *her* doubtful privilege, i.e., how much a person would pay for the benefit of her enjoyment,[20] this being what they [the false witnesses] will not pay, and they will pay the remainder [up to the face value of the contract]? Because the remainder is the loss they would have caused by their testimony, since even now, when their testimony has been discredited, he would gladly give her this amount for her rights.

Or, does the Mishna say that we should estimate *his* doubtful privilege, and this is the amount that the witnesses will pay? Thus we would not obligate the witnesses to pay so much, i.e., the face value less the benefit of her enjoyment ..." (Makkoth 3a, in the Rashi).

Rashi is rarely so expansive; usually his writing is terse and to the point. His expansiveness in this instance may indicate that he is feeling his way on unfamiliar ground, or at the least, expounding an unfamiliar viewpoint. Be that as it may, it appears that Rashi may have come fairly close to the interpretation in the body of this paper.

Dedication

This article is dedicated to Mordecai Kurz, a great economist and a wonderful friend, colleague, and coworker. In addition to his many fundamental contributions to economic theory, Mordecai will be remembered for the twenty magnificent years during which he ran the Institute for Mathematical Studies in the Social Sciences (Economics) – the unforgettable IMSSS – at Stanford University. Every summer during the seventies and eighties, the cream of Economic Theory, worldwide, gathered for two tremendously exciting months at the frontiers of knowledge. These were not ordinary conferences. Though there were presentations, they took up less than half the working time; most of the time was taken up by work between the participants. Mordecai selected the participants and speakers with great care, and imposed an iron discipline, which forged the group into a major force in the development of Economic Theory during those formative decades.

As mentioned in the title footnote, this paper was first presented at the IMSSS in the summer of 1981 at a special session on Talmudic economics for which Mordecai provided the initiative. It is a very special pleasure and privilege to publish it, for the first time, in this issue of *Economic Theory* dedicated to Mordecai Kurz.

---

[19] By having the husband write off the face value of the contract immediately (at the time of the marriage), Rashi has normalized so that all payoffs to all protagonists are non-negative. If the wife dies before the husband, the amount of the contract appears as a credit to him; he "inherits" it. This normalization is like in the body of the paper.

[20] "Tovat Hana'a."

Annuities and retirement*

Eytan Sheshinski

Department of Economics, The Hebrew University of Jerusalem, Jerusalem, ISRAEL
(e-mail: mseytan@mscc.huji.ac.il)

Received: November 21, 2002; revised version: February 1, 2003

Summary. This paper examines the interaction between the market for annuities and retirement and consumption decisions in the presence of lifetime uncertainty. We focus on two aspects of the demand for annuities: the *timing* of annuitization and the *information* available to the issuers of annuities with regard to purchasers' survival probabilities. The *First-Best* is attained by continuous annuitization of savings, with a guaranteed lump-sum payment to beneficiaries upon death. Annuitization of savings at retirement, and *a-fortioti* no annuitization, are inferior and lead to distortions in retirement and consumption decisions. Applying a '*Stochastic Dominance*' approach, we show how these decisions depend on individuals' degree of risk aversion. Under imperfect information, we analyze *Pooling Equilibria* and compare them with the *First-Best*.

Keywords and Phrases: Annuitization, Survival function, Information arrival, Optimal retirement age, Pooling and separating equilibrium.

JEL Classification Numbers: D6, H55, J26.

1 Introduction

Retirement plans involve numerous uncertainties. Foremost is the uncertainty with respect to lifetime. Not only is lifetime duration uncertain but, particularly early in life, survival probabilities are difficult to predict. These depend largely on health conditions which only unfold over time. With these uncertainties, lifetime consumption and savings plans depend crucially on the availability of insurance markets.

---

 * This paper was written while I was visiting Princeton University in the Fall of 1998. I wish to thank my colleagues and friends at Princeton for their hospitality. I also wish to acknowledge useful comments received from Kenneth Arrow, Peter Diamond and Sergiu Hart.

The market most relevant for the length of life uncertainty is the market for *annuities*. These financial instruments provide payments to annuity holders starting on a specified date and contingent on survival.[1]

By pooling mortality risks of many individuals, issuers of annuities can fully insure against individual lifetime uncertainty. The terms at which annuities are offered in competitive markets depend on the *information* available to issuers. Purchasers belong to one of a number of 'risk classes', each consisting of individuals with the same survival function. When suppliers can identify the 'risk class' to which customers belong, competitive annuity prices differ by 'risk class' and a *First Best* allocation is attained (Yaari, 1965).

When information on survival probabilities is private, or only imperfectly available to firms, annuity prices are based on 'pooled' (average) survival rates of several 'risk classes.' Pooling leads to *adverse selection*, that is, annuities are purchased primarily by groups with high survival probabilities while those with low survival probability prefer to purchase fewer annuities.[2] In extreme cases, some 'risk classes' may choose not to purchase any annuities. In a *Separating Equilibrium* of this kind, the actions of individuals reveal (partially or fully) their characteristics (Rothschild and Stiglitz, 1976).

Whenever annuity prices are based on pooling of some 'risk classes' the allocation is not *First Best*. These prices imply transfers ('cross-subsidization') across groups and these entail, in turn, distortions in consumption and in retirement decisions. The objective of this paper is to explore the interaction between these decisions and the characteristics of the annuity market.

At an optimum retirement age, the benefits and costs of a marginal postponement of retirement are equal. The instantaneous benefits of such postponement are the wage rate times the marginal utility of consumption. The level of consumption depends, in turn, on the structure of the insurance markets, in particular the market for annuities. Costs are the instantaneous disutility of labor. When benefits decrease and costs increase with age, there is a unique optimum retirement age.

This paper focuses on the *timing of the annuitization of savings*. When the purpose of savings is the provision of consumption during retirement whose duration is uncertain, it is most efficient to annuitize savings continuously, i.e. to purchase *'deferred annuities'* which start payments, contingent on survival, upon retirement. Age-dependent annuity prices are cheaper for young purchasers, reflecting the probability that the annuity holder may die prior to retirement. Annuitization of savings at retirement is inferior to continuous annuitization (the same amount of cumulative savings yields a lower post-retirement flow of payments), because insurance firms are not able to pull mortality risks prior to retirement. Consequently, retirement is delayed and consumption during retirement is lower, compared to the *First-Best*. This assumes that individuals place no (or low) value on non-annuitized assets left behind when death occurs before retirement.

---

[1] These are called *'single annuities.'* *'Joint annuities'* provide, in addition, specified payments to spouses or children after the death of the annuity holder. Section 6 below analyzes this type of annuities.

[2] Convincing evidence on the different mortality probabilities of annuity purchasers compared to the population average is presented in Mitchell, Poterba, Warshawsky, and Brown (1999).

When individuals have a *bequest motive,* the *First-Best* allocation is attained with continuous annuitization and a guaranteed lump-sum payment upon death to designated beneficiaries (which entails correspondingly lower retirement benefits). Annuitization at retirement leads to excessive (compared to the *First-Best*) accumulation of assets during the working phase, whereby part of these assets is annuitized upon retirement and part is left for bequest. As with no bequest motive, retirement is delayed, post-retirement consumption and bequests are lower, compared to the *First-Best*.

All individuals are shown to purchase some annuities at any price. Hence, no *Separating Equilibrium* exists. A *Pooling Equilibrium* distorts consumption and retirement decisions. Specifically, if individuals are sufficiently risk-averse (coefficient of relative risk aversion larger than one), retirement ages and consumption are 'pooled' together compared to the *First-Best*.

It has been argued that annuity markets in the U.S. are expensive and inefficient (e.g. Friedman and Warshawsky, 1990; Mitchell et al., 1999; Walliser, 1998). It is not clear how much of this inefficiency can be attributed to the existence of a mandatory Social Security system which provides (real) annuities at pooled rates and leaves little room for residual demand. We do not discuss directly the merits of different reform proposals for the U.S. Social Security system. Our underlying assumption is that a competitive, decentralized and voluntary annuity market is feasible and we analyze its potential efficiency under alternative informational structures.

The use of annuities to insure against lifetime uncertainty has been introduced by Yaari (1965). A number of papers (Merton, 1981; Sheshinski and Weiss, 1981; Abel, 1993) excluded by assumption private annuity markets and focused on insurance offered by the government. The first paper that examined the different types of annuity market equilibria is Eckstein, Eichenbaum and Peled (1985). The issue of the timing of annuity purchases was analyzed by Brugiavini (1993).[3] These papers used a two or three-period model and did not examine endogenous retirement and its interaction with the annuity market.

2 Optimum consumption and retirement with lifetime uncertainty

Consider an individual who has to decide on his or her optimum consumption at different ages and the age of retirement in the presence of uncertainty about the length of life. Suppose that this is the only uncertainty that the individual faces. Later we analyze the effects of additional uncertainties with respect to future income and survival rates.

Let T, $T > 0$, be the planning horizon, i.e. maximum lifetime (it is possible to allow $T = \infty$). Lifetime uncertainty is represented by a *survival distribution function* $F(z)$, which is non-increasing in age, z, $0 \leq z \leq T$, with $F(0) = 1$, $F(T) = 0$.

Denote consumption at age z by $c(z)$, $c(z) \geq 0$. Utility of consumption at different ages is separable and independent of age and there is no subjective discount

---

[3] I became aware of Brugiavini's paper and of the timing of annuitization from a lecture note by Peter Diamond.

rate. Instantaneous utility, $u(c)$, displays risk-aversion $(u'(c) > 0, u''(c) < 0)$ and, to ensure an interior solution, satisfies the end-conditions $u'(0) = \infty$ and $u'(\infty) = 0$.

When working, the individual provides one unit of labor. Disutility of work, $a(z) > 0, 0 < z < T$, is independent of consumption and nondecreasing with age, $a'(z) \geq 0$. Again, to ensure an interior solution, assume that $a(0) = 0$ and $a(T) = \infty$. Contingent on survival, the individual works between ages 0 and R, $0 < R < T$, i.e. *retirement* occurs at age R.

The individual's objective is to maximize *expected utility*, V,

$$V = \int_0^T F(z)u(c(z))dz - \int_0^R F(z)a(z)dz. \tag{1}$$

The choice of the optimum consumption path and age of retirement will depend on the insurance options available in the market.

2.1 No annuities

Consider first the case when no insurance is available. Let wages at age z be $w(z)$, $w(z) \geq 0$. Savings, $w(z) - c(z)$, earn a zero rate of interest.[4] With no initial assets, the individual's assets at age z, $B(z)$, are equal to cumulative savings:

$$B(z) = W(z) - \int_0^z c(x)dx, 0 \leq z \leq T, \tag{2}$$

where $W(z) = \int_0^{\min(z,R)} w(x)dx$, $0 \leq z \leq T$, are cumulative wages. Feasible consumption plans must have non-negative assets at all ages:[5]

$$B(z) \geq 0. \tag{3}$$

Maximization of (1) subject to (3) with respect to $c(z)$ and R yields F.O.C. for an interior solution $(\hat{c}(z),\hat{R})$

$$F(z)u'(\hat{c})(z)) - \lambda = 0, 0 \leq z \leq T, \tag{4}$$

$$a(\hat{R}) - \psi(\hat{R}) = 0, \tag{5}$$

where $\psi(R) = u'(\hat{c}(R))w(R)$ for all $R, 0 < R < T$, and $\lambda, \lambda > 0$, is a constant. Condition (4) states that optimum expected marginal utility is constant over age.

---

[4] It is well-known how the results are modified when the rate of interest and the subjective discount rate are positive.

[5] Since optimum consumption is shown below to strictly decrease with age, (3) implies that $B(z) > 0, R \leq z < T$. The constraint $B(z) \geq 0$ may bind the optimum consumption path for $0 \leq z < R$.

Denote the solution to (4) by $\hat{c}(z)$ and write $\hat{c}(z) = c_0 h(z)$, where $h(0) = 1$ and, from (4), $u'(c_0) = \lambda$. Optimum consumption is seen to decrease with age,

$$\frac{\hat{c}'(z)}{\hat{c}(z)} = \frac{h'(z)}{h(z)} = -\frac{1}{\sigma}\frac{f(z)}{F(z)}, \tag{6}$$

where $\sigma = \sigma(z) = -\frac{u''(c)c}{u'(c)} > 0$, is the elasticity of marginal utility, termed *Coefficient of Relative Risk Aversion*, $f(z)$, $f \geq 0$, is the density of $1 - F(z)$ and $\frac{f(z)}{F(z)}$ is the conditional probability of dying at age z, termed the *Hazard-Rate*. By (4) and the assumption that $u'(0) = \infty$, $\hat{c}(T) = h(T) = 0$.

With no bequest motive, constraint (3) is binding at T. This condition is used to solve for initial consumption, c_0,

$$c_0 = c_0(R) = \frac{\int_0^R w(z)dz}{\int_0^T h(z)dz} \tag{7}$$

The condition which determines the optimum retirement age, (5), has an intuitive interpretation. A marginal postponement of retirement yields an additional income of $w(R)$. It's value in utility terms is $\psi(R) = u'(\hat{c}(R))w(R)$. The costs of postponement in utility terms are the disutility of labor, $a(R)$. At the optimum, costs and benefits are equalized.

The sign of $\psi'(R)$ cannot be established in general and hence there is a possibility of multiple solutions to (5).[6] An example is provided below. We shall not dwell on this question and assume a unique solution, denoted $\hat{R}$.[7]

Optimum expected utility, denoted $\hat{V}$, is given by

$$\hat{V} = \int_0^T F(z)u(\hat{c}(z))dz - \int_0^{\hat{R}} F(z)a(z)dz \tag{8}$$

2.2 Continuous annuitization

An annuity is a financial instrument which promises certain payments at specified ages, contingent on survival of the holder of the annuity.[8] The issuer of annuities can

---

[6] Differentiating $\psi(R)$ with respect to R,

$$\frac{\psi'(R)}{\psi(R)} = \frac{f(R)}{F(R)} - \sigma(R)\frac{w(R)}{\int_0^R w(z)dz} + \frac{w'(R)}{w(R)}$$

[7] $\hat{R}$ satisfies second order conditions if $a(R) - \psi(R) < (>)0$ for $R < (>) \hat{R}$ whenever (5) is satisfied.

[8] Annuities that also provide payments to beneficiaries after the death of the annuity holder are discussed in Section 6.

promise certain payments by pooling the mortality risks of many individuals. When individuals save in order to finance consumption during retirement, it is efficient to convert savings continuously during the working phase into annuities which will start payments upon retirement. The purchase of such '*deferred-annuities*' provides perfect insurance against lifetime uncertainty and the individual takes advantage of the lower prices of early annuity purchases which reflect the probability that the purchaser does not survive to retirement.[9]

All individuals are assumed to have the same survival distribution function, $F(z)$. With a zero rate of interest, lending and borrowing is made at a discount rate equal to this common rate. Hence, without loss of generality, we can take annuity payments to be synchronized with consumption during retirement.

With continuous conversion of savings, $w(z) - c(z)$, into annuities which start payments at R, competition implies a zero expected profits condition:

$$\int_R^T F(z)c(z)dz - \int_0^R F(z)(w(z) - c(z))dz = 0$$

or

$$\int_0^T F(z)c(z)dz - \int_0^R F(z)w(z)dz = 0. \tag{9}$$

Maximization of (1) subject to (9) with respect to $c(z)$ and R, yields F.O.C. for an interior solution $(c^*(z), R^*)$:

$$u'(c^*(z)) - \lambda = 0, 0 \leq z \leq T \tag{10}$$

$$a(R^*) - \phi(R^*) = 0 \tag{11}$$

where $\phi(R) = u'(c^*(R))w(R)$ for all $R, 0 \leq R \leq T$, and $\lambda, \lambda > 0$, is a constant.[10]

It is seen from (10) that optimum consumption is constant: $c^*(z) = c^*, 0 \leq z \leq T$. By (9), for all $R, 0 \leq R \leq T$,

$$c^* = c^*(R) = \frac{\overline{W}(R)}{\overline{z}} \tag{12}$$

where $\overline{z} = \int_0^T F(z)dz$ is *life-expectancy*,[11] and

$$\overline{W}(R) = \int_0^R F(z)w(z)dz = \int_0^R f(z)W(z)dz + (1 - F(R))W(R) \tag{13}$$

---

[9] Age-dependent prices assume, of course, full infomration on the purchasers age. Purchasing deferred annuities is equivalent to investing in a large pension fund which distributes the assets of deceased members of the same age cohort.

[10] The functional form of ϕ, (11), and ψ, (5), is the same, but their dependence on R differs because of the different consumption paths, $c^*(z)$ and $\hat{c}(z)$, respectively. We use thoughout the same notation for the Lagrangan constants, λ, but their values differ, of course, for each maximization.

[11] Integrating by parts: $\int_0^T F(z)dz = \int_0^T zf(z)dz$ where $f(z)$ is the density of $1 - F(z)$, when it exists.

is *expected total wages* until retirement.

Condition (11) determines the optimum retirement age, R^*. As before, it balances the benefits and costs of a marginal postponement in retirement. When $w'(R) \leq 0$, $\phi(R)$ decreases with R and (11) determines a unique solution.[12]

Optimum expected utility, denoted V^*, is

$$V^* = u(c^*) \int_0^T F(z)dz - \int_0^{R^*} F(z)a(z)dz =$$

$$u\left(\frac{\overline{W}(R^*)}{\overline{z}}\right)\overline{z} - \int_0^{R^*} F(z)a(z)dz. \tag{14}$$

Fully insuring against lifetime uncertainty, continuous annuitization is the *First-Best* allocation. Hence,

Proposition 1. $V^* > \hat{V}$.

Proof. By concavity, $u(c) < u(c^*) + u'(c^*)(c - c^*)$, for any (c, c^*), $c \neq c^*$. Let $c = c(z)$ be a function of z. Multiplying by $F(z)$ and integrating,

$$\int_0^T F(z)u(c(z))dz < u(c^*) \int_0^T F(z)dz$$

$$+u'(c^*)\left[\int_0^T F(z)c(z)dz - c^* \int_0^T F(z)dz\right]$$

$$= u(c^*) \int_0^T F(z)dz + u'(c^*)\left[\int_0^T F(z)c(z)dz - \int_0^R F(z)w(z)dz\right], \tag{15}$$

where R corresponds to $c^* = c^*(R)$ by (12). Now take $c(z) = \hat{c}(z)$ and $R = \hat{R}$, the optimum with no annuities. Setting $w(z) = 0$ for $\hat{R} < z \leq T$, write $\int_0^{\hat{R}} F(z)w(z)dz = \int_0^T F(z)w(z)dz$. Integrating by parts, using condition (3), $\int_0^T F(z)(w(z) - \hat{c}(z))dz \geq 0$, or $\int_0^T F(z)\hat{c}(z)dz \leq \int_0^{\hat{R}} F(z)w(z)dz$. It now follows from (15) that

$$\int_0^T F(z)u(\hat{c}(z))dz < u(c^*(\hat{R})) \int_0^T F(z)dz. \tag{16}$$

---

[12] For all R, $0 \leq R \leq T$,

$$\frac{\phi'(R)}{\phi(R)} = \frac{w'(R)}{w(R)} - \sigma(R)\frac{w(R)}{W(R)}F(R), \text{ where } \sigma(R) = -\frac{u''(c^*(R))c^*(R)}{u'(c^*(R))} > 0$$

Subtracting $\int_0^{\hat{R}} F(z)a(z)dz$ on both sides of (16), noting that V^* is maximum at $R = R^*$, it follows from (8) and (14) that $\hat{V} < V^*$.   □

2.3 Annuitization at retirement

Suppose that savings are converted to annuities only upon retirement.[13] The zero expected profits condition is now

$$\int_R^T F(z)c(z)dz - F(R) \int_0^R (w(z) - c(z))dz = 0 \qquad (17)$$

Maximization of (1) subject to (17) yields F.O.C. for an interior solution $(\hat{c}_R(z), \hat{R}_R)$:

$$F(z)u'(\hat{c}_R(z)) - \lambda F(R) = 0, \quad 0 \le z \le R \qquad (18)$$

$$u'(\hat{c}_R(z)) - \lambda = 0, \quad R < z \le T \qquad (19)$$

and

$$a(\hat{R}_R) - \phi_R(\hat{R}_R) = 0 \qquad (20)$$

where $\phi_R(R) = u'(\hat{c}_R(R))w(R)$, for all R, $0 \le R \le T$, and λ, $\lambda > 0$, is a constant. It is seen from (18) and (19) that optimum consumption decreases for $0 \le z \le R$ and is constant thereafter (Figure 1). Again, write $\hat{c}_R(z) = c_{0R}h(z)$, $0 \le z \le R$, where $h(0) = 1$ and, by (18), $h(z)$ satisfies differential Equation (6).

Initial consumption, c_{0R}, is solved from (17),

$$c_{0R} = c_{0R}(R) = \frac{\int_0^R w(z)dz}{\frac{h(R)}{F(R)} \int_R^T F(z)dz + \int_0^R h(z)dz}. \qquad (21)$$

Optimum expected utility with annuitization at retirement, denoted $\hat{V}_R$, is thus

$$\hat{V}_R = \int_0^R F(z)u(\hat{c}_R(z))dz + u(\hat{c}_R(R)) \int_{\hat{R}_R}^T F(z)dz - \int_0^{\hat{R}_R} F(z)a(z)dz \qquad (22)$$

Annuitization at retirement is inferior to continuous annuitization and superior to no annuitization:

Proposition 2. $V^* > \hat{V}_R > \hat{V}$.

Proof is straightforward.

---

[13] This is the case in many of the mandatory privately managed pension schemes, following the example of Chile (see Diamond and Valdez-Prieto, 1994). Assets of deceased members in each fund are transferred to designated beneficiaries. It is assumed here that individuals put no value on these transfers. In Section 6 below, we introduce a bequest motive.

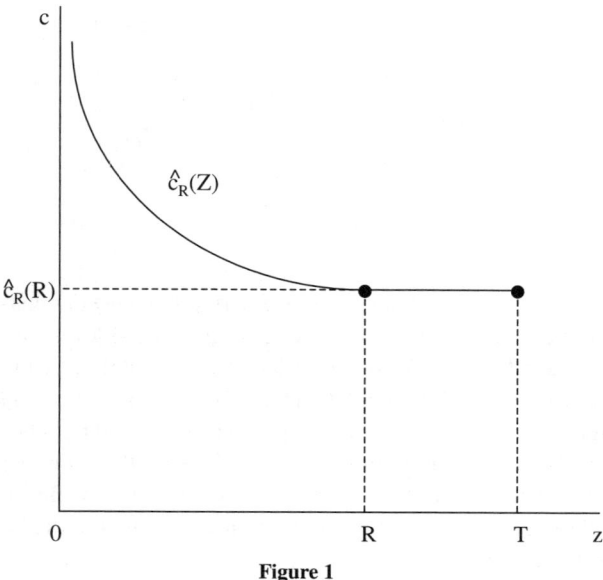

Figure 1

3 Comparison of optimum retirement ages

We want to compare optimum retirement ages and levels of consumption during retirement under the different insurance regimes. Without annuities, optimum consumption during retirement decreases with age. In this case, therefore, optimum retirement age is uniquely related only to the level of consumption *upon* retirement.

3.1 Continuous annuitization vs. annuitization at retirement

First, we show that optimum retirement age is lower with continuous annuitization than with annuitization at retirement:

Proposition 3. $R^* < \hat{R}_R$ and $c^* > \hat{c}_R(\hat{R}_R)$.

Proof. From conditions (11) and (20) it is seen that $R^* < \hat{R}_R$ if $c^*(R) > \hat{c}_R(R)$ for all R, $0 \leq R \leq T$. From (3),

$$F(R) \int_0^R (w(z) - \hat{c}_R(z)) dz < \int_0^R F(z)(w(z) - \hat{c}_R(z)) dz \qquad (23)$$

for all R, $0 \leq R \leq T$. Hence, by (17) and (24)

$$\int_0^R F(z)\hat{c}_R(z) dz + \hat{c}_R(R) \int_R^T F(z) dz \leq \int_0^R F(z)w(z) dz \qquad (24)$$

Since, by (18), $\hat{c}_R(z) > \hat{c}_R(R), 0 \leq z < R$, it follows from (24) and (12) that

$$\hat{c}_R(R) < \frac{\int\limits_0^R F(z)w(z)dz}{\int\limits_0^T F(z)dz} = c^*(R). \tag{25}$$

□

This result is intuitive. Without insurance during the working phase and facing decreasing survival rates, the individual chooses decreasing levels of consumption so as to keep expected marginal utility constant. Consequently consumption during retirement is lower than under complete insurance, i.e. continuous annuitization.

In the absence of annuitization, consumption is uninsured both before and after retirement. The decision on optimum retirement depends therefore both on the survival distribution function and on the individual's risk aversion. Before we analyze this case it will be useful for subsequent discussions to link it to the familiar notion of 'Stochastic Dominance'.

3.2 Ranking of survival functions

Consider two survival distribution functions, $F_1(z)$ and $F_2(z)$, $0 \leq z \leq T$. By assumption, $F_i(0) = 1$, $F_i(T) = 0$ and $F_i(z)$ non-increases in z, $i = 1, 2$.

Definition 1. ('Single Crossing' or 'Stochastic Dominance'): The function $F_1(z)$ is said to (strictly) stochastically dominate $F_2(z)$ if the 'Hazard-Rates' satisfy

$$\frac{f_2(z)}{F_2(z)} > \frac{f_1(z)}{F_1(z)}, \quad 0 \leq z \leq T. \tag{26}$$

In words, the rate of decrease of survival probabilities, $\frac{d\ell n F(z)}{dz} = -\frac{f(z)}{F(z)}$, is smaller at all ages with distribution 1 than with 2.

Two implications of this definition are important. First, consider the functions $\frac{F_i(z)}{\int\limits_0^T F_i(z)dz}$, $0 \leq z \leq T$, $i = 1, 2$. Being positive and their integral over $(0, T)$ equal to one, they must intersect (cross) at least once over this range. At any such crossing, when $\frac{F_1(z)}{\int\limits_0^T F_1(z)dz} = \frac{F_2(z)}{\int\limits_0^T F_2(z)dz}$, condition (26) implies that

$\frac{d}{dz}\left(\frac{F_1(z)}{\int\limits_0^T F_1(z)dz}\right) > \frac{d}{dz}\left(\frac{F_2(z)}{\int\limits_0^T F_2(z)dz}\right)$. Hence, there can be only a 'single crossing'.

That is, there exists an age z_c, $0 < z_c < T$, such that (Figure 2),

$$\frac{F_1(z)}{\int\limits_0^T F_1(z)dz} \underset{>}{\overset{\leq}{\lessgtr}} \frac{F_2(z)}{\int\limits_0^T F_2(z)dz} \quad \text{as } z \underset{>}{\overset{\leq}{\lessgtr}} z_c. \tag{27}$$

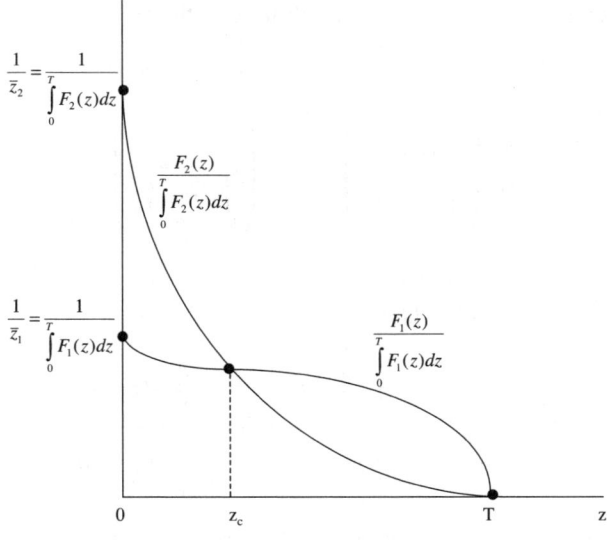

Figure 2

Intuitively, (27) means that the dominant distribution has higher (lower) survival rates, relative to life expectancy, at older (younger) ages.

Second, since $F_i(0) = 1$, $i = 1, 2$, it follows from (27) that

$$\bar{z}_1 = \int_0^T F_1(z)dz > \int_0^T F_2(z)dz = \bar{z}_2, \tag{28}$$

i.e., stochastic dominance implies *higher life expectancy*.

3.3 Effects of annuitization

We are now ready to compare the no-annuities case with those where annuitization is available, either continuously or at retirement.

With no insurance against lifetime uncertainty, the individual's consumption and retirement plans depend on his or her attitude towards risk, represented by the *coefficient of (relative) risk-aversion*, σ.

Proposition 4. *If $\sigma \geq 1$ for all z, $0 \leq z \leq T$, then $R^* < \hat{R}_R < \hat{R}$ and $c^* > \hat{c}_R > \hat{c}$.*

Proof. Recall that with no annuities, optimum consumption is $\hat{c}(z) = c_0 h(z)$, where c_0 is given by (7), $h(0) = 1$ and $h(z)$ obeys the differential equation (6). Note that, by (6), if $\sigma = 1$ for all z, $0 \leq z \leq T$, then $h(z) = F(z)$. We now want to show that, when $\sigma > 1$, $h(z)$ stochastically dominates $F(z)$.

When $\sigma > 1$, then by (6),

$$\frac{h'(z)}{h(z)} = -\frac{1}{\sigma}\frac{f(z)}{F(z)} > -\frac{f(z)}{F(z)} \tag{29}$$

Hence, whenever $\dfrac{h(z)}{\int_0^T h(z)dz} = \dfrac{F(z)}{\int_0^T F(z)dz}$, (29) implies that

$$\frac{d}{dz}\left(\frac{h(z)}{\int_0^T h(z)dz}\right) > \frac{d}{dz}\left(\frac{F(z)}{\int_0^T F(z)dz}\right),$$

which implies a 'single-crossing'.

Since $h(z)$ is non-increasing, by (7),

$$\hat{c}(R) = c_0 h(R) = \frac{\int_0^R w(z)dz}{\int_0^T h(z)dz} h(R) < \frac{\int_0^R h(z)w(z)dz}{\int_0^T h(z)dz}, \tag{30}$$

for all $0 \le R \le T$. An implication of 'single-crossing' is that

$\int_0^R \left[\dfrac{h(z)}{\int_0^T h(z)dz} - \dfrac{F(z)}{\int_0^T F(z)dz} \right] dz < 0$, for all $R < T$ (because, for $R = T$ the integral

is equal to zero and hence, by (27), it is negative for all $R < T$).

From (30) and the identity $h(z) = F(z)$ for $\sigma = 1$, we conclude that whenever $\sigma \ge 1$

$$\hat{c}(R) < \frac{\int_0^R h(z)w(z)dz}{\int_0^T h(z)dz} \le \frac{\int_0^R F(z)w(z)dz}{\int_0^T F(z)dz} = c^*(R) \tag{31}$$

for all $0 \le R \le T$. Thus, by (5) and (11) $\psi(R) > \phi(R)$, $0 \le R \le T$. Since $a'(R) \ge 0$, it follows from these conditions that $\hat{R} > R^*$.

To show that $\hat{c}_R(R) > \hat{c}(R)$, for all $0 \le R \le T$, compare c_{0R}, given by (22), to c_0, given by (7). We see that $c_{0R} \lessgtr c_0$ as $\int_R^T \dfrac{F(z)}{F(R)}dz \gtrless \int_R^T \dfrac{h(z)}{h(R)}dz$. By (6), if $\sigma \ge 1$,

$$\frac{d\ell n h(z)}{dz} = \frac{h'(z)}{h(z)} = -\frac{1}{\sigma}\frac{f(z)}{F(z)} \le -\frac{f(z)}{F(z)} = \frac{d\ell n F(z)}{dz}, \text{ or } \frac{h(z)}{h(R)} \ge \frac{F(z)}{F(R)} \text{ for all}$$

$R \le z \le T$.

Thus, $c_{0R} \ge c_0$ (with equality when $\sigma = 1$), and the result follows (Figure 3).
□

4 Income effects on optimum retirement

Write wages at age z as $\theta w(z)$, where θ, $\theta > 0$, is a constant. A change in θ is a proportionate change in wages at all ages.[14]

---

[14] In Section 7 we study a more general case, where θ is uncertain and occurs sometime in the future.

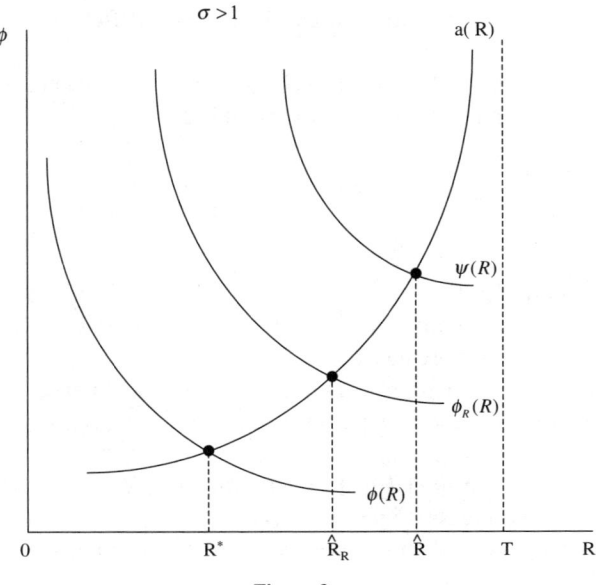

Figure 3

Using (12), (7) and (22), it can be shown that the elasticities of $c^*(R)$, $\hat{c}_R(R)$ with respect to θ and $\hat{c}_R(R)$ are all equal to one. Furthermore, from the definitions of $\phi(R)$, $\psi(R)$, and $\phi_R(R)$, (11), (5) and (20), respectively, it can be shown that

$$\frac{\theta}{\phi}\frac{\partial\phi}{\partial\theta} = \frac{\theta}{\psi}\frac{\partial\psi}{\partial\theta} = \frac{\theta}{\phi_R}\frac{\partial\phi_R}{\partial\theta} = 1 - \sigma \tag{32}$$

for all $0 \le R \le T$. Consequently, R^*, $\hat{R}$ and $\hat{R}_R$ all decrease (increase) when wages increase provided $\sigma > (<)1$. Intuitively, a large *coefficient of risk aversion* implies that the negative income effect due to increased wages (labor being a normal good) outweighs the positive substitution effect (due to a larger opportunity cost to retire), and *vice-versa*.

5 Example

Consider the survival function

$$F(z) = \frac{e^{-\alpha z} - e^{-\alpha T}}{1 - e^{-\alpha T}}, \alpha > 0 \text{ a constant}, 0 \le z \le T \tag{33}$$

Expected lifetime is $\bar{z} = \frac{1}{\alpha}\left(\frac{e^{\alpha T}-1-\alpha T}{e^{\alpha T}-1}\right)$, which is inversely related to α. The simplest case to explore is when $T \to \infty$ (and $\bar{z} \to \frac{1}{\alpha}$).[15]

Let $u(c) = \frac{c^{1-\sigma}}{1-\sigma}$, $\sigma \ne 1$ or $u(c) = \ell n(c)$. Assume that $w(z) = w, 0 \le z \le T$.

---

[15] All subsequent results hold also for finite T.

With continuous annuitization, it can be calculated from (12) that $c^*(R) = w(1 - e^{-\alpha R})$.

With no annuities, by (4) and (7), $\hat{c}(R) = \alpha' Rwe^{-\alpha' R}$, where $\alpha' = \frac{\alpha}{\sigma}$.

With annuitization at retirement, by (18) and (22),

$$\hat{c}_R(z) = \frac{\alpha' Rwe^{-\alpha' z}}{1 + e^{-\alpha' R}\left(\frac{\alpha' - \alpha}{\alpha}\right)}, 0 \leq z \leq R$$

and $\hat{c}_R(z) = \hat{c}_R(R)$ for $R \leq z$.

It can be demonstrated that for $\alpha' < \alpha$ (i.e. $\sigma > 1$), we have $c^*(R) > \hat{c}_R(R) > \hat{c}(R)$ for all $R \geq 0$, implying that $R^* < \hat{R}_R < \hat{R}$.

Notice that when $\alpha' > \alpha$ ($\sigma < 1$), $\hat{c}_R(R) < \hat{c}(R)$ and hence $\hat{R}_R > \hat{R}$. It can also be seen that for small σ it is possible that $\hat{c}(R) > c^*(R)$, which implies that $R^* > \hat{R}$.

When $\alpha' = \alpha$, i.e. $\sigma = 1$ ($u(c) = \ell n c$), then $c^*(R) > \hat{c}_R(R) = \hat{c}(R)$ and $R^* < \hat{R}_R = \hat{R}$. For this case (Figure 4), $\phi(R) = \frac{w}{c^*} = \frac{1}{1 - e^{-\alpha/R}}$ is the inverse of the 'replacement-ratio', i.e., the ratio of post-retirement consumption to pre-retirement income.

$\psi(R) = \phi_R(R) = \frac{1}{\alpha R}e^{\alpha R}$, is downward (upward) sloping for $R < (>)\frac{1}{\alpha}$. Thus, if it is assumed that retirement age is earlier than life expectancy, $\frac{1}{\alpha}$, uniqueness is guaranteed.

Comparative statics also yield expected results. For example, for $\sigma = 1$ the elasticity of optimum retirement with respect to life expectancy is

$$\frac{\alpha}{R^*}\frac{dR^*}{d\alpha} = \frac{\frac{\alpha R^* e^{-\alpha R^*}}{1 - e^{-\alpha R^*}}}{\frac{R^*\alpha'(R^*)}{\alpha(R^*)} + \frac{\alpha R^* e^{-\alpha R^*}}{1 - e^{-\alpha R^*}}} \tag{34}$$

Hence, $-1 \leq \frac{d}{R^*}\frac{dR^*}{d\alpha} \leq 0$.

6 Bequest motive

It was assumed above that individuals have no bequest motive. Consequently, savings are used solely to finance consumption during retirement. With continuous annuitization, since all wealth is in the form of (deferred) annuities, no resources are left behind when individuals die. In the absence of a market for annuities, or when savings are annuitized at retirement and death occurs earlier, some resources held to finance consumption during retirement are left behind when death occurs, their level depending on the age of the deceased. So far, we assumed that individuals' consumption and retirement plans place no value on these uncertain resources left behind.

Suppose now that individuals have a bequest motive, that is, they have a utility, $v(B)$, from assets left behind as bequests $B(\geq 0)$. Consider first the complete insurance case, i.e. continuous annuitization of savings with payments, contingent

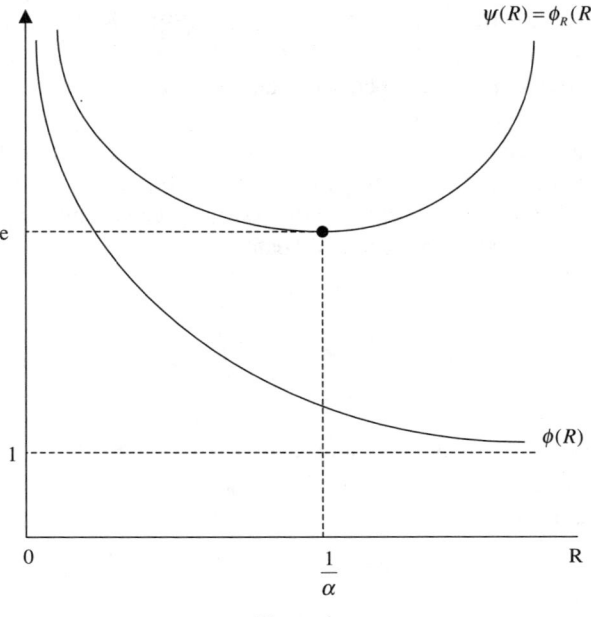

Figure 4

on survival, starting after retirement and, *in addition*, a certain amount provided to beneficiaries upon death.[16]

The zero expected profits condition is

$$B + \int_0^T F(z)u(c(z))dz - \int_0^R F(z)w(z)dz - B_0 = 0 \qquad (35)$$

where B_0 is initial endowment. Maximization of expected utility

$$V = \int_0^T F(z)u(c(z))dz - \int_0^R F(x)a(z) + v(B) \qquad (36)$$

subject to (34) determines optimum consumption, c^*, retirement age, R^*, and bequests B^*, satisfying F.O.C.

$$u'(c^*) - v'(B^*) = 0 \qquad (37)$$

$$a(R^*) - u'(c^*)w(R^*) = 0 \qquad (38)$$

and

$$c^* = \frac{\overline{W}(R^*) + B_0 - B^*}{\overline{z}} \qquad (39)$$

---

[16] This is similar to the '*fixed payments certain*' type of annuities which ensure payments to annuitants' beneficiaries for a fixed period. This annuity, however, does not provide complete insurance if the amount that beneficiaries receive depends on the time of death. For a description of the type of annuities available in the U.S. see Poterba (1997).

where $\overline{W}(R)$ is given by (13). Differentiating (37)-(39), it is seen that consumption and bequests are normal goods, $\frac{dc^*}{dB_0} > 0$ and $0 < \frac{dB^*}{dB_0} < 1$.[17]

Thus, in a dynamic setting, bequests can be viewed as converging to a fixed-point or *steady state*, $B_0 = B^*$.

When savings are annuitized at retirement, the terms of the annuities can again include a certain level of bequests (reflected, correspondingly, with a lower level of payments). If however, an individual dies during the working phase, the level of assets left behind depends on the age of death.

Expected utility is given by

$$V = \int_0^T F(z)u(c(z))dz - \int_0^R F(z)a(z)dz + \int_0^R f(z)v(B(z))dz + F(R)v(B_R)$$

(40)

where B_R is the (certain) level of bequest after retirement.

The level of assets, $B(z)$, during the working phase is given by (2). Assets change is equal to savings,

$$B'(z) = w(z) - c(z), \quad 0 \le z \le R$$

(41)

with $B(0) = B_0$.

The zero expected profits condition is

$$\int_R^T F(z)c(z)dz - F(R)\left[\int_0^R (w(z) - c(z))dz - B_R + B_0\right] = 0.$$

(42)

Maximization of (40) subject to (41) and (42) yields optimum consumption, $\hat{c}_R(z)$, the corresponding path of assets prior to retirement, $\hat{B}(z)$, $0 \le z \le R$, retirement age, $\hat{R}_R$, and bequest after retirement, $\hat{B}_R$. As before, $\hat{c}(z) = \hat{c}(R)$, $R \le z \le T$, i.e. optimum level of consumption after retirement is constant. All the optimum values depend on initial endowment, B_0.

We want to compare continuous annuitization with annuitization at retirement when both are in a steady-state. That is, when $B_0 = B^*$ and $B_0 = \hat{B}_R$, respectively.

Proposition 5. *In a steady-state with a bequest motive,* $R^* < \hat{R}_R$, $\hat{c}_R(R) < c^*$ *and* $\hat{B}_R < B^*$.

Proof. See Appendix A.

The comparison of steady states eliminates, as seen for the budget constraints (35) and (42), *income effects*.

Qualitatively, the previous results are seen to carry-over in the presence of a bequest motive. Of course, without annuitization, lifetime uncertainty makes bequests random at all ages and the appropriate model to analyze optimum consumption and retirement is via the stationary asset distributions generated by bequests.[18]

---

[17] $\frac{dB^*}{dB_0} = \dfrac{u''(c^*)}{u''(c^*)+v''(B^*)\left[\bar{z} - \frac{u''(c^*)w(R^*)^2F(R^*)}{a'(R^*)-u'(c^*)w'(R^*)}\right]}$ and, by second-order conditions, $a'(R^*) - u'(c^*)w'(R^*) > 0$.

[18] We plan to analyze this case in a companion paper.

7 Lifetime and income uncertainty

Suppose that individuals face lifetime uncertainty and are also uncertain about their future income. Income uncertainty may reflect, for example, the possibility of future *disability*. It is assumed that income uncertainty cannot be insured, presumably due to '*Moral-Hazard*'. We model income uncertainty by assuming that beyond a certain age M, prior to retirement, wages become $\theta w(z)$, where θ, $\theta > 0$, is a constant which can assume different values, $0 < \theta < \bar{\theta}$, with probability $p(\theta)$, $0 \leq p(\theta) \leq 1$, i.e. $\int_0^{\bar{\theta}} p(\theta)d\theta = 1$.

Consumption after the realization of θ (at M), and the chosen retirement age, depend on θ. Denote these $c(\theta, z)$ and $R(\theta)$, respectively.

Expected utility is now

$$V = \int_0^M F(z)u(c(z))dz +$$

$$\int_0^{\bar{\theta}} p(\theta) \left[\int_M^T F(z)u(c(\theta, z))dz - \int_0^{R(\theta)} F(z)a(z)dz \right] d\theta \qquad (43)$$

Under continuous annuitization, a zero expected profits condition holds for each θ,

$$\int_0^M F(z)c(z)dz + \int_M^T F(z)c(\theta, z)dz - \int_0^M F(z)w(z)dz -$$

$$\theta \int_M^{R(\theta)} F(z)w(z)dz = 0, \quad 0 \leq \theta \leq \bar{\theta}. \qquad (44)$$

Maximization of (43) subject to (44) yields constant optimum consumption prior to M, c^*, and constant optimum consumption after age M dependent on θ, $c^*(\theta)$. These contingent optimum consumption levels satisfy the condition,

$$u'(c^*) = \int_0^{\bar{\theta}} p(\theta)u'(c^*(\theta))d\theta \qquad (45)$$

Optimum marginal utility of consumption prior to M is equal to *expected* optimum marginal utility of consumption after M.

For each θ, optimum retirement, $R^*(\theta)$, is determined by the condition

$$a(R^*(\theta)) - \phi(\theta, R^*(\theta)) = 0 \qquad (46)$$

where for any R, $\phi(\theta, R) = \theta w(R) u'(c^*(\theta, R))$. The level of $c^*(\theta, R)$ depends on R through (44),

$$c^*(\theta, R) = \frac{\int\limits_0^M F(z)(w(z) - c^*)dz + \theta \int\limits_M^R F(z)w(z)dz}{\int\limits_M^T F(z)dz} \tag{47}$$

To find the effect of θ, differentiate ϕ with respect to θ, given R,

$$\frac{\theta}{\phi} \frac{\partial \phi}{\partial \theta} = 1 - \sigma \varepsilon \tag{48}$$

where, by (44) and (47),

$$\varepsilon = \frac{\theta}{c^*(\theta, R)} \frac{\partial c^*(\theta, R)}{\partial \theta} = \frac{\theta \int\limits_M^R F(z)w(z)dz}{\int\limits_0^M F(z)(w(z) - c^*)dz + \theta \int\limits_M^R F(z)w(z)dz} \tag{49}$$

The effect of θ on ϕ depends on the product of the coefficient of risk aversion, σ, and the elasticity of optimum consumption after M with respect to θ, ε. The value of ε depends on the *timing* of the change in wages, M, relative to R.

By (47), $\varepsilon > 0$ for all $0 \leq M \leq R$, $\varepsilon = 1$ when $M = 0$ and $\varepsilon = 0$ when $M = R$.[19] If annuitized savings prior to M are negative, $\int\limits_0^M F(z)(w(z) - c^*)dz < 0$ (in expectation of high wages afterward), ε may be larger than 1 for some M (Figure 5a).

The effect of θ on ϕ and, by (46), on the optimum R, is seen to depend on the sign of $1 - \sigma \varepsilon$ (Figure 5b). If $\sigma > 1$ for all z, then $\frac{\partial \phi}{\partial \theta} < 0$ for small M and $\frac{\partial \phi}{\partial \theta} > 0$ for M close to R. This implies, by (46), that for small (large) M, an increase in wages decreases (increases) the optimum retirement age. The explanation is that the earlier the wage increase occurs the larger is the income effect, which, with $\sigma > 1$, outweighs the (opposite) substitution effect. When $\sigma < 1$ the substitution effect works in the same direction as the income effect to increase optimum retirement age.

8 Different 'risk classes': full information

Suppose that the population consists of two homogeneous groups, $i = 1, 2$. Individuals in each group, called 'risk-class' by insurance firms, have the same survival function, $F_i(z)$.

---

[19] In (47), c^* also depends on M through (46). Furthermore, as $M \to R$, $\int\limits_0^M F(z)(w(z) - c^*)dz > 0$ since $c^*(\theta, R) \to c^*$. The case $M = 0$ has been discussed in Section 4.

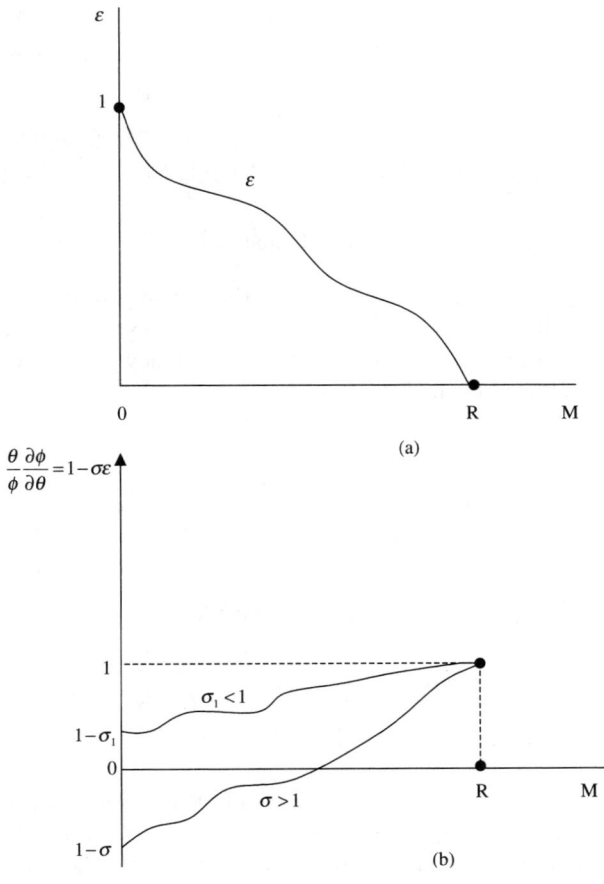

Figure 5a,b

Assume that groups 1's survival function *stochastically dominates*, according to (26), that of group 2. In particular, group 1 has a higher life expectancy.

Assume further that firms can identify annuity purchasers according to the group to which they belong. In a perfectly competitive market, annuities are then priced differently for each group, the analysis in Section 2 applying to each group separately.

We focus on the effect of heterogeneity in survival rates and hence assume that wages are the same for both groups.

Proposition 6. *With full information and continuous annuitization,* $R_1^* > R_2^*$.

Proof. Applying (12) to each group, $c_i^* = c_i^*(R) = \dfrac{\int\limits_0^R F_i(z)w(z)dz}{\int\limits_0^T F_i(z)dz}$, $i = 1, 2$.

By (27),

$$c_1^*(R) = \frac{\int\limits_0^R F_1(z)w(z)dz}{\int\limits_0^T F_1(z)dz} < \frac{\int\limits_0^R F_2(z)w(z)dz}{\int\limits_0^T F_2(z)dz} = c_2^*(R), \tag{50}$$

for all R, $0 < R < T$. It follows from (11) that $R_1^* > R_2^*$. $\square$

Individuals with higher life expectancy partially compensate for longevity by retiring later, but their optimum consumption remains lower throughout.

Without insurance, the effect of higher life expectancy on optimum retirement is indeterminate. Recall from (7) that

$$\hat{c}_i(R) = \frac{\int\limits_0^R w(z)dz}{\int\limits_0^T h_i(z)dz} h_i(R) \quad i = 1, 2 \tag{51}$$

where $h_i(R)$ is defined by (6): $h_i(0) = 1$ and $\frac{h_i'(z)}{h_i(z)} = -\frac{1}{\sigma}\frac{f_i(z)}{F_i(z)}$. It now follows from (27) that

$$\frac{h_1(R)}{\int\limits_0^T h_1(z)dz} \lesseqgtr \frac{h_2(R)}{\int\limits_0^T h_2(z)dz} \quad \text{as } R \lesseqgtr R_c \tag{52}$$

for some R_c, $\theta < R_c < T$. Recall that the benefit from a marginal postponement of retirement, (5), for a group i individual is $\psi_i(R) = w(R)u'(\hat{c}_i(R))$. In view of (51) and (52), if $\hat{R}_1 < R_c$ then $\hat{R}_1 > \hat{R}_2$, while if $\hat{R}_1 > R_c$, then $\hat{R}_1 < \hat{R}_2$.

The explanation of this indeterminacy is as follows. In the absence of insurance and with the same age of retirement, individuals with higher life expectancy consume less at early ages and more at advanced ages, compared to individuals with lower life expectancy. Hence, the benefit from a marginal postponement of retirement is higher for the former group, provided that retirement takes place when their consumption is lower. The opposite holds if retirement occurs when their consumption is higher.

This argument is easily demonstrated with the help of a simple case from Section 5. With $u(c) = \ell nc$, $F_i(z) = e^{-\alpha_i z}$, $i = 1, 2$, and $w(z) = w$, one finds that $\psi_i(R) = \frac{1}{\alpha_i R}e^{\alpha_i R}$. Figure 6 clearly displays the reversal possibility and the critical switch point, R_c, for these functions.

9 Different 'risk classes': pooling equilibrium

Social security systems, pension funds and pension schemes offered by employers, trade-unions and professional organizations offer annuities at prices which are based on the 'pooling' of survival rates of annuity purchasers who belong to different 'risk-classes'. Sometimes the non-discriminatory pricing of annuities is due to a legal

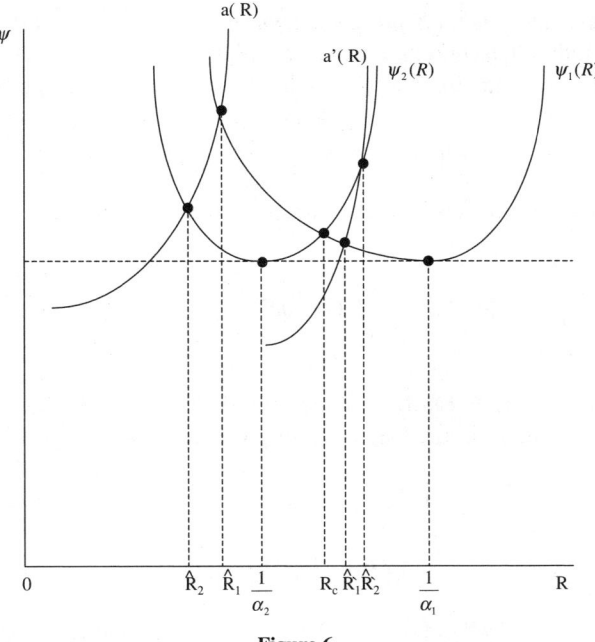

Figure 6

constraint and often it reflects the imperfect information available to insurers on purchasers' risk characteristics. Presumably, the costs of additional information are higher than the benefits from selective pricing.

When a variety of annuities are offered in terms of the time pattern of payment (over age), different 'risk classes' will typically choose different annuities, thereby potentially revealing their type. However, such 'self-selection' is not transparent when individuals purchase annuities from many insurers. Rather than question the feasibility of *pooling equilibria*, we want to explore their implications.

Assume again that the population consists of two groups, such that all individuals who belong to a certain group who have the same survival function. We continue to assume that the survival function of group 1 *stochastically dominates*, according to (26), that of group 2. The relative size of group 1 in the population is $g, 0 < g < 1$.

Annuity prices are based on a uniform age survival function, $G(z)$, which is a weighted average of the two groups' survival functions, $F_1(z)$ and $F_2(z)$. Clearly individuals in group 1 benefit from being pooled with the shorter-lived group 2. In contrast, individuals in group 2 pay higher prices than those based on their own survival function. However, it can be shown that they always prefer to annuitize *some* of their savings (see Appendix B). In particular, with continuous annuitization and the ability to adjust consumption at the pooled rate, group 2 individuals choose to annuitize *all* their savings. When annuities provide a constant flow of payments and consumption may differ from these payments by using non-annuitized resources, group 2 individuals will choose to have a lump-sum available upon retirement in order to supplement consumption over a certain period after retirement (Appendix

B). In any case, there is no *Separating Equilibrium*, i.e. an equilibrium in which *only* group 1 individuals purchase annuities. With everybody purchasing annuities, the pooled survival function has as weights the relative size of the two groups in the population:

$$G(z) = gF_1(z) + (1 - g)F_2(z) \tag{53}$$

Each group's annuities satisfy a zero expected profits condition:

$$\int_0^T G(z)c_i(z)dz - \int_0^{R_i} G(z)w(z)dz = 0, \quad i = 1, 2,. \tag{54}$$

Maximization of expected utility subject to (54) yields F.O.C. for optimum consumption, $\widetilde{c}_i(z)$, and retirement age, $\widetilde{R}_i$, of group i:

$$F_i(z)u'(c_i(z)) - \lambda_i G(z) = 0 \tag{55}$$

$$a(\widetilde{R}_i) - \widetilde{\phi}_i(\widetilde{R}_i) = 0 \tag{56}$$

where $\widetilde{\phi}_i(R) = u'(\widetilde{c}_i(R))w(R)$ for all R, $0 \le R \le T$, and λ_i, $\lambda_i > 0$, constant. As before, write $\widetilde{c}_i(z) = c_{0i}h_i(z)$, $h_i(0) = 1$.

Solving for c_{0i} from (54),

$$c_{0i} = c_{0i}(R) = \frac{\int_0^R G(z)w(z)dz}{\int_0^T G(z)h_i(z)dz} \tag{57}$$

Differentiating (55), $h_1(z)$ and $h_2(z)$ are seen to satisfy:

$$\frac{h_1'(z)}{h_1(z)} = \frac{1-\beta}{\sigma}\left[\frac{f_2(z)}{F_2(z)} - \frac{f_1(z)}{F_1(z)}\right], \tag{58}$$

and

$$\frac{h_2'(z)}{h_2(z)} = \frac{\beta}{\sigma}\left[\frac{f_1(z)}{F_1(z)} - \frac{f_2(z)}{F_2(z)}\right], \tag{59}$$

where $\beta = \frac{gF_1(z)}{G(z)} = \frac{gF_1(z)}{gF_1(z)+(1-g)F_2(z)}, 0 < \beta < 1$.

Since $F_1(z)$ stochastically dominates $F_2(z)$, (58) and (59) show that optimum consumption of group 1 increases, $h_1'(z) > 0$, while that of group 2 decreases, $h_2'(z) < 0, 0 \le z \le T$.

Pooling distorts retirement decisions compared with the *First-Best*. Specifically, optimum retirement ages of the two groups are 'pooled together'.

Proposition 7. *In a Pooling Equilibrium with continuous annuitization, if $\sigma \ge 1$ for all z, $0 \le z \le T$, then $R_1^* > \widetilde{R}_1$ and $R_2^* < \widetilde{R}_2$.*

Proof. By (58), since $h'_1(z) > 0$,

$$\widetilde{c}_1(R) = c_{01} h_1(R) = \frac{\int\limits_0^R G(z)w(z)dz}{\int\limits_0^T G(z)h_1(z)dz} h_1(R) > \frac{\int\limits_0^R G(z)h_1(z)w(z)dz}{\int\limits_0^T G(z)h_1(z)dz} \tag{60}$$

By (58) and (59)

$$\frac{(G(z)h_1(z))'}{G(z)h_1(z)} < -\frac{f_1(z)}{F_1(z)} \tag{61}$$

Inequality (62) implies that $\frac{G(z)h_1(z)}{\int\limits_0^T G(z)h_1(z)dz}$ and $\frac{F_1(z)}{\int\limits_0^T F_1(z)dz}$ satisfy the 'single-crossing' condition (the latter crossing 'from below') and hence, by (27),

$$\int\limits_0^R \left[\frac{G(z)h_1(z)}{\int\limits_0^T G(z)h_1(z)dz} - \frac{F_1(z)}{\int\limits_0^T F_1(z)dz} \right] dz > 0 \tag{62}$$

for all R, $0 < R < T$. In view of (60) and (63),

$$\widetilde{c}_1(R) > \frac{\int\limits_0^R G(z)h_1(z)w(z)dz}{\int\limits_0^T G(z)h_1(z)dz} > \frac{\int\limits_0^R F_1(z)w(z)dz}{\int\limits_0^T F_1(z)dz} = c_1^*(R) \tag{63}$$

This inequality and conditions (56) and (11) imply that $R_1^* > \widetilde{R}_1$. A similar argument can be used to show that $R_2^* < \widetilde{R}_2$. □

Note that it has not been shown whether $\widetilde{R}_1$ is larger or smaller than $\widetilde{R}_2$. Indeed, observe that, by (58) and (59), $\frac{(G(z)h_1(z))'}{G(z)h_1(z)} > \frac{(G(z)h_2(z))'}{G(z)h_2(z)}$ implying that $\frac{G(z)h_i(z)}{\int\limits_0^T G(z)h_i(z)dz}$, $i = 1, 2$, satisfy the 'single-crossing' condition. Hence, there exists an age R_c, $0 < R_c < T$, such that

$$\frac{G(R)h_1(R)}{\int\limits_0^T G(z)h_1(z)dz} \lesseqgtr \frac{G(R)h_2(R)}{\int\limits_0^T G(z)h_2(z)dz}, \text{ as } R \lesseqgtr R_c \tag{64}$$

Since $\widetilde{c}_i(R) = \frac{\int\limits_0^R G(z)w(z)dz}{\int\limits_0^T G(z)h_i(z)dz} h_i(R)$, it follows from (50) that $\widetilde{c}_1(R) \lesseqgtr \widetilde{c}_2(R)$ and $\widetilde{\phi}_1(R) \gtreqless \widetilde{\phi}_2(R)$, as $R \lesseqgtr R_c$. Thus, if $\widetilde{R}_1 < R_c$ then $\widetilde{R}_1 > \widetilde{R}_2$, while if $\widetilde{R}_1 > R_c$ then $\widetilde{R}_1 < \widetilde{R}_2$.

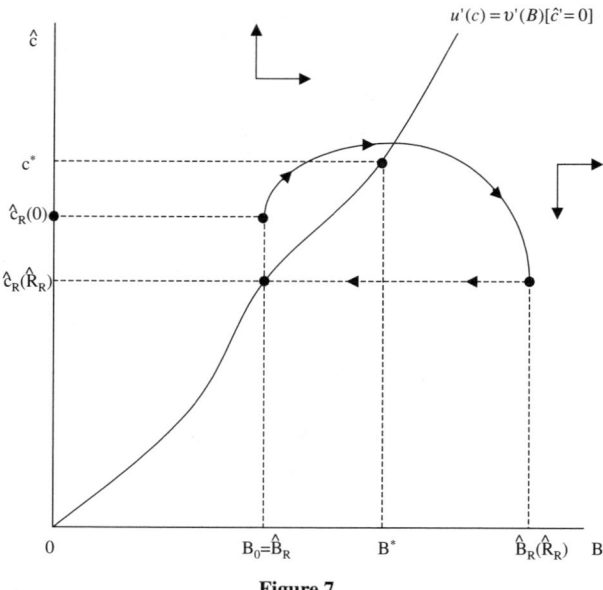

Figure 7

Appendix A

Maximization of (40) subject to (42) with respect to $c(z)$, $R \leq z \leq T$, R and B_R yields F.O.C. for the optimum, $(\hat{c}(z), \hat{R})$:

$$u'(\hat{c}_R(z)) - v'(\hat{B}_R) = 0, \quad R \leq z \leq T \tag{A.1}$$

$$a(\hat{R}_R) - \phi_R(\hat{R}_R) = 0 \tag{A.2}$$

where $\phi_R(R) = u'(\hat{c}_R(R))w(R)$, for all R, $0 \leq R \leq T$. It is seen from (A.1) that consumption after retirement is constant, $\hat{c}_R(z) = \hat{c}_R(R)$, $R \leq z \leq T$.

The (Hamiltonian) dynamic optimization condition with respect to $c(z)$, $0 \leq z \leq R$, is

$$\frac{\hat{c}'_R(z)}{\hat{c}_R(z)} = \frac{1}{\sigma} \frac{f(z)}{F(z)} \left[\frac{v'(\hat{B}_R(z)) - u'(\hat{c}_R(z))}{u'(\hat{c}_R(z))} \right] \tag{A.3}$$

Note that equation (6) in the paper is a special case of (A.3) when $v'(B) = 0$.

The two differential equations (A.3) and (41) in the paper, with initial condition $\hat{B}_R(0) = B_0$ and the end-condition $\hat{c}_R(R)$ satisfying (A.1) and (A.2), determine the optimum path $(\hat{c}_R(z), \hat{B}_R(z))$, $0 \leq z \leq R$.

The budget constraint, (42) in the paper, and (A.2) are used jointly to solve for $\hat{R}_R$ and $\hat{c}_R(\hat{R}_R)$. Figure 7 displays the phase diagram for the optimum path. A higher R corresponds to a higher $\hat{c}_R(0)$. The path drawn in the Figure is for $\hat{R}_R$ satisfying (A.2), with $\hat{B}_R = B_0$, i.e. *steady state*.

Unlike the model with no bequest motive, optimum consumption first increases and then decreases with age. The reason is that the marginal utility of bequests initially exceeds the marginal utility of consumption, leading to accelerated accumulation of assets and slowly increasing consumption. As assets increase, eventually this relation is reversed. At point A (at age $\hat{R}_R$), assets have accumulated to $\hat{B}_R(\hat{R})$, where $\hat{B}_R(\hat{R}_R) = \int_0^R (w(z) - \hat{c}_R(z))dz + \hat{B}_R$. At this point, cumulative savings $\int_0^{\hat{R}_R} (w(z) - \hat{e}_R(z))dz > 0$ (i.e. $\hat{B}_R(\hat{R}_R) > \hat{B}_R$) are annuitized to a flow of $\hat{c}_R(\hat{R}_R)$, reducing assets to $\hat{B}_R$.

The proof that $\hat{c}_R(\hat{R}_R) < c^*$ (and hence, by (A.1), $\hat{B}_R < B^*$), is the same as the proof of Proposition 3 in the paper. The assumption that $B_0 = \hat{B}_R$, i.e. *steady-state*, is needed to establish that $\hat{c}_R(z) > \hat{c}_R(\hat{R}_R)$ for all z, $0 \leq z \leq \hat{R}_R$, which is used in that proof. It may not hold for an arbitrary $B_0 < \hat{B}_R$.

Appendix B

Groups 2's survival function, $F_2(z)$, is stochastically dominated by that of group 1, $F_1(z)$. First we want to show that when annuities are priced at the pooled survival function, $G(z)$, and the payment of benefits can be adjusted optimally, group 2 individuals annuitize *all* their savings.

Let $e_2(z)$ be consumption at age z, $R \leq z \leq T$, from *non-annuitized* resources. The zero-expected profits condition is

$$\int_0^T G(z)c(z)dz + G(R) \int_{R_2}^T e_2(z)dz - \int_0^{R_2} G(z)w(z)dz = 0 \qquad \text{(B.1)}$$

Maximization of expected utility

$$V_2 = \int_0^{R_2} F_2(z)u(c_2(z))dz + \int_{R_2}^T F_2(z)u(c_2(z)+e_2(z))dz - \int_0^{R_2} F_2(z)a(z)dz \quad \text{(B.2)}$$

subject to (B.1) yields F.O.C. with respect to $c(z)$, $R \leq z \leq T$,

$$F_2(z)u'(c_2(z) + e(z)) - \lambda_2 G(z) = 0 \qquad \text{(B.3)}$$

where λ_2, $\lambda_2 > 0$, is a constant. Hence, for any z, $z > R_2$,

$$\frac{\partial V_2}{\partial e_2(z)} = \lambda_2[G(z) - G(R_2)] < 0, \qquad \text{(B.4)}$$

which means that, at the optimum, *all* savings are annuitized.

Suppose now that annuities provide a constant flow of payments, denoted by b. The zero expected profits condition is now

$$b_2 \int_{R_2}^T G(z)dz + G(R_2) \int_{R_2}^T e_2(z)dz - \int_0^{R_2} G(z)(w(z) - c_2(z))dz = 0 \qquad \text{(B.5)}$$

and expected utility is

$$V_2 = \int\limits_{0}^{R_2} F_2(z)u(c_2(z))dz + \int\limits_{R_2}^{T} F_2(z)u(b_2 + e_2(z))dz - \int\limits_{0}^{R_2} F_2(z)a(z)dz = 0$$

(B.6)

Maximization of V_2 subject to (B.5) with respect to $e_2(z)$ and b_2 yields F.O.C.

$$F_2(z)u'(b_2 + e_2(z)) - \lambda_2 G(R_2) = 0, \quad R \le z \le T \qquad (B.7)$$

and

$$\int\limits_{R_2}^{T} F_2(z)u'(b_2 + e_2(z))dz - \lambda_2 \int\limits_{R_2}^{T} G(z)dz = 0 \qquad (B.8)$$

Suppose that at the optimum, $e(z) = 0$ for all $R_2 \le z \le T$. Then, from (B.6) and (B.7),

$$\frac{\partial V_2}{\partial e_2}\Big|_{e_2=0} = \lambda_2 \left[\frac{F_2(R_2)}{\int\limits_{R_2}^{T} F_2(z)dz} - \frac{G(R_2)}{\int\limits_{R_2}^{T} G(z)dz} \right] \int\limits_{R_2}^{T} G(z)dz \qquad (B.9)$$

which, by (53) and (26) in the paper, is positive. Hence, $e_2(z) > 0$ for some z, $R \le z \le T$. Next, we want to show that $b_2 > 0$ at the optimum. Suppose $b_2 = 0$ and hence (B.7) holds for *all* z, $R_2 \le z \le T$. Using (B.8),

$$\frac{\partial V}{\partial b_2}\Big|_{b_2=0} = \lambda_2 \left[G(R_2)(T - R_2) - \int\limits_{R_2}^{T} G(z)dz \right] \qquad (B.10)$$

which is positive. Hence, at the optimum $b_2 > 0$.

References

Abel, A. B.: Capital accumulation and uncertain lifetimes with adverse selection. Econometrica **54**, 1079–1097 (1986)

Brugiavini, A.: Uncertainty resolution and the timing of annuity purchases. Journal of Public Economics **50**, 31–62 (1993)

Diamond, P., Valdez-Prieto, S.: Social security reforms. In: Bosworth, B., Dornbush, R., Labàn, R. (eds.) The Chilean economy: policy lessons and challenges. Brookings Institution (1994)

Eckstein, Z., Eichenbaum, M., Peled, D.: Uncertain lifetimes and the welfare enhancing properties of annuity markets and social security. Journal of Public Economics **26**, 303–326 (1985)

Friedman, B. M., Warshawsky, M.: The cost of annuities: implications for saving behavior and bequest. Quarterly Journal of Economics **420**, 135–154 (1990)

Merton, R. C.: On the role of social security as a means for efficient risk-bearing in an economy where human capital is not tradeable. In: Bodie, Z., Shoven, J. (eds.) Financial aspects of the United States pension system, ch. 12, pp. 325–358. Chicago: University of Chicago Press (1981)

Mitchell, O. S., Poterba, J. R., Warshawsky, M. J., Brown, J. R.: New evidence on money's worth of individual annuities. American Economic Review **89**, 1299–1318 (1999)

Poterba, J. R.: The history of annuities in the United States. N.B.E.R. Working Paper No. 6001 (1997); also in: Brown, J., Mitchell, O., Poterba, J., Warshawsky, M.: The role of annuity markets in financing retirement, ch. 2. Cambridge, MA London: MIT Press 2001

Rothschild, M., Stiglitz, J.: Equilibrium in competitive insurance markets: an essay on the economics of incomplete information. Quarterly Journal of Economics **90**, 624–649 (1976)

Sheshinski, E., Weiss, Y.: Uncertainty and optimal social security systems. Quarterly Journal of Economics **95**, 189–206 (1981)

Walliser, J.: Understanding adverse selection in the annuities market and the impact of privatizing social security. Congressional Budget Office 1998

Yaari, M. E.: Uncertain lifetime, life insurance and the theory of the consumer. Review of Economic Studies **32**, 137–150 (1965)

Claims problems and weighted generalizations of the Talmud rule*

Toru Hokari[1] and William Thomson[2]

[1] Institute of Social Sciences, University of Tsukuba, 1-1-1 Ten'no-dai,
   Tsukuba, Ibaraki 306-8571, JAPAN (e-mail: hokari@social.tsukuba.ac.jp)
[2] Department of Economics, University of Rochester, Rochester, NY 14627, USA
   (e-mail: wth2@troi.cc.rochester.edu)

Received: April 8, 2002; revised version: June 26, 2002

Summary. We investigate the existence of consistent rules for the resolution of conflicting claims that generalize the Talmud rule but do not necessarily satisfy equal treatment of equal. The first approach we follow starts from the description of the Talmud rule in the two-claimant case as "concede-and-divide", and an axiomatic characterization for the rule. When equal treatment of equals is dropped, we obtain a one-parameter family, "weighted concede-and-divide rules". The second approach starts from the description of the Talmud rule as a hybrid of the constrained equal awards and constrained equal losses rules, and weighted generalizations of these rules. We characterize the class of consistent rules that coincide with weighted concede-and-divide rules in the two-claimant case or with weighted hybrid rules. They are defined by partitioning the set of potential claimants into "priority classes" or "half-priority classes" respectively, and selecting reference weights for all potential claimants. For the first approach however, and in each class with more than two claimants, equal treatment is actually required

Keywords and Phrases: Claims problems, Weighted generalizations of Talmud rule, Consistency, Converse consistency.

JEL Classification Numbers: C71, D71.

1 Introduction

When a firm goes bankrupt, how should its liquidation value be divided among its creditors? This problem is an example of a general class that we call "claims

---

* Thomson acknowledges support from NSF under grant SBR-9731431. We thank Jean-Pierre Benoît for his comments, and the referee for several useful suggestions.

Correspondence to: W. Thomson

problems". In general, the question is how to allocate a resource among agents having incompatible claims on it, the goal being to identify well-behaved methods of making a recommendation for each problem. We call such methods division "rules".[1] In the standard specification of a claims problem, claimants only differ in their claims, and the requirement is imposed on rules that two agents with equal claims should receive equal amounts. This is the property of "equal treatment of equals". However, in applications, in addition to their claims, agents may have rights, needs, obligations and so on, that could, or should, be taken into account when performing the division. Then, two agents with equal claims need not receive equal amounts.[2] In fact, fairness requires that this be the case.

The possibility of treating differently two agents with equal claims can be formally accommodated in two ways. First, claimants can be sorted into priority classes, precedence being given to each class before any consideration is given to any lower class. This is actual practice. An extreme case is when a strict order is specified on the set of claimants and they are compensated one after the other until money runs out. We call such a rule a "priority rule". Alternatively, and somewhat less radically, we can choose "weights" reflecting the relative importance to be given to each claimant. Using weights so as to gear the social choice towards agents who are perceived as more deserving is standard in cooperative game theory and in various branches of the theory of fair allocation. (How to use these weights in the present context is made explicit later on.) Finally, rules can be defined by mixing both procedures, partitioning the set of claimants into priority classes and within each class, assigning them weights.

Most of the rules that are central to the literature can be redefined in this way. This is in particular the case for two important rules that appear in Medieval writings, the constrained equal awards and constrained equal losses rules. The former assigns equal amounts to all claimants subject to no one receiving more than his claim; the latter assigns amounts so that all claimants experience equal losses subject to no one receiving a negative amount. However, a rule whose introduction has been very importantly responsible for the considerable development of the axiomatic literature on the subject has not: it is the rule defined by Aumann and Maschler (1985) to rationalize the resolutions proposed in the Talmud for certain numerical examples described there, the "Talmud rule". Our objective is to investigate the existence of generalizations of the rule that would allow recognizing ways in which claimants may differ besides their claims.

We follow two strategies. To explain them, we recall two important facts concerning the Talmud rule. First, in the two-claimant case, the rule assigns to each claimant the sum of (i) the difference between the amount to divide and the claim of the other claimant (or 0 if this difference is negative), and (ii) half of the remainder. We call this method "concede-and-divide" because one of its justifications is the following natural and simple scenario: the difference in (i) is interpreted as the amount conceded to each agent by the other, and in a "first-round" it certainly makes sense to award him that amount; the dispute is really about the remainder,

---

[1] For a survey of the literature devoted to this subject, see Thomson (1995).

[2] For another study of a model of claims resolution that dispenses with the equal treatment assumption, see Moulin (2000). We discuss this contribution in the concluding section.

and (ii) says that in a "second-round", this remainder should be divided equally (Aumann and Maschler, 1985, give arguments in favor of equal division in the second round). Seen from a different perspective, the difference in (i) can be described as a "minimal" amount to which the agent is entitled, his "minimal right". A second important fact concerning the Talmud rule is that it is "consistent": when some agents leave with their awards and the situation is re-evaluated, the rule makes the same recommendation for the remaining agents as before. This property has recently played an important role in the axiomatic study of a variety of classes of problems. A central justification for the Talmud rule is that, although many ways of looking at the problem coincide with concede-and-divide in the two-claimant case, the rule is the unique consistent one to do so (Aumann and Maschler, 1985; Benoît, 1997).

Our first strategy starts from the observation just made concerning the coincidence of many rules with concede-and-divide in the two-claimant case. Here are important examples. For the random arrival rule, imagine claimants arriving one at a time, and fully compensate them until money runs out; then, take the average of the awards vectors so obtained under the assumption that all orders of arrival are equally likely.[3] For the minimal overlap rule, imagine that each agent claims a specific part of the amount to divide; position these claims so as to maximize in the lexicographic maximin order the part claimed by exactly one claimant, then the part claimed by exactly two claimants, and so on; finally, divide each part equally among all agents claiming it.[4] Two other rules that coincide with concede-and-divide in the two-claimant case are the version of the constrained equal awards rule obtained by first assigning to each claimant his minimal right and then applying the rule to divide the remainder, and the version of the constrained equal losses rule obtained by first truncating claims by the amount available. For each of the rules just enumerated, a natural asymmetric version of the scenario that underlies it can be defined. It turns out that all the rules so obtained coincide. We call them "weighted" concede-and-divide rules.

For the passage from two claimants to larger populations, we appeal to consistency. Our first main result is a characterization of the consistent rules extending the weighted concede-and-divide rules. They are defined as follows: an ordered partition of the set of potential claimants into "priority classes" is defined, a "reference partition", and a positive weight is chosen for each potential claimant belonging to a two-claimant class. For each specific problem, the reference partition induces an ordered partition of the set of claimants who are present; for a component of this partition that coincides with a two-claimant reference priority class, apply the weighted concede-and-divide rule relative to the weights assigned to its two members; otherwise, apply the Talmud rule itself.

For example, take as reference partition $\{\ldots, 11, 9, 7, 5, 3, 1, 0\}$, $\{2, 4, 6\}$, $\{8, 10\}$, $\{12, 14, 16\}$, $\{18, \ldots\}$, with the left-to-right direction corresponding to lower and lower priorities. For a problem with claimant set $\{7, 5, 3, 1, 0, 2, 6, 8, 10, 12, 18, 20\}$, the induced partition is $\{7, 5, 3, 1, 0\}$, $\{2, 6\}$, $\{8, 10\}$, $\{12\}$, $\{18, 20\}$.

---

[3] This rule is based on the scenario underlying the Shapley value (Shapley, 1953).

[4] This rule is defined by O'Neill (1982).

For the component $\{7, 5, 3, 1, 0\}$, apply the Talmud rule because this component is induced from a reference class that has more than two claimants. The next component, $\{2, 6\}$, has two members, but apply the Talmud rule to it as well because this component is induced from a reference class that has more than two claimants. The next component, $\{8, 10\}$, coincides with a two-claimant reference class, so a weighted concede-and-divide rule can now be used. The next component consists of a single agent, and all rules coincide in that trivial case. The final component is $\{18, 20\}$, and since it is induced from a reference class that has more than two claimants, return to the Talmud rule itself.

Our second strategy takes as point of departure an alternative description of the Talmud rule as a hybrid of the constrained equal awards and constrained equal losses rules. Since both have straightforward "weighted" generalizations to the sort of situations we have in mind here, we use those as ingredients in constructing generalizations of the Talmud rule. Interestingly, in the two-claimant case, the resulting "weighted hybrids" rules are not the same as the weighted concede-and-divide rules. We then ask about their consistency and show that when the weights are all positive, and if they are chosen in a "consistent" manner across populations, consistency holds. This conclusion parallels the aforementioned characterization of Aumann and Maschler's. Moreover, the limit case where some of the weights are 0 is also allowed. The choice of weighted hybrid rules or of their limits in the two-claimant case, and consistency, give us weighted hybrid rules in general: a "reference ordered partition" of the set of potential claimants is chosen and each potential claimant is assigned a positive "reference weight". This time and for a reason to be explained shortly, we refer to its components as "half-priority classes". For each specific problem, the reference partition induces an ordered partition on the set of claimants actually present. The awards vector chosen by the rule can then be described as a function of the amount available as follows: divide the first units among the members of the first component of the induced partition by applying the weighted hybrid rule with weights proportional to the weights of its members, and do so until the amount to divide is equal to the half-sum of their claims. Then, turn to the second component of the induced partition and divide the next units among its members by applying the weighted hybrid rule with weights proportional to their weights. Proceed in this way until each claimant has received his half-claim. At that point, return to the first component of the induced partition and divide additional units among its members by picking up the application of the first weighted hybrid rule where it was interrupted, and do so until all of its members are fully compensated. Then, return to the second component of the induced partition and to the second weighted hybrid rule, and divide additional units until its members are fully compensated, and so on.

The need to select the half-claims vectors when the amount to divide is equal to the half-sum of the claims limits the extent to which some claimants can be given precedence over others. Indeed, the weighted hybrid rules have the following feature: focusing on the two-claimant case, as it allows for a simple statement, it is that, as we increase the parameter describing the extent to which a claimant is favored at the expense of the other, the weighted hybrid rule does not get closer to the priority rule in which the favored agent is first. Our first strategy provides this limit behavior, which one may consider desirable, but at the price of restricting the choice of weights, that is, the flexibility with which one can favor certain claimants at the expense of others. This flexibility is present only for two-claimant priority classes.

We will leave it to the reader to decide which of the two classes of rules we identify is more suitable. In applications, it certainly will be a matter of circumstances how much of an asymmetric treatment of claimants is needed. It is clear however that in generalizing the Talmud rule so as to accommodate the need to favor particular claimants, consistency places constraints that do not exist for other rules, and that rarely exist in other contexts. We say rarely because we know of at least one other model, for which a similar phenomenon occurs (Section 3). Another revealing study in this regard is discussed in the concluding section (Moulin, 2000). There, we also show how to incorporate exogenously given information about priority classes and weights, and we present an alternative to consistency that accommodates an asymmetric treatment of claimants (Dagan and Volij, 1997).

2 The problem of resolving conflicting claims

There is an infinite set of "potential" claimants, indexed by the natural numbers $\mathbb{N}$. Each given problem, however, only involves a finite number of them. Let $\mathcal{N}$ denote the class of non-empty finite subsets of $\mathbb{N}$. A *claims problem* is a pair $(c, E) \in \mathbb{R}_+^N \times \mathbb{R}_+$, where $N \in \mathcal{N}$, such that $\sum_N c_i \geq E$. Let $\mathcal{C}^N$ denote the class of all problems with claimant set N. A division *rule* is a function defined on $\cup_{N \in \mathcal{N}} \mathcal{C}^N$ that associates with each $N \in \mathcal{N}$ and each $(c, E) \in \mathcal{C}^N$ an *awards vector of* (c, E), namely a vector $x \in \mathbb{R}^N$ such that $0 \leqq x \leqq c$ and satisfying the efficiency condition $\sum_N x_i = E$.[5]

Let R be a rule. For each $N \in \mathcal{N}$ and each $c \in \mathbb{R}_+^N$, the *path of awards of R for c* is the locus of the awards vector R selects as E ranges from 0 to $\sum_N c_i$:
$$x \in P^R(N, c) \Longleftrightarrow \text{there exists } E \in [0, \sum_N c_i] \text{ such that } x = R(c, E).$$

Here are important rules. For the constrained equal awards rule, awards are equal subject to no-one receiving more than his claim. For the constrained equal losses rule, at the chosen awards vector, the losses agents experience are equal subject to no-one receiving a negative amount.

Constrained equal awards rule, CEA. For each $N \in \mathcal{N}$, each $(c, E) \in \mathcal{C}^N$, and each $i \in N$, $CEA_i(c, E) \equiv \min\{c_i, \lambda\}$, where $\lambda \in \mathbb{R}_+$ is chosen so as to achieve efficiency.

Constrained equal losses rule, CEL. For each $N \in \mathcal{N}$, each $(c, E) \in \mathcal{C}^N$, and each $i \in N$, $CEL_i(c, E) \equiv \max\{c_i - \lambda, 0\}$, where $\lambda \in \mathbb{R}_+$ is chosen so as to achieve efficiency.

The Talmud rule (Aumann and Maschler, 1985) can be described in several ways. Its description as a hybrid of the constrained equal awards and constrained equal losses rules is our point of departure in our second attempt at defining generalizations of it that accommodate a desired preferential treatment of some agents.

---

[5] The notation $x \leqq y$ means that for each $i \in N$, $x_i \leq y_i$.

Although this description has almost always been adopted in the literature, an important lesson of our work is that which description is used considerably influences the way in which one goes about searching for such generalizations.

Talmud rule, T. For each $N \in \mathcal{N}$, each $(c, E) \in \mathcal{C}^N$, and each $i \in N, T_i(c, E) \equiv$ $\min\{\frac{c_i}{2}, \lambda\}$ if $\sum \frac{c_j}{2} \geq E$, and $T_i(c, E) \equiv \max\{\frac{c_i}{2}, c_i - \lambda\}$ otherwise, where in each case, $\lambda \in \mathbb{R}_+$ is chosen so as to achieve efficiency.

The following requirement on a rule is central to our analysis: starting from some problem, and having applied the rule, we imagine some of the claimants leaving with their awards, and we re-evaluate the situation from the viewpoint of the remaining ones. We ask that for this "reduced problem", the rule should assign to them the same awards as it did initially.[6]

Consistency. For each $N \in \mathcal{N}$, each $(c, E) \in \mathcal{C}^N$, and each $N' \subset N$, if $x \equiv$ $R(c, E)$, then $x_{N'} = R(c_{N'}, E - \sum_{N \backslash N'} x_i)$.[7]

Bilateral consistency is obtained by adding the restriction $|N'| = 2$. In deriving necessary conditions for our main characterizations, we only use this weaker property, but the rules we end up with are *consistent*.

3 A first approach

We begin with a discussion of the case of two claimants. In order to generalize a given rule so as to favor one of them, we introduce a parameter representing the extent of this desired bias. The parameter is specified by applying the rule to a problem in which claims are equal. The choice made then reflects society's preferences in treating the two claimants. In principle, the parameter could depend on the common value of their claims and on the amount to divide. We limit ourselves to rules for which asymmetry is "uniform", as they can be described by means of a single parameter in the one-dimensional simplex.

We observed earlier that in the two-claimant case, the Talmud rule coincides with a number of other important rules. We listed concede-and-divide, the random arrival and minimal overlap rules, the rule obtained from the constrained equal awards rule by first assigning to each agent his minimal right, and the rule obtained from the constrained equal losses rule by first truncating any claim greater than the amount to divide by that amount. Our first approach starts from asymmetric versions of the scenarios underlying these rules. In spite of their great diversity, they all result in the same family of rules. We then give an axiomatic justification for this family. Finally, we extend its members to arbitrary populations by means of consistency.

The first part of the program just outlined simply consists in reconsidering the various scenarios that we have seen lead to rules that all coincide in the two-claimant case, but this time allowing for an asymmetric treatment of the two claimants.

---

[6] For a survey, see Thomson (1996).

[7] Since rules are such that for each $i \in N$, $x_i \in [0, c_i]$, then the sum of the claims of the remaining claimants is still greater than the amount that the remainder, so that the reduced problem is well-defined.

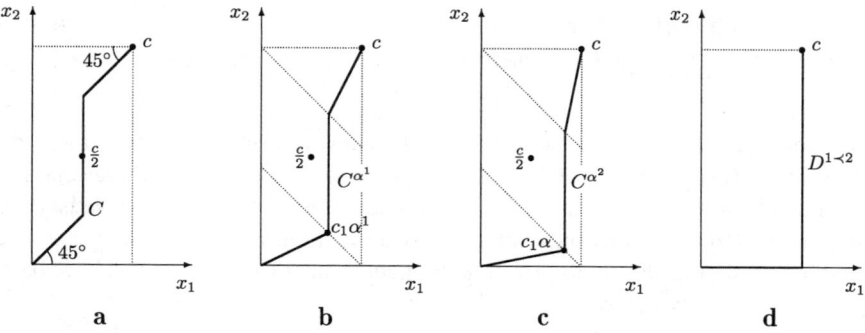

Figure 1a–d. Weighted versions of concede-and-divide. Starting from concede-and-divide, we give progressively more weight to agent 1 than to agent 2. The kinks in the paths of awards still occur when $E = c_1$ or $E = c_2$. **a** Concede-and-divide. **b** and **c**: Weighted versions of weights $\alpha^1 \equiv (\frac{2}{3}, \frac{1}{3})$ and $\alpha^2 \equiv (\frac{5}{6}, \frac{1}{6})$. **d** At the limit, we obtain the priority rule relative to the ordering $1 \prec 2$, $D^{1 \prec 2}$.

For "concede-and-divide", the concession step is unchanged but the remainder of the second step is divided proportionally to the weights assigned to the two claimants. For the random arrival rule, we place a greater weight on the order for which one of them is first. For the minimal overlap rule, when two agents claim the same part of the amount available, we perform the division proportionally to their weights. The constrained equal awards rule operated from minimal rights and the constrained equal losses rule operated from truncated claims can be similarly redefined. Remarkably, all of these generalizations coincide with the rules defined as follows. Given pair $\{i, j\} \in \mathcal{N}$, let $\Delta^{\{i,j\}}$ denote the unit simplex in $\mathbb{R}_+^{\{i,j\}}$ and $int \Delta^{\{i,j\}}$ its relative interior. Let $\alpha \equiv (\alpha_{ij}, \alpha_{ji}) \in \Delta^{\{i,j\}}$:

Weighted concede-and-divide rule of weights $\alpha \in \Delta^{\{i,j\}}, C^\alpha$. For each $(c, E) \in \mathcal{C}^{\{i,j\}}$,

$$
\begin{cases}
C_i^\alpha(c, E) \equiv \max\{E - c_j, 0\} + \alpha_{ij}\big[E - \max\{E - c_j, 0\} - \max\{E - c_i, 0\}\big], \\
C_j^\alpha(c, E) \equiv \max\{E - c_i, 0\} + \alpha_{ji}\big[E - \max\{E - c_j, 0\} - \max\{E - c_i, 0\}\big].
\end{cases}
$$

Next, we enquire about a possible axiomatic justification for these rules. A natural starting point is the characterization of concede-and-divide itself offered by Dagan (1996). It involves three requirements. First, two claimants with equal claims should receive equal amounts. Second, one should be able to ignore any part of a claim that is over the amount to divide. Third, one should be able to solve each problem in either one of the following two ways: (i) directly, or (ii) by first assigning to each claimant his minimal right, decreasing claims by the amounts received in this first round, and then applying the rule to divide what is left.

Equal treatment of equals. For each $(c, E) \in \mathcal{C}^N$ and each pair $\{i, j\} \subseteq N$, if $c_i = c_j$, then $R_i(c, E) = R_j(c, E)$.

Invariance under claims truncation. For each $(c, E) \in \mathcal{C}^N$, $R(c, E) = R(t(c, E), E)$, where for each $i \in N$, $t_i(c, E) \equiv \min\{c_i, E\}$.

Minimal rights first. For each $(c, E) \in \mathcal{C}^N$, $R(c, E) = m(c, E) + R(c - m(c, E), E - \sum m_i(c, E))$, where for each $i \in N$, $m_i(c, E) \equiv \max\{E - \sum_{j \in N \setminus \{i\}} c_j, 0\}$.[8]

Concede-and-divide is the only two-claimant rule satisfying these three requirements (Dagan, 1996). If *equal treatment of equals* is dropped, a large class of additional rules become admissible, but a small subclass of rules also satisfy the requirement that if claims and amount to divide are multiplied by the same positive number, so should all awards. This requirement is met by most rules that have been discussed in the literature.

Homogeneity. For each $(c, E) \in \mathcal{C}^N$ and each $t > 0$, $R(tc, tE) = tR(c, E)$.

Our characterization, whose proof does not depart much from Dagan's own argument, is the following:

Proposition 1. *For $|N| = 2$, the weighted concede-and-divide rules are the only rules on $\mathcal{C}^N$ satisfying* homogeneity, invariance under claims truncation, *and* minimal rights first.

Proof. We omit the proof that the weighted concede-and-divide satisfy the three properties. Conversely, let $N \equiv \{i, j\}$, R be a rule satisfying these properties, and $(c, E) \in \mathcal{C}^N$. Without loss of generality, suppose $c_i \leq c_j$.

Case 1: $E \leq c_i$. Then, $\min\{c_i, E\} = \min\{c_j, E\} = E$. By *invariance under claims truncation*, $R(c, E) = R(E, E; E)$. Let $\alpha \equiv R(E, E; E) / \|R(E, E; E)\| = R(E, E; E) / E$. Let $E' \leq c_i$. By *invariance under claims truncation*, $R(c, E') = R(E', E'; E')$ and by *homogeneity*, $R(c, E') = E'\alpha$. Altogether, we conclude that for each $E \leq c_i$, $R(c, E) = C^\alpha(c, E)$.

Case 2: $c_i < E \leq c_j$. Then, minimal rights are $\max\{E - c_j, 0\} = 0$ and $\max\{E - c_i, 0\} = E - c_i$. Claims revised down by the minimal rights are c_i and $c_j - (E - c_i)$. There remains $E - (E - c_i) = c_i$ to divide. After truncation, claims are $\min\{c_i, c_i\} = c_i$ and $\min\{c_j - (E - c_i), c_i\} = c_i$. Since, by Case 1, $R(c_i, c_i; c_i) = c_i \alpha$, we obtain, by *minimal rights first*, $R(c, E) = (0, E - c_i) + c_i \alpha = C^\alpha(c, E)$.

Case 3: $c_j < E$. Then, minimal rights are $E - c_j$ and $E - c_i$. Claims revised down by the minimal rights are $c_i - (E - c_j)$ and $c_j - (E - c_i)$, which are equal. There remains $E - (E - c_j) - (E - c_i) = c_i + c_j - E$ to divide. We have already calculated that $R(c_i + c_j - E, c_i + c_j - E, c_i + c_j - E) = (c_i + c_j - E)\alpha$. By *minimal rights first*, $R(c, E) = (E - c_i, E - c_j) + (c_i + c_j - E)\alpha = C^\alpha(c, E)$.
$\square$

Equipped with both a definition of a family of rules that constitute attractive weighted generalizations of the two-claimant version of the Talmud rule as well as with an axiomatic justification for them, we now turn to general populations. We appeal to *consistency*. Our next result is a characterization of the class of rules satisfying all our requirements. First, define an ordered partition of the set

---

[8] Note that $(c - m(c, E), E - \sum m_i(c, E))$ is a well-defined problem.

of potential claimants into priority classes. Also, assign a positive weight to each potential claimant, with the weights assigned to all members of a class that contains more than two agents being equal. Without loss of generality, they can be chosen to be 1. Then, for each specific group of claimants and each specific problem this group may face, identify the ordered partition of this group induced by the reference partition. Do not give anything to a class until all classes with higher priorities have been fully satisfied. For a component of the partition induced by a two-claimant reference class, divide between them what remains available when its turn comes by applying the weighted concede-and-divide rule with weights proportional to their reference weights. For a component induced by any larger reference class, (for which all weights are equal,) apply the Talmud rule itself.

Formally, let $\preccurlyeq$ be a complete and transitive binary relation on $\mathbb{N}$, with $\prec$ and $\sim$ denoting its asymmetric and symmetric parts. Let $w \in \mathbb{R}_{++}^{\mathbb{N}}$ be such that for each $i \in \mathbb{N}$, if $|\{j \in \mathbb{N} \mid i \sim j\}| \neq 2$, then $w_i = 1$.

First definition of the sequential Talmud rule relative to $\preccurlyeq$ and w, $T^{C,\preccurlyeq,w}$. Let $N \in \mathcal{N}$ and $(c, E) \in \mathcal{C}^N$. Let $(S_1, S_2, \ldots, S_K)$ be the ordered partition of N such that for each pair $\{k, \ell\} \subseteq \{1, 2, \ldots, K\}$, each $i \in S_k$, and each $j \in S_\ell$, (i) if $k = \ell$, then $i \sim j$, and (ii) if $k < \ell$, then $i \prec j$. Let $E_1 \equiv \min\{E, \sum_{i \in S_1} c_i\}$, $E_2 \equiv \min\{E - E_1, \sum_{i \in S_2} c_i\}$, $E_3 \equiv \min\{E - E_1 - E_2, \sum_{i \in S_3} c_i\}$, and so on

For each pair $\{i, j\} \in \mathcal{N}$, let $\alpha_{ij} \equiv \frac{w_i}{w_i + w_j}$. Then, let $T^{C,\preccurlyeq,w}(c, E)$ be the awards vector x of (c, E) such that for each $k \in \{1, 2, \ldots, K\}$,

(i)   if $S_k = \{i\}$, then $x_i \equiv E_k$;
(ii) if $S_k = \{i, j\}$, then $(x_i, x_j) \equiv C^{(\alpha_{ij}, \alpha_{ji})}(c_i, c_j, E_k)$;
(iii) if $|S_k| \geq 3$, then $x_{S_k} \equiv T(c_{S_k}, E_k)$.

Our first main result is the following characterization:

Theorem 1. *A rule R on $\mathcal{C}$ is such that*

(∗) *for each pair $\{i, j\} \in \mathcal{N}$, there exists a vector of weights $(\alpha_{ij}, \alpha_{ji}) \in \Delta^{\{i,j\}}$ such that on $\mathcal{C}^{\{i,j\}}$, $R = C^{(\alpha_{ij}, \alpha_{ji})}$,*

and satisfies consistency *only if there exists a complete and transitive binary relation $\preccurlyeq$ on the set of potential claimants such that for each pair $\{i, j\} \in \mathcal{N}$,*

(i) $(\alpha_{ij}, \alpha_{ji}) = (1, 0)$ *if and only if $i \prec j$;*
(ii) $(\alpha_{ij}, \alpha_{ji}) \in int\Delta^{\{i,j\}}$ *if and only if $i \sim j$;*
(iii) *if there exists $k \in \mathbb{N} \backslash \{i, j\}$ such that $i \sim j \sim k$, then $\alpha_{ij} = \alpha_{ji}$.*

Then, R coincides with the sequential Talmud rule $T^{C,\preccurlyeq,w}$, where $w \in \mathbb{R}_{++}^{\mathbb{N}}$ is defined by setting for each $i \in \mathbb{N}$,

$$w_i \equiv \begin{cases} \alpha_{ij} & \text{if there exists } j \in \mathbb{N} \backslash \{i\} \text{ such that } \{k \in \mathbb{N} \mid k \sim i\} = \{i, j\}, \\ 1 & \text{otherwise.} \end{cases}$$

Conversely, each sequential Talmud rule $T^{C,\preccurlyeq,w}$ satisfies condition (∗) and consistency. *Here, for each pair $\{i, j\}$ of claimants, the weights $(\alpha_{ij}, \alpha_{ji})$ are* *constructed from $\preccurlyeq$ and w by*

$$(\alpha_{ij}, \alpha_{ji}) \equiv \begin{cases} (1, 0) & \text{if } i \prec j, \\ \left(\frac{w_i}{w_i + w_j}, \frac{w_j}{w_i + w_j}\right) & \text{if } i \sim j. \end{cases}$$

Clearly the rules $T^{C, \preccurlyeq, w}$ satisfy condition $(*)$. Also:

Lemma 1. *The sequential Talmud rules $T^{C, \preccurlyeq, w}$ are* consistent.

Proof. Let $N \in \mathcal{N}$, $(c, E) \in \mathcal{C}^N$, and $x \equiv T^{C, \preccurlyeq, w}(c, E)$. We show that for each $N' \subset N$, $x_{N'} = T^{C, \preccurlyeq, w}(c_{N'}, E - \sum_{N \setminus N'} x_i)$. By the definition of *consistency*, we can suppose that $N' \equiv N \setminus \{j\}$ for some $j \in N$.

Let $(S_1, S_2, \ldots, S_K)$ be the ordered partition of N induced by $\preccurlyeq$. Let $\ell \in \{1, 2, \ldots, K\}$ be such that $j \in S_\ell$. For each $k \in \{1, 2, \ldots, K\}$, let E_k be the amount to divide among the members of S_k in (c, E). For each $k \in \{1, 2, \ldots, K\}$ with $k \neq \ell$, let E'_k be the amount to divide among the members of S_k in $(c_{N'}, E - x_j)$. Also, if $S_\ell \setminus \{j\} \neq \emptyset$, let E'_ℓ be the corresponding amount for $S_\ell \setminus \{j\}$ in $(c_{N'}, E - x_j)$.

Clearly, for each $k < \ell$, $E'_k = E_k$. Thus, for each $k < \ell$, $x_{S_k} = T^{C, \preccurlyeq, w}_{S_k}(c_{N'}, E - x_j)$.

First, suppose that $S_\ell = \{j\}$. Then $x_j = E_\ell$ and

$$E'_{\ell+1} = \min\left\{ E - x_j - \left(\sum_{k=1}^{\ell} E_k - E_\ell\right), \sum_{i \in S_{\ell+1}} c_i \right\} = E_{\ell+1}.$$

Similarly, for each $k \geq \ell + 1$, $E'_k = E_k$. Thus, for each $k > \ell$, $x_{S_k} = T^{C, \preccurlyeq, w}_{S_k}(c_{N'}, E - x_j)$.

Next, suppose that $|S_\ell| \geq 2$. By definition,

$$E'_\ell = \begin{cases} \min\{E - x_j, \sum_{i \in S_\ell \setminus \{j\}} c_i\} & \text{if } \ell = 1, \\ \min\{E - x_j - \sum_{k=1}^{\ell-1} E_k, \sum_{i \in S_\ell \setminus \{j\}} c_i\} & \text{if } \ell > 1. \end{cases}$$

Case 1: $x_j = c_j$. Then $E'_\ell = E_\ell - x_j$ and $E'_{\ell+1} = \min\{E - x_j - (\sum_{k=1}^{\ell} E_k - x_j), \sum_{i \in S_{\ell+1}} c_i\} = E_{\ell+1}$. Similarly, for each $k \geq \ell + 1$, $E'_k = E_k$.

Case 2: $x_j < c_j$. Then, $E_\ell = E$ if $\ell = 1$, and $E_\ell = E - \sum_{k=1}^{\ell-1} E_k$ if $\ell > 1$. Also, for each $k > \ell$, $E_k = 0$. Recall that for each $i \in N$, $x_i \leq c_i$. Thus,

$$\sum_{i \in S_\ell \setminus \{j\}} c_i \geq \sum_{i \in S_\ell \setminus \{j\}} x_i = E_\ell - x_j = \begin{cases} E - x_j & \text{if } \ell = 1, \\ E - x_j - \sum_{k=1}^{\ell-1} E_k & \text{if } \ell > 1. \end{cases}$$

This implies that $E'_\ell = E_\ell - x_j \leq \sum_{i \in S_\ell \setminus \{j\}} c_i$, and hence, for each $k > \ell$, $E'_k = E_k = 0$.

In both cases, $E'_\ell = E_\ell - x_j$ and for each $k > \ell$, $E'_k = E_k$. If $|S_\ell| \geq 3$, then by *consistency* of the Talmud rule, for each $i \in S_\ell \setminus \{j\}$, $x_i = T_i(c_{S_\ell}, E_\ell) = T_i(c_{S_\ell \setminus \{j\}}, E_\ell - x_j) = T^{C, \preccurlyeq, w}_i(c_{N'}, E - x_j)$. If $S_\ell = \{i, j\}$, then $x_i = E_\ell - x_j = T^{C, \preccurlyeq, w}_i(c_{N'}, E - x_j)$. Finally, for each $k > \ell$, since $E'_k = E_k$, we have $x_{S_k} = T^{C, \preccurlyeq, w}_{S_k}(c_{N'}, E - x_j)$. $\square$

Clearly, the weighted concede-and-divide rules satisfy the following property:

Resource-monotonicity. For each $(c, E) \in \mathcal{C}^N$ and each $E' \in \mathbb{R}_+$, if $E < E' \leq \sum c_i$, then for each $i \in N$, $R_i(c, E') \geq R_i(c, E)$.

Moreover, if a rule is *resource-monotonic* in the two-claimant case and *consistent*, it is *resource-monotonic* in general (Dagan, Serrano, and Volij, 1996; Hokari and Thomson, 2000). Thus, if a rule satisfies condition $(*)$ and *consistency*, it is *resource-monotonic*.

Lemma 2. *Let R be a rule on $\mathcal{C}$ satisfying* condition $(*)$ *and* consistency. *Let $N \equiv \{i, j, k\}$. Then the α's appearing in* condition $(*)$ *satisfy the following relations*:

(i)   *If $\alpha_{ji}, \alpha_{jk} \notin \{0, 1\}$, then $\alpha_{ji} = \alpha_{jk}$.*
(ii) *If $\alpha_{ji} < 1$ and $\alpha_{jk} = 1$, then $\alpha_{ik} = 1$.*

Proof. **(i)** Suppose that $\alpha_{ji}, \alpha_{jk} \notin \{0, 1\}$. First, we show that $\alpha_{ji} \geq \alpha_{jk}$. Suppose, by contradiction, that $\alpha_{ji} < \alpha_{jk}$. Let $(c_i, c_j, c_k) \equiv (1, 2, 1)$. Since R is *resource-monotonic*, $R(c, E)$ is non-decreasing and continuous in E. Thus, for some $E_1 > 0$, $R_i(c, E_1) + R_j(c, E_1) = 1$. By condition $(*)$ and *consistency*, $\big(R_i(c, E_1), R_j(c, E_1)\big) = R(c_i, c_j; 1) = (\alpha_{ij}, \alpha_{ji})$. By *consistency*, $\big(R_j(c, E_1), R_k(c, E_1)\big) \in P^R(\{j, k\}, c_j, c_k)$. Thus, as indicated in Figure 2a, $R_k(c, E_1) = \frac{\alpha_{ji} \cdot \alpha_{kj}}{\alpha_{jk}}$. Similarly, since $R(c, E)$ is non-decreasing and continuous in E, then for some $E_2 > 0$, $R_j(c, E_2) + R_k(c, E_2) = 1$. By *consistency* and condition $(*)$, $\big(R_j(c, E_2), R_k(c, E_2)\big) = R(c_j, c_k; 1) = (\alpha_{jk}, \alpha_{kj})$. By *consistency*, $\big(R_i(c, E_2), R_j(c, E_2)\big) \in P^R(\{i, j\}, c_i, c_j)$. Since $\alpha_{ji} < \alpha_{jk} < 1$, it can be seen from Figure 2(b) that $R_i(c, E_2) = \alpha_{ij}$.

By *resource-monotonicity*, the line segment in $\mathbb{R}_+^{\{i,k\}}$ connecting $(\alpha_{ij}, \frac{\alpha_{ji} \cdot \alpha_{kj}}{\alpha_{jk}})$ and $(\alpha_{ij}, \alpha_{kj})$ is a subset of $P^R(\{i, k\}, c_i, c_k)$. Since $\alpha_{ji} < \alpha_{jk}$, then $\frac{\alpha_{ji} \cdot \alpha_{kj}}{\alpha_{jk}} < \alpha_{kj}$. Thus, $P^R(\{i, k\}, c_i, c_k)$ contains a non-degenerate line segment along which claimant i's award is constant and strictly between 0 and c_1 (Fig. 2b). Since $c_i = c_k = 1$, this contradicts condition $(*)$. Thus, $\alpha_{ji} \geq \alpha_{jk}$.

By a similar argument, it can be shown that $\alpha_{ji} \leq \alpha_{jk}$. Thus, $\alpha_{ji} = \alpha_{jk}$.

(ii) Suppose that $\alpha_{ji} < 1$ and $\alpha_{jk} = 1$. Let $(c_i, c_j, c_k) \equiv (1, 1, 1)$. Since $R(c, E)$ is non-decreasing and continuous in E, then for some $E_1 > 0$, $R_i(c, E_1) + R_j(c, E_1) = 1$. By *consistency* and condition $(*)$, $\big(R_i(c, E_1), R_j(c, E_1)\big) = R(c_i, c_j; 1) = (\alpha_{ij}, \alpha_{ji})$. By *consistency*, $\big(R_j(c, E_1), R_k(c, E_1)\big) \in P^R(\{j, k\}, c_j, c_k)$. Thus, as indicated in Figure 3, $R_k(c, E_1) = 0$. By *consistency*, $(\alpha_{ij}, 0) = \big(R_i(c, E_1), R_k(c, E_1)\big) \in P^R(\{i, k\}, c_i, c_k)$. By $\alpha_{ij} > 0$ and condition $(*)$, $\alpha_{ik} = 1$. □

Lemma 2 essentially implies the existence of a complete and transitive binary relation on $\mathbb{N}$ that is used in defining each sequential Talmud rule. The next lemma shows how to construct this relation.

Lemma 3. *Let R be a rule on $\mathcal{C}$ satisfying condition $(*)$ and* consistency. *Define a binary relation $\preccurlyeq$ on $\mathbb{N}$ as follows: for each pair $\{i, j\} \in \mathcal{N}$, $i \preccurlyeq j$ if and only if $\alpha_{ij} > 0$. Let $\sim$ denote the symmetric part of $\preccurlyeq$. Then the following two conditions hold*:

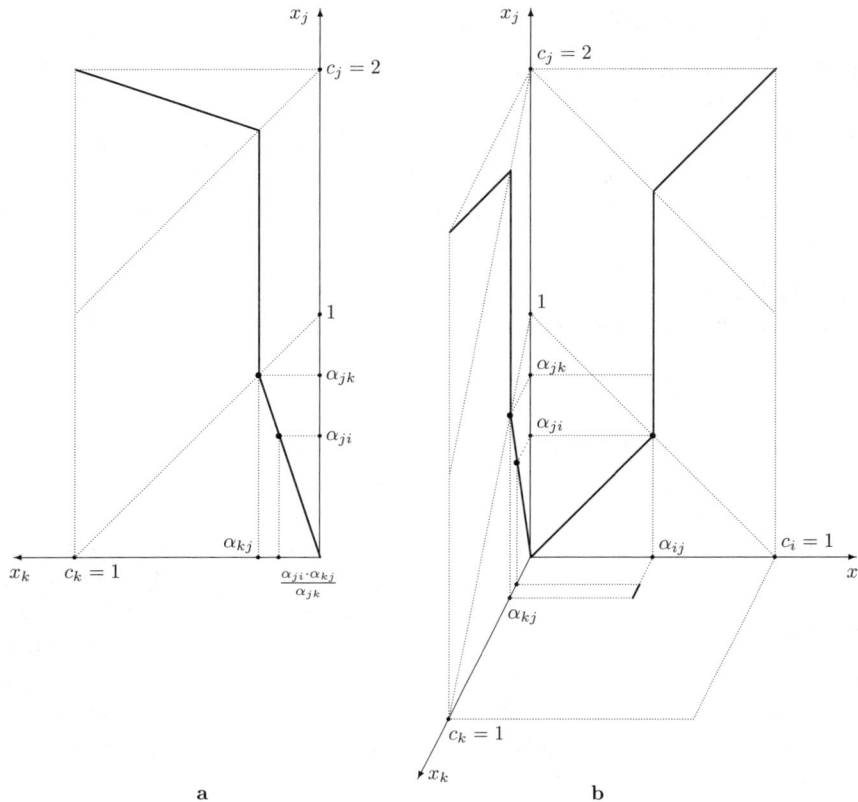

Figure 2a,b. *Proof of Lemma 2(i).* Panel **a** represents a side-view of the three-claimant problem in panel **b**

(i) $\preceq$ *is complete and transitive.*

(ii)   *If $i, j, k \in \mathbb{N}$ are distinct and $i \sim j \sim k$, then*

$$\alpha_{ij} = \alpha_{ji} = \alpha_{ik} = \alpha_{ki} = \alpha_{jk} = \alpha_{kj} = \frac{1}{2}.$$

Proof. Clearly, $\preceq$ is complete. We show that $\preceq$ is transitive. Let $i, j, k \in \mathbb{N}$ be such that $i \preceq j$ and $j \preceq k$. Then $\alpha_{ij} > 0$ and $\alpha_{jk} > 0$. We want to show that $\alpha_{ik} > 0$. Suppose, by contradiction, that $\alpha_{ik} = 0$. Then $\alpha_{ki} = 1$. By condition (ii) of Lemma 2, $\alpha_{kj} < 1$ and $\alpha_{ki} = 1$ imply $\alpha_{ji} = 1$, which contradicts $\alpha_{ij} > 0$. Thus, $\alpha_{ik} > 0$.

Next, let $i, j, k \in \mathbb{N}$ be distinct and $i \sim j \sim k$. Then, by the definition of $\preceq$, $\alpha_{ij}, \alpha_{ji}, \alpha_{ik}, \alpha_{ki}, \alpha_{jk}, \alpha_{kj} \notin \{0, 1\}$.

By condition (i) of Lemma 2, $\alpha_{ij} = \alpha_{ji} = \alpha_{ik} = \alpha_{ki} = \alpha_{jk} = \alpha_{kj} = \frac{1}{2}$. $\square$

So far, we have shown that if a rule satisfies condition ($*$) and *consistency*, then it coincides with a sequential Talmud rule for the two-claimant case. The following property is useful to show that this coincidence holds in general.

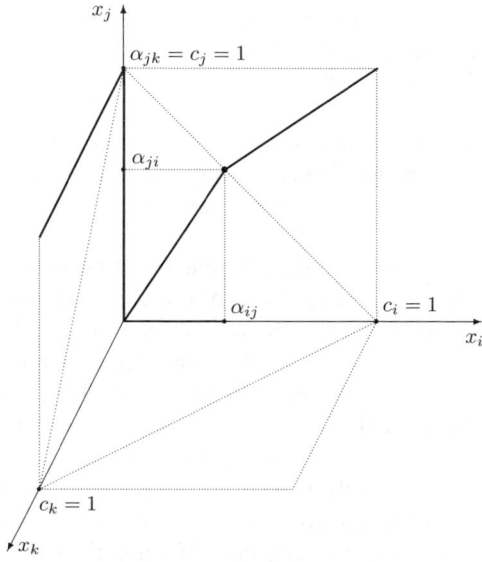

Figure 3. *Proof of Lemma 2(ii).* Relating the claimants' weights

Converse consistency. For each $N \in \mathcal{N}$ with $|N| \geq 3$, each $(c, E) \in \mathcal{C}^N$, and each $x \in \mathbb{R}_+^N$ with $\sum_N x_i = E$, if for each $N' \subset N$ with $|N'| = 2$, $x_{N'} = R(c_{N'}, E - \sum_{N \setminus N'} x_i)$, then $x = R(c, E)$.

Lemma 4. *The sequential Talmud rules* $T^{C, \preccurlyeq, w}$ *are conversely consistent.*

Proof. Clearly, the rules $T^{C, \preccurlyeq, w}$ satisfy condition (∗). By Lemma 1, they are *consistent*. As mentioned earlier, condition (∗) and *consistency* imply *resource-monotonicity*. *Resource-monotonicity* and *consistency* imply *converse consistency* (Chun, 1999). Thus, they are *conversely consistent*. ◻

Proof of Theorem 1. As mentioned before, each rule $T^{C, \preccurlyeq, w}$ satisfies condition (∗) and by Lemma 1, *consistency*. Conversely, let R be a rule on $\mathcal{C}$ satisfying condition (∗) and *consistency*. Let $\preccurlyeq$ denote the complete and transitive binary relation defined as in Lemma 3. For each $i \in \mathbb{N}$, let

$$w_i \equiv \begin{cases} \alpha_{ij} & \text{if there exists } j \in \mathbb{N} \setminus \{i\} \text{ such that } \{k \in \mathbb{N} \mid k \sim i\} = \{i, j\}, \\ 1 & \text{otherwise.} \end{cases}$$

Then, by Lemma 3, R coincide with $T^{C, \preccurlyeq, w}$ in the two-claimant case. Let $N \in \mathcal{N}$ with $|N| > 2$, $(c, E) \in \mathcal{C}^N$, and $x \equiv R(c, E)$. Since R is *consistent*, for each $N' \subset N$ with $|N'| = 2$,

$$x_{N'} = R(c_{N'}, E - \sum_{N \setminus N'} x_i) = T^{C, \preccurlyeq, w}(c_{N'}, E - \sum_{N \setminus N'} x_i).$$

Finally, by *converse consistency* of $T^{C, \preccurlyeq, w}$, $x = T^{C, \preccurlyeq, w}(c, E)$.[9] ◻

---

[9] This proof is an instance of a very general lemma, "Elevator Lemma" (Thomson, 1996), which asserts that if a solution correspondence is *consistent* (*bilateral consistency* would suffice), and coincides with a *conversely consistent* solution correspondence in the two-agent case, then coincidence holds for an arbitrary number of agents.

Since each rule $T^{C,\preccurlyeq,w}$ satisfies *homogeneity, invariance under claims truncation*, and *minimal rights first*, we obtain the following result as a corollary of Proposition 1 and Theorem 1.

Corollary 1. *The sequential Talmud rules* $T^{C,\preccurlyeq,w}$ *are the only rules satisfying* homogeneity, invariance under claims truncation, minimal rights first, *and* consistency.

In the theory of coalitional games, a similar question to the one we addressed here has arisen concerning the possibility of defining and justifying asymmetric versions of the "standard solution", the two-player solution that divides equally between the players the surplus above their individual rationality utilities. Such operations are in general possible, and rich classes of solutions have emerged that have been very useful in applications, the primary example being the weighted versions of the Shapley value. One solution however has not been so generalized, the nucleolus (Schmeidler, 1969). Indeed, if the counterpart of the condition that we referred to as *consistency* is imposed on a solution to coalitional games, no weighted generalization exists (Orshan, 1994; Hokari, 2000). Given the correspondence between the nucleolus and the Talmud rule (Aumann and Maschler, 1985),[10] one may think that our result could be obtained by somehow adapting these authors' proofs. This is indeed the case, although there is no logical relation between these earlier results and ours because we work on a smaller domain on which the axioms lose force. At the same time, and precisely because claims problems constitute a considerably simpler class of problems, a much more direct proof is available. That is the proof we have presented.

4 A second approach

If we think of the Talmud rule as a hybrid of the constrained equal awards and constrained equal losses rules, our search for generalizations of the rule naturally passes by first defining weighted generalizations of these rules. This is easily done, even for an arbitrary number of agents. Given $N \in \mathcal{N}$, let Δ^N denote the unit simplex in $\mathbb{R}_+^N$ and $int\Delta^N$ its relative interior.

Weighted constrained equal awards rule of weights $\alpha \in int\Delta^N, CEA^\alpha$. For each $(c, E) \in \mathcal{C}^N$ and each $i \in N$, $CEA_i^\alpha(c, E) \equiv \min\{c_i, \alpha_i\lambda\}$, where $\lambda \in \mathbb{R}_+$ is chosen so as to achieve efficiency.

Weighted constrained equal losses rule of weights $\alpha \in int\Delta^N, CEL^\alpha$. For each $(c, E) \in \mathcal{C}^N$ and each $i \in N$, $CEL_i^\alpha(c, E) \equiv \max\left\{c_i - \frac{\lambda}{\alpha_i}, 0\right\}$, where $\lambda \in \mathbb{R}_+$ is chosen so as to achieve efficiency.

---

[10] If, as suggested by O'Neill (1983), one associates to a claims problem a coalitional game by assigning to each coalition a "worth" equal to the difference between the amount to divide and the sum of the claims of the members of the complementary coalition (or 0 if this difference is negative), and then calculate the nucleolus of the game, one obtains the awards vector produced by the Talmud rule for the problem.

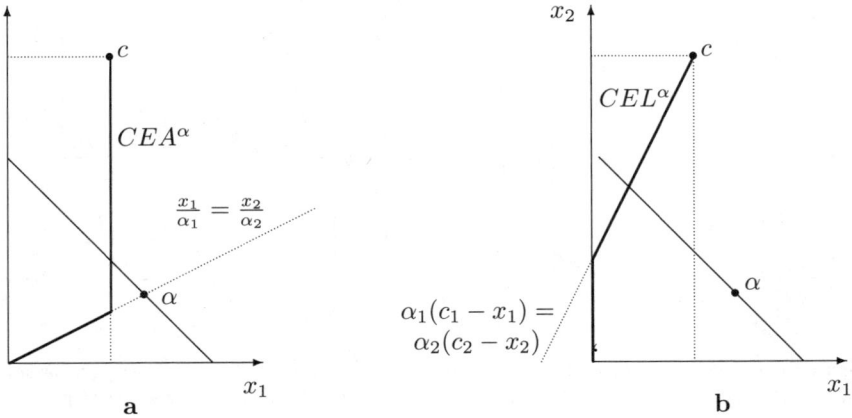

Figure 4a,b. Weighted versions of the constrained equal awards and constrained equal losses rules. **a** Weighted constrained equal awards rule of weights $\alpha \equiv (\frac{2}{3}, \frac{1}{3})$. **b** Weighted constrained equal losses rule of weights α

Note that the limit case, where α has coordinates equal to 0, is excluded, as the formulae would then not produce a well-defined rule. Choosing weights equal to 0 amounts to giving priority to some agents over the others, a possibility that we discuss a little later. In order to get an idea of the behavior of the rules just introduced, we plot representative paths of awards in the two-claimant case (Fig. 4).

We now combine these rules so as to obtain a rule in the spirit of the Talmud rule but that incorporates society's bias towards one or the other of the two claimants. We give the definition for a fixed N, and again, for positive weights.

Weighted hybrid rule of weights $\alpha \in int\Delta^N$, H^α. For each $(c, E) \in \mathcal{C}^N$ and each $i \in N$, $H_i^\alpha(c, E) \equiv \min\{\frac{c_i}{2}, \alpha_i\lambda\}$ if $\sum \frac{c_i}{2} \geq E$, and $\max\{\frac{c_i}{2}, c_i - \frac{\lambda}{\alpha_i}\}$ otherwise, where in each case, λ is chosen so as to achieve efficiency.

Figure 5 shows for the two-claimant case how the weighted hybrid rule behaves as the relative weight placed on the claim of agent 1 increases. Also represented is the limit case, which is naturally associated with the limit weight vector $(1, 0)$. It is of interest that for each $\alpha \in \Delta^N$ and each claims vector, the path of awards of H^α goes through the half-claims vector.[11]

We just wrote the three definitions above for a fixed population but when population varies, the weights assigned to a given claimant could in principle vary depending upon the identity of his fellow claimants. However, in order to obtain *consistency*, we will see that weight vectors assigned to different populations should satisfy certain relations: given two claimants and two groups related by inclusion and containing both, the relative weights assigned to these two claimants should be equal in both groups. One way to achieve this property is to assign a "reference" weight to each potential claimant, and for each actual problem, to use the weights that have been assigned to whoever is present. If all reference weights are positive,

---

[11] They satisfy the property called "midpoint property", formulated by Chun, Schummer, and Thomson (1999).

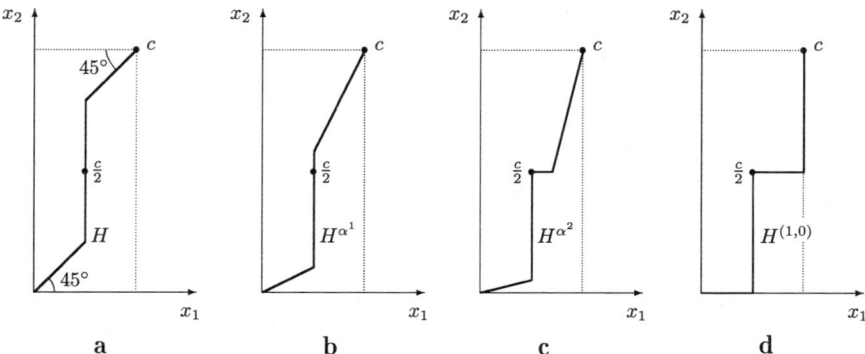

Figure 5a–d. Weighted versions of the Talmud rule, seen as a hybrid of the constrained equal awards and constrained equal losses rules. Here, $N \equiv \{1, 2\}$. In each panel, the slope of the segment emanating from the claims point is the inverse of that of the segment emanating from the origin. (a) The symmetric case. In the following panels, we give progressively more weight to claimant 1 relative to claimant 2. (b) Here, $\alpha^1 \equiv (\frac{2}{3}, \frac{1}{3})$, (c) Here, $\alpha^2 \equiv (\frac{4}{5}, \frac{1}{5})$. (d) At the limit, we obtain a rule that is composed of "twice" the priority rule in which claimant 1 has priority over claimant 2, the half-claims vector being used instead of the claims vector itself. We denote this "half-priority" rule $H^{(1,0)}$

the weighted generalizations of the constrained equal awards and constrained equal losses rules so obtained are *consistent*. So are the weighted generalizations of the hybrid rules.[12]

The limit cases can be described as follows. First, define an ordered partition of the set of potential claimants. For each problem, identify the partition of the set of claimants actually present induced by the reference partition. We naturally refer to its components as "priority classes". Handle each component of the induced partition in succession. If the starting point is the constrained equal awards rule, apply the weighted constrained equal awards rule of weights proportional to the reference weights assigned to its members until each of them is fully compensated. If the starting point is the constrained equal losses rule, follow a similar procedure. If the starting point is the hybrid of the two rules, let us refer to the components of the reference partition of the set of potential claimants as "half-priority classes". For each problem, once again, identify the partition of the set of claimants actually present induced by the reference partition; then, apply the weighted hybrid rule of weights proportional to the reference weights assigned to its members until each of them has received his half-claim; next, turn to the component of the induced partition with the next highest half-priority and apply the weighted hybrid rule of weights proportional to the reference weights assigned to these claimants until each of them has received his half-claim, and so on. When each claimant has received his half-claim, revisit each component of the partition, in the same order, and return to the corresponding hybrid rule (the rule relative to weights proportional to the reference weights assigned to its members), and do so until each of them is fully compensated.

Formally, let $\preccurlyeq$ be a complete and transitive binary relation on the set of potential claimants $\mathbb{N}$. Let $\prec$ and $\sim$ denote the asymmetric and symmetric parts of $\preccurlyeq$, respectively. Let $w \in \mathbb{R}^{\mathbb{N}}_{++}$ be a list of positive weights.

---

[12] They also all admit parametric representations (Young, 1987).

Second definition of the sequential Talmud rule relative to $\precsim$ and w, $\precsim$ and w, $T^{H,\precsim,w}$. Let $N \in \mathcal{N}$ and $(c, E) \in \mathcal{C}^N$. Let $(S_1, S_2, \ldots, S_K)$ be the ordered partition of N such that for each pair $\{k, \ell\} \subseteq \{1, 2, \ldots, K\}$, each $i \in S_k$, and each $j \in S_\ell$, (i) if $k = \ell$, then $i \sim j$, and (ii) if $k < \ell$, then $i \prec j$. Let $E_1 \equiv \min\{E, \frac{1}{2} \sum_{i \in S_1} c_i\}$, $E_2 \equiv \min\{E - E_1, \frac{1}{2} \sum_{i \in S_2} c_i\}$, $E_3 \equiv \min\{E - E_2 - E_3, \frac{1}{2} \sum_{i \in S_1} c_i\}$, and so on. For each $E \in \,]\frac{1}{2} \sum_N c_i, \sum_N c_i[$ and each $k \in \{1, 2, \ldots, K\}$, let F^k be defined as E^k was by replacing E by $E - \frac{1}{2} \sum_{S_k} c_i$, and each E_ℓ by F_ℓ.

Now, for each $N \in \mathcal{N}$ and each $i \in N$, let $\alpha_i \equiv \frac{w_i}{\sum_N w_j}$ and $\alpha \equiv (\alpha_i)_{i \in N}$. Then let $T^{H,\precsim,w}(c, E)$ be the awards vector x obtained as follows: for each $k \in \{1, 2, \ldots, K\}$,

(i) if $E \leq \frac{1}{2} \sum_N c_i$, then $x_{S_k} \equiv CEA^\alpha\left(\frac{1}{2} c_{S_k}, E_k\right)$;
(ii) if $E \in \,]\frac{1}{2} \sum_N c_i, \leq \sum_N c_i]$, then $x_{S_k} \equiv \frac{1}{2} c_{S_k} + CEL^\alpha\left(\frac{1}{2} c_{S_k}, F_k\right)$.

We have the following theorem: (The proof is omitted because of the overlap with that of Theorem 1.)

Theorem 2. *A rule R on $\mathcal{C}$ is such that*

$(**)$ *for each pair $\{i, j\} \in \mathcal{N}$, there exists a vector of weights $(\alpha_{ij}, \alpha_{ji}) \in \Delta^{\{i,j\}}$ such that $R = H^{(\alpha_{ij}, \alpha_{ji})}$ on $\mathcal{C}^{\{i,j\}}$.*

and satisfy consistency *only if there exist a complete and transitive binary relation $\precsim$ on the set of potential claimants and a list of positive weights $w \in \mathbb{R}^{\mathbb{N}}_{++}$ such that for each pair $\{i, j\} \in \mathcal{N}$, the following relations hold:*

(i) *$(\alpha_{ij}, \alpha_{ji}) = (1, 0)$ if and only if $i \prec j$;*
(ii) *$(\alpha_{ij}, \alpha_{ji}) \in int\Delta^{\{i,j\}}$ if and only if $i \sim j$ and $(\alpha_{ij}, \alpha_{ji})$ is proportional to (w_i, w_j).*

Then, R coincides with the sequential hybrid rule $T^{H,\precsim,w}$.

Conversely, each sequential hybrid rule $T^{H,\precsim,w}$ satisfies condition $(**)$ and consistency*. Here, for each pair $\{i, j\}$ of claimants, the weights $(\alpha_{ij}, \alpha_{ji})$ are constructed from $\precsim$ and w by invoking conditions* (i) *and* (ii)*.*

The condition relating the weights chosen for the two-claimant populations is necessary for *consistency*. Another way to express the condition is as follows: if agent i belongs to a higher half-priority class than agent j, then $(\alpha_{ij}, \alpha_{ji}) = (1, 0)$, and if agents i, j, and k belong to the same half-priority class, then the weights $(\alpha_{ij}, \alpha_{ji})$ and $(\alpha_{jk}, \alpha_{kj})$ determine the weights $(\alpha_{ik}, \alpha_{ki})$.

Note that in our definition, the bias in favor of an agent is independent of the amount to divide. One could argue that a distinction should be made about who should be favored depending upon how much is available. This two-regime idea is part of the Talmud rule, which trades-off a particularly favorable treatment of the agents with the smaller claims over half the range of the amount to divide for a particularly favorable treatment of the agents with the larger claims over the complementary range. However, it does so in a way that respects *equal treatment of equals*. If this property is dropped, we have the option of choosing weights that depend on the regime. However, for each regime, the weights should satisfy consistency conditions of the kind expressed in the theorem.

A special case is when a reversal of the weights occurs between the two regimes. Let $N \equiv \{i, j\}$. The suggestion here is to apply the weighted constrained equal awards rule of weights (α_i, α_j) for an amount to divide at most as large as the half-sum of the claims (of course, using the half-claims instead of the claims themselves) and the weighted constrained equal losses rule of weights (α_j, α_i) otherwise (once again, using the half-claims).[13]

If some weights are equal to 0, *consistency* and this reversal have an interesting implication. Let $N \equiv \{1, 2, 3\}$. Suppose that when the amount to divide is less than the half-sum of the claims, the weights are $(\alpha_{12}, \alpha_{21}) = (1, 0), (\alpha_{13}, \alpha_{31}) = (1, 0)$, and $(\alpha_{23}, \alpha_{23}) = (1, 0)$. Suppose also that the weights are reversed when the amount to divide is greater than the half-sum of the claims. Then, a *consistent* rule that coincides with these hybrid rules for two-claimant problems behaves as follows. As the amount to divide first increases from 0, agent 1 gets everything until his award is his half-claim; agent 2 receives any additional amount available until his award is his half-claim; then agent 3 receives any additional amount available until his award is his half-claim. At that point, *consistency* forces us to reverse the half-priority relation: agent 3 receives any additional amount until he is fully compensated; agent 2 comes next until he too is fully compensated; and it is only then that agent 1 starts receiving more.

5 Concluding comments

1. Several rules have been discussed in the literature that are closely related to the Talmud rule, and it is natural to enquire about the possibility of defining weighted generalizations of them. One such rule is introduced by Piniles (1861) in an attempt to explain the numerical examples in the Talmud. This rule can be described as consisting of a "double application" of the constrained equal awards rule, the half-claims being used instead of the claims themselves. Also, a "reverse" of Piniles' rule and a "reverse" of the Talmud rule can be defined by exchanging the order in which the constrained equal awards and constrained equal losses rules are applied. It is easy to construct *consistent* extensions of the weighted versions of Piniles' rule and of its reverse. On the other hand, for the "reverse Talmud rule", two approaches can be taken to define weighted generalizations. These approaches lead to counterparts of the rules we identified in Theorems 1 and 2 . Like the rules we describe in Theorem 1, the "sequential reverse Talmud rules" allow a flexible differential treatment of claimants only in the two-claimant case.

2. Instead of letting priority classes and weights emerge from the axioms, one could enrich the model by adding information of this kind. This possibility is discussed by Thomson (1995), who defines a *claims problem with priority classes* as a list $(c, \prec, E)$ where $(c, E) \in \mathcal{C}^N$ and $\prec$ is a complete and transitive binary relation

---

[13] We thank Jean-Pierre Benoît for the suggestion.

on N whose equivalence classes constitute a partition $\mathcal{P}^N$ of N. Also, a *claims problem with weights* is a list (c, w, E), where $(c, E) \in \mathcal{C}^N$ and $\alpha \in int\Delta^N$ is a point in the interior of the $(|N| - 1)$-dimensional simplex indicating the *relative* importance that should be given to claimants. A richer formulation would include priority classes and weights.[14] Let us then consider rules defined over a domain of such problems when N runs over $\mathcal{N}$. A rule should reflect the value judgments incorporated in the choice of classes and weights: for each problem, it should assign nothing to a class until all classes with higher priority are fully compensated; within each class, it should gear the awards vector towards the members of the class to whom higher weights have been assigned: if they have equal claims, the ratio of their awards should be equal to the ratio of their weights unless the agent with the greater weight is fully compensated.

In the variable population version of the model, one would have to specify classes and weights for each population of claimants, and require the rule to behave accordingly for each population. However, *consistency* will imply that the exogenous data be "consistent" across populations: if in some population, an agent is assigned to a higher priority class than some other agent, so should he in any other population to which they both belong. Also, if in some population, two agents are assigned to the same priority class, so should they in any other population to which they both belong; also, the ratios of the weights assigned to them in the two populations should be equal. These requirements will imply two basic restrictions on the class and weight assignments exogenously chosen for the various populations: these classes should be induced from a reference partition of the set of potential agents into priority classes, and for each induced class, the exogenous weights assigned to its members should be proportional to a reference weight vector chosen for the reference class of which this induced class is a subset. If our first approach is taken, Theorem 1 says more. It implies that in any reference priority class containing more than two agents, the reference weights should be equal. On the other hand, if our second approach is taken, Theorem 2 would imply no additional restrictions on classes and weights.

3. In a recent paper, Moulin (2000) has considered a similar issue to the one we address and the similarities between his conclusions and the ones we reached by following our first approach should help shed light on both studies. His starting point is a rich family of two-claimant rules, to which he had arrived by imposing a certain list of axioms (*homogeneity*, and two "composition" properties, expressing the invariance of the choice with respect to two alternative ways of dealing with possible increases, or decreases respectively, in the amount to divide). He asked how such two-claimant rules could be extended to general populations, and showed that any rule passing all of his tests can be described as follows: the population of potential claimants is partitioned into reference priority classes; for each problem, this reference partition induces an ordered partition of the set of claimants actually present, and these classes are handled in succession (this is as we discovered for our problem); within each two-claimant induced class, a rule of the kind he had obtained for the two-claimant case can be used (again, this conclusion parallels

---

[14] Alternatively, one could have a lexicographic system of weights.

ours), but within each induced class containing three or more claimants, the choice is narrowed down to one in the small subclass consisting of the proportional, weighted constrained equal awards, and weighted constrained equal losses rules; finally, the weights used to modify the constrained equal awards and constrained equal losses rules have to be proportional to weights assigned to the members of the reference class of which this class is a subset (we too obtain an important narrowing, but of course to a different rule, the Talmud rule).

Incidentally, we note that a simple proof of Moulin's characterization along the lines of the proof we develop here can be devised (Thomson, 2001b), thereby showing the usefulness of our approach beyond our specific application. For another presentation of the approach, see Thomson (2001a).

4. Given the somewhat limiting conclusions uncovered in Theorem 1, one may wonder whether a weaker version of *consistency* would be available to help guide the passage from two-claimant populations to general populations. The answer is yes. Dagan and Volij (1997) define a rule to be "average consistent" if for each problem, and for each claimant involved in this problem, this agent's award is equal to the average of his awards in the associated reduced problems relative to all the subgroups to which he belongs. They also propose a version in which the average is limited to subgroups of two claimants. They show that each *resource monotonic* rule has an *average consistent* extension, and this extension is unique. Since the weighted concede-and-divide rules are *resource monotonic*, they are covered. Therefore *average consistency* offers a reasonable alternative to *consistency* that allows significantly broader opportunities for extending to general populations rules that have been found desirable in the two-claimant case.

References

Aumann, R., Maschler, M.: Game theoretic analysis of a bankruptcy problem from the Talmud. Journal of Economic Theory **36**, 195–213 (1985)
Benoît, J-P.: The nucleolus is contested-garment–consistent: a direct proof. Journal of Economic Theory **77**, 192-196 (1997)
Chun, Y.: Equivalence of axioms for bankruptcy problems. International Journal of Game Theory **28**, 511–520 (1999)
Chun, Y., Schummer, J., Thomson, W.: Constrained egalitarianism: a new solution to claims problems. Seoul Journal of Economics **14**, 269–297 (2001)
Dagan, N.: New characterizations of old bankruptcy rules. Social Choice and Welfare **13**, 51–59 (1996)
Dagan, N., Serrano, R., Volij, O.: A non-cooperative view of consistent bankruptcy rules. Games and Economic Behavior **18**, 55–72 (1997)
Dagan, N., Volij, O.: Bilateral comparisons and consistent fair division rules in the context of bankruptcy problems. International Journal of Game Theory **26**, 11–25 (1997)
Hokari, T.: Axiomatic analysis of TU coalitional games: population monotonicity and consitency. Ph.D. Dissertation (2000)
Hokari, T., Thomson, W.: Lifting lemmas for bankruptcy rules. Mimeo (2000)
Moulin, H.: Priority rules and other asymmetric rationing methods. Econometrica **68**, 643–684 (2000)
O'Neill, B.: A problem of rights arbitration from the Talmud. Mathematical Social Sciences **2**, 345–371 (1982)
Orshan, G.: Non-symmetric prekernels. Hebrew University of Jerusalem, Mimeo (1994)
Piniles, H.M.: Darkah shel Torah. Vienna: Forester 1861

Schmeidler, D.: The nucleolus of a characteristic function game. SIAM Journal on Applied Mathematics **17**, 1163–1170 (1969)

Shapley, L. S.: A value for n-person games. In: Kuhn, H., Tucker, A. W. (eds.) Contributions to the theory of games, Vol. 2, pp. 307–317. Princeton: Princeton University Press 1953

Thomson, W.: Axiomatic analysis of bankruptcy and taxation problems: a survey. Mimeo (1995)

Thomson, W.: Consistent allocation rules. Mimeo (1996)

Thomson, W.: Claims problems and the non-existence of a consistent compromise between the constrained equal awards and constrained equal losses rules. Mimeo (2001a)

Thomson, W.: A characterization of a family of rules for the problems of claims resolution. Mimeo (2001b)

Young, P.: On dividing an amount according to individual claims or liabilities. Mathematics of Operations Research **12**, 398–414 (1987)

Speculative trading with rational beliefs and endogenous uncertainty[*]

Ho-Mou Wu[1] and Wen-Chung Guo[2]

[1] Department of International Business and Department of Economics, National Taiwan University, Academia Sinica and Chung-Hua Institution for Economic Research, Taipei, TAIWAN
 (e-mail: homouwu@ccms.ntu.edu.tw)
[2] Department of Finance, Yuan Ze University, Chung-Li, TAIWAN

Received: March 15, 2001; revised version: April 26, 2002

Summary. This paper introduces the framework of rational beliefs of Kurz (1994), which makes the assumptions of heterogeneous beliefs of Harrison and Kreps (1978) and Morris (1996) more plausible. Agents hold diverse beliefs that are "rational" in the sense of being compatible with ample observed data. In a non-stationary environment the agents only learn about the stationary measure of observed data, but their beliefs can remain non-stationary and diverse. Speculative trading then stems from disagreements among traders. In a Markovian framework of dividends and beliefs, we obtain analytical results to show how the speculative premium depends on the extent of heterogeneity of beliefs. In addition, we demonstrate that there exists a unique Rational Belief Equilibrium (RBE) generically with endogenous uncertainty (as defined by Kurz and Wu, 1996) and that the RBE price is higher than the rational expectation equilibrium price (REE) under some general conditions.

Keywords and Phrases: Speculation, Asset pricing, Rational beliefs, Endogenous uncertainty.

JEL Classification Numbers: D84, G12.

---

 * We are deeply grateful to Mordecai Kurz for his constant encouragement and inspiring guidance over the years. We wish to express our gratitude to an anonymous referee for the very valuable comments provided. We also thank Kenneth Arrow, Peter Hammond, Roko Aliprantis and Nicholas Yannelis for their helpful suggestions and Academia Sinica and the National Science Council of the R.O.C. for their indispensable support.
Correspondence to: H.-M. Wu

1 Introduction

Speculation has been a major research topic for economists, especially in light of recent financial crises and speculative attacks on currency and stock markets. Faced with a similarly turbulent world sixty years ago, Lord Keynes brought to public's attention the relationship between speculation and subjective expectations, comparing the determination of stock prices to a "beauty contest." Investors are "concerned, not with what an investment is really worth to a man who buys it for keeps, but with what the market will value it at, under the mass psychology, three months or a year hence."(Keynes, 1936). According to Kaldor (1939), speculation may be defined as "the purchase (or sale) of goods with a view to resale (or repurchase) at a later date . . . ." Such speculative behavior cannot exist in a world of complete markets or rational expectations (see Arrow, 1953; Feiger, 1976; Tirole, 1982), where investors do not change their asset holding even when markets reopen later. So the appropriate framework for the study of speculative trading is that of incomplete financial markets with sequential trading; speculative trading can then stem from disagreements among investors. The purpose of this paper is to probe further into the relationship between speculation and subjective valuation and to provide a rigorous foundation for a theory on asset pricing with speculative trading.

There are at least three approaches to modeling disagreements and speculation.[1] The first is found in a wide literature based on the presence of private information and noise (liquidity) investors (see, for example, Grossman and Stiglitz, 1980; De-Long et al., 1990). Then the difference-of-opinion approach by Varian (1985, 1989) and Harris and Raviv (1993) dispenses with the noise investors and obtains diverse posterior beliefs from the differences in the way investors interpret common information. A third method also exists to explain diverse posterior beliefs by relaxing the assumption of common prior, as in Harrison and Kreps (1978) and Morris (1996). Harrison and Kreps studied a model in which agents have different beliefs about the stochastic process of future dividends. They established the existence of the "minimal consistent price scheme" which is a partial equilibrium version of the Radner's "equilibrium of plans, prices and price expectations" (Radner, 1972). They also demonstrated the existence of positive speculative premiums, but only by providing some numerical examples. Morris (1996) adopted a simplified framework in which dividends follow an independently and identically distributed (i.i.d.) binomial distribution. The agents may have different prior distributions, but the difference of beliefs will disappear as agents learn from observing realized dividends. So the price of a risky asset can be greater than its fundamental value, but only initially, and the difference will converge to zero as investors' beliefs converge with Bayesian learning when more information becomes available. Such a framework may be more appropriate for modeling asset pricing during initial public offerings, but not for other speculative phenomena. Besides, the framework of i.i.d. binomial distribution for dividends is also quite limited.

According to Morris (1996), the result of Harrison and Kreps "has apparently been largely ignored, presumably because of the assumption of (unmodeled) hetero-

---

[1] Other related studies include Hirshleifer (1975), Kreps (1977), Milgrom and Stokey (1982), Leach (1991) and Detemple and Murphy (1994).

geneity of expectations." (Morris, 1996, p. 1112). We intend to correct the weakness of Harrison and Kreps (1978) by providing a theory to justify the continued presence of diverse beliefs in a general Markovian framework. Since the heterogeneity of beliefs will be sustained in our model, our theory is also different from that of Morris (1996). This paper proposes a new framework for the study of speculative phenomena without the shortcomings mentioned above, through the introduction of the rational beliefs of Kurz (1994) (see also Kurz and Schneider, 1996; Kurz and Wu, 1996; Kurz and Beltratti, 1997; Nielsen, 1996, for further development and applications). The theory of rational beliefs assumes that agents have ample data and that an empirical distribution exists which is commonly known to all agents. The theory then shows that the empirical distribution can be uniquely extended to a probability measure on an infinite sequence of observations and that relative to that measure, the process of observed variables is stationary. We call that probability measure the "stationary measure" of the dynamics. The stationary measure may be different from the original measure, possibly non-stationary, under which the data was generated, but the stationary measure is the common empirical knowledge on which all agents agree.

Investors have diverse beliefs which are "rational" in the sense of being compatible with observed data. In a stable environment the investors can learn only about the stationary measure of observed data. Although the stationary measures of investors' beliefs will become the same as those of the data with complete learning, these beliefs may stay non-stationary and diverse. In other words, the set of rational beliefs compatible with data (having the same stationary measure) can be quite large, including those non-stationary beliefs which may differ on the timing of some rare events (such as structural changes). Therefore, investors may disagree even when they are allowed to learn from a substantial number of observations (see Kurz, 1994). Our model provides a foundation for the continued presence of heterogeneous expectations in speculative trading, which was not discussed by Harrison and Kreps or Morris. The framework of rational beliefs enables us to study many interesting phenomena. In particular, we demonstrate in Section 3 of this paper the emergence of endogenous uncertainty (see Kurz and Wu, 1996; Huang and Wu, 1999) in a Rational Belief Equilibria (RBE) and the continued deviation of asset prices from the agents' valuation if obliged to hold on to the asset forever. We show that positive speculative premiums will persist in a Markovian model of speculative trading. We also demonstrate how the Rational Belief equilibrium (RBE) price may differ from the Rational Expectations Equilibrium (REE) price.

Before introducing rational beliefs in Section 3, we will discuss the basic model and analyze the properties of asset prices with a general Markovian belief system in Section 2. We demonstrate that the equilibrium asset prices can be determined on the basis of a "representative belief", which is constructed systematically from the heterogeneous beliefs of agents in the economy. The equilibrium asset prices with speculative trading were demonstrated by Harrison and Kreps (1978) to be no less than any investor's valuation. Morris (1996) showed that asset prices are strictly greater than any investor's valuation under certain conditions, but in a simplified framework of i.i.d. binomial distribution for dividends. Unlike Morris (1996), we adopt a Markovian framework to model dividends and agents' beliefs. In our framework of Markovian dividend processes and Markovian beliefs, we find the conditions for the emergence of positive speculative premiums by utilizing the technique of constructing a "representative belief" for the economy. In addition, we provide analytical results to show how the premium depends upon the extent of heterogeneity of beliefs, whereas Morris (1996) obtained results only from numerical simulation. All proofs are provided in the Appendix. Section 4 concludes.

2 Heterogeneous Markovian beliefs

We consider an economy with a finite number of types of investors ($i = 1, \cdots, K$), each type having different expectations about the future values of a risky asset. Following Harrison and Kreps (1978) and Morris (1996), we also assume that investors are risk neutral, that each type of investor has infinite collective wealth, and that no investor can sell the asset short. As discussed by Harrison and Kreps (1978, Section VI), such a model is a good approximation to a world of risk averse investors with finite wealth. As for the no short sales assumption, allowing some finite amount of short selling would not change the main results (see Morris, 1996, p. 1122) in this type of model.[2]

2.1 The basic model

All investors have access to the same information set, and future dividends $\{d_t\}$ of the risky asset are believed to follow a specific exogenous stochastic process. Harrison and Kreps (1978) allowed for a general functional dependence of dividends on the current information set while Morris (1996) considered the case of an i.i.d. dividend process. We adopt the assumption that the dividends follow a finite-state stationary Markov chain, which is similar to Section III of Harrison and Kreps. Suppose the transition matrix of investors' beliefs is an $S \times S$ matrix,

$$Q^i = \begin{bmatrix} q_{11}^i & q_{12}^i & \cdots & q_{1S}^i \\ \vdots & \vdots & \cdots & \vdots \\ q_{s1}^i & q_{s2}^i & \cdots & q_{sS}^i \\ \vdots & \vdots & \cdots & \vdots \\ q_{S1}^i & q_{S2}^i & \cdots & q_{SS}^i \end{bmatrix} = \begin{bmatrix} q_1^i \\ \vdots \\ q_s^i \\ \vdots \\ q_S^i \end{bmatrix}, \tag{1}$$

where q_s^i is the sth row vector of Q^i, and $q_{ss'}^i$ represents the probability of state s' occurring in the next period, given that the current state is s, $s = 1, \cdots, S$. The elements of q_s^i should be between 0 and 1, and the sum of all elements is equal to one. The investors also believe that the dividends follow a Markov process. As in Kaldor or Harrison and Kreps, we say that investors exhibit "speculative trading

---

[2] The assumption of infinite wealth can be relaxed without much change to our results so long as the class of investors has enough wealth to purchase the security they have chosen. Harrison and Kreps (1978) also discussed the alternative of introducing a bond that makes unnecessary the infinite wealth assumption. For further development, please see Wu and Guo (2002).

behavior" if they are willing to pay more for the risky asset with a right to resell, than what they would pay if obliged to hold on to it forever.[3] Investors are willing to pay a "speculative premium" for the anticipated gains from speculative trading. That is, investors of type i will buy the asset when the price of the security in state s is lower than his willingness to pay in that state,

$$p_s \leq \gamma E_s^i(\vec{p} + \vec{d}) = \gamma q_s^i(\vec{p} + \vec{d}) , s = 1, \cdots S,$$

where p_s represents the price of the asset in state s, $\vec{p} = (p_1, \cdots, p_s, \cdots, p_S)'$, $\vec{d} = (d_1, \cdots, d_s, \cdots, d_S)'$, $E_s^i(\cdot)$ represents expectation of investors of type i when state s occurs, and γ denotes the common discount rate, $\gamma < 1$. As the investors purchase the asset in state s, the price p_s goes up until the market price is equal to the highest willingness to pay among investors.

We can define a stationary consistent price scheme p_s^* as follows:

$$p_s^* = \max_i \gamma E_s^i(\vec{p^*} + \vec{d}) = \max_i \gamma q_s^i(\vec{p^*} + \vec{d}) , s = 1, \cdots S . \tag{2}$$

Harrison and Kreps (1978) showed that there are many non-stationary consistent price schemes. In Proposition 3 we will demonstrate that there exists a unique stationary consistent price scheme, which is also the equilibrium price.

Let $\vec{p^i}$ be the expected present value of subjectively evaluated dividends to an investor of type i. With the risk neutrality assumption we can derive the "subjective valuation" $\vec{p^i}$ of an investor of type i if obliged to hold the asset forever: $\vec{p^i} = \gamma Q^i(\vec{p^i} + \vec{d})$. Such a "subjective valuation" will be used to define "speculative premium" later. By applying the following Lemma, $\vec{p^i}$ can be solved uniquely:

$$\vec{p^i} = (I - \gamma Q^i)^{-1} \gamma Q^i \vec{d}. \tag{3}$$

Lemma 1 (McKenzie, 1960). *For any $S \times S$ transition matrix Q and $\gamma < 1$, $I - \gamma Q$ is invertible.*

2.2 The case of two states

Before considering the general case of S states in the next subsection, we will illustrate some basic ideas in the framework of 2×2 Markov chains ($S = 2$). Suppose the belief of investors of type i is

$$Q^i = \begin{bmatrix} 1 - a^i & a^i \\ 1 - b^i & b^i \end{bmatrix}, i = 1, \cdots, K, \tag{4}$$

where a^i, b^i are all between 0 and 1. Suppose there are two possible values of dividends $\vec{d} = \begin{bmatrix} d_1 \\ d_2 \end{bmatrix}$, with $d_2 > d_1$.

---

[3] It is difficult to find a satisfactory definition of speculation (see the discussion in Hart and Kreps, 1986). The definition adopted here follows the lead of Keynes (1936), Kaldor (1939), Feiger (1976), Harrison and Kreps (1978) and Hart and Kreps (1986). Our rational belief model introduces a new element into the concept of speculation in this literature by allowing agents to speculate that the beliefs of other agents may change in the future and choose their portfolios accordingly. We are grateful to the anonymous referee for helpful comments on this point.

When there is only one type of investor in the economy or when there are no disagreements among different types of investors ($Q^i = Q$, $\forall i$), the market equilibrium price can be represented by (3). However, when we have heterogeneous beliefs, the phenomenon of speculative trading occurs as shown in the following example.

Example 1 (Harrison and Kreps, 1978). As in Harrison and Kreps (1978), we assume $K = 2$, $\gamma = 0.75$, $d_1 = 0$, $d_2 = 1$,

$$Q^1 = \begin{bmatrix} \frac{1}{2} & \frac{1}{2} \\ \frac{2}{3} & \frac{1}{3} \end{bmatrix}, \ Q^2 = \begin{bmatrix} \frac{2}{3} & \frac{1}{3} \\ \frac{1}{4} & \frac{3}{4} \end{bmatrix}.$$

By applying equation (3) we can find

$$\vec{p^1} = \begin{bmatrix} \frac{4}{3} \\ \frac{11}{9} \end{bmatrix} \approx \begin{bmatrix} 1.33 \\ 1.22 \end{bmatrix}, \ \vec{p^2} = \begin{bmatrix} \frac{16}{11} \\ \frac{21}{11} \end{bmatrix} \approx \begin{bmatrix} 1.45 \\ 1.91 \end{bmatrix}.$$

With the infinite wealth and no short sale assumptions, it can be conjectured that investors of type 2 will hold the asset and the market price will be $\vec{p^2}$. However, investors of type 1 can "speculate" by buying the asset in $s = 1$ with the intention of selling it when $s = 2$ occurs. Such a speculative plan can generate a revenue of

$$[\frac{1}{2}(0.75) + (\frac{1}{2})^2(0.75)^2 + \cdots] \cdot (1 + \frac{21}{11}) \approx 1.75,$$

which is greater than the purchase cost of $\frac{16}{11}$ in $s = 1$. So the market price in $s = 1$ should be at least 1.75. That is, the market price should become $\begin{bmatrix} 1.75 \\ 1.91 \end{bmatrix}$, due to the speculative behavior of type 1 investors. However, 1.91 cannot be the final price in $s = 2$ since investors of type 2 can "speculate" by buying the asset in $s = 2$ with the intention of selling it when $s = 1$ occurs. Then there exists another speculation plan and so on. The speculation process will continue until it converges. The limit in this case is $\vec{p^*} \approx \begin{bmatrix} 1.85 \\ 2.08 \end{bmatrix}$.

Harrison and Kreps demonstrate that the infinite progression as in the above example finally stops and achieves a "minimal consistent price scheme", which is also the market price. In the following proposition, we obtain a characterization of the market price inspired by the treatment in Harrison and Kreps. We introduce a concept of "representative belief" which is constructed from the beliefs of all investors. Instead of getting an approximate limit, we can determine the market price precisely and efficiently on the basis of the "representative beliefs."

Proposition 1. *For the $S = 2$ case, there exists a unique (stationary) market price $\vec{p^*}$ and a representative belief Q^*, such that*

$$\vec{p^*} = (I - \gamma Q^*)^{-1} \gamma Q^* \vec{d}, \tag{5}$$

$$Q^* = \begin{bmatrix} 1 - \max_i a^i & \max_i a^i \\ 1 - \max_i b^i & \max_i b^i \end{bmatrix}. \tag{6}$$

We can apply this result to Example 1:

Example 1 (Extended). We can construct a "representative belief" Q^* by equation (6) of Proposition 1, and use equation (5) in Proposition 1 to derive $\vec{p^*}$:

$$Q^* = \begin{bmatrix} \frac{1}{2} & \frac{1}{2} \\ \frac{1}{4} & \frac{3}{4} \end{bmatrix}, \quad \vec{p^*} = \begin{bmatrix} \frac{24}{13} \\ \frac{27}{13} \end{bmatrix} \approx \begin{bmatrix} 1.85 \\ 2.08 \end{bmatrix},$$

which is the same as the "minimal consistent price scheme" derived by Harrison and Kreps. Note that our method efficiently generates quite a precise result (in fractions) in just one step, without having to go through an infinite sequence of finding various speculative plans.

Next we consider whether positive speculative premiums would exist in this economy. Speculative premiums are defined as the differences between the market price $\vec{p^*}$ and the subjective valuation $\vec{p^i}$ by investors of type i ($i = 1, \cdots, K$) if investors are obliged to hold the asset forever. Only when the market price is strictly greater than the subjective valuation of all investors, can we say that speculative premiums are positive. In Example 1, the speculative premiums $\vec{p^*} - \max\{\vec{p^1}, \vec{p^2}\}$ are positive; however, if there is a "dominant" investor j who has the highest valuation in all states, i.e., $a^j = \max_i a^i$ and $b^j = \max_i b^i$, this investor becomes the representative investor in the market with his belief Q^j being selected as Q^*. Then the speculative premium is zero. Morris (1996) demonstrated the existence of positive speculative premiums with an i.i.d. dividend process. In the following proposition, we provide conditions for the existence of positive speculative premiums in a Markovian framework, which was also adopted by Harrison and Kreps (1978) in their Section III. We also study how the size of premiums relates to the extent of heterogeneity of beliefs. Without loss of generality, we assume that $a^1 = \max_i a^i$, and $b^2 = \max_i b^i$ (no dominant investor).

Proposition 2. *Suppose there is no dominant investor ($a^1 = \max_i a^i$ and $b^2 = \max_i b^i$) in the market for the $S = 2$ case.*

(a) If $b^2 \neq 1$, then strictly positive speculative premiums exist in each state.

(b) The speculative premium $\vec{p^*} - \max\{\vec{p^1}, \vec{p^2}\} = \min\{\vec{p^*} - \vec{p^1}, \vec{p^*} - \vec{p^2}\}$, where

$$\vec{p^*} - \vec{p^1} = \frac{\gamma}{1-\gamma} \cdot \frac{1}{1 + (a^1 - b^2)\gamma} \cdot \frac{1}{1 + (a^1 - b^1)\gamma} \tag{7}$$
$$\cdot (d_2 - d_1) \cdot (b^2 - b^1) \cdot \left[\frac{a^1 \gamma}{1 - \gamma + a^1 \gamma} \right],$$

$$\vec{p^*} - \vec{p^2} = \frac{\gamma}{1-\gamma} \cdot \frac{1}{1 + (a^1 - b^2)\gamma} \cdot \frac{1}{1 + (a^2 - b^2)\gamma} \tag{8}$$
$$\cdot (d_2 - d_1) \cdot (a^1 - a^2) \cdot \left[\frac{1 - b^2 \gamma}{(1 - b^2)\gamma} \right].$$

(c) The size of speculative premiums increases as a^1 or b^2 increases and as a^2 or b^1 decreases:

$$\frac{\partial(\vec{p^*} - \vec{p^1})}{\partial b^1} < 0 < \frac{\partial(\vec{p^*} - \vec{p^1})}{\partial b^2}, \quad \frac{\partial(\vec{p^*} - \vec{p^2})}{\partial a^2} < 0 < \frac{\partial(\vec{p^*} - \vec{p^2})}{\partial a^1}. \tag{9}$$

This proposition demonstrates that strictly positive speculative premiums exist in a Markovian framework unless there is a dominant investor or the investors who hold the asset do not expect to sell it in the next period with probability 1, i.e., they do not expect to speculate at all. The later exception is described mathematically by the condition $b^2 \neq 1$. Since $a^1 - a^2$ and $b^2 - b^1$ measure the extent of heterogeneity of beliefs in the economy, changing a^i or b^i, while holding the other parameter constant in Proposition 2(c) tells us that the size of speculative premiums is an increasing function of the heterogeneity of beliefs.

2.3 The case of S states

Next we extend the results in the 2×2 Markov chains to the general case of $S \times S$ Markov chains. With Lemma 1, the subjective valuation by investors of type i can also be solved for the general $S \times S$ case,

$$\vec{p^i} = (I - \gamma Q^i)^{-1} \gamma Q^i \vec{d}. \tag{3}$$

Following the same reasoning as Proposition 1, we can find a type $i(s)$ of investors who have the highest valuation of the asset for any given state s. We will show that the market equilibrium price for the general $S \times S$ case can also be written as follows:

$$\vec{p^*} = \gamma \begin{bmatrix} q_1^{i(1)} \\ \cdot \\ q_s^{i(s)} \\ \cdot \\ q_S^{i(S)} \end{bmatrix} (\vec{p^*} + \vec{d}) = \gamma Q^*(\vec{p^*} + \vec{d}),$$

where Q^* is the "representative belief" of a fictitious investor. In contrast to the 2×2 case with a unique Q^* as in Proposition 1, the representative belief is not necessarily unique in the general case. However, there exists a unique equilibrium price.

Example 2. In a 3×3 Markovian framework, there may exist multiple representative beliefs:

$$Q^1 = \begin{bmatrix} 0.5\ 0.4\ 0.1 \\ 0.5\ 0.4\ 0.1 \\ 0.3\ 0.3\ 0.4 \end{bmatrix}, Q^2 = \begin{bmatrix} 0.3\ 0.2\ 0.5 \\ 0.3\ 0.5\ 0.2 \\ 0.7\ 0.2\ 0.1 \end{bmatrix}, Q^3 = \begin{bmatrix} 0.4 & 0.4 & 0.2 \\ 0.4 & 0.5 & 0.1 \\ 0.22\ 0.43\ 0.35 \end{bmatrix}.$$

Given the dividend vector $\vec{d} = [0, 0.5, 1]'$ and discount factor $\gamma = 0.75$, there are two representative beliefs Q^{*1} and Q^{*2}, but a unique equilibrium price $\vec{p^*}$ still exists.

$$\vec{p^*} = \begin{bmatrix} 1.6544 \\ 1.5221 \\ 1.6103 \end{bmatrix}, Q^{*1} = \begin{bmatrix} 0.3\ 0.2\ 0.5 \\ 0.3\ 0.5\ 0.2 \\ 0.3\ 0.3\ 0.4 \end{bmatrix}, Q^{*2} = \begin{bmatrix} 0.3 & 0.2 & 0.5 \\ 0.3 & 0.5 & 0.2 \\ 0.22\ 0.43\ 0.35 \end{bmatrix}.$$

In fact, the third row of Q^{*1} is constructed from Q^1 and the third row of Q^{*2} is constructed from Q^3. When $s = 3$, both rows satisfy

$$p_3^* = \gamma q_3^i (\vec{p}^* + \vec{d}) = 1.6103 \, , \, i = 1, 3.$$

Now we can study the general "representative belief" by considering the set of possible "combined beliefs" of some fictitious investor f, $f = 1, \cdots, K^S$:

$$\Psi = \{ Q^f | q_s^f = q_s^i \, , i = 1, \cdots K, s = 1, \cdots, S \} \, .$$

The "representative belief" Q^* can be found as an element in Ψ. In Harrison and Kreps (1978, pp.331), the equilibrium price was calculated by iterated speculation of different investors until the price converges, as shown in Example 1. A shortcoming of this procedure is that it needs an infinite number of steps to find the solution, which is an approximation at best. Introducing "representative beliefs," as done in the last subsection, provides both a precise and efficient solution. Since the number of elements in Ψ is equal to K^S, just going through every element member of Ψ to find Q^* cannot be undertaken easily when the number of types of investors K becomes large (for example, where $K = 100$ and $S = 4$, it may need 100^4 steps). However, in what follows, we provide an efficient algorithm to discover equilibrium price $\vec{p}^*$ and representative belief Q^* in just a few steps. Furthermore, the number of steps depends only on the number of state S, not the number of types of investors K. Our experiences with numerical simulation suggest that, in a typical 4×4 case, it only needs at most 4 steps even when the number of types of investors is very large. The algorithm is as follows:

Step 1: Compute the subjective valuation $\vec{p}^i = (I - \gamma Q^i)^{-1} \gamma Q^i \vec{d}$ if investors of type i are obliged to hold forever, $i = 1, \cdots, K$. For each s, find the highest valuation $p_s^{i0(s)}$ and its corresponding type $i0(s)$. Set

$$Q^{f0} = \begin{bmatrix} q_1^{i0(1)} \\ \vdots \\ q_S^{i0(S)} \end{bmatrix} \in \Psi \, .$$

Step 2: Compute the corresponding price $\vec{p}^{f0} = (I - \gamma Q^{f0})^{-1} \gamma Q^{f0} \vec{d}$ for the fictitious belief Q^{f0} constructed in Step 1. Then compute the "willingness to pay" $\vec{W}^{i0} = \gamma Q^i (\vec{p}^{f0} + \vec{d})$ associated with $\vec{p}^{f0}$ for each type i. Find the highest willingness to pay for each state s and its corresponding type $i1(s)$. Set

$$Q^{f1} = \begin{bmatrix} q_1^{i1(1)} \\ \vdots \\ q_S^{i1(S)} \end{bmatrix} \in \Psi \, .$$

Let $Q^{f1} = F(Q^{f0})$, which defines a mapping $F : \Psi \to \Psi$.

Step 3: Compute the corresponding price $\vec{p}^{f1} = (I - \gamma Q^{f1})^{-1}\gamma Q^{f1}\vec{d}$, for the fictitious belief Q^{f1} constructed in Step 2. If $\vec{p}^{f1} = \vec{p}^{f0}$, stop the algorithm and list the price $\vec{p}^{f1}$ and belief Q^{f1} as the equilibrium values. If $\vec{p}^{f1} \neq \vec{p}^{f0}$, repeat Step 2 until $\vec{p}^{f(n+1)} = \vec{p}^{fn}(= \vec{p}^*)$. The corresponding $Q^{f(n+1)}$ and Q^{fn} are exactly the "representative belief" Q^* discussed earlier. Note that while the equilibrium price is unique the representative beliefs are not.

This algorithm searches through the elements of Ψ for a candidate for the representative belief. A fixed point of the mapping F is shown to exist in Proposition 3. Before presenting our formal results, we can illustrate the procedure for finding $\vec{p}^*$ and Q^* in the following examples:

Example 2 (continued). Given the beliefs Q^i for $i = 1, 2, 3$, we first compute their subjective valuation if obliged to hold the asset forever:

$$\vec{p^1} = \begin{bmatrix} 0.9726 \\ 0.9726 \\ 1.2145 \end{bmatrix}, \vec{p^2} = \begin{bmatrix} 1.4069 \\ 1.3288 \\ 1.1762 \end{bmatrix}, \vec{p^3} = \begin{bmatrix} 1.2208 \\ 1.1690 \\ 1.3589 \end{bmatrix}.$$

We can find that $i0(1) = 2$, $i0(2) = 2$ and $i0(3) = 3$ are the types $i0(s)$ with highest valuation in state $s = 1, 2, 3$. Then we can construct Q^{f0} and compute $\vec{p}^{f0}$:

$$Q^{f0} = \begin{bmatrix} 0.3 & 0.2 & 0.5 \\ 0.3 & 0.5 & 0.2 \\ 0.22 & 0.43 & 0.35 \end{bmatrix}, \quad \vec{p}^{f0} = \begin{bmatrix} 1.6544 \\ 1.5221 \\ 1.6103 \end{bmatrix}.$$

In fact, this algorithm converges in one step: $\vec{p}^{f0} = \vec{p}^*$. In all examples, the algorithm converges in a small number of steps. In the proof of Proposition 3, we provide such an efficient and finite algorithm, which is in contrast to the infinite algorithm of Harrison and Kreps (1978).

Proposition 3. *For the $S \times S$ Markovian beliefs, there exists a finite algorithm to find a unique market price $\vec{p}^*$ and at least one fictitious investor whose "representative belief" Q^* satisfies the following equation:*

$$\vec{p}^* = (I - \gamma Q^*)^{-1}\gamma Q^*\vec{d}, \tag{10}$$

and $Q^* \in \Psi = \{Q^f | q_s^f = q_s^i, i = 1, \cdots K, s = 1, \cdots, S\},$ \tag{11}

where Ψ is the set of possible "combined beliefs" of a fictitious investor.

In Proposition 3, we follow Proposition 2 of Harrison and Kreps (1978) and present an alternative way of demonstrating the existence and uniqueness of stationary equilibrium price $\vec{p}^*$ (the proof is in the appendix). In contrast to Harrison and Kreps (1978), the result is demonstrated by introducing "representative belief" in a specialized framework of Markovian beliefs. In doing so, we obtain a sharpened characterization of the properties of equilibrium prices, as summarized by equations (10) and (11). Note that Harrison and Kreps (1978) adopt a general framework and formulate an infinite algorithm to find the "minimal consistent price scheme" while there are many non-stationary equilibrium prices. In fact, they allow the price to be infinite and do not prove the uniqueness explicitly. In our specialized (Markovian) framework, we provide a different formulation to obtain by a finite algorithm a unique stationary equilibrium price, which is the same as their "minimal consistent price scheme".

Next we demonstrate that there are positive speculative premiums except in two extreme cases. The first is the presence of a "dominant" investor. An investor i is defined to be "dominant" if [4]

$$\vec{p^i} \geq \vec{p^j} \text{ and } \vec{p^i} \geq Q^j(\vec{p^i} + \vec{d}) , \forall j. \tag{12}$$

The second exceptional case is when investors with probability one do not expect to sell the security on any subsequent date. A general condition $g^i_{ss'} > 0, \forall i, s, s'$ is sufficient to ensure that this case will not occur. Both conditions will be stated in Proposition 4.

With the general Markov chains, it is possible to get multiple "representative beliefs" in equilibrium, as illustrated in Example 2. Hence it is not straightforward to generalize the measure of the extent of heterogeneity of Proposition 2. However, we can use the difference between equilibrium price and investors' "willingness to pay" associated with $\vec{p^*}$ to measure the heterogeneity in the economy. We can then show that the size of speculative premium is an increasing function of this heterogeneity measure of the economy. Define

$$m^i_s = p^*_s - \gamma q^i_s(\vec{p^*} + \vec{d}) = \gamma(q^*_s - q^i_s)(\vec{p^*} + \vec{d}) , \tag{13}$$

which is dependent on the difference of beliefs $q^*_s - q^i_s$ and is evaluated with respect to $\vec{p^*} + \vec{d}$. It measures how the subjective valuation of the asset $\gamma q^i_s(\vec{p^*} + \vec{d})$ by investors of type i deviates from the market valuation p^*_s. From equation (2) we know that $m^i_s \geq 0$. Define $\vec{m^i}$ to be $[m^i_1, \cdots, m^i_S]'$. We can then demonstrate the relationship between the size of speculative premiums and the extent of heterogeneity $\vec{m^i}$ in the following proposition.

Proposition 4. *Suppose that there is no dominant investor and that $q^i_{ss'} > 0$, $\forall i, s, s'$. Then speculative premium of state s, as denoted as $\pi_s = p^*_s - \max_i\{p^i_s\}$, must be positive. In addition, the speculative premium is an increasing function of the extent of heterogeneity measured by $\vec{m^i}$, that is,*

$$\vec{p^*} - \vec{p^i} = (I - \gamma Q^i)^{-1} \vec{m^i} . \tag{14}$$

---

[4] One might think that

$$q^i_s \vec{d} \geq q^j_s \vec{d} , \forall j, s ,$$

could be one possible definition for agent i to be a "dominant" investor since he is the most optimistic with regard to dividends. It is equivalent to (12) only in the case of 2 states. There exist counter-examples such that investor i is dominant ($\vec{p^*} = \vec{p^i}$) but is not the most optimistic with regard to dividends when S is greater than 2. We are grateful to the referee for helpful comments on the concept of the "dominant investor."

3 Endogenous uncertainty with rational Markovian beliefs

In this section we will study how endogenous uncertainty may emerge with speculative trading. We also provide reasons to explain why heterogeneous beliefs can
persist in a Markovian framework with rational beliefs.

3.1 The meaning of endogenous uncertainty

In the previous section we found that positive speculative premiums exist with
sufficiently diverse beliefs. In the analysis we need not require the dividends to be
distinct in all S states. The states $s = 1, \cdots, S$ can be used to represent possible
values of market prices, of which only some are affected by the exogenously given
values of dividends. We say that "endogenous uncertainty" is present in a Markovian
economy if the endogenously determined equilibrium prices are distinct even when
the exogenous variables (dividends) take the same value (see Kurz and Wu, 1996;
Huang and Wu, 1999, for a formal definition and general discussion).

Utilizing the technique of constructing a "representative belief" as in the previous section, we find conditions under which market equilibrium prices can be
different even when dividends are the same. In the following example there are two
possible values of dividends ($d_L = 0$, $d_H = 1$), but three distinct market equilibrium prices are present for $S = 3$. We define this phenomenon as the emergence
of endogenous uncertainty, which can occur in a specialization of the model of
Harrison and Kreps (1978).

Example 3. Suppose $S = 3$, $\gamma = 0.75$. The market equilibrium prices can be
determined by the "representative beliefs" Q^* as in the last section. In equation
(15) there are three distinct prices while there are only two prices in equation (16).

$$\vec{d} = \begin{bmatrix} 0 \\ 1 \\ 1 \end{bmatrix}, Q^* = \begin{bmatrix} 0.4\ 0.1\ 0.5 \\ 0.2\ 0.3\ 0.5 \\ 0.3\ 0.3\ 0.4 \end{bmatrix}, \vec{p^*} = \begin{bmatrix} 1.995 \\ 2.171 \\ 2.089 \end{bmatrix}. \tag{15}$$

$$\vec{d} = \begin{bmatrix} 0 \\ 1 \\ 1 \end{bmatrix}, Q^* = \begin{bmatrix} 0.4\ 0.1\ 0.5 \\ 0.3\ 0.2\ 0.5 \\ 0.3\ 0.3\ 0.4 \end{bmatrix}, \vec{p^*} = \begin{bmatrix} 1.946 \\ 2.027 \\ 2.027 \end{bmatrix}. \tag{16}$$

A closer examination reveals one possible reason for the differences between
equation (15) and (16). In (15), the probability $Pr^*(d_{s^{t+1}} = 1 | s^t = s)$ of getting
high dividend ($d_H = 1$) given the current state s, according to the representative
belief Q^*, is equal to $0.3 + 0.5 = 0.8$ when $s = 2$ and is equal to $0.3 + 0.4 = 0.7$ when $s = 3$. These two conditional probabilities are different. In (16), the two
conditional probabilities given $s = 2$ and $s = 3$ are both equal to 0.7.

Example 3 represents a general phenomenon not only valid for representative
belief Q^*, but also true for some given belief Q. In the following lemma, we state
the conditions for the subjective valuation $\vec{p}^Q$ of an investors with belief Q to be
the same in all states where the dividends are the same, i.e.,

$$\text{if } d_s = d_{s'}, \text{ then } p_s^Q = p_{s'}^Q.$$

Note that such an investor can be a fictitious investors as described in Proposition 3.

Lemma 2. *Given S states and 2 possible values of dividends ($d = d_L, d_H$), the necessary and sufficient condition for the subjective price $\vec{p}^Q$ of an investor with belief Q to be the same in all states where the dividends are the same is that the conditional probabilities of getting high (low) dividends are the same for all current states with high (low) dividends. This condition can be written as*

$$Pr^Q(d_{s^{t+1}} = d_H | s^t) = k_H \text{ , for all } s^t \text{ such that } d_{s^t} = d_H , \tag{17a}$$

and

$$Pr^Q(d_{s^{t+1}} = d_L | s^t) = k_L \text{ , for all } s^t \text{ such that } d_{s^t} = d_L . \tag{17b}$$

Lemma 2 applies to the case of two possible values of dividends ($d_t \in D = \{d_L, d_H\}$), which is assumed in this section. We also assume that the stationary measure of the dividend process follows a Markov chain which can be represented by a transition matrix of the following form as in Kurz and Schneider (1996) and Kurz (1998):

$$\Lambda = \begin{bmatrix} \phi & 1 - \phi \\ 1 - \phi & \phi \end{bmatrix} . \tag{18}$$

Consider the rational expectations equilibrium (REE) approach in economic modeling where all agents are assumed to be able to carry out the necessary calculations to deduce the equilibrium price map $p_t = P(s_t)$ given the knowledge of the exogenous state s_t. The agents are then said to have "structural knowledge."[5] If all investors possess structural knowledge including equation (18), the investors will not choose any belief other than the one in equation (18). Then from Proposition 1 the REE price should be represented by equation (19). We will use the REE price $\vec{p}^\Lambda$ as a reference point for comparison later.

$$\vec{p}^\Lambda = \begin{bmatrix} p_L^\Lambda \\ p_H^\Lambda \end{bmatrix} \tag{19}$$

$$= \frac{\gamma}{(1 - \gamma)(1 + \gamma - 2\gamma\phi)} \cdot \begin{bmatrix} (\gamma - 2\gamma\phi + \phi)d_L + (1 - \phi)d_H \\ (1 - \phi)d_L + (\gamma - 2\gamma\phi + \phi)d_H \end{bmatrix} .$$

However, investors have neither structural knowledge nor information about the beliefs of other investors in reality. Therefore, they may form different subjective beliefs about the process of dividends and prices. As the equilibrium price depends on the state of beliefs y_t of all agents, so

$$p_t = P(s_t, y_t) .$$

---

[5] One could interpret structural knowledge as common knowledge by all agents of the objectives of all investors and the law of motion for $\{d_t\}$. In general, structural knowledge also includes precise information of demand or supply functions and probability laws, which are usually unobservable.

Note that p_t now fluctuates when the state of beliefs varies. The component of economic fluctuations due to the agents' beliefs hence represents an important kind of uncertainty faced by all agents. This is referred to as "endogenous uncertainty" by Kurz and Wu (1996). In the rest of this paper we will study the emergence of endogenous uncertainty in a rational belief framework.

3.2 The rational belief framework

We now introduce the framework of rational beliefs of Kurz (1994) where agents do not have structural knowledge of the economy. If agents can form expectations without restrictions, we are in the framework of temporary equilibrium, which is criticized for its lack of rational utilization of information in learning. We impose a set of rationality conditions to place greater restrictions on the system of heterogeneous Markovian beliefs we have analyzed so far. Agents are said to have "rational beliefs" if their beliefs cannot be refuted by data. In a non-stationary environment, agents can learn only about the stationary measure of the observed data. Agents agree only on the stationary measure of the environment, but they can still disagree on the likelihood of some important and rare events or on the timing of regime changes even after exhausting all possibilities of learning.

Similarly to the overlapping generation framework adopted by Kurz (1998), we think of each investor as a "dynasty" consisting of a sequence of short-lived decision making agents. All investors live for a short finite life and those who belong to the same dynasty are linked by bequest motives. The theory of rational beliefs is based on the observation that our economy is a non-stationary environment with a sequence of changing regimes and the length of decision life of any agent is relatively short. During his economic life each agent has too few observations to determine if his private theory is correct or not. The private theory of each agent will be represented by an "assessment variable." The little learning which can be done during the lifetime of the agents cannot be used to refute agents' private theories or change their "assessment variables." This provides a theoretical foundation for the continued presence of heterogeneous expectation even when learning is allowed. In the rest of this paper we will study the properties of rational belief equilibria (RBE).

We assume that there are two types $i = 1, 2$ and many investors of each type.[6] The individual investor n of type i adopts a "state of belief" or an "assessment variable" $y^{i,n} \in Y^i = \{0, 1\}$, which is an i.i.d. random variable representing an investor's private signal or state of mind. Investors of type i receives at date t a random signal $y_t^{i,n}$ which determines his one-period ahead conditional probability as represented by a Markov matrix $Q_t^{i,n}$. They adopt matrix $Q_t^{i,0}$ as their belief when $y_t^{i,n} = 0$, and matrix $Q_t^{i,1}$ as their belief when $y_t^{i,n} = 1$. The probabilities of assessment variables for each agent are represented by $\text{Prob}\{y_t^{i,n} = 0\} = \alpha^i$ and $\text{Prob}\{y_t^{i,n} = 1\} = 1 - \alpha^i$. The vector $y = (y^{1,1}, \cdots, y^{1,N}, y^{2,1}, \cdots, y^{2,N})$ is a collection of individual states in an economy with $2N$ agents ($i = 1, 2$, $n = 1, \cdots, N$). Agents of type i who have the same assessment will have the same

---

[6] The results in this section hold for any finite number of types.

demand behavior. However, each agent knows his own assessment variables but not the assessment variables of any of the other agents in the economy, past or present. Within our framework, it is not possible to learn about others' private theories. It is also not possible to assess whether his private theory is wrong since the length of each economic regime is short and the agent does not have enough data.

Following Cass, Chichilnisky and Wu (1996) and Kurz (1998), we can define a "social state" to include all those collections of individual states which yield the same aggregate composition of investors in the economy. For example, all collections of individual states with half of type i investors being optimistic ($y^{i,n} = 1$) and half of type i investors being pessimistic ($y^{i,n} = 0$) yield the same aggregate composition of investors, so they are all represented by one social state for type i, denoted by its distribution $F^i = (0.5, 0.5)$. To be more precise, the distribution of type i investors can be represented by $F^i = (f^i, 1 - f^i)$, $0 < f^i < 1$, where f^i is the proportion of type i investors with assessment variables taking the value $y^{i,n} = 1$ (the rest $1 - f^i$ have assessment variables with the value $y^{i,n} = 0$). Let $\mathcal{F}^i$ be the space of possible distributions of type i investors, then distributions $F^i \in \mathcal{F}^i$.

Note that such a space of distributions $\mathcal{F}^i$ allows for possible correlation within the type. For example, $\mathcal{F}^i = \{(0.5, 0.5)\}$ corresponds to the case when individual assessments of type i investors are independently distributed with $\alpha^i = 0.5$. When individual assessments are perfectly correlated for all agents of type i, the possible distributions for type i investors can be represented by $\mathcal{F}^i = \{(0, 1), (1, 0)\}$, with two element. Other forms of correlation result in a non-degenerate representation of $\mathcal{F}^i$. For example, the element $F^i = (0.8, 0.2)$ of $\mathcal{F}^i = \{(0.8, 0.2), (0.2, 0.8)\}$ represents the case with a proportion 0.8 of type i investors adopting belief $Q^{i,1}$ and a proportion 0.2 of type i investors adopting belief $Q^{i,0}$, and similarly for $F^i = (0.2, 0.8)$. Although each investor of type i still adopts the belief $Q^{i,0}$ with probability α^i, for $F^i = (0.8, 0.2)$, 80% of them act together in choosing belief $Q^{i,1}$ while the rest of them act together in choosing the other belief $Q^{i,0}$. As for $F^i = (0.2, 0.8)$, 80% of them now choose $Q^{i,0}$ together and the rest choose $Q^{i,1}$ together.[7] This gives rise to the space of distributions $\mathcal{F}^i = \{(0.8, 0.2), (0.2, 0.8)\}$. So it must have two elements. It is a case of partial correlation.

Let there be J elements of $\mathcal{F}^i$ and $D = \{d_L, d_H\}$, then the state space considered in this section is $\hat{S} = D \times \mathcal{F}^1 \times \mathcal{F}^2$ with $S = 2 \cdot J^2$ elements (two values for dividends, and two types of agents are assumed). Each of the S elements will be called a "social state" for the whole economy.

The stationary measure on $\hat{S} = D \times \mathcal{F}^1 \times \mathcal{F}^2$ can be represented by the following transition matrix

$$\Gamma = \begin{bmatrix} \phi A & (1 - \phi)A \\ (1 - \phi)A & \phi A \end{bmatrix}, \tag{20}$$

---

[7] Since we adopt the assumptions of infinite wealth and risk-neutrality as in Harrison and Kreps (1978), only the optimists affect the equilibrium. The number of investors who are optimists or pessimists does not matter for the equilibrium price. To incorpoate a genuine influence of the distribution of investors, we need to study a model with finite wealth, as in Wu and Guo (2002).

where A is a $(J^2 \times J^2)$ transition matrix and the marginals of Γ on the dividend states D are represented by equation (18). We can represent the stationary transition matrix and belief matrices as

$$
\Gamma = \begin{bmatrix} \Gamma_1 \\ \vdots \\ \Gamma_s \\ \vdots \\ \Gamma_S \end{bmatrix} = \begin{bmatrix} \Gamma_{11} & \cdots & \Gamma_{1S} \\ \vdots & \ddots & \vdots \\ \Gamma_{s1} & \cdots & \Gamma_{sS} \\ \vdots & \ddots & \vdots \\ \Gamma_{S1} & \cdots & \Gamma_{SS} \end{bmatrix} , \tag{21}
$$

$$
Q^{i,j} = \begin{bmatrix} q_1^{i,j} \\ \vdots \\ q_s^{i,j} \\ \vdots \\ q_S^{i,j} \end{bmatrix} = \begin{bmatrix} q_{11}^{i,j} & \cdots & q_{1S}^{i,j} \\ \vdots & \ddots & \vdots \\ q_{s1}^{i,j} & \cdots & q_{sS}^{i,j} \\ \vdots & \ddots & \vdots \\ q_{S1}^{i,j} & \cdots & q_{SS}^{i,j} \end{bmatrix} , \quad \begin{array}{l} i = 1, 2, \\ j = 0, 1. \end{array}
$$

The stationary measure Γ is computed from data, and the investors' beliefs will be made consistent with Γ. Then the rationality constraint of Kurz and Schneider (1996) in a Markovian framework requires that for each individual investor

$$
\alpha^i Q^{i,0} + (1 - \alpha^i) Q^{i,1} = \Gamma \text{ for } i = 1, 2, \tag{22}
$$

where $\alpha^i = \text{Prob}\{y^{i,n} = 0\}$ and $1 - \alpha^i = \text{Prob}\{y^{i,n} = 1\}$. Note that in the rationality restriction we only use the ex-ante probability of private signals α^i, which may be different from the actual proportion f^i in the social state F^i. The equilibrium achieved in such a framework with a state space described above and beliefs satisfying (22) is defined to be a "Rational Belief Equilibrium (RBE) with social states of beliefs" (first introduced by Kurz, 1998).

Before proceeding with our analysis, we consider the benchmark case of rational expectation equilibrium (REE) when all investors' beliefs coincide with the stationary measure, i.e. $Q^{i,j} = \Gamma$, $\forall i, j$. One can show that there exists no endogenous uncertainty and there are only two possible values of prices. The market equilibrium price vector achieved in this benchmark case is reduced to $\vec{p}^A$, which is an REE price represented by equation (19). We can now write $\vec{p}^A = [p_L^A, p_L^A, p_L^A, p_L^A, \ p_H^A, p_H^A, p_H^A, p_H^A]'$. Note that the equilibrium price $\vec{p}^A$ has only two possible values p_L^A, p_H^A, associated with dividends $d_s = d_L, d_H$, respectively.

With a rational belief structure as specified by (22), investors may use different beliefs from time to time. Let the current belief at date t be $Q_t^{i,j}$ by an investor of type i with $y^{i,n} = j$, $j = 0, 1$. The agent knows his current belief $Q_t^{i,j}$, but before receiving private signals $y_{t+1}^{i,n}$ he does not know his future beliefs $Q_{t+1}^{i,j}$ except that it is one of the two Markovian matrices in equation (22) with probabilities α^i and $1 - \alpha^i$. The expectation at date t of his future beliefs at date $t + 1$ and beyond is thus equal to the weighted average of these two Markovian matrices, which is equal to Γ as required by (22). We can show in the following lemma that the expected revenue from sales of the asset in the next period is no less than the expected revenue from sales of the asset in a later period. (the right hand side of equation (23)):

Lemma 3. *For any investor of type i the following holds in an RBE:*

$$\gamma Q^{i,j}(\vec{p^*} + \vec{d}) \geq \gamma Q^{i,j}(\gamma \Gamma(\vec{p^*} + \vec{d}) + \vec{d}) \tag{23}$$

From Lemma 3, it is clear that selling in later periods will not provide any higher revenues than selling in the next period, as shown by the following inequalities:

$$\begin{aligned}
&\gamma Q^{i,j}(\vec{p^*} + \vec{d}) \tag{24}\\
&\geq \gamma Q^{i,j}(\gamma \Gamma(\vec{p^*} + \vec{d}) + \vec{d})\\
&\geq \gamma Q^{i,j}(\gamma \Gamma(\gamma \Gamma(\vec{p^*} + \vec{d}) + \vec{d}) + \vec{d})
\end{aligned}$$

Hence, in equilibrium the valuation of any investor can be represented just by his expected revenue from sales in the next period. With this lemma, the rational belief framework developed here can accommodate for the model of Harrison and Kreps (1978).

Lemma 3 performs an important role in the dynamic decision of speculation. As investors of type i adopt beliefs of $Q^{i,0}, Q^{i,1}$, Lemma 3 allows us only to consider simple trading strategies consisting of selling the security in the next period instead of selling in later periods. So we can apply the technique and results developed in the previous sections even when the beliefs adopted by investors are not constant over time.

3.3 The properties of rational belief equilibrium

Now we can show that endogenous uncertainty may exist in RBE, using the technique of constructing a "representative belief", developed in the previous sections. We also demonstrate that the RBE prices can deviate from the REE prices $\vec{p^A}$ of equation (19). The following example illustrates the possibility of the presence of endogenous uncertainty. Agents may not have structural knowledge, but they still try to learn from the data until the stationary measures of their beliefs coincide with Γ, i.e., their beliefs satisfy the rationality restrictions of the theory of rational beliefs.

Example 4. Suppose that $D = \{1, 2\}$ and each $\mathcal{F}^i = \{(0.8, 0.2), (0.2, 0.8)\}$ has two elements. The state space is $\hat{\mathcal{S}} = D \times \mathcal{F}^1 \times \mathcal{F}^2$ with eight elements. To construct a Γ consistent with equation (20), let A be a 4×4 matrix with all elements equal to 0.25, $\phi = 0.5$ and Γ be a 8×8 matrix with all elements equal to 0.125. Let $\alpha^1 = \alpha^2 = 0.5$, $\gamma = 0.75$ and B be a 1×4 matrix with all elements equal to 0.25. This is the case of i.i.d. type states with no correlation for the stationary distribution. Suppose

$$
Q^{1,0} = \begin{bmatrix} 0.30B\ 0.70B \\ 0.20B\ 0.80B \\ 0.40B\ 0.60B \\ 0.10B\ 0.90B \\ 0.60B\ 0.40B \\ 0.55B\ 0.45B \\ 0.68B\ 0.32B \\ 0.58B\ 0.42B \end{bmatrix}, \ Q^{1,1} = \begin{bmatrix} 0.70B\ 0.30B \\ 0.80B\ 0.20B \\ 0.60B\ 0.40B \\ 0.90B\ 0.10B \\ 0.40B\ 0.60B \\ 0.45B\ 0.55B \\ 0.32B\ 0.68B \\ 0.42B\ 0.58B \end{bmatrix},
$$

$$
Q^{2,0} = \begin{bmatrix} 0.60B\ 0.40B \\ 0.80B\ 0.20B \\ 0.55B\ 0.45B \\ 0.60B\ 0.40B \\ 0.25B\ 0.75B \\ 0.35B\ 0.65B \\ 0.40B\ 0.60B \\ 0.45B\ 0.55B \end{bmatrix}, \ Q^{2,1} = \begin{bmatrix} 0.40B\ 0.60B \\ 0.20B\ 0.80B \\ 0.45B\ 0.55B \\ 0.40B\ 0.60B \\ 0.75B\ 0.25B \\ 0.65B\ 0.35B \\ 0.60B\ 0.40B \\ 0.55B\ 0.45B \end{bmatrix}.
$$

we can check that these belief matrices satisfy the rationality restrictions (22) since $\alpha^1 = 0.5$ and $\alpha^2 = 0.5$: $\frac{1}{2}Q^{1,0} + \frac{1}{2}Q^{1,1} = \Gamma$, $\frac{1}{2}Q^{2,0} + \frac{1}{2}Q^{2,1} = \Gamma$. At any of the eight social states, all types of beliefs $\{Q^{1,1}, Q^{1,0}, Q^{2,1}, Q^{2,0}\}$ are present. Utilizing the techniques of constructing "representative beliefs" of Proposition 3, we can find $\vec{p}^*$ and Q^* by equation (10) and the algorithm presented in Section 2.3:

$$
Q^* = \begin{bmatrix} 0.30B\ 0.70B \\ 0.20B\ 0.80B \\ 0.40B\ 0.60B \\ 0.10B\ 0.90B \\ 0.25B\ 0.75B \\ 0.35B\ 0.65B \\ 0.32B\ 0.68B \\ 0.42B\ 0.58B \end{bmatrix},
$$

$$
\vec{p}^* = \begin{bmatrix} 5.0799 \\ 5.1504 \\ 5.0094 \\ 5.2209 \\ 5.1152 \\ 5.0447 \\ 5.0658 \\ 4.9953 \end{bmatrix} > \vec{p}^\Lambda = \begin{bmatrix} 4.5 \\ 4.5 \\ 4.5 \\ 4.5 \\ 4.5 \\ 4.5 \\ 4.5 \\ 4.5 \end{bmatrix}, \ \vec{p}^* - \max_i\{\vec{p}^i\} = \begin{bmatrix} 0.3300 \\ 0.3412 \\ 0.3188 \\ 0.3524 \\ 0.3789 \\ 0.4029 \\ 0.4713 \\ 0.4115 \end{bmatrix}.
$$

Note that in Example 4 speculative premiums $p_s^* - \max_i\{p_s^i\}$ are positive since the conditions of Proposition 4 are satisfied. Since the results of Proposition 4 also apply in the framework of rational beliefs, the focus of this section will go beyond finding conditions for positive speculative premiums. In Example 4, also note that the RBE prices $\vec{p}^*$ exhibit the phenomena of "endogenous uncertainty". The agents in the economy form their beliefs conditional on the social states and may end up in equilibrium with distinct prices for different social states. However, there are cases in which there exists RBE without endogenous uncertainty. In the following proposition, we demonstrate the generic presence of endogenous uncertainty in a Rational Belief Equilibrium with social states of beliefs.

In the proof of Proposition 5, we employ that fact that the necessary and sufficient conditions for the nonexistence of endogenous uncertainty for RBE are

$$\sum_{k=1}^{\hat{s}} q^*_{sk} = k_L, \text{ for } s=1 \text{ to } \hat{s}, \tag{25}$$

$$\sum_{k=\hat{s}+1}^{S} q^*_{sk} = k_H, \text{ for } s = \hat{s}+1 \text{ to } S, \tag{26}$$

where $s = 1, \cdots \hat{s}$ correspond to the case of low dividend, $s = \hat{s}+1, \cdots S$ correspond to the case of high dividend, $\hat{s} = \frac{S}{2}$ and Q^* with elements q^*_{sk} is the representative belief. When these conditions are not satisfied, there exists endogenous uncertainty. We identify the space of possible rational beliefs of all investors as $\mathcal{Q}$ with

$$\vec{Q} = \{Q^{1,0}, Q^{1,1}, \cdots, Q^{I,0}, Q^{I,1}\} \in \mathcal{Q} \subset \Delta^{2SI},$$

$$Q^{i,j} \in \Delta^S = \{q_s^{i,j} \in R^S \mid \sum_{s'} q_{ss'}^{i,j} = 1, 0 \leq q_{ss'}^{i,j} \leq 1\}$$

$$\text{and } \alpha^i Q^{i,0} + (1 - \alpha^i)Q^{i,1} = \Gamma.$$

Given a uniform measure on $\mathcal{Q}$, we can show that there exists generally (on $\mathcal{Q}$) RBE with endogenous uncertainty. Note that $(Q^{i,0}, Q^{i,1}) \in \Delta^{2S}$ implies that $\dim(\Delta^{2S}) = 2S - 2$. With S rationality constraints for each i, rational beliefs $(Q^{i,0}, Q^{i,1})$ is in the space of dimension $(2S-2) - S = S-2$. Hence the space of rational beliefs of all agents $\mathcal{Q}$ has a dimension of $(S-2)I$. From Proposition 3, we can represent the equilibrium price $\vec{p}^*$ by some representative belief Q^*. Then we apply Lemma 2 to such a Q^* to obtain the result in Proposition 5(b):

Proposition 5. *(a). Given the beliefs of agents there is a unique RBE with social states of beliefs.*
(b). There exists generically RBE with endogenous uncertainty. Such an RBE price $\vec{p}^*$ will not be equal to the REE price $\vec{p}^\Lambda$.
(c). When there is no endogenous uncertainty, the RBE price $\vec{p}^*$ may not be equal to the REE price $\vec{p}^\Lambda$. The RBE price $\vec{p}^*$ is equal to the REE price $\vec{p}^\Lambda$ if, and only if, $k_L = \phi$ and $k_H = \phi$ hold, in addition to equations (25) and (26).

Note that in Harrison and Kreps (1978) even when agents have diverse beliefs, $\vec{p}^\Lambda$ could still be the equilibrium price, for instance, when a dominant investor has rational expectations. However, under rational beliefs this is only possible when all agents have rational expectations. This is another way, in which the rational beliefs case differentiates itself from the Harrison and Kreps model.

The other interesting phenomenon (besides endogenous uncertainty) demonstrated in Example 4 is that the RBE price $\vec{p^*}$ is higher than the REE price $\vec{p^A}$. In fact, we can show later that the RBE price $\vec{p^*}$ can be higher than $\vec{p^A}$ as long as the beliefs of agents remain sufficiently diverse, which will be the focus of the rest of this section. However, the equilibrium price may be lower than $\vec{p^A}$ of REE if the agents' beliefs become perfectly correlated. We will first discuss the case of perfect correlation, then present the results without perfect correlation in Proposition 6.

If there are perfect correlations among the individual assessment variables $y^{i,n}$ within type i, then $y^{i,n} = y^i \in Y^i = \{0, 1\}$, $\mathcal{F}^i = \{(0, 1), (1, 0)\}$ for all i. We refer to the equilibrium of this case as "RBE with perfect correlation within types." The state space can be represented by the following index mapping Φ, with $s = 1, \cdots, 8$:

$$
\begin{bmatrix} 1 \\ 2 \\ 3 \\ 4 \\ 5 \\ 6 \\ 7 \\ 8 \end{bmatrix} = \Phi \begin{bmatrix} d = d_L \ y^1 = 0 \ y^2 = 0 \\ d = d_L \ y^1 = 0 \ y^2 = 1 \\ d = d_L \ y^1 = 1 \ y^2 = 0 \\ d = d_L \ y^1 = 1 \ y^2 = 1 \\ d = d_H \ y^1 = 0 \ y^2 = 0 \\ d = d_H \ y^1 = 0 \ y^2 = 1 \\ d = d_H \ y^1 = 1 \ y^2 = 0 \\ d = d_H \ y^1 = 1 \ y^2 = 1 \end{bmatrix}.
\tag{27}
$$

The structure of the economy as represented by the above equation is not known to investors in the economy, but the stationary measure Γ on $\hat{S} = D \times \mathcal{F}^1 \times \mathcal{F}^2$ can be learned by agents:

$$
\Gamma = \begin{bmatrix} \phi A & (1 - \phi)A \\ (1 - \phi)A & \phi A \end{bmatrix},
\tag{20}
$$

where

$$
A = [A_{ij}] = \begin{bmatrix} A_1 \\ A_2 \\ A_3 \\ A_4 \end{bmatrix} = \begin{bmatrix} a_1 \ \alpha^1 - a_1 \ \alpha^2 - a_1 \ 1 + a_1 - \alpha^1 - \alpha^2 \\ a_2 \ \alpha^1 - a_2 \ \alpha^2 - a_2 \ 1 + a_2 - \alpha^1 - \alpha^2 \\ a_3 \ \alpha^1 - a_3 \ \alpha^2 - a_3 \ 1 + a_3 - \alpha^1 - \alpha^2 \\ a_4 \ \alpha^1 - a_4 \ \alpha^2 - a_4 \ 1 + a_4 - \alpha^1 - \alpha^2 \end{bmatrix}
\tag{28}
$$

is a 4×4 transition matrix in which a_k, k=1,2,3,4, measure correlation across types. Note that (28) implies that Prob$\{y^i = 0\} = \alpha^i$ for $i = 1, 2$, which is compatible with our specification for individual assessment variables. Type i investors adopt beliefs $Q^{i,0}$ with an ex-ante probability α^i and $Q^{i,1}$ with ex-ante probability of $1 - \alpha^i$. In general, investors of type i adopt one of the two beliefs independently, depending on their assessment variable y^i, so that both beliefs are present in the economy. However, with perfect correlation as described by equation (27), there is only one possible type of belief for each type of investors at any state. For example, at state $s = 1$, the investors of type 1 all have belief $Q^{1,0}$ ($y^1 = 0$) and investors of type 2 all have belief $Q^{2,0}$ ($y^2 = 0$). In the following examples, we show that the RBE price with perfect correlation within types can be either higher or lower than REE price $\vec{p^A}$.

Example 5. Given Γ, γ, A and B as specified in Example 4. Suppose

$$Q^{1,0} = \begin{bmatrix} 0.40B & 0.60B \\ 0.55B & 0.45B \\ 0.40B & 0.60B \\ 0.80B & 0.20B \\ 0.60B & 0.40B \\ 0.15B & 0.85B \\ 0.75B & 0.25B \\ 0.30B & 0.70B \end{bmatrix}, Q^{1,1} = \begin{bmatrix} 0.60B & 0.40B \\ 0.45B & 0.55B \\ 0.60B & 0.40B \\ 0.20B & 0.80B \\ 0.40B & 0.60B \\ 0.85B & 0.15B \\ 0.25B & 0.75B \\ 0.70B & 0.30B \end{bmatrix},$$

$$Q^{2,0} = \begin{bmatrix} 0.60B & 0.40B \\ 0.65B & 0.35B \\ 0.45B & 0.55B \\ 0.30B & 0.70B \\ 0.30B & 0.70B \\ 0.45B & 0.55B \\ 0.65B & 0.35B \\ 0.90B & 0.10B \end{bmatrix}, Q^{2,1} = \begin{bmatrix} 0.40B & 0.60B \\ 0.35B & 0.65B \\ 0.55B & 0.45B \\ 0.70B & 0.30B \\ 0.70B & 0.30B \\ 0.55B & 0.45B \\ 0.35B & 0.65B \\ 0.10B & 0.90B \end{bmatrix}.$$

Given perfect correlation as described by equation (27), at state $s = 1$ only the first row of $Q^{1,0}$ will become "effective" since state $s = 1$ is described by $y^1 = 0$. Hence we can construct the beliefs $Q^{i,e}$ of these two types of agents, $i = 1, 2$, that become effective for states $s = 1, \cdots, 8$ and the corresponding "representative beliefs" Q^*:

$$Q^{1,e} = \begin{bmatrix} 0.40B & 0.60B \\ 0.55B & 0.45B \\ 0.60B & 0.40B \\ 0.20B & 0.80B \\ 0.60B & 0.40B \\ 0.15B & 0.85B \\ 0.25B & 0.75B \\ 0.70B & 0.30B \end{bmatrix}, Q^{2,e} = \begin{bmatrix} 0.60B & 0.40B \\ 0.35B & 0.65B \\ 0.45B & 0.55B \\ 0.70B & 0.30B \\ 0.30B & 0.70B \\ 0.55B & 0.45B \\ 0.65B & 0.35B \\ 0.10B & 0.90B \end{bmatrix}, Q^* = \begin{bmatrix} 0.40B & 0.60B \\ 0.35B & 0.65B \\ 0.45B & 0.55B \\ 0.20B & 0.80B \\ 0.30B & 0.70B \\ 0.15B & 0.85B \\ 0.25B & 0.75B \\ 0.10B & 0.90B \end{bmatrix}.$$

Applying the technique developed in the last section, we can find equilibrium price $\vec{p^*} > \vec{p^A}$.

$$\vec{p^*} = \begin{bmatrix} 5.1549 \\ 5.1972 \\ 5.1127 \\ 5.3239 \\ 5.2394 \\ 5.3662 \\ 5.2817 \\ 5.4085 \end{bmatrix} > \vec{p^A} = \begin{bmatrix} 4.5 \\ 4.5 \\ 4.5 \\ 4.5 \\ 4.5 \\ 4.5 \\ 4.5 \\ 4.5 \end{bmatrix}.$$

Note that the RBE price $\vec{p^*}$ demonstrates endogenous uncertainty and $p_s^* > p_s^A$ in all states. We can also find cases in which the RBE prices p_s^* can be lower than REE price p_s^A in all states, as in the following example.

Example 6. Given Γ, γ, A and B as specified in Example 4. Suppose

$$
Q^{1,0} = \begin{bmatrix} 0.50B & 0.50B \\ 0.80B & 0.20B \\ 0.20B & 0.80B \\ 0.10B & 0.90B \\ 0.80B & 0.20B \\ 0.65B & 0.35B \\ 0.45B & 0.55B \\ 0.20B & 0.80B \end{bmatrix}, \quad
Q^{1,1} = \begin{bmatrix} 0.50B & 0.50B \\ 0.20B & 0.80B \\ 0.80B & 0.20B \\ 0.90B & 0.10B \\ 0.20B & 0.80B \\ 0.35B & 0.65B \\ 0.55B & 0.45B \\ 0.80B & 0.20B \end{bmatrix},
$$

$$
Q^{2,0} = \begin{bmatrix} 0.80B & 0.20B \\ 0.40B & 0.60B \\ 0.70B & 0.30B \\ 0.20B & 0.80B \\ 0.75B & 0.25B \\ 0.20B & 0.80B \\ 0.80B & 0.20B \\ 0.55B & 0.45B \end{bmatrix}, \quad
Q^{2,1} = \begin{bmatrix} 0.20B & 0.80B \\ 0.60B & 0.40B \\ 0.30B & 0.70B \\ 0.80B & 0.20B \\ 0.25B & 0.75B \\ 0.80B & 0.20B \\ 0.20B & 0.80B \\ 0.45B & 0.55B \end{bmatrix}.
$$

Then the beliefs $Q^{i,e}$ become effective for states $s = 1, \cdots, 8$ and the representative belief Q^* can be found:

$$
Q^{1,e} = \begin{bmatrix} 0.50B & 0.50B \\ 0.80B & 0.20B \\ 0.80B & 0.20B \\ 0.90B & 0.10B \\ 0.80B & 0.20B \\ 0.65B & 0.35B \\ 0.55B & 0.45B \\ 0.80B & 0.20B \end{bmatrix}, \quad
Q^{2,e} = \begin{bmatrix} 0.80B & 0.20B \\ 0.60B & 0.40B \\ 0.70B & 0.30B \\ 0.80B & 0.20B \\ 0.75B & 0.35B \\ 0.80B & 0.20B \\ 0.80B & 0.20B \\ 0.45B & 0.55B \end{bmatrix}, \quad
Q^* = \begin{bmatrix} 0.50B & 0.50B \\ 0.60B & 0.40B \\ 0.70B & 0.30B \\ 0.80B & 0.80B \\ 0.75B & 0.25B \\ 0.65B & 0.35B \\ 0.55B & 0.45B \\ 0.45B & 0.55B \end{bmatrix}.
$$

Then we have equilibrium prices $\vec{p^*} < \vec{p^\Lambda}$.

$$
\vec{p^*} = \begin{bmatrix} 4.2078 \\ 4.1299 \\ 4.0519 \\ 3.9740 \\ 4.0130 \\ 4.0909 \\ 4.1688 \\ 4.2468 \end{bmatrix} < \vec{p^\Lambda} = \begin{bmatrix} 4.5 \\ 4.5 \\ 4.5 \\ 4.5 \\ 4.5 \\ 4.5 \\ 4.5 \\ 4.5 \end{bmatrix}.
$$

Besides Example 5 and 6, we can also find cases in which the equilibrium prices p_s^* may be higher than p_s^Λ in some states, but lower than p_s^Λ in other states. These examples all involve social states with perfect correlation within types. Since the rationality restrictions (22) guarantee that agents are sometimes pessimistic and sometimes optimistic, when all four belief matrices are present, we can get Q^* to be the most optimistic. Then RBE will have $\vec{p^*} > \vec{p^A}$. But for RBE with perfect correlation within types, only two belief matrices are present at any time, i.e., one of the four possible cases $\{Q^{1,0}, Q^{2,0}\}$, $\{Q^{1,0}, Q^{2,1}\}$, $\{Q^{1,1}, Q^{2,0}\}$, $\{Q^{1,1}, Q^{2,1}\}$ must occur. Then there is no definite relationship between $\vec{p^*}$ and $\vec{p^A}$, since agents can become all optimistic or all pessimistic with perfect correlation within types, as demonstrated in Examples 5 and 6.

Now we study the general relationship between $\vec{p^*}$ and $\vec{p^A}$ besides those exceptional cases, when four belief matrices $\{Q^{1,0}, Q^{1,1}, Q^{2,0}, Q^{2,1}\}$ are all present. It also holds for the cases with three belief matrices $\{Q^{1,0}, Q^{1,1}, Q^{2,0}\}$ and $\{Q^{1,0}, Q^{2,0}, Q^{2,1}\}$, so long as one type has both belief matrices present in the economy. We call these cases "social states without perfect correlation within types." The properties of RBE in such a case were illustrated in Example 4.

Proposition 6. *For social states without perfect correlation within types, the RBE prices $\vec{p^*}$ are strictly greater than the REE prices $\vec{p^A}$ if all elements of Γ are positive and $k_L \neq \phi$ or $k_H \neq \phi$.*

This proposition applies to RBE with or without endogenous uncertainty. It helps to distinguish the Harrison and Kreps model from the present model, since in the Harrison and Kreps model no definite relation between REE prices and the minimal consistent price scheme can be proved. In our framework there is such a generic relationship. One implication of this proposition is that for RBE without endogenous uncertainty (its necessary and sufficient condition was provided in (25) and (26)), the difference $\vec{p^*} - \vec{p^A}$ can still be positive under the conditions specified in Proposition 6. This phenomenon, also called the "amplification effect," has been studied by Kurz (1998) and Wu and Guo (2002). Note that the relationship between $\vec{p^*}$ and $\vec{p^A}$ is the focus of this proposition while the properties of speculative premiums $\pi_s = p_s^* - \max_i p_s^i$, $s = 1, \cdots S$, were studied in Proposition 4.

4 Concluding remarks

Using unmodeled heterogeneity of expectations in economics does not have much support in the modern literature on economic research. However, the continued presence of heterogeneous expectations observed in the real world remains an important phenomenon for us to understand. It is particularly necessary when we try to study the volatile behavior of speculative trading in financial markets. This paper proposes a research framework for modeling speculative trading with diverse beliefs which are rational and compatible with the data. Even with complete learning, when all agents learn about the stationary measure, their beliefs can still remain diverse and non-stationary.

The framework of rational beliefs introduced by this paper makes the assumption of heterogeneous beliefs of Harrison and Kreps (1978) more plausible. It provides a theory of belief formation in a stable environment such that a stationary measure can be learned by all agents, but there is still left a formally defendable room for disagreement among agents (see Kurz, 1994). We demonstrate that a Rational Belief Equilibrium (RBE) exists. We also demonstrate that such a framework with rational belief can help us to understand the functioning of speculative trading in the market place. In particular, we find that Keynes' insight on speculation and subjective expectations can be supported in a rigorous framework and our paper demonstrates that heterogeneous beliefs can be the major reason for speculative trading. The extent of belief heterogeneity affects the size of speculative premiums. In such a framework, investors evaluate the asset according to its resale value and not just its dividend streams, just as described by M. Keynes. Hence, it provides a framework for pricing assets when speculative trading is allowed.

This paper also studies the phenomenon of endogenous uncertainty where equilibrium prices are distinct even when exogenous dividend states remain the same. In a framework of rational beliefs, it is essentially the uncertainty about the future beliefs of other agents. The concepts of endogenous uncertainty and rational beliefs, although formally distinct, are brought together in the last section of the paper. Our framework is in line with the thinking of M. Keynes as formulated in his "beauty contest" example. We demonstrate that endogenous uncertainty may emerge with speculative trading in a Rational Belief Equilibrium (RBE). In addition, the RBE prices are shown to be generally higher than the REE prices and speculative premiums are positive when beliefs are sufficiently diverse. These results shed light on our understanding of belief, speculation and uncertainty. Although we adopt a simple framework, the basic results can be shown to be robust (see Wu and Guo, 2002) in a more general environment with finite wealth and limited short selling.

Appendix: Proofs

To prove Propositions 1, 2 and 3, we first establish the following lemmas:

Lemma A 1. *For $S = 2$,*

$$\vec{p^i} = \frac{\gamma}{(1 - \gamma)(1 + \gamma(a^i - b^i))} \cdot \begin{bmatrix} (1 - a^i + a^i\gamma - b^i\gamma)d_1 + a^i d_2 \\ (1 - b^i)d_1 + (b^i + a^i\gamma - b^i\gamma)d_2 \end{bmatrix} \quad (A.1)$$

Proof. It is easily shown by completing the following calculation:

$$\vec{p^i} = (I - \gamma Q^i)^{-1}\gamma Q^i \vec{d}$$
$$= \left[\begin{bmatrix} 1 & 0 \\ 0 & 1 \end{bmatrix} - \gamma \begin{bmatrix} 1 - a^i & a^i \\ 1 - b^i & b^i \end{bmatrix} \right]^{-1} \gamma \begin{bmatrix} 1 - a^i & a^i \\ 1 - b^i & b^i \end{bmatrix} \begin{bmatrix} d_1 \\ d_2 \end{bmatrix}$$
$$= \frac{\gamma}{(1-\gamma)(1+\gamma(a^i-b^i))} \cdot \begin{bmatrix} (1 - a^i + a^i\gamma - b^i\gamma)d_1 + a^i d_2 \\ (1 - b^i)d_1 + (b^i + a^i\gamma - b^i\gamma)d_2 \end{bmatrix}.$$

□

Lemma A 2. *If $(I - \gamma Q) \cdot \vec{x} \geq 0$ is satisfied, then each element of $\vec{x}$ must be non-negative.*

Proof. Suppose some component of $\vec{x}$ is negative. Let $x_1 < 0$ and $x_1 \leq x_s$ for all s. Then we have

$$(1 - \gamma q_{11})(x_1) + (-\gamma q_{12})(x_2) + \cdots + (-\gamma q_{1S})(x_S)$$
$$= (1 - \gamma)(x_1) + \gamma(q_{12}(x_1 - x_2) + \cdots + q_{1S}(x_1 - x_S)) < 0 , \quad (A.2)$$

which leads to a contradiction. Hence all components of $\vec{x}$ must be non-negative.

$\square$

Proof of Lemma 1. Since $I - \gamma Q$ has a dominant diagonal, it is invertible (see McKenzie, 1960).

Proof of Proposition 1. At market equilibrium, for any state s there always exists a type of investor who has the highest subjective valuation for the asset and gets hold of it:

$$p_s^* = \max_i \gamma q_s^i (\vec{p^*} + \vec{d}), \ s = 1, 2. \tag{A.3}$$

The stationary market equilibrium price p_s^* should satisfy $p_s^* = \gamma q_s^{i(s)}(\vec{p^*} + \vec{d})$ for some types of investors $i(s)$. By applying Lemma A1 it is easily shown that

$$p_2^* - p_1^* = \frac{\gamma(b^{i(2)} - a^{i(1)})}{1 - \gamma(b^{i(2)} - a^{i(1)})}(d_2 - d_1).$$

Since $d_2 > d_1$, we have $p_2^* - p_1^* + d_2 - d_1 > 0$. Suppose state 1 occurs, the difference of willingness to pay between investors of type i and j is computed from (A1):

$$\gamma q_1^i(\vec{p^*} + \vec{d}) - \gamma q_1^j(\vec{p^*} + \vec{d}) = \gamma(a^i - a^j)((p_2^* - p_1^*) + (d_2 - d_1)),$$

which is positive if $a^i > a^j$. So the investors with maximal a^i will have the highest valuation of the asset when state 1 occurs. Similarly, when state 2 occurs the investors with maximal b^i will have the highest valuation. Substituting these back into equation (A1) and applying Lemma 1, we can derive equations (5) and (6).

Proof of Proposition 2. From Proposition 1

$$Q^* = \begin{bmatrix} 1 - \max_i a^i & \max_i a^i \\ 1 - \max_i b^i & \max_i b^i \end{bmatrix} = \begin{bmatrix} 1 - a^1 & a^1 \\ 1 - b^2 & b^2 \end{bmatrix}, \tag{A.4}$$

then by Lemma A1 we have the formula for equilibrium prices

$$\vec{p^*} = \frac{\gamma}{(1-\gamma)(1+\gamma(a^1 - b^2))} \cdot \begin{bmatrix} (1 - a^1 + a^1\gamma - b^2\gamma)d_1 + a^1 d_2 \\ (1 - b^2)d_1 + (b^2 + a^1\gamma - b^2\gamma)d_2 \end{bmatrix}. \tag{A.5}$$

By further calculation we can obtain equations (7) and (8) as in Proposition 2. We can obtain the results in (9) by differentiating equation (7) and (8).

Proof of Proposition 3. The subject valuation of a fictitious investor is

$$\vec{p^f} = (I - \gamma Q^f)^{-1}\gamma Q^f \vec{d} = \vec{p^f}(Q^f), \ f = 1, \cdots, K^S.$$

Given $p^{\vec{f}}$, we can find the representative investor $i(s, p^{\vec{f}}(Q^f))$ who has the highest valuation at state s. This defines the following mapping from Ψ into itself:

$$F(Q^f) = \begin{bmatrix} q_1^{i(1,p^{\vec{f}}(Q^f))} \\ \cdot \\ q_s^{i(s,p^{\vec{f}}(Q^f))} \\ \cdot \\ q_S^{i(S,p^{\vec{f}}(Q^f))} \end{bmatrix} .$$

A fixed point of this mapping implies that the associated prices also have a fixed point $p^{\vec{f}} = p^{\vec{*}}$, which is the equilibrium or minimal consistent price scheme. There are only finite elements in Ψ. If there is no fixed point of F, then there must exist a cycle $Q^{f1} \cdots Q^{fM}$ such that $F(Q^{fm}) = F(Q^{f(m+1)})$ for $m \leq M-1$ and $F(Q^{fM}) = Q^{f1}$. The corresponding prices are $p^{\vec{f}1} \cdots p^{\vec{f}M}$. We will show in the following that $p^{\vec{f}m} = p^{\vec{*}}$, $m = 1, \cdots, M$.

Since $F(Q^{f1}) = Q^{f2}$, as the price is equal to $p^{\vec{f}1} = p^{\vec{f}}(Q^f)$, the belief $q_s^{f2} = q_s^{i(s,p^{\vec{f}1}(Q^f))}$ gives the highest valuation $\gamma q_s^{f2}(p^{\vec{f}1} + \vec{d})$ of the asset and

$$\gamma q_s^{f2}(p^{\vec{f}1} + \vec{d}) \geq \gamma q_s^{f1}(p^{\vec{f}1} + \vec{d}) = p_s^{f1} \text{ for } s = 1, \cdots, S.$$

Hence we have

$$\gamma Q^{f2}(p^{\vec{f}1} + \vec{d}) \geq p^{\vec{f}1},$$
$$\text{or, } (I - \gamma Q^{f2})p^{\vec{f}1} \leq \gamma Q^{f2}(\vec{d}).$$

In addition,

$$(I - \gamma Q^{f2})p^{\vec{f}2} = \gamma Q^{f2}(\vec{d}).$$

By combining them we have

$$(I - \gamma Q^{f2})(p^{\vec{f}2} - p^{\vec{f}1}) \geq 0.$$

Then by applying Lemma A2, all components of $p^{\vec{f}2} - p^{\vec{f}1}$ must be non-negative. Therefore, $p^{\vec{f}2} \geq p^{\vec{f}1}$ and $p^{\vec{f}m}$ increases with m. Since $p^{\vec{f}m}$ forms a cycle, all these prices must be equal to a constant $p^{\vec{f}}$. Then fictitious beliefs $\{Q^{fm}\}$ all come with the same price $p^{\vec{f}} = p^{\vec{*}}$, which is the equilibrium or minimal consistent price scheme.

The proof above helps us to understand the working of the algorithm mentioned in text. It can also be shown by applying Proposition 1 of Harrison and Kreps (1978) to get

$$p_s^* = \max_i \gamma q_s^i(p^{\vec{*}} + \vec{d}) = \gamma q^{i(s)}(p^{\vec{*}} + \vec{d}) \ s = 1, \cdots, S.$$

Then by combining them we obtain that

$$p^{\vec{*}} = \gamma Q^*(p^{\vec{*}} + \vec{d}),$$

where

$$q_s^* = q_s^{i(s)}, \ s = 1, \cdots, S.$$

Certainly Q^* is one of the members in Ψ.

Next we prove the uniqueness of equilibrium prices by contradiction. Suppose that there exist two equilibrium prices $\vec{p}^{*1}$ and $\vec{p}^{*2}$. Q^{*1} and Q^{*2} are the representative beliefs, respectively. Combining two equations $\vec{p}^{*2} = \gamma Q^{*2}(\vec{p}^{*2} + \vec{d})$ and $\vec{p}^{*1} \geq \gamma Q^{*2}(\vec{p}^{*1} + \vec{d})$, we find that

$$(I - \gamma Q^{*2})(\vec{p}^{*2} - \vec{p}^{*1}) \geq 0. \tag{A.6}$$

By applying Lemma A2 it follows that all components of $\vec{p}^{*2} - \vec{p}^{*1}$ must be non-negative. It can also be shown that all components of $\vec{p}^{*1} - \vec{p}^{*2}$ must be non-negative by the same argument. Then $\vec{p}^{*1} = \vec{p}^{*2}$, and so the uniqueness is proved. $\square$

Proof of Proposition 4. We first prove the first part by two steps.

(a). No investor holds the asset in all states. If this were the case, $\vec{p}^* = \vec{p}^i$ for some i. However, since there is no dominant investor, $p_s^i < p_s^j$ or $p_s^i < q_s^j(\vec{p}^* + \vec{d})$ for some s and j, a contradiction.

(b). $\vec{p}^* > \vec{p}^i$

By (a), for any i we may find $p_s^* > p_s^i$ for some s. It follows that for any given state t

$$p_t^* - p_t^i \geq \gamma q_t^i(\vec{p}^* + \vec{d}) - \gamma q_t^i(\vec{p}^i + \vec{d}) = \gamma q_t^i(\vec{p}^* - \vec{p}^i) > 0,$$

where the last inequality holds since $q_{tt'} > 0$, $\forall\, t, t'$, and $p_s^* - p_s^i$ must be positive for some s. So there exist strictly positive premiums.

Next prove the second part. Suppose that state s occurs,

$$p_s^* - p_s^i = \gamma q_s^*(\vec{p}^* + \vec{d}) - \gamma q_s^i(\vec{p}^i + \vec{d}) = m_s^i + \gamma q_s^i(\vec{p}^* - \vec{p}^i), \quad \text{for } s = 1, \cdots, S.$$

It follows that

$$\vec{p}^* - \vec{p}^i = \vec{m}^i + \gamma Q^i(\vec{p}^* - \vec{p}^i),$$

and then the second part of Proposition 4 is proved.

Proof of Lemma 2. Without loss of generality, let states $s = 1, \cdots \hat{s}$ correspond to the case of low dividend ($d_s = d_L$) and states $s = \hat{s} + 1, \cdots S$, high dividend ($d_s = d_H$). The subjective price $\vec{p}^Q$ of an investor with belief Q will satisfy

$$\begin{bmatrix} p_1 \\ \vdots \\ p_{\hat{s}} \\ p_{\hat{s}+1} \\ \vdots \\ p_S \end{bmatrix} = \gamma \begin{bmatrix} q_{1,1}^Q & \cdots & q_{1,\hat{s}}^Q & q_{1,\hat{s}+1}^Q & \cdots & q_{1,S}^Q \\ & & \vdots & & & \\ q_{\hat{s},1}^Q & \cdots & q_{\hat{s},\hat{s}}^Q & q_{\hat{s},\hat{s}+1}^Q & \cdots & q_{\hat{s},S}^Q \\ q_{\hat{s}+1,1}^Q & \cdots & q_{\hat{s}+1,\hat{s}}^Q & q_{\hat{s}+1,\hat{s}+1}^Q & \cdots & q_{\hat{s}+1,S}^Q \\ & & \vdots & & & \\ q_{S,1}^Q & \cdots & q_{S,\hat{s}}^Q & q_{S,\hat{s}+1}^Q & \cdots & q_{S,S}^Q \end{bmatrix} \left(\begin{bmatrix} p_1 \\ \vdots \\ p_{\hat{s}} \\ p_{\hat{s}+1} \\ \vdots \\ p_S \end{bmatrix} + \begin{bmatrix} d_L \\ \vdots \\ d_L \\ d_H \\ \vdots \\ d_H \end{bmatrix} \right) \tag{A.7}$$

The necessary part is proved first. Suppose $p_s^Q = p_L^Q$ for $s = 1, \cdots \hat{s}$ and $p_s^Q = p_H^Q$ for $s = \hat{s} + 1, \cdots S$. Then equation (A.7) is equivalent to

$$\begin{bmatrix} p_L^Q \\ p_H^Q \end{bmatrix} = \gamma \begin{bmatrix} \sum_{j=1}^{\hat{s}} q_{s_1,j}^Q & \sum_{j=\hat{s}+1}^S q_{s_1,j}^Q \\ \sum_{j=1}^{\hat{s}} q_{s_2,j}^Q & \sum_{j=\hat{s}+1}^S q_{s_2,j}^Q \end{bmatrix} \begin{bmatrix} p_L^Q + d_L \\ p_H^Q + d_H \end{bmatrix}, \tag{A.8}$$

for any pair s_1 and s_2 where $s_1 = 1, \cdots \hat{s}$ and $s_2 = \hat{s}+1, \cdots S$. From our discussion on 2×2 Markovian case, the subjective price in (A.8) has a unique solution. Hence we have the conditions in Lemma 2:

$$\sum_{j=1}^{\hat{s}} q_{s,j}^{Q} = k_L \text{, for all } s = 1, \cdots \hat{s},$$
$$\text{and } \sum_{j=\hat{s}+1}^{S} q_{s,j}^{Q} = k_H \text{, for all } s = \hat{s}+1, \cdots S. \tag{A.9}$$

Next we prove the sufficiency part. Given that equation (17) is satisfied, then (A.7) can be reduced to

$$\begin{bmatrix} p_L^Q \\ p_H^Q \end{bmatrix} = \gamma \begin{bmatrix} k_L & 1 - k_L \\ 1 - k_H & k_H \end{bmatrix} \begin{bmatrix} p_L^Q + d_L \\ p_H^Q + d_H \end{bmatrix},$$

whose unique solution is also the solution to equation (A.8). It follows that $p_s^Q = p_L^Q$ for $s_1 = 1, \cdots \hat{s}$ and $p_s^Q = p_H^Q$ for $s_2 = \hat{s}+1, \cdots S$.

Proof of Lemma 3. In equilibrium

$$p_s^* \geq \gamma q_s^{i,j}(\vec{p^*} + \vec{d}) \text{, for } s = 1, \cdots S \text{ and } i = 1, 2, \; j = 0, 1 , \tag{A.10}$$

which follows from optimizing behavior. From the rationality constraint (22) we have

$$p_s^* \geq \gamma \Gamma_s(\vec{p^*} + \vec{d}), \text{ and } \vec{p^*} \geq \gamma \Gamma(\vec{p^*} + \vec{d}).$$

Since Q^{ij} has no-negative elements, it follows that

$$\gamma Q^{i,j}(\vec{p^*} + \vec{d}) \geq \gamma Q^{i,j}(\gamma \Gamma(\vec{p^*} + \vec{d}) + \vec{d}) .$$

Proof of Proposition 5. (a). In general, four types of beliefs $\{Q^{1,1}, Q^{1,0}, Q^{2,1}, Q^{2,0}\}$ are all present. Proposition 3 holds for any given set of beliefs, so by Lemma 3 there exists a unique equilibrium price $\vec{p^*}$ with a fictitious representative belief Q^*.

(b). Proposition 3 provides a characterization of the equilibrium price $\vec{p^*}$ in terms of some representative belief Q^*. Then we apply Lemma 2 to such a Q^* to help to prove Proposition 5(b): Now we prove that endogenous uncertainty is generic for RBE. For any $\vec{i} = (i_s)_{s=1}^{S}, i_s \in \{1, \cdots, 2I\}$, we define a set $\mathcal{B}(\vec{i})$, as a subset of $\mathcal{Q} \subset \Delta^{2SI}$ with $\vec{Q} \in \mathcal{Q}$ as its elements satisfying

$$\sum_{k=1}^{\hat{s}} q_{sk}^{i_s} = \sum_{k=1}^{\hat{s}} q_{s'k}^{i_{s'}}, \text{ for any } s, s'=1 \text{ to } \hat{s} ,$$

$$\sum_{k=\hat{s}+1}^{S} q_{sk}^{i_s} = \sum_{k=\hat{s}+1}^{S} q_{s'k}^{i_{s'}}, \text{ for any } s, s' = \hat{s} + 1 \text{ to } S ,$$

where $s = 1, \cdots \hat{s}$ correspond to the case of low dividend, $s = \hat{s} + 1, \cdots S$ correspond to the case of high dividend, $\hat{s} = \frac{S}{2}$. For any $\vec{i}$, the set $\mathcal{B}(\vec{i})$ is the subset of $\mathcal{Q}$ with elements being consistent with the $2(S-1)$ linear constraints listed above.

Since it has a lower dimension, $\mathcal{B}(\vec{i})$ becomes a set with zero measure on $\mathcal{Q}$, in terms of the uniform measure $m_{\mathcal{Q}}()$:

$$m_{\mathcal{Q}}(\mathcal{B}(\vec{i})) = 0 .$$

We collect all possible $\mathcal{B}(\vec{i})$ to define a set $\mathcal{B}$, i.e.,

$$\mathcal{B} = \cup_{\vec{i}} \mathcal{B}(\vec{i}) .$$

Since there are only finite possible $\vec{i}$, a finite union of measure-zero sets also has a zero measure. Therefore, $\mathcal{B}$ is a set of zero measure, i.e.,

$$m_{\mathcal{Q}}(\mathcal{B}) = 0 .$$

Next we show that for any $\vec{Q} \in \mathcal{Q}$, if there is no endogenous uncertainty in RBE, then $\vec{Q} \in \mathcal{B}$. By the proof in Proposition 3 we can find a fixed point (for the mapping F) $Q^{f*} \in \Psi$, which is a representative belief Q^*. We define i^* as the collection of i_s^* such that

$$q_s^{i^*} = q_s^{f*} = q_s^* .$$

Since Q^* satisfies (25) and (26), such a $\vec{Q}$ should be included in $B(i^*)$. Then we can conclude that such a set of exogenous beliefs which does not lead to endogenous uncertainty in RBE has a zero measure in $\mathcal{Q}$.

(c). When there is no endogenous uncertainty,

$$\begin{bmatrix} p_L^* \\ p_H^* \end{bmatrix} = \gamma \begin{bmatrix} k_L & 1 - k_L \\ 1 - k_H & k_H \end{bmatrix} \begin{bmatrix} p_L^* + d_L \\ p_H^* + d_H \end{bmatrix} ,$$

which may not have the same solution as $\vec{p}^\Lambda$, which solves the following equation

$$\begin{bmatrix} p_L^* \\ p_H^* \end{bmatrix} = \gamma \begin{bmatrix} \phi & 1 - \phi \\ 1 - \phi & \phi \end{bmatrix} \begin{bmatrix} p_L^* + d_L \\ p_H^* + d_H \end{bmatrix} ,$$

unless $k_L = \phi, k_H = \phi$.

Proof of Proposition 6. For a given type i where both belief matrices are present in the economy,

$$\gamma Q^{i,j}(\vec{p^*} + \vec{d}) \le \vec{p^*}, \text{ for } i = 0, 1. \tag{A.11}$$

The above inequality and the rationality constraints of equation (22) imply that

$$\gamma \Gamma(\vec{p^*} + \vec{d}) \le \vec{p^*} .$$

Combining it with $\gamma \Gamma(\vec{p^\Lambda} + \vec{d}) = \vec{p^\Lambda}$ we have

$$\vec{p^*} - \vec{p^\Lambda} \ge \gamma \Gamma(\vec{p^*} + \vec{d}) - \gamma \Gamma(\vec{p^\Lambda} + \vec{d}) = \gamma \Gamma(\vec{p^*} - \vec{p^\Lambda}) , \tag{A.12}$$

and then $(I - \gamma \Gamma)(\vec{p^*} + \vec{p^\Lambda}) \ge 0$. Then by Lemma A2 it leads to $\vec{p^*} - \vec{p^\Lambda} \ge 0$. Since $k_L \ne \phi$ or $k_H \ne \phi$ holds, $p_s^* - p_s^\Lambda$ must be positive for some state s by Proposition 5(c). Then by the condition that all elements of Γ are positive and (A.12), it follows that for any state s,

$$p_s^* - p_s^\Lambda \ge \gamma \Gamma_s(\vec{p^*} - \vec{p^\Lambda}) > 0 .$$

So $\vec{p^*}$ is strictly higher than $\vec{p^\Lambda}$.

References

Arrow, K. J.: Le Rôle des valeurs boursières pour la repartition la Meilleure des risques. Économetrie C.N.R.S. **40**, 41-47 (1953). Translated as: The role of securities in the optimal allocation of risk bearing. Review of Economic Studies **31**, 91–96 (1964)

Cass, D., Chichilnisky, G., Wu, H. M.: Individual risk and mutual insurance. Econometrica **64**, 331–341 (1996)

DeLong, J. B., Shiller, A., Summers, L. H., Waldman, R. J.: Noise trader risk in financial markets. Journal of Political Economy **98**, 703–738 (1990)

Detemple J., Murthy, S.: Intertemporal asset pricing with heterogeneous beliefs. Journal of Economic Theory **62**, 294–320 (1994)

Feiger, G.: What is speculation. Quarterly Journal of Economics **90**, 677–687 (1976)

Grossman S. J., Stiglitz, J. E.: On the impossibility of informationally efficient markets. American Economic Review **70**, 393–408 (1980)

Harris, M., Raviv, A.: Differences of opinion make a horse race. Review of Financial Studies **6**, 473–506 (1993)

Harrison, M., Kreps, D. M.: Speculative investor behavior in a stock market with heterogeneous expectation. Quarterly Journal of Economics **92**, 323–336 (1978)

Hart, O. D., Kreps, D.: Price destabilizing speculation. Journal of Political Economy **94**, 927–952 (1986)

Hirshleifer, J.: Foundations of the theory of speculation: information, risk and markets. Quarterly Journal of Economics **89**, 519–542 (1975)

Huang, P., Wu, H. M.: Market equilibrium with endogenous price uncertainty and options. In: Chichilnisky, G. (ed.) Markets, information and uncertainty: essays in honor of Kenneth Arrow, Chapter 11. New York: Cambridge University Press 1999

Kaldor, N.: Speculation and economic stability. Review of Economic Studies **7**, 1–27 (1939)

Keynes, J. M.: The general theory of employment, interest and money. London: Macmillan 1936

Kreps, D. M.: A note on "fulfilled expectations" equilibria. Journal of Economic Theory **14**, 32–43 (1977)

Kurz, M.: On the structure and diversity of rational beliefs. Economic Theory **4**, 877–900 (1994)

Kurz, M.: Social states of belief and the determinants of the equity risk premium in a rational beliefs equilibrium. In: Abramovich, Y., Avgerinos, E., Yannelis, N. C. (eds.) Functional analysis and economic theory, pp. 171–220. Berlin Heidelberg New York: Springer 1998

Kurz, M., Beltratti, A.: The equity premium is no puzzle. In: Kurz, M. (eds.) Endogenous economic fluctuations: studies in the theory of rational belief. Berlin Heidelberg New York: Springer 1997

Kurz, M., Schneider, M.: Coordination and correlation in Markov rational belief equilibria. Economic Theory **8**, 489–520 (1996)

Kurz, M., Wu, H. M.: Endogenous uncertainty in a general equilibrium model with price contingent contracts. Economic Theory **8**, 461–488 (1996)

Leach, J.: Rational speculation. Journal of Political Economy **99**, 131–144 (1991)

McKenzie, L.: Matrices with dominant diagonals and economic theory. In: Arrow, K., Karlin, S., Suppes, P. (eds.) Mathematical methods in the social sciences. Stanford: Stanford University Press 1960

Milgrom, P., Stokey, N.: Information, trade and common knowledge. Journal of Economic Theory **26**, 17–27 (1982)

Morris, S.: Speculative investor behavior and learning. Quarterly Journal of Economics **111**, 1113–1133 (1996)

Nielsen, C. K.: Rational belief structures and rational belief equilibria. Economic Theory **8**, 399–422 (1996)

Radner, R.: Existence of equilibrium plans, prices and price expectations in a sequence of markets. Econometrica **40**, 289–303 (1972)

Tirole, J.: On the possibility of speculation under rational expectation. Econometrica **50**, 1163–1181 (1982)

Varian, H. R.: Divergence of opinion in complete markets: A note. Journal of Finance **90**, 309–317 (1985)

Varian, H. R.: Differences of opinion in financial markets. In: Stone, C. C. (eds.) Financial risk: theory, evidence and implications-proceedings of the Eleventh Annual Economic Policy Conference of the Federal Reserve Bank of St. Louis 1989

Wu, H. M., Guo, W. C.: Asset market speculation with a large number of rational agents. Working Paper (2002)

Floating exchange rates versus a monetary union under rational beliefs: the role of endogenous uncertainty*

Carsten Krabbe Nielsen

Department of Statistics and Operations Research, Institute of Mathematical Sciences,
University of Copenhagen, Universitetsparken 5, 2100 Copenhagen Ø, DENMARK
(e-mail: CarstenKNielsen@excite.com)

Received: September 1, 2001; revised version: 24 June 2002

Summary. Using the concept of ex-post optimality, we compare different exchange rate regimes, including floating exchange rates and fixed exchange rates with a Monetary Union in a two country OLG model with stochastic endowments. The emphasis of this comparison is on the welfare consequences of agents having incorrect beliefs. We do not assume that agents can hold any beliefs, but rather that their beliefs are rational that is consistent with the observed empirical behavior of the economy. We study a large set of possible policies, but two of them have our particular interest. The first policy implies devaluations in reaction to a negative shock, while the other implies a fixed exchange rate. These policies have very different consequences. The first will for generic beliefs not result in an ex-post optimal allocation. The other policy is on the other hand always feasible and results in an ex-post optimal allocation. When the two countries form a Monetary Union, the ex-post optimal allocation is also achieved. The meaning of "endogenous uncertainty" as an institutionally induced uncertainty is illustrated.

Keywords and Phrases: Endogenous uncertainty, Exchange rates, Monetary union, OLG model, Rational beliefs.

JEL Classification Numbers: D51, D84, F31, F33.

---

* I would like to thank Horace W. Brock, Gianluca Cassese, Paula Orlando, Ho-Mou Wu as well as seminar participants at Copenhagen Business School, ESEM98, Keio University, Kyoto University, Osaka University, SITE (Stanford) and University of Copenhagen for many useful comments on the paper. I am also grateful to Mark J. Garmaise, Takako Fujiwara-Greve, and an anonymous referee for many helpful suggestions for improving the paper. Without the many discussions about Rational Beliefs and related issues I have had with Mordecai Kurz over the years, the research presented here would not have been possible. Financial support from The Carlsberg Foundation, Danish Social Research Council, University of Copenhagen and SITE is gratefully acknowledged.

1 Introduction

This study is concerned with the choice of an optimal exchange rate regime. The starting point for our analysis is that in general agents have incorrect expectations when they try to forecast future exchange rate movements. This starting point makes our contribution differ from other studies of the optimal exchange rate regime (f.i. Obstfeld and Rogoff, 2000), in which agents are assumed to have rational expectations. When the exchange rate is constant, agents know the future rate, and they will make decisions that are ex-post optimal. However, when the exchange rate is stochastic and when agents do not use the objectively correct distribution when choosing their portfolios of currencies, these portfolios are not ex-post optimal. Based on this argument, our study concludes in favor of a monetary policy that leaves the exchange rate constant or, alternatively, of a monetary union, where there is only one currency and, consequently, no exchange rate to forecast.

In our formalization of the above heuristic argument, we have to deal with several theoretical issues. The first is how to evaluate the outcomes of various regimes. In welfare economics if one allocation is Pareto dominated by another it is generally agreed that the latter should be chosen. This makes sense when agents have rational expectations. But when the expectations of different agents are mutually inconsistent this is no longer a plausible rule, as has been argued for instance by Hammond (1981). We adopt ex-post optimality (see Hammond,1981) as the criterion by which to compare different regimes. As a consequence, the performance of any exchange rate regime (or a Monetary Union) will be judged by whether it will lead to a particular ex-post optimal allocation C^* (to be defined below). This allocation will in general be Pareto dominated by other allocations, but such dominance would be based on inconsistent subjective expectations, and as such, should not be of concern. The bottom line is that the Pareto criterion for optimality cannot account for the mistakes that agents with subjective beliefs make, while ex-post optimality can.

Another issue is what expectations we are to assume that agents have. It would be hard to defend that these expectations are arbitrary and have no relation to what is observed in equilibrium. We propose in stead that they, like rational expectations, are consistent with empirical observations. Such expectations, called rational beliefs, were proposed by Kurz (1994a,b)[1]. Rational beliefs are expectations that are consistent with empirical data, but may none the less be incorrect. If one assumes that agents have rational expectations, a stricter condition is being imposed, namely that beliefs are correct[2]. Even though rationality rules out many beliefs, there is still much room left for disagreement about expectations. We argue that typically there will be disagreement, and under most exchange rate regimes such disagreement results in an allocation that is different from C^*.

When putting ourselves in the place of decision makers, we realize that when they have to decide on what regime to choose they do not know what beliefs agents will have in the future. One presumption made in the present study is that they do

---

[1] For a review, see Kurz (1997a).

[2] Rational expectations are in an environment where an empirical distribution exists (f.i. a stationary environment) rational beliefs.

know that these beliefs will be rational. This means that the decision makers know that agents will learn from observations and adapt their beliefs accordingly. For primarily technical reasons, we go further and assume that they know that these beliefs will lie in a restricted set, $\mathcal{H}$ of rational beliefs . This set of possible rational beliefs is equivalent to an open subset of a Euclidean space. As a result, it is possible to use Lebesgue measure to represent the likelihood of different beliefs (from the viewpoint of the decision makers).

The model we study is a two-country model in which each country experiences idiosyncratic real shocks. Allowing the agents in the two countries to hold portfolios of both currencies may help them insure against these shocks (in particular one should not expect that autarky is optimal in this model). The allocation C^* is precisely chosen because it is fair and involves optimal insurance against these shocks. The central question then is: For which exchange rate regimes (if any) is it the case, that for most (in terms of Lebesgue measure) beliefs in $\mathcal{H}$ the ex-post optimal allocation C^* is achievable as an equilibrium allocation? The answer is that there are only two possibilities (within a large class of policies): Either a Monetary Union with a coordinated fiscal policy or a fixed exchange rate regime, where the monetary policies of the two countries is coordinated in such a way as to implement optimal insurance against the country specific shocks. One alternative in which the exchange rate is floating and moves in response to the idiosyncratic shocks *does* achieve C^* in case agents have rational expectations. But rational expectations have measure zero in $\mathcal{H}$ and we show that for most (rational) beliefs in $\mathcal{H}$, C^* is not an equilibrium allocation under this floating exchange rate regime. This observation then serves to demonstrate that the introduction of rational beliefs does have a bearing on policy conclusions. In particular, under some exchange rate regimes mutual inconsistency of beliefs may result in ex-post suboptimal equilibrium allocations, something that would not happen when agents have rational expectations (which are necessarily mutually consistent).

It has been argued that a regime with floating exchange rates provides governments with an effective instrument for macroeconomic policy (for a formalization, see Obstfeld and Rogoff, 2000). However, in recent years, a monetary union has been formed in Europe, and local currencies have been tied to the dollar in other places. Why is that? Two answers have been proposed. One is that such an arrangement saves on transaction costs, the other is that it enhances trade by reducing risk. The first answer only makes sense, when transaction costs are large, something one would doubt is the case in the sophisticated financial markets of Europe. The other answer does not explain why there are not sufficiently many financial instruments available to insure against risk, if it is so costly. The present study argues that there is another reason to form a monetary union: Under such an arrangement agents do not, based on their subjective beliefs, choose currency positions that are ex-post inoptimal. In particular, speculation in currencies is absent. From the viewpoint of society this is an important advantage to forming a monetary union.

Our analysis of different exchange rate regimes, based on the role of subjective expectations and on ex-post optimality, provides a methodological framework that could potentially be used with regard to many other issues in public policy. The fundamental question could be phrased as follows: When we take into account that agents have subjective beliefs, how should we compare institutions and determine which are optimal? For instance, when it comes to the stock markets it has been argued that investment strategies based on 'irrational' beliefs have resulted in serious misallocations of funds. If we believe that mistaken beliefs result in realized welfare that is severely suboptimal, public policy that seeks to reduce the extent to which expectations are wrong (for instance by reducing volatility) becomes interesting.

Related literature

The model we use is a simple OLG model in the Kareken-Wallace (1981) tradition. This model was extended to the stochastic case by Manuelli and Peck (1994) who showed in an OLG model with two currencies that the same Pareto optimal allocation is compatible with many different patterns of exchange rate behavior. These patterns range from a constant rate to complicated non-stationary stochastic processes. But no matter the degree of complexity, the agents are assumed to know the true distribution of the exchange rate process and take it into account when making their decisions. The results of Manuelli and Peck can be replicated in our model if we assume rational expectations (see Appendix 5). The RE assumption is widely used in theoretical studies of the exchange rate – for instance in Neumeyer (1998): In a model with incomplete markets and cash in advance constraints the trade-off between floating exchange rates and a monetary union is explained as one between better spanning when there are many currencies and less uncertainty (since a monetary union removes erratic central bank behavior).[3]

The empirical evidence, as for instance surveyed in Taylor (1995), goes against the rational expectations hypothesis. One source of rejection, reported in Taylor, is macroeconometric studies showing that observed departures from the market efficiency hypothesis can only be explained by rejecting the RE hypothesis. Surveys of the expectations of individual traders in the market are another source. These surveys have been able to categorize traders according to the type of theory (and thus beliefs) they were using, some being chartists, others being fundamentalists (see f.i. Frankel and Froot, 1990). This very diversity of theories/beliefs is incompatible with the RE assumption, unless one is willing to postulate that some traders somehow obtain important information that others cannot access about future market movements. The theory of rational beliefs is a simple way to formalize the idea that beliefs may be diverse and thus incorrect even though they do conform with observations. This theory also suggests that volatility in short term beliefs may be the root of the observation of many empirical studies (see Taylor, 1995), that the degree of volatility in endogenous variables is beyond what can be explained by fundamentals.

This is an issue taken up in Black (1997) and Kurz (1997b) in which models of the markets for foreign exchange, with cash-in-advance constraints, are calibrated and simulated for the case where agents have rational expectations and rational

---

[3] In Neumeyer's model, forming a monetary union is simply a way to keep central banks from this erratic unexplained behavior. Unlike in this model and models by Nickelsburg (1984) and Sibert (1989) we do not assume unexplained behavior by any agent in the economy.

beliefs respectively. They demonstrate that the assumption of rational beliefs is considerably more satisfactory when it comes to generating price movements resembling the observed ones. Hauswald (1999) uses the Lucas two-country model to study volatility of foreign exchange rates and shows that excess volatility, seemingly a puzzle, is really due to the assumption of RE, but can be accounted for by using the theory of rational beliefs. McKinnon and Pill (1998) employ rational beliefs in a theoretical study of the overborrowing syndrome on capital markets in South East Asia. As in our model, they use RB as a way to describe the importance of mistaken beliefs for the behavior of international markets. The idea that agents use different theories to make forecasts is also exploited in Goldberg and Frydman (1996). They conclude that the introduction of heterogeneous beliefs, which are qualitatively consistent with the true model, can explain the observed behavior of the exchange rate.

In the next section we present the model. Section 3 provides a heuristic introduction to the theory of rational beliefs and introduces a major tool for our analysis, a rational belief structure (RBS), that describes the rational beliefs and the exogenous environment without referring to endogenous variables. In the following section, we investigate the effects of different policies when agents have rational beliefs. Section 5 discusses different forms of evaluations under rational beliefs, in particular ex-post versus ex-ante optimality, as well as the concept of endogenous uncertainty, and Section 6 concludes.

2 The model

In each of the two countries, called A and B, reside two (representative) agents with the same name as their country, one young and one old, providing us with the classical OLG framework. Each country also has a government, which can issue fiat money, the proceeds of which is used to subsidize its citizenry.

Agents in each country and across time are ex-ante identical. At date t the young's endowment of the single consumption good is e while the old have e_a^t, $a = A, B$. Here e_a^t is stochastic, taking values in $\{e_1, e_2\}$ with $e_1 > e_2$ and probabilities, $\overline{\pi}^E = (\overline{\pi}_1^E, \overline{\pi}_2^E)$. Consequently, at any date t the exogenous state of the economy is described by (e_{Am}, e_{Bn}) where $e_{ak} = e_k$, $a = A, B$, $k = 1, 2$. We let the total endowments of the old generation be $e_{mn} = e_{Am} + e_{Bn}$ and the difference be $\Delta e_{mn} = e_{Am} - e_{Bn}$. $\{e_A^t\}$ and $\{e_B^t\}$ are both assumed to be independently distributed over time (more precisely SIDS (i.i.d) – as defined below) and mutually independent.

Since we are concerned with savings decisions we assume that agents only consume when they are old, and we describe their preferences by a utility function f, defined on $\mathbb{R}_+$, which is C^2, strictly monotone, and strictly concave. Money, which can only be held in positive amounts, is the only store of value. This means that the only decision made by the young agents is how to invest their first period endowments. It also means that we only study the allocation of risk; the issue of intertemporal allocation is simply assumed away. In light of the identical risk averse utility functions of the agents it is not surprising that we are interested in the allocation, where each agent consumes $C_{mn}^* \equiv e + (e_{Am} + e_{Bn})/2$, so that the effects of a shock to one country is shared with the rest of the world. We assume that $e > |e_A^t - e_B^t|$ with probability 1 implying, as we shall see, that there is a monetary policy which attains C^*.

We intend to study two institutional set-ups. In the first, each country has its own currency, a for country A and b for country B. There are no capital controls so both agents are free to use both currencies as a means of savings. This set-up is characterized by the exchange rate being determined endogenously by the demands of the agents and the policies (to be defined shortly) of the governments, and by the agents having to decide what portfolio of the two currencies to hold between the two periods. In the second institutional set-up there is one common currency, which both agents have to use as a means of savings. This set-up will be interpreted as a monetary union.

The money stock of country A and B at the *beginning* of period t is M_{At} and M_{Bt} respectively, and $M_{A1} > 0$ and $M_{B1} > 0$. During period t the governments in the two countries issue money (in positive or negative amounts) to finance the real expenditures S_{At} and S_{Bt} – they do not themselves consume anything. Thus if we let the prices of the commodity be P_{At} in currency a and P_{Bt} measured in currency b, the two governments issue the amounts $\Delta M_{At} = P_{At} S_{At}$ and $\Delta M_{Bt} = P_{Bt} S_{Bt}$ respectively during period t.

S_{At} and S_{Bt}, which we call transfer schemes, are assumed to depend only on the current exogenous "shocks," e_A^t and e_B^t. We write their realizations as S_{Amn} and S_{Bmn}. Our main result is then that within the large class of *all* such policies there is only one candidate for achieving the ex-post optimal allocation (C^*, C^*), called Policy 2 below. Another Policy, 1 has a natural interpretation as an institution with a floating exchange rate and devaluations in response to adverse economic shocks. Because of this interpretation and because that when agents have rational expectations C^* is achieved under Policy 1, it has our particular interest.

Policy 1: $S_{Amn}^1 = \max\{0, e_{Bn} - e_{Am}\}$, $S_{Bmn}^1 = \max\{0, e_{Am} - e_{Bn}\}$.

Policy 2: $S_{Amn}^2 = \dfrac{c}{2(c-1)}(e_{Bn} - e_{Am})$, $S_{Bmn}^2 = \dfrac{2-c}{2(c-1)}(e_{Bn} - e_{Am})$ for $c \in (0, 2) \setminus \{1\}$.

The first policy involves only positive transfers to the citizens of the two countries, consequently the governments only issue positive amounts of their currency and inflation is always non-negative. If the realized endowment of one country is lower than that of the rest of the world, i.e., it realizes what we call a negative shock, it issues money to finance a real transfer to its own citizen. If the citizens of both countries hold its currency the result is a net transfer from foreign to domestic citizens, although all of them pay an inflation tax. In this simple setting the issuance of money amounts to an expansionary monetary policy with a resulting devaluation. This policy resembles the devaluation policy (in a context of sticky prices) which some proponents of floating exchange rates suggest should be pursued. Under a monetary union Policy 1 amounts to a coordinated subsidy program for countries which experience negative shocks financed by a seigniorage tax on the common currency.

With two currencies the second policy involves positive or negative transfers, i.e. subsidies or taxes. Here, either the two governments both issue money (an "expansionary" policy) or they both buy money. In the last case the purchase of money which is financed by a tax on the citizens of each country (so that the policy is "contractionary") results in a negative rate of inflation i.e. a positive return on currency holdings. It turns out that with this policy one is able to achieve a fixed exchange rate whether agents have rational expectations or rational beliefs. Note that the set of policies we study also includes no active policy in either of the countries, in which case $S_{Amn} = S_{Bmn} \equiv 0$.

The problem of the young agent A in period t, where he has a belief B_{At} about prices, endowments, and real government transfers in period $t + 1$, is:

$$\underset{\alpha \in [0,1]}{\text{Max}} \ E_{At} f(\alpha(P_{At}/P_{At+1})e + (1 - \alpha)(P_{Bt}/P_{Bt+1})e + e_A^{t+1} + S_{At+1}) \ .$$

Here E_{At} denotes expectation about period $t + 1$ stochastic variables. $C_{At+1} = \alpha(P_{At}/P_{At+1})e + (1 - \alpha)(P_{Bt}/P_{Bt+1})e + e_A^{t+1} + S_{At+1}$ is the consumption of agent A in period $t+1$ if he invests a fraction α of his initial endowment in currency A and the rest, $1 - \alpha$, in currency B. Let the solution to his problem be α_t^*. The young agent B has a problem similar to that of the young agent A, but possibly different expectations. We denote the solution to his problem, β_t^* and C_{Bt+1} is his consumption level in period $t + 1$.

In equilibrium where all outstanding money of both countries is held by the agents, the demand for the commodity by the old agents is $M_{At}/P_{At} + M_{Bt}/P_{Bt} + e_A^t + e_B^t + S_{At} + S_{Bt}$ so that equilibrium in the market for the commodity good means $M_{At}/P_{At} + M_{Bt}/P_{Bt} + e_A^t + e_B^t + S_{At} + S_{Bt} = 2e + e_A^t + e_B^t$. Equilibrium in the market for currency a means $\alpha_t^* P_{At} e + \beta_t^* P_{At} e = M_{At} + P_{At} S_{At} = M_{At+1}$ and similarly, equilibrium for currency b means $(1 - \alpha_1^*) P_{Bt} e + (1 - \beta_t^*) P_{Bt} e = M_{Bt} + P_{Bt} S_{Bt} = M_{Bt+1}$.

Consequently, in equilibrium we have for the gross returns on currency a and currency b respectively:

(2.1)
$$\frac{P_{At}}{P_{At+1}} = \left(\frac{M_{At+1}}{\alpha_t^* e + \beta_t^* e} \right) \left(\frac{M_{At+1}}{\alpha_{t+1}^* e + \beta_{t+1}^* e - S_{At+1}} \right)^{-1}$$
$$= \frac{\alpha_{t+1}^* e + \beta_{t+1}^* e - S_{At+1}}{\alpha_t^* e + \beta_t^* e} \quad \text{and}$$

(2.2)
$$\frac{P_{Bt}}{P_{Bt+1}} = \frac{(1 - \alpha_{t+1}^*)e + (1 - \beta_{t+1}^*)e - S_{Bt+1}}{(1 - \alpha_t^*)e + (1 - \beta_t^*)e} \ .$$

As is clear from these formulas there are, for a given policy, two sources of uncertainty for an agent, namely the future exogenous shocks and the future investment behavior of the next generation which in turn depends on the future beliefs among agents of that generation. The RE framework not only assumes that all agents have rational expectations but also that this is common knowledge. We assume neither of these here. Thus there are two effects of any policy that induces a fixed exchange rate (or of a Monetary Union). Firstly, the policy removes the volatility in the exchange rate induced by volatile beliefs and thus removes subjective uncertainty. Secondly, the policy removes the consequences of mistaken beliefs. It is the second effect that drives our results. However, the first effect may also be very important, and when markets are incomplete a fixed exchange rate regime may Pareto dominate a regime with floating exchange rates for that reason (see Nielsen, 2001).

Markets are incomplete and we assume that governments try to design a policy that makes up for this incompleteness and avoids losses due to the mistaken beliefs of agents. Note, that without free capital mobility the investment policies of agents are trivial and not influenced by beliefs. This framework then accomplishes the second of these two goals of the government, but because there are no insurance possibilities, not the first. Before turning to a definition of a rational beliefs equilibrium in this model we need to specify the beliefs of the agents.

3 Rational beliefs and rational belief structures

The theory of rational beliefs first and foremost formulates a type of environment where much but not everything can be learned from observations, and states which beliefs agents can have about this environment. The specified environment is characterized by the stochastic process of observable variables being *stable*. Stability of a stochastic process means that the empirical properties of the process are well defined (something that stationarity implies[4]). In other words, assuming that all theoretical moments exist, the empirical mean as well as all other empirical moments converge as the number of observations increase. One crucial property of a stable process is that its empirical properties do not identify it in a unique way. Even when endowed with all possible empirical information about the process an agent is not able to ascertain its true nature. The empirical properties of the stable process, only identify a subset of all stable processes to which the true process must belong. A rational belief based on observations of a stable stochastic process is then defined to be a member of this subset. Consequently, the definition tells what agents *cannot* believe and what they *can* believe, given what the true stochastic process is. This definition leaves the door open for disagreement among agents, who have at their disposal the *same empirical data*, but interpret it in terms of *different statistical models*. We refer the reader to Kurz (1994a) and Kurz (1997a) for detailed accounts of the theory of rational beliefs. The following simple example is meant to capture the idea of the type of rational beliefs we consider here.

Suppose the empirical distribution of an infinite series of coin tosses is known and i.i.d. with probability (of heads and tails) being $\bar{\pi} = (2/3, 1/3)$. If the observer is furthermore told that the true distribution of the series of coin tosses is stationary, he can deduce (using the law of large numbers) that this true distribution is equal to the empirical distribution. If, instead, he is only told that the true distribution is stable, he can deduce less. Specifically, he cannot a priori rule out that the true distribution is, say, $(5/6, 1/6)$, at a particular set of dates consisting of half of the dates (we say that the frequency of these dates is one half) and $(3/6, 3/6)$ at the other half

---

[4] The set of stationary processes is a subset of the set of stable processes but there is no way empirically to judge whether a particular process is stationary or just stable.

of the dates, where the two (known) sets of dates are dispersed in a non-systematic way over time (note, we have $1/2(5/6, 1/6) + 1/2(3/6, 3/6) = (2/3, 1/3)$) or more generally that the (one-period) probability is $(B_1^k, 1 - B_1^k)$ on a particular set of dates whose frequency is Q_k, where $\sum_{k=1}^K Q_k(B_1^k, 1 - B_1^k) = (2/3, 1/3)$. If he believes this is the case, he has an SIDS (i.i.d.) belief (introduced in Nielsen, 1996) with K one-period beliefs, $(B_1^k, 1 - B_1^k)$, $k = 1, 2 \ldots K$.[5] Two important features of an SIDS (i.i.d.) measure are that it is stable and that the empirical distribution it generates is i.i.d., so that we only need describe this distribution in terms of a one-period probability.

When there are several possible interpretations of the empirical distribution there is implicitly an added *endogenous* component to the uncertainty each individual agent faces, namely about how other future actors in his environment interpret the data (and act on their interpretations). The concept of a Rational Belief Structure (RBS), introduced in Nielsen (1996), is a way to describe the beliefs of all agents in terms of primitives taking this added endogenous component into account.

With an eye to simplifying the exposition and derivation of the results below we perform our analysis using a very simple set of RBS. We assume that the two agents both have SIDS (i.i.d.) beliefs with two possible one-period beliefs each. At any given date t, what matters, i.e., what constitutes the *primitives* of the economy, is the one-period beliefs held by the two young agents and the realization of e_A^t and e_B^t. Consequently the *index set of primitives* is $S = \{e_{A1}, e_{A2}\} \times \{e_{B1}, e_{B2}\} \times \{1, 2\} \times \{1, 2\}$. An equilibrium is a map from members of the index set of primitives to endogenous and exogenous variables (so we observe at most 16 different endogenous variables in equilibrium). Not only the exogenous variables but also the state of beliefs of the agents may influence the realized equilibrium values of endogenous variables and this is precisely the source of endogenous uncertainty. In equilibrium the beliefs of the agents are beliefs about endogenous and exogenous variables, but because of the equilibrium relation between the set of primitives and the latter variables, we can think of the beliefs as being defined on the set of primitives. In this way we can analytically distinguish between beliefs and exogenous variables on the one hand and the equilibrium map associated with any particular economic institution (or policy) on the other. Thus, in the definition of an RBS provided below we let the one-period beliefs be defined on S. Let $\Delta^N = \{x \in \mathbb{R}_+^N : \sum_{n=1}^N x_n = 1\}$ be the $N - 1$ dimensional simplex. In the following definition Q refers to frequency of one-period beliefs, B refers to beliefs, and $\bar{\pi}$ refers to empirical distribution.

Definition 1. *SIDS (i.i.d.) Rational Belief Structure (RBS) given $\bar{\pi}^E$*
A collection $(Q^D, (B^{Ak})_{k=1}^2, (B^{Bl})_{l=1}^2)$, where

(i) $Q^D = (Q_{kl}^D)_{k,l=1,2} \in \Delta^4$

---

[5] Technically speaking, letting heads correspond to 0 and tails to 1, the belief has the form of an infinite product measure, $\otimes_{t=1}^\infty \beta_t$, on the infinite product space, $\times_{t=1}^\infty \{0, 1\}$, where $\beta_t \in \{(B_1^1, 1 - B_1^1), \ldots, (B_1^K, 1 - B_1^K)\}$ and the frequency of dates at which $\beta_t = (B_1^k, 1 - B_1^k)$ is Q_k (so $\sum_k Q_k = 1$).

(ii) $B^{Ak} = (\{B^{Ak}_{ijmn}\}_{i,j,m,n=1,2}) \in \Delta^{16}$ $(k = 1, 2)$

(iii) $B^{Bl} = (\{B^{Bl}_{ijmn}\}_{i,j,m,n=1,2}) \in \Delta^{16}$ $(l = 1, 2)$

is a RBS if, when we define $\bar{\pi} \in \Delta^{16}$ *by*

(3.1) $\bar{\pi}_{ijmn} = Q^D_{ij}\bar{\pi}^E_m\bar{\pi}^E_n$, $(i, j, m, n = 1, 2)$

then

(3.2) $\displaystyle \bar{\pi}_{ijmn} = \sum_{k=1}^{2}\sum_{l=1}^{2} Q^D_{kl}B^{Ak}_{ijmn} = \sum_{k=1}^{2}\sum_{l=1}^{2} Q^D_{kl}B^{Bl}_{ijmn}$ $(i, j, m, n = 1, 2)$
                                                                                                             $\square$

The SIDS (i.i.d.) beliefs of the two agents are described in (i)–(iii). In (ii) and (iii) we have the two possible one-period beliefs of agent A and agent B respectively, all 16-dimensional probability vectors. The frequency of these beliefs is described in (i) as a joint distribution, i.e. a $(2 \times 2 =)$ 4-dimensional probability vector. So we allow for correlation between the processes of the one-period beliefs of the two agents. To obtain the one-period distribution for the (i.i.d.) empirical process of primitives we combine the empirical one-period distribution of the two exogenous processes with Q^D, the product of which gives us $\bar{\pi}$, a 16-dimensional probability vector, described in (3.1).[6] Finally, in (3.2), the *rationality condition* states that the one-period distribution of the empirical process of primitives, $\bar{\pi}$, is equal to the one-period distribution of the empirical process that agents expect according to their individual beliefs. In other words, the frequency at which agent A (or B) expects to observe the primitive state $(e_A, e_B, B^{Ak}, B^{Bl})$ is equal to the frequency at which this state appears, and in this sense, the beliefs of the agents are consistent with observations and thus confirmed.

In this paper we consider an OLG model where each agent is short lived. But we assume that he is a member of a dynasty where all members have the same belief, a probability measure on infinite sequences of primitives.[7] As noted, these beliefs about primitives will be carried over to beliefs about exogenous and endogenous variables. If the one-period beliefs of the agents vary over time (so that their beliefs are not stationary, but only stable)[8], this causes the *true* process of endogenous

---

[6] Note that in (3.2) we implicitly make the assumption of *structural independence* (defined in Nielsen (1994)) between the exogenous process and the beliefs in that we write the joint frequency of e_{Am}, e_{Bn}, B^{Ak}, and B^{Bl} as a product of the marginal frequencies, $\bar{\pi}^E_m$, $\bar{\pi}^E_n$, and Q^D_{kl}. Structural independence is an assumption that the agents do not know more than what is implied by the rationality condition (3.2).

[7] We could also interpret the assumption in a slightly different way. Each of the countries have their own distribution of young agents. At any date t, a young agent (identified by his belief) is picked at random in each country, the belief of this agent is now an SIDS (i.i.d.) measure on infinite sequences, but the only thing which matters is what his one-period belief is at date t. In any case, the individual agent living at date t is assumed either not to think about the connection between future beliefs and future prices or not to know future beliefs even of his own dynasty.

[8] We can interpret the possession of a non-stationary rational belief by an agent as an expression of overconfidence (see Nielsen, 1994, for more on this point). The agent believes that he has a model of his environment which is consistent with, but tells more than, the empirical distribution. This overconfidence could also be interpreted as "animal spirit" [Kurz (1997a) suggests a connection between the thinking of Keynes and the theory of rational beliefs].

variables to be non-stationary but stable. In this sense the beliefs of the agents, that their environment is non-stationary, is self-fulfilling.

The set of RBS should be thought of as the a priori set of *possible* rational beliefs in a situation with interacting agents, i.e., in a situation with both endogenous and exogenous uncertainty. The concept of the set of RBS can in this sense be seen as a natural extension of the concept of the set of rational beliefs. Whereas the latter describes what an individual agent can rationally believe when informed about all empirical properties of a given stable process, the former describes the set of possible distributions of rational beliefs in a society of *interacting agents* for a given set of empirical data about the exogenous process. In an RBS the beliefs of the agents have to be *jointly* consistent, meaning that the beliefs of each agent has to be rational given the empirical data *and* the distribution of all agents' beliefs.

In the analysis in the following two sections we propose that a government may use the set of RBS in its assessment of the consequences of different economic institutions and policies. Ultimately the government is interested in the outcome in terms of a rational belief equilibrium (RBE). But this RBE depends on the beliefs of the agents as described in an RBS, beliefs which, however, are not observable ex ante and most likely not even ex post. When the governments make a coordinated decision about which policy to implement they do not know what RBS and hence RBE will come about. We suggest that the governments instead reason in terms of the likely outcome of their policy, i.e. the likely RBE obtained, which in turn will depend on the likely underlying RBS. Thus we suggest that the governments reason in terms of a distribution on the set of RBS, which are consistent with the observed empirical properties of the exogenous process. In view of the linear restrictions in the definition it is not surprising that this set can be parametrized as a bounded subset (of $\mathbb{R}^{33}$) with open interior (see Nielsen, 1997a, for a formal statement and proof and Appendix 3 for a parametrization). In the following we assume that the governments reason in terms of the Lebesgue measure, λ, (or any distribution absolutely continuous w.r.t. to it) restricted to the set of RBS. Note that this implies that the set of rational expectations has zero measure.

4 Comparing regimes under rational beliefs

The issue to be considered is under what institutional arrangements there will for λ-almost all RBS be an RBE where the consumption of both agents is C^*. In Section 5 our interest in this particular allocation is explained.

Though in an RBE, agents observe prices and transfers and not the more fundamental choice variables, it simplifies the analysis of the model to concentrate on the former. Following Kurz (1994b) and Nielsen (1996) we define an RBE for two particular transfer schemes S_A and S_B for this economy as follows:

Definition 2. *Rational Belief Equilibrium (RBE) for an RBS*
A set $\{\alpha_{kl}, \beta_{kl}\}_{k,l=1,2}$ constitutes a Rational Belief Equilibrium if

(i) $\quad \alpha_{kl} + \beta_{kl} \in (0, 2)$ $\hspace{5cm}$ $(k, l = 1, 2)$

(ii) $[\alpha_{kl} + \beta_{kl}]e > S_{Amn}$, $[2 - \alpha_{kl} - \beta_{kl}]e > S_{Bmn}$, $C_{Am} \geq 0$, and $C_{Bn} \geq 0$
$(k, l, m, n = 1, 2)$

(iii) $\forall k, l : \alpha_{kl}$ solve: $\text{Max}_{\alpha \in [0,1]} \sum_{i=1}^{2} \sum_{j=1}^{2} \sum_{m=1}^{2} \sum_{n=1}^{2} B_{ijmn}^{Ak}$
$\times f \left(\alpha \frac{[\alpha_{ij} + \beta_{ij}]e - S_{Amn}}{[\alpha_{kl} + \beta_{kl}]e} e + (1 - \alpha) \frac{[2 - \alpha_{ij} - \beta_{ij}]e - S_{Bmn}}{[2 - \alpha_{kl} - \beta_{kl}]e} e + e_{Am} + S_{Amn} \right)$
and β_{kl} solve: $\text{Max}_{\beta \in [0,1]} \sum_{i=1}^{2} \sum_{j=1}^{2} \sum_{m=1}^{2} \sum_{n=1}^{2} B_{ijmn}^{Bl}$
$\times f \left(\beta \frac{[\alpha_{ij} + \beta_{ij}]e - S_{Amn}}{[\alpha_{kl} + \beta_{kl}]e} e + (1 - \beta) \frac{[2 - \alpha_{ij} - \beta_{ij}]e - S_{Bmn}}{[2 - \alpha_{kl} - \beta_{kl}]e} e + e_{Bn} + S_{Bmn} \right)$.

$\square$

Condition (i) guarantees that both currencies are in demand by the agents in equilibrium. Condition (ii) is a feasibility condition stating that the transfers of each government is less than the real savings in its currency.

Suppose that the governments use Policy 1 and that we have an RBS where both agents have stationary beliefs i.e. that $B^{A1} = B^{A2}$ and $B^{B1} = B^{B2}$. Then we have an RBE where $\alpha_{kl} = \beta_{kl} = 1/2, \forall k, l$ (we can simply apply the proof of existence of an REE from Appendix 5). In Nielsen (1997)) it is also shown that for a neighborhood of any such RBS there exists an RBE. This implies that for a set of RBS of positive measure there will be an RBE under Policy 1. In this sense this policy is at least a priori interesting. The following proposition identifies two important consequences of Policy 2.

Proposition 1. *For $c \in (0, 1 - (e_{A1} - e_{A2})/2e] \cup [1 + (e_{A1} - e_{A2}/2e, 2)$ and for all RBS there is an RBE under policy 2 such that the exchange rate is constant and such that the resulting allocation is (C^*, C^*).*

Proof. See Appendix 1.

In the following three lemmatas we consider *all* possible transfer schemes which depend on the current shock and ask for which of them the allocation (C^*, C^*) can be attained for a generic set of RBS.

Lemma 1. *A necessary condition for obtaining (C^*, C^*) is that one of the following two conditions holds:*

(i) $\exists c$ s.t. $\alpha_{kl} + \beta_{kl} = c$ $(\forall k, l)$

(ii) $\alpha_{kl} = \beta_{kl}$ $(\forall k, l)$.

Proof. See Appendix 2.

Next we show that if (C^*, C^*) is to be obtained then case (i) from Proposition 4 can be ruled out, unless we use Policy 2.

Lemma 2. *Let the transfer scheme (S_A, S_B) be given. If for a set of RBS of positive measure we have an RBE where there is a constant c such that $\alpha_{kl} + \beta_{kl} = c, \forall k, l$ and where the allocation (C^*, C^*) is attained it follows that (S_A, S_B) is the transfer scheme used under Policy 2.*

Proof. See Appendix 3.

The idea of the proof is that we have for any transfer scheme other than Policy 2 for most RBS that if the allocation (C^*, C^*) is obtained under this RBS there is a perturbation of that RBS so that we can no longer have an RBE where the allocation is (C^*, C^*). In the next proposition we use a similar argument to establish that $\alpha_{kl} = \beta_{kl}, \forall\, k, l$ for an RBE that attains (C^*, C^*) is generically, in the set of RBS, not possible.

Lemma 3. *Suppose the transfer scheme (S_A, S_B) is given and different from transfer scheme 2. Then only for a set of RBS of zero measure do we have an RBE where $\alpha_{kl} = \beta_{kl}, \forall\, k, l$ and where the allocation (C^*, C^*) is attained.*

Proof. See Appendix 4 □

We summarize the findings of Proposition 1 and Lemmatas 1–3 in the following proposition.

Proposition 2. *For Policy 2 there is for all RBS an RBE with constant exchange rate and consumption (C^*, C^*). For all other feasible transfer schemes, the set of RBS for which (C^*, C^*) can be attained has Lebesgue measure 0.*

Note in particular, that for almost all RBS, (C^*, C^*) is not obtained under Policy 1.

If the governments implement a monetary union together with Policy 1, the beliefs of the agents will not matter and as under rational expectations they will each consume (C^*, C^*). In that case the subsidies are financed by printing the common currency. Thus $C_A = e - \frac{1}{2}\max(0, e_B - e_A) - \frac{1}{2}\max(0, e_A - e_B) + e_A + \max(0, e_B - e_A) = e + \frac{1}{2}[\max(0, e_B - e_A) - \max(0, e_A - e_B)] = C^*$. Note, that the uncertainty the agents then face is different from the case, where the exchange rate is floating. While they will observe as many as 16 different returns on savings under a floating exchange rate, they will, under a monetary union, only observe 3 different returns, no matter what the underlying RBS is. Similarly, the empirical distribution of consumption will in general exhibit larger variance under floating exchange rates than under a Monetary Union. In this sense the uncertainty that agents face depends on the institution implemented by the government and therefore we say that part of this uncertainty is *endogenous*.

5 Evaluation under rational beliefs

Why may the two governments be interested in implementing a policy which results in the allocation (C^*, C^*)? We start our discussion of this issue by presenting the notion of *ex-post optimality* as an alternative to the notion of Pareto optimality when agents have subjective (rational) beliefs.[9] (C^*, C^*) is ex-post optimal but in general not Pareto optimal.

It is instructive first to consider a possible Pareto optimal allocation. Suppose that at some date the agents have beliefs B^{Ai} and B^{Bj}. Then the allocation $(2e + e_{Am} + e_{Bn} - C^{ij}_{mn}, C^{ij}_{mn})_{(mn) \in \{1,2\}^2}$ would be Pareto optimal if it

---

[9] See Kurz (1997a) for another perspective on the problems of employing the notion of Pareto optimality in the context of rational beliefs.

solved $\max_{\{C_{mn}\}} U(\{C_{mn}\}) = \sum_{m=1}^{2} \sum_{n=1}^{2} B_{\cdots mn}^{Ai} f(2e + e_{Am} + e_{Bn} - C_{mn}) +$ $\sum_{m=1}^{2} \sum_{n=1}^{2} B_{\cdots mn}^{Bj} f(C_{mn})$. Since the frequency of dates at which the beliefs are (B^{Ai}, B^{Bj}) is Q_{ij}^{D} the time average of the sum (in each period) of subjective utilities would be $U^{PO} = \sum_{i=1}^{2} \sum_{j=1}^{2} Q_{ij}^{D} U(\{C_{mn}^{ij}\})$.

One notable exception to the Pareto (ex-ante) optimality criterion is the notion of ex-post optimality as for instance discussed in Hammond (1981). In this approach one does not evaluate a (weighted) sum of (subjectively) expected utilities but instead an expected weighted welfare of state by state consequences. As is shown in Hammond (1981), if the two forms of evaluations agree then, among other conditions, each agent must use the same belief in constructing expected utility (which is not the case in our model), and this belief must in turn be used when calculating the expected, state by state, weighted welfare of consequences. If in our model the decision makers somehow agree in each period t to maximize the sum of expected utilities of the two agents according to *one* probability measure, B_{t}^{g}, on the state space of endowments the result would be ex-post optimal. They would then at date t solve $\max_{C_{mn}} \sum_{m=1}^{2} \sum_{n=1}^{2} B_{tmn}^{g} [f(2e + e_{Am} + e_{Bn} - C_{mn} + f(C_{mn})]$ with solution C^{*}.[10] Let us assume, that the common rational (hence subjective!) belief of the two governments, B_{t}^{g}, is SIDS (i.i.d.) with two one-period probabilities B^{g1} and B^{g2} and frequency Q_{r}^{g}, $r = 1, 2$, as was the case for the agents. The resulting average sum of utilities will according to this belief then be $U^{g} = \sum_{r=1}^{2} Q_{r}^{g} \sum_{m=1}^{2} \sum_{n=1}^{2} B_{mn}^{gr} \{2f(C_{mn}^{*})\} = \sum_{m=1}^{2} \sum_{n=1}^{2} \overline{\pi}_{m}^{A} \overline{\pi}_{n}^{B} \{2f(C_{mn}^{*})\}$, the equality stemming from the rationality of the belief. It is important to notice that this equality will hold for *all* rational beliefs that the governments may have and in particular if the belief being used happens to be *correct*. This fact is known by everyone, so the governments know that if they implement (C^{*}, C^{*}) the *realized* average sum of utilities will be U^{g} and the agents agree on this. Moreover, everyone agrees that no other sequence, $\{C_{tmn}\}_{t=1}^{\infty}$ of contingent consumptions of agent B (with agent A consuming the residual) – including the Pareto optimal sequence considered before – will generate a higher realized or expected (average) sum of utilities.

This agreement about evaluations despite the disagreements about beliefs is an important consequence of the diverse beliefs being rational beliefs. The rationality of beliefs implies that agents still agree on some core facts (including the period-by-period constraints on total available resources). Even with this objective component of their beliefs we still have that $U^{OP} \geq U^{g}$ with strict inequality when for some (i, j) with $Q_{ij}^{D} > 0$ we have $B^{Ai} \neq B^{Bj}$, a generic feature of an RBS. Furthermore, under the same generic condition there is an allocation $\{C_{mn}^{ij}\}$ which is Pareto optimal and strictly Pareto dominates (C^{*}, C^{*}). In this sense, the incompatibility of the subjective beliefs of the agents results in "unrealistic" expectations in the aggregate. In conclusion, if the governments rejected the (ex-ante) Pareto optimality criterion and instead used ex-post optimality as a criterion and accepted equality (in each state) in the weighing of utilities they would implement (C^{*}, C^{*}).

---

[10] Note the importance of the identical weights and utility functions for the independence of the solution from B_{t}^{g}.

This would also be the case if they formed a union and ignored the beliefs, but not the preferences, of their citizens. Presumably, the common government for all citizens would have a unified view on the exogenous environment, i.e., a single subjective belief, which, if it was furthermore egalitarian, would again lead to (C^*, C^*).[11]

Policy 2 is designed to keep exchange rates constant and this is the way the endogenous uncertainty is avoided. The fact that one country is called upon to tax its own citizens and to (indirectly) subsidize the citizens of the other country (by buying its money from them) raises the questions as to whether such an arrangement will be politically stable. The fragility of this institution may explain why the governments would opt for a monetary union where supposedly the room for deviations is smaller. In that case, in order to obtain the allocation (C^*, C^*), an *integrated* part of the arrangement is that the union of the two countries subsidizes the region (formerly country) which suffers an adverse real shock, funding this subsidy with seigniorage. In other words, a monetary union without an integrated financial policy does not work.[12]

6 Concluding remarks

An underlying theme of the analysis presented here is the relationship between institutions and subjective expectations. Note, for instance, that under Policy 2 agents' beliefs about the future exchange rate will be correct while this will not be the case under Policy 1. To our mind, this illustrates an important sense in which uncertainty is endogenous in any society: The uncertainty that agents face is intimately connected with the institution with which they live and this individual uncertainty is not just the result of an allocation of a given exogenous uncertainty. Consequently, it becomes an important task to study the connection between market structure and beliefs as has been the aim here.

Kurz (1974) introduced endogenous uncertainty[13] as an uncertainty (about endogenous variables) that cannot be reduced to uncertainty about exogenous variables, but which ultimately is about the actions of other agents. The uncertainty that agents face in the economy studied here is partly endogenous in this sense, and obviously, the way we used the term in the previous paragraph is intimately connected with Kurz' definition.

The purpose of the paper has been to formalize how the choice of exchange rate regime may influence the extent to which agents make objectively suboptimal decisions. Important for our analysis is a deviation from the traditional approach to evaluating economic policy. While under rational expectations it seems entirely justifiable to use the preferences (including beliefs) of the agents in such an evaluation, the case of diverse beliefs suggests a new methodological approach. This

---

[11] In Nielsen (1997a) other reasons why the government may be interested in the allocation $\{C^*_{mn}\}$ are discussed.

[12] The absence of such a financial policy among member states of the European Union has been used by proponents of floating exchange rates to argue against the implementation of a monetary union in Europe.

[13] Endogenous uncertainty is further studied in Kurz and Wu (1996) and Nielsen (1994).

approach takes into account that the complexity of the economic environment and thus the degree to which the beliefs of agents are incorrect depends on the economic institution. While we do not assume that governments are better informed about fundamentals of the economy than any other agent, we suggest that they may have a unified approach to evaluating the outcome of a particular economic institution, which rectifies some inconsistencies among the beliefs on the individual level of society.

Appendix 1

Proof of Proposition 1. We check that the FOC for (iii) of Definition 2 holds for any $\{\alpha_{kl}, \beta_{kl}\}_{(k,l)\in(1,2)^2}$ s.t. $\alpha_{kl} \geq 0$ and $\beta_{kl} \geq 0$ with $\alpha_{kl} + \beta_{kl} = c, \forall\, k, l$. In that case we have that

$$\frac{(\alpha_{ij} + \beta_{ij})e - \frac{c}{2(1-c)}\Delta e_{mn}}{\alpha_{kl} + \beta_{kl}} - \frac{(2 - \alpha_{ij} - \beta_{ij})e - \frac{2-c}{2(1-c)}\Delta e_{mn}}{2 - \alpha_{kl} - \beta_{kl}}$$

$$= \frac{ce - \frac{c}{2(c-1)}\Delta e_{mn}}{c} - \frac{(2-c)e - \frac{2-c}{2(c-1)}\Delta e_{mn}}{2 - c} = 0, \quad (\forall\, k, l, m, n, i, j),$$

so the FOC does hold. Furthermore the quotient between the exchange rates of two consecutive periods is

$$\frac{P_{Bt+1}/P_{At+1}}{P_{Bt}/P_{At}} = \frac{ce - \frac{c}{2(c-1)}\Delta e_{mn}}{c} \cdot \left(\frac{(2-c)e - \frac{2-c}{2(c-1)}\Delta e_{mn}}{2 - c}\right)^{-1} = 1.$$

The consumption of agent A is $\alpha_{kl}(e - \frac{1}{2(c-1)}\Delta e_{mn}) + (1 - \alpha_{kl})(e - \frac{1}{2(c-1)}\Delta e_{mn}) + e_{Am} - \frac{c}{2(c-1)}\Delta e_{mn} = e + \frac{1}{2}e_{mn} = C^*_{mn}$ and so the consumption of agent B is the same. Note also, that the restrictions on α and β and c guarantee that (i) and (ii) of Definition 2 are fulfilled, i.e. that we have an RBE. $\square$

Appendix 2

Proof of Lemma 1. We require that

$$\alpha_{kl}\frac{[\alpha_{ij} + \beta_{ij}]e - S_{Amn}}{[\alpha_{kl} + \beta_{kl}]} + (1 - \alpha_{kl})\frac{[2 - \alpha_{ij} - \beta_{ij}]e - S_{Bmn}}{[2 - \alpha_{kl} - \beta_{kl}]}$$

$$+ e_{Am} + S_{Amn} = e + \frac{1}{2}e_{mn} \quad (\forall\, k, l, i, j, m, n)$$

which can be rewritten as

$$\left(\frac{\alpha_{kl}}{[\alpha_{kl} + \beta_{kl}]} - \frac{(1 - \alpha_{kl})}{[2 - \alpha_{kl} - \beta_{kl}]}\right)[\alpha_{ij} + \beta_{ij}]e = -\frac{1 - \alpha_{kl}}{[2 - \alpha_{kl} - \beta_{kl}]}2e$$

$$+ e + \frac{1}{2}e_{mn} - e_{Am} - S_{Amn}\left(1 - \frac{\alpha_{kl}}{\alpha_{kl} + \beta_{kl}}\right) + S_{Bmn}\frac{1 - \alpha_{kl}}{2 - \alpha_{kl} - \beta_{kl}}$$

and similarly for country B:

$$\left(\frac{\beta_{kl}}{[\alpha_{kl}+\beta_{kl}]} - \frac{(1-\beta_{kl})}{[2-\alpha_{kl}-\beta_{kl}]}\right)[\alpha_{ij}+\beta_{ij}]e = -\frac{1-\beta_{kl}}{[2-\alpha_{kl}-\beta_{kl}]}2e$$

$$+e+\frac{1}{2}e_{mn}-e_{Bn}-S_{Bmn}\left(1-\frac{1-\beta_{kl}}{[2-\alpha_{kl}-\beta_{kl}]}\right)+S_{Amn}\frac{\beta_{kl}}{\alpha_{kl}+\beta_{kl}}.$$

The right hand side of these equations do not depend on $\alpha_{ij}+\beta_{ij}$. It follows that if for some $(i',j'),(i'',j'')\alpha_{i'j'}+\beta_{i'j'} \neq \alpha_{i''j''}+\beta_{i''j''}$ then $\alpha_{kl}/[\alpha_{kl}+\beta_{kl}] = (1-\alpha_{kl})/[2-\alpha_{kl}-\beta_{kl}]$ and $\beta_{kl}/[\alpha_{kl}+\beta_{kl}] = (1-\beta_{kl})/[2-\alpha_{kl}-\beta_{kl}], \forall\, k,l$ and so $\alpha_{kl}/\beta_{kl} = (1-\alpha_{kl})/(1-\beta_{kl}), \forall\, k,l$ i.e. $\alpha_{kl} = \beta_{kl}, \forall\, k,l$. $\square$

Appendix 3

Proof of Lemma 2. Let some RBS and some RBE for this RBS be given. Under the stated assumption we have either

(a) $\exists\, c \in (0,2), K \in [0,1]$ with $c - K \in [0,1]$ s.t. $\alpha_{kl} = K \& \beta_{kl} = c - K, \forall\, k,l$
or
(b) $S_{Bmn}/(2-c) - S_{Amn}/c = 0, \forall\, m,n$.

To see this note that for (C^*,C^*) to be attained we require, $\forall\, k,l,m,n$ that $-[\alpha_{kl}S_{Amn}/c+(1-\alpha_{kl})S_{Bmn}/(2-c)]+e_{Am}+S_{Amn} = e_{mn}/2$ or equivalently $\alpha_{kl}[S_{Bmn}/(2-c) - S_{Amn}/c] = S_{Bmn}/(2-c) - S_{Amn} - \Delta e_{mn}/2, \forall\, k,l,m,n$. This equality can however only hold if either α_{kl} is constant (so (a) holds) or $S_{Bmn}/(2-c) - S_{Amn}/c = 0, \forall\, m,n$ (i.e. (b) holds).

Now, if (b) holds $-S_{Amn}/c + S_{Amn} = -\Delta e_{mn}/2$ implying that $c \neq 1$ so we get $S_{Amn} = -c\Delta e_{mn}/2(c-1)$ and thus $S_{Bmn} = -(2-c)\Delta e_{mn}/2(c-1)$, i.e. scheme 2.

Suppose next that (a) holds together with the conditions of the Proposition. We first show that the FOC of agent A holds with equality. If not then say $\sum_{m=1}^{2}\sum_{n=1}^{2}B_{\cdot\cdot mn}^{A1}f'(e+e_{mn}/2)[S_{Bmn}/(2-c)-S_{Amn}/c] < 0$, where $B_{\cdot\cdot mn}^{Ak} \equiv \sum_{i=1}^{2}\sum_{j=1}^{2}B_{ijmn}^{Ak}$. Then, $\alpha_{kl} = K = 0$ implying that $\sum_{m=1}^{2}\sum_{n=1}^{2}B_{\cdot\cdot mn}^{A2}f'(e+e_{mn}/2)[S_{Bmn}/(2-c)-S_{Amn}/c] \leq 0$ so that, using the rationality condition (3.2), $\sum_{i=1}^{2}\sum_{j=1}^{2}\sum_{m=1}^{2}\sum_{n=1}^{2}\bar{\pi}_{ijmn}f'(e + e_{mn}/2)[S_{Bmn}/(2 - c) - S_{Amn}/c] = \sum_{k=1}^{2}\sum_{l=1}^{2}Q^{D}(k,l)\sum_{m=1}^{2}\sum_{n=1}^{2}B_{\cdot\cdot mn}^{Ak}f'(e + e_{mn}/2)[S_{Bmn}/(2 - c) - S_{Amn}/c] < 0$. But as a consequence for some l we have that $\sum_{m=1}^{2}\sum_{n=1}^{2}B_{\cdot\cdot mn}^{Bl}f'(e + e_{mn}/2)[S_{Bmn}/(2 - c) - S_{Amn}/c] < 0$, implying that $c - K = \beta(k,l) = 0$, which is not possible. The symmetric argument applies to agent B's FOC.

We now introduce a specific parametrization, $\mathcal{H}$ of most of the set of RBS. More precisely, there is a set $\mathcal{H}'$ with $\mathcal{H} \subset \mathcal{H}' \subset \overline{\mathcal{H}}$ and a function $G : \mathcal{H}' \to \{RBS\}$ which is one-to-one and onto. Next we prove a lemma. $\mathcal{H} \subset \mathbb{R}_{++}^{33}$ is open, consisting of vectors, $h = (Q_{11}^{D}, Q_{12}^{D}, Q_{21}^{D}, B_{1111}^{A1}, \dots, B_{2212}^{A1}, B_{2221}^{A1}, B_{1111}^{B1}, \dots, B_{2212}^{B1}, B_{2221}^{B1})$ with the following properties:

(i) $Q_{ij}^D \in (0,1)$ for $(i,j) = (1,1),(1,2),(2,1), Q_{11}^D + Q_{12}^D < 1/2, Q_{11}^D + Q_{21}^D < 1/2$

(ii) Defining first $Q_{22}^D = 1 - Q_{11}^D - Q_{12}^D - Q_{21}^D$ and then $\overline{\pi} \in \Delta^{16}$ by $\overline{\pi}(o,p,m,n) = Q_{op}^D \overline{\pi}_m^E \overline{\pi}_n^E, o, p, m, n = 1, 2$ we have for $M = A, B$:

a) $B_{mnop}^{M1} \in (0,1)$ for $m, n, o, p = 1, 2$, where $B_{2222}^{M1} \equiv 1 - \sum_{(m,n,o,p) \neq (2,2,2,2)} B_{mnop}^{M1}$,

b) $B^{M1} \neq \overline{\pi}$, and

c) $1 - \sum_{l=1}^2 Q_{1l}^D > \overline{\pi}(o,p,m,n) - \sum_{l=1}^2 Q_{1l}^D B_{opmn}^{A1} > 0$ and $1 - \sum_{k=1}^2 Q_{k1}^D > \overline{\pi}(o,p,m,n) - \sum_{k=1}^2 Q_{k1}^D B_{opmn}^{B1} > 0$ for $o, p, m, n = 1, 2$.

Note: We can then define B_{mnop}^{M2} s.t. (3.2) holds. For any $h \in \mathcal{H}$ we then get a matrix

$$B(h) = \begin{pmatrix} B_{\cdot\cdot 11}^{A1} & B_{\cdot\cdot 12}^{A1} & B_{\cdot\cdot 21}^{A1} & B_{\cdot\cdot 22}^{A1} \\ B_{\cdot\cdot 11}^{A2} & B_{\cdot\cdot 12}^{A2} & B_{\cdot\cdot 21}^{A2} & B_{\cdot\cdot 22}^{A2} \\ B_{\cdot\cdot 11}^{B1} & B_{\cdot\cdot 12}^{B1} & B_{\cdot\cdot 21}^{B1} & B_{\cdot\cdot 22}^{B1} \\ B_{\cdot\cdot 11}^{B2} & B_{\cdot\cdot 12}^{B2} & B_{\cdot\cdot 21}^{B2} & B_{\cdot\cdot 22}^{B2} \end{pmatrix}$$

in the obvious way. Since (3.2) holds, rank $B(h) < 4$.

Lemma. Let $x, y \in \mathbb{R}^4$ be given.

Letting $A = \{h \in \mathcal{H} : B(h)[x + ty] = 0 \Rightarrow [x + ty] = 0\}$ we have that $\lambda(\mathcal{H} \backslash A) = 0$.

Proof. (i) Suppose $z \in \mathbb{R}^4 \backslash \{0\}$. Then for generic h, $B(h)z \neq 0$. If $B(h)z = 0$ then since all elements of $B(h)$ are > 0 we have for some i, j, k, l that $z_{ij} > 0$ and $z_{kl} < 0$. Consequently if we add $\delta > 0$ to $B_{\cdot\cdot ij}^{A1}$ and subtract it from $B_{\cdot\cdot kl}^{A1}$ and adjust the second row to retain the rationality constraints, obtaining $B(h)_\delta$, we do not have $B(h)_\delta z = 0$. Genericity is then a consequence of Fubini's theorem.

(ii) If $x = 0, y \neq 0$ then since by (i) for generic h, $B(h)y \neq 0$ the result follows. If $x \neq 0, y = 0$ then also by (i) the result holds. Suppose then that $x \neq 0, y \neq 0$ and (as is by the proof of (i) the case for generic h) that $B^{B1}(h)x \neq 0$ and $B^{B1}(h)y \neq 0$. $B(h)[x + ty] = 0$ implies $B^{A1}(h)[x + ty] = 0$ and $B^{B1}(h)[x + ty] = 0$. There is then a unique $t = \overline{t}$ s.t. $B^{B1}(h)[x + \overline{t}y] = 0 (\overline{t} = -\frac{B^{B1}(h)x}{B^{B1}(h)y})$. Using the same argument as in the proof of (i) we can add some $\delta > 0$ to $B_{\cdot\cdot ij}^{A1}$ and subtract the same δ from $B_{\cdot\cdot kl}^{A1}$ to obtain a perturbed $B(h_\delta)$ with $B^{B1}(h_\delta) = B^{B1}(h)$ and $B^{A1}(h_\delta)[x + \overline{t}y] \neq 0$. Thus by Fubini's theorem the lemma follows. $\square$

We introduce some notation. Letting $f'_{mn} = f'(e + e_{mn}/2)$, define

$$
A = \begin{pmatrix}
B^{A1}_{\cdot\cdot 11} f'_{11} & B^{A1}_{\cdot\cdot 12} f'_{12} & B^{A1}_{\cdot\cdot 21} f'_{21} & B^{A1}_{\cdot\cdot 22} f'_{22} \\
B^{A2}_{\cdot\cdot 11} f'_{11} & B^{A2}_{\cdot\cdot 12} f'_{12} & B^{A2}_{\cdot\cdot 21} f'_{21} & B^{A2}_{\cdot\cdot 22} f'_{22} \\
B^{B1}_{\cdot\cdot 11} f'_{11} & B^{B1}_{\cdot\cdot 12} f'_{12} & B^{B1}_{\cdot\cdot 21} f'_{21} & B^{B1}_{\cdot\cdot 22} f'_{22} \\
B^{B2}_{\cdot\cdot 11} f'_{11} & B^{B2}_{\cdot\cdot 12} f'_{12} & B^{B2}_{\cdot\cdot 21} f'_{21} & B^{B2}_{\cdot\cdot 22} f'_{22}
\end{pmatrix}
= \begin{pmatrix}
B^{A1}_{\cdot\cdot 11} & B^{A1}_{\cdot\cdot 12} & B^{A1}_{\cdot\cdot 21} & B^{A1}_{\cdot\cdot 22} \\
B^{A2}_{\cdot\cdot 11} & B^{A2}_{\cdot\cdot 12} & B^{A2}_{\cdot\cdot 21} & B^{A2}_{\cdot\cdot 22} \\
B^{B1}_{\cdot\cdot 11} & B^{B1}_{\cdot\cdot 12} & B^{B1}_{\cdot\cdot 21} & B^{B1}_{\cdot\cdot 22} \\
B^{B2}_{\cdot\cdot 11} & B^{B2}_{\cdot\cdot 12} & B^{B2}_{\cdot\cdot 21} & B^{B2}_{\cdot\cdot 22}
\end{pmatrix}.
$$

$$
\begin{pmatrix}
f'_{11} & 0 & 0 & 0 \\
0 & f'_{12} & 0 & 0 \\
0 & 0 & f'_{21} & 0 \\
0 & 0 & 0 & f'_{22}
\end{pmatrix}, \quad
S_i = \begin{pmatrix}
S_{i11} \\
S_{i12} \\
S_{i21} \\
S_{i22}
\end{pmatrix}, i = A, B
$$

$e_d = (1/2)(0, e_{B2} - e_{A1}, e_{B1} - e_{A2}, 0)^T$ and $S(c) = S_B/(2 - c) - S_A/c\, (\neq 0$ unless scheme 2 is being used). Then K and c fulfill:

(A3.1) $KS(c) = S_B/(2 - c) + e_d - S_A$ (since (C^*, C^*) is obtained)

(A3.2) $A \cdot S(c) = 0$ (the FOC of agent A).

(A3.1) and (A3.2) imply $A(\frac{S_B}{2-c} + e_d - S_A) = 0$ which for generic h implies that $\frac{S_B}{2-c} + e_d - S_A = 0$. But then from (A3.1) $S(c) = 0$ $\square$

Appendix 4

Proof of Lemma 3. We let the common value of $\alpha_{kl} = \beta_{kl}$ be $\gamma_{kl} \in (0, 1)$. Note that when (C^*, C^*) is attained we have that

(A4.1) $e - \dfrac{1}{2}(S_{Amn} + S_{Bmn}) + e_{Am} + S_{Amn} = e + \dfrac{1}{2}e_{mn} = C^*_{mn}$

$e - \dfrac{1}{2}(S_{Amn} + S_{Bmn}) + e_{Bn} + S_{Bmn} = e + \dfrac{1}{2}e_{mn} = C^*_{mn}$

the FOC for A holds with equality and is: $\sum_{i=1}^{2} \sum_{j=1}^{2} \sum_{m=1}^{2} \sum_{n=1}^{2} B^{Ak}_{ijmn} f'(e + e_{mn}/2) \cdot [(2\gamma_{ij}e - S_{Amn})/2\gamma_{kl} - (2(1 - \gamma_{ij})e - S_{Bmn})/2(1 - \gamma_{kl})] = 0$ or $\sum_{i=1}^{2} \sum_{j=1}^{2} \sum_{m=1}^{2} \sum_{n=1}^{2} B^{Ak}_{ijmn} f'(e + e_{mn}/2) \cdot \{[\gamma_{ij}/\gamma_{kl} - (1 - \gamma_{ij})/(1 - \gamma_{kl})]e + [S_{Bmn}/(1 - \gamma_{kl}) - S_{Amn}/\gamma_{kl}]/2\} = 0$ and similarly for B. Multiplying through by $\gamma_{kl}(1 - \gamma_{kl})$ and rearranging we then get for A and B:

(A4.2) $\gamma_{kl} \displaystyle\sum_{m=1}^{2} \sum_{n=1}^{2} B^{Ak}_{\cdot\cdot mn} f'(e + e_{mn}/2)(\{S_{Amn} + S_{Bmn}\}/2 - e) =$

$- \displaystyle\sum_{i=1}^{2} \sum_{j=1}^{2} \sum_{m=1}^{2} \sum_{n=1}^{2} B^{Ak}_{ijmn} f'(e + e_{mn}/2)(\gamma_{ij}e - S_{Amn}/2)$ $(k, l \in \{1, 2\})$

and

$$(A4.3) \qquad \gamma_{kl} \sum_{m=1}^{2} \sum_{n=1}^{2} B^{Bl}_{\cdots mn} f'(e + e_{mn}/2)(\{S_{Amn} + S_{Bmn}\}/2 - e) =$$

$$-\sum_{i=1}^{2} \sum_{j=1}^{2} \sum_{m=1}^{2} \sum_{n=1}^{2} B^{Bl}_{ijmn} f'(e + e_{mn}/2)(\gamma_{ij} e - S_{Amn}/2) \quad (k, l \in \{1, 2\})$$

Now if the following conditions

$$(A4.4) \quad \sum_{m=1}^{2} \sum_{n=1}^{2} B^{Ak}_{\cdots mn} f'(e + e_{mn}/2)(\{S_{Amn} + S_{Bmn}\}/2 - e) \neq 0 \quad (k = 1, 2)$$

$$(A4.5) \quad \sum_{m=1}^{2} \sum_{n=1}^{2} B^{Bl}_{\cdots mn} f'(e + e_{mn}/2)(\{S_{Amn} + S_{Bmn}\}/2 - e) \neq 0 \quad (l = 1, 2)$$

hold (A4.2) and (A4.3) imply that $\gamma_{k1} = \gamma_{k2}$ for $k = 1, 2$ and $\gamma_{1l} = \gamma_{2l}$ for $l = 1, 2$, i.e. that γ_{kl} is constant, i.e. $\alpha_{kl} + \beta_{kl}$ is constant. But from Proposition 5 we know that generically we can only attain the allocation (C^*, C^*) when $\alpha_{kl} + \beta_{kl}$ is constant if (S_A, S_B) is of type 2. So if we prove that (A4.4) and (A4.5) hold generically under the stated assumptions we have proved the proposition. A necessary condition for (A4.4) and (A4.5) is that for some m, n:

$$(A4.6) \qquad K_{mn} = \frac{1}{2}\{S_{Amn} + S_{Bmn}\} - e \neq 0$$

But if we obtain the allocation (C^*, C^*) (A4.6) must hold for some m' and n'. Else (from (A4.1)) we would have that $S_{Amn} = e + (e_{Bn} - e_{Am})/2$ and $S_{Bnm} = e + (e_{Am} - e_{Bn})/2$ which contradict the requirement that $[\alpha_{kl} + \beta_{kl}]e - S_{Amn} > 0$ and $[2 - \alpha_{kl} - \beta_{kl}]e - S_{Bmn} > 0, \forall k, l, m, n$ (the feasibility of the transfer scheme).

To finish the proof we show that if for some m' and n' (4.9) does hold then for almost all RBS (A4.4) and (A4.5) hold. Suppose to the contrary for a given RBS in $\mathcal{H}$ and some $k \in \{1, 2\}$ that, say, (A4.4) does not hold and that $K_{m', n'} \neq 0$. It then follows that for some $(m'', n'') \neq (m', n')$, $K_{m'', n''} \neq 0$ and we can assume that $K_{m', n'} > 0$ and $K_{m'', n''} < 0$.

Let $\varepsilon > 0$. For $\delta \in (0, \varepsilon)$ define the perturbed belief, $B^{\delta A1}$ as follows: $B^{\delta A1}_{11m'n'} = B^{A1}_{11m'n'} + \delta, B^{\delta A1}_{11m''n''} = B^{A1}_{11m''n''} - \delta$ and $B^{\delta A1}_{ijmn} = B^{A1}_{ijmn}$ for $(i, j, m, n) \neq (1, 1, m', n'), (1, 1, m'', n'')$. Let $Q^A = Q^D_{11} + Q^D_{12}$. Then let $B^{\delta A2}_{11m'n'} = B^{\delta A2}_{11m'n'} - \delta Q^A/(1 - Q^A), B^{\delta A2}_{11m''n''} = B^{\delta A2}_{11m''n''} + \delta Q^A/(1 - Q^A)$ and $B^{\delta A2}_{ijmn} = B^{A2}_{ijmn}$, for $(i, j, m, n) \neq (1, 1, m', n'), (1, 1, m'', n'')$. By picking ε sufficiently small we can ensure that A's perturbed beliefs, $(B^{\delta A1}, B^{\delta A2})$ belong to a new RBS which is, with the exception of these beliefs, identical to the original one and also that for $\delta \in (0, \varepsilon) : \sum_{m=1}^{2} \sum_{n=1}^{2} B^{\delta A1}_{\cdots mn} f'(e + e_{mn}/2) K_{mn} > 0$ and $\sum_{m=1}^{2} \sum_{n=1}^{2} B^{\delta A2}_{\cdots mn} f'(e + e_{mn}/2) K_{mn} < 0$. It then follows from Fubini's theorem that for $\lambda-$ a.a. h (A4.4) holds. $\qquad \square$

Appendix 5

Definition. (Stationary) *Rational Expectations Equilibrium with two currencies.*

(Stationary) stochastic processes $\{\alpha_t^*, \beta_t^*, S_{At}, S_{Bt}\}_{t=1}^{\infty}$ s.t., $\forall t$

(i) $0 < \alpha_t^* + \beta_t^* < 2$ almost surely
(ii) $\alpha_{t+1}^* e + \beta_{t+1}^* e - S_{At+1} > 0$ and $(1 - \alpha_{t+1}^*)e + (1 - \beta_{t+1}^*)e - S_{Bt+1} > 0$
$C_{At+1} \geq 0, C_{Bt+1} \geq 0$ almost surely.
(iii) α_t^* and β_t^* solve respectively

$$\text{Max}_{\alpha \in [0,1]} E \left\{ f \left(\alpha \frac{\alpha_{t+1}^* e + \beta_{t+1}^* e - S_{At+1}}{\alpha_t^* e + \beta_t^* e} e \right. \right.$$
$$\left. \left. + (1 - \alpha) \frac{(1 - \alpha_{t+1}^*)e + (1 - \beta_{t+1}^*)e - S_{Bt+1}}{(1 - \alpha_t^*)e - (1 - \beta_t^*)e} e + e_{At+1} + S_{At+1} \right) \Big| I_t \right\}$$

and

$$\text{Max}_{\beta \in [0,1]} E \left\{ f \left(\beta \frac{\alpha_{t+1}^* e + \beta_{t+1}^* e - S_{At+1}}{\alpha_t^* e + \beta_t^* e} e \right. \right.$$
$$\left. \left. + (1 - \beta) \frac{(1 - \alpha_{t+1}^* e) + (1 - \beta_{t+1}^*)e - S_{Bt+1}}{(1 - \alpha_t^*)e - (1 - \beta_t^*)e} e + e_{Bt+1} + S_{Bt+1} \right) \Big| I_t \right\}.$$

$\square$

Remark. (i) guarantees that there will be positive (private) demand for both currencies at all dates. (ii) is a feasibility requirement. We require that the real subsidy of each government does not exceed its proceeds from printing money. (iii) is the requirement of individual optimality. (I_t is the information at date t.)

With an equilibrium as described above we have for any initial level of nominal money stock via (1.1) and (1.2) determined a stochastic process of currency prices and the amount of the two currencies in circulation. Note, that in this equilibrium the two agents choose the same portfolio in each period. Of course, whether an equilibrium exists or not will depend on the stochastic processes $\{S_{At}\}$ and $\{S_{Bt}\}$ chosen by the two governments. We consider Policy 1.

Proposition. *Under Policy 1 there exists a stationary REE with $\alpha_t^* = \beta_t^* = \frac{1}{2}$ and for any REE we have $C_{At} = C_{Bt} = C^*(e_{At}, e_{Bt}), \forall t$.*

Proof. Note, that under Policy 1, $e_{At+1} + S_{At+1} = \max(e_{At+1}, e_{Bt+1}) = e_{Bt+1} + \max(0, e_{At+1} - e_{Bt+1})$ so the problems of the two agents become identical. Because of the strict concavity of f there is then a unique (identical) solution to the two problems. So in equilibrium $\alpha_t^* = \beta_t^*$ and both agents hold positive amounts of money. This implies that if there is an equilibrium $C_{At} = C^*(e_{At}, e_{Bt}) = C_{Bt}, \forall t$.

Next note, that if we have an equilibrium where both agents hold positive amounts of both currencies, the following first order conditions hold:

$$E\left\{ f'\left(\alpha_t^* \frac{\alpha_{t+1}^* e + \beta_{t+1}^* e - S_{At+1}^1}{\alpha_t^* e + \beta_t^* e} e \right. \right.$$
$$+ (1 - \alpha_t^*) \frac{(1 - \alpha_{t+1}^*)e + (1 - \beta_{t+1}^*)e - S_{Bt+1}^1}{(1 - \alpha_t^*)e + (1 - \beta_t^*)e} e$$
$$+ \max(e_{At+1}, e_{Bt+1})) \cdot \left(\frac{\alpha_{t+1}^* e + \beta_{t+1}^* e - S_{At+1}^1}{\alpha_t^* e + \beta_t^* e} e \right.$$
$$\left. \left. - \frac{(1 - \alpha_{t+1}^*)e + (1 + \beta_{t+1}^*)e - S_{Bt+1}^1}{(1 - \alpha_t^*)e + (1 - \beta_t^*)e} e \right) \middle| I_t \right\} = 0.$$

A similar condition holds for agent B. We check that $\alpha_t^* = \beta_t^* = 1/2, \forall\, t$ is a solution. Inserting in the first order condition we get:
$Ef'(e + (e_{At+1} + e_{Bt+1})/2)(e_{At+1} - e_{Bt+1}) = 0$ which holds by the assumption that $\{e_{At}\}$ and $\{e_{Bt}\}$ are mutually independent, identically distributed and i.i.d. as shown by the following argument: $\mu((e_A, e_B) \in B) = \mu((e_B.e_A) \in B)$ for measurable B, so $E(f'(e + (e_A + e_B)/2)(e_A - e_B)) = E(f'(e + (e_A + e_B)/2)(e_B - e_A)) = -E(f'(e + (e_A + e_B)/2)(e_A - e_B))$, implying that $E(f'(e + (e_A + e_B)/2)(e_A - e_B)) = 0$ □

The allocation obtained is Pareto optimal no matter the process of price movements, in particular also for sunspot equilibria (see Nielsen, 1997 for more on this). Note also, that in this model free capital mobility plays an important role, since it allows a scheme which insures the agents against variations in their second period consumption. One easily shows that also under Policy 2 do we get an REE with the same Pareto Optimal allocation.

It turns out that a monetary union, combined with transfer scheme 1, invoked by some central authority and financed by issuing the common currency, will also result in this allocation. Under a monetary union there is only one currency, so the problem of the agents is completely trivial. Thus under this institution the return on savings in the single currency is $1/2e[2e - \max(0, e_{Bt} - e_{At}) - \max(0, e_{At} - e_{Bt})]$ and consequently consumption of agent A is $1/2[2e - \max(0, e_{Bt} - e_{At}) - \max(0, e_{At} - e_{Bt})] + e_{At} + \max(0, e_{Bt} - e_{At})] = e + (e_{At} + e_{Bt})/2$ at date t and similarly for agent B. In summary, if we used the rational expectations hypothesis to analyze this model, we would not detect any difference between an institution with floating exchange rates and a Monetary Union.

References

Black, S.: The forward discount puzzle in a rational beliefs framework. Manuscript, Dept. of Economics, Stanford University (1997)

Frankel, J.A., Froot, K.A.: Chartists, fundamentalists and trading in the foreign exchange market. American Economic Review **80**, 24–38 (1990)

Goldberg, M.D., Frydman, R.: Imperfect knowledge and behavior in the foreign exchange markets. The Economic Journal **106**, 869–893 (1996)

Hammond, P.: Ex-ante and ex-post welfare optimality under Uncertainty. Economica **48**, 235–250 (1981)

Hauswald, R.B.H.: The excess volatility of foreign exchange rates: statistical puzzle or theoretical artifact? Kelley School of Business, Indiana University (1999)

Kareken, J., Wallace, N.: On the indeterminacy of equilibrium exchange rates. Quarterly Journal of Economics **96**, 207–222 (1981)

Kurz, M.: The Kesten-Stigum model and the treatment of uncertainty in equilibrium theory. In: Balch, M.S., McFadden, D.L., Wu, S.Y. (eds.) Essays on economic behavior under uncertainty, pp. 389–399. Amsterdam: North-Holland 1974

Kurz, M.: On the structure and diversity of rational beliefs. Economic Theory **4**, 977–900 (1994a). Reprinted as chapter 2 in: Kurz, M. (ed.) Endogenous economic fluctuations: studies in the theory of rational beliefs. Berlin Heidelberg New York: Springer 1997

Kurz, M.: On rational belief equilibria. Economic Theory **4**, 859–976 (1994b). Reprinted as chapter 5 in: Kurz M. (ed.) Endogenous economic fluctuations: studies in the theory of rational beliefs. Berlin Heidelberg New York: Springer 1997

Kurz, M.: Endogenous economic fluctuations and rational beliefs: a general perspective. In: Kurz, M. (ed.) Endogenous economic fluctuations: studies in the theory of rational beliefs, pp. 1–37. Berlin Heidelberg New York: Springer 1997a

Kurz, M.: On the volatility of foreign exchange rates. In: Kurz, M. (ed.) Endogenous economic fluctuations: studies in the theory of rational beliefs, pp. 337–352. Berlin Heidelberg New York Springer 1997b

Kurz, M.: Heterogenous forecasting and federal reserve information. Manuscript, Stanford University (2001)

Kurz, M., Wu, H.-M. Endogenous uncertainty in a general equilibrium model with price contingent contracts. Economic Theory **8**, 461–488 (1996). Reprinted as chapter 7 in: Kurz, M. (ed.) Endogenous economic fluctuations: studies in the theory of rational beliefs. Berlin Heidelberg New York: Springer 1997

Manuelli, R., Peck, J.: Exchange rate volatility in an equilibrium asset pricing model. International Economic Review **31**, no. 3 (1990)

McKinnon, R.I., Pill, H.: International overborrowing. A decomposition of credit and currency risks. World Development **7**, 1267–1282 (1998)

Nickelsburg, G.: Dynamic exchange rate equilibria with uncertain government policy. Review of Economic Studies **51**, 509–520 (1984)

Neumeyer, P.A.: Currencies and the allocation of risk: the welfare effects of a monetary union. American Economic Review **88**, 245–259 (1998)

Nielsen, C.K.: Weakly rational beliefs, structural independence and rational belief structures. Ph.D. Dissertation, Stanford University, unpublished (1994)

Nielsen, C.K.: Rational belief structures and rational belief equilibria. Economic Theory **8**, 399–422 (1996). Reprinted as chapter 6 in: Kurz, M. (ed.) Endogenous economic fluctuations: studies in the theory of rational beliefs. Berlin Heidelberg New York: Springer 1997

Nielsen, C.K.: Floating exchange rates versus a monetary union under rational beliefs: the role of endogenous uncertainty. Discussion Paper, University of Copenhagen (1997)

Nielsen, C.K.: Stabilizing, pareto improving policies in an OLG model with incomplete markets: the rational expectations and rational beliefs case. Manuscript, University of Copenhagen (2001)

Obstfeld, M., Rogoff, K.: New directions for stochastic open economy models. Journal of International Economics **50**, 117–153 (2000)

Sibert, A.: The risk premium in the foreign exchange market. Journal of Money, Credit, and Banking **21**, 49–65 (1989)

Taylor, M.P.: The economics of exchange rates. Journal of Economic Literature **33**, 13–47 (1995)

Endogenous uncertainty
and the non-neutrality of money[*]

Maurizio Motolese

Istituto di Politica Economica, Università Cattolica di Milano, via Necchi 5, 20123 Milano, ITALY
(e-mail: maurizio.motolese@mi.unicatt.it)

Received: January 14, 2002; revised version: April 5, 2002

Summary. We study some implications of the Theory of Rational Beliefs to monetary policy. We show that monetary policy in a Rational Beliefs environment can have an important effect on the characteristics of economic fluctuations. In Rational Beliefs Equilibria money is generically non-neutral unlike Rational Expectations Equilibria in which money is neutral and monetary policy is ineffective. Under Rational Beliefs Equilibria nominal prices and real output change not only in response to changes in the exogenous growth rate of money but also in response to changes in the state of beliefs. In Rational Beliefs Equilibria monetary shocks have real effects even when they are observed but are not fully anticipated. Furthermore, the non-neutrality of money results in a short run Phillips curve. When money *"flutters, real output sputters"* [8]. We show that *Endogenous Uncertainty* and the distribution of market beliefs are the major explanatory variables of such fluctuations. Under Rational Expectations monetary policy is ineffective because agents neutralize it by predicting correctly the effect of the policy. Under Rational Beliefs it is shown instead that inflation and recessions can be substantially aggravated by the distribution of market beliefs.

Keywords and Phrases: Money non-neutrality, Monetary policy, Rational expectations, Rational beliefs, Rational belief equilibrium, Endogenous uncertainty, States of belief, Phillips curve.

JEL Classification Numbers: D5, D84, E52.

---

   * I would like to thank Mordecai Kurz for his constant help and support. Most of the ideas developed hereby have been inspired by innumerable and fruitful discussions with him. I have also greatly benefited from helpful comments by Stanley Black, Luigi Campiglio, Carsten Nielsen and Ho-Mou Wu. I also received valuable remarks from participants at the V meeting of "The Society for the Advancement of Economic Theory" held in Ischia, Italy, on July 2-8, 2001, where an initial draft of the present work was presented.

1 Introduction

The work reported in this paper is a positive application of the Theory of Rational Belief Equilibrium (in short RBE) to investigate and understand the nature of money non-neutrality and its effects on the real economy. Given the expectational perspective proposed by the Theory of RBE, we show that one of the most important role in the emerging of money non-neutrality is played by endogenous uncertainty, the internally propagated uncertainty about endogenous variables such as beliefs and actions of other agents and prices. We also show that the heterogeneity of beliefs together with the distribution and intensity of agents' states of optimism/pessimism can expand or reduce the real effect of monetary policy and generate *Endogenous Fluctuations of Output*. This, in contrast to the Rational Expectations results of money neutrality and policy ineffectiveness, leads to a scenario in which monetary policy has an impact on the real economy by affecting motives and decisions. This view is not new to economists. J.M. Keynes already rejected the money neutrality presumption and wrote:

> "...An economy, which uses money but uses it merely as a neutral link between transactions in real things and real assets and does not allow it to enter into motives or decisions, might be called - for want of a better name - a *real-exchange economy*. The theory which I desiderate would deal, in contradistinction to this, with an economy in which money plays a part of its own and affects motives and decisions and is, in short, one of the operative factors in the situation, so that the course of events cannot be predicted, either in the long period or in the short, without a knowledge of the behavior of money between the first state and the last. And it is this which we ought to mean when we speak of a *monetary economy*". (J.M. Keynes [10] pp. 408, 409).

The New Classical Economics discarded this perspective and insisted on the neutrality property of money. One can find early assertions of the neutrality of money in writings regarding monetary issues several hundred years ago. Untill the late 1960's the neutrality proposition was merely regarded as a general principle following from the *Quantity Theory of Money*. No precise account of the role of expectations was given to support such a proposition. With Lucas [25] money neutrality began to be viewed as an "expectational" phenomenon. Under Rational Expectations money is neutral because it is common knowledge that all agents expect the equilibrium price function to exhibit neutrality. The neutrality theory, developed by Lucas [25], Sargent and Wallace [36] often referred to as LSW proposition, states that any anticipated monetary shock would have no effect on real economic variables neither in the short run nor in the long run. Based on their theory of Rational Expectations, New Classical economists believe in the idea of policy irrelevance. In other words, since individuals hold Rational Expectations, they conclude that fully anticipated monetary policies are ineffective in the short run, as well as in the long run; only unanticipated policy shocks can influence real variables in the short run while in the long run the classical neutrality proposition holds. Empirical evidence has not supported such a conclusion: Mishkin [29], using quarterly data for the United States for the period 1954-1976, conclude that both anticipated and unanticipated monetary policy have a long lasting effect on output. Similar conclusions on United States and five additional countries are reached by Hoffman and Schlagenhauf [9]. Aware of the empirical evidence of money non-neutrality, early studies of monetary policy under Rational Expectations Equilibrium (in short REE) focused on informational imperfections. Lucas' [25] argued that money is not neutral because, due to asymmetric information, people confuse changes in absolute price level for changes in relative prices. Lucas' [25] turned out to be one of the most celebrated papers about monetary theory. It received widespread attention from economists: it has been regarded as the flagship of the Rational Expectations revolution. One of the main results that drew the attention of economists in Lucas [25] has been the emergence of

"the Phillips curve not as an unexplained empirical fact, but as a central feature of the solution to a general equilibrium system."(Lucas [25] p. 122)

However he also showed that the non zero slope of this curve has no use for policy purposes. That is, monetary policy (of his k-percent type) is neutral in his model. To strengthen this result he also claimed that the equilibrium allocation of his model with a *Friedman-like* monetary policy is Pareto-optimal.

The Rational Expectations approach dominated almost all economic thought in the 1970's and early 1980's. However, some concerns were raised: Chiappori and Guesnerie [6] proves the existence of a new class of non-linear solutions to the Lucas' [25] model which exhibit the non-neutrality property. In fact, under those solutions the labor-supply process is non-stationary and strongly affected by the money stock. Concern about the validity of the LSW proposition has also been raised by the theoretical *sunspots* literature. Azariadis [2], Azariadis and Guesnerie [3], Cass and Shell [5] and others show that money can be non-neutral in a wide class of models in which equilibrium depends on the realization of such random variables as sunspots which have no inherent relevance to the fundamentals of the economy. Some more concerns have as well been raised in a recent paper by Woodford [37] who shows that money non-neutrality can emerge as a simple consequence of the failure of common knowledge of monetary shocks to hold at any level of the chain of reasoning. This shows even further how much expectations are at the foundation of any theory of non-neutrality.

We note that when agents hold diverse beliefs the LSW neutrality proposition need not to hold in equilibrium. An equilibrium in which rational agents hold diverse beliefs is one where they have *subjective theories about the dynamics of the economy*. In particular, in an RBE agents have a non-stationary perspective on the economy hence they hold different opinions about the true probability distributions of the equilibrium stochastic process of exogenous and endogenous variables (i.e. they hold different conditional probabilities). It follows agents arrive at different conclusions when they condition their probability beliefs on current information. That is, *agents interpret current information differently*. Hence, even fully observed monetary shocks could be interpreted differently by agents and money non-neutrality is implied (for earlier results see Motolese [30], [31] and Kurz, Jin and Motolese [21]). Furthermore, the non-neutrality of money under RBE results in the emergence of a short run Phillips curve. The non-zero slope of this curve in the monetary equilibria with diverse beliefs, in spite of Lucas' [25] conclusion, opens up the door to a crucial role for monetary policy as explained in Kurz, Jin and Motolese [21].

The rest of the paper is organized as follows. Section 2 presents a brief review of Lucas'[25] model and results. Section 3 formalizes the OLG models at study, defines both the REE and the RBE and reports and discusses the simulated results of money neutrality/non-neutrality and the real effects of monetary policy and state of beliefs. Section 4 examines the endogenous GNP fluctuations and the Phillips curve results of our model. Section 5 concludes.

2 Review of Lucas model

Lucas' [25] model of money neutrality was, for a time, very influential and contributed to the development of the Rational Expectations literature on economic policy. However, its empirical and theoretical relevance, as we mentioned above, has been later subjected to strong criticism. We use it in the present paper, which is methodological in nature, only insofar as it simplifies the analysis and allows a better understanding of how a monetary RBE works. Nevertheless, as we explain in Section 3, we depart from Lucas' [25] model in three fundamental ways: (i) trading occurs on one single competitive market thus we relax the assumption of separate isolated markets; (ii) monetary policy is observed by all agents hence we relax the Lucas' [25] assumption of asymmetric information; (iii) agents' do not hold Rational Expectations but heterogeneous conditional Rational Beliefs. We now turn to a brief review of Lucas' [25] model.

The structure of the model is very simple. It is an Overlapping Generation model where agents live for two periods. At each date t, N identical agents are born and hence at each date $2N$ (divisible) agents are present in the economy. When young they make consumption and labor supply decisions over the two periods of their life. They supply n_t units of labor (per capita) only when young producing the amount $y_t = n_t$ of consumption good. Since agents are only productive in the first period of their life (when young) and the consumption good is not storable, in order to consume in the second period of their life (when old) they have to hold money. All agents have identical preferences and rational expectations in the sense of Muth [32]. Trading occurs on two competitive geographically isolated markets[1] without any trading occurring between them and without information exchange between them as well. At each date t the young agents are allocated randomly across the two markets in the proportions $\theta_t/2$ and $(1 - \theta_t/2)$, respectively, where $\{\theta_t, t = 1, 2, ...\}$ are independently and identically distributed random variables with continuous symmetric density function g on $(0, 2)$. The old generation is allocated in such a way that each market possesses at all times one-half the aggregate supply of money. The two markets are treated symmetrically hence an equilibrium can be computed by looking at just one of them.

---

[1] In the following literature this type of model has been explicitly referred to as the two-island model (see Lucas [26]). In Azariadis [2] the two markets are named, in a very colorful way, *Pacifica* and *Atlantica*

Fiat money is issued by a government and is transferred to the old agents at the beginning of each period. The transfer is assumed to be proportional to the pre-transfer holdings of the agents. Let $\overline{m}_t$ denote the pre-transfer money holdings, per old agent, at date t. All agents know $\overline{m}_t$ as well as the ruling price p_t on their own island and all past prices. Post transfer money holdings, per old agent, are given by the equation

$$\overline{m}_{t+1} = \overline{m}_t x_t \tag{1}$$

where $\{x_t, t = 1, 2, ...\}$ are independently and identically distributed random variables with continuous density function f on $(0, \infty)$. To rule out wealth redistribution effect, it is assumed that x_t is common to all old agents. The state of the economy at any date t is then completely described by the three variables $(\overline{m}_t, x_t, \theta_t)$ and the motion from state to state is independent of decisions made by agents and completely determined by equation (1), and by the density functions g and f. There is no inheritance. Old agents do not have any labor to sell and money does not yield any direct utility. At the start of trading at date t young agents do not observe the monetary policy variable x_t as well as the real shock random variable θ_t. Since the observed price p_t responds to both nominal and real shocks, all the additional information about the actual state of the economy hide in a noisy price message. Asymmetry of information is assumed as at the beginning of trading at each date t all old agents possess more information than young agents. In fact, by the mere fact of receiving a transfer of money proportional to their pre-transfer holdings, they do observe x_t. Some monetary policy x_t is implemented and yet it is not public information. Furthermore, such valuable information for young agents is completely valueless to old agents who simply give up their cash balances in return for as much consumption as they can buy at market price.

The young agent's utility, measured by a Von Neumann-Morgestern utility function which has the form

$$U\left(c_t, n_t\right) + V\left(c_{t+1}\right) \tag{2}$$

depends on consumption c_t, supply of labor n_t at date t and consumption c_{t+1} at date $t+1$. Let λ_t denote the young agent's demand for money at date t, which is the money holdings he is carrying over into next period and which will then be affected by the monetary shock at date $t + 1$. Then his consumption at date $t + 1$ will be $c_{t+1} = (x_{t+1}\lambda_t/p_{t+1})$. Let x_{t+1} and p_{t+1} have the joint conditional distribution function $F\left(x_{t+1}, p_{t+1} \mid \overline{m}_t, p_t\right)$, then the decision problem of the young agent is to choose a non-negative triple (c_t, n_t, λ_t) to maximize his expected utility function:

$$U\left(c_t, n_t\right) + \int V\left(\frac{x_{t+1}\lambda_t}{p_{t+1}}\right) dF(x_{t+1}, p_{t+1} \mid \overline{m}_t, p_t) \tag{3}$$

subject to the budget constraint:

$$p_t c_t \leq p_t n_t - \lambda_t. \tag{4}$$

In the market where the fraction $\theta_t/2$ of young agents is allocated at date t, if the state of the economy is $(\overline{m}_t, x_t, \theta_t)$, the total demand for consumption good by the old agents is $(N\overline{m}_t x_t/2p_t)$. Young agents choose their net good supply functions $(n_t - c_t) = \xi\,(p_t; F\,(x_{t+1}, p_{t+1} \mid \overline{m}_t, p_t))$. The total net good supply is then $N\theta_t\xi(\ )/2$ and the consumption good market will clear if

$$(N\overline{m}_t x_t/2p_t) = N\theta_t\xi\,(p_t; F\,(x_{t+1}, p_{t+1} \mid \overline{m}_t, p_t))\,/2, \qquad (5)$$

that is

$$\overline{m}_t x_t/\theta_t = p_t\xi\,(p_t; F\,(x_{t+1}, p_{t+1} \mid \overline{m}_t, p_t)). \qquad (6)$$

By Walras' law, equilibrium on the money market is also obtained once the market clearing condition (6) is satisfied. On the money market the total money supplied by the old agents is $(N\overline{m}_t x_t/2)$, hence the money supplied per demander is $(N\overline{m}_t x_t/2)\,/\,(N\theta_t/2) = \overline{m}_t x_t/\theta_t$ and in equilibrium $\lambda_t = \overline{m}_t x_t/\theta_t$.

The equilibrium price is a function, that is, a time invariant rule of the form $p = \Psi\,(\overline{m}, x, \theta)$ that relates the price level to the exogenous shocks (x, θ) and to the pre-transfer money holdings $\overline{m}$. The price function Ψ is known to all agents, although they do not observe x and θ. The true probability distribution of next period's price, $p' = \Psi\,(\overline{m}', x', \theta') = \Psi\,(\overline{m}x, x', \theta')$ is known, conditional on $\overline{m}$, from the known distribution of x, x' and θ'. Agents, treating $\overline{m}$ as parameter, take expectations with respect to the well-defined joint distribution of (x, x', θ') conditional on $\Psi\,(\overline{m}, x, \theta)$ denoted in Lucas' paper by $G\,(x, x', \theta' \mid \Psi\,(\overline{m}, x, \theta))$, hence dispensed with unspecified distribution F. Returning to the exogenous shocks (x, θ), from the form of the market clearing condition (6) and in view of Lemma 1 in Lucas ([25], p.111), it is easy to see that they affect the economy only through their ratio $z = x/\theta$. Hence the market clearing price function $p = \Psi\,(\overline{m}, x, \theta)$ can be rewritten as $p = \Psi\,(\overline{m}, z)$. Under the a priori neutrality-like hypothesis that the state variable $\overline{m}$ affects equilibrium prices in proportion and quantities are not affected at all, the equilibrium price function is then rewritten as $p = \Psi\,(\overline{m}, x, \theta) = \overline{m}\varphi\,(z)$. Lucas [25] assumes that $\overline{m}$ and $\varphi\,(z)$ are known to the agents. He then proves that $\varphi\,(z)$ is monotonic in $z = x/\theta$ (see Lemma 1 in Lucas [25], p. 111)[2]. Hence if agents observe the price p they can deduce the value of the ratio $z = x/\theta$. If neither x nor θ are observed by agents, knowledge of the ratio $z = x/\theta$ still leads to confusion between relative and absolute prices. Indeed, under these conditions, *money is non-neutral* in Lucas' model. On the other hand, when either x or θ are observed by agents, price signals are not noisy and consequently all nominal movements are only reflected in the absolute level of prices. That is, *money is neutral*. In the case where the exogenous shock to the economy is not a purely monetary one, Lucas finds a positive price function with elasticity between zero and one. This implies that the equilibrium employment function has the form $n\,(x/\theta)$ and that $n'\,(x/\theta) > 0$. That is, increases in demand induce increases in real output, weather the initial increase in demand is monetary (an increase in x) or real (a reduction in θ). At this point the relevance to the Phillips curve debate is clear. In Lucas' model there is no usable trade-off between inflation and employment and yet in equilibrium inflation and employment appear to have positive correlation.

---

[2] Following Lucas' erratum [27] and relaxing the a priori neutrality-like hypothesis, Chiappori and Guesnerie [6] found a broader class of *non-neutral* solutions to the price function $p = \Psi\,(\overline{m}, x, \theta) = \psi\,(\overline{m}, z)$ with $z = x/\theta$.

3 Money non-neutrality under rational beliefs

3.1 The model and its stochastic structure

We now consider the same monetary economy studied by Lucas [25] and compute equilibria with the perspective of the theory of RBE and show how money non-neutrality arises from the perspective of this theory. The economy is studied by use of an Overlapping Generation model where agents live for two periods. At each date t, N agents are born within each of the K agent-types present in the economy and hence at each date $2NK$ agents (N young and N old for each one of the K agent-types) are present in the economy. Generally within each agent-type the N agents might have the same utility and hold different conditional beliefs. This is the case of *social states of beliefs* studied in Kurz [18]. In the present study we concentrate on the extreme case in which the beliefs of the agents are perfectly correlated within each agent-type hence agents of the same type hold the same conditional beliefs.[3]

We now turn to our monetary model of perfectly correlated beliefs within each agent type. Note that this economy has the same results of an economy where only $2K$ agents (K young and K old) are present. We then proceed as if only K agents are born at each date t. To describe our model we first introduce some notation.

l_t^k - consumption of leisure by the young agent k at t,

C_t^k - consumption of commodities by the old agent k at t,

m_t^k - money holdings of the young agent k at t,

P_t - price of consumption good at date t,

M_t - money supply at date t,

g_t - production of consumption good by the young agent k at t,

I_t - information available at t.

When young, agents make consumption and labor supply decisions over the two period of their life. It is assumed as in Azariadis [2] that in the first period of their life (when young) agents do not consume any consumption good. They just supply $(1 - l_t)$ units of labor (per capita) producing the amount $g_t = (1 - l_t)$ of consumption good. We do not follow Lucas [25] in creating informationally isolated markets. We assume that trading occurs on a single competitive market. Young agents supply $(1 - l_t)$ units of labor (per capita). At each date t an exogenous shock θ_t impacts on young agents' productivity, where $\{\theta_t, t = 1, 2, ...\}$ is a stationary Markov process that will be specified below. Hence, at each date t each young agent produces the amount $g_t = \theta_t (1 - l_t)$ of consumption good. Since agents are only productive in the first periods of their life (when young) and the consumption good is not storable, in order to consume in the second period of their life (when old) they have to hold money.

Fiat money is issued by a government and is transferred to the old agents at the beginning of each period. Let M_{t-1} denote the aggregate pre-transfer money

---

[3] For a better understanding of the notation and terminology used here see chapter 3 in Motolese [30] or Kurz [14].

holdings, that is the aggregate money purchased by the agents at $t-1$ (when young) and carried over next period at date t (when old). Then the aggregate post-transfer money holdings at date t are given by the equation

$$M_t = M_{t-1}x_t \tag{7}$$

where $\{x_t, t = 1, 2, ...\}$ is a stationary Markov process that will be specified below. The money increase/decrease $\Delta M_t = M_t - M_{t-1} = M_{t-1}(x_t - 1)$ is hence transferred to the old agents. We adopt *helicopter money* as it is in Lucas [25], hence the transfer is proportional to the pre-transfer holdings of the agents. Diverging from Lucas [25], no asymmetry of information is assumed in our model; indeed we assume that, at date t, x_t and θ_t are observed by all agents. This is equivalent to say that once the monetary policy at date t is announced it is known to all agents. Under these assumptions in Lucas' model with proportional transfers money is neutral. This is not the case under Rational Beliefs where, though agents observe x_t and θ_t, money is generically non-neutral.

At all dates the aggregate stock of money, the total demand for consumption by the old agents is M_t $t = 1, 2, ...$ Young agents choose their net good supply functions

$$g_t^k = \theta_t \left(1 - l_t^k\right) \quad t = 1, 2, ...$$

The aggregate good supply is then

$$\theta_t \left(K - \sum_{k=1}^{K} l_t^k \right) \quad t = 1, 2, ...$$

Hence in equilibrium we shall have

$$M_t = P_t \theta_t \left(K - \sum_{k=1}^{K} l_t^k \right) \quad t = 1, 2, ... \tag{8}$$

To make the equilibrium solutions independent of the initial money supply at date 0, M_0, define $p_t = P_t/M_{t-1}$ and $p_{t+1} = P_{t+1}/M_t$ which by equation (7) can be written as $p_{t+1} = P_{t+1}/M_{t-1}x_t$.[4] This allows us to satisfy the stability conditions in the sense of Kurz [12].

In equilibrium inflation rates for all transitions from t to $t + 1$ are defined by

$$1 + i_{t+1} = \frac{P_{t+1}}{P_t} \quad t = 1, 2, ...$$

and using the definition, given above, of p_t and equation (7) we redefine them by

$$1 + i_{t+1} = \frac{x_t p_{t+1}}{p_t} \quad t = 1, 2, ... \tag{9}$$

At each date t the young agents make an allocative decision with full knowledge of their labor endowment, the price level, the real exogenous shock θ_t and the observed monetary policy x_t. However they are uncertain about the future price level P_{t+1}

---

[4] To simplify notation in the rest of the section we call the ratios $p_t = P_t/M_{t-1}$ prices.

that will be realized when they are old. Therefore they are uncertain about the future rate of return on their labor which is defined by

$$1 + r_{t+1} = \frac{P_t x_{t+1}}{P_{t+1}} \quad t = 1, 2, ...$$

and using the definition, given above, of p_t and equation (7) it can be redefined by

$$1 + r_{t+1} = \frac{p_t x_{t+1}}{x_t p_{t+1}} \quad t = 1, 2, ... \tag{10}$$

Note that the interest factor $1 + r$ can be also interpreted as a *wage* rate.

The stochastic structure of the model consists of: the process of money growth $\{x_t, t = 1, 2, ...\}$ where $x_t \in X$, the process of exogenous productivity shocks $\{\theta_t, t = 1, 2, ...\}$ where $\theta_t \in \Theta$ and of the process of the agents' assessment variables denoted by $y_t = \left(y_t^1, y_t^2, ..., y_t^k\right) \in Y \equiv Y^1 \times Y^2 \times ... \times Y^k$. The state of the economy at any date t is then entirely described by (x_t, θ_t, y_t).

Assumption 1 X *is a finite subset in* $\mathbb{R}_{++}$ *with* $|X|$ *elements; the exogenous process* $\{x_t, t = 1, 2, ...\}$ *is a stable* [5] *Markov process on* X *with probability measure* Π_X *defined on* $(X^\infty, \mathbb{B}(X^\infty))$ *with stationary measure* m_X.

Assumption 2 Θ *is a finite subset in* $(0, 2)$ *with* $|\Theta|$ *elements; the exogenous process* $\{\theta_t, t = 1, 2, ...\}$ *is a stable Markov process on* Θ *with probability measure* Π_Θ *defined on* $(\Theta^\infty, \mathbb{B}(\Theta^\infty))$ *with stationary measure* m_Θ.

Agents do not have structural knowledge and hence do not know the true equilibrium map between states (x_t, θ_t, y_t) and prices. Agents take as known at each date t the past observed prices $(p_1, p_2, ..., p_t)$, the past realizations of the exogenous shocks $\{x_\tau, \tau = 1, 2, ..., t\}$ and $\{\theta_\tau, \tau = 1, 2, ..., t\}$ and they do observe past realizations of $\{y_\tau, \tau = 1, 2, ..., t\}$ which occur within their own life. They do not observe the assessment variables of any other agent or of their own "parents". For this reason they will never learn the true probability distribution of x_t and θ_t and the true structure of the equilibrium map. Agents then form probability beliefs about prices and exogenous states knowing that the exogenous state space is a partition of the price state space. We can then write the beliefs Q^k as probabilities on $(P \times Y^k)$ and then state the following

---

[5] Agents try to learn something about the true probability distribution Π by observing the data generated by the economy. Let $v^t = (v_{t+1}, v_{t+2}, v_{t+3}, ...) \in V^\infty$ and $v = (v_1, v_2, v_3, ...) \in V^\infty$ then the shift transformation is defined by $v^t = Tv^{t-1}$. Agents compute $m^n(B)(v) = \frac{1}{n}\sum_{k=0}^{n-1} 1_B(T^k v)$, the relative frequency at which the dynamical system visits each measurable set $B \in \mathbb{B}(V^\infty)$ given that it started at v, where $1_B(z) = \begin{cases} 1 \text{ if } z \in B \\ 0 \text{ if } z \notin B \end{cases}$.

Learning occurs only if, given sufficient data, $m^n(B)(v)$ converges. It hence follows that a dynamical system $(V^\infty, \mathbb{B}(V^\infty), \Pi, T)$ is said to be *stable* if for all finite-dimensional sets, or cylinders, $B \in \mathbb{B}(V^\infty)$, $\lim_{n\to\infty} m^n(B)(v) = \overline{m}(B)(v)$ Π a.e. The system is said to be *strongly stable* if the limit of $m^n(B)(v)$ exists Π a.e. for all $B \in \mathbb{B}(V^\infty)$. If the additional assumption of ergodicity is made (as we do later) then the limit above is independent of v. (For more details see Kurz [12] and Kurz [14].)

Assumption 3 *For all k, under the belief Q^k the process $\left\{ \left(p_t, y_t^k \right), t = 1, 2, \ldots \right\}$ is jointly a Markov process and the dynamical system*

$$\left(\left(P \times Y^k \right)^\infty, \mathbb{B} \left(\left(P \times Y^k \right)^\infty \right), Q^k, T \right)$$

(where T is the shift transformation) is stationary and ergodic. Y^k is a finite subset in $\mathbb{R}$ with $\left| Y^k \right|$ elements. The non-stationarity induced by each assessment variable y^k is a selection, at each date, of a Markov transition function (a matrix if the set of prices is finite) which is determined by the value taken by y_t^k.

Further explanation of assumption 3 will be given later.

Assumption 4 *The utility function of the kth agent is $u^k \left(l^k, C^k \right)$ which is C^2, strictly increasing and strictly concave.*

Using the definition, given above, of p_t we now write the optimization problem of the young agent k:

$$\max_{(l^k, C^k)} E_{Q_t^k} \left\{ u^k \left(l_t^k, C_{t+1}^k \right) \mid \left(p_t, x_t, \theta_t, y_t^k \right) \right\} \tag{11}$$

subject to

$$p_{t+1} C_{t+1}^k = \frac{m_t^k x_{t+1}}{M_t}$$

$$m_t^k = \theta_t (1 - l_t^k) p_t M_{t-1}$$

which imply that

$$C_{t+1}^k = \frac{\theta_t (1 - l_t^k) p_t x_{t+1}}{p_{t+1} x_t}. \tag{12}$$

The market clearing conditions are:

$$x_t = p_t \theta_t \left(K - \sum_{k=1}^{K} l_t^k \right) \quad t = 1, 2, \ldots \tag{13}$$

Under the Markov assumption, the demand function for leisure of all generations take the form

$$l_t^k = \xi^k \left(p_t, x_t, \theta_t, y_t^k \right)$$

and in equilibrium

$$x_t = p_t \theta_t \left(K - \sum_{k=1}^{K} \xi^k \left(p_t, x_t, \theta_t, y_t^k \right) \right). \tag{14}$$

We can solve (14) and write the equilibrium map in the form

$$p_t = \Phi^* \left(x_t, \theta_t, y_t \right) \quad t = 1, 2, \ldots \tag{15}$$

Solution of the form (15) are also derived by Nielsen [33], Kurz and Schneider [23], Kurz and Wu [24], Kurz and Beltratti [20], Kurz [16], [18] and by Kurz and Motolese [22]. The state space for equilibrium analysis is $(X \times \Theta \times Y)$. Note that the number of distinct equilibrium prices cannot exceed $M = |\Theta| |X| \prod_{k=1}^{K} |Y^k|$. We then conclude that in any RBE only a finite number of equilibrium prices such that

$$p_s = \Phi^* (x_s, \theta_s, y_s) \quad \text{for} \quad s = 1, 2, ..., M \qquad (16)$$

is observed.

Now we have to specify the true joint distribution of private assessment variables y_t, x_t and θ_t as a probability $\Pi_{X\Theta Y}$ on the measurable space

$$((X \times \Theta \times Y)^\infty, \mathbb{B}((X \times \Theta \times Y)^\infty)).$$

Some restriction apply to the probability measure $\Pi_{X\Theta Y}$. First consider the vector y_t of assessment variables. Each agent k determines his own probability of y_t^k and knows his own distribution. The probability of y_t^k is $Q_{Y^k}^k$ the marginal measure of Q^k on Y^k. Given $\Pi_{X\Theta Y}$ the implied probability of y_t^k is $\Pi_{(X\Theta Y)_{Y^k}}$ and the following must be satisfied:

$$\Pi_{(X\Theta Y)_{Y^k}} = Q_{Y^k}^k \quad \text{for} \quad k = 1, 2, ..., K. \qquad (17)$$

Assumption 5 *Under $\Pi_{X\Theta Y}$ the process $\{(x_t, \theta_t, y_t), t = 1, 2, ...\}$ is a Markov process and the dynamical system*

$$((X \times \Theta \times Y)^\infty, \mathbb{B}((X \times \Theta \times Y)^\infty), \Pi_{X\Theta Y}, T)$$

is stable and ergodic with stationary measure $m_{X\Theta Y}$.

Lemma 6 *The price process $\{p_t, t = 1, 2, ...\}$ is a stable and ergodic process on the finite state space $X \times \Theta \times Y$ with probability Π_P and a stationary measure m_P. The probability Π_P on $((P), \mathbb{B}((P)^\infty))$ is defined by the probability $\Pi_{X\Theta Y}$ together with the equilibrium map (16). The measure m_P is also obtained from $m_{X\Theta Y}$ and the map Φ^* in (16).*

Now, for any set $A \in \mathbb{B}((X \times \Theta)^\infty)$ define $\Phi_{X\Theta}^* (A) = \{p \in (P)^\infty : p_t = \Phi^* (x_t, \theta_t, y_t)$ all t, for $(x_t, \theta_t) \in A$, some $y_t \in Y^\infty\}$. It then follows from the map (16) that in equilibrium we must have

$$\Pi_{X\Theta} (A) = \Pi_P (\Phi_{X\Theta}^* (A)) \quad \text{for all } A \in \mathbb{B}((X \times \Theta)^\infty) \qquad (18)$$

and therefore

$$m_{X\Theta} (A) = m_P (\Phi_{X\Theta}^* (A)) \quad \text{for all } A \in \mathbb{B}((X \times \Theta)^\infty). \qquad (19)$$

(16), (17), (18) and (19) provide the tools for stating the rationality conditions of the agents.

The belief Q^k is a probability on the space $((P \times Y^k)^\infty, \mathbb{B}((P \times Y^k)^\infty))$ since the agent is not assumed to know the map Φ^*. However, the data reveals that the empirical distribution of prices, money growth and productivity shocks must conform to (19) and this condition must be satisfied by Q^k as required by the rationality axioms. The following is then implied by the Conditional Stability Theorem:

Lemma 7 *Under the assumptions of Lemma 6 Q^k is a rational belief relative to Π_P if:*

1. $\Pi_{(X \ominus Y)_{Y^k}} = Q^k_{Y^k}$,
2. $Q^k_{X\Theta}(A) = Q^k_P(\Phi^*_{X\Theta}(A)) = m_{X\Theta}(A)$ *for all* $A \in \mathbb{B}((X \times \Theta)^\infty)$,
3. $Q^k_P = m_P$.

Using Lemma 7 we can define a RBE as follows:

Definition 8 $\left\{ \Pi_P, \left\{ (Q^k, l^k_s, m^k_s) \text{ for } k = 1, 2, ..., K \text{ and } s = 1, 2, ..., M \right\} \text{ and } (p_s) \text{ for } s = 1, 2, ..., M \right\}$ *constitute a RBE of the monetary OLG economy if:*

1. Q^k *is a rational belief relative to Π_P for $k = 1, 2, ..., K$ and Π_P is defined by $\Pi_{X\Theta Y}$ and by the equilibrium map induced by $(Q^1, Q^2, ..., Q^k)$,*
2. $(l^k_1, l^k_2, ..., l^k_M)$ *are optimal agent allocations for $k = 1, 2, ..., K$,*
3. $x_s = p_s \theta_s \left(K - \sum_{k=1}^K l^k \right)$ *for all t and s.*

Theorem 9 *Under assumptions 4 and 5 there exist a RBE.*

Proof. See proof of existence from Kurz and Schneider [23] in Kurz [14] pp. 278–282.

3.2 Simulations of the monetary economy

Our main tool of analysis here is economic simulations. Such tool enables us to study dynamical systems which are complex mathematical objects for which it would be very difficult to provide a complete characterization by using only mathematical tools. The objective of all simulations is to exhibit the workings of economic principles. For instance, the simulations in the present work focus on the conditions under which the Markov monetary economy under study exhibits money non-neutrality and endogenous fluctuations. We numerically solve for equilibria and through MonteCarlo simulations we generate time series to study the real effect of monetary policy.

Now, let $K = 2$. Assume a CRRA time separable utility function common to all agents with common relative risk aversion (i.e. $\gamma^1 = \gamma^2 = \gamma$). The optimization problem of agent k is then stated as follows for $(k = 1, 2)$:

$$\max_{(l^k, C^k)} E_{Q_{k_t}} \left\{ \frac{1}{1-\gamma} (l^k_t)^{1-\gamma} + \frac{\beta}{1-\gamma} (C^k_{t+1})^{1-\gamma} \mid I_t \right\} \tag{20}$$

subject to

$$C^k_{t+1} = \frac{\theta_t(1 - l^k_t) p_t x_{t+1}}{p_{t+1} x_t}. \tag{21}$$

Inserting (21) into (20) we can write down agent-type 1's first order conditions for optimization:

$$\left(\frac{1 - l^1_t}{l^1_t} \right)^\gamma = \left(\frac{\theta_t p_t}{x_t} \right)^{1-\gamma} \beta E_{Q_1} \left(\left(\frac{x_{t+1}}{p_{t+1}} \right)^{1-\gamma} \mid I_t \right) \tag{22}$$

and agent-type 2's first order conditions for optimization:

$$\left(\frac{1 - l_t^2}{l_t^2}\right)^\gamma = \left(\frac{\theta_t p_t}{x_t}\right)^{1-\gamma} \beta E_{Q_2}\left(\left(\frac{x_{t+1}}{p_{t+1}}\right)^{1-\gamma} \mid I_t\right). \tag{23}$$

In equilibrium we must have

$$p_t \theta_t \left(2 - l_t^1 - l_t^2\right) = x_t. \tag{24}$$

Under assumptions 3 and 4 the standard theorems of dynamic programming apply to the optimization of the agents. The demand function of agent-type k $(k = 1, 2)$ for leisure depends upon $(p_t, x_t, \theta_t, y_t^k)$. We can then rewrite the market clearing condition as

$$p_t \theta_t \left(2 - l_t^1\left(p_t, x_t, \theta_t, y_t^1\right) - l_t^2\left(p_t, x_t, \theta_t, y_t^2\right)\right) = x_t. \tag{25}$$

Condition (25) implies that the equilibrium map of the economy takes the form:

$$p_t = \Phi^*\left(x_t, \theta_t, y_t^1, y_t^2\right). \tag{26}$$

Assumption 10 $Y^1 = Y^2 = \{0, 1\}$, $X = \{x^H, x^L\}$ and $\Theta = \{\theta^H, \theta^L\}$. The marginal measure of $\Pi_{X\Theta Y}$ on $(X^\infty, \mathbb{B}(X^\infty))$ and on $(\Theta^\infty, \mathbb{B}(\Theta^\infty))$ specify these processes to be stationary and ergodic Markov processes with transition matrices

$$\begin{bmatrix} \chi & 1-\chi \\ 1-\chi & \chi \end{bmatrix} \text{ and } \begin{bmatrix} \vartheta & 1-\vartheta \\ 1-\vartheta & \vartheta \end{bmatrix} \tag{27}$$

respectively. Similarly, the marginal measures of $\Pi_{X\Theta Y}$ on $\left(Y^{1\infty}, \mathbb{B}\left(Y^{1\infty}\right)\right)$ and on $\left(Y^{2\infty}, \mathbb{B}\left(Y^{2\infty}\right)\right)$ specify these processes to be i.i.d. with the probability of $y_t^1 = 1$ being α_1 and the probability of $y_t^2 = 1$ being α_2.

The state space is $\left(X \times \Theta \times Y^1 \times Y^2\right)$ but we can consider this space to be the index set $S = \{1, 2, \ldots, 16\}$. We thus write down a new equilibrium map Φ between the *indices* of prices and the states of monetary policy, of productivity shock and of assessment variables[6]:

---

[6] Note that the map Φ is not the equilibrium map Φ^* which defines the values prices take in equilibrium but rather a map between the *indices* of the price states 1,2,...,16 and the vectors of exogenous states and states of beliefs.

$$
\begin{bmatrix} 1 \\ 2 \\ 3 \\ 4 \\ 5 \\ 6 \\ 7 \\ 8 \\ 9 \\ 10 \\ 11 \\ 12 \\ 13 \\ 14 \\ 15 \\ 16 \end{bmatrix} = \Phi
\begin{bmatrix}
x = x^H & \theta = \theta^H & y^1 = 1 & y^2 = 1 \\
x = x^H & \theta = \theta^H & y^1 = 1 & y^2 = 0 \\
x = x^H & \theta = \theta^H & y^1 = 0 & y^2 = 1 \\
x = x^H & \theta = \theta^H & y^1 = 0 & y^2 = 0 \\
x = x^L & \theta = \theta^H & y^1 = 1 & y^2 = 1 \\
x = x^L & \theta = \theta^H & y^1 = 1 & y^2 = 0 \\
x = x^L & \theta = \theta^H & y^1 = 0 & y^2 = 1 \\
x = x^L & \theta = \theta^H & y^1 = 0 & y^2 = 0 \\
x = x^H & \theta = \theta^L & y^1 = 1 & y^2 = 1 \\
x = x^H & \theta = \theta^L & y^1 = 1 & y^2 = 0 \\
x = x^H & \theta = \theta^L & y^1 = 0 & y^2 = 1 \\
x = x^H & \theta = \theta^L & y^1 = 0 & y^2 = 0 \\
x = x^L & \theta = \theta^L & y^1 = 1 & y^2 = 1 \\
x = x^L & \theta = \theta^L & y^1 = 1 & y^2 = 0 \\
x = x^L & \theta = \theta^L & y^1 = 0 & y^2 = 1 \\
x = x^L & \theta = \theta^L & y^1 = 0 & y^2 = 0
\end{bmatrix}. \tag{28}
$$

Given the map (28) we refer to states $\theta^H (\theta^L)$ for $s = 1, 2, ..., 8$ $(s = 9, 10, ..., 16)$ as states of *high (low) exogenous productivity shock*. We refer to states $x^H (x^L)$ for $s = 1, ..., 4, 9, ..., 12$ $(s = 5, ..., 8, 13, ..., 16)$ as states of *loose (tight) monetary policy*.

We proceed to construct the RBE by specifying the Markov transition matrix representing the stationary measure as follows

$$
\Gamma = \begin{bmatrix}
\vartheta\chi A & \vartheta(1-\chi)A & (1-\vartheta)\chi A & (1-\vartheta)(1-\chi)A \\
\vartheta(1-\chi)B & \vartheta\chi B & (1-\vartheta)(1-\chi)B & (1-\vartheta)\chi B \\
(1-\vartheta)\chi C & (1-\vartheta)(1-\chi)C & \vartheta\chi C & \vartheta(1-\chi)C \\
(1-\vartheta)(1-\chi)D & (1-\vartheta)\chi D & \vartheta(1-\chi)D & \vartheta\chi D
\end{bmatrix} \tag{29}
$$

which has $\pi = (\pi_1, ..., \pi_{16})$ as its stationary distribution and where (A, B, C, D) are all 4×4 matrices characterized by the 18 parameters $(\alpha_1, \alpha_2, a, b, c, d)$, where $a = (a_1, a_2, a_3, a_4)$, $b = (b_1, b_2, b_3, b_4)$, $c = (c_1, c_2, c_3, c_4)$, $d = (d_1, d_2, d_3, d_4)$, and of the following type:

$$
A = \begin{bmatrix}
a_1 \alpha_1 - a_1 & a_2 - a_1 & 1 + a_1 - \alpha_1 - \alpha_2 \\
a_2 \alpha_1 - a_2 & a_2 - a_2 & 1 + a_2 - \alpha_1 - \alpha_2 \\
a_3 \alpha_1 - a_3 & a_2 - a_3 & 1 + a_3 - \alpha_1 - \alpha_2 \\
a_4 \alpha_1 - a_4 & a_2 - a_4 & 1 + a_4 - \alpha_1 - \alpha_2
\end{bmatrix}. \tag{30}
$$

From (29) and (30) it follows that the marginals of Γ are indeed as specified in assumption 10. In fact, the marginal measures Γ_{Y^k} specify y_t^k to be i.i.d. with $P\{y_t^k = 1\} = \alpha_k$ for $k = 1, 2$ and the marginal measures Γ_X and Γ_Θ are specified by the stationary Markov processes (27). Also, although each process $\{y_t^k, t = 1, 2, ...\}$ for $k = 1, 2$ is very simple, the joint process may be complex and allow joint correlation among the y_t^k over time.

3.2.1 Equilibrium conditions in terms of price states

Given the price state space defined in (28) we can then write the system of Euler equations, the budget constraints and the market clearing conditions in the form used in the computations below (for $s = 1, 2, ..., 16$, $j = 1, 2, ..., 16$ and $k = 1, 2$). Equations (21)-(23) in terms of price states are respectively written as follows[7]:

$$c_{s,j}^k = \frac{\theta_s(1 - l_s^k)p_s x_j}{p_j x_s} \tag{31}$$

$$\left(\frac{1 - l_s^1}{l_s^1}\right)^\gamma = \left(\frac{\theta_s p_s}{x_s}\right)^{1-\gamma} \beta \sum_{j=1}^{16} \left(\frac{x_j}{p_j}\right)^{1-\gamma} Q_{1(s,j)}^s \tag{32}$$

$$\left(\frac{1 - l_s^2}{l_s^2}\right)^\gamma = \left(\frac{\theta_s p_s}{x_s}\right)^{1-\gamma} \beta \sum_{j=1}^{16} \left(\frac{x_j}{p_j}\right)^{1-\gamma} Q_{2(s,j)}^s. \tag{33}$$

And the market clearing conditions become:

$$p_s \theta_s \left(2 - l_s^1 - l_s^2\right) = x_s. \tag{34}$$

We now turn to specify the probabilities Q_1^s and Q_2^s of the two agent-types to complete the specification of the equilibria described by the system of equations (31)-(34).

3.3.2 Rational beliefs

By assumption 3 the beliefs of the agents are probabilities of joint Markov process on prices and individual assessment variables. Since by assumption the marginals on the assessment variables are i.i.d. with probabilities α_1 and α_2, if we denote the two matrices for agent-type 1 by (F_1, F_2) and for agent-type 2 by (G_1, G_2), it then follows that the rationality conditions of beliefs are:

$$\alpha_1 F_1 + (1 - \alpha_1) F_2 = \Gamma, \quad \alpha_2 G_1 + (1 - \alpha_2) G_2 = \Gamma. \tag{35}$$

Then the following probabilities (where $F_1^{s,j}$ is the (s, j) element of F_1)

$$Q_{1(s,j)}^s = \begin{cases} F_1^{s,j} \text{ if } y_s^1 = 1 \\ F_2^{s,j} \text{ if } y_s^1 = 0 \end{cases} \quad Q_{2(s,j)}^s \begin{cases} G_1^{s,j} \text{ if } y_s^2 = 1 \\ G_2^{s,j} \text{ if } y_s^2 = 0 \end{cases} \tag{36}$$

define Q_1^s and Q_2^s in (32)-(33).[8] To complete the definition of $Q_{1(s,j)}^s$ and $Q_{2(s,j)}^s$ in (36) it is left to specify matrices (F_1, F_2, G_1, G_2). Select 2 parameters λ and μ which will be interpreted later, and define the row vectors of A with the notation

$$A^j = (a_j, \alpha_1 - a_j, \alpha_2 - a_j, 1 + a_j - (\alpha_1 + \alpha_2)) \quad j = 1, 2, 3, 4. \tag{37}$$

---

[7] Note that to compute equilibrium solutions the system of equations (31)-(34) has been solved using standard Newton method as supplied by the software package TENSOLVE, a suite of FORTRAN 77 subroutines. For a complete overview of the software package TENSOLVE see Bouaricha A. and Schnabel R. B. [4]. All computations have been implemented in FORTRAN 77 on DECStations Sun Ultra Enterprise 5000/200.

[8] The superscript in Q_1^s and Q_2^s stress the dependence on y_s^1 and y_s^2.

Similar notation is used for B, C and D. With this notation we define eight matrix functions of a real number z as follows:

$$A_1(z) = \begin{bmatrix} zA^1 \\ zA^2 \\ zA^3 \\ zA^4 \end{bmatrix}, \quad A_2^\vartheta(z) = \begin{bmatrix} \dfrac{1-\vartheta z}{1-\vartheta}A^1 \\ \dfrac{1-\vartheta z}{1-\vartheta}A^2 \\ \dfrac{1-\vartheta z}{1-\vartheta}A^3 \\ \dfrac{1-\vartheta z}{1-\vartheta}A^4 \end{bmatrix} \tag{38}$$

$$B_1(z) = \begin{bmatrix} zB^1 \\ zB^2 \\ zB^3 \\ zB^4 \end{bmatrix}, \quad B_2^\vartheta(z) = \begin{bmatrix} \dfrac{1-\vartheta z}{1-\vartheta}B^1 \\ \dfrac{1-\vartheta z}{1-\vartheta}B^2 \\ \dfrac{1-\vartheta z}{1-\vartheta}B^3 \\ \dfrac{1-\vartheta z}{1-\vartheta}B^4 \end{bmatrix} \tag{39}$$

$$C_1(z) = \begin{bmatrix} zC^1 \\ zC^2 \\ zC^3 \\ zC^4 \end{bmatrix}, \quad C_2^\vartheta(z) = \begin{bmatrix} \dfrac{1-(1-\vartheta)z}{\vartheta}C^1 \\ \dfrac{1-(1-\vartheta)z}{\vartheta}C^2 \\ \dfrac{1-(1-\vartheta)z}{\vartheta}C^3 \\ \dfrac{1-(1-\vartheta)z}{\vartheta}C^4 \end{bmatrix} \tag{40}$$

$$D_1(z) = \begin{bmatrix} zD^1 \\ zD^2 \\ zD^3 \\ zD^4 \end{bmatrix}, \quad D_2^\vartheta(z) = \begin{bmatrix} \dfrac{1-(1-\vartheta)z}{\vartheta}D^1 \\ \dfrac{1-(1-\vartheta)z}{\vartheta}D^2 \\ \dfrac{1-(1-\vartheta)z}{\vartheta}D^3 \\ \dfrac{1-(1-\vartheta)z}{\vartheta}D^4 \end{bmatrix}. \tag{41}$$

Hence given the definitions (38)-(41) we define the matrices $(F_1(\lambda), F_2(\lambda))$ by

$$F_1(\lambda) = \begin{bmatrix} \vartheta\chi A_1(\lambda) & \vartheta(1-\chi)A_1(\lambda) & (1-\vartheta)\chi A_2^\vartheta(\lambda) & (1-\vartheta)(1-\chi)A_2^\vartheta(\lambda) \\ \vartheta(1-\chi)B_1(\lambda) & \vartheta\chi B_1(\lambda) & (1-\vartheta)(1-\chi)B_2^\vartheta(\lambda) & (1-\vartheta)\chi B_2^\vartheta(\lambda) \\ (1-\vartheta)\chi C_1(\lambda) & (1-\vartheta)(1-\chi)C_1(\lambda) & \vartheta\chi C_2^\vartheta(\lambda) & \vartheta(1-\chi)C_2^\vartheta(\lambda) \\ (1-\vartheta)(1-\chi)D_1(\lambda) & (1-\vartheta)\chi D_1(\lambda) & \vartheta(1-\chi)D_2^\vartheta(\lambda) & \vartheta\chi D_2^\vartheta(\lambda) \end{bmatrix} \tag{42}$$

and $F_2(\lambda)$ is then defined by the usual condition

$$F_2(\lambda) = \frac{1}{1-\alpha_1}\left(\Gamma - \alpha_1 F_1(\lambda)\right). \tag{43}$$

Finally given the definitions (38)-(41) we define the matrices $(G_1(\mu), G_2(\mu))$ by

$$
G_1(\mu) = \begin{bmatrix}
\vartheta\chi A_1(\mu) & \vartheta(1-\chi)A_1(\mu) & (1-\vartheta)\chi A_2^\vartheta(\mu) & (1-\vartheta)(1-\chi)A_2^\vartheta(\mu) \\
\vartheta(1-\chi)B_1(\mu) & \vartheta\chi B_1(\mu) & (1-\vartheta)(1-\chi)B_2^\vartheta(\mu) & (1-\vartheta)\chi B_2^\vartheta(\mu) \\
(1-\vartheta)\chi C_1(\mu) & (1-\vartheta)(1-\chi)C_1(\mu) & \vartheta\chi C_2^\vartheta(\mu) & \vartheta(1-\chi)C_2^\vartheta(\mu) \\
(1-\vartheta)(1-\chi)D_1(\mu) & (1-\vartheta)\chi D_1(\mu) & \vartheta(1-\chi)D_2^\vartheta(\mu) & \vartheta\chi D_2^\vartheta(\mu)
\end{bmatrix}
\tag{44}
$$

and $G_2(\mu)$ is then defined by the usual condition

$$
G_2(\mu) = \frac{1}{1-\alpha_2}\left(\Gamma - \alpha_2 G_1(\mu)\right). \tag{45}
$$

Given the equilibrium map (28), the Markovian Rational Beliefs structure constructed above allows any revision of the conditional probabilities of states of high exogenous productivity shock $(1,2,3,4,5,6,7,8)$ relative to the stationary measure Γ, to be offset by an opposite direction revision of the conditional probabilities of states of low exogenous productivity shock $(9,10,11,12,13,14,15,16)$.

In order to ensure non-negative probability entries in (F_1, F_2, G_1, G_2), the selection of the parameters λ and μ is restricted by 10 inequality constraints which define the feasible region. These constraints are as follows:

$$
\lambda \le \frac{1}{\vartheta}, \lambda \le \frac{1}{1-\vartheta}, \lambda \le \frac{1}{\alpha_1}, \lambda \ge \frac{\alpha_1+\vartheta-1}{\vartheta\alpha_1}, \lambda \ge \frac{\alpha_1-\vartheta}{(1-\vartheta)\alpha_1},
$$

$$
\mu \le \frac{1}{\vartheta}, \mu \le \frac{1}{1-\vartheta}, \mu \le \frac{1}{\alpha_2}, \mu \ge \frac{\alpha_2+\vartheta-1}{\vartheta\alpha_2}, \mu \ge \frac{\alpha_2-\vartheta}{(1-\vartheta)\alpha_2}.
\tag{46}
$$

To motivate these construction note that the parameters λ and μ are multiplied by the rows of A, B, C and D and hence are proportional changes of the conditional probabilities of the four sets of four states $(1,2,3,4)$, $(5,6,7,8)$, $(9,10,11,12)$ and $(13,14,15,16)$ relative to the stationary measure represented by the matrix Γ in (29). Since λ and μ are the factors of proportionality by which agents' conditional probability beliefs deviate from the stationary probabilities in Γ, we refer to them as *intensity* parameters. So far agents' assessment variables have been used to endogenously enlarge the price state space and no actual economic meaning has been attached to them. They attain meaning only when the agents specify how they interpret these variables in generating their conditional probability beliefs. For example, $\lambda > 1$ implies increased probabilities of states $(1,2,3,4)$ and $(5,6,7,8)$ in F_1 relative to Γ of agent-type 1 given an initial state s and by the rationality conditions (35) the probabilities of the same states are decreased in F_2. This means that the assessment variables induce more *optimism* or *pessimism* about the occurrence of prices $(1,2,3,4)$ and $(5,6,7,8)$ at $t+1$ relative to Γ. Suppose that at some date t state $s=1$ occurs so that the price p_1 is realized. If $\lambda > 1$ then agent-type 1 with assessment variable $y_t^1 = 1$ uses matrix F_1 to forecast prices at $t+1$ and by (38)-(41) he is more optimistic (relative to Γ) about the probabilities of $(p_1, p_2, p_3, p_4, p_5, p_6, p_7, p_8)$ at $t+1$. The converse applies when $y_t^1 = 0$. Suppose agent 1 is an optimist using $F1$. As $\lambda > 1$ increases probabilities in $F1$, the rationality conditions (35) $\alpha_1 F_1 + (1 - \alpha_1) F_2 = \Gamma$ require a downward adjustment of the probabilities of $(p_1, p_2, p_3, p_4, p_5, p_6, p_7, p_8)$ in the pessimistic matrix $F2$.[9]

The simulations focus on the factors and the conditions under which the Markov monetary economy at study exhibits endogenous uncertainty hence money non-neutrality. There are two such factors:

1. *Deviations over time of the intensity parameters* (λ, μ) *from* 1 reflecting the non-stationarity of beliefs of the agents.
2. *Correlation among agents* represented by the vectors (a, b, c, d) of parameters inducing a joint distribution of the assessments which is Markov and not i.i.d.

The results reported below are focused on studying the characteristics of the joint effects of monetary policy and beliefs on inflation, labor supply and real output.

For any transition of the economy from state s to state j the inflation rates defined by (9) and the rates of return on labor defined by (10) are respectively written as $1 + i_{s,j} = \frac{x_s p_j}{p_s}$, $1 + r_{s,j} = \frac{x_j p_s}{p_j x_s}$ for all $s, j = 1, 2, ..., 16$ and respectively approximated by $i_{s,j} \approx \ln(1 + i_{s,j})$ and $r_{s,j} \approx \ln(1 + r_{s,j})$. Let us note that, because the economy at study has a finite number of prices, the long term rate of return on labor r both under REE and RBE is equal to zero and the long term inflation rate i both under REE and RBE is equal to the long term growth rate of money $(E_\Gamma x_t)$.

Before reporting the results of our simulations, we first note that in all states $s = 1, 2, ..., 16$ nominal and real variables are jointly affected by the exogenous monetary shocks $x_s \in X = \{x^H, x^L\}$, the endogenous states of beliefs $(y_s^1, y_s^2) \in Y \times Y = \{(1, 1), (1, 0), (0, 1), (0, 0)\}$ and the exogenous productivity shocks $\theta_s \in \Theta = \{\theta^H, \theta^L\}$. Due to our Markov assumption any transition from $\theta_s = \theta^L$ to $\theta_j = \theta^H$ for $s, j = 1, 2, ..., 16$ or viceversa represents a productivity regime switch. When such a regime switch occurs prices and real variables are jointly affected by the productivity shock, the monetary policy and the endogenous states of beliefs. We want then to decompose the standard deviation of each variable under study and subtract from it the effect of the productivity exogenous shocks in order to isolate the compound effect of monetary forces and states of beliefs. In the case of inflation rates, for example, we first compute the following two long term conditional averages with fixed money supply (i.e. $x_s = 1$ for $s = 1, 2, ..., 16$):

$$\bar{i}_{HL} = \ln \left(E_\Gamma \left(1 + i_{s,j} \mid \theta_s = \theta^H, \theta_j = \theta^L \right) \right)$$

and

$$\bar{i}_{LH} = \ln \left(E_\Gamma \left(1 + i_{s,j} \mid \theta_s = \theta^L, \theta_j = \theta^H \right) \right).$$

---

[9] Many more complex different structures can be constructed and rational belief equilibria studied. Different patterns of optimism/pessimism may be constructed. Motolese [30] shows that in the case of a markov *helicopter* type monetary policy there is a particular belief structure that when adopted by all agents it causes no endogenous uncertainty to be present and money to be dynamically neutral. Such a structure implies a perfect correlation between states of optimism/pessimism and monetary policy states. This beliefs structure would be justified if agents learned from the data such a perfect correlation. In other words it would be justified if the data told them that money is neutral. Which empirically is a questionable result. But again, this conclusion is valid only when the monetary policy takes place through proportional transfers.

Now let $\varpi_{s,j}^{HL} = 1$ when $\theta_s = \theta^H, \theta_j = \theta^L$ and 0 otherwise, and let $\varpi_{s,j}^{LH} = 1$ when $\theta_s = \theta^L, \theta_j = \theta^H$ and 0 otherwise. Hence the inflation rates volatility, which is the solely joint effect of state of beliefs and monetary policy, will be defined by the standard deviation of the following random variable (for all $s, j = 1, 2, ..., 16.$):

$$\iota_{s,j} = i_{s,j} - \bar{i}_{HL}\varpi_{s,j}^{HL} - \bar{i}_{LH}\varpi_{s,j}^{LH}. \tag{47}$$

Similarly we compute the two long term conditional rates of return on labor ($\bar{r}_{HL}$ and $\bar{r}_{LH}$) and study the impact of states of beliefs and monetary policy on expected real wage rates by the statistics of the following random variable (for all $s, j = 1, 2, ..., 16.$):

$$\rho_{s,j} = i_{s,j} - \bar{r}_{HL}\varpi_{s,j}^{HL} - \bar{r}_{LH}\varpi_{s,j}^{LH}. \tag{48}$$

We can now state the following:

Definition 11 *Money is dynamically neutral if, for any transition of the economy from state s to state j ($s, j = 1, 2, ..., 16$) any monetary shock $x \in X = \{x^H, x^L\}$ leaves all real variables unaffected and leads to the following:*

$$1 + \iota_{s,j} = x_s \text{ for all } s, j = 1, 2, ..., 16.$$

(i.e. within any given real exogenous shock regime the growth rate of money is the sole determinant of the rate of inflation).

To measure the pure effect of monetary policy and states of beliefs on output fluctuations we also compute the output deviations from the conditional means $\overline{\Omega}_H = E_\Gamma(\Omega_s \mid \theta_s = \theta^H)$ and $\overline{\Omega}_L = E_\Gamma(\Omega_s \mid \theta_s = \theta^L)$. This component is uncorrelated with the exogenous productivity shocks and represents the compound effect of Endogenous Uncertainty and monetary policy. To measure this effect let $\zeta_s^H = 1$ when $\theta_s = \theta^H$ and 0 otherwise, and let $\zeta_s^L = 1$ when $\theta_s = \theta^L$ and 0 otherwise. Now we can define the output deviations by the following random variable (for all $s = 1, 2, ..., 16.$):

$$\omega_s = \Omega_s - \overline{\Omega}_H\zeta_s^H - \overline{\Omega}_L\zeta_s^L. \tag{49}$$

3.3 Results of the simulations

3.3.1 The parameterization

Utility function. In all simulations we assume that $\beta = 0.92$, the coefficients of relative risk aversion are assumed to be $\gamma = 3.5$.

The Exogenous stochastic processes. In all simulations we set $\vartheta = 0.43$ and $\chi = 0.5$. The exogenous productivity shock state space is set to be $\Theta = \{\theta^H, \theta^L\} = \{1.3, 0.7\}$ and the money growth state space is set to be $X = \{x^H, x^L\} = \{1.02, 0.98\}$.

The parameters above will remain unchanged across the different simulation cases below unless explicitly specified.

The benchmark case: REE. We define the benchmark case when all agents hold rational expectations. Note that the joint process $\left\{ (x_t, \theta_t, y_t^1, y_t^2), t = 1, 2, ... \right\}$

allows correlation among the four variables over time and such correlation effect plays a central role in the simulations. However, if we set $\alpha_1 = \alpha_2 = 0.5$ and $a_i = b_i = c_i = d_i = 0.25$ for $i = 1, 2, 3, 4$ then all correlations among the four variables of the joint process are eliminated. If, in addition, we assume that agents ignore their assessment variables and set their beliefs by selecting $Q_{1_{(s,j)}} = Q_{2_{(s,j)}} = \Gamma_{(s,j)} \quad \forall s, j$, then we have exactly the REE of Lucas [25]. In the benchmark case of rational expectations the intensity parameters are required to be $\lambda = \mu = 1$. It is easy to see that in such case there is no endogenous uncertainty and that in equilibrium only four possible values of prices, p_{HH}^{Γ}, p_{LH}^{Γ}, p_{HL}^{Γ} and p_{LL}^{Γ}, respectively associated with the exogenous states regimes $\left\{ \left(x^H \theta^H \right), \left(x^L \theta^H \right), \left(x^H \theta^L \right), \left(x^L \theta^L \right) \right\}$, will be realized.

RBE. In all simulations we set $\alpha_1 = \alpha_2 = 0.57$. The vectors of correlation parameters are set to $a = b = c = d = (0.20, 0.14, 0.14, 0.14)$ while the intensity parameters λ_s and μ_s for $s = 1, 2, ..., 16$ and will be accordingly specified later.

Note that we are not attempting to calibrate the model so to simulate the behavior of any given country economy. We are only trying to show how money non-neutrality arises in a RBE and how this differs from a REE. In the tables below we report statistics on inflation rates (ι), rates of return on labor (ρ), level of output (Ω) for the economy and output deviations from conditional means (ω). Additionally we report expected aggregate output, inflation rate, wage rate and output deviations conditional upon the occurrence of both a *loose* and a *tight* monetary policy (i.e. respectively conditional upon $x_s = x^H$ and $x_s = x^L$).

3.3.2 Characteristics of money neutrality in a REE

Under any REE of the model money is neutral. The choice of the exogenous monetary shock parameters $\left\{ x^H, x^L \right\} = \{1.02, 0.98\}$ imply the long term growth rate of money ($E_\Gamma x_t$) to be equal to zero. It follows that long run neutrality of money is achieved under the REE and the RBE of the model. In other words RBE are neutral on average. However, money is generically non-neutral in the short run.

We report in columns 2 and 3 of Table 1 the REE results which provide a reference point for the study of the characteristics of inflation and non-neutrality of money under RBE. From Table 1 one can see the well known result of money neutrality under the Rational Expectations hypothesis. In fact, definition 11 is satisfied: the average inflation rate is equal to the average growth rate of money (i.e. zero), for all states $s, j = 1, 2, ..., 16$. $1 + \iota_{s,j} = x_s$ and its standard deviation exactly matches the one implied by the growth rate of money x_s. All real variables in the economy are not correlated with monetary policy. Their conditional expected values do not depend upon $x \in \left\{ x^H, x^L \right\}$. In the REE any monetary fluctuation leaves the real economy unaffected. In fact, the average level of output is not responsive to monetary forces and exhibits a long term standard deviation exclusively induced by the labor productivity shocks θ_t. The last conclusion we can draw from the Rational Expectations results is that monetary fluctuations, due to the structural knowledge implied in such equilibrium, do not offer any opportunity for positive returns on labor. In fact, the conditional expected rates of return are equal to zero. Indeed, all the results reported in the REE columns of Table 1 are well known and constitute the hallmark of money neutrality.

Table 1. Characteristics of money (non)-neutrality

	REE		RBE (Case 1)	
	avg	σ	avg	σ
ι_t	0.0000	0.0200	0.0000	0.0503
$\iota_t \mid x_t = x^H$	0.0000	0.0200	0.0165	0.0344
$\iota_t \mid x_t = x^L$	0.0000	0.0200	-0.0169	0.0577
ρ_t	0.0000	0.0000	0.0000	0.0529
$\rho_t \mid x_t = x^H$	0.0000	0.0000	-0.0167	0.0381
$\rho_t \mid x_t = x^L$	0.0000	0.0000	0.0167	0.0600
Ω_t	1.0248	0.2623	1.0188	0.2699
$\Omega_t \mid x_t = x^H$	1.0248	0.2623	1.0334	0.2655
$\Omega_t \mid x_t = x^L$	1.0248	0.2623	1.0041	0.2735
ω_t	0.0000	0.0000	0.0000	0.0330
$\omega_t \mid x_t = x^H$	0.0000	0.0000	0.0147	0.0024
$\omega_t \mid x_t = x^L$	0.0000	0.0000	-0.0147	0.0417

3.3.3 Characteristics of money non-neutrality in a RBE

We now turn to the results obtained in the RBE simulations. Our main aim is to explore if heterogeneous rational beliefs can expand(contract) the effect of monetary policy causing endogenous fluctuations in output and employment and affecting price volatility in contrast to the neutrality property of REE of Table 1. To characterize the structure of beliefs we have to choose the intensity parameters λ and μ. We thus start considering the following pattern of intensity parameters which we name as *Case 1* or x^H-*consensus*:

$$\lambda_s = \mu_s = 1.0000 \text{ for } s = 1, 2, 3, 4, 9, 10, 11, 12$$
$$\lambda_s = \mu_s = 1.7542 \text{ for } s = 5, 6, 7, 8, 13, 14, 15, 16. \tag{50}$$

The results obtained under this pattern of beliefs are reported in the RBE columns of Table 1. From that table we can immediately see that RBE are drastically different than REE. Monetary forces are not neutral and have a non trivial impact on the real sector of the economy as well as on inflation. Monetary shocks have an effect on expectations of agents and although they observe the current growth rate of money x_t, their demand for money and their labor supply (which depend upon their beliefs) change according to their interpretation of the monetary signal. This results in a percentage change in the price level not equal to the growth rate of money (i.e. money is dynamically non-neutral). Monetary policy under RBE is perceived as a signal that carries information about future price levels hence inflation rates and wage rates which is not the case under REE.

Before further discussing the results reported in Table 1 we need to get some intuition about the pattern of intensity parameters in (50). Recall the equilibrium map (28) in which the states $x_s^H \left(x_s^L \right)$ for $s = 1, ..., 4, 9, ..., 12$ ($s = 5, ..., 8, 13, ..., 16$) are states of *loose (tight) monetary policy*. Bear in mind that $\alpha_1 = \alpha_2 = 0.57$ means that both agents are optimistic in 57% of the dates and that $\lambda = \mu = 1.7542$

is the maximal ratio, implied by the feasibility conditions (46), by which, when optimistic, agents can adjust probabilities of states $s = 1, \ldots, 8$ at $t + 1$ which are the states of low price and high output levels. On the other hand $\lambda = \mu = 1$ means that agents believe in the stationary measure Γ thus realizing an overall consensus among them with no states of optimism or pessimism. Note that the intensity parameters $\lambda_s = \mu_s = 1$ from (50) are associated with states of *loose* monetary policy (i.e. $x_t = x^H$) while $\lambda_s = \mu_s = 1.7542$ are associated with states of *tight* monetary policy (i.e. $x_t = x^L$). It follows that when $x_t = x^H$ agents' expectations are in accord with the stationary measure resulting in a perfect *consensus*. When $x_t = x^L$ states of disagreement among agents occur. Notice that the parameter choice in (50) along with the choice of the correlation parameters $a = b = c = d = (0.20, 0.14, 0.14, 0.14)$ and $\alpha_1 = \alpha_2 = 0.57$ imply that states of optimism and pessimism are asymmetric.

The implications of the x^H-*consensus* structure are such that $\Omega_t \mid x_t = x^H$ is higher than $\Omega_t \mid x_t = x^L$. In fact when $x_t = x^L$ agents' expectations are pushed to the extreme and with high probability they forecast less inflation and high wage rates and so their labor supply is higher. However, the pessimists in the economy are almost sure that high inflation and low wage rates will occur at date $t + 1$ and work less and push down the total output $\Omega_t \mid x_t = x^L$. The intensity level of pessimism dominates and hence aggravates the occurrence of crashes or recession periods[10]. Monetary policy under Rational Beliefs does have an impact on the real variables as it influences the terms of trade between the present and the future. It is also the case under Rational Beliefs that the rational mistakes of the agents can amplify or reduce the effect of monetary policy leading to a higher standard deviation of inflation rates than under Rational Expectations. Also the long term rates of return on labor under Rational Beliefs exhibit higher standard deviations than under Rational Expectations.

We report in Figures 1 and 2 respectively samples of 300 observations of equilibrium inflation rates (ι) and rates of return (ρ) generated by Montecarlo simulations under the x^H-*consensus* structure.

We now consider a pattern of rational beliefs which is symmetric in the choice of intensity parameters to the one examined above and exhibits quite the opposite behavior in the statistics. The following choice of beliefs parameters identifies *Case 2* or x^L-*consensus*:

$$\begin{aligned} \lambda_s = \mu_s &= 1.7542 \text{ for } s = 1, 2, 3, 4, 9, 10, 11, 12 \\ \lambda_s = \mu_s &= 1.0000 \text{ for } s = 5, 6, 7, 8, 13, 14, 15, 16. \end{aligned} \tag{51}$$

The results obtained under this pattern of beliefs are reported in Table 2.

From Table 2 we can immediately notice the reversed behavior in the reported statistics of the x^L-*consensus* belief structure with respect to that of *Case 1*. In fact, conditional on a *loose* monetary policy the expected rate of inflation is below average as well as the expected output while the wage rate is above average. The contrary occurs when conditioning on a *tight* monetary policy. Also notice that,

---

[10] The parameters choice and the interpretation of the implied dynamics is similar to that of Kurz and Motolese [22] in explaining the equity premium and stock prices dynamics.

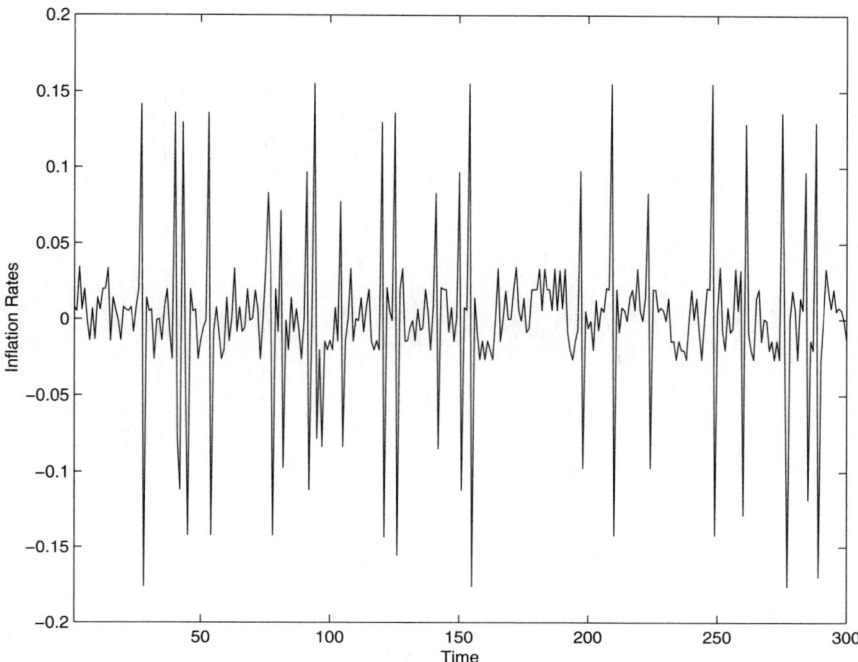

Figure 1. Inflation rates under rational beliefs (Case 1)

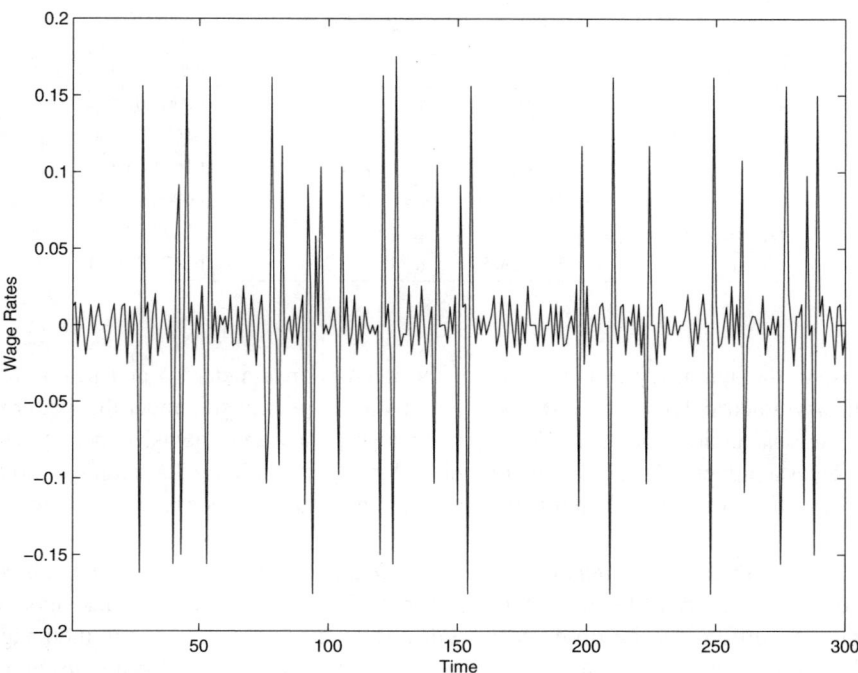

Figure 2. Wage rates under rational beliefs (Case 1)

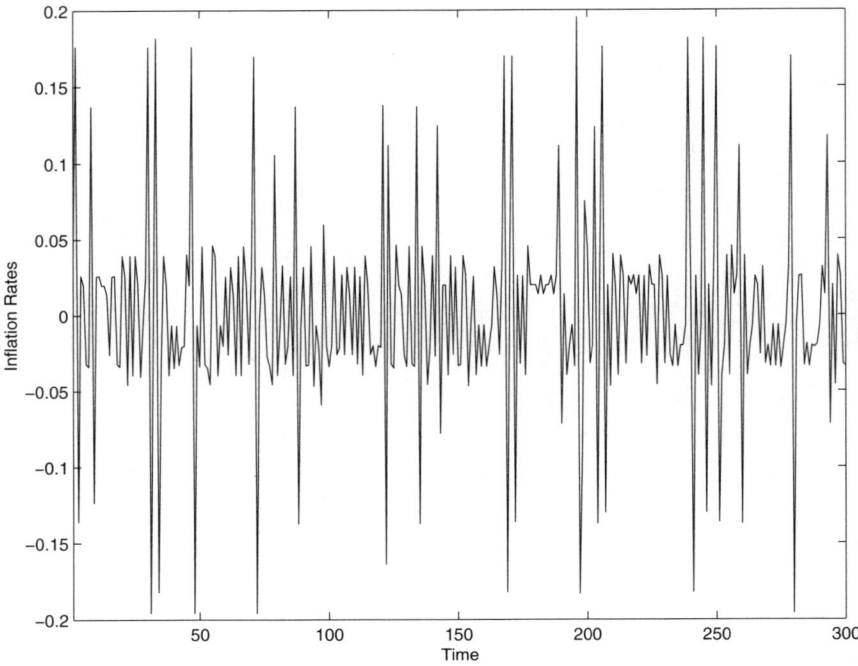

Figure 3. Inflation rates under rational beliefs (Case 2)

Table 2. Characteristics of money non-neutrality

				avg	σ
	RBE (Case 2)				
	avg	σ		avg	σ
ι_t	0.0000	0.0622	ρ_t	0.0000	0.0529
$\iota_t \mid x_t = x^H$	−0.0169	0.0683	$\rho_t \mid x_t = x^H$	0.0167	0.0600
$\iota_t \mid x_t = x^L$	0.0165	0.0502	$\rho_t \mid x_t = x^L$	−0.0167	0.0381
Ω_t	1.0188	0.2699	ω_t	0.0000	0.0330
$\Omega_t \mid x_t = x^H$	1.0041	0.2735	$\omega_t \mid x_t = x^H$	−0.0147	0.0417
$\Omega_t \mid x_t = x^L$	1.0334	0.2655	$\omega_t \mid x_t = x^L$	0.0147	0.0024

despite the symmetry in the choice of the intensity parameters λ and μ, due to the asymmetric dynamics of states of optimism and pessimism, under this pattern of rational beliefs the joint effect of monetary policy and agents' expectations induces a higher volatility in inflation rates. We report in Figure 3 a sample of 300 observations of equilibrium inflation rates (ι) under the x^L-*consensus* structure.

Why is money non neutral under RBE?   The real effect of money in an RBE arises from the diversity of beliefs. Monetary shocks have a real effect because agents interpret differently the observed growth rate of money at date t thus inducing diverse forecasts of all variables in the economy at date $t + 1$. Agents disagree about the effects of monetary shocks on future inflation, wage rates and output levels. Agent hold diverse forecast of the real variables in the economy and once they disagree on the effects of monetary shocks on future inflation, monetary shocks have real effects. This is analogous to Lucas' [25] case justifying money non-neutrality by agents *confusion* between monetary and real shocks. However, in an RBE there is no confusion about shocks and all the observables, only diversity of opinions about forecasting future variables and this is sufficient for money to be non neutral. In agreement with the results achieved by Kurz, Jin and Motolese [21] we also assert that *once agents perceive money to have real effects, it is rational for them to incorporate monetary shocks as a component affecting their own beliefs about all variables in the economy* (i.e. for x_t to affect agents' perceptions).

In Tables 1 and 2 we have reported results from two different Rational Beliefs structures. Both of them exhibit money non-neutrality but its behavior goes in two opposite directions. The difference between Table 1 and Table 2 is the quantitative impact of the monetary shock x_t: the effect of x_H in Table 1 is equal to the effect of x_L in Table 2 and vice versa. The first structure (x^H-*consensus*) presents a *pro-cyclical* behavior: it shows positive correlation between the growth rate of money and output. The second structure x^L-*consensus*, instead, presents a *counter-cyclical* behavior: it shows negative correlation between the growth rate of money and output[11].

The empirical evidence is compatible with the *pro-cyclical* rational belief structure, that is, we observe a positive correlation between the growth rate of money and output (see the highly debated issue outlined by Friedman and Schwartz [7] and recently reestablished by Aksoy and Piskorski [1]). Analogous modeling and results have been studied by Kurz, Jin and Motolese [21].

The above conclusion leads to the natural question of whether the structure of beliefs under *Case 1* is the actual pattern of beliefs observed in the market. We can not give an answer yet. Further research and availability of market data on expectations is crucial to address this issue. Within the RBE theory a recent paper by Kurz [19] started tackling such a topic following earlier issues raised by Romer and Romer [35].

4 Real GNP fluctuations and the Phillips curve

Tables 1 and 2 show a very important feature of the RBE under study: some components of real GNP fluctuations in the economy are explained by an endogenous mechanism induced by the distribution of beliefs among agents. Furthermore, GNP fluctuations under the two monetary regimes x^H and x^L are different thus implying an effective role for monetary policy in the settling of cycles. This is a drastically different scenario compared to the one obtained under Rational Expectations where the monetary forces play no role at all to explain GNP fluctuations and cycles. The distribution of beliefs in the economy at any date is a crucial factor which constitutes an *endogenous components of output fluctuations*. This is one of the most important conclusions we can draw from the results shown above. To support the argument

---

[11] The sign of such a correlation under both cases is also respectively confirmed by the regression coefficients of the growth rate of money x_t in the Phillips curves (52) and (53).

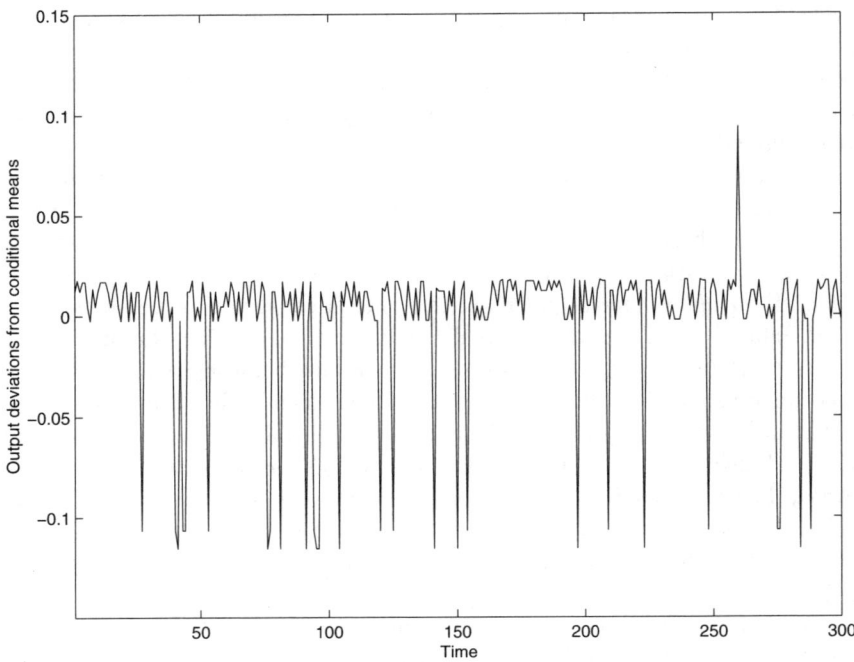

Figure 4. Endogenous output fluctuations under rational beliefs

above we have also reported in Tables 1 and 2 statistics for output deviations from the conditional means (ω) within each monetary regime. This component is uncorrelated with the exogenous productivity shocks and represents the compound effect of Endogenous Uncertainty and monetary policy. In Figure 4, a time series realization of 300 observations of output deviations is plotted.

From Figure 4 we can immediately see that in a RBE the distribution of beliefs creates *Endogenous Fluctuations of Output* which are not generated under REE. Such fluctuations take the form of *crashes and booms* or of *recession and expansion* periods. From Figure 4 as well as from Tables 1 and 2 we notice that negative deviations (*crashes*) have larger standard deviations then positive deviations (*booms*). In other words they are not symmetric. Once again this is due to the asymmetric behavior of states of optimism and pessimism as explained above. The intensity of the pessimists dominates and it is just counterbalanced by a less intense level of optimism and this pushes the level of output down to *recession*. We thus conclude that under Rational Beliefs *recessions* can be substantially aggravated by the distribution of market beliefs.

To strengthen the RBE result of money non-neutrality we also simulated the economy and generated 10,000 observations and estimated the following basic and simple regression model:

$$\Omega_t = \delta_0 + \delta_1 E_{t,\Gamma|p_t}\left(\iota_{t+1}\right) + \delta_2 x_t + \delta_3 \theta_t + \epsilon_t.$$

where $E_{t,\Gamma|p_t}(\iota_{t+1})$ is the $t+1$ expected inflation rate conditional upon the state price p_t observed at t measured by the stationary probability Γ. We obtained the following estimation for the RBE *Case 1*:

$$\Omega_t = \underset{(0.0050)}{0.0940} + \underset{(0.0026)}{0.8399} \; E_{t,\Gamma|p_t}(\iota_{t+1})$$

$$+ \underset{(0.0050)}{0.0319} \; x_t + \underset{(0.0003)}{0.8930} \; \theta_t + \epsilon_t, \;\; R^2 = 0.999, \tag{52}$$

and the following estimation for the RBE *Case 2*:

$$\Omega_t = \underset{(0.0049)}{0.1556} + \underset{(0.0026)}{0.8358} \; E_{t,\Gamma|p_t}(\iota_{t+1})$$

$$- \underset{(0.0049)}{0.0296} \; x_t + \underset{(0.0003)}{0.8931} \; \theta_t + \epsilon_t, \;\; R^2 = 0.999. \tag{53}$$

In both regressions the Phillips curve coefficient of $E_{t,\Gamma|p_t}(\iota_{t+1})$ is positive. In the regression of *Case 1* the sign of the coefficient of the growth rate of money x_t is also positive which stresses the positive correlation between monetary regimes and GNP we already highlighted in Table 1. On the contrary in the regression of *Case 2* the sign of the coefficient of the growth rate of money x_t is negative as expected given the negative correlation between monetary regimes and GNP we already found in Table 2.

We thus conclude that the non-zero slope of the Phillips Curves above is due to the endogenous distribution of beliefs in the economy. Furthermore, such a result is consistent with those of Kurz, Jin and Motolese [21] where also further implications of the role of monetary policy under RBE are discussed.

5 Conclusions

In this paper we have demonstrated that dynamic money non-neutrality is a generic property of any RBE. Endogenous uncertainty is the main force generating money non-neutrality. Endogenous uncertainty is the dominant uncertainty in an RBE. Such a form of uncertainty is that component of market volatility which is generated by the diversity of agent's beliefs. Hence monetary forces do have an impact on the real variables of the economy and such an impact can be amplified or reversed by the beliefs of the agents. We studied a two-agent OLG economy similar to the two islands model of Lucas' [25]. We assumed all transaction occur on one single market and no informational asymmetry exists. We showed that the heterogeneity of beliefs together with the distribution and intensity of agents' states of optimism/pessimism can expand or reduce the real effect of monetary policy and generate *Endogenous Fluctuations of Output* and give rise to higher volatility of inflation rates and wage rates.

We found that under RBE, when agents perceive monetary policy as a signal that carries information about future prices and wage rates, different patterns of volatility are observed in the economy. Such patterns depend upon agents' subjective interpretation of the observed monetary signal and upon the induced optimism/pessimism in their conditional probability beliefs. Furthermore, the non-neutrality of money results in a short run Phillips curve and its non-zero slope is due to the endogenous distribution of beliefs in the economy.

It is clear that the proposed paradigm of the Theory of Rational Beliefs has very important implications to the study of economic fluctuations and to the formulation of monetary policy. For this reason we think that the distribution of market beliefs constitutes useful information for the conduct of monetary policy. The results obtained in the present study and Nielsen [34] represent a first step in addressing the implications of the theory proposed by Kurz [12], [13] to monetary policy. Further implications have already been addressed by Kurz, Jin and Motolese [21] who show under an infinite horizon RBE money non-neutrality, the existence of a short-run Phillips curve and of "hump shaped" impulse response functions with respect to monetary shocks. They also address the very debated issue about the role of monetary policy which under RBE is aimed to reducing the excess endogenously created volatility.

References

1. Aksoy, Y., Piskorski, T.: Domestic money and US output and inflation. CFS Working Paper 2001/08, Center for Financial Studies, Johann Wolfgang Goethe-Universität (2001)
2. Azariadis, C.: A reexamination of natural rate theory. American Economic Review **71**, 944–960 (1981)
3. Azariadis, C., Guesnerie, R.: Sunspots and cycles. Review of Economic Studies **53**, 725–737 (1986)
4. Bouaricha, A., Schnabel, R. B.: TENSOLVE: A software package for solving systems of nonlinear equations and nonlinear least squares problems using tensor methods. Argonne National Laboratories, Illinois (1994)
5. Cass, D., Shell, K.: Do sunspots matter? Journal of Political Economy **91**, 193–227 (1983)
6. Chiappori, P. A., Guesnerie, R.: Anticipations, indétermination et non-neutralité de la monnaie. Annales d'Economie et de Statistique **19**, 1–25 (1990)
7. Friedman, M., Schwartz, A.: A Monetary history of the United States: 1867–1960. Princeton, NJ: Princeton University Press 1963
8. Gurley, J. G.: Review of a program of monetary stability by Milton Friedman. Review of Economics and Statistics **43**, 307–308 (1961)
9. Hoffman, D. L., Schlagenhauf, D. E.: An econometric investigation of the monetary neutrality and rationality propositions from an international perspective. Review of Economics and Statistics **64**, 562–571 (1982)
10. Keynes, J. M.: A monetary theory of production. First published in 1933; reprinted in: Moggridge, D. (ed.) The collected writings of John Maynard Keynes, vol. 13. London: Macmillan 1973
11. Kurz, M.: The Kesten-Stigum model and the treatment of uncertainty in equilibrium theory. In: Balch, M.S., McFadden, D.L., Wu, S.Y. (eds.) Essays on economic behavior under uncertainty, pp. 389–399. Amsterdam: North Holland 1974
12. Kurz, M.: On the structure and diversity of rational beliefs. Economic Theory **4**, 877–900 (1994)
13. Kurz, M.: On rational belief equilibria. Economic Theory **4**, 859–876 (1994)
14. Kurz, M. (ed): Endogenous economic fluctuations: studies in the theory of rational belief. Studies in economic theory, No. 6. Berlin Heidelberg New York: Springer 1997
15. Kurz, M.: Asset prices with rational beliefs. In: Kurz, M. (ed.) Endogenous economic fluctuations: studies in the theory of rational belief, Ch. 9. Studies in economic theory, No. 6. Berlin Heidelberg New York: Springer 1997
16. Kurz, M.: On the volatility of foreign exchange rates. In: Kurz, M. (ed.) Endogenous economic fluctuations: studies in the theory of rational belief, Ch. 12. Studies in economic Theory, No. 6. Berlin Heidelberg New York: Springer 1997
17. Kurz, M.: Endogenous uncertainty: a unified view of market volatility. Working Paper 97–027, Department of Economics, Stanford University (1997)

18. Kurz, M.: Social states of belief and the determinant of the equity risk premium in a rational belief equilibrium. In: Abramovich, Y. A., Avgerinos, E., Yannelis, N.C. (eds.) Functional analysis and economic theory. Springer Series in Applied mathematics and Operation Research, pp. 171–220. Berlin Heidelberg New York: Springer 1998
19. Kurz, M.: Heterogeneous forecasting and federal reserve information. Working Paper 02–002, Department of Economics, Stanford University (2002)
20. Kurz, M., Beltratti, A.: The equity premium is no puzzle. In: Kurz, M. (ed.) Endogenous economic fluctuations: studies in the theory of rational belief, Ch. 11. Studies in economic theory, No. 6. Berlin Heidelberg New York: Springer 1997
21. Kurz, M., Jin, H., Motolese, M.: Endogenous fluctuations and the role of monetary policy. In: Aghion, P., Frydman, R., Stiglitz, J., Woodford, M. (eds.) Knowledge, information and expectations in modern macroeconomics: Essays in honor of Edmund S. Phelps. Princeton, NJ: Princeton University Press 2002
22. Kurz, M., Motolese, M.: Endogenous uncertainty and market volatility. Economic Theory **8**, 461–488 (1996)
23. Kurz, M., Schneider, M.: Coordination and correlation in Markov rational belief equilibria. Economic Theory **8**, 489–520 (1996)
24. Kurz, M., Wu, H. M.: Endogenous uncertainty in a general equilibrium model with price contingent contracts. Economic Theory **17**, 497–544 (2001)
25. Lucas, Jr., R.: Expectations and the neutrality of money. Journal of Economic Theory **4**, 103–124 (1972)
26. Lucas, Jr., R.: Some international evidence on output inflation trade-offs. American Economic Review **63**, 326–334 (1973)
27. Lucas, Jr., R.: Expectations and the neutrality of money: Corrigendum. Journal of Economic Theory **31**, 197–199 (1983)
28. Lucas, Jr., R.: Nobel lecture: monetary neutrality. Journal of Political Economy **104**, 661–682 (1996)
29. Mishkin, F.S.: Does anticipated monetary policy matter? An econometric investigation. Journal of Political Economy **90**, 22–51 (1982)
30. Motolese, M.: Dynamic non-neutrality of money under rational beliefs: the role of endogenous uncertainty. A Ph.D. dissertation, University of Bologna (submitted 1998)
31. Motolese, M.: Money non-neutrality in a rational belief equilibrium with financial assets. Economic Theory **18**, 97–126 (2001)
32. Muth, J. F.: Rational expectations and the theory of price movements. Econometrica **29**, 315–335 (1961)
33. Nielsen, C. K.: Rational belief structures and rational belief equilibria. Economic Theory **8**, 399–422 (1996)
34. Nielsen, C. K.: Floating exchange rates versus a monetary union under rational beliefs: the role of endogenous uncertainty. Mimeo, Institute of Economics, University of Copenhagen (1997)
35. Romer, C. D., Romer, D. H.: Federal reserve information and the behavior of interest rates. American Economic Review **90**, 429–457 (2000)
36. Sargent, T. J., Wallace, N.: Rational expectations, the optimal monetary instrument and the optimal money supply rule. Journal of Political Economy **83**, 241–254 (1975)
37. Woodford, M.: Imperfect common knowledge and the effects of monetary policy. In: Aghion, P., Frydman, R., Stiglitz, J., Woodford, M. (eds.) Knowledge, information and expectations in modern macroeconomics: Essays in honor of Edmund S. Phelps. Princeton, NJ: Princeton University Press 2002

Inside and outside fiat money, gains to trade, and IS-LM

Pradeep Dubey[1] and John Geanakoplos[2]

[1] Center for Game Theory, Department of Economics, SUNY, Stony Brook, NY 11794-4384, USA
(e-mail: pradeepkdubey@yahoo.com)
[2] Cowles Foundation for Research in Economics, Yale University, New Haven, CT 06520-8281, USA
(e-mail: john.geanakoplos@yale.edu)

Received: January 2, 2002; revised version: April 8, 2002

Summary. We build a one-period general equilibrium model with money. Equilibrium exists, and fiat money has positive value, as long as the ratio of outside money to inside money is less than the gains to trade available at autarky. We show that the nominal effects of government fiscal and monetary policy can be completely described by a diagram identical in form to the IS-LM curves introduced by Hicks to describe Keynes' general theory. IS-LM analysis is thus not incompatible with full market clearing, multiple commodities, and heterogeneous households. We show that as the government deficit approaches a finite threshold, hyperinflation sets in (prices converge to infinity and real trade collapses). At the other extreme, if the government surplus is too large, the economy enters a liquidity trap in which nominal GNP sinks and monetary policy is ineffectual.

Keywords and Phrases: Central bank, Gains to trade, Inside money, IS-LM, Outside money.

JEL Classification Numbers: D50, E40, E50, E58.

1 Introduction

Fiat money is a creature of the state, since nobody else can create it. When the state injects it into the private sector in exchange for assets promising the future delivery of money, its arrival foreshadows its departure, and it is called inside money. Money injected into the private sector as a transfer, or in exchange for a commodity (which gives no claim on future repayment), is called outside money.[1]

Correspondence to: P. Dubey

[1] Fiat money as a creature of the state is taken from the title of an article by Lerner [26]. Our definitions of inside and outside money are taken from Gurley-Shaw [21], and stand in contrast to the "inside money" used by some authors to describe private IOU notes which can be further circulated.

There is a longstanding puzzle about how to guarantee that outside money has positive value, often called the Hahn paradox. We argue in this paper that if fiat money is the sole medium of exchange, and if the ratio of outside money to inside money is less than the gains to trade available at autarky, then money must have positive value and a full-fledged monetary equilibrium must exist.

The Hahn paradox arises because households do not want to hold money at the end.[2] Equilibrium models usually rely on one of two devices to overcome the paradox. The first device is to assume there is no last period (see Samuelson [31] or Grandmont-Younes [20] for infinite horizon models, or Grandmont-Younes [19] or Hool [24] for temporary equilibrium models where the last period is really not the last period, since agents have expectations there about the future value of money). The second device is to *oblige* some agent, either the government or the households themselves, to sell something valuable for money. This device is used for example when the government or some external agent is postulated to sell commodities for money at prearranged prices. Lerner [26], and later Heller [23]and Balasko-Shell [2], assumed that the government is owed in taxes (payable only in money) precisely the sum of the cash balances of all the households. The government is thus obliged to offer relief from taxes in exchange for money. Finally, Lucas [28], [29], and a long literature following him, assumes that in each period all agents must sell their entire endowment of commodities for money. Magill-Quinzii [30] make a similar assumption, as do Karatzas-Shubik-Sudderth [25].

In the language we shall develop, one can reinterpret this involuntary trade as voluntary, but conducted in the presence of *infinite* gains to trade. For example, as Lucas himself more or less points out, in a representative agent model with a single good, one can paint the endowment of every individual agent with a different color, and then assume that no agent wants to consume his own color. This gives rise to an equivalent economy with multiple commodities and infinite gains to trade.[3]

The purpose of money is to facilitate trade. If there are no potential gains to trade, then fiat money cannot have value. This point has long been recognized, but never quantified. How big must the gains to trade be?[4]

We construct a scalar measure $\gamma(x) = \gamma_u(x)$ of the real gains to trade available in an economy, with utilities u, at an arbitrary allocation of goods x. The measure is an alternative to Debreu's coefficient of resource allocation $\delta(x)$: $\delta(x)$ quantifies the global gains to trade at x, while $\gamma(x)$ quantifies the local gains to trade. We characterize $\gamma(x)$ in terms of a maximal cycle of trades. Though defined entirely by the real sector of the economy, $\gamma(x)$ is ideally suited to our study of monetary equilibrium.

---

[2] Thus models that have only inside money (see Shubik-Wilson [32], Cass [5], Balasko-Cass [1], Geanakoplos-Mas-Colell [17], and Dubey-Shapley [14]), steer clear of the Hahn paradox, since there is no money at the end.

[3] If agents were obliged to sell a fraction $\alpha > 0$ of their endowment for money, we could still reinterpret the situation as infinite gains to trade (painting only the portion that must involuntarily be sold).

[4] Bewley [3], [4] and Levine [27] link the existence of monetary equilibrium in an infinite horizon model to the precautionary demand for money and therefore to the inefficiency of the allocation without money. They do not, however, offer a measure of gains to trade.

Consider a multiple-good economy with heterogeneous utilities u and endowments e, as in the theory of general equilibrium. Suppose that agents are endowed with "outside fiat money" m, and can borrow "inside fiat money" M from a central bank. Suppose also that fiat money is the sole medium of exchange, so that agents must pay fiat money in advance for their purchases of goods. But all actions are purely voluntary: no agent is forced to sell his endowment, or to borrow from the bank.

We prove that a monetary equilibrium exists in which money has value whenever $\gamma(e) > m/M$. If utilities are separable, the condition is tight: monetary equilibrium fails to exist if $\gamma(e) \leq m/M$. This makes precise the link between monetary equilibrium and gains to trade.

Money has value in our model because the assets (bonds), exchanged for inside money when it is injected into the system, sell for endogenous prices (interest rates). In equilibrium they will promise more than they cost (the interest rate will be positive). When their payoffs are discharged, more money leaves the system than entered, and so the outside money is pulled out along with the inside money. Indeed, the interest rate rises to the level m/M to make this happen. As long as the gains to trade are large enough, households will be anxious to get the money, and will voluntarily agree to pay back more than they borrowed.

In this paper we confine ourselves to a one-period setting. This has two advantages. First, it shows that it is possible to develop a theory of fiat money with a finite horizon. This is important because one can build truly computable models with heterogeneous agents and multiple commodities. In stochastic infinite horizon models this is nearly impossible, without resorting to heroic assumptions such as stationarity, a representative consumer, and one or two consumption goods. The curse of dimensionality becomes especially severe when one considers international trade and multiple currencies. Our approach provides a tractable alternative. (See Geanakoplos-Tsomocos [18], Geanakoplos-Kubler [16] for applications of this methodology.)

The second advantage of the one-period model is that several macroeconomic phenomena already emerge which are valid with multiple periods and uncertainty, but can be brought out most cleanly by avoiding the added complications of intertemporal trade. These phenomena include: (1) the link between gains to trade and the ratio of outside to inside money;[5] (2) generic finiteness of equilibrium (in contrast to models with only inside money, such as Cass [5]; Balasko-Cass [1], Geanakoplos-Mos-Colell [17]); (3) the non-neutrality of inside and outside money, when each is varied by itself, in contrast to the "classical dichotomy"; (4) the Pareto inefficiency of monetary equilibrium; (5) hyperinflation with bounded stock of money; (6) liquidity trap; (7) IS-LM. All these phenomena are established here for the one-period model, to the same degree of generality as in Arrow-Debreu. They carry over to stochastic finite horizon models, and even to stochastic OLG models, with the exception that IS-LM becomes much subtler and generic finiteness fails in stochastic OLG.

---

[5] Many authors emphasize one of outside or inside money at the expense of the other. But it is the *interplay* between them that gives rise to many of the phenomena that form the focus of this paper.

In our forthcoming work we use the one-period model as a building block of a general approach that in turn considers multiple periods without uncertainty, then with uncertainty, and finally infinite horizon stochastic overlapping generations models of money (Dubey-Geanakoplos [9], [10], [11], [12]). It is true that the nominal rate of interest is independent of the real sector in the one-period setting. Evidently there can be no precautionary or speculative demand for money, only a transactions demand. In equilibrium, the interest rate must be $r = m/M$, as we already pointed out. But once we connect many one-period models in a stochastic multiple-period model, these other demands for money are reinstated and a term structure of interest rates naturally emerges.

But even in the one-period setting of our model, prices reflect a subtle interaction between the real sector (u, e) and the monetary sector (m, M) of the economy. We prove that the price level is not a monotonic function of the money supply: it is U-shaped in M, fixing m. As M goes to infinity, price levels go to infinity, eventually giving merely nominal inflation with little effect on real trades and price ratios. However, as M diminishes (holding m fixed), prices stop declining and start to rise. As M declines still further to some finite threshold $M^*(m) > 0$, price levels accelerate to infinity and *real* trade crashes, in what we call a hyperinflation. This means that when money is much too tight, easing the bank money M will increase output (trade) and reduce the price level. Tightening even further will have the paradoxical consequence of starting a hyperinflation. Many models almost axiomatically assume that output and inflation must go in the same direction, as in the Phillips curve. That regime occurs in our model only when inside money is sufficiently abundant.

The same phenomenon can be expressed in a dual manner: keeping M fixed, we find that an increase in m eventually produces a hyperinflation at a *finite* level $m^*(M)$.

We first present these results in a stripped-down one-period model in which a central bank injects a fixed stock M of inside money into the economy in exchange for bonds promising money at the end of the period. The Treasury branch of the government does nothing else but give a fixed stock m of outside money free and clear to households, who treat it as part of their endowment.

In our second model, we flesh out the Treasury, giving the government five policy instruments: the stock of bank money, the supply of government bonds, lump sum transfers to households, expenditures on inputs for the production of public goods, and ad valorem taxes. We can now reinterpret the hyperinflation in terms of the government deficit. As the Treasury deficit approaches a *finite* threshold, hyperinflation sets in where prices go to infinity and trade crashes. If the Treasury runs a large budget surplus, it will push the economy into a liquidity trap where the interest rate is zero, and where small changes in government monetary policy have no effect whatsoever.

The effects of the government's policy tools on aggregate nominal variables can be completely described by a graphical framework nearly identical in form to the IS-LM diagram used by Hicks in formalizing Keynes' model. This shows that there is nothing incompatible between IS-LM and full market clearing and rational expectations. We also describe the effects of government policy on welfare, consumption, and on price levels, and show that in terms of real variables, the five policy instruments achieve nothing more than is available by using any two of them: e.g., open market operations and government spending on commodities by printing money.

Our IS-LM diagram gives a macroeconomic picture of nominal income and the nominal interest rate in a genuine microeconomic model. It is remarkable that an economy with many heterogeneous consumers and commodities can be faithfully summarized by a two-dimensional diagram. This is possible because we use nominal income, not real income, in our IS-LM equations, and because there is only one period in our model. The heterogeneity of consumer tastes is suppressed, because all consumers want to spend their money income at the same time.

The subtlety of monetary equilibrium is attenuated in our model by the restriction to one period without uncertainty. Market clearing in the goods markets, namely that aggregate expenditure equals aggregate income (which Hicks called the IS equation), becomes much less interesting when it loses the investment and savings components from which its name derives. Similarly money market clearing (which Hicks called the LM equation) loses much of its complexity because in a one-period model we can retain only the transactions demand for money, necessarily ignoring the precautionary demand and speculative demand. In companion papers we describe monetary equilibrium in time [10], which introduces savings and investment motives and endogenizes the velocity of money; and equilibrium with uncertainty [9], in which the speculative and precautionary demands for money reappear, and in which a liquidity trap can arise without government surplus. We also combine time and uncertainty in an infinite horizon setting in [11].

A crucial ingredient of our model is the Clower [6] cash-in-advance constraint we put on all transactions.[6] In our model of money and time [10], where the trading rounds can come every nanosecond, this is a much less restrictive assumption, and monetary equilibria can still be shown to exist. Obviously many transactions in the real world are carried out by credit cards and by checks, as well as via money. One of the most important virtues of our model is that by making the transactions technology explicit (rather than subsuming it in a reduced-form utility of money), it becomes straightforward to add credit cards to the model, which we do in [8]. There we find that credit cards do not destroy the value of money (indeed equilibrium with a positive value of money is more likely to exist). On the other hand, credit cards do reduce the value of money, i.e., its purchasing power, leading to inflation.

The idea of introducing inside fiat money into general equilibrium via a bank came to our attention in the work of M. Shubik and C. Wilson (see [32]). Our contribution is to combine the inside money of the bank with outside money. We first did so in Dubey-Geanakoplos [7]. Theorems 2 and 3 from this paper, connecting the gains to trade with the existence of monetary equilibrium and showing the local uniqueness of monetary equilibrium, already appeared there. The other nine theorems, as well as a simpler existence proof for Theorem 2, appear here for the

---

[6] This constraint was used approximately, but not exactly, by Grandmont and Younes [19, 20]. They were unable to prove the existence of equilibrium in their infinite horizon model. Their model had only outside money, and established the existence of a quasi-equilibrium in which some agents did not optimize.

first time, including the characterization of the measure $\gamma(x)$ of the gains to trade at an arbitrary allocation x.

2 The model

Consider an economy in which money is the *sole* medium of exchange. Furthermore, suppose that there is just one round of trade between money and commodities. Since the money receipts from commodity sales come after the round is over, let us add the possibility of borrowing money prior to the trading round and repaying it after. Thus the period is divided into three time intervals: borrowing, trading, and repaying.[7]

2.1 The underlying economy

We first analyze a pure exchange economy which has only private goods (commodities) $L = \{1, ..., L\}$. (Later we shall add a government sector and public goods.) The agents in the economy are households $H = \{1, ..., H\}$. Each $h \in H$ has an endowment of commodities $e^h \in \mathbb{R}_+^L$ and a utility of consumption $u^h : \mathbb{R}_+^L \to \mathbb{R}$. We assume: (a) $e^h \neq 0$ for all $h \in H$, i.e., every household has at least some endowment (e.g., its own labor); (b) $\sum_{h \in H} e^h \gg 0$, i.e., every named commodity is present in the aggregate; (c) u^h is continuous, concave, and strictly increasing in each[8] variable, for all $h \in H$. The underlying economy, which constitutes the real sector of our model, is denoted $\mathcal{E} \equiv (u^h, e^h)_{h \in H}$.

2.2 Money and bank loans

Money is fiat and gives no direct utility of consumption to the households; they value money only insofar as it enables them to acquire commodities for consumption. Money enters the economy in two ways: as private endowment $m^h \geq 0$ of household $h \in H$ and as a stock $M > 0$ at a (central) bank. Apart from households, the bank is the only other agent in our model, but it has a passive role. It stands ready to lend M to households at an interest rate that is determined endogenously in equilibrium. Both $m \equiv \{m^h\}_{h \in H}$ and M are exogenously fixed as part of the data of the model. The sum $\bar{m} \equiv \sum_{h \in H} m^h$ constitutes the stock of *outside money*, which households own free and clear of debt, at the start of the economy. The bank stock M is *inside money* and is always accompanied by debt when it comes into households' hands. We denote the monetary economy by $(\mathcal{E}, m, M) \equiv ((u^h, e^h, m^h)_{h \in H}, M)$; and its private sector by $(\mathcal{E}, m) \equiv (u^h, e^h, m^h)_{h \in H}$.

The period, as was said, is divided into three time intervals. In the first interval, households borrow money from the bank. In effect, households sell IOU notes or bonds to the bank in exchange for money. In the second interval, they sell

---

[7] In our companion papers we take up multiple trading rounds, uncertainty, and infinite horizon [9], [10], [11].

[8] *Strict* monotonicity is assumed for ease of presentation, and will be weakened (see the first remark after the proof of Theorem 2 in the Appendix).

commodities for money and simultaneously buy (other) commodities with money. In the third interval, they repay bank loans with money and consume. Default is not permitted.

All commodity markets meet simultaneously in the second interval. Households are required to pay money to purchase commodities at the different markets.[9] It is only in the third interval, after these markets close, that revenue from the sales of commodities comes into households' hands, by which time it is too late to use this revenue for purchases. Those households who find their endowment m^h of money insufficient will need to borrow money from the bank to finance purchases, and will defray the loan out of their sales revenue.[10]

2.3 Macrovariables: prices and quantities

Let $p_\ell > 0$ denote the price of commodity $\ell \in L$ in terms of money, and let $r \geq 0$ denote the money rate of interest on the bank loan. Money is borrowed by selling bonds to the bank. Each bond constitutes a promise to pay 1 dollar after commodity trade. Thus the price before commodity trade of a bond is $1/(1+r)$.

The vector $(p, r) \in \mathbb{R}_{++}^L \times \mathbb{R}_+$ will be referred to as "market prices." The price of money is $1/p_\ell$ in terms of commodity ℓ, and $(1+r)$ in terms of the bond. The value of money is reflected by these prices. As $p \to \infty$, money loses all value (in terms of commodities). As $r \to -1$, money-now loses all value (in terms of money-later) and as $r \to \infty$ money-later loses all value (in terms of money-now). In this paper our interest is on p, since it is determined by the interaction of the real sector $\mathcal{E}$ and monetary sector (m, M) of the economy, and not so much on r, which is determined entirely by the monetary sector.[11]

We denote money by m (without confusing it with the vector $m \equiv (m^1, \ldots, m^H)$ of household endowments) and bonds by b. Since money is the sole medium of exchange, the vector q^h of market actions of household h has $2L+1$ components (where $\ell \in L$):

$q_{bm}^h \equiv$ quantity of bonds sold by h to the bank for money
$q_{m\ell}^h \equiv$ money spent by h to purchase ℓ
$q_{\ell m}^h \equiv$ quantity of ℓ sold by h for money

(It is evident, on account of their being just one period, that no household would improve its consumption by depositing money at the bank to earn interest. So, we suppress deposits, i.e., the purchase of bonds q_{mb}^h.)

By real income q we mean the vector of aggregate commodity sales, with components $q_\ell = \sum_{h \in H} q_{\ell m}^h$. By nominal income we mean the value of real

---

[9] In [8], we allow households to buy on credit, as well as with cash.

[10] The loans are purely short-term, intraperiod transactions loans. This is on account of the fact that there is only one consumption period in the model. Elsewhere ([10], [11]) we consider long-term, interperiod loans in a multiperiod model. In general, both kinds of loans involve money and carry weight in a modern-day economy.

[11] In the multiperiod setting, which we study in [10], there is a term structure of interest rates determined by the interaction of the real and monetary sectors, and our focus shifts to both p and r.

income

$$Y = p \cdot q \equiv \sum_{\ell \in L} \sum_{h \in H} p_\ell q_{\ell m}^h.$$

Notice that income corresponds to sales and not to endowments. Since households are not obliged to sell their endowments, real income is genuinely endogenous. Nominal income appears doubly endogenous, since both p and q are endogenous, but often it can be deduced from monetary considerations alone.

Irving Fisher introduced a famous formula for the velocity of money, v, which in our context becomes

$$(M + \bar{m})v = p \cdot q \equiv Y.$$

In a one-period model the velocity of money is not very interesting. If all the money is spent, then $v = 1$ and nominal income is determined. If some of the money is unspent, v may be less than 1 and Y becomes endogenous. This happens in our model in a liquidity trap (Section 11.2).

2.4 The budget set of a household

We consider the case of a perfectly competitive household sector. Each $h \in H$ regards market prices $(p, r) \in \mathbb{R}_{++}^L \times \mathbb{R}_+$ as fixed (uninfluenced by its own actions). The *budget set* $B(p, r, e^h, m^h)$ consists of all market actions and consumptions $(q^h, x^h) \in \mathbb{R}_+^{2L+1} \times \mathbb{R}_+^L$ that satisfy the budget constraints (1), (2), (5), and (3ℓ), (4ℓ), (6ℓ) for all $\ell \in L$. The residual variables $\tilde{x}^h = \tilde{x}^h(q^h, p)$ and $\tilde{m}^h = \tilde{m}^h(q^h, r)$ are determined automatically by q^h, p, r.

$$\tilde{m}^h \equiv \frac{q_{bm}^h}{1 + r} \tag{1}$$

$$\sum_{\ell \in L} q_{m\ell}^h \leq m^h + \tilde{m}^h \tag{2}$$

$$q_{\ell m}^h \leq e_\ell^h \tag{3ℓ}$$

$$\tilde{x}_\ell^h \equiv \frac{q_{m\ell}^h}{p_\ell} \tag{4ℓ}$$

$$q_{bm}^h \leq \Delta(2) + \sum_{\ell \in L} p_\ell q_{\ell m}^h \tag{5}$$

$$x_\ell^h \leq (\Delta 3\ell) + \tilde{x}_\ell^h. \tag{6ℓ}$$

Here $\Delta(\alpha)$ is the difference between the right and left sides of inequality (α). The interpretation is clear: (1) says that household h borrows $\tilde{m}^h$ dollars by promising to pay $q_{bm}^h = (1 + r)\tilde{m}^h$ dollars after commodity trade, i.e., by selling q_{bm}^h bonds; (2) says that total money spent on purchases cannot exceed the money on hand, i.e., money endowed plus money borrowed; (3ℓ) says that no household can sell more of any commodity than it is endowed with; (4ℓ) says that households purchase commodities $\tilde{x}^h$ with money at market prices p; (5) says that we are not

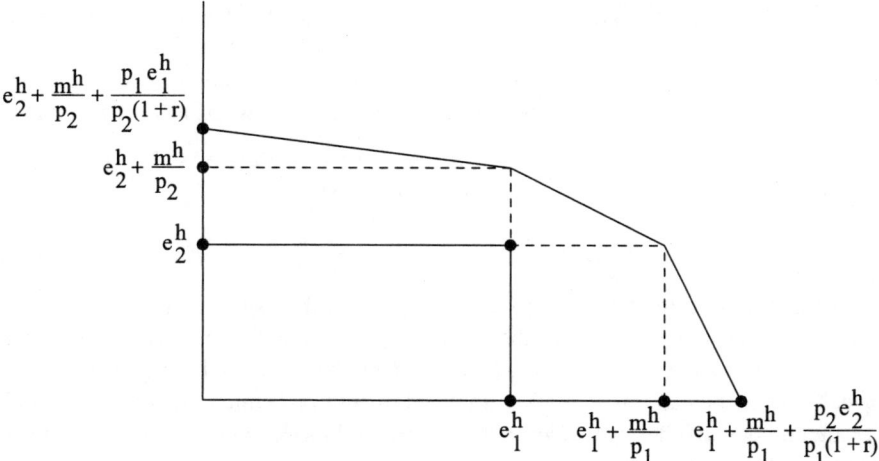

Figure 1. Budget-feasible consumptions

permitting default, i.e., every household must fully deliver on its bonds; (6ℓ) says that consumption cannot exceed what a household winds up with after trade.

The budget set describes constraints on the flows of money and commodities that a household may send to market. Implicitly, these flows define changes in the household stocks of money and commodities after trade. The budget set ensures that the stocks are always nonnegative.

2.5 Another view of the budget set

We denote the set of *budget-feasible consumptions* for household h by

$$B_C(p, r, e^h, m^h) = \{x^h \in \mathbb{R}_+^L : \exists q^h \in \mathbb{R}_+^{2L+1} \text{ with } (q^h, x^h) \in B(p, r, e^h, m^h)\}.$$

Note that B_C is homogeneous of degree zero in p, m^h; i.e., for any $\lambda > 0$,

$$B_C(\lambda p, r, e^h, \lambda m^h) = B_C(p, r, e^h, m^h).^{12}$$

We can picture the budget-feasible consumptions for a household h, endowed with money and both goods 1 and 2 (see Fig. 1).

For any trade vector $\tau \in \mathbb{R}^L$, and prices $(p, r) \in \mathbb{R}_{++}^L \times \mathbb{R}_+$, define the *cost* $C_r(p, \tau)$ of τ by

$$C_r(p, \tau) = p \cdot {}^*\tau - \frac{1}{1+r} p \cdot {}_*\tau$$

where $\cdot$ denotes dot product, and

$$ {}^*\tau_\ell = \max\{\tau_\ell, 0\}, \qquad {}_*\tau_\ell = -\min\{\tau_\ell, 0\}$$

give the purchases and sales in $\tau = {}^*\tau - {}_*\tau$.

---

[12] Indeed, if $(q^h, x^h) \in B(p, r, e^h, m^h)$, then $(\tilde{q}^h, x^h) \in B(\lambda p, r, e^h, \lambda m^h)$ where $\tilde{q}_{\ell m}^h = q_{\ell m}^h$, $\tilde{q}_{bm}^h = \lambda q_{bm}^h$ and $\tilde{q}_{m\ell}^h = \lambda q_{m\ell}^h$.

The cost $C_r(p, \tau)$ discounts the revenue from sales by the interest rate, since money-later (arriving after commodity trade) is worth less to the household than money-now, which can be used for commodity purchases. Using this cost function we can replace the six inequalities describing the budget set with just one, as Lemma 1 shows.

Lemma 1 *Let* $(p, r) \in \mathbb{R}^L_{++} \times \mathbb{R}_+$. *Then for any* $x^h \in \mathbb{R}^L_+$

$$x^h \in B_C(p, r, e^h, m^h) \Leftrightarrow C_r(p, x^h - e^h) \leq m^h.$$

Any finite horizon model with outside money, i.e., with positive endowments of fiat money, must somehow dissipate the money through trade. In our model the banking system extracts money every time a household purchases beyond its purely financial wealth m^h. One can imagine other financial institutions which facilitate trade and extract money in other ways. We examine a general model, via an abstract financial cost function $C(p, r, \tau)$, in [13].

2.6 Monetary equilibrium

A vector of prices and household actions

$$\langle p, r, (q^h, x^h)_{h \in H} \rangle \in \mathbb{R}^L_{++} \times \mathbb{R}_+ \times (\mathbb{R}^{2L+1}_+ \times \mathbb{R}^L_+)^H$$

is a *pre-monetary equilibrium* (preME) of $(\mathcal{E}, m, M)$ if all household actions are in their budget sets, i.e.,

$$(q^h, x^h) \in B(p, r, e^h, m^h) \tag{7}$$

and demand equals supply for the loan market and for all commodity markets, i.e.,

$$\text{(a)} \sum_{h \in H} \tilde{m}^h(q^h, r) = M.$$

$$\text{(b)} \sum_{h \in H} \tilde{x}^h_\ell(q^h, p) = \sum_{h \in H} q^h_{\ell m}, \, \ell \in L \tag{8}$$

It is worth noting that in a pre-monetary equilibrium, the total stock of money and commodities held collectively in the hands of the bank and the households is conserved in all three time intervals into which the period is divided. At the start, the bank holds M and households hold $\bar{m}$ of money. Money market clearing (8a) guarantees that the bank stock M flows to households at the end of the first interval. Commodity market clearing (8b) guarantees that the total stock of commodities is conserved and redistributed among the households during the second time interval. And (8b), multiplied by p_ℓ, shows that the total stock of money is conserved and redistributed among the households during the second time interval. Thus at the end of the first and second intervals, all of $M + \bar{m}$ is with households. The no-default condition (5) implies that the total bonds sold by households do not exceed $M + \bar{m}$. At the end of the third interval in a preME, the bank holds $(1 + r)M \leq M + \bar{m}$, and households hold the balance $\bar{m} - rM$.

A preME $\langle p, r, (q^h, x^h)_{h \in H} \rangle$ is a *monetary equilibrium* (ME) iff

$$u^h(x^h) \geq u^h(\underline{x}^h) \text{ for all } (\underline{q}^h, \underline{x}^h) \in B(p, r, e^h, m^h).$$

In any ME, at the end of the third interval, after repaying the bank, no household will be left with unowed cash, otherwise it should have spent more money earlier to purchase commodities, or else curtailed its sale of commodities, improving its utility. Hence at least $M + \bar{m}$ is owed to the bank. But no more could be owed, since default is not permitted. Thus $(1 + r)M = M + \bar{m}$ at any ME, i.e., $r = \bar{m}/M$.

This shows that the rate of interest r in (our one-period) monetary equilibrium is determined solely by the stocks of inside and outside money, and is unaffected by the real sector $\mathcal{E}$. In a multiperiod setting [13], there would be a genuine interaction between the real and monetary sectors that determines the interest rates.

In contrast, even with one period, p is determined by a genuine interaction between the real and monetary sectors. Notice that since the components of p at any ME must be finite by definition, money will have positive value at an ME. Thus the existence of an ME is tantamount to a resolution of the Hahn paradox.

3 Gains to trade

At first glance the cash-in-advance constraint (embodied in (2)) and the presence of the bank seem to provide a way out of the Hahn paradox: the bank, as was said, is an agent that demands money for its own sake, and households will need to hold money at the end in order to repay their loans to the bank. This argument would be fine if we could guarantee that households took out bank loans in the first place. But, unless money already has value to begin with, why should anyone want to take out loans? In a representative agent economy, for instance, nobody would take out loans and money would have no value. Thus the bank, while necessary, does not in and of itself ensure that money will have value. Something more is needed. We show in Theorem 6 that money fails to have value if nothing is added.

One device is to oblige households to put up some positive fraction of their endowment for sale against money (i.e., require in condition (3) of the budget set that $\alpha e_\ell^h \leq q_{\ell m}^h \leq e_\ell^h$ for some $0 < \alpha \leq 1$). Indeed the case when the entire endowment must be put up for sale (i.e., $\alpha = 1$) is considered by Lucas [28], [29], and Magill-Quinzii [30]. Such forced sales, of course, ensure that money will buy something of value in equilibrium (i.e., an ME exists, see Remark 2). But the trouble is that some of these sales *must* be forced. With even the tiniest transactions cost, households would strictly prefer not to sell and buy back the same commodities. For any $\alpha > 0$, if any $e^h \gg 0$, household h would not voluntarily undertake to sell αe^h, for then there would be a commodity ℓ which h would be buying as well as selling. In our model there is no transaction cost; but there is a positive rate of interest at any ME if $\bar{m} > 0$. Households are loath to indulge in wash sales, because they would lose the interest float.

Another device is to introduce a government, ready to defend the sanctity of its fiat money by putting up some exogenous stock of commodities (e.g., gold) for sale against money. By this device we could again get ME without much ado: government sales back the fiat money and guarantee its purchasing power.

We do not have to take recourse to such extraneous and drastic measures as forced sales of commodities, or gold-backed money, in order to guarantee that money has value. What is required is an intrinsic "gains to trade hypothesis."

Fiat money is wanted only for trading commodities. It follows that the value of money should depend on households' motivation to trade commodities with each other. We develop a measure of this motivation called gains to trade and show that, whenever they are strong enough, monetary equilibrium exists. Money is valued and used to move commodities through markets.

Let $\tau^h \in \mathbb{R}^L$ be a trade vector of h (with positive components representing purchases and negative components representing sales). For any scalar $\gamma \geq 0$, define

$$\tau_\ell^h(\gamma) = \min\{\tau_\ell^h, \tau_\ell^h/(1+\gamma)\}$$

Note $\tau_\ell^h(\gamma) = \tau_\ell^h$ if $\tau_\ell^h < 0$, $\tau_\ell^h(\gamma) = \tau_\ell^h/(1+\gamma)$ if $\tau_\ell^h > 0$. Thus $\tau^h(\gamma)$ entails a diminution of purchases in τ^h by the fraction $\gamma/(1+\gamma)$.

We say that there are *gains to γ-diminished trade* at $x \equiv (x^h)_{h \in H} \in (\mathbb{R}_+^L)^H$ if there exist trades $(\tau^h)_{h \in H}$ such that:

(a) $\sum_{h \in H} \tau^h = 0$
(b) $x^h + \tau^h \in \mathbb{R}_+^L$ for all $h \in H$
(c) $u^h(x^h + \tau^h(\gamma)) > u^h(x^h)$ for all[13] $h \in H$.

In other words, it should be possible – in spite of the "γ-handicap" on trade – for households to Pareto-improve on x. We define $\gamma(x)$ as the supremum of all handicaps that permit Pareto improvement.

Definition The **gains to trade at x** are given by

$$\gamma(x) = \sup\{\gamma : \text{there are gains to } \gamma\text{-diminished trade at } x\}$$
$$= \min\{\gamma : \text{there are } not \text{ gains to } \gamma\text{-diminished trade at } x\}.$$

To clarify the definition of gains to trade at e, define the utility functions

$$v_\gamma^h(x) \equiv u^h(e^h + (x - e^h)(\gamma))$$

for $h \in H$. Observe that for fixed γ, every component of $(x - e^h)(\gamma)$ is concave in x. Since u^h is also concave and increasing, it follows that v_γ^h inherits both these properties. Thus, by standard arguments, the economy $(v_\gamma^h, e^h)_{h \in H}$ has a Walras equilibrium.

Lemma 2 *There are no gains to γ-diminished trade at e if and only if the economy $(v_\gamma^h, e^h)_{h \in H}$ has a no-trade Walras equilibrium.*

---

[13] Since utilities are strictly monotonic, this is equivalent to requiring that some household is strictly better off and none are worse off.

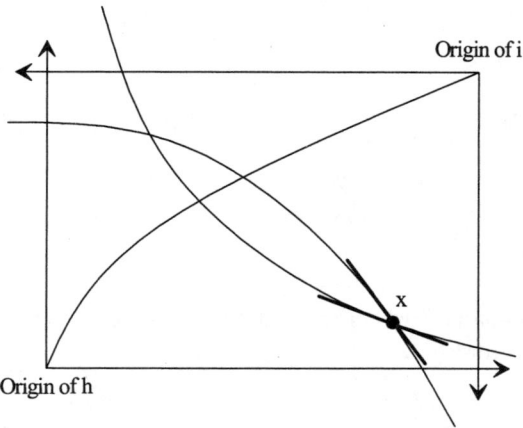

Figure 2. $[1 + \gamma(x)]^2 =$ ratios of slopes (not the angle between them)

For smooth economies we can give an explicit formula for $\gamma(x)$. Suppose $x^h \gg 0$ and u^h is continuously differentiable at x^h, for all $h \in H$. With two agents and two goods, $[1 + \gamma(x)]^2$ is the ratio of the agents' marginal rates of substitution for the two goods (see Fig. 2). It would be incorrect to connect $\gamma(x)$ with the angle between the indifference curves at x in an Edgeworth Box. An important property of $\gamma(x)$ is that rescaling the units of a commodity, say from pounds to ounces, leaves $\gamma(x)$ invariant, though it changes the angle. One can check that (if $H = 2$ or $L = 2$),

$$[1 + \gamma(x)]^2 = \max_{\substack{h \neq i \\ k \neq \ell}} \left[\left(\frac{\frac{\partial u^h(x^h)}{\partial x_k}}{\frac{\partial u^h(x^h)}{\partial x_\ell}} \right) \left(\frac{\frac{\partial u^i(x^i)}{\partial x_\ell}}{\frac{\partial u^i(x^i)}{\partial x_k}} \right) \right].$$

More generally, Pareto improvement may require trade involving more than two households and two commodities. Define a trading cycle c_n as a sequence of distinct commodities $(\ell_1, ..., \ell_n)$ and agents $(h_1, ..., h_n)$ where h_i sells ℓ_i and buys ℓ_{i+1} (where $\ell_{n+1} \equiv \ell_1$). Although each agent h_i trades only a pair of goods, the trading cycle may require many goods and agents because there may be no double coincidence of wants. The next theorem and its corollary show that in calculating the gains to trade available at an allocation x, it suffices to examine trading cycles. Since there are only a finite number of trading cycles, the following formula for the gains to trade can in principle be explicitly calculated.

Theorem 1 *If u^h is continuously differentiable and $x^h \gg 0 \; \forall h \in H$, then*

$$1 + \gamma(x) = \max_{2 \leq n \leq L} \; \max_{c_n \in C_n} \left\{ \prod_{i=1}^{n} \left(\frac{\frac{\partial u^{h_i}(x^{h_i})}{\partial x_{\ell_{i+1}}}}{\frac{\partial u^{h_i}(x^{h_i})}{\partial x_{\ell_i}}} \right) \right\}^{1/n}$$

where $\ell_{n+1} \equiv \ell_1$ and the second max is taken over the finite set C_n of all trading cycles $c_n = (\ell_1, ..., \ell_n, h_1, ..., h_n)$ of length n.[14]

Corollary 1 Suppose $(u^h)_{h \in H}$ are continuously differentiable and $x^h \gg 0 \ \forall h \in H$. Suppose there are trades $(\tau^h)_{h \in H}$, $\sum_{h \in H} \tau^h = 0$, such that $v_\gamma^h(x^h + \tau^h) > v_\gamma^h(x^h) \ \forall h \in H$. Then there is a cycle $(\ell_1, ..., \ell_n)$, $(h_1, ..., h_n)$ and trades $\tilde{\tau}^{h_i}$ on the cycle (i.e., $\tilde{\tau}_{\ell_{i+1}}^{h_i} = -\tilde{\tau}_{\ell_{i+1}}^{h_{i+1}} > 0$ for all $i = 1, ..., n$, and $\tilde{\tau}_\ell^{h_i} = 0$ if $\ell \neq \{\ell_i, \ell_{i+1}\}$) such that $v_\gamma^{h_i}(x^h + \tilde{\tau}^h) > v_\gamma^{h_i}(x^h) \ \forall i = 1, ..., n$.

If utilities are not continuously differentiable, or if some $x_\ell^h = 0$, we can give lower and upper bounds for $\gamma(x)$.

Corollary 2 Drop the differentiability and interiority assumptions of Theorem 1. Define $\gamma_*(x)$ (or $\gamma^*(x)$) in exactly the same manner as $\gamma(x)$, but with the right-hand derivative in the numerator (or denominator) and the left-hand derivative in the denominator (or numerator). Then $\gamma_*(x) \leq \gamma(x) \leq \gamma^*(x)$.

The gains to trade are not given by the distance from x to the Pareto frontier, but as Theorem 1 shows, by the ratio of the slopes of the two indifference curves at x. The number $\gamma(x)$ can be viewed as a local measure of the departure from Pareto optimality. It represents how easy it is to make a *small* Pareto improvement starting from x. If x is Pareto optimal, then $\gamma(x) = 0$; otherwise $\gamma(x) > 0$. Our measure is to be contrasted with a global measure of inefficiency suggested by Debreu, which represents how far from *full* Pareto optimality the allocation is. His coefficient of resource allocation $\delta(x)$ is given by:

$$1 + \delta(x) = \sup \left\{ \lambda : \text{there exists a reallocation of } \frac{1}{\lambda} \sum_{h \in H} x^h \right.$$

$$\left. \text{which leaves each } h \text{ at utility level at least } u^h\left(x^h\right) \right\}.$$

Notice that perturbations of utility functions u^h only around x^h will alter our measure $\gamma(x)$, but will have little effect on Debreu's measure $\delta(x)$.

An example makes this local/global distinction transparent. Take $u^h(x_1, x_2) = u^i(x_1, x_2) = \min\{x_1 + 2x_2, 2x_1 + x_2\}$. For any $x = (x^h, x^i)$, where $x_1^h + 2x_2^h < 2x_1^h + x_2^h$ and $x_1^i + 2x_2^i > 2x_1^i + x_2^i$, utilities are differentiable and the local gains to trade are quite steep. By our formula, $\gamma(x) = 1$, since $[1 + \gamma(x)]^2 = \frac{2}{1} \cdot \frac{2}{1} = 4$.

If aggregate endowments are equal in both goods, i.e., $x_1^h + x_1^i = x_2^h + x_2^i$, then the Pareto surface consists of the diagonal of the Edgeworth box drawn below in Figure 3. The point x in the picture has $\gamma(x) = 1$, no matter how close it is to the diagonal. In contrast, $\delta(x)$ falls to zero as x approaches the diagonal. The formula shows that the function $\gamma(x)$ is continuous when the utilities are continuously differentiable and $x \gg 0$. The example shows that, without differentiability, $\gamma(x)$ may not be continuous. But it is always lower semi-continuous.

---

[14] When the maximum is achieved only on cycles of length at least $n > 2$, there is a failure of double coincidence of wants in the economy. We might even use the minimum length n which achieves the maximum defining $\gamma(x)$ as a measure of this failure. We shall not pursue this point here.

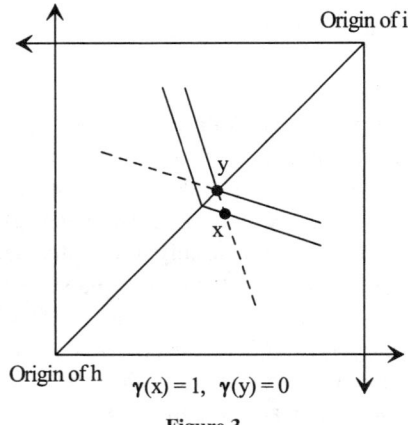

Figure 3

Lemma 3 *Let $(u^h)_{h \in H}$ be concave and continuous. Let the allocations $x(k) \to x$ as $k \to \infty$. Then $\liminf_k \gamma(x(k)) \geq \gamma(x)$.*

We are now ready to state our key condition for the existence of ME.

Gains to trade hypothesis $\gamma(e) > \bar{m}/M$; i.e., there are gains to γ-diminished trade at the initial endowment, where $\gamma = \bar{m}/M$.

4 Existence of monetary equilibrium

Theorem 2 *Consider a monetary economy $(\mathcal{E}, m, M)$ in which $\bar{m} \equiv \sum_{h \in H} m^h > 0$ and the Gains to Trade Hypothesis holds, i.e., $\gamma(e) > \bar{m}/M$. Then a monetary equilibrium exists and, at any monetary equilibrium, the interest rate $r = \bar{m}/M$.*

Corollary 3 *(a) Suppose $\bar{m} = 0$. Then monetary equilibria exist and coincide (in allocations and price ratios) with the Walras equilibria of the underlying economy. (b) Suppose $(p(k), r(k), q(k), x(k))$ is an ME of $(\mathcal{E}, m(k), M(k))$, with $\bar{m}(k) > 0$, and $x(k) \to_{k \to \infty} x$, and $\bar{m}(k)/M(k) \to_{k \to \infty} 0$. Then x is Walrasian for $\mathcal{E}$.*

According to the theorem, increasing the stock of inside money M must eventually guarantee the orderly functioning of markets, if the initial endowment is not Pareto-optimal. Equilibrium exists in $(\mathcal{E}, m, M)$ once M exceeds the finite threshold $\bar{m}/\gamma(e)$. In our model, inside money is indeed "the grease that turns the wheels of commerce."

Part (b) of the corollary assures us that although all trades are conducted via money, and although all prices are quoted in terms of money, as bank money M approaches infinity, the final allocation of goods becomes essentially no different from the Walrasian allocation obtained in an idealized world without any money at all, and in which prices really only have meaning as exchange rates between pairs of commodities.

Levels of bank money beyond $\bar{m}/\gamma(e)$, but short of infinity, give a large domain in which the real sector $\mathcal{E} = (u^h, e^h)_{h \in H}$ and the financial sector (m, M) influence each other, as we show in Section 6.

4.1 Outline of the proof of Theorem 2

The intuition behind the proof is as follows. Imagine an external agent, say the government, that commits to activating all markets by selling ε units on each side of every market (except for the money side of the bond market, where the bank already commits $M > 0$). In particular, the government sells ε units of each real good for money. Money would then naturally have value because any holder of it could get something real in exchange for it. We quickly show that, with such an external agent, a full blown ε-monetary equilibrium (ε-ME), including commodity prices $p(\varepsilon) < \infty$ and an interest rate $r(\varepsilon)$, necessarily exists. In an ε-ME, as in an ME, the total stock of money and commodities held by households, the bank, and the external agent is conserved in all three time intervals.

We remove the external agent by letting $\varepsilon \to 0$, and examine a limit of ε-ME to see if the households themselves are imputing positive value to money. If $r(\varepsilon)$ and $p(\varepsilon)$ stay bounded as $\varepsilon \to 0$, we can pass to a convergent subsequence which will be a bona fide ME. Clearly $r(\varepsilon) \geq 0$, for otherwise households could arbitrage the bank. Thus money-now must have value in terms of money-later. Note that no $p_\ell(\varepsilon) \to 0$, for otherwise any household h with $m^h > 0$ would be buying more of good ℓ than there is, contradicting the feasibility of ε-ME.

The interest rate $r(\varepsilon)$ must be no bigger than $(\bar{m} + L\varepsilon)/M$, no matter how small $\varepsilon > 0$ is, since default is not allowed. Otherwise, some household would necessarily default, because it would not be able to find the money to pay its debts. Thus $0 \leq \lim r(\varepsilon) \leq \bar{m}/M$.

It remains to show that $p(\varepsilon) \nrightarrow \infty$. Suppose $p(\varepsilon) \to \infty$. Then, since the total money in the system is bounded and since money is the sole medium of exchange, trade in goods $\to 0$ as $p(\varepsilon) \to \infty$. Hence households end up consuming their initial endowment e in the limit. At the same time, notice that with $p(\varepsilon) \to \infty$, the purchasing power of the endowed money m goes to zero and may be ignored. Consider now the limiting price *ratios* (on some subsequence) given by p, where $p_\ell = \lim_{\varepsilon \to 0} p_\ell(\varepsilon)/\sum_{k \in L} p_k(\varepsilon)$. The trading opportunity for any household (at the limit) is effectively to purchase goods solely out of borrowed money and to pay the loan back, at the interest rate $r \equiv \lim r(\varepsilon) \leq \bar{m}/M$ out of his sales revenue (conducting all trade via money, of course, at the prices p). A little reflection reveals that this is tantamount to doing standard Walrasian trades at prices p but consuming only the fraction $1/(1+r)$ of purchases, which in turn may be viewed as consuming the whole Walrasian trade via modified utilities v_r^h. Thus e is a Walras allocation for $(v_r^h)_{h \in H}$ at prices p, and must be Pareto-optimal with respect to $(v_r^h)_{h \in H}$. Since $r \leq \bar{m}/M < \gamma(e)$, e is also Pareto-optimal with respect to $(v_{\gamma(e)}^h)_{h \in H}$. This contradicts the gains-to-trade hypothesis. So $p(\varepsilon) \nrightarrow \infty$, finishing the proof.

In short, once the external agent has given households the confidence that money-now has value (i.e., that $p(\varepsilon) < \infty$), and that money-later has enough value (i.e., that $r(\varepsilon)$ is not too high), they themselves will offer large amounts of goods, and small enough amounts of bonds, for sale against money. Their actions will fulfill their own prophecies, propping up the value of money-now and money-later, if there are gains to $\bar{m}/M$-diminished trade.

Our proof in effect uses the gold-backed device of supporting the value of money, but shows that in the end the gold is not needed, if there are enough potential gains to trade in the underlying real economy. We could have given almost the same proof, using the alternative device of forced commodity sales against money, showing again in the end that such sales are unnecessary.

The proof in the appendix also shows that we can drop the hypothesis of strict monotonicity, which becomes important when we consider a multiperiod model with overlapping generations.

5 Determinacy of monetary equilibrium and the value of money

If outside money $\bar{m} \equiv \sum_{h \in H} m^h = 0$, the ME are Walrasian, as pointed out in the Corollary to Theorem 2. In this event, it is clear that there is great indeterminacy of the commodity price levels. Households can borrow, hoard, and return money to the bank without spending it and without incurring any interest cost. Hence ME prices can be scaled down arbitrarily (with households hoarding increasing amounts of bank money M) without disturbing the ME.

But the moment $\bar{m} > 0$ we must have the interest rate positive, indeed equal to $\bar{m}/M$. Consequently there is no hoarding at any ME and the above indeterminacy abruptly disappears. In particular, the value of money (given by the price levels) is determinate.

We can state this intuition formally as follows. Fix utilities $(u^h)_{h \in H}$. Let $\mathcal{U}$ be an (suitably small) open set of linear perturbations of the utilities. More precisely for each vector $c \in \mathbb{R}^L$, let $u_c^h(x) = u^h(x) + c \cdot x$ (where $\cdot$ denotes inner product). Take $\mathcal{U}$ to be a (suitably small) open set in $\mathbb{R}^L$, including the origin, and (for household h) identify $c \in \mathcal{U}$ with the utility u_c^h.

Let $\mathbb{R}_{++}^L$ denote the set of possible endowment vectors e^h for any household $h \in H$. Similarly, let $\mathbb{R}_{++}$ denote the set of possible monetary endowments m^h for any household $h \in H$. Finally, also let $\mathbb{R}_{++}$ denote the set of possible levels of bank money. Then we may think of any monetary economy $((c^h, e^h, m^h)_{h \in H}, M)$ as a point in $\Xi \equiv (\mathcal{U} \times \mathbb{R}_{++}^L \times \mathbb{R}_{++})^H \times \mathbb{R}_{++}$.

Theorem 3 *For an open and full measure set Ξ' of vectors $\xi \in \Xi$, the set of ME for the economy defined by ξ is finite in number; and this set varies continuously on Ξ'.*

Theorem 3 tells us that as the data of the economy change, equilibrium moves differentiably. We are particularly interested in showing that perturbations of (m, M) cause real changes in $(x^h)_{h \in H}$ and in relative prices p_ℓ/p_k, as well as in nominal price levels and in interest rates. We will also show that changes in the real sector $(u^h, e^h)_{h \in H}$ affect the general price level, as well as price ratios and final consumption.

6 Welfare in a monetary equilibrium

As was said, at any monetary equilibrium the interest rate r is the ratio of outside to inside money in the economy, i.e., $r = \bar{m}/M$ (see Theorem 2). Thus, if $\bar{m} > 0$, households who borrow money lose the interest-float on their marginal purchases, which discourages some trade. The upshot is that ME allocations are not Pareto-optimal, leaving room for further gains to trade.

A simple example will clarify the picture. Suppose that $e^h = e^i = (50, 50)$, $m^h = m^i = 5$, $M = 90$, and $u^h(x_1, x_2) = 10 \log x_1 + 3 \log x_2$ and $u^i(x_1, x_2) = 3 \log x_1 + 10 \log x_2$. In equilibrium (i) household i sells part of its endowment of commodity 1 and buys commodity 2; while household h sells part of its endowment of commodity 2 and buys commodity 1, (ii) both households borrows money from the bank. Since the utilities are differentiable, let

$$\nabla_\ell^j(y) \equiv \text{the partial derivative of } u^j \text{ w.r.t. the variable } x_\ell, \text{ evaluated at } y \in \mathbb{R}_+^L.$$

Then, if x^i and x^h are the households' consumptions at an ME with prices (p, r), we must have

$$\frac{\nabla_2^i(x^i)}{p_2} = (1 + r)\frac{\nabla_1^i(x^i)}{p_1}$$

and

$$\frac{\nabla_1^h(x^h)}{p_1} = (1 + r)\frac{\nabla_2^h(x^h)}{p_2}.$$

(Such an equality holds in general for any "*active*" household j, i.e., any j that borrows money and purchases good ℓ and sells only a *part* of his endowment of k.[15]) Therefore gradients of households i and h tilt away from the price ratio p_1/p_2 in opposite directions (see Fig. 4 below). The misalignment implies that from the ME it is possible for households i and h to trade further for joint gains. The expansion of trade would no doubt occur in a Walrasian world in which all trade is free. In Walrasian equilibrium, final consumption would be approximately $x^h = (77, 23)$ and $x^i = (23, 77)$. But, in a monetary equilibrium, the cost imposed by the interest-float hampers the possibility.

One can check that in monetary equilibrium $p_1 = p_2 = 2$, $r = 1/9$, $x^h = (75, 25)$, $x^i = (25, 75)$. Agent h spends his $5 and buys 2.5 units of good 1. He also borrows $45 from the bank, promising to repay $50. This is spent to buy 22.5 units more of good 1. To repay the bank he must sell 25 units of good 2, which is purchased by agent i.

The example also reveals that the misalignment is no more than the wedge factor $(1 + r)$, putting a bound on the inefficiency of ME allocations.

To describe the general situation, we first introduce a condition which will ensure that households never sell all of any commodity they are positively endowed with.

---

[15] The reason for the equality is clear. If $\nabla_\ell^j(x^j)/p_\ell > (1 + r)\nabla_k^j(x^j)/p_k$, then j could improve its utility by borrowing δ more dollars from the bank and spending it to purchase ℓ, while defraying the loan by selling $(1 + r)\delta$ dollars' worth more of good k; if the reverse inequality holds, j would benefit by reducing slightly its bank loan and purchase of ℓ, while simultaneously curtailing the concomitant sale of k. There is a "wedge" of size $(1 + r)$ between the buying and selling prices of any commodity.

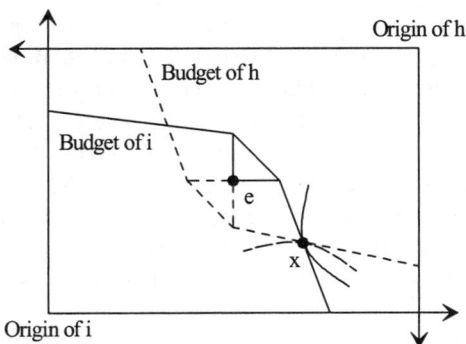

Figure 4. ME allocation x in the Edgeworth box

Regularity condition Assume that each u^h is continuously differentiable, and that

$$\left.\begin{array}{r} u^h(y^h) \geq u^h(e^h) \\ y^h \leq \sum_{i \in H} e^i \\ e_\ell^h > 0 \end{array}\right\} \Rightarrow y_\ell^h > 0$$

The "social welfare" at an ME is summed up in:

Theorem 4 *Let (p, r, q, x) be an ME of the monetary economy $(\mathcal{E}, m, M)$ with $\bar{m} > 0$. Then $\gamma(x) \leq r$. If the regularity condition holds, $\gamma(x) = r = \bar{m}/M$.*

Remark 1 Even without regularity, if utilities $(u^h)_{h \in H}$ are differentiable, there are allocations y arbitrarily close to x with $\gamma(y) \geq r$.

7 Non-neutrality of money

There is no money illusion in our model. If we scale M and the vector $m \equiv (m^h)_{h \in H}$ by the same factor, then clearly the allocation and the interest rate at an ME remain unaffected, and all commodity prices are scaled by the same factor. The scaling is tantamount to a change in units, and it would be surprising indeed if a switch from dollars to cents caused rational agents to behave differently.

But for the above circumstance, money is never neutral. Indeed let $\mathcal{A}(\mathcal{E}, m, M)$ denote the set of all allocations of commodities achieved at monetary equilibria of $(\mathcal{E}, m, M)$. Then we have

Theorem 5 *Consider two monetary economies $(\mathcal{E}, m, M)$ and $(\mathcal{E}, m^*, M^*)$ with the same underlying real sector $\mathcal{E}$, and suppose that the regularity condition holds for $\mathcal{E}$. If $\bar{m}/M \neq \bar{m}^*/M^*$, then the sets $\mathcal{A}(\mathcal{E}, m, M)$ and $\mathcal{A}(\mathcal{E}, m^*, M^*)$ are disjoint.*

Proof. Take any $x \in \mathcal{A}(\mathcal{E}, m, M)$ and any $x^* \in \mathcal{A}(\mathcal{E}, m^*, M^*)$. By Theorem 4, the gains to trade at x and at x^* are $\gamma(x) = \bar{m}/M$ and $\gamma(x^*) = \bar{m}^*/M^*$. Since $\bar{m}/M \neq \bar{m}^*/M^*$, x and x^* are distinct. $\qquad\square$

A change in M alone, or in m alone, or in both but in different proportions, will (by Theorem 5) invariably affect real trades. The injection of bank money (with private endowments of money held fixed) corresponds to a form of elementary monetary policy in our model. It is evident that this policy will lower the interest rate (since $r = \bar{m}/M$) and alter commodity allocations, moving them "closer" to Pareto-efficiency since the unexploited gains to trade "left on the table" become smaller. However, as we shall see in Section 9, increases in M will also eventually raise equilibrium price levels p. Households that began with relatively large endowments m^h of money will be hurt, since their cash endowments lose purchasing power. These households could be expected to use their influence on the central bank to resist such expansionary monetary policy. Gifts of fiat money to households constitute fiscal policy. They will cause interest rates to rise, and the ensuing ME allocations are bound to be affected, becoming less (locally) efficient in the process. Of course households that were the primary recipients of the fiscal gifts may be better off than before.

The welfare-reducing impact of fiscal injections is most pronounced in the setting of exchange economies with private goods and complete markets. When there is production and incomplete markets, fiscal injections may be Pareto improving. But we deal with this important issue elsewhere [12]. Fiscal injections can also be Pareto improving when there are public goods. We defer our analysis of monetary and fiscal policy to Section 10.

8 The necessity of gains to trade

We have claimed that the presence of a bank (which "demands" money for its own sake, at the end of commodity trade) is not sufficient for money to have value. We now make this claim precise via Theorems 6 and 7, showing that monetary equilibrium does not exist unless there are sufficient gains to trade.

Let $\tilde{\nabla}_\ell^h(y)$ ($\bar{\nabla}_\ell^h(y)$) denote the left-hand (right-hand) derivative of u^h, with respect to the variable x_ℓ, at the point $y \in \mathbb{R}_+^L$. And let $\mathcal{A}(e)$ be the set of all individually rational and feasible allocations, i.e.,

$$\mathcal{A}(e) = \{ y \equiv (y^h)_{h \in H} \in (\mathbb{R}_+^L)^H : \sum_{h \in H} y^h$$
$$= \sum_{h \in H} e^h, \text{ and } u^h(y^h) \geq u^h(e^h) \text{for } h \in H \}$$

Put

$$\Gamma(e) = \sup_{y \in \mathcal{A}(e)} \gamma^*(y)$$

i.e., $\Gamma(e)$ is an upper bound on the gains to trade at any point in $\mathcal{A}(e)$.

Theorem 6

$$(\bar{m}/M) > \Gamma(e) \Rightarrow \textit{no ME exists.}$$

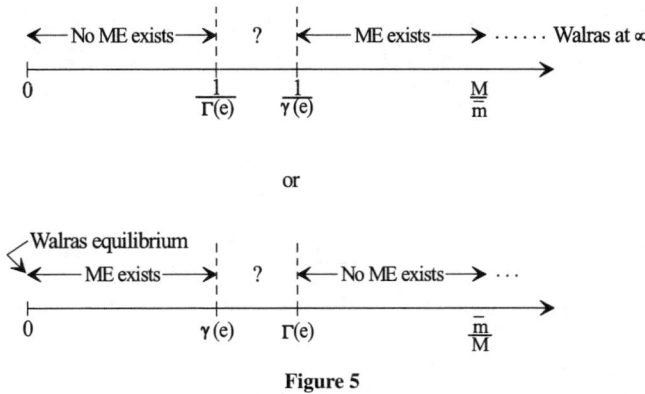

Figure 5

Theorem 6 shows that the value of money depends on household heterogeneity. If all households were identical, $\Gamma(e) = 0$, and monetary equilibrium could not exist. In fact, if e were Pareto efficient, then $\Gamma(e) = 0$ and again money would have no value.

Putting together Theorems 2 and 6, we obtain Figure 5. In general, we cannot say whether equilibrium exists in the gap $\gamma(e) \le \bar{m}/M \le \Gamma(e)$. But when all utilities u^h are separable, the question mark disappears: the region in question in Figure 5 falls entirely into the domain of nonexistence.

The function $u : \mathbb{R}_+^L \to \mathbb{R}$ is said to be separable if there exist functions $u_\ell : \mathbb{R}_+ \to \mathbb{R}$, for $\ell \in L$, such that $u(x) = \sum_{\ell \in L} u_\ell(x_\ell)$ for all $x \in \mathbb{R}_+^L$. (Cobb-Douglas utilities have this property.)

First we establish the following Theorem, which may be of some interest in its own right. It asserts that for separable utilities, an ME strictly uses up some of the gains to trade available at the initial endowment.

Theorem 7 *Consider a monetary economy* $((u^h, e^h, m^h)_{h \in H}, M)$ *with* $\bar{m} > 0$, *and suppose that* u^h *is strictly concave and separable for all* $h \in H$. *Suppose there exists an ME* $\langle p, r, q, x \rangle$ *with interest rate* r. *Then* $\gamma(e) > r \ge \gamma(x)$.

(The second inequality in Theorem 7 was shown in Theorem 4.) From Theorem 7 we immediately get

Theorem 8 *Assume* u^h *is strictly concave and separable for* $h \in H$. *Then an ME exists if and only if* $\gamma(e) > \bar{m}/M$.

In other words, for strictly concave and separable utilities, the sufficient condition for existence (in Theorem 2) is also necessary (and this incidentally shows that Theorem 2 cannot be strengthened).

9 Hyperinflation

Let us fix $\bar{m}$ and start with M so large that $M/\bar{m}$ is well to the right of $1/\gamma(e)$. By Theorem 2 equilibrium exists, and by Corollary 3 it is nearly Walrasian. Increasing

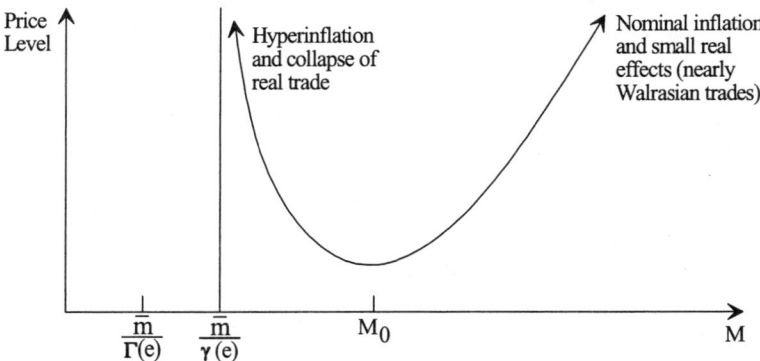

Figure 6. m fixed

M still further has little real effect; nearly exactly the same real trades are conducted. Since all of $M + \bar{m}$ is spent on practically the same purchases, the price level[16] rises linearly with M, and we have the nominal inflation depicted in Figure 6.

As M falls toward $\bar{m}/\gamma(e)$, what happens to price levels? The decline in money suggests that price levels will continue to fall. But as M falls, $r = \bar{m}/M$ rises, discouraging trade and moving us to less efficient allocations. (By Theorem 4, at any ME allocation x of a regular economy, $\gamma(x) = r$.) Smaller volumes of trade Q make for higher price levels. Which effect dominates?

Suppose the economy has strictly concave and separable utilities. Consider a sequence of ME with bank money $M(n)$ and equilibrium prices $p(n)$. Suppose $M(n)$ converges to $\bar{m}/\gamma(e)$ from above. If $p(n)$ remains bounded, then by passing to a convergent subsequence, we could show the existence of ME at $M = \bar{m}/\gamma(e)$, contradicting Theorem 8. Hence $p(n) \to \infty$ and the price level must look something like in Figure 6.

Our analysis has the paradoxical feature that there is some stock M_0 of bank money which minimizes the price level. If the bank eases, and lends more money, inflation will creep in, though the equilibrium allocation will improve somewhat. If the bank tightens its policy, lending less than M_0, inflation will again occur, and eventually price levels will *rise* much more rapidly (i.e., much faster than linearly, since they reach infinity over a finite move $M_0 - \bar{m}/\gamma(e)$). We call this explosion of prices, a hyperinflation.

We can also attribute the hyperflation to changes in the stock of outside money $\bar{m}$, occasioned perhaps by a government eager to distribute money to particular projects.

Too big a fiscal injection $\bar{m}$ must result in an explosion of prices, and collapse of trade, as we approach the finite threshold $\gamma(e)M$. Note that we cannot be sure that the price level falls for increases of $\bar{m}$ near 0, since price levels at the Walras equilibrium are now finite (see Fig. 7).

It is very useful to compare monetary injections ΔM with fiscal injections $\Delta \bar{m}$. As $M \to \infty$ (holding $\bar{m}$ fixed), price levels rise to infinity at the same rate, and

---

[16] I.e., the price of some fixed good in terms of money. (All price ratios will be bounded, since the interest rate is bounded across all ME.)

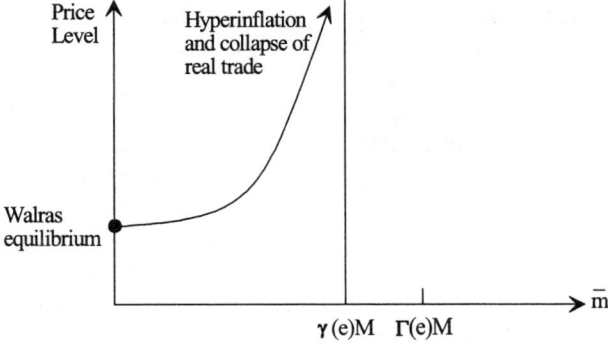

Figure 7. M fixed

we have *nominal* inflation. As $\bar{m}$ increases (holding M fixed), price levels rise at an accelerating pace, reaching infinity when $\bar{m}$ is still finite, and thus giving *hyperinflation*.

We summarize our discussion of hyperinflation in the following theorem.

Theorem 9 (Hyperinflation) *Fix the real sector* $\mathcal{E} \equiv (u^h, e^h)_{h \in H}$. *Assume* u^h *is strictly concave and separable for* $h \in H$. *Consider a sequence* $\{M(n),$ $(m^h (n))_{h \in H}\}_{n=1}^{\infty}$, *indexed by* n, *such that* $M(n)$ *is bounded away from 0 and* $\sum_{h \in H} m^h(n)/M(n)$ *increases monotonically to* $\gamma(e)$. *Every* n *defines an economy which, by Theorem 2, has monetary equilibria. Let* $p(n)$ *be an ME commodity price vector of the* n*th economy. Then* $p(n) \to \infty$ *as* $n \to \infty$.

We conjecture that hyperinflation occurs, without the separability hypothesis on utilities, at some point m^*/M^* between $\gamma(e)$ ansd $\Gamma(e)$.

10 Government: treasury and central bank

We extend our one-period model by adding a government. Fiat money, after all, is money by government decree, and brought into being by government issue.

The government produces public goods denoted $P = \{1, ..., P\}$. These goods are not marketed, but give utility to households. For simplicity we imagine a production function $F : \mathbb{R}_+^L \to \mathbb{R}_+^P$ mapping inputs of private goods into a unique output of public goods. We suppose F is continuous and $F(0) = 0$.

Public goods have impact on households' utilities. For any vector z of public goods present in the economy, $u_z^h : \mathbb{R}_+^L \to \mathbb{R}$ is the utility of $h \in H$ over his consumption of private goods. We assume that, for every fixed z, $u_z^h(x)$ is concave and strictly monotonic in x; and also that $u_z^h(x)$ is continuous jointly in z, x. (Notice that we do not need to assume that $u_z^h(x)$ is monotonic in z.)

We distinguish the government Treasury department, which chooses $(Q_{bm}, \Delta m, Q_m, \sigma)$, from the government central bank, which sets M. The complete government policy is given by a vector

$$\pi \equiv (M, Q_{bm}, \Delta m, Q_m, \sigma) \in \mathbb{R}_{++} \times \mathbb{R}_+ \times \mathbb{R}^H \times \mathbb{R}_+^L \times [0, 1).$$

We now describe the four policy instruments of the Treasury.

The Treasury buys all inputs for production, such as labor, from the private sector. Denote these expenditures by $Q_m = (Q_{m\ell})_{\ell \in L}$.

The Treasury also raises taxes and transfers wealth. For simplicity we assume the same ad valorem tax rate on the sale of every good and denote it by $\sigma \in [0, 1)$. An agent who sells a vector q of goods at prices p must pay the Treasury $\sigma p \cdot q$ out of his sales revenue $p \cdot q$. We denote by $\Delta m \equiv (\Delta m^h)_{h \in H}$ the transfer of money (to household h if $\Delta m^h > 0$, and from household h if $\Delta m^h < 0$). We assume these lump sum transfers occur after bank loans are made and before commodity trade. Lump sum transfers Δm^h are rarely observed in reality, especially if $\Delta m^h < 0$. We include them for theoretical reasons.

To finance its expenditures $\bar{Q}_m \equiv \sum_{\ell \in L} Q_{m\ell}$ on inputs for production and on money transfers $\Delta \bar{m} \equiv \sum_{h \in H} \Delta m^h$, the Treasury can borrow money from the bank, in competition with households, by issuing its own bonds Q_{bm}. Let

$$m_\beta \equiv \Delta \bar{m} + \bar{Q}_m - Q_{bm}/(1 + r). \tag{9.1}$$

A shortfall $m_\beta > 0$ must be covered by printing money, and a surplus $m_\beta < 0$ must be inventoried.

The Treasury is not allowed to default on its bank loan. Let

$$m_\alpha \equiv Q_{bm} - \text{tax revenue} - \max\{-m_\beta, 0\}. \tag{9.2}$$

If it cannot repay the loan out of its tax revenue and inventory, it must print additional $m_\alpha > 0$; and must dispose of excess money if $m_\alpha < 0$.

The budget deficit of the Treasury is defined by the amount of money it must print to meet its spending

$$\text{deficit} \equiv \max\{m_\beta, 0\} + m_\alpha \equiv -\text{surplus}.$$

If there is no deficit (or surplus), the budget is *balanced*.

We think of the Treasury making a budget plan at the beginning of the period. This plan consists of its issue of bonds Q_{bm}, its transfers $(\Delta m^h)_{h \in H}$ and expenditures $(Q_{m\ell})_{\ell \in L}$, and the tax rate σ. The Treasury can always carry out its plan, no matter how households behave, so long as it is free to print money before trade $(m_\beta > 0)$, or hoard money across trading time $(m_\beta < 0)$, and print money $(m_\alpha > 0)$ or destroy money $(m_\alpha < 0)$ at the end. Policies for which the government borrows precisely what it needs to spend, and taxes precisely to cover its debt, i.e., $m_\beta = 0 = m_\alpha$, are called *totally balanced* budget policies. In an uncertain world, however, the Treasury might not be sure of the proceeds from its sale of bonds, or of its tax revenue. There would need to be a mechanism for covering shortfalls and disposing of excess money. Though there is no uncertainty in our model, we wish to investigate whether monetary equilibrium can be maintained even when the Treasury faces imbalances. The residual variables (m_β, m_α), which adjust for imbalances, are determined in our model uniquely from $M, (Q_{bm}, \Delta m, Q_m, \sigma), r$ and tax revenue by equations (9.1) and (9.2). We investigate whether equilibrium exists for arbitrary complete policies $(M, Q_{bm}, \Delta m, Q_m, \sigma)$, even if they entail printing money or destroying money.

In American law, the Treasury cannot literally print money, but must borrow it from the Federal Reserve. But the Federal Reserve can print the money, giving it to the Treasury in exchange for an IOU note. By not redeeming the IOU note, or equivalently by rolling it over in perpetuity, the Treasury prints money by proxy.

10.1 Monetary equilibrium with government

We shall denote by $\Pi(m)$ the set of all (complete) policies π that are consistent with the private sector $(\mathcal{E}, m)$. Denote $\bar{m} \equiv \sum_{h \in H} m^h$, $\Delta \bar{m} \equiv \sum_{h \in H} \Delta m^h$, $\bar{Q}_m \equiv \sum_{\ell \in L} Q_{m\ell}$. Then

$$\Pi(m) = \left\{ (M, Q_{bm}, \Delta m, Q_m, \sigma) \in \mathbb{R}_{++} \times \mathbb{R}_+ \times \mathbb{R}_+^H \times \mathbb{R}_+^L \times [0,1) : \right.$$
$$\left. m + \Delta m \in \mathbb{R}_+^H, \bar{m} + \Delta \bar{m} + \bar{Q}_m > 0 \right\}.$$

Thus the government cannot take more money from any household than it is endowed with $(m + \Delta m \in \mathbb{R}_+^H)$, or wipe out all the outside money in the system $(\bar{m} + \Delta \bar{m} + \bar{Q}_m > 0)$. *Throughout, when we consider an economy $(\mathcal{E}, m, \pi)$, it will be assumed that $\pi \in \Pi(m)$.*

The budget set $B(p, r, e^h, m^h + \Delta m^h, \sigma)$ of household h is defined exactly as $B(p, r, e^h, m^h)$, but with two amendments. First, replace m^h with $m^h + \Delta m^h$. Second, tax must be deducted from sales revenue (i.e., $\sum_{\ell \in L} p_\ell q_{\ell m}^h$ must be replaced by $\sum_{\ell \in L} (1 - \sigma) p_\ell q_{\ell m}^h$).

The vector $\langle p, r, (q^h, x^h)_{h \in H} \rangle$ is a *monetary equilibrium* for the economy $(\mathcal{E}, m, \pi \equiv (M, Q_{bm}, \Delta m, Q_m, \sigma))$ iff

(i) $(q^h, x^h) \in B(p, r, e^h, m^h + \Delta m^h, \sigma)$ and $u_z^h(x^h) \geq u_z^h(\underline{x}^h)$ for all $(\underline{q}^h, \underline{x}^h)$
$\in B(p, r, e^h, m^h + \Delta m^h, \sigma)$ where $z = F(Q_{m1}/p_1, ..., Q_{mL}/p_L)$

(ii) $p_\ell \sum_{h \in H} q_{\ell m}^h = Q_{m\ell} + \sum_{h \in H} q_{m\ell}^h, \forall \ell \in L$

(iii) $Q_{bm} + \sum_{h \in H} q_{bm}^h = (1 + r)M$

Thus households are optimal in their amended budget sets ((i)) and all markets clear ((ii), (iii)), taking the government's actions into account.

10.2 Gains to trade and existence of monetary equilibrium

Given private money $(m^h)_{h \in H} \equiv m$ and government policy $\pi \equiv (M, Q_{bm}, \Delta m, Q_m, \sigma)$, consider the equation:

$$g(x) \equiv (1 - \sigma) \left[\frac{xM}{Q_{bm} + x} + \sum_{h \in H} (m^h + \Delta m^h) + \sum_{\ell \in L} Q_{m\ell} \right] = x. \qquad (10)$$

Since $\bar{m} + \Delta \bar{m} + \bar{Q}_m > 0$ and $\sigma < 1$, we have $g(0) > 0$. Moreover, since $M > 0$ and the function $\delta(x) \equiv x/(Q_{bm} + x)$ is strictly concave[17] and monotonic and bounded above (by 1), all these properties are inherited by g. It follows that there exists a unique positive scalar which solves the equation. We denote it by $x(m, \pi)$.

---

[17] Unless $Q_{bm} = 0$, in which case the graph of g is horizontal, still yielding a unique intersection with the 45° line.

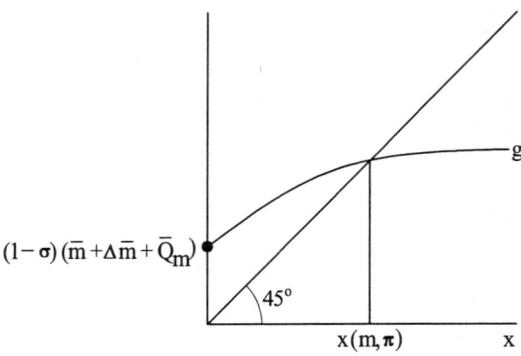

Figure 8

As we shall see in the proof of Theorem 10, the variable x gives the total number of bonds issued by households, and $g(x)$ gives the total money in their hands after commodity trade, assuming $r > 0$.

Define

$$\rho(m, \pi) = \max \left\{ 0, \frac{Q_{bm} + x(m, \pi)}{M} - 1 \right\}. \tag{11}$$

For any z in $\mathbb{R}_+^P$, define the gains to trade $\gamma_z(x)$ in the private sector in the presence of public goods z, exactly as before, using the utilities $(u_z^h)_{h \in H}$. Then $\gamma_0(e)$ represents the gains to trade before any trade or government production occurs.

Theorem 10 *Consider a monetary economy* $(\mathcal{E}, m, \pi)$*. Suppose* $\gamma_0(e) > r^* \equiv r^*(m, \pi) \equiv (\sigma + \rho(m, \pi))/(1 - \sigma)$*. Then an ME exists. Moreover if* $\langle p, r, q, x \rangle$ *is an ME of* $(\mathcal{E}, m, \pi)$*, then* $r = \rho(m, \pi)$*.*

The existence theorem for monetary equilibrium with the government differs in two respects from our previous existence Theorem 2. First, the gains to trade $\gamma_0(e)$ must exceed a threshold r^* that is greater than just the interest rate r. Second, the equilibrium interest rate r is no longer a simple ratio of exogenous stocks of outside and inside money.

As in monetary equilibrium without the Treasury, all the money issued by the bank and all the money printed by the Treasury (and not destroyed) must be owed and returned to the bank. Thus the interest rate must be

$$r = \frac{\bar{m} + \max\{m_\beta, 0\} + m_\alpha}{M}.$$

The Treasury can inject money into the system by running a deficit and printing it $(\max\{m_\beta, 0\} + m_\alpha > 0)$ or take it out of the system by running a surplus and destroying it $(\max\{m_\beta, 0\} + m_\alpha < 0)$. Previously only the bank could inject or withdraw money. The values m_β and m_α are solved endogenously, as part of the monetary equilibrium along with r. Without the Treasury, the equilibrium interest rate r was a simple ratio of the stock of outside money to the stock of inside money. With the Treasury endogenously printing and destroying money (depending on its budget deficit which depends on taxes raised) the interest rate cannot be so easily calculated from the exogenous parameters $(\mathcal{E}, m, \pi)$.

Nevertheless, Theorem 10 shows that the equilibrium interest rate $r = \rho(m, \pi)$ and the impediment to trade $r^* = r^*(m, \pi)$ do not depend on the underlying physical economy $(u^h, e^h)_{h \in H}$ in any way. We emphasize that this is because there is only one time period. With two or more time periods, nominal interest rates would depend on the underlying physical economy [10].

Theorem 10 enables a partition of the policy space $\Pi(m)$ into two regions $\Pi_0(m)$ and $\Pi_+(m)$ such that, if $\pi \in \Pi_0(m)$, the interest rate (at every ME of $(\mathcal{E}, m, \pi)$) is zero; and if $\pi \in \Pi_+(m)$, the interest rate is positive. Define

$$\Pi_0(m) \equiv \{(M, Q_{bm}, \Delta m, Q_m, \sigma) \in \Pi(m) : (\bar{m} + \Delta\bar{m} + \bar{Q}_m)$$
$$\leq (\sigma/(1 - \sigma))(M - Q_{bm})\}$$

and
$$\Pi_+(m) \equiv \Pi(m) \backslash \Pi_0(m).$$

Corollary 4 *Suppose* $\langle p, r, (q^h, x^h)_{h \in H} \rangle$ *is an ME of* $(\mathcal{E}, m, \pi)$. *Then*

$$r = 0 \Leftrightarrow \pi \in \Pi_0(m)$$
$$(and\ so\ r > 0 \Leftrightarrow \pi \in \Pi_+(m)).$$

10.3 Welfare and gains to trade

Without taxation, we saw that the interest rate created an impediment to trade. Taxation adds another impediment, and exacerbates the interest rate impediment. To see why, consider a household that wishes to sell enough goods to finance purchases worth one dollar. The household must incur a debt of $1 + r$ dollars, and then sell goods on which it will pay the tax σ, bringing its total cost to $(1 + r)/(1 - \sigma)$. This is tantamount to paying an effective interest rate of $r^* = [(1 + r)/(1 - \sigma)] - 1 = (r + \sigma)/(1 - \sigma)$. All our previous theorems on welfare hold (by the same proofs) with this threshold r^* in place of the bank interest rate r. For regular economies (in which utilities are smooth and equilibrium consumption $x_\ell^h > 0$ whenever $e_\ell^h > 0$), r^* measures precisely the unexploited gains to trade at any monetary equilibrium allocation $((x^h)_{h \in H}, z)$.

Government policies which increase r^* move the equilibrium allocation to $((x^h)_{h \in H}, z)$ at which there are more unexploited gains to trade. But welfare might nevertheless be increased if the policy leads to more public goods $z = F(Q_{m1}/p_1, ..., Q_{mL}/p_L)$, and if consumers value these public goods.

11 Monetary and fiscal policy

11.1 Nominal comparative statics in an "IS-LM" framework

Macroeconomics seeks to understand aggregate variables such as total output without paying careful attention to the microeconomic details. A complete description of equilibrium in our model would require knowledge of all microeconomic particulars. Changing some household u^h or e^h would typically change equilibrium prices p and consumption $(x^h)_{h \in H}$ for all households. Raising $Q_{m\ell}$ by \$1 and lowering some other $Q_{m\ell'}$ by \$1 would do the same.

We shall show, however, that the equilibrium interest rate r and nominal GNP Y can be calculated without knowledge of microeconomic details. Indeed, write *aggregate* fiscal and monetary policy as the four-dimensional vector

$$\bar{\pi} \equiv (M, Q_{bm}, \mu \equiv \bar{m} + \Delta\bar{m} + \bar{Q}_m, \sigma).$$

We think of the first two coordinates as aggregate monetary policy, and the latter two coordinates as aggregate fiscal policy. We shall show that r and Y can be calculated from $\bar{\pi}$ alone via a graphical analysis identical to the Hicksian IS-LM framework. Each curve corresponds to the locus of points (Y, r) describing goods market clearing, or money market clearing, or bond market clearing.

In this section we assume that government policy is consistent with the existence of monetary equilibrium with a positive interest rate, i.e. (from Theorem 10)

$$\pi \in \Pi_+^*(m) \equiv \left\{ \pi \in \Pi_+(m) : \gamma_0(e) > r^*(m, \pi) \equiv \frac{\sigma + \rho(m, \pi)}{1 - \sigma} \right\}.$$

Define nominal GNP Y as aggregate spending on commodities: $Y = \sum_{\ell \in L} p_\ell \sum_{h \in H} q_{\ell m}^h$. Agents will always spend all their cash $\bar{m} + \Delta\bar{m}$. Anticipating revenue $(1 - \sigma)Y$ from their sale of commodities, and knowing that there is only one consumption period, consumers will borrow and spend an additional $(1 - \sigma)Y/(1 + r)$. Since government spends $\bar{Q}_m$, market clearing for commodities requires

$$\left[\frac{(1 - \sigma)Y}{1 + r} + \bar{m} + \Delta\bar{m} \right] + \bar{Q}_m = Y, \text{ or}$$

$$Y = \frac{1 + r}{\sigma + r}(\bar{m} + \Delta\bar{m} + \bar{Q}_m) = \frac{1 + r}{\sigma + r}\mu. \qquad (10)$$

Note that for $\sigma < 1$, Y declines in r, for any fixed $\mu > 0$. We call (10) the income = spending equation, or IS for short.

Anticipating that they will be spending $Y - \bar{Q}_m$, households will demand precisely this same amount of money if $r > 0$. Since government demand for money is $Q_{bm}/(1 + r)$, money market clearing requires that

$$(Y - \bar{Q}_m) + \frac{Q_{bm}}{1 + r} = M + \bar{m} + \Delta\bar{m}, \text{ or}$$

$$Y = M + \mu - \frac{Q_{bm}}{1 + r} \qquad (11)$$

Anticipating again that their revenue from commodity sales will be $(1 - \sigma)Y$, households will offer to sell exactly that many bonds (assuming $r > 0$ so they do not inventory any money). Bond market clearing therefore requires that

$$1 + r = \frac{Q_{bm} + (1 - \sigma)Y}{M}, \text{ or}$$

$$Y = \frac{1 + r}{1 - \sigma}M - \frac{Q_{bm}}{1 - \sigma}. \qquad (12)$$

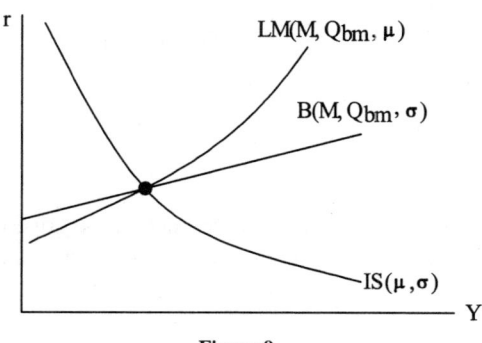

Figure 9

These three equations are drawn in the familiar (Y, r) plane in Figure 9.

Note that the commodity market clearing equation (10) depends only on fiscal parameters (μ, σ), and is independent of M and Q_{bm}. Similarly bond market clearing does not depend on μ, and money market clearing does not depend on σ. The slopes of the curves are derived from equations (10)–(12). In particular, since $Q_{bm}/(1 + r)^2 < M$, the money market clearing curve slopes up faster than the bond market clearing curve near the equilibrium (r, Y).

All the usual Keynesian comparative statics hold. Increasing M by \$1 shifts the LM curve \$1 to the right and the B curve $(1 + r)/(1 - \sigma) > 1$ dollars to the right, leaving the IS curve unchanged. Thus r declines and Y rises by less than \$1. Increasing government spending by \$1, financed by printing money, moves the IS curve $(1 + r)/(\sigma + r) > 1$ dollars to the right, and shifts the LM curve exactly \$1 to the right, while leaving the B curve unchanged. Thus r rises and Y rises by more than \$1. Increasing government borrowing by selling $1 + r$ more bonds shifts the LM curve \$1 to the left and the B curve $(1 + r)/(1 - \sigma) > 1$ dollar to the left. Thus r increases and Y decreases by less than \$1. In all three cases r^* moves in the same direction as r, since σ is fixed.

Increasing σ by 1% moves the B curve and the IS curve down by more than 1%, while leaving the LM curve unchanged. Therefore r goes down by more than 1%, and Y goes down. From (10) we have that $Y[(1 - \sigma)/(1 + r)] = \mu/r^*$. From (12) we know that $Y(1 - \sigma)$ drops by a bigger percentage than $1 + r$, since Q_{bm} and M are fixed. Hence $Y[(1 - \sigma)/(1 + r)]$ declines and so r^* increases when σ increases.

At any budget-balanced policy, $r = \bar{m}/M$ and $Y \leq \bar{m} + M$. At any totally budget-balanced policy, $Y = \bar{m} + M$, exactly as was the case without the Treasury. Budget-balanced increases in government expenditures $\bar{Q}_m$ do not raise nominal income Y. In the Keynesian model they do increase Y, because households' marginal propensity to consume is assumed to be less than one (while the government spends all its tax revenue). In our one-period model, households have no reason to save, and their marginal propensity to consume is therefore one. In a totally budget-balanced policy, the government's propensity to spend is also one, while in a balanced-budget policy, the Treasury may borrow and hoard the money, paying the interest by raising taxes, and thus its propensity to spend out of tax revenue may be less than one.

11.2 Liquidity trap

One new phenomenon introduced by the presence of the government is the possibility that interest rates become zero even when there are initial stocks of outside money $\bar{m} > 0$. If the government surplus, $- \max\{m_\beta, 0\} - m_\alpha = \bar{m}$, then r becomes zero.

At first glance this seems like a knife-edge case, but in fact the region where $r = 0$ is quite robust. Define

$$\Pi_0^*(m) \equiv \left\{ \pi \in \Pi_0(m) : \gamma_0(e) > \frac{\sigma}{1-\sigma} \right\}.$$

Corollary 5 *Fix* $(\mathcal{E}, m)$. *Then for any government policy* $\pi = (M, Q_{bm}, \Delta m, Q_m, \sigma)$ *in the nonempty open set* $T \equiv \mathrm{int}\ \Pi_0^*(m)$, *monetary equilibrium exists, and at every monetary equilibrium the interest rate is zero.*

For any policy in the liquidity trap, $\pi \in T$, equilibrium leaves $\sigma/(1-\sigma)$ unexploited gains to trade, and leaves $r = 0$. No small policy change that maintains the tax rate σ will be able to budge the interest rate or improve the gains to trade. In the liquidity trap, government monetary policy (M, Q_{bm}) is powerless to improve household trade if the Treasury dares not reduce taxes.

In the liquidity trap households borrow money and hoard some of it. By equation (10), with $r = 0$, total expenditure must be μ/σ. The remaining money must be borrowed and hoarded. As the tax rate σ is increased, households hoard more, $Y = $ GNP declines, and tax revenue $= \sigma Y = \sigma\mu/\sigma = \mu$ stays the same, as does the budget surplus. This explains why the liquidity trap region is robust, and how an open set of government policies can leave the budget surplus unchanged.

If the government insists on further and further increases in the tax rate σ, it will depress nominal GNP. As $\sigma/(1-\sigma)$ approaches $\gamma_0(e)$, real GNP will collapse to zero (if utilities are separable), by Theorem 7.

In the liquidity trap, increasing fiscal expenditures $\Delta\bar{m} + \bar{Q}_m$, without raising taxes, has a somewhat surprising multiplier effect. If the government prints one extra dollar, increasing $\Delta\bar{m} + \bar{Q}_m$ by 1, then aggregate spending Y must increase by $1/\sigma$.

Unfortunately, this remarkable stimulus to GNP springing from government spending is mostly inflation driven. Consider the special case that $1 = \sum_{h \in H} \Delta m^h$ and $\Delta m^h = m^h / \sum_{i \in H} m^i$ for all h, and $\bar{Q}_m = 0 = Q_{bm}$. Then the effect is purely from inflation, with consumption $(x^h)_{h \in H}$ unchanged.

It is worth noting that when the government takes no actions $\Delta\bar{m} = \bar{Q}_m = Q_{bm} = 0$, except to tax $\sigma > 0$, the ME correspond to Walras equilibrium with ad valorem tax σ and lump sum redistributions of the tax revenue $m^h / \sum_{i \in H} m^i$ to each agent $h \in H$. Our existence Theorem 10 thus proves the existence of Walras equilibrium with taxes.

11.3 Hyperinflation

Public goods expenditures $\bar{Q}_m > 0$ and transfers $\Delta\bar{m} > 0$ may be of great value to the economy. But if the Treasury becomes too ambitious by spending or borrowing too much, it will necessarily engender a self-defeating hyperinflation: $p \to \infty$, $x^h \to e^h$, $z \to 0$ as $\bar{Q}_m$ or Q_{bm} approach finite thresholds.

We argue this in the context of separable utilities, as in Section 9. There we showed that if the impediment to trade approached $\gamma(e)$, then hyperinflation necessarily set in. The fact that no money was being hoarded by households was crucial in that argument.

We will suppose $\pi \in \Pi_+(m)$ so that the interest rate is positive and households again do not hoard money. Recall that the impediment to trade is now $r^* \equiv r^*(m, \pi) \equiv [\sigma + \max\{0, (Q_{bm} + x(m, \pi))/M) - 1\}]/(1 - \sigma)$. Clearly $r^*(m, \pi)$ approaches $\gamma_0(e)$ at finite thresholds Q_{bm}^* and $\bar{Q}_m^*$. Indeed let $Q_{bm}^* = M(\gamma_0(e) + 1)$. Then, for $Q_{bm} \geq Q_{bm}^*$, the impediment $r^*(m, \pi) \geq \gamma_0(e)$ no matter what Δm, Q_m, or σ may be. To compute $\bar{Q}_m^*$, recall $x(m, \pi) = g(x(m, \pi)) \geq (1 - \sigma)\bar{Q}_m$; so let $\bar{Q}_m^* = M(\gamma_0(e) + 1)/(1 - \sigma)$. Then if $\bar{Q}_m \geq \bar{Q}_m^*$, we get $r^*(m, \pi) \geq \gamma_0(e)$ no matter what Δm and Q_{bm} may be (but this second threshold $\bar{Q}_m^*$ does depend on σ).

By its own profligacy, borrowing or spending too much, the Treasury destroys the value of its fiat money, and also its power to produce any public goods or to transfer any real wealth.

11.4 Real comparative statics: the Ricardian equivalence between policies

In this section we investigate the real consequences of monetary and fiscal policy. This gives a very different picture from the nominal effects described in Section 11.1. For example, we shall show that government expenditures on public goods have the same *real* effect (up to scale) whether they are budget-balanced financed, by debt repaid later out of tax revenue, or deficit financed by printing money. Yet we saw in Section 11.1 that balanced-budget financing does not increase the interest rate r or nominal income Y, whereas printing money does increase both r and Y. This real "Ricardian equivalence" turns out to be delicate to prove. If the Treasury finances its purchases by printing money, it will cause an inflation of commodity prices, which will force it to plan proportionately higher expenditures to maintain the same real purchases. In addition, in order to achieve all the same real effects as the budget-balanced expenditures, the Treasury will be obliged to print still more money to make transfers to compensate the holders of money who will be hurt by the inflation. We describe the precise equivalence between government policies in the Corollary to Theorem 11. The key to the equivalence is the observation that real trade is influenced by r^*, and not by r. Printing and spending money $\Delta\bar{m}$ increases r, and therefore also r^*, while budget-balanced expenditures leave r fixed, but increase σ, and hence also r^*.

By directing government spending toward one public good as opposed to another, or by transferring wealth from one group to another, the government can influence real outcomes without changing the nominal values calculated in the last section. Ricardian equivalence does not apply in these cases. However, we wish to concentrate on the aggregate real effects of aggregate policy. So we restrict attention to aggregate policies for which all Δm^h move in the same proportion, and all Q_m move in the same proportion. Thus we are back to five policy instruments $(M, Q_{bm}, \Delta m, Q_m, \sigma)$.

The Treasury can always create real changes by changing its mix of expenditures between transfers to households, and spending on public goods. If we fix this mix, then Theorem 11 demonstrates that the Treasury, with its three policy tools $(Q_{bm}, (\Delta m, Q_m), \sigma)$, can achieve no more and no less than the same real effects achievable by the central bank with its single policy instrument M. In the nominal IS-LM framework, the Treasury and the central bank had complementary policy tools, one controlling the LM curve, and the other the IS curve (and perhaps the LM curve). In real terms, it turns out that whatever the Treasury can do by printing money (and increasing all expenditures by the same proportion) or by raising commodity taxes (uniformly across goods), the central bank can do by reducing the money stock. Treasury power becomes distinct from the central bank only when it targets a part of the economy, for example by shifting resources from private production to public production.

Let $\langle p, r, q, x \rangle$ be an ME of $(\mathcal{E}, m, \pi)$ where $\pi \equiv (M, Q_{bm}, \Delta m, Q_m, \sigma)$. What are "equivalent" policies $\tilde{\pi} \equiv (\tilde{M}, \tilde{Q}_{bm}, \Delta\tilde{m}, \tilde{Q}_m, \tilde{\sigma})$ for which x remains an ME allocation?

We shall vary $(\Delta m, Q_m)$ linearly with a single parameter $\lambda \in \mathbb{R}_{++}$ as follows:

$$\Delta m^h(\lambda) = \lambda(m^h + \Delta m^h) - m^h, \text{ for } h \in H$$
$$Q_{m\ell}(\lambda) = \lambda Q_{m\ell}, \text{ for } \ell \in L.$$

This enables us to think of a policy as a four-dimensional vector $(\tilde{M}, \tilde{Q}_{bm}, \lambda, \tilde{\sigma})$, where $(\Delta\tilde{m}, \tilde{Q}_m) = (\Delta m(\lambda), Q_m(\lambda))$.

Equivalent policies can be pictured as a smooth surface in four dimensions. In particular this picture reveals that an arbitrary small change in any three policy variables, can be compensated by adjusting the fourth variable, to retain x as an ME allocation.

We describe the surface as a function $\Lambda : D \to \mathbb{R}_+$ where $D \subset \mathbb{R}_+^3$. For any $(\tilde{M}, \tilde{Q}_{bm}, \tilde{\sigma}) \in D$, setting $\tilde{\lambda} = \Lambda(\tilde{M}, \tilde{Q}_{bm}, \tilde{\sigma})$ will yield an equivalent policy. Needless to say, the domain D and the function Λ depend upon both $\langle p, r, q, x \rangle$ and $\pi \equiv (M, Q_{bm}, \Delta m, Q_m, \sigma)$.

Denote $r^* \equiv (r + \sigma)/(1 - \sigma)$. Then the domain D is given by:

$$D = \left\{ (\tilde{M}, \tilde{Q}_{bm}, \tilde{\sigma}) \in \mathbb{R}_{++} \times \mathbb{R}_+^2 : 0 \leq \tilde{\sigma} \leq \frac{r^*}{r^* + 1}, \right.$$
$$\left. \tilde{Q}_{bm} < (r^* + 1)(1 - \tilde{\sigma})\tilde{M} \right\}.$$

We will suppose $B \equiv M - (Q_{bm}/(1+r)) > 0$, i.e., the government is not borrowing *all* the bank money at the ME $\langle p, r, q, x \rangle$ of $(\mathcal{E}, m, M, Q_{bm}, \Delta m, Q_m, \sigma)$. With this proviso, define

$$\Lambda(\tilde{M}, \tilde{Q}_{bm}, \tilde{\sigma}) = \frac{1}{B} \left(\tilde{M} - \frac{\tilde{Q}_{bm}}{(r^* + 1)(1 - \tilde{\sigma})} \right).$$

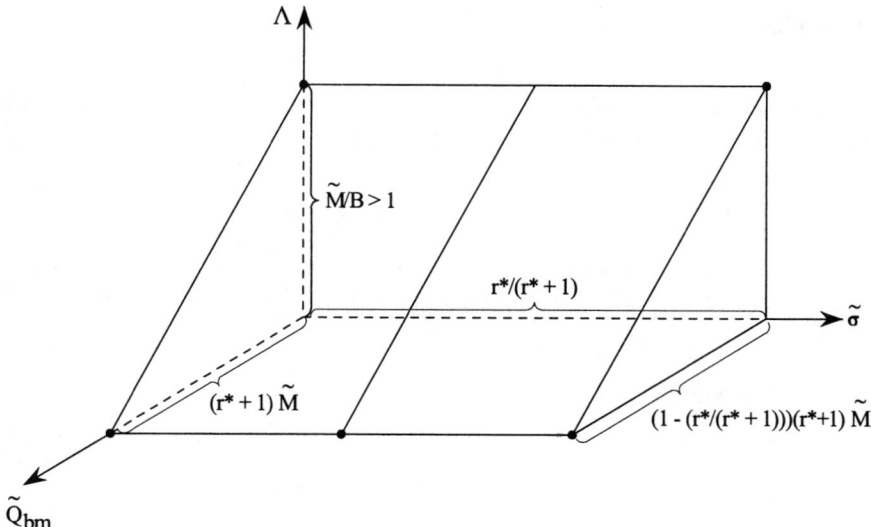

Figure 10. The graph of Λ for fixed $\tilde{M}$

We are ready to state

Theorem 11 *Let $\langle p, r, q, x \rangle$ be an ME of $(\mathcal{E}, m, M, Q_{bm}, \Delta m, Q_m, \sigma)$ and assume $M - (Q_{bm}/(1+r)) > 0$. Let $\Lambda : D \to \mathbb{R}_{++}$ be defined as above. Then, for any policy $\tilde{\pi} \in \{(\tilde{M}, \tilde{Q}_{bm}, \Lambda(\tilde{M}, \tilde{Q}_{bm}, \tilde{\sigma}), \tilde{\sigma}) : (\tilde{M}, \tilde{Q}_{bm}, \tilde{\sigma}) \in D\}$, there exists an ME $\langle \tilde{p}, \tilde{r}, \tilde{q}, \tilde{x} \rangle$ of $(\mathcal{E}, m, \tilde{\pi})$ such that $\tilde{x} = x$; and $Q_{m\ell}/p_\ell = \tilde{Q}_{m\ell}/\tilde{p}_\ell$ for all $\ell \in L$, where $\tilde{Q}_{m\ell} = \Lambda(\tilde{M}, \tilde{Q}_{bm}, \tilde{\sigma})Q_{m\ell}$ is government's expenditure in $\tilde{\pi}$.*

Corollary *Let $\langle p, r, q, x \rangle$ be an ME of $(\mathcal{E}, m, M, Q_{bm}, \Delta m, Q_m, \sigma)$ and assume $M - (Q_{bm}/(1+r)) > 0$. Then there exist policies π_1, π_2, π_3 which are equivalent to $(M, Q_{bm}, \Delta m, Q_m, \sigma)$ such that the government finances its expenditures solely by*

(1) *printing money $\tilde{m}_\beta$ prior to trade in π_1 (i.e., $\tilde{Q}_{bm}$, $\tilde{\sigma}$ and $\tilde{m}_\alpha$ are zero under π_1)*
(2) *borrowing money and repaying it by printing $\tilde{m}_\alpha$ after trade in π_2 (i.e., $\tilde{\sigma}$ and $\tilde{m}_\beta$ are zero under π_2)*
(3) *borrowing money and repaying it out of tax revenue in π_3 (i.e., via a totally balanced policy where $\tilde{m}_\beta = \tilde{m}_a = 0$ under π_3)*

(Here $\tilde{m}_\beta$, $\tilde{m}_\alpha$ are the endogenously determined quantities of money printed in the ME achieved under the relevant policy π_i.)

Appendix

Proof of Lemma 1. Let $C_r(p, x^h - e^h) \leq m^h$. We write $\tau \equiv x^h - e^h$. We will construct q^h so that $(q^h, x^h) \in B(p, r, e^h, m^h)$. Let

$$q^h_{bm} = (1 + r) \max\{p \cdot {}^*\tau - m^h, 0\}$$
$$q^h_{\ell m} = {}_*\tau_\ell \text{ for all } \ell \in L$$
$$q^h_{m\ell} = p_\ell {}^*\tau_\ell \text{ for all } \ell \in L.$$

Observe that with $\tilde{m}^h \equiv q^h_{bm}/(1 + r)$, $p \cdot {}^*\tau \leq m^h + \tilde{m}^h$, so

$$\sum_{\ell \in L} q^h_{m\ell} = p \cdot {}^*\tau \leq m^h + \tilde{m}^h$$

and inequality (2) of the budget set is verified. Since $x^h \in \mathbb{R}^L_+$, ${}_*\tau_\ell \leq e^h_\ell$, verifying inequality (3ℓ). Observe next that if $q^h_{bm} = 0$, then (5) is automatic. Otherwise,

$$\begin{aligned} q^h_{bm} &= (1 + r)[p \cdot {}^*\tau - m^h] \\ &= (1 + r)\left[p \cdot {}^*\tau - \frac{1}{1+r}p \cdot {}^*\tau - m^h\right] + p \cdot {}^*\tau \\ &= (1 + r)[C_r(p, \tau) - m^h] + p \cdot {}^*\tau \\ &\leq p \cdot {}^*\tau = \sum_{\ell \in L} p_\ell q^h_{\ell m} \end{aligned}$$

verifying inequality (5). Finally, letting $\tilde{x}^h_\ell \equiv q^h_{m\ell}/p_\ell = {}^*\tau_\ell$, we get

$$\begin{aligned} x^h_\ell &= (e^h_\ell - {}_*\tau_\ell) + {}^*\tau_\ell \\ &= (\Delta 3\ell) + \tilde{x}^h_\ell \end{aligned}$$

verifying inequality (6ℓ). Hence $(q^h, x^h) \in B(p, r, e^h, m^h)$, and so $x^h \in B_C(p, r, e^h, m^h)$.

Conversely, suppose there exists $q^h \in \mathbb{R}^{2L+1}_+$ with $(q^h, x^h) \in B^h(p, r, e^h, m^h)$. Let $\tau \equiv x^h - e^h$, and recall that $\tilde{x}^h_\ell = q^h_{m\ell}/p_\ell$ and (by (3ℓ) and (6ℓ))

$$\tilde{x}^h_\ell - q^h_{\ell m} \geq x^h_\ell - e^h_\ell = {}^*\tau_\ell - {}_*\tau_\ell.$$

Hence $\tilde{x}^h_\ell \geq {}^*\tau_\ell$ and $q^h_{\ell m} \geq {}_*\tau_\ell$.

Now,

$$
\begin{aligned}
C_r(p, x^h - e^h) &= p \cdot {}^*\tau - \frac{1}{1+r} p \cdot {}^*\tau \\
&= \sum_{\ell \in L} p_\ell \tilde{x}_\ell^h - \frac{1}{1+r} \sum_{\ell \in L} p_\ell q_{\ell m}^h \\
&\quad - \left[\sum_{\ell \in L} p_\ell(\tilde{x}_\ell^h - {}^*\tau_\ell) - \frac{1}{1+r} \sum_{\ell \in L} p_\ell(q_{\ell m}^h - {}^*\tau_\ell) \right] \\
&\leq \sum_{\ell \in L} p_\ell \tilde{x}_\ell^h - \frac{1}{1+r} \sum_{\ell \in L} p_\ell q_{\ell m}^h
\end{aligned}
$$

(by the preceding three inequalities)

$$
\begin{aligned}
&= \sum_{\ell \in L} q_{m\ell}^h - \frac{1}{1+r} \sum_{\ell \in L} p_\ell q_{\ell m}^h \\
&\leq \sum_{\ell \in L} q_{m\ell}^h - \frac{1}{1+r} \left[(1+r)\tilde{m}^h - (m^h + \tilde{m}^h - \sum_{\ell \in L} q_{m\ell}^h \right]
\end{aligned}
$$

(by the budget set inequalities (5), (2), and (1))

$$
\begin{aligned}
&= \frac{r}{1+r} \sum_{\ell \in L} q_{m\ell}^h - \frac{r}{1+r} (m^h + \tilde{m}^h) + m^h \\
&\leq \frac{r}{1+r} (m^h + \tilde{m}^h) - \frac{r}{1+r} (m^h + \tilde{m}^h) + m^h
\end{aligned}
$$

(by the budget set inequality (2))

$$
= m^h.
$$

$\square$

Proof of Lemma 2.   If no-trade is a Walras equilibrium, then, by the first welfare theorem, e is Pareto optimal with respect to $(v_\gamma^h)_{h \in H}$, and hence there are not gains-to-γ-diminished-trade at e (with respect to $(u^h)_{h \in H}$). Conversely, if there are not gains-to-γ-diminished-trade at e (with respect to $(u^h)_{h \in H}$), then e is Pareto optimal with respect to the utilities $(v_\gamma^h)_{h \in H}$. By the second welfare theorem, no-trade is a Walras equilibrium for $(v_\gamma^h)_{h \in H}$. $\square$

Proof of Lemma 3.   If $\gamma(x) = 0$, the lemma is obvious, since $\gamma(y) \geq 0$ for any y. So suppose $\gamma(x) > \gamma > 0$. Then we can find trades $(\tau^h)_{h \in H}$ such that $\sum_{h \in H} \tau^h = 0$ and for all $h \in H$, $x^h + \tau^h \in \mathbb{R}_+^L$ and $u^h(x^h + \tau^h(\gamma)) > u^h(x^h)$. It follows immediately from the continuity of u^h that for large enough k and $\lambda < 1$ but sufficiently close to 1, $x^h(k) + \lambda \tau^h(\gamma) \geq 0$ and $u^h(x^h(k) + \lambda \tau^h(\gamma)) > u^h(x^h(k))$.
$\square$

Proof of Theorem 1.   We first prove the theorem assuming that each u^h is strictly concave.

Let $(\gamma_k)_{k=1}^\infty$ be a monotonically increasing sequence $\gamma_k \to_k \gamma(x)$. Let $\langle p(k),$ $(x^h(k))_{h \in H} \rangle$ be a Walras equilibrium of $(v_{\gamma_k}^h, x^h)_{h \in H}$ for each k. Since w.l.o.g.

$\sum_{\ell=1}^{L} p_\ell(k) = 1$ for all k, by passing to convergent subsequences we may suppose that[18] $p(k) \to p \gg 0$ and $x^h(k) \to \tilde{x}^h$ for all $h \in H$. Since the functions $v_{\gamma_k}^h$ converge uniformly to $v_{\gamma(x)}^h$ on compact sets, $\langle p, (\tilde{x}^h)_{h \in H} \rangle$ must be a Walras equilibrium for the economy $(v_{\gamma(x)}^h, x^h)_{h \in H}$. By strict concavity and Lemma 2, $\tilde{x}^h = x^h$ for all $h \in H$.

Furthermore, by Lemma 2 we know that for each k, $\tau^h(k) \equiv \tilde{x}^h(k) - x^h \neq 0$ for at least one $h \in H$ (otherwise there are no more than $\gamma_k < \gamma(x)$ gains-to-trade at x). Hence for each k we can find a trading cycle $(\ell_1(k), \ell_2(k), ..., \ell_{n_k}(k))$ and $(h_1(k), h_2(k), ..., h_{n_k}(k))$ such that each $h_i(k)$ sells $\ell_i(k)$ and buys $\ell_{i+1}(k)$. Since the set of all possible cycles is finite, we may assume the same cycle obtains on our subsequence of k. We denote it $(\ell_1, ..., \ell_n), (h_1, ..., h_n)$.

For each k, and each trader $h \in \{h_1, ..., h_n\}$, define the net trades $\tilde{\tau}^h(k) \in \mathbb{R}^L$ by

$$\tilde{\tau}_\ell^h(k) = \begin{cases} -\varepsilon^h(k) p_{\ell_{i+1}}(k) & \text{if } \ell = \ell_i \\ \varepsilon^h(k) p_{\ell_i}(k) & \text{if } \ell = \ell_{i+1} \\ 0 & \text{otherwise} \end{cases}$$

where $\varepsilon^h(k) < \min\{-\tau_{\ell_i}^{h_i}(k)/p_{\ell_i}(k), \tau_{\ell_{i+1}}^{h_i}(k)/p_{\ell_{i+1}}(k)\}$. From utility maximization at Walras equilibrium,

$$v_{\gamma_k}^{h_i}(x^{h_i}(k) - \tilde{\tau}^{h_i}(k)) < v_{\gamma_k}^{h_i}(x^{h_i}(k)) \text{ for all } k.$$

It follows that

$$\varepsilon^{h_i}(k) p_{\ell_i}(k) \frac{\partial u^{h_i}(x^{h_i}(k) - \tilde{\tau}^{h_i}(k))}{\partial x_{\ell_{i+1}}}$$
$$> (1 + \gamma_k) \varepsilon^{h_i}(k) p_{\ell_{i+1}}(k) \frac{\partial u^{h_i}(x^{h_i}(k) - \tilde{\tau}^{h_i}(k))}{\partial x_{i_\ell}}.$$

Rearranging terms,

$$\frac{\dfrac{\partial u^{h_i}(x^{h_i}(k) - \tilde{\tau}^{h_i}(k))}{\partial x_{\ell_{i+1}}}}{\dfrac{\partial u^{h_i}(x^{h_i}(k) - \tilde{\tau}^{h_i}(k))}{\partial x_{i_\ell}}} > (1 + \gamma_k) \frac{p_{\ell_{i+1}}(k)}{p_{\ell_i}(k)}.$$

Taking products over $i = 1, ..., n$, and passing to the limit as $k \to \infty$,

$$\prod_{i=1}^{n} \frac{\dfrac{\partial u^{h_i}(x^{h_i})}{\partial x_{\ell_{i+1}}}}{\dfrac{\partial u^{h_i}(x^{h_i})}{\partial x_{i_\ell}}} > (1 + \gamma(x))^n.$$

---

[18] All price ratios are bounded since utilities are strictly monotonic. Then since $p \neq 0$ we have $p \gg 0$.

Conversely, appealing again to Lemma 2, let $\langle p, (x^h)_{h\in H}\rangle$ be a Walras equilibrium for the economy $(v^h_{\gamma(x)}, x^h)_{h\in H}$. Then for each $h \in H$,

$$\frac{\dfrac{\partial u^h(x^h)}{\partial x_i}}{\dfrac{\partial u^h(x^h)}{\partial x_j}} \le (1 + \gamma(x))\frac{p_i}{p_j}.$$

Hence on any chain,

$$\overset{n}{\underset{i=1}{\times}} \frac{\dfrac{\partial u^{h_i}(x^{h_i})}{\partial x_{\ell_{i+1}}}}{\dfrac{\partial u^{h_i}(x^{h_i})}{\partial x_{i_\ell}}} \le (1 + \gamma(x))^n.$$

This concludes the proof assuming the u^h are strictly concave. But if u^h is not strictly concave, we can replace each u^h with $\tilde{u}^h$ defined by $\tilde{u}^h(x) = u^h(x) + \varepsilon \sum_{\ell \in L} \sqrt{x_\ell}$. Taking limits as $\varepsilon \to 0$ gives our result. $\quad\square$

Proof of Corollary 1. If $(\tau^h)_{h\in H}$ improves on $(x^h)_{h\in H}$, for the utilities v^h_γ, then any Walras equilibrium $\langle p, (\tilde{x}^h)_{h\in H}\rangle$ for the economy $(v^h_\gamma, x^h)_{h\in H}$ must involve some nonzero trade. From these excess demands at equilibrium we can extract a cycle, as in the proof of Theorem 1. $\quad\square$

Proof of Corollary 2. Proceed exactly as in the proof of Theorem 1, noting that for concave utilities, left-hand derivatives are continuous from the left and exceed right-hand derivatives, which are continuous from the right. $\quad\square$

Proof of Theorem 2. For any $\varepsilon > 0$, we establish the existence of an ε-monetary equilibrium (ε-ME) whose limit (as $\varepsilon \to 0$) will yield an ME.

An ε-ME may be thought of as the strategic equilibrium of the following generalized game $\mathcal{G}_\varepsilon$. Replace each $h \in H$ by a continuum $(h-1, h]$ of identical households. Each t in the interval $(h-1, h]$ has the characteristics

$$(e^t, m^t) \equiv (e^h, m^h)$$
$$u^t \equiv u^h.$$

The *ambient* strategy-set of each $t \in [h-1, h]$ is $B(\varepsilon) = \{(q^t, x^t) \in \mathbb{R}^{2L+1}_+ \times \mathbb{R}^L_+ : \text{every component is} \le 1/\varepsilon\}$. Throughout we shall focus on *type-symmetric* strategies. (This permits the use of the notation (q^h, x^h) in three different senses: as the *vector* in $\mathbb{R}^{2L+1}_+ \times \mathbb{R}^L_+$, which is the common individual strategy chosen by each household $t \in (h-1, h]$ of type h; as the constant *function* which maps each $t \in (h-1, h]$ to the vector (q^h, x^h) and describes the symmetric strategy-selection of households of type h; and as the *integral* of this constant function on the unit interval $(h-1, h]$, which gives the aggregate strategy by households of type h. The sense in which (q^h, x^h) is used will be indicated, or else will be clear from the context.)

Given a strategy-selection $(q, x) \equiv (q^h, x^h)_{h \in H}$ by all households, market prices $p(\varepsilon, q, x)$, $r(\varepsilon, q, x)$ form according to the rule:

$$p_\ell(\varepsilon, q, x) = \frac{\varepsilon + \sum\limits_{h \in H} q^h_{m\ell}}{\varepsilon + \sum\limits_{h \in H} q^h_{\ell m}}$$

$$1 + r(\varepsilon, q, x) = \frac{\varepsilon + \sum\limits_{h \in H} q^h_{bm}}{M}.$$

(In the above formulae, read q^h as integral and (q, x) as a function.) In the game $\mathcal{G}_\varepsilon$, we imagine an external agent who puts up ε units of goods, money and bonds as indicated in the formulae. Prices form to clear all markets (taking the external agent into account). In other words, all of M is disbursed to households and the external agent in proportion to their bonds; and at each commodity-money market all the money (or, commodity) received is disbursed to households and the external agent in proportion to the commodity (or, money) sent by them. Of course, given an arbitrary selection (q, x), it may well happen that at the emergent prices households do not balance their budgets. So we are led to consider a generalized game with strategy sets that depend on others' choices. We define the *feasible* strategy-set of each $t \in (h - 1, h]$ of type h by:

$$B^h_\varepsilon(q, x) \equiv B(\varepsilon) \cap B(p(\varepsilon, q, x), r(\varepsilon, q, x), e^h, m^h).$$

(Here $(q, x) \equiv$ strategy selection; (e^h, m^h) are t's individual characteristics; $B(p(\varepsilon, q, x), r(\varepsilon, q, x), e^h, m^h)$ is just the budget-set defined earlier.) Given the joint strategies (q, x), each $t \in [h - 1, h)$ gets the payoff $u^h(x^h)$.

We define an ε-ME to be a type-symmetric strategic equilibrium of this generalized game $\mathcal{G}_\varepsilon$. We shall shortly prove that ε-ME exists.

Notice that, at any ε-ME, (1) we have a physically closed system, in which all the money or goods sent to market are conserved and redistributed among households and the external agent; (2) all households view $p(\varepsilon, q, x)$, $r(\varepsilon, q, x)$ as fixed, since their individual vector q^h does not affect the integral q^h involved in forming prices; (3) each household chooses optimal strategies (q^h, x^h) in his truncated budget-set (which just consists of those vectors in his standard budget-set that are of size $\leq 1/\varepsilon$ in each component).

Fix $\mu > M + \bar{m}$ and $\eta > \sum_{h \in H} \sum_{\ell \in L} e^h_\ell$, and choose ε small enough to ensure that $M + \bar{m} + L\varepsilon < \mu < 1/\varepsilon$ and $\eta < 1/\varepsilon$. Define

$$u^h_* = u^h(\eta, ..., \eta)$$

and let η^* be chosen to guarantee that

$$u^h(0, ..., 0, \eta^*, 0, ..., 0) > u^h_*$$

for η^* in any component. (W.l.o.g.[19] we may suppose that such a η^* exists.)

---

[19] Let $\square$ be the cube in $\mathbb{R}^L_+$ with sides of length η. Recall $u^h : \square \to \mathbb{R}$ is strictly increasing in the variables x_ℓ for $\ell \in L$. Define $\tilde{u}^h : \mathbb{R}^L_+ \to R$ by $\tilde{u}^h(y) = \inf\{L_x(y) : x \in \square, L_x \text{ is an affine}$

Let $B^*(\varepsilon) = ((B(\varepsilon))^H$.

Define the individual "best reply" ("demand") correspondence $\psi_\varepsilon^h : B^*(\varepsilon) \rightrightarrows B(\varepsilon)$ by

$$\psi_\varepsilon^h(q, x) = \arg\max\{u^h(\bar{x}^h) : (\bar{q}^h, \bar{x}^h) \in B_\varepsilon^h(q, x)\}$$

for $h \in H$; and then define ψ_ε from $B^*(\varepsilon)$ to itself by

$$\psi_\varepsilon = \psi_\varepsilon^1 \times \cdots \times \psi_\varepsilon^H.$$

Clearly $B_\varepsilon^h(q, x) \equiv B(\varepsilon) \cap B(p(\varepsilon, q, x), r(\varepsilon, q, x), e^h, m^h)$ is non-empty, compact, and convex. On account of the external agent's ε, prices $p(\varepsilon, q, x)$ are always positive, and since $e^h \neq 0$, $B_\varepsilon^h(q, x)$ is a continuous correspondence in (q, x). Hence each ψ^h is non-empty, convex, and upper semi-continuous.

By Kakutani's Theorem ψ_ε has a fixed point $(q(\varepsilon), x(\varepsilon)) \equiv (q^h(\varepsilon), x^h(\varepsilon))_{h \in H}$, with induced prices $p(\varepsilon)$, $r(\varepsilon)$. The vector $\langle p(\varepsilon), r(\varepsilon), q(\varepsilon), x(\varepsilon) \rangle$ will be called an ε-ME. Select a subsequence of ε-ME as $\varepsilon \to 0$ to ensure that all its components and all ratios of all components converge (possibly to zero or infinity).

Note that $q_{bm}^h(\varepsilon) < \mu < 1/\varepsilon$ (being budget feasible, h must return $q_{bm}^h(\varepsilon)$ to the bank, and there is at most μ units of money in the economy), $q_{m\ell}^h(\varepsilon) \leq$ total expenditure across markets $< \mu$, $q_{\ell m}^h(\varepsilon) \leq e_\ell^h < \eta < 1/\varepsilon$ and $x_\ell^h(\varepsilon) < \eta < 1/\varepsilon$ for $\ell \in L$. Since the limits on their optimal actions are not binding, and their utilities are concave, we obtain

Step 1. For sufficiently small ε, $(q^h(\varepsilon), x^h(\varepsilon))$ is optimal in $B(p(\varepsilon), r(\varepsilon), e^h, m^h)$, not just in $B_\varepsilon^h(q(\varepsilon), x(\varepsilon))$.

Step 2. $r(\varepsilon) \geq 0$, for sufficiently small ε.

Proof. Suppose $r(\varepsilon) < 0$. Then let h increase $q_{bm}^h(\varepsilon)$ by a positive δ obtaining $\delta(1 + r(\varepsilon))^{-1} > \delta$ units of bank money. Let him inventory δ to repay this additional loan and spend the surplus to buy $\ell \in L$. This improves his utility, contradicting Step 1.

Step 3. $r(\varepsilon) \leq (\bar{m} + L\varepsilon + \varepsilon)/M < (1 + \bar{m})/M \equiv \tilde{r}$, for small enough ε.

Proof. Since all households are budget feasible, all their debts to the bank are honored. So no more than $M + \bar{m} + L\varepsilon$ bonds could have been sold by households. And the external agent sells only ε bonds, so $1 + r \leq (M + \bar{m} + L\varepsilon + \varepsilon)/M$.

Step 4. For small enough ε, $p_k(\varepsilon)/p_\ell(\varepsilon) \leq \eta^*(1 + \tilde{r})/e^*$ for all $k, \ell \in L$, where $e^* \equiv \min_{i \in L} \max_{h \in H} e_i^h$, and $\tilde{r}$ is as in Step 3.

---

function representing a supporting hyperplane to the graph of u^h at the point $(x, u^h(x))\}$. Then it is clear that (a) $\tilde{u}^h$ is concave and strictly monotonic, and coincides with u^h on $\square$, hence ME of our economy are unaltered if we replace u^h by $\tilde{u}^h$; and (b) there exists a η^* such that $\tilde{u}^h(0, ..., 0, \eta^*, 0, ..., 0) > u_*^h$ for η^* in any component.

Proof. Consider h with $e_k^h = \max\{e_k^i : i \in H\} > 0$. Let $q_{bm}^h = p_k(\varepsilon)e_k^h$, $q_{km}^h = e_k^h$, $q_{m\ell}^h = p_k(\varepsilon)e_k^h/(1 + r(\varepsilon))$ and all other components of $q^h = 0$. From his sale of k, h obtains $p_k(\varepsilon)e_k^h$ units of money, and is able to repay the loan q_{bm}^h. So this action is in his untruncated budget set, and (by Step 1) cannot improve his payoff. But his consumption of ℓ via this action is at least $p_k(\varepsilon)e_k^h/((1+r(\varepsilon))p_\ell(\varepsilon))$ which must be less than η^* (otherwise h gets more than u_*^h utility, a contradiction). Recalling from Step 3 that $r(\varepsilon) \leq \tilde{r}$, Step 4 follows.

Step 5. Let $m^* = \max\{m^h : h \in H\}$. Then, for small enough ε,

$$\frac{m^*}{\eta^*} \leq p_\ell(\varepsilon).$$

Proof. It is clear that u_*^h is an upper bound on the utility of h at an ε-equilibrium (for small enough ε). But if $p_\ell(\varepsilon) < m^*/\eta^*$, then any agent h with $m^h = m^*$ can spend all his private endowment of money to purchase ℓ, consuming at least η^* of ℓ, and thus obtaining more than u_*^h utiles, a contradiction.

Step 6. $p_\ell(\varepsilon) \not\to \infty$ for any $\ell \in L$.

Proof. Suppose some $p_\ell(\varepsilon) \to \infty$. By Step 4, $p_k(\varepsilon) \to \infty$ for all $k \in L$. Since $p_k(\varepsilon) < \mu/(\sum_{h \in H} q_{km}^h(\varepsilon))$ we obtain $q_{km}^h(\varepsilon) \to 0$ for all $h \in H$ and $k \in L$, and hence $x^h(\varepsilon) \to e^h$ for all $h \in H$. Let $\hat{p}_\ell(\varepsilon) = p_\ell(\varepsilon)/\sum_{k \in L} p_k(\varepsilon)$ for $\ell \in L$ and $\hat{p} = \lim \hat{p}(\varepsilon)$. By Step 4, $\hat{p} \gg 0$. We also know that (see Step 3) $r \equiv \lim r(\varepsilon) \leq \bar{m}/M$.

Consider the consumption feasible budget sets $B_C(p(\varepsilon), r(\varepsilon), e^h, m^h)$. By homogeneity,

$$B_C(p(\varepsilon), r(\varepsilon), e^h, m^h) = B_C\left(\hat{p}(\varepsilon), r(\varepsilon), e^h, \frac{m^h}{\sum_{\ell \in L} p_k(\varepsilon)}\right).$$

Since $\hat{p}(\varepsilon) \to \hat{p} \gg 0$, $e^h \neq 0$, and $m^h/\sum_{\ell \in L} p_k(\varepsilon) \to 0$, we have (by a standard argument) the set convergence

$$B_C(p(\varepsilon), r(\varepsilon), e^h, m^h) \to B_C(\hat{p}, r, e^h, 0).$$

Hence, since $x^h(\varepsilon)$ is u^h-optimal in $B_C(p(\varepsilon), r(\varepsilon), e^h, m^h)$, $\lim x^h(\varepsilon) = e^h$ must be u^h-optimal in $B_C(\hat{p}, r, e^h, 0)$. Therefore $(\hat{p}, e)$ is Walrasian for the economy $(v_r^h, e^h)_{h \in H}$, where $v_r^h(x) \equiv u^h(e^h + (x - e^h)(r))$. By Lemma 3, there are not gains-to-r-diminished trade at e, with respect to the utilities $(u^h)_{h \in H}$. Hence $\gamma(e) \leq r$, but by Step 3, $r \leq \bar{m}/M$, and so $\gamma(e) \leq \bar{m}/M$, contradicting the gains-to-trade hypothesis.

Step 7. All markets clear at $\langle p, r, q, x \rangle \equiv \lim_{\varepsilon \to 0} \langle p(\varepsilon), r(\varepsilon), q(\varepsilon), x(\varepsilon) \rangle$.

Proof. By Steps 2 and 3, $0 \leq r \leq \lim(\bar{m} + L\varepsilon/M) = \bar{m}/M < \infty$, and by Steps 5 and 6, $0 \ll p \ll \infty$. By definition, all markets clear at an ε-ME, once we include the actions of the external agent. But since his actions go to zero, and all prices are finite and positive, it follows that markets clear without the external agent at the limit $\langle p, r, q, x \rangle$.

Step 8. An ME exists.

Proof. Utility maximization at $\langle p, r, q, x \rangle$ follows from the continuity of u^h and the upper semi-continuity and lower semi-continuity of B_C. □

Step 9. At any ME $\langle p, r, q, x \rangle$ of $(\mathcal{E}, m, M)$, $r = \bar{m}/M$.

Proof. This was proved just after the definition of ME in Section 1.6. □

Proof Corollary 3. First consider case (b). By Theorem 1, $r(k) = \bar{m}(k)/M(k)$, hence $r(k) \to 0$. Define $\hat{p}_\ell(k) \equiv p_\ell(k)/\sum_{j \in L} p_j(k)$ for $\ell \in L$. It is easily verified that x is Walrasian with prices $\hat{p} \equiv \lim \hat{p}(k)$.

Next consider case (a). If $\langle p, r, q, x \rangle$ is an ME, then r must be zero (by an argument analogous to Step 2), hence $\langle p, x \rangle$ is Walrasian. On the other hand if $\langle p, x \rangle$ is Walrasian, choose $\lambda > 0$ so that $\lambda p \cdot \sum_{h \in H} e^h = M$. Then $(\lambda p, x)$ is achieved at an ME by letting each h borrow and spend $\lambda p \cdot e^h$ to buy the goods x^h. These actions clearly constitute an ME. □

Remarks

Dropping Strict Monotonicity: the Having–Wanting Chain. It is important to note that strict monotonicity of u^h in every commodity can be dropped from the hypotheses of our model. What is needed is a version of resource relatedness. We formulate this in terms of a "having-wanting" chain. If $e_\ell^h > 0$, say that "agent h has ℓ"; and if $u^h(x)$ is everywhere strictly increasing in the variable x_ℓ, say that "agent h wants ℓ." Consider a directed graph on node-set L with arc (ℓ, k) if there exists an agent h who has ℓ and wants k. Our existence theorem holds if we assume for every (ℓ, k) in $L \times L$, with $\ell \neq k$, that there is a directed path (chain) from ℓ to k. The only change required is in the proof of Step 4, which we now indicate.

Exactly as in the proof of Step 4, if arc (ℓ, k) exists then the upper bound of Step 4 is valid. Since any two goods ℓ and k are connected by a chain of length at most $L - 1$, $(\eta^*(1 + \tilde{r})/e^*)^{L-1}$ will be an upper bound for $p_\ell(\varepsilon)/p_k(\varepsilon)$ (for every (ℓ, k) in $L \times L$).

Forced Sales of Commodities. For $0 < \alpha \leq 1$, define an α-ME exactly like an ME, but with condition (3) of the budget set amended to read

$$\alpha e_\ell^h \leq q_{\ell m}^h \leq e_\ell^h.$$

Then an α-ME exists *without* the gains-to-trade hypothesis. To see this, repeat the proof of Theorem 1 up to Step 5 (but with households being forced to obey the above inequality) and then notice that $p_\ell(\varepsilon) \leq \mu/\sum_{h \in H} \alpha e_\ell^h$, so $p(\varepsilon) \twoheadrightarrow \infty$, hence the limit is an α-ME. The case $\alpha = 1$ corresponds to the hypothesis made in Lucas [28] and Magill-Quinzii [30]. Theorem 2 shows that money has value even without such α-forced sales.

Proof of Theorem 3. See Dubey-Geanakoplos [7]. □

Proof of Theorem 4. The proof that $\gamma(x) \leq r$ relies on properties of the cost function C_r which are of interest on their own. Recall that for any trade vector $\tau \in \mathbb{R}^L$, $^*\tau_j = \max\{0, \tau_j\}$ denotes purchases, and $_*\tau_j = -\min\{0, \tau_j\}$ denotes sales; and that we defined the present-value cost function $C_r : \mathbb{R}_+^L \times \mathbb{R}_+ \to \mathbb{R}$ by

$$C_r(p, \tau) \equiv p \cdot {}^*\tau - \frac{1}{1+r} p \cdot {}_*\tau = \frac{1}{1+r}[p \cdot \tau + rp \cdot {}^*\tau].$$

Recall also that we defined $\tau(\gamma) \equiv \frac{1}{1+\gamma} {}^*\tau - {}_*\tau \leq \tau$, for any $\gamma \geq 0$.

Lemma 4 *The cost function $C_r(p, \tau)$ is continuous, convex in τ, homogenous of degree one in p and τ separately, and satisfies*

$$C_r(p, \tau + \tilde{\tau}) \leq C_r(p, \tau) + C_r(p, \tilde{\tau})$$
$$C_r(p, \tau(r)) = \frac{p \cdot \tau}{1+r} \leq C_r(p, \tau)$$

for any trade vectors $\tau, \tilde{\tau} \in \mathbb{R}^L$, and any $p \geq 0$, $r \geq 0$.

Proof of Lemma 4. Continuity and homogeneity are evident, as is the second displayed inequality. Convexity in τ and homogeneity in τ guarantee the first inequality. To verify convexity in τ, write

$$C_r(p, \tau) = \frac{1}{1+r} \sum_{\ell \in L} p_\ell[\tau_\ell + r \max\{\tau_\ell, 0\}].$$

The ℓth term is convex in τ_ℓ, for each $\ell \in L$, hence C_r is convex in τ. Finally, the equality is straightforward from the definitions.   □

We are now ready to prove Theorem 4.

If $\bar{m} = 0$, the theorem follows from Case (a) of Corollary 3. So assume $\bar{m} > 0$. Note that this implies, by Theorem 2, that $r > 0$.

We show first that $\gamma(x) \leq r$. If $\gamma(x) > r$, then there exist gains-to-r-diminished-trade at x, i.e., there exist $(\tau^h)_{h \in H}$ such that $\sum_{h \in H} \tau^h = 0$, $x^h + \tau^h \in \mathbb{R}_+^L$ and $u^h(x^h + \tau^h(r)) > u^h(x^h)$ for $h \in H$. It follows that for some h, $p \cdot \tau^h \leq 0$, where p is the ME price vector. From Lemma 4 we must then have $C_r(p, \tau^h(r)) \leq 0$. Let $\bar{\tau}^h = x^h - e^h$ be the ME trade of household h. From Lemma 1, $C_r(p, \bar{\tau}) \leq m^h$. By Lemma 4,

$$C_r(p, \bar{\tau}^h + \tau^h(r)) \leq C_r(p, \bar{\tau}^h) + C_r(p, \tau^h(r)) \leq m^h,$$

hence by Lemma 1 again, $x^h + \tau^h(r) = e^h + \bar{\tau}^h + \tau^h(r) \in B_C(p, r, e^h, m^h)$. But this contradicts the optimization behavior of h in the ME.

To show that $\gamma(x) = r$ under the regularity condition, it is useful to concentrate on "active" households h, namely households that borrow money, sell goods, and buy other goods. By the regularity condition, no active household consumes zero units of any commodity he is positively endowed with.

Since all of the bank money $M > 0$ is borrowed at the ME (see condition (8)), there exists at least one household h_1 with borrowed money $\tilde{m}^{h_1} > 0$. Since the interest rate $r = \bar{m}/M > 0$, every household h spends all the money on hand (i.e., $\tilde{m}^h + m^h$) on purchases at the ME. Hence h_1 could not be hoarding money, and so he must be selling goods to repay the bank, i.e., h_1 is an active seller. Suppose h_1 buys ℓ_2 and sells ℓ_1. Since $r > 0$, h_1 is not indulging in "wash sales," i.e., buying and selling the same commodity. Then some other household h_2 must be selling ℓ_2, and in this case he too must be borrowing money from the bank (otherwise, why sell?) and spending it on purchases of some commodity other than ℓ_2.

Consider a directed graph with a node for each commodity; and arc $(\ell, k) \in L \times L$ if there exists an active household who buys k and sells ℓ.

We have shown that there exists at least one arc; and that if there is an incoming arc at any node, there must also be an outgoing arc at that node. Since L is finite, there is a cycle $(\ell_1, \ell_2), ..., (\ell_i, \ell_{i+1}), ..., (\ell_n, \ell_1)$ and active households $h_1, ..., h_n$ such that h_i buys ℓ_{i+1} and sells ℓ_i (with $\ell_{n+1} \equiv \ell_1$).

The regularity condition implies (as discussed in Section 5)

$$\frac{\nabla^{h_i}_{\ell_{i+1}}(x^{h_i})}{p_{\ell_{i+1}}} = \frac{(1+r)\nabla^{h_i}_{\ell_i}(x^{h_i})}{p_{\ell_i}}$$

for $i = 1, ..., n$. Consider any scalar $\tilde{\gamma} < r$. Then for small enough ε, household h_i would benefit by selling $\varepsilon p_{\ell_{i+1}}$ more units of commodity ℓ_i, and buying εp_{ℓ_i} more units of commodity ℓ_{i+1} but consuming only $\varepsilon p_{\ell_i}/(1+\tilde{\gamma})$ more units of ℓ_{i+1} (since $\tilde{\gamma} < r$, and at r he is indifferent).

So define trades accordingly on the cycle, i.e., let

$$(\tilde{t}^{h_i})_\ell = \begin{cases} -\varepsilon p_{\ell_{i+1}} & \text{if } \ell = \ell_i \\ \varepsilon p_{\ell_i} & \text{if } \ell = \ell_{i+1} \\ 0 & \text{otherwise} \end{cases}$$

for $i = 1, ..., n$. For small enough ε, the $(\tilde{t}^{h_i})_{i=1}^n$ constitute a feasible trade at $x = (x^h)_{h \in H}$ and we have

$$u^{h_i}(x^{h_i} + \tilde{t}^{h_i}(\tilde{\gamma})) > u^{h_i}(x^{h_i})$$

for $i = 1, ..., n$. This shows that there are gains-to-$\tilde{\gamma}$-diminished trades at x, i.e., $\gamma(x) > \tilde{\gamma}$ for all $\tilde{\gamma} < r$, proving Theorem 4. $\qquad\square$

Proof of Theorem 6. Suppose $\langle(p, r), (q^h, x^h)_{h \in H}\rangle$ is an ME. Then $r = \bar{m}/M > 0$. Consider a cycle $(\ell_1, \ell_2), ..., (\ell_n, \ell_1)$ in which household h_i buys commodity ℓ_{i+1} and sells commodity ℓ_i (with $\ell_{n+1} \equiv 1$). As shown in the proof of Theorem 4, such a cycle always exists. Then, letting $\tilde{\nabla}$ (or, $\bar{\nabla}$) denote left (or, right) hand derivative, we must have

$$\frac{\tilde{\nabla}^{h_i}_{\ell_{i+1}}(x^{h_i})}{p_{\ell_{i+1}}} \geq (1+r)\frac{\bar{\nabla}^{h_i}_{\ell_i}(x^{h_i})}{p_{\ell_i}}$$

for $i = 1, ..., n$; otherwise h_i would do better to reduce both his purchase of ℓ_{i+1} and the concomitant sale of ℓ_i (by a little). Hence

$$\frac{\tilde{\nabla}^{h_i}_{\ell_{i+1}}(x^{h_i})}{\bar{\nabla}^{h_i}_{\ell_i}(x^{h_i})} \geq (1+r)\frac{p_{\ell_{i+1}}}{p_{\ell_i}}$$

for $i = 1, ..., n$. Taking products of the left and right sides, we obtain

$$1 + \Gamma^*(e) \geq \left\{ (1+r)^n \bigtimes_{i=1}^{n} \frac{p_{\ell_i}}{p_{\ell_{i+1}}} \right\}^{1/n} \equiv (1+r)1$$

which implies

$$\Gamma^*(e) \geq r$$

a contradiction, since $r = \bar{m}/M$ by Theorem 2. This proves Theorem 6.   □

Proof of Theorem 7. Let $x \equiv (x^h)_{h \in H}$ denote the ME allocation. W.l.o.g., rescaling units of commodities if necessary, suppose $p_\ell = 1$ for all $\ell \in L$. (Recall that $\gamma(e)$ remains unaltered by rescaling.) Also, relabelling households and commodities if necessary, there exists (as shown in the proof of Theorem 3) a "cycle" of households such that household k sells commodity k and buys commodity $k+1$ for $k = 1, ..., n$ (where $n + 1 \equiv 1$). We can *reduce* trade on this cycle by a small $\delta > 0$ without affecting other households' trades. Define trade vectors T^k for $k = 1, ..., n$ by

$$T^k_\ell = \begin{cases} \delta & \text{if } \ell = k \\ -\delta & \text{if } \ell = k+1 \\ 0 & \text{otherwise} \end{cases}$$

Also define $\tilde{T}^k \equiv -T^k$ (see Fig. 11.) Notice that the bundle $\tilde{x}^k$ given by

$$\tilde{x}^k_\ell = \begin{cases} x^k_\ell + \delta & \text{if } \ell = k \\ x^k_\ell - \frac{\delta}{1+r} & \text{if } \ell = k+1 \\ x^k_\ell & \text{otherwise} \end{cases}$$

is feasible in the budget set of $k = 1, ..., n$; and also (for small enough δ)

$$\tilde{x}^k_k < e^k_k, \tilde{x}^k_{k+1} > e^k_{k+1} \tag{13}$$

for $k = 1, ..., n$, since no agent conducts wash sales at any ME.

By the strict concavity of u^k, the convexity of budget sets and the fact that x^k maximizes u^k on k's budget set, we get

$$u^k(\tilde{x}^k) < u^k(x^k) \equiv u^k(\tilde{x}^k + \tilde{T}^k(r))$$

for $k = 1, ..., n$; hence (since u^k is continuous), for some $\tilde{\gamma} > r$,

$$u^k(\tilde{x}^k) < u^k(\tilde{x}^k + \tilde{T}^k(\tilde{\gamma})) \tag{14}$$

for $k = 1, ..., n$.

Now we invoke separability of the u^h. Denote $u^k(x) \equiv \sum_{\ell=1}^{L} u^k_\ell(x_\ell)$. Then

$$u^k_{k+1}\left(e^k_{k+1} + \frac{\delta}{1+\tilde{\gamma}}\right) - u^k_{k+1}(e^k_{k+1}) \equiv \Delta_{k+1}$$

$$u^k_k(e^k_k) - u^k_k(e^k_k - \delta) \equiv \Delta_k$$

$$u^k_{k+1}\left(\tilde{x}^k_{k+1} + \frac{\delta}{1+\tilde{\gamma}}\right) - u^k_{k+1}(\tilde{x}^k_{k+1}) = \tilde{\Delta}_{k+1}$$

$$u^k_k(\tilde{x}^k_k) - u^k_k(\tilde{x}^k_k - \delta) = \tilde{\Delta}_k$$

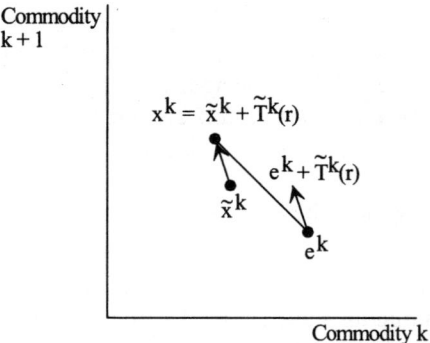

Figure 11. By separability and concavity, the increase in utility from e^k to $e^k + \tilde{T}^k(r)$ is at least as high as from $\tilde{x}^k$ to $\tilde{x}^k + \tilde{T}^k(r)$

By (13), and the concavity of the u_ℓ^h,

$$\Delta_{k+1} \geq \tilde{\Delta}_{k+1}$$
$$\Delta_k \leq \tilde{\Delta}_k$$

Now since u^k is separable,

$$u^k(\tilde{x}^k + \tilde{T}^k(\tilde{\gamma})) - u^k(\tilde{x}^k) = \tilde{\Delta}_{k+1} - \tilde{\Delta}_k$$

and

$$u^k(e^k + \tilde{T}^k(\tilde{\gamma})) - u^k(e^k) = \Delta_{k+1} - \Delta_k$$

So

$$u^k(e^k + \tilde{T}^k(\tilde{\gamma})) - u^k(e^k) \geq u^k(\tilde{x}^k + \tilde{T}^k(\tilde{\gamma})) - u^k(\tilde{x}^k)$$
$$> 0$$

for $k = 1, ..., n$ (the strict inequality coming from (14)).

But $\sum_{k=1}^n \tilde{T}^k = 0$ by construction, so we have shown that there are gains-to-$\tilde{\gamma}$-diminished trades at e. Since $\tilde{\gamma} > r$, this proves the lemma. $\square$

Proof of Remark 1. First assume that utilities are strictly concave. Let $\langle p, r, q, x \rangle$ be an ME of $(\mathcal{E}, m, M)$. Define $\tilde{x}^h$ and $\tilde{\tau}^h$ as in the proof of Lemma 5. Then, as shown in that proof, there exist gains-to-$\tilde{\gamma}$-diminished-trade at $\tilde{x} \equiv (\tilde{x}^h)_{h \in H}$ for some $\tilde{\gamma} > r$. Moreover, $\tilde{x}$ can be made arbitrarily close to x by choosing δ small. This proves that there exist y arbitrarily close to x with $\gamma(y) > r$, provided the u^h are strictly concave.

If u^h is not strictly concave, consider $\tilde{u}^h$ defined by $\tilde{u}^h(z) = u^h(z) + \varepsilon \sum_{\ell \in L} (z_\ell)^{1/2}$ and take limits.

Proof of Theorem 8. Suppose there exists a monetary equilibrium with interest rate r. Then, by Theorem 7, $\gamma(e) > r$; and by Theorem 1, $r = \bar{m}/M$. Thus $\gamma(e) > \bar{m}/M$. (This proves "only if"; "if" follows from Theorem 1.) $\square$

Proof of Theorem 9.   Since $r(n) \equiv \sum_{h \in H} m^h(n)/M(n)$ is bounded above as $n \to \infty$, all price ratios $p_\ell(n)/p_k(n)$ (for $\ell, k \in L$) are bounded as $n \to \infty$ (by the argument in Step 4 of the proof of Theorem 1). Suppose $p(n) \not\to \infty$. Then take a subsequence along which the price vectors converge to some limit p^*. A standard argument shows that p^* is an ME price of the limit economy. But the interest rate in the limit economy is $r = \lim_{n \to \infty} \sum_{h \in H} m^h(n)/M(n) = \gamma(e)$. However, by Theorem 8, $r < \gamma(e)$, a contradiction. $\square$

Proof of Theorem 10.   Define an ε-ME $\langle p(\varepsilon), r(\varepsilon), q(\varepsilon), x(\varepsilon) \rangle$ as before, with the external agent putting up ε at each market and with the government's market actions fixed according to Q. It exists as before. The interest rate $r(\varepsilon) \geq 0$, otherwise households would arbitrage as in Step 2 (of the proof of Theorem 2). Moreover the government prints at most $\sum_{\ell \in L} Q_{m\ell} + \sum_{h \in H} \Delta m^h$ before commodity trade at any ε-ME; and at most Q_{bm} after commodity trade. Thus the total money in the system, as also the interest rate $r(\varepsilon)$, is bounded above independent of ε. This implies that price-ratios $p_\ell(\varepsilon)/p_k(\varepsilon)$ are bounded as in Step 4. If $\sum_{h \in H}(m^h + \Delta m^h) > 0$, then Step 5 holds as before. Otherwise, $\bar{Q}_m > 0$. But, since the government spends at least $\bar{Q}_m/L > 0$ on some commodity, the price of some $\ell \in L$ is bounded from below by $\bar{Q}_m/(L(\sum_{h \in H} e_\ell^h + \varepsilon)) > \bar{Q}_m/(L(\sum_{h \in H} e_\ell^h + 1)) > 0$. Hence Step 5 also holds, bounding prices away from 0.

Let $\langle p, r, q, x \rangle \equiv \lim_{\varepsilon \to 0} \langle p(\varepsilon), r(\varepsilon), q(\varepsilon), x(\varepsilon) \rangle$ with possibly $p = \infty$. First we claim that $r \leq \max\{0, ([Q_{bm} + x(m, \pi)]/M) - 1\} = \rho(m, \pi)$. If $r = 0$, there is nothing to show. So suppose $r > 0$. Denote $\sum_{h \in H} q_{bm}^h \equiv \bar{q}_{bm}$ and $\sum_{h \in H} q_{bm}^h(\varepsilon) \equiv \bar{q}_{bm}(\varepsilon)$. The external agent borrows $\varepsilon/(1 + r(\varepsilon)) \leq \varepsilon$ units of bank money, since he sells only ε units of the bond and since $r(\varepsilon) \geq 0$. Hence households borrow $(\bar{q}_{bm}(\varepsilon)/(Q_{bm} + \bar{q}_{bm}(\varepsilon)))(M - \varepsilon/(1 + r(\varepsilon))) \to \bar{q}_{bm} M/(Q_{bm} + \bar{q}_{bm})$ units of bank money. Since $r(\varepsilon) \to r > 0$, households spend all the money at hand on commodity trade, i.e., they collectively spend $\delta(\bar{q}_{bm})M + m^h + \Delta m^h$ in the limit (where $\delta(x) \equiv x/(Q_{bm} + x)$). The external agent spends $L\varepsilon \to 0$, which we ignore. The government spends $\bar{Q}_m \equiv \sum_{\ell \in L} Q_{m\ell} \geq 0$. Thus (letting $\bar{m} \equiv \sum_{h \in H} m^h$ and $\Delta \bar{m} \equiv \sum_{h \in H} \Delta m^h$) total expenditure on commodity trade converges to $\delta(\bar{q}_{bm})M + \bar{m} + \Delta \bar{m} + \bar{Q}_m$. Of this, a fraction σ is taxed away by the government. Hence only the amount $(1 - \sigma)[\delta(\bar{q}_{bm})M + \bar{m} + \Delta \bar{m} + \bar{Q}_m] = g(\bar{q}_{bm})$ accrues to households *and* the external agent. Since households have spent all their cash, this is their only source of funds to repay the bank. From Figure 11, we know $\bar{q}_{bm} \leq g(\bar{q}_{bm})$ implies that $\bar{q}_{bm} \leq x(m, \pi)$. Hence

$$1 + r = \frac{Q_{bm} + \bar{q}_{bm}}{M} \leq \frac{Q_{bm} + x(m, \pi)}{M},$$

establishing the claim on r.

Now suppose $p(\varepsilon) \to \infty$. Then all commodity sales $\to 0$ since the money in the system is bounded above. Hence $x^h(\varepsilon) \to e^h$ for all $h \in H$, and public goods produced $F((Q_{m\ell}/p_\ell(\varepsilon))_{\ell \in L}) \to F(0) = 0$.

Let $\tilde{p} = \lim(p(\varepsilon)/\sum_{\ell \in L} p_\ell(\varepsilon))$. Since all price ratios stay bounded in $p(\varepsilon)$, $\tilde{p} \gg 0$.

Let $\hat{r} \equiv (\sigma + r)/(1 - \sigma)$. Notice that, in the limit, all trades τ such that $e^h + \tau \in \mathbb{R}_+^L$, and

$$\tilde{p} \cdot {}^*\tau = \frac{1}{1 + \hat{r}} \hat{p} \cdot {}_*\tau$$

are feasible for household h. As before, this implies that no trade is a Walras equilibrium for utilities $(v_{\hat{r}}^h)_{h \in H}$ (where $v_{\hat{r}}^h$ is defined using u_0^h), and hence there are *no* gains-to-$\hat{r}$-diminished-trade at e for the utilities $(u_0^h)_{h \in H}$, contradicting the gains-to-trade hypothesis that $\gamma_0(e) > [\sigma + \rho(m, \pi)]/(1 - \sigma)$, since $r \leq \rho(m, \pi)$ as we saw.

We conclude that $p(\varepsilon) \nrightarrow \infty$. It is now straightforward to verify that $\langle p, r, q, x \rangle$ is a bona-fide ME.

We still must show that $r = \rho(m, \pi)$ at any ME $\langle p, r, q, x \rangle$ of $(\mathcal{E}, m, \pi)$. Suppose there is no hoarding at the ME by households and that they spend all their borrowed and privately held money $\delta(\bar{q}_{bm})M + \bar{m} + \Delta\bar{m}$ on purchases of commodities, where $\bar{q}_{bm} \equiv \sum_{h \in H} q_{bm}^h$. Government spends $\bar{Q}_m$. Thus exactly $g(\bar{q}_{bm}) \equiv (1 - \sigma)(\delta(\bar{q}_{bm})M + \bar{m} + \Delta\bar{m} + \bar{Q}_m)$ is in the hands of households after trade. Since they will not default, or be left holding worthless cash in equilibrium, $g(\bar{q}_{bm}) = \bar{q}_{bm}$. This shows $r = \rho(m, \pi)$.

If there is hoarding by households at the ME, then $r = 0$ and spending is strictly less than $\bar{q}_{bm} = \delta(\bar{q}_{bm})M$. Households will clearly spend at least as much as their privately held money. (If they spend less, they could do so exclusively out of their private money, thus avoiding all debt, and freeing the rest of their private money for more purchases.) So let $\alpha\delta(\bar{q}_{bm})M$ be the fraction of borrowed money that households collectively spend where $0 \leq \alpha < 1$. We must have

$$(1 - \sigma)(\alpha\delta(\bar{q}_{bm})M + \bar{m} + \Delta\bar{m} + \bar{Q}_m) = \alpha\bar{q}_{bm},$$

since the hoarded money $(1 - \alpha)\delta(\bar{q}_{bm})M$ pays off an equal amount of bonds $(1 - \alpha)\bar{q}_{bm}$. This implies

$$(1 - \sigma)(\bar{m} + \Delta\bar{m} + \bar{Q}_m) = \sigma\alpha\bar{q}_{bm} < \sigma\bar{q}_{bm} = \sigma(M - Q_{bm}).$$

(The first equality holds since $\bar{q}_{bm} = \delta(\bar{q}_{bm})M$; the inequality holds since $\alpha < 1$; and the last equality holds since $\bar{q}_{bm} + Q_{bm} = M$ on account of $r = 0$.) We conclude that if

$$\bar{m} + \Delta\bar{m} + \bar{Q}_m \geq (\sigma/(1 - \sigma))(M - Q_{bm})$$

then there is no hoarding of money by households, and hence $r = \rho(m, \pi)$.

Suppose households hoard money. Then $r = 0$ and $q_{bm} + Q_{bm} = M$ and

$$\bar{m} + \Delta\bar{m} + \bar{Q}_m < (\sigma/(1 - \sigma))(M - Q_{bm}).$$

Put $z \equiv M - Q_{bm}$. Rewriting the above inequality,

$$(1 - \sigma)[\bar{m} + \Delta\bar{m} + \bar{Q}_m] < \sigma z$$
$$\Leftrightarrow (1 - \sigma)[z + \bar{m} + \Delta\bar{m} + \bar{Q}_m] < z$$
$$\Leftrightarrow g(z) < z$$
$$\Leftrightarrow x(m, \pi) < z = M - Q_{bm}.$$

(The second "⇔" follows since $zM/(Q_{bm} + z) = z$ for $z \equiv M - \bar{Q}_{bm}$; the third "⇔" follows from Figure 11 for g.) But then $\rho(m, \pi) = 0$ by definition (see (11) of Section 10.2) and again $r = \rho(m, M, \pi)$. $\qquad\qquad\qquad\qquad$ □

Proof of Corollary to Theorem 10. (This follows the last three implications in the Proof of Theorem 10.) Indeed

$$
\begin{aligned}
r = 0 \ &\Leftrightarrow\ x(m, \pi) \leq M - Q_{bm} \\
&\Leftrightarrow\ g(M - Q_{bm}) \leq M - Q_{bm} \\
&\Leftrightarrow\ (1 - \sigma)(M - Q_{bm} + \bar{m} + \Delta\bar{m} + \bar{Q}_m) \leq M - Q_{bm} \\
&\Leftrightarrow\ \bar{m} + \Delta\bar{m} + \bar{Q}_m \leq (\sigma/(1 - \sigma))(M - Q_{bm}).
\end{aligned}
$$

(The first "⇔" comes from Theorem 10; the second "⇔" from Figure 10 for g; the third "⇔" from substituting $M - Q_{bm}$ for x in $g(x)$.) $\qquad\qquad\qquad$ □

Proof of Theorem 11. We shall construct $\langle \tilde{p}, \tilde{r}, \tilde{q}, \tilde{x} \rangle$. First define $\tilde{r} \equiv (1 - \tilde{\sigma})r^* - \tilde{\sigma}$, where $r^* \equiv (r + \sigma)/(1 - \sigma)$. Then $\tilde{r} \geq 0$ since $\tilde{\sigma} \leq r^*/(r^* + 1)$. By the definition of $\tilde{r}$

$$
\frac{\tilde{r} + \tilde{\sigma}}{1 - \tilde{\sigma}} = \frac{r + \sigma}{1 - \sigma} \tag{i}
$$

Next denote $M_1 \equiv Q_{bm}/(1 + r)$, $M_2 \equiv \tilde{Q}_{bm}/(1 + \tilde{r})$ and observe

$$
\Lambda(\tilde{M}, \tilde{Q}_{bm}, \tilde{\sigma}) = (\tilde{M} - M_2)/(M - M_1) > 0.
$$

(We omit the straightforward algebra that checks this.) For brevity, denote $\Lambda(\tilde{M}, \tilde{Q}_{bm}, \tilde{\sigma}) \equiv \tilde{\lambda}$. Now define, for $h \in H$ and $\ell \in L$

$$
\Delta\tilde{m}^h = \tilde{\lambda}(m^h + \Delta m^h) - m^h \tag{ii}
$$

$$
\tilde{q}_{m\ell}^h = \tilde{\lambda}q_{m\ell}^h \tag{iii}
$$

$$
\tilde{q}_{\ell m}^h = q_{\ell m}^h \tag{iv}
$$

$$
\tilde{q}_{bm}^h = \tilde{\lambda}\frac{(1 + \tilde{r})}{(1 + r)}q_{bm}^h \tag{v}
$$

$$
\tilde{x}_{\ell m}^h = x_{\ell m}^h \tag{vi}
$$

$$
\tilde{Q}_{m\ell} = \tilde{\lambda}Q_{m\ell} \tag{vii}
$$

$$
\tilde{p}_\ell = \tilde{\lambda}p_\ell \tag{viii}
$$

Notice that $(\Delta\tilde{m}^h, \tilde{Q}_{m\ell}) = (\Delta m^h(\tilde{\lambda}), Q_{m\ell}(\tilde{\lambda}))$ as required.

We submit that $\langle \tilde{p}, \tilde{r}, \tilde{q}, \tilde{x} \rangle$ is an ME of $(\mathcal{E}, m, \tilde{M}, \tilde{Q}_{bm}, \Delta\tilde{m}, \tilde{Q}_{m\ell}, \tilde{\sigma})$. To verify this, notice (by (ii) and (viii)) that $m^h + \Delta\tilde{m}^h$ has the same purchasing power at $\tilde{p}$, as $m^h + \Delta m^h$ has at p. Moreover, by (i), the impediment to trade is the same for each household at the first scenario $\langle p, r, q, x \rangle$ and at the second scenario $\langle \tilde{p}, \tilde{r}, \tilde{q}, \tilde{x} \rangle$. Hence households face the same budget set in the two scenarios and their optimal consumptions are invariant.

All sales of commodities remain the same (by (iv)), and expenditures are scaled by $\tilde{\lambda}$ in the second scenario (by (iii) and (vii)), hence the definition of prices $\tilde{p}$

in (viii) shows that commodity markets clear in the second scenario (since they already cleared in the first scenario by assumption). Finally $\sum_{h \in H} \tilde{q}_{bm}^h = \tilde{\lambda}(1 + \tilde{r}) \sum_{h \in H} q_{bm}^h / (1 + r) = ((\tilde{M} - M_2)/(M - M_1))(1 + \tilde{r})(\sum_{h \in H} q_{bm}^h / (1 + r)) = ((\tilde{M} - M_2)/(M - M_1))(1 + \tilde{r})(M - M_1) = (\tilde{M} - M_2)(1 + \tilde{r})$. (The third equality holds because the bond market clears in the first scenario by assumption; i.e., households borrow $M - M_1$ with their bonds $\sum_{h \in H} q_{bm}^h$ at interest rate r, given that the government is borrowing $M_1 = Q_{bm}/(1+r)$.) Hence $\tilde{Q}_{bm} + \sum_{h \in H} \tilde{q}_{bm}^h = (1 + \tilde{r})M_2 + (1 + \tilde{r})(\tilde{M} - M_2) = (1 + \tilde{r})\tilde{M}$ showing that the bond market clears in the second scenario as well.

This proves that $\langle \tilde{p}, \tilde{r}, \tilde{q}, \tilde{x} \rangle$ is an ME of $(\mathcal{E}, m, \tilde{M}, \tilde{Q}_{bm}, \Delta\tilde{m}, \tilde{Q}_{m\ell}, \tilde{\sigma})$. But by (vii) and (viii), it is evident that $\tilde{Q}_{m\ell}/\tilde{p}_\ell = Q_{m\ell}/p_\ell$ for all $\ell \in L$. □

Proof of Corollary to Theorem 11. Fix $\tilde{M} = M$, and consider Figure 10. Let $D(M) = \{(\tilde{Q}_{bm}, \tilde{\sigma}) : (M, \tilde{Q}_{bm}, \tilde{\sigma}) \in D\}$, and consider the restriction of Λ to $D(M)$. The policy π_1 is clearly given by the point $(0, 0)$ in the restricted domain $D(M)$ of Λ. (If $\bar{Q}_m = 0$ then π_1 suffices for π_2 and π_3, so we assume $\bar{Q}_m > 0$.) Now fix $\tilde{\sigma} \in [0, r^*/(r^* + 1)]$, and go up the "$\tilde{\sigma}$-vertical line" in $D(M)$ consisting of points $(\tilde{Q}_{bm}, \tilde{\sigma})$, where $\tilde{Q}_{bm}$ increases in $[0, (r^* + 1)(1 - \tilde{\sigma})M]$. It is evident from our formulae that (on the $\tilde{\sigma}$-vertical line):

(i) the interest rate $\tilde{r} = r^*(1 - \tilde{\sigma}) - \tilde{\sigma}$ is constant (It falls linearly from r^* to 0 as we shift the vertical line, by raising $\tilde{\sigma}$ from 0 to $r^*/(r^* + 1)$.)

(ii) government expenditure falls linearly from $(M/B)\bar{Q}_m$ to 0

(iii) total expenditure falls linearly to 0

(iv) money borrowed by the government rises linearly from 0 to M (not hitting M since the points $((r^* + 1)(1 - \tilde{\sigma})M, \tilde{\sigma})$, on the sloping boundary of the trapezium, are excluded from $D(M)$).

The two straight lines, given by (ii) and (iv) for fixed $\tilde{\sigma}$, intersect when $\tilde{Q}_{bm} = Q_{bm}^*(\tilde{\sigma}) \equiv M\bar{Q}_m (r^* + 1)(1 - \tilde{\sigma})/(B + \bar{Q}_m)$; and therefore the policy $(Q_{bm}^*(\tilde{\sigma}), \tilde{\sigma}, \Lambda(Q_{bm}^*(\tilde{\sigma}), \tilde{\sigma}))$ has $\tilde{m}_\beta = 0$ for all $\tilde{\sigma} \in [0, r^*/(r^* + 1)]$. Let π_2 correspond to $\tilde{\sigma} = 0$. It is easy to check that it satisfies the requirements of part (b) of the Corollary.

Again consider the points $(\tilde{Q}_{bm}, \tilde{\sigma})$ on the $\tilde{\sigma}$-vertical line for fixed $\tilde{\sigma} \in (0, r^*/(r^* + 1))$. As $\tilde{Q}_{bm} \downarrow 0$, the total expenditure and tax revenue converge to positive numbers by (ii), (iii); but the amount borrowed from, and owed to, the bank converges to 0. Hence $\tilde{m}_\alpha < 0$ for small $\tilde{Q}_{bm}$. On the other hand, as $\tilde{Q}_{bm} \uparrow (r^* + 1)(1 - \tilde{\sigma})M$ the amount borrowed by the government converges to M, while expenditures and tax revenue converge to 0 (by (ii), (iii)). Since the interest rate is positive (by (i)), the only way for the government to meet its interest payment is to print money at the end. Thus $\tilde{m}_\alpha > 0$ as $\tilde{Q}_{bm} \uparrow (r^* + 1)(1 - \tilde{\sigma})M$. Since clearly $\tilde{m}_\alpha$ varies linearly on the $\tilde{\sigma}$-vertical line, there is a unique $\tilde{Q}_{bm} = Q_{bm}^{**}(\tilde{\sigma})$ at which $\tilde{m}_\alpha = 0$. Any of the policies $(Q_{bm}^{**}(\tilde{\sigma}), \tilde{\sigma}, \Lambda(Q_{bm}^{**}(\tilde{\sigma}), \tilde{\sigma}))$ for $\tilde{\sigma} \in (0, r^*/(r^* + 1))$ suffices for π_3.

The locus of points $(Q_{bm}^*(\tilde{\sigma}), \tilde{\sigma})$ and $(Q_{bm}^{**}(\tilde{\sigma}), \tilde{\sigma})$, for which $\tilde{m}_\beta = 0$ and $\tilde{m}_\alpha = 0$ respectively, form lines in $D(M)$. We shall show that $Q_{bm}^{**}(\tilde{\sigma}) < Q_{bm}^*(\tilde{\sigma})$ for $\tilde{\sigma}$ close to 0, and $Q_{bm}^{**}(\tilde{\sigma}) > Q_{bm}^*(\tilde{\sigma})$ for $\tilde{\sigma}$ close to $r^*/(r^* + 1)$, so that the two lines intersect yielding a totally balanced policy which is equivalent to the original policy.

To this end, recall that $\tilde{m}_\alpha(\tilde{Q}_{bm}, \tilde{\sigma})$ varies linearly in $\tilde{Q}_{bm}$ for fixed $\tilde{\sigma} \in (0, r^*/(r^*+1))$. Denote $\tilde{m}_\alpha^0(\tilde{\sigma}) \equiv \tilde{m}_\alpha(0, \tilde{\sigma})$, $\tilde{m}_\alpha^+(\tilde{\sigma}) \equiv \tilde{m}_\alpha((r^*+1)(1-\tilde{\sigma})M, \tilde{\sigma})$. For $\tilde{Q}_{bm} = 0$, the government finances its expenditure by printing $\tilde{m}_\beta > 0$ since it borrows nothing from the bank, hence $\tilde{m}_\alpha^0(\tilde{\sigma}) = -$tax revenue $\downarrow 0$ as $\tilde{\sigma} \downarrow 0$. But interest rate $\tilde{r} \uparrow r^*$ as $\tilde{\sigma} \downarrow 0$ by (i); and by (iii) and (iv) expenditures/tax revenue $\downarrow 0$, while money borrowed by the government $\uparrow M$, as $\tilde{Q}_{bm} \uparrow (r^*+1)(1-\tilde{\sigma})M$, so $\tilde{m}_\alpha^+(\tilde{\sigma})$ is approximately r^*M for small $\tilde{\sigma}$. Since $Q_{bm}^{**}(\tilde{\sigma})$ is defined as the zero point of the line joining $\tilde{m}_\alpha^0(\tilde{\sigma})$ and $\tilde{m}_\alpha^+(\tilde{\sigma})$, we see that $Q_{bm}^{**}(\tilde{\sigma}) \downarrow 0$ as $\tilde{\sigma} \downarrow 0$. But $Q_{bm}^*(\tilde{\sigma}) \to M\bar{Q}_m(r^*+1)/(B+\bar{Q}_m)$ as $\tilde{\sigma} \downarrow 0$. This proves $Q_{bm}^{**}(\tilde{\sigma}) < Q_{bm}^*(\tilde{\sigma})$ for $\tilde{\sigma}$ close to 0.

Finally as $\tilde{\sigma} \uparrow r^*/(r^*+1)$ and $\tilde{Q}_{bm} = 0$, by (i) tax revenue converges to $(r^*/(r^*+1)) \times$ expenditures $\geq (r^*/(r^*+1))(M/B)\bar{Q}_m \equiv K$ (say) for positive K, i.e., $\tilde{m}_\alpha^0(\tilde{\sigma}) \leq -K$ as $\tilde{\sigma} \uparrow r^*/(r^*+1)$. However $\tilde{r} \downarrow 0$ as $\tilde{\sigma} \uparrow r^*/(r^*+1)$ by (i). So, for $\tilde{Q}_{bm}$ close to its upper bound $(r^*+1)(1-\tilde{\sigma})M$ (where expenditures/tax revenue are nearly zero by (iii), and the government has borrowed almost all of M by (iv)), $\tilde{m}_\alpha$ is approximately $\tilde{r}M \to 0$ as $\tilde{\sigma} \uparrow r^*/(r^*+1)$. Thus $\tilde{m}_\alpha^+(\tilde{\sigma})$ is close to zero as $\tilde{\sigma} \uparrow r^*/(r^*+1)$. We conclude that $Q_{bm}^{**}(\tilde{\sigma}) \uparrow (r^*+1)(1-\tilde{\sigma})M$ as $\tilde{\sigma} \uparrow r^*/(r^*+1)$ and becomes bigger than $Q_{bm}^*(\tilde{\sigma}) = M\bar{Q}_m(r^*+1)(1-\tilde{\sigma})/(B+\bar{Q}_m)$. $\square$

References

1. Balasko, Y., Cass, D.: The structure of financial equilibrium with exogenous yields: the case of incomplete markets. Econometrica **57**, 135–162 (1989)
2. Balasko, Y., Shell, K.: Lump-sum taxation: the static case. CARESS discussion paper (1983)
3. Bewley, T.: The optimum quantity of money. In: Hahn, F. H., Brechling, F. P. R. (eds.) Models of monetary economies. Federal Reserve Bank of Minneapolis (1980)
4. Bewley, T.: A difficulty with the optimum quantity of money. Econometrica **51**, 1485–1504 (1983)
5. Cass, D.: On the 'number' of equilibrium allocations with incomplete financial markets. CARESS Working Paper 85–16, University of Pennsylvania (1985)
6. Clower, R.: A reconsideration of the microfoundations of monetary theory. Western Economic Journal **6**(1), 1–8 (1967)
7. Dubey, P., Geanakoplos, J.: The value of money in a finite horizon economy: a role for banks. In: Dasgupta, P., Gale, D., Hart, O., Maskin, E. (eds.) Economic analysis of markets: Essays in Honor of Frank Hahn, pp. 407–444. Cambridge, MA; MIT Press 1992
8. Dubey, P., Geanakoplos, J.: Credit cards and money. Mimeo, Yale University (2000)
9. Dubey, P., Geanakoplos, J.: Monetary equilibrium with missing markets. Mimeo, Yale University (2000)
10. Dubey, P., Geanakoplos, J.: Money and time. Mimeo, Yale University (2000)
11. Dubey, P., Geanakoplos, J.: A general equilibrium model of financial markets with infinite horizon. Mimeo, Yale University (2000)
12. Dubey, P., Geanakoplos, J.: Firms, incomplete markets and money. Mimeo, Yale University (2000)
13. Dubey, P., Geanakoplos, J.: Monetary equilibrium with general financial costs of transactions. Mimeo, Yale University (2000)
14. Dubey, P., Shapley, L. S.: Noncooperative general exchange with a continuum of traders: two models. Journal of Mathematical Economics **23**, 253–293 (1994)
15. Friedman, M.: The optimum quantity of money and other essays. Aldine Publishing Company 1969
16. Geanakoplos, J., Kubler, F.: Currencies, incomplete markets, and crises. Yale mimeo (1999)
17. Geanakoplos, J., Mas-Colell, A.: Real indeterminacy with financial assets. Journal of Economic Theory **47**(1), 22–38 (1985)

18. Geanakoplos, J., Tsomocos, D.: International finance in general equilibrium. Cowles Foundation Discussion Paper 1313 (2001); Research in Economics (forthcoming)
19. Grandmont, J.-M., Younes, Y.: On the role of money and the existence of a monetary equilibrium. Review of Economic Studies **39**, 355–372 (1972)
20. Grandmont, J.-M., Younes, Y.: On the efficiency of a monetary equilibrium. Review of Economic Studies **40**: 149–165 (1973)
21. Gurley, J. G., Shaw, E. S.: Money in a theory of finance. Washington, DC: The Brookings Institution 1960
22. Hahn, F. H.: On some problems of proving the existence of an equilibrium in a monetary economy. In: Hahn, F. H., Brechline, F. P. R. (eds.) The theory of interest rates. London: Macmillan 1965
23. Heller, W.: The holding of money balances in general equilibrium. Journal of Economic Theory **7**, 93–108 (1974)
24. Hool, R. B.: Money, expectations and the existence of a temporary equilibrium. Review of Economic Studies **43**, 439–445 (1976)
25. Karatzas, I., Shubik, M., Sudderth, W. D.: Construction of stationary Markov equilibria of a strategic market game. Mathematics of Operations Research **19**(4), 975–1006 (1994)
26. Lerner, A.: Money as a creature of the state. American Economic Review Special Proceedings **37** (Proceedings Suppl.), 312–317 (1947)
27. Levine, D. K.: Efficiency and the value of money. Review of Economic Studies **56**, 77–88 (1989)
28. Lucas, R. E., Jr.: Equilibrium in a pure currency economy. Economic Enquiry **18**, 203–220 (1980)
29. Lucas, R. E., Jr.: Liquidity and interest rates. Journal of Economic Theory **50**, 237–264 (1990)
30. Magill, M., Quinzii, M.: Real effects of money in general equilibrium. Journal of Mathematical Economics **21**, 302–342 (1992)
31. Samuelson, P.: An exact consumption loans model of interest with or without the social contrivance of money. Journal of Political Economy **LXVI**(6), 467–482 (1958)
32. Shubik, M., Wilson, C.: The optimal bankruptcy rule in a trading economy using fiat money. Zeitschrift für Nationalökonomie **37**, 3–4 (1977)

The economic effects of restrictions on government budget deficits: imperfect private credit markets[*]

Christian Ghiglino[1] and Karl Shell[2]

[1] Department of Economics, Queen Mary, University of London, Mile End Road, London E1 4NS, UK (e-mail: c.ghiglino@qmul.ac.uk)
[2] Department of Economics, 402 Uris Hall, Cornell University, Ithaca, NY 14853-7601, USA (e-mail: ks22@cornell.edu)

Received: August 28, 2001; revised version: March 25, 2002

Summary. The present paper is an extension of Ghiglino and Shell [7] to the case of imperfect consumer credit markets. We show that with constraints on individual credit and only anonymous (i.e., non-personalized) lump-sum taxes, strong (or "global") irrelevance of government budget deficits is not possible, and weak (or "local") irrelevance can hold only in very special situations. This is in sharp contrast to the result for perfect credit markets. With credit constraints and anonymous consumption taxes, weak irrelevance holds if the number of tax instruments is sufficiently large and at least one consumer's credit constraint is not binding. This is an extension of the result for perfect credit markets.

Keywords and Phrases: Balanced-budget amendment, Consumption taxes, Credit constraints, Government budget deficit irrelevance, Lump-sum taxes, Overlapping generations.

JEL Classification Numbers: D50, D90, E52, E60, H62, H63.

1 Introduction and summary

In Ghiglino and Shell [7], we analyzed the economic effects of constitutional or other restrictions on the government budget deficit. We assumed that private agents have access to perfect markets for borrowing and lending. We allowed for the possibility that the full range of personalized taxes might not be available to the government. Our leading case was that of anonymous taxation, in which the government must choose taxes to be the same for every member of a given generation. The main results of Ghiglino and Shell [7] are:

---

[*] We thank Todd Keister, Bruce Smith, and two referees for helpful comments.
Correspondence to: C. Ghiglino

(1) When lump-sum taxes are available, deficit restrictions are (strongly or "glob-
     ally") irrelevant in the sense that the set of allocations the government is able
     to achieve as competitive equilibria is independent of the deficit restrictions.
     This result holds even if taxes cannot be perfectly personalized or even if they
     must be anonymous [1]. The intuition behind this is that the government is able to
     avoid any effects of the deficit restrictions by replacing explicit borrowing with
     what is in effect a "social security scheme" in which individuals are taxed when
     young and promised offsetting transfers when old. Since taxes are lump-sum
     and consumer credit markets are perfect, only net lifetime taxation matters.
     Individuals offset their early tax bills by borrowing in the credit markets. Es-
     sentially, consumers do the explicit borrowing "on behalf" of the government.
(2) When only anonymous consumption taxes are available, the situation is more
     complicated. Strong (or global) irrelevance of deficit restrictions is impossible
     because of the constraint that before-tax and after-tax prices must be non-
     negative. Nonetheless, if there is a sufficient number of tax instruments, the
     government is able to offset local changes in the deficit restrictions while hold-
     ing constant (normalized) after-tax prices and varying before-tax prices so as to
     leave lifetime incomes unchanged. In particular, there is weak (or "local") irrel-
     evance of the deficit restrictions when the number of consumers per generation
     does not exceed the number of goods per period [2].

The assumption of perfect private borrowing and lending markets is very strong in
this context. In the real world, some consumers do face binding credit constraints or
other imperfections in the borrowing market. If government borrowing is restricted
but private borrowing is unconstrained, then the government can ease the effects of
its own borrowing restrictions by in effect "borrowing off the books" by increasing
the early-life taxes on some individuals while at least partially offsetting this by
increasing late-life subsidies to the same individuals. It is natural therefore to ask
how private credit constraints affect the government's ability to avoid restrictions
on its deficit.

In the present paper, we assume that the government *and* individuals face credit
restrictions. The restrictions on the government are from the constitution or other
law, or from international borrowing agreements. The reasons for private credit
constraints include imperfect collateral and other "moral hazards". The underlying
sources of private credit rationing will not be analyzed here. We simply assume
that there are (exogenously given) private credit constraints which possibly differ
across individuals.

Following Ghiglino and Shell [7], we employ a pure-exchange overlapping-
generations model with several consumers per generation and several commodities
per period. We allow for lump-sum taxes and consumption taxes. We also allow for
the fact that tax schedules cannot be made perfectly individual-specific. In this paper,
we focus for simplicity on the perfectly anonymous case in which each consumer
from the same generation faces the same tax situation. Finally, we also assume for
simplicity that each commodity can be taxed at its own rate. In Ghiglino and Shell

---

[1] See Ghiglino and Shell [7], Proposition 5.
[2] See Ghiglino and Shell [7], Proposition 12.

[7], the situation is somewhat more general: The number of commodity tax classes is less than or equal to the number of commodities, while the number of consumer tax classes is less than or equal to the number of consumers per generation.

We use the classic economic definitions [3] of *relevance* and *irrelevance* applied in this case to government-budget-deficit restrictions. The government-budget-deficit restriction is said to be *irrelevant* if the set of achievable equilibrium allocations is unaffected by the restriction. Otherwise, the restriction is said to be *relevant*. Of course, saying that the restriction is irrelevant is not saying that the restriction does not matter. If the restriction either directly or indirectly affects expectations in such a manner that it affects the selection of the equilibrium, then the restriction does matter.

As one would expect, irrelevance of deficit restrictions is less likely with individual credit constraints than in the case with perfect consumer credit markets. We show in the present paper:

(1) If private credit is constrained but these constraints are not binding on any individual and the only tax instruments are anonymous lump-sum taxes, then there is weak (local) irrelevance of the government budget restrictions. If some credit constraints are binding, then with only anonymous lump-sum taxes even local irrelevance is unlikely or impossible.

(2) If there are private credit constraints, the case with only anonymous consumption taxes is more interesting. Surprisingly, consumption taxes, although distorting, are more likely to provide (at least some form of) irrelevance than lump-sum taxes. With credit constraints and anonymous consumption taxes, there is weak (local) irrelevance if the number of tax instruments is sufficiently large and at least one consumer's credit constraint is not binding. This generalizes the result for the case with no private credit constraints. In particular, we show that if in all periods the number of commodities is no less than the sum of the number of individuals *plus* the number of individuals for whom credit rationing is binding, then the deficit restriction is weakly (locally) irrelevant [4]. In moving from the unconstrained credit case to the constrained case, the minimal number of tools for weak irrelevance is increased by the number of constrained consumers. To undo the effects of the local change in the deficit restriction, the government has to maintain not only the lifetime incomes of all consumers but also the first period incomes of the constrained consumers.

Why is weak (i.e. local) budget-deficit irrelevance more likely with anonymous consumption taxes than with anonymous lump-sum taxes? With only anonymous lump-sum taxation, if taxes must be increased on the young in order to reduce the deficit, constrained consumers will typically have to decrease their early-life consumptions. This means that the deficit restriction is relevant. With anonymous consumptions taxes, the government will be able to accommodate local changes in the deficit restriction if there are sufficiently many types of commodities to tax. This is because, by altering tax rates, the government is able to affect early-life incomes,

---

[3] See Barro [6]. See also Ghiglino and Shell [7] and the references therein.

[4] If none of the private credit constraints are binding, this inequality reduces (as it should) to the one in Proposition 12 of Ghiglino and Shell [7].

late-life incomes, and the rates of commodity substitution, while generating the necessary revenue.

We choose, because we consider it most natural, to work within the monetary, overlapping-generations framework. If we had adopted real taxes, none of our analysis would have been altered. We see no reason that our basic results would be affected by going to infinite-lived agents, but since agent heterogeneity is essential to our analysis, the infinite-lifetime model would have been more clumsy to work with.

2 The model

We employ a pure-exchange overlapping-generations model in which there are n different consumers per generation and ℓ perishable commodities per period. We suppose that consumers live for two periods. The government collects taxes, distributes transfers (negative taxes), and finances government consumption. We focus on two types of government instruments: lump-sum taxes and consumption taxes. We assume that lump-sum taxes and consumption tax rates must be the same for every member of a given generation, but that consumption taxes can vary freely over the l commodities. For the general case, see Ghiglino and Shell [7], which allows for more general consumer tax classes and more general commodity tax classes. We assume that government consumption of commodities is exogenously determined. It is denoted by the sequence $g = (g^1, \ldots, g^t, \ldots)$ with $g^t \in \mathbb{R}^l_+$ for $t = 1, 2, \ldots$. It is assumed that use of capital markets is constrained, viz. each individual faces exogenously given constraints on his borrowing.

Our set-up is based on the Samuelson [10] overlapping-generations model presented in Balasko and Shell [2,3,4], but new tax instruments and the individual credit constraints must be defined. As in Balasko and Shell [3], let $m^s_{th} \in \mathbb{R}$ be the lump-sum money transfer to consumer h of generation t in period s; if m^s_{th} is negative, then the consumer is paying a lump-sum tax. Following Ghiglino and Shell [7], we let $\tau^{si}_{th} \in \mathbb{R}$ be the present tax rate levied on consumer h of generation t on his consumption of commodity i in period s.

Let $x^s_{th} = (x^{s1}_{th}, \ldots, x^{si}_{th}, \ldots, x^{sl}_{th}) \in \mathbb{R}^\ell_{++}$ be the vector of consumption in period s by individual h of generation t and $\omega^s_{th} = (\omega^{s1}_{th}, \ldots, \omega^{si}_{th}, \ldots, w^{s\ell}_{th}) \in \mathbb{R}^\ell_{++}$ be the vector of commodity endowments in period s of individual h from generation t for $t = 0, 1, \ldots, s = 1, 2, \ldots$, and $h = 1, \ldots, n$. Let $m^s_t \in \mathbb{R}$ be the money transfer in period s to each consumer from generation t, and $\tau^s_t = (\tau^{s1}_t, \ldots, \tau^{si}_t, \ldots, \tau^{sl}_t) \in \mathbb{R}^\ell$ be the vector of anonymous consumption tax rates in period s for consumers from generation t. Consumers from generation 0 are alive in period 1, while consumers from generation t ($t = 1, 2, \ldots,$) are alive in periods t and $t + 1$. Hence it is convenient to define the following vectors:

$$x_{0h} = x^1_{0h} \in \mathbb{R}^\ell_{++}, \quad x_{th} = (x^t_{th}, x^{t+1}_{th}) \in \mathbb{R}^{2\ell}_{++},$$

$$\omega_{0h} = \omega^1_{0h} \in \mathbb{R}^\ell_{++}, \quad \omega_{th} = (\omega^t_{th}, \omega^{t+1}_{th}) \in \mathbb{R}^{2\ell}_{++},$$

$$m_0 = m^1_0 \in \mathbb{R}, \quad m_t = (m^t_t, m^{t+1}_t) \in \mathbb{R}^2,$$

and

$$\tau_0 = \tau_0^1 \in \mathbb{R}^\ell, \quad \tau_t = (\tau_t^t, \tau_t^{t+1}) \in \mathbb{R}^{2\ell}.$$

Let $p^s = (p^{s1}, \ldots, p^{si}, \ldots, p^{s\ell}) \in \mathbb{R}_{++}^\ell$ be the vector of present (before-tax) prices for commodities available in period s and let

$$q_t^s = (q_t^{s1}, \ldots, q_t^{si}, \ldots, q_t^{s\ell}) \in \mathbb{R}_{++}^\ell$$

be the present after-tax vector of commodity prices facing consumers of generation t in period s. Define the after-tax present price vectors facing consumers of generation $t = 0, 1, 2, \ldots$, by

$$q_0 = q_0^1 = p^1 + \tau_0 \in R_{++}^\ell \text{ for } t = 0$$

and

$$q_t = (q_t^t, q_t^{t+h}) = (p^t, p^{t+1}) + (\tau_t^t, \tau_t^{t+1}) \in R_{++}^{2\ell} \text{ for } t = 1, 2, \ldots . \quad (2.1)$$

Then define the following quantity and price sequences: $x = ((x_{0h})_{h=1}^{h=n}, \ldots, (x_{th})_{h=1}^{h=n}, \ldots)$, $\omega = ((\omega_{0h})_{h=1}^{h=n}, \ldots, (\omega_{th})_{h=1}^{h=n}, \ldots)$, $p = (p^1, \ldots, p^t, \ldots)$, $m = (m_0, \ldots, m_t, \ldots)$, $\tau = (\tau_0, \ldots, \tau_t, \ldots)$ and $q = (q_0, \ldots, q_t, \ldots)$.

We assume that the preferences of consumer h from generation t can be described by the utility function u_{th} defined over the consumption set of all strictly positive x_t's (i.e. $\mathbb{R}_{++}^\ell$ or $\mathbb{R}_{++}^{2\ell}$) with the properties:

(i) u_{th} is twice differentiable with strictly positive first-order derivatives and with corresponding negative definite Hessian

and

(ii) the closure of every indifference surface of u_{th} is in the consumption set (i.e. $\mathbb{R}_{++}^\ell$ or $R_{++}^{2\ell}$).

These rather standard assumptions simplify the comparative statics [5]. Note that we have also assumed that the endowment of the consumer lies in his consumption set, i.e. we have ω_{th} is in $\mathbb{R}_{++}^\ell$ or $\mathbb{R}_{++}^{2\ell}$.

Let $b_{th}^s \in \mathbb{R}_+$ be the maximum credit in money units available in period s to consumer h from generation t. The behavior of consumer h $(h = 1, 2, \ldots, n)$ from generation t $(t = 1, 2, \ldots)$ is then described by

$$\text{maximize } u_{th}(x_{th}^t, x_{th}^{t+1})$$

subject to

$$q_{th}^t \cdot x_{th}^t + x_{th}^{tm} = p^t \cdot \omega_{th}^t + m_{th}^t,$$

$$q_{th}^{t+1} \cdot x_{th}^{t+1} + x_{th}^{t+1,m} = p^{t+1} \cdot \omega_{th}^{t+1} + m_{th}^{t+1}, \quad (2.2)$$

$$x_{th}^{tm} \geq -b_{th}^t,$$

and

$$x_{th}^{tm} + x_{th}^{t+1,m} = 0,$$

[5] See Balasko [1]. See Balasko and Shell [2,3] for their application in overlapping generations models.

where $x_{th}^{sm} \in \mathbb{R}$ is the gross addition to money holding in period s by consumer h of generation t. The last equation in (2.2) is the requirement that the consumer's indebtedness be zero in his final period of life. The borrowing constraint is not binding on consumer th if in equilibrium $x_{th}^{tm} > -b_{th}^t$. The inequality in (2.2) is the credit constraint. We have implicitly assumed in writing (2.2) that the borrowing constraint of at least one consumer is not binding, so that we can use the usual no-arbitrage argument to establish that the present price of money is constant, i.e.,

$$p^{t,m} = p^{t+1,m} = p^m \in \mathbb{R}_+ \tag{2.3}$$

where $p^{s,m} \in \mathbb{R}_+$ is the present price of money in period $s = 1, 2, \cdots$. Assuming that the economy is in proper monetary equilibrium, we can set $p^m = 1$.[6]

The nominal (coupon) rate of interest on money is assumed without loss of generality to be zero.[7] Hence the only return on holding money is the capital gain relative to commodities. Condition (2.3) is thus that money appreciate in value relative to any commodity at the commodity rate of interest. For consumers for which the credit restriction is not binding, Condition (2.3) allows us to rewrite (2.2) somewhat as Balasko and Shell [3] to yield

$$\text{maximize } u_{th}(x_{th}^t, x_{th}^{t+1})$$

subject to

$$q_t^t \cdot x_{th}^t + q_t^{t+1} \cdot x_{th}^{t+1}$$

$$= p^t \cdot \omega_{th}^t + p^{t+1} \cdot \omega_{th}^{t+1} + m_t^t + m_t^{t+1}$$

\begin{flushright}(2.4)\end{flushright}

for $h = 1, 2, \ldots, n$ and $t = 1, 2, \ldots$, where by choice of numeraire we set $q_{01}^1 = 1$. The transfers $m_t = (m_t^t, m_t^{t+1}) \in \mathbb{R}^2$ affect the behavior of the consumer only through the lifetime transfer $\mu_t = m_t^t + m_t^{t+1} \in \mathbb{R}$.

It remains to describe the behavior of the older generation ($t = 0$) in period 1. Consumer $0h$ maximizes his utility subject to his one-period budget constraint:

$$\text{maximize } u_{0h}(x_{0h}^1)$$

subject to

$$q_0^1 \cdot x_{0h}^1 + x_{0h}^{1m} = p^1 \cdot \omega_{0h}^1 + m_0^1$$

$$x_{0h}^{1m} = 0,$$

\begin{flushright}(2.5)\end{flushright}

and

$$x_{0h}^1 \in \mathbb{R}_{++}^\ell.$$

---

[6] Strictly speaking, setting $p^m = 1$ is not without loss of generality. We know, however, that we can reconstruct the full set of perfect-foresight equilibria by using the absence-of-money-illusion property.

[7] This is because the super-neutrality of money.

3 Fiscal policy

We assume in this paper that the government has at its disposal either anonymous lump-sum taxation *or* anonymous consumption taxation. Thus, the government's fiscal policy is either the sequence of anonymous lump-sum transfers m or the sequence of the consumption tax rates τ.

Let d^t be the present commodity value (and also the dollar value) of the government budget deficit incurred in period t. Hence we have for the case of lump-sum taxation

$$d^t = p^t g^t + n \left(m^t_{t-1} + m^t_t \right)$$

for $t = 1, 2, \ldots$, where n is the number of consumers per generation. For the case of consumption taxes

$$d^t = p^t g^t - \sum_{h=1}^{n} \sum_{i=1}^{l} \left(\tau^{ti}_{t-1} x^{ti}_{t-1,h} + \tau^{ti}_t x^{ti}_{th} \right)$$

for $t = 1, 2, \ldots$ Let d denote the sequence $(d^1, \ldots, d^t, \ldots)$. Let δ^t be the present value (and money value) of the constitutionally imposed deficit restriction (assumed for convenience in the form of an equality) in period t. Let δ denote the sequence $(\delta^1, \ldots, \delta^t, \ldots)$. The budget deficit restriction is then

$$d = \delta.$$

According to the previous definition, the deficit is denominated in money, or equivalently in Arrow-Debreu units of accounts. Our results do not depend on this convention: They still hold if the deficit restrictions are expressed in real terms.

4 Equilibrium

We maintain throughout this paper some strong assumptions. We suppose perfect-foresight on the part of consumers and the government. We also suppose that the government is able to perfectly commit to its announced fiscal policy.

Next we define equilibrium in the economy with taxes.

Definition *Given the sequence of endowments ω, the feasible fiscal policy m or τ, the exogenous consumption g, the behavior of consumers described by the systems (2.2), (2.4) and (2.5), the numeraire choice yielding $p^{11} = 1$, the (further) monetary normalization yielding $p^m = 1$ and the deficit-restriction sequence δ, a constitutional competitive equilibrium is defined by a positive price sequence p and the allocation sequence x such that markets clear, so that we have*

$$g^t + \sum_{h=1}^{h=n} (x^t_{t-1,h} + x^t_{t,h}) = \sum_{h=1}^{h=n} (\omega^t_{t-1,h} + \omega^t_{t,h})$$

for $t = 1, 2, \ldots$, and the deficit restriction $d = \delta$ is satisfied.

From Balasko and Shell [2], one might expect that the existence of competitive equilibrium to be guaranteed in "nice" overlapping-generation models, but this does not extend to our Definition. There are three reasons that competitive equilibrium as defined above could fail to exist. The first reason is because we are seeking a *proper* monetary equilibrium, one for which the price of money is strictly positive. For a proper monetary equilibrium to exist the fiscal policy must be bonafide[8]. The second reason applies only to commodity taxation. It might not be possible to equilibrate supply and demand while maintaining the positivity of the two price sequences p and q. The third reason is that equilibrium may fail to exist because of excessive government consumption.

When the model is stationary, i.e., preferences, endowments, and government consumption are constant across generations, one is tempted to focus on equilibria in which allocations are constant across periods. We provide separate definitions of the steady state for the two tax regimes.

Definition L (Steady state with lump-sum taxes). *Let $p = (p^1, \cdots, p^t, \cdots) \in (\mathbb{R}^l_{++})^\infty$ be the equilibrium sequence of commodity prices when the fiscal policy is given by the sequence of lump-sum transfers $(m^1_0, m^1_1, m^2_1, \cdots, m^t_t, m^{t+1}_t, \cdots)$. These describe a steady-state equilibrium if there is a scalar $\beta \in \mathbb{R}_{++}$ such that*

$$p^t = \beta^{t-1}\mathbf{p}$$
$$m^t_t = \beta^{t-1}\mathbf{m}^1 \quad \text{and}$$
$$m^{t+1}_t = \beta^t\mathbf{m}^0$$

for $t = 1, 2, \cdots$, where $\mathbf{p} = p^1, \mathbf{m}^1 = m^1_1$ and $\mathbf{m}^0 = m^1_0$.

Definition C (Steady state with consumption-taxes). *Let $p = (p^1, \cdots, p^t, \cdots) \in (\mathbb{R}^l_{++})^\infty$ be an equilibrium vector of before-tax commodity prices when the fiscal policy is given by the sequence of consumption taxes $(\tau^1_0, \tau^1_1, \tau^2_1, \cdots, \tau^t_t, \tau^{t+1}_t, \cdots) \in (\mathbb{R}^l)^\infty$. These describe a steady-state equilibrium if there is a scalar $\beta \in \mathbb{R}_{++}$ such that*

$$p^t = \beta^{t-1}\mathbf{p}$$
$$\tau^t_t = \beta^{t-1}\tau^1 \quad \text{and}$$
$$\tau^{t+1}_t = \beta^t\tau^0$$

for $t = 1, 2, \cdots$, where $\mathbf{p} = p^1, \tau^1 = \tau^1_1$ and $\tau^0 = \tau^1_0$.

When focusing on steady states it makes sense to focus on budget deficits $d = (d^1, \cdots, d^t, \cdots)$ that are constant in current terms, so that we have for a scalar $\mathbf{d} \in \mathbb{R}$

$$d^t = \beta^{t-1}\mathbf{d}$$

for $t = 1, 2, \ldots$ From Walras's law and market clearing, we have the two steady-state relations:

$$(\beta - 1)\left[\sum_{h=1}^n \left(\mathbf{p}\left(x^1_h - \omega^1_h\right)\right) - n\mathbf{m}^1\right] + \mathbf{d} = 0 \qquad \text{(SS-L)}$$

---

[8] See Balasko and Shell [3,4,5] and Ghiglino and Shell [7].

for lump-sum taxation, and

$$(\beta - 1) \sum_{h=1}^{n} \left[\mathbf{p} \left(x_h^1 - \omega_h^1 \right) + \sum_{i=1}^{l} \tau^{1i} x_h^{1i} \right] + \mathbf{d} = 0 \qquad \text{(SS-C)}$$

for consumption taxation, where $\tau^{1i} \in \mathbb{R}$ is the ith component of τ^1. The steady-state conditions (SS-L) and (SS-C) do not directly involve government consumption, but steady-state $\mathbf{g} = g^t$ for $t = 1, 2, \cdots$ is implied through the equilibrium allocations and prices. If $d = 0$, from (SS-L) and (SS-C), we have the familiar OG steady-state result that either the interest rate is zero ($\beta = 1$) or aggregate savings is zero. Our aim is to find conditions under which the government is able to "avoid" the restrictions on its deficit with changing neither its own consumption nor the consumption of any private consumer. When this is possible the deficit restriction is said to be *irrelevant*. We recall the formal definitions given in Ghiglino and Shell [7].

Definition (Irrelevance of the deficit restriction). *Let g be government consumption sequences and let x be an allocation that can be implemented as a competitive equilibrium with some feasible fiscal policy m (resp. τ) and with the resulting deficits given by the sequence d. The deficit restriction $d = \delta$ is said to be irrelevant if for any other deficit restriction sequence δ' there exists a feasible fiscal policy m (resp. τ) that implements the allocation x as a competitive equilibrium and is compatible with g, but with the resulting deficit given by the sequence δ'.*

The above notion of irrelevance is very strong because it involves any possible deficit sequence other than the pre-reform, or baseline, deficit d. In many situations, this type of irrelevance does not obtain. Following Ghiglino and Shell [7], we employ a weaker notion of irrelevance. The first characteristic of the weaker deficit restriction is that it is based on finite but arbitrarily long time horizons. Consider the time horizon $T = 1, 2, \ldots$. Then define $\delta(T) = (\delta^1, \delta^2, \ldots, \delta^t, \ldots, \delta^T) \in \mathbb{R}^T$ as a deficit restriction of (finite) length T. For a competitive equilibrium to be weakly constitutionally feasible the deficit in period t, d^t, must be equal to δ^t if $t = 1, 2, \ldots, T$, while for $t > T$, the deficit is unrestricted. The second characteristic of the weaker deficit restriction is that only restrictions "near" the base-line deficit are considered, i.e., only period-by-period deficits that are not too different from the baseline deficits are considered. In other words, only a neighborhood (in any topology, since T is finite) of the original sequence is considered. According to the weaker notion of irrelevance only restrictions of finite length, $\delta(T)$, that belong to a first T-period neighborhood of the base deficit vector $d = (d^1, d^2, \ldots, d^t, \ldots, d^T, \ldots)$, denoted $\mathcal{D}^T(d)$, are considered.

Definition (Weak irrelevance of the deficit restriction). *Let g be the sequence of government consumptions and let x be an allocation that can be implemented as a competitive equilibrium with some feasible fiscal policy m (resp. τ) and with the resulting deficits given by the sequence d. A deficit restriction is said to be weakly irrelevant if for any positive integer T there is a set $\mathcal{D}^T(d)$ such that for all $\delta \in \mathcal{D}^T(d)$ there is a fiscal policy m' (resp. τ') that implements the allocations x and g, but with the resulting deficit given by the sequence δ.*

Note that the time horizon of the deficit specification is arbitrary: For every T (no matter how large) there must be a neighborhood $\mathcal{D}^T(d)$. These neighborhoods will typically depend on T. The limit of these neighborhood as T goes to infinity might be empty.

5 Relevance of government budget deficit restrictions with lump-sum taxes

The relevance of government deficit restrictions is first investigated in economies in which some consumers face credit constraints and only lump-sum taxation is available. It is shown that deficit restrictions are likely to be relevant unless the government can use non-anonymous taxes. In the leading example considered below, the government uses only anonymous lump-sum taxes and transfers and thus has no way to treat differently the consumers, so that the likely outcome is that some consumers are hurt or benefited by the change to a fiscal scheme that respects the government deficit restrictions.

For overlapping-generations economies with perfect borrowing markets and lump-sum taxes and transfers, restrictions on the government budget have no impact on the set of equilibrium allocations (see Ghiglino and Shell [7], Proposition 5). The reason for this is that in these economies only the present value of taxes and transfers, not their timing, matters to consumers. In this case, the government can "borrow off the books" from taxpayers by adjusting the timing of individual taxes and transfers.

When credit restrictions are included, weak (or local) irrelevance of the budget deficit restriction obtains if non-anonymous taxes can be personalized to some consumer whose constraint is not binding. In this case, the government respects the market constraints using personalized taxes. Being able to personalize taxes is not always possible. If this is not possible, then matters dramatically change. This is illustrated in the following example. For simplicity, a stationary equilibrium is considered. This amounts to ignoring the transition path. In other words, in this example we will assume that a suitable money transfer is made so that the economy "starts" at the steady state and only deficit specifications from $t = 2$ onward are considered.[9]

Example (Relevance of government deficit restrictions when consumer credit is constrained). Let the economy be stationary with one commodity per period and two consumers. Perfectly anonymous lump-sum taxation is available. No other tax instruments are available. The two consumers, 1 and 2, have log-linear utility functions

$$u_{th} = 1/2 \log x_{th}^t + 1/2 \log x_{th}^{t+1}$$

for $h = 1, 2$ and $t = 1, 2, \ldots$. Endowments are given by

---

[9] Another, equivalent, way to view steady states is to consider a model with no beginning as well as no end (see Ghiglino and Tvede [8]).

$$\omega_{t1} = (\omega_{t1}^t, \omega_{t1}^{t+1}) = (1, 20),$$

and

$$\omega_{t2} = (\omega_{t2}^t, \omega_{t2}^{t+1}) = (0.75, 1).$$

Consider generation t. Suppose that the borrowing of consumer 1 is unconstrained, $b_{t1}^t = \infty$, but that consumer 2 cannot borrow, $b_{t2}^t = 0$. At a steady state the individual demands of the consumers depend on the interest factor β. When the credit constraint is not binding, the demands must satisfy

$$x_{t1}^t = \frac{1 + 20\beta}{2},$$

$$x_{t1}^{t+1} = \frac{1 + 20\beta}{2\beta},$$

$$x_{t2}^t = \frac{0.75 + \beta}{2},$$

and

$$x_{t2}^{t+1} = \frac{0.75 + \beta}{2\beta}.$$

For $\beta \geq 0.75$, we have

$$\left(x_{t2}^t, x_{t2}^{t+1}\right) = (0.75, 1)$$

because then the credit restriction is binding.

Suppose that $g^t = 1$ for $t = 1, 2, \ldots$ Without credit restrictions $\beta = 0.89498$ and $\beta = 0.093112$ solve the equilibrium equations, but $\beta = 0.89498$ is not an equilibrium interest factor because the borrowing constraint for consumer 2 is violated. The steady-state equilibrium interest factor is $\beta = 0.89408$ and the corresponding equilibrium allocations are

$$x_{t1} = (x_{t1}^t, x_{t1}^{t+1}) = (9.4408, 10.5592)$$

and

$$x_{t2} = (x_{t2}^t, x_{t2}^{t+1}) = (0.75, 1).$$

Since the government is not taxing any consumer, the associated deficit is $d^t = p^t g^t = p^t = (0.89408)^{t-1}$ in present real or money units.

Suppose now that the government is required to balance its budget in every period, so that $d^t = \delta^t = 0$ for $t = 2, 3, \ldots$. We will show that the new restriction on the deficits leads to a modification of the existing allocation. First, note that in order to keep unchanged the consumption of consumer 1, β should be unchanged at

$\beta = 0.89408$. Now, the government can either tax the young or tax the old. Suppose first that consumers are taxed in their youth and receive transfers in their old age. The procedure is similar to that used in the proof of Proposition 5 in Ghiglino and Shell [7]. Since the consumers of the same generation are perfectly anonymous for tax purposes, suppose that we tax each young equally with a lump-sum tax $-m_2^2 > 0$ and no tax on the consumers born in the first period, $m_1^2 = 0$. The government budget constraint is then $\beta^{t-1} g^t + 2m_2^2 = 0$ so that $m_{21}^2 = m_{22}^2 = -\beta/2$ (or $1/2$ in current terms) in order that the deficit be zero, $d^2 = 0$. Note in the next period, these same consumers have to be compensated by a positive transfer of $\beta/2$ in present terms, or $1/(2\beta)$ in current terms. After the transfer, the endowments (in current terms) of consumer 2 are $(0.75 - 0.5, 1 + 0.5\beta)$. At $\beta = 0.89408$, consumer 2 would still like to borrow. However, due to the borrowing constraint his first period consumption is now 0.25. The equilibrium allocation has been affected by the fiscal policy. The other possibility is to subsidize the young and tax the old. A similar reasoning shows that also in this case the fiscal policy affects the equilibrium allocation. Therefore, the deficit sequence is relevant.   □

The previous example suggests that when some consumers are credit-constrained, anonymous lump-sum taxes are not powerful enough to achieve irrelevance of the government budget deficit. This is generalized in the following

Proposition (Relevance of the government deficit restrictions with credit constraints). *Let the allocation x be implemented as a constitutional competitive equilibrium with a fiscal policy consisting only of lump-sum taxes and transfers compatible with the deficit restriction δ. If at least one consumer's credit constraint is binding then the deficit sequence δ is weakly (and strongly) relevant. Otherwise, it is weakly irrelevant.*

Proof. If no consumer's credit constraint is binding, then Proposition 5 in Ghiglino and Shell [7] applies. However, in general since the government is employing only anonymous taxation, any transfer changes the consumer's actual borrowings (or savings) and therefore affects his demand for commodities. Indeed, assume there are two consumers and that $h = 2$ is the consumer whose credit constraint is binding (if the credit constraint of consumer 1 is also binding then the deficit restriction is obviously relevant). Consider consumer 2 first. His demands are the solutions to the problem

$$\text{maximize } u_{t2}(x_{t2}^t, x_{t2}^{t+1})$$

subject to

$$
\begin{aligned}
p^t \cdot x_{t2}^t + x_{t2}^{tm} &= p^t \cdot \omega_{t2}^t + m_t^t, \\
p^{t+1} \cdot x_{t2}^{t+1} + x_{t2}^{t+1,m} &= p^{t+1} \cdot \omega_{t2}^{t+1} + m_t^{t+1}, \\
x_{t2}^{tm} &= -b_{t2}^t,
\end{aligned}
\tag{5.1}
$$

and

$$x_{t2}^{tm} + x_{t2}^{t+1,m} = 0.$$

The sum of the transfers made to the two consumers of type 2 in period t are

$$m_{t-1}^t + m_t^t = p^t \cdot (x_{t2}^t - \omega_{t2}^t) + p^t \cdot (x_{t-1,2}^t - \omega_{t-1,2}^t) - b_{t2}^t + b_{t-1,2}^t$$

On the other hand, because transfers are anonymous the government budget deficit is

$$d^t = 2(m_t^t + m_{t-1}^t) + p^{t,1} g^t.$$

Then we obtain

$$\delta^t = d^t = 2p^t \cdot (x_{t2}^t - \omega_{t2}^t + x_{t-1,2}^t - \omega_{t-1,2}^t) - 2b_{t2}^t + 2b_{t-1,2}^t + p^{t,1} g^t.$$

However, in order to keep consumer 1 unaffected by the fiscal policy, $p^t/p^{t-1,1}$ should be unchanged for all t. Since $p^{1,1} = 1$, the entire sequence of prices is predetermined as is the deficit sequence δ.   ☐

Remark. If the government were able to use personalized lump-sum taxes, then the deficit sequence δ would be *weakly irrelevant*. To show this, renumber the consumers so that consumer 1 is the unconstrained consumer and reproduce the proof of Proposition 5 in Ghiglino and Shell [7].

6 Restoring irrelevance with consumption taxes

In this section we assume that only anonymous taxes on consumption are available. The question is then whether the government is able to "avoid" the deficit restriction with these instruments even though some consumers are credit constrained. As in the case with unconstrained borrowing and lending, the answer depends on the number of tax instruments compared to the number of goals (consumers) and on the duration (in periods) of the restriction. We start with an example.

Example (Irrelevance of deficit restrictions in an economy with several tax instruments). Consider a stationary, overlapping-generations economy with four commodities per period ($\ell = 4$) and two consumers per generation ($n = 2$). Assume that the second consumer faces credit restrictions while the other has free access to the credit market. The government has a constant consumption of (only) the first good, $g^t = (g^{t1}, g^{t2}, g^{t3}, g^{t4}) = (3, 0, 0, 0)$. Preferences and endowments of consumer th are given by:

$$u_{th}(x_{th}^t, x_{th}^{t+1}) = \sum_{k=1}^{4} \alpha_{kh} \log x_{th}^{tk} + \sum_{k=1}^{4} \beta_{kh} \log x_{th}^{t+1k}$$

and

$$\omega_{th} = (\omega_h^{0k}, \omega_h^{1k})_{k=1}^4,$$

where

$$(\alpha_{kh})_{h=1,2,}^{k=1,\ldots,4}=\begin{bmatrix}1/8 & 5/8 & 1/8 & 1/8 \\ 1/8 & 4/7 & 1/7 & 9/56\end{bmatrix}, (\beta_{kh})_{h=1,2}^{k=1,\ldots,4}=\begin{bmatrix}1/4 & 1/4 & 1/5 & 6/20 \\ 1/5 & 1/4 & 1/4 & 6/20\end{bmatrix}$$

and

$$w_1^{01} = 300, w_1^{13} = w_1^{14} = w_2^{13} = 200, w_2^{14} = 230, w_2^{12} = 120, w_2^{01} = 250,$$
$$w_1^{12} = 500, w_2^{02} = 1000, \text{ and } w_i^{jk} = 100 \text{ for all other } h, j, k.$$

With anonymous consumption taxes, we have

$$\tau_{t-1,1}^{tk} = \tau_{t-1,2}^{tk} = \tau_{t-1}^{tk} \text{ and } \tau_{t1}^{tk} = \tau_{t2}^{tk} = \tau_t^{tk} \text{ for } k = 1, .., 4 \text{ and } t = 1, 2, \ldots .$$

For convenience, we look at steady state competitive equilibrium. As noted earlier, we will only consider the periods from $t = 2$ onward. First, we assume that the government finances its consumption by running a deficit, i.e. we look at a steady state with $\tau_t^{tk} = 0$ and $\tau_{t-1}^{tk} = 0$. By restricting our attention to prices of the form $p^{tk} = (\beta)^{t-1}p^k$, $k = 1, \ldots, 4$, it can be shown that the following set of allocations and prices represents a steady state

$$\mathbf{p}^2 = 1.06975, \mathbf{p}^3 = 1.30350, \mathbf{p}^4 = 1.51932, \beta = 1.03351$$

and

$$(x_1^{0k})_{k=1}^4 = (120.5557, 563.4745, 92.4864, 79.3485),$$
$$(x_1^{1k})_{k=1}^4 = (233.2935, 218.0816, 143.1801, 184.2615),$$
$$(x_2^{0k})_{k=1}^4 = (154.2905, 659.3371, 135.2762, 130.5673),$$
$$(x_2^{1k})_{k=1}^4 = (238.8603, 279.1068, 229.0573, 235.8228).$$

The associated deficit is $p^{t1} = 3(\beta)^{t-1}\mathbf{p}^1 = 3(\beta)^{t-1} = 3(1.03351)^{t-1}$ in present real (and money) terms. In current units, the savings are -275.1889 for consumer 1, 367.7109 for consumer 2 producing aggregate savings of 92.5221.

The issue is whether $(\tau_{t-1}^t, \tau_t^t)_{k=1}^4$ can be used in order to meet the deficit requirement $\delta^t = 0$ in period t without disturbing these allocations. Such a tax scheme must at least satisfy for each t $(t = 2, 3, \ldots)$ the following equations

$$x_{t-1,1}^{tk} = (1 - \alpha_1)\frac{\beta_{k1}W_{t-11}}{p^{tk} + \tau_{t-1}^{tk}} = x_1^{0k},$$

$$x_{t1}^{tk} = \alpha_1\frac{\alpha_{k1}W_{t1}}{p^{tk} + \tau_t^{tk}} = x_1^{1k},$$

$$(6.1)$$

$$x_{t-1,2}^{tk} = (1 - \alpha_2)\frac{\beta_{k2}W_{t-12}}{p^{tk} + \tau_{t-1}^{tk}} = x_2^{0k},$$

and

$$x_{t2}^{tk} = \alpha_2\frac{\alpha_{k2}W_{t2}}{p^{tk} + \tau_t^{tk}} = x_2^{1k},$$

where

$$W_{th} = \sum_{k=1}^{4} p^{tk} \omega_h^{0k} + \sum_{k=1}^{4} p^{t+1k} \omega_h^{1k}$$

and

$$\sum_{k=1}^{4}(x_1^{1k} + x_2^{1k})\tau_{t-1}^{tk} + \sum_{k=1}^{4}(x_1^{0k} + x_2^{0k})\tau_t^{tk} + 3p^{t1} = 0.$$

A natural candidate for a solution to the first four equations of (6.1) is of the form $p^{tk} = (\beta)^{t-1}\mathbf{p}^k$, $\tau_t^{tk} = (\beta)^{t-1}\tau^{0k}$, and $\tau_{t-1}^{tk} = (\beta)^{t-1}\tau^{1k}$, where $\beta \in \mathbb{R}$ is the interest factor, $\tau^{0k} \in \mathbb{R}$ is the present and nominal value of the tax rate on the young and $\tau^{1k} \in \mathbb{R}$ is the present and nominal value of the tax rate on the old.

In this example, consumer 2 is assumed to face a binding credit constraint while the government cannot personalize his taxes. Then it is required that his saving or borrowing should remain exactly as it was in the untaxed situation. The budget equation for this consumer when young is

$$\sum_{k=1}^{4}(p_t^k + \tau_{t-1}^{tk})x_2^{0k} - x_{t2}^m = \sum_{k=1}^{4} p_t^k \omega_2^{0k}.$$

In the initial situation with no taxes, this yields at the steady state

$$\sum_{k=1}^{4}(\beta^{t-1}\mathbf{p}^k x_2^{0k}) - x_{t2}^m = \sum_{k=1}^{4} \beta^{t-1}\mathbf{p}^k \omega_2^{0k},$$

which can be rewritten as

$$\beta^{t-1}\left(\sum_{k=1}^{4} \mathbf{p}^k(x_2^{0k} - \omega_2^{0k})\right) = x_{t2}^m$$

or as

$$x_{t2}^m = \beta^{t-1}x_2^m.$$

We should point out that in the absence of taxes, a strictly positive government consumption is financed through a permanent deficit (see Equation (SS-C)) implying that both aggregate and individual savings are non-zero and β is different from unity.

Since we assume that consumer 2's savings are not affected by the fiscal policy, if $\widehat{\beta}$ is the value of the interest factor obtained with taxes we should have

$$\sum_{k=1}^{4}(\widehat{\beta}^{t-1}\mathbf{p}^k + \widehat{\beta}^{t-1}\tau^{0k})x_2^{0k} - \widehat{\beta}^{t-1}\mathbf{p}^k\omega_2^{0k} = \beta^{t-1}x_2^m \text{ for } t = 1, 2, \dots .$$

These equations imply that $\widehat{\beta}$ should remain unaffected by the introduction of taxes, $\widehat{\beta} = \beta$. By replacing the demand functions, savings are unaltered for consumer 2 if

$$\frac{1}{\beta^{t-1}}\left(\alpha_2 \sum_{k=1}^4 \alpha_{k2}W_{t2} - \sum_{k=1}^4 \beta^{t-1}\mathbf{p}^k\omega_2^{0k}\right) = \alpha_2 \sum_{k=1}^4 (\mathbf{p}^k\omega_2^{0k} + \beta\mathbf{p}^k\omega_2^{1k})$$

$$-\sum_{k=1}^4 \mathbf{p}^k\omega_2^{0k} = x_2^m \qquad \text{(CS)}$$

In this example the total system is composed of 11 equations; 7 normalized prices, 2 normalized incomes, the government budget deficit equation and Equation (CS) concerning the individual borrowing/saving constraint. On the other hand, there are 3 prices (β is fixed) and 8 taxes. Hence a solution to the set of equations is likely to exist. Indeed, the following set of values represents the steady state equilibrium

$$\mathbf{p}^2 = 1.2399, \mathbf{p}^3 = 1.24845, \mathbf{p}^4 = 2.64413$$

and

$$(t_1^{0k})_{k=1}^4 = (t_1^{1k})_{k=1}^4 = (0.22449, 0.07003, 0.3477, -0.7837).$$

Finally, saving is now -367.7109 for consumer 1 and 367.7109 for consumer 2, so that aggregate savings are zero. This last result agrees with Equation (SS-C) as in the new situation a balanced budget ($\delta = 0$) co-exists with β different from unity, implying that the aggregate savings are zero. $\qquad \square$

The previous example illustrates the mechanism for irrelevance. Starting from a steady state with non-zero aggregate savings, there is a tax scheme which keeps the interest rate unchanged while achieving a zero government budget deficit. In the new situation, the government is taxing the consumers just enough to balance its budget. Moreover, this tax-transfer is such that aggregate, after tax, savings become zero. In this game, the change in the aggregate savings is completely done through the unrestricted consumer. A further important fact is that the taxes required to obtain irrelevance are age-anonymous, i.e. the tax on the young is the same as the tax on the old.

The role played by the consumer whose credit constraint is not binding is crucial. Indeed, if all consumers would be kept at their initial levels of borrowing or saving, achieving irrelevance would be impossible. This is clearly seen by considering an economy consisting of only one consumer.

The budget constraint for the generation $t-1$ consumer during his old age can be rearranged as

$$\tau_{t-1}^t x_{t-1}^t = -p_t x_{t-1}^t + p_t \omega_{t-1}^t - x_{t-1}^{tm},$$

while for the consumer in generation t, his first period constraint is

$$\tau_t^t x_t^t = -p_t x_t^t + p_t \omega_t^t - x_t^{tm}.$$

Adding these two, we obtain

$$\tau_t^t x_t^t + \tau_{t-1}^t x_{t-1}^t = p_t \left[\omega_t^t + \omega_{t-1}^t - x_t^t - x_{t-1}^t\right] - x_{t-1}^{tm} - x_t^{tm}.$$

Considering now the government budget deficit equation

$$
\begin{aligned}
d^t &= p^t g^t - \left[\tau_t^t x_t^t + \tau_{t-1}^t x_{t-1}^t \right] \\
&= p^t \left[g^t + x_t^t + x_{t-1}^t - \omega_t^t - \omega_{t-1}^t \right] + x_{t-1}^{tm} + x_t^{tm} \\
&= x_{t-1}^{tm} + x_t^{tm},
\end{aligned}
$$

we see that the deficit is completely determined by the aggregate borrowing and saving decisions of the consumers. If these are unaffected by the fiscal policy, the deficit will remain unchanged.

The example above shows that when consumption tax instruments are sufficiently diversified, irrelevance of deficit restrictions may hold. However, in general as was the case with no credit restrictions (see Ghiglino and Shell [7], Example 11), having several tax instruments is not sufficient for irrelevance, which also depends on the length and the magnitude of the deficit restriction. Indeed, even when there are enough instruments, it is not assured that $q_{th}^{sk} - \tau_{th}^{sk}$ is positive, i.e. we could have for some s ($s = 1, 2, \dots$) and some k ($k = 1, \dots, \ell$) that $\mathbf{p}^{sk} < 0$. This would be consistent with the formal model, but is, of course, inconsistent with free disposal of endowments. As a consequence, the next proposition which generalizes the former example, gives only a *sufficient* condition for *relevance* and a *sufficient* condition for *weak irrelevance*. This proposition holds only generically–i.e. for an open and dense set of economies. In this way, degenerate cases–principally those in which individual endowments are co-linear–are excluded.

Proposition (Relevance of deficit restrictions in economies with consumption taxes and consumer credit restrictions). *Suppose that only anonymous consumption taxes are available and that the credit constraint on at least one consumer is not binding. Let x be an allocation that can be implemented as a constitutional equilibrium with a fiscal policy and deficit restriction δ and let $r_t, 0 \le r_t < n$, be the number of consumers of generation t for which the credit constraint is binding.*

Then, if $n + r_t > \ell$ for all t the deficit restriction is weakly (and strongly) relevant. On the other hand, if $n + r_t \le \ell$ for all t then the deficit restriction is weakly irrelevant.

Proof. When consumers are potentially credit constrained, demand for commodities may depend on the individual borrowings or lendings, so that these must be kept constant when the policy changes. Formally, x_{th}^{tm}, with $x_{th}^{tm} = p^t \cdot \omega_{th}^t - q_t^t \cdot x_{th}^t$, is kept constant for constrained consumers. Denote this quantity by $\overline{x}_{th}^{tm}$. Furthermore, since there is some consumer whose credit constraint is not binding, prices in successive periods are linked. Therefore, in period t the relevant system consists of $2\ell - 1$ conditions on prices and $r_t + n$ conditions on individual wealths. Let the consumers whose credit constraint is binding be denoted by $h = 1, \cdots, r_t$ while the remaining $h = r_t + 1, \cdots, n$ have non-binding credit restrictions. Taking into account the restriction on the deficit, the system of $2\ell + r_t + n$ equations can be written as

$$\hat{p}^t + \hat{\tau}_t^t = (p^{t1} + \tau_t^{t1})R_t^t,$$

$$p^{t+1} + \tau_t^{t+1} = (p^{t1} + \tau_t^{t1})R_t^{t+1},$$

$$p^t \cdot \omega_{th}^t - \bar{b}_{th}^t = (p^{t1} + \tau_t^{t1})W_{th}^t, \quad h = 1, \cdots, r_t$$

$$p^t \cdot \omega_{th}^t + p^{t+1} \cdot \omega_{th}^{t+1} = (p^{t1} + \tau_t^{t1})W_{th}, \quad h = 1, \cdots, n$$

and

$$\sum_{h=1}^{h=n} \sum_{i=1}^{i=\ell} \tau_t^{ti} f_{th}^{ti}(R_t^t, R_t^{t+1}, W_{th}) + \tau_{t-1}^{ti} f_{t-1,h}^{ti}(R_{t-1}^{t-1}, R_{t-1}^t, W_{t-1,h}) = -\delta^t,$$

for $i = 1, \ldots, \ell$, where the rates of commodity substitution R, the normalized incomes W, and the government deficits δ are fixed. Suppose that $n \leq \ell$ and that $r_t = r$ is constant across time. In the Appendix, it is shown that in this case it is useful to consider as "free" variables the last $\ell - n$ prices of period t: $p^{t,n+1}, \ldots$,$p^{t\ell}$ and the first n prices of period $t + 1$: $p^{t+1,1}, \ldots, p^{t+1,n}$. This system, which is linear in 3ℓ unknowns, has a solution if and only if $n + r \leq \ell$. The usual sign restrictions on the p's apply so this condition is not sufficient for irrelevance. The proof can easily be generalized to the case in which r_t is not constant. $\square$

The above result does not concern paths such that there exist periods t' and t'' such that $n + r_{t'} \leq \ell$ and $n + r_{t''} > \ell$. In this case, $n + r_t$ should be compared to the number of available degrees of freedom in period t keeping in mind that this number depends also on the number of variables used in the other periods. The relevant quantity is $v_t + w_t$, where v_t and w_t are sequences defined by $v_t = \min(\ell, v_{t-1} + \ell - r_{t-1} - n)$ and $w_t = \min(\ell, w_{t+1} + \ell - r_{t+1} - n)$. The condition $n + r_t \leq \ell + v_t + w_t$ then indicates that the number of available instruments in period t is sufficient. However, in order to prove irrelevance the rank of the relevant matrices should also be checked. Unfortunately, there is no general result concerning these ranks and a case-by-case analysis is necessary.

7 Concluding remarks

We consider an OG exchange economy with anonymous taxes and transfers and constraints on individual borrowings. We ask whether or not the set of equilibrium allocations is affected by constitutional restrictions on the government's budget deficits.

Consumer credit constraints are important. With credit constraints and only anonymous lump-sum taxes, global irrelevance of government deficit restrictions is impossible and local irrelevance can obtain only in uninteresting circumstances. This contrasts with the case of unconstrained consumer credit, in which case deficit restrictions are globally (and locally) irrelevant. However, with credit constraints on individuals and only anonymous consumption taxes, global deficit irrelevance is impossible just as it is for the case without credit constraints. If there are a sufficient number of tax instruments and at least one consumer's credit constraint is not binding, then there is local irrelevance of the deficit restriction. This generalizes a similar result for the model without consumer credit constraints. More tools are needed for local irrelevance in the credit-constrained economy: For each consumer whose credit constraint is binding, there must be another tax tool. Hence the requirement for local irrelevance becomes: The number of commodities cannot be less then the number of consumers plus the number of consumers for whom the borrowing constraint is binding. (Of course, the number of consumers with binding credit restrictions might vary over time. In this case, the irrelevance result would be based on the maximum over time of the number of consumers with binding credit constraints.)

Consumption taxes are better for avoiding deficit restrictions than are lump-sum taxes in the case with constraints on consumer credit. With only anonymous lump-sum taxation, there is no way to increase the taxes on the young without reducing the early expenditures of the constrained youth. On the other hand, with a sufficient number of consumption tax rates, the government can increase locally the taxes on the young while leaving unchanged the consumptions of the credit-constrained young.

Compare our approach to taxation in the credit-constrained economy to that of Sargent and Smith [11]. Both Sargent and Smith [11] and we recognize that the tax powers of the government are limited. For us, this is captured by the assumption that all taxes must be anonymous. Sargent and Smith do allow for personalized lump-sum taxes but only as long as they do no not alleviate the credit constraint of any consumer. We make the strong assumption that the credit constraints (the b's) are unaffected by taxes and hence reducing early-life taxes can alleviate these constraints. Sargent and Smith make the strong assumption that the government cannot alleviate credit constraints by reducing early-life (lump-sum) taxes.

The implicit goal of the government in this paper (and other "irrelevance" papers) is to leave consumptions unchanged in the face of changed deficit restrictions. It is natural to ask how our results would be affected if the goal of the government were instead to maximize social welfare. To do this, one needs to characterize how taxes affect the equilibrium path of the economy. This is not so simple, at least for OG models. We leave this for further research.

A simpler question that can be addressed within our present approach is whether or not the government can free each consumer from his credit constraint. The government has the power to alleviate credit constraints by making transfers. Of course, the government might not be able to target the transfers at only credit-constrained individuals because of the anonymity requirement for taxes and transfers. Because of the need for anonymity, the government's power to alleviate binding credit constraints is limited, but this power is not zero. It turns out that the conditions for making each consumer liquid are similar to those for irrelevance of deficit restrictions. When only anonymous lump-sum transfers are used, freeing all consumer credit constraints is not possible in general, unless the deficit restriction does not include the first period of the economy. In the latter case, the government might be able to run a large first-period deficit and inject enough liquidity so that no generation faces binding credit constraints. When consumption taxes are available, it is likely that the same conditions that ensure weak irrelevance of deficit restrictions also ensure that all credit constraints can be weakened locally.

The present paper along with Ghiglino and Shell [7] indicates that there can be limits on the government's ability to avoid the restrictions on its deficit. Of course, as Kotlikoff [9] and others have argued, there is still ample scope for the government to evade (as opposed to avoid) the deficit restrictions by altering the timing of receipts and disbursements, by guaranteeing private loans, and so forth.

We stress again that to say the deficit restriction is irrelevant is not to say that the deficit does not matter. It is likely to matter if individuals condition their expectations on the deficit. For the analysis to be complete we must consider the role of expectations in "selecting" equilibria and we must also extend perfect-foresight expectations to rational expectations (that include sunspots). We believe that this is important.

We chose to use the OG model as the vehicle for our analysis because we think that it is both more tractable and more appropriate for this problem than the model with infinite-lived consumers. Nonetheless we do not believe that our basic results would be altered if we had employed the model with infinite-lived dynasties. (Of course, the present problem requires a model with many commodities and many consumers. The dynasties model has only recently been generalized to accommodate these heterogeneities.) On the other hand, the infinite time-horizon in either the OG or the dynasties model does make the analysis of local irrelevance much more complicated. Hence the finite version of this model deserves immediate investigation, to serve at least as a benchmark. The finite version of the model would also relate better to the existing optimal taxation literature. Finally, the model needs to be extended to include not only many consumption goods and many consumers but also to include production with many inputs and many outputs. Heterogeneity is crucial in this problem.

Appendix: Rank computations

For notational convenience, we focus attention on the case $n \geq 2$. Assume also that the number of consumers for whom the constraint is binding is constant though their identity might change from one generation to the next. Let the consumers for whom their credit constraint is binding be $h = 1, \ldots, r$ while for the remaining, $h = r+1, \ldots n$, their credit constraint is not binding. First, consider a consumer of generation 0. It is clear from Ghiglino and Shell [7] that the set of free parameters left after imposing the condition of constant individual demands to these consumers is a set of dimension $l - n$. Let us then consider as free the last $l - n$ prices $p^{1,n+1}, \ldots, p^{1,l}$.

Second, consider the consumers of generation t, $(t = 1, 2, \ldots)$, with the constraint that the prices $p^{t1}, \ldots, p^{tn}$ are already fixed (from previous-period conditions).

Using the relationship between the prices in periods t and $t+1$, the system of equations associated to a given demand can be written as

$$\hat{p}^t + \hat{\tau}_t^t = (p^{t1} + \tau_t^{t1})R_t^t$$

$$p^{t+1} + \tau_t^{t+1} = (p^{t1} + \tau_t^{t1})R_t^{t+1}$$

$$p^t \omega_{th}^t - \bar{b}_{th}^t = (p^{t1} + \tau_t^{t1})W_{th}^t \quad \text{for } h = 1, \ldots, r$$

$$p^t \omega_{th}^t + p^{t+1}\omega_{th}^{t+1} = (p^{t1} + \tau_t^{t1})W_{th} \quad \text{for } h = 1, \ldots, n$$

$$\sum_{h=1}^{n}\sum_{i=1}^{l} \tau_t^{ti} f_{th}^{ti}(R_t^t, R_t^{t+1}, W_{th}) +$$
$$+\tau_{t-1}^{ti} f_{t-1,h}^{ti}(R_{t-1}^{t-1}, R_{t-1}^t, W_{t-1,h}) = \delta^t,$$

where the quantities $R_t^t \in \mathbb{R}^{l-1}$, $R_{t-1}^t \in \mathbb{R}^l$, W_{th}^t, W_{th} and δ^t are fixed. The system of $2l - 1 + n + r + 1 = 2l + n + r$ equations becomes linear in $3l$ unknowns, $p^{t,n+1}, \ldots, p^{t,l}, p^{t+1,1}, \ldots, p^{t+1,n}, \tau_t^t$ and τ_t^{t+1}.

Introduce the vectors $P_0^t \in \mathbb{R}^{l-n}$, $P_1^{t+1} \in \mathbb{R}^n$ and $J_s \in \mathbb{R}^s$ defined by

$$P_0^t = \begin{bmatrix} p^{t,n+1} \\ p^{t,n+2} \\ \vdots \\ p^{t,l} \end{bmatrix}, \quad P_1^{t+1} = \begin{bmatrix} p^{t+1,1} \\ p^{t+1,2} \\ \vdots \\ p^{t+1,n} \end{bmatrix} \quad \text{and } J_s = \begin{bmatrix} 1 \\ 1 \\ \vdots \\ 1 \end{bmatrix}$$

In matrix form, the system can be written as $A_t z_t = b_t$ with

$$A_t = \begin{bmatrix} 0 & 0 & -R_t^t & I_{n-1} & 0 & 0 & 0 \\ I_{l-n} & 0 & -\overline{R}_t^t & 0 & I_{l-n} & 0 & 0 \\ 0 & I_n & -\underline{R}_t^{t+1} & 0 & 0 & I_n & 0 \\ 0 & 0 & -\overline{R}_t^{t+1} & 0 & 0 & 0 & I_{l-n} \\ \varpi_t^t & 0 & -W_t^t \cdot J_r & 0 & 0 & 0 & 0 \\ \omega_t^t & \omega_t^{t+1} & -W_t \cdot J_n & 0 & 0 & 0 & 0 \\ 0 & 0 & \sum_{h=1}^n f_{th}^{t1} & \sum_{h=1}^n \hat{f}_{th}^t & 0 & 0 \end{bmatrix}_{2l+n+r \times 3l},$$

$$\underline{R}_t^t = \begin{bmatrix} R_t^{t2} \\ R_t^{t3} \\ \vdots \\ R_t^{tn} \end{bmatrix}_{n-1\times 1}, \quad \overline{R}_t^t = \begin{bmatrix} R_t^{t,n+1} \\ R_t^{t,n+2} \\ \vdots \\ R_t^{tl} \end{bmatrix}_{l-n\times 1}, \quad \underline{R}_t^{t+1} = \begin{bmatrix} R_t^{t+1,1} \\ R_t^{t+1,2} \\ \vdots \\ R_t^{t+1,n} \end{bmatrix}_{n\times 1},$$

$$\overline{R}_t^{t+1} = \begin{bmatrix} R_t^{t+1,n+1} \\ R_t^{t+1,n+2} \\ \vdots \\ R_t^{t+1,l} \end{bmatrix}_{l+n\times 1}, \quad \varpi_t^t = \begin{bmatrix} \omega_{t1}^{t,n+1} & \omega_{t1}^{t,n+2} & \cdots & \omega_{t1}^{tl} \\ \omega_{t2}^{t,n+1} & \omega_{t2}^{t,n+2} & \cdots & \omega_{t2}^{tl} \\ \vdots & \vdots & \vdots & \vdots \\ \omega_{tr}^{t,n+1} & \omega_{tr}^{t,n+2} & \cdots & \omega_{tr}^{tl} \end{bmatrix}_{r\times l-n},$$

$$
\omega_t^t = \begin{bmatrix} \omega_{t1}^{t,n+1} & \omega_{t1}^{t,n+2} & \cdots & \omega_{t1}^{tl} \\ \omega_{t2}^{t,n+1} & \omega_{t2}^{t,n+2} & \cdots & \omega_{t2}^{tl} \\ \vdots & \vdots & \vdots & \vdots \\ \omega_{tn}^{t,n+1} & \omega_{tn}^{t,n+2} & \cdots & \omega_{tn}^{tl} \end{bmatrix}_{n \times l-n} , \quad \omega_t^{t+1} = \begin{bmatrix} \omega_{t1}^{t+1,1} & \omega_{t1}^{t+1,2} & \cdots & \omega_{t1}^{t+1,n} \\ \omega_{t2}^{t+1,1} & \omega_{t2}^{t+1,2} & \cdots & \omega_{t2}^{t+1,n} \\ \vdots & \vdots & \vdots & \vdots \\ \omega_{tn}^{t+1,1} & \omega_{tn}^{t+1,2} & \cdots & \omega_{tn}^{t+1,n} \end{bmatrix}_{n \times n}
$$

and
$$
z_t = \begin{bmatrix} P_1^t \\ P_0^{t+1} \\ \tau_t^{t1} \\ \tau_t^{t2} \\ \vdots \\ \tau_t^{t+1,l} \end{bmatrix}.
$$

Let also $W_t^t \in \mathbb{R}^n$ and $W_t \in \mathbb{R}^n$ be the vectors of individual wealths. The rank of the matrix A_t is equal to the rank of the matrix

$$
\begin{bmatrix} 0 & 0 & -\underline{R}_t^t & I_{n-1} & 0 \\ I_{l-n} & 0 & -\overline{R}_t^t & 0 & I_{l-n} \\ \varpi_t^t & 0 & -W_t^t \cdot J_n & 0 & 0 \\ \omega_t^t & \omega_t^{t+1} & -W_t \cdot J_n & 0 & 0 \\ 0 & 0 & \sum_{h=1}^n f_{th}^{t1} & \sum_{h=1}^n \hat{f}_{th}^{\hat{t}} & \end{bmatrix}_{l+n+r \times 2l}
$$

plus l. Some tedious manipulations similar to those performed in Ghiglino and Shell [7], show that generically the above matrix has maximal rank. Then, for $l = n + r$ the A_t matrix has full rank $3l$. In this case the system has always a solution. The same can be said for $n + r < l$.

Suppose now that $l + 1 = n + r$. Then the A_t matrix is a $3l + 1 \times 3l$ matrix which has generically maximal rank $3l$. Consider the square $3l + 1$ matrix associated to the augmented system, (A_t, b_t) and let us prove that Rank $(A_t, b_t) = 3l + 1$. Indeed, the last coordinate of b_t is a function of δ_t that can be written as

$$
\delta^t - \sum_{h=1}^n \sum_{i=1}^l \tau_{t-1}^{ti} f_{it-1}^{ti}.
$$

The determinant of (A_t, b_t) is a first degree polynomial expression in δ^t. Therefore, to prove that the relevant matrix has full rank for an open and dense set of values of δ^t it is enough that the coefficient of δ^t in the polynomial expression is nonzero, which can be seen to be generically true. Since $Rank(A) < Rank(A, b)$, the solution set is empty. This is the borderline case so the same result holds also whenever $n + r > l + 1$.

References

1. Balasko, Y.: Foundations of the theory of general equilibrium. Boston, MA: Academic Press 1988
2. Balasko, Y., Shell, K.: The overlapping-generations model. I. The case of pure exchange without money. Journal of Economic Theory 23, 281–306 (1980)

3. Balasko, Y., Shell, K.: The overlapping-generations model. II. The case of pure exchange with money. Journal of Economic Theory **24**, 112–142 (1981)
4. Balasko, Y., Shell, K.: Lump-sum taxes and transfers: public debt in the overlapping generation model. In: Heller, W., Starr, R., Starrett, D. (eds.) Equilibrium analysis: Essays in honor of Kenneth J. Arrow, Vol. II. Equilibrium analysis. New York: Cambridge University Press 1986
5. Balasko, Y., Shell, K.: Lump-sum taxation: the static economy. In: Becker, R., Boldrin, M., Jones, R., Thomson, W. (eds.) General equilibrium, growth and trade. Essays in honor of Lionel McKenzie, Vol. II. New York: Academic Press 1993
6. Barro, R.: Are government bonds net wealth? Journal of Political Economy **82**, 1095–1118 (1974)
7. Ghiglino, C., Shell, K.: The economic effects of restrictions on government budget deficits. Journal of Economic Theory **94** (1), 106–137 (2000)
8. Ghiglino, C., Tvede, M.: Endowments, stability and fluctuations in OG models. Journal of Economic Dynamics and Control **19**, 621–653 (1995)
9. Kotlikoff, L. J.: From deficit delusion to the fiscal balance rule. Journal of Economics (Suppl.) **7**, 17–41 (1993)
10. Samuelson, P. A.: An exact consumption-loan model of interest with or without the social contrivance of money. Journal of Political Economy **66**, 467–482 (1958)
11. Sargent, T. J., Smith, B.: Irrelevance of open market operations in some economies with government currency being denominated in rate of return. American Economic Review **77** (1), 78–92 (1987)

Speculative trade, asset prices and investment levels[*]

Alvaro Sandroni[1,2]

[1] Department of Economics, University of Rochester, 238 Harkness Hall, Rochester, NY 14627, USA
(e-mail: alsn@troi.cc.rochester.edu)
[2] J.L. Kellogg Graduate School of Management, Department of Managerial Economics and Decision Sciences, 2001 Sheridan Road, Evanston, IL 60208, USA
(e-mail: sandroni@nwu.edu)

Received: January 8, 2001; revised version: April 11, 2002

Summary. In this paper I consider a dynamically complete market model without intrinsic uncertainty. Agents' beliefs are different, but correct in the limit. Some agents are more patient than others. I show that infinitely often share prices are low and the economy stagnates. Also, infinitely often share prices are high and the economy grows. The changes from growth to stagnation and from stagnation to growth are not caused by exogenous shocks. They are caused by speculative trade among agents with different propensities to save and invest.

Keywords and Phrases: Economic fluctuations, Speculative trade.

JEL Classification Numbers: D83.

1 Introduction

There is plenty of evidence that people give different interpretations to the same facts. Analogously, agents' forecasts may be different even if all agents observe the same data. Differences in forecasts may induce trade and, consequently, changes in agents' relative wealth. Therefore, if in addition to having heterogenous beliefs agents also differ in other dimensions like discount factors then aggregate economic variables such as asset prices and investment levels may behave as if agents were sometimes more patient and sometimes less patient although each individual agent has a constant discount factor. This is, of course, in direct contrast with models in which agents do not trade (e.g. a representative agent economy).

---

  * I thank an anonymous referee for helpful comments. I gratefully acknowledge financial support from the National Science Foundation.

A systematic approach to the economic implications of speculative trade can be found in Kurz (1997) who shows how agents' beliefs should comply with the data and also how differences in beliefs may help explain the predictability of excess returns and the excess volatility of stock prices. In this paper, I consider a different, but related, methodology. I show how small differences in beliefs may have a significant impact on asset prices and investment levels. The basic idea is that even vanishingly small differences in beliefs may lead to speculative trade that causes the wealth to bounce back and forth between patient and impatient agents. When the wealth is concentrated in the hands of patient agents then both asset prices and investment levels are high because these agents are more willing to trade current consumption for future consumption. Conversely, when the wealth is concentrated in the hands of impatient agents then both asset prices and investment levels are low because impatient agents are less willing to trade current for future consumption. This raises the possibility that there exists a link between investment levels and asset prices caused by speculative trade.

I consider a dynamically complete market model in which there are long-lived trees and a risk-free asset. Two long-lived agents maximize an expected discounted logarithmic utility function according to their beliefs and discount factor. There are two states of nature h and l. The probability of state l vanishes over time slowly enough so that l occurs infinitely often. Each tree delivers the same amount of fruits and seeds in both states. Hence, the only uncertainty is extrinsic (i.e. over share prices). A new tree can be produced with one seed and γ fruits. The assumption that seeds are needed for the production of trees implies that capital is not entirely substitutable for consumption goods. That is, *if* there were no seeds in the model then trees produce fruits and trees could be produced by fruits. Share prices would be fixed and determined by a no-arbitrage condition. Other relevant economic variables such as investment levels would also remain fixed. Hence, the assumption that capital is not entirely substitutable for consumption goods is critical in this model. The economy *grows* when new trees are produced. The economy *stagnates* when new trees are not produced.

In this model, *if* agents' beliefs were identical then agents would differ only in discount factor and initial wealth. Eventually, the most patient agent will hold all wealth. In the absence of exogenous shocks, share prices would converge to a fixed level. However, assume that both agents believe that state l has vanishing probability. In the limit, both agents' beliefs are exactly correct, but at any point in time there is a vanishingly small difference in beliefs. These small differences in beliefs suffice to produce "runs" that sometimes favor the patient agent and sometimes the impatient agent. The most patient agent do not end up with all wealth. Infinitely often the wealth will be concentrated in the hands of the patient agent and infinitely often the wealth will be concentrated in the hands of the impatient agents.[1]

Agents' discount factors are assumed to be such that if the wealth is sufficiently concentrated in the hands of the impatient agent then share prices will be lower than γ. Hence, it is cheaper to buy a tree in the market than to produce one even

---

[1] In a simpler model in which no trees could be produced Sandroni (1998) shows that these small differences in beliefs may cause market crashes (defined by an arbitrarily low return on equity). After the crash asset prices slowly rebound but, infinitely often, crash again.

if the seeds are free. There is no demand for seeds and the economy stagnates. Conversely, if the wealth is sufficiently concentrated in the hands of patient agents then share prices will be above γ. The seeds must be sold at a strictly positive price otherwise it would be cheaper to produce a tree than to buy one. This would cause an excess supply of trees which is not possible in equilibrium. So, all seeds are sold and the economy grows. I show that both scenarios, one in which share prices are low and the economy stagnates and one in which share prices are high and the economy grows, reoccur infinitely often. These changes from growth to stagnation and from stagnation to growth are not caused by exogenous shocks in the economy. They are caused by speculative trade among agents with different propensities to save and invest.

2 The model

There are two long-lived agents, long-lived trees, a risk-free asset in zero supply, a single consumption good c, and two states of nature given by the set $\Sigma = \{h, l\}$. Each tree gives d units of consumption and s seeds in every state of nature. Agents are born with k_0^i trees, $i \in \{1, 2\}$, and with no credit or debt on the risk-free asset. A new tree can be produced with one seed and γ units of consumption. Seeds and consumption are perishable and last only one period. I assume that $d > \gamma s$.

Let $\sum^t$, $1 \le t \le \infty$, be the set of all t-histories. Let $\Im_1 \subset ...\Im_t \subset ... \subset \Im$ be the filtration on $\sum^\infty$ where $\Im_t$ is the σ-algebra generated by all t-histories, and $\Im$ is the σ-algebra generated by the algebra $\Im^0 \equiv \bigcup_{t \ge 1} \Im_t$.

At period t, agent i sells all his assets, buys b_t^i units of the risk-free asset, c_t^i units of consumption, k_t^i of the existing trees, and plants v_t^i new trees (with v_t^i seeds with γv_t^i units of consumption bought in the market). Share prices, seed prices and interest rates are given by p_t, q_t, and r_t, respectively. Gross interest rates are defined by $i_t \equiv 1 + r_t$. The variables c_t^i, v_t^i, k_t^i, b_t^i, p_t, q_t, and i_t are assumed to be $\Im_t$−measurable.

The market value of agent i's assets before consumption takes place is

$$w_t^i \equiv (p_t + d + q_t s)\left(k_{t-1}^i + v_{t-1}^i\right) + b_{t-1}^i.$$

I refer to w_t^i as agent i's wealth.

At period t, agent i's observed and anticipated budget constraints are

$$c_{t+j}^i + (q_{t+j} + \gamma)\, v_{t+j}^i + p_{t+j} k_{t+j}^i$$
$$+ \frac{b_{t+j}^i}{i_{t+j}} = (p_{t+j} + d + q_{t+j}s)(v_{t+j-1}^i + k_{t+j-1}^i) + b_{t+j-1}^i;$$

$$c_{t+j}^i \ge 0;\; v_{t+j}^i \ge 0;\; w_{t+j}^i \ge 0.$$

The total supply of trees is given by K_t, where

$$K_t = K_{t-1} + v_{t-1}^1 + v_{t-1}^2,\; K_0 = k_0^1 + k_0^2,\text{ and } v_0^1 = v_0^2 = 0.$$

Markets clear at period t if

$$c_t^1 + c_t^2 + \gamma \left(v_t^1 + v_t^2 \right) = K_t d, \; b_t^1 + b_t^2 = 0, \; k_t^1 + k_t^2 = K_t, \text{ and}$$

$$\left(v_t^1 + v_t^2 \right) \leq K_t s, \text{ with equality if } q_t > 0.$$

Let P and P^i be probability measures on $(\Sigma^\infty, \Im)$ representing the true probability measure and agent i's belief about the histories of states of nature, respectively. At period t, agent i's expected discounted utility function is given by

$$E^{P^i} \left\{ \sum_{j=0}^{\infty} (\beta^i)^j \log(c_{t+j}^i) \mid \Im_t \right\},$$

where β^i is agent i's discount factor, and E^{P^i} is the expectations operator associated with agent i's belief P^i.

In equilibrium, agents maximize expected discounted utility subject to the budget constraints, and markets clear in every period.

2.1 Agents' beliefs

Given $0 < \delta < 1$, let $y(t)$ be defined by

$$y(t) \equiv \frac{\delta + \ln \left(\frac{\ln t - t^{-\delta}}{\ln t - 1} \right)}{\delta \ln t + \ln \left(\frac{\ln t - t^{-\delta}}{\ln t - 1} \right)} \ln t.$$

It is easy to check that $\lim_{t \to \infty} y(t) = 1$. The results would be the same if $y(t)$ were defined as 1, but the complex expression for $y(t)$ actually simplifies some calculations. If $t \geq 3$ then $\frac{y(t)}{\ln(t)} \in (0, 1)$; $\frac{1}{\ln t} \in (0, 1)$; and $\frac{1}{t^\delta \ln t} \in (0, 1)$.

The true probability of the state of nature l, conditional on all information available at period $t - 1$, $t \geq 3$, is $\frac{y(t)}{\ln(t)}$. Agent 1 believes that this probability is $\frac{1}{\ln(t)}$, whereas agent 2 believes that it is $\frac{1}{t^\delta \ln t}$.[2]

Both agents correctly believe that state l has a vanishingly small probability. The differences in beliefs become vanishingly small as time progresses. In the limit, both agents' beliefs are identical and correct. In particular, both agents' beliefs are rational in the sense of Kurz (1997) since the asymptotic distribution of agents' beliefs is identical to the true distribution. However, at any point in time agents' beliefs are different. These differences in beliefs are essential for the main results. In this model, markets are dynamically complete. Hence, it follows from Cass and Shell (1983) that in equilibrium if agents' beliefs were identical then asset prices would be the same in both states of nature. There would be no uncertainty over asset prices. The most patient agent would eventually hold all wealth and the relevant economy variables would be determined by this agent.

---

[2] Agents' beliefs and the true probabilities at period 1 are not relevant and, therefore, are not defined.

2.2 Comments on the model

1. A novelty in this model is the use of seeds in the production of trees. As mentioned in the introduction, the role of seeds is as follows: In the model, consumption comes from trees. So, if consumption were the only input needed to produce trees then share prices would be determined by a no-arbitrage condition (between consumption goods and trees). In the absence of intrinsic uncertainty, the remaining uncertainty is over prices, but there would be no such uncertainty if share prices were fixed.

2. If the price of a seed is strictly positive then, in equilibrium, the demand for seeds must be equal to the supply. On the other hand, if the demand for seeds does not meet the supply even when seeds are costless then, in equilibrium, there must be an excess supply of seeds. Under free disposal, the price of seeds cannot become negative.

3. The logarithmic utility function simplifies many calculations, but it is not an essential feature of this model. Under close inspection, the proofs reveal that the main results would still hold with other utility functions.

4. The technology in this model is so simple that I felt no need to model firms explicitly. Agents buy seeds and produce trees "at home." The results would remain the same if firms were explicitly modelled.

5. The assumption of a linear technology in the production of trees implies that, in equilibrium, either all seeds are used to produce trees or no trees are produced. I conjecture that in more general models where this technology is not linear, intermediary levels of investments are possible in equilibrium.

6. The assumption that $d > \gamma s$ ensures that all seeds might be used to produce new trees. To see this assume that $d \leq \gamma s$. If there are K_t trees and all available seeds are used to produce new trees then $K_t \gamma s$ units of consumptions are needed for this purpose. Given that there are $K_t d$ units of consumption available, this is either impossible or leaves nothing to be consumed.

3 Main results

If a seed is costly (i.e. $q_t > 0$) then, in equilibrium, all seeds must be used to produce new trees (i.e. $\left(v_t^1 + v_t^2\right) = K_t s$). The economy grows. By no-arbitrage, a tree must cost a seed plus γ units of consumption (i.e. $p_t = q_t + \gamma$). So, a tree costs more than γ units of consumption. On the other hand, assume that a tree costs less than γ units of consumption. So, it is cheaper to buy a tree than to produce one. Hence, in equilibrium, the seeds are free (i.e. $q_t = 0$) and no new trees are produced (i.e. $v_t^1 = v_t^2 = 0$). The economy stagnates. I define

Definition 1. *Given an infinite sequence of states of nature $\sigma \in \sum^\infty$, the economy grows in period t if $p_t(\sigma) > \gamma$ and $v_t^1(\sigma) + v_t^2(\sigma) = K_t(\sigma)s$. The economy stagnates in period t if $p_t(\sigma) < \gamma$ and $v_t^1(\sigma) = v_t^2(\sigma) = 0$.*

This connection between asset prices and economic growth cannot be obtained in a representative agent economy. To see this assume that $k_0^1 = 0$. Given that agent 1 has no wealth, agent 2 can be thought of as a representative agent. The proposition below shows that, for a sufficiently patient representative agent, the economy always grows. For a less patient representative agent, the economy always stagnates. The critical discount factor $\bar{\beta}$ is

$$\bar{\beta} \equiv \frac{\gamma(1+s)}{d+\gamma} < 1.$$

Proposition 1. *Assume that $k_0^1 = 0$ and $k_0^2 > 0$. Consider an arbitrary infinite sequence of states of nature. If $\beta^2 > \bar{\beta}$ then, in equilibrium, the economy grows in every period. If $\beta^2 < \bar{\beta}$ then, in equilibrium, the economy stagnates in every period.*

Proof. See Appendix.

On the other hand, assume that both agents have strictly positive endowments (i.e. $k_0^1 > 0$ and $k_0^2 > 0$). Given that agents 1 and 2 have different beliefs and discount factors, they will trade. That is, they will make different choices over consumption, investments and portfolio allocations. Hence, the relative wealth of these two agents will fluctuate according to the stochastic realizations of the states of nature.

Assume that $\beta^2 > \bar{\beta} > \beta^1$. By Proposition 1, agent 1 is not patient enough so that if 1 were a representative agent the economy would stagnate. Agent 2 is sufficiently patient so that if 2 were a representative agent the economy would grow. Therefore, if agent 2 has a sufficiently high fraction of the total wealth then asset prices will be high (i.e. above γ) and the economy grows. On the other hand, if agent 1 has a sufficiently high fraction of the total wealth then asset prices will be low and the economy stagnates.[3] The main result in this paper shows that if $\ln \beta^2 = \ln \beta^1 + \delta$ then the relative wealth of these two agents will fluctuate to the point that, almost surely, infinitely often both agents will have a high fraction of the total wealth. Hence, infinitely often, the economy grows and, infinitely often, it stagnates.

Proposition 2. *Assume that $k_0^1 > 0$ and $k_0^2 > 0$. Also assume that $\beta^2 > \bar{\beta} > \beta^1$ and $\ln \beta^2 = \ln \beta^1 + \delta$. For $P-$almost all sequences of states of nature, the economy grows infinitely often and stagnates infinitely often.*

Proof. See Appendix.

An outsider observing aggregate variables may find the behavior of this economy puzzling. There are no significant exogenous shocks to the economy. The trees gives d units of consumption and s seeds every period. However, suddenly share prices become sufficiently low so that it is cheaper to buy a tree than to produce one. As a result, the economy stagnates. After some periods, share prices rebound and the economy grows again. These cycles continue forever. They are caused by trade between agents with different propensities to save and invest.

Remark 1. The restriction $\ln \beta^2 = \ln \beta^1 + \delta$ makes this example special, but it must be satisfied so that no long-lived agent eventually obtains all wealth. In an overlapping generation model agents' characteristics do not have to be precisely

---

[3] In this economy, share prices are often quite different from γ.

chosen in order for meaningful interactions among different agents to continue eternally. However, introducing natural death may not add a relevant insight to the ideas presented in this example.

Remark 2. Assume that dividends per tree depends on the states of nature. That is, assume that each tree gives d_h units of consumption of the state is h and d_l units of consumption of the state is l, $d_h \neq d_l$. Also assume that this difference is small, i.e. d_h and d_l are close to d. There exists a unique equilibrium in this economy. If d_h and d_l are sufficiently close to d then this unique equilibrium approaches the equilibrium, described in this paper (see Manuelli and Peck, 1992, for additional details on this issue).

4 Conclusion

This paper shows a positive correlation between asset prices and investment levels. A drop in asset prices and investment levels is associated with an increase in the relative wealth of agents who are not inclined to save and invest. Conversely, an increase in asset prices and investment levels is associated with an increase in the relative wealth of agents who are inclined to save and invest. These fluctuations in relative wealth are caused by vanishingly small differences in beliefs. In a dynamically complete market model with no intrinsic uncertainty, these fluctuations would not occur if agents' beliefs were exactly identical.

The model presented here is counterfactual. It may take an arbitrarily long time for the economy to switch from stagnation to growth and conversely from growth to stagnation. This is partially caused by the assumption that it may take an arbitrarily long time for one of the two states of nature to occur. This unrealistic assumption was made so that fluctuations in economic activity reoccur even if agents' beliefs are eventually identical to the truth. If agents' beliefs are not required to become identical then the economy may switch more frequently between stagnation and growth. In this expanded model, I conjecture that economic activity may change even if the distribution of wealth remains the same.

The contribution of this paper is the presentation of a mechanism for fluctuations in economic activity that differs from standard models in which these fluctuations are caused by exogenous productivity shocks. However, it is not clear at this point how this mechanism can be part of a more realistic model of economic fluctuations. This is a subject for future research.

5 Appendix

Let $\rho_t^i \equiv 1 - \frac{c_t^i}{w_t^i}$ be agent i's saving ratio. It is known that if agents have log utility then their optimal savings ratio is identical to their discount factor (see Sandroni, 1998, Lemma 1). So, in equilibrium,

$$c_t^i = (1 - \beta^i)w_t^i. \tag{5.1}$$

Let $w_t \equiv w_t^1 + w_t^2$ be the aggregate wealth. Let $\varpi_t^i \equiv \frac{w_t^i}{w_t}$ be agent i's fraction of the aggregate wealth. Let $r_t^2 \equiv \frac{w_t^2}{w_t^1}$ be agent 2's relative wealth.

By definition, $\varpi_t^1 = \frac{1}{1+r_t^2}$ and $\varpi_t^2 = \frac{r_t^2}{1+r_t^2}$. In equilibrium, $w_t = K_t(p_t + d + q_t s)$. By (5.1),

$$(1 - \beta^1)w_t^1 + (1 - \beta^2)w_t^2 + \gamma \left(v_t^1 + v_t^2\right) = K_t d.$$

If $q_t > 0$ then

$$(1 - \beta^1)w_t^1 + (1 - \beta^2)w_t^2 = K_t (d - \gamma s) \Rightarrow$$

$$p_t + d + q_t s = \frac{d - \gamma s}{(1 - \beta^1)\varpi_t^1 + (1 - \beta^2)\varpi_t^2}.$$

Moreover, $p_t = q_t + \gamma$. So,

$$p_t + d + (p_t - \gamma) s = \frac{d - \gamma s}{(1 - \beta^1)\varpi_t^1 + (1 - \beta^2)\varpi_t^2} \Rightarrow$$

$$p_t = \frac{d - \gamma s}{(1 + s)} \left(\frac{\beta^1 + \beta^2 r_t^2}{(1 - \beta^1) + (1 - \beta^2)r_t^2} \right).$$

On the other hand if $q_t = 0$ then

$$(1 - \beta^1)w_t^1 + (1 - \beta^2)w_t^2 = K_t d \Rightarrow$$

$$p_t + d = \frac{d}{(1 - \beta^1)\varpi_t^1 + (1 - \beta^2)\varpi_t^2} \Rightarrow$$

$$p_t = d \left(\frac{\beta^1 + \beta^2 r_t^2}{(1 - \beta^1) + (1 - \beta^2)r_t^2} \right).$$

These results are summarized in Lemma 1 below.

Lemma 1. *In equilibrium, if $q_t > 0$ then*

$$p_t = \frac{d - \gamma s}{(1 + s)} \left(\frac{\beta^1 + \beta^2 r_t^2}{(1 - \beta^1) + (1 - \beta^2)r_t^2} \right) > \gamma. \tag{5.2}$$

If $q_t = 0$ then

$$p_t = d \left(\frac{\beta^1 + \beta^2 r_t^2}{(1 - \beta^1) + (1 - \beta^2)r_t^2} \right) \leq \gamma. \tag{5.3}$$

Proof of Proposition 1. By definition,

$$d \left(\frac{\bar{\beta}}{(1 - \bar{\beta})} \right) > \frac{d - \gamma s}{(1 + s)} \left(\frac{\bar{\beta}}{(1 - \bar{\beta})} \right) = \gamma.$$

By assumption, the wealth of agent 1 is zero. So, $r_t^2 = \infty$. Hence, if $\beta^2 > \bar{\beta}$ then Eq. (5.3) cannot be satisfied. Hence, $q_t > 0$. If $\beta^2 < \bar{\beta}$ then Eq. (5.2) cannot be satisfied. Hence, $q_t = 0$. $\qquad \square$

Lemma 2. *In equilibrium,*

$$\limsup_{t\to\infty} r_t^2 = \infty \text{ and } \liminf_{t\to\infty} r_t^2 = 0 \qquad P-a.s.$$

Proof. The proof of this lemma is essentially identical to the proof of Lemma 3 in Sandroni (1998). I reproduce the argument here for completeness. Let x_t be a $\Im_t$−measurable random variable defined by

$$x_t(\sigma) = \ln\left(\frac{P^2_{\sigma_{t-1}}(\sigma_t))}{P^1_{\sigma_{t-1}}(\sigma_t))}\right) + \delta. \tag{5.4}$$

where $\sigma = (\sigma_t, ...) \in \Sigma^\infty$, $\sigma_t \in \Sigma^t$, $\sigma_t = (\sigma_{t-1}, a)$, $\sigma_{t-1} \in \Sigma^{t-1}$, $a \in \Sigma$, and $P^i_{\sigma_{t-1}}(\sigma_t))$, $i = 1, 2$, is the conditional probability of σ_t, given σ_{t-1} (according to agent i's beliefs).

By the definition of agents beliefs P^1 and P^2, $x_t = -\delta \ln t + \delta$ if l occurs at period t, and $x_t = \ln\left(\frac{\ln t - t^{-\delta}}{\ln t - 1}\right) + \delta$ if h occurs at period t. Hence, by the definition of $y(t)$ and the true probability measure P, if $t \geq 3$ then

$$E^P\{x_t \mid \Im_{t-1}\} = -\delta y(t) + (1 - \frac{y(t)}{\ln t}) \ln\left(\frac{\ln t - t^{-\delta}}{\ln t - 1}\right) + \delta = 0.$$

Let $Var^P\{x_t \mid \Im_{t-1}\}$ be the conditional variances of x_t according to the probability P. By definition,

$$Var^P\{x_t \mid \Im_{t-1}\} = \delta^2 y(t) \ln t + (1 - \frac{y(t)}{\ln t})\left(\ln\left(\frac{\ln t - t^{-\delta}}{\ln t - 1}\right)\right)^2 - \delta^2.$$

Hence, $Var^P\{x_t \mid \Im_{t-1}\} \geq \delta^2(\ln t - 1)$. Therefore,

$$\sum_{t=3}^{\infty} Var^P\{x_t \mid \Im_{t-1}\} = \infty.$$

Let T_m be $\sum_{t=3}^{m} Var^P\{x_t \mid \Im_{t-1}\}$. Then, $T_m \geq \sum_{t=3}^{\infty} \delta^2(\ln t - 1)$. Moreover, $\ln t$ goes to infinity as t goes to infinity. Hence, there exists $\bar{m}$ such that if $m \geq \bar{m}$ then $T_m \geq m$.

By L'Hopital theorem, the ratio $\frac{m^{0.25}}{\delta \ln(m+1)+\delta}$ goes to infinity as m goes to infinity. Moreover, $\ln\left(\frac{\ln(m+1) - (m+1)^{-\delta}}{\ln(m+1) - 1}\right) + \delta$ goes to δ as m goes to infinity. Hence, the ratio $\frac{m^{0.25}}{\ln\left(\frac{\ln(m+1) - (m+1)^{-\delta}}{\ln(m+1) - 1}\right)+\delta}$ goes to infinity as m goes to infinity. Therefore, there exists a positive constant ζ such that

$$\zeta(T_m)^{0.25} \geq \max\left\{\ln\left(\frac{\ln(m+1) - (m+1)^{-\delta}}{\ln(m+1) - 1}\right) + \delta, \ \delta \ln(m+1) + \delta\right\}$$

for every $m \geq 3$.

Let $\epsilon(b) \equiv \zeta b^{-0.25}$. Let $\phi(b) \equiv b^{0.5}$. Clearly, $\epsilon(b) \equiv \zeta b^{-0.25}$ tends to zero as b goes to infinity and $\epsilon(b)\phi(b)$ is a non-decreasing function of b. By construction, $|x_{m+1}^2| \leq epsilon(T_m)\phi(T_m)$.

Let S_m be $\sum_{t=3}^{m} x_t$. By Freedman (75), Proposition 2.6,

$$\lim_{m\to\infty} \sup S_m = \infty \text{ and } \lim_{m\to\infty} \inf S_m = -\infty \ P - a.s. \tag{5.5}$$

Agents' first order conditions imply that, in equilibrium, for every path $\sigma = (\sigma_m, ...) \in \Sigma^\infty, \sigma_m \in \Sigma^m$,

$$\frac{(\beta^2)^m P^2(\sigma_m)(c_m^2(\sigma))^{-1}}{(\beta^1)^m P^1(\sigma_m)(c_m^1(\sigma))^{-1}} = \frac{\lambda^2}{\lambda^1},$$

where λ^i is agent i's Lagrange multiplier.
By (5.1),

$$\frac{(\beta^2)^m P^2(\sigma_m)}{(\beta^1)^m P^1(\sigma_m)} = r_m^2(\sigma)\frac{\lambda^2}{\lambda^1}\frac{1-\beta^2}{1-\beta^1}.$$

By assumption, $\ln \beta^2 = \ln \beta^1 + \delta$. So, taking logs on both sides,

$$\ln r_m^2(\sigma) = m\delta + \ln\frac{P^2(\sigma_m)}{P^1(\sigma_m)} - \ln\frac{\lambda^2}{\lambda^1}\frac{1-\beta^2}{1-\beta^1}.$$

By Bayes' rule,

$$\frac{P^2(\sigma_m)}{P^1(\sigma_m)} = \frac{\prod_{t=1}^{m} P^2_{\sigma_{t-1}}(\sigma_t)}{\prod_{t=1}^{m} P^1_{\sigma_{t-1}}(\sigma_t)}$$

where $\sigma_m = (\sigma_t, ...)$.
Substituting in (5.4),

$$\ln r_m^2(\sigma) = \sum_{t=3}^{m} x_t + \ln\frac{P^2(\sigma_2)}{P^1(\sigma_2)} - \ln\frac{\lambda^2}{\lambda^1}\frac{1-\beta^2}{1-\beta^1}$$

The proof is concluded by (5.5).   □

Proof of Proposition 2. If r_t^2 is sufficiently large then

$$d\left(\frac{\beta^1 + \beta^2 r_t^2}{(1-\beta^1) + (1-\beta^2)r_t^2}\right) \approx d\left(\frac{\beta^2}{(1-\beta^2)}\right) > d\left(\frac{\bar{\beta}}{(1-\bar{\beta})}\right) = \gamma.$$

Hence, Eq. (5.3) cannot be satisfied. By Lemma 1, if r_t^2 is sufficiently large then $q_t > 0$.
If r_t^2 is sufficiently close to zero then

$$\frac{d-\gamma s}{(1+s)}\left(\frac{\beta^1 + \beta^2 r_t^2}{(1-\beta^1) + (1-\beta^2)r_t^2}\right) \approx \frac{d-\gamma s}{(1+s)}\left(\frac{\beta^1}{(1-\beta^1)}\right) < \gamma.$$

Hence, Eq. (5.2) cannot be satisfied. By Lemma 1, if r_t^2 is sufficiently close to zero then $q_t = 0$. The proof is concluded by Lemma 2.   □

References

Cass, D., Shell, K.: Do sunspots matter? Journal of Political Economy **91** (2), 193–227 (1983)

Freedman, D.: On tail probabilities for martingales. Annals of Probability **3**, 100–118 (1975)

Kurz, M.: Endogenous economic fluctuations, studies in the theory of rational beliefs. Berlin Heidelberg New York: Springer 1997

Manuelli, R., Peck, J.: Sunspot-like effects of random endowments. Journal of Economic Dynamics and Control **16**, 193–206 (1992)

Sandroni, A.: Learning, rare events, and recurrent market crashes in frictionless economies without intrinsic uncertainty. Journal of Economic Theory **82**, 1–18 (1998)

Indeterminacy of equilibrium in stochastic OLG models[*]

Michael Magill[1] and Martine Quinzii[2]

[1] Department of Economics, University of Southern California, Los Angeles, CA 90089-0253, USA
(e-mail: magill@usc.edu)
[2] Department of Economics, University of California, Davis, CA 95616-8578, USA
(e-mail: mmquinzii@ucdavis.edu)

Received: November 19, 2001; revised version: March 22, 2002

Summary. This paper studies the equilibria of a stochastic OLG exchange economies consisting of identical agents living for two periods, and having the opportunity to trade a single infinitely-lived asset in constant supply. The agents have uncertain endowments and the stochastic process determining the endowments is Markovian. For such economies, the literature has focused on studying strongly stationary equilibria in which quantities and prices are functions of the exogenous states of nature which describe the uncertainty: such equilibria are generalizations of deterministic steady states, and this paper investigates if they have the same special status as asymptotic limits of other equilibrium paths. The difficulty in extending the analysis of equilibria beyond the class of strongly stationary equilibria comes from the presence of indeterminacy: we propose a procedure for overcoming this difficulty which can be decomposed into two steps. First backward induction arguments are used to restrict the domain of possible prices; then if some indeterminacy is left, expectation functions are introduced to make the forward equilibrium equations determinate. The properties of the resulting trajectories, in particular their asymptotic properties, can then be studied. For the class of models that we study this procedure provides a justification for focusing on strongly stationary equilibria. For the model with positive dividends (equity or land) the justification is complete, since we show that the strongly stationary equilibrium is the unique equilibrium. For the model with zero dividends (money) there is a continuum of self-fulfilling expectation functions resulting in a continuum of equilibrium paths starting from any admissible initial condition: under conditions given in the paper, these equilib-

---

[*] We are grateful for the stimulating environment and research support provided by the Cowles Foundation at Yale University during the Fall 2000 when this paper was first conceived. We are also grateful to the participants of the SITE Workshop at Stanford University and the Incomplete Markets Workshop at SUNY Stony Brook during the summer 2001 for helpful discussions.

Correspondence to: M. Magill

rium paths converge almost surely to one of the strongly stationary equilibria-either autarchy or the stochastic analogue of the Golden Rule.

Keywords and Phrases: Stochastic overlapping generations model, Stationary rational expectations equilibrium, Indeterminacy, Expectation functions, Martingale convergence theorem.

JEL Classification Numbers: D50, D84, C62.

1 Introduction

The overlapping generations model has proved to be a useful model for exploring properties of equilibria over time. It draws on the fact that one of the simplest sources of heterogeneity of agents comes from the fact they are born at different dates and the resulting intergenerational trade reflects the differences in agents' needs over their life cycle. While the model can be studied at various levels of generality (see Geanakoplos and Polemarchakis, 1991, for a survey), many of its basic properties and insights can be derived within the simplest class consisting of identical agents living for two periods, and having the opportunity to trade a single infinitely-lived asset in constant supply. The deterministic model was analyzed with great clarity by Gale (1973). For stochastic economies the theoretical literature has focused on studying strongly stationary equilibria in which quantities and prices are functions of the exogenous states of nature which describe the uncertainty. Existence of non-trivial equilibria of this type has been proved by Cass et al. (1992) and Gottardi (1996), while their normative properties, which are similar to those of the Golden Rule in the deterministic case, have been analyzed by Peled (1984), Aiyagari and Peled (1991), Demange and Laroque (1999) and Chattopadhyay and Gottardi (1999).

In the deterministic model, steady-state equilibria have a special status not only because they are simple, but also because in standard cases they are the asymptotic limits of all other equilibrium paths. Strongly stationary equilibria of the stochastic model are the generalizations of deterministic steady states, and the goal of this paper is to investigate if they have the same special status as asymptotic limits of other equilibrium paths.

The difficulty in extending the analysis of equilibria beyond the class of strongly stationary equilibria comes from the presence of indeterminacy. If uncertainty is modeled by the occurrence of S possible shocks at each date, then at each node of the associated event-tree the equilibrium equation is a first-order condition relating the price of the asset at this node to the S prices at the immediate successors: if the equilibrium equation is read forward then there is an $(S-1)$-dimensional manifold of "candidate prices" for the next period. If the model takes place over a finite horizon T and if there is a natural terminal condition (typically that the price is zero) then the equations can be read (solved) backwards to obtain a (typically) determinate equilibrium. If the horizon is infinite and there is no natural terminal condition, this way of obtaining determinacy cannot be applied. However backward induction can still be useful for finding restrictions on the equilibrium prices: at each date T in the future the prices have to satisfy certain inequalities (e.g. be positive and affordable by young agents) and drawing the consequences of these inequalities by backward induction can substantially restrict the indeterminacy of the prices. In the case where the asset pays a positive dividend (the asset is "land" or "equity") we show that this procedure has a rather dramatic outcome, for it eliminates all price sequences except the strongly stationary equilibrium, which, under our assumptions, is thus the unique equilibrium. However if the asset pays zero dividends (i.e is "money"), then this procedure only eliminates prices which lie above the stationary equilibrium prices, thus leaving an indeterminacy of dimension $S - 1$ at each node.

The hypothesis of rational expectations requires that agents correctly anticipate both the support and the probabilities of future prices. In a series of papers, Kurz (1997) has argued that the latter hypothesis is unreasonable because agents can not be expected to know the stochastic process driving the uncertainty: learning about frequencies from past data still leaves room for differences in beliefs about the nature of the stochastic process.[1] However in the examples of equilibria with rational beliefs, Kurz still retains the assumption that agents anticipate the same prices, which is natural given the strong stationarity of the equilibria. In the stochastic model with money in which there is a continuum of prices which can be self fulfilling, it is the former assumption which might seem more restrictive, namely that agents anticipate the same future prices.

To resolve this difficulty we assume that agents co-ordinate their expectations through an expectation function, and the set of expectation functions parametrizes the $(S - 1)$-dimensions of next period prices at each node. Since we retain a form of stationarity – the expectations are stationary in that they depend only on the current price and the current shock – it is perhaps not unreasonable to assume that agents could learn such an expectation function. Once introduced, the expectation function leads to a determinate stochastic difference equation for the equilibrium prices: the price equation at each node can be read forward as in the deterministic case, and the asymptotic properties of the equilibrium paths can again be studied. Using a martingale convergence argument we prove, under conditions spelled out in Section 2, that the equilibrium paths converge almost surely to one of the strongly stationary equilibria–either autarchy or the stochastic analogue of the Golden Rule.

Thus the general procedure that we propose for extending the analysis beyond the class of strongly stationary equilibria can be decomposed into two steps. First backward induction arguments are used to restrict the domain of prices. Then if some indeterminacy is left, expectation functions are introduced to make the forward

---

[1] More precisely Kurz introduces the concept of a "rational belief equilibrium" in which agents have differing conditional probabilities for prices at the immediate successors, even though they agree on the probabilities of tail (or asymptotic or long-run) events. Keynes (1930, ch.15) has emphasized the importance of differences in investors' opinions for a proper understanding of the functioning of financial markets–in the language of Wall Street, it is the changing proportion of the population of investors between bulls (optimists) and bears (pessimists) which accounts for much of the volatility of asset prices. The theory of rational beliefs is a way of formalizing this view of the functioning of financial markets based on short-run differences in beliefs, which are however compatible with the long-run behavior of prices.

equations determinate: the properties of the resulting trajectories, in particular their asymptotic properties, can then be studied. For the class of models that we are studying this procedure provides a justification for focusing on strongly stationary equilibria. For the model with positive dividends (equity or land) the justification is complete in that the the strongly stationary equilibrium is the only equilibrium; for the model with zero dividends (money) the justification is less complete–indeed the analysis suggests that it is something of an act of faith to focus attention on the stationary Golden Rule since many (and for some expectation functions, all) equilibrium paths converge to autarchy.

2 The model with money

Consider a one-good overlapping generations exchange economy in which agents live for two periods and have random endowments. The uncertainty is modeled by the occurrence of one of a finite number of shocks at each date, $s \in S = \{1, \ldots, S\}$, s_0 being the initial shock. Let $\sigma_t = (s_0, \ldots, s_t)$ denote the history of the shocks from date 0 to date t: let $\Sigma_t = S \times \cdots \times S$ denote the set of all such histories up to date t and let $\Sigma = \cup_{t=0}^{\infty} \Sigma_t$ denote the collection of all such histories for all dates, $\sigma = (s_0, \ldots, s_t, \ldots)$ denoting a typical path of the event-tree Σ. We assume that the shocks follow a first-order Markov process and denote by P the induced probability on the event-tree Σ.

Assumption 1. (*Markov structure*): There exists a Markov transition matrix $\rho = [\rho_{ss'}]_{s,s' \in S}$ with $\rho_{ss'} > 0$, $\forall s, s' \in S$ such that $P(s_{t+1} = s' | s_t = s) = \rho_{ss'}$, $\forall s, s' \in S$.

At each date-event $\sigma_t \in \Sigma$, n identical agents enter the economic stage: since we do not consider growth or fluctuations in the cohort size, we may set $n = 1$. The representative agent lives for two periods and has the random endowment stream $\omega(\sigma_t) = (\omega^1(\sigma_t), \omega^2(\sigma_t, s'))_{s' \in S}$ which depends only on the shock s_t realized when the agent is young, i.e. if $\sigma_t = (s_0, \ldots, s_t)$, then $\omega(\sigma_t) = \left(\omega^1_{s_t}, (\omega^2_{s_t, s'})_{s' \in S}\right)$.

Assumption 2. (*Positive endowments*): $\omega(s) = (\omega^1_s, (\omega^2_{ss'})_{s' \in S}) \in \mathbf{R}^{S+1}_{++}$, $\forall s \in S$.

All agents maximize the expected utility of their lifetime consumption streams, with the same utility indices. The representative agent born at node σ_t ranks the possible consumption streams $x(\sigma_t) = \left(x^1(\sigma_t), (x^2(\sigma_t, s'))_{s' \in S}\right) \in \mathbf{R}^{S+1}_+$ according to a utility function $U_{\sigma_t} : \mathbf{R}^{S+1}_+ \to \mathbf{R}$ satisfying:

Assumption 3. (*Preferences*): There exist increasing, concave, differentiable functions $u_1, u_2 : \mathbf{R}_{++} \to \mathbf{R}$ such that

$$U_{\sigma_t}(x(\sigma_t)) = u_1(x^1(\sigma_t)) + \sum_{s' \in S} \rho_{s_t s'} u_2(x^2(\sigma_t, s'))$$

where for $i = 1, 2$, $\lim_{c \to 0} u'_i(c) = +\infty$ and u_2 has a coefficient of relative risk aversion less than or equal to 1: $\forall c > 0, -c\dfrac{u''_2(c)}{u'_2(c)} \leq 1$.

To study the consequences of intergenerational trade, we assume that there is an infinitely-lived asset available in positive supply, normalized to 1, which pays no dividends (usually called money). The asset is initially held by the representative old agent at date 0 and is then exchanged (if prices are non zero) at each date between the old and the young. Let $q(\sigma_t)$ denote the price of the asset at node σ_t. The young agent at node σ_t faces the budget constraints

$$x^1(\sigma_t) = \omega^1(\sigma_t) - q(\sigma_t)z, \ z \in \mathbf{R} \tag{1}$$
$$x^2(\sigma_t, s') = \omega^2(\sigma_t, s') + q(\sigma_t, s')z, \ s' \in \mathbf{S} \tag{2}$$

and chooses z to maximize $U_{\sigma_t}(x(\sigma_t))$. Under Assumption 3 the optimal choice of z is defined by the FOC for maximizing U_{σ_t} under the budget constraints (1), (2): since the equilibrium condition is $z(\sigma_t) = 1$, $\forall \sigma_t \in \Sigma$, the definition of an equilibrium takes the simple form:

Definition 1. $(q(\sigma_t))_{\sigma_t \in \Sigma}$ is an *equilibrium price process* if

$$u'_1(\omega^1_{s_t} - q(\sigma_t))q(\sigma_t) = \sum_{s' \in \mathbf{S}} \rho_{s_t s'} u'_2(\omega^2_{s_t s'} + q(\sigma_t, s'))q(\sigma_t, s'), \ \forall \sigma_t \in \Sigma \ (\mathcal{E})$$

It will sometimes be convenient to write $(\mathcal{E})$ in stochastic process notation as

$$u'_1(\omega^1_t - q_t)q_t = E\left(u'_2(\omega^2_{t+1} + q_{t+1})q_{t+1} \,|\, \mathcal{F}_t\right), \ \forall t \geq 0 \tag{$\mathcal{E}'$}$$

where $\mathcal{F}_t$ is the information available at date t.

In the deterministic case the stochastic difference equation $(\mathcal{E}')$ reduces to a simple difference equation which, under Assumption 3, defines q_{t+1} as a function of q_t: an equilibrium is a solution which satisfies some initial condition $q_0 = \bar{q}_0$ and respects the non-negativity of consumption at every date. In the stochastic case $(\mathcal{E})$ gives a single equation at each node σ_t for determining the S prices $(q(\sigma_t, s'))_{s' \in \mathbf{S}}$ at the immediate successors, and this suggests that the equilibria will be indeterminate – unless the assumption of rational expectations which requires that agents' expectations be fulfilled at all dates along a trajectory introduces further restrictions which eliminate this indeterminacy. Most of the current literature on stochastic OLG models sidesteps the indeterminacy problem by studying stationary equilibria which depend only on the current shock. Since we will be led to study equilibria which are stationary on a larger state space, we will refer to equilibria which depend only on the exogenous shocks as strongly stationary equilibria.

Definition 2. $(q^*_s)_{s \in \mathbf{S}}$ is a *strongly stationary* equilibrium (SE^*) price vector[2] if

$$u'_1(\omega^1_s - q^*_s)q^*_s = \sum_{s' \in \mathbf{S}} \rho_{ss'} u'_2(\omega^2_{ss'} + q^*_{s'})q^*_{s'}, \quad \forall s \in \mathbf{S} \tag{$\mathcal{E}^*$}$$

---

[2] We use the short hand SE^* for strongly stationary equilibria to avoid confusion with the stationary sunspot equilibria which are often referred to as SSE.

$(\mathcal{E}^*)$ is a system of S equations to determine the S unknowns $(q_s^*)_{s \in \boldsymbol{S}}$, which typically has a finite number of solutions. It clear that the trivial (no-trade) equilibrium $q_s^* = 0$ for all s is always a solution of $(\mathcal{E}^*)$. Cass et al. (1992) and Gottardi (1996) have proved existence of non-trivial SE^* in a more general class of models in which there is intergenerational trade with the same infinitely-lived asset as that considered here and, in addition, the young at each node are heterogeneous and trade short-lived assets with members of the same cohort to share their risks. The model studied here is simplified to focus attention on the intergenerational trade of the long-lived asset. Cass et al. concentrate on the case in which there is a positive solution to the system of equations $(\mathcal{E}^*)$, while Gottardi studies the general case in which there is either a positive or a negative solution, a negative solution corresponding to the case where the supply of the asset is negative. We follow the former authors and focus on the positive case – namely where agents want to transfer income forward and the welfare of all agents can be improved by trading an asset in positive supply. To express the conditions under which this occurs, consider the matrix of present-value vectors of the representative agents born in the S possible states, at their initial endowments

$$\Pi^0 = \left[\pi_{ss'}^0\right]_{s,s' \in \boldsymbol{S}} = \left[\frac{\rho_{ss'}\, u_2'(\omega_{ss'}^2)}{u_1'(\omega_s^1)}\right]_{s,s' \in \boldsymbol{S}}$$

Since Π^0 is a matrix with positive coefficients, by the Frobenius theorem (Gantmacher, 1959; Takayama, 1974) it has a unique positive eigenvalue (its Frobenius root) associated with a positive eigenvector. Let $\lambda_f(\Pi^0)$ denote this eigenvalue.

Assumption 4. $\lambda_f(\Pi^0) > 1$.

When Assumption 4 is satisfied, by the Frobenius theorem, there exists[3] a vector of transfers $dx = (dx_s)_{s \in \boldsymbol{S}} \gg 0$ such that

$$\Pi^0 dx = \lambda_f(\Pi^0) dx \gg dx$$

which can be expressed as

$$-u_1'(\omega_s^1)dx_s + \sum_{s' \in \boldsymbol{S}} \rho_{ss'}\, u_2'(\omega_{ss'}^2)dx_{s'} > 0, \ \forall s \in \boldsymbol{S}$$

Thus if an infinitely-lived benevolent planner were to transfer the amount dx_s from the young to the old at each date-event when the shock is s, $(s \in \boldsymbol{S})$, the welfare of all agents would be improved. In short when $\lambda_f(\Pi^0) > 1$, agents need to transfer income to their old age, and an asset in positive supply permits such transfers to occur.

Proposition 1. *Under Assumptions A1–A4, there exists a unique positive strongly stationary equilibrium.*

---

[3] We use the following notation for vector inequalities. For $x \in \mathrm{R}^S$, $x \geq 0$ implies $x_s \geq 0, \forall s$ (x non-negative); $x > 0$ implies $x_s \geq 0, \forall s$, and $x_{s'} > 0$ for some s' (x semi-positive); $x \gg 0$ implies $x_s > 0, \forall s$ (x positive).

Proof. For a price vector $q \in \mathbf{R}^S$ for which $w_s^1 - q_s > 0, w_{ss'}^2 + q_{s'} > 0, \forall s, s' \in \mathbf{S}$, define

$$\Pi(q) = [\pi_{ss'}(q)]_{s,s' \in \mathbf{S}} = \left[\frac{\rho_{ss'} \, u_2'(w_{ss'}^2 + q_{s'})}{u_1'(w_s^1 - q_s)} \right]_{s,s' \in \mathbf{S}}$$

Equation $(\mathcal{E}^*)$ for a SE^* can be written as

$$\Pi(q^*)q^* = q^* \tag{3}$$

which implies that the Frobenius root $\lambda_f(\Pi(q^*)) = 1$. It can be deduced from Gottardi (1996) that if $\lambda_f(\Pi^0) \neq 1$ there exists either a positive or a negative solution to (3). By concavity of u_1 and u_2, if $q^* \ll 0$, $\pi_{ss'}(q^*) > \pi_{ss'}^0$ so that, by the Frobenius theorem, $\lambda_f(\Pi(q^*)) > \lambda_f(\Pi^0)$. Thus if the initial endowments of the economy satisfy $\lambda_f(\Pi^0) > 1$, (3) cannot have a negative solution: it follows that it has at least one positive solution.

To prove uniqueness of this solution, construct by induction the following sequence of prices: let $q^{(0)} = \omega^1$ where $\omega^1 = (\omega_s^1)_{s \in \mathbf{S}}$. Note that, for any SE^* price vector q^*, $u'(\omega_s^1 - q_s^*)$ must be well defined so that $q^* \ll q^{(0)}$. Define the next price $q^{(1)}$ by

$$u_1'(\omega_s^1 - q_s^{(1)})q_s^{(1)} = \sum_{s' \in \mathbf{S}} \rho_{ss'} u_2'(\omega_{ss'}^2 + q_{s'}^{(0)})q_{s'}^{(0)}, \ \forall s \in \mathbf{S}$$

By concavity of the function u_1, the function $y \to u_1'(\omega_s^1 - y)y$ is increasing. By the Inada condition it increases from 0 to $+\infty$ when y increases from 0 to ω_s^1: thus $q^{(1)}$ is well defined and, since $u_1'(\omega_s^1 - q_s^{(1)})$ is also well defined for all s, $q^{(1)} \ll q^{(0)}$. By Assumption 3, the functions $h_{ss'}$ defined by $h_{ss'}(y) = u_2'(\omega_{ss'}^2 + y)y$ are increasing for $y > 0$ since

$$h_{ss'}'(y) = u_2'(\omega_{ss'}^2 + y)\left(1 + \frac{u_2''(\omega_{ss'}^2 + y)y}{u_2'(\omega_{ss'}^2 + y)}\right)$$

$$> u_2'(\omega_{ss'}^2 + y)\left(1 + \frac{u_2''(\omega_{ss'}^2 + y)(\omega_{ss'}^2 + y)}{u_2'(\omega_{ss'}^2 + y)}\right) \geq 0$$

Since, for any SE^*, $(\mathcal{E}^*)$ holds, $q^* \ll q^{(0)}$ implies $q^* \ll q^{(1)}$. It is now easy to see that the sequence $(q^{(n)})_{n \geq 1}$ defined by

$$u_1'(\omega_s^1 - q_s^{(n)})q_s^{(n)} = \sum_{s' \in \mathbf{S}} \rho_{ss'} u_2'(\omega_{ss'}^2 + q_{s'}^{(n-1)})q_{s'}^{(n-1)}, \ \forall s \in \mathbf{S} \tag{4}$$

is a decreasing sequence such that $q^* \ll q^{(n)}$ for all n. It converges to $\bar{q}$ which, since the functions in (4) are continuous, is a SE^*.

Suppose that there is a SE^* price $q^* < \bar{q}$. Then by the monotonicity properties shown above $\pi_{ss'}(q^*) \geq \pi_{ss'}(\bar{q})$, with at least one strict inequality. By the Frobenius theorem, $\lambda_f(\Pi(q^*)) > \lambda_f(\Pi(\bar{q}))$, which contradicts $\lambda_f(\Pi(q^*)) = \lambda_f((\Pi(\bar{q})) = 1$. Thus $\bar{q}$ is the unique positive SE^*. $\qquad\square$

Enlarging the class of equilibrium solutions. Our objective is to study a broader class of solutions of the equilibrium equations $(\mathcal{E})$ than the strongly stationary solutions defined by $(\mathcal{E}^*)$. The equilibrium conditions $(\mathcal{E})$ assert that at any given node σ_t of the event-tree Σ, given a current price $y \in \mathbf{R}_+$, there is a-priori an $(S-1)$-manifold $\Phi_{\sigma_t}(y)$ of prices of the asset at the S immediate successors of σ_t

$$\Phi_{\sigma_t}(y) = \left\{ q' \in \mathbf{R}^S \ \middle|\ \sum_{s' \in S} \rho_{s_t s'}\, u_2'(\omega_{s_t s'}^2 + q_{s'}')q_{s'}' = u_1'(\omega_{s_t}^1 - y)y \right\}$$

which justify an agent paying y for the asset at node σ_t. There are thus two possible ways of studying a broader class of solutions of $(\mathcal{E})$: the first is to study the solutions of the system of stochastic inclusions $q(\sigma_t^+) \in \Phi_{\sigma_t}(q(\sigma_t))$, $\forall\, \sigma_t \in \Sigma$, where $q(\sigma_t^+)$ denotes the vector of prices at the S immediate successors of σ_t; the second is to pick selections $\phi_{\sigma_t}(q(\sigma_t)) \in \Phi_{\sigma_t}(q(\sigma_t))$ and then study the stochastic difference equation $q(\sigma_t^+) = \phi_{\sigma_t}(q(\sigma_t))$, $\forall\, \sigma_t \in \Sigma$. We follow the latter approach.

Given the Markov structure of the endowments, Φ only depends on the current price y and the current shock s_t: $\Phi_{\sigma_t}(y) = \Phi_{s_t}(y)$. Although the selection in $\Phi_{\sigma_t}(q(\sigma_t))$ could depend on properties of the trajectory before date t, we restrict attention to selections which depend only on the current variables: $\phi_{\sigma_t}(q(\sigma_t)) = \phi_{s_t}(q(\sigma_t))$. Under this stationarity requirement a selection of the correspondence Φ is a family of S functions $q_s \to \phi_s(q_s)$ such that $\phi_s(q_s) \in \Phi_s(q_s)$ for all $s \in S$, and such a family of functions will be called an expectation function. This function must be restricted to prices which are feasible, i.e. it must be restricted to a domain of prices affordable by the young.

Definition 3. An *expectation function for state s* is a function $\phi_s : Q_s \to \mathbf{R}_+^S$, where $Q_s \subset [0, \omega_s^1)$, which associates with every current price q_s a vector of prices $(\phi_{s1}(q_s), \ldots, \phi_{sS}(q_s))$ which justifies the purchase of the asset at price q_s, i.e. which satisfies

$$u_1'(\omega_s^1 - q_s)q_s = \sum_{s' \in S} \rho_{ss'}\, u_2'(\omega_{ss'}^2 + \phi_{ss'}(q_s))\phi_{ss'}(q_s), \ \forall\, q_s \in Q_s \qquad (5)$$

If $(\phi_s)_{s \in S}$ is a family of expectation functions for the S states then the function

$$\phi : Q \times S \to \mathbf{R}_+^S \quad \text{defined by}\ \ \phi(q, s) = \phi_s(q_s), \quad \text{where}\ Q = \prod_{s \in S} Q_s$$

is called an *expectation function* for the economy.

Definition 4. $(q(\sigma_t))_{\sigma_t \in \Sigma}$ is a *rational expectations equilibrium with expectation function ϕ* if at each node $\sigma_t \in \Sigma$, with $\sigma_t = (s_0, \ldots, s_t)$, the prices at the S successors satisfy

$$q(\sigma_t, s') = \phi_{s_t s'}(q(\sigma_t)), \quad \forall s' \in S \qquad (\mathcal{E}_\phi)$$

Note that the assumption of rational expectations requires that all agents have the same expectation function ϕ: since the function ϕ is stationary on the state space

$Q \times S$ it is perhaps not unreasonable to think that agents could learn to coordinate on a function ϕ. To generate a solution to $(\mathcal{E}_\phi)$, at each node $\sigma_t \in \Sigma$ and for each of its successors $s' \in S$, the expectation function $\phi_{s_t s'}(q(\sigma_t))$ must select a feasible price to which the function $\phi_{s'}$ can be applied to form the expectation for the next date, i.e. $\phi_{s_t s'}(q(\sigma_t)) \in Q_{s'}$, and so on indefinitely. Thus to show that a rational expectations equilibrium in the sense of Definition 4 exists, one must exhibit an expectation function which is defined on a *self-justified domain*[4] $\bar{Q}$ i.e. a domain such that $\phi(\bar{Q} \times S) \subset \bar{Q}$.

For any a, $b \in \mathbf{R}^S$ such that $a \leq b$ let $[a, b] = \prod_{s \in S}[a_s, b_s]$ and $(a, b) = \prod_{s \in S}(a_s, b_s)$.

Proposition 2. *Let q^* denote the positive strongly stationary equilibrium of Proposition 1. (i) The maximal domain $\bar{Q}$ for which there exists an expectation function ϕ such that $\phi(\bar{Q} \times S) \subset \bar{Q}$ is $\bar{Q} = [0, q^*]$. (ii) There is a continuum of candidate expectation functions $\phi : \bar{Q} \times S \to \bar{Q}$, and they all satisfy $\phi_s(0) = 0$, $\phi_s(q_s^*) = q^*$ for all $s \in S$.*

Proof. (i) Consider the sequence $q^{(0)}, q^{(1)}, \ldots, q^{(n)}, \ldots$ constructed in the proof of Proposition 1, which converges to the unique SE^* price vector q^*. The domain Q of an expectation function must satisfy $Q \subset [0, \omega^1]$. Since by Assumption 3 the functions $y \to u_2'(\omega_{ss'}^2 + y)y$ are increasing, in order that the next period price expectations can be chosen in the domain Q, the price q_s observed in state s must be such that

$$
\begin{aligned}
u_1'(\omega_s^1 - q_s)q_s &\leq \sum_{s' \in S} \rho_{ss'} u_2'(\omega_{ss'}^2 + \omega_{s'}^1)\omega_{s'}^1 \\
&= \sum_{s' \in S} \rho_{ss'} u_2'(\omega_{ss'}^2 + q_{s'}^{(0)})q_{s'}^{(0)}, \ \forall s \in S
\end{aligned}
$$

Thus $Q \subset [0, q^{(1)}]$. By the same reasoning, since $\phi_s(q_s)$ needs to be in $[0, q^{(1)}]$ for all s, q_s must be less than $q^{(2)}$. Thus

$$
\phi(Q \times S) \subset Q \Longrightarrow Q \subset \bigcap_{n \geq 0}[0, q^{(n)}] = [0, q^*]
$$

To show that there is an expectation function defined on $\bar{Q} = [0, q^*]$ whose range is in $\bar{Q}$, define $\phi_{ss'}(q_s) = \lambda q_{s'}^*$ where λ is the solution to

$$
u_1'(\omega_s^1 - q_s)q_s = \sum_{s' \in S} \rho_{ss'} u_2'(\omega_{ss'}^2 + \lambda q_{s'}^*)\lambda q_{s'}^* \tag{6}
$$

The function $h_s(\lambda) = \sum_{s' \in S} \rho_{ss'} u_2'(\omega_{ss'}^2 + \lambda q_{s'}^*)\lambda q_{s'}^*$ is increasing (by Assumption 3), and such that $h_s(0) = 0 \leq u_1'(\omega_s^1 - q_s)q_s \leq u_1'(\omega_s^1 - q_s^*)q_s^* = h_s(1)$ for $q_s \in [0, q_s^*]$. Thus for any $q_s \in [0, q_s^*]$, there is a solution $\lambda \in [0, 1]$ and the function $\phi = (\phi_s)_{s \in S}$ maps $\bar{Q}$ into itself.

---

[4] Using the terminology of Duffie et al. (1994).

(ii) Note that if $0 < q_s < q_s^*$, then the solution λ of (6) satisfies $0 < \lambda < 1$ and the intersection of $\Phi_s(q_s)$ and the S-dimensional open set $(0, q^*)$ is a $(S-1)$-dimensional manifold of possible expectations $q' \in \bar{Q}$. Thus for each $s \in S$ and each $q_s \in (0, q_s^*)$ there is a continuum of possible choice for $\phi_s(q_s)$. However since $q' \leq q^*$ implies

$$\sum_{s' \in S} \rho_{ss'} \, u_2'(\omega_{ss'}^2 + q_{s'}') q_{s'}' \leq \sum_{s' \in S} \rho_{ss'} \, u_2'(\omega_{ss'}^2 + q_{s'}^*) q_{s'}^* = u_1'(\omega_s^1 - q_s^*) q_s^*$$

the only expectation in $\bar{Q}$ which justifies the price q_s^* in state s is q^*. Since it is clear that the only price expectation which justifies a zero price is zero, the properties (ii) hold.   □

Thus the global requirement that agents must anticipate prices which can be continued to form an infinite horizon equilibrium forces the feasible initial conditions to lie in $[0, q_{s_0}^*]$ and constrains agents' expectations of next period prices to lie in $\bar{Q}$ at every date. This property leads to a minimal stability property for the positive SE^* : if at some node where the current shock is s, the observed price of the asset is q_s^*, then the only expectation compatible with equilibrium is q^*, and the equilibrium coincides with the strongly stationary equilibrium forever after. The same stability property clearly holds for the no-trade equilibrium. However if the initial condition lies in the intermediate range, $0 < q_0 < q_{s_0}^*$, then there is a continuum of possible price expectations for the next period which can be fulfilled in equilibrium, so that there is a continuum of equilibrium trajectories starting at q_0.

In many deterministic OLG models the equilibrium conditions do not determine the initial conditions (prices and consumption at date 0) and these initial conditions must be specified exogenously: in these models, the dimension of indeterminacy is equal to the number of initial conditions which must be exogenously specified. For the deterministic version of the model that we are studying the dimension of indeterminacy is 1, since there is an equilibrium trajectory starting from any initial price q_0 for money such that $0 \leq q_0 \leq q^*$, where q^* is the Golden Rule price. In the stochastic case, not only is there an equilibrium starting at $q_0 \in [0, q_{s_0}^*]$, but if $0 < q_0 < q_{s_0}^*$ there is a continuum of such equilibrium paths. Thus the indeterminacy associated with the choice of the expectation function is over and above the indeterminacy present in the deterministic model.

The indeterminacy studied here has interesting connexions with the indeterminacy created by sunspots. The literature on sunspots has analyzed the conditions under which an economy with deterministic fundamentals (preferences and endowments) admits stochastic equilibria in which agents co-ordinate their beliefs on an exogenous stochastic process (see e.g. Chiappori and Guesnerie, 1991, or Guesnerie and Woodford, 1992, for surveys). The exogenous state space (extrinsic uncertainty) is introduced as a device to show that when agents believe that exogenous events can influence the economic outcome, such beliefs can be self-fulfilling – as a result the exogenous events end up influencing the equilibrium. The sunspot literature has focused on conditions under which economies with deterministic fundamentals have strongly stationary sunspot equilibria in which the equilibrium variables are functions of the extrinsic state. Because of this strong stationarity requirement no explicit expectation function ϕ is needed to describe the equilibria.

Our motivation for introducing a state space is quite different – we study equilibria of economies in which the fundamentals are stochastic, and this brings with it the natural state space required to describe the uncertainty. If agents' endowments are stochastic, then agents' beliefs must depend on the state of nature and this "intrinsic uncertainty" creates a continuum of possible beliefs. However the argument in the proof of Proposition 2 does not rely on the property that agents' endowments are random. Thus in the case where endowments are non-random ($w_s^1 = a, \forall s$ and $w_{s,s'}^2 = b, \forall s, s'$) Proposition 2 shows that there is a continuum of sunspot equilibria, stationary on the state space $Q \times S$. However there is no strongly stationary sunspot equilibrium i.e. one for which the price is a non-trivial function of the exogenous state.[5]

The choice of an expectation function ϕ is less crucial if the asymptotic properties of the equilibria do not depend too much on the particular choice of ϕ. To study these asymptotic properties we need to introduce some additional notation.

Any vector $q \in \bar{Q}$ can be written as $q = \lambda \circ q^*$ where $q^* = (q_1^*, \ldots, q_S^*)$ is the positive SE^* price vector, $\lambda = (\lambda_1, \ldots, \lambda_S) \in [0, 1]^S$ is the vector of scale factors relative to this price vector, and $\circ$ denotes the component-wise multiplication: $\lambda \circ q^* = (\lambda_1 q_1^*, \ldots, \lambda_S q_S^*)$. An expectation function $\phi : \bar{Q} \times S \to \bar{Q}$ has associated with it a unique function ψ which determines the scale factors expected for next period. We define the *scale function for state s*, $\psi_s : [0, 1] \to [0, 1]^S$, induced by the expectation function ϕ, by:

$$\psi_s(\lambda_s) \circ q^* = \phi_s(\lambda_s q_s^*) = \phi_s(q_s)$$

where $\psi_s(\lambda_s) = (\psi_{s1}(\lambda_s), \ldots, \psi_{sS}(\lambda_s))$. The family of functions $(\psi_s)_{s \in S}$ can be summarized by the function $\psi : [0, 1]^S \times S \to [0, 1]^S$ defined by:

$$\psi(\lambda, s) = \psi_s(\lambda_s), \quad \forall \lambda \in [0, 1]^S, \ s \in S$$

The requirement that ϕ satisfies the equilibrium condition (5) is equivalent to the requirement that ψ satisfies the condition

$$u_1'(w_s^1 - \lambda_s q_s^*)\lambda_s q_s^* = \sum_{s' \in S} \rho_{ss'} \, u_2'(w_{ss'}^2 + \psi_{ss'}(\lambda_s)q_{s'}^*)\psi_{ss'}(\lambda_s)q_{s'}^*, \quad (7)$$

$$\forall \lambda_s \in [0, 1], \ \forall s \in S$$

If, for $s, s' \in S$, we define the function

$$\Gamma_{ss'}(x, y) = \frac{u_2'(w_{ss'}^2 + x q_{s'}^*)q_{s'}^*}{u_1'(w_s^1 - y q_s^*)q_s^*}, \quad \forall x, y \in [0, 1]$$

---

[5] It is well known that under Assumption A3 there is no non-trivial strongly stationary sunspot equilibrium. This can be seen as a simple consequence of Proposition 1. For if y^* is the Golden Rule price of the deterministic model, namely the positive solution of $u_1'(a - y^*)y^* = u_2'(b + y^*)y^*$, then $q_s^* = y^*, \forall s \in S$, is a stationary equilibrium (which is trivial) and by Proposition 1 it is unique.

then (7) can be written as

$$\lambda_s = \sum_{s' \in S} \rho_{ss'} \Gamma_{ss'}(\psi_{ss'}(\lambda_s), \lambda_s)\psi_{ss'}(\lambda_s), \quad \forall \lambda_s \in [0,1], \ \forall s \in S, \quad (8)$$

A rational expectations equilibrium generated by an expectation function ϕ: $[0, q^*] \times S \to [0, q^*]$ can thus be equivalently described by its associated scale function ψ as:

$$q(\sigma_t) = \lambda(\sigma_t)q^*_{s_t}, \quad (9)$$
$$\lambda(\sigma_t, s') = \psi_{s_t s'}(\lambda(\sigma_t)),$$
$$\forall \sigma_t = (s_0, \ldots, s_t) \in \Sigma_t, \ \forall t \geq 0$$

where ψ satisfies (8).

Proposition 3. *Let $(q(\sigma_t))_{t \geq 0}$ be a rational expectations equilibrium associated with an expectation function ϕ. (i) If ϕ is continuous and such that the associated scale function ψ satisfies*

$$\lambda_s \geq \sum_{s' \in S} \rho_{ss'} \psi_{ss'}(\lambda_s), \quad \forall \lambda_s \in [0,1], \ \forall s \in S \quad (10)$$

then for almost all $\sigma \in \Sigma$ the equilibrium path $(q_t(\sigma))_{t \geq 0}$ converges to a strongly stationary equilibrium. (ii) If $0 \leq q_0 < q^*_{s_0}$, equilibrium paths converge to the no-trade equilibrium with positive probability, i.e. there exists $\Sigma' \subset \Sigma$ with $P(\Sigma') > 0$ such that $q_t(\sigma) \to 0$ for all $\sigma \in \Sigma'$.

Proof. Let q_0 be the initial condition with $q_0 = \lambda_0 q^*_{s_0}$, $0 \leq \lambda_0 \leq 1$ and let $\lambda(\sigma_t)$ be the stochastic process defined by (9), which can also be written in stochatic process notation

$$\lambda_{t+1}(\sigma) = \psi_{s_t s_{t+1}}(\lambda_t(\sigma))$$

Note that the two strongly stationary equilibria of the economy, the no-trade and the positive stationary equilibrium, are characterized respectively by $\lambda_t = 0$ and $\lambda_t = 1$ for all $t \geq 0$. Thus to prove convergence of an equilibrium path to a SE^* we need to prove that the sequence $(\lambda_t(\sigma))_{t \geq 0}$ converges either to 0 or 1. Condition (10) implies that for all $t \geq 0$, $\lambda_t \geq E(\lambda_{t+1}|\mathcal{F}_t)$, i.e. that the process $(\lambda_t)_{t \geq 0}$ is a supermartingale. Since it is bounded below by 0, for almost all $\sigma \in \Sigma$ the sequence $(\lambda_t(\sigma))_{t \geq 0}$ converges. Let σ be a path on which the sequence converges and let $\bar{\lambda} \in [0,1]$ denote the limit.

Let us show that for any s, s', $\psi_{ss'}(\bar{\lambda}) = \bar{\lambda}$. Note that

$$|\psi_{ss'}(\bar{\lambda}) - \bar{\lambda}| \leq |\psi_{ss'}(\bar{\lambda}) - \psi_{ss'}(\lambda_t(\sigma))| + |\psi_{ss'}(\lambda_t(\sigma)) - \bar{\lambda}|$$

Since $\rho_{ss'} > 0$, with probability 1 the succession of states s and s' occurs an infinite number of times on a trajectory. Thus w.l.o.g we can assume that the trajectory σ is such that for any $T > 0$ there exists $\tau \geq T$ such that $s_\tau = s$, $s_{\tau+1} = s'$. For any such τ

$$|\psi_{ss'}(\bar{\lambda}) - \bar{\lambda}| \leq |\psi_{ss'}(\bar{\lambda}) - \psi_{ss'}(\lambda_\tau(\sigma))| + |\lambda_{\tau+1}(\sigma) - \bar{\lambda}| \quad (11)$$

Since ψ is continuous and $\lambda_t(\sigma) \to \bar{\lambda}$, for any $\varepsilon > 0$, there exist $T > 0$ such that if $\tau \geq T$ each term of the right side of (11) is less than ε. Since this is true for any $\varepsilon > 0$ it follows that $\psi_{ss'}(\bar{\lambda}) = \bar{\lambda}$.

Since ψ satisfies (8)

$$\bar{\lambda} = \sum_{s' \in S} \rho_{ss'} \Gamma_{ss'}(\bar{\lambda}, \bar{\lambda})\bar{\lambda} \Longleftrightarrow \bar{\lambda}\left(1 - \sum_{s' \in S} \rho_{ss'} \Gamma_{ss'}(\bar{\lambda}, \bar{\lambda})\right) = 0, \forall s \in S \quad (12)$$

If $0 < \bar{\lambda} < 1$, then $\Gamma_{ss'}(\bar{\lambda}, \bar{\lambda}) > \Gamma_{ss'}(1, 1) = \frac{u_2'(\omega_{ss'}^2 + q_{s'}^*)q_{s'}^*}{u_1'(\omega_s^1 - q_s^*)q_s^*}$ and, since $\sum_{s' \in S} \rho_{ss'} \Gamma_{ss'}(1, 1) = 1$, $1 - \sum_{s' \in S} \rho_{ss'} \Gamma_{ss'}(\bar{\lambda}, \bar{\lambda}) < 0$. Thus the only solutions to (12) are $\bar{\lambda} = 0$ or $\bar{\lambda} = 1$, which proves that the equilibrium path converges either to the no-trade equilibrium or to the positive SE^*.

(ii) Let $\bar{\lambda}(\sigma)$ denote the random variable (defined almost everywhere) which is the limit of $(\lambda_t(\sigma))_{t \geq 0}$. If $\lambda_0 = 0$, then $\lambda_t(\sigma) = 0$ for all t and all $\sigma \in \Sigma$, so that (ii) clearly holds. If $0 < \lambda_0 < 1$, by property (10), $\lambda_t \geq E(\lambda_{t+1}|\mathcal{F}_t)$: it follows that the sequence $E(\lambda_t)$ is non-increasing and $E(\lambda_t) \leq \lambda_0$. Since $\lambda_t(\sigma)$ converges almost surely to $\bar{\lambda}(\sigma)$ and is dominated by the integrable constant function 1, by the dominated convergence theorem, $E(\lambda_t) \to E(\bar{\lambda})$. Thus $E(\bar{\lambda}) \leq \lambda_0$, which implies that $\bar{\lambda}(\sigma)$ can not be equal to 1 on a set of probability 1, and thus must be equal to zero on a set of positive probability. $\qquad \square$

In order to use the martingale theorem to prove convergence of the equilibrium paths, we had to impose condition (10) on the expectation function.[6] The strength of Proposition 3 depends on whether or not condition (10) seriously restricts the admissible expectation functions. First note that this condition is not vacuous. Condition (8), which is satisfied by any admissible function ψ, can be written as

$$\begin{aligned}
\lambda_s &= E_s\left(\Gamma_s(\psi_s(\lambda_s), \lambda_s)\psi_s(\lambda_s)\right) \\
&= E_s\left(\Gamma_s(\psi_s(\lambda_s), \lambda_s)\right) E_s\left(\psi_s(\lambda_s)\right) \quad (13) \\
&\quad + \text{cov}_s\left(\Gamma_s(\psi_s(\lambda_s), \lambda_s), \psi_s(\lambda_s)\right)
\end{aligned}$$

where $\psi_s(\lambda_s)$ denotes the S-vector $(\psi_{ss'}(\lambda_s))_{s' \in S}$, $\Gamma_s(\psi_s(\lambda_s), \lambda_s)$ denotes the vector $(\Gamma_{ss'}(\psi_{ss'}(\lambda_s), \lambda_s))_{s' \in S}$ and E_s is the expectation with respect to the probability conditional on state s. Since λ_s and $\psi_{ss'}(\lambda_s)$ are less or equal to 1 and $\Gamma_{ss'}$ is decreasing in both components, $E_s(\Gamma_s(\psi_s(\lambda_s), \lambda_s)) \geq E_s(\Gamma_s(1, 1)) = 1$ with a strict inequality if either λ_s or some component of $\psi_{ss'}$ is strictly less than 1. Thus if the covariance term is either non-negative or negative and small, the inequality

---

[6] For any expectation function, the price process (q_t) is a supermartingale with respect to the measure $\tilde{P}$ on Σ induced by the Markov chain $\tilde{\rho}_{ss'} = \rho_{ss'} \Gamma_{ss'}(1, 1)$. Thus for a set of measure 1 with respect to $\tilde{P}$, the equilibrium paths $q_t(\sigma)$ converge. Unfortunately in the infinite horizon case a change of conditional probability does not result in an "equivalent martingale measure" since the measure $\tilde{P}$ is not absolutely continuous with respect to P: convergence on a set of measure 1 for $\tilde{P}$ essentially proves nothing for the typical trajectory $\sigma \in \Sigma$ under the measure P. We are indebted to Jean Francois Mertens for pointing out this mistake in an earlier version of this paper.

(10) will hold. In particular if $\psi_{ss'}(\lambda_s)$ is constant as in the expectation function constructed in the proof of Proposition 2 then (10) is satisfied. (10) essentially places restrictions on how different the scale factors for each future state can be. When the scale factors are not constant, the greater the term $E_s\Big(\Gamma_s(\psi_s(\lambda_s),\lambda_s)\Big)$, the more negative the covariance term in (13) can be without violating (10). Intuitively since for given x, y, the function $\Gamma_{ss'}(x, y)$ increases when $\omega_{ss'}^2$ decreases (or ω_s^1 increases), condition (10) will tend to be satisfied if the endowments of the old are sufficiently small relative to that of the young. Although we do not have an argument which applies to the general case, we can show that for log and power utilities, if the endowments of the old are sufficiently small relative to those of the young, then condition (10) is satisfied by any expectation function ϕ.

Example. Let $S = 2$, $\mathbf{S} = \{a, b\}$, and assume that the shocks are i.i.d with conditional probabilities ρ_a and ρ_b. The preferences of the representative agent born at any node are defined by the utility function

$$U(x) = \log(x^1) + \rho_a \log(x_a^2) + \rho_b \log(x_b^2)$$

We assume that the endowments $(\omega_a^1, \omega_b^1, \omega_a^2, \omega_b^2)$ satisfy Assumption 4 which, in this case, reduces to $\rho_a \dfrac{\omega_a^1}{\omega_a^2} + \rho_b \dfrac{\omega_b^1}{\omega_b^2} > 1$: if the endowments of the young are on average greater that those of the old, then agents want to transfer forward. A strongly stationary equilibrium is a solution of the system of equations

$$\frac{q_a}{\omega_a^1 - q_a} = \frac{q_b}{\omega_b^1 - q_b} = \frac{\rho_a q_a}{\omega_a^2 + q_a} + \frac{\rho_b q_b}{\omega_b^2 + q_b}$$

which is equivalent to the system

$$\frac{q_a}{\omega_a^1} = \frac{q_b}{\omega_b^1}, \quad q_a\left(\frac{1}{\omega_a^1 - q_a} - \frac{\rho_a}{\omega_a^2 + q_a} - \frac{\rho_b}{\frac{\omega_b^2\omega_a^1}{\omega_b^1} + q_a}\right) = 0 \qquad (14)$$

which has the solution $q = 0$ and a positive solution q^* whose analytical expression is too complicated to be interesting. We will however use the fact that when $\omega^2 \to 0$ this solution tends to $(\omega_a^1/2, \omega_b^1/2)$.

An expectation function $\phi = (\phi_a, \phi_b)$ is such that, for $q_a \in [0, q_a^*]$, ϕ_a selects a point in the set

$$\Phi_a(q_a) \cap [0, q^*] = \left\{(q_a', q_b') \in [0, q^*] \,\middle|\, \frac{\rho_a q_a'}{\omega_a^2 + q_a'} + \frac{\rho_b q_b'}{\omega_b^2 + q_b'} = \frac{q_a}{\omega_a^1 - q_a}\right\}$$

and, for $q_b \in [0, q_b^*]$, ϕ_b selects a point in the set

$$\Phi_b(q_b) \cap [0, q^*] = \left\{(q_a', q_b') \in [0, q^*] \,\middle|\, \frac{\rho_a q_a'}{\omega_a^2 + q_a'} + \frac{\rho_b q_b'}{\omega_b^2 + q_b'} = \frac{q_b}{\omega_b^1 - q_b}\right\}$$

From the analysis above we know that the sets $\Phi_a(q_a) \cap [0, q^*]$ and $\Phi_b(q_b) \cap [0, q^*]$ are non empty if $0 \le q_a \le q_a^*$ and $0 \le q_b \le q_b^*$.

When the price vectors in the cube $[0, q^*]$ are expressed in terms of the scale factors, $q = \lambda \circ q^*$ with $\lambda \in [0, 1]^2$, a function ϕ is equivalent to a scale function ψ where, for each $\lambda_a \in [0, 1]$, ψ_a selects a point in

$$\Psi_a(\lambda_a) \cap [0, 1]^2 = \left\{ (\lambda'_a, \lambda'_b) \in [0, 1]^2 \;\middle|\; \frac{\rho_a \lambda'_a}{e_a^2 + \lambda'_a} + \frac{\rho_b \lambda'_b}{e_b^2 + \lambda'_b} = \frac{\lambda_a}{e_a^1 - \lambda_a} \right\}$$

and for each $\lambda_b \in [0, 1]$ ψ_b selects a point in

$$\Psi_b(\lambda_b) \cap [0, 1]^2 = \left\{ (\lambda'_a, \lambda'_b) \in [0, 1]^2 \;\middle|\; \frac{\rho_a \lambda'_a}{e_a^2 + \lambda'_a} + \frac{\rho_b \lambda'_b}{e_b^2 + \lambda'_b} = \frac{\lambda_b}{e_b^1 - \lambda_b} \right\}$$

where e_s^i denote the normalized endowments: $e_s^i = \omega_s^i / q_s^*$, $i = 1, 2$, $s = a, b$. Note that by (14) $e_a^1 = e_b^1$, so that $\Psi_a(\lambda) = \Psi_b(\lambda)$ for any $\lambda \in [0, 1]$. Furthermore since the function $x \to \frac{x}{\alpha + x}$ is concave for all positive values of α, the level curves $\Psi_s(\lambda)$ ($s = a, b$) have the shape of standard indifference curves. A selection ψ_s is represented by a curve in the box $[0, 1] \times [0, 1]$, $\psi_s(\lambda)$ being at the intersection of the curve with the level curve $\Psi_s(\lambda)$. Figure 1 represents one of these possible selections. While the level curves $\Psi_a(\lambda)$ and $\Psi_b(\lambda)$ are the same, the selections ψ_a and ψ_b can differ.

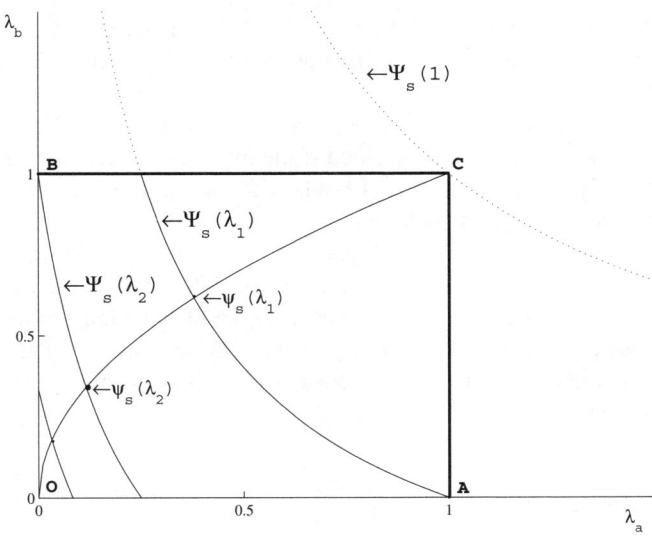

Figure 1

When the selection ψ_s is given by the intersection of the level curves with the diagonal of the box OABC then the covariance term in (13) is zero and (10) is satisfied. The further the selection ψ_s is from the diagonal, the greater the variance of ψ_s and the greater the likelihood that inequality (10) is violated. The maximum variance is obtained when the (scale) expectation function selects the point A or B of Figure 1. The point A with coordinates $(1, 0)$ is on the level curve corresponding to $\lambda_1 = \frac{\rho_a e_a^1}{1+\rho_a+e_a^2}$, while the point B with coordinates $(1,0)$ is on the level curve

corresponding to $\lambda_2 = \frac{\rho_b e_b^1}{1+\rho_b+e_b^2}$. If the inequality (10) is to be satisfied when $\lambda_s = \lambda_1$ and the expectation function selects A, or when $\lambda_s = \lambda_2$ and the expectation function selects B, then the following inequalities must be satisfied

$$e_s^1 \geq 1 + \rho_s + e_s^2, \quad s = a, b \tag{15}$$

When these inequalities are satisfied any expectation function ψ satisfies inequality (10).

Lemma 1. *If conditions (15) are satisfied, then (10) holds for any expectation function* ψ.

Proof. See Appendix.

Since $e_s^i = \omega_s^i/q_s^*$, inequalities (15) involve the SE^* prices which are difficult to express in closed form. However when the endowments of the old tend to zero, $q^* \to (\omega_a^1/2, \omega_b^1/2)$, so that $e_s^1 \to 2 > 1 + \rho_s$ and in the limit (15) is satisfied with strict inequalities. Thus there exist ε such that if $\omega_s^2 \leq \varepsilon, s = a, b$, then (15) hold. In this case Proposition 3 implies that for any admissible expectation function almost all equilibrium paths converge to an SE^*. Note that the equilibrium path may not converge with probability 1 to the no-trade equilibrium. If the expectation function selects scale factors on the segment AC or BC and if the initial condition satisfies $\lambda_0 \geq \lambda_1$ or $\lambda_0 \geq \lambda_2$ then, with positive probability, some paths converge to the positive SE^* $(1,1)$.

For power utilities, conditions analogous to (15) can be derived, expressing the condition that inequality (10) is satisfied if the expectation function selects point A and/or B of Figure 1 and Lemma 1 holds with this appropriate version of (15). However it is no longer possible to compute explicitly the limit of the price vector q^* when $\omega^2 \to 0$. For all numerical examples that we considered we found that the appropriate version of (15) holds in the limit with strict inequalities. Thus it seems that, at least when ω^2 is small, convergence of the equilibrium path to an SE^* is independent of the expectation function chosen–although the probability of converging to the no-trade or to the positive SE^* may depend on the choice of the expectation function.

3 The model with equity

The problem of indeterminacy disappears if the asset used for intergenerational transfer gives positive instead of zero dividends, i.e., if the asset is "equity"[7] instead of being "money". To see this, consider the same economy as in Section 2, with agents' characteristics satisfying Assumptions 1–3, the only change being that the infinitely-lived asset gives a dividend of $D_s \geq 0$ units of good at every date when the state of nature is s, for all $s \in S$, with at least one strict inequality. At date 0 the asset, whose supply is normalized to 1, belongs to the representative agent of the

---

[7] It can be the equity of any productive asset like "land" or a Lucas "tree".

old generation and is then exchanged at each date between the old and the young agent. The budget constraint of the representative agent born at node σ_t is now

$$x^1(\sigma_t) = \omega^1(\sigma_t) - q(\sigma_t)z, \ z \in R \tag{16}$$

$$x^2(\sigma_t, s') = \omega^2(\sigma_t, s') + (D_{s'} + q(\sigma_t, s'))z, \ s' \in S \tag{17}$$

where we retain the notation $q(\sigma_t)$ for the price of the asset. We assume free disposal of the (property right to the) asset so that the price $q(\sigma_t)$ must be non-negative. An equilibrium price process for the economy with equity is a process $(q(\sigma_t))_{\sigma_t \in \Sigma}$ such that $q(\sigma_t) \geq 0$ for all $\sigma_t \in \Sigma$ and

$$u_1'(\omega_{s_t}^1 - q(\sigma_t))q(\sigma_t) = \sum_{s' \in S} \rho_{s_t s'} u_2'(\omega_{s_t s'}^2 + D_{s'} + q(\sigma_t, s')) \tag{$\mathcal{E}_Q$}$$

$$(D_{s'} + q(\sigma_t, s')), \forall \sigma_t \in \Sigma$$

A strongly stationary equilibrium price vector is a non-negative vector $(q_s^*)_{s \in S}$ such that

$$u_1'(\omega_s^1 - q_s^*)q_s^* = \sum_{s' \in S} \rho_{ss'} u_2'(\omega_{ss'}^2 + D_{s'} + q_{s'}^*)(D_{s'} + q_{s'}^*), \ \forall s \in S \tag{$\mathcal{E}_Q^*$}$$

Proposition 4. *Under Assumptions 1-3 the economy with equity has a unique equilibrium which is a positive strongly stationary equilibrium.*

Proof. Let $(q(\sigma_t))_{\sigma_t \in \Sigma}$ be an equilibrium price process. Let σ_t be a node where the current shock is s. To be affordable by the young the price of the asset at a successor node (σ_t, s') of σ_t must be strictly less than $\omega_{s'}^1$. As before let $q^{(0)} = \omega^1$ be the vector of the initial endowments of the young. Then by monotonicity of the functions $y \to u_1'(\omega_s^1 - y)y$ and $y \to u_2'(\omega_{ss'}^2 + y)y$, $q(\sigma_t)$ must be less than $q_s^{(1)}$, where $q_s^{(1)}$ is defined by

$$u_1'(\omega_s^1 - q_s^{(1)})q_s^{(1)} = \sum_{s' \in S} \rho_{ss'} u_2'(\omega_{ss'}^2 + D_{s'} + q_{s'}^{(0)})(D_{s'} + q_{s'}^{(0)}), \ \forall s \in S$$

Note that $q^{(1)} \ll q^{(0)}$. If we define the sequence $(q^{(n)})_{n \geq 0}$ by induction

$$u_1'(\omega_s^1 - q_s^{(n+1)})q_s^{(n+1)} = \sum_{s' \in S} \rho_{ss'} u_2'(\omega_{ss'}^2 + D_{s'} + q_{s'}^{(n)})(D_{s'} + q_{s'}^{(n)}), \ \forall s \in S$$

then the equilibrium price process must be such that $q(\sigma_t) \leq q_{s_t}^{(n)}$ for all n and all $\sigma_t \in \Sigma$. Since $q^{(1)} \ll q^{(0)}$ the sequence is decreasing, and since it is bounded below by zero it converges to a vector q^* satisfying $(\mathcal{E}_Q^*)$. Since a vector with zero components cannot satisfy $(\mathcal{E}_Q^*)$, it follows that $q^* \gg 0$ and q^* is a positive SE^*. Since an equilibrium price process must satisfy $q(\sigma_t) \leq q_{s_t}^{(n)}$ for all n and all $\sigma_t \in \Sigma$, it must be such that $q(\sigma_t) \leq q_{s_t}^*$ for all $\sigma_t \in \Sigma$.

Since an equilibrium price process $(q(\sigma_t))_{\sigma_t \in \Sigma}$ must satisfy $(\mathcal{E}_Q)$ at each node and the right side of $(\mathcal{E}_Q)$ is strictly positive, the price must be positive at each node. Let $\tilde{q}^{(0)} = 0$ (the zero vector of $\mathbf{R}^S$). Since the prices at the successor nodes (σ_t, s') of σ_t must be strictly positive, by monotonicity, $q(\sigma_t)$ must be greater than $\tilde{q}_s^{(1)}$, where $\tilde{q}_s^{(1)}$ is defined by

$$u_1'(\omega_s^1 - \tilde{q}_s^{(1)})\tilde{q}_s^{(1)} = \sum_{s' \in S} \rho_{ss'} u_2'(\omega_{ss'}^2 + D_{s'} + \tilde{q}_{s'}^{(0)})(D_{s'} + \tilde{q}_{s'}^{(0)}), \ \forall s \in S$$

Note that $\tilde{q}^{(1)} \gg \tilde{q}^{(0)}$. If we define the sequence $(\tilde{q}^{(n)})_{n \geq 0}$ by induction

$$u_1'(\omega_s^1 - \tilde{q}_s^{(n+1)})\tilde{q}_s^{(n+1)} = \sum_{s' \in S} \rho_{ss'} u_2'(\omega_{ss'}^2 + D_{s'} + \tilde{q}_{s'}^{(n)})(D_{s'} + \tilde{q}_{s'}^{(n)}), \ \forall s \in S$$

then the equilibrium price process must be such that $q(\sigma_t) \geq \tilde{q}_{s_t}^{(n)}$ for all n and all $\sigma_t \in \Sigma$. Since $\tilde{q}^{(1)} \gg \tilde{q}^{(0)}$ the sequence is increasing, and since it is bounded above by $q^{(0)}$ it converges to a SE^* price vector $\tilde{q}^* \gg 0$. Any equilibrium price process must be such that $q(\sigma_t) \geq \tilde{q}_{s_t}^*$ for all $\sigma_t \in \Sigma$. Combining the two previous steps we find that all the equilibria must satisfy

$$\tilde{q}_{s_t}^* \leq q(\sigma_t) \leq q_{s_t}^*, \quad \forall \sigma_t \in \Sigma \tag{18}$$

For a vector $q \in \mathbf{R}^S$ such that $0 \leq q \ll \omega^1$ define the matrix

$$\hat{\Pi}(q) = [\hat{\pi}_{ss'}(q)]_{s,s' \in S} = \left[\frac{\rho_{ss'} u_2'(\omega_{ss'}^2 + D_{s'} + q_{s'})}{u_1'(\omega_s^1 - q_s)} \right]_{s,s' \in S}$$

If q is a SE^*, it satisfies

$$(I - \hat{\Pi}(q))q = \hat{\Pi}(q)D$$

Since $\hat{\Pi}(q)$ is a positive matrix and $D > 0$, $\hat{\Pi}(q)D \gg 0$. Thus for the vector $q \gg 0$, $(I - \hat{\Pi}(q))q \gg 0$. It follows that the matrix $I - \hat{\Pi}(q)$ is diagonal dominant, invertible, and $(I - \hat{\Pi}(q))^{-1} = I + \hat{\Pi}(q) + \ldots + \hat{\Pi}^n(q) + \ldots$ (see McKenzie, 1960; Takayama, 1974). Thus for any SE^* vector q

$$q = (\hat{\Pi}(q) + \hat{\Pi}^2(q) + \ldots + \hat{\Pi}^n(q) + \ldots)D \tag{19}$$

Consider the stationary equilibria q^* and $\tilde{q}^*$. Since $q^* \geq \tilde{q}^*$, and since the functions $q \to \hat{\pi}_{ss'}(q)$ are decreasing, $\hat{\Pi}(q^*) \leq \hat{\Pi}(\tilde{q}^*)$. By (19), $q^* \leq \tilde{q}^*$. Combining with (18) gives $q^* = \tilde{q}^*$, so that the equilibrium is unique. $\square$

Note that, for an economy with equity, the existence of a positive SE^* does not require Assumption 4, or rather its modified version $\lambda_f(\hat{\Pi}(0)) > 1$. Even if the agents have greater endowments in old age than in their youth, there is always a positive price – perhaps small – at which young agents will want to buy an asset yielding positive dividends.

From the positive point of view, the model with positive dividends is better be-haved than the model with money, since it does not exhibit the same indeterminacy.[8] It would be interesting to know if there are assumptions under which the uniqueness result extends to the more complicated models with heterogenous agents and sev-eral securities – typically bonds and equity – which are used in financial economics. This is left for future research.

Appendix

Proof of Lemma 1. Since $e_a^1 = e_b^1$ and for all $\lambda \in [0, 1]$, $\Psi_a(\lambda) = \Psi_b(\lambda)$ we omit the subscript a or b from e^1, Ψ and λ. Because of the shape of the level curves, if the 2 points at which a level curve $\Psi(\lambda)$ intersects the boundary of the box $OABC$ are below the line with equation $\rho_a \lambda_a' + \rho_b \lambda_b' = \lambda$, i.e. if these points have coordinates such that $\rho_a \lambda_a' + \rho_b \lambda_b' \leq \lambda$, then the same inequality will hold for all the points on the level curve $\Psi(\lambda)$ which are inside the box.

The level curve $\Psi(\lambda)$ intersects OB at the point with coordinates $\lambda_a'(\lambda) = 0$, $\lambda_b'(\lambda) = \frac{\lambda e_b^2}{\rho_b e^1 - \lambda(1+\rho_b)}$, if λ is such that $\frac{\lambda e_b^2}{\rho_b e^1 - \lambda(1+\rho_b)} \leq 1$ or equivalently

$$\lambda \leq \lambda_2 = \frac{\rho_b e^1}{e_b^2 + 1 + \rho_b} \tag{20}$$

Note that when (15) is satisfied, $\lambda_2 \geq \rho_b$. When $\lambda \geq \lambda_2$ then $\Psi(\lambda)$ does not interset OB and it intersects BC at the point $(\lambda_a'(\lambda), 1)$ where $\lambda_a'(\lambda)$ is the solution of the equation

$$\frac{\rho_a \lambda_a'}{e_a^2 + \lambda_a'} = \frac{\lambda}{e^1 - \lambda} - \frac{\rho_b}{e_b^2 + 1} \tag{21}$$

Let us show that in both cases the inequality $\rho_a \lambda_a'(\lambda) + \rho_b \lambda_b'(\lambda) \leq \lambda$ holds. If $\lambda \leq \lambda_2$, then $\rho_a \lambda_a'(\lambda) + \rho_b \lambda_b'(\lambda) = \rho_b \lambda_b'(\lambda)$ and

$$\rho_b \lambda_b'(\lambda) = \frac{\lambda \rho_b e_b^2}{\rho_b e^1 - \lambda(1 + \rho_b)} \leq \frac{\lambda \rho_b e_b^2}{\rho_b e^1 - \lambda_2(1 + \rho_b)} = \frac{\lambda \rho_b e_b^2}{\lambda_2 e_b^2} = \frac{\lambda \rho_b}{\lambda_2} \leq \lambda$$

since $\rho_b/\lambda_2 \leq 1$. If $\lambda \geq \lambda_2$, $\rho_a \lambda_a'(\lambda) + \rho_b \lambda_b'(\lambda) = \rho_a \lambda_a'(\lambda) + \rho_b$. Differentiating (21) gives $\frac{d}{d\lambda}(\lambda_a'(\lambda)) = \frac{e^1(e_a^2 + \lambda_a'(\lambda))^2}{e_a^2(e^1 - \lambda)^2}$, which, since $\lambda_a'(\lambda)$ is an increasing function of λ, is increasing. Thus the function $\lambda \to \rho_a \lambda_a'(\lambda) + \rho_b - \lambda$ is convex and if it is non-positive for $\lambda = \lambda_2$ and for $\lambda = 1$ it is non-positive in the interval $[\lambda_2, 1]$. $\lambda_a'(\lambda_2) = 0$ and $\lambda_2 \geq \rho_b$ implies that it is non-positive for $\lambda = \lambda_2$, while $\lambda_a'(1) = 1$ implies that it is non-positive for $\lambda = 1$. The analysis for the intersection

---

[8] However the positive SE^* of the model with equity has less good normative properties than that of the model with money. Since $\lambda_f(\hat{\Pi}(q^*)) < 1$, it can be deduced from Aiyagari and Peled (1991) that the equilibrium is Pareto optimal, or dynamically efficient, but is not optimal in the set of all feasible stationary allocations. This lack of the "Golden Rule property" can be deduced from Peled (1984) or seen directly by considering a transfer $dx = (dx_1, \ldots, dx_S) \gg 0$ from the old to the young, where dx is an eigenvector associated with the eigenvalue $\lambda_f(\hat{\Pi}(q^*))$.

of $\Psi(\lambda)$ with either OA or AC is similar and leads to the result that the inequality (10) holds for any function ψ which selects points on the boundary or in the interior of $OABC$.   □

References

Aiyagari, S. R., Peled, D.: Dominant root characterization of Pareto optimality and the existence of optimal equilibria in stochastic overlapping generations models. Journal of Economic Theory **54**, 69–83 (1991)

Cass, D., Green, R. C., Spear, S. E.: Stationary equilibria with incomplete markets and overlapping generations. International Economic Review **33**, 495–512 (1992)

Chattopadhyay, S., Gottardi, P.: Stochastic OLG models, market structure and optimality. Journal of Economic Theory **89**, 21–67 (1999)

Chiappori, P. A., Guesnerie, R.: Sunspot equilibria in sequential markets models. In: Hildenbrand, W., Sonnenschein, H. (eds.) Handbook of mathematical economics, Vol. IV. Amsterdam: North Holland 1991

Demange, G., Laroque, G.: Social security and demographic shocks. Econometrica **67**, 527–542 (1999)

Duffie, D., Geanakoplos, J., Mas-Colell, A., McLennan, A.: Stationary Markov equilibria. Econometrica **62**, 745–781 (1994)

Gale, D.: Pure exchange equilibrium of dynamic economic models. Journal of Economic Theory **6**, 12–36 (1973)

Gantmacher, F. R.: The theory of matrices. New York: Chelsea 1959

Geanakoplos, J. D., Polemarchakis, H. M.: Overlapping generations. In: Hildenbrand, W., Sonnenschein, H. (eds.) Handbook of mathematical economics, Vol. IV. Amsterdam: North Holland 1991

Gottardi, P.: Stationary monetary equilibria in overlapping generations models with incomplete markets. Journal of Economic Theory **71**, 75–89 (1996)

Guesnerie, R., Woodford, M.: Endogenous fluctuations. In: Laffont, J. J. (ed.) Advances in economic theory. Sixth World Congress, Vol. 2. Cambridge: Cambridge University Press 1992

Keynes, J. M.: A treatise on money, Vol. I. The pure theory of money. London: Macmillan 1936

Kurz, M.: Endogenous economic fluctuations: studies in the theory of rational beliefs. Berlin Heidelberg New York: Springer 1997

McKenzie, L. W.: Matrices with dominant diagonal and economic theory. In: Arrow, K., Karlin S., Suppes, P. (eds.) Mathematical methods in the social sciences. Stanford: Stanford University Press 1959

Peled, D.: Stationary Pareto optimality of stochastic asset equilibria with overlapping generations. Journal of Economic Theory **34**, 396–403 (1984)

Takayama, A.: Mathematical economics. Hinsdale. IL: The Dryden Press 1974

Existence and uniqueness of 'money' in general equilibrium: natural monopoly in the most liquid asset[*]

Ross M. Starr

Economics Department, University of California, San Diego, La Jolla, CA 92093-0508, USA
(e-mail: rstarr@weber.ucsd.edu)

Received: February 15, 2002; revised version: December 27, 2002

Summary. The monetary character of trade, use of a common medium of exchange, is shown to be an outcome of economic general equilibrium in the presence of transaction costs and market segmentation (in trading posts with a separate budget constraint at each transaction). Commodity money arises endogenously as the most liquid (lowest transaction cost) asset. Scale economies in transaction cost account for uniqueness of the (fiat or commodity) money in equilibrium, creating a natural monopoly. Trading posts using a medium of exchange create a network externality inducing others' adoption of the same medium. Bertrand monetary equilibria (among competing trading posts) and uniqueness of 'money' are robust to threats of entry. Government-issued fiat money has a positive equilibrium value from its acceptability for tax payments and sustains its natural monopoly through the scale of government economic activity.

Keywords and Phrases: Commodity money, Fiat money, Transaction cost, Scale economy, Double coincidence of wants.

JEL Classification Numbers: E40, D50.

---

* This paper has benefited from seminars and colleagues' helpful remarks at the University of California - Santa Barbara, University of California - San Diego, NSF-NBER Conference on General Equilibrium Theory at Purdue University, Society for the Advancement of Behavioral Economics at San Diego State University, Econometric Society at the University of Wisconsin - Madison, SITE at Stanford University-2001, Federal Reserve Bank of Kansas City, Federal Reserve Bank of Minneapolis, Midwest Economic Theory Conference at the University of Illinois - Urbana Champaign, University of Iowa, Southern California Economic Theory Conference at UC - Santa Barbara, Midwest Macroeconomics Conference at University of Iowa, University of California - Berkeley, European Workshop on General Equilibrium Theory at University of Paris I, Society for Economic Dynamics at San Jose Costa Rica, World Congress of the Econometric Society at University of Washington, Cowles Foundation at Yale University. It is a pleasure to acknowledge comments of Henning Bohn, Harold Cole, James Hamilton, Walter P. Heller, Mukul Majumdar, Harry Markowitz, Herbert Newhouse, Joseph Ostroy, Chris Phelan, Meenakshi Rajeev, Wendy Shaffer, Bruce Smith, and Max Stinchcombe.

"[An] important and difficult question...[is] not answered by the approach taken here: the integration of money in the theory of value..."
— Gerard Debreu, Theory of Value (1959)

1 Introduction

One of the oldest issues in economics is to explain the use of money, preferably in elementary terms based on the theory of value. There are contributions extending from Aristotle's *Politics* and Smith's *Wealth of Nations* to the present. The superiority of monetary trade to barter explains why monetary trade is efficient but not why monetary trade is a market equilibrium. No economic agent can individually decide to monetize; monetary exchange should be the equilibrium outcome of interaction among optimizing agents. *Money*, like *written language*, is one of the fundamental discoveries of civilization. Despite the evident superiority of monetary trade, it is puzzling; monetary trade involves one party to a transaction giving up something desirable (labor, his production, a previous acquisition) for something useless (a fiduciary token or a commonly traded commodity for which he has no immediate use) in the hope of advantageously retrading it. The foundations of monetary theory should include elementary economic conditions that allow this paradox to be sustained as an individually rational market equilibrium. Is there a (parsimonious) model of an economy where existence of a common medium of exchange is a result of the optimizing behavior of individual firms and households? Does the price system create money? The solution proposed in this paper focuses on transaction costs and their scale economies. The monetary character of trade, use of a common medium of exchange, is shown to be an outcome of an economic general equilibrium. Markets are assumed to be segmented[1] in trading posts, with a separate budget constraint at each transaction creating demand for a carrier of value between trading posts. Commodity money arises endogenously as the most liquid (lowest transaction cost) asset. Scale economies in transaction cost account for uniqueness of the (fiat or commodity) money in equilibrium, creating a natural monopoly. Trading posts using a medium of exchange create a network externality inducing others' adoption of the same medium. Bertrand monetary equilibria (among competing trading posts) and uniqueness of 'money' are robust to threats of entry. Government-issued fiat money has a positive equilibrium value from its acceptability for tax payments (a notion attributable to Adam Smith) and it sustains its natural monopoly due to the scale of government economic activity.

2 Money in Walrasian general equilibrium

Consider four commonplace observations on the character of trade in virtually all economies:

(i) Trade is monetary. One side of almost all transactions is the economy's common medium of exchange.

---

[1] The notion of market segmentation is essential to monetization (Alchian, 1977).

(ii) Money is (virtually) unique. Though each economy has a 'money' and the 'money' differs among economies, almost all the transactions in most places most of the time use a single common medium of exchange.

(iii) 'Money' is government-issued fiat money, trading at a positive value though it conveys directly no utility or production.

(iv) Even transactions displaying a double coincidence of wants are transacted with money.[2]

Where economic behavior displays such uniformity, a general elementary economic theory should be able to account for the universal usages. But (i), (ii), and (iii) contradict the implications of a frictionless Walrasian general equilibrium model, and (iv) contradicts the conventional view of the role of money (with regard to the double coincidence of wants). This essay presents a class of examples with a slight modification of the Arrow-Debreu general equilibrium model sufficient to derive points (i)-(iv) as outcomes. In doing so, this essay responds to a challenge expressed by Tobin (1980):

> Social institutions like money are public goods ... General equilibrium theory is not going to explain the institution of a monetary ... common means of payment.

Thus the examples below are intended to show that a general equilibrium model can explain endogenously from price theory the institution of a common monetary means of payment.[3] The price system itself designates 'money' and guides transactors to trade using 'money.' The model emphasizes complete markets and complete information. Points (ii) and (iv) involve scale economies, nonconvex transaction costs; it will typically be difficult to develop general existence of equilibrium theorems – hence the use of examples.

It is well known that a frictionless Arrow-Debreu model cannot accomodate a role for money. The single budget constraint facing transactors in the model precludes a carrier of value between transactions. This essay is intended as a partial counterexample, demonstrating that minimal friction in trade is sufficient to induce the existence of money as a result, not an assumption. The monetary structure of the economy is derived from elementary price theory in a class of examples. Use of a common medium of exchange, a commodity money, is an outcome of the market

---

[2] University of California faculty whose children are enrolled at the University pay fees in money, not in kind; Ford employees buying a Ford car pay in money, not in kind; Albertson's supermarket checkout clerks acquiring groceries pay in money, not in kind. This observation suggests that the focus on the absence of double coincidence of wants – as distinct from transaction costs – as an explanation for the monetization of trade may miss a significant part of the underlying causal mechanism.

[3] A bibliography of the issues involved in this inquiry appears in Ostroy and Starr (1990). In addition, note particularly Banerjee and Maskin (1996), Hellwig (2000), Howitt (2000), Howitt and Clower (2000), Iwai (1996), Kiyotaki and Wright (1989), Marimon, McGrattan and Sargent (1990), Rajeev (1999), Rey (2001), Trejos and Wright (1995), and Young (1998). The treatment of transaction costs in this essay (as opposed to the recent focus in the literature on search and random matching equilibria) resembles the general equilibrium models with transaction cost developed in Foley (1970), Hahn (1971), Starrett (1973), and Kurz (1974). The structure of bilateral trade here however is more detailed, with a budget constraint enforced on each transaction separately, so that the Foley, Hahn, and Starrett models do not immediately translate to the present setting.

equilibrium. Starting from a (non-monetary) Arrow-Debreu model, the monetary quality of the economic equilibrium is derived through the addition of market segmentation (with a separate budget constraint in each segment) and transaction costs. Multiplicity of budget constraints – requiring that goods acquired be paid for by delivery of equal value at each trade separately (Ostroy, 1973) – creates a demand for media of exchange. Transaction costs imply differing bid and ask prices for each good. Liquidity is priced: its price is the bid/ask spread. The most liquid asset, the instrument that provides liquidity at lowest cost, will be chosen as the medium of exchange. Thus, the choice of a 'money' is the outcome of optimizing behavior of economic agents in a market equilibrium. Fiat money – issued by government – derives its positive value from acceptability in payment of taxes; it becomes the medium of exchange from its low transaction cost. Uniqueness of (fiat or commodity) money follows from scale economy in transaction costs.

Section 4 of the paper presents the model of segmented markets with linear transaction costs without double coincidence of wants. Commodity money arises endogenously in market equilibrium. Section 5 demonstrates that the absence of double coincidence of wants is essential to monetization of trade in a linear model by considering the same problem with full double coincidence of wants. The result is a nonmonetary equilibrium. Section 7 considers a (nonconvex) transaction technology with scale economies. The examples there demonstrate that uniqueness of money (uniqueness of the endogenously chosen medium of exchange) results from scale economies in transaction costs. Further, Section 7 demonstrates that scale economies in transaction cost account for monetization of trade with a unique 'money' even when there is full double coincidence of wants. Section 8 presents the same issues in an oligopolistic setting, as a Bertrand equilibrium. Section 9 considers government-issued fiat money whose value is supported by acceptability in payment of taxes. Scale economies in transaction cost and government's large scale ensure that fiat money is the unique common medium of exchange.[4]

---

[4] It is useful to distinguish search/random matching models of money, e.g. Kiyotaki and Wright (1989), Trejos and Wright (1995), from general equilibrium models with transaction cost, e.g. Foley (1970), Hahn (1971), Starrett (1973), Ostroy and Starr (1974), Iwai (1996), Howitt (2000), and this essay. Search models emphasize very imperfect markets with limited ability of traders to locate desirable trades and with limited price flexibility. That approach is consistent with Smith (1776), v.I, book I, ch. 4. General equilibrium models typically model complete markets and a fully articulated price system. Using the general equilibrium approach allows us to pursue a parsimonious theory: What is a minimal set of market imperfections so that money arises endogenously?

The random matching/search formalization of the friction in trade has a very classical implication: in the rare case where two agents have a double coincidence of wants and meet to trade, they will trade their goods or services directly for one another (Kiyotaki and Wright, 1991; Trejos and Wright, 1993). This is a distinctive feature, distinguishing the random matching/search models from general equilibrium with transaction cost models. In the present model, direct trade between agents with reciprocal demands will take place only when that arrangement provides the lowest available transaction cost (Example 5.1). Hence, even in the rare instance of double coincidence of wants, general equilibrium models with transaction cost need not predict direct trade between parties with reciprocal demands and supplies.

In actual monetary economies, in those comparatively rare instances where double coincidence of wants occurs, it is seldom resolved by barter exchange. Trade between agents – even with a double coincidence of wants – usually takes a monetary form. This is typified by the examples above of a University of California professor's child's University fees, a supermarket checkout clerk's payment for groceries, and an autoworker's purchase of a car. Even in the setting most propitious for barter, those instances where

3 Formalizing Menger's 'origin of money'

Over a century ago, Carl Menger presented the paradox of monetary trade as a challenge to monetary theory and proposed an outline of its solution, a theory of liquidity as the basis of monetary theory (Menger, 1892):

> It is obvious ... that a commodity should be given up by its owner ...for another more useful to him. But that every[one] ... should be ready to exchange his goods for little metal disks apparently useless as such...or for documents representing [them]...is...mysterious... why...is...economic man ...ready to accept a certain kind of commodity, *even if he does not need it*, ... in exchange for all the goods he has brought to market[?] [Call] goods ... *more or less saleable*, according to the ... facility with which they can be disposed of ... at current purchasing prices or with less or more diminution... Men ... exchange goods ... for other goods ... more saleable....[which] become generally acceptable media of exchange [emphasis in original].

Menger's proposed solution focused on the liquidity of commodities. A good is very *saleable* (liquid) in Menger's definition above, if the price at which a household can sell it (the market's prevailing bid price) is very near the price at which it can buy (the market's prevailing ask price). In this setting, price theory includes a theory of liquidity. The segmented market creates a demand for a carrier of value between transactions. Separate bid and ask prices represent transaction costs and put a price on liquidity: a good's bid/ask spread is the price of using it as a medium of exchange. Hence, a good with a uniformly narrow bid/ask spread is highly liquid – in Menger's word 'saleable' – and constitutes a natural 'money.' Price theory implies monetary theory. Liquidity creates monetization. This is the insight that will be formalized in the examples below.

Starting from the non-monetary Arrow-Debreu model, two additional structures are sufficient to give endogenous monetization in equilibrium: multiple budget constraints (one at each transaction, not just on net trade) and transaction costs. One way of formalizing multiple budget constraints is a trading post model. Thus, if there are N goods actively traded, there are $N(N-1)/2$ possible trading posts. That is the starting point of the examples below. The choice of which trading posts a typical household will trade at is part of the household optimization. The equilibrium structure of exchange is the array of trading posts that actually host active trade. The determination of which trading posts are active in equilibrium is endogenous to the model and characterizes the monetary character of trade. The equilibrium is monetary with a unique money if only $(N-1)$ trading posts are active, those trading all goods against 'money.'

The examples below derive monetary equilibrium as a market equilibrium of optimizing agents based on elementary considerations of transaction cost. Household optimization includes deciding at which trading posts the household will trade. For a given mix of goods, trade is drawn to the lowest transaction cost trading posts.

---

double coincidence of wants occurs, monetary trade prevails. This usage contradicts the predictions of the random matching/search models. It is consistent however with Ostroy and Starr (1974, Theorem 4), and it is precisely the behavior Examples 7.2, 7.3, 7.4, 8.2, 8.3, and 9.1 below would predict.

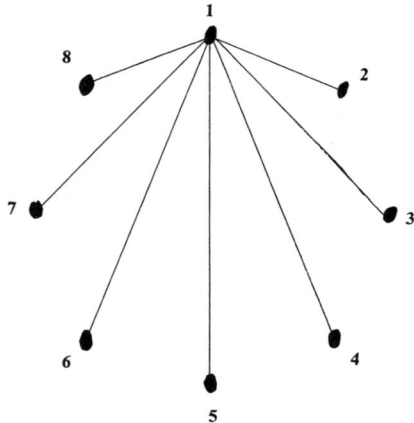

Figure 1. Monetary equilibrium with unique money

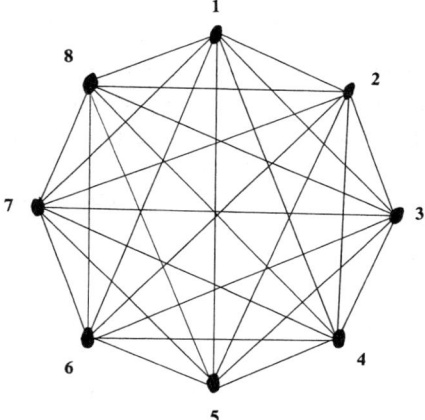

Figure 2. Barter equilibrium for H^D

The question *Why is there money?* can then be answered by presenting sufficient conditions so that an equilibrium trading array has N-1 active trading posts, those trading in a common medium of exchange versus the N-1 other goods. This is illustrated in Figures 1 and 2. Each node in the figures represents a commodity. Active trade is represented by a chord between nodes. A barter economy will have chords among a wide variety of goods – one for each pair of goods where there is a household with a matching demand and supply (Figure 2). A monetary economy with a unique money will be a sparser array. There will be one good so that the only chords are those linking that good to all others (Figure 1). The question *why is there money?* is then reduced to asking for sufficient conditions so that the array of active trading posts in equilibrium looks like Figure 1 (spider-shaped) instead of Figure 2 (star-shaped).

4 Monetization comes from liquidity:
monetary competitive equilibrium with linear transaction costs

The distinctive features of the model are (i) transactions exchange pairs of goods, (ii) budget constraints are enforced at each transaction separately, generating a role for a carrier of value between transactions (a medium of exchange), (iii) transaction costs are assumed to be linear in Sections 4 and 5 and nonconvex (displaying scale economies) in Sections 6, 7, 8 and 9. In the linear transaction cost case without double coincidence of wants, the most liquid (lowest transaction cost) good becomes the common medium of exchange. There may be multiple media of exchange when there is a tie for lowest cost.

Let there be $N+1$ commodities, numbered 0,1,2,...N. They are traded in pairs – good i for good j – at specialized trading posts. The trading post for trade of good i versus good j (and vice versa) is designated $\{i, j\}$; trading post $\{i,j\}$ is the same trading post as $\{j, i\}$. Trading post $\{i, j\}$ is a business firm, the market maker in trade between goods i and j. $\{i, j\}$ actively buys and (re)sells both i and j. Trade as a resource using activity is modeled by describing the post's transaction costs. The notion of transaction cost summarizes costs that in an actual economy are incurred by retailers, wholesalers, individual firms and households. The bid/ask spread summarizes these costs to the model's transactors. Thus, part of transaction cost represents the (non-marketed) time and resources used by households in arranging their transactions, summarized here imprecisely as a price spread.[5]

Specify a transaction cost function for these pairwise trading posts so that all transaction costs accrue in good 0. This is obviously a restrictive convention, but it simplifies accounting for transaction costs. It is simplest to think of good 0 as the labor used in the transaction technology. Trading post $\{i,j\}$ buys good 0 as an input to its transaction costs. The typical transactions of trading post $\{i,j\}$ will consist of purchases $y_i^{\{i,j\}B}$, $y_j^{\{i,j\}B}$, $y_0^{\{i,j\}B} \geq 0$ of i, j, and 0 respectively and sales $y_i^{\{i,j\}S}$, $y_j^{\{i,j\}S} \geq 0$ of i and j. In this section, we use the further simplifying assumption of linear transaction costs. The cost structure is generalized to non-convex costs in Sections 6, 7, 8, and 9.

The transaction cost function[6] for trading post $\{i, j\}$ is

$$C^{\{i,j\}} = y_0^{\{i,j\}B} = \delta^i y_i^{\{i,j\}B} + \delta^j y_j^{\{i,j\}B} \tag{TCL}$$

where $\delta^i, \delta^j > 0$. In words, the transaction technology looks like this: Trading post $\{i, j\}$ makes a market in goods i and j, buying each good in order to resell it. It incurs transaction costs in good 0. These costs vary directly (in proportions δ^i, δ^j) with volume of trade. The transaction cost structure is separable in the two principal traded goods. The trading post $\{i, j\}$ buys good 0 to cover the transaction costs it incurs, paying for 0 in good i and j. The transaction cost function $C^{\{i,j\}}$ is sufficiently flexible to distinguish transaction costs differing among commodities, including differences in durability, portability, recognizibility, divisibility.

---

[5] An alternative more explicit treatment of household non-market transaction cost decisions is embodied in Kurz (1974).

[6] (TCL) is intended as a mnemonic for linear transaction cost.

The population of households is denoted H, consisting of a mix of subpopulations (with different tastes and endowments). A typical household $h \in H$, has an endowment $r^h \in R_+^N$; r_n^h is h's endowment of good n. For simplicity in the examples below, each household is endowed with only one commodity. This is obviously inessential. h's utility function is $u^h(x) = u^h(x_0, x_1, \ldots, x_N)$.

It is convenient to arrange a subpopulation H^0 to provide good 0 (transaction labor). H^0's endowment of good 0 is characterized as $\sum_{h \in H^0} r_0^h > \sum_{h \in H} \sum_{i=1}^N \delta^i r_i^h$. For typical $h \in H^0$, h's utility function is

$$u^h(x) = \sum_{i=0}^N x_i. \tag{U0}$$

That is, a subpopulation H^0 owns all of the good 0 in sufficient quantity to cover all the transaction costs in the economy that are likely to be incurred; h's tastes, for $h \in H^0$, treat all goods as perfect substitutes with MRS equal to unity. This unrealistic assumption is designed to make accounting for transaction costs particularly easy.

A typical household outside of H^0 may be denoted $h = [m, n]$ where m and n are integers between 1 and N (inclusive). m denotes the good with which h is endowed. n denotes the good h prefers. $[m, n]$'s utility function can then be taken to be

$$u^{[m,n]}(x) = \sum_{i=0, i \neq n}^N x_i + 3x_n. \tag{U1}$$

$[m, n]$'s endowment, $r_m^{[m,n]}$, is specified as part of the description of the subpopulation.

Households formulate their trading plans deciding how much of each good to trade at each pairwise trading post. This leads to the rather messy notation:

$b_\ell^{[m,n]\{i,j\}}$ = planned purchase of good ℓ by household $[m, n]$ at trading post $\{i, j\}$.

$s_\ell^{[m,n]\{i,j\}}$ = planned sale of good ℓ by household $[m, n]$ at trading post $\{i, j\}$. The bid prices (the prices at which the trading post will buy from households) at $\{i, j\}$ are $q_i^{\{i,j\}}$, $q_j^{\{i,j\}}$ for goods i and j respectively. The price of i is in units of j. The price of j is in units of i. The ask price (the price at which the trading post will sell to households) of j is the inverse of the bid price of i (and vice versa). That is, $(q_i^{\{i,j\}})^{-1}$ and $(q_j^{\{i,j\}})^{-1}$ are the ask prices of j and i at $\{i, j\}$. The trading post $\{i, j\}$ covers its costs by the difference between the bid and ask prices of i and j, that is, by the spread $(q_j^{\{i,j\}})^{-1} - q_i^{\{i,j\}}$ and the spread $(q_i^{\{i,j\}})^{-1} - q_j^{\{i,j\}}$. Transaction costs at the trading post are incurred in good 0. Post $\{i, j\}$ pays for 0 in i and j, acquired in trade through the difference in bid and ask prices. The bid price of 0 in terms of i is $q_{(i)0}^{\{i,j\}}$. The bid price of 0 in terms of j is $q_{(j)0}^{\{i,j\}}$.

Given $q_i^{\{i,j\}}$, $q_j^{\{i,j\}}$, for all $\{i, j\}$, household h then forms its buying and selling plans, in particular deciding which trading posts to use to execute his desired trades. Household $h \in H$ faces the following constraints on its transaction plans:

(T.i) $b_n^{h\{i,j\}} > 0$ only if $n = i, j$; $s_n^{h\{i,j\}} > 0$ only if $n = i, j, 0$.

(T.ii) $b_i^{h(i,j)} \leq q_j^{\{i,j\}} \cdot s_j^{h\{i,j\}}$, $b_j^{h\{i,j\}} \leq q_i^{\{i,j\}} \cdot s_i^{h\{i,j\}}$ for each $\{i,j\}$. There is a slightly distinct version of (T.ii), (T.ii'), applying to households in H^0.

(T.ii') For $h \in H^0$, decompose $s_0^{h\{i,j\}}$ into nonnegative elements $s_{(i)0}^{h\{i,j\}}$ and $s_{(j)0}^{h(i,j)}$, so that $s_{(i)0}^{h\{i,j\}} + s_{(j)0}^{h\{i,j\}} = s_0^{h\{i,j\}}$, then we have $b_i^{h\{i,j\}} \leq q_{(i)0}^{\{i,j\}} \cdot s_{(i)0}^{h\{i,j\}}$, and $b_j^{h\{i,j\}} \leq q_{(j)0}^{\{i,j\}} \cdot s_{(j)0}^{h\{i,j\}}$ for each $\{i,j\}$.

(T.iii) $x_n^h = r_n^h + \sum_{\{i,j\}} b_n^{h\{i,j\}} - \sum_{\{i,j\}} s_n^{h\{i,j\}} \geq 0$, $0 \leq n \leq N$.

Note that condition (T.ii)[and (T.ii')] defines a budget balance requirement at the transaction level, implying the decentralized character of trade. Since the budget constraint applies to each pairwise transaction separately, there may be a demand for a carrier of value to move purchasing power between distinct transactions. h faces the array of bid prices $q_i^{\{i,j\}}, q_j^{\{i,j\}}$ and chooses $s_n^{h\{i,j\}}$ and $b_n^{h\{i,j\}}$, $n = i, j$ (and $n = 0$ for $h \in H^0$), to maximize $u^h(x^h)$ subject to (T.i), (T.ii), (T.iii). That is, h chooses which pairwise markets to transact in and a transaction plan to optimize utility, subject to a multiplicity of pairwise budget constraints.

The trading posts in Sections 4 and 5 have linear transaction technologies. A competitive equilibrium is an appropriate solution concept resulting in zero profits for the typical trading post (with the additional benefit that no account need be taken of distribution of profits). The threat of entry (by other similar trading post firms) rationalizes the competitive model, but for simplicity we take there to be a unique trading post firm making a market in goods i and j, denoted indiscriminately $\{i,j\}$ $= \{j, i\}$.

A *competitive equilibrium* under (TCL) consists of $q_{(i)0}^{o\{i,j\}}, q_{(j)0}^{o\{i,j\}}, q_i^{o\{i,j\}}$, $q_j^{o\{i,j\}}$, $1 \leq i, j \leq N$, so that :

- For each household $h \in H$, there is a utility optimizing plan $b_n^{oh\{i,j\}}, s_n^{oh\{i,j\}}$, (subject to T.i, T.ii [or T.ii' for $h \in H^0$], T.iii) so that $\sum_h b_n^{oh\{i,j\}} = y_n^{o\{i,j\}S}$, $\sum_n s_n^{oh\{i,j\}} = y_n^{o\{i,j\}B}$, $n = i, j$, for each $\{i,j\}$, each n, where
- $y_n^{o\{i,j\}S} \leq y_n^{o\{i,j\}B}$, $n = i, j$.
- $y_0^{o\{i,j\}B}$ can be divided into two parts, $y_{(i)0}^{o\{i,j\}B} \geq 0$, $y_{(j)0}^{o\{i,j\}B} \geq 0$, so that $y_{(i)0}^{o\{i,j\}B} + y_{(j)0}^{o\{i,j\}B} = y_0^{o\{i,j\}B} = C^{\{i,j\}}$
- $q_{(i)0}^{o\{i,j\}} y_{(i)0}^{o\{i,j\}B} \leq y_i^{o\{i,j\}B} - q_j^{o\{i,j\}B} y_j^{o\{i,j\}B} \cdot q_{(j)0}^{o\{i,j\}} y_{(j)0}^{o\{i,j\}B} \leq y_j^{o\{i,j\}B} - q_i^{o\{i,j\}} y_i^{o\{i,j\}B}$.
- $\delta^i + \delta^j q_i^{o\{i,j\}} = (q_{(i)0}^{o\{i,j\}})^{-1}(1 - q_i^{o\{i,j\}} q_j^{o\{i,j\}})$, $\delta^j + \delta^i q_j^{o\{i,j\}}$ $= (q_{(j)0}^{o\{i,j\}})^{-1}(1 - q_i^{o\{i,j\}} q_j^{o\{i,j\}})$.

The expression in the last bullet is a marginal cost pricing condition: the transaction cost (in good 0) of buying one unit of i and enough j to pay for it (pricing the 0 in good i) is equal to the amount of i left over after completing the trade in i and j. Similarly for trade in j.

An equilibrium is said to be *monetary* with a unique money, μ, if – for all households – good μ is the only good that a household will both buy and sell.

An equilibrium will be monetary with multiple moneys, $\mu^1, \mu^2, \ldots$, if – for all households – $\mu^1, \mu^2, \ldots$ are the only goods that a household will both buy and sell.

Jevons (1875) reminds us that monetization of trade follows in part from the absence of a double coincidence of wants. In the present model, that logic is particularly powerful. Absence of coincidence of wants means that the typical traded good will be traded more than once in moving from endowment to consumption. Barter trade successfully rearranging the allocation to an equilibrium will transact an endowment first at the trading post where it is supplied and again at a distinct post where it is demanded. Hence monetary trade as an alternative (substituting retrade of money for the retrade of nonmonetary goods) can be undertaken without increasing total trading volume or transaction cost, even without scale economies. Conversely, when there is a full double coincidence of wants *and linear transaction cost*, equilibrium will be non-monetary even in the presence of a natural money (Sect. 5).

We now formalize the notion of the absence of double coincidence of wants. Let N be an integer, $N \geq 3$. For $m = 1, 2, \ldots, N$ and positive integers i, $1 \leq i \leq N-1$, let

$$m \oplus i = \begin{cases} m + i & \text{if } m + i \leq N, \\ m + i - N & \text{if } m + i > N \end{cases}$$

That is, $m \oplus i$ denotes $m + i$ mod N, skipping 0 (since good 0 is used primarily as an input to the transaction process). Recall that $[m, n]$ denotes a household endowed with good m, strongly preferring good n. Using the notation above, let $H^1 = \{[m, m \oplus 1] | m = 1, 2, \ldots, N; r_m^{[m,m\oplus 1]} = A > 0\}$. H^1 characterizes a population of N households with the same size of initial endowment, so that no pair of them have reciprocal matching endowments and preferences but so that their endowments in aggregate can be reallocated to make each one significantly better off (roughly by arranging the households clockwise in a circle ordered by endowment good and having each household $[m, m \oplus 1]$ send his endowment one place counterclockwise).

Example 4.1 (Existence of monetary equilibrium with a most liquid asset, absent double coincidence of wants): Let the population of households be $H = H^0 \cup H^1$. Let $C^{\{i,j\}}$ be described by (TCL). Let $0 < \delta^i < 1/3$ and $0 < \delta^1 < \delta^i$, for $i = 2, 3, \ldots, N$. Transaction costs are constant and non-trivial for all goods; they are significantly lower in good 1. Then there is a unique competitive equilibrium allocation (though a range of prices may support the *unique* real allocation of trades and consumptions). The equilibrium is a monetary equilibrium with good 1 as the unique 'money'.

Demonstration of Example 4.1: Using marginal cost pricing and market clearing, we have for each $\{i, j\}$, $i \neq j$, $1 \leq i, j \leq N$, $q_{(i)0}^{\{i,j\}} = q_{(j)0}^{\{i,j\}} = 1$, $q_{i\oplus 1}^{\{i,i\oplus 1\}} = 1$, $q_i^{\{i,i\oplus 1\}} = \frac{1-\delta^i}{1+\delta^{i\oplus 1}}$, and for $j \neq 1, i \oplus 1$, $q_i^{\{i,j\}} = 1 - \delta^i$; $q_i^{\{i,1\}} = \frac{1-\delta^i}{1+\delta^1}$, $q_1^{\{i,1\}} = 1$. $s_i^{[i,i\oplus 1]\{i,1\}} = A$, $b_1^{[i,i\oplus 1]\{i,1\}} = q_i^{\{i,1\}} A = s_1^{[i,i\oplus 1]\{i\oplus 1,1\}}$, $b_{i\oplus 1}^{[i,i\oplus 1]\{i\oplus 1,1\}} = q_i^{\{i,1\}} q_1^{\{i\oplus 1,1\}} A$.

What's happening in Example 4.1? At first household $[i, i \oplus 1]$ goes to trading post $\{i, i \oplus 1\}$ offering i in exchange for $i \oplus 1$. But no one is coming to the trading post offering $i \oplus 1$. So good i is priced at a large discount at the post, reflecting the transaction costs of both i and $i \oplus 1$. On all other markets $\{i, j\}$ goods are priced to reflect their transaction costs, $q_i^{\{i,j\}} = 1 - \delta^i$. But at that pricing, since $\delta^1 < \delta^i$, it is advantageous for $[i, i \oplus 1]$ to trade through 1 as an intermediary. This follows since $(1 - \delta^i) \cdot (1 - \delta^1) > (1 - \delta^i) \cdot (1 - \delta^{i \oplus 1})$. This pricing creates a small shortage of 1 at each trading post (since small quantities of 1 are being retained at the post to cover 1's transaction costs) so prices are readjusted so that all of the discount in bid prices at $\{i,1\}$ appears in the bid price of i. This results in $q_i^{\{i,1\}} = \frac{1-\delta^i}{1+\delta^1}$, $q_1^{\{i,1\}} = 1$. All trade of i for $i \oplus 1$ now goes through 1. Good 1 has become 'money,' the unique low transaction cost common medium of exchange.

In actual monetary economies we usually see a single 'money' as in Example 4.1. We'll argue in Sections 6 through 9 that the reason for uniqueness of 'money' is scale economy. Does there have to be a reason for uniqueness? Yes. US dollars, pounds sterling, and euros, all have similar low transaction costs but in their separate markets they are virtually unique in use. Economic theory should have an explanation for this uniqueness. Example 4.2 below emphasizes, by counterexample, that the nonconvexity in Section 6 is important. In Example 4.2, absent the nonconvexity, when there's a tie for lowest transaction cost, there are many media of exchange in use. Is a tie realistic; isn't it a singularity? The example of dollars, sterling, and euros suggests that on the contrary, the notion of a tie for lowest transaction cost is a non-trivial event, so that uniqueness requires an explanation.

Example 4.2 (Multiple 'moneys''s in equilibrium): Let the population of households be $H = H^0 \cup H^1$. Let $C^{\{i,j\}}$ be described by (TCL). Let $0 < \delta^1 = \delta^2 = \delta^3 < \delta^i < 1/3$, $i = 4, 5, ...N$. Then there is a continuum of competitive equilibrium allocations with 1,2,3 acting as 'money' in proportions from 0% to 100%. Consumptions and utilities of all households are the same as in the equilibrium of Example 4.1.

Demonstration of Example 4.2: The marginal cost market-clearing pricing is identical to that in Example 4.1 with goods 2 and 3 priced simlarly to good 1. The exception is trade between 'money''s where $q_1^{\{1,2\}} = 1 - \delta^1$, and similarly for 2,3, all of these bid prices being equal.

The trading posts $\{i,1\}$, $\{i,2\}$, and $\{i,3\}$, $i=4,5,...,N$, (for trade in good i versus goods 1,2,3) are the trading posts with narrow bid/ask spreads since 1,2,3 have low transaction costs. Households can now divide their transactions among trading posts for goods 1, 2, and 3 versus all other goods in any proportion (though in equilibrium they will be the same proportions for all households). Markets clear. The logic of Example 4.2 is merely the multi-money version of 4.1. Goods 1, 2, 3 are equally liquid and become media of exchange. They can be used however in any proportionate combination from 0% to 100% since absent economies of scale there is no reason further to specialize.

5 Absence of double coincidence of wants is essential to monetization in a linear model

Let $H^D = \{[m, n] | m, n = 1, 2, 3, ..., N, m \neq n\}$. H^D is distinctive in creating a population of households with fully complementary demands and supplies, full double coincidence of wants. We can use this population to illustrate the importance of the absence of double coincidence of wants to monetization in a linear model. Under the same conditions where monetary equilibria existed – and indeed were the only equilibria – in Examples 4.1 and 4.2 in the absence of double coincidence of wants, we can show that for H^D, with full double coincidence of wants, a barter equilibrium is the unique competitive equilibrium. Hence the classical focus on the absence of double coincidence of wants is confirmed; it is essential to monetization in a linear model. Note that this result depends on the linearity (or convexity) of transaction costs; if scale economies are present, then even with full double coincidence of wants, it may be more economical to use a common medium of exchange with resulting high trading volumes.

Example 5.1 (Barter equilibrium with full double coincidence of wants): Let the population of households be $H = H^0 \cup H^D$. Let $C^{\{i,j\}}$ be described by (TCL). Let $0 < \delta^i < 1/3$ and $0 < \delta^1 < \delta^i$, for all $i, i = 2, 3, ...N$. Transaction costs are constant and non-trivial for all goods but 1. Then there is a unique competitive equilibrium allocation. The equilibrium is non-monetary with active trade in all trading posts $\{i, j\}$, $1 \leq i, j \leq N$.

Demonstration of Example 5.1: For each $i, j, 1 \leq i, j \leq N$, $q_i^{\{i,j\}} = (1 - \delta^i)$, $q_j^{\{i,j\}} = (1 - \delta^j)$. $s_i^{[i,j]\{i,j\}} = A, b_j^{[i,j]\{i,j\}} = q_j^{\{i,j\}} A, s_j^{[j,i]\{i,j\}} = A, b_i^{[j,i]\{i,j\}} = q_j^{\{i,j\}} A$. Markets clear. The allocation is an equilibrium.

What's happening in Example 5.1? Direct barter trade works successfully in the presence of double coincidence of wants. For each household $[i, j]$ with a supply of one good and a demand for another, there is a precise mirror image $[j, i]$ in the population. They each go to the trading post $\{i, j\}$ where their common demands and supplies are traded. They trade, each incurring the cost of trading one good. Monetary trade is not advantageous since it requires twice the transactions volume – with corresponding cost – of direct barter trade (similar volumes for each non-monetary good and an equal volume of trade in the medium of exchange). Monetization of trade in equilibrium in a *linear model* depends on absence of double coincidence of wants.

6 Uniqueness of the medium of exchange: scale economies in transaction cost

Monetary trade is typically characterized by a unique medium of exchange or a small number of related media (e.g. currency, credit cards, travelers' checks, all denominated in $US). How does this come about? Professor Tobin (1980) suggests that scale economies in transaction costs are essential:

The use of a particular language or a particular money by one individual increases its value to other actual or potential users. Increasing returns to scale
... explains the tendency for one basic language or money to monopolize
the field.

When monetization takes place, households supplying good i and demanding good
j are induced to trade in a monetary fashion, first trading i for 'money' and then
'money' for j, by discovering that transaction costs are lower in this indirect trade
than in direct trade of i for j. But as Example 4.2 points out, monetization of trade
is no guarantee of uniqueness of the medium of exchange. Scale economies in
transaction costs induce specialization in the medium of exchange function. High
volume leads to low unit transaction costs (see also Howitt, 2000; Rey, 2001; Starr
and Stinchcombe, 1999). Scale economy is not a necessary condition for uniqueness
of the medium of exchange in equilibrium (Example 4.1), but scale economy helps
to ensure uniqueness (Example 7.1, below). If there are many equally low cost
candidates for the medium of exchange, then scale economy in transaction costs
will allow one to be endogenously chosen as the unique medium of exchange.

The transaction cost structure of Sections 7 through 9 with large scale economies
is unsuitable for competitive equilibrium. Competitive equilibria typically cannot
exist in the unbounded scale economy environment. In Section 7, instead of competitive equilibria, average cost pricing equilibria are developed. The use of average
cost pricing is subject to interpretation. A literal interpretation is that the there is a
natural monopoly market maker pricing at average cost to discourage new entry. An
alternative is that the operation of the market is in the nature of a public good; the
nonconvex technology is a summary of the interactions of many individual agents
sharing an economy of scale, and hence average cost pricing reflects the common
benefit from the level of activity in the market (a Marshallian externality). In Section
8 the scale economy allows a Bertrand equilibrium with monopoly trading posts to
form. In Section 9 government provides fiat money; government's large scale combined with the scale economy in transaction costs assures that government-issued
fiat money becomes the common medium of exchange. Scale economy implies a
cost saving resulting from uniqueness of 'money,' since only N (in the case of
fiat money) or $N - 1$ (commodity money) trading posts need to operate, incurring significantly lower costs than $N(N - 1)/2$ (under barter). Scale economies
make it cost-saving to concentrate transactions in a few firms and one intermediary
instrument.

7 Monetization comes from liquidity again:
monetary general equilibrium with unique money
under average cost pricing of non-convex transaction costs

Scale economies in the transaction cost structure induce uniqueness of the equilibrium medium of exchange. 'Money' is a natural monopoly. As Professor To-
bin(1959) tells us, "Why are some assets selected by a society as generally acceptable media of exchange while others are not? This is not an easy question, because
the selection is self-justifying." Thus gold and dollar bills may have low transaction costs and be excellent candidates for medium of exchange, but if (despite high

transaction cost) Yap Island stones are already the commonly chosen medium of exchange with high trading volume, then stones may have the lowest average transaction cost. The choice of Yap Island stones as the common medium of exchange is then 'self-justifying.'

The nonconvex (scale economy) cost function[7] for trading post $\{i,j\}$ is

$$C^{\{i,j\}} = y_0^{\{i,j\}B} = \min[\delta^i y_i^{\{i,j\}B}, \gamma^i] + \min[\delta^j y_j^{\{i,j\}B}, \gamma^j] \qquad \text{(TCNC)}$$

where $\delta^i, \delta^j, \gamma^i, \gamma^j > 0$. In words, the transaction technology looks like this: Trading post $\{i,j\}$ makes a market in goods i and j, buying each good in order to resell it. It incurs transaction costs in good 0. These costs vary directly (in proportions δ^i, δ^j) with volume of trade at low volume and then hit a ceiling after which they do not increase with trading volume. The specification in (TCNC) is an extreme case: zero marginal transaction cost beyond the ceiling. Adding additional linear terms would represent a more general case.

Since the trading posts in this economy have nonconvex transaction technologies, a competitive equilibrium is not an appropriate solution concept. The equilibrium notion used is an average cost pricing equilibrium resulting in zero profits for the typical trading post firm. The rationale for this choice of equilibrium concept may be the threat of entry (by other similar firms) if any economic rent is actually earned. The presence of potential entrants and their actions is not explicitly modeled. An *average cost pricing equilibrium* consists of $q_{(i)0}^{o\{i,j\}}, q_{(j)0}^{o\{i,j\}}, q_i^{o\{i,j\}}$, $q_j^{o\{i,j\}}, 1 \le i, j \le N$, so that :

- For each household h, there is a utility optimizing plan $b_n^{oh\{i,j\}}, s_n^{oh\{i,j\}}$, (subject to T.i, T.ii [or T.ii' for $h \in H^0$], T.iii) so that $\sum_h b_n^{oh\{i,j\}} = y_n^{o\{i,j\}S}$, $\sum_h s_n^{oh\{i,j\}} = y_n^{o\{i,j\}B}$, for each $\{i,j\}$, each n, where
- $y_n^{o\{i,j\}S} \le y_n^{o\{i,j\}B}, n = i, j$.
- $y_0^{o\{i,j\}B}$ can be divided into two parts, $y_{(i)0}^{o\{i,j\}B} \ge 0, y_{(j)0}^{o\{i,j\}B} \ge 0$, so that $y_{(i)0}^{o\{i,j\}B} + y_{(j)0}^{o\{i,j\}B} = y_0^{o\{i,j\}B} = C^{\{i,j\}}$.
- $q_{(i)0}^{o\{i,j\}} y_{(i)0}^{o\{i,j\}B} = y_i^{o\{i,j\}B} - q_j^{o\{i,j\}} y_j^{o\{i,j\}B}$. $q_{(j)0}^{o\{i,j\}} y_{(i)0}^{o\{i,j\}B} = y_j^{o\{i,j\}B} - q_i^{o\{i,j\}} y_i^{o\{i,j\}B}$.

Let κ be a positive integer, $2 \le \kappa < (N/2)$. Let $H^\kappa = \{[m, m \oplus i]|m = 1, 2, ...N; i = 1, 2, ..., \kappa; r_m^{[m,m\oplus1]} = A > 0\}$. H^κ is a set of κN households without double coincidence of wants. One way to visualize H^κ's situation is to think of the households arrayed in a circle clockwise, each one's position designated by endowment. They can arrange a Pareto improving redistribution by each taking his endowment and sending it i places counterclockwise. However, reflecting the absence of double coincidence of wants, if each of the housheholds in H^κ goes to the trading post where his endowment is traded against his desired good, he finds himself alone. He's dealing on a thin market. The following Example 7.1

---

[7] (TCNC) is intended as a mnemonic for non-convex transaction cost.

demonstrates that, with scale economies in transaction cost, virtually any good can become money; the designation is self-confirming.

Example 7.1 (Monetary equilibrium absent double coincidence of wants with scale economy in transaction costs): Let the population of households be $H = H^0 \cup H^\kappa$. Let $C^{\{i,j\}}$ be described by (TCNC). Let $0 < \delta^i < 1$ for all $i = 1, 2, ...N$. Let $\frac{\gamma^i + \gamma^j}{\kappa A} < \frac{2}{3}$ and $\left(1 - \frac{\gamma^i + \gamma^j}{\kappa A}\right) > (1 - \delta^j)$ for all $i \neq j, i, j = 1, 2, ..., N$. Then for each $i = 1, 2, ..., N$ there is a monetary average cost pricing equilibrium with good i as the unique 'money'.

Demonstration of Example 7.1: Choose an arbitrary $i = 1, 2, ..., N$ as 'money.' For all $j \neq i, j = 1, 2, ..., N$, let $q_i^{\{i,j\}} = 1$, $q_j^{\{i,j\}} = 1 - \frac{\gamma^i + \gamma^j}{\kappa A}$. For all j, and $k = 1, 2, ..., N, j \neq k \neq i$, $q_j^{\{j,k\}} = 1 - \delta^j$, $q_k^{\{j,k\}} = 1 - \delta^k$. For $1 \leq \ell \leq \kappa$, let $s_m^{[m,m\oplus\ell]\{i,m\}} = A$, $b_i^{[m,m\oplus\ell]\{i,m\}} = q_m^{\{i,m\}} A$, $s_i^{[m,m\oplus\ell]\{i,m\oplus\ell\}} = q_m^{\{i,m\}} A$, $b_{m\oplus\ell}^{[m,m\oplus\ell]\{i,m\oplus\ell\}} = q_m^{\{i,m\}} A$.

What's happening in Example 7.1? Virtually any good i can become money. Monetization comes from liquidity and – with scale economies – liquidity comes from trading volume. The economy is focusing on good i as its common medium of exchange. Since there are scale economies in transaction costs, high trading volume means low average cost with concommitant narrow bid/ask spread. The narrow bid/ask spread is the way the price system confirms and reinforces the choice of i as the medium of exchange. Trader $[m, m \oplus \ell]$ wants to trade good m for good $m \oplus \ell$. He could do so directly, but the transaction costs are heavy, reducing his return on the trade to $A(1 - \delta^m)(1 - \delta^{m\oplus\ell})$ units of $m \oplus \ell$ after starting with A units of good m. The alternative is to trade good m for good i and then trade i for $m \oplus \ell$. This results in $A(1 - [(\gamma^i + \gamma^{m\oplus\ell})/\kappa A])$ units of $m \oplus \ell$. When κ is sufficiently large, that's a much greater return. Because of the narrow bid/ask spread on trade through i, every market with good i on one side attracts high trading volume, κ traders on each side of the market, the high trading volume needed to maintain good i's low bid/ask spreads. The scale economy means that the choice of good i as the common medium of exchange is self-confirming.

The difference between barter and monetary exchange is the contrast between a complex of many thin high transaction cost markets and an array of a smaller number of thick low transaction cost markets dealing in each good versus a unique common medium of exchange. The choice of medium of exchange is self-justifying. Any good i with sufficient scale economy in its transaction technology (with γ^i, the ceiling on its transaction costs, sufficiently low) can become the unique medium of exchange in equilibrium when trading volume κA is sufficiently high. Mint-standardized gold coins (with a low cost transaction technology) or Yap Island stones (high cost technology) may be 'money' depending on which is well established. Sufficient trading volume can confirm either choice.

Recall $H^D = \{[m, n] | m, n = 1, 2, 3, .., N, \ m \neq n, \ r_m^{[m,n]} = A > 0\}$. H^D is a set of $N(N-1)$ households with full double coincidence of wants. The following Example 7.2 demonstrates that even in the presence of double coincidence of wants, sufficient scale economies in transaction costs can lead to monetization of trade, the use of a common medium of exchange.

Example 7.2 (Monetary equilibrium with full double coincidence of wants and scale economy in transaction costs): Let the population of households be $H = H^0 \cup H^D$. Let $C^{\{i,j\}}$ be described by (TCNC). Let $0 < \delta^i < 1$ all $i = 1, 2, ..., N$. For some i and all j, $1 \le i, j \le N$, $i \ne j$, let $\frac{\gamma^i + \gamma^j}{(N-1)A} < \frac{2}{3}$ and $(1 - \frac{\gamma^i + \gamma^j}{(N-1)A}) > (1 - \delta^j)$, $(1 - \frac{\gamma^i + \gamma^j}{(N-1)A}) > (1 - \delta^i)$. Then there is a monetary average cost pricing equilibrium with good i as the unique 'money.'

Demonstration of Example 7.2: For all $j \ne i$, $j = 1, 2, ..., N$, let $q_i^{\{i,j\}} = 1$, $q_j^{\{i,j\}} = 1 - \frac{\gamma^i + \gamma^j}{(N-1)A}$. For all j, and $k = 1, 2, ..., N$, $j \ne k \ne i$, $q_j^{\{j,k\}} = 1 - \delta^j$, $q_k^{\{j,k\}} = 1 - \delta^k$. Let $s_m^{[m,n]\{i,m\}} = A$, $b_i^{[m,n]\{i,m\}} = q_m^{\{i,m\}} A$, $s_i^{[m,n]\{i,n\}} = q_m^{\{i,m\}} A$, $b_n^{[m,n]\{i,m\}} = q_m^{\{i,m\}} A$.

What's happening in Example 7.2? Monetization comes from liquidity and – with scale economies – liquidity comes from trading volume. But how can monetization of trade occur where there is double coincidence of wants? The answer is scale economies. Trader $[m, n]$ wants to trade good m for good n. He could do so directly at post $\{m, n\}$, and he'd find a willing trading counterpart at the trading post, so he'd only have to pay for the transaction costs on one side of the trade. But the transaction costs are still substantial, reducing his return on the trade to $A(1 - \delta^m)$ units of n after starting with A units of good m. The alternative is to trade good m for good i and then trade i for n. This results in $A(1 - [(\gamma^i + \gamma^n)/(N-1)A])$ units of n. When N is sufficiently large, that's a much greater return. Because of the narrow bid/ask spread on trade through i, every market with good i on one side attracts high trading volume, N-1 traders on each side of the market, the high trading volume needed to maintain good i's low bid/ask spreads. The scale economy means that the choice of good i as the common medium of exchange is self-confirming.[8]

7.1 Convergence to a unique 'money'

Einzig (1966, p. 345), suggests "Money tends to develop automatically out of barter, through the fact that favourite means of barter are apt to arise ... object[s] ... widely accepted for direct consumption." That is, Einzig suggests those goods with high trading volumes are the most liquid (presumably reflecting scale economy in transaction cost), and evolve into common media of exchange. That medium is unique because scale economies lead to 'money' as a natural monopoly. The following example demonstrates this process.

As monetization takes place, households supplying good i and demanding good j start by trading directly. They may also consider monetary trade, first trading i for 'money' and then 'money' for j. When they discover that transaction costs are lower in this indirect trade than in direct trade of i for j, they adopt monetary trade. Starting

---

[8] For a network externality interpretation see Hahn(1997) which notes that in the presence of market set-up costs, each transactor in the market benefits from the participation of others. "If the number who can gain from trade is ... sufficiently [large] ..., the Pareto improving trade will take place. There is thus an externality induced by set-up costs." Young (1998) assumes the externality without additional explanation. Rey (2001) denotes this interaction the "thick markets externality."

from a barter array consisting of $N(N-1)/2$ active trading posts, the allocation evolves through price and quantity adjustments to a monetary array where only N-1 trading posts are active. The impetus for the concentration of the trading function in a few trading posts (those specializing in trade that includes the commodity that is endogenously designated as 'money') in the monetary equilibrium comes from pricing the scale economies in transaction technology.

Example 7.3, below, starts with an economy of diverse endowments and demands and with a double coincidence of wants. The demand structure is arranged at the outset positing some goods most "widely accepted for direct consumption." With scale economies in the transaction technology, these high volume goods will also be those with the lowest unit transaction cost. Thus they are, in Menger's view, the most saleable, and excellent candidates for "*generally* acceptable media of exchange." As they are so adopted by some households, their trading volumes increase, reducing their average transaction costs, and making them more saleable still. This process converges to an equilibrium with a unique medium of exchange, reflecting the interaction of scale economy and liquidity. As households discover that some pairwise markets (those with high trading volumes) have lower transaction costs, they rearrange their trades to take advantage of the low cost. That leads to even higher trading volumes and even lower costs at the most active trading posts. The process converges to an equilibrium where only the high volume trading posts dealing in a single intermediary good ('money') are in use. Under nonconvex transaction costs, this implies a cost saving, since only N-1 trading posts need to operate, incurring significantly lower costs than $N(N-1)/2$ posts.

Scale economies make it cost-saving to concentrate transactions in a few trading posts and a unique 'money'. Scale economies in the transactions technology generate a strong tendency to multiple equilibria. This creates an interest in determining which of the several equilibria the economy will actually select. One solution to this problem is to posit an adjustment process to equilibrium that makes the choice. Hence we use the following

Tatonnement adjustment process for average cost pricing equilibrium: Prices will be adjusted by an average cost pricing auctioneer. Specify the following adjustment process for prices.

STEP 0: The starting point is somewhat arbitrary. In each pairwise market the bid-ask spread is set to equal average costs at low trading volume.

CYCLE 1 STEP 1: Households compute their desired trades at the posted prices and report them for each pairwise market

STEP 2: Average costs (and average cost prices) are computed for each pairwise market based on the outcome of STEP 1. Prices are adjusted upward for goods in excess demand at a trading post, downward for goods in excess supply, with the bid-ask spread adjusted to average cost. A market's (market making firm's) nonzero prices are specified only for those goods where the firm has the technical capability of being active in the market; other prices are unspecified, indicating no available trade.

CYCLE 2 Repeat STEP 1 (at the new posted prices) and STEP 2.

CYCLE 3, CYCLE 4, ... repeat until the process converges.

Einzig encourages us to look for favorite means of barter as latent money; we'll define a population with some favorite means of barter. Define a household population H^F as follows: Let N be an integer, $N \geq 3$. Without loss of generality, designate goods 1 and 2 for distinctive roles: 1 is widely heavily traded, particularly in exchange for 2. Let $H^F = \{[m,n] \mid 1 \leq m, n \leq N, m \neq n; r_m^{[m,n]} = A > 0,$ except $r_m^{[m,1]} = 2A = r_1^{[1,m]}$ for $m \neq 2, r_2^{[2,1]} = 3A = r_1^{[1,2]}\}$. That is, there is a distinctively high desired net trade volume in good 1, particularly in exchange for good 2 (the numerical designation is inessential).

Example 7.3 (High trading volume with scale economy designates 'money'): Let the population be $H^F \cup H^0$ Let transactions costs be characterized by (TCNC) with $\delta^i = \frac{1}{2}, \gamma^i = (.6)A$, all i. That is, there is full double coincidence of wants. All goods have the same transaction technology but there is higher desired net trading volume in good 1. Scale economies in transaction costs are evident at trading volumes slightly higher than the desired trade size of most traders but well within the size of traders desiring net trades in good 1, particularly in exchange for 2. Then the tatonnement process converges to a monetary equilibrium where 1 is the unique money.

Demonstrating Example 7.3: The economy has a full double coincidence of wants. For most pairs of goods m, n, the desired net trade is uniformly distributed; the desired trade between them is A. For pairs $1, n$ the desired trading volume is $2A$ except for the pair 1,2 where the desired volume is $3A$. This structure of preferences and endowments creates a desire for relatively high trading volumes among households trading in good 1.

The scale economy in transactions costs begins to be apparent at trading volumes just slightly larger than the endowment of most households. The scale economy is manifest well within the desired trading volumes of households endowed with or desiring good 1. The progression from barter to money is then the movement from a diffuse array of many active low volume markets to the concentration on a connected family of high volume (low average cost) markets. The tatonnement proceeds as follows:

STEP 0: For all $1 \leq i, j \leq N, i \neq j, q_{(i)0}^{\{i,j\}} = q_{(j)0}^{\{i,j\}} = 1, q_i^{\{i,j\}} = q_j^{\{i,j\}} = \frac{1}{2}$.

CYCLE 1, STEP 1:

- For $[m,n] \in H^F, m \neq 1 \neq n, b_n^{[m,n]\{m,n\}} = \left(\frac{1}{2}\right)A = q_m^{\{m,n\}}A, s_m^{[m,n]\{m,n\}} = A$; all other purchases and sales are nil.
- For $[m,1] \in H^F, m \neq 2, b_1^{[m,n]\{m,n\}} = A = q_m^{\{m,1\}}2A, s_m^{[m,n]\{m,1\}} = 2A$; all other purchases and sales are nil. For $[1,n] \in H^F, n \neq 2, b_n^{[1,n]\{1,n\}} = A = q_1^{\{1,n\}}2A, s_1^{[1,n]\{1,n\}} = 2A$; all other purchases and sales are nil.
- For the two remaining elements of $H^F, [1,2]$ and and $[2,1], b_2^{[1,2]\{2,1\}} = \left(\frac{3}{2}\right)A = q_1^{\{2,1\}}3A, s_1^{[1,2]\{2,1\}} = 3A; b_1^{[2,1]\{2,1\}} = \left(\frac{3}{2}\right)A = q_2^{\{2,1\}}3A, s_2^{[2,1]\{2,1\}} = 3A$; all other purchases and sales are nil.
- For $h \in H^0$, for $i \neq 1 \neq j, b_i^{h\{i,j\}} = b_j^{h\{i,j\}} = A/2, s_0^{h\{i,j\}} = A$; for i or $j = 1, b_i^{h\{i,j\}} = b_j^{h\{i,j\}} = \gamma = (.6)A, s_0^{h\{i,j\}} = 2\gamma = (1.2)A$.

STEP 2:

- For $\{m,n\}$ where $m \neq 1 \neq n$, $1 = q^{\{m,n\}}_{(m)0} = q^{\{m,n\}}_{(n)0}$, $q^{\{m,n\}}_m = q^{\{m,n\}}_n = \left(\frac{1}{2}\right)$.
- For $\{m,1\}$, $m \neq 2$, $1 = q^{\{m,1\}}_{(1)0} = q^{\{m,1\}}_{(m)0}$, $q^{\{m,1\}}_m = \frac{2A-\gamma}{2A} = .70$.
- For $\{2,1\}$, $1 = q^{\{2,1\}}_{(2)0} = q^{\{2,1\}}_{(1)0}$, $q^{\{2,1\}}_2 = q^{\{2,1\}}_1 = \frac{3A-\gamma}{3A} = .80$.

At this stage we can see the initial effect of the scale economy. At STEP 0 prices started essentially equivalent in all pairwise markets. But the prices announced at the end of CYCLE 1 STEP 2 show that the bid prices of goods are much higher in the high volume markets; the bid/ask spread is lower there. The high volume markets are more liquid.

On entering CYCLE 2 STEP 1 households recalculate their desired trades. Those who have been trading on $\{2, 1\}$ and on $\{m,1\}$ find that trade on these markets has become even more attractive since the bid-ask spreads have narrowed. Those who had been trading on $\{2,m\}$ face a quandary: goods 2 and m are the goods that they want to trade, but trading indirectly through good 1 in $\{2, 1\}$ and $\{m,1\}$ may be a lower cost alternative. In order to make that decision the household compares $q^{\{2,m\}}_m$ to the product $q^{\{2,1\}}_1 \cdot q^{\{m,1\}}_m$. The former is the value of m in terms of 2 in direct trade, the latter through trade mediated by good 1. $q^{\{m,1\}}_m \cdot q^{\{2,1\}}_1 = .56 > .5 = q^{\{m,2\}}_m$. Household $[m,2]$ can get more 2 for his m by trading indirectly through the markets with good 1, and household $[2, m]$ can get more m for his 2 by trading indirectly through the markets with good 1. They decide to trade through good 1. Good 1 is beginning to take on the character of money.

The transformation of good 1 into money is not complete however. Household $[m, n]$ for $m \neq 2 \neq n$ considers but does not adopt indirect trade through good 1. He calculates $q^{\{m,1\}}_m \cdot q^{\{n,1\}}_1 = .49 < .5 = q^{\{m,n\}}_m$. Household $[m, n]$ still gets a better deal trading directly good m for n.

CYCLE 2, STEP 1:

- For $[m,n] \in H^F$, $m, n \neq 2$, $m, n \neq 1$, $s^{[m,n]\{m,n\}}_m = A$, $b^{[m,n]\{m,n\}}_n = Aq^{\{m,n\}}_m$; all other purchases and sales are nil.
- For $[m,2]$, $m \neq 1$, $s^{[m,2]\{m,1\}}_m = A$, $b^{[m,2]\{m,1\}}_1 = Aq^{\{m,1\}}_m$, $s^{[m,2]\{1,2\}}_1 = Aq^{\{m,1\}}_m$, $b^{[m,2]\{1,n\}}_1 = Aq^{\{m,1\}}_m q^{\{2,1\}}_1$; all other purchases and sales are nil.
- For $[2,n]$, $n \neq 1$, $s^{[2,n]\{2,1\}}_2 = A$, $b^{[2,n]\{2,1\}}_2 = Aq^{\{2,1\}}_2$, $s^{[2,n]\{1,n\}}_1 = Aq^{\{2,1\}}_2$, $b^{[2,n]\{1,n\}}_n = Aq^{\{2,1\}}_2 q^{\{1,n\}}_1$; all other purchases and sales are nil.
- For $[m,1]$, $m \neq 2$, $s^{[m,1]\{m,1\}}_m = 2A$, $b^{[m,1]\{m,1\}}_1 = 2Aq^{\{m,1\}}_m$; all other purchases and sales are nil. For $[1,n]$, $n \neq 2$, $s^{[1,n]\{1,n\}}_1 = 2A$, $b^{[1,n]\{1,n\}}_n = 2Aq^{\{n,1\}}_1$; all other purchases and sales are nil.
- For $[2,1]$, $s^{[2,1]\{2,1\}}_2 = 3A$, $b^{[2,1]\{2,1\}}_1 = 3Aq^{\{2,1\}}_2$. For $[1,2]$, $s^{[1,2]\{2,1\}}_1 = 3A$, $b^{[1,2]\{2,1\}}_1 = 3Aq^{\{2,1\}}_1$.
- For $h \in H^0$, for each $\{1, j\}$, $b^{h\{1,j\}}_j = \gamma = s^{h\{1,j\}}_0$; for each $\{i, j\}$ so that $1 \neq j \neq 2 \neq i \neq 1$, $b^{h\{i,j\}}_j = A/2 = s^{h\{i,j\}}_0$; all other $b^{h\{i,j\}}_j$ and $s^{h\{i,j\}}_j$ are nil. In particular $b^{h\{i,2\}}_i$ and $s^{h\{i,2\}}_0$ are nil.

STEP 2:

- For $\{m,n\}$ where $m \neq 1 \neq n$, $1 = q_{(m)0}^{\{m,n\}} = q_{(n)0}^{\{m,n\}}$, $q_m^{\{m,n\}} = q_n^{\{m,n\}} = (1/2)$.
- For $\{m,1\}$, $m \neq 2$, $1 = q_{(1)0}^{\{m,j\}} = q_{(m)0}^{\{m,j\}} = q_1^{\{m,1\}} = \frac{3A-2\gamma}{3A} = 0.60$.
- For $\{2,1\}$, $1 = q_{(2)0}^{\{2,1\}} = q_{(1)0}^{\{2,1\}} = q_1^{\{2,1\}}$, $q_2^{\{2,1\}} = \frac{(N+2)A-2\gamma}{(N+2)A} \geq 0.76$.

As CYCLE 2 STEP 1 is completed, trade has become partially monetized. All trade in good 2 goes through good 1 as a medium of exchange. As STEP 2 is completed, prices reflect the higher trading volumes on markets including 1. For convenience, pricing at trading posts $\{1,m\}$ dealing in good 1 is characterized by setting $q_1^{\{1,m\}}$ (the bid price of 1) at 1 and discounting only $q_m^{\{1,m\}}$ to reflect transaction cost. Going into CYCLE 3 STEP 1, typical $[m,n]$ for $1 \neq m \neq 2 \neq n \neq 1$, can reconsider whether to trade in goods m and n directly or to trade through good 1 as a medium of exchange. In order to make that decision he compares $q_m^{\{m,n\}}$ to the product $q_1^{\{n,1\}} \cdot q_m^{\{m,1\}}$. The former is the value of m in terms of n in direct trade, the latter through trade mediated by good 1. This is the same comparison $[m,n]$ made at CYCLE 2 STEP 1, and decided to continue to trade directly. But at the new posted prices we have $.5 = q_m^{\{m,n\}} < 0.60 = q_1^{\{n,1\}} \cdot q_m^{\{m,1\}}$. It is more advantageous to trade indirectly. The outcome of CYCLE 3 STEP 1 will be full monetization; all trade will go through good 1.

CYCLE 3, STEP 1:

- For $[m,n] \in H^F$, $s_m^{[m,n]\{m,1\}} = A$, $b_1^{[m,n]\{m,1\}} = Aq_m^{\{m,1\}}$, $s_1^{[m,n]\{1,n\}} = Aq_m^{\{m,1\}}$, $b_n^{[m,n]\{1,n\}} = A(q_m^{\{m,1\}}q_1^{\{n,1\}})$; all other purchases and sales are nil.
- For $[m,n] \in H^F$, $m \neq 1$, $s_m^{[m,1]\{m,1\}} = 2A$, $b_1^{[m,1]\{m,1\}} = 2Aq_m^{\{m,1\}}$; all other purchases and sales are nil. For $[1,n] \in H^F$, $n \neq 1$, $s_1^{[1,n]\{1,n\}} = 2A$, $b_n^{[1,n]\{1,n\}} = 2Aq_1^{\{1,n\}}$; all other purchases and sales are nil.
- For $[2,1]$, $s_2^{[2,1]\{2,1\}} = 3A$, $b_1^{[2,1]\{2,1\}} = 3Aq_2^{\{2,1\}}$. For $[1,2]$, $s_1^{[1,2]\{2,1\}} = 3A$, $b_1^{[1,2]\{2,1\}} = 3Aq_1^{\{2,1\}}$.
- For $h \in H^0$, for each $\{i,j\}$ with $i \neq 1 \neq j$, all transactions are nil. For $\{1,j\}$, $2 \leq j \leq N$, $b_j^{h\{1,j\}} = \gamma = s_0^{h\{i,j\}}$.

STEP 2:

- For $\{m,n\}$ where $m \neq 1 \neq n$, $1 = q_{(m)0}^{\{m,n\}} = q_{(n)0}^{\{m,n\}}$, $q_m^{\{m,n\}} = q_n^{\{m,n\}} = (\frac{1}{2})$.
- For $\{m,1\}$, $m \neq 2$, $1 = q_{(1)0}^{\{m,1\}} = q_{(m)0}^{\{m,1\}} = q_1^{\{m,1\}}$, $q_m^{\{m,1\}} = \frac{NA-2\gamma}{(N+2)A} \geq 0.60$.
- For $\{2,1\}$, $1 = q_{(2)0}^{\{2,1\}} = q_{(1)0}^{\{2,1\}} = q_1^{\{2,1\}}$, $q_2^{\{2,1\}} = \frac{(N+2)A-2\gamma}{(N+2)A} \geq 0.76$.

CYCLE 4, STEP 1: Repeat Cycle 3, Step 1

STEP 2: Repeat Cycle 3, Step 2

CONVERGENCE.

What's happening in Example 7.3? Preferences and endowments are structured so that at roughly the same prices for all goods, there is a balance between supply and demand. Some pairs of goods are more actively traded than others. Good 1 has approximately twice as much active demand (and supply) as most other goods. Good 2 has slightly more active trade than most other goods, and that active trade is concentrated in a supplier who demands good 1 and a demander endowed with good 1.

Here's how trade takes place. The starting point is a barter economy, the full array of $N(N-1)/2$ trading posts. For every pair of goods i, j, where $1 \leq i, j \leq N$, there is a post where that pair can be traded. The starting prices are chosen (somewhat arbitrarily) to cover average costs at low trading volume. The bid-ask spread is uniform across trading posts so trade at each post is as attractive as anywhere else. Then each household computes its demands and supplies at those prices. It figures out what it wants to buy and sell and to which trading posts it should go to implement the trades. Since all bid-ask spreads start out equal, each household just goes to the post that trades in the pair of goods that the household wants to exchange for one another; demanders of good j who are endowed with good i go to $\{i, j\}$. Because of the distribution of demands and supplies, there is twice the trading volume on posts $\{1, j\}$ as on most $\{i, j\}$ and three times as much on $\{1,2\}$.

Then the average cost pricing auctioneer responds to the planned transactions. He prices bid/ask spreads in all markets to cover the costs of the trade on them. Since there is a scale economy in the transactions technology, this leads to slightly narrower bid/ask spreads on the $\{1, j\}$ markets and an even narrower spread on the $\{1, 2\}$ market. The auctioneer announces his prices.

Households respond to the new prices. Households who want to buy or sell good 2 discover that the bid/ask spread on market $\{1, 2\}$ is lower than on any other market trading 2. It makes sense to channel transactions through this low cost market, even if the household has to undertake additional transactions to do so. Ordinarily households $[i, 2]$ and $[2, i]$ would have gone directly to the market $\{i, 2\}$ to do their trading. But the combined transaction costs on $\{i, 1\}$ and on $\{1,2\}$ are lower than those on $\{i, 2\}$. Households $[i, 2]$ and $[2, i]$ find that they incur lower transaction costs by trading through good 1 as an intermediary. They exchange i for 1 and 1 for 2 (or 2 for 1 and 1 for i) rather than trade directly. The market makers on the many different $\{i, 1\}$ markets, $2 \leq i \leq N$, find their trading volumes increased as the $[i, 2]$ and $[2, i]$ traders move their trades to $\{i, 1\}$ and $\{2,1\}$.

The average cost pricing auctioneer responds to the revised trading plans once again. Bid-ask spreads narrow on $\{i,1\}$, $2 \leq i \leq N$. Now the discounts incurred through bid-ask spreads in trading for $i \neq 1 \neq j$ indirectly – through $\{i,1\}$ and $\{1,j\}$ – are significantly smaller than those trading directly at $\{i, j\}$ (particularly when N is large). The auctioneer announces his prices. Households respond to the new prices. For all households $[i, j]$, it is now less expensive to trade through good 1 as an intermediary than to trade directly i for j or j for i. All $[i, j]$ now trade on $\{i,1\}$ and $\{j,1\}$; none trade on $\{i, j\}$, for $i \neq 1 \neq j$. Trade is fully monetized with good 1 as the 'money.'

The average cost pricing auctioneer re-prices the markets. Inactive markets, $\{i, j\}$ for $i \neq 1 \neq j$, necessarily continue to post their starting prices (which reflected anticipated low trading volume). The active markets $\{i,1\}$ get posted prices reflecting their high trading volumes, with narrow bid-ask spreads.

Households review the newly posted prices. The narrow bid-ask spreads on the $\{i,1\}$ markets reinforce the attractiveness of their previous plans, which called for trading through good 1 as an intermediary. They leave their monetary trading plans in force. At current prices, it is much more economical to trade i for j by first trading i for 1 and then 1 for j than to trade i for j directly. High trading volumes on the $\{i,1\}$ and $\{j,1\}$ markets ensure low transaction costs and keep them attractive. All trade takes place at $\{i,1\}$, $i = 2, 3, 4, ..., N$. Good 1 has become the unique 'money'.

Example 7.3 demonstrates price and trading adjustment to the property that scale economies in the transactions technology mean that high volume markets will be low average cost markets. The transition from barter to monetary exchange is the transition from a complex of many thin markets – one for trade of each pair of goods for one another – to an array of a smaller number of thick markets dealing in each good versus a unique common medium of exchange. This transition is resource saving when scale economies in transactions technology are large enough.

Example 7.3 shows that the transition progresses through individually rational decisions when prices reflect the scale economy and the initial condition includes a commodity (the latent 'money') with a relatively high transaction volume (hence low average transaction cost). Then, as Einzig notes, "favourite means of barter are apt to arise" and a barter economy thus converges incrementally to a monetary economy. Menger (1892) describes this transition:

> when any one has brought goods not highly saleable to market, the idea up-permost in his mind is to exchange them, not only for such as he happens to be in need of, but...for other goods...more saleable than his own...By...a mediate exchange, he gains the prospect of accomplishing his purpose more surely and economically than if he had confined himself to direct exchange...Men have been led...without convention, without legal compulsion,...to exchange...their wares...for other goods...more saleable...which ...have ...become generally acceptable media of exchange.

Thus, Menger argues that starting from a relatively primitive market setting, some goods will be more liquid than others. As they are adopted as media of exchange, markets for trade in them versus other goods become increasingly liquid. Eventually they become the common media of exchange in equilibrium. Example 7.3 formalizes this argument emphasizing that the increasing liquidity develops endogenously as a result of scale economy in the transaction process.

7.2 A large pure trade economy with average cost pricing monetary equilibrium

Since scale economies enter into this argument in an essential way, we'd now like to consider a large economy. This class of examples starts with the same structure as in Example 7.2, but we allow the economy to be large in the sense that there are G (positive integer) households of each type $[m, n]$. Let $H^{D \times G}$ denote the G-fold replication of H^D with typical element $[m, n, g]$ where m and n are integers between 1 and N (inclusive), $m \neq n$, and g is an integer between 1 and G. m denotes the good with which h is endowed. n denotes the good he prefers. g is a serial number for the agent of type $[m, n]$.

Example 7.4 (Average Cost Pricing Monetary Equilibrium in a Large Economy): Let $H = H^{D \times G} \cup H^0$. Let transaction technology be characterized by (TCNC). For all $1 \leq i, j \leq N$, let $\delta^i > 0$, $A - [(\gamma^i + \gamma^j)/G(N-1)] > A(1 - \delta^i)$ and $A - [(\gamma^i + \gamma^j)/G(N-1)] > (1/3)A$. Without loss of generality, distinguish any single good, $1, \ldots, N$, as μ. Then the economy has a monetary average cost pricing equilibrium with good μ as 'money'.

Demonstration of Example 7.4:

- For $j \neq \mu$, $q_\mu^{\{\mu, j\}} = 1 = q_0^{\{\mu, j\}}$. $q_j^{\{\mu, j\}} = 1 - [(\gamma^i + \gamma^j)/GA(N-1)]$.
- For all other i, j, combinations, $q_i^{\{i, j\}} = (1 - \delta^i)$, $q_j^{\{i, j\}} = (1 - \delta^j)$.
- For $h = [m, n, g]$ (where $m, n \neq \mu$), we have
$b_n^{[m,n,g]\{\mu, n\}} = Aq_m^{\{\mu, m\}}$, $s_\mu^{[m,n,g]\{\mu, n\}} = Aq_m^{\{\mu, m\}}$,
$b_\mu^{[m,n,g]\{\mu, m\}} = Aq_m^{\{\mu, m\}}$, $s_m^{[m,n,g]\{\mu, m\}} = A$.
- For $h = [m, n, g]$ (where $m = \mu$) we have
$b_n^{[m,n,g]\{n, \mu\}} = A$. $s_\mu^{[m,n,g]\{i, \mu\}} = A$.
- For $h = [m, n, g]$ (where $n = \mu$), we have
$b_\mu^{[m,n,g]\{\mu, m\}} = Aq_m^{\{\mu, m\}}$. $s_m^{[m,n,g]\{\mu, m\}} = A$.
- For some elements $h'' \in H^0$, $\sum_{h''} b_m^{h''\{\mu, m\}} = \gamma^\mu + \gamma^m$, $\sum_{h''} s_0^{h''\{\mu, m\}} = \gamma^\mu + \gamma^m$.

The examples of Section 7 demonstrate Tobin's (1959) argument: the choice of the medium of exchange is self-justifying. There is a significant resource saving – and a competitive pricing advantage to the market-maker – in moving from a barter to a monetary equilibrium, but the choice of what is 'money' is (under these assumptions) essentially arbitrary.[9] Once the choice is made, the equilibrium, including the designation of 'money,' is stable against small perturbations and entry by alternative media of exchange. These characteristics of the monetary equilibrium reflect the underlying transactions technology: the complementarity among pairwise goods markets implicit in the structure of the problem and the scale economies in transaction costs encourage concentration of trading activity in a few market-makers and a single medium of exchange.[10] Conversely, Examples 4.1 and 4.2 suggest that scale economies are essential to monetization of the economy. Without assuming properties peculiar to the designated 'money' as in Example 4.1 (that 'money' is the single good so that trades that include it are achieved at the lowest possible transaction cost) there seems to be no impetus in a convex model driving the equi-

---

[9] This arbitrariness is in contrast to the example of Banerjee and Maskin (1996) where, without explicit transaction costs, in a convex model, the choice of 'money' is fully determined by the parameters of the model as the unique good whose quality is most easily recognized.

[10] The notion of scale economy is consistent with the models of Iwai (1995) and Kiyotaki and Wright (1989) where concentrating trading activity on a single transaction medium reduces waiting times for the completion of trades.

librium toward a single distinguished medium of exchange. Unique monetization results from scale economies in the transaction technology.

8 Barter and monetary Bertrand equilibria with scale economies

Under nonconvex transaction costs, competitive equilibria are unlikely to exist. Hence the focus on average cost pricing equilibria in Sections 6 and 7. Imperfectly competitive equilibria, Bertrand equilibria among competing trading posts, may exist and be monetary. A single good will be distinguished in equilibrium as the medium of exchange common to virtually all transactions. Monetary Bertrand equilibria with nonconvex transaction costs share the strong stability property of the average cost pricing equilibria investigated in Section 7. They are stable against entry by a new post offering an alternative medium of exchange. This will typically be true even if the alternative medium is superior in the sense that total transaction costs would be lower if it were generally adopted. A superior (lower cost) medium will typically not be adopted in preference to the prevailing medium, precisely because it is not prevailing. Markets using the alternative medium are thin, displaying high average transaction costs. In order to be attractive to individual buyers and sellers the markets must become thicker; there must be additional posts available where the alternative medium is in use.

The transaction cost structure with scale economies is a very suitable setting for Bertrand equilibrium. Each potential market-making firm operating a trading post can survey prevailing prices and demand functions and decide to enter by determining prices to post. This section develops a class of examples of barter and monetary Bertrand equilibrium in a pure trade economy with pairwise goods markets and nonconvex technology. A *Bertrand equilibrium* is a Nash equilibrium of best responses in price based on imputed demand functions from the households.

In order to model oligopoly, expand the set of market making firms in the following way. For each pair of goods i, j, let there be several firms capable of making the market in i and j. Denote the firms $\{i, j; 1\}, \{i, j; 2\}..., \{i, j; \ell\}, ...\{i, j; \Lambda\}$, where $\Lambda \geq 2$ is the number of potential entrants into making the market in i and j, indexed by ℓ. Most of the firms $\{i, j, \ell\}$ will remain inactive, but their potential to enter the market affects equilibrium prices. We'll focus on the population structure $H = H^D \cup H^0$ as in Example 5.1. The reason for focusing on this setting with full double coincidence of wants is not that double coincidence is essential, but that it gives us a fair sized economy, with enough symmetry that the algebra is relatively simple. The economy is large enough that if traders concentrate their transactions on a few trading posts, there may be scope for scale economies. It is small enough that if trading activity is dispersed, then markets are thin and no scale economy is experienced (at the cost of greater complexity we could alternatively use $H^\kappa \cup H^0, 1 \leq \kappa \leq N - 1$). We'll take the market makers' cost functions to display scale economies following the specification (TCNC). The scale economies become active only at relatively high trading volumes.

We'll consider a range of cases. First, Example 8.1, we'll suppose that market makers price for low trading volume and that turns out to be an equilibrium: a barter economy with thin markets is a Bertrand equilibrium. Then we'll consider the opposite tack, Example 8.2. The interesting case, reflecting a network externality, shows up where most active trading post market makers have adopted a common medium of exchange but others are deciding whether to enter in active market-making. We'll show that once most active market makers have adopted the plan of making a market using a common medium of exchange, the remaining market makers will find that demand facing them makes using the common medium of exchange compelling as well. Once a common medium has been widely adopted, the pressure to adopt universally is decisive.

Example 8.1 (A Bertrand barter equilibrium; When you're not [hot], you're not): Let the population of households be $H = H^D \cup H^0$. Let $C^{\{i,j\}}$ be described by (TCNC). Let $0 < \delta^i < 1/3$ for all $i = 1, 2, ..., N$. Let $(\delta^i + \delta^j)A < \gamma^i + \gamma^j$ for all $i \neq j$, $i, j = 1, 2, ...N$. Let $\Lambda \geq 2$. Note that this setting implies that scale economies in transaction costs are available, but that they are inactive when each trading post deals in quantities of each good comparable to the endowment of each household. Then there is a Bertrand barter equilibrium. For all i, j commodity pairs there is an active market where that pair is transacted.

Demonstration of Example 8.1: There are no active scale economies in this example, but it is convenient to concentrate on a single trading post for each commodity pair. For $i, j = 1, 2, ...N$, $i \neq j$, let $1 = q_0^{\{i,j,\ell\}}$, $q_i^{\{i,j,\ell\}} = (1 - \delta^i)$, $q_j^{\{i,j,\ell\}} = (1 - \delta^j)$, For $i = m$ or n and $j = n$ or m respectively and (without loss of generality) for and $\ell = 1$, let $b_n^{[m,n]\{i,j,\ell\}} = q_m^{\{i,j,\ell\}}A$, $s_m^{[m,n]\{i,j,\ell\}} = A$. Suppose all other trading posts $\{i, j, \ell\}$, $\ell > 1$, are inactive. Then markets clear. Each trading post covers its transaction costs. The pricing and allocation is a Bertrand equilibrium.

The equilibrium in Example 8.1 is essentially Walras's (1874) trading post model. For each pair of distinct goods in active trade, $i, j = 1, 2, 3, ..., N$, there is a market maker dealing in the pair. The volume of trade at each trading post is modest; A units (a single endowment) of each of two goods is traded at each post. For each good there is a bid price and a higher ask price, so that the market maker retains a surplus (in the proportions δ^i and δ^j) from each transaction. The market maker incurs transaction costs (in the proportion, δ^i and δ^j) in good 0; to provide for the transaction costs the market maker buys 0 from agents in H^0 in exchange for the surplus i and j left over from the direct transactions. A zero profit condition is fulfilled, enforced by the threat of entry of other identical market makers $\ell > 1$.

The starting point of the following Example 8.2 is the same as the previous Example 8.1, but the result is a monetary, not a barter equilibrium. This demonstrates the notion that, reflecting the scale economies at the level of individual trading posts, use of a single common medium of exchange in trade of one good encourages trade in that medium for all goods. There are multiple equilibria; any good can become 'money.' The same initial conditions can result in a barter equilibrium or a monetary equilibrium. Reflecting the scale economy, a common usage of monetary trade encourages monetary trade in all goods. Common usage of barter trade discourages monetary trade in all goods.

Example 8.2 (When you're hot, you're hot; A Bertrand Monetary Equilibrium): Let the population of households be $H = H^D \cup H^0$. Let $C^{\{i,j\}}$ be described by (TCNC). Let $0 < \delta^i < 1/2$ for $i = 1, 2, ..., N$. Let $(\delta^i + \delta^j)A < \gamma^i + \gamma^j$, $(N-1)A - (\gamma^i + \gamma^j) > (N-1)A(1 - \delta^i)$ for all $i \neq j$, $i, j = 1, 2, ..., N$. Let $\Lambda \geq 2$. Note that this setting implies that scale economies in transaction costs are available, and active when each active trading post deals in quantities of each good comparable to the total endowment of that good. Without loss of generality, distinguish any single good, $1, ..., N$, as μ. Then there is a Bertrand monetary equilibrium with good μ acting as 'money'. For all i, μ commodity pairs there is an active market where that pair is transacted and these are the only active markets in equilibrium.

Demonstration of Example 8.2: The following prices and allocations constitute a Bertrand equilibrium.

- For $\ell = 1$, all $j \neq \mu$: $q_0^{\{\mu,j,1\}} = 1$.
  $q_j^{\{\mu,j,1\}} = [1 - (\gamma^j + \gamma^\mu)/(N-1)A]$, $q_\mu^{\{\mu,j,1\}} = 1$.
- For $h = [m, n]$ (where $m, n \neq \mu$) and $\ell = 1$ we have
  $b_n^{[m,n]\{\mu,n,1\}} = Aq_m^{\{\mu,m,1\}}$, $s_\mu^{[m,n]\{\mu,n,1\}} = Aq_m^{\{\mu,m,1\}}$, $b_\mu^{[m,n]\{\mu,m,1\}} = Aq_m^{\{\mu,m,1\}}$, $s_m^{[m,n]\{\mu,m,1\}} = A$
- For all m, n so that neither $m, n = \mu$, and for all $\ell \neq 1$, $q_m^{\{m,n,\ell\}} = (1 - \delta^m)$ and $b_n^{[m,n]\{m,n,\ell\}} = 0$, $s_m^{[m,n]\{m,n,\ell\}} = 0$, $b_m^{[m,n]\{m,n,\ell\}} = 0$, $s_n^{[m,n]\{m,n,\ell\}} = 0$.

The only trading firms active in equilibrium in Example 8.2 are those trading good μ for other goods $m = 1, 2, ..., \mu-1, \mu+1, ..., N-1$. Only one trading post in each pair μ, n is active, reflecting the scale economy. Without loss of generality, that trading post is designated number 1. Other potential entering trading posts in μ and n are inactive, but their threat of entry keeps the active post's pricing at average cost. Household $h \in H^D$, $h = [m, n]$, goes to the trading post $\{m, \mu, 1\}$ dealing in good m, his endowment, and sells his endowment at the bid price in exchange for μ. Household h then takes the proceeds of the sale to $\{n, \mu, 1\}$ the trading post dealing in n, his desired good, and buys n for μ at the ask price. Buyers and sellers are evenly matched so all demands are fulfilled. Since ask prices of good m exceed bid prices, post $\{m, \mu, 1\}$ accumulates net stocks of m. Post $\{m, \mu, 1\}$ incurs transaction costs in good 0. Households $h \in H^0$ supply good 0, the needed input to the transaction process, receive payment in μ, and spend the μ on good m, absorbing post $\{m, \mu, 1\}$'s net accumulation of m(prices for these transactions are unity). The allocation and market structure constitute an equilibrium since no firm finds it profitable, taking other firms' announced prices as given, to change its prices.

Example 8.2 is the strategic counterpart of 7.2 (a price-taking average cost pricing equilibrium). Note that the preferences and endowments in this example fullfil 'double coincidence of wants.' Nevertheless, the structure of transaction costs keeps agents from trading directly with those whose endowments and preferences are reciprocal to their own (as they do in Example 5.1), but encourages them to use monetary trade. Firms trading in i and j where $i \neq \mu \neq j$, find that entry is unprofitable because their markets are thin. Since most trade goes through μ, transaction volumes in markets μ, j are much higher than in i, j. The i, j firms cannot successfully (at positive trading volumes) charge wide enough margins between bid and ask prices to cover costs. It is in this sense that the choice of μ as 'money' is self-justifying.

Example 8.2 emphasizes as well that there are multiple equilibria. Virtually any of the N goods can become μ, the common medium of exchange. The only ones ruled out are those with insufficient scale economies (excessively high γ^i). There is no assurance that any single equilibrium designation of μ will be the best choice. On the contrary, there are monetary equilibria with a choice of 'money' that is dominated by possible alternatives. A best choice would be one with the lowest γ^i, but there is no mechanism posited in Example 8.2 to seek out the lowest cost medium. On the contrary, once a common medium of exchange has been selected (by chance, history, an invisible hand), its costs become *locally* lowest by scale economy. Hence the choice is sustained despite the availability of superior alternatives. Their superiority is clear to us as observing economists globally, but it is not locally evident, because (with nonconvex transaction costs) only global changes, not incremental local changes, result in a cost saving. Once the equilibrium with μ as 'money' is established, the markets for the firms $\{i, j, \ell\}$ for $i \neq \mu \neq j$ are thin and unprofitable, even if they have a lower set-up cost.

There is a network externality associated with the choice of a common medium of exchange. As additional markets and traders use a particular medium of exchange, that medium becomes more attractive for others, reflecting complementarities among markets. The strategic situation facing a market-making firm as the economy approaches the Bertrand equilibrium in Example 8.2 reflects the power of the network externality in the common medium of exchange. As the economy approaches a Bertrand equilibrium with good μ as common medium of exchange, consider the situation of a trading post firm deciding to enter the market. Suppose that almost all goods except one, good n^*, are already traded for μ. That is, there are active trading posts $\{\mu, n, 1\}$ for all goods $n = 1, 2, ...n^* - 1, n^* + 1, ..., N$, but the market for n^* is still unsettled. Trade in all goods but n^* is already monetized. What is the demand situation confronting trading posts entering the market in n^*? All the households $[m, n^*]$ who want to trade good m for n^* face low transaction costs in trading their endowments, m, for the prevailing medium of exchange μ. All the traders $[n^*, m]$ who want to trade their endowment n^* for a variety of other goods m face low transaction costs in buying m, for the prevailing medium of exchange μ. Potential entrants to trading post activity in n^* and μ, $\{\mu, n^*, \ell\}$ see this immense latent demand. There are N-1 buyers and N-1 sellers who will find it advantageous (other things being equal) – based on the low costs at complementary trading posts $\{\mu, n^*, \ell\}$, $n \neq n^*$ – to trade at $\{\mu, n^*, \ell\}$ if the price is right. The demand facing $\{\mu, n^*, \ell\}$ promises a thick market at the trading post. Conversely, the (barter) trading posts $\{m, n^*, \ell\}$, $m \neq \mu$, face the prospect of trading in a thin market. At break-even prices, these trading posts face low volume. The only traders interested in their markets are those with precisely matching demands $[m, n^*]$ and $[n^*, m]$. The message to potential entrant trading posts is clear. There is immense demand for trading post $\{\mu, n^*, \ell\}$ providing transactions in the prevailing medium of exchange, μ. There is little demand for trading post $\{m, n^*, \ell\}$ providing transactions to a thin barter market.

The (network) externality here follows these lines: High trading volume at active trading posts using the prevailing medium of exchange, μ, leads to low average transaction costs at those posts, implying low Bertrand pricing of transaction services. That leads to high demand for potential entrant trading posts (in goods not currently served) trading in the prevailing medium of exchange μ. When those posts enter and find their average costs are low, their Bertrand prices are also low. Use of μ as a common medium of exchange builds on itself. Each trading post making a market in μ adds to the demand for other trading posts making a market in μ and other goods. Each additional active market $\{m, n, \ell\}$ increases trading volumes and reduces average costs for the complementary markets.

8.1 Bertrand monetary equilibrium in a large Economy[11]

As in Example 6.4, we now consider a large economy.

Example 8.3 (Bertrand Monetary Equilibrium in a Large Economy): Let $\Lambda \geq 2$. Let $H = H^{D \times G} \cup H^0$. Let transaction technology be characterized by (TCNC). For all $1 \leq i, j \leq N$, let $A - [(\gamma^i + \gamma^j)/G(N-1)] > (1 - \delta^i)A$, and $A - [(\gamma^i + \gamma^j)/G(N-1)] > (1/3)A$. Without loss of generality, distinguish any single good, $1, ..., N$, as μ. Then the economy has a Bertrand monetary equilibrium with good μ as 'money'.

Demonstration of Example 8.3: The following prices and quantities constitute a Bertrand equilibrium. Without loss of generality designate the active trading post firm in each μ, n market as $\ell = 1$.

- For $j \neq \mu$, $\ell = 1$, $q_\mu^{\{\mu,j,1\}} = 1 = q_0^{\{\mu,j,1\}}$.
  $q_j^{\{\mu,j,1\}} = 1 - [(\gamma^i + \gamma^j)/GA(N-1)]$. For all other i, j, ℓ, combinations, $q_i^{\{i,j,\ell\}} = (1 - \delta^i)$, $q_j^{\{i,j,\ell\}} = (1 - \delta^j)$.
- For $h = [m, n, g]$ (where $m, n \neq \mu$) we have
  $b_n^{[m,n,g]\{\mu,n,1\}} = Aq_m^{\{\mu,m\}}$, $s_\mu^{[m,n,g]\{\mu,n,1\}} = Aq_m^{\{\mu,m\}}$, $b_\mu^{[m,n,g]\{\mu,m,l\}} = Aq_m^{\{\mu,m\}}$, $s_m^{[m,n,g]\{\mu,m,l\}} = A$.
- For $h = [m, n, g]$ (where $m = \mu$) we have
  $b_n^{[m,n,g]\{n,\mu,1\}} = A$. $s_\mu^{[m,n,g]\{i,\mu,1\}} = A$.
- For $h = [m, n, g]$ (where $n = \mu$) we have
  $b_\mu^{[m,n,g]\{\mu,m,1\}} = Aq_m^{\{\mu,m\}}$. $s_m^{[m,n,g]\{\mu,m,1\}} = A$.
- For some elements $h'' \in H^0$, $\sum_{h''} b_m^{h''\{\mu,m,1\}} = \gamma^\mu + \gamma^m$, $\sum_{h''} s_0^{h''\{\mu,m,1\}} = \gamma^\mu + \gamma^m$.

As in Example 8.3, the firms $\ell > 1$ are potential market entrants and their threat of entry affects price determination. The firms with positive levels of actual transactions are $\ell = 1$. There are zero profits. The allocation and market structure constitute

---

[11] A version of this example appeared in Starr and Stinchcombe (1998). See also Howitt (2000), for a large economy with a Bertrand monetary equilibrium in a trading post model.

an equilibrium since no firm finds it profitable, taking other firms' announced prices as given, to change its price offers. Again as in Example 8.2, firms $\{i, j, \ell\}$ where $i \neq \mu \neq j$, find that entry is unprofitable because their markets are thin. They post prices $q_i^{\{i,j,\ell\}} = (1 - \delta^i)$, $q_j^{\{i,j,\ell\}} = (1 - \delta^j)$ that cover their operating costs at low volume, but these prices are unattractive to active traders, dominated by the posted prices of firms trading through μ. It is in this sense that the choice of μ as 'money' is self-justifying. Since most trade goes through μ, transaction volumes in markets $\{\mu, j, 1\}$ are much higher than in $\{i, j, 1\}$ for $i \neq \mu \neq j$. Hence $\{\mu, j, 1\}$ firms can successfully operate at narrower bid/ask spreads than $\{i, j, 1\}$ firms can.

The distinction between Examples 8.2 and 8.3 is that the large numbers of traders in 8.3 allow the economy to overcome larger set-up costs on the transactions technology than would otherwise be possible. This shows up as the assumption that $A - [(\gamma^i + \gamma^j)/G(N-1)] > (1 - \delta^i)A$ in Example 8.3 versus $(N-1)A - (\gamma^i + \gamma^j) > (N - 1)A(1 - \delta^i)$ in Example 8.2.

The examples of Section 8 demonstrate Tobin's (1959) argument: the choice of the medium of exchange is 'self-justifying.' There is a significant resource saving – and a competitive pricing advantage to the market-maker – in moving from a barter to a monetary equilibrium, but the choice of what is 'money' is (under these assumptions) essentially arbitrary. Once the choice is made, the equilibrium, including the designation of 'money,' is stable against small perturbations and entry by competing market-makers or by alternative media of exchange. These characteristics of the monetary equilibrium reflect the underlying transactions technology: the complementarity among pairwise goods markets implicit in the structure of the problem and the scale economies in transaction costs encourage concentration of trading activity in a few market-makers and a single medium of exchange.[12] Conversely, Example IV.2 shows that scale economies are essential to unique monetization of the economy. Without assuming properties peculiar to the designated 'money' (e.g. that 'money' is the single good so that trades that include it are achieved at the lowest possible transaction cost, as in Example 4.1) there seems to be no impetus in a convex model driving the equilibrium toward a single distinguished medium of exchange. Unique monetization of trade results from scale economies in the transaction technology.

9 Government-issued fiat money

In order to study fiat money we introduce a government with the unique power to issue fiat money. Fiat money is intrinsically worthless; it enters no one's utility function. But government is uniquely capable of declaring it acceptable in payment of taxes. Adam Smith (1776) notes "A prince, who should enact that a certain proportion of his taxes be paid in a paper money of a certain kind, might thereby give a certain value to this paper money." (v. I, book II, ch. 2). Abba Lerner (1947) comments, "The modern state can make anything it chooses generally acceptable

---

[12] The notion of scale economy is consistent with the models of Iwai (1995) and Kiyotaki and Wright (1989) where concentrating trading activity on a single transaction medium reduces waiting times for the completion of trades.

as money...if the state is willing to accept the proposed money in payment of taxes."
Taxation – and fiat money's guaranteed value in payment of taxes – explains the
positive equilibrium value of fiat money.[13] Scale economies explain its uniqueness
as the medium of exchange.

As an economic agent, government is denoted G. Government sells tax receipts,
the $N+1^{st}$ good. It also sells good $N+2$, an intrinsically worthless instrument, (latent)
fiat money, that government undertakes to accept in payment of taxes, that is, in
exchange for $N + 1$. The typical household $[m, n]$ in H^1 or H^κ desires to purchase
tax receipts to the extent it prefers not to have a quarrel with the government's tax
authorities. Government sets a target tax receipt purchase by the taxayer of $\tau^{[m,n]}$.
Then we rewrite $[m, n]$'s utility function as

$$u^{[m,n]}(x) = \sum_{i=0,i\neq n}^{N} x_i + 3x_n - 10[\max[(\tau^{[m,n]} - x_{N+1}^{[m,n]}), 0]) \qquad \text{(UT)}$$

That is, household $[m, n]$ values paying his taxes with a positive marginal utility
up to his tax bill $\tau^{[m,n]}$ and with zero marginal utility for tax payments thereafter.
Government uses its revenue to purchase a variety of goods $n = 1, ..., N$, in the
amount x_n^G.

Good $N + 2$ represents latent fiat money. Government, G, sells $N+1$ (tax receipts) for $N+2$ at a fixed ratio of one-for-one. The trading post $\{N+1, N+2\}$ where
tax receipts are traded for $N+2$ operates with zero transaction cost. Acceptability
in payment of taxes ensures $N+2$'s positive value. If, in addition, $N+2$ is assumed
to have sufficiently low transaction cost, then it becomes the common medium
of exchange. Thus if we assume a low linear transaction cost, the existence of a
fiat money equilibrium is merely an application of Example 4.1 and need not be
repeated here.

Government-issued fiat money is typically the unique common medium of
exchange: in the US virtually all transactions are denominated in US dollars; in
the UK virtually all (nonfinancial) transactions are denominated in pounds sterling.
The virtual uniqueness of the monetary instrument is not merely a possibility; it
seems to be a general fact. Dollars, euros, pounds sterling, and other government-
issued fiat money's all seem to have similar low transaction costs. But in any
single market economy precisely one of these instruments is likely to be the unique
common medium of exchange. Example 9.1 harnesses scale economy to explain
why fiat money is (almost universally) the unique common medium of exchange.

Particularly in the case of scale economies in the transactions technology, there
is a strong tendency to multiple equilibria (recall Example 7.1). This creates an
interest in determining which of the several equilibria the economy will actually
select. Hence we posit the same tatonnement adjustment process for average cost
pricing equilibrium as in Section 7. That plausible adjustment process explains why
government-issued fiat money becomes the unique common medium of exchange
– and would do so even in the absence of legal tender rules. Government has two
distinctive characteristics: it has the power to support the value of fiat money by

---

[13] See also Li and Wright (1998) and Starr (1974).

making it acceptable in payment of taxes; it is a large economic presence undertaking a high volume of transactions in the economy. Hence, government can make its fiat money the common medium of exchange merely by using it as such. The scale economies implied will make fiat money the low transaction cost instrument and hence the most suitable medium of exchange, not just for government but for all transactors.

Example 9.1: Let the population of households be $H = H^0 \cup H^\kappa$. Let $u^{[m,n]}$ be described by (UT). Let $\tau^0 > 0$ be a constant. Let $0 < \tau^{[m,n]} = \tau^0 < A(1 - \delta^{N+2})(1 - \delta^m)$, all $[m,n] \in H^\kappa$. Let $x_n^G = \kappa\tau^0 q_{N+2}^{\{N+2,n\}}$ all $n = 1, 2, ...N$. Let $C^{\{i,j\}}$ be described by (TCNC). Let $(\gamma^{N+2}/\kappa\tau^0) < \delta^i < 1/3$ all $i = 1, 2, ...N$. Then there exists a monetary average cost pricing equilibrium with taxation with good $N+2$ as the unique 'money.' That monetary equilibrium is the unique limit point of the tatonnement adjustment.

Demonstration of Example 9.1: Step 0: For $n \neq m$ set $q_n^{\{m,n\}} = (1 - \delta^n)$.
Cycle 1, Step 1:

- For $i = 1, 2, ..., \kappa$, let $s_n^{[n,n\oplus i]\{n,n\oplus i\}} = A - (\tau^0/q_n^{\{N+2,n\}})$, $b_{n\oplus i}^{[n,n\oplus i]\{n,n\oplus i\}} = (A - (\tau^0/q_n^{\{N+2,n\}}))q_n^{\{n,n\oplus i\}}$, $s_{N+2}^{[n,n\oplus i]\{N+2,N+1\}} = \tau^0 = b_{N+1}^{[n,n\oplus i]\{N+2,n\}}$; $b_{N+2}^{[n,n\oplus i]\{N+2,n\}} = \tau^0$, $s_n^{[n,n\oplus i]\{N+2,n\}} = \tau^0/q_n^{\{N+2,n\}}$
- For $n = 1, 2, ..., N$, let $s_{N+2}^{G\{N+2,n\}} = \kappa\tau^0$, $b_n^{G\{N+2,n\}} = \kappa\tau^0 q_{N+2}^{\{N+2,n\}}$.

Cycle 1, Step 2:

- For $n, m \neq N + 2, n \neq m$, set $q_n^{\{m,n\}} = (1 - \delta^n)$. $q_n^{\{N+2,n\}} = (1 - \min[\delta^n, \gamma^n/\kappa\tau^0])(1 - \gamma^{N+2}/\kappa\tau^0)$, $q_{N+2}^{\{N+2,n\}} = 1$.

Cycle 2, Step 1:

- For $n = 1, 2, ...N$, let $s_{N+2}^{G\{N+2,n\}} = \kappa\tau^0$, $b_n^{G\{N+2,n\}} = \kappa\tau^0 q_{N+2}^{\{N+2,n\}}$; $s_{N+1}^{G\{N+1,N+2\}} = N\kappa\tau^0$, $b_{N+2}^{G\{N+1,N+2\}} = N\kappa\tau^0$; $b_{N+1}^{[n,n\oplus i]\{N+2,N+1\}} = \tau^0$, $s_{N+2}^{[n,n\oplus i]\{N+2,N+1\}} = \tau^0$; $s_n^{[n,n\oplus i]\{N+2,n\}} = A$, $b_{N+2}^{[n,n\oplus i]\{n,N+2\}} = Aq_n^{\{N+2,n\}}$; $s_{N+2}^{[n,n\oplus i]\{n\oplus i,N+2\}} = Aq_n^{\{N+2,n\}} - \tau^0$, $b_{n\oplus i}^{[n,n\oplus i]\{n\oplus i,N+2\}} = (Aq_n^{\{N+2,n\}} - \tau^0)q_{N+2}^{\{n\oplus i,N+2\}}$.

Cycle 2, Step 2:

- For $n, m \neq N + 2$, set $q_n^{\{m,n\}} = (1 - \delta^n)$. $q_n^{\{N+2,n\}} = (1 - \min[\delta^n, \gamma^n/\kappa A])(1 - \gamma^{N+2}/\kappa A)$, $q_{N+2}^{\{N+2,n\}} = 1$.

Cycle 3, Step 1: Repeat Cycle 2, Step 1. Cycle 3, Step 2: Repeat Cycle 2, Step 2. Convergence.

What's happening in Example 9.1? Scale economies are taking their course! Government expenditures in all goods markets in exchange for $N+2$ (and large household demand to acquire $N+2$ to finance tax payments) result in a large trading volume on the trading posts for good $N+2$ versus $n = 1, ..., N$. Volume is large enough that scale economies kick in. The average cost pricing auctioneer adjusts

Table 1. Equilibrium monetary structure returns to scale in transaction technology

Demand structure	Linear transaction technology	Increasing returns transaction technology
Absence of double coincidence of wants	Monetary equilibrium where the low transaction cost instrument becomes 'money' (Example 4.1); possibly multiple 'moneys' (Example 4.2)	Monetary average cost pricing equilibrium with unique 'money' (Example 7.1)
Absence of double coincidence of wants with fiat money	Fiat money equilibrium if fiat money is the low transaction cost instrument (apply Example 4.1)	Fiat money equilibrium ('money' is unique) when tax payments and government purchases are sufficiently large (Example 9.1)
Full double coincidence of wants	Nonmonetary equilibrium (Example 5.1)	Monetary equilibria (average cost pricing and Bertrand) with unique 'money' (Examples 7.2, 8.2)

prices, the bid/ask spread, to reflect the scale economies. The bid/ask spreads incurred on trading m for $m \oplus i$ by way of good $N+2$ become considerably narrower than on trading m for $m \oplus i$ directly. The price system then directs each household to the market $\{m, N+2\}$ where its endowment is traded against good $N+2$. The household sells all its endowment there for $N+2$ and trades $N+2$ subsequently for tax payments and desired consumption. Scale economy has turned $N+2$ from a mere tax payment coupon into 'money,' the unique universally used common medium of exchange.

10 Conclusion

The monetary structure of trade in general equilibrium, the uniqueness of money, and the existence of a fiat money equilibrium can be demonstrated as the outcome of a market general equilibrium with transaction costs. The monetary character of trade, the existence of a common medium of exchange in economic equilibrium, is logically derived from price theory. Starting from a (non-monetary) Arrow-Debreu Walrasian model the addition of two constructs is sufficient: segmented markets with multiple budget constraints (one at each transaction) and transaction costs. The multiplicity of budget constraints creates a demand for a carrier of value (medium of exchange) between transactions. Money (the common medium of exchange) arises endogenously as the most liquid (lowest transaction cost) asset. Government-issued fiat money derives its value from acceptability in payment of taxes. Uniqueness of the monetary instrument (fiat or commodity money) in equilibrium comes from scale economies in transaction cost. The taxonomy of cases developed is depicted in Table 1. Absent double coincidence of wants, with linear transaction costs, a low transaction cost instrument is endogenously chosen as a medium of exchange. In the case of linear transaction costs, absence of double coincidence of wants is essential to monetary equilibrium. Alternatively scale economies in transaction cost (nonconvex transaction costs) lead to a corner solution, uniqueness of the common medium of exchange. Fiat money derives its positive equilibrium value from acceptability in payment of taxes. Fiat money becomes the unique common medium of exchange when government taxation and purchases are sufficiently large that scale economies in transaction costs make it the low (average) transaction cost instrument.

References

Alchian, A.: Why money? Journal of Money, Credit, and Banking **9**, 131–140 (1977)

Aristotle (350 BCE): Politics. Jowett translation

Banerjee, A., Maskin, E.: A Walrasian theory of money and barter. Quarterly Journal of Economics **CXI**(4), 955–1005 (1996)

Clower, R.: A reconsideration of the microfoundations of monetary theory. Western Economic Journal **6**, 1–8 (1967)

Clower, R.: On the origin of monetary exchange. Economic Inquiry **33**, 525–536 (1995)

Debreu, G.: Theory of value. New York: Wiley 1959

Einzig, P.: Primitive money. Oxford: Pergamon Press 1966

Foley, D.K.: Economic equilibrium with costly marketing. Journal of Economic Theory **2**(3), 276–291 (1970)

Hahn, F.H.: Equilibrium with transaction costs. Econometrica **39**(3), 417–439 (1971)

Hahn, F.H.: Fundamentals. Revista Internazionale di Scienze Sociali **CV**, 123–138 (1997)

Hellwig, C.: Money, intermediaries and cash-in-advance constraints. London School of Economics, July 27, 2000, pdf duplicated (2000)

Howitt, P.: Beyond search: fiat money in organized exchange. Ohio State University, pdf duplicated, December (2000)

Howitt, P., Clower, R.: The emergence of economic organization. Journal of Economic Behavior and Organization **41**, 55–84 (2000)

Iwai, K.: The bootstrap theory of money: a search theoretic foundation for monetary economics. Structural Change and Economic Dynamics **7**, 451–477 (1996)

Jevons, W.S.: Money and the mechanism of exchange. London: Macmillan 1875

Jones, R.A.: The origin and development of media of exchange. Journal of Political Economy **84**, 757–775 (1976)

Kiyotaki, N., Wright, R.: On money as a medium of exchange. Journal of Political Economy **97**, 927–954 (1989)

Kurz, M.: Equilibrium in a finite sequence of markets with transaction cost. Econometrica **42**(1), 1–20 (1974)

Lerner, A.P.: Money as a creature of the state. Proceedings of the American Economic Association **37**, 312–317 (1947)

Li, Y., Wright, R.: Government transaction policy, media of exchange, and prices. Journal of Economic Theory **81**, 290–313 (1998)

Marimon, R., McGrattan, E., Sargent, T.: Money as a medium of exchange in an economy with artificially intelligent agents. Journal of Economic Dynamics and Control **14**, 329–373 (1990)

Menger, C.: On the origin of money. Economic Journal **II**, 239–255 (1892). Translated by Foley., C.A. Reprinted in: Starr, R. (ed.) General equilibrium models of monetary economies. San Diego: Academic Press 1989

Ostroy, J.M.: The informational efficiency of monetary exchange. American Economic Review **LXIII**(4), 597–610 (1973)

Ostroy, J., Starr, R.: The transactions role of money. In: Friedman, B., Hahn, F. (eds.) Handbook of monetary economics. New York: Elsevier North Holland 1990

Ostroy, J., Starr, R.: Money and the decentralization of exchange. Econometrica **42**, 597–610 (1974)

Rajeev, M.: Marketless set-up vs trading posts: a comparative analysis. Annales d'Economie et de Statistique **53**, 197–211 (1999)

Rey, H.: International trade and currency exchange. Review of Economic Studies **68**(2)(235), 443–464 (2001)

Smith, A.: Wealth of nations, Vol. I, Book II, Chap. II (1776)

Starr, R.M.: The price of money in a pure exchange monetary economy with taxation. Econometrica **42**, 45–54 (1974)

Starr, R.M., Stinchcombe, M.B.: Exchange in a network of trading posts. In: Chichilnisky, G. (ed.) Markets, information, and uncertainty: essays in economic theory in honor of Kenneth Arrow. Cambridge: Cambridge University Press 1999

Starr, R.M., Stinchcombe, M.B.: Monetary equilibrium with pairwise trade and transaction costs. University of California, San Diego, duplicated (1998)

Starrett, D.A.: Inefficiency and the demand for money in a sequence economy. Review of Economic Studies **XL**(4), 437–448 (1973)

Tobin, J.: Discussion. In: Kareken, J., Wallace, N. (eds.) Models of monetary economies. Minneapolis: Federal Reserve Bank of Minneapolis 1980

Tobin, J.: The Tobin manuscript. New Haven, Yale University, mimeographed (1959)

Tobin, J., Golub, S.: Money, credit, and capital. Boston: Irwin/McGraw-Hill 1998

Trejos, A., Wright, R.: Search, bargaining, money and prices. Journal of Political Economy **103**(1), 118–141 (1995)

Walras, L.: Elements of pure economics (Jaffe translation 1954). Homewood, IL: Irwin 1874

Young, H.P.: Individual strategy and social structure: an evolutionary theory of institutions. Princeton: Princeton University Press 1998

Is assortative matching efficient?[*]

S. N. Durlauf and A. Seshadri

Department of Economics, 1180 Observatory Drive, University of Wisconsin, Madison, WI 53706-1393, USA (e-mail: {sdurlauf,aseshadr}@ssc.wisc.edu)

Received: September 25, 2001; revised version: February 26, 2002

Summary. This paper develops some general conditions under which complementarities between individual agents imply that assortative matching is efficient. Our analysis has four main findings. First, when agents are organized into equal-sized groups, just as in Becker (1973), the presence of within-group complementarities is sufficient for stratification to be efficient. Second, if group sizes vary, assortative matching may not be efficient even though complementarities are present, unless particular functional form assumptions are imposed. Third, the connection between assortative matching, complementarities and efficiency reemerges if one considers sequences of replications of the economy in which individual coalitions are uniformly bounded in size. Fourth, the presence of feedbacks from the composition of group memberships has important effects on efficient allocations and breaks any simple link between assortative matching and efficiency. Together, these results suggest that the characterization of the cross-section evolution of an efficiently sorted economy is likely to be highly complex.

Keywords and Phrases: Assortative matching, Matching models.

JEL Classification Numbers: J41.

1 Introduction

Recent work in economic theory, ranging from models of economic development to income inequality has begun to focus on the question of stratification, defined as the

---
[*] We thank William Brock for many helpful conversations and Scott Page for detailed comments on an earlier draft of this paper. The National Science Foundation, John D. and Catherine T. MacArthur Foundation and Center for Urban Land Economic Research have generously provided financial support.
Correspondence to: S. N. Durlauf

tendency of agents with similar characteristics to interact with one another in isolation of others. Two natural examples within a given economy are the assignment of workers to firms (Kremer, 1993; Kremer and Maskin, 1996) and the assignment of families to neighborhoods (Bénabou, 1993, 1996a,b; Durlauf, 1996a,b). When the productivity of each worker in an organization depends positively on the productivity of coworkers, incentives will exist for relatively high skilled workers to form firms that exclude their lower skilled counterparts. Similarly, when the quality and or quantity of a public good, such as education, within a neighborhood is an increasing function of the income distribution of the neighborhood due to tax base, role model influences or other effects, incentives exist for wealthier or better educated families to isolate themselves from poorer or less educated ones. Underlying these different environments is the common assumption that interactions between agents exhibit positive spillover effects, usually in the form of complementarities. In turn, each of these models can be thought of as a solution to a sorting problem, whose details are embedded in the economic environment of interest.

While the conditions under which stratification will emerge as an equilibrium allocation have been studied in a number of contexts, there has been less attention given to the efficiency of such allocations. Any relationship between equilibrium and efficiency in these models is not self-evident, of course, due to the many spillover effects and associated incomplete markets that typically are built into the economies under study. Neighborhood models, for example, typically exhibit role model and peer group effects between students attending a common school that are not adjudicated through market mechanisms. The major exception to this lack of attention is Becker's seminal work (1973) on the marriage problem. Becker considers the allocation of men and women into marriages in which the "productivity" of each marriage is assumed to depend on the ability levels of each of the partners. Becker then provides conditions under which the efficient allocation of partners results in "assortative matching," i.e. the most able male is matched with the most able female, etc. Specifically, the efficiency of assortative matching is shown to depend on the presence of positive cross-partial derivatives between the abilities of the partners in the output of a marriage. The generalization of this assumption - strategic complementarities between individuals - is typically assumed in describing interactions between agents in the more recent literature on interactions and stratification; as a result, Becker's proof is often used as evidence that in the presence of strategic complementarities, efficiency will induce stratification, suggesting that the stratification found in equilibrium models will at least qualitatively generalize when markets are complete.

The purpose of this paper is to determine whether the presence of complementarities between individuals is sufficient to determine whether stratification of agents by an attribute such as ability is efficient. We do this by employing a particular formalization of the notion of complementarity, increasing differences, which allows one to work with fairly general payoff structures. Basic properties of functions exhibiting increasing differences lead to four general conclusions. First, when agents are organized into equal-sized groups, increasing differences is a sufficient condition for stratification. This shows how the assortative matching results for the marriage problem may be generalized. Second, when group sizes vary, stratification may not be efficient even in the presence of increasing differences. Third, we show that the connection between stratification and efficiency will reemerge when one considers t-fold replications of the economy. Fourth, when there are incentive effects associated with the allocation of agents across coalitions, then stratification by ability may be inefficient even in a large economy limit.

These results in turn have two implications for the existing literatures on inequality. First, from the theory side, these results clarify the importance of market imperfections or specific functional forms, as opposed to complementarities *per se*, in generating stratified equilibria. In this sense, stratification is not a primitive feature of complementarities-driven economies. From the empirical side, these results indicate that stratification is not, as asserted by Herrnstein and Murray (1994), a logical consequence of the breakdown of barriers to mobility in society, and therefore cannot be treated as a self-evident explanation for persistent income inequality. Further, by clarifying the relationship between individual characteristics and group assignments, the analysis is important for determining the identifiability of group spillover effects in different economic and social contexts, as is clear from the work of Manski (1993) and Brock and Durlauf (2000, 2001).

Section 2 of the paper specifies the basic model under study. Section 3 develops the relationship between stratification and efficiency for economies exhibiting strategic complementarities. Equal and variable coalition-sizes are considered. Section 4 analyzes economies in which some additional restrictions are placed either on the coalition-specific production functions or on the cross-section distribution of abilities. Section 5 introduces incentive effects into the analysis of efficient allocations of agents. Section 6 discusses some examples that illustrate the general claims of the paper. Section 7 provides a summary and conclusions.

2 Model specification

Consider an economy consisting of N agents denoted by n. Each agent is associated with a nonnegative ability level a_n, which may be interpreted as any scalar attribute that relates to productivity. The collection of the ability levels for the agents, $\{a_1, \ldots, a_N\}$, is denoted by $\mathcal{A}$.

Agents are organized into coalitions. Coalition k, of size I can be represented as the vector $\underset{\sim}{a}_k = a_{n_1}, \ldots, a_{n_I}$ when it is comprised of agents n_1 to n_I. There exists a sequence of production functions indexed by the number of agents that interact within a coalition. Total output of the coalition equals

$$\Phi_I\left(\underset{\sim}{a}_k\right). \tag{1}$$

Abilities are measured so that $\Phi_I(\underset{\sim}{0}) = 0 \forall I$. Any permutation of the ability levels within a coalition is assumed to leave output unchanged. Throughout, we will be interested in allocations of individuals across coalitions that maximize aggregate output.

We will employ two definitions in the subsequent discussion. The first definition formalizes the notion of strategic complementarities for functions that are not required to be continuous or differentiable, thereby generalizing the positive cross-partial derivative assumption of Becker to arbitrary coalition-specific payoff functions.

Definition 1. *Increasing differences*
An I-size coalition-specific production function exhibits increasing differences if for any pair of J-length ability vectors $\underset{\sim}{b}$ and $\underset{\sim}{b}'$ and $I - J$ length vector $\underset{\sim}{c}$,

$$\Phi_I\left(\underset{\sim}{b}, \underset{\sim}{c}\right) - \Phi_I\left(\underset{\sim}{b}', \underset{\sim}{c}\right) \text{ is strictly increasing in } \underset{\sim}{c} \text{ if } \underset{\sim}{b} > \underset{\sim}{b}'.$$

When $\Phi_I(\cdot)$ is twice-differentiable, strict increasing differences is equivalent to the condition that all cross-partial derivatives of this function are positive, and thus corresponds exactly to the notion of strategic complementarities studied by Cooper and John (1988) and many others.

Topkis (1978) shows that if a function (with domain and range defined on the reals) exhibits increasing differences, it will also exhibit the property of supermodularity. A given size-specific coalition production function is strictly supermodular if for any nonscalar vectors $\underset{\sim}{a}$ and $\underset{\sim}{b}$ such that $\underset{\sim}{a} \neq \underset{\sim}{b}$,

$$\Phi_I\left(\underset{\sim}{a} \vee \underset{\sim}{b}\right) + \Phi_I\left(\underset{\sim}{a} \wedge \underset{\sim}{b}\right) > \Phi_I\left(\underset{\sim}{a}\right) + \Phi_I\left(\underset{\sim}{b}\right), \tag{2}$$

where for any two vectors $\underset{\sim}{a}$ and $\underset{\sim}{b}$, $\underset{\sim}{a} \vee \underset{\sim}{b} = (\max(a_1, b_1), \ldots, \max(a_I, b_I))$ and $\underset{\sim}{a} \wedge \underset{\sim}{b} = (\min(a_1, b_1), \ldots, \min(a_I, b_I))$.

This equivalence between increasing differences and supermodularity will prove to be very useful below.[1]

The second definition formalizes what we mean by a stratified allocation.

Definition 2. *Stratified allocations*
An allocation of agents is said to be stratified if
  A. Agents are allocated to at least 2 distinct coalitions.
  B. For any pair of coalitions $\underset{\sim}{a}$ and $\underset{\sim}{b}$ either $\min_n a_n \in \{\underset{\sim}{a}\} \geq \max_n b_n \in \{\underset{\sim}{b}\}$ or $\max_n a_n \in \{\underset{\sim}{a}\} \leq \min_n b_n \in \{\underset{\sim}{b}\}$.

3 Efficient allocations

i) Fixed coalition size

We first consider the problem of the efficient allocation of agents across coalitions when the size of each coalition is fixed at some I. Assume that N/I equals some

---

[1] Milgrom and Roberts (1990) provide a comprehensive introduction to increasing differences and supermodularity as well as a survey of its use in studying a wide variety of economic environments; see Milgrom and Shannon (1994) for many additional results. Cooper and John (1988) provide an excellent overview of the macroeconomic implications of strategic complementarities.

integer K. An efficient allocation of agents across coalitions is one that maximizes

$$\sum_{k=1}^{K} \Phi_I \left(\underset{\sim}{a}_k \right),$$ (3)

subject to the conditions that all agents are allocated, i.e.

$$\bigcup_{k=1}^{K} \left\{ \underset{\sim}{a}_k \right\} = \mathcal{A}.$$ (4)

Our first result indicates how strict increasing differences induces stratification in this case. The result is a slight generalization of the Becker condition for assortative mating in marriages, extending the assumptions of that paper to include cases where 1) more than two agents are matched, 2) agents within a coalition are of a common type, (as opposed to distinguishable between men and women), and 3) the coalition payoff functions are not differentiable.

Proposition 1. *Efficiency of stratification under increasing differences and fixed coalition size.*

Suppose that all coalitions must be of the same size I and that the associated payoff function exhibits strict increasing differences. Then stratification is necessary for output maximization.

Proof. Since N is finite, there always exists at least one optimal coalition configuration. Within such an output maximizing allocation, let $\underset{\sim}{b}$ and $\underset{\sim}{c}$ denote allocations to any particular pair of coalitions; associated with these vectors let $\underset{\sim}{b}^{(I)}$ denote the reordering of $\underset{\sim}{b}$ in ascending order of ability levels and $\underset{\sim}{c}_{(I)}$ denote the reordering of $\underset{\sim}{c}$ in descending order of abilities. In order for $\underset{\sim}{b}$ and $\underset{\sim}{c}$ to be part of an optimal allocation, it must be the case that $\Phi_I(\underset{\sim}{b}) + \Phi_I(\underset{\sim}{c})$ maximizes output among all possible coalitions of agents in $\{\underset{\sim}{b}\} \cup \{\underset{\sim}{c}\}$. This implies, since the ordering of agents within a coalition has no effect on output, that

$$\Phi_I \left(\underset{\sim}{b}^{(I)} \vee \underset{\sim}{c}_{(I)} \right) + \Phi_I \left(\underset{\sim}{b}^{(I)} \wedge \underset{\sim}{c}_{(I)} \right) \le \Phi_I \left(\underset{\sim}{b}^{(I)} \right)$$
$$+ \Phi_I \left(\underset{\sim}{c}_{(I)} \right) = \Phi_I \left(\underset{\sim}{b} \right) + \Phi_I \left(\underset{\sim}{c} \right).$$ (5)

This latter condition can hold given (2), only if either $\underset{\sim}{b}^{(I)} \vee \underset{\sim}{c}_{(I)} = \underset{\sim}{b}^{(I)}$ or $\underset{\sim}{b}^{(I)} \vee \underset{\sim}{c}_{(I)} = \underset{\sim}{c}_{(I)}$, which implies that across the coalitions, the agents are stratified. $\square$

ii) Variable coalition size

Proposition 1 indicates how stratification will be efficient for models of such organizations as marriages or athletic teams in which social norms have established a fixed set of coalition sizes. On the other hand, when one considers organizations such as firms or neighborhoods, the assumption of fixed coalition sizes is no longer sensible. When the coalition size is a choice variable, then the property of increasing differences no longer implies that multiple stratified coalitions are efficient. The efficiency of multiple stratified coalitions depends critically on whether large coalitions with low ability levels among some agents are less efficient than groups of small coalitions. Proposition 2 formalizes this by providing a condition under which stratification is never efficient - namely that an additional agent can never reduce the payoff of a coalition.

Proposition 2. *Condition for efficiency of integration under increasing differences and variable coalition size.*

Suppose that all production functions exhibit increasing differences and that coalitions may vary in size. If

$$\Phi_{I+J}\left(\underset{\sim}{a}, \underset{\sim}{0}\right) \geq \Phi_I\left(\underset{\sim}{a}\right) \forall I, \tag{6}$$

then output is maximized by a single coalition of all agents, regardless of the distribution of abilities.

Proof. If output is maximized under multiple coalitions, then

$$\Phi_{I+J}\left(\underset{\sim}{b}, \underset{\sim}{c}\right) < \Phi_I\left(\underset{\sim}{b}\right) + \Phi_J\left(\underset{\sim}{c}\right), \tag{7}$$

for any coalitions $\underset{\sim}{b}$ and $\underset{\sim}{c}$ which are part of the optimal allocation of agents. Given (2) this implies

$$\Phi_{I+J}\left(\underset{\sim}{b}, \underset{\sim}{c}\right) + \Phi_{I+J}\left(\underset{\sim}{0}, \underset{\sim}{0}\right)$$
$$< \Phi_{I+J}\left(\underset{\sim}{b}, \underset{\sim}{0}\right) + \Phi_{I+J}\left(\underset{\sim}{0}, \underset{\sim}{c}\right). \tag{8}$$

But $(\underset{\sim}{b}, \underset{\sim}{c}) = (\underset{\sim}{b}, \underset{\sim}{0}) \vee (\underset{\sim}{0}, \underset{\sim}{c})$ and $(\underset{\sim}{0}, \underset{\sim}{0}) = (\underset{\sim}{b}, \underset{\sim}{0}) \wedge (\underset{\sim}{0}, \underset{\sim}{c})$, which means that (8) contradicts (2).   □

We now consider the case where

$$\Phi_{I+J}\left(\underset{\sim}{a}, \underset{\sim}{0}\right) < \Phi_I\left(\underset{\sim}{a}\right), \tag{9}$$

which, by Proposition 2, is necessary for multiple coalitions to be efficient under some conditions. The existence of a link between increasing differences and stratification can now be considered. In fact, the existence of multiple coalition sizes means that no such link exists. To see this, suppose that at an efficient allocation, agents are allocated into two coalitions of sizes $N - K$ and K respectively, where $N - K > K$. Suppose that the agents in these coalitions are ordered in terms of increasing ability and are divided into three non-overlapping ability groups, $\underset{\sim}{a}_{\text{low}}, \underset{\sim}{a}_{\text{mid}}$ and $\underset{\sim}{a}_{\text{high}}$ such that low and high groups each have K members. Segregation will be efficient if and only if either $\Phi_{N-K}(\underset{\sim}{a}_{\text{low}}, \underset{\sim}{a}_{\text{mid}}) + \Phi_K(\underset{\sim}{a}_{\text{high}})$ or $\Phi_{N-K}(\underset{\sim}{a}_{\text{mid}}, \underset{\sim}{a}_{\text{high}}) + \Phi_K(\underset{\sim}{a}_{\text{low}})$ maximizes output relative to all possible configurations of agents. Consider the latter case; symmetric reasoning applies to the former. Let $\gamma_K(\cdot) = \Phi_{N-K}(\underset{\sim}{a}_{\text{mid}}, \cdot)$. For segregation to be efficient, such an allocation must be necessary to maximize the sum of two distinct K-size increasing differences functions, $\gamma_K(\cdot) + \Phi_K(\cdot)$. Such a requirement is not implied by, and is quite different from, the supermodularity condition that characterizes the individual functions. Following this logic, under variable coalition size, the assumption of increasing differences no longer implies any necessary link between stratification and efficiency.

Proposition 3. *Possible efficiency of integration under increasing differences and variable coalition size.*

There exist sets of size-specific payoff functions such that output is maximized by integrated coalitions even though each payoff function exhibits increasing differences.

Proof. We verify the proposition by example. Suppose that there are three agents with ability levels $a_1 = 1, a_2 = 1.5$ and $a_3 = 2$ respectively. The three size-specific payoff functions are:

$$\Phi_1(a_i) = 0.001 a_i^2 + 1.1 \max(a_i - 1, 0), \tag{10}$$

$$\Phi_2(a_i, a_j) = 1.5(a_i \cdot a_j), \tag{11}$$

and

$$\Phi_3(a_i, a_j, a_k) = .1(a_i \cdot a_j \cdot a_k). \tag{12}$$

These functions exhibit increasing differences yet the output-maximizing configuration of agents would place those with abilities a_1 and a_3 in one coalition, leaving the agent with ability a_2 isolated. □

While the example in the proposition's proof is ungainly, it illustrates some general ideas of interest. The high ability of agent 3 is productive in conjunction with another agent, requiring at least one multiple agent coalition. The isolation of the least able agent will leave him totally unproductive. In turn, his integration with agent 3 represents the most productive allocation. This example indicates how the integration of very skilled and unskilled workers in an organization may be more efficient than the integration of very skilled and moderately skilled workers even when each coalition exhibits complementarities.

The example in Proposition 3 possesses an additional feature of interest. Suppose that the ability of agent 1 is increased from 1 to 1.4. In this case, the efficient allocation of agents places 2 and 3 together, and leaves 1 in isolation. Therefore, an inequality-decreasing change in the cross-section distribution of abilities leads to an increase in the degree of stratification. To see why this can hold more generally, suppose that at some initial distribution of abilities, it is efficient to stratify into coalitions $\underset{\sim}{b}$ and $\underset{\sim}{c}$, i.e. $\Phi_I(\underset{\sim}{b}) + \Phi_J(\underset{\sim}{c}) > \Phi_{I+J}(\underset{\sim}{b}, \underset{\sim}{c})$. Assuming that $\underset{\sim}{b}$ comprises the higher ability agents, a monotonic decrease in ability from $\underset{\sim}{c}$ to $\underset{\sim}{c}'$, would necessarily preserve this stratification, relative to an integration of both coalitions only if

$$\Phi_{I+J}\left(\underset{\sim}{b}, \underset{\sim}{c}\right) - \Phi_{I+J}\left(\underset{\sim}{b}, \underset{\sim}{c}'\right) + \Phi_J\left(\underset{\sim}{c}\right) - \Phi_J\left(\underset{\sim}{c}'\right) > 0. \qquad (13)$$

This condition, in turn, is equivalent to the condition of increasing differences for the composite payoff function $\Phi_I(\underset{\sim}{b}) + \Phi_J(\underset{\sim}{c}) - \Phi_{I+J}(\underset{\sim}{b}, \underset{\sim}{c})$, which is not an implication of the fact that each function exhibits increasing differences individually.[2] These considerations are summarized in Proposition 4.

Proposition 4. *Lack of relationship between degree of cross-section inequality and stratification*

Consider an economy in which all production functions exhibit increasing differences and in which all agent allocations are efficient.

A. For an initial distribution of abilities such that in the efficient allocation no agents with abilities above $\bar{a}$ form coalitions with agents with abilities below $\underline{a}$, this stratification will not necessarily be preserved if all abilities above $\bar{a}$ are increased whereas all abilities below $\underline{a}$ are decreased.

B. A rightward shift in the distribution of abilities has no necessary implication for the degree of efficient stratification.

Propositions 3 and 4 illustrate that stratification cannot be treated as a generic property of economies exhibiting complementarities and some minimum degree of cross-section inequality without additional restrictions on the payoff functions under study.[3]

4 Stratification in restricted economic environments

A link between stratification and efficiency can be re-established with additional restrictions on the economic environment under study. One possibility concerns restrictions on the coalition-specific production functions. Alternatively, one can restrict the distribution of abilities across agents.

---

[2] We thank Paul Milgrom for discussion of this point.

[3] This result is related to Becker's (1974) analysis of assortative mating in polygamous societies. Becker argues that efficiency in the marriage market may be achieved equally well by matching a relatively able male with either one able or two less able females. Our analysis shows how this type of result may be strictly efficiency-enhancing in a number of alternative contexts; for example, we do not require any necessary link between the efficiency of integration and coalition size.

i) Restrictions on coordination costs

One approach to delimiting the class of production functions so as to restore a link between stratification and efficient allocations is to parameterize the costs of coordination in larger coalitions. Proposition 5 indicates how a restriction on coordination costs, namely the representation of these costs as additive and a function only of coalition size, leads to the efficiency of stratification. The proposition additionally shows that this sort of restriction on costs produces a relationship between the relative ability distributions of any two coalitions and their relative sizes.

Proposition 5. *Efficiency of stratification under increasing differences and variable coalition size when increased coalition size induces fixed costs.*

If

$$\Phi_{I+J}\left(\underset{\sim}{a}, 0\right) = \Phi_I\left(\underset{\sim}{a}\right) - C\left(I, J\right), \tag{14}$$

where $C\left(I, J\right)$ is positive, then

A. Any output maximizing configuration is stratified.

B. When coalitions are ordered by ability, if one coalition contains abilities which are greater than another, then that coalition will be at least as large as the other.

Proof. At an equilibrium it must be the case that any pair of coalitions $\underset{\sim}{b}$ and $\underset{\sim}{c}$ must maximize $\Phi_I\left(\underset{\sim}{b}\right) + \Phi_J\left(\underset{\sim}{c}\right)$, relative to any reallocation of agents across coalitions which preserves their size. Without loss of generality, take $I > J$. Using eq. (14) this sum can be rewritten as

$$\Phi_{I+J}\left(\underset{\sim}{b}, \underset{\sim}{0}_J\right) + \Phi_{I+J}\left(\underset{\sim}{c}, \underset{\sim}{0}_I\right) + C\left(I, J\right) + C\left(J, I\right), \tag{15}$$

where the subscripts on the 0 vectors denote length. It is clear that eq. (15) can only be maximized if $\left(\underset{\sim}{b}, \underset{\sim}{0}_J\right) = \left(\underset{\sim}{b}, \underset{\sim}{0}_J\right) \vee \left(\underset{\sim}{c}, \underset{\sim}{0}_I\right)$ and $\left(\underset{\sim}{c}, \underset{\sim}{0}_I\right) = \left(\underset{\sim}{b}, \underset{\sim}{0}_J\right) \wedge \left(\underset{\sim}{c}, \underset{\sim}{0}_I\right)$, which can only occur when the elements of $\underset{\sim}{b}$ are at least as large as those of $\underset{\sim}{c}$, i.e. the coalitions are stratified. This proves A.

To prove B, observe that if $I < J$, then the argument still goes through, only in this case $\underset{\sim}{c} > \underset{\sim}{b}$ as the larger coalition must have more nonzero elements, given the argument in A. $\square$

ii) Multiplicative interactions

A second type of production function which always produces stratification in efficient allocations is the so-called O-ring production function, introduced by Kremer

(1993). Kremer's production function is a special case of coalition-specific payoff functions of the form

$$\Phi_I\left(\underset{\sim}{a}\right) = \Gamma\left(I\right) \underset{n}{\Pi} \varphi\left(a_n\right). \tag{16}$$

In this expression, $\Gamma\left(\cdot\right)$ is a scaling factor which reflects economies (or diseconomies) of scale for coalitions of different sizes. The individual ability component $\varphi\left(\cdot\right)$ is assumed to be nonnegative and increasing.[4] As Proposition 6 states, such production functions will always produce stratification for efficient allocations.

Proposition 6. *Relationship between O-ring production functions and efficiency of stratification*

If Eq. (1) characterizes the set of coalition-specific production functions, then any pair of distinct coalitions will be stratified.

Proof. Since any pair of coalitions of equal size must be stratified by Proposition 1, we only need to consider any pair of coalitions with sizes $I > J$ and associated allocations $\underset{\sim}{a}$ and $\underset{\sim}{b}$. Rewrite $\underset{\sim}{a}$ as $\left(\underset{\sim}{c}, \underset{\sim}{d}\right)$, where $\underset{\sim}{c}$ contains the smallest J elements of $\underset{\sim}{a}$. Total production from the two coalitions is

$$\frac{\Gamma\left(I\right)}{\Gamma\left(J\right)} \overset{I-J}{\underset{n=1}{\Pi}} \varphi\left(d_n\right) \Phi_J\left(\underset{\sim}{c}\right) + \Phi_J\left(\underset{\sim}{b}\right) = K\Phi_J\left(\underset{\sim}{c}\right) + \Phi_J\left(\underset{\sim}{b}\right) \tag{17}$$

for some nonnegative K. If $K < 1$, maximization requires that $\underset{\sim}{c} = \underset{\sim}{c} \wedge \underset{\sim}{b}$. Further, since these coalitions are output maximizing, then they must be invariant under any alternative partition of $\underset{\sim}{a}$, as alternative partitions will lower the value of K. Similarly, if $K = 1$, then we can arbitrarily assign $\underset{\sim}{c} = \underset{\sim}{c} \wedge \underset{\sim}{b}$ and repeat the same argument. When $K > 1$, it is clear that this expression can only be maximized if $\underset{\sim}{c} = \underset{\sim}{c} \vee \underset{\sim}{b}$. Since the elements of $\underset{\sim}{d}$ exceed those of $\underset{\sim}{c}$, they must exceed those of $\underset{\sim}{b}$, and the coalitions are again stratified. $\square$

iii) Limiting behavior of replicated economies

While the above results illustrate how, in a finite economy, integrated economies may be efficient even in the presence of complementarities, these results can disappear when one considers replications of the economy. In particular, one can consider t-fold replications of the agents of an economy comprised of a finite number of agents with ability set $\mathcal{A}$, i.e. increasing collections of agents with associated abilities $\mathcal{A}_t$ such that

$$\mathcal{A}_t = \mathcal{A}_{t-1} \cup \mathcal{A}. \tag{18}$$

---

[4] Kremer (1993) studies a particular parameterization of this function, $\Phi_I(\underset{\sim}{a}) = \Gamma\left(I\right) \underset{n}{\Pi} a_n$ where $\Gamma\left(\cdot\right)$ is increasing and a_n is bounded between 0 and 1.

In such sequences, the number of agents with identical abilities becomes unbounded for each element of the support of abilities, while the number of agent types will remain unchanged. This turns out to have a critical effect on the limiting behavior of the economy, as it affects the capacity for completely homogeneous coalitions to emerge.

Proposition 7. *Asymptotic stratification with fixed upper coalition bound on coalition size.*

Suppose that the maximum coalition size never exceeds some finite I. Then as t becomes large in a sequence of t-fold replications of a finite collection of agents, the fraction of agents located in completely homogeneous coalitions will approach 1.

Proof. At an optimal allocation of agents across coalitions for the t'th replication, suppose that $C_{i,t}$, agents are located in coalitions of size i, so that $\sum_{i=1}^{I} C_{i,t} = tN$, the number of agents in the economy at that replication. For agents in coalitions of size i, let $K_{i,t}$ be the largest integer such that $C_{i,t} - i \cdot K_i$ is non-negative. For the $i \cdot K_{i,t}$ agents in coalitions of size i to be optimally assigned, Proposition 1 implies those agents must be in homogeneous coalitions, since such coalitions are feasible within the i-size class. Therefore, at most $\sum_{i=1}^{I} (C_{i,t} - i \cdot K_{i,t}) < I^2$ agents can lie in nonhomogeneous coalitions, given the upper bound on the number of types. Therefore, the fraction of agents in homogeneous coalitions can be no smaller than $1 - I^2/tN$, which converges to 1 as tN becomes large. $\qquad\square$

Observe that this result does not depend on the assumption that the maximum coalition size is exogenously limited; the proposition will still hold if, along any sequence of efficient allocations corresponding to increasing the number of replications of the original economy, the size of the largest coalition is uniformly bounded.

The applicability of Proposition 7 will depend, of course on the environment under study. In the case of firms, for example, it seems plausible that difficulties in the coordination of activities can impose an upper bound on the size of individual coalitions that is small relative to the pool of available workers.[5] On the other hand, for the problem of determining the number and size of school districts, it is clearly plausible that returns to scale and/or politically imposed restrictions on the number of districts will mean that the size of districts is large relative to the size of the population of families which is to be allocated.

5 Interactions between sorting allocations and productive inputs

The discussion thus far has focused on the efficiency properties of sorting when the determinants of output, namely ability, are fixed. When effort is integrated into production, the relationship between sorting and efficiency becomes much more complicated.

In order to generalize the earlier discussion, we focus on the case in which all coalitions must be of size I. Individuals supply effort e_i as well as ability. For ease

---

[5] See Becker and Murphy (1992) for an analysis of this issue.

of exposition, we take the effort level to be binary so that $e_i \in \{\underline{e}, \bar{e}\}$. Each member of a coalition receives a payoff of the form

$$\phi_{I,i}\left(a_i, \underset{\sim}{a}_{-i}, e_i, \underset{\sim}{e}_{-i}\right) \tag{19}$$

Total coalition output may be written

$$\Phi_I\left(\underset{\sim}{a}, \underset{\sim}{e}\right) = \sum_{i=1}^{I} \phi_{I,i}\left(a_i, \underset{\sim}{a}_{-i}, e_i, \underset{\sim}{e}_{-i}\right)^6 \tag{20}$$

Milgrom and Shannon (1994) verify that the coalition output functions exhibit increasing differences so long as the individual payoff functions do so, given the additive form of (20). Notice that by allowing the individual payoff functions to differ within a coalition, agents with different abilities can receive different levels of compensation.

For any allocation of agents across coalitions, Milgrom and Roberts verify the following. Given an ability vector $\underset{\sim}{a}$ within a coalition, there will exist at least one vector $\underset{\sim}{e}$ such that each e_i solves

$$\max_{e_i \in \{\underline{e}, \bar{e}\}} \phi_{I,i}\left(a_i, \underset{\sim}{a}_{-i}, e_i, \underset{\sim}{e}_{-i}\right) \tag{21}$$

We now consider the allocation of individuals across coalitions. Observe that conditional on common effort levels across all agents, output maximization will require that all coalitions are stratified by ability. Further, if the effort level of an agent with given ability is always at least as high as the effort level of an agent with lesser ability, then stratification will also be efficient. However, when effort is endogenous, stratification may be inefficient due to its effects on the set of efforts across agents.

In particular, suppose $I = 2$ and that $N/2$ agents have high ability $\bar{a}$ and $N/2$ agents have low ability $\underline{a}$. Suppose as well that the following inequalities hold:

$$\phi_{2,i}(\bar{a}, \underline{a}, \bar{e}, \underline{e}) > \phi_{2,i}(\bar{a}, \underline{a}, \underline{e}, \underline{e}) \tag{22}$$

$$\phi_{2,i}(\underline{a}, \underline{a}, \bar{e}, \underline{e}) < \phi_{2,i}(\underline{a}, \underline{a}, \underline{e}, \underline{e}) \tag{23}$$

In this case, it is clear stratification will always be output maximizing if

$$|\Phi_2(\bar{a}, \bar{a}, \bar{e}, \bar{e}) - \Phi_2(\bar{a}, \underline{a}, \bar{e}, \bar{e})| > |\Phi_2(\bar{a}, \underline{a}, \bar{e}, \bar{e}) - \Phi_2(\underline{a}, \underline{a}, \underline{e}, \underline{e})| \tag{24}$$

Increasing differences imply only that

$$|\Phi_2(\bar{a}, \bar{a}, \bar{e}, \bar{e}) - \Phi_2(\bar{a}, \underline{a}, \bar{e}, \bar{e})| > |\Phi_2(\bar{a}, \underline{a}, \bar{e}, \bar{e}) - \Phi_2(\underline{a}, \underline{a}, \bar{e}, \bar{e})| \tag{25}$$

---

[6] It might seem more natural to assume that each individual chooses an effort level which maximizes a utility function whose arguments include effort as well as compensation from the coalition. In fact, our current formulation can be rewritten this way without any qualitative change of results. The current formulation, by folding any utility aspects of effort into the coalition output function, avoids any ambiguity in what is meant by "efficient."

which makes clear the effect of incentives on efficiency. Proposition 8 verifies that these incentive effects can be powerful enough to render integrated coalitions efficient, even for a fixed-coalition size economy.

Proposition 8. *Efficiency of integration under increasing differences and constant coalition size in the presence of endogenous effort.*

There exist fixed-size coalition and individual payoff functions such that
A. Total payoffs are maximized by integrated coalitions even though the payoff functions exhibit increasing differences jointly in effort and ability.
B. The efficiency of integration holds for arbitrary replications of the economy.

Proof. We prove by example. Suppose that $\underline{a} = 1$, $\bar{a} = 6$ and $e_i \in \{1, 2\}$. Let the individual payoff functions within each coalition obey

$$\phi_{2,i}\left(a_i, \underline{a}_{-i}, e_i, \underline{e}_{-i}\right) = a_i \cdot a_j \cdot e_i \cdot e_j - 5e_i + 100e_j^2 \tag{26}$$

It is easy to verify that high ability agents will always choose $\bar{e}$ whereas low ability agents will choose $\underline{e}$ when matched with one another versus $\bar{e}$ when paired with high ability agents. A coalition of two high ability agents will produce a payoff of 534 for each, a coalition of two low ability agents will produce a payoff of 0 for each, and a mixed coalition will produce a payoff of 414 for each, which implies that integration will maximize total payoffs, which proves A. Part B is immediate since replications of the agents, keeping the distribution of abilities constant, do not change the set of feasible coalition types in the economy. □

One interesting feature of the efficient allocation of agents across coalitions in this example is that once the effort levels are set, the allocation is *ex post* inefficient; as in fact must always be true given Proposition 1. Durlauf (1996c) argues that this feature makes it difficult to assess the efficiency of programs such as affirmative action which may have desirable effects on unobservable variables such as effort by altering the way agents are sorted.

Finally, observe that if the coalition payoff function is replaced by $(a_i - 5) e_i + 100e_i e_j$ and the effort support is $\{0,1\}$, then while any coalition with a high ability agent will be associated with high effort by both members, a coalition with two low ability members will exhibit multiple Nash equilibria in effort as $\{0,0\}$ and $\{1,1\}$ are both self-reinforcing choices. This suggests that integrated allocations can help overcome coordination problems. This basic idea has application far beyond this simple example. For example, Brock and Durlauf (2001) show how the presence of multiple equilibria due to coordination failure in binary choice environments depends on the interplay of a relatively strong interdependence of choices on "social utility" effects (the large coefficient on $100e_i e_j$ in this case) with a relatively weak dependence of private utility on choices (the negative coefficient in $(a_i - 5) e_i$ for low ability agents). Alterations of the cross-section characteristics of agents within groupings, each of which obeys that model, can eliminate the presence of multiple equilibria. Durlauf (1995c) shows that this type of argument suggests that certain classes of affirmative action policies may be efficiency enhancing.

6 Examples

i) Endogenous growth

One variant of endogenous growth models of the type pioneered by Romer (1986) and Lucas (1988) may be thought of as positing individual production functions of the form[7]

$$\phi\left(a_n, F_{N_k}, \mu\left(N_k\right)\right),\qquad(27)$$

where N_k denotes the interaction neighborhood of agent n, F_{N_k} denotes the empirical distribution function of agent abilities in the neighborhood and $\mu\left(N_k\right)$ denotes the population of the neighborhood. These models typically assume a large population of individuals so that no single agent's actions affect the characteristics of the whole population. The function $\phi\left(\cdot,\cdot,\cdot\right)$ is usually taken to exhibit increasing differences with respect to the ability levels of other agents (in the sense that a rightward shift in F_{N_k} increases the marginal product of a_n), and to exhibit increasing returns in all arguments whereas $\partial^2\phi\left(\cdot,\cdot,\cdot\right)/\partial a_n^2 < 0$, so that the function is concave in its first argument. The aggregate production function Φ is simply the sum of the individual production functions so that

$$\Phi\left(F_{N_1},\ldots,F_{N_K},\mu\left(N_1\right),\ldots,\mu\left(N_k\right)\right)=\sum_n\phi\left(a_n,F_{N_k},\mu\left(N_k\right)\right).\qquad(28)$$

As before, conditional on any distribution of firms across neighborhoods, the aggregate production function inherits the concavity and increasing differences properties of the individual production functions.

One possible form for the individual production function (27) is

$$\phi\left(a_n,\mu\left(N_k\right)^\rho\cdot\int_{N_k}a\cdot dF_{N_k}\left(a\right)\right),\qquad(29)$$

so that the population and ability density terms interact multiplicatively.

Different choices of the function ρ allow one to distinguish the extent to which spillovers depend on total versus average ability levels as well as the output-maximizing configuration of firms across neighborhoods for a given distribution of abilities if one generalizes these models to allow agents to choose with whom they interact.[8] In particular, two extreme cases exist. When $\rho = 1$, individual productivity depends only on the magnitude of the ability aggregate. In this case, the requirement of Proposition 2 holds, so that if the interaction range is endogenous, then all agents will choose to interact together. This configuration corresponds to

---

[7] To be precise, this exercise should be thought of as describing sorting behavior in a model with capital externalities in which firm-specific capital stocks are given.

[8] The appropriateness of endogenizing the spillover environments in such models will of course depend upon the spillover in question. For example, it certainly seems reasonable *a priori* that firms will organize and locate themselves to account for human capital spillovers both internally and externally, as occurs, for example in the Silicon Valley. Similarly, the voluntary allocation of families into neighborhoods and associated house price and rental barriers is dependent on the feedback from neighborhood characteristics into offspring outcomes.

the case studied by Romer and Lucas, in which all agents spillover symmetrically onto one another. Of course, the same fully integrated outcome occurs when $\rho = 0$.

On the other hand, if $\rho = -1$, then individual productivity depends on the ability mean and Proposition 2's requirements are violated. In this case, the economy will completely stratify. This is intuitively obvious and can be verified as follows. Observe that for any ability distribution where a fraction λ of all agents have ability $\bar{a}$ and a fraction $1 - \lambda$ have ability $\underline{a}$, concavity in the first argument of the individual production function implies (supposing other arguments of the function)

$$
\phi \left(\lambda \bar{a} + (1 - \lambda) \underline{a}, \lambda \bar{a} + (1 - \lambda) \underline{a} \right) > \\
\lambda \phi \left(\bar{a}, \lambda \bar{a} + (1 - \lambda) \underline{a} \right) + (1 - \lambda) \phi \left(\underline{a}, \lambda \bar{a} + (1 - \lambda) \underline{a} \right). \tag{30}
$$

Further note that by social increasing returns

$$
\lambda \phi \left(\bar{a}, \bar{a} \right) + (1 - \lambda) \phi \left(\underline{a}, \bar{a} \right) > \phi \left(\lambda \bar{a} + (1 - \lambda) \underline{a}, \lambda \bar{a} + (1 - \lambda) \underline{a} \right). \tag{31}
$$

Together these inequalities imply

$$
\lambda \phi \left(\bar{a}, \bar{a} \right) + (1 - \lambda) \phi \left(\underline{a}, \underline{a} \right) > \lambda \phi \left(\bar{a}, \lambda \bar{a} + (1 - \lambda) \underline{a} \right) \\
+ (1 - \lambda) \phi \left(\underline{a}, \lambda \bar{a} + (1 - \lambda) \underline{a} \right) \tag{32}
$$

which means that integration is never efficient for any mixture of types. Repeated use of this argument can be used to show that the economy will always break up into isolated individuals when the spillover effects are based on mean ability. By extension, since the efficiency of the economy is unaffected whenever individuals of equal ability inhabit the same coalition or different coalitions, the economy will completely stratify whenever the number of feasible coalitions equals the number of different types.[9] For $-1 < \rho < 0$, the model can exhibit stratification or integration, depending on the distribution of abilities.

ii) Classroom size

To further illustrate the connection, or lack thereof, between the efficiency of stratification and complementarity, we consider the question of allocating students across classrooms when teachers are of differing qualities. The issue of class size, and the manner in which it influences individual learning has been a contentious one in the education literature. To model this problem, one can employ the same basic framework as in the endogenous growth case. However, the introduction of matching between students and teachers introduces some important differences.

We assume that the production function that determines a student's human capital h_n is, instead of (29)

$$
\Psi \left(a_n, \mu \left(N_k \right)^\rho \cdot \int_{N_k} a \cdot dF_{N_k} \left(a \right), H_n \right) \tag{33}
$$

---

[9] This follows from the fact that the average product of two identical agents in separate size-1 coalitions will equal their average product when they combine since the spillover associated with the mean ability level does not change.

where H_n denotes the human capital of the teacher assigned to a student with ability a_n.

Given a vector of student abilities, $\underset{\sim}{a}$ and a vector of teacher human capital levels $\underset{\sim}{H}$, the question under consideration is the matching of teachers and students so as to maximize total human capital among students. (Classrooms are implicitly defined by sets of students assigned to a common teacher.) This problem turns out to be complicated because of the possibilities that classroom sizes differ. For ease of exposition, we consider a special case of (33),

$$a_n^\alpha \cdot \mu \left(N_k \right)^\rho \cdot H_n^\gamma \left(\int_{N_k} a \cdot dF_{N_k} \left(a \right) \right)^\beta \tag{34}$$

Here, γ is the elasticity of a student's human capital with respect to his teacher's human capital. Observe that if $\alpha, \gamma > 0$, then student ability a_n and H_n are complementary in determining the student's human capital. When $\gamma = 0$, we are back to the endogenous growth case.

As before, ρ determines whether classmates are complementary inputs in human capital production. In particular, regardless of the value of γ, if $\rho > 0$, then the conditions of Proposition 2 go through, and integration is efficient. When $\rho < 0$, efficient allocations are again difficult to characterize. Hence, we use some numerical examples to illustrate the main workings of the model.

Imagine that there are three students with ability levels, 1, 1.5 and 2.0 and two teachers with human capital levels, 0.5 and 0.6. Further, assume that $\alpha = \beta = 0.5$. Finally, assume that teachers can go unassigned. Table 1 depicts the various efficient allocations for alternative values of ρ and γ.

Table 1 illustrates some basic ideas. First, consider allocations for different values of ρ when γ is fixed. Higher values of ρ imply larger negative effects of group size on individual achievement. This in turns renders stratification relatively efficient. Further, there is substitutability between group size and teacher human capital when ρ is negative. This implies that efficient allocations assign low ability students to high human capital teachers. When γ is small, the effect of substitutability between group size and individual achievement dominates the effects of complementarity between individual ability and teacher human capital. Again, lower ability students are assigned to better teachers. As γ increases, the latter effect dominates and more able students are assigned to better teachers.

What happens as ρ varies for a fixed γ? Unsurprisingly, higher values of ρ induce stratification. Further, notice that when $-1.6 < \rho < -1.2$, the efficient allocation assigns weaker students to stronger teachers. As in the case of varying γ, there are two forces at work. When ρ is sufficiently negative, the direct negative effect of group size on individual achievement is so strong that efficiency requires matching more able students and teachers. When γ and ρ are both high, stratification of students across classrooms and assignment of more able students to better teachers is required for efficiency.

Taken as a whole, these results suggest that the determination of efficient assignments when groups sizes are endogenous is potentially quite complex, and that the link between the efficiency of stratification and the degree of complementarity

Table 1. Effect of stratification on efficiency

$\rho\backslash\gamma$	0.1	0.2	1.5
-1	$\{a_1 \Leftrightarrow H_1\}, \{(a_2,a_3) \Leftrightarrow H_2\}$	$\{(a_1,a_2,a_3) \Leftrightarrow H_2\}$	$\{(a_1,a_2,a_3) \Leftrightarrow H_2\}$
-1.2	$\{(a_1,a_2) \Leftrightarrow H_2\}, \{a_3 \Leftrightarrow H_1\}$	$\{(a_1,a_2) \Leftrightarrow H_2\}, \{a_3 \Leftrightarrow H_1\}$	$\{a_1 \Leftrightarrow H_1\}, \{(a_2,a_3) \Leftrightarrow H_2\}$
-1.4	$\{(a_1,a_2) \Leftrightarrow H_2\}, \{a_3 \Leftrightarrow H_1\}$	$\{(a_1,a_2) \Leftrightarrow H_2\}, \{a_3 \Leftrightarrow H_1\}$	$\{a_1 \Leftrightarrow H_1\}, \{(a_2,a_3) \Leftrightarrow H_2\}$
-1.6	$\{(a_1,a_2) \Leftrightarrow H_2\}, \{a_3 \Leftrightarrow H_1\}$	$\{(a_1,a_2) \Leftrightarrow H_2\}, \{a_3 \Leftrightarrow H_1\}$	$\{a_1 \Leftrightarrow H_1\}, \{(a_2,a_3) \Leftrightarrow H_2\}$
-1.8	$\{(a_1,a_2) \Leftrightarrow H_1\}, \{a_3 \Leftrightarrow H_2\}$	$\{(a_1,a_2) \Leftrightarrow H_1\}, \{a_3 \Leftrightarrow H_2\}$	$\{(a_1,a_2) \Leftrightarrow H_1\}, \{a_3 \Leftrightarrow H_2\}$

The rows depict varying values for ρ, while the columns stand for different values of γ, the elasticity of a student's human capital with respect to his teacher's human capital. Each cell indicates the efficient allocation of students to teachers. For instance, $\{a_1 \Leftrightarrow H_1\}, \{(a_2,a_3) \Leftrightarrow H_2\}$ means that the student with the lowest ability a_1 is matched to the teacher with the lowest human capital H_1, while the other two students are assigned to the better teacher H_2.

will depend on a range of factors. And of course, this analysis does not consider the question of efficiency in dynamic contexts, where additional considerations arise. For example, Seshadri (2000) shows how the assignment of students to teachers has subtle implications for both intergenerational mobility and growth.

7 Conclusions

This paper has examined conditions under which the efficient allocation of agents produced stratification by ability. The assortative mating solution proposed by Becker (1973) was shown to generalize in two senses. First, assortative mating will hold for all economies comprised of identical sized coalitions. Second, replications of the population will lead to asymptotic stratification when the number of agents per coalition is bounded. On the other hand, when agents are free to choose coalition size, strategic complementarities were shown to be compatible with the efficiency of integrated equilibria. Similar results were obtained when the configuration of individuals across coalitions influenced the degree of effort made by each individual. The implication of these results is that the link between stratification and efficiency, which in the marriage problem holds whenever strategic complementarities exist between spouses, does not generalize to a wide range of sorting problems of interest. Therefore the wide range of stratified equilibria which have emerged as a robust implication of the new theoretical literature on inequality carry no presumption of efficiency. By extension, stratification as an equilibrium phenomenon must therefore depend on the presence of market incompleteness, except in relatively special cases. For example, it suggests that various assertions in Herrnstein and Murray (1994) on the role of economic forces in the rise of cognitive stratification are incorrect. Finally, neither the evidence of increasing stratification of firms by skill (Kremer and Maskin, 1995) nor of neighborhoods by income (Jargowsky, 1997) is self-evidently explained by efficiency considerations. In fact, the implication of models such as those of Bénabou (1993) and Durlauf (1995a,b), is that such increased stratification may be explained by the interaction of changes in the cross-section distribution of individual characteristics with the presence of either direct externalities or restrictions on the compensation rules available to coalitions.

One important extension of the current analysis is suggested by the possibility that the contemporaneous allocation of agents by coalition affects the distribution of abilities next period, as would occur in the neighborhood-based human capital models or in models in which worker ability is influenced by learning-by-doing in an environment conditioned by coworkers. In this context, there will exist dynamic efficiency considerations beyond those which we have explored. Bénabou (1995) provides an interesting analysis of dynamic efficiency of this type in a comparison of completely stratified versus completely integrated economies; a useful complement to that analysis would consider the dynamics of efficient stratification.

Finally, our analysis suggests the importance of developing a metric for identifying features of heterogeneous economies which are robust with respect to changes in the distribution of cross-section characteristics as well as with respect to functional form specifications. One fundamental difference between models in economic science and models in the natural sciences is the relative lack of guidance provided by economic theory on the specifics of individual agent behavior. Yet it is precisely these details which will determine the cross-section allocation of a heterogeneous population. Research in complex systems has already identified, in many contexts, equivalence classes of dynamic processes with similar limiting behavior. A similar research program represents an important ingredient in the development of a complex systems approach to endogenous groupings of individuals.

References

Becker, G.: A theory of marriage: Part I. Journal of Political Economy **81**, 813–846 (1973)

Becker, G.: A theory of marriage: Part II. Journal of Political Economy **82** (2 pt. 2), S11–S26 (1974)

Becker, G., Murphy, K.: The division of labor, coordination costs, and knowledge. Quarterly Journal of Economics, **107**, 1137–1160 (1992)

Bénabou, R.: Workings of a city: location, education, and production. Quarterly Journal of Economics **108**, 619–652 (1993)

Bénabou, R.: Equity and efficiency in human capital investment: the local connection. Review of Economic Studies **62**, 237–264 (1996a)

Bénabou, R.: Heterogeneity, stratification, and growth: macroeconomic implications of community structure and school finance. American Economic Review **86**, 584–609 (1996b)

Brock, W., Durlauf, S.: Discrete choice with social interactions. Review of Economic Studies **68**, 235–260 (2001a)

Brock, W., Durlauf, S.: Interactions-based models. Mimeo, University of Wisconsin. In: Heckman, J., Leamer, E. (eds.) Handbook of econometrics, Vol. 5. Amsterdam: North-Holland 2001b

Cooper, R., John, A.: Coordinating coordination failures in Keynesian models. Quarterly Journal of Economics **103**, 441–464 (1988)

Durlauf, S.: A Theory of persistent income inequality. Journal of Economic Growth **1**, 75–93 (1996a)

Durlauf, S.: Neighborhood feedbacks, endogenous stratification, and income inequality. In: Barnett, W., Gandolfo, G., Hillinger, C. (eds.) Dynamic disequilibrium modelling: Proceedings of the Ninth International Symposium on Economic Theory and Econometrics. Cambridge: Cambridge University Press 1996b

Durlauf, S.: Associational redistribution: a defense. Politics and Society **24** (4), 391–401 (1996c)

Herrnstein, R., Murray, C.: The bell curve. New York: Basic Books 1994

Jargowsky, P.: Poverty and place. New York: Russell Sage Foundation 1997

Kremer, M.: The O-ring theory of economic development. Quarterly Journal of Economics **108**, 551–575 (1993)

Kremer, M., Maskin, E.: Wage inequality and segregation by skill. NBER Working Paper no. 5718 (1996)

Loury, G.: A dynamic theory of racial income differences. In: Wallace, P., LaMond, A. (eds.) Women minorities and employment discrimination. Lexington: Lexington Books 1977

Lucas, R.: On the mechanics of economic development. Journal of Monetary Economics **22**, 3–42 (1988)

Manski, C.: Identification problems in the social sciences. In: Marsden, P. (ed.) Sociological methodology, Vol. 23. Cambridge: Basil Blackwell 1993a

Manski, C.: Identification of endogenous social effects: the reflection problem. Review of Economic Studies **60**, 531–542 (1993b)

Milgrom, P., Roberts J.: Rationalizability, learning, and equilibrium in games with strategic complementarities. Econometrica **58**, 1255–1277 (1990)

Milgrom, P., Shannon C.: Monotone comparative statics. Econometrica **62**, 157–180 (1994)

Romer, P.: Increasing returns and long run growth. Journal of Political Economy **94**, 1002–1037 (1986)

Seshadri, A.: Specialization in education. Mimeo, University of Wisconsin (2000)

Topkis, D.: Minimizing a submodular function on a lattice. Operations Research **26**, 305–321 (1978)

On extensive form implementation of contracts in differential information economies[*]

Dionysius Glycopantis[1], Allan Muir[2], and Nicholas C. Yannelis[3]

[1] Department of Economics, City University, Northampton Square, London EC1V 0HB, UK
(e-mail: d.glycopantis@city.ac.uk)
[2] Department of Mathematics, City University, Northampton Square, London EC1V 0HB, UK
(e-mail: a.muir@city.ac.uk)
[3] Department of Economics, University of Illinois at Urbana-Champaign, IL 61820, USA
(e-mail: nyanneli@uiuc.edu)

Received: November 19, 2001; revised version: April 17, 2002

Summary. In the context of differential information economies, with and without free disposal, we consider the concepts of Radner equilibrium, rational expectations equilibrium, private core, weak fine core and weak fine value. We look into the possible implementation of these concepts as perfect Bayesian or sequential equilibria of noncooperative dynamic formulations. We construct relevant game trees which indicate the sequence of decisions and the information sets, and explain the rules for calculating ex ante expected payoffs. The possibility of implementing an allocation is related to whether or not it is incentive compatible. Implementation through an exogenous third party or an endogenous intermediary is also considered.

Keywords and Phrases: Differential information economy, Private core, Radner equilibrium, Rational expectations equilibrium, Weak fine core, Weak fine value, Free disposal, Coalitional Bayesian incentive compatibility, Game trees, Perfect Bayesian equilibrium, Sequential equilibrium, Contracts.

JEL Classification Numbers: 020, 226.

1 Introduction

An economy with differential information consists of a finite set of agents each of which is characterized by a random utility function, a random consumption set,

---

[*] This paper comes out of a visit by Nicholas Yannelis to City University, London, in December 2000. We are grateful to Dr A. Hadjiprocopis for his invaluable help with the implementation of Latex in a Unix environment. We also thank Leon Koutsougeras and a referee for several, helpful comments.
Correspondence to: N. C. Yannelis

random initial endowments, a private information set defined on the states of nature, and a prior probability distribution on these states. For such an economy there are a number of cooperative and non-cooperative equilibrium concepts.

We have the noncooperative concepts of the generalized Walrasian equilibrium ideas of Radner equilibrium and rational expectations equilibrium (REE) defined in Radner (1968), Allen (1981) and Einy, Moreno, and Shitovitz (2000, 2001).[1] We also have the cooperative concepts of the private core (Yannelis, 1991), of the weak fine core, defined in Yannelis (1991) and Koutsougeras and Yannelis (1993), and that of the weak fine value (Krasa and Yannelis, 1994). The last two concepts allow the agents to pool their information.[2]

In a comparison of the equilibrium concepts we note that contrary to the private core any rational expectations Walrasian equilibium notion will always give zero quantities to an agent whose initial endowments are zero in each state. This is so irrespective of whether his private information is the full partition or the trivial partition of the states of nature. Hence the Radner as well as the REE do not register the informational superiority of an agent.

In Glycopantis, Muir, and Yannelis (2001) we provided a noncooperative interpretation of the private core for a three persons economy without free disposal. We constructed game trees which indicate the sequence of decisions and the information of the agents, and explained the rules for calculating ex ante, expected payoffs, through the reallocation of initial endowments. We showed that the private core can be given a dynamic interpretation as a perfect Bayesian equilibrium (PBE) of a noncooperative extensive form game.

The term implementation is used in the sense of realization of an allocation and not in the formal sense of implementation theory or mechanism design. Implementation or support of an allocation is sought through the PBE concept, described in Tirole (1988), which is a variant of the Kreps-Wilson (1982) idea of sequential equilibrium.

A PBE consists of a set of players' optimal behavioral strategies, and consistent with these, a set of beliefs which attach a probability distribution to the nodes of each information set. Consistency requires that the decision from an information set is optimal given the particular player's beliefs about the nodes of this set and the strategies from all other sets, and that beliefs are formed from updating, using the available information. If the optimal play of the game enters an information set then updating of beliefs must be Bayesian. Otherwise appropriate beliefs are assigned arbitrarily to the nodes of the set. This equilibrium concept is further looked at in Appendix I.

Our main observation in Glycopantis, Muir, and Yannelis (2001) was that Bayesian incentive compatible concepts, like the private core, can be implemented as a PBE of a noncooperative, extensive form game. Moreover we provided a counter example which demonstrates that core concepts which are not necessarily Bayesian incentive compatible, as for example the weak fine core, cannot be supported, under reasonable rules, in a dynamic framework. In the present paper we

---

[1] Kurz (1994) has provided the alternative idea of rational belief equilibria.

[2] See also Allen and Yannelis (2001) for additional references.

examine further the issue of extensive form implementation and obtain additional results.

Firstly, we consider cooperative and noncooperative solution concepts *with* and *without* free disposal. To our surprise, as it was not intuitively obvious, we found that solution concepts which are Bayesian incentive compatible without free disposal, do not retain this property under free disposal. In particular, not only free disposal *destroys* incentive compatibility but a problem also appears in verifying that an agent has actually destroyed part of his initial endowment.

Secondly, we provide examples which demonstrate that with free disposal cooperative and noocooperative solution concepts are not implementable as a PBE. However implementation becomes possible by introducing a third party, such as a court which has perfect knowledge in order to be able to penalize the lying agents.

Thirdly, for the purpose of implementation of the (non-free) disposal private core, we follow an alternative approach. We consider the (non-free) disposal private core example of the one-good, three-agent economy discussed in Glycopantis, Muir, and Yannelis (2001). The introduction of a third party results in the implementation of the private core allocation as a PBE. We show here that it can also be implemented as a sequential equilibrium (Kreps and Wilson, 1982).

Finally we provide a full characterization of our Bayesian incentive compatibility concept in the case of one good per state.

The analysis suggests that if an allocation is not incentive compatible, i.e. the agents do not find that it is in accordance with their interests, then there is a difficulty in implementing it in a dynamic framework. On the other hand incentive compatible allocations are implementable through contracts with reasonable conditions. We note that the implementation analysis is independent of the equilibrium notion. It applies to contracts in general which can be analysed by a similar tree structure.

Parts of the investigation fall into the area of the Nash programme the purpose of which has been, as explained in Glycopantis, Muir, and Yannelis (2001), to provide support and justification of cooperative solutions through noncooperative formulations. On the other hand we extend here the investigation into more general areas by discussing explicitly the possible implementation of noncooperative concepts such as Radner equilibrium and REE. It appears that in general the issue is the relation between dynamic and static considerations, not necessarily between cooperative and noncooperative formulations.

The paper is organized as follows. Section 2 defines a differential information exchange economy. Section 3 contains the equilibrium concepts discussed in this paper. Section 4 describes ideas of incentive compatibility. Section 5 discusses the non-implementation of free disposal private core allocations and Section 6 the implementation of private core and Radner equilibria through the courts. Section 7 discusses the implementation of non-free disposal private core allocations through an endogenous intermediary. Section 8 offers concluding remarks. Appendix I contains further remarks on PBE.

2 Differential information economy

We define the notion of a finite-agent economy with differential information, confining ourselves to the case where the set of states of nature, Ω, is finite and there is a finite number of goods, l, per state. $\mathcal{F}$ is a σ-algebra on Ω, I is a set of n players and $\mathbb{R}^l_+$ will denote the positive orthant of $\mathbb{R}^l$.

A *differential information exchange economy* $\mathcal{E}$ is a set $\{((\Omega, \mathcal{F}), X_i, \mathcal{F}_i, u_i, e_i, q_i) : i = 1, \ldots, n\}$ where

1. $X_i : \Omega \rightarrow 2^{\mathbb{R}^l_+}$ is the set-valued function giving the *random consumption set* of Agent (Player) i, who is denoted also by Pi;
2. $\mathcal{F}_i$ is a partition of Ω, denoting the *private information*[3] of Pi;
3. $u_i : \Omega \times \mathbb{R}^l_+ \rightarrow \mathbb{R}$ is the *random utility* function of Pi;
4. $e_i : \Omega \rightarrow \mathbb{R}^l_+$ is the *random initial endowment* of Pi, assumed to be constant on elements of $\mathcal{F}_i$, with $e_i(\omega) \in X_i(\omega)$ for all $\omega \in \Omega$;
5. q_i is an $\mathcal{F}$-measurable probability function on Ω giving the *prior* of Pi. It is assumed that on all elements of $\mathcal{F}_i$ the aggregate q_i is positive. If a common prior is assumed it will be denoted by μ.

We will refer to a function with domain Ω, constant on elements of $\mathcal{F}_i$, as $\mathcal{F}_i$-*measurable*, although, strictly speaking, measurability is with respect to the σ-algebra generated by the partition. We can think of such a function as delivering information to Pi which does not permit discrimination between the states of nature belonging to any element of $\mathcal{F}_i$.

In the first period agents make contracts in the ex ante stage. In the interim stage, i.e., after they have received a signal[4] as to what is the event containing the realized state of nature, one considers the incentive compatibility of the contract.

For any $x_i : \Omega \rightarrow \mathbb{R}^l_+$, the *ex ante expected utility* of Pi is given by

$$v_i(x_i) = \sum_{\omega \in \Omega} u_i(\omega, x_i(\omega)) q_i(\omega). \tag{1}$$

Denote by $E_i(\omega)$ the element in the partition $\mathcal{F}_i$ which contains the realized state of nature, $\omega \in \Omega$. It is assumed that $q_i(E_i(\omega)) > 0$ for all $\omega \in \Omega$. The *interim expected utility* function of Pi is given by

$$v_i(\omega, x_i) = \sum_{\omega' \in \Omega} u_i(\omega', x_i(\omega')) q_i(\omega' | E_i(\omega)), \tag{2}$$

where

$$q_i(\omega' | E_i(\omega)) = \begin{cases} 0 & \text{for} \quad \omega' \notin E_i(\omega) \\ \dfrac{q_i(\omega')}{q_i(E_i(\omega))} & \text{for} \quad \omega' \in E_i(\omega). \end{cases}$$

---

[3] Following Aumann (1987) we assume that the players' information partitions are common knowledge. Sometimes $\mathcal{F}_i$ will denote the σ-algebra generated by the partition, in which case $\mathcal{F}_i \subseteq \mathcal{F}$, as it will be clear from the context.

[4] A *signal* to Pi is an $\mathcal{F}_i$-measurable function from Ω to the set of the possible distinct observations specific to the player; that is, it induces the partition $\mathcal{F}_i$, and so gives the finest discrimination of states of nature directly available Pi.

3 Private core, weak fine core, Radner equilibrium, REE and weak fine value

We define here the various equilibrium concepts in this paper, distinguishing between the free disposal and the non-free disposal case. A comparison is also made between these concepts. All definitions are in the context of the exchange economy $\mathcal{E}$ in Section 2.

We begin with some notation. Denote by $L_1(q_i, \mathbb{R}^l)$ the space of all equivalence classes, with respect to q_i, of $\mathcal{F}$-measurable functions $f_i : \Omega \to \mathbb{R}^l$.

L_{X_i} is the set of all $\mathcal{F}_i$-measurable selections from the random consumption set of Agent i, i.e.,

$$L_{X_i} = \left\{ x_i \in L_1(q_i, \mathbb{R}^l) : \ x_i : \Omega \to \mathbb{R}^l \text{ is } \mathcal{F}_i\text{-measurable} \right.$$
$$\left. \text{and } x_i(\omega) \in X_i(\omega) \ q_i\text{-a.e.} \right\}$$

and let $L_X = \prod_{i=1}^{n} L_{X_i}$.

Also let

$$\bar{L}_{X_i} = \left\{ x_i \in L_1(q_i, \mathbb{R}^l) : \ x_i(\omega) \in X_i(\omega) \ q_i\text{-a.e.} \right\}$$

and let $\bar{L}_X = \prod_{i=1}^{n} \bar{L}_{X_i}$.

An element $x = (x_1, \ldots, x_n) \in \bar{L}_X$ will be called an *allocation*. For any subset of players S, an element $(y_i)_{i \in S} \in \prod_{i \in S} \bar{L}_{X_i}$ will also be called an allocation, although strictly speaking it is an allocation to S.

We note that the above notation is employed also for purposes of comparisons with the analysis in Glycopantis, Muir, and Yannelis (2001). In case there is only one good, i.e. $l = 1$, we shall use the notation $L_{X_i}^1$, $\bar{L}_{X_i}^1$ etc. When a common prior is also assumed $L_1(q_i, \mathbb{R}^l)$ will be replaced by $L_1(\mu, \mathbb{R}^l)$.

First we define the notion of the (ex ante) private core[5] (Yannelis, 1991).

Definition 3.1. An allocation $x \in L_X$ is said to be a *private core allocation* if

(i) $\sum_{i=1}^{n} x_i = \sum_{i=1}^{n} e_i$ and
(ii) there do not exist coalition S and allocation $(y_i)_{i \in S} \in \prod_{i \in S} L_{X_i}$ such that

$$\sum_{i \in S} y_i = \sum_{i \in S} e_i \text{ and } v_i(y_i) > v_i(x_i) \text{ for all } i \in S.$$

Notice that the definition above does not allow for free disposal. If the feasibility condition (i) is replaced by (i)' $\sum_{i=1}^{n} x_i \leq \sum_{i=1}^{n} e_i$ then *free disposal* is allowed.

Example 3.1. Consider the following three agents economy, $I = \{1, 2, 3\}$ with one commodity, i.e. $X_i = \mathbb{R}_+$ for each i, and three states of nature $\Omega = \{a, b, c\}$.

We assume that the initial endowments and information partitions of the agents are given by

---

[5] The private core can also be defined as an interim concept. See Yannelis (1991) and Glycopantis, Muir, and Yannelis (2001).

$$e_1 = (5, 5, 0), \quad \mathcal{F}_1 = \{\{a, b\}, \{c\}\};$$
$$e_2 = (5, 0, 5), \quad \mathcal{F}_2 = \{\{a, c\}, \{b\}\};$$
$$e_3 = (0, 0, 0), \quad \mathcal{F}_3 = \{\{a\}, \{b\}, \{c\}\}.$$

It is also assumed that $u_i(\omega, x_i(\omega)) = x_i^{\frac{1}{2}}$, which is a typical strictly concave and monotone function in x_i, and that every player expects that each state of nature occurs with the same probability, i.e. $\mu(\{\omega\}) = \frac{1}{3}$, for $\omega \in \Omega$. For convenience, in the discussion below expected utilities are multiplied by 3.

It was shown in Appendix II of Glycopantis, Muir, and Yannelis (2001) that, without free disposal, a private core allocation of this economy is $x_1 = (4, 4, 1)$, $x_2 = (4, 1, 4)$ and $x_3 = (2, 0, 0)$. It is important to observe that in spite of the fact that Agent 3 has zero initial endowments, his superior information allows him to make a Pareto improvement for the economy as a whole and he was rewarded for doing so. In other words, Agent 3 traded his superior information for actual consumption in state a. In return Agent 3 provided insurance to Agent 1 in state c and to Agent 2 in state b. Notice that if the private information set of Agent 3 is the trivial partition, i.e., $\mathcal{F}_3' = \{a, b, c\}$, then no-trade takes place and clearly in this case he gets zero utility. Thus the private core is sensitive to information asymmetries.

Next we define another core concept, the weak fine core (Yannelis, 1991; Koutsougeras and Yannelis, 1993). This is a refinement of the fine core concept of Wilson (1978). Recall that the fine core notion of Wilson as well as the fine core in Koutsougeras and Yannelis may be empty in well behaved economies. It is exactly for this reason that we are working with a different concept.

Definition 3.2. An allocation $x = (x_1, \ldots, x_n) \in \bar{L}_X$ is said to be a *weak fine core allocation* if

(i) each $x_i(\cdot)$ is $\bigvee_{i=1}^{n} \mathcal{F}_i$-measurable [6]

(ii) $\sum_{i=1}^{n} x_i = \sum_{i=1}^{n} e_i$ and

(iii) there do not exist coalition S and allocation $(y_i)_{i \in S} \in \prod_{i \in S} \bar{L}_{X_i}$ such that $y_i(\cdot) - e_i(\cdot)$ is $\bigvee_{i \in S} \mathcal{F}_i$-measurable for all $i \in S$, $\sum_{i \in S} y_i = \sum_{i \in S} e_i$ and $v_i(y_i) > v_i(x_i)$ for all $i \in S$.

Existence of private core and weak fine core allocations is discussed in Glycopantis, Muir, and Yannelis (2001). The weak fine core is also an ex ante concept. As with the private core the feasibility condition can be relaxed to (ii)' $\sum_{i=1}^{n} x_i \leq \sum_{i=1}^{n} e_i$. Notice however that now coalitions of agents are allowed to pool their own information and all allocations will exhaust the resource. The example below illustrates this concept.

Example 3.2. Consider the Example 3.1 without Agent 3. Then if Agents 1 and 2 pool their own information a possible allocation is $x_1 = x_2 = (5, 2.5, 2.5)$. Notice

---

[6] $\bigvee_{i=1}^{n} \mathcal{F}_i$ denotes the smallest σ-algebra containing each $\mathcal{F}_i$.

that this allocation is $\bigvee_{i=1}^{2} \mathcal{F}_i$-measurable and cannot be dominated by any coalition of agents using their pooled information. Hence it is a weak fine core allocation.[7]

Next we shall define a Walrasian equilibrium notion in the sense of Radner. In order to do so, we need the following. A *price system* is an $\mathcal{F}$-measurable, non-zero function $p : \Omega \to \mathbb{R}^l_+$ and the *budget set* of Agent i is given by

$$B_i(p) = \left\{ x_i : x_i : \Omega \to \mathbb{R}^l \text{ is } \mathcal{F}_i\text{-measurable } x_i(\omega) \in X_i(\omega) \right.$$

$$\left. \text{and } \sum_{\omega \in \Omega} p(\omega) x_i(\omega) \le \sum_{\omega \in \Omega} p(\omega) e_i(\omega) \right\}.$$

Notice that the budget constraint is across states of nature.

Definition 3.3. A pair (p, x), where p is a price system and $x = (x_1, \ldots, x_n) \in L_X$ is an allocation, is a *Radner equilibrium* if

(i) for all i the consumption function maximizes v_i on B_i
(ii) $\sum_{i=1}^{n} x_i \le \sum_{i=1}^{n} e_i$ (free disposal), and
(iii) $\sum_{\omega \in \Omega} p(\omega) \sum_{i=1}^{n} x_i(\omega) = \sum_{\omega \in \Omega} p(\omega) \sum_{i=1}^{n} e_i(\omega)$.

Radner equilibrium is an ex ante concept. We assume free disposal, for otherwise it is well known that a Radner equilibrium with non-negative prices might not exist. This can be seen through straightforward calculations in Example 3.1.

Next we turn our attention to the notion of REE. We shall need the following. Let $\sigma(p)$ be the smallest sub-σ-algebra of $\mathcal{F}$ for which $p : \Omega \to \mathbb{R}^l_+$ is measurable and let $\mathcal{G}_i = \sigma(p) \vee \mathcal{F}_i$ denote the smallest σ-algebra containing both $\sigma(p)$ and $\mathcal{F}_i$. We shall also condition the expected utility of the agents on $\mathcal{G}_i$ which produces a random variable.

Definition 3.4. A pair (p, x), where p is a price system and $x = (x_1, \ldots, x_n) \in \bar{L}_X$ is an allocation, is a *rational expectations equilibrium* (REE) if

(i) for all i the consumption function $x_i(\omega)$ is $\mathcal{G}_i$-measurable.
(ii) for all i and for all ω the consumption function maximizes

$$v_i(x_i|\mathcal{G}_i)(\omega) = \sum_{\omega' \in E_i^{\mathcal{G}_i}(\omega)} u_i(\omega', x_i(\omega')) \frac{q_i(\omega')}{q_i\left(E_i^{\mathcal{G}_i}(\omega)\right)}, \tag{3}$$

(where $E_i^{\mathcal{G}_i}(\omega)$ is the event in $\mathcal{G}_i$ which contains ω and $q_i(E_i^{\mathcal{G}_i}(\omega)) > 0$) subject to

$$p(\omega) x_i(\omega) \le p(\omega) e_i(\omega)$$

i.e. the budget set at state ω, and
(iii) $\sum_{i=1}^{n} x_i(\omega) = \sum_{i=1}^{n} e_i(\omega)$ for all ω.

---

[7] See Koutsougeras and Yannelis (1993).

This is an interim concept because we condition expectations on information received from prices as well. In the definition, free disposal can easily be introduced. The idea of conditioning on the σ-algebra, $v_i(x_i|\mathcal{G}_i)(\omega)$, is rather well known.

REE can be classified as (i) *fully revealing* if the price function reveals to each agent all states of nature, (ii) *partially revealing* if the price function reveals some but not all states of nature and (iii) *non-revealing* if it does not disclose any particular state of nature.

Finally we define the concept of *weak fine value allocation* (see Krasa and Yannelis, 1994). As in the definition of the standard value allocation concept, we must first define a transferable utility (TU) game in which each agent's utility is weighted by a factor λ_i $(i = 1, \ldots, n)$, which allows interpersonal comparisons. In the value allocation itself no side payments are necessary.[8] A game with side payments is then defined as follows.

Definition 3.5. A game with side payments $\Gamma = (I, V)$ consist of a finite set of agents $I = \{1, \ldots, n\}$ and a superadditive, real valued function V defined on 2^I such that $V(\emptyset) = 0$. Each $S \subset I$ is called a coalition and $V(S)$ is the 'worth' of the coalition S.

The Shapley value of the game Γ (Shapley, 1953) is a rule that assigns to each Agent i a 'payoff', Sh_i, given by the formula[9]

$$Sh_i(V) = \sum_{\substack{S \subseteq I \\ S \supseteq \{i\}}} \frac{(|S|-1)!(|I|-|S|)!}{|I|!} [V(S) - V(S\backslash\{i\})]. \tag{4}$$

The Shapley value has the property that $\sum_{i \in I} Sh_i(V) = V(I)$, i.e. it is Pareto efficient.

We now define for each economy with differential information, $\mathcal{E}$, and a common prior, and for each set of weights, $\lambda_i : i = 1, \ldots, n$, the associated game with side payments (I, V_λ) (we also refer to this as a 'transferable utility' (TU) game) as follows:

For every coalition $S \subset I$, let

$$V_\lambda(S) = \max_x \sum_{i \in S} \lambda_i \sum_{\omega \in \Omega} u_i(\omega, x_i(\omega))\mu(\omega) \tag{5}$$

subject to

(i) $\sum_{i \in S} x_i(\omega) = \sum_{i \in S} e_i(\omega)$, μ–a.e.,
(ii) $x_i - e_i$ is $\bigvee_{i \in S} \mathcal{F}_i$–measurable.

We are now ready to define the weak fine value allocation.

---

[8] See Emmons and Scafuri (1985, p. 60) for further discussion.

[9] The Shapley value measure is the sum of the expected marginal contributions an agent can make to all the coalitions of which he/she is a member (see Shapley, 1953).

Definition 3.6. An allocation $x = (x_1, \ldots, x_n) \in \bar{L}_X$ is said to be a *weak fine value allocation* of the differential information economy, $\mathcal{E}$, if the following conditions hold

(i) Each net trade $x_i - e_i$ is $\bigvee\limits_{i=1}^{n} \mathcal{F}_i$-measurable,

(ii) $\sum_{i=1}^{n} x_i = \sum_{i=1}^{n} e_i$ and

(iii) There exist $\lambda_i \geq 0$, for every $i = 1, \ldots, n$, which are not all equal to zero, with
$$\sum_{\omega \in \Omega} \lambda_i u_i(\omega, x_i(\omega)) \mu(\omega) = Sh_i(V_\lambda) \text{ for all } i, \text{ where } Sh_i(V_\lambda) \text{ is the Shapley}$$
value of Agent i derived from the game (I, V_λ), defined in (5) above.

Condition (i) requires the pooled information measurability of net trades, i.e. net trades are measurable with respect to the "join". Condition (ii) is the market clearing condition and (iii) says that the expected utility of each agent multiplied by his/her weight, λ_i, must be equal to his/her Shapley value derived from the TU game (I, V_λ).

An immediate consequence of Definition 3.6 is that

$$Sh_i(V_\lambda) \geq \lambda_i \sum_{\omega \in \Omega} u_i(\omega, e_i(\omega)) \mu(\omega)$$

for every i, i.e. the value allocation is individually rational. This follows immediately from the fact that the game (V_λ, I) is superadditive for all weights λ. Similarly, efficiency of the Shapley value for games with side payments immediately implies that the value allocation is weak-fine Pareto efficient.

On the basis of the definitions and the analysis of Example 3.1 of an exhange economy with 3 agents and of Example 3.2 with 2 agents we make comparisons between the various equilibrium notions. The calculations of all, cooperative and noncooperative, equilibrium allocations are straightforward.

Contrary to the private core any rational expectation Walrasian equilibium notion, such as Radner equilibrium or REE, will always give zero to an agent who has no initial endowments. For example, in the 3-agent economy of Example 3.1, Agent 3 receives no consumption since his budget set is zero in each state. This is so irrespective of whether his private information is the full information partition $\mathcal{F}_3 = \{\{a\}, \{b\}, \{c\}\}$ or the trivial partition $\mathcal{F}_3' = \{a, b, c\}$. Hence the Walrasian, competitive equilibrium ideas do not take into account the informational superiority of an agent.

The set of Radner equilibrium allocations, with and without free disposal, are a subset of the corresponding private core allocations. Of course it is possible that a Radner equilibrium allocation might not exist. In the two-agent economy of Example 3.2, assuming non-free disposal the unique private core is the initial endowments allocation while no Radner equilibrium exists. On the other hand, assuming free disposal, for the same example, the REE coincides with the initial endowments allocation which does not belong to the private core. It follows that *the REE allocations need not be in the private core.*

We also have that *a REE need not be a Radner equilibrium.* In Example 3.2, without free disposal no Radner equilibrium with non-negative prices exists but REE does. It is unique and it implies no-trade.

As for *the comparison between private and weak fine core allocations* the two sets could intersect but there is no definite relation. Indeed the measurability requirement of the private core allocations separates the two concepts. In Example 3.2 the allocation $(5, 2.5, 2.5)$ to Agent 1 and $(5, 2.5, 2.5)$ to Agent 2, as well as $(6, 3, 3)$ and $(4, 2, 2)$ belong to the weak fine core but not to the private core. There are many weak fine core allocations which do not satisfy the measurability condition.

For $n = 2$ one can easily verify that the weak fine value belongs to the weak fine core. However it is known (see, for example, Scafuri and Yannelis, 1984) that for $n \geq 3$ *a value allocation may not be a core allocation, and therefore may not be a Radner equilibrium.*

Also, in Example 3.1 *a private core allocation is not necessarily in the weak fine core.* Indeed the division $(4, 4, 1)$, $(4, 1, 4)$ and $(2, 0, 0)$, to Agents 1, 2 and 3 respectively, is a private core but not a weak fine core allocation. The first two agents can get together, pool their information and do better. They can realize the weak fine core allocation, $(5, 2.5, 2.5)$, $(5, 2.5, 2.5)$ and $(0, 0, 0)$ which does not belong to the private core.

Finally notice that even with free disposal no allocation which does not distribute the total resource could be in the weak fine core. The three agents can get together, distribute the surplus and increase their utility.

In the next section we shall discuss whether core and Walrasian type allocations have certain desirable properties from the point of view of incentive compatibility. Following this, we shall turn our attention in later sections to the implementation of such allocations.

4 Incentive compatibility

The basic idea is that an allocation is incentive compatible if no coalition can misreport the realized state of nature to the complementary set of agents and become better off.

Let us suppose we have a coalition S, with members denoted by i, and the complementary set $I \setminus S$ with members j. Let the realized state of nature be ω^*. A member $i \in S$ sees $E_i(\omega^*)$. Obviously not all $E_i(\omega^*)$ need be the same, however all Agents i know that the actual state of nature could be ω^*.

Consider now a state of nature ω' with the following property. For all $j \in I \setminus S$ we have $\omega' \in E_j(\omega^*)$ and for at least one $i \in S$ we have $\omega' \notin E_i(\omega^*)$ (otherwise ω' would be indistinguishable from ω^* for all players and, by redefining utilities appropriately, could be considered as the same element of Ω). Now the coalition S decides that each member i will announce that she has seen her own set $E_i(\omega')$ which, of course, definitely contains a lie. On the other hand we have that $\omega' \in \bigcap_{j \notin S} E_j(\omega^*)$, (we also denote $j \in I \setminus S$ by $j \notin S$).

Now the idea is that if all members of $I \setminus S$ believe the statements of the members of S then each $i \in S$ expects to gain. For *coalitional Bayesian incentive compatibility* (CBIC) of an allocation we require that this is not possible. This is the incentive compatibility condition used in Glycopantis, Muir, and Yannelis (2001) where we gave a formal definition.

We showed there that in the three-agent economy without free disposal the private core allocation $x_1 = (4, 4, 1)$, $x_2 = (4, 1, 4)$ and $x_3 = (2, 0, 0)$ is incentive compatible. This follows from the fact that Agent 3 who would potentially cheat in state a has no incentive to do so. It has been shown in Koutsougeras and Yannelis (1993) that if the utility functions are monotone and continuous then private core allocations are *always* CBIC.

On the other hand the weak fine core allocations are not always incentive compatible, as the proposed redistribution $x_1 = x_2 = (5, 2.5, 2.5)$ in the two-agent economy shows. Indeed, if Agent 1 observes $\{a, b\}$, he has an incentive to report c and Agent 2 has an incentive to report b when he observes $\{a, c\}$.

CBIC coincides in the case of a two-agent economy with *Individually Bayesian Incentive Compatibility* (IBIC) which corresponds to the case in which S is a singleton.

The concept of *Transfer Coalitionally Bayesian Incentive Compatible* (TCBIC) allocations, used in this paper,[10] allows for transfers between the members of a coalition, and is therefore a strengthening of the concept of Coalitionally Bayesian Incentive Compatibility (CBIC).

Definition 4.1. An allocation $x = (x_1, \ldots, x_n) \in \bar{L}_X$, with or without free disposal, is said to be *Transfer Coalitionally Bayesian Incentive Compatible* (TCBIC) if it is not true that there exists a coalition S, states ω^* and ω', with ω^* different from ω' and $\omega' \in \bigcap_{i \notin S} E_i(\omega^*)$ and a random net-trade vector, z, among the members of S,

$$(z_i)_{i \in S}, \sum_S z_i = 0$$

such that for all $i \in S$ there exists $\bar{E}_i(\omega^*) \subseteq Z_i(\omega^*) = E_i(\omega^*) \cap (\bigcap_{j \notin S} E_j(\omega^*))$,

for which

$$\sum_{\omega \in \bar{E}_i(\omega^*)} u_i(\omega, e_i(\omega) + x_i(\omega') - e_i(\omega') + z_i) q_i(\omega|\bar{E}_i(\omega^*)) \qquad (6)$$

$$> \sum_{\omega \in \bar{E}_i(\omega^*)} u_i(\omega, x_i(\omega)) q_i(\omega|\bar{E}_i(\omega^*)).$$

Notice that the z_i' s above are not necessarily measurable. The definition is cast in terms of all possible z_i' s. It follows that $e_i(\omega) + x_i(\omega') - e_i(\omega') + z_i(\omega) \in X_i(\omega)$ is not necessarily measurable. The definition means that no coalition can form with the possibility that by misreporting a state, every member will become better off if the announcement is believed by the members of the complementary set.

Returning to Definition 4.1, one then can define CBIC to correspond to $z_i = 0$ and then IBIC to the case when S is a singleton. Thus we have (not IBCI) $\Rightarrow$ (not CBIC) $\Rightarrow$ (not TCBIC). It follows that TCBIC $\Rightarrow$ CBIC $\Rightarrow$ IBIC.

We now provide a *characterization* of TCBIC:

---

[10] See Krasa and Yannelis (1994) and Hahn and Yannelis (1997) for related concepts.

Proposition 4.1. *Let $\mathcal{E}$ be a one-good differential information economy as described above, and suppose each agent's utility function, $u_i = u_i(\omega, x_i(\omega))$ is monotone in the elements of the vector of goods x_i, that $u_i(., x_i)$ is $\mathcal{F}_i$-measurable in the first argument, and that an element $x = (x_1, \ldots, x_n) \in \bar{L}_X^1$ is a feasible allocation in the sense that $\sum_{i=1}^n x_i(\omega) = \sum_{i=1}^n e_i(\omega) \ \forall\omega$. Consider the following conditions:*

(i) $x \in L_X^1 = \prod_{i=1}^n L_{X_i}^1$ *and*

(ii) x *is TCBIC.*

Then (i) is equivalent to (ii).

Proof. First we show that (i) implies (ii) by showing that (i) and the negation of (ii) lead to a contradiction.

Let $x \in L_X$ and suppose that it is not TCBIC. Then, varying the notation for states to emphasize that Definition 4.1 does not hold, there exists a coalition S, states a and b, with $a \neq b$ and $b \in \bigcap_{i \notin S} E_i(a)$ and a net-trade vector, z, among the members of S,

$$(z_i)_{i \in S}, \quad \sum_S z_i = 0$$

such that for all $i \in S$ there exists $\bar{E}_i(a) \subseteq Z_i(a) = E_i(a) \cap (\bigcap_{j \notin S} E_j(a))$, for which

$$\sum_{c \in \bar{E}_i(\alpha)} u_i(c, e_i(c) + x_i(b) - e_i(b) + z_i) q_i(c | \bar{E}_i(a)) \qquad (7)$$

$$> \sum_{c \in \bar{E}_i(a)} u_i(c, x_i(c)) q_i(c | \bar{E}_i(a)).$$

For $c \in \bar{E}_i(a), e_i(c) = e_i(a)$ since e_i is $\mathcal{F}_i$-measurable, so

$$e_i(c) + x_i(b) - e_i(b) + z_i = e_i(a) + x_i(b) - e_i(b) + z_i$$

and hence also

$$u_i(c, e_i(c) + x_i(b) - e_i(b) + z_i) = u_i(a, e_i(a) + x_i(b) - e_i(b) + z_i),$$

by the assumed $\mathcal{F}_i$-measurability of u_i.

Since, by (i), $x_i(c) = x_i(a)$ for $c \in \bar{E}_i(a)$, we similarly have $u_i(c, x_i(c)) = u_i(a, x_i(a))$. Thus in equation (7) the common utility terms can be lifted outside the summations giving

$$u_i(a, e_i(a) + x_i(b) - e_i(b) + z_i) > u_i(a, x_i(a))$$

and hence $e_i(a) + x_i(b) - e_i(b) + z_i > x_i(a)$, by monotonicity of u_i.

Consequently,

$$\sum_{i \in S} (x_i(b) - e_i(b)) > \sum_{i \in S} (x_i(a) - e_i(a)). \qquad (8)$$

On the other hand for $i \notin S$ we have $x_i(b) - e_i(b) = x_i(a) - e_i(a)$ from which we obtain

$$\sum_{i \notin S} (x_i(b) - e_i(b)) = \sum_{i \notin S} (x_i(a) - e_i(a)). \tag{9}$$

Taking equations (8),(9) together we have

$$\sum_{i \in I} (x_i(b) - e_i(b)) > \sum_{i \in I} (x_i(a) - e_i(a)), \tag{10}$$

which is a contradiction since both sides are equal to zero, by feasibility.[11]

We now show that (ii) implies (i). For suppose not. Then there exists some Agent j and states a, b with $b \in E_j(a)$ such that $x_j(a) \neq x_j(b)$. Without loss of generality, we may assume that $x_j(a) > x_j(b)$. Since $e_j(.)$ is $\mathcal{F}_j$-measurable $e_j(b) = e_j(a)$ and therefore

$$x_j(a) - e_j(a) > x_j(b) - e_j(b). \tag{11}$$

Let $S = I \backslash \{j\}$. From the feasibility of x and (11) it follows that

$$\sum_{i \in S} (x_i(a) - e_i(a)) = -(x_j(a) - e_j(a)) < -(x_j(b) - e_j(b)) \tag{12}$$

$$= \sum_{i \in S} (x_i(b) - e_i(b)).$$

From (12) we have that

$$\delta = \sum_{i \in S} (e_i(a) + x_i(b) - e_i(b) - x_i(a)) > 0. \tag{13}$$

For each $i \in S$, let

$$z_i = x_i(a) - e_i(a) - x_i(b) + e_i(b) + \frac{\delta}{n-1}$$

so that $\sum_{i \in S} z_i = 0$ and

$$e_i(a) + x_i(b) - e_i(b) + z_i > x_i(a).$$

By monotonicity of u_i, we can conclude that

$$u_i(a, e_i(a) + x_i(b) - e_i(b) + z_i) > u_i(a, x_i(a)), \tag{14}$$

for all $i \in S$, a contradiction to the fact that x is TCBIC as the role of $\bar{E}_i$ in the definition can be played by $\{a\}$.

Finally note that a particular case of $\mathcal{F}_i$-measurability of u_i is when it is independent of ω. This completes the proof of Proposition 4.1. $\square$

---

[11] Koutsougeras and Yannelis (1993) and Krasa and Yannelis (1994) show that (i) implies (ii) for any number of goods, but for ex post utility functions. This means that the contract is made ex ante and after the state of nature is realized we see that we have incentive compatibility. Hahn and Yannelis (1997) show that (i) implies (ii) for any number of goods and for interim utility functions. Notice that since the non-free disposal Radner equilibrium is a subset of the non-free disposal ex ante private core, it follows from Hahn and Yannelis that the non-free disposal Radner equilibrium is TCBIC.

In the lemma that follows we refer to CBIC, as TCBIC does not make much sense since z_i is not available. CBIC is obtained when all z_i's are set equal to zero.

Lemma 4.1. *Under the conditions of the Proposition, if there are only two agents then (ii) x is CBIC, which is the same as IBIC, implies (i).*

Proof. For suppose not. Then lack of $\mathcal{F}_i$-measurability of the allocations implies that there exist Agent j and states a, b, where $b \in E_j(a)$, such that $x_j(b) < x_j(a)$ and therefore

$$x_j(b) - e_j(b) < x_j(a) - e_j(a). \tag{15}$$

Feasibility implies

$$x_i(b) - e_i(b) + x_j(b) - e_j(b) = x_i(a) - e_i(a) + x_j(a) - e_j(a) \tag{16}$$

from which we obtain

$$x_i(b) - e_i(b) > x_i(a) - e_i(a). \tag{17}$$

By monotonicity and the one-good per state assumption it follows that,

$$u_i(a, e_i(a) + x_i(b) - e_i(b)) > u_i(a, x_i(a)). \tag{18}$$

This implies that we have

$$u_i(a, e_i(c) + x_i(b) - e_i(b)) > u_i(a, x_i(c)) \tag{19}$$

which contradicts the assumption that x is CBIC. This completes the proof of the lemma. □

The above results characterize TCBIC and CBIC in terms of private individual measurability, i.e. $\mathcal{F}_i$-measurability, of allocations. These results will enable us to conclude whether or not, in case of non-free disposal, any of the solution concepts, i.e. Radner equilibrium, REE, private core, weak fine core and weak fine value will be TCBIC whenever feasible allocations are $\mathcal{F}_i$-measurable.

It follows from the lemma that the redistribution shown in the matrix below, which is a weak fine core allocation of Example 3.2, where the ith line refers to Player i and the columns from left to right to states a, b and c,

$$\begin{pmatrix} 5 & 2.5 & 2.5 \\ 5 & 2.5 & 2.5 \end{pmatrix}$$

is not CBIC as it is not $\mathcal{F}_i$-measurable. Thus, *a weak fine core allocation may not be CBIC.*

On the other hand the proposition implies that, in Example 3.2, the no-trade allocation

$$\begin{pmatrix} 5 & 5 & 0 \\ 5 & 0 & 5 \end{pmatrix}$$

is incentive compatible. This is a non-free disposal REE, and a private core allocation.

We note that the Proposition 4.1 refers to non-free disposal. As a matter of fact Proposition 4.1 is not true if we assume free disposal. Indeed if free disposal is allowed $\mathcal{F}_i$-measurability PBE *does not* imply incentive compatibility.

In the case with *free disposal, private core and Radner equilibrium need not be incentive compatible*. In order to see this we notice that in Example 3.2 the (free disposal) Radner equilibrium is $x_1 = (4, 4, 1)$ and $x_2 = (4, 1, 4)$. The above allocation is clearly $\mathcal{F}_i$-measurable and it can easily be checked that it belongs to the (free disposal) private core. However it is not TBIC since if state a occurs Agent 1 has an incentive to report state c and gain.

Now in employing game trees in the analysis, as it is done below, we will adopt the definition of IBIC. The equilibrium concept employed will be that of PBE. The definition of a play of the game is a directed path from the initial to a terminal node.

In terms of the game trees, a core allocation will be IBIC if there is a profile of optimal behavioral strategies and equilibrium paths along which no player misreports the state of nature he has observed. This allows for the possibility, as we shall see later, that such strategies could imply that players have an incentive to lie from information sets which are not visited by an optimal play.

In view of the analysis in terms of game trees we comment again on the general idea of CBIC. First we look at it once more, in a similar manner to the one in the beginning of Section 4.

Suppose the true state of nature is $\bar{\omega}$. Any coalition can only see that the state lies in $\bigcap_{i \in S} E_i(\bar{\omega})$ when they pool their observations. If they decide to lie they must first guess at what is the true state and they will do so at some $\omega^* \in \bigcap_{i \in S} E_i(\bar{\omega})$. Then of course we have $\bigcap_{i \in S} E_i(\bar{\omega}) = \bigcap_{i \in S} E_i(\omega^*)$. Having decided on ω^* as a possible true state, they now pick some $\omega' \in \bigcap_{j \notin S} E_j(\omega^*)$ and (assuming the system is not CBIC) they hope, by announcing (each of them) that they have seen $E_i(\omega')$ to secure better payoffs.

This is all contingent on their being believed by $I \setminus S$. This, in turn, depends on their having been correct in their guessing that $\omega^* = \bar{\omega}$, in which case they might be believed. If $\omega^* \neq \bar{\omega}$, i.e they guess wrongly, then since $\bigcap_{j \notin S} E_j(\omega^*) \neq \bigcap_{j \notin S} E_j(\bar{\omega})$ they may be detected in their lie, since possibly $\omega' \notin \bigcap_{j \notin S} E_j(\bar{\omega})$.

This is why the definition of CBIC can only be about possible existence of situations where a lie might be beneficial. It is not concerned with what happens if the lie is detected. On the other hand the extensive form forces us to consider that alternative. It requires statements concerning earlier decisions by other players to lie or tell the truth and what payoffs will occur whenever a lie is detected, through observations or incompatibility of declarations. Only in this fuller description can players really make a decision whether to risk a lie, since only then can they balance the gains from not being caught against a definitely declared payoff if they are.

The issue is whether cooperative and noncooperative static solutions can be obtained as perfect Bayesian or sequential equilibria. That is whether such allocations can also be supported through an appropriate noncooperative solution concept. The analysis below shows that CBIC allocations can be supported by a PBE while lack of incentive compatibility implies non-support, in the sense that the two agents, left on their own, do not sign the contract. It is also shown how implementation of allocations becomes possible through the introduction in the analysis of an exogenous third party or an endogenous intermediary.

5 Non-implementation of free disposal private core and Radner equilibria, and of weak fine core allocations

The main point here is that lack of IBIC implies that the two agents based on their information cannot sign a proposed contract because both of them have an incentive to cheat the other one and benefit. Indeed PBE leads to no-trade. This so irrespective of whether in state a the contract specifies that they both get 5 or 4.

Note that to impose free disposal in state a causes certain problems, because the question arises as to who will check that the agents have actually thrown away 1 unit. In general, free disposal is not always a very satisfactory assumption in differential information economies with monotone preferences.

We shall investigate the possible implementation of the allocation

$$\begin{pmatrix} 4 & 4 & 1 \\ 4 & 1 & 4 \end{pmatrix}$$

in Example 3.2, contained in a contract between P1 and P2 when no third party is present. For the case with free disposal, this is both a private core and a Radner equilibrium allocation.

This allocation is not IBIC because, as we explained in the previous section, if Agent 1 observes $\{a, b\}$, he has an incentive to report c and Agent 2 has an incentive to report b when he observes $\{a, c\}$.

We construct a game tree and employ reasonable rules for describing the outcomes of combinations of states of nature and actions of the players. In fact we look at the contract

$$\begin{pmatrix} 5 & 4 & 1 \\ 5 & 1 & 4 \end{pmatrix}$$

in which the agents get as much per state as under the private core allocation above. The latter can be obtained by invoking free disposal in state a.

The investigation is through the analysis of a specific sequence of decisions and information sets shown in the game tree in Figure 1. Notice that vectors at the terminal nodes of a game tree will refer to payoffs of the players in terms of quantities. The first element will be the payoff to P1, etc.

The players are given strategies to tell the truth or to lie, i.e., we model the idea that agents truly inform each other about what states of nature they observe, or deliberately aim to mislead their opponent. The issue is what type of behavior is optimal and therefore whether a proposed contract will be signed or not. We find that the optimal strategies of the players imply no-trade.

Figures 1 and 2 show that the allocation $(5, 4, 1)$ and $(5, 1, 4)$ will be rejected by the players. They prefer to stay with their initial endowments and will not sign the proposed contract as it offers to them no advantage.

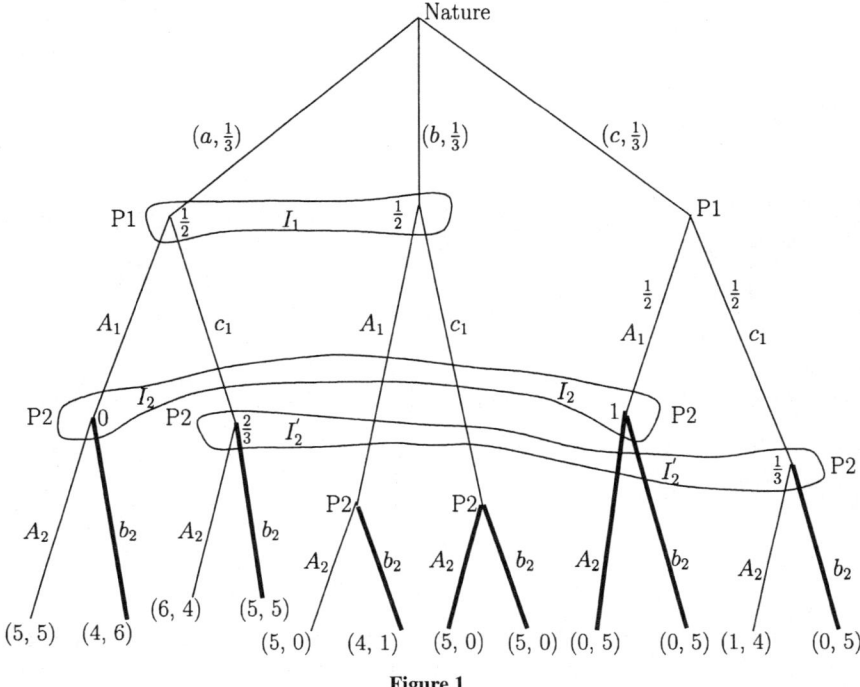

Figure 1

In Figure 1, nature chooses states a, b or c with equal probabilities. This choice is flashed on a screen which both players can see. P1 cannot distinguish between a and b, and P2 between a and c. This accounts for the information sets I_1, I_2 and I_2' which have more than one node. A player to which such an information set belongs cannot distinguish between these nodes and therefore his decisions are common to all of them. A behavioral strategy of a player is to declare which choices he would make, with what probability, from each of his information sets. Indistinguishable nodes imply the $\mathcal{F}_i$-measurability of decisions.

P1 moves first and he can either play $A_1 = \{a, b\}$ or $c_1 = \{c\}$, i.e., he can say "I have seen $\{a, b\}$ or "I have seen c". Of course only one of these declarations will be true. Then P2 is to respond saying that the signal he has seen on the screen is $A_2 = \{a, c\}$ or that it is $b_2 = \{b\}$. Obviously only one of these statements is true.

Strictly speaking the notation for choices should vary with the information set but there is no danger of confusion here. Finally notice that the structure of the game tree is such that when P2 is to act he knows exactly what P1 has chosen.

Next we specify the *rules* for calculating the payoffs, i.e. the terms of the contract:

(i) If the declarations by the two players are incompatible, that is (c_1, b_2) then no-trade takes place and the players retain their initial endowments. That is the case when either state c, or state b occurs and Agent 1 reports state c and Agent 2 state b. In state a both agents can lie and the lie cannot be detected by either of them. They are in the events $\{a, b\}$ and $\{a, c\}$ respectively, they get

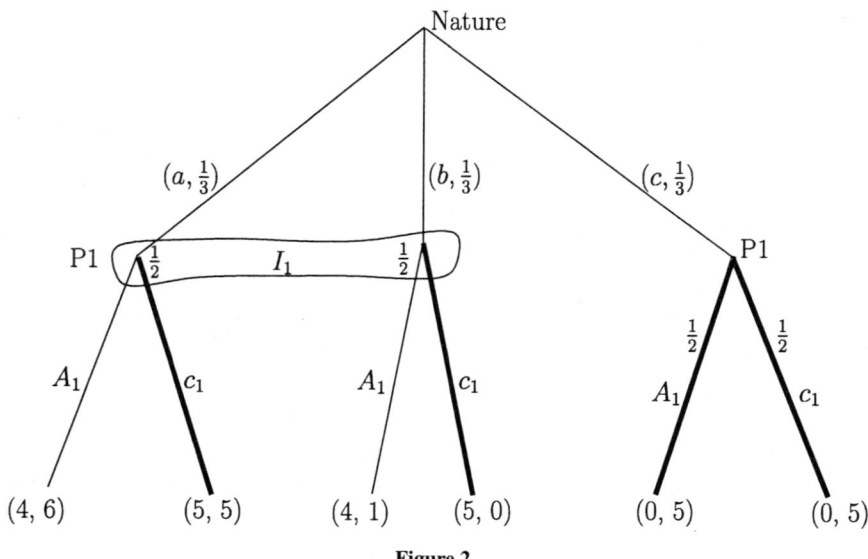

Figure 2

5 units of the initial endowments and again they are not willing to cooperate. Therefore whenever the declarations are incompatible, no trade takes place and the players retain their initial endowments.

(ii) If the declarations are (A_1, A_2) then even if one of the players is lying, this cannot be detected by his opponent who believes that state a has occured and both players have received endowment 5. Hence no-trade takes place.

(iii) If the declarations are (A_1, b_2) then a lie can be beneficial and undetected. P1 is trapped and must hand over one unit of his endowment to P2. Obviously if his initial endowment is zero then he has nothing to give.

(iv) If the declarations are (c_1, A_2) then again a lie can be beneficial and undetected. P2 is now trapped and must hand over one unit of his endowment to P1. Obviously if his initial endowment is zero then he has nothing to give.

The calculations of payoffs do not require the revelation of the actual state of nature. Optimal decisions will be denoted by a heavy line. We could assume that a player does not lie if he cannot get a higher payoff by doing so.

Assuming that each player chooses optimally from his information sets, the game in Figure 1 folds back to the one in Figure 2. Inspection of Figure 1 reveals that from the information set I_2 agent P2 can play b_2 with probability 1. (A heavy line A_2 indicates that this choice also would not affect the analysis). This accounts for the payoff $(4, 6)$ and the first payoff $(0, 5)$ from left to right in Figure 2. Similarly by considering the optimal decisions from all other information sets of P2 we arrive at Figure 2. Analyzing this figure we obtain the optimal strategies of P1.

In conclusion, the optimal behavioral strategy for P1 is to play c_1 with probability 1 from I_1, i.e to lie, and from the singleton to play any probability mixture of options, and we have chosen $(A_1, \frac{1}{2}; c_1, \frac{1}{2})$. The optimal strategy of P2 is to play b_2 from both I_2 and I_2', i.e. to lie, and from the second singleton he can either tell the truth or lie, or spin a wheel, divided in proportions corresponding to A_1 and c_1, to decide what to choose.

In Figures 1 and 2, the fractions next to the nodes in the information sets correspond to beliefs of the agents obtained, wherever possible, through Bayesian updating. I.e., they are consistent with the choice of a state by nature and the optimal behavioral strategies of the players. This means that strategies and beliefs satisfy the conditions of a PBE.

These probabilities are calculated as follows. From left to right, we denote the nodes in I_1 by j_1 and j_2, in I_2 by n_1 and n_2 and in I_2' by n_3 and n_4. Given the choices by nature, the strategies of the players described above and using the Bayesian formula for updating beliefs we can calculate, for example, the conditional probabilities

$$Pr(n_1/A_1) = \frac{Pr(A_1/n_1) \times Pr(n_1)}{Pr(A_1/n_1) \times Pr(n_1) + Pr(A_1/n_2) \times Pr(n_2)} \tag{20}$$

$$= \frac{1 \times 0}{1 \times 0 + 1 \times \frac{1}{3} \times \frac{1}{2}} = 0$$

and

$$Pr(n_3/c_1) = \frac{Pr(c_1/n_3) \times Pr(n_3)}{Pr(c_1/n_3) \times Pr(n_3) + Pr(c_1/n_4) \times Pr(n_4)} \tag{21}$$

$$= \frac{1 \times \frac{1}{3}}{1 \times \frac{1}{3} + 1 \times \frac{1}{2} \times \frac{1}{3}} = \frac{2}{3}.$$

In Figure 3 we indicate, through heavy lines, plays of the game which are the outcome of the choices by nature and the optimal behavioral strategies by the players. The interrupted heavy lines signify that nature does not take an optimal decision but simply chooses among three alternatives, with equal probabilities. The directed path (a, c_1, b_2) with payoffs $(5, 5)$ occurs with probability $\frac{1}{3}$. The paths (b, c_1, A_2) and (b, c_1, b_2) lead to payoffs $(5, 0)$ and occur with probability $\frac{1}{3}(1 - q)$ and $\frac{1}{3}q$, respectively. The values $(1 - q)$ and q denote the probabilities with which P2 chooses between A_2 and b_2 from the singleton node at the end of (b, c_1). The paths (c, A_1, b_2) (c, c_1, b_2) lead to payoffs $(0, 5)$ and occur, each, with probability $\frac{1}{3} \times \frac{1}{2}$.

For all choices by nature, at least one of the players tells a lie on the optimal play. The players by lying avoid the possibility of having to make a payment to their opponent and stay with their initial endowments. The PBE obtained above confirms the initial endowments. The decisions to lie imply that the players will not sign the contract $(5, 4, 1)$ and $(5, 1, 4)$.

We have constructed an extensive form game and employed reasonable rules for calculating payoffs and shown that the proposed allocation $(5, 4, 1)$ and $(5, 1, 4)$ will not be realized. A similar conclusion would have been reached if we investigated the allocation $(4, 4, 1)$ and $(4, 1, 4)$ which would have been brought about by considering free disposal.

Finally suppose we were to modify (iii) and (iv) of the *rules* and adopt those in Section 5 of Glycopantis, Muir, and Yannelis (2001):

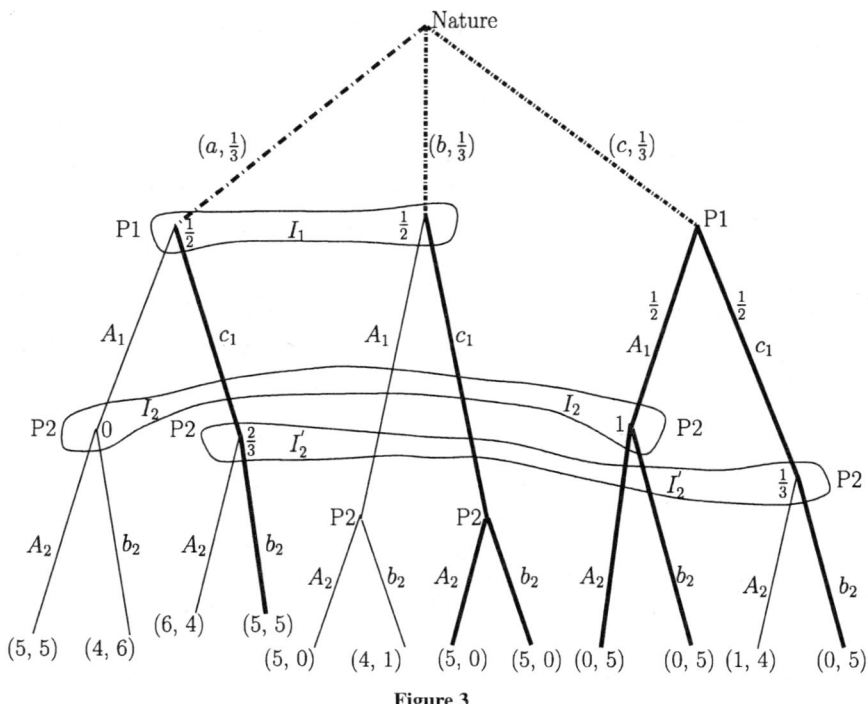

Figure 3

(iii) If the declarations are (A_1, b_2) then a lie can be beneficial and undetected, and P1 is trapped and must hand over half of his endowment to P2. Obviously if his endowment is zero then he has nothing to give.

(iv) If the declarations are (c_1, A_2) then again a lie can be beneficial and undetected. P2 is now trapped and must hand over half of his endowment to P1. Obviously if his endowment is zero then he has nothing to give.

The new rules would imply, starting from left to right, the following changes in the payoffs in Figure 1. The second vector would now be $(2.5, 7.5)$, the third vector $(7.5, 2.5)$, the sixth vector $(2.5, 2.5)$ and the eleventh vector $(2.5, 2.5)$. The analysis in Glycopantis, Muir, and Yannelis (2001) shows that the weak fine core allocation in which both agents receive $(5, 2.5, 2.5)$ cannot be implemented as a PBE. Again this allocation is not IBIC.

Since we have two agents, the weak fine value belongs to the weak fine core. We can also check through routine calculations that the non-implementable allocation $x_1 = x_2 = (5, 2.5, 2.5)$ belongs to the weak fine value, with the two agents receiving equal weights.

Finally we note that, in the context of Figure 1, the perfect Bayesian equilibrium implements the initial endowments allocation

$$\begin{pmatrix} 5 & 5 & 0 \\ 5 & 0 & 5 \end{pmatrix}.$$

In the case of non-free disposal, no-trade coincides with the REE and it is implementable. However as it is shown in Glycopantis, Muir, and Yannelis (2002) a REE is not in general implementable.

6 Implementation of private core and Radner equilibria through the courts; implementation of weak fine core

We shall show here how the free disposal private core and also Radner equilibrium allocation

$$\begin{pmatrix} 4 & 4 & 1 \\ 4 & 1 & 4 \end{pmatrix}$$

of Example 3.2 can be implemented as a PBE by invoking an exogenous third party, which can be interpreted as a court which imposes penalties when the agents lie.

We shall assume that the agents do not hear the choice announced by the other player or that they do not pay much attention to each other because the court will verify the true state of nature.

It should be noted that now if the two players see the events (A_1, A_2) the exogenous agent will not allow them to misreport the state of nature by imposing a penalty for lying. Therefore the contract will be enforced exogenously.

The analysis is through the figures below. Figure 4 contains the information sets of the two agents, P1 and P2, their sequential decisions and the payoffs in terms of quantities. Each agent can choose either to tell the truth about the information set he is in, or to lie.

Nature chooses states a, b and c with equal probabilities. P1 acts first and cannot distinguish between a and b. When P2 is to act he has two kinds of ignorance. Not only he cannot distinguish between a and c but also he does not know what P1 has chosen before him. This is an assumption about the relation between decisions. The one unit that the courts take from a lying agent can be considered to cover the costs of the court.

Next given the sequence of decisions of the two players, shown on the tree, we specify the rules for calculating payoffs in terms of quantities, i.e we specify the terms of the contract. They will, of course include the penalties that the court would impose to the agents for lying.

The *rules* are:

(i) If a player lies about his observation, then he is penalized by 1 unit of the good. If both players lie then they are both penalized. For example if the declarations are (c_1, b_2) and state a occurs both are penalized. If they choose (c_1, A_2) and state a occurs then the first player is penalized. If a player lies and the other agent has a positive endowment then the court keeps the quantity substracted for itself. However, if the other agent has no endowment, then the court transfers to him the one unit subtracted from the one who lied.

(ii) If the declarations of the two agents are consistent, that is (A_1, A_2) and state a occurs, (A_1, b_2) and state b occurs, (c_1, A_2) and state c occurs, then they divide equally the total endowments in the economy.

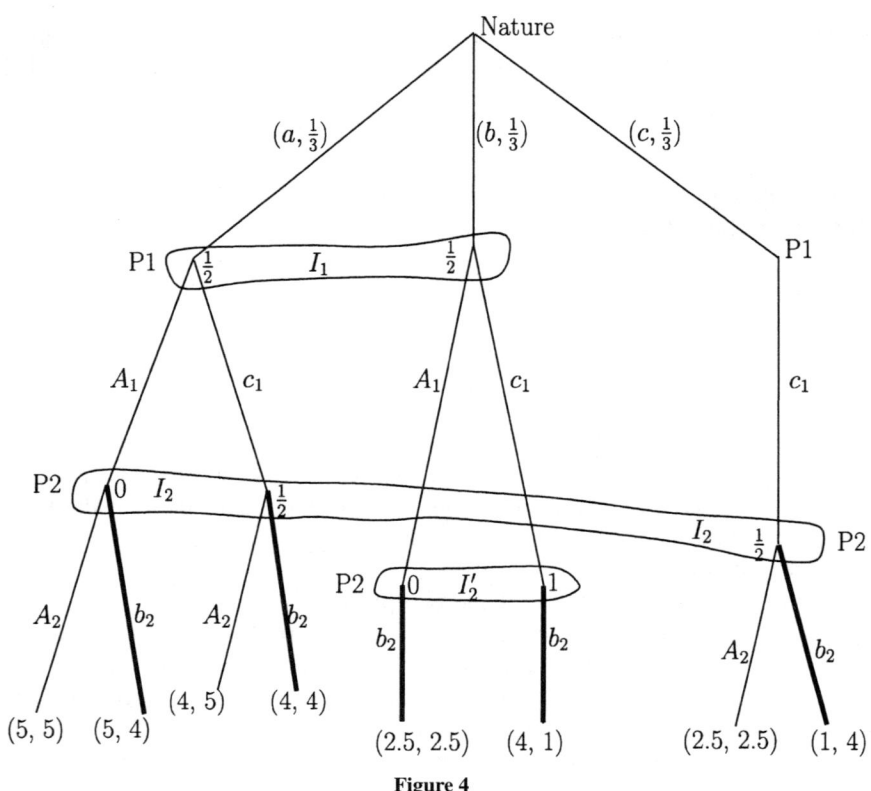

Figure 4

One explanation of the size of the payoffs is that if the agents decide to share, they do so voluntarily. On the other hand the court feel that they can punish them for lying but not to the extent of forcing them to share their endowments.

Assuming that each player chooses optimally, given his stated beliefs, from the information sets which belong to him, P2 chooses to play b_2 with probability 1 from both I_2 and I'_2 and the game in Figure 4 folds back to the one in Figure 5. The choice of b_2 is justified as follows. We ignore for the moment the specific conditional probabilities attached to the nodes of I_2. On the other hand, starting from left to right, the sum of the probabilities of the first two nodes must be equal to $\frac{1}{2}$, and this implies that strategy b_2 overtakes, in utility terms, strategy A_2, as $\frac{1}{2}5^{\frac{1}{2}} + \frac{1}{2}2.5^{\frac{1}{2}} < 4^{\frac{1}{2}}$. It follows that P2 chooses to play the behavioral strategy b_2 with probability 1. Now inspection of Figure 5 implies that P1 will choose c_1 from I_1. The conditional probabilities on the nodes of I_1 follow from the fact that nature chooses with equal probabilities and the optimal choice of c_1 with probability 1 follows again from the fact that $\frac{1}{2}5^{\frac{1}{2}} + \frac{1}{2}2.5^{\frac{1}{2}} < 4^{\frac{1}{2}}$.

Figure 6 indicates, through heavy lines, plays of the game which are the outcome of choices by nature and the optimal strategies of the players. The fractions next to the nodes of the information sets are obtained through Bayesian updating. I.e.

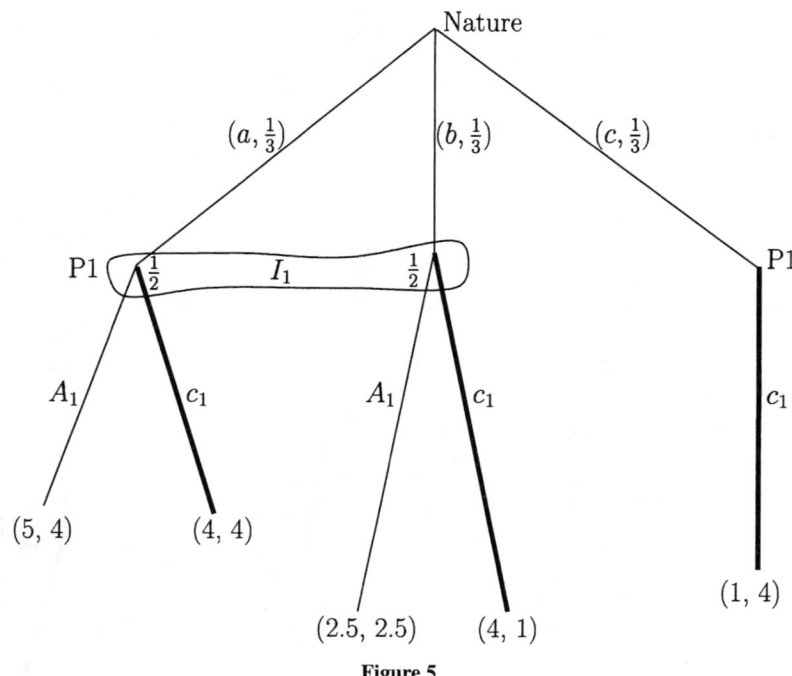

Figure 5

they are consistent with the choice of a state by nature and the optimal behavioral strategies of the players. We have thus obtained a PBE and the above argument implies that it is unique.

The free disposal private core allocation that we are concerned with is implemented, always, by at least one of the agents lying. The reason is that they make the same move from all the nodes of an information set and the rules of the game imply that they are not eager to share their endowments. They prefer to suffer the penalty of the court.

Finally notice the following. Suppose that the penalties are changed as follows. The court is extremely severe when an agent lies while the other agent has no endowment. It takes all the endowment from the one who is lying and transfers it to the other player. Everything else stays the same. Then the game is summarized in a modified Figure 4. Numbering the end points from left to right, the 2nd vector will be replaced by $(5, 0)$, the 3rd by $(0, 5)$, the 4th by $(0, 0)$, the 6th by $(0, 5)$ and the 8th one by $(5, 0)$.

The analysis of the game implies now that P2 will play A_2 from I_2 and P1 will play A_1 from I_1. Therefore invoking an exogenous agent implies that the PBE will now implement the weak fine core allocation

$$\begin{pmatrix} 5 & 2.5 & 2.5 \\ 5 & 2.5 & 2.5 \end{pmatrix}.$$

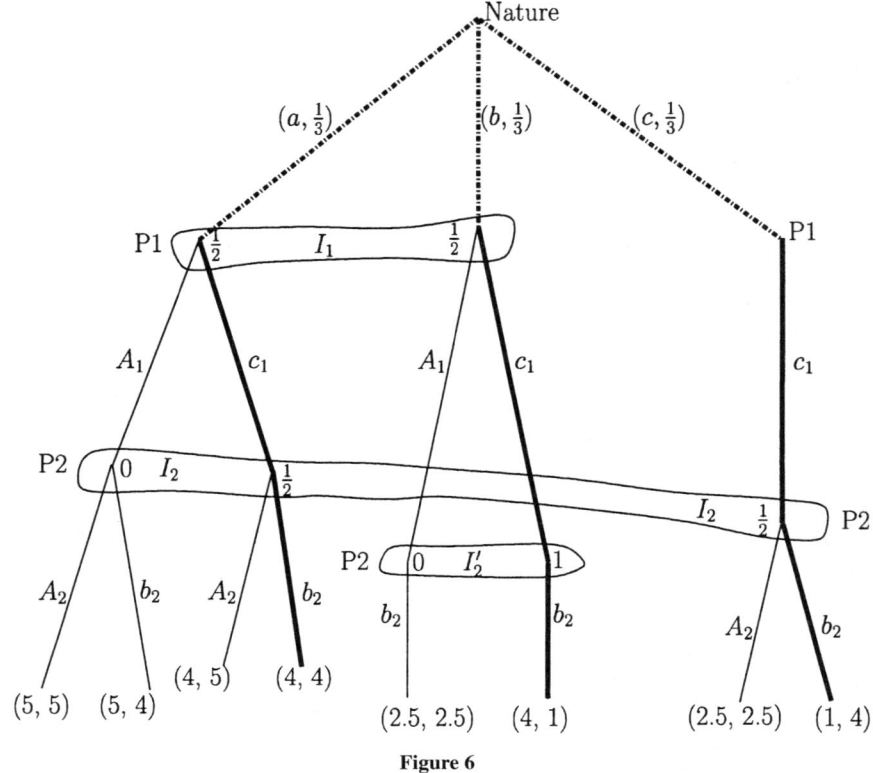

Figure 6

7 Implementation of non-free disposal private core through an endogenous intermediary

Here we draw upon the discussion in Glycopantis, Muir, and Yannelis (2001) but we add the analysis that the optimal paths obtained are also part of a sequential equilibrium. Hence we obtain a stronger conclusion, in the sense that we implement the private core allocation as a sequential equilibrium, which requires more conditions than PBE.

In the case we consider now there is no court and the agents in order to decide must listen to the choices of the other players before them. The third agent, P3, is endogenous and we investigate his role in the implementation, or realization, of private core allocations.

Private core without free disposal seems to be the most satisfactory concept. The third agent who plays the role of the intermediary implements the contract and gets rewarded in state a. We shall consider the private core allocation, of Example 3.1,

$$\begin{pmatrix} 4 & 4 & 1 \\ 4 & 1 & 4 \\ 2 & 0 & 0 \end{pmatrix}.$$

We know that such core allocations are CBIC and we shall show now how they can be supported as a perfect Bayesian equilibrium of a noncooperative game.

P1 cannot distinguish between states a and b and P2 between a and c. P3 sees on the screen the correct state and moves first. He can either announce exactly what he saw or he can lie. Obviously he can lie in two ways. When P1 comes to decide he has his information from the screen and also he knows what P3 has played. When P2 comes to decide he has his information from the screen and he also knows what P3 and P1 played before him. Both P1 and P2 can either tell the truth about the information they received from the screen or they can lie.

We must distinguish between the announcements of the players and the true state of nature. The former, with the players' temptations to lie, cannot be used to determine the true state which is needed for the purpose of making payoffs. P3 has a special status but he must also take into account that eventually the lie will be detected and this can affect his payoff.

The *rules* of calculating payoffs, i.e. the terms of the contract, are as follows:

If P3 tells the truth we implement the redistribution in the matrix above which is proposed for this particular choice of nature.

If P3 lies then we look into the strategies of P1 and P2 and decide as follows:

(i) If the declaration of P1 and P2 are incompatible we go to the initial endowments and each player keeps his.

(ii) If the declarations are compatible we expect the players to honour their commitments for the state in the overlap, using the endowments of the true state, provided these are positive. If a player's endowment is zero then no transfer from that agent takes place as he has nothing to give.

The extensive form game is shown in Figure 7, in which the heavy lines can be ignored in the first instance. We are looking for a PBE, i.e. a set of optimal behavioral strategies consistent with a set of beliefs. The beliefs are indicated by the probabilities attached to the nodes of the information sets, with arbitrary r, s, q, p and t between 0 and 1. The folding up of the game tree through optimal decisions by P2, then by P1 and subsequently by P3 is explained in Glycopantis, Muir, and Yannelis (2001).

In Figure 7 we indicate through heavy lines the equilibrium paths. The interrupted heavy lines at the beginning of the tree signify that nature does not take an optimal decision but simply chooses among three alternatives, with equal probabilities. The directed paths (a, a, A_1, A_2) with payoffs $(4, 4, 2)$, (b, b, A_1, b_2) with payoffs $(4, 1, 0)$ and (c, c, c_1, A_2) with payoffs $(1, 4, 0)$ occur, each, with probability $\frac{1}{3}$. It is clear that nobody lies on the optimal paths and that the proposed reallocation is incentive compatible and hence it will be realized.

Along the optimal paths nobody has an incentive to misrepresent the realized state of nature and hence the private core allocation is incentive compatible. However even optimal strategies can imply that players might have an incentive to lie from information sets which are not visited by the optimal play of the game. For example, P1, although he knows that nature has chosen a or b, has an incentive to declare c_1 from I_1^3, trying to take advantage of a possible lie by P3. Similarly P2, although he knows that nature has chosen a or c, has an incentive to declare b_2 from

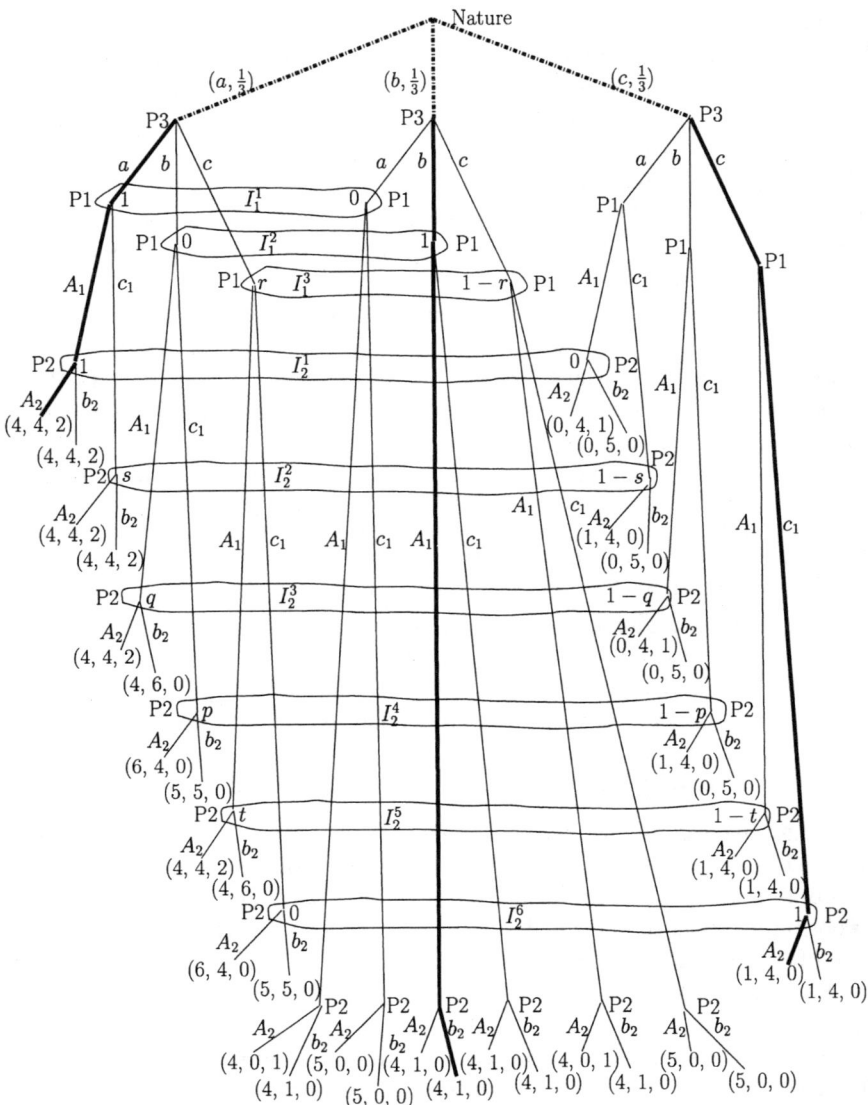

Figure 7

I_2^2, I_2^3, I_2^4 and I_2^5, trying to take advantage of possible lies by the other players. Incentive compatibility has now been defined to allow that the optimal strategies can contain lies, while there must be an optimal play which does not.

We also note that the same payoffs, i.e. (4, 4, 2), (4, 1, 0) and (1, 4, 0), can be confirmed as a PBE for all possible orders of the players.

Next we turn our attention to obtaining a sequential equilibrium. This adds further conditions to those of a PBE. Now, it is also required that the optimal behavioral strategies and the beliefs consistent with these are the limit of a sequence consisting of completely stochastic behavioral strategies, that is all choices are played with positive probability, and the implied beliefs. Throughout the sequence it is only required that beliefs are consistent with the strategies. The latter are not expected to be optimal.

We discuss how the PBE shown in Figure 7 can also be obtained as a sequential equilibrium in the sense of Kreps and Wilson (1982). Therefore, we are looking for a sequence of positive probabilities attached to all the choices from each information set and beliefs consistent with these such that their limits are the results given in Figure 7.

First we specify the positive probabilities, i.e. the completely stochastic strategies, with which the players choose the available actions. The sequence is obtained through $\{n = 2, 3, \ldots\}$.

In the first instance we consider the singletons from left to right belonging to P3. At the first one the positive probabilities attached to the various actions are given by $(a, 1 - \frac{2}{n}; \ b, \frac{1}{n}; \ c, \frac{1}{n})$, at the second one by $(a, \frac{1}{n}; \ b, 1 - \frac{2}{n}; \ c, \frac{1}{n})$ and at the third one by $(a, \frac{1}{n}; \ b, \frac{1}{n}; \ c, 1 - \frac{2}{n})$.

Then we come to the probabilities with which P1 chooses his actions from the various information sets belonging to him. From I_1^1 and I_1^2 the choices and the probabilities attached to these are $(A_1, 1 - \frac{1}{n}; \ c_1, \frac{1}{n})$, and from I_1^3, as well as from all the singletons, they are $(A_1, \frac{1}{n}; \ c_1, 1 - \frac{1}{n})$.

With respect to P2 choices and probabilities are given as follows. From I_2^1 and I_2^6 they are $(A_2, 1 - \frac{1}{n}; \ b_2, \frac{1}{n})$ and from I_2^2, I_2^3, I_2^4 and I_2^5 they are $(A_2, \frac{1}{n}; \ b_2, 1 - \frac{1}{n})$. With respect to the singletons belonging to P2 we have for all of them $(A_2, \frac{1}{n}; b_2, 1 - \frac{1}{n})$.

Beliefs are indicated by the probabilities attached to the nodes of the information sets. Below by the left (right) probability we mean the consistent with the above behavioral strategies belief that the player attaches to being at the left (right) corner node of an information set. We also give the limit of these beliefs as n tends to ∞.

In I_1^1 the left probability is $\dfrac{1 - \frac{2}{n}}{1 - \frac{1}{n}}$ and the right probability is $\dfrac{\frac{1}{n}}{1 - \frac{1}{n}}$. The limit is $(1, 0)$.

In I_1^2 the left probability is $\dfrac{\frac{1}{n}}{1 - \frac{1}{n}}$ and the right probability is $\dfrac{1 - \frac{2}{n}}{1 - \frac{1}{n}}$. The limit is $(0, 1)$.

In I_1^3 the left probability is $\dfrac{1}{2}$ and the right probability is $\dfrac{1}{2}$. The limit is $(\frac{1}{2}, \frac{1}{2})$.

In I_2^1 the left probability is $\dfrac{(1 - \frac{1}{n})(1 - \frac{2}{n})}{(1 - \frac{2}{n})(1 - \frac{1}{n}) + (\frac{1}{n})^2}$ and the right probability is $\dfrac{(\frac{1}{n})^2}{(1 - \frac{1}{n})(1 - \frac{2}{n}) + (\frac{1}{n})^2}$. The limit is $(1, 0)$.

In I_2^2 the left probability is $\dfrac{(1 - \frac{2}{n})\frac{1}{n}}{(1 - \frac{2}{n})\frac{1}{n} + (1 - \frac{1}{n})\frac{1}{n}}$ and the right probability is $\dfrac{(1 - \frac{1}{n})\frac{1}{n}}{(1 - \frac{2}{n})\frac{1}{n} + (1 - \frac{1}{n})(\frac{1}{n})}$. The limit is $(\frac{1}{2}, \frac{1}{2})$.

In I_2^3 the left probability is $\dfrac{(1-\frac{1}{n})\frac{1}{n}}{(1-\frac{1}{n})\frac{1}{n}+(\frac{1}{n})^2}$ and the right probability is

$\dfrac{(\frac{1}{n})^2}{(1-\frac{1}{n})\frac{1}{n}+(\frac{1}{n})^2}$. The limit is $(1,0)$.

In I_2^4 the left probability is $\dfrac{(\frac{1}{n})^2}{(1-\frac{1}{n})\frac{1}{n}+(\frac{1}{n})^2}$ and the right probability is

$\dfrac{(1-\frac{1}{n})\frac{1}{n}}{(1-\frac{1}{n})\frac{1}{n}+(\frac{1}{n})^2}$. The limit is $(0,1)$.

In I_2^5 the left probability is $\dfrac{(\frac{1}{n})^2}{(1-\frac{2}{n})\frac{1}{n}+(\frac{1}{n})^2}$ and the right probability is

$\dfrac{(1-\frac{2}{n})\frac{1}{n}}{(1-\frac{2}{n})\frac{1}{n}+(\frac{1}{n})^2}$. The limit is $(0,1)$.

In I_2^6 the left probability is $\dfrac{(1-\frac{1}{n})\frac{1}{n}}{(1-\frac{1}{n})\frac{1}{n}+(1-\frac{1}{n})(1-\frac{2}{n})}$ and the right proba-

bility is $\dfrac{(1-\frac{1}{n})(1-\frac{2}{n})}{(1-\frac{1}{n})\frac{1}{n}+(1-\frac{1}{n})(1-\frac{2}{n})}$. The limit is $(0,1)$.

The belief attached to each singleton is that it has been reached with probability 1.

The limits of the sequence of strategies and beliefs confirm a particular Bayesian equilibrium as a sequential one. In an analogous manner, sequential equilibria can also be obtained for the models analyzed in the previous sections.

8 Concluding remarks

As we have already emphasized in Glycopantis, Muir, and Yannelis (2001), we consider the area of incomplete and differential information and its modelling important for the development of economic theory. We believe that the introduction of game trees, which gives a dynamic dimension to the analysis, helps in the development of ideas.

The discussion in that paper is in the context of one-good examples without free disposal. The conclusion was that core notions which may not be CBIC, such as the weak fine core, cannot easily be supported as a PBE. On the other hand, in the presence of an agent with superior information, the private core which is CBIC can be supported as a PBE. The discussion provided a noncooperative interpretation or foundation of the private core while making, through the game tree, the individual decisions transparent. In this way a better understanding of how incentive compatible contracts are formed is obtained.

In the present paper we investigate, in a one-good, two-agent economy, with and without free disposal, the implementation of private core, of Radner equilibrium, of weak fine core and weak fine values allocations. We obtain, through the construction of a tree with reasonable rules, that free disposal private core allocations, to which also the Radner equilibrium belongs, are not implementable. A brief comparison of the idea of CBIC in the static presentation with the case when the analysis is in terms of game trees is made.

It is surprising that free disposal destroys incentive compatibility and creates problems for implementation. Implementation in this case can be achieved by invoking an exogenous third party which can be thought of as a court that penalizes lying agents. It is of course possible that rational agents, once they realize that they can be cheated, might decide not to trade rather than rely on a third party which has to prove that he has perfect knowledge and can execute the correct trades. Notice that the third, exogenous party, in this case the court, plays the role of the mechanism designer in the relevant implementation literature (see Hahn and Yannelis, 2001, and the references there).

Similarly, implementation of a private core allocation becomes possible through the introduction of an endogenous third party with zero endowments but with superior information. In this case the third party is part of the model, i.e. an agent whose superior information allows him to play the role of an intermediary. The analysis overlaps with the one in Glycopantis, Muir, and Yannelis (2001). On the other hand we show here that implementation can also be achieved through a sequential equilibrium. It should be noted that the endogenous third agent is rewarded for his superior information by receiving consumption in a particular state, in spite of the fact that he has zero initial endowments in each state. However, both Radner equilibria and REE would not recognize a special role to such an agent. These Walrasian type notions would award to him zero consumption in all states of nature.

In summary, the analysis here considers the relation between, cooperative and noncooperative, static equilibrium concepts and noncooperative, game theoretic dynamic processes in the form of game trees. We have examined the possible support and implementation as perfect Bayesian equilibria of the cooperative concepts of the private core and the weak fine core, and the noncooperative generalized, Walrasian type equilibrium notions of Radner equilibrium and REE. In effect what we are doing is to look directly into the Bayesian incentive compatibility of the corresponding allocations, as if they were contracts, and then consider their implementability.

Appendix I: A note on PBE

In this note we look briefly at equilibrium notions when sequential decisions are taken by the players, i.e. in the context of game trees. For strategies we shall employ the idea of a *behavioral strategy* for a player being an assignment to each of his information sets of a probability distribution over the options available from that set. For a game of perfect recall, Kuhn (1953) shows that analysis of the game in terms of behavioral strategies is equivalent to that in terms of, the more familiar, mixed strategies. In any case, behavioral strategies are more natural to employ with an extensive form game. Sometimes we shall refer to them simply as *strategies*.

Consider an extensive form game and a given profile of behavioral strategies

$$s = \{s_i : i \in I\}$$

where I is the set of players.

When s is used each node of the tree is reached with probability obtained by producting the option probabilities given by s along the path leading to that node. In particular, there is a probability distribution over the set of terminal nodes so the expected payoff E_i to each player Pi may be expressed in terms of option probabilities from each information set.

Consider any single information set J owned by Pi, with corresponding option probabilities $(1-\pi_J, \pi_J)$, where for simplicity of notation we assume binary choice. The dependence of E_i on π_J is determined only by the paths which pass through J. Taking any one of these paths, on the assumption that the game is of perfect recall, the term it contributes to E_i will only involve π_J once in the corresponding product of probabilities. Thus, on summing over all such paths, the dependence of E_i on π_J is seen to be linear, with coefficients depending on the remaining components of s.

This allows the formation of a reaction function expressing π_J in terms of the remaining option probabilities, by optimizing π_J while holding the other probabilities constant; hence the Nash equilibria are obtained, as usual, as simultaneous solutions of all these functional relations. We are here adopting an *agent* form for a player, where optimization with respect to each of his decisions is done independently from all the others. A solution is guaranteed by the usual proof of existence for Nash equilibria.

For example, consider the tree in Figure 4, denoting the option probabilities from I_1, I_2 by $(1 - \alpha, \alpha), (1 - \beta, \beta)$ respectively. The payoff functions are then (apart from the factor $\frac{1}{3}$ expressing the probability of Nature's choice, and leaving out terms not involving α which come from paths not passing through I_1, I_2

$$E_1 = 5(1 - \alpha)(1 - \beta) + 5(1 - \alpha)\beta$$
$$+4\alpha(1 - \beta) + 4\alpha\beta + 2.5(1 - \alpha) + 4\alpha + \ldots$$
$$= 7.5 + 0.5\alpha + \ldots;$$
$$E_2 = 5(1 - \alpha)(1 - \beta) + 4(1 - \alpha)\beta$$
$$+5\alpha(1 - \beta) + 4\alpha\beta + 2.5(1 - \beta) + 4\beta + \ldots$$
$$= 7.5 + 0.5\beta + \ldots.$$

Since the coefficient of α in E_1 is positive, the optimal choice of α, i.e. the reaction function of Agent 1 is 1. Similarly for β in E_2 we obtain the value 1, and this is the reaction function of Agent 2.

Note that in any such calculation, only the coefficient of each π_J is important for the optimization – the rest of E_i is irrelevant. We may similarly treat the 21 option probabilities in Figure 7, obtaining 21 conditions which they must satisfy. These are quite complex and there are, probably, many solutions but it may be checked that the one given satisfies all conditions.

When an equilibrium profile is used, it is possible that some nodes are visited with zero probability. This means that the restriction of the strategy profile to subsequent nodes has no effect on the expected payoffs, so may be chosen arbitrarily. To eliminate this redundancy in the set of Nash equilibria, a refinement of the equilibrium concept to that of perfect equilibrium, was introduced for games of perfect information – that is, games in which each information set is a singleton.

This requires an equilibrium strategy also to be a Nash equilibrium for *all* sub-games of the given game. In other words, the strategy profile should be a Nash equilibrium for the game which might be started from any node of the given tree, not just the nodes actually visited in the full game.

Any attempt to extend this notion to general games encounters the problem that sub-trees might start from nodes which are not in singleton information sets. In such a case, the player who must move first cannot know for certain at which node he is located within that set. He can only proceed if he adopts *beliefs* about where he might be, in the form of a probability distribution over the nodes of the information set. Moreover, these beliefs must be common knowledge, for the other players to be able to respond appropriately, so the desired extension of the equilibrium concept must take into account both strategies and beliefs of the players. The game will be played from any information set as if the belief probabilities had been realised by an act of nature.

We need, therefore, to consider pairs (s, μ), consisting of a behavioral strategy profile s and a belief profile

$$\mu = \{\mu_J : J \in \mathcal{J}\}.$$

Here, $\mathcal{J}$ denotes the set of information sets and μ_J is a probability distribution over the nodes of information set J, expressing the beliefs of the player who might be required to play from that set. Given the belief profile, we then require that the strategy profile give a perfect equilibrium, in the sense of being optimal for each player starting from every information set. But we need also to consider the source of the beliefs.

Given any behavioral strategy profile s denote the probability of reaching any node a, using s, by $\nu(a)$. Consider first an information set, J, not all of whose nodes are visited with zero probability when using s. We may calculate the conditional probability of being at a node $a \in J$ given that it is in J by

$$\nu(a|J) = \frac{\nu(\{a\} \cap J)}{\nu(J)} = \frac{\nu(a)}{\nu(J)}$$

since $a \in J \Rightarrow \{a\} \cap J = \{a\}$. Thus the belief probabilities $\mu_J(a) = \nu(a|J)$ for J are just the relative probabilities of reaching the nodes of J.

For example, returning to Figure 4 and employing the only Nash solution $\alpha = \beta = 1$ noted above, the probabilities of reaching the nodes of I_2 are $0, \frac{1}{3}, \frac{1}{3}$ which relativises, given the condition that we reach I, to $0, \frac{1}{2}, \frac{1}{2}$ as stated.

Thus for a *PBE*, the behavioral strategy-belief profile pair (s, μ) should satisfy two conditions:

(i) For the given belief profile μ, the strategy profile s should be a perfect equilibrium, as defined above;
(ii) For the given strategy profile s, the belief profile μ should be calculated at each information set for which $\nu(I) \neq 0$ by the formula above.

Justifications of the concept of perfect equilibrium in games of perfect information will argue that the players need to have good strategies to employ, even were something to go wrong with the intended play so that the game accidentally enters sub-trees which ought not to be accessed. One way to argue this is through the notion of a trembling hand which makes errors, so possibly choosing the wrong move. Employing this same idea in the context of perfect Bayesian equilibria, we can allow small perturbations in the strategies, such that all information sets are visited with non-zero probability. Then the relation determining beliefs from strategies is well posed and we may consider only beliefs which arise as limiting cases of such perturbations. This more restrictive definition of equilibrium is called a *sequential equilibrium*.

References

Allen, B.: Generic existence of completely revealing equilibria with uncertainty, when prices convey information. Econometrica **49**, 1173–1199 (1986)

Allen, B., Yannelis, N. C.: Differential information economies. Economic Theory **18**, 263–273 (2001)

Aumann, R. J.: Correlated equilibria as an expression of Bayesian rationality. Econometrica **55**, 1–18 (1987)

Emmons, D., Scafuri, A. J.: Value allocations – An exposition. In: Aliprantis, C. D., Burkinshaw, O., Rothman, N. J. (eds.) Advances in economic theory. Lecture notes in economics and mathematical systems, pp. 55–78. Berlin Heidelberg New York: Springer 1985

Einy, E., Moreno, D., Shitovitz, B.: Rational expectations equilibria and the ex post core of an economy with asymmetric information. Journal of Mathematical Economics **34**, 527–535 (2000)

Einy, E., Moreno, D., Shitovitz, B.: Competitive and core allocations in large economies with differential information. Economic Theory **34**, 321–332 (2001)

Glycopantis, D., Muir, A., Yannelis, N. C.: An extensive form interpretation of the private core. Economic Theory **18**, 293–319 (2001)

Glycopantis, D., Muir, A., Yannelis, N. C.: On the extensive form implementation of REE. Mimeo (2002)

Hahn, G., Yannelis, N. C.: Efficiency and incentive compatibility in differential information economies. Economic Theory **10**, 383–411 (1997)

Hahn, G., Yannelis, N. C.: Coalitional Bayesian Nash implementation in differential information economies. Economic Theory **18**, 485–509 (2001)

Koutsougeras, L., Yannelis, N. C.: Incentive compatibility and information superiority of the core of an economy with differential information. Economic Theory **3**, 195–216 (1993)

Krasa, S., Yannelis, N. C.: The value allocation of an economy with differential information. Econometrica **62**, 881–900 (1994)

Kreps, M. D., Wilson, R.: Sequential equilibrium. Econometrica **50**, 889–904 (1982)

Kuhn, H. W.: Extensive games and the problem of information. In: Kuhn, H. W., Tucker, A. W. (eds.) Contributions to the theory of games, Vol. II. Annals of mathematical studies, Vol. 28, pp. 193–216. Princeton, NJ: Princeton University Press 1953

Kurz, M.: On rational belief equilibria. Economic Theory **4**, 859–876 (1994)

Radner, R.: Competitive equilibrium under uncertainty. Econometrica **36**, 31–58 (1968)

Scafuri, A. J., Yannelis, N. C.: Non-symmetric cardinal value allocations. Econometrica **52**, 1365–1368 (1984)

Shapley, L. S.: A value for n-person games. In: Kuhn, H. W., Tucker, A. W. (eds.) Contributions to the theory of games, Vol. II. Annals of mathematical studies, Vol. 28, pp. 307–317. Princeton, NJ: Princeton University Press 1953

Tirole, J.: The theory of industrial organization. Cambridge, MA London: The MIT Press 1988

Wilson, R.: Information, efficiency, and the core of an economy. Econometrica **46**, 807–816 (1978)

Yannelis, N. C.: The core of an economy with differential information. Economic Theory **1**, 183–198 (1991)

Vickrey auctions with reserve pricing[*]

Lawrence M. Ausubel and Peter Cramton

Department of Economics, University of Maryland, College Park, MD 20742-7211, USA
(e-mail: ausubel@econ.umd.edu; peter@cramton.umd.edu)

Received: December 31, 2002; revised version: May 5, 2003

Summary. We generalize the Vickrey auction to allow for reserve pricing in a multi-unit auction with interdependent values. In the Vickrey auction with reserve pricing, the seller determines the quantity to be made available as a function of the bidders' reports of private information, and then efficiently allocates this quantity among the bidders. Truthful bidding is a dominant strategy with private values and an *ex post* equilibrium with interdependent values. If the auction is followed by resale, then truthful bidding remains an equilibrium in the auction-plus-resale game. In settings with perfect resale, the Vickrey auction with reserve pricing maximizes seller revenues.

Keywords and Phrases: Auctions, Vickrey auctions, Multi-unit auctions, Reserve price, Resale.

JEL Classification Numbers: D44, C78, D82.

1 Introduction

A Vickrey auction has the distinct advantage of assigning goods efficiently – putting the goods in the hands of those who value them most. However, one critique of a Vickrey auction is that it may yield low revenues for the seller. Indeed, Vickrey expressed this concern in his seminal article (Vickrey, 1961). When competition

---

[*] The authors gratefully acknowledge the generous support of National Science Foundation Grants SES-97-31025, SES-01-12906 and IIS-02-05489. We appreciate valuable comments from Ilya Segal. Special thanks go to Mordecai Kurz, who served as Larry's dissertation advisor and who introduced both authors to the economics profession back at IMSSS at Stanford. Congratulations and best wishes are extended to Mordecai and his family on the happy occasion of the publication of this Festschrift in Mordecai's honor.

Correspondence to: L.M. Ausubel

is weak and the bidders are asymmetric, revenues from a Vickrey auction may be small. A vivid example was the 1990 New Zealand sale of spectrum licenses by second-price auction. In one case, the winner bid $100,000, but paid only $6; in another, the winner bid $7,000,000, but paid only $5,000 (McMillan, 1994). Reserve pricing is a simple and effective device to avoid such disasters. The seller may charge the reserve price or reduce the quantity sold if the bids are too low. Reserve pricing is also an effective device for mitigating collusion, since it limits the maximum gain collusion can reap.

Reserve pricing is especially important in auctions, such as electricity auctions, spectrum auctions or Treasury auctions, where participants bid for multiple items. Then the largest market participant may be so large that removing this bidder may lead to no excess demand. In a Vickrey auction, prices are based on the opportunity cost of winning; that is, a winner pays the value that the goods would have in their best use without the winner. If a bidder's winnings are greater than the excess demand in the auction with the bidder removed, then some of the Vickrey prices are undefined (or zero). In auctions to supply electricity during peak periods, it is common for the capacity of the largest generator to be far greater than the excess capacity in the system. In such a setting, a Vickrey auction must involve reserve pricing.

We generalize the Vickrey auction to allow for reserve pricing in a multi-unit auction with interdependent values. In the Vickrey auction with reserve pricing, the seller determines the quantity to be made available as a function of the bidders' reports of their private information, and then efficiently allocates this quantity among the bidders. We prove in Theorem 1 that truthful bidding by all bidders is an *ex post* equilibrium in a model with interdependent values (a bidder's value also depends on the private information of other bidders) and that truthful bidding is a dominant strategy in a model with private values (a bidder's value depends only on its own private information). Thus, reserve pricing does not interfere with many of the desirable features of a Vickrey auction.

An important motivation for this article is the possibility of resale after an auction. The "optimal auctions" literature requires the seller to misassign items, that is, to put the items in hands other than those who value them the most, with positive probability (except in symmetric models). However, the seller's ability to do this may be undermined when resale cannot be prevented; bidders will anticipate the resale market and adjust their bids accordingly. In Ausubel and Cramton (1999), we prove that whenever resale markets are *fully* efficient, a seller *cannot* increase revenues by misassigning the items among the bidders. The revenue-maximizing auction is simply for the seller to decide on an optimal quantity to sell based on the bidders' reports and to assign these units efficiently among the bidders. Thus, faced with a perfect resale market, the best that the seller can do is to withhold some of the supply, but then to sell the remaining supply efficiently.

This immediately raises the question of how to construct a mechanism that limits the quantity sold but allocates efficiently whatever quantity is sold. The Vickrey auction with reserve pricing defined in the current article performs precisely this job, and thus does exactly what is required for an optimizing seller facing a perfect resale market.[1] In particular, we prove in Theorem 2 that truthful bidding in the Vickrey auction with reserve pricing (and hence an efficient assignment of the goods that are sold) is an *ex post* equilibrium of the two-stage game consisting of the auction followed by resale. Indeed, for this result, we do not need the resale procedure to be perfect. Truthful bidding will be an *ex post* equilibrium whenever the resale game is such that no bidder expects to get more than 100% of the gains from trade that it brings to the table.

The current article is related to three strands of literature. First, a number of articles extend the Vickrey auction to settings where bidders have interdependent values. Crémer and McLean (1985) construct a mechanism through which the full surplus can be extracted from bidders and, as a step along the way, they construct a mechanism for discrete types that yields an efficient assignment as an *ex post* equilibrium. Maskin (1992) defined a modified second-price auction, which yields an efficient assignment in a single-good setting with interdependent values. Ausubel (1999) extends Maskin's approach by defining a "generalized Vickrey auction" for multiple identical items with interdependent values. Dasgupta and Maskin (2000), Jehiel and Moldovanu (2001) and Perry and Reny (2002) also define auction mechanisms that, for the case of multiple identical objects, are outcome-equivalent to the generalized Vickrey auction. None of these papers explore reserve pricing or the implications of resale markets.

The second strand of literature considers multi-unit auctions with variable supply. Back and Zender (2001) show that in a uniform-price auction the seller can eliminate low-price equilibria (Back and Zender, 1993) by restricting supply after the bids are in. Lengwiler (1999), in a model allowing two possible price levels, considers the effects of variable supply on seller revenues in both uniform-price and pay-your-bid auctions. McAdams (2002) also examines variations on the uniform-price auction in which the auctioneer is able to increase or decrease quantity after receiving the bids. None of these papers consider Vickrey pricing or resale.

The third strand of literature considers auctions with resale. Haile (1999, 2003) demonstrates that, in auctions followed by resale, bidders will anticipate the resale market and adjust their bids accordingly. Furthermore, Haile (1999) considers Vickrey auctions with resale in the case of auctioning a single item. Ausubel and Cramton (1999) consider multi-unit optimal auctions with efficient resale. Here we consider how to implement these optimal auctions. Zheng (2002) and Calzolari and Pavan (2002) consider single-item optimal auctions with alternative resale games.[2]

Section 2 presents a general model for the auction of a divisible good. Bidders' demands for the items may be interdependent. Section 3 defines the Vickrey auction with reserve pricing, and demonstrates that truthful bidding is an *ex post*

---

[1] The current article does not concern itself with the determination of the optimal quantity based on bidders' reports, and instead takes as given the quantity as a function of bidders' reports. For a treatment of the problem of determining the optimal quantity, see Ausubel and Cramton (1999).

[2] The principal difference between the approaches of Ausubel and Cramton (1999) and Zheng (2002) is that we assume that the resale market is fully efficient, whereas Zheng assumes that the winner of the auction has full monopoly power in the resale game. Calzolari and Pavan (2002) assume that the resale market comprises a single take-it-or-leave-it offer by a seller (with probability λ) or a buyer (with probability $1-\lambda$).

equilibrium, despite the fact that the bidding affects the quantity sold. Section 4 analyzes an auction followed by resale. It is shown that the possibility of resale does not distort the Vickrey auction with reserve pricing. Truthful bidding remains an *ex post* equilibrium, despite the presence of a resale market following the auction. Section 5 concludes.

2 The general divisible good model

A seller has a quantity 1 of a divisible good to sell to n bidders, $N \equiv \{1,\dots,n\}$. The seller's valuation for the good equals zero. Each bidder i can consume any quantity $q_i \in [0,1]$. We can interpret q_i as bidder i's share of the total quantity. Let $q \equiv (q_1,\dots,q_n)$, and let $Q \equiv \{q \mid \sum_i q_i \leq 1\}$ be the set of all feasible assignments. Each bidder's value for the good depends on the private information of all the bidders. Let $t_i \in T_i \equiv [0, t_i^{max}]$ be bidder i's type (i's private information), $t \equiv (t_1,\dots,t_n) \in T \equiv T_1 \times \cdots \times T_n$, and $t_{-i} \equiv t\backslash t_i = (t_1,\dots,t_{i-1},t_{i+1},\dots,t_n)$. A bidder's value $V_i(t, q_i)$ for the quantity q_i depends on its own type t_i and the other bidders' types t_{-i}. A bidder's utility is its value less the amount it pays: $V_i(t, q_i) - X_i$. Let $v_i(t, q_i)$ denote the marginal value for bidder i, given the vector t of types and quantity q_i. Then $V_i(t, q_i) = \int_0^{q_i} v_i(t, y)\, dy$.

We assume that $v_i(t, q_i)$ satisfies the following assumptions:

Continuity. For all i, t, and q_i, $v_i(t, q_i)$ is jointly continuous in (t, q_i).

Value monotonicity. For all i, t and q_i, $v_i(t,q_i)$ is nonnegative, strictly increasing in t_i, and weakly decreasing in q_i.

Single-crossing property. For all $i, j \neq i, q_i, q_j, t_{-i}$, and $t_i' > t_i, v_i(t, q_i) > v_j(t, q_j) \Rightarrow v_i(t_i', t_{-i}, q_i) > v_j(t_i', t_{-i}, q_j)$ and $v_i(t_i', t_{-i}, q_i) < v_j(t_i', t_{-i}, q_j) \Rightarrow v_i(t, q_i) < v_j(t, q_j)$.

Value monotonicity implies that types are naturally ordered, and that the bidders have weakly downward-sloping demand curves. The single-crossing property implies that, if a fixed quantity is assigned efficiently among the bidders, then bidder i's quantity $q_i(t)$ may be chosen to be weakly increasing in t_i. The single-crossing property holds if an increase in bidder i's type raises bidder i's marginal value at least as much as any other bidder's.

Three special cases of the general model are particularly useful.

Private values. A bidder's value $V_i(t_i, q_i)$ depends only on its own type.

Common value. The bidders' values are the same: $V_i(t, q_i) = V_j(t, q_i)$.

Independent types. The bidders' types are drawn independently from the distribution functions F_i with positive and finite density f_i on T_i.

The private values assumption enables us to strengthen many of the results. In particular, truthful bidding becomes a dominant strategy, rather than simply a best response. Also, value monotonicity automatically implies the single-crossing property in the private value setting.

The common value assumption often is made in models of oil lease auctions and in models of Treasury and other financial auctions.

Independent types is needed in the optimal auction analysis (our final result). Expected revenues depend on the probability distribution of types, and independence is needed for a general revenue equivalence theorem. However, most of our analysis is based on *"ex post"* arguments, which do not require any assumptions about the distribution of types.

Our starting point for describing a Vickrey auction with reserve pricing is to specify the aggregate quantity $\bar{q}(t) \equiv \Sigma_i\, q_i(t)$ that the seller assigns to the bidders, as a function of the vector of reported types. The description of the Vickrey auction is only guaranteed to make sense if the aggregate quantity $\bar{q}(t)$ is weakly increasing. We therefore require

Monotonic aggregate quantity. The aggregate quantity rule $\bar{q}(t)$ is weakly increasing in each bidder's type.

This assumption, together with the single-crossing property, guarantees that the quantity $\bar{q}(t)$ can be assigned efficiently among the bidders in such a way that each bidder i's quantity $q_i(t)$ is weakly increasing in t_i.

3 Vickrey auction with reserve pricing

The Vickrey auction with reserve pricing can be thought of as a three-step procedure. First, the bidders simultaneously and independently report their types t to the seller, and the seller determines the aggregate quantity $\bar{q}(t)$ that it wishes to assign to bidders. Second, the seller determines an efficient assignment of this aggregate quantity; that is, the seller solves for $q^*(t) \equiv (q_1^*(t), \cdots, q_n^*(t))$ that maximizes $\Sigma_i\, V_i(t, q_i^*(t))$ subject to $\Sigma_i\, q_i^*(t) = \bar{q}(t)$. When the efficient assignment is not unique due to flat regions in the aggregate demand curve, $q_i^*(t)$ is chosen so that it is weakly increasing in t_i. Third, the seller determines a payment $X_i^*(t)$ for each bidder i associated with the assignment of $q_i^*(t)$, where $q_i^*(t)$ and $X_i^*(t)$ must be specified so that truthful bidding is incentive compatible and individually rational for every type of every bidder.

The determination of the payment rule is most easily understood in an environment with discrete units. Hold the reports t_{-i} of bidders other than bidder i fixed, and consider the quantity $q_i^*(t_i, t_{-i})$ assigned to i as a function of t_i. Let $t_i^1(t_{-i})$ denote the minimum type such that i is awarded at least one unit, let $t_i^2(t_{-i})$ denote the minimum type such that i is awarded at least two units, etc. More precisely, for every $k \geq 1$, let $t_i^k(t_{-i}) = \inf \{t_i : q_i^*(t_i, t_{-i}) \geq k\}$, the minimum type such that bidder i is awarded at least k units. By hypothesis, $q_i^*(t)$ is weakly increasing in t_i. Therefore, by value monotonicity and the single-crossing property, $t_i^k \leq t_i^{k+1}$ for all $k \geq 1$.

Discrete payment rule. If bidder i is assigned K units, then for every $k (1 \leq k \leq K)$, bidder i is charged a price of $v_i(t_i^k(t_{-i}), t_{-i}, k)$ for the kth unit.

Vickrey pricing is best thought of in terms of opportunity costs. The winner pays the opportunity cost of its winnings. In a standard Vickrey auction, the opportunity cost is always the value to the other bidder that would receive the good if the winner did not participate. In a Vickrey auction with reserve pricing, the opportunity cost can come instead from the seller. This occurs for a good that the seller would withhold were it not for the winner's bids. Critical to the analysis, observe that bidder i's value is evaluated at the *minimal* type at which i receives the kth unit. This specification has the effect of subsuming the proper pricing rule both for the case where the kth unit of bidder i comes from another bidder as well as for the case where the kth unit of bidder i comes from the seller's reserve. If the kth unit for bidder i is assigned to bidder i from another bidder j, then bidder i is charged the other bidder's value $v_j(t_i^k(t_{-i}), t_{-i}, q_j)$, assuming i's type is just high enough to receive k units, as by definition, $t_i^k(t_{-i})$ is the minimal type of bidder i such that bidder i receives this unit, so $v_i(t_i^k(t_{-i}), t_{-i}, k) = v_j(t_i^k(t_{-i}), t_{-i}, q_j)$. Meanwhile, if the kth unit for bidder i is assigned to bidder i out of the seller's reserve, then the seller's implicit "reserve price" for this unit also equals $v_i(t_i^k(t_{-i}), t_{-i}, k)$, since all types of bidder i greater than $t_i^k(t_{-i})$ are receiving this unit while all types of bidder i less than $t_i^k(t_{-i})$ are not.

Returning to the case of continuous quantity, let $q_{-i}^*(t) \equiv \bar{q}(t) - q_i^*(t)$ denote the aggregate quantity allocated to bidders other than i (bidders $N\backslash i$) following reports t. Furthermore, for any quantity y, let $v_{-i}(t, y)$ denote the marginal value to bidders $N\backslash i$ if the quantity y is allocated *efficiently* among bidders $N\backslash i$. Observe that, for any aggregate quantity rule $\bar{q}(t)$ and for any reports t, an efficient assignment rule $q^*(t)$ satisfies

$$v_i(t, q_i^*(t)) \begin{cases} \leq v_{-i}(t, q_{-i}^*(t)), & \text{for } i \quad \text{such that } q_i^*(t) = 0 \\ = v_{-i}(t, q_{-i}^*(t)), & \text{for } i \quad \text{such that } 0 < q_i^*(t) < \bar{q}(t) \\ \geq v_{-i}(t, q_{-i}^*(t)), & \text{for } i \quad \text{such that } q_i^*(t) = \bar{q}(t). \end{cases} \quad (1)$$

Otherwise, from continuity and value monotonicity, if $0 < q_i^*(t) < \bar{q}(t)$ and $v_i(t, q_i^*(t)) > v_{-i}(t, q_{-i}^*(t))$, then there exists $\varepsilon > 0$ such that allocating $q_i^*(t) + \varepsilon$ to bidder i and $q_{-i}^*(t) - \varepsilon$ to bidders $-i$ would generate social improvement, and similarly if $v_i(t, q_i^*(t)) < v_{-i}(t, q_{-i}^*(t))$.

From Eq. (1) and the single-crossing property, for any monotonic aggregate quantity rule $\bar{q}(t)$, there exists an associated efficient assignment rule $q_i^*(t)$ that is weakly increasing in t_i. To see this, note that the single-crossing property implies that, in an efficient assignment, any quantity that must go to i when t_i is reported must still go to i when $t_i' > t_i$ is reported, and any quantity that cannot go to i when $t_i' > t_i$ is reported still cannot go to i when t_i is reported. This would guarantee that if aggregate demand were strictly downward sloping, then $q_i^*(t)$ would be uniquely defined, and it would be weakly increasing in t_i. However, when the aggregate demand curve has a flat region and the flat portion includes more than one bidder, then $q_i^*(t)$ is no longer unique, and indeed some efficient assignment rules may not be monotonic. In this case, the seller must choose a tie-breaking rule that is consistent with a monotonic efficient assignment. For example, in the flat portion of aggregate demand, award the good first to the bidder with the higher type, and split the quantity equally among bidders with the same type.

Also observe that, although $\bar{q}(t)$ is monotonic, $\bar{q}(t)$ need not be continuous in t_i, so it is useful to define limits of $\bar{q}(t)$ from above and below in t_i:

$$\bar{q}_i^+(\hat{t}_i, t_{-i}) = \lim_{t_i \downarrow \hat{t}_i} \bar{q}(t_i, t_{-i}) \text{ and } \bar{q}_i^-(\hat{t}_i, t_{-i}) = \lim_{t_i \uparrow \hat{t}_i} \bar{q}(t_i, t_{-i}).$$

We can now define the generalized Vickrey auction with reserve pricing.

Definition. *Vickrey auction with reserve pricing.* Given any monotonic efficient assignment rule $q^*(t)$, and for reports t_{-i} of bidders other than bidder i and for any quantity z such that $0 \le z \le q_i^*(t_i^{\max}, t_{-i})$, define:

$$\hat{t}_i(t_{-i}, z) = \inf_{t_i} \{t_i \mid q_i^*(t_i, t_{-i}) \ge z\}. \tag{2}$$

Following reports t, bidder i is assigned $q_i^*(t)$ units and is charged a payment $X_i^*(t)$ computed by:

$$X_i^*(t) = \int_0^{q_i^*(t)} v_i(\hat{t}_i(t_{-i}, z), t_{-i}, z) \, dz. \tag{3}$$

Note that the payment formula of Eq. (3) is well defined, since the value monotonicity assumption assures that, for any reports t and for any quantity $z \in [0, q_i^*(t)]$, we have $0 \le v_i(\hat{t}_i(t_{-i}, z), t_{-i}, z) \le v_i(t, 0)$.

In the Vickrey auction, a bidder pays the opportunity cost of its winning for each incremental quantity won. Hence, the marginal payment made at each quantity z is determined by the bidder's marginal value assuming the bidder makes the *lowest possible* report consistent with winning a quantity z. This marginal value may be determined either by the opportunity to sell to another bidder or by the opportunity to withhold the good. In this way, the bidder receives 100 percent of the gains from trade that it brings to the table. The fact that the bidder receives 100 percent of its incremental contribution is what gives the bidder the incentive for truthful bidding.

Theorem 1. *For any monotonic aggregate quantity rule $\bar{q}(t)$ and associated monotonic efficient assignment rule $q_i^*(t)$, and for any valuation functions $v_i(t, q_i)$ satisfying continuity, value monotonicity and the single-crossing property, the Vickrey auction with reserve pricing has truthful bidding as an ex post equilibrium.*

Proof. By continuity, value monotonicity and the single-crossing property, we can choose $q_i^*(t)$ to be weakly increasing in t_i. Then $\hat{t}_i(t_{-i}, z)$ defined by Eq. (2) is weakly increasing in z. Substituting Eq. (3) into the expression, $V_i(t, q_i) - X_i$, for bidder i's utility yields the following integral for bidder i's utility from reporting its type as t_i' when its true type is t_i and the other bidders' true and reported types are t_{-i}:

$$U_i(t_i'|t) = \int_0^{q_i^*(t_i', t_{-i})} \left[v_i(t, z) - v_i(\hat{t}_i(t_{-i}, z), t_{-i}, z) \right] dz. \tag{4}$$

Observe that the integrand of Eq. (4) is independent of t_i', bidder i's reported type; t_i' enters into Eq. (4) *only* through the upper limit on the integral. Moreover, by value monotonicity, the integrand of Eq. (4) is nonnegative for all $z \le q_i^*(t)$ and is nonpositive for all $z \geq q_i^*(t)$. Hence, $U_i(t_i'|t)$ is maximized for every t when the upper limit on the integral equals $q_i^*(t)$, which is attained by truthful bidding. $\square$

For the special case of private values, truthful bidding is a dominant strategy. Then truthful bidding is a best response for *any* reports by the other bidders. Without private values, the dominant strategy result is lost, since a bidder's value depends on the types of the other bidders, and so the bidder cares whether the reports of the others are truthful. Truthful bidding is only a best response if the other bidders are truthful; but it remains a best response after the bidder learns the opposing bidders' (truthful) reports. Hence, the *ex post* equilibrium property of truthful bidding always holds.

4 Auction followed by resale

A main motivation for assigning goods efficiently is the possibility of resale (Ausubel and Cramton, 1999). Resale undermines the seller's incentive to misassign the goods, since the misassignment may be undone in the resale market. The bidders anticipate the possibility of resale, which alters their incentives and distorts the bidding in the initial auction. Hence, an equilibrium in the auction game typically is not an equilibrium in the auction-plus-resale game.

Here we wish to show that a Vickrey auction with reserve pricing is not distorted by the possibility of resale. To prove this, we need to show that a bidder i with type t_i does not wish to misreport type t_i' in a Vickrey auction with reserve pricing followed by resale. Let $\Delta_i(t_i'|t)$ denote the optimal quantity of resale between bidder i and the coalition $N \backslash i$ if bidder i misreports its type as t_i' when its true type is t_i and the other bidders' true and reported types are t_{-i}, and let $\mathrm{GFT}_i(t_i'|t)$ denote the gains from trade available via resale between bidder i and the coalition $N \backslash i$ if bidder i misreports its type as t_i' when its true type is t_i and the other bidders' true and reported types are t_{-i}.

Lemma 1. *If bidder i misreports its type as t_i' when its true type is t_i and the other bidders' true and reported types are t_{-i}, the (minimum) optimal quantity of resale between bidder i and the coalition $N \backslash i$ is given by*

$$\Delta_i(t_i'|t) = \begin{cases} \min\{z \geq 0 | v_{-i}(t, q_{-i}^*(t_i', t_{-i}) + z) \leq v_i(t, q_i^*(t_i', t_{-i}) - z)\}, & \text{if } t_i' > t_i, \\ \min\{z \geq 0 | v_i(t, q_i^*(t_i', t_{-i}) + z) \leq v_{-i}(t, q_{-i}^*(t_i', t_{-i}) - z)\}, & \text{if } t_i' < t_i, \end{cases}$$
(5)

and the gains from trade available via resale between bidder i and the coalition $N \backslash i$ are given by

$$\mathrm{GFT}_i(t_i'|t) = \int_0^{\Delta_i(t_i'|t)} \left[v_{-i}(t, q_{-i}^*(t_i', t_{-i}) + z) - v_i(t, q_i^*(t_i', t_{-i}) - z) \right] dz .$$
(6)

Proof. Observe that the integrand of Eq. (6) gives the marginal gains of the zth unit transferred from coalition $N \backslash i$ to bidder i. By value monotonicity, if $z' < z$, then $v_{-i}(t, q_{-i}^*(t_i', t_{-i}) + z) > v_i(t, q_i^*(t_i', t_{-i}) - z)$ implies $v_{-i}(t, q_{-i}^*(t_i', t_{-i}) + z') >$

$v_i(t, q_i^*(t_i', t_{-i}) - z')$ and $v_i(t, q_{-i}^*(t_i', t_{-i}) + z) > v_{-i}(t, q_i^*(t_i', t_{-i}) - z)$ implies $v_i(t, q_{-i}^*(t_i', t_{-i}) + z') > v_{-i}(t, q_i^*(t_i', t_{-i}) - z')$. Thus, $\Delta_i(t_i'|t)$ defined by Eq. (5) provides the (minimal) upper limit for the integral in Eq. (6) which maximizes the value of the integral.   □

The following calculation will be helpful in what follows:

Lemma 2. *For any monotonic aggregate quantity rule $\bar{q}(t)$ and associated monotonic efficient assignment rule $q_i^*(t)$, for any valuation functions $v_i(t, q_i)$ satisfying continuity, value monotonicity and the single-crossing property, for any bidder i, for any true type t_i, for any overreport $t_i' > t_i$, for any vector t_{-i} of other bidders' reported and true types, and for any z such that $0 \le z \le \Delta_i(t_i'|t)$,*

$$v_{-i}(t, q_{-i}^*(t_i', t_{-i}) + z) \le v_i(\hat{t}_i(t_{-i}, q_i^*(t_i', t_{-i}) - z),\ t_{-i},\ q_i^*(t_i', t_{-i}) - z). \quad (7)$$

Proof. Consider any z such that $0 \le z \le \Delta_i(t_i'|t)$, and define $\hat{t}_i^z \equiv \hat{t}_i(t_{-i}, q_i^*(t_i', t_{-i}) - z) \ge t_i$. By the definition of $\hat{t}_i^z$, for every $\tilde{t}_i > \hat{t}_i^z$, it is the case that $q_i^*(\tilde{t}_i, t_{-i}) \ge q_i^*(t_i', t_{-i}) - z$; therefore, $v_{-i}(\tilde{t}_i, t_{-i}, \bar{q}(\tilde{t}_i, t_{-i}) - q_i^*(t_i', t_{-i}) + z) \le v_i(\tilde{t}_i, t_{-i}, q_i^*(t_i', t_{-i}) - z)$, for every $\tilde{t}_i > \hat{t}_i^z$, and so taking the limit as $\tilde{t}_i \downarrow \hat{t}_i^z$ implies that $v_{-i}(\hat{t}_i^z, t_{-i}, \bar{q}^+(\hat{t}_i^z, t_{-i}) - q_i^*(t_i', t_{-i}) + z) \le v_i(\hat{t}_i^z, t_{-i}, q_i^*(t_i', t_{-i}) - z)$. Note that $v_{-i}(t, q_{-i}^*(t_i', t_{-i}) + z) \equiv v_{-i}(t, \bar{q}(t_i', t_{-i}) - q_i^*(t_i', t_{-i}) + z) \le v_{-i}(\hat{t}_i^z, t_{-i}, \bar{q}^+(\hat{t}_i^z, t_{-i}) - q_i^*(t_i', t_{-i}) + z)$, since $\hat{t}_i^z \le t_i'$ implies $\bar{q}(t_i', t_{-i}) \ge \bar{q}^+(\hat{t}_i^z, t_{-i})$, and since $\hat{t}_i^z \ge t_i$. Combining inequalities, we conclude that $v_{-i}(t, q_{-i}^*(t_i', t_{-i}) + z) \le v_i(\hat{t}_i^z, t_{-i}, q_i^*(t_i', t_{-i}) - z)$, as desired.   □

To prove Theorem 2, we need some structure on the resale game. In particular, we need a constraint on how much a misreporting bidder can gain in the resale game. With two bidders, individual rationality is all that is required. In the resale game, a bidder cannot get a surplus that is greater than the available gains from trade, for to do so the other bidder would have to strictly lose from resale. In this case, the other bidder would simply refuse to participate in resale. With more than two bidders and interdependent values, we must extend the definition of individual rationality. This is because one bidder's misreport in the auction may create gains from trade among the other bidders. These other bidders, then, should consider the gains from trade they can secure among themselves in deciding whether to participate in resale with the misreporting bidder.

Coalitional rationality. For any initial allocation a of the good among bidders, for any vector t of types and for any subset S of the set N of bidders, let $v(S|a, t)$ denote the available gains from trade if the bidders in subset S trade only amongst themselves (starting at allocation a and evaluated at types t). Further, let s_i denote the surplus from the resale process realized by bidder i. The resale process is *coalitionally rational* if, for every subset S of the set N of bidders, the bidders in subset S obtain no more surplus s_i than they bring to the table:

$$\sum_{i \in S} s_i \leq v(N|a,t) - v(N \backslash S|a,t). \tag{8}$$

The resale process is *coalitionally-rational against individual bidders* if, for every element i of the set N of bidders, bidder i obtains no more surplus s_i than it brings to the table:

$$s_i \leq v(N|a,t) - v(N \backslash i|a,t). \tag{9}$$

The intuition behind this assumption is that, in the bargaining process underlying resale, the bidders in coalition $N \backslash S$ always have the outside option of excluding the bidders in the complementary set, S, from the bargaining and only trading amongst themselves. Hence, the bidders in S cannot deprive the bidders in $N \backslash S$ of the gains from trade that they could still obtain by trading amongst themselves.

We should remark that the assumption of coalitional rationality is quite natural and quite weak. It is implied, for example, by the requirement in the definition of the core that no coalition can improve upon an allocation. All we will need for our resale theorem is the still-weaker assumption of coalitional rationality against individual bidders. This is the requirement that any individual bidder i not receive any higher payoff than its marginal contribution to the set $N \backslash i$ of bidders. Observe that this is trivially implied by coalitional rationality. With superadditive values (which is always the case when value reflects potential gains from trade), it is also satisfied by standard solution concepts such as the Shapley value, which has every bidder i receiving its expected marginal contribution to the set S of bidders (the expectation taken over all subsets $S \subseteq N \backslash i$).

In the private values case, the definition of coalitional rationality reduces to individual rationality. With private values, if all bidders except bidder i report truthfully in the auction, then observe that in the resale round, $v(N \backslash i|a,t) = 0$, since the objects distributed to the coalition $N \backslash i$ are already assigned efficiently. Thus, individual rationality, $s_i \geq 0$, and feasibility, $\sum_{j \in N} s_j \leq v(N|a,t)$, imply that $s_i \leq v(N|a,t) - v(N \backslash i|a,t)$, which is coalitional rationality.

We now can prove our second theorem, which concerns the game with resale.

Theorem 2. *For any monotonic aggregate quantity rule $\bar{q}(t)$ and associated monotonic efficient assignment rule $q_i^*(t)$, and for any valuation functions $v_i(t, q_i)$ satisfying continuity, value monotonicity and the single-crossing property, truthful bidding followed by no resale is an ex post equilibrium of the two-stage game consisting of the Vickrey auction with reserve pricing followed by any resale process that is coalitionally-rational against individual bidders.*

Proof. Let $\pi_i(t_i'|t)$ denote the combined payoff to bidder i in the Vickrey auction and the resale market from misreporting t_i', when its true type is t_i and the other bidders' reported and true types are t_{-i}. By coalitional rationality against individual bidders, $\pi_i(t_i'|t) \leq U_i(t_i'|t) + \text{GFT}_i(t_i'|t)$, since $\text{GFT}_i(t_i'|t)$ is defined to be the gains from trade available via resale between bidder i and the coalition $N \backslash i$. By Eqs. (4)

and (6),

$$\pi_i(t_i'|t) \leq \int_0^{q_i^*(t)} [v_i(t,z) - v_i(\hat{t}_i(t_{-i},z), t_{-i}, z)]dz \tag{10}$$

$$+ \int_0^{q_i^*(t_i',t_{-i})-q_i^*(t)} [v_i(t, q_i^*(t_i',t_{-i})-z) - v_i(\hat{t}_i(t_{-i}, q_i^*(t_i',t_{-i})-z), t_{-i}, q_i^*(t_i',t_{-i})-z]dz$$

$$+ \int_0^{\Delta_i(t_i'|t)} [v_{-i}(t, q_{-i}^*(t_i',t_{-i})+z) - v_i(t, q_i^*(t_i',t_{-i})-z)]dz .$$

Since $t_i \leq \hat{t}_i(t_{-i}, q_i^*(t_i',t_{-i})-z)$, for all z between 0 and $q_i^*(t_i',t_{-i}) - q_i^*(t)$, the second integrand of Eq. (10) is weakly negative. Since $0 \leq \Delta_i(t_i'|t) \leq q_i^*(t_i',t_{-i}) - q_i^*(t)$, we further have

$$\pi_i(t_i'|t) \leq \int_0^{q_i^*(t)} [v_i(t,z) - v_i(\hat{t}_i(t_{-i},z), t_{-i}, z)]dz \tag{11}$$

$$+ \int_0^{\Delta_i(t_i'|t)} [v_i(t, q_i^*(t_i',t_{-i})-z) - v_i(\hat{t}_i(t_{-i}, q_i^*(t_i',t_{-i})-z), t_{-i}, q_i^*(t_i',t_{-i})-z]dz$$

$$+ \int_0^{\Delta_i(t_i'|t)} [v_{-i}(t, q_{-i}^*(t_i',t_{-i})+z) - v_i(t, q_i^*(t_i',t_{-i})-z)]dz .$$

But, then, using Eq. (4), we can simplify this as

$$\pi_i(t_i'|t) \leq U_i(t_i|t) + \int_0^{\Delta_i(t_i'|t)} [v_{-i}(t, \, q_{-i}^*(t_i',t_{-i})+z) \tag{12}$$

$$- v_i(\hat{t}_i(t_{-i}, q_i^*(t_i',t_{-i})-z), t_{-i}, q_i^*(t_i',t_{-i})-z)] \, dz.$$

Finally, observe by Lemma 2 that the integrand of Eq. (12) is nonpositive for all z such that $0 \leq z \leq \Delta_i(t_i'|t)$; consequently the integral is nonpositive whenever $\Delta_i(t_i'|t) \geq 0$. By the single-crossing property and the monotonicity of $\bar{q}(t)$, $t_i' > t_i$ implies $\Delta_i(t_i'|t) \geq 0$. This allows us to conclude that $\pi_i(t_i'|t) \leq U_i(t_i|t)$, for all $t_i' > t_i$, and for all t_{-i}. Analogous reasoning applies for all underreports $t_i' < t_i$. $\qquad \square$

Finally, consider the problem of a seller that seeks to maximize revenues, but cannot prevent resale. Ausubel and Cramton (1999) show that a seller faced with a perfect resale market cannot gain by misassigning goods. The best the seller can hope to do is to assign the goods efficiently, perhaps withholding quantity. This result requires independent types, so that the optimal auction program is well specified and a general revenue equivalence theorem holds.

Theorem 2 states that any monotonic aggregate quantity rule and associated monotonic efficient assignment rule can be implemented with a Vickrey auction with reserve pricing. This suggests that a revenue-maximizing seller then can optimize over all monotonic aggregate quantity rules to attain the upper bound on revenues given by the resale-constrained auction program in Ausubel and Cramton (1999). Indeed, this is the case provided the Vickrey auction with reserve pricing holds the lowest type ($t_i = 0$) of every bidder to a payoff of zero. To see this, note that $\hat{t}_i(t_{-i}, y) = 0$ for all t_{-i} and $y \in [0, q_i^*(0, t_{-i})]$, so that the lowest type's payment $X_i^*(0, t_{-i})$ is exactly equal to the value it gets from $q_i^*(0, t_{-i})$. Hence, we have:

Corollary. *With independent types, the Vickrey auction with reserve pricing attains the upper bound on revenues in the resale-constrained auction program.*

5 Conclusion

A Vickrey auction with reserve pricing has two main advantages. First, it assigns goods efficiently. Efficiency is especially important in auction markets with resale, since the apparent revenue benefits from misassignment are undermined by resale. Second, it allows the seller to withhold supply and set reserve prices to improve revenues. The use of reserve prices is especially important when competition is weak and the bidders are asymmetric. It is also important in auctions of multiple identical items, where one or more of the bidders purchases a significant share of the goods.

We have extended the Vickrey auction to include reserve pricing in a multi-unit setting with interdependent values. Truthful bidding remains an *ex post* equilibrium despite the fact that the seller varies the quantity based on the bids. This efficient outcome is robust to the possibility of resale. So long as the resale game satisfies a natural extension of individual rationality, truthful bidding followed by no resale is an equilibrium in the auction-plus-resale game. Moreover, if resale is efficient, then the Vickrey auction with appropriate reserve pricing is the optimal auction. No alternative auction can yield higher revenues.

A practical difficulty of using Vickrey pricing when auctioning multiple items is that identical items sell for different prices. Worse, large winners tend to pay lower average prices than small winners. This fact is an unavoidable implication of achieving efficiency. Large bidders have a greater incentive to reduce demands than small bidders. Hence, efficient pricing must reward large bidders for bidding their true demands by letting large bidders win the efficient quantity at lower average prices. In contrast, uniform pricing necessarily leads to an inefficient assignment (Ausubel and Cramton, 2002), and hence to suboptimal revenues when resale is efficient.

Participants in many actual markets voice a strong preference for uniform pricing (Wilson, 2002). Often the case for uniform pricing is made on efficiency grounds, and the case against Vickrey pricing is based on examples of lost revenue. With diminishing marginal valuations, these arguments have little merit. On either efficiency or revenue grounds, a Vickrey auction with reserve pricing should be preferred.

References

Ausubel, L.M.: A mechanism generalizing the Vickrey auction. University of Maryland Working Paper (1999)

Ausubel, L.M., Cramton, P.: The optimality of being efficient. University of Maryland Working Paper (1999)

Ausubel, L.M., Cramton, P.: Demand reduction and inefficiency in multi-unit auctions. University of Maryland Working Paper 96-07 (2002)

Back, K., Zender, J.F.: Auctions of divisible goods: on the rationale for the treasury experiment. Review of Financial Studies **6**, 733–764 (1993)

Back, K., Zender, J.F.: Auctions of divisible goods with endogenous supply. Economics Letters **73**, 29–34 (2001)

Calzolari, G., Pavan, A: Monopoly with resale. Northwestern University Working Paper (2002)

Crémer, J., McLean, R.P.: Optimal selling strategies under uncertainty for a discriminating monopolist when demands are interdependent. Econometrica **53**, 345–361 (1985)

Dasgupta, P., Maskin, E.: Efficient auctions. Quarterly Journal of Economics **115**, 341–388 (2000)

Haile, P.A., Auctions with resale. University of Wisconsin Working Paper (1999)

Haile, P.A., Auctions with private uncertainty and resale opportunities. Journal of Economic Theory **108**, 72–110 (2003)

Jehiel, P., Moldovanu, B.: Efficient design with interdependent valuations. Econometrica **69**, 1237–1259 (2001)

Lengwiler, Y.: The multiple unit auction with variable supply. Economic Theory **14**, 373–392 (1999)

Maskin, E.: Auctions and privatization. In: Siebert, H. (ed.) Privatization: Symposium in Honor of Herbert Giersch, pp. 115–136. Tübingen: Mohr (Siebeck) 1992

McAdams, D.: Modifying the uniform-price auction to eliminate 'collusive-seeming equilibria'. MIT Working Paper (2002)

McMillan, J.: Selling spectrum rights. Journal of Economic Perspectives **8**, 145–162 (1994)

Perry, M., Reny, P.J., An efficient auction. Econometrica **70**, 1199–1212 (2002)

Vickrey, W.: Counterspeculation, auctions, and competitive sealed tenders. Journal of Finance **16**, 8–37 (1961)

Wilson, R.: Architecture of power markets. Econometrica **70**, 1299–1340 (2002)

Zheng, C.Z.: Optimal auction with resale. Econometrica **70**, 2197–2224 (2002)

Incentives in market games
with asymmetric information: the core[*]

Beth Allen[**]

Department of Economics, University of Minnesota, 1035 Heller Hall, 271 – 19th Avenue South, Minneapolis, MN 55455, USA (e-mail: assist@econ.umn.edu)

Received: December 26, 2001; revised version: June 11, 2002

Summary. This paper examines the *ex ante* core of a pure exchange economy with asymmetric information in which state-dependent allocations are required to satisfy incentive compatibility. This restriction on players' strategies in the cooperative game can be interpreted as incomplete contracts or partial commitment. An example is provided in which the incentive compatible core with nontransferable utility is empty; the game fails to be balanced because convex combinations of incentive compatible net trades can violate incentive compatibility. However, randomization of such strategies leads to *ex post* allocations which satisfy incentive compatibility and are feasible on average. Hence, convexity is preserved in such a model and the resulting cooperative games are balanced. In this framework, an incentive compatible core concept is defined for NTU games derived from economies with asymmetric information. The main result is nonemptiness of the incentive compatible core.

Keywords and Phrases: Incentive compatability, Core, NTU cooperative games, General equilibrium with asymmetric information, Balancedness.

JEL Classification Numbers: D82, D71, D51, D79.

---

    * This work was financed, in part, by contract No 26 of the programme "Pôle d'attraction interuniversitaire" of the Belgian government, and, in part, by research grant SBR93-09854 from the U.S. National Science Foundation. Much of my thinking about this topic was developed during a wonderful visit to CORE for the 1991–1992 academic year (on sabbatical from the University of Pennsylvania). This paper was originally circulated in December 1991 as CARESS Working Paper #91-38, Center for Analytic Research in Economics and the Social Sciences, Department of Economics, University of Pennsylvania and in February 1992 as CORE Discussion Paper 9221, Center for Operations Research and Econometrics, Université Catholique de Louvain, Louvain-la-Neuve, Belgium.

    ** At the very start of my research, Jean-François Mertens was almost a co-author. François Forges provided detailed comments at a later stage, during my visit to THEMA, Université Cergy-Pontoise, in Spring 1997. They are entitled to the customary disclaimer.

1 Introduction

Incentive considerations are interesting in cooperative games because they provide an argument that agreements made within coalitions are self-enforcing. For example, if contracts are incomplete,[1] a fully credible verification and enforcement arrangement may be lacking. More generally, commitments may be only partially enforceable. Games with partial commitment power define an interesting territory between cooperative and noncooperative game theory.

The introduction of information into cooperative game theory was initiated by Wilson (1978) and extended, in the standard Harsanyi framework of games with incomplete information, by Myerson (1984) and Rosenmüller (1990). A series of more recent papers – i.e., as initiated by Yannelis (1991) and Allen (1991) – has examined the (NTU) core with asymmetric information, but without incentive compatibility requirements.

On the other hand, Myerson (1984) and Rosenmüller (1990) do discuss incentives, but do not analyze any core concept. Allen (1993) examines verification – the notion that a player's or coalition's strategy must be measurable with respect to the information available to other players – but this is very different from incentive compatibility. Krasa and Yannelis (1994) define the (NTU) value with incentive compatibility for a simple class of economies with private information sharing and (essentially) finitely many states of the world. Koutsougeras and Yannelis (1995) find core allocations (with private, fine, and the balanced cover of coarse information sharing) and check that private information sharing allocations satisfy incentive compatibility. However, they do not require incentive compatibility for blocking allocations. [Allen (1995 and 1999, respectively) demonstrates positive existence results for the (NTU) incentive-compatible core in large economies under either a relaxation to an approximate core concept or a dispersion hypothesis.]

Recent attention has focused on a variety of definitions of coalitional incentive compatibility. For an overview of part of this large and growing literature, see Allen and Yannelis (2001) and the references cited there.

For a class of taxation problems having a very special structure, Berliant (1992) defines and analyzes the incentive compatible core; he argues that it may, in general, be empty. In the context of financial intermediation, Boyd and Prescott (1986) and Kahn (1987) find incentive compatible core allocations directly. Marimon (1989) examines an asymmetric information core with adverse selection.

However, all of this work has failed to define the incentive compatible core in general economies with asymmetric information and to analyze when the incentive compatible core is nonempty. The work of Berliant (1992) suggests that problems can be expected to arise in general, but his model is sufficiently different from (for instance) the general exchange economy paradigm that such a conclusion is not fully justified.

In this paper, I formulate the incentive compatible core for exchange economies in which agents with asymmetric information have state-dependent utilities and initial endowments. Incentive compatibility is taken to mean that players' strategies (in

---

[1] See, for example, Section 3 of Hart and Holmstrom (1977) and the article by Hart and Moore (1988).

the induced cooperative game with nontransferable utility) satisfy a self-selection constraint that the state-dependent net trade must give rise to allocations that are at least as good as those for the state-dependent net trade of another state of the world. Here "at least as good" refers to the true state-dependent utility function if the player is perfectly informed and to conditional expected utility given his information otherwise. I focus on games with nontransferable utility because this is the appropriate setting for the study of incentives. Incentive compatibility or truthful revelation does not pose a problem when the goal of all coalition members is to maximize a single objective function.

My notion of incentive compatibility should be interpreted as a restriction on players' strategies rather than a specification of what happens if a lie occurs. Indeed, I do not emphasize mechanisms to define allocations when one or more players lie about their information. Instead, one performs the thought experiment in which the player receives the net trade corresponding to whatever information he announces, even though all of these net trades may not be feasible for the whole economy (unless all players lie in an internally consistent way). Any state-dependent net trade commitments that give anyone a strict incentive to lie are eliminated from the set of strategies that players may use. For example, a contract may detail outcomes based on the information provided by a party to the contract but may be independent of the information given by others in similar contracts. Any strategy that is not self-enforceable in the sense described above is simply not believed to be a credible commitment by other members of a coalition.

Existence of allocations in the incentive compatible core requires a weakening of the usual solution concept. Convex combinations of incentive compatible net trades may violate incentive compatibility, so that the usual argument to show balancedness of cooperative games derived from exchange economies fails, even though traders are assumed to have concave utilities. To bypass this obstacle, I introduce a randomization.[2] Incentive compatible core allocations thus consist of probabilities over incentive compatible state-dependent net trades such that the actual state-dependent allocation is, for almost all states of the world, feasible on average with respect to the randomization.

The next section furnishes an example of a pure exchange economy with three traders, five commodities, and five states of the world in which incentive compatibility leads to nonconvex strategy sets. As a result, the game is not balanced and its (NTU) incentive compatible core is empty. This demonstrates that the randomization is indeed necessary. The example further shows that resource feasibility must be expressed in terms of averages over the randomization within every state rather than with probability one with respect to the randomization.

The necessity for weakening the resource feasibility requirement suggests that perhaps the phenomenon should be viewed as a negative result. While this argument indeed has its rationale, asset markets, inventories and the law of large numbers could be used to justify my average feasibility condition. Nevertheless, the existence

---

[2] This is reminiscent of the work of Prescott and Townsend (1984a,b), who analyze competitive equilibria and Pareto optimality in large economies with asymmetric information and lotteries or many independent risks.

of such incentive compatible core allocations should be interpreted with caution. On the other hand, my analysis has the virtue of providing a cooperative game-theoretic explanation of the endogenous and strategic choice of a mechanism in the presence of asymmetric information. The choice of a cooperative solution point can be interpreted as the choice of a mechanism. While the core cannot be expected to contain, in general, a unique incentive-compatible state-dependent allocation (and hence to correspond to a unique equivalence class of mechanisms), allocations or mechanisms outside of my core can be blocked with incentive compatible outcomes so that points not in the core can be eliminated from consideration when strategic agents can cooperate within coalitions.

The remainder of this paper is organized as follows: Section 2 gives the example illustrating that incentives can destroy convexity, balancedness, and nonemptiness of the set of incentive compatible core allocations. Section 3 presents the model. Incentive constraints and randomizations are discussed in Sections 4 and 5 respectively, while Section 6 presents and proves the main result. Section 7 concludes with some remarks.

2 The example

This section presents an example which demonstrates the main ideas of the paper. The example illustrates a pure exchange economy under uncertainty in which the induced strategy sets of the cooperative game fail to be convex due to incentive compatibility constraints. The set of utility vectors attainable by the grand coalition is also nonconvex, although the nonconvexities in payoff space occur where they are inessential for the existence of points in the core. The nonconvexity of incentive-compatible strategy sets leads to a game with nontransferable utility which is not balanced and has an empty (NTU) core.

However, the introduction of randomizations over allocations restores convexity, balancedness, and nonemptiness of the core at the expense of weakening feasibility requirements. Almost sure resource feasibility is replaced by feasibility, in each state of the world, of (state-dependent) allocations on average with respect to the randomization. While randomization over almost surely feasible allocations would convexify the sets of attainable payoffs, it does not guarantee either balancedness or the existence of core allocations. Hence the original game with incentive constraints derived from my example fails to have convex feasible payoff sets, has convex strategy sets, is not balanced, and has an empty core. Randomization with almost sure feasibility is sufficient to ensure convexity of feasible payoffs, but does not solve any of the other problems. Randomization with mean feasibility enlarges the attainable utility sets further and suffices to convexify the underlying strategy sets so as to give a balanced game which therefore has a nonempty core. Yet removal of the incentive compatibility constraints would alter the game still more and lead to larger core payoffs so that my randomization operation is definitely not equivalent to the elimination of incentive compatibility.

The example features three traders, five states of the world, and five commodities, of which the first two matter in the first two states while each of the remaining goods is of consequence in exactly one state. All states are equally likely, so that total utility is proportional to expected utility (which slightly simplifies the calculations). Let $\Omega = \{a, b, c, d, f\}$ denote the set of states of the world. Use subscripts $i = 1, 2, 3$ to distinguish players, and let commodities be indicated by x, y, z, r, and t. Nonnegative amounts of all goods can be consumed in every state $s \in \Omega$.

Traders' initial endowment vectors do not depend on the state of the world in this pure exchange example. Write $e_i(s) \in \mathbb{R}^5_+$ for player i's initial endowment in state $s \in \Omega$ and suppose that $e_1(s) = (1, 1, 3, 0, 0)$, $e_2(s) = (0, 4, 0, 1, 0)$, and $e_3(s) = (4, 0, 0, 0, 1)$ for all $s = a, b, c, d, f$.

The second and third economic agents have almost identical state-dependent cardinal utilities. They are given by $u_2(x, y, z, r, t; a) = \sqrt{x}$, $u_3(x, y, z, r, t; a) = \sqrt{y}$, $u_2(x, y, z, r, t; b) = -\infty I(y < 4)$, $u_3(x, y, z, r, t; b) = -\infty I(x < 4)$, $u_2(x, y, z, r, t; c) = u_3(x, y, z, r, t; c) = z$, $u_2(x, y, z, r, t; d) = r$, $u_3(x, y, z, r, t; f) = t$ and $u_3(x, y, z, r, t; d) = u_2(x, y, z, r, t; f) = 0$.[3]

Trader 1 has rather complicated utility functions in states a and b, but straightforward utilities in the remaining three states. Let $u_1(x, y, z, r, t; c) = z/3$, $u_1(x, y, z, r, t; d) = r/2$, $u_1(x, y, z, r, t; f) = t/2$, and $u_1(x, y, z, r, t; b) = (x + y)/2$ if $x + y \leq 2$, $u_1(x, y, z, r, t; b) = 1$ if $x = 0$ and $y \geq 2$ or if $y = 0$ and $x \geq 2$, and set $u_1(x, y, z, r, t; b) > 1$ if $x > 0$, $y > 0$, and $x + y > 2$ so as to be concave and continuous. Finally, for state a, define[4] $u_1(x, y, z, r, t; a) = \frac{1}{4}\min(x + y - 1 + \sqrt{(x + y - 1)^2 + 8x}, x + y - 1 + \sqrt{(x + y - 1)^2 + 8y})$. Note that, in particular, utility along each axis in the plane is given by $(y - 1)/2$ if $x = 0$ or $(x - 1)/2$ if $y = 0$. This utility function is more easily described by its utility level m indifference surfaces as the union of the line segment joining $(0, 2m + 1)$ and (m, m) and the line segment joining (m, m) and $(2m + 1, 0)$ in the $x - y$ plane, so that the points $(3, 0)$, $(0, 3)$ and $(1, 1)$ are associated with a utility level of 1, $(4, 0)$, $(0, 4)$ and $(3/2, 3/2)$ have utility $3/2$, $(5, 0)$, $(0, 5)$ and $(2, 2)$ give utility 2, and $(7, 0)$, $(0, 7)$, and $(3, 3)$ give utility 3 regardless of trader 1's allocation of commodities z, r, and t. Note that this utility function is concave and, moreover, is the least concave representation for these indifference curves because it is linear along the diagonal.

Writing payoffs as total utilities (summed over the five states) gives 3 as the individually rational utility level for player 1 and 1 as the individual rationality constraint for players 2 and 3; these are the total utilities associated with each player's initial endowment. By interchanging their initial endowments in state a, the coalition composed of players 2 and 3 can achieve utilities of 3 each; this presumes that they can distinguish state a from state b.

With complete information and in the absence of any incentive compatibility considerations, the (total utility) imputation $(5, 3.5, 3.5)$ belongs to the core. It can be achieved with the state-dependent allocations $((4, 4, 3, 0, 0), (1, 1, 3, 0, 0), (1, 1, 0, 0, 0), (1, 1, 3, 0, 0), (1, 1, 3, 0, 0))$ for player 1, $((1, 0, 0, 1, 0), (0, 4, 0, 1, 0),$

---

[3] Define $\infty \cdot 0 = 0$ and $\infty \cdot 1 = \infty$, so that $-\infty \cdot I(y < 4) = -\infty$ if $y < 4$, $-\infty \cdot I(y < 4) = 0$ if $y \geq 4$, $-\infty \cdot I(x < 4) = -\infty$ if $x < 4$ and $-\infty \cdot I(x < 4) = 0$ if $x \geq 4$, where $I(\cdot)$ denotes the indicator function of a set.

[4] A discussion with Heraklis Polemarchakis helped me to transform the desired indifference curves into this utility function.

$(0, 4, 1.5, 1, 0)$, $(0, 4, 0, 1, 0)$, $(0, 4, 0, 1, 0))$ for player 2, and $((0, 1, 0, 0, 1)$, $(4, 0, 0, 0, 1)$, $(4, 0, 1.5, 0, 1)$, $(4, 0, 0, 0, 1)$, $(4, 0, 0, 0, 1))$ for player 3. These outcomes are individually rational and cannot be blocked by the coalition $\{2, 3\}$. Clearly the allocation is Pareto optimal and hence cannot be blocked by the grand coalition. Finally, notice that it cannot be blocked by $\{1, 2\}$ or $\{1, 3\}$, as these coalitions can, at best, achieve utility imputations such as $(3.94, 3.5)$, $(5.19, 0)$ or $(4.95, 0.7)$.

Incentive compatibility implies that the above core allocation (as well as the latter two imputations for coalitions $\{1, 2\}$ or $\{1, 3\}$) cannot be sustained. Indeed, player 1 prefers his allocation in state a to his allocation in state b when the true state of the world is b. The role of the particular indifference curve specified for player 1 in state b is to demonstrate that incentives destroy convexity of sets of feasible net trades. Note that $(5, 0, 0, 0, 0)$ or $(0, 5, 0, 0, 0)$ in state a with $(1, 1, 0, 0, 0)$ in state b is an incentive compatible allocation for trader 1, but the convex combination of $(2.5, 2.5, 0, 0, 0)$ in state a violates incentive compatibility if trader 1 receives $(1, 1, 0, 0, 0)$ in state b, as he would then always claim that the state is a even when b is true. In this way, incentive considerations change the cooperative game in characteristic function form that is derived from this economy.

An easy way to see why this is true in general is to observe that if there were only one commodity and if all agents' state-dependent utilities were strictly increasing in that commodity, then the only incentive compatible allocations are those that give, to each trader, identical amounts of the good in each state that occurs with positive probability. Clearly this reduces the possibilities for risk sharing and for gains from trade based on different preferences or subjective probabilities.

To construct the cooperative game $V: 2^I \to \mathbb{R}^3$ in characteristic function form (where $I = \{1, 2, 3\}$ is the player set) derived from my economic example with incentive compatibility constraints, examine efficient, incentive compatible and feasible state-dependent commodity allocations and their associated total utility payoff vectors for each coalition. In order to do so, I must first specify the information of each player. To set notation, let $v: 2^I \to \mathbb{R}$ be the associated game with transferable utility, so that $v(S)$ is the total worth of coalition S [set $v(\emptyset) = 0$], and if T is any subset of some Euclidean space, let $\text{comp}(T)$ denote its comprehensive hull, where $\text{comp}(T) = T - \mathbb{R}_+^n$ if T is considered as a subset of $\mathbb{R}^n$.

Information will be defined for each agent so as to make player 2 or player 3 need another player in order to be able to distinguish state a from state b. Player 1 knows the precise state of the world – i.e., his information is specified by the partition $\{\{a\}, \{b\}, \{c\}, \{d\}, \{f\}\}$ of S. Players 2 and 3 each receive a signal drawn randomly, with equal probabilities, from the set $\{0, 1, c, d, f\}$. These signals are correlated with states of the world as follows: If the signal received by a player is c, d, or f, then the same signal is received by the other player and that (common) signal equals the state of the world. Conditional on the state (and signal) not being equal to c, d, or f, the combinations $00, 01, 10$ and 11 are equally likely. If both players receive the same signal, the true state is a, while if they receive different signals, the state is b. Thus, player 2 or player 3 has the information partition $\{\{a, b\}, \{c\}, \{d\}, \{f\}\}$ while all other coalitions, including $\{2, 3\}$, have the complete information partition $\{\{a\}, \{b\}, \{c\}, \{d\}, \{f\}\}$.

Since incentive compatibility does not affect initial endowments – indeed, agents' endowments in the example do not depend on the state of the world – the worth of any singleton equals its worth in the game without incentive constraints. Hence $v(\{1\}) = \sum_{s\in\Omega} u_1(1,1,3,0,0;s) = 1+1+1+0+0 = 3$, $v(\{2\}) = \sum_{s\in\Omega} u_2(0,4,0,1,0;s) = 0+0+0+1+0 = 1$, and $v(\{3\}) = \sum_{s\in\Omega} u_3(4,0,0,0,1;s) = 0+0+0+0+1 = 1$ in the TU game while $V(\{1\}) = \text{comp}(v(\{1\})) \times \mathbb{R}^2$, $V(\{2\}) = \mathbb{R} \times \text{comp}(v(\{2\})) \times \mathbb{R}$, and $V(\{3\}) = \mathbb{R}^2 \times \text{comp}(v(\{3\}))$ in the NTU game. Alternatively, $V(\{1\}) = \{p \in \mathbb{R}^3 \mid p_1 \leq 3\}$, $V(\{2\}) = \{p \in \mathbb{R}^3 \mid p_2 \leq 1\}$, and $V(\{3\}) = \{p \in \mathbb{R}^3 \mid p_3 \leq 1\}$. Clearly these are nonempty closed convex comprehensive cylinder sets.

The efficient feasible allocations for $\{2,3\}$ are defined by giving all of good x to player 2 in state a (giving him a utility of $\sqrt{4} = 2$ in state a) and all of good y to player 3 in state a (for a utility of $\sqrt{4} = 2$ in state a). Goods x and y are not traded in state b, nor are r and t traded in states d and f. The resulting allocations are incentive compatible and assign three units of total utility to each player in the coalition $\{2,3\}$. Therefore, $v(\{2,3\}) = 6$ and $V(\{2,3\}) = \mathbb{R} \times \text{comp}((3,3)) = \{p \in \mathbb{R}^3 \mid p_2 \leq 3$ and $p_3 \leq 3\}$. Note that this coalition uses its combined (complete) information and neither player can gain from cheating in information revelation.

Now consider the remaining two-player coalitions, $\{1,2\}$ and $\{1,3\}$. As players 2 and 3 play symmetric roles here, it suffices to examine the incentive compatible utility imputations of one of these coalitions. Individual rationality precludes trades in state b. Therefore the (unique) efficient and incentive compatible allocation in state a gives all five units of y to player 1 and one unit of x to player 2. In state c, the three units of commodity z can be divided arbitrarily between the two players and similarly for the one unit of good r in state d. Satiation in utilities implies that state-dependent allocations of all other goods do not matter providing that they do not cause violations of incentive compatibility. Hence $V(\{1,2\}) = \{p \in \mathbb{R}^3 \mid p_1 \leq 4.5 - z/3$ and $p_2 \leq 1 + z$ for $z \in [0,3]$ or $p_1 \leq 3.5 - r/2$ and $p_2 \leq 4 + r$ for $r \in [0,1]\}$. Similarly, $V(\{1,3\}) = \{p \in \mathbb{R}^3 \mid p_1 \leq 4.5 - z/3$ and $p_3 \leq 1 + z$ for $z \in [0,3]$ or $p_1 \leq 3.5 - t/2$ and $p_3 \leq 4 + t$ for $t \in [0,1]\}$ Therefore, in the game with transferable utility, $v(\{1,2\}) = \max\{p_1 + p_2 \mid p \in V(\{1,2\})\} = 8$ and $v(\{1,3\}) = \max\{p_1 + p_3 \mid p \in V(\{1,3\})\} = 8$. Note that both $V(\{1,2\})$ and $V(\{1,3\})$, like $V(\{2,3\})$, are convex sets.

Finally, the grand coalition $\{1,2,3\} = I$ is also restricted by incentive compatibility. Individual rationality for players 2 and 3 implies that no trades occur in state b. This forces player 1 in state a to receive either 1 unit each of x and y or to receive an allocation consisting of zero units of either x or y and a quantity between three units and five units (inclusive) of the other good. If player 1 keeps his one unit of x and one unit of y while players 2 and 3 interchange their initial endowments of x and y in state a, the sum total utilities received by players is maximized. This defines worth $v(I) = 11$ of the grand coalition in the game with transferable utility. The corresponding nontransferable utility imputations are specified by $(2 + z_1/3 + r_1/2 + t_1/2, 2 + z_2 + r_2, 2 + z_3 + t_3)$ where $z_1 + z_2 + z_3 = 3$, $r_1 + r_2 = 1$, $t_1 + t_3 = 1$, and $z_1, z_2, z_3, r_1, r_2, t_1, t_3$ are all nonnegative. However, the imputations of this form that are individually rational for player 1 $[v(\{1\}) = 3]$ are dominated by those based on giving all five units of x or y to player 2 or 3 respectively and the remainder to player 1 in state a. This yields imputations of the form $(3 + z_1/3 + r_1/2 + t_1/2, 2.236 + z_2 + r_2, 0 + z_3 + t_3)$ and $(3 + z_1/3 + r_1/2 + t_1/2, 0 + z_2 + r_2, 2.236 + z_3 + t_3)$ where $z_1 + z_2 + z_3 = 3$, $r_1 + r_2 = 1$, $t_1 + t_3 = 1$, and $z_1, z_2, z_3, r_1, r_2, t_1, t_3$ are all nonnegative. Inspection of the imputations $(5, 2.236, 0)$ and $(5, 0, 2.236)$ shows that $V(I)$ is not convex; the utility vector $(5, \frac{1}{2}\sqrt{5}, \frac{1}{2}\sqrt{5})$ cannot be realized by incentive compatible allocations.

Recapitulation of the game with transferable utility gives $v(\{1\}) = 3, v(\{2\}) = v(\{3\}) = 1, v(\{1,2\}) = v(\{1,3\}) = 8, v(\{2,3\}) = 6$, and $v(\{1,2,3\}) = 11$. This game is balanced and (equivalently, by the theorem of Bondareva (1962) and Shapley (1967)) has a nonempty core. The easiest way to see this is to observe that $(5, 3, 3)$ is the unique core imputation for the TU game derived from my example with incentive compatibility constraints. However, notice that $(5, 3, 3) \notin V(I)$; this utility vector cannot be attained in the game with nontransferable utility unless incentive compatibility is violated (in which case $(5, 3.5, 3.5)$ can be achieved); the total utility of 11 in the NTU game uses the imputation $(2, 4.5, 4.5)$, which fails individual rationality for player 1.

A sufficient (see Scarf, 1967) but not necessary condition for nonemptiness of the (NTU) core of a cooperative game with nontransferable utility is balancedness. This means that for any balanced collection $\mathcal{B}$ of subsets S of I with balancing weights $\gamma_S \geq 0$ [where $\sum_{\substack{S \in \mathcal{B} \\ S \ni i}} \gamma_S = 1$ for all $i \in I$], $V(I) \supseteq \sum_{T \subseteq I} \gamma_T V(T)_T$, where $V(T)_T = \{w \in V(T) \mid w_i = 0 \text{ if } i \notin T\}$. A weaker condition which suffices for nonemptiness of the NTU core is quasibalancedness, or $\bigcap_{S \in \mathcal{B}} V(S) \subseteq V(I)$ for every balanced collection $\mathcal{B}$ of coalitions with nonnegative balancing weights. To see that my example does not generate a balanced game,[5] consider the collection $\{\{1,2\}, \{1,3\}, \{2,3\}\}$ of two-player coalitions with balancing weights $\frac{1}{2}$ each. If one takes $(3, 5, 0) \in V(\{1,2\})_{\{1,2\}}, (3, 0, 5) \in V(\{1,3\})_{\{1,3\}}$ and $(0, 3, 3) \in V(\{2,3\})_{\{2,3\}}$, the sum $\frac{1}{2}(3, 5, 0) + \frac{1}{2}(3, 0, 5) + \frac{1}{2}(0, 3, 3) = (3, 4, 4) \notin V(I)$, so that the game is not balanced. However, these vectors do not show that the game fails to be quasibalanced, as $(3, 3, 3) \in V(I)$. Alternatively, checking $(3\frac{5}{6}, 3, 0) \in V(\{1,2\})_{\{1,2\}}, (3\frac{5}{6}, 0, 3) \in V(\{1,3\})_{\{1,3\}}$, and $(0, 3, 3) \in V(\{2,3\})_{\{2,3\}}$ gives $\frac{1}{2}(3\frac{5}{6}, 3, 0) + \frac{1}{2}(3\frac{5}{6}, 0, 3) + \frac{1}{2}(0, 3, 3) = (3\frac{5}{6}, 3, 3) \notin V(I)$ but $(3\frac{5}{6}, 3, 3) \in V(\{1,2\}) \cap V(\{1,3\}) \cap V(\{2,3\})$ so that the game is neither balanced nor quasibalanced.[6] Unfortunately, this need not prove that the NTU core is empty.

Perhaps the easiest way to verify that the core actually is empty is to examine the maximum payoffs that players can obtain in $V(I)$ subject to certain other coalitions

---

[5] Contrast this to the standard case of a pure exchange economy without uncertainty, which necessarily generates a balanced game, as demonstrated by Scarf (1971).

[6] Almost by definition, the (NTU) game is superadditive: $V(T) \cap V(T') \subseteq V(T \cup T')$ whenever $T \cap T' = \emptyset$. This is true because resources are additive in an exchange economy and incentive compatibility is a restriction on an individual's allocation which is independent of the coalition, so that the union of disjoint coalitions can always choose any allocations that were permitted in the smaller coalitions.

being unable to block. The logic here is to observe that if the core is nonempty, then there are strictly Pareto optimal core allocations, where strict optimality means that no single agent can be made better off without making someone else worse off. (Contrast this to the definition of blocking by the grand coalition, which requires strict improvements for every player.) Requiring individual rationality for players 1 and 3 leads to the problem of maximizing w_2 subject to $(w_1, w_2, w_3) \in V(I)$ with $w_1 \geq 3$ and $w_3 \geq 1$. Its solution is $(3, 6.236, 1) \in V(I)$. However, this imputation can be blocked by the coalition $\{1, 3\}$, which can attain 4 for player 1 and 2 for player 3. The argument works the same way if the roles of players 2 and 3 are reversed. Alternatively, the condition that coalition $\{2, 3\}$ cannot block combined with individual rationality for players 2 and 3 yields the problem of maximizing w_1 subject to $(w_1, w_2, w_3) \in V(I)$ and $w_2 \geq 3$, $w_3 \geq 1$ (or, equivalently, $w_2 \geq 1$ and $w_3 \geq 3$). The solution is $(4.413, 3, 1) \in V(I)$, which can be blocked by $\{1, 3\}$ with the imputation of 4.45 to player 1 and 1.15 to player 3. This proves that all strictly Pareto optimal feasible allocations for the grand coalition can be blocked by some smaller coalition. Therefore, my example has an empty core.

Randomization over almost surely feasible state-dependent allocations satisfying incentive compatibility would convexify the sets $V(S)$ for all $S \subseteq I$. In my example, this randomization operation is thus equivalent to replacing $V(I)$ by its (closed) convex hull in the definition of the NTU game with incentive constraints. However, this does not suffice to give a balanced game. In fact, the solutions to the two optimization problems examined above would be the same if $V(I)$ were replaced by its convex hull. Hence, almost surely feasible randomization does not insure nonemptiness of the core.

On the other hand, randomization with feasibility on average in every state of the world does lead to the existence of an incentive compatible core. Indeed, for my example, giving probability $\frac{1}{2}$ to the allocation $((0, 5, 0, 1, 1), (1, 1, 0, 1, 1),$ $(1, 1, 0, 1, 1), (1, 1, 0, 1, 1), (1, 1, 0, 1, 1))$ to player 1, $((4, 0, 1, 0, 0), (0, 4, 1, 0, 0),$ $(0, 4, 1, 0, 0), (0, 4, 1, 0, 0), (0, 4, 1, 0, 0))$ to player 2 and $((0, 1, 2, 0, 0), (4, 0, 2, 0,$ $0), (4, 0, 2, 0, 0), (4, 0, 2, 0, 0), (4, 0, 2, 0, 0))$ to player 3 (so that total utilities are $4, 3$ and 3 while the total resource allocation is $(4, 6, 3, 1, 1)$ in state a and $(5, 5, 3, 1, 1)$ in all other states) and probability $\frac{1}{2}$ to the allocation $((5, 0, 0, 1, 1),$ $(1, 1, 0, 1, 1), (1, 1, 0, 1, 1), (1, 1, 0, 1, 1), (1, 1, 0, 1, 1))$ to player 1, $((1, 0, 2, 0, 0),$ $(0, 4, 2, 0, 0), (0, 4, 2, 0, 0), (0, 4, 2, 0, 0), (0, 4, 2, 0, 0))$ to player 2 and $((0, 4, 1,$ $0, 0), (4, 0, 1, 0, 0), (4, 0, 1, 0, 0), (4, 0, 1, 0, 0), (4, 0, 1, 0, 0))$ to player 3 (for total utilities of $4, 3$, and 3 again with a total allocation of $(6, 4, 3, 1, 1)$ in state a and $(5, 5, 3, 1, 1)$ in all other states) is an incentive compatible core allocation.

Note that this randomization is not equivalent to removal of incentive compatibility constraints. Without incentive compatibility, the above randomized core allocation could be blocked (and strictly Pareto improved) by the nonrandomized state-dependent allocation of $((2.5, 2.5, 0, 1, 1), (1, 1, 0, 1, 1), (1, 1, 0, 1, 1),$ $(1, 1, 0, 1, 1), (1, 1, 0, 1, 1))$ to player 1, $((2.5, 0, 1.5, 0, 0), (0, 4, 1.5, 0, 0), (0, 4,$ $1.5, 0, 0), (0, 4, 1.5, 0, 0), (0, 4, 1.5, 0, 0))$ to player 2 and $((0, 2.5, 1.5, 0, 0), (4, 0,$ $1.5, 0, 0), (4, 0, 1.5, 0, 0), (4, 0, 1.5, 0, 0), (4, 0, 1.5, 0, 0))$ to player 3 which yields total utilities of $4.5, 3.08$, and 3.08 respectively. Of course, this allocation violates incentive compatibility for player 1, who prefers his state a allocation to his state b

allocation when the true state of the world is b; the problem is clearly nonconvexity of player 1's set of incentive compatible allocations for state a given the allocation which is forced by the utilities of players 2 and 3 in state b.

It is relatively easy to find examples in which incentive compatibility violates convexity of strategy sets and balancedness, but the economy nevertheless admits nonrandom incentive compatible core allocations. Indeed, if one were to delete the last two states and two commodities from the example analyzed above, the resulting game is not balanced but has a nonempty core. In this case, $V(\{1\}) = (-\infty, 3] \times \mathbb{R}^2$, $V(\{2\}) = \mathbb{R} \times (-\infty, 0] \times \mathbb{R}$, $V(\{3\}) = \mathbb{R}^2 \times (-\infty, 0]$, and $V(\{2, 3\}) = \mathbb{R} \times \text{comp}((2, 2))$. Coalition $\{1, 2\}$ can trade to the incentive compatible allocation $((0, 5, 3), (1, 1, 3), (1, 1, 3))$ for player 1 and $((1, 0, 0), (0, 4, 0), (0, 4, 0))$ for player 2 (with total utilities of 4 and 1 respectively). Similarly, $\{1, 3\}$ can attain 4 and 1 using the allocations $((5, 0, 3), (1, 1, 3), (1, 1, 3))$ and $((0, 1, 0), (4, 0, 0), (4, 0, 0))$. However, placing balancing weights of one-half each on the balanced collections of two-player coalitions requires $\frac{1}{2}(4, 1, 0) + \frac{1}{2}(4, 0, 1) + \frac{1}{2}(0, 2, 2) = (4, 1.5, 1.5) \in V(\{1, 2, 3\})$, for balancenedness, which is false. Nor is the game quasi-balanced $[(4, 1, 1) \notin V(\{1, 2, 3\})]$, although it is superadditive. The efficient and individually rational points in $V(\{1, 2, 3\})$ are of the form $(3 + z_1/3, \sqrt{5} + z_2, z_3)$ or $(3 + z_1/3, z_2, \sqrt{5} + z_3)$ where $z_i \geq 0$ for all i and $z_1 + z_2 + z_3 = 3$. (This also shows that $V(\{1, 2, 3\})$ is not convex. Convexification of this set via extending strategy sets to include almost surely feasible random allocations does not lead to balanced game.) The imputations $(4, 2.236, 0)$ and $(4, 0, 2.236)$ belong to the incentive compatible core. They can be achieved by the incentive compatible (nonrandom) allocations $((0, 5, 3), (1, 1, 3), (1, 1, 3))$ to player 1, $((5, 0, 0), (0, 4, 0), (0, 4, 0))$ to player 2, and $((0, 0, 0), (4, 0, 0), (4, 0, 0))$ to player 3 or $((5, 0, 3), (1, 1, 3), (1, 1, 3))$, $((0, 0, 0), (0, 4, 0), (0, 4, 0))$, and $((0, 5, 0), (4, 0, 0), (4, 0, 0))$ respectively. Recall that $\{1, 2\}$ or $\{1, 3\}$ can obtain exactly 4 and 1. These two core points cannot be blocked in the convexified game with almost surely feasible random allocations, but its core also includes closed line segments with these endpoints; more precisely, the core contains all total utility vectors of the form $(4, \lambda\sqrt{5}, (1 - \lambda)\sqrt{5})$ for $\lambda \in [0, 2/\sqrt{5}] \cup [1 - 2/\sqrt{5}, 1]$. If one allows randomizations that are only feasible on average, then the imputation $(3, 3, 3)$ belongs to the core. It can be obtained from the allocation $((0, 5, 0), (1, 1, 0), (1, 1, 0))$, $((4, 0, 0), (0, 4, 0), (0, 4, 0))$, and $((0, 1, 3), (4, 0, 3), (4, 0, 3))$ with probability $\frac{1}{2}$ (for a total allocation of $(4, 6, 3)$ in state a and $(5, 5, 3)$ in states b and c) and $((5, 0, 0), (1, 1, 0), (1, 1, 0))$, $((1, 0, 3), (0, 4, 3), (0, 4, 3))$, and $((0, 4, 0), (4, 0, 0), (4, 0, 0))$ with probability $\frac{1}{2}$ (for a total allocation of $(6, 4, 3)$ in state a and $(5, 5, 3)$ in states b and c). The (nonrandomized) transferable utility version of this example with three states has $v(\{1\}) = 3$, $v(\{2\}) = v(\{3\}) = 0$, $v(\{1, 2\}) = v(\{1, 3\}) = 7$, $v(\{2, 3\}) = 4$, and $v(\{1, 2, 3\}) = 9$. Its core equals the imputation $(5, 2, 2)$ which cannot be achieved in the nontransferable utility game with incentive compatibility constraints.

For a somewhat more dramatic example of the difference between balancedness and the existence of core allocations with incentive compatibility, remove the third state and third commodity from the preceding example – i.e., take only the first two states and two good in the original example. Then $V(\{1\}) = (-\infty, 2] \times \mathbb{R}^2$,

$V(\{2\}) = \mathbb{R} \times \mathbb{R}_{-} \times \mathbb{R}$, $V(\{3\}) = \mathbb{R}^2 \times \mathbb{R}_{-}$, $V(\{2,3\}) = \mathbb{R} \times (-\infty, 2] \times (-\infty, 2]$, $V(\{1,2\}) = (-\infty, 3] \times (-\infty, 1] \times \mathbb{R}$, and $V(\{1,3\}) = (-\infty, 3] \times \mathbb{R} \times (-\infty, 1]$. The game is not balanced because $\frac{1}{2}(3,1,0) + \frac{1}{2}(3,0,1) + \frac{1}{2}(0,2,2) = (3, 1.5, 1.5) \notin V(\{1,2,3\})$, nor is it quasibalanced as $(3, 1, 1) \notin V(\{1, 2, 3\})$. On the other hand, it is superadditive; to check this directly verify that the points $(2,2,2)$, $(3,1,0)$, $(3,0,1)$, and $(2,0,0)$ all belong to $V(\{1,2,3\})$. The transferable utility version of the game has $v(\{1\}) = 2$, $v(\{2\}) = v(\{3\}) = 0$, $v(\{1,2\}) = v(\{1,3\}) = v(\{2,3\}) = 4$, and $v(\{1,2,3\}) = 6$. The TU core consists of the (unique) point $(2,2,2)$, which can be obtained under an incentive compatible allocation $[((1,1),(1,1)), ((4,0),(0,4)), ((0,4)(4,0))]$ which belongs to the NTU core. However, every Pareto optimal and individually rational imputation of the NTU game with incentive compatibility constraints belongs to its core, including also the imputations $(2.5, \sqrt{5}, 1), (2.5, 1, \sqrt{5}), (2, \sqrt{5}, \sqrt{2}), (2, \sqrt{2}, \sqrt{5})$, $(3, \sqrt{5}, 0)$, and $(3, 0, \sqrt{5})$.

Finally, a further remark regarding the original example is in order. It can be perturbed slightly so as to give a pure exchange economy having strictly positive initial endowments (which are constant across the world) and strictly concave and continuous – or even smooth – utility functions for which the incentive compatible core is empty.

3 The model

To begin, let Ω be a finite set of states of the world (with typical element ω) and let μ be a probability on $(\Omega, \mathcal{F})$ where $\mathcal{F} = 2^{\Omega}$. For convenience, assume that the (subjective) probability measure μ is the same for all agents and that $\mu(\omega) > 0$ for all $\omega \in \Omega$ (otherwise reduce Ω by a μ-null set). Interpret Ω as a description of all of the relevant uncertainty in the economy, where points in Ω represent systematic risk or states of the world common to all agents. Alternatively, think of Ω as the set of all possible profiles of agents' types. Note that if Ω is infinite but each agent's information consists of a finite partition of Ω, one could redefine a finite set of states of the world by events in the pooled information partition.

Assume that there is a finite number, ℓ, of commodities potentially available in each state of the world. Take $\mathbb{R}_{+}^{\ell}$ to be the consumption set of each consumer in each state of the world.

Finitely many economic agents are present in my pure exchange economy. Let I be the set of traders (or players in the induced games), write #I for its cardinality, and use subscript i ($i \in I$) to signify a typical individual agent.

Each trader is endowed with a nonnegative vector $e_i \in \mathbb{R}_{+}^{\ell}$ of commodities. For simplicity, these initial allocations are assumed to be constant as states of the world vary. Thus, an agent's initial endowment is always incentive compatible and does not contain any information. Nor does the economy's total resource endowment vector contain any information. If incentive compatibility were violated for state-dependent endowments, the worth of a singleton might not be well defined as the utility level that an agent can guarantee itself should be attainable with an incentive compatible allocation if incentive compatibility is required for all other coalitions.

Preferences are described by state-dependent cardinal utility functions. Expected utilities define payoffs. For $i \in I$, write $u_i \colon \mathbb{R}_+^\ell \times \Omega \to \mathbb{R}$ and assume that, for all $\omega \in \Omega$, $u_i(\cdot; \omega)$ is a continuous and concave function on $\mathbb{R}_+^\ell$. (For the generalization to infinitely many basic states of the world, $\mathcal{F}$-measurability in ω and continuity on $\mathbb{R}_+^\ell$ imply joint measurability of this mapping.)

The data of the pure exchange economy under uncertainty generates a cooperative game. The correspondence $V \colon 2^I \to \mathbb{R}^{\#I}$ defines a cooperative game with nontransferable utility if $V(\emptyset) = \{0\}$ and for all $S \subseteq I, S \neq \emptyset, V(S)$ is a nonempty closed comprehensive cylinder set. In the absence of incentive compatibility constraints, my economic model generates a balanced game with a nonempty core.[7]

4 Information and incentives

With asymmetric information, the incentive compatible core of a pure exchange economy under uncertainty should consist of exactly those feasible incentive compatible state-dependent allocations that cannot be blocked by any coalition. Here blocking requires strict improvement in expected utility using state-dependent allocations that are incentive compatible and feasible for the coalition. Equivalently, one could define the NTU game generated by the economy with incentive compatibility constraints and examine the core of the induced game. Analysis of either concept requires stating the information that each trader possesses and formulating the appropriate incentive compatibility restriction.

To specify traders' information, for each $i \in I$, let S_i be a finite set and let $\mathbf{s}_i \colon \Omega \to S_i$ be a function. Then $\mathbf{s}_i$ generates a finite partition P_i of Ω and a finite sub-σ-field $\mathcal{F}_i$ of $\mathcal{F}$. Interpret S_i as the set of signals that i can receive about the state of the world and P_i or $\mathcal{F}_i$ as i's initial information. Think of the sets S_i and the maps $\mathbf{s}_i \colon \Omega \to S_i$ for all $i \in I$ as common knowledge for all agents (and to the planner or mechanism designer). Take S_i also to be the set of messages that agents can implicitly communicate. In other words, an agent can convey a (true or false) subset of his actual information partition. Note that the realizations $\mathbf{s}_i(\omega)$ of i's signal are not observable to agents other than i (or to the planner or mechanism designer). Note also that "random signals" are allowed in this model in that otherwise identical "copies" of $\omega \in \Omega$ could be mapped into different elements of S_i, so that this is equivalent to expanding Ω to a larger finite set. Let $S = S_1 \times \cdots \times S_{\#I}$ with typical element $s = (s_1, \ldots, s_{\#I})$ and write $(s_i', s_{-i}) = (s_1, \ldots, s_{i-1}, s_i', s_{i-1}, \ldots s_{\#I}) \in S$ and $s_{-i} \in S_{-i} = \prod_{j \neq i} S_j$. Let $\mathbf{s} \colon \Omega \to S$ be defined by $\mathbf{s}(\omega) = (\mathbf{s}_1(\omega), \ldots, \mathbf{s}_{\#I}(\omega))$ and define $\mathbf{s}_{-i}$ in the obvious way. Write $\mu(\cdot \mid s_i)$ and $\mu(\cdot \mid s)$ for the conditional probabilities on Ω given $s_i \in S_i$ or $s \in S$ respectively. The distribution μ on Ω and the map $\mathbf{s} \colon \Omega \to S$ induce a distribution ν on S. For $i \in I$, denote its conditional distribution on S_{-i} given $s_i \in S_i$ by $\nu_i(\cdot \mid s_i)$.

A state-dependent allocation $x_i \colon \Omega \to \mathbb{R}_+^\ell$ for trader $i \in I$ is *strongly incentive compatible* if $x_i(\cdot)$ is $\sigma(\bigcup_{i \in I} \mathcal{F}_i)$-measurable [i.e., if $x_i(\omega) = x_i(\omega')$ whenever $\mathbf{s}(\omega) = \mathbf{s}(\omega')$] and if for all $\omega \in \Omega$, $u_i(x_i(\omega); \omega) \geq u_i(x_i(\omega'); \omega)$ for all $\omega' \in \Omega$.

---

[7] Hildenbrand and Kirman (1976) is a useful reference.

It is *Bayesian incentive compatible* if $x_i(\cdot)$ is $\sigma(\bigcup_{i \in I} \mathcal{F}_i)$-measurable and if, for all $s_i \in S_i$ and all $s_i' \in S_i$, one has

$$\sum_{\omega \in \Omega} u_i(x_i(\mathbf{s}(\omega)); \omega)\mu_i(\omega \mid s_i) \geq \sum_{\omega \in \Omega} u_i(x_i(s_i', \mathbf{s}_{-i}(\omega)); \omega)\mu(\omega \mid s_i)$$

where $x_i(\mathbf{s}(\omega)) = x_i(\omega)$, $x_i(s_i', \mathbf{s}_{-i}(\omega)) = x_i(\omega')$ if $\mathbf{s}(\omega') = (s_i', \mathbf{s}_{-i}(\omega))$, and $x_i(s_i', \mathbf{s}_{-i}(\omega)) = 0$ if there is no $\omega' \in \Omega$ for which $\mathbf{s}(\omega') = (s_i', \mathbf{s}_{-i}(\omega))$. Both definitions require that allocations depend only on signals; each agent's state-dependent allocation must be measurable with respect to the joint information received – or reported – by all agents. Both formulations also capture the notion that agent i can do no better by reporting s_i' rather than s_i when his true signal is $s_i \in S_i$. Strong incentive compatibility clearly implies Bayes incentive compatibility, but not conversely. The difference is that strong incentive compatibility applies to each possible realization of $\omega \in \Omega$ separately, while the Bayes incentive compatibility requirement is stated in terms of trader i's expected utility given the signal that he has received. Strong incentive compatibility is more appropriate if players do not know the probabilities affecting their opponents. Either definition can be applied (consistently) in the remainder of this paper.

Say that an allocation belongs to the (strong or Bayes) *incentive compatible core* if it is feasible and (strongly or Bayesian) incentive compatible and if it cannot be blocked by any coalition using an allocation that is (strongly or Bayesian) incentive compatible and feasible for the coalition. More formally, the *strongly incentive compatible core* (respectively, *Bayesian incentive compatible core*) of a pure exchange economy with asymmetric information consists of $((x_i(\cdot), \ldots, x_{\#I}(\cdot)): \Omega \to \mathbb{R}_+^{\#I\ell}$ such that $\sum_{i \in I} x_i(\omega) = \sum_{i \in I} e_i(\omega)$ for (almost) all $\omega \in \Omega$, $x_i(\cdot)$ is strongly incentive compatible (Bayesian incentive compatible) for all $i \in I$, and there does not exist $T \subseteq I$, $T \neq \emptyset$, and $x_i'(\cdot): \Omega \to \mathbb{R}_+^\ell$ for $i \in T$ such that $\sum_{i \in T} x_i'(\omega) = \sum_{i \in T} x_i(\omega)$ for (almost) all $\omega \in \Omega$ with $x_i'(\cdot)$ strongly incentive compatible (Bayesian incentive compatible) for all $i \in T$ and $\int_\Omega u_i(x_i'(\omega); \omega) \, d\mu(\omega) > \int_\Omega u_i(x_i(\omega); \omega) \, d\mu(\omega)$ for every $i \in T$. Allocations in the strongly incentive compatible core (Bayesian incentive compatible core) of an economy correspond to imputations in the (NTU) core of the strongly incentive compatible game (Bayesian incentive compatible game) $V^S: 2^I \to \mathbb{R}^{\#I}(V^B: 2^I \to \mathbb{R}^{\#I})$ with nontransferable utility defined by $V(\emptyset) = \{0\}$ and for $T \subseteq I$, $T \neq \emptyset$, $V(T) = \{(w_1, \ldots, w_{\#I}) \in \mathbb{R}^{\#I} \mid$ there exist strongly incentive compatible (Bayesian incentive compatible) $x_i: \Omega \to \mathbb{R}_+^\ell$ for $i \in T$ with $\sum_{i \in T} x_i(\omega) = \sum_{i \in T} e_i(\omega)$ for (almost) all $\omega \in \Omega$ and $w_i \leq \int_\Omega u_i(x_i(\omega); \omega) \, d\mu(\omega)\}$.

Any Bayesian incentive compatible allocation (and hence any strongly incentive compatible allocation) can be achieved as a Nash equilibrium of a mechanism in which each player's message space is S_i. However, to do so may require the mechanism to waste resources outside of (this) equilibrium. The reason is that initial endowment vectors may dominate the desired incentive compatible allocation for some signal realization, so that the usual device of forcing all agents to keep their endowments whenever the messages are inconsistent may not cause truthful revelation. This requires, of course, the extremely weak monotonicity assumption on state-dependent preferences that for all $i \in I$ and all $\omega \in \Omega$, $u_i(0; \omega) \leq u_i(x_i; \omega)$ for all $x_i \in \mathbb{R}_+^\ell$.

5 Randomization

To demonstrate existence of incentive compatible core allocations, one might enlarge players' strategy spaces to as to ensure balancedness of the induced game. Since balancedness is a type of convexity condition – albeit involving convex combinations of objects taken from different sets being required to belong to yet another set – extension to "mixed strategies" in the cooperative game intuitively appears to be a promising approach.

However, as the example of Section 2 proves, randomization over state-dependent allocations which are feasible with probability one for the economy serves to convexify $V(I)$ [and also $V(S)$ for $S \subseteq I$ if needed] but does not guarantee that this modified game has a nonempty core. Balancedness relates to convexification over allocations that are feasible for different coalitions in the balanced family. Because different players can belong to these coalitions, that total resource requirement may therefore also be random, although feasibility can be preserved on average in each state of the world.

My story is as follows: Agents reveal their messages, perhaps strategically, and a commonly observed and verifiable random device selects an allocation according to a known, verifiable, and agreed upon probability over allocations that are state-dependent and incentive compatible. The *ex ante* random allocation is incentive compatible (in terms of its expected utility, not the utility of its expectation) as is its *ex post* realization. In this sense, the sequencing of communication and the randomization does not matter. Messages can be sent before, during, or after the random drawing occurs in my model. When averaged over the randomization, the (state-dependent) allocations satisfy mean resource feasibility in each state of the world, although the given realization of the allocation need not be feasible.

The law of large numbers could perhaps justify mean feasibility. For instance, the deviation from feasibility vanishes if the state-dependent resource allocation problem with incentives is repeated many times or if there are many independent "copies" (i.e., "islands") of the economy or the game. Inventories could also play a role.

However, perhaps a better interpretation is to appeal to asset markets.[8] Imagine that the grand coalition (or planner or mechanism designer) buys and sells risky commodity contracts so as to offset the discrepancy between the total incentive compatible state-dependent resource allocation and the group's total initial endowment. The randomization could even be defined to depend on the outcome in the risky asset market so that after the outside commodity market's random addition to or subtraction from total resources, feasibility is exactly satisfied.

To summarize this discussion more formally, first consider the case of almost surely feasible randomizations. Define the *convexified strongly incentive compatible game* $V^{CS}: 2^I \to \mathbb{R}^{\#I}$ by $V^{CS}(T) = \text{conv}(V^S(T))$ for $T \subseteq I$. Sim-

---

[8] It's a pleasure to acknowledge a helpful discussion with Jacques Drèze on this point.

ilarly, define the *convexified Bayesian incentive compatible game* $V^{CB}: 2^I \rightarrow \mathbb{R}^{\#I}$ by $V^{CB}(T) = \text{conv}(V^B(T))$ for $T \subseteq I$. [If S is a closed subset of $\mathbb{R}^n$, $\text{conv}(S)$ denotes its (closed) convex hull.] The *convexified strongly incentive compatible core* is the core of the convexified strongly incentive compatible game and the *convexified Bayesian incentive compatible core* is the core of the convexified Bayesian incentive compatible game. Not surprisingly, core allocations in the respective incentive-constrained economies give rise to core imputations in the corresponding games. Hence, the convexified strongly (Bayesian) incentive compatible core of an economy includes precisely those randomizations over strongly (Bayesian) incentive compatible allocations that are resource feasible with probability one and that cannot be blocked by any coalition using an almost surely feasible randomization over strongly (Bayesian) incentive compatible allocations for the coalition. Somewhat more formally, these core concepts are defined by the statement that the convexified strongly (Bayesian) incentive compatible core of a pure exchange economy with asymmetric information consists of probability measures α on state-dependent allocations with $\alpha(\{(x_1(\cdot), \ldots, x_{\#I}(\cdot)): \Omega \rightarrow \mathbb{R}_+^{\#I\ell} | \sum_{i \in I} x_i(\omega) = \sum_{i \in I} e_i(\omega)$ for (almost) all $\omega \in \Omega$ and for all $i \in I$, $x_i(\cdot)$ is strongly (Bayesian) incentive compatible$\}) = 1$ such that there does not exist $T \subseteq I$, $T \neq \emptyset$, and a probability measure β on state-dependent allocations for T with $\beta(\{(x_i'(\cdot))_{i \in T}: \Omega \rightarrow \mathbb{R}_+^{\#T\ell} | \sum_{i \in T} x_i(\omega) = \sum_{i \in T} e_i(\omega)$ for (almost) all $\omega \in \Omega$ and for all $i \in T$, $x_i'(\cdot)$ is strongly (Bayesian) incentive compatible$\}) = 1$ such that $\int \int u_i(x_i(\omega); \omega) \, d\mu(\omega) \, d\alpha(x_i(\cdot)) < \int \int u_i(x_i'(\omega); \omega) \, d\mu(\omega) \, d\beta(x_i'(\cdot))$ for all $i \in T$.

It remains to define these concepts when the randomization is only required to satisfy resource feasibility on average. The term "modified" signifies this distinction. The *modified strongly (Bayesian) incentive compatible game* $V^{MS}: 2^I \rightarrow \mathbb{R}^{\#I}(V^{MB}: 2^I \rightarrow \mathbb{R}^{\#I})$ is defined by $V^{MS}(\emptyset) = V^{MB}(\emptyset) = \{0\}$ and, for $T \subseteq I, T \neq \emptyset$, $V^{MS}(T)$ (respectively $V^{MB}(T)$) equals the set $\{(w_1, \ldots, w_{\#I}) \in \mathbb{R}^{\#I} |$ there exists a probability measure α on state-dependent allocations for coalition T, with $\alpha(\{(x_i(\cdot))_{i \in T}: \Omega \rightarrow \mathbb{R}_+^{\#T\ell} | x_i(\cdot)$ is strongly (Bayesian) incentive compatible$\}) = 1$ and $\sum_{i \in T} \int x_i(\omega) \, d\alpha(x_i(\cdot)) = \sum_{i \in T} e_i(\omega)$ for (almost) every $\omega \in \Omega$, for which $w_i \leq \int \int u_i(x_i(\omega); \omega) \, d\mu(\omega) \, d\alpha(x_i(\cdot))$ for all $i \in T\}$. Then the modified strongly (Bayesian) incentive compatible core of the economy equals the set of randomized strongly (Bayesian) incentive compatible allocations, with resource feasibility on average, that yield imputations in the (NTU) core of the modified strongly (Bayesian) incentive compatible game. To state this more explicitly, the *modified strongly (Bayesian) incentive compatible core* of a pure exchange economy with asymmetric information consists of those probability measures α on state-dependent allocations such that $\alpha(\{(x_1(\cdot), \ldots, x_{\#I}(\cdot)): \Omega \rightarrow \mathbb{R}_+^{\#I\ell} | x_i$ is strongly (Bayesian) incentive compatible for all $i \in I\}) = 1$, $\sum_{i \in I} \int x_i(\omega) \, d\alpha(x_i(\cdot)) = \sum_{i \in I} e_i(\omega)$ for (almost) all $\omega \in \Omega$, and there is no coalition $T \subseteq I$ ($T \neq \emptyset$) with probability measure β on state-dependent allocations for T such that $\beta(\{(x_i'(\cdot))_{i \in T}: \Omega \rightarrow \mathbb{R}_+^{\#T\ell} | x_i'(\cdot)$ is strongly (Bayesian) incentive compatible for each $i \in T\}) = 1$, $\sum_{i \in T} \int x_i'(\omega) \, d\beta(x_i'(\cdot)) = \sum_{i \in T} e_i(\omega)$ for (almost) all $\omega \in \Omega$, and $\int\int u_i(x_i(\omega);\omega)d\mu(\omega)\,d\alpha(x_i(\cdot)) < \int\int u_i(x_i'(\omega);\omega)\,d\mu(\omega)\,d\beta(x_i'(\cdot))$ for all $i \in T$.

Since my main result does not depend on the difference between strong incentive compatibility and Bayes incentive compatibility, call either concept the modified game and the modified incentive compatible core. To simplify notation, write $\tilde{V}\colon 2^I \to \mathbb{R}^{\#I}$ for either V^{MS} or V^{MB}. Call $\tilde{V}$ the modified NTU game with incentive compatibility constraints and randomization with feasibility on average.

6 Nonemptiness of the modified incentive compatible core

The purpose of this section is to state and prove the main result that there are (state-dependent) allocations in the modified incentive compatible core (as defined in Section 5), or, equivalently, that the modified game has a nonempty core. As discussed above, this holds because balancedness follows from the extension of strategy sets to permit randomization over state-dependent and incentive compatible allocations that are feasible on average for every state of the world.

Theorem. *The modified NTU game* $\tilde{V}\colon 2^I \to \mathbb{R}^{\#I}$ *with incentive compatibility constraints (and randomization with feasibility on average) is balanced and its (modified incentive compatible) core is nonempty.*

Proof. Let $\mathcal{B}$ be a balanced family of coalitions with (nonnegative) balancing weights $\gamma_T \geq 0$ for $T \in \mathcal{B}$. I need to show that $\sum_{T\in\mathcal{B}}\gamma_T\tilde{V}(T)_T \subseteq \tilde{V}(I)$. First choose $w^T \in V(T)_T$ for $T \in \mathcal{B}$. Recall that this means $w^T \in V(T) \subseteq \mathbb{R}^{\#I}$ and $w_i^T = 0$ if $i \notin T$. By definition, there are (nonrandom) incentive compatible (state-dependent) allocations $x_i^T\colon \Omega \to \mathbb{R}_+^\ell$ such that, for (almost) all $\omega \in \Omega$,

$$\sum_{i\in T} x_i^T(\omega) = \sum_{i\in T} e_i(\omega)$$

and for all $i \in T$

$$\sum_{\omega\in\Omega} u_i(x_i^T(\omega);\omega)\mu(\omega) \geq w_i^T.$$

Assign $i \in I$ the random allocation which equals $x_i^T(\cdot)$ with probability γ_T if $i \in T$. Since

$$\sum_{\substack{T\in\mathcal{B}\\T\ni I}} \gamma_T = 1 \qquad \text{for all} \qquad i \in I,$$

this defines a probability measure λ on state-dependent allocations for the grand coalition. Feasibility holds on average because

$$\sum_{i\in I}\sum_{T\ni i} x_i^T(\omega)\lambda(\{x_1'(\cdot),\ldots,x_{\#I}'(\cdot))|x_i'(\omega) = x_i^T(\omega)\})$$

$$= \sum_{T\in\mathcal{B}}\sum_{i\in T}\gamma_T x_i^T(\omega) = \sum_{T\in\mathcal{B}}\gamma_T\sum_{i\in T} x_i^T(\omega)$$

$$= \sum_{T\in\mathcal{B}}\gamma_T\sum_{i\in T} e_i(\omega) = \sum_{i\in I} e_i(\omega).$$

If, in place of $w_T \in V(T)_T$, we had $w_T \in \tilde{V}(T)_T$, then there would exist finitely many (nonrandom) incentive compatible allocations $\tilde{x}_i^T : \Omega \to \mathbb{R}_+^\ell$ such that the resulting random allocation is feasible on average and always generates expected utility w_i^T. Repetition of the above argument completes the proof that $\tilde{V} : 2^I \to \mathbb{R}^{\#I}$ is balanced. By Scarf's (1967) theorem, $\tilde{V}$ therefore has a nonempty core. Hence there exist (random) allocations in the modified incentive compatible core. $\qquad\square$

Remark. The same argument shows that $\tilde{V}$ is totally balanced and hence that all subgames have nonempty modified incentive compatible cores.

7 Remarks

1. If for all agents $i \in I$ and all pairs $\omega, \omega' \in \Omega$ of states of the world, the state-dependent preferences represented by the utilities $u_i(\cdot; \omega)$ and $u_i(\cdot; \omega')$ on $\mathbb{R}_+^\ell$ satisfy the "single crossing" property, then the randomization is not necessary for the existence of state-dependent allocations in the incentive compatible core. In this case, incentive compatibility may require the disposal of some resources. As one would expect, even without randomization the core and the incentive compatible core are (generally) distinct.
2. If there are infinitely many states of the world (and especially if Ω is an uncountable set and the support of μ fails to be at most countable), closedness of the $V(S)$ sets may prove problematic since incentive compatibility destroys the convexity of strategy sets. Technically, the problem is that the proof in Allen (1991) that the NTU game $V : 2^I \to \mathbb{R}^{\#I}$ is well defined (based on the Theorem of Dunford and Pettis) relies on the fact that strongly closed *convex* subsets of an $\mathcal{L}^1$ space are also weakly closed or, equivalently, that strongly continuous and concave utilities must be weakly upper semicontinuous functions.
3. Communication requirements or communication restrictions could be modeled as exogenous information sharing rules. Such limits to the endogeneity of information revelation constraints should satisfy the boundedness condition for information sharing rules to avoid destroying balancedness of the resulting NTU game. See Allen (1991).
4. The incentive compatibility constraints in the definition of the (modified) incentive compatible core could be altered to reflect the specific asymmetric information situation. For instance, particular restrictions may not be desirable if they involve states of the world that agents know. However, such modifications move one away from the mechanism story.

References

Allen, B.: Market games with asymmetric information and nontransferable utility: Representation results and the core. CARESS Working Paper #91–09 (1991)

Allen, B.: Market games with asymmetric information: Verification and the publicly predictable information core. Hitotsubashi Journal of Economics (Special issue: Proceedings of Tokyo Conference in Honor of Lionel McKenzie) **34**, 101–122 (1993)

Allen, B.: Incentives in market games with asymmetric information: Approximate NTU cores in large economies. In: Barnett, W.A., Moulin, H., Salles, M., Schofield, N.J. (eds) Social choice, welfare and ethics: Proceedings of the Eighth International Symposium on Economic Theory and Econometrics, Caen, France, (1992). Cambridge: Cambridge University Press 1995

Allen, B.: On the existence of core allocations in a large economy with incentive compatibility constraints. In: Wooders, M.H. (ed) Topics in game theory and mathematical economics in honor of Robert Aumann. Fields Institute Communications, American Mathematical Society (1999)

Allen, B., Yannelis, N.C.: Differential information economies: introduction. Economic Theory 18, 263–273 (2001)

Berliant, M.: On income taxation and the core. Journal of Economic Theory 56, 121–141 (1992)

Bondareva, O.N.: Theory of core in an n-person game. Vestnik Leningradskogo Universiteta, Seriia Matematika, Mekhaniki i Astronomii 13, 141–142 (1962)

Boyd, J.H., Prescott, E.C.: Financial intermediary-coalitions. Journal of Economic Theory 38, 211–232 (1986)

Hart, O., Holmstrom, B.: The theory of contracts. In: Bewley, T. (ed) Advances in economic theory. Cambridge: Cambridge University Press 1985

Hart, O., Moore, J.: Incomplete contracts and renegotiation. Econometrica 56, 755–785 (1988)

Hildenbrand, W., Kirman, A.P.: Introduction to equilibrium analysis. Amsterdam: North Holland 1976

Kahn, J.A.: Endogenous financial structure in an economy with private information. Rochester Center for Economic Research Working Paper No. 96 (1987)

Koutsourgeras, L., Yannelis, N.C.: Incentive compatibility and information superiority of the core of an economy with differential information. Economic Theory 3, 195–216 (1995)

Krasa, S., Yannelis, N.C.: The value allocation of an economy with differential information. Econometrica 62, 881–900 (1994)

Marimon, R.: The core of private information economies. UAB/IAE Discussion Papers W.P. 131.90, Universitat Autónoma de Barcelona (1989)

Myerson, R.B.: Cooperative games with imcomplete [sic] information. International Journal of Game Theory 13, 69–96 (1984)

Prescott, E.C., Townsend, R.M.: Pareto optima and competitive equilibria with adverse selection and moral hazard. Econometrica 52, 21–45 (1984a)

Prescott, E.C., Townsend, R.M.: General competitive analysis in an economy with private information. International Economic Review 25, 1–20 (1984b)

Rosenmüller, J.: Fee games: (N)TU games with incomplete information. Institut für Mathematische Wirtschaftsforschung Working Paper No. 190, Universität Bielefeld (1990)

Scarf, H.E.: On the existence of a cooperative solution for a general class of n-person games. Journal of Economic Theory 3, 169–181 (1971)

Scarf, H.E.: The core of an n-person game. Econometrica 35, 50–69 (1967)

Shapley, L.S.: On balanced sets and cores. Naval Research Logistics Quarterly 14, 453–460 (1967)

Wilson, R.: Information, efficiency and the core of an economy. Econometrica 46, 807–816 (1978)

Yannelis, N.C.: The core of an economy with differential information. Economic Theory 1, 183–198 (1991)

The cheapest hedge[*]

Charalambos D. Aliprantis[1], Yiannis A. Polyrakis[2], and Rabee Tourky[3]

[1] Department of Economics, Purdue University, West Lafayette, IN 47907–1310, USA
(e-mail: aliprantis@mgmt.purdue.edu)
[2] Department of Mathematics, National Technical University, 157 80 Athens, GREECE
(e-mail: ypoly@math.ntua.gr)
[3] Department of Economics, University of Melbourne, Melbourne, VIC 3010, AUSTRALIA
(e-amil: rtourky@unimelb.edu.au)

Received: June 15, 2001; revised version: April 19, 2002

Summary. Investors often wish to insure themselves against the payoff of their portfolios falling below a certain value. One way of doing this is by purchasing an appropriate collection of traded securities. However, when the derivatives market is not complete, an investor who seeks portfolio insurance will also be interested in the cheapest hedge that is marketed. Such insurance will not exactly replicate the desired insured-payoff, but it is the cheapest that can be achieved using the market.

Analytically, the problem of finding a cheapest insuring portfolio is a linear programming problem. The present paper provides an alternative *portfolio dominance* approach to solving the minimum-premium insurance portfolio problem. This affords remarkably rich and intuitive insights to determining and describing the minimum-premium insurance portfolios.

Keywords and Phrases: Super-replicating portfolio, Portfolio insurance, Hedging, Portfolio dominance, Ordered vector spaces.

JEL Classification Numbers: D52, G11.

1 Introduction

Portfolio insurance guarantees a minimum payoff or floor on the downside while capturing the upside. The desired insured payoff can be replicated by holding a

---

* We thank Bruce Grundy and Yvan Lengwiler for their helpful advice. We also thank an anonymous referee for constructive comments. The research of C. D. Aliprantis is supported by the Purdue University E-Business Research Center and the 21st Century Fund and by the NSF Grants EIA-0075506 and SES-0128039. The research of R. Tourky is funded by the Australian Research Council Grant A00103450. This paper is reprinted from the *Journal of Mathematical Economics* [**37**, 269–295 (2002)], with the permission of the publisher.
Correspondence to: C.D. Aliprantis

riskless asset and fiduciary call options. Alternatively, it can be replicated by holding the portfolio and protective put options.

When derivative markets are not complete, the desired insured payoff need not be marketed and a perfectly insuring portfolio may not be available. However, there always exist tradable portfolios that pay at least as much in every state of the world as the desired payoff. These portfolios are candidates for portfolio insurance when markets are not complete. The price of such a *super insuring* portfolio is its insurance-premium. Therefore, an investor who seeks portfolio insurance would be interested in the cheapest hedge that combines available securities, even though it need not exactly replicate the desired insured payoff. That is, an investor will strive to purchase a portfolio whose payoff dominates the desired insured payoff and which has the lowest insurance-premium. Such a portfolio is termed a *minimum-premium insurance portfolio*.

The problem of finding a minimum-premium insurance portfolio is a standard linear programming problem. This paper presents an alternative approach to solving the minimum-premium insurance portfolio problem in a general setting. This is done by taking advantage of the order theoretic structure of portfolio dominance— whereby a portfolio dominates another portfolio if it pays at least as much in each state of the world. As we shall see, the portfolio dominance approach affords remarkably rich and intuitive insights to determining and describing minimum-premium insurance portfolios.

The principal insight of this paper is that we can always obtain a minimum-premium insurance portfolio by looking at portfolio dominance over a restricted number of states of the world. In particular, its analysis focuses on the structure of portfolio dominance over as many uncertain states as available securities.

Technically, the argument goes as follows. When markets are complete, it is easy to determine the portfolio that replicates a desired insured payoff, since in such a setting there are as many states of the world as the available (non-redundant) securities. In terms of portfolio dominance, this portfolio is the least upper bound of the underlying portfolio and the floor.[1]

In contrast, when markets are not complete there are more states in the world than available securities and the desired insured payoff need not be marketed. In such a case, we construct a number of different notions of portfolio dominance by discarding enough states of the world. For instance, if there are J securities, then we can say that a portfolio dominates another portfolio if it pays at least as much in the first J states of the world. Likewise, a portfolio dominates another portfolio if it pays at least as much in the last J states of the world. Now for every such restricted notion of portfolio dominance, we can calculate the least upper bound of the underlying portfolio and the floor giving us a finite number of candidate portfolios. The main result of this paper asserts the following:

– *One of the finite number of the least upper bounds or candidate portfolios of the underlying portfolio and the floor must be a minimum-premium insurance portfolio.*

---

[1] Here the matrix of non-redundant contingent claims is non-singular and the replicating portfolio can be calculated by taking the inverse value of the payoff matrix at the desired insured-payoff.

A characterization of investors that demand portfolio insurance has been presented in the classical article of Leland [12]; where it is assumed that option markets are complete and therefore any desired insured payoff can be perfectly replicated through the purchase of traded securities. Clearly, an investor that demands insurance in a complete market also demands insurance in the case of an incomplete derivatives market. However, if she cannot perfectly replicate the desired insured payoff, then why would she be interested in the "exotic" insurance studied in the present paper? Why characterize the cheapest hedge? Why the minimum cost criterion?

The replication of derivatives in constrained markets using the minimum cost criterion has been the subject of many articles in the literature. For instance, Naik and Uppal [14] use the minimum cost criterion to construct optimal hedging strategies in the presence of leverage constraints. Their work determines the strategy that minimizes the initial cost of hedging given leverage constraints. They argue that the criterion of minimum cost has several advantages. First, for constrained institutions that need to hedge liabilities, this approach is equivalent to maximizing profit. Second, the minimum cost approach also determines the maximum price that a constrained investor would be willing to pay for a contingent claim for exact portfolio insurance. That is, it is the maximum price that an investor is willing to pay for a non-traded Over-The-Counter portfolio insurance. Third, the authors show how the minimum cost criterion is related to utility maximization in the presence of leverage constraints.

The minimum cost criterion is by now a well studied in the literature on hedging and option pricing under constraints. For example, Edirisinghe, Naik, and Uppal [6] study minimum-premium hedging in the presence of transactions costs. Moreover, Broadie, Cvitanic, and Soner [4] use the minimum cost criterion to determine the cheapest portfolio that dominates an option in the presence of extremely general constraints. Our analysis is motivated by the issues considered in these papers.

The cheapest hedge problems under portfolio constraints lend themselves comfortably to the realm of linear optimization (see for instance Naik and Uppal [14]) and to convex as well as to stochastic optimization approaches (see for instance Edirisinghe, Naik, and Uppal [6] and Karatzas and Kou [9, 10]).

The order theoretic approach taken in this paper affords a new and intuitively appealing characterization of the cheapest hedge. The idea is simple: Portfolio dominance captures an important mathematical aspect of options—the building blocks of hedging strategies. Indeed, under the portfolio dominance approach an option is simply a vector lattice operation in the portfolio space.[2] Generally, when markets are not complete portfolio dominance has no lattice structure.[3] Fortunately, every "pseudo-complete" market defines a coarser ordering that has a vector lattice

---

[2] For an underlying security with replicating portfolio θ, the call option at strike price k is replicated by the portfolio $(\theta - \mathbf{k})^+$, where $\mathbf{k}$ is the riskless portfolio paying k in each state of the world and the lattice operation $(\theta - \mathbf{k})^+$ is taken in the space of portfolios. Similarly, the put option at strike price k is replicated by the portfolio $(\mathbf{k} - \theta)^+$.

[3] In fact, it is shown in [3] that generically when there are less than half as many assets as states of the world not a single non-trivial option can be replicated.

structure;[4] and generates its own "put options" and "call options." The main idea of this paper is that the cheapest hedge can always be constructed using these new "options."

The portfolio dominance approach has yielded several other results on portfolio trading. See for instance the work of Brown and Ross [5] who extend Ross' classical result [15] on the role of options in completing markets (see also Green and Jarrow [7]). Hedging in the non-generic case in which portfolio dominance has a vector lattice structure is studied in [1]. Possibly the most successful use of vector ordering methods in economics is in general equilibrium theory. Indeed, the order theoretic properties of commodity spaces—in particular their vector lattice properties—have proven crucial for the development of the Arrow–Debreu–McKenzie general equilibrium model.

The structure of this paper is as follows. The model is in Section 2. The main result regarding the minimum-premium insurance portfolio is stated in Section 3. Section 4 illustrates the results with several examples. The mathematical background needed for establishing the main result is presented in Section A1 of the Appendix. Section A2 studies the concept of portfolio dominance, while the proof of the main result of this work is in Section A3.

2 Minimum-premium insurance portfolio

This section begins with a brief exposition of portfolio insurance in the standard state-space assets markets model, see for example the models in [15,13]. We then look at hedging when markets are complete. Using the insights gained from the case of complete markets we extend the analysis to the case of incomplete markets.

We consider the two-period securities model. There is a finite number S of states of the world. Agents trade $J \leq S$ non-redundant securities $r_1, r_2, \ldots, r_J$ in period-zero whose period-one payoffs are state contingent claims. Therefore, we allow for incomplete markets in which the number of no-redundant securities J is smaller than the number of states S. As usual, the *asset returns matrix* (or the *payoff matrix*) R is the $S \times J$ matrix whose columns are the available no-redundant (i.e., linearly independent) security vectors:

$$
R = \begin{bmatrix} r_1(1) & r_2(1) & \ldots & r_J(1) \\ r_1(2) & r_2(2) & \ldots & r_J(2) \\ \vdots & \vdots & \ddots & \vdots \\ r_1(S) & r_2(S) & \ldots & r_J(S) \end{bmatrix}
$$

Portfolios are linear combinations of the available securities. A portfolio is therefore represented by a vector in $\mathbb{R}^J$. Portfolios are considered as column vectors and the payoff of a portfolio θ is $R\theta$.

A state contingent claim, which is a vector in $\mathbb{R}^S$, is said to be a *marketed payoff* if it lies in the asset span (i.e., the range) $M = \langle R \rangle$ of the asset returns matrix R in $\mathbb{R}^S$; in which case, there is a unique portfolio (called the *replicating portfolio*) of

---

[4] Under this ordering the portfolio space is called a *minimal lattice subspace*.

the available securities whose payoff is the state contingent claim. We shall assume that the riskless bond $\mathbf{1} = (1, 1, \ldots, 1)$ is marketed.

We shall also say that a portfolio θ *super replicates* a state contingent claim $x \in \mathbb{R}^S$ if $R\theta \geq x$. That is, θ pays at least as much in each state as x. A portfolio θ (*perfectly*) *replicates* a state contingent claim $x \in \mathbb{R}^S$ over a set of states I if $R\theta(s) = x(s)$ for every $s \in I$.

If the asset span equals the entire space of contingent claims (i.e., if $J = S$), then markets are *complete*. When $J < S$ the markets are *incomplete* in which case some state contingent claims cannot be replicated by a portfolio.

We shall restrict our study to *arbitrage-free* security prices. That is, we restrict our attention to vectors $q \in \mathbb{R}^J$ of security prices that give a non-zero positive value $q \cdot \theta > 0$ to any non-zero portfolio θ with a positive payoff $R\theta \geq 0$. A price $q \in \mathbb{R}^J$ is *arbitrage free* (resp. *weakly arbitrage free*) if $q \cdot \theta > 0$ (resp. $q \cdot \theta \geq 0$) whenever the portfolio θ satisfies $R\theta > 0$ (resp. $R\theta \geq 0$).

Portfolio insurance: The *insured payoff* of a portfolio $\theta = (\theta_1, \theta_2, \ldots, \theta_J)$ at a *floor* $k \in \mathbb{R}$ is a state contingent claim that captures the upside of the portfolio and insures against any downside below the floor. In other words, the insured payoff is the state contingent claim

$$
\max \{R\theta, \mathbf{k}\} = \begin{bmatrix} \max \left\{ \sum_{j=1}^{J} r_j(1)\theta_j, k \right\} \\ \max \left\{ \sum_{j=1}^{J} r_j(2)\theta_j, k \right\} \\ \vdots \\ \max \left\{ \sum_{j=1}^{J} r_j(S)\theta_j, k \right\} \end{bmatrix},
$$

where $\mathbf{k} = k\mathbf{1}$ is the riskless bond paying k in each state of the world. In a complete market the insured payoff of a portfolio is the contingent claim that can be replicated by holding the payoff of the portfolio and a put option with a strike price k; or it can be replicated by holding $\mathbf{k}$ and a call option on the portfolio with a strike price k. The basic problem is that when markets are incomplete the insured payoff need not be a marketed payoff.

Minimum–premium insurance portfolios: Once again we consider a portfolio θ and a floor k. Any portfolio η whose payoff $R\eta$ dominates the insured payoff $\max \{R\theta, \mathbf{k}\}$ in each state is viewed as an *insurance portfolio*. There are many such portfolios. The cost of such a portfolio is the *insurance-premium*. So, if q is a securities price, then the insurance-premium associated with an insurance portfolio η is $q \cdot \eta$. We are, therefore, interested in a *minimum–premium insurance portfolio* (or a *cheapest hedge portfolio*) of θ at the floor k, which is the least costly portfolio whose payoff dominates the insured payoff of θ and the floor k. That is, a minimum–premium insurance portfolio is a solution to the following minimization problem:

$$(MP) \quad \min q \cdot \eta$$
$$\text{s.\,t.:}\ \eta \in \mathbb{R}^J,\ R\eta \geq R\theta,\ \text{and}\ R\eta \geq \mathbf{k}$$

A solution to this minimization problem always exists. As a matter of fact:

– *The solution set of the minimization problem* (MP) *is a non-empty, convex and compact subset of* $\mathbb{R}^J$.

3 Portfolio dominance and the cheapest hedge solution

In this section we shall sketch briefly the basic ideas behind our solution to the hedging problem. As mentioned before, our solution is based on the notion of portfolio dominance that is related to the lattice structures of the spaces.

We shall say that a portfolio θ *dominates* a portfolio η if $R\theta \geq R\eta$, in which case we write $\theta \succeq \eta$. The portfolio dominance relation $\succeq$ makes $\mathbb{R}^J$ a partially ordered vector space. We shall denote by C the (pointed convex) cone generated by $\succeq$, i.e,

$$C = \left\{ \theta \in \mathbb{R}^J : \theta \succeq 0 \right\}.$$

Now for any two portfolios θ and η we write $\theta \vee_C \eta$ to mean the least upper bound of the set $\{\theta, \eta\}$ relative to $\succeq$. That is, the portfolio $\theta \vee_C \eta$, if it exists, has the property that $\theta \vee_C \eta \succeq \theta$ and $\theta \vee_C \eta \succeq \eta$ and if $\mu \succeq \theta$ and $\mu \succeq \eta$, then $\mu \succeq \theta \vee_C \eta$.

Whenever markets are complete, one can calculate a unique portfolio that is a minimum–premium insurance portfolio for any arbitrage free securities price. However, when markets are not complete the minimum–premium insurance portfolio depends on the prevailing price. Nevertheless, as we shall see, the incomplete markets case is quite similar to the case of complete markets. The details follow.

Complete markets: Assume for now that markets are complete. That is, assume that the payoff matrix R is a $J \times J$ matrix. Recall that we have fixed a portfolio θ and a floor k. When markets are complete, it is easy to calculate a perfect hedge, or a portfolio that replicates the insured payoff of θ at floor k.

Indeed, if the portfolio κ replicates $\mathbf{k}$ (i.e., if $R\kappa = \mathbf{k}$), then since R is invertible the insured payoff is replicated by the portfolio:

$$\theta^* = \theta \vee_C \kappa = R^{-1} \max \{R\theta, \mathbf{k}\}.$$

The portfolio θ^* is clearly a minimum–premium insurance portfolio for any arbitrage free price. In particular, it is independent of the prevailing arbitrage free security prices. That is, we have the following result.

Theorem 3.1 *If markets are complete, then for any arbitrage free price the unique minimum–premium insurance portfolio is replicated by the portfolio* $\theta \vee_C \kappa$*, which exists* (*and is the call option on the portfolio* θ *at strike price* k *and* k *bonds* $\mathbf{1}$.)

Incomplete markets: Assume now that the market is incomplete. We shall see that discarding some $S - J$ states of the world allows us to use a procedure for calculating a minimum–premium portfolio insurance as though the market is complete. We shall describe this method next.

For any collection I of J elementary states let R_I be the $J \times J$ matrix whose rows are the rows of the payoff matrix R corresponding to the states of I. For instance, if there are three securities and four states then

$$R_{(1,3,4)} = \begin{bmatrix} r_1(1) & r_2(1) & r_3(1) \\ r_1(3) & r_2(3) & r_3(3) \\ r_1(4) & r_2(4) & r_3(4) \end{bmatrix}.$$

If R_I is invertible, then we say that R_I (or even I) defines a *pseudo-complete market*. Since the rank of R is J there always exists at least one pseudo-complete market.

Before proceeding further, let us introduce some further notation. If a set of states $I = \{s_1 < s_2 < \cdots < s_J\}$ defines a pseudo-complete market and θ is a portfolio, then we let $\theta_I = (\theta_{s_1}, \theta_{s_2}, \ldots, \theta_{s_J})$. If we view θ_I as a column vector, then we shall denote $R_I\theta_I$ by $R_I\theta$, that is, $R_I\theta = R_I\theta_I$.

Now each *pseudo-complete market* R_I generates a new notion of portfolio dominance $\succeq_I$ by defining $\theta \succeq_I \eta$ whenever $R_I\theta \geq R_I\eta$. It turns out that not only this portfolio dominance relation $\succeq_I$ partially orders the portfolio space $\mathbb{R}^J$ but it also induces a lattice ordering. That is, for every *pseudo-complete market* R_I its portfolio dominance cone

$$C_I = \{\theta \in \mathbb{R}^J : \theta \succeq_I 0\}$$

is a lattice cone—which is also a super-cone of C, i.e., $C \subseteq C_I$. This means that if η and θ are two portfolios, then the $\succeq_I$-supremum of the two portfolios $\theta \vee_I \eta$ exists and is given by $\theta \vee_I \eta = R_I^{-1} \max\{R_I\theta, R_I\eta\}$. Assuming that $R\kappa = \mathbf{k}$, for each pseudo-complete market R_I we let

$$\eta_I = \theta \vee_I \kappa = R_I^{-1} \max\{R_I\theta, \mathbf{k}\}.$$

If θ is any portfolio and k is a floor price, then a *potentially insuring portfolio* is any portfolio of the form η_I satisfying $R\eta_I \geq \max\{R\theta, \mathbf{k}\}$. We shall denote the finite collection of all potentially insuring portfolios of θ at the floor k by $\mathcal{P}_{\theta,k}$, i.e.,

$$\mathcal{P}_{\theta,k} = \{\eta \in \mathbb{R}^J : \eta = \eta_I \text{ for a pseudo-complete market } R_I \text{ and } R\eta \geq R\theta \vee \mathbf{k}\}.$$

Clearly, there is a finite number of potentially insuring portfolios that are calculated independently of the arbitrage free security price.

The remarkable property is that one of the potentially insuring portfolios is a minimum-insurance premium portfolio. This is the main result of this paper and it will be stated next. Its proof is quite involved and it will be presented in Section A3 of the Appendix.

Theorem 3.2 (The Cheapest Hedge Theorem) *For any portfolio θ, any arbitrage price q, and any floor k we have the following:*

1. *There exists at least one potentially insuring portfolio $\theta \vee_I \kappa$ that is a minimum-premium insurance portfolio for θ at floor k.*
2. *A minimum-premium insurance portfolio $\theta \vee_I \kappa$ is the cheapest potentially insuring portfolio. That is, $q \cdot (\theta \vee_I \kappa) \leq q \cdot \eta$ for all $\eta \in \mathcal{P}_{\theta,k}$.*

3. *The portfolio $\eta^* = \theta \vee_C \kappa$ exists if and only if $\mathcal{P}_{\theta,k}$ consists of one portfolio η^*, which is automatically a minimum-premium insurance portfolio for any arbitrage free price.*

The third statement in the theorem is an extension of the main result in [1], which shows that a price independent minimum-premium insurance portfolio insurance exists for any portfolio–floor pair if and only if the portfolio dominance cone is a lattice cone; i.e., it generates a vector lattice on the portfolio space.

Let us conclude this section with a final remark. There is an intuitively appealing way of identifying the potentially insuring portfolios:

  – *A portfolio is a potentially insuring portfolio if and only if it super replicates the insured payoff and perfectly replicates the insured payoff over a set I of J states for which R_I is a pseudo-complete market.*

4 Illustrative examples

The Cheapest Hedge Theorem 3.2 can be reformulated as follows.

Theorem 4.1 *For any portfolio θ, any arbitrage free price q, and any floor k we have the following:*

1. *There exists at least one potentially insuring portfolio that is a minimum-premium insurance portfolio for θ at floor k.*
2. *A minimum-premium insurance portfolio can be obtained by solving the finite minimization problem:*

$$(\mathcal{F}MP) \qquad \min \, q \cdot \eta$$
$$s.\,t.\!: \eta \in \mathcal{P}_{\theta,k}$$

3. *If $\mathcal{P}_{\theta,k}$ consists of one portfolio, say η^*, then η^* is automatically a minimum-premium insurance portfolio.*

That is: *For any arbitrage free price, the cheapest potentially insuring portfolio is a minimum-premium insurance portfolio.* In other words, we have reduced the minimum-premium insurance portfolio problem (MP) to the following minimization problem over a finite set:

$$\min \, q \cdot \eta$$
$$s.\,t.\!: \eta \text{ is a potentially insuring portfolio.}$$

This section presents some illustrative examples of the preceding result. With this in mind, let θ be a portfolio, k a floor, and q an arbitrage free price. Moreover, for each set I of J states, let R_I be the $J \times J$ matrix whose rows are the rows of R determined by I. Now consider the following steps:

1. For each invertible R_I find the portfolio

$$\eta_I = R_I^{-1} \max \{R\theta, \mathbf{k}\},$$

and form the collection $\mathcal{P}_{\theta,k}$ of all potentially insuring portfolios of θ at the floor k.

2. If $\mathcal{P}_{\theta,k}$ consists of one portfolio, say η^*, then we are done. The portfolio η^* is the only minimum-premium insurance portfolio for any arbitrage free price.
3. If $\mathcal{P}_{\theta,k}$ contains more than one portfolio, then the least costly portfolio η in $\mathcal{P}_{\theta,k}$ with respect to the price q is a minimum-premium insurance portfolio.

We are now ready to present three examples. The first example is an example of a complete market.

Example 1 (A complete market). Suppose that there are four states of the world and that the market has the following non-redundant securities:

1. A treasury bond with payoff $\mathbf{1} = (1, 1, 1, 1)$.
2. A corporate bond with payoff $(0, 1, 1, 1)$.
3. A share with payoff $(0, 1, 2, 4)$.
4. A call option on the share with a strike price of 3. That is, the security $\max\{(0, 1, 2, 4) - \mathbf{3}, 0\} = (0, 0, 0, 1)$.

Therefore, the asset returns matrix R is

$$R = \begin{bmatrix} 1 & 0 & 0 & 0 \\ 1 & 1 & 1 & 0 \\ 1 & 1 & 2 & 0 \\ 1 & 1 & 4 & 1 \end{bmatrix}.$$

Keep in mind that the payoff of any portfolio θ is $R\theta$.

Now consider the portfolio $\theta = (1, 2, 3, 0)$. The insured payoff on a portfolio θ at a floor $k = 10$ is the contingent claim

$$\max\{R\theta, \mathbf{10}\} = \max\left\{ \begin{bmatrix} 1 \\ 6 \\ 9 \\ 15 \end{bmatrix}, \begin{bmatrix} 10 \\ 10 \\ 10 \\ 10 \end{bmatrix} \right\} = \begin{bmatrix} 10 \\ 10 \\ 10 \\ 15 \end{bmatrix}.$$

This contingent claim is obviously marketed and is the payoff of the portfolio

$$\theta^* = R^{-1}\max\{R\theta, \mathbf{10}\} = \begin{bmatrix} 1 & 0 & 0 & 0 \\ -1 & 2 & -1 & 0 \\ 0 & -1 & 1 & 0 \\ 0 & 2 & -3 & 1 \end{bmatrix} \begin{bmatrix} 10 \\ 10 \\ 10 \\ 15 \end{bmatrix} = \begin{bmatrix} 10 \\ 0 \\ 0 \\ 5 \end{bmatrix}.$$

Clearly, for any arbitrage free securities price q the portfolio θ^* is the unique minim-premium insurance portfolio. Figure 1 provides a graphical illustration of this example.

Example 2 (Incomplete markets with only one potentially insuring portfolio). We consider the market in the previous example. But now we suppose that the call option is not available. Thus, the market is described by the returns matrix

$$R = \begin{bmatrix} 1 & 0 & 0 \\ 1 & 1 & 1 \\ 1 & 1 & 2 \\ 1 & 1 & 4 \end{bmatrix}.$$

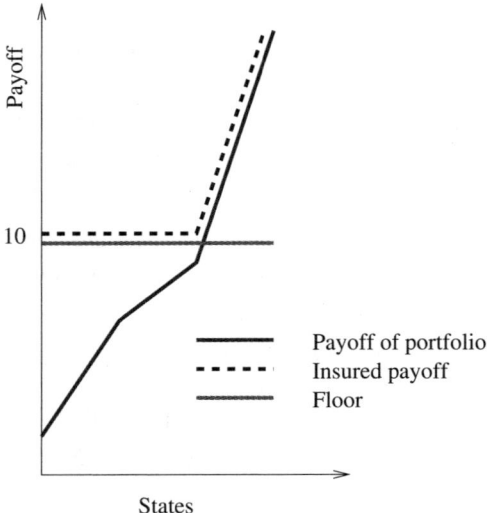

Figure 1. When markets are complete the insured payoff can be replicated by a portfolio containing ten treasury bonds and five call options

Consider the portfolio $\theta = (1, 2, 3)$. The insured payoff on the portfolio θ at a floor $k = 10$ is once again the contingent claim

$$\max\{R\theta, \mathbf{10}\} = \max\left\{\begin{bmatrix} 1 \\ 6 \\ 9 \\ 15 \end{bmatrix}, \begin{bmatrix} 10 \\ 10 \\ 10 \\ 10 \end{bmatrix}\right\} = \begin{bmatrix} 10 \\ 10 \\ 10 \\ 15 \end{bmatrix}.$$

This contingent claim is not marketed since as we saw in the previous example it is the payoff of a portfolio using the unavailable call option.

However, we can calculate (at most) four important portfolios by looking at the four 3×3 matrices whose rows are taken from R. These are the matrices:

$$R_{(1,2,3)} = \begin{bmatrix} 1 & 0 & 0 \\ 1 & 1 & 1 \\ 1 & 1 & 2 \end{bmatrix}, \quad R_{(1,2,4)} = \begin{bmatrix} 1 & 0 & 0 \\ 1 & 1 & 1 \\ 1 & 1 & 4 \end{bmatrix},$$

$$R_{(1,3,4)} = \begin{bmatrix} 1 & 0 & 0 \\ 1 & 1 & 2 \\ 1 & 1 & 4 \end{bmatrix}, \quad R_{(2,3,4)} = \begin{bmatrix} 1 & 1 & 1 \\ 1 & 1 & 2 \\ 1 & 1 & 4 \end{bmatrix}.$$

Notice that $R_{(2,3,4)}$ is a singular matrix. Therefore, we restrict our attention to the remaining three pseudo-complete markets and obtain the following three

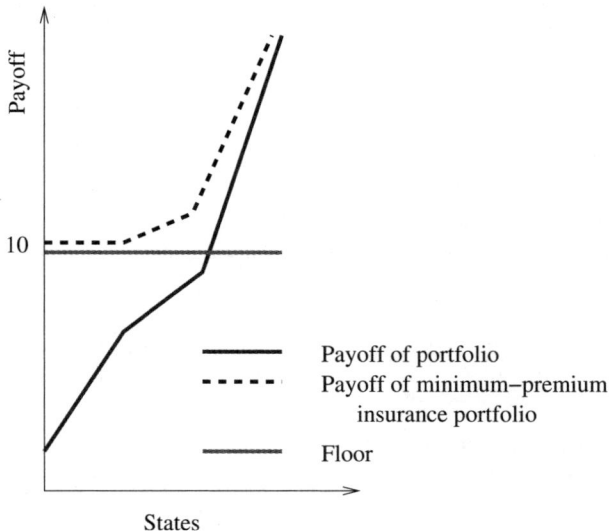

Figure 2. When the call option is not available the insured payoff cannot be replicated. However, the unique minimum-premium insurance portfolio contains ten treasury bonds, a short sale of one and two thirds of the corporate bond, and one and two thirds of the share

portfolios:

$$\eta_{(1,2,3)} = R_{(1,2,3)}^{-1} \max\left\{R_{(1,2,3)}\theta, \mathbf{10}\right\} = \begin{bmatrix} 10 \\ 0 \\ 0 \end{bmatrix},$$

$$\eta_{(1,2,4)} = R_{(1,2,4)}^{-1} \max\left\{R_{(1,2,4)}\theta, \mathbf{10}\right\} = \begin{bmatrix} 10 \\ -\frac{5}{3} \\ \frac{5}{3} \end{bmatrix},$$

$$\eta_{(1,3,4)} = R_{(1,3,4)}^{-1} \max\left\{R_{(1,3,4)}\theta, \mathbf{10}\right\} = \begin{bmatrix} 10 \\ -5 \\ \frac{5}{2} \end{bmatrix}.$$

From these portfolios only $\eta_{(1,2,4)}$ has a payoff greater than the insured payoff of θ with floor 10. That is,

$$\theta^* = R\eta_{(1,2,4)} = \begin{bmatrix} 10 \\ 10 \\ \frac{35}{3} \\ 15 \end{bmatrix} \geq \begin{bmatrix} 10 \\ 10 \\ 10 \\ 15 \end{bmatrix}.$$

Therefore, for any arbitrage free securities price q the portfolio $\eta_{(1,2,4)}$ is the only minimum-premium insurance portfolio. Therefore, we have found a solution that is independent of the arbitrage free security prices. This example is illustrated in Figure 2.

Example 3 (Incomplete markets with price dependent insurance). Consider a market
with the payoff matrix

$$R = \begin{bmatrix} 1 & 2 & 1 \\ 1 & 1 & 5 \\ 1 & 0 & 3 \\ 1 & 0 & 0 \end{bmatrix},$$

and, once again we consider the portfolio $\theta = (1, 2, 3)$.

The insured payoff on the portfolio θ at a floor $k = 10$ is the contingent claim

$$\max\{R\theta, \mathbf{10}\} = \max\left\{ \begin{bmatrix} 8 \\ 18 \\ 10 \\ 1 \end{bmatrix}, \begin{bmatrix} 10 \\ 10 \\ 10 \\ 10 \end{bmatrix} \right\} = \begin{bmatrix} 10 \\ 18 \\ 10 \\ 10 \end{bmatrix}.$$

This contingent claim is not marketed.

Next, we can calculate (at most) four portfolios by looking at the four 3×3
matrices whose rows are taken from R. These are the matrices:

$$R_{(1,2,3)} = \begin{bmatrix} 1 & 2 & 1 \\ 1 & 1 & 5 \\ 1 & 0 & 3 \end{bmatrix}, \quad R_{(1,2,4)} = \begin{bmatrix} 1 & 2 & 1 \\ 1 & 1 & 5 \\ 1 & 0 & 0 \end{bmatrix},$$

$$R_{(1,3,4)} = \begin{bmatrix} 1 & 2 & 1 \\ 1 & 0 & 3 \\ 1 & 0 & 0 \end{bmatrix}, \quad R_{(2,3,4)} = \begin{bmatrix} 1 & 1 & 5 \\ 1 & 0 & 3 \\ 1 & 0 & 0 \end{bmatrix}.$$

All four matrices are invertible. So, we consider the portfolios:

$$\eta_{(1,2,3)} = R_{(1,2,3)}^{-1} \max\{R_{(1,2,3)}\theta, \mathbf{10}\} = \begin{bmatrix} 2 \\ \frac{8}{3} \\ \frac{8}{3} \end{bmatrix},$$

$$\eta_{(1,2,4)} = R_{(1,2,4)}^{-1} \max\{R_{(1,2,4)}\theta, \mathbf{10}\} = \begin{bmatrix} 10 \\ -\frac{8}{9} \\ \frac{16}{9} \end{bmatrix},$$

$$\eta_{(1,3,4)} = R_{(1,3,4)}^{-1} \max\{R_{(1,3,4)}\theta, \mathbf{10}\} = \begin{bmatrix} 10 \\ 0 \\ 0 \end{bmatrix},$$

$$\eta_{(2,3,4)} = R_{(2,3,4)}^{-1} \max\{R_{(2,3,4)}\theta, \mathbf{10}\} = \begin{bmatrix} 10 \\ 8 \\ 0 \end{bmatrix}.$$

Notice that the portfolios $\eta_{(1,2,4)}$ and $\eta_{(2,3,4)}$ have a payoff greater than the insured
payoff of θ at floor 10. (see Figure 3). That is,

$$R\eta_{(1,2,4)} = \begin{bmatrix} 10 \\ 18 \\ 15\frac{1}{3} \\ 10 \end{bmatrix} \geq \begin{bmatrix} 10 \\ 18 \\ 10 \\ 10 \end{bmatrix} \quad \text{and} \quad R\eta_{(2,3,4)} = \begin{bmatrix} 26 \\ 18 \\ 10 \\ 10 \end{bmatrix} \geq \begin{bmatrix} 10 \\ 18 \\ 10 \\ 10 \end{bmatrix}.$$

Now let us take three arbitrage free prices.

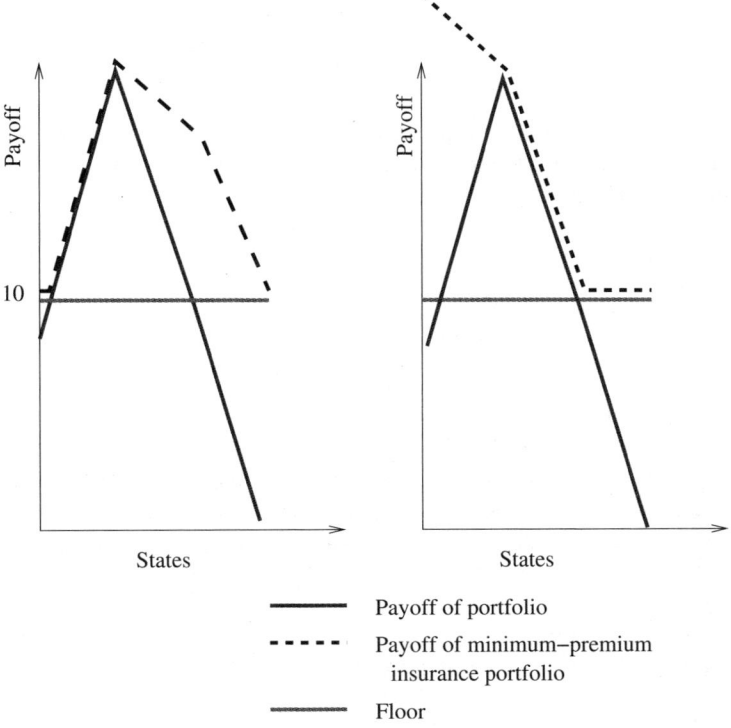

	Payoff of portfolio
- - - -	Payoff of minimum–premium insurance portfolio
	Floor

Figure 3. In this example the insured payoff which is a butterfly-spread cannot be replicated. However, there are two choices for portfolio insurance; and the choice depends on the prevailing securities prices

1. Let $q = (1, 1, 1) = \frac{4}{9}(1, 2, 1) + \frac{1}{9}(1, 1, 5) + \frac{4}{9}(1, 0, 0)$. From

$$q \cdot \eta_{(1,2,4)} = 10\frac{8}{9} < q \cdot \eta_{(2,3,4)} = 18\,,$$

   we see that the minimum-premium insurance portfolio for the price q is $\eta_{(1,2,4)}$.
2. For the arbitrage free securities price

$$q = (4, 1, 12) = \frac{1}{3}(1, 2, 1) + \frac{1}{3}(1, 1, 5) + \frac{10}{3}(1, 0, 3)\,,$$

   we have

$$q \cdot \eta_{(1,2,4)} = 60 + \frac{4}{9} \quad \text{and} \quad q \cdot \eta_{(2,3,4)} = 48\,.$$

   Thus, $q \cdot \eta_{(1,2,4)} > q \cdot \eta_{(2,3,4)}$, and so $\eta_{(2,3,4)}$ is the minimum-premium insurance portfolio for the price $q = (4, 1, 12)$.
3. For the price $q = (11, 5, 25) = 2(1, 2, 1) + 6(1, 0, 3) + 2(1, 0, 0) + (1, 1, 5)$, we get $q \cdot \eta_{(1,2,4)} = q \cdot \eta_{(2,3,4)} = 150$. Therefore, both portfolios are minimum-premium insurance portfolios for this price q.

Appendix: Background and proofs

A1. Mathematical preliminaries

We present here the basic concepts and results concerning cones in finite dimensional spaces that are needed to prove the main theorem of this paper. The generic finite dimensional vector space will be $\mathbb{R}^J$.

Recall that a *pointed convex cone*, or simply a *cone*, is a non-empty subset K of $\mathbb{R}^J$ such that:

1. $K + K \subseteq K$,
2. $\alpha K \subseteq K$ for each $\alpha \geq 0$, and
3. $K \cap (-K) = \{0\}$.

Every cone K induces a vector space order $\geq_K$ (or $\leq_K$) on $\mathbb{R}^J$ by defining $x \geq_K y$ (or $y \leq_K x$) whenever $x - y \in K$. The vectors of K are precisely the vectors satisfying $x \geq_K 0$ and (if there is no other cone under consideration) they are referred to as *positive vectors*. We also write $x >_K 0$ to mean $x \geq_K 0$ and $x \neq 0$. For each vector $x \in K$, the K-order interval $\{y \in \mathbb{R}^J : 0 \leq_K y \leq_K x\}$ will be denoted $[0, x]_K$, i.e., $[0, x]_K = \{y \in \mathbb{R}^J : 0 \leq_K y \leq_K x\}$.

A cone K is said to be *generating* if $\mathbb{R}^J = K - K$, i.e., if every vector in $\mathbb{R}^J$ can be written as a difference of two vectors in K. The following result is well known and we state it for completeness.

Lemma A1.1 *A cone in $\mathbb{R}^J$ is generating if and only if it has an interior point.*

The *dual cone* K' of a cone K is defined by

$$K' = \left\{ q \in (\mathbb{R}^J)' = \mathbb{R}^J : \ q \cdot x \geq 0 \text{ for all } x \in K \right\}.$$

The members of K' are called *positive linear functionals*.

Regarding dual cones, we have the following basic duality result.

Theorem A1.2 (Duality Theorem) *If K is a closed generating cone in $\mathbb{R}^J$, then:*

1. *The dual cone K' is also a closed and generating cone.*
2. *The dual cone of K' coincides with K, i.e., $K = K'' = (K')'$.*

In particular, we have:

a. *$x \geq_K y$ if and only if $q \cdot x \geq q \cdot y$ for each $q \in K'$, and*
b. *$q_1 \geq_{K'} q_2$ if and only if $q_1 \cdot z \geq q_2 \cdot z$ for each $z \in K$.*

Proof. It should be clear that $K' + K' \subseteq K'$, $\alpha K' \subseteq K'$ for each $\alpha \geq 0$, and that K' is a closed subset of $\mathbb{R}^J$. To see that K' is a cone, let $q \in K' \cap (-K')$. Then, $q \cdot x \geq 0$ and $q \cdot x \leq 0$ both hold for all $x \in K$. That is, $q \cdot x = 0$ for each $x \in K$. Since K is generating, it follows that $q \cdot x = 0$ for all $x \in \mathbb{R}^J$, i.e., $q = 0$.

Clearly, $K \subseteq K''$. To see that $K = K''$ is indeed true, assume by way of contradiction that K is a proper subset of K''. So, there exists some $x \in K''$ such that $x \notin K$. Since K is closed and convex, it follows (from the separation theorem) that there exist some $q \in \mathbb{R}^J$ and some real number c such that $q \cdot y \geq c > q \cdot x$

for each $y \in K$. Since K is a cone, we get $c \leq 0$ and $q \cdot y \geq 0$ for all $y \in K$. This implies $q \in K'$, and so $q \cdot x \geq 0$, which contradicts $q \cdot x < c \leq 0$. Hence, $K = K''$.

Finally, we show that K' is generating, i.e., that $\mathbb{R}^J = K' - K'$. To see this, assume that some $q \in \mathbb{R}^J$ satisfies $q \cdot y = 0$ for all $y \in K' - K'$. This implies $q \in K'' \cap (-K'') = K \cap (-K) = \{0\}$, i.e., $q = 0$. Thus, the closed vector subspace $K' - K'$ is dense in $\mathbb{R}^J$, and consequently $\mathbb{R}^J = K' - K'$.

A vector $q \in (\mathbb{R}^J)' = \mathbb{R}^J$ is said to be K-*strictly positive* (or simply *strictly positive*), denoted $q \gg_K 0$, if $x >_K 0$ implies $q \cdot x > 0$. The strictly positive vectors will play the role of the arbitrage free prices.

There are two more notions related to strict positivity. If K is a cone in a vector space X, then a vector $x \in K$ is said to be:

a. *internal*, if for each $y \in X$ there exists some $\alpha_0 > 0$ such that $x + \alpha y \in K$ for all $|\alpha| \leq \alpha_0$, and

b. *an order unit*, or simply a *unit*, if for each $y \in X$ there exists some $\alpha > 0$ such that $y \leq_K \alpha x$.

For the dual of a closed and generating cone in R^J all these notions coincide.

Lemma A1.3 *For a closed and generating cone K and some $q \in K'$ the following statements are equivalent.*

1. *q is K-strictly positive.*
2. *q is an interior point of K'.*
3. *q is an internal point of K'.*
4. *q is an order unit of K'.*

Moreover, the interior of K' is non-empty—and so the collection $(K')^\circ$ of all strictly positive vectors is dense in K'.

Proof. Notice first that from Theorem A1.2 and Lemma A1.1 we know that $(K')^\circ$ (the interior of K') is non-empty. This easily implies that $(K')^\circ$ is dense in K'.

$(1) \implies (2)$ Let q be a strictly positive vector and assume by way of contradiction that $q \notin (K')^\circ$. Since $(K')^\circ$ in non-empty and convex, there exists (in view of the separation theorem) some non-zero vector $x \in \mathbb{R}^J$ such that $q \cdot x \leq p \cdot x$ for all $p \in (K')^\circ$. Since $(K')^\circ$ is dense in K', it follows that $q \cdot x \leq p \cdot x$ holds for all $p \in K'$. Taking into account that K' is a cone, we see that $q \cdot x \leq 0 \leq p \cdot x$ for all $p \in K'$. This implies $x \in K'' = K$, and so $x >_K 0$. But then, the strict positivity of q implies $q \cdot x > 0$, contrary to $q \cdot x \leq 0$. Thus, $q \in (K')^\circ$.

$(2) \implies (3)$ This is obvious.

$(3) \implies (4)$ Assume that q in internal point of K' and let $p \in \mathbb{R}^J$. Pick some $\alpha > 0$ such that $q + \alpha(-p) \in K'$. This implies $p \leq_{K'} \frac{1}{\alpha}q$, and so q is an order unit.

$(4) \implies (1)$ Fix an interior vector p in the dual cone K'. Also, choose a symmetric neighborhood V of zero such that $p + V \subseteq K'$. From $p \pm v \in K'$ for each $v \in V$, it follows that $-p \leq_{K'} v \leq_{K'} p$ for each $v \in V$, i.e., $V \subseteq [-p, p]_{K'}$.

Since q is an order unit, there exists some $\alpha > 0$ such that $\alpha q \geq_{K'} \pm p$, and hence $\frac{1}{\alpha}[-p,p]_{K'} \subseteq [-q,q]_{K'}$. So, if we let $W = \frac{1}{\alpha}V$, then $W \subseteq [-q,q]_{K'}$, and thus $q + W \subseteq [0,2p]_{K'} \subseteq K'$. This shows that q is an interior point of K'.

Now let $x >_K 0$ and assume by way of contradiction that $q \cdot x = 0$. If $r \in K'$ is arbitrary, then there exists some $\lambda > 0$ such that $\pm \lambda r \in W$. This yields $q \pm \lambda r \in K'$, and so $0 \leq (q \pm \lambda r) \cdot x = \pm \lambda r \cdot x$. This implies $r \cdot x = 0$ for all $r \in K'$, and consequently $r \cdot x = 0$ for all $r \in \mathbb{R}^J$. Therefore, $x = 0$, which is impossible. This contradiction shows that $q \cdot x > 0$, and so q is strictly positive.

Lemma A1.4 *Let K be a closed and generating cone in $\mathbb{R}^J$. If q is a strictly positive vector, then a closed subset A of K is compact if and only if the set of real numbers $q \cdot A = \{q \cdot a : a \in A\}$ is bounded.*

Proof. Let A be a closed subset of K such that $q \cdot A$ is bounded, where q is a strictly positive vector. Since (according to Lemma A1.3) q is an interior point of K', there exists an open neighborhood V of zero such that $q + V \subseteq K'$. Now let $p \in \mathbb{R}^J$ be an arbitrary vector. Choose some $\lambda > 0$ such that $\pm \frac{1}{\lambda}p \in V$, and so $q \pm \frac{1}{\lambda}p \in K'$. Therefore, $q \pm \frac{1}{\lambda}p \geq_{K'} 0$ or $-\lambda q \leq_{K'} p \leq_{K'} \lambda p$. This—and the fact that $q \cdot A$ is bounded—imply that the set $p \cdot A$ is bounded for each $p \in \mathbb{R}^J$. Consequently, A is a bounded subset of $\mathbb{R}^J$. Since A is also closed, it must be a compact set.

Corollary A1.5 *If K is a closed and generating cone in $\mathbb{R}^J$, then the K-order intervals of $\mathbb{R}^J$ are compact.*

Proof. Let $[0,x]_K = K \cap (x + K)$ be an order interval. Since K is closed, it should be obvious that $[0,x]_K$ is also closed. Now fix some vector $q \in (K')^\circ$ and note that $0 \leq q \cdot y \leq q \cdot x$ for each $y \in [0,x]_K$, i.e., the set $q \cdot [0,x]_K$ is bounded. By Lemma A1.4, the order interval $[0,x]_K$ is compact.

Now let K be a cone in $\mathbb{R}^J$. The K-supremum of two points $x,y \in \mathbb{R}^J$, if it exists, will be denoted $x \vee_K y$. We shall say that K is a *lattice cone* if for any two points $x,y \in \mathbb{R}^J$ the supremum $x \vee_K y$ exits. An immediate consequence of the basic duality Theorem A1.2 is the following.

Lemma A1.6 *A closed and generating cone in $\mathbb{R}^J$ is a lattice cone if and only if its dual cone K' is likewise a lattice cone.*

A non-zero vector x in a cone K is called a *K-extremal vector* if $0 \leq_K y \leq_K x$ implies $y = \alpha x$ for some $\alpha \geq 0$. The half-line $L(x) = \{\alpha x : \alpha \geq 0\}$ generated by a K-extremal vector x is called a *K-extremal ray* (or simply an *extremal ray*) of K.

Lemma A1.7 *For a cone K in $\mathbb{R}^J$ we have the following.*

1. *If K is a lattice cone, then K has (aside of scalar multiples) exactly J extremal vectors (which are necessarily linearly independent) that generate the cone K.*
2. *If K is generated by J linearly independent vectors of K, then K is a lattice cone and (aside of scalar multiples) these linearly independent vectors are the only extremal vectors of K.*

In other words, K is a lattice cone if and only if there exist J linearly independent vectors $e_1, e_2, \ldots, e_J$ in K that generate K, i.e.,

$$K = \left\{ \sum_{i=1}^{J} \lambda_i e_i : \ \lambda_i \geq 0 \ \text{for all} \ i = 1, 2, \ldots, J \right\}.$$

Moreover, when K is a lattice cone, the half rays $L(e_1), L(e_2), \ldots, L(e_J)$ are the only extremal rays of K and for each pair of vectors $x = \sum_{i=1}^{J} \lambda_i e_i$ and $x = \sum_{i=1}^{J} \mu_i e_i$ we have

$$x \vee_K y = \sum_{i=1}^{J} \max\{\lambda_i, \mu_i\} e_i \quad \text{and} \quad x \wedge_K y = \sum_{i=1}^{J} \min\{\lambda_i, \mu_i\} e_i.$$

Recall that a non-empty convex subset B of a cone K is said to be a *base* for K if for each non-zero $x \in K$ with $x \neq 0$ there exist a unique vector $b \in B$ and a unique scalar $\lambda > 0$ such that $x = \lambda b$. The following simple result follows easily from the definitions.

Lemma A1.8 *If B is a base for a cone K, then (aside from scalar multiples) the extremal vectors of K are precisely the extreme points of the convex set B.*

Regarding the existence of bases we have the following result of V. Klee [11]. (For a proof see [8, Theorem 3.12.8, p. 144 and Corollary 3.12.9, p. 145].)

Lemma A1.9 (Klee) *If K is a closed cone in $\mathbb{R}^J$, then:*

a. *K has a compact base, and*
b. *K coincides with the convex hull of its extremal vectors.*

The proof of the existence of our cheapest hedge will be based upon the following duality result that is a special case of a result in [2].

Theorem A1.10 *Let K be a closed and generating cone in $\mathbb{R}^J$. Then, for any $x, y \in \mathbb{R}^J$ and any $q \in K'$ we have*

$$\inf_{z \geq_K x, \, z \geq_K y} q \cdot z = \max_{p \in [0,q]_{K'}} \left[p \cdot (x - y) + q \cdot y \right] = \max_{p \in [0,q]_{K'}} \left[p \cdot x + (q - p) \cdot y \right].$$

Proof. Fix $q \in K'$, and let $x, y \in \mathbb{R}^J = (R^J)''$. By Corollary A1.5, the K'-order intervals of $\mathbb{R}^J = (R^J)'$ are norm compact. Now the desired formula follows from [2, Theorem 7.6] applied to the partially ordered vector space $L = (R^J, K')$ whose order dual is $L^\sim = (R^J, K)$.

A2. Portfolio dominance

In this section we shall discuss the two-period securities model when there are S states and $J \leq S$ non-redundant securities. The only information needed for our analysis is the payoff matrix

$$
R = \begin{bmatrix}
r_1(1) & r_2(1) & \ldots & r_J(1) \\
r_1(2) & r_2(2) & \ldots & r_J(2) \\
\vdots & \vdots & \ddots & \vdots \\
r_1(S) & r_2(S) & \ldots & r_J(S)
\end{bmatrix},
$$

where $r_1, r_2, \ldots, r_J$ are the J non-redundant securities. As mentioned before, the sth row of the matrix R will be denoted q_s, i.e., $q_s = \big(r_1(s), r_2(s), \ldots, r_J(s)\big)$.

We shall consider the matrix R as a linear operator $R \colon \Theta = \mathbb{R}^J \to \mathbb{R}^S$, where Θ is viewed as the *portfolio space* and $\mathbb{R}^S$ as the asset space. Since the rank of the matrix R is J, the matrix R as an operator from R^J to R^S is one-to-one.

The *asset span* or the *marketed space* is the range of the operator R, and is denoted M or $\langle R \rangle$. Clearly, the operator $R \colon \mathbb{R}^J \to M$ is one-to-one and surjective. We always consider the marketed space M partially ordered by the closed cone $M_+ = \mathbb{R}^S_+ \cap M$. When M_+ is lattice cone of M, then M is called a *lattice-subspace* of $\mathbb{R}^S$.

Although the non-redundant securities $r_1, r_2, \ldots, r_J$ are not assumed to be positive vectors, we shall impose the following technical condition on M.

– Assumption: *The cone M_+ is generating in M, i.e., $M = M_+ - M_+$.*

If the riskless bond is marketed, then it should be clear that M_+ is generating. Also, if each security r_i is positive, then M_+ is automatically generating. We are now ready to define the portfolio cone.

Definition A2.1 *The **portfolio cone** is the cone in the portfolio space defined by*

$$
C = \{\theta \in \Theta = \mathbb{R}^J \colon R\theta \geq 0\} = \{\theta \in \mathbb{R}^J \colon q_s \cdot \theta \geq 0 \text{ for each } s = 1, 2, \ldots, J\}.
$$

That is, the portfolio cone C consists of all portfolios in $\mathbb{R}^J$ with non-negative payoff and is the inverse image of the standard cone in $\mathbb{R}^S$ under the operator R, i.e., $C = R^{-1}(\mathbb{R}^S_+) = R^{-1}(M_+)$. This easily implies that C is a closed cone in Θ, and our basic assumption shows that we have following.

Lemma A2.2 *The portfolio cone C is closed and generating.*

Recall that the vectors in $\Theta' = (\mathbb{R}^J)'$ are also known as *security prices*. If $p \in \Theta'$ and $\theta \in \Theta$, then $p \cdot \theta$ represents the value of the portfolio θ ate prices p. The prices in the dual cone of C are known as weakly arbitrage prices.

Definition A2.3 *A **weakly arbitrage free price** is a price lying in the dual cone of the portfolio cone C. That is, the weakly arbitrage free prices are the prices in*

$$
C' = \{q \in \Theta' = \mathbb{R}^J \colon q \cdot \theta \geq 0 \text{ for all } \theta \in C\}.
$$

A price $q \in C'$ is said to be *arbitrage free* if $\theta \in C$ and $\theta \neq 0$ imply $q \cdot \theta > 0$. That is, the arbitrage free prices are the C-strictly positive vectors—which, according to Lemma A1.3, they are precisely the vectors in $(C')^\circ$. Since $(C')^\circ$ is dense in C', we have the following important property.

Lemma A2.4 *The cone of weakly arbitrage free prices C' is closed and generating and is precisely the closure of the convex set $(C')^\circ$ of all arbitrage free prices.*

Specializing Lemma A1.6 to C and C' we have the following.

Lemma A2.5 *The three statements below are equivalent.*

1. *The portfolio cone C is a lattice cone.*
2. *The cone of weakly arbitrage free prices C' is a lattice cone.*
3. *The marketed space M is a lattice-subspace of $\mathbb{R}^S$.*

We now come to the notions of dominance by portfolios and prices.

Definition A2.6 *A portfolio θ is said to **dominate** another portfolio η if $\theta \geq_C \eta$, i.e., if $R\theta \geq R\eta$.*
*Similarly, a weakly arbitrage free price q **dominates** another weakly arbitrage free price p if $q \geq_{C'} p$, that is, if for any portfolio $\theta \in C$ we have $q \cdot \theta \geq p \cdot \theta$.*

Since $R\theta \geq 0$ is equivalent to $R\theta \cdot y \geq 0$ for all $y \in \mathbb{R}_+^S$ and $R\theta \cdot y = \theta \cdot R^t y$ holds (where R^t denotes the transpose of the matrix R), it follows that $R^t y$ belongs to C' for each $y \in \mathbb{R}_+^J$. That is, we have the inclusion $\{R^t y\colon y \in \mathbb{R}_+^J\} \subseteq C'$, where $\{R^t y\colon y \in \mathbb{R}_+^J\}$ is clearly the (closed) cone generated by the rows of the payoff matrix R. The next results informs that, in fact, we have equality.

Lemma A2.7 *The cone of weakly arbitrage free prices C' is precisely the cone generated by the rows of the payoff matrix R. That is,*

$$C' = \left\{R^t y\colon y \in \mathbb{R}_+^J\right\} = \left\{\sum_{s=1}^{S} \lambda_s q_s\colon \lambda_s \geq 0 \text{ for each } s = 1, 2, \ldots, S\right\}.$$

Proof. Let $C_1 = \{R^t y\colon y \in \mathbb{R}_+^J\}$. As noticed above, C_1 is the closed (convex) subcone of C' that is generated by the rows of the payoff matrix R. If $C_1 \neq C'$, then there exists some $q \in C'$ such that $q \notin C_1$. So, by the Separation Theorem, there exists some $\theta \in \mathbb{R}^J$ such that $r \cdot \theta \geq 0 > q \cdot \theta$ holds for all $r \in C_1$. In particular, we have $q_s \cdot \theta \geq 0$ for each s, and so $\theta \in C$. This implies $q \cdot \theta \geq 0$, which contradicts $q \cdot \theta < 0$. This contradiction establishes that $C_1 = C'$. ∎

The next result presents a connection between the extremal rays of C' and the rows of the payoff matrix R. This is a basic result for our work.

Theorem A2.8 *The cone of weakly arbitrage free prices C' enjoys the following properties.*

1. *Every extremal ray of C' coincides with the half ray generated by some row of R (and so C' has a finite number of extremal rays).*
2. *The number ℓ of all extremal rays of C' satisfies $J \leq \ell \leq S$. In particular, C' is a lattice cone if and only if $\ell = J$.*

Proof. (1) Let q be an extremal vector of C' and let $L(q)$ be its half-ray. By Lemma A2.7, there exist row vectors $q_{s_1}, \ldots, q_{s_k}$ of the payoff matrix R and positive constants $\alpha_1, \ldots, \alpha_k$ such that $q = \sum_{i=1}^{k} \alpha_i q_{s_i}$. From $0 \leq_{C'} \alpha_1 q_{s_1} \leq_{C'} q$ and the extremality of q, there exists some $\lambda > 0$ such that $\alpha_1 q_{s_1} = \lambda q$. Hence, $q = \mu q_{s_1}$ holds for some $\mu > 0$, and so $L(q) = L(q_{s_1})$. This shows that C' has a finite number of extremal rays.

(2) Let ℓ be the number of extremal rays of C'. By part (1), it follows that $\ell \leq S$. Also, let $q_{s_1}, \ldots, q_{s_\ell}$ be ℓ rows of R that generate all extremal rays of C'.

By Lemma A1.9, we know that C' is the convex hull of its extremal vectors. This implies that C' is generated by the row vectors $q_{s_1}, \ldots, q_{s_\ell}$. In particular, from $\mathbb{R}^J = C' - C'$ it follows that $\ell \geq J$. Otherwise, if $\ell < J$ were true, then the vector space $C' - C'$ could not be of dimension J.

For the last part, notice first that if $\ell = J$, then the vectors $q_{s_1}, \ldots, q_{s_\ell}$ must be linearly independent. This implies that the cone C' must be a lattice cone. On the other hand, if C' is a lattice cone, then it must have exactly J extremal rays, in which case we infer that $\ell = J$.

We are now ready to discuss the existence of cheapest hedging portfolios.

Theorem A2.9 *For any portfolio θ and any arbitrage free price q there exists a portfolio θ^* such that: its payoff is positive, it is dominating θ, and*

$$q \cdot \theta^* = \min_{\eta \geq_C \theta, \, \eta \geq_C 0} q \cdot \eta = \max_{0 \leq_{C'} p \leq_{C'} q} p \cdot \theta .$$

Proof. Fix a portfolio θ and let q be an arbitrage free price. Since C has interior points, there exists some $\eta_1 \in C$ such that $\eta_1 \geq_C \theta$. Now consider the convex set

$$A = \{\eta \in C : \ \eta \geq_C \theta \text{ and } q \cdot \eta \leq q \cdot \eta_1\} .$$

Clearly, A is a closed subset of C and $q \cdot A$ is bounded. By Lemma A1.4, the set A is compact. Now, from $\eta_1 \in C$ and $\eta_1 \geq_C \theta$, we see that $\eta_1 \in A$. To complete the proof notice that

$$\inf_{\eta \in A} q \cdot \eta = \inf_{\eta \geq_C \theta, \, \eta \geq_C 0} q \cdot \eta ,$$

and then use Theorem A1.10 and the compactness of A.

Corollary A2.10 *Let θ_1 and θ_2 be two portfolios, and let q be an arbitrage free price. Then there exists a portfolio θ^* dominating θ_1 and θ_2 such that*

$$q \cdot \theta^* = \min_{\eta \geq_C \theta_1, \, \eta \geq_C \theta_2} q \cdot \eta = \max_{0 \leq_{C'} p \leq_{C'} q} \left[p \cdot \theta_1 + (q - p) \cdot \theta_2 \right] .$$

Proof. By Theorem A2.9 there exists some portfolio ϵ such that

$$q \cdot \epsilon = \min_{\eta \geq_C \theta_1 - \theta_2, \, \eta \geq_C 0} q \cdot \eta = \max_{0 \leq_{C'} p \leq_{C'} q} p \cdot (\theta_1 - \theta_2) .$$

Now if we let $\theta^* = \epsilon + \theta_2$, then it is easy to check that θ^* satisfies the desired properties.

Any portfolio θ^* dominating θ_1 and θ_2 satisfying the optimality equation of Corollary A2.10 is known as a *cheapest hedging portfolio* (or a *minimum-premium insurance portfolio*) for θ_1 and θ_2 with respect to the arbitrage free price q.

In [1] it was shown that a unique minimum-premium insurance portfolio exists for any pair of portfolios that is independent of the arbitrage free price if and only if C is a lattice cone. We can prove that result easily from our analysis here.

Lemma A2.11 (Aliprantis–Brown–Werner) *The following are equivalent:*

1. *Each pair of portfolios θ_1 and θ_2 admits a unique minimum-premium insurance portfolio θ^* that is independent of the arbitrage free price. That is, for each pair θ_1 and θ_2 of portfolios there exists a unique portfolio θ^* dominating θ_1 and θ_2 such that for each arbitrage free price q we have*

$$q \cdot \theta^* = \min_{\eta \geq_C \theta_1, \, \eta \geq_C \theta_2} q \cdot \eta.$$

2. *The portfolio cone C is a lattice cone in $\mathbb{R}^J$ or, equivalently, the marketed space M is a lattice-subspace of $\mathbb{R}^S$.*

In particular, if C is a lattice cone, then the unique portfolio θ^* that satisfies property (1) is the portfolio $\theta^* = \theta_1 \vee_C \theta_2$.

Proof. (1) $\implies$ (2) Assume that θ^* has the stated uniqueness property. If some portfolio η satisfies $\eta \geq_C \theta_1$ and $\eta \geq_C \theta_2$, then we have $q \cdot \eta \geq q \cdot \theta^*$ for each arbitrage free price q. Since the arbitrage free prices are dense in C', we see that $q \cdot \eta \geq q \cdot \theta^*$ for each $q \in C'$. By Theorem A1.2, we get $\eta \geq \theta^*$, and this shows that $\theta^* = \theta_1 \vee_C \theta_2$.

(1) $\implies$ (2) If C is a lattice cone, then it is easy to see that the portfolio $\theta^* = \theta_1 \vee_C \theta_2$ satisfies the properties stated in (1).

A3. The proof of Theorem 3.2

For any non-empty subset I of the index set of states $\{1, 2, \ldots, S\}$, let H_I be the vector subspace generated in $\mathbb{R}^J$ by the collection of the row vectors $\{q_s : s \in I\}$. Clearly, there is a finite number of distinct vector subspaces of the form H_I. Let

$$\mathcal{H} = \bigcup_{\{I : \, \dim H_I < J\}} H_I.$$

Thus, the set $\mathcal{H}$ is a (finite) union of vector subspaces. As expected, the closed set $\mathcal{H}$ has an empty interior.

Lemma A3.1 *The set $\mathcal{H}$ is closed and has no interior points. In particular, the set of arbitrage free prices not in $\mathcal{H}$ is open and dense in the set of arbitrage free prices.*

Proof. Clearly, each H_I is a closed subspace of $\mathbb{R}^J$. Since $\dim H_I < J$ implies $H_I^\circ = \emptyset$, it follows that $\mathcal{H}$ is a finite union of closed sets with empty interior. The conclusion now follows from the following topological fact.

(•) *If $C_1, C_2, \ldots, C_k$ are closed subsets of a topological space such that $C_i^\circ = \emptyset$ holds for each i, then the closed set $C = \bigcup_{i=1}^k C_i$ has an empty interior.*

A proof of the preceding claim goes as follows. Assume that x is an interior point of $C = \bigcup_{i=1}^k C_i$. Pick an open neighborhood N of x such that $N \subseteq C$. Since x is not an interior point of C_1, there exists some point $x_1 \in N$ such that $x_1 \notin C_1$. Thus, x belongs to the open set C_1^c, and so there exists an open neighborhood N_1 of x_1 such that $N_1 \cap C_1 = \emptyset$. Replacing N_1 by $N \cap N_1$, we can assume that $N_1 \subseteq N$.

Similarly, since x_1 is not an interior point of C_2 there exists some point $x_2 \in N_1$ and an open neighborhood N_2 of x_2 satisfying $N_2 \subseteq N_1$ and $N_2 \cap C_2 = \emptyset$. Proceeding this way, we see that there exist points $x_1, x_2, \ldots, x_k$ and open sets $N_1, N_2, \ldots, N_k$ such that $x_i \in N_i$ and $N_i \cap C_i = \emptyset$ for each $1 \le i \le k$, and

$$N_k \subseteq N_{k-1} \subseteq \cdots \subseteq N_2 \subseteq N_1 \subseteq N \subseteq C.$$

Now notice that

$$\emptyset \ne N_k = N_k \cap C = N_k \cap \left(\bigcup_{i=1}^k C_i \right) = \bigcup_{i=1}^k N_k \cap C_i \subseteq \bigcup_{i=1}^k N_i \cap C_i = \emptyset,$$

which is impossible. This contradiction completes the proof of (•).

For the last claim observe that the set of arbitrage free prices is $(C')^\circ$ satisfies

$$(C')^\circ = (C')^\circ \cap \overline{\mathcal{H}^c} \subseteq \overline{(C')^\circ \cap \mathcal{H}^c} \subseteq \overline{(C')^\circ} = C'.$$

Since $(C')^\circ$ is dense in C', we infer that $\overline{(C')^\circ \cap \mathcal{H}^c} = C'$.

Recall that a subset $I = \{s_1, s_2, \ldots, s_J\}$ of the set of states $\{1, 2, \ldots, S\}$ defines a *pseudo-complete market* if the $J \times J$ matrix R_I with rows the vectors $q_{s_1}, q_{s_2}, \ldots, q_{s_J}$ is invertible. In this case, we also say that R_I is a *pseudo-complete market*.

The basic result needed to prove Theorem 3.2 is the following.

Lemma A3.2 *If θ is an arbitrary portfolio and q is an arbitrage free price, then there exists a portfolio θ^* such that:*

1. *θ^* dominates θ and has positive payoff, i.e., $\theta^* \ge_C \theta$ and $\theta^* \ge_C 0$.*
2. *θ^* solves the optimization problem*

$$q \cdot \theta^* = \min_{\eta \ge_C \theta, \, \eta \ge_C 0} q \cdot \eta.$$

3. *$\theta^* = R_I^{-1} \max\{R_I \theta, 0\}$ for some pseudo-complete market R_I.*

Proof. If $\theta \in -C$, i.e., if $\theta \le_C 0$, then the conclusion should be obvious; the portfolio $\theta^* = 0$ does the job. So, we can suppose that $\theta \notin -C$. We shall assume first that the arbitrage free price q does not belong to $\mathcal{H}$, i.e., $q \notin \mathcal{H}$.

By Theorem A2.9 there exists a portfolio θ^* that satisfies (1) and

$$q \cdot \theta^* = \min_{\eta \ge_C \theta, \, \eta \ge_C 0} q \cdot \eta = \max_{0 \le_{C'} p \le_{C'} q} p \cdot \theta.$$

Since $\theta \notin -C$, it follows from $\theta^* \geq_C \theta$ and $\theta^* \geq_C 0$ that $\theta^* >_C 0$. Consequently, the strict positivity of q implies $q \cdot \theta^* > 0$. Under the assumption $q \notin \mathcal{H}$ we shall verify next that this θ^* also satisfies (3).

Start by observing that since the order interval of security prices $[0, q]_{C'}$ is compact, there exists some $p^* \in [0, q]_{C'}$ such that

$$p^* \cdot \theta = \max_{0 \leq_{C'} p \leq_{C'} q} p \cdot \theta . \tag{$\star$}$$

¿From $p^* \leq_{C'} q$, $\theta \leq_C \theta^*$ and (2), we get $p^* \cdot \theta^* \leq q \cdot \theta^* = p^* \cdot \theta \leq p^* \cdot \theta^*$. Therefore,

$$p^* \cdot \theta^* = p^* \cdot \theta = q \cdot \theta^* > 0 . \tag{$\star\star$}$$

In particular, we have $p^* \neq 0$.

Since $p^* \in C'$, there exist (in view of Lemma A2.7) a non-empty set of states I_1 and positive constants $\{\alpha_s : s \in I_1\}$ such that $p^* = \sum_{s \in I_1} \alpha_s q_s$. We claim that $q_s \cdot \theta \geq 0$ holds for each $s \in I_1$. To see this, assume that for some $s_0 \in I_1$ we have $q_{s_0} \cdot \theta < 0$. From $\sum_{s \in I_1} \alpha_s (q_s \cdot \theta) = p^* \cdot \theta = q \cdot \theta^* > 0$, it follows that I_1 must have at least two states. Now notice that the inequalities

$$\left[\sum_{s \in I_1 \setminus \{s_0\}} \alpha_s q_s \right] \cdot \theta = \sum_{s \in I_1 \setminus \{s_0\}} \alpha_s (q_s \cdot \theta) > \sum_{s \in I_1} \alpha_s (q_s \cdot \theta) = p^* \cdot \theta ,$$

and $0 \leq_{C'} \sum_{s \in I_1 \setminus \{s_0\}} \alpha_s q_s \leq_{C'} \sum_{s \in I_1} \alpha_s q_s = p^* \leq_{C'} q$ contradict $(\star)$. So, $q_s \cdot \theta \geq 0$ for each $s \in I_1$.

From $(\star\star)$ we have $\sum_{s \in I_1} \alpha_s (q_s \cdot \theta) = \sum_{s \in I_1} \alpha_s (q_s \cdot \theta^*)$. Taking into account that $\theta \leq_C \theta^*$ is equivalent to $q_s \cdot \theta \leq q_s \cdot \theta^*$ for each $s = 1, 2, \ldots, S$, it follows that $q_s \cdot \theta = q_s \cdot \theta^* \geq 0$ for each $s \in I_1$. Therefore,

$$q_s \cdot \theta^* = \max\{q_s \cdot \theta, 0\} \text{ for each } s \in I_1 . \tag{$\dagger$}$$

Next, notice that $p^* \in [0, q]_{C'}$ implies $q - p^* \in [0, q]_{C'}$. If $q - p^* = 0$, let $I_2 = \emptyset$. If $q - p^* >_{C'} 0$, let I_2 be a non-empty subset of $\{1, 2, \ldots, S\}$ for which there exist positive scalars $\{\beta_s : s \in I_2\}$ such that $q - p^* = \sum_{s \in I_2} \beta_s q_s$. From $(\star\star)$, it follows that $\sum_{s \in I_2} \beta_s (q_s \cdot \theta^*) = (q - p^*) \cdot \theta^* = 0$. Since $\theta^* \geq_C 0$ is equivalent to $q_s \cdot \theta^* \geq 0$ for each s, the latter implies $q_s \cdot \theta^* = 0$ for each $s \in I_2$. In particular, from $\theta \leq_C \theta^*$ we infer that $q_s \cdot \theta \leq q_s \cdot \theta^* = 0$ holds for all $s \in I_2$. This shows that

$$q_s \cdot \theta^* = \max\{q_s \cdot \theta, 0\} \text{ for each } s \in I_2 . \tag{$\dagger\dagger$}$$

By assumption $q \notin \mathcal{H}$. So, from $q = p^* + (q - p^*) \in H_{I_1 \cup I_2}$, it follows that $\dim H_{I_1 \cup I_2} = J$. This guarantees the existence of J linearly independent row vectors in $\{q_s : s \in I_1 \cup I_2\}$. Let $I = \{s_1, s_2, \ldots, s_J\} \subseteq I_1 \cup I_2$ be such a set of J states for which the set of vectors $\{q_s : s \in I\}$ is linearly independent. From $(\dagger)$

and (††), we see that

$$R_I \theta^* = \begin{bmatrix} q_{s_1} \cdot \theta^* \\ q_{s_2} \cdot \theta^* \\ \vdots \\ q_{s_J} \cdot \theta^* \end{bmatrix} = \begin{bmatrix} \max\{q_{s_1} \cdot \theta, 0\} \\ \max\{q_{s_2} \cdot \theta, 0\} \\ \vdots \\ \max\{q_{s_J} \cdot \theta, 0\} \end{bmatrix}$$

$$= \max \left\{ \begin{bmatrix} q_{s_1} \cdot \theta \\ q_{s_2} \cdot \theta \\ \vdots \\ q_{s_J} \cdot \theta \end{bmatrix}, \begin{bmatrix} 0 \\ 0 \\ \vdots \\ 0 \end{bmatrix} \right\} = \max\{R_I \theta, 0\} .$$

Finally, notice that the $J \times J$ square matrix R_I has rank J and so it is invertible. Consequently, $\theta^* = R_I^{-1} \max\{R_I \theta, 0\}$, and the validity of (3) has been established.

Next, we consider the case $q \in \mathcal{H}$. By Lemma A3.1 there exists a sequence $\{q_n\}$ of arbitrage free prices such that $q_n \to q$ and $q_n \notin \mathcal{H}$ for each n. By Theorem A2.9, for each n there exists a portfolio θ_n^* dominating θ with positive payoff satisfying

$$q_n \cdot \theta_n^* = \min_{\eta \geq_C \theta, \eta \geq_C 0} q_n \cdot \eta = \max_{0 \leq_{C'} p \leq_{C'} q_n} p \cdot \theta .$$

By the preceding case, for each n there exists a set I_n of J states such that

$$\theta_n^* = R_{I_n}^{-1} \max\{R_{I_n} \theta, 0\} .$$

Since there is only a finite number of subsets of the set of states $\{1, 2, \ldots, S\}$, we can assume (by passing to a subsequence if necessary) that there exists a fixed subset I of J indices such that $I_n = I$ for each n. This implies

$$\theta_n^* = R_I^{-1} \max\{R_I \theta, 0\} = \theta^*$$

for each n. We shall show that θ^* satisfies properties (1), (2), and (3). Clearly, (1) and (3) are satisfied automatically. So, to finish the proof, we must prove the validity of (2).

To this end, take any $\eta \geq_C 0$ satisfying $\eta \geq_C \theta$. Then, we have $q_n \cdot \eta \geq q \cdot \theta_n^* = q \cdot \theta^*$ for all n. Taking limits yields $q \cdot \eta \geq q \cdot \theta^*$. This shows that θ^* is a solution to the optimization problem

$$\min_{\eta \geq_C \theta, \eta \geq_C 0} q \cdot \eta ,$$

and the proof is finished.

Corollary A3.3 *If θ_1 and θ_2 are arbitrary portfolios and q is an arbitrage free price, then there exists a portfolio θ^* such that:*

1. θ^* dominates θ_1 and θ_2.
2. θ^* solves the optimization problem

$$q \cdot \theta^* = \min_{\eta \geq_C \theta_1, \eta \geq_C \theta_2} q \cdot \eta .$$

3. $\theta^* = R_I^{-1} \max\{R_I\theta_1, R_I\theta_2\}$ *for some pseudo-complete market R_I.*

Proof. Consider the portfolio $\theta = \theta_1 - \theta_2$. According to Lemma A3.2 there exists a portfolio ϵ^* such that:

a. ϵ^* dominates θ and has positive payoff.
b. ϵ^* solves the optimization problem

$$q \cdot \epsilon^* = \min_{\eta \geq_C \theta, \, \eta \geq_C 0} q \cdot \eta = \max_{0 \leq_{C'} p \leq_{C'} q} p \cdot \theta.$$

c. $\epsilon^* = R_I^{-1} \max\{R_I\theta, 0\}$ for some pseudo-complete market R_I.

Now let $\theta^* = \epsilon^* + \theta_2$ and note that θ^* satisfies properties (1), (2), and (3).

Finally, we are ready to prove the Cheapest Hedge Theorem 3.2. Start by observing that since the bond $\mathbf{k}$ is marketed, there exists some portfolio $\theta_1 \in \mathbb{R}^J$ such that $R\theta_1 = \mathbf{k}$. By Corollary A3.3 there exists some portfolio θ^* such that:

i. θ^* dominates θ and θ_1.
ii. θ^* solves the optimization problem

$$q \cdot \theta^* = \min_{\eta \geq_C \theta, \, \eta \geq_C \theta_1} q \cdot \eta.$$

iii. $\theta^* = R_I^{-1} \max\{R_I\theta, R_I\theta_1\}$ for some pseudo-complete market R_I.

Next, consider the finite minimization problem:

$$(\mathcal{FMP}) \qquad \min q \cdot \eta$$
$$\text{s.t.: } \eta \in \mathcal{P}_{\theta,k},$$

where $\mathcal{P}_{\theta,k}$ is the set of all potentially insuring portfolios of θ at the floor k, i.e.,

$$\mathcal{P}_{\theta,k} = \{\eta \in \mathbb{R}^J : \eta = \eta_I \text{ for some pseudo-complete market } R_I \text{ and } R\eta \geq R\theta \vee \mathbf{k}\}.$$

From (i), (ii), and (iii), we see that the portfolio θ^* is a solution of the minimization problem $(\mathcal{FMP})$, and that any solution of $(\mathcal{FMP})$ satisfies (i), (ii), and (iii). Now the validity of all statements in Theorem 3.2 follow from this equivalence.

References

1. Aliprantis, C.D., Brown, D.J., Werner, J.: Minimum-cost portfolio insurance. Journal of Economic Dynamics and Control **24**, 1703–1719 (2000)
2. Aliprantis, C.D., Tourky, R.: The super order dual of an ordered vector space and the Riesz–Kantorovich formula. Trans. American Mathematical Society **354**, 2055–2077 (2002)
3. Aliprantis, C.D., Tourky, R.: Markets that don't replicate any options. Economics Letters **76**, 437–442 (2002)
4. Broadie, M., Cvitanic, J., Soner, H.M.: Optimal replication of contingent claims under portfolio constraints. Review of Financial Studies **11**, 59–81 (1998)
5. Brown, D.J., Ross, S.A.: Spanning, valuation and options. Economic Theory **1**, 3–12 (1991)

6. Edirisinghe, C., Naik, V., Uppal, R.: Optimal replication of options with transaction costs and trading restrictions. Journal of Financial and Quantitative Analysis **28**, 117–139 (1993)
7. Green, R., Jarrow, R.A.: Spanning and completeness in markets with contingent claims. Journal Economic Theory **41**, 202–210 (1987)
8. Jameson, G.J.O.: Ordered linear spaces. Lecture Notes in Mathematics, **141**. Berlin Heidelberg New York: Springer 1970
9. Karatzas, I., Kou, S.G.: On the pricing of contingent claims under constraints. The Annals of Applied Probability **6**, 321–369 (1996)
10. Karatzas, I., Kou, S.G.: Hedging American contingent claims with constrained portfolios. Finance and Stochastics **2**, 215–258 (1998)
11. Klee, Jr., V.L.: Extremal structure of convex sets. Archive of Mathematics **8**, 234–240 (1957)
12. Leland, H.: Who should buy portfolio insurance. Journal of Finance **35**, 581–594 (1980)
13. Magill, M.M., Quiinzi, M.: The theory of incomplete markets. Cambridge, MA: MIT Press 1995
14. Naik, V., Uppal, R.: Leverage constraints and the optimal hedging of stock and bond options. Journal of Financial and Quantitative Analysis **29**,199–223 (1994)
15. Ross, S.A.: Options and efficiency. Quarterly Journal of Economics **90**, 75–89 (1976)

The informational efficiency of finite price mechanisms

Leonid Hurwicz and Thomas Marschak

[1] Department of Economics, University of Minnesota, Minneapolis, MN 55455, USA
(e-mail: hurwicz@tc.umn.edu)
[2] Walter A. Haas School of Business, University of California, Berkeley, CA 94720, USA
(e-mail: marschak@socrates.berkeley.edu)

Received: March 20, 2002; revised version: July 26, 2002

Summary. This paper obtains finite counterparts of previous results that showed the informational efficiency of the Walrasian mechanism among all mechanisms yielding Pareto-optimal individually rational trades in exchange economies while using a continuum of possible messages. Such "continuum" mechanisms lack realism, even when the space of environments (characteristics) is a continuum, since it is not possible to transmit or announce all elements of a message continuum, and it generally takes infinite time to find an exact equilibrium message among all messages in such a continuum. Accordingly the paper studies finite approximations (having a finite number of messages) of the continuum Walrasian mechanism and of other continuum allocation mechanisms. An approximation's *overall error* for a class of exchange economies, is the largest distance, over all economies in the class, between the continuum mechanism's final allocation and the approximation's final allocation. In particular, we develop finite counterpart s of the superiority, with respect to message-space dimension, of the Walrasian mechanism over Direct Revelation (DR). We measure a finite mechanism's *cost* by the number of its (equilibrium) messages.

Our two main results are as follows: (1) For exchange economies we find that the overall error of a (sufficiently fine) approximate Walrasian mechanism is less than the overall error of a not-more-costly approximation of a continuum DR mechanism whose equilibrium outcomes are trades that are Pareto optimal and individually rational. More generally, approximate Walrasian mechanisms are superior, in the same sense, to approximations of any continuum mechanism whose equilibrium outcomes (like those of the continuum Walrasian mechanism) are Pareto optimal individually rational trades and whose message space has higher dimension than that of the Walrasian mechanism. (2) As we increase without limit the dimension of the set of environments (characteristics) defining our class of exchange economies, we find that the extra cost of such DR approximations relative to Walrasian approximations, when both achieve the same overall error, also grows without limit.

Thus the informational superiority of the Walrasian mechanism emerges again when we approximate it and take the finite number of messages in the approximation as our cost measure.

Keywords and Phrases: Mechanisms, Mechanism design, Approximating smooth mechanisms, Information processing, Exchange economies.

JEL Classification Numbers: D20, D50, D80, D83.

1 Introduction[1]

This paper obtains finite counterparts of previous results that showed the informational efficiency of the Walrasian mechanism among all mechanisms that yield Pareto-optimal individually rational trades in exchange economies while using a continuum of possible messages. Such mechanisms lack realism, even when the set of possible environments (characteristics) is also a continuum, since one cannot transmit or announce all points of a message continuum. Moroever finding an exact equilibrium message, among all the messages in such a continuum, generally takes an infinite time. It is therefore of considerable interest to see whether the informational superiority of the Walrasian mechanism again emerges when we approximate it, using a finite set of messages. In our study of this question we apply general results from another paper [6]. That paper considered finite approximations of continuum mechanisms in general organizations, with exchange economies as a particular example.

To set the stage, let us recall the puzzle which gave rise to the previous literature.[2] While the optimality of the Walrasian equilibrium and the conditions under which it exists were well understood, it remained true that many other mechanisms could be constructed, which also yield, at their equilibrium positions, trades that are Pareto optimal and individually rational. Such mechanisms might not use prices at all. One of the alternative mechanisms is Direct Revelation, where a Center gathers complete information about the traders' preferences and endowments, and then calculates and imposes an optimal allocation. Other mechanisms might pool part of the traders' private information but not all of it.

It was natural to conjecture that the Walrasian mechanism had a special property: the information transmitted is no "more" than the task of finding an optimal allocation requires. The alternative mechanisms, on the other hand, generally require the transmission of more information than necessary. That conjecture certainly agreed with the informal but widely shared view that the performance of price mechanisms cannot be matched by "command" mechanisms. It also agreed with more

---

[1] This paper is an expanded version of [7].

[2] A survey of the literature (up to the mid 1980s) is found in [5]. Among the papers that dealt with the informational efficiency (specifically the message-space minimality) of the Walrasian mechanism are [4,9,14–17].

explicit statements in classic writings, by Hayek and others, as to the merits of price mechanisms, with regard to information.[3]

The difficulty was that some precise modelling had to be done before the conjecture could be checked. "A mechanism achieving the task" and "amount of information transmitted" had to be given exact meanings.

The literature that we seek to supplement took an approach that may be characterized in the following way.

- It considered an n-person organization (an n-trader exchange economy for example), where person i privately observes e_i, which lies in a set E_i; we call e_i person i's *characteristic* or *local environment*. In the exchange-economy case, e_i describes $i's$ endowment and preferences.

- It defined a *mechanism on* $E \equiv E_1 \times \cdots \times E_n$ *with action space* A as a triple $\langle M, (g_1, \ldots, g_n), h \rangle$. Here M is the *message space*; g_i, which we call *person i's agreement function*, is a function from $M \times E_i$ to a finite-dimensional Euclidean space; and h is a function, called the *outcome function*, from M to A. In an important special case, M is the Cartesian product $A \times Y$ and h is a projection operator, i.e., for $m = (a, y) \in M$, we have $h(m) = a$. We shall then say that the mechanism is a *projection mechanism*. An example is the Walrasian mechanism (treated in more detail below), where a message is a trade/price pair (a, p).

- It interpreted the operation of a mechanism in several ways. In one of them, sometimes called the "verification scenario", a sequence of messages m in M is announced. Person i responds to a given m by computing $g_i(m, e_i)$ and signals "agreement" if the result is zero.[4] In the case of the Walrasian mechanism, with non-numeraire commodities $1, \ldots, L$ and a numeraire $L + 1$, the function g_i has real-valued components $g_{i\ell}$, one for each non-numeraire commodity. A message m consists of an allocation a and a price vector $p = (p_1, \ldots, p_L)$, and

$$g_{i\ell}((a, p), e_i)$$
$$= \frac{i\text{'s marginal utility for commodity } \ell \text{ if the trades in } a \text{ were carried out}}{i\text{'s marginal utility for commodity } L + 1 \text{ if the trades in } a \text{ were carried out}} - p_\ell.$$

- It defined an *equilibrium message for* $e = (e_1, \ldots, e_n)$ to be a message m to which all agree at e, i.e., $g_i(m, e_i) = 0, i = 1, \ldots, n$, and supposed that when an equilibrium message m has been found, then the action $h(m)$, called *an equilibrium action for* e, is taken. In the Walrasian case, h is the projection operator, i.e., $h(a, p) = a$. It confined attention to mechanisms which *cover* E, i.e., for every e in E an equilibrium message exists.

- It defined a mechanism as *realizing the goal function* $\gamma : E \to A$ if $h(m) = \gamma(e)$ whenever m is an equilibrium message for e. For the exchange-economy case, we have a goal function γ^*, where $\gamma^*(e)$ is a balanced n-tuple of trade vectors which is Pareto-optimal and individually rational for e.

---

[3] While the paper by Hayek ([3]) is frequently cited as an articulate statement of the superiority of price mechanisms, an extensive discussion of the issue is found earlier, in the debate about the possibility of socialism. See, for example, Lange [11].

[4] An alternative interpretation, called the "iterative-adjutment-process scenario" is given below, near the start of Section 2.

- It chose the *size of the message space* M as a natural measure of a mechanism's informational cost. If M is a differentiable manifold, then its *dimension* is the appropriate size measure.
- It found that for a class of exchange economies with classic properties (convex preferences) there is no mechanism which obeys certain regularity conditions, realizes γ^*, and has a message space of smaller dimension than that of the Walrasian mechanism.

Thus the approach yielded a rigorous version of the informational claims long made informally for the Walrasian mechanism. But the claim dealt only with mechanisms whose message spaces are continua.

In seeking a version of the claim when we require mechanisms to have a *finite* message space, we start by defining a *finite approximation* of a continuum mechanism, say $\Lambda = \langle M, (g_1, \ldots, g_n), h \rangle$, whose message space M is a D-dimensional subset of $\mathbb{R}^D$ and whose action space is a subset of $\mathbb{R}^\alpha$, where $D \geq 1, \alpha \geq 1$. The agreement function g_i has real-valued components g_{ik}. Suppose M is compact. Then, in our approach, the message space of a *mesh-ϵ* finite approximation of Λ, where $\epsilon > 0$, is denoted M_ϵ. It is the intersection of M with an unbounded lattice of points in $\mathbb{R}^D$, called the ϵ-*lattice*. The points of the lattice are spaced 2ϵ apart in each coordinate. We can interpret ϵ as determined by the number of digits to which we round off each of the D real message variables when we turn from the continuum mechanism Λ to its finite approximation.

We obtain the kth component of person i's agreement function in our finite approximation by "softening" the original function g_{ik}. The softened function takes the value zero for a given e_i and a message m^* in M_ϵ if and only if $|g_{ik}(m^*, e_i)| \leq \eta$, where η is a positive number called the approximating mechanism's *tolerance*.

Our new "softened" agreement function may be interpreted as the agreement function of the original continuum mechanism, but now computed to a specified accuracy.

As for the finite approximation's outcome function, we shall consider two variants. In the *exact-outcome* variant, the outcome function is simply the restriction of h to M_ϵ. In defining *the round-outcome* variant, we confine attention to an action set A which is an open set of $\mathbb{R}^\alpha$. We specify an *action mesh*, say ν. Then the rounded-outcome function assigns to a message m in M_ϵ that point of the ν-lattice in $\mathbb{R}^\alpha$ which lies closest to the action $h(m) \in A$ selected by the original continuum mechanism. (Distance is defined as the maximum difference over all coordinates.)

The *overall error* of a finite approximation of Λ is the largest distance, over all e in E, between an equilibrium action for e in the original mechanism Λ and an equilibrium action for e in the approximation. We shall take the *cost* of a finite approximation to be the finite number of its equilibrium messages. We shall say that the first of two approximations is *not-more-costly-than* the second if the number of equilibrium messages in the first is not more than the number of equilibrium messages in the second.[5] We shall say that a class of approximations of a given continuum mechanism is *complete* if for some $\epsilon^\# > 0$ the class contains a mesh-ϵ approximation for every ϵ with $0 < \epsilon \leq \epsilon^\#$. Given two regular continuum

---

[5] In [6] the term "cost-equivalent to" is used instead of "not-more-costly than".

mechanisms, we shall say that a complete class A_1 of finite approximations of the first is *superior* to a complete class A_2 of finite approximations of the second if and only if the following holds:

If the mesh of an approximation $a_1 \in A_1$ is sufficiently fine, and $a_2 \in A_2$ is not-more-costly than a_1, then the overall error of a_1 is less than the overall error of a_2.

In the present paper we state three general propositions about approximations and we then apply them to allocation mechanisms in a class of exchange economies. Propositions A and B lead to Exchange-economy Propositions 1 and 2, which are our main results. Proposition C deals with the relatively minor issue of approximating two continuum mechanisms whose message spaces have exactly the *same* dimension; it leads to Exchange-economy Proposition 3.

Proposition A considers two regular continuum mechanisms on an environment set E. Each has an action space which is a subset of $\mathbb{R}^\alpha$ and the first has a lower message-space dimension than the second. The Proposition asserts that a complete class of approximations of the first is superior to a complete class of approximations of the second in the following cases:

- the approximations in both classes are of the rounded-outcome type, with the ratio of action mesh to message mesh meeting a common upper bound and a common and lower bound for all members of both classes
- the approximations in the first class are of the rounded-outcome type (with a common upper bound and a common positive lower bound to the action-mesh/message-mesh ratio), the approximations in the second class are of the exact-outcome type, and the second continuum mechanism has the projection property
- the approximations in the first class are of the rounded-outcome type (with a common upper bound and a common positive lower bound to the action-mesh/message-mesh ratio), the approximations in the second class are of the exact-outcome type, and the second continuum mechanism has the Direct Revelation (DR) property: its message space is the environment set E, and each person agrees to a message if and only if it describes her local environment correctly.

Proposition B deals with some general issues raised by the striking virtue which the continuum Walrasian mechanism displays in a convex economy: its message-space dimension remains the same no matter how large we make the environment set, i.e., no matter what the dimension of the set of the traders' possible individual characteristics may be. The Walrasian mechanism is superior to the DR mechanism with regard to message-space dimension as long as the environment set's dimension exceeds the dimension of the Walrasian mechanism, and the Walrasian mechanism's superiority over the DR mechanism grows without bound as we increase without bound the dimension of the environment set, since the DR mechanism's message space is (by our definition of DR) the environment set itself.

In Proposition B we study this matter in a very general setting, with the Walrasian mechanism in an exchange economy as an example. We consider an infinite sequence of environment sets of increasing dimension. For each set in the sequence we consider a continuum DR mechanism and an *indirect* continuum mechanism, both realizing the same goal function. In all the indirect continuum mechanisms the message space has *the same* dimension and is contained in a cube which remains the same for all the mechanisms. But while the cost of the indirect continuum mechanism (its message space dimension) remains the same for all environment sets, the cost of the DR mechanism is the dimension of the evironment set itself, and so it goes to infinity as the environment set's dimension increases without bound. We seek finite analogues of that continuum statement. Two types of analogues can be studied. The first approach maintains the viewpoint of Proposition A and studies, for each environment set in our sequence, the *indirect-versus-DR* **error** *differential*: the difference in overall error between an approximation of the indirect mechanism and a not-more-costly approximation of the DR mechanism. The second approach is to adopt a viewpoint that is "dual" to that of Proposition A and to study, for each environment set in our sequence, the *indirect-versus-DR* **cost** *differential*: the difference in cost between an approximation of the indirect mechanism and an approximation of a *no-larger-error* DR mechanism, i.e., an approximation whose overall error is at least as small as the overall error of the indirect approximation.

Pursuing the first approach, we see that if the possible values of the goal-fulfilling actions satisfy a bound that stays unchanged for all environment sets, then the indirect-versus-DR error differential cannot go to infinity as the environment-set dimension goes to infinity.[6] Proposition B pursues the second approach and obtains an analogue of the continuum statement: we find that under certain further conditions, the cost differential indeed goes to infinity as the environment set's dimension goes to infinity, even though the possible values of the goal-fulfilling actions satisfy a bound that stays the same for all environment sets.

In Proposition C we return to the case of a fixed environment set E. It deals with approximations of two continuum mechanisms whose message spaces are of the *same* dimension, say D. We shall say that a complete class of approximations of one continuum mechanism is *indifferent* to a complete class of approximations of another if we can find two approximations, one in each class, whose overall errors are as close as we wish and at the same time one approximation is not-more-costly than the other. Proposition C says that if the goal function realized by each of the two D-dimensional continuum mechanisms has some regularity properties, then a complete class of rounded-outcome approximations of the first is indifferent to a complete class of rounded-outcome approximations of the second. Similarly, the Proposition says that a complete class of exact outcome approximations of the first is indifferent to a complete class of exact-outcome approximations of the second.

We apply Proposition A,B, and C to a class of *convex* exchange economies, each defined by an element $e = (e_1, \ldots, e_n)$ of an environnment set $E = E_1 \times \cdots \times E_n$. Each e_i in E_i is a vector of parameters defining a concave quasi-linear utility function for Trader i.

---

[6] In the exchange-economy example with fixed endowments, the actions are trades and those are bounded by the fixed endowments for all environment sets. So both the overall error of an approximation of the Walrasian mechanism, and the overall error of an approximation of a DR mechanism realizing the same goal as the Walrasian mechanism, are bounded as the environment set dimension grows without bound.

Our Exchange-economy Proposition 1 has two parts. In Part I, we apply a portion of Proposition A, letting the low-dimensional continuum mechanism be the Walrasian mechanism and letting the high-dimensional continuum mechanism be a Direct Revelation mechanism which has E itself as its message space and has trade vectors as its actions. We show that the Walrasian mechanism indeed has the regularity required if we are to apply Proposition A. Part I says that if (as is typically the case) the message-space dimension of the continuum Walrasian mechanism is less than the dimension of the environment set E (the message space of the continuum DR mechanism), then a sufficiently fine finite exact-outcome approximation of the DR mechanism has higher overall error than a not-more-costly rounded-outcome approximation of the continuum Walrasian mechanism. It is readily checked that if we have two mesh-ϵ approximations of any given continuum mechanism and one has exact outcomes while the other has rounded outcomes, then the exact-outcome approximation has a not-larger overall error than the rounded-outcome approximation. Thus Part I shows the superiority of the approximate Walrasian mechanisms over the approximate DR mechanisms even when we are "harsh" to the Walrasian approximations by insisting that the outcomes be rounded while we are "kind" to the DR approximations by letting the outcomes be exact.

In Part II of Exchange-economy Proposition 1 we apply another portion of Proposition A. Part II says that a sufficiently fine rounded-outcome approximation of the Walrasian mechanism has lower overall error than a not-more-costly rounded-outcome approximation of *any* continuum mechanism whose outcomes are trade vectors and whose message-space dimension is higher than that of the continuum Walrasian mechanism.

In Exchange-economy Proposition 2, we apply Proposition B. Proposition 2 provides a finite counterpart of the continuum Walrasian mechanism's increasing superiority over the continuum DR mechanism as the size of the environment set E grows. Proposition 2 considers three sequences: a sequence of environment sets of increasing dimension, the associated sequence of continuum Walrasian mechanisms, and an associated sequence of continuum DR mechanisms whose equilibrium outcomes are Pareto optimal individually rational allocations. It says that (1) for every triple in the three sequences, there is an exact-outcome approximation of the DR mechanism with a not-larger overall error than any given exact-outcome approximation of the Walrasian mechanism, but (2) as the three sequences grow without limit so does the amount by which the cost of the DR approximation exceeds the cost of the Walrasian approximation.

Thus for *convex* economies[7] we indeed obtain a finite analogue of the continuum Walrasian mechanism's striking informational virtue relative to a continuum DR mechanism that also yields Pareto optimal individually rational allocations: as the dimension of the environment set (parameter space) grows without bound, the continuum DR mechanism's message-space dimension goes to infinity but the continuum Walrasian mechanism's message-space dimension remains unchanged. For

---

[7] I.e., economies in which all preferences and production sets are convex.

certain *non-convex* economies[8], however, the situation is different. As Calsamiglia showed [1], in an economy with convex preferences but increasing returns in production (hence a non-convex economy), even with the number of agents and goods constant one finds that as the number of parameters in the production functions goes to infinity, so must the dimension of any "smooth" mechanism yielding Pareto-optimal allocations. A similar situation arises in economies with externalities that result in non-convex transformation sets (Hurwicz [8], especially Proposition 1), and it seems likely that analogous impossibility results can be obtained for exchange economies with one convex trader and one non-convex trader.

Thus in those non-convex economies all continuum mechanisms share with DR the unpleasant feature that as the environment-set grows without bound, so must the message space. In the presence of non-convexities there may be no mechanism with the message-space boundedness property possessed by the continuum Walrasian mechanism in convex economies. That does not mean, however, that in non-convex economies one cannot do better, with regard to message-space size, than DR. In fact for any given (finite) dimension of the environment set, there typically exist continuum mechanisms realizing the goal function of interest but using a lower message-space dimension than DR. *Parameter transfer* is one example of such a mechanism.[9]

We conjecture that there are finite counterparts of the preceding statements about continuum mechanisms in non-convex economies.

Finally we apply Proposition C to obtain Exchange-economy Proposition 3. That proposition returns to the case of a fixed environment set E. Proposition 3 considers the (possibly empty) class of regular continuum mechanisms that realize the same goal function as the continuum Walrasian mechanism (for our class of exchange economies), are distinct from the Walrasian mechanism, and have a message-space dimension equal to that of the continuum Walrasian mechanism. We find that nothing is to be gained by considering approximations of such a continuum mechanism: a complete class of approximations of such a continuum mechanism is indifferent (in our sense) to a complete class of approximaitions of the continuum Walrasian mechanism.

Exchange-economy Propositions 1 and 3 recapture in a new way the "winning" status of the Walrasian mechanism among all regular mechanisms that yield Pareto-optimal individually rational trade-vector n-tuples for all the exchange economies defined by the elements of our environment set E. If one seeks finite mechanisms for an exchange economy, if one confines attention to finite mechanisms that are approximations of regular continuum mechanisms, and if one wants the overall

---

[8] I.e., economies in which *not all* preferences or production sets are convex.

[9] In a parameter-transfer mechanism a message contains a proposed description of the local environment of persons $1, \ldots, n-1$ and also proposes an action. Persons $1, \ldots, n-1$ agree to a message if and only if their local environments are correctly described, while person n agrees if and only if she finds (given her own local environment) that the proposed action meets a specified goal were the other $n-1$ local environments to have the values give in the message. Any goal function can be realized by such a mechanism. For certain goal functions, moreover, one cannot improve on the parameter-transfer mechanism: its message-space dimension is a lower bound to the message-space dimension of all (regular) continuum mechanisms that realize the goal function. Inner product is an example of such a goal function. See [5, Section 3.2].

error of any sufficiently fine approximation to be as small as possible for a given cost, then one cannot do better than to approximate the Walrasian mechanism.

The remainder of this paper is organized in the following way. In Section 2 we present the concepts and definitions that we need, including the regularity requirement that we impose on the continuum mechanisms to be considered. The discusssion includes an extensive remark (Remark R1) on conditions that imply regularity. In Section 3 we summarize our three general propositions A-C. They are obtained from our companion paper [6]. Section 4 describes our exchange economy with quasi-linear utilities and presents the continuum Walrasian mechanism for that economy. We show (following the program presented in Remark R1) that the mechanism obeys our regularity conditions, so that the three propositions of Section 3 can indeed be applied. We illustrate our concepts for the special case of a two-trader one-nonnumeraire-commodity economy, in which the nonlinear part of each trader's quasi-linear utility function is quadratic.

In Section 5 we apply the propositions of Section 3 and obtain the three propositions about approximations of the continuum Walrasian mechanism that were just sketched. Section 6 offers brief concluding remarks.

2 Continuum mechanisms and their finite approximations: concepts and definitions

We suppose throughout that Person i's environment set $E_i, i = 1, \ldots, n$ is a compact set in $\mathbb{R}^{J_i}$. Its typical element is

$$e_i = (e_{i1}, \ldots, e_{iJ_i}).$$

We shall suppose throughout that E_i contains an open set in $\mathbb{R}^{J_i}$. We define

$$J \equiv \sum_{i=1}^{n} J_i.$$

We shall also suppose that the action space, denoted A, is a subset of $\mathbb{R}^\alpha$, where α is a positive integer.

A *mechanism on* $E = E_1 \times \cdots \times E_n$ *with action space* A is a triple

$$\Lambda = \langle M, (g_1, \ldots, g_n), h \rangle,$$

where M is the *message space*; $g_1, \ldots, g_n$ are the *individual agreement functions*; and h is the *outcome function*.

They have the following properties:

- For some D, with $2 \leq D \leq J$, M is a subset of $\mathbb{R}^D$. Its typical element is the message $m = (m_1, \ldots, m_D)$.
- $g_i : M \times E_i \to \mathbb{R}^{D_i}, i = 1, \ldots, n$, where $D_1 + \cdots + D_n = D$.
- For $i = 1, \ldots, n$, the function g_i has components $g_{i1}, \ldots, g_{iD_i}$, where $g_{ik} : M \times E_i \to \mathbb{R}, k = 1, \ldots, D_i$.

- ("Coverage") For every $e = (e_1, \ldots, e_n) \in E$, there exists $m \in M$ such that $g_1(m, e_1) = g_2(m, e_2) = \cdots = g_n(m, e_n) = 0$.
- $h : M \to A$.

We shall sometimes use the statement[10] $g(m) = 0$ as an abbreviation for the statement $g_i(m, e_i) = 0, i = 1, \ldots, n$.

If M is a continuum, then Λ is a *continuum mechanism*. If M is finite, then Λ is a *finite mechanism*.

Definition D1 *We shall call the mechanism* $\Lambda^* = \langle M^*, g_1^*, \ldots, g_n^*, h^* \rangle$ *on* E *with action space* A *a* **Direct Revelation mechanism** *if*

- $M^* = E$ *and each* $m^* \in M^*$ *is a J-tuple*

$$
\left(m_1^*, \ldots, m_{J_1}^*, \ldots, m_{\left(\sum_{k=1}^{i-1} J_k\right)+1}^*, \ldots, m_{\left(\sum_{k=1}^{i} J_k\right)}^*, \ldots, m_J^* \right)
$$

- *For* $i = 1, \ldots, n$, *we have* $g_i^* : M^* \times E_i \to \mathbb{R}^{J_i}$ *and*

$$
g_i^*(m^*, e_i) = \left(m_{\left(\sum_{k=1}^{i-1} J_k\right)+1}^*, \ldots, m_{\left(\sum_{k=1}^{i} J_k\right)}^* \right) - e_i.
$$

Thus, for a DR mechanism, we have $D_i = J_i, i = 1, \ldots, n$, and $D = J_1 + \cdots + J_n = J$.

An important class of mechanisms has the property that a portion of each message is a proposed action, and the outcome function is a projection operator, which assigns that proposed action to the message.

Definition D2 *We shall call* Λ *a* **projection mechanism** *if there is a set* Y *such that* $M = A \times Y$ *and, for every* $m = (a, y)$ *with* $a \in A, y \in Y$, *we have* $h(m) = a$.

Definition D3 *A message* $m \in M$ *is said to be an* **equilibrium message of** Λ **for the environment** $e = (e_1, \ldots, e_n) \in E$ *if* $g_i(m, e_i) = 0, i = 1, \ldots, n$. *An outcome (action)* $a \in A$ *is said to be an* **equilibrium outcome (action) of** Λ *for*

---

[10] We may interpret the D_i-tuple $(g_{i1}, \ldots, g_{iD_i})$ using an *iterative adjustment-process* scenario rather than the "agreement" (or "verification") scenario of the Introduction. In the iterative adjustment-process scenario, there is a sequence of messages, each belonging to M. Person i responds to the tth message, denoted

$$
m(t) = (m_1(t), \ldots, m_D(t)),
$$

by broadcasting to everyone else the message variables

$$
m_{\left(\sum_{j=1}^{i-1} D_j\right)+1}(t+1) = f_{i1}(m(t), e_i), \ldots, m_{\left(\sum_{j=1}^{i} D_j\right)}(t+1) = f_{iD_i}(m(t), e_i),
$$

where

$$
f_{i1} - m_{\left(\sum_{j=1}^{i-1} D_j\right)+1} = g_{i1}, f_{i2} - m_{\left(\sum_{j=1}^{i-1} D_j\right)+2} = g_{i2}, \ldots, f_{iD_i} - m_{\left(\sum_{j=1}^{i} D_j\right)} = g_{iD_i}.
$$

Thus an equilibrium message for e for the functions $g_1, \ldots, g_n$ is a *stationary* message for e in the iterative adjustment process defined by the functions $\{f_{ij}\}_{i=1,\ldots,n, j=1,\ldots,D_i}$.

the environment $e \in E$ *if* $a = h(m)$ *and* m *is an equilibrium message for* e. *The* **equilibrium message set for** Λ *is the set*

$$\{m \in M \; : \; m \text{ is an equilibrium message of } \Lambda \text{ for some } e \in E\}.$$

The **equilibrium outcome (action) set for** Λ *is the set*

$$\{a = h(m) \; : \; m \text{ is an equilibrium message of } \Lambda \text{ for some } e \in E\}.$$

Definition D4 *The mechanism* Λ *is said to* **realize the goal function** $\gamma : E \to A$ *if we have*

$$a = \gamma(e)$$

whenever the following two statements hold:

- $a = h(m)$
- m *is an equilibrium message for* e.

Note that the goal function realized by a DR mechanism $\Lambda^* = \langle M^*, (g_1^*, \ldots, g_n^*), h^* \rangle$ (where $M^* = E$) is identical to the outcome function h^*.

We next state regularity conditions that we shall henceforth impose on the continuum mechanis that we consider.

Definition D5 *We shall call a continuum mechanism* $\Lambda = \langle M, (g_1, \ldots, g_n), h \rangle$ *on the environment set* E **regular** *if it satisfies the following conditions:*

(a) M *is compact.*

(b) *For each* $i \in \{1, \ldots, n\}$ *and each* $k \in \{1, 2, \ldots, D_i\}$, *the function* g_{ik} *is* $\mathcal{C}^2$.

(c) *There exists a real number* $\bar{\delta} > 0$ *and a* $\mathcal{C}^2$ *function*

$$\bar{t} : E \times \underbrace{[-\bar{\delta}, \bar{\delta}] \times [-\bar{\delta}, \bar{\delta}] \times \cdots \times [-\bar{\delta}, \bar{\delta}]}_{D \text{ times}} \to M,$$

called an **approximate-solution function,** *such that for all* $e \in E$, *and all*

$$\delta_{11}, \ldots, \delta_{1D_1}, \ldots, \delta_{n1}, \ldots, \delta_{nD_n} \in [-\bar{\delta}, \bar{\delta}],$$

the message

$$m = \bar{t}(e, \delta_{11}, \ldots, \delta_{1D_1}, \ldots, \delta_{n1}, \ldots, \delta_{nD_n})$$

is the only message in M *satisfying the equation system*

$$g_{i1}(m, e_i) = \delta_{i1}, g_{i2}(m, e_i) = \delta_{i2}, \ldots, g_{iD_i}(m, e_i) = \delta_{iD_i}, i = 1, \ldots, n.$$

In particular, for any $e \in E$, *the only message in* M *satisfying* $g(m, e) = 0$ *is* $m = t(e)$, *where the function* $t : E \to M$, *is called the* **thread** *of the mechanism* Λ, *and is defined by:*

$$t(e) \equiv \bar{t}(e, \underbrace{0, 0, \ldots, 0}_{D \text{ times}}).$$

(d) *The image set* $t(E)$ *includes an open set of* $\mathbb{R}^D$.

Note that a regular mechanism realizes the goal function

$$\gamma = h \circ t$$

(the composition of h and t) and no other goal function.

Note also that if Λ is a DR mechanism, then t is the identity function.

Remark R1 (concerning Regularity Condition (c))

Regularity Condition (c) appears to be demanding. In particular, while one might expect a mechanism to have a smooth thread t and a smooth approximate-solution function $\bar{t}$, the *uniqueness* of those functions may be difficult to guarantee.

Hence it is natural to ask whether, once we impose Regularity Conditions (a) and (b), we can conclude that (c) holds as well if the Jacobian

$$Q(m, e) \equiv \left(\left(\frac{\partial g_{ik_i}(m, e_i)}{\partial m_j} \right) \right)_{i = 1, \ldots, n; \, k_i = 1, \ldots, D_i; \, j = 1, \ldots, D}$$

has appropriate properties. In particular, one might require that for every e in E and every m in M, the matrix $Q(m, e)$ has nonzero determinant. That would imply that the agreement rules g_i have no redundancy.

It turns out that we can indeed obtain all of Condition (c), including uniqueness, if our mechanism has some further properties, including some restrictions on $Q(m, e)$. Those properties will be exhibited, in particular, by the exchange-economy mechanism that we construct in Section 4 below. The further properties are as follows.

(1) The compact message space M contains *in its interior* a set $\bar{M}$ such that for every $e \in E$, there is a message $m_e \in \bar{M}$ which solves $g(m, e) = 0$. (Since the mechanism covers E we already know that M contains such a solution).

(2) The functions g_i obey the following *extension condition*:
There exist

   (i) open sets $\hat{E}_i \subset \mathbb{R}^{J_i}$ such that E_i is in the interior of $\hat{E}_i$, i=1,...,n;
   (ii) an open set $\hat{M} \subset \mathbb{R}^D$ such that M is in the interior of $\hat{M}$ and $\hat{M}$ is a *rectangular region* [11]
   (iii) an extension of g, namely a C^2 function $\hat{g} : \hat{M} \times \hat{E} \to \mathbb{R}^D$ (where $\hat{E} \equiv \hat{E}_1 \times \cdots \times \hat{E}_n$) such that $\hat{g}(m, e) = g(m, e)$ for all $m \in M, e \in E$.

(3) For every $(m, e) \in \hat{M} \times \hat{E}$, the Jacobian

$$\hat{Q}(m, e^*) \equiv \left(\left(\frac{\partial \hat{g}_{ik_i}(m, e_i^*)}{\partial m_j} \right) \right)_{i = 1, \ldots, n; \, k_i = 1, \ldots, D_i; \, j = 1, \ldots, D}$$

has nonzero determinant.

---

[11] That means that $\hat{M} = \{m = (m_1, \ldots, m_D) : \bar{m}_j < m_j < \bar{\bar{m}}_j, \text{ all } j \in \{1, \ldots, D\}\}$, where $\bar{m}_j < \bar{\bar{m}}_j$.

(4) For every (m, e) in $\hat{M} \times \hat{E}$, the Jacobian $\hat{Q}(m, e)$ obeys a suitable "univalence" condition, selected from the work of Gale and Nikaido. Such a condition guarantees, for any e^* in $\tilde{E}$, that if m belongs to the open rectangular region $\hat{M}$ and solves an equation system for which $\hat{Q}(m, e^*)$ is the associated Jacobian, then $\hat{M}$ contains no other solution to that system.

Now property **(1)** tells us that for every e in E, the message m_e, which solves $g(m, e) = 0$, lies in the interior of M. Then properties **(2)** and **(3)**, together with the Implicit Function Theorem, tell us that for every e^* in E, there is an open neighborhood $U(e^*) \subset \hat{E}$, a neighborhood $V(m_{e^*}) \subset M$, and a C^2 function $\rho_{e^*} : U(e^*) \to M$, such that for all $e \in U(e^*)$:

(i)   the message $\rho_{e^*}(e) = m_{e^*}$ solves $g(m, e) = 0$
(ii) $m_{e^*} \in V(m_{e^*})$
(iii) for every $\bar{m} \neq m_{e^*}$ in $V(m_{e^*})$, we have $g(\bar{m}, e) \neq 0$.

Moreover, since m_{e^*} is in the interior of M and solves $g(m, e^*) = 0$, the Implicit Function Theorem has a broader implication.[12] For $i = 1, \ldots, n$, let $\overrightarrow{\delta}_i$ denote the vector $(\delta_{i1}, \ldots, \delta_{iD_i})$. There is a positive number r_{e^*}, and a C^2 function

$$\bar{\rho}_{e^*} : U(e^*) \times \underbrace{[-r_{e^*}, r_{e^*}] \times \cdots \times [-r_{e^*}, r_{e^*}]}_{D \text{ times}} \to M,$$

such that for all $\overrightarrow{\delta}_1, \ldots, \overrightarrow{\delta}_n$ with components in $[-r_{e^*}, r_{e^*}]$, and all $e \in U(e^*)$, the message

$$m_{e^* \overrightarrow{\delta}_1, \ldots, \overrightarrow{\delta}_n} = \bar{\rho}_{e^*}(e, \overrightarrow{\delta}_1, \ldots, \overrightarrow{\delta}_n)$$

solves the equation system

$$(*) \qquad g_{ik}(m, e_i) = \delta_{ik}, k = 1, \ldots, D_i, i = 1, \ldots, n,$$

and no other message in a neighborhood $\bar{V}\left(m_{e^* \overrightarrow{\delta}_1, \ldots, \overrightarrow{\delta}_n}\right)$ of $m_{e^* \overrightarrow{\delta}_1, \ldots, \overrightarrow{\delta}_n}$ solves that system. (Thus $\bar{\rho}_{e^*}(e, \underbrace{0, \ldots, 0}_{D \text{ times}}) = \rho_{e^*}(e)$ and $\bar{V}(m_{e^* \underbrace{0, \ldots, 0}_{D \text{ times}}}) = V(m_{e^*})$).

The Jacobian associated with the equation system $(*)$, for $(m, e) \in M \times E$, is again $\hat{Q}(m, e)$. By **(4)**, that Jacobian satisfies a Gale/Nikaido condition. Such a condition, together with the fact that $\hat{M}$ is an open rectangular region, assures us that $m_{e^* \overrightarrow{\delta}_1, \ldots, \overrightarrow{\delta}_n}$ is the only solution to the equation system $(*)$ to be found in all of $\hat{M}$ and, *a fortiori*, in all of M.

To summarize what we have so far obtained:

---

[12] To obtain the broader implication, one uses the facts that (1) for every $m \neq m_{e^*}$ in $V(m_{e^*})$, we have $g(m, e^*) \neq 0$, and (2) the set $g(V(m_{e^*})) \subset \mathbb{R}^D$ has dimension D (since $Q(m, e^*)$ has full rank for all $m \in M$).

for all $e^* \in \hat{E}$ and for all $\vec{\delta}_1, \ldots, \vec{\delta}_n$ with components in $[-r_{e^*}, r_{e^*}]$, the compact set M contains just one solution to the equation system

$$(*) \qquad \hat{g}_{ik}(m, e_i^*) = \delta_{ik}, k = 1, \ldots, D_i, i = 1, \ldots, n,$$

namely the message $m_{e^* \vec{\delta}_1, \ldots, \vec{\delta}_n}$.

Finally, we can construct the positive constant $\bar{\delta}$ of Regularity Condition (c). The procedure is as follows. The collection of open sets $\{U(e^*) : e^* \in \hat{E}\}$ covers the compact set E (including its boundary points). By the Heine-Borel theorem, that collection contains a *finite* sub-collection, say Σ, which also covers E. Now for each set $U(e^*)$ in Σ, consider the associated positive number r_{e^*} in (+). Take the smallest such positive number, over all the sets in the finite collection Σ. In view of (+), that smallest positive number meets the requirements of $\bar{\delta}$ in Condition (c).

Now define the function

$$\bar{t} : E \times \underbrace{[-\bar{\delta}, \bar{\delta}] \times \cdots \times [-\bar{\delta}, \bar{\delta}]}_{D \text{ times}} \to M,$$

required in Condition (c), by

$$\bar{t}(e^*, \vec{\delta}_1, \ldots, \vec{\delta}_n) \equiv m_{e^* \vec{\delta}_1, \ldots, \vec{\delta}_n}.$$

Since the function $\bar{\rho}_{e^*}$ is a C^2 function on $U(e^*) \times \underbrace{[-r_{e^*}, r_{e^*}] \times \cdots \times [-r_{e^*}, r_{e^*}]}_{D \text{ times}}$, it follows that $\bar{t}$ is also a C^2 function, as Condition (c) requires.

We conclude that **if our continuum mechanism obeys Regularity Conditions (a) and (b), and has properties (1)–(4), then it also obeys Regularity Condition (c).**

The message space of a finite approximation to a continuum mechanism $\Lambda = \langle M, g_1, g_2, h \rangle$ is the intersection of the continuum $M \subset \mathbb{R}^D$ with an unbounded lattice of separated and evenly spaced points in $\mathbb{R}^D$. Each agreement function of the approximation is a "softening" of the corresponding agreement function in the continuum mechanism: it equals zero if and only if the absolute value of the continuum agreement function does not exceed a positive number called the *tolerance*.

Definition D6

(i)   Given a continuum mechanism $\Lambda = \langle M, (g_1, \ldots, g_n), h \rangle$ on E, we shall call the mechanism $\Lambda_{\eta\epsilon} = \langle M_\epsilon, (g_1^{\eta\epsilon}, \ldots, g_n^{\eta\epsilon}), h^0 \rangle$, with $\epsilon > 0, \eta > 0$, **the finite exact-outcome approximation of Λ with message mesh ϵ and tolerance η** if

- $M_\epsilon = M \cap S_\epsilon^D$, where S_ϵ^D is called the ϵ-**lattice** and is defined as follows:

$$S_\epsilon^D \equiv \text{ the } D \text{-fold Cartesian product of}$$

$$S_\epsilon \equiv \{\ldots, -2(\ell+1)\epsilon, -2\ell\epsilon, \ldots, -4\epsilon, -2\epsilon, 0, 2\epsilon, 4\epsilon, \ldots, 2\ell\epsilon, 2(\ell+1)\epsilon, \ldots\};$$

- for $i = 1, \ldots, n$ we have $g_i^{\eta\epsilon} : M_\epsilon \times E_i \to \mathbb{R}^{D_i}$;
- the function $g_i^{\eta\epsilon}$ has components $g_{i1}^{\eta\epsilon}, \ldots, g_{iD_i}^{\eta\epsilon}$, where, for $m \in M_\epsilon, e = (e_1, \ldots, e_n) \in E, k \in \{1, 2, \ldots, D_i\}$,

$$g_{ik}^{\eta\epsilon}(m, e_i) = \begin{cases} 0 \text{ if } |g_{ik}(m, e_i)| \le \eta \\ 1 \text{ otherwise} \end{cases}$$

- the coverage requirement is met, i.e., for every $e \in E$, there exists $m \in M_\epsilon$ such that $g_i^{\eta\epsilon}(m, e_i) = 0, i = 1, \ldots, n$;
- for $m \in M_\epsilon$ we have $h^0(m) = h(m)$.

(ii) A **finite rounded-outcome approximation of** Λ **with message mesh** ϵ, **action mesh** ν **and tolerance** η is a triple $\Lambda_{\eta\epsilon}^\nu = \langle M_\epsilon, (g_1^{\eta\epsilon}, \ldots, g_n^{\eta\epsilon}), h^\nu \rangle$, where $M_\epsilon, g_1^{\eta\epsilon}, \ldots, g_n^{\eta\epsilon}$ have the definitions just given, while the outcome function $h^\nu : M_\epsilon \to A$ is defined as follows: [13]

$$h^\nu(m) = \begin{cases} \text{the element } a = (a_1, \ldots, a_\alpha) \text{ of } A \cap S_\nu^\alpha \text{ which is closest to } h(m), \\ \text{where distance is measured by } \max_{r \in \{1, \ldots, \alpha\}} \left(|h(m) - a_r| \right) \text{ and} \\ \text{ties are broken downward.} \end{cases}$$

It is shown in [6] that for the approximations $\Lambda_{\eta\epsilon}, \Lambda_{\eta\epsilon}^\nu$, there exists, for a given mesh ϵ, a *minimal tolerance* $\hat{\eta}(\epsilon)$ among all tolerances that achieve coverage, and that we minimize overall error when we use the minimal tolerance. Henceforth we shall confine attention to *minimal-tolerance approximations* and shall omit the tolerance symbol η. **It will be understood that the tolerance used is** $\hat{\eta}(\epsilon)$.

We now define an approximation's error.

Definition D7 *Consider a continuum mechanism* $\Lambda = \langle M, (g_1, \ldots, g_2), h \rangle$ *on the environment set* $E = E_1 \times \cdots \times E_n$ *with action set* $A \in \mathbb{R}^\alpha$. *Consider a (minimal-tolerance) approximation of* Λ *whose mesh is* ϵ. *Let the type of outcome function be unspecified: it is either of the exact type or of the rounded type with some action mesh* ν. *To avoid repetition, denote the mechanism* $\mathcal{L} = \langle M_\epsilon, (g_1^{\epsilon\eta}, \ldots, g_n^{\epsilon\eta}), h^\# \rangle$, *where* $h^\#$ *stands for* h^0 *if the mechanism is of the exact-outcome type and stands for* h^ν *if the mechanism is of the rounded-outcome type with action mesh* ν. *The domain of* h *and of* $h^\#$ *is* M_ϵ *and the range of both functions is* $\mathbb{R}^\alpha$. *So* h *has* α *real-valued component functions, denoted* $h_1, \ldots, h_\alpha$ *and* $h^\#$ *has* α *real-valued component functions, denoted* $h_1^\#, \ldots, h_\alpha^\#$.

For the environment $e = (e_1, \ldots, e_n) \in E$, *the* **error at** e **of** $\mathcal{L}$ *is defined to be*

$$\sup \{ |h_j^\#(\bar{m}) - h_j(m)| : m \in M, \bar{m} \in M_\epsilon; g_i(m, e_i) = 0, g_i^{\epsilon\eta}(\bar{m}, e_i) = 0, i = 1, \ldots, n; j \in \{1, \ldots, \alpha\} \}.$$

The **overall error of** $\mathcal{L}$, *denoted* $\mathcal{E}(\mathcal{L})$, *is defined to be* $\sup_{e \in E}$ *(error at* e *of* $\mathcal{L}$ *)*.

---

[13] "Ties broken downward" means that if the actions a', a'', with $a' \neq a''$, are equally close to $h(m)$, then $h^\nu(m)$ equals the action whose kth component is lower, where the kth component is the first one in which a' and a'' differ.

It is readily verified that if $\mathcal{L}, \mathcal{L}'$ are both mesh-ϵ approximations of the same regular mechanism Λ, and if $\mathcal{L}$ has exact outcomes while $\mathcal{L}'$ has rounded outcomes, then $Œ(\mathcal{L}) \leq Œ(\mathcal{L}')$.

3 Three propositions about the message-space dimension of a continuum mechanism and the error of its finite approximations

We now present three general propositions that we shall be applying to the Walrasian mechanism in a class of exchange economies.

Proposition A, the first of our general Propositions, considers a regular continuum mechanism $\Lambda^{(1)}$. The mechanism's message space has dimension $D(1)$.[14] Its action space is a subset of $\mathbb{R}^\alpha$ for some $\alpha \geq 1$. The Proposition also considers three other regular continuum mechanisms. Their action spaces are also subsets of $\mathbb{R}^\alpha$, but their message spaces have dimension higher than $D(1)$. The three other mechanisms are denoted $\Lambda^{(2)}, \Lambda^{(3)}$, and $\Lambda^{(4)}$. The mechanism $\Lambda^{(2)}$ has no further restrictions. The mechanism $\Lambda^{(3)}$ has the projection property and the mechanism $\Lambda^{(4)}$ is a DR mechanism. We shall use the term "superior" defined in the Introduction. The Proposition asserts the superiority of finite approximations of $\Lambda^{(1)}$ over finite approximations of the other three mechanisms: if the mesh used to approximate any of the three is sufficiently fine, then a not-more-costly approximation of $\Lambda^{(1)}$ has smaller overall error, where cost is measured by the number of equilibrium messages. The strongest superiority result concerns approximations of $\Lambda^{(1)}$ versus approximations of $\Lambda^{(3)}$ and $\Lambda^{(4)}$. We obtain the superiority of $\Lambda^{(1)}$ even if we "handicap" it by insisting that its approximations have rounded-outcome functions, while we "favor" the other two by permitting their approximations to have exact outcome functions. On the other hand rounded-outcome approximations of $\Lambda^{(1)}$ are superior to rounded-outcome approximations of the unrestricted mechanism $\Lambda^{(2)}$.

To state the Proposition formally, we first review several terms that were presented in the Introduction and we introduce several new terms.

We shall say that a finite approximation to a continuum mechanism is *nontrivial* if it has two or more equilibrium messages.

We view the cost of a finite approximation as the number of its equilibrium messages. If $\mathcal{L}$ is a finite approximation of a continuum mechanism Λ, the symbol $C(\mathcal{L})$, called *the cost of* $\mathcal{L}$, denotes the number of equilibrium messages in $\mathcal{L}$.

Given a continuum mechanism Λ, we shall call a class $\mathcal{B}$ of finite approximations of Λ *complete*, if there exists $\epsilon^{\#} > 0$ such that for every ϵ with $0 < \epsilon \leq \epsilon^{\#}$, $\mathcal{B}$ contains at least one ϵ-mesh approximation of Λ.

We shall say that a complete class $\mathcal{S}$ of rounded-outcome approximations Λ_ϵ^ν is *controlled* if the set of action-mesh/message-mesh ratios, i.e., $\left\{ \frac{\nu}{\epsilon} : \Lambda_\epsilon^\nu \in \mathcal{S} \right\}$, is (i) bounded from above, and (ii) bounded from below by a positive number.

Finally we shall say that a goal function $\gamma : E \to A$, where $A \subseteq \mathbb{R}^\alpha$, is *well-behaved* if (i) γ is $\mathcal{C}^1$, and (ii) the image set $\gamma(E)$ contains an open set of $\mathbb{R}^\alpha$.

---

[14] More accurately, the message space is a subset of $\mathbb{R}^{D(1)}$ and therefore, by regularity, it includes an open set of $\mathbb{R}^{D(1)}$. Recall that if the message space M of a regular continuum mechanism on E is a subset of $\mathbb{R}^D$, then it has a thread t such that the image set $t(E)$ contains an open set of $\mathbb{R}^D$ (Regularity Condition (d)). Since $t(E) \subseteq M$, the message space also contains an open set of $\mathbb{R}^D$.

Our first proposition now follows. It is proved in [6], our companion paper.[15]

Proposition A. *For $i = 1, \ldots, n$, let*

$$E_i = \{e_i = (e_{i1}, \ldots, e_{iJ_i}) : \bar{e}_{ij} \le e_{ij} \le \bar{\bar{e}}_{ij}, j = 1, \ldots, J_i\},$$

where $\bar{e}_{ij} < \bar{\bar{e}}_{ij}, j = 1, \ldots, J_i$.

Suppose $\Lambda^{(1)}, \Lambda^{(2)}, \Lambda^{(3)}, \Lambda^{(4)}$ are all regular continuum mechanisms on $E \equiv E_1 \times \cdots \times E_n$. Suppose each realizes a well-behaved goal function whose range is $\mathbb{R}^\alpha$. Suppose:

- *the message space of $\Lambda^{(1)}$ is a subset of $\mathbb{R}^{D(1)}$, where $D(1) < J_1 + \cdots + J_n$.*
- *the message space of $\Lambda^{(2)}$ is a subset of $\mathbb{R}^{D(2)}$, where $D(2) > D(1)$.*
- *$\Lambda^{(3)}$ has the projection property and its message space is a subset of $\mathbb{R}^{D(3)}$ where $D(3) > D(1)$.*
- *$\Lambda^{(4)}$ is a DR mechanism (and hence its message space is E).*

Then any complete class of nontrivial controlled rounded-outcome approximations of $\Lambda^{(1)}$ is superior to:

- *any complete class of nontrivial controlled rounded-outcome approximations of $\Lambda^{(2)}$*
- *any complete class of nontrivial exact-outcome approximations of $\Lambda^{(3)}$*
- *any complete class of nontrivial exact-outcome approximations of $\Lambda^{(4)}$.*

We now turn to a proposition that concerns the informational performance of DR mechanisms relative to indirect mechanisms when the environment set grows.

As we remarked in the Introduction, the informational deficiency of continuum DR mechanisms, as compared to continuum "indirect" mechanisms that realize the same goal function, reveals itself most strikingly when we vary the dimension of the environment set E. That is illustrated by the convex exchange economy with n traders and a fixed number of commodities, which we will study in Part IV below. The environment set E is the set of possible individual characteristics (endowments and utility-function parameters). Each environment e in E defines a convex exchange economy. We are interested in a goal function which assigns to every e in E a trade vector that is Pareto optimal and individually rational for the economy described by e. We can realize the goal function by a continuum DR mechanism, whose message space is E itself, and we can also realize it by an indirect continuum mechanism, namely the Walrasian mechanism.

As the dimension of E rises, so does the message-space dimension of the associated DR mechanism (E itself). But no matter how large the dimension of E, the Walrasian mechanism retains the same message-space dimension, since that dimension depends only on the number of traders and commodities, and those are fixed.

---

[15] This proposition combines Propositions 1,2, and 3 of [6]. Those are proved for two persons and a single action variable (and an arbitrary number of message variables), but, as remarked in [6], the proof is easily extended to many persons and many action variables.

Thus the "informational differential" (dimensional difference) between the DR and the indirect mechanisms goes to infinity as the environment set's dimension grows without bound. That remains the case if, for all sets E, the endowments stay the same, or satisfy the same bounds, so that only the dimension of the set of possible utility-function parameters grows without bound.

Generalizing from the exchange-economy example, we have the following statement about continuum mechanisms:

($§$) For each environment set E in a family of environment sets, consider a continuum DR mechanism on E and a continuum indirect mechanism on E, where both realize the same goal function, with α real-valued components, and the indirect mechanism has a message space that is a subset of $\mathbb{R}^D$ for every E. As we increase without bound the dimension of the environment set E, we also increase without bound the "informational differential" (difference in message-space dimension) between the two mechanisms. That is true, in particular, if (a) the indirect mechanism's message space is not all of $\mathbb{R}^D$ but is contained, for all E, in the same closed (nondegenerate) "message cube" in $\mathbb{R}^D$, or (b) for both mechanisms, the equilibrium actions are contained, for all E, in the same closed (nondegenerate) "action cube" in $\mathbb{R}^\alpha$.

We ask whether there is a finite analogue of ($§$). In Proposition A, we found that higher message-space dimension for continuum mechanisms implied higher overall error for not-more-costly finite approximations. It is therefore natural to consider first a finite analogue of ($§$) which says

($§§$) If the environment set's dimension grows without bound, then so does the DR-versus-indirect "error differential", i.e., the difference in overall error between an approximation of the indirect continuum mechanism and a not-more-costly exact-outcome approximation of the corresponding continuum DR mechanism.

But it is immediately clear that ($§§$) cannot hold if the possible values of the goal function that both continuum mechanisms realize, or (equivalently) the possible values of an equilibrium action of either continuum mechanism, are indeed contained in the same "action cube" for all environment sets. For then the overall error of an approximation of either mechanism cannot exceed the difference between the upper and lower action bounds defined by the cube and so the overall error cannot go to infinity as the dimension of the environment set increases without bound.

In the Walrasian example, the messages are price/trade pairs and the outcome function is a projection operator which selects the trade portion of each message. If the messages are contained in the same closed message cube for all environment sets E, then the possible trades are bounded from above and from below by bounds which remain the same for all E, and hence the equilibrium actions (trades) are contained in an unchanging action cube.[16] (A special case is that in

---

[16] More generally, if the indirect mechanism has, for every E, the projection property (as the Walrasian mechanism does), then the existence of an unchanging message cube implies the existence of an unchanging action cube.

which endowments never change). So neither the overall error of approximations of the Walrasian mechanism, nor the overall error of approximations of continuum DR mechanisms that realize the same goal function as the Walrasian mechanism, can go to infinity as we vary E.

Thus (§§), our first attempt at a finite analogue of (§), fails to hold when there is an unchanging action cube for all environment sets.

Since (§§) fails when there is an unchanging action cube, it is natural to make a second attempt at a finite analogue of (§) by investigating a "dual" of (§§). We require, in the dual, that the overall error of the DR approximation be no larger than that of the indirect approximation. We again vary the environment set's dimension but now we study, for each environment set, the "DR-versus-indirect cost differential", i.e., **the amount by which the DR approximation's** *cost* **(number of equilibrium messages) exceeds the indirect approximation's cost when we require the DR approximation to have an overall error not larger than that of the indirect approximation**. Our next Proposition, [17] Proposition B, formally establishes the following informally stated "dual" finite analogue of (§).

(§§§)
> As the environment sets' dimension grows without bound, the DR-versus-indirect cost differential also grows without bound. It does so, in particular, if the continuum indirect mechanism has a message space that is contained, for every environment set E, in the same closed message cube in $\mathbb{R}^D$ and, for both mechanisms, the equilibrium actions are contained, for all E, in the same closed action cube in $\mathbb{R}^\alpha$.

To state Proposition B, we consider a *sequence* of environment sets, $\{E^v : v = 1, 2, \dots \}$. The environment set E^v has a larger dimension than E^{v-1} and the dimension goes to infinity as v increases without bound. For each v, we consider an indirect continuum mechanism on E^v, denoted Λ^v, and a DR continuum mechanism on E^v, denoted Λ^{*v}. For all v, both mechanisms have actions in the same Euclidean space $\mathbb{R}^\alpha$. In particular, it may be the case that for all v, the mechanisms Λ^v and Λ^{*v} realize the same goal function, but (as in Proposition A) that will not be needed for the Propositions we obtain. If the action sets lie in a common Euclidean space $\mathbb{R}^\alpha$, then that is enough to allow us to compare the overall error of an approximation of the indirect mechanism relative to the continuum indirect mechanism itself with the overall error of an approximation of the DR mechanism relative to the continuum DR mechanism itself.

We now impose conditions on the environment-set sequence $\{E^v = E_1^v \times \cdots \times E_n^v : v = 1, 2 \dots \}$, and on the continuum-mechanism sequences $\{\Lambda^v\}, \{\Lambda^{*v}\}$.

(i)   *For $i = 1, \dots, n$, $E_i^v = \{e_i^v = (e_{i1}^v, \dots, e_{iK_i(v)}^v) : \bar{e}_{ij}^v \le e_{ij}^v$*

$\le \bar{\bar{e}}_{ij}^v, j = 1, \dots, K_i(v)\}, v \ge 1$

(ii)   *For $i = 1, \dots, n$, $\bar{e}_{ij}^v, \bar{\bar{e}}_{ij}^v \in \mathbb{R}$ and $\bar{e}_{ij}^v < \bar{\bar{e}}_{ij}^v, j = 1, \dots, K_i(v), v \ge 1$.*

(iii) $K_1(v+1) + K_2(v+1) + \cdots + K_n(v+1) > K_1(v) + K_2(v)$

$+ \cdots + K_n(v), v \ge 1$.

---

[17] Proposition B, which follows, corresponds to Part B of Proposition 4 in [6].

Thus E^v has dimension $J^v = K_1(v) + K_2(v) + \cdots + K_n(v)$, where $J^{v+1} > J^v$, and hence

(iv)
$$\lim_{v \to \infty} J^v = \infty.$$

In Proposition B we shall make the following two further assumptions on the environment-set sequence.

(v)
there exist $W, X \in \mathbb{R}$, with $W < X$, such that
$$\bar{e}^v_{ij} \leq W, X \leq \overset{=v}{e}_{ij}, i = 1, \ldots, n, j = 1, \ldots, K_i(v), v \geq 1.$$

(vi)
there exist $Y, Z \in \mathbb{R}$, with $Y < Z$, such that $Y \leq \bar{e}^v_{ij}, \overset{=v}{e}_{ij} \leq Z$,
$i = 1, \ldots, n, j = 1, \ldots, K_i(v), v \geq 1.$

Next we consider a sequence of indirect regular mechanisms $\{\Lambda^v = \langle M^v, (g^v_1, \ldots, g^v_n), h^v \rangle\}$, where

(vii)
for $v \geq 1$, Λ^v is a mechanism on E^v
and its action space is a subset of $\mathbb{R}^\alpha, v \geq 1,$

(viii)
for $v \geq 1$, M^v is a subset of $\mathbb{R}^D$,

and

(ix)
there exist $a, b, r, s \in \mathbb{R}$, with $a < b, r < s$, such that for $M = \underbrace{[a, b] \times \cdots \times [a, b]}_{D \text{ times}}$, $M' = \underbrace{[r, s] \times \cdots \times [r, s]}_{D \text{ times}}$, and for all $v \geq 1$, we have $M' \subset M^v \subset M \subset \mathbb{R}^D$.

Condition *(viii)* and the regularity of Λ^v imply that M^v includes an open set of $\mathbb{R}^D$. Thus every mechanism in the indirect sequence has a message space of dimension D and (by condition *(ix)*) all of the message spaces are proper subsets of the same D-dimensional cube M and all of them properly include the D-dimensional cube M'.

Next we consider a sequence of regular continuum DR mechanisms $\{\Lambda^{*v} = \langle M^{*v}, (g^{*v}_1, \ldots, g^{*v}_n), h^{*v} \rangle\}$, where

(x)
Λ^{*v} is a DR mechanism on $E^v, i.e. M^{*v} = E^v, v \geq 1,$

and

(xi)
for $v \geq 1$, the action space of Λ^{*v} is a subset of $\mathbb{R}^\alpha$.

Note that *(vi)* implies that for all $v \geq 1$ there is a "universal" upper bound to the mesh of an approximation of Λ^{*v}, namely $\frac{Z-Y}{2}$. For any mesh larger than that, the set E^v contains no lattice points.

The proof of Proposition B requires that exact-outcome approximations of the DR mechanisms Λ^{*v} have the following property: if we fix the mesh, and if it is sufficiently small, then there is a uniform lower bound, valid for all v, to the ratio of overall error to mesh. One can show that this condition follows if there is a uniform lower bound, valid for all v, to the sensitivity of the goal function realized by Λ^{*v} to some environment component, where sensitivity is measured by the absolute value of the derivative of the goal function with respect to that component.[18] Accordingly we shall assume that the α real-valued components of the outcome function h^{*v}, denoted $h_1^{*v}, \ldots, h_\alpha^{*v}$, satisfy the following condition:

(xii) there exists $\tau > 0$ such that for every $v \geq 1$, there is some triple (i, j, k), with $i \in \{1, \ldots, n\}, j \in \{1, \ldots, K_i(v)\}, k \in \{1, \ldots, \alpha\}$ such that for all $e^v \in E^v$, the absolute value of $\frac{\partial h_k^{*v}}{\partial e_{ij}}$ is not less than τ.

It can then be shown that – as the proof of Proposition B requires – there exists $\tau^* > 0$ and $\tilde{\epsilon}$, with $0 < \tilde{\epsilon} \leq \frac{X-W}{2}$, such that for all v and for $0 < \epsilon < \tilde{\epsilon}$ we have $Œ(\Lambda_\epsilon^{*v}) \geq \tau^* \cdot \epsilon$.

Finally, Proposition B uses a condition on the derivatives associated with the indirect-mechanism sequence and the DR sequence $\{\Lambda^v\}$. Since we require this property for both the indirect sequence and the DR sequence.[19], we state it for an arbitrary sequence, $\{\bar{\Lambda}^v\}$, where $\bar{\Lambda}^v = \langle \bar{M}^v, (\bar{g}_1^v, \ldots, \bar{g}_n^v), \bar{h}^v \rangle$ is a regular continuum mechanism on E^v with actions in $\mathbb{R}^\alpha$. We shall say that the sequence $\{\bar{\Lambda}^v\}$ has *uniformly bounded derivatives* if there exist $\bar{\delta} > 0, \xi > 0, \rho > 0, \lambda > 0$ such that for every $v \geq 1$ the following holds:

- The mechanism Λ^v satisfies Regularity Condition (c), concerning the approximate-solution function of Λ^v, which we denote $\bar{t}^v = (\bar{t}_{11}^v, \ldots, \bar{t}_{1,K_1(v)}^v, \ldots, \bar{t}_{n1}^v, \ldots, \bar{t}_{n,K_n(v)}^v)$, where the domain of each $\bar{t}_{ij}^v$ is $[-\bar{\delta}, \bar{\delta}]$.
- For each e^v in E^v, and each $(\delta_{11}, \ldots, \delta_{1,K_1(v)}, \ldots, \delta_{n1}, \ldots, \delta_{n,K_n(v)})$ with every component in $[-\bar{\delta}, \bar{\delta}]$, the derivative of each $\bar{t}_{ij}^v$ with respect to each component of e^v does not exceed ρ in absolute value.

---

[18] The required argument is suggested by an argument found in the proof of Part 2 of Lemma 3 of [6].

[19] To interpret the following four parts of the "uniform boundedness of derivatives" condition when we apply them to each continuum DR mechanism Λ^{*v}, recall that for that mechanism and for $i \in \{1, \ldots, n\}, j \in \{1, \ldots, K_i(v)\}$, the thread is the identity function and for the typical real-valued component of the approximate-solution function $\bar{t}^v$, we have $\bar{t}_{ij}^v(e^v, \delta_{11}, \ldots, \delta_{1K_1(v)}, \ldots, \delta_{n1}, \ldots, \delta_{nK_n(v)}) = e_{ij} - \delta_{ij}$. The derivative of the latter expression with respect to any component of e^v is either zero or one. Every component of each agreement function is the difference between a message component and an environment component; hence its derivative with respect to any message component is either zero or one. Thus the first three parts of the uniform-boundedness-of-derivatives condition are automatically satisfied. The fourth part requires that there exist $\lambda > 0$ such that for every v, the absolute value of the derivative of every component of the outcome function with respect to any of its arguments does not exceed λ, or (equivalently) the absolute value of the derivative of every component of the goal function realized by Λ^{*v} with respect to any environment component does not exceed λ in absolute value.

- For each (m, e^v) in $M^v \times E^v$ the derivative of each real-valued component of each agreement function g_i^v with respect to each component of m does not exceed ξ in absolute value.
- For each m in M^v, the derivative of h^v with respect to each component of m does not exceed λ in absolute value.

We shall consider exact-outcome approximations for both of the continuum-mechanism sequences.[20]

We now state Proposition B.

Proposition B. *Let the environment sequence $\{E^v\}$ satisfy $(i) - (vi)$. Let the continuum indirect-mechanism sequence $\{\Lambda^v\}$ satisfy $(vii) - (ix)$. Let the continuum DR sequence $\{\Lambda^{*v}\}$ satisfy $(x) - (xii)$. Let both sequences have uniformly bounded derivatives. Then there exists $\hat\epsilon > 0$ and a function $\bar r$ from $\mathbb{R}^+ \times I^+$ to $\mathbb{R}^+$ (where I^+ denotes the positive integers) such that*

$(\dagger)$ *for $0 < \epsilon < \hat\epsilon$, the approximations $\Lambda_\epsilon^v, \Lambda_{\bar r(\epsilon, v)}^{*v}$ exist for all v.*

and

$(\dagger\dagger)$ *for $0 < \epsilon < \hat\epsilon$, $Œ(\Lambda_{\bar r(\epsilon, v)}^{*v}) \leq Œ(\Lambda_\epsilon^v)$.*

(The DR approximation has a not-larger overall error than the indirect approximation.) Moreover, for any $\hat\epsilon, \bar r$ satisfying $(\dagger), (\dagger\dagger)$, we have

there exists $\epsilon^* \leq \hat\epsilon$ such that for all ϵ with

$(\dagger\dagger\dagger)$ $0 < \epsilon < \epsilon^*, \displaystyle\lim_{v\to\infty} \left(C(\Lambda_{\bar r(\epsilon, v)}^{*v}) - C(\Lambda_\epsilon^v) \right) = \infty.$

(The DR-versus-indirect cost differential goes to infinity as v goes to infinity.)

We now turn to Proposition C. Like proposition A, it concerns a fixed environment set E. It considers approximations of distinct continuum mechanisms whose message spaces are of *equal* dimension.

To state it, we shall need to consider two distinct complete classes of approximations, say C and D, such that the difference between the C-error and the D-error can be made arbitrarily close to zero by choosing the C-mesh sufficiently small and the D-mesh not more costly, and also by choosing the D-mesh sufficiently small and the C-mesh not more costly. More formally:

Given two regular continuum mechanisms, a complete class C of finite approximations of the first will be called *indifferent* to a complete class D of finite approximations to the second if the following holds:

---

[20] Uniform boundedness of derivatives guarantees the existence of a number $\bar G > 0$ such that for all $v \geq 1$, and for any sufficiently small mesh ϵ, we have $Œ(\bar\Lambda_\epsilon^v) \leq \bar G \cdot \epsilon$, a fact used in the Proof of proposition B. To demonstrate that fact, one applies an argument found in the proof of Lemma 2 of [6].

Let Γ_ϵ denote the ϵ-mesh approximation in C and let Δ_ϵ denote the ϵ-mesh approximation in $\mathcal{D}$. Then there exist $V > 0, \epsilon^\# > 0$ such that for any ϵ with $0 < \epsilon \leq \epsilon^\#$:

(i) $\mathcal{D}$ contains an approximation Γ', such that Γ' is not-more-costly than Γ_ϵ and

$$|\textrm{Œ}(\Gamma_\epsilon) - \textrm{Œ}(\Gamma')| \leq V \cdot \epsilon.$$

(ii) C contains an approximation Δ', such that Δ' is not-more–costly than Δ_ϵ and

$$|\textrm{Œ}(\Delta_\epsilon) - \textrm{Œ}(\Delta')| \leq V \cdot \epsilon.$$

Proposition C. *Suppose that Λ and Λ' are regular continuum mechanisms on the environment set E defined in Proposition A. Suppose that the message space of each is a subset of $\mathbb{R}^D$. Suppose that each realizes a well-behaved goal function whose range is $\mathbb{R}^\alpha$. Then*

- *any complete class of nontrivial controlled rounded-outcome approximations of Λ is indifferent to any complete class of nontrivial controlled rounded-outcome approximations of Λ'*
- *any complete class of nontrivial exact-outcome approximations of Λ is indifferent to any complete class of nontrivial exact-outcome approximations of Λ'*

The proof of Proposition C is obtained by modifying certain steps in the proofs of our companion paper [6].[21]

4 The continuum Walrasian mechanism in an exchange economy with quasi-linear utilities: regularity and goal realization

The agenda of this section. In this section we consider an exchange economy with n traders, a numeraire commodity, and L other commodities, where n and L are arbitrary. An environment e will identify the n traders' utility functions. We shall construct a version of the continuum Walrasian mechanism, whose typical message is a point in $\mathbb{R}^{nL}$. The message consists of L prices and an L-component trade vector for each of $n - 1$ traders, say Traders $1, \ldots, n - 1$. For convenient reference we call this the *condensed* Walrasian mechanism. In this mechanism trader n's trade vector is not an explicit part of the typical messsage, but is understood (by Trader n) to equal the negative of the sum of the other $n - 1$ traders' trade vectors.

It will turn out to be useful to define as well the *full* Walrasian mechanism, whose typical message has $L(n + 1)$ real components. The message explicitly describes

---

[21] In the proof of Lemma 1 of that paper, one modifies the chain of inequalities following statement (i) of Step 3, by letting $D = D^*$ and then dividing the chain into two parts. The first part provides a lower bound to the first overall error appearing in the chain and the second part provides an upper bound to the second overall error. One then considers the absolute value of the difference between the two bounds. The absolute value is bounded from above by a constant times the mesh. That fact can then be used to obtain a version of Lemma 1 which deals with the case $D^* = D$ and becomes part of the proof of Proposition C.

a trade vector for each of the n traders, and those trade vectors sum to zero if the message is an equilibrium message. For every environment e, an equilibrium message of one mechanism uniquely identifies an equilibrium message of the other.

For every e, the condensed mechanism's equilibrium outcome will be an individually rational and Pareto-optimal vector of trades.[22]

Our main task in this section is to show that under appropriate assumptions on the set of environments and on the utility functions, the condensed Walrasian continuum mechanism meets our regularity conditions. That will allow us to apply Propositions A-C, which require regularity, and to study finite approximations of the Walrasian mechanism as well as finite approximations of other regular continuum mechanisms that also realize Pareto optimality and individual rationality. Those other mechanisms will include a DR mechanism.

While Regularity Conditions (a),(b),(d) are quickly established for the condensed mechanism (under the assumptions that we make about the exchange economy), that turns out not to be true for Condition (c).

It turns out, in fact, that a straightforward way of establishing Condition (c) for the condensed mechanism is to establish it first for the full mechanism. We do so by following, for the full mechanism, the program given in Remark R1. We shall show that the full mechanism indeed has the four properties presented in that Remark. Recall, in particular, that the fourth of those properties requires the Jacobian associated with the agreement functions to satisfy a suitable Gale/Nikaido condition. The task of verifying such a condition occupies a major part of the argument. Once the four properties are established, Regularity Condition (c) will follow for the full mechanism. It is then easily argued that this implies satisfaction of Regularity Condition (c) by the condensed mechanism as well.

As for Regularity Condition (d), the full mechanism *fails* to satisfy it [23], but the condensed mechanism does satisfy it.

We conclude that the condensed Walrasian continuum mechanism indeed satisfies all four regularity conditions. Hence we may proceed to study its finite approximations by applying Propositions A-C, which concern approximations of regular continuum mechanisms.

The exchange economy to be studied. We consider an economy with traders $1, \ldots, i, \ldots, n$ and commodities $1, \ldots, \ell, \ldots, L, L + 1$. Commodity $L + 1$ will be a numeraire.

In the course of the discussion, we shall occasionally make a temporary departure from the n-trader $(L+1)$-commodity case in order to visit a special illustration: a two-commodity economy with two traders whose quasl-linear utility function has a quadratic function as its nonlinear part.

Trader i's initial endowment of each of the $L + 1$ commodities is positive and fixed. For commodity $\ell, \ell \in \{1, \ldots, L\}$, i's endowment is $W_\ell^i > 0$; for commodity $L + 1$ it is $V^i > 0$. Let $v^i, w_1^i, \ldots, w_L^i \in \mathbb{R}$ denote i's net trades (additions to

---

[22] The same is true of the full mechanism.

[23] The message space of the full mechanism is a subset of $\mathbb{R}^{L(n+1)}$. Condition (d) requires that for the mechanism's unique thread (exact-solution function), denoted t^F, the image set $t^F(E)$ includes an open set of $\mathbb{R}^{L(n+1)}$. But that is not the case, since at any equilibrium message, say $t^F(e)$, the n trade vectors sum to zero.

endowment or subtractions from endowment) for the numeraire commodity and the L other commodities, respectively. We shall use the notation

$$w^i \equiv (w_1^i, \ldots, w_L^i), \ (v^i, w^i) \equiv (v^i, w_1^i, \ldots, w_L^i).$$

We shall consider certain sets in the commodity space and the environment space. Some of these sets are open and others are closed. The caret symbol ($\wedge$), when placed over a symbol denoting a set, will always indicate that the set is open.

Trader i has a utility function defined on the commodity bundles that i would hold if the trade (v^i, w^i) lies in the open set

$$\hat{S}_i \equiv \left\{ (v^i, w^i) \ : \ -V^i < v^i < \sum_{j \neq i} V^j + \gamma; \ \text{for all} \right.$$

(4.1) $$\left. \ell \in \{1, \ldots, L\}, -W_\ell^i < w_\ell^i < \sum_{j \neq i} W_\ell^j + \gamma \right\},$$

where $\gamma > 0$.

We shall also be considering a subset of $\hat{S}^i$, namely the closed set

$$S_i \equiv \left\{ (v^i, w^i) \ : \ -V^i + \sigma \leq v^i \leq \sum_{j \neq i} V^j; \ \text{for all} \right.$$

(4.2) $$\left. \ell \in \{1, \ldots, L\}, -W_\ell^i + \sigma \leq w_\ell^i \leq \sum_{j \neq i} W_\ell^j \right\},$$

where σ is a real number satisfying

(4.3) $$0 < \sigma < \min(V^i, W_\ell^i), \ \text{for all} \ i \in \{1, \ldots, n\}, \ell \in \{1, \ldots, L\}.$$

Thus for every trade vector in S_i, trader i's post-trade holding of every commodity is at least $\sigma > 0$ but is at least γ less than the upper bound given in the definition of $\hat{S}_i$ in (4.1).

It will be convenient to define the nonnumeraire projections

$$S_i^\# \equiv \{ w^i \ : \ \text{for some} \ v^i \ \text{we have} \ (v^i, w^i) \in S_i \}$$

$$\hat{S}_i^\# \equiv \{ (w^i \ : \ \text{for some} \ v^i \ \text{we have} \ (v^i, w^i) \in \hat{S}_i \}.$$

Finally, we fix a real number Δ such that

(4.4) $$0 < \Delta < \min(\sigma, \gamma)$$

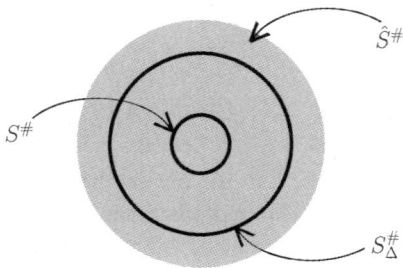

Figure 1

and we define the following closed set of nonnumeraire[24] trade vectors for trader i:

$$(4.5) \qquad S^{\#}_{i\Delta} \equiv \left\{ w^i \ : \ -W^i_\ell + \Delta \leq w^i_\ell \leq \sum_{j \neq i} W^i_\ell + \Delta, \ell = 1, \ldots, L \right\}.$$

We write:

$$S = S_1 \times \cdots \times S_n, S^{\#} = S^{\#}_1 \times \cdots \times S^{\#}_n, S^{\#}_\Delta = S^{\#}_{1\Delta} \times \cdots \times S^{\#}_{n\Delta}.$$

Note that since $0 < \Delta < \sigma$, the most that i gives up of nonnumeraire commodity ℓ, among the trade vectors in $S^{\#}_i$, is his endowment minus σ. But the most that i gives up among the trade vectors in $S^{\#}_{i\Delta}$ is a larger quantity, namely his endowment minus Δ. We have

$$(4.6) \qquad\qquad\qquad S^{\#}_i \text{ is in the interior of } S^{\#}_{i\Delta}.$$

Moreover, since (by ("IV)-1) $\Delta < \gamma$

$$(4.7) \qquad\qquad\qquad S^{\#}_{i\Delta} \text{ is in the interior of } \hat{S}^{\#}_i.$$

Figure 1 schematically indicates the relation between three sets in the space of nonnumeraire-commodity trade-vector n-tuples: the open set $\hat{S}^{\#}$ and the closed sets $S^{\#}_\Delta$ and $S^{\#}$.

It will turn out that in our full Walrasian mechanism the inner closed set $S^{\#}$ contains the trade portion of the mechanism's equilibrium-message set, the larger closed set $S^{\#}_\Delta$ will be the trade portion of the message space, and the still larger open set $\hat{S}^{\#}$ (the entire shaded area) will be the trade portion of an extended message space, which is needed in establishing one of the properties of Remark R1. It follows from (4.3) and (4.4) that for all i and all ℓ, we have $-W^i_\ell + \Delta < 0$. That, together with (4.5) and our assumption that all W^i_ℓ are positive, implies

$$(4.8) \qquad\qquad \text{for all } i, \text{ the no-trade vector 0 belongs to } S^{\#}_{i\Delta}.$$

---

[24] We again use the superscript # to indicate that the set contains only nonnumeraire trade vectors.

Since i's endowments are assumed fixed, we can write i's utility as a function of i's trade vector. The function belongs to a family of real-valued functions, each of them uniquely identified by a vector of parameters $e_i = (e_{i1}, \ldots, e_{iJ_i}) \in \mathbb{R}^{J_i}$. The vector e_i is i's local environment (individual characteristic). We consider two local-environment sets: (i) an open set

(4.9) $\hat{E}_i = \{e_i = (e_{i1}, ..., e_{iJ_i}) : \bar{e}_{ij} - \zeta_i < e_{ij} < \bar{\bar{e}}_{ij} + \zeta_i, j = 1, \ldots, J_i\},$

where $\zeta_i > 0$ and $\bar{e}_{ij} < \bar{\bar{e}}_{ij}, i = 1, \ldots, n; j = 1, \ldots, J_i$; and (ii) a subset of $\hat{E}_i$, namely the closed set

$$E_i = \{e_i = (e_{i1}, ..., e_{iJ_i}) : \bar{e}_{ij} \le e_{ij} \le \bar{\bar{e}}_{ij}, j = 1, \ldots, J_i\}.$$

We now restrict our class of exchange economies, in the following Assumptions $(\alpha 1)$ and $(\alpha 2)$, so that it becomes narrower than the class usually studied. Each of our traders has preferences represented by a quasi-linear utility function. The non-linear part of that function, i.e., the "valuation" function, is thrice continuously differentiable and strictly concave. That implies the existence of a Walrasian equilibrium. (As we shall see when we construct and study the full Walrasian mechanism, $(\alpha 2)$ in fact implies that there is exactly one). We shall also require (Assumption $(\alpha 3)$) that in any given economy of our class, every equilibrium nonnumeraire trade-vector n-tuple lies in $S^{\#}$, which means that after the trades, every trader has a positive amount of every nonnumeraire commodity. Finally, we require (Assumption $(\alpha 4)$) that the set of all equilibrium trade/marginal-utility nL-tuples which we obtain as we pass over all economies in our class, includes an open set of $\mathbb{R}^{nL}$.

Formally, for every local environment $e_i \in \hat{E}_i$ and every trade vector (v^i, w^i) with $w^i \in \hat{S}_i$, trader i's utility is given by the following quasi-linear function:

$$u^i = v^i + \phi^i(w^i; e_i).$$

We assume:

$(\alpha 1)$ Every function $\phi^i : \hat{S}_i \times \hat{E}_i \to \mathbb{R}$ is C^3.

$(\alpha 2)$ For every $e_i \in \hat{E}_i$, and every (nonnumeraire) trade vector $w^i = (w_1^i, \ldots, w_L^i)$ such that $-W_\ell^i < w_\ell^i < \sum_{j \neq i} W_\ell^j + \gamma, \ell = 1, \ldots, L$, the function ϕ^i:
- is strictly increasing in $w_\ell^i, \ell = 1, \ldots, L$
- is *strictly differentiably concave*, i.e., letting

$$\phi_{rs}^i(w^i; e_i) \equiv \frac{\partial \phi^i}{\partial w_r^i \partial w_s^i},$$

the following Hessian is negative definite:

$$\phi_{ww}^i \equiv \begin{pmatrix} \phi_{11}^i & \cdots & \phi_{1L}^i \\ .. & \ddots & .. \\ \phi_{L1}^i & \cdots & \phi_{LL}^i \end{pmatrix}.$$

Assumption $(\alpha 2)$ implies[25] that a Walrasian equilibrium exists, i.e., letting

(4.10),
$$\phi_\ell^i(w^i; e_i) \equiv \frac{\partial \phi^i(w^i; e_i)}{\partial w_\ell^i}.$$

for every $e \in \hat{E}$, there exists a (nonnumeraire) trade-vector n-tuple $(w^1, \ldots, w^n)$ which satisfies

(4.11) $$\phi_\ell^1(w^1; e_1) = \cdots = \phi_\ell^n(w^n; e_n), \sum_{i=1}^n w_\ell^i = 0, \ell = 1, \ldots, L.$$

We shall assume[26]

$(\alpha 3)$ For every $e \in \hat{E}$, every trade-vector n-tuple satisfying the equalities in (4.11) belongs to $S^\#$.

To state our final assumption, we first define, for $e = (e_1, \ldots, e_n) \in E$, the set

$$\Phi(e) \equiv \left\{ \left((w^1, \ldots, w^{n-1}), p) \right) : \left(w^1, \ldots, w^{n-1}, -\sum_{i=1}^{n-1} w^i \right) \in \hat{S}_{1\Delta}^\# \times \cdots \right.$$
$$\left. \times \hat{S}_{n\Delta}^\#; p = (\phi_1^1(w_1^1; e_1), \ldots, \phi_L^1(w_L^1; e_1)) \right\}$$

and the set $\Phi(E) \equiv \{\Phi(e) : e \in E\}$. Our assumption is:[27]

$(\alpha 4)$ The set $\Phi(E)$ includes an open set of $\mathbb{R}^{nL}$.

To illustrate the preceding concepts, we now turn to a two-person two-commodity exchange economy, in which each person's valuation function (the non-linerar part of the quasi-linear utility function) is quadratic.

---

[25] Mas-colell, Whinston, Green, ([13]), Proposition 17.B.2, p.581.

[26] The role played by Assumption $(\alpha 3)$ is as follows. We will show that the full Walrasian mechanism obeys Regularity Condition (c), by following the program given in Remark R1. To do so we will need to show (in establishing Property 2 of Remark R1) that the full mechanism's compact message space is extendable to an open set that includes the message space, in such a way that the extended agreement functions and the Jacobian associated with them, have the properties which are used to establish Property 4. The domain of the ith extended agreement function is the Cartesian product of the extended message space and the open set $\hat{E}_i$. The trade portion of our compact message space will be the set $S_\Delta^\#$, in which every trade leaves every trader with at least $\Delta > 0$ of every nonnumeraire commodity. Assumption $(\alpha 3)$ implies, in view of the definition of $S^\#$, that at an equilibrium message of the full mechanism, the trades leave every trader with at least $\sigma > \Delta$ of every nonnumeraire commodity. The fact that $S^\# \subset S_\Delta^\#$ will permit us to argue that our compact message space achieves coverage. We can then extend that space while still leaving every trader with a positive amount of every nonnumeraire commodity. The trade portion of the extended space will be the open set $\hat{S}$.

[27] We impose this requirement on the environment set E in order that the condensed Walrasian mechanism on E, which we construct below, obey Regularity Condition (d). The set $\Phi(e)$ will be the set of equilibrium messages for e in the condensed mechanism, and $\Phi(E)$ will be the mechanism's entire set of equilibrium messages.

An illustration: a two-person economy with quadratic valuation functions. We have two traders ($n = 2$), a numeraire commodity, one non-numeraire commodity ($L = 1$), and the following linear-quadratic utility function:

$$U_i = Y_i + \alpha_i X_i - \frac{1}{2}\beta_i X_i^2,$$

where X_i, Y_i are $i's$ total holdings of the non-numeraire commodity and the numeraire commodity respectively; $\alpha_i > 0, \beta_i > 0$; and the function gives i's utility for bundles (X_i, Y_i) such that $X_i < \frac{\alpha}{\beta}$. (At such a bundle the function is strictly increasing in X_i).

Now for $i = 1, 2$, we shall fix $V^i > 0, W^i > 0$, trader i's endowments of the numeraire and non-numeraire commodities, respectively. We define

$$\theta_i \equiv \alpha_i - \beta_i W^i .$$

and we let trader i's local environment be the pair $e_i = (e_{i1}, e_{i2}) = (\theta_i, \beta_i)$. Then we can write i's utility as

$$u_i = v^i + \phi^i(w^i; e_i),$$

where

$$\phi^i : \hat{S}_i \times \hat{E}_i \rightarrow \mathbb{R}; \phi^i(w^i; e_i) = (\theta_i + \beta_i W^i)(W^i + w^i) - \frac{1}{2}\beta_i(W^i + w^i)^2.$$

We have

$$\frac{d\phi^i}{dw^i} = \theta_i + \beta_i W^i - \beta_i(W^i + w^i) = \theta_i - \beta_i w^i.$$

So ϕ^i is strictly increasing as long as

$$w^i < \frac{\theta_i}{\beta_i}.$$

In the Appendix, we define the sets E_i, $\hat{E}_i$, S_i, $\hat{S}_i$, $S_{i\Delta}$ in such a way that each ϕ^i is indeed strictly increasing and in addition Assumptions ($\alpha1$)–($\alpha4$) are met. We provide a numerical example of those sets.

The full Walrasian continuum mechanism. In the full mechanism a message specifies a nonnumeraire trade vector for every trader $i, i = 1, \dots, n$ and also specifies a price for each of the L nonnumeraire commodities. The message space's possible values for i's nonnumeraire trade vector are those in the compact set $\hat{S}_{i\Delta}^{\#}$, defined in (4.5).

To construct the set of "price" components of the mechanism's possible messages, we first need to identify, for each non-numeraire commodity ℓ, a smallest and largest marginal utility. We again use the marginal-utility notation introduced in (4.10). Define

$$p_\ell^* \equiv \min_{w^i \in S_i^\#, e_i \in E_i, i \in \{1,\dots,n\}} \phi_\ell^i(w^i; e_i)$$

and

$$p_\ell^{**} \equiv \max_{w^i \in S_i^\#, e_i \in E_i, i \in \{1,\ldots,n\}} \phi_\ell^i(w^i; e_i).$$

These minima and maxima exist since the sets $S_i^\#, E_i$ are compact and, by Assumption $(\alpha 1)$ and the fact that $S_i^\# \subset \hat{S}_i^\#$, the function ϕ_ℓ^i is continuous on $S_i^\#$. Moreover, by Assumption $(\alpha 2)$, and the definition of the sets $S_i^\#$, every marginal utility is positive when the trade n-tuple lies in $S^\#$. We can therefore choose a real number Ψ so that

$$(4.12) \qquad\qquad 0 < \Psi < \min_{\ell \in \{1,\ldots,L\}} p_\ell^*.$$

For each non-numeraire commodity ℓ, our set of possible prices will be $\{p_\ell : p_\ell^* - \Psi \le p_\ell \le p_\ell^{**} + \Psi\}$. All members of that set are positive.

We are now ready to combine the above statements into a definition of the compact message space of the full continuum Walrasian mechanism, and to specify the mechanism's remaining elements. The mechanism is the triple $\Lambda^F = \langle M^F, (\bar{g}_1, \ldots, \bar{g}_n), h^F \rangle$, with the following elements. The message space is the compact set

$$M^F = \big\{ m = ((w^1, \ldots, w^n), (p_1, \ldots, p_L)) : w^i \in S_{i\Delta}^\#; i = 1, \ldots, n;$$

$$\cdot p_\ell^* - \Psi \le p_\ell \le p_\ell^{**} + \Psi, \ell \in \{1, \ldots, L\} \big\}.$$

For Trader i, with $i \in \{2, \ldots, n\}$, the agreement function is $\bar{g}_i = (\bar{g}_{i1}, \ldots, \bar{g}_{iL})$, where, for $\ell \in \{1, \ldots, L\}$, we have $\bar{g}_{i\ell}(m, e_i) = \phi_\ell^i(w^i; e_i) - p_\ell$. Trader 1 has the additional task of checking whether or not the proposed trades balance and accordingly has the agreement function $\bar{g}_1 = (\bar{g}_{11}, \ldots, \bar{g}_{1L}, g_1^*, \ldots, g_L^*)$, where, for $\ell \in \{1, \ldots, L\}$, we have $\bar{g}_{1L}(m, e_1) = \phi_\ell^1(w^1; e_i) - p_\ell$ and $g_\ell^*(m, e_1) = \sum_{i=1}^n w_\ell^i$. Turning to the outcome function, for $m = ((w^1, \ldots, w^n), (p_1, \ldots, p_L)) \in M^F$, we have $h^F(m) = (w^1, \ldots, w^n)$.

The condensed mechanism. This mechanism is the triple $\Lambda = \langle M, (g_1, \ldots, g_n), h \rangle$, with the following elements.

The message space is an nL-dimensional set in $\mathbb{R}^{nL}$, namely the compact set

$$M = \big\{ m = ((w^1, \ldots, w^{n-1}), (p_1, \ldots, p_L)) : w^i \in S_{i\Delta}^\#; i = 1, \ldots, n-1;$$

$$p_\ell^* - \Psi \le p_\ell \le p_\ell^{**} + \Psi, \ell \in \{1, \ldots, L\} \big\}.$$

For Trader i, with $i \in \{1, \ldots, n-1\}$, the agreement function is $g_i = (g_{i1}, \ldots, g_{iL})$, where, for $\ell \in \{1, \ldots, L\}$, we have $g_{i\ell}(m, e_i) = \phi_\ell^i(w^i; e_i) - p_\ell$. Trader n does not explicitly find a proposed trade vector for himself among the components of the message m. Rather n views the negative of the sum of the other traders' vectors as his own proposed trade vector and accordingly has the agreement function $g_n = (g_{n1}, \ldots, g_{nL})$, where $g_{n\ell}(m, e_n) = \phi_\ell^n(-\sum_{i=1}^{n-1} w_\ell^i; e_n) - p_\ell, \ell = 1, \ldots, L$. Turning to the outcome function, for $m = ((w^1, \ldots, w^n), (p_1, \ldots, p_L)) \in M$, we have $h(m) = (w^1, \ldots, w^{n-1})$.

Results to be established for the two mechanisms. We now show that under Assumptions $(\alpha 1) - (\alpha 4)$:

- Both mechanisms cover E.
- Both mechanisms satisfy Regularity Conditions (a) and (b).
- The full mechanism satisfies Regularity Condition (c), and that implies that the condensed mechanism also satisfies Regularity Condition (c).
- While the full mechanism does not satisfy Regularity Condition (d), the condensed mechanism does. Thus the condensed mechanism satisfies all four regularity conditions.
- For every e in E, the trade-vector $(n-1)$-tuple which is the condensed mechanism's unique equilibrium outcome (when combined with the balancing trade vector for trader n) is individually rational and Pareto-optimal in the set of all trade-vector n-tuples that sum to zero.

Coverage. To deal with the coverage issue, recall first that Assumption $(\alpha2)$ implies the existence of a Walrasian equilibrium, i.e., a nonnumeraire trade n- tuple which sums to zero and has the property that all traders have the same marginal utility for each non-numeraire commodity ℓ. Assumption $(\alpha3)$ guarantees that at this equilibrium each trader holds a bundle with at least $\sigma > 0$ of each non-numeraire commodity. At those bundles the common marginal utility for the nonnumeraire commodity ℓ is a positive number, which we may call p_ℓ, satisfying $p_\ell^* \leq p_\ell \leq p_\ell^{**}$. Using the definition of the message space M^F, we have:

$(\dagger)$
$$\left[\begin{array}{l} \text{for every } e \in E, \text{ the equation system} \\[2mm] \bar{g}_{i\ell}(m, e_i) = 0, g_\ell^*(m, e_1) = 0, i \in \{1, \ldots, n\}, \ell \in \{1, \ldots, L\} \\[2mm] \text{of the full mechanism has a solution} \\[2mm] \qquad m = ((w^1, \ldots, w^n), (p_1, \ldots, p_L)) \in M^F, \\[2mm] \text{with } w^i \in S_i^\# \text{ for all } i \text{ and } p_\ell^* \leq p_\ell \leq p_\ell^{**} \text{ for all } \ell. \end{array}\right.$$

So the full mechanism covers E. Similarly, using the definition of M:

$(\dagger\dagger)$
$$\left[\begin{array}{l} \text{for every } e \in E, \text{ the equation system} \\[2mm] g_{i\ell}(m, e_i) = 0, i \in \{1, \ldots, n-1\}, \ell \in \{1, \ldots, L\} \\[2mm] \text{of the condensed mechanism has a solution} \\[2mm] \qquad m = ((w^1, \ldots, w^{n-1}), (p_1, \ldots, p_L)) \in M, \\[2mm] \text{with } w^i \in S_i^\# \text{ for } i = 1, \ldots, n-1, \text{ and } p_\ell^* \leq p_\ell \leq p_\ell^{**} \text{ for all } \ell. \end{array}\right.$$

So the condensed mechanism also covers E

Regularity Conditions (a),(b). By construction, the message space M^F of the full mechanism is compact and, by Assumption $(\alpha1)$ (which says that every valuation function ϕ^i is C^3), every component of every agreement function $\bar{g}_i$ is C^2 at every (m, e_i) in $M^F \times E_i$. The message space M of the condensed mechanism is also compact and every component of every agreement function g_i is also C^2 at every (m, e_i) in $M \times E_i$. So Regularity Conditions (a) and (b) are satisfied for both mechanisms.

The full mechanism satisfies Regularity Condition(c). We shall establish this by verifying that the four properties in Remark R1 hold for the full mechanism.

Property (1). The full mechanism has this property if M^{F} includes in its interior a set $\bar{M}^{\mathsf{F}}$, which contains an equilibrium message for every e in E. In view of the above statement (†), and the fact that $S_\Delta^{\#}$ contains $S^{\#}$ in its interior, that is indeed the case for the set

$$\bar{M}^{\mathsf{F}} = \{ m = ((w^1, \ldots, w^n), (p_1, \ldots, p_L)) : w^i \in S_i^{\#}; i$$

$$= 1, \ldots, n; p_\ell^* \le p_\ell \le p_\ell^{**}, \ell \in \{1, \ldots, L\}\}.$$

Property (2). This property requires the agreement functions to obey extension conditions.

The open set $\hat{E}$ – the extended environment set – was defined in (4.9). We now define the open set $\hat{M}^{\mathsf{F}}$, which will be our extended message space. In interpreting the definition, recall that $\hat{S}^{\#}$ is an open set. We have:

$$\hat{M}^{\mathsf{F}} \equiv \left\{ m = (w^1, \ldots, w^n), (p_1, \ldots, p_L)) : w^i \in \hat{S}_i^{\#}; i \right.$$

$$= 1, \ldots, n; p_\ell^* - \tau < p_\ell < p_\ell^{**} + \tau, \ell \in \{1, \ldots, L\}\},$$

where, for the number Ψ introduced in (4.12), we have

$$\Psi < \tau < p_\ell^*, \text{ all } \ell.$$

By assumption $(\alpha 1)$, each utility function $u_i = v^i + \phi^i(\ \cdot \ ; \ \cdot \ )$ is defined on $\hat{S}_i \times \hat{E}_i$ and is $\mathcal{C}^3$. So each agreement function $\bar{g}_i$ is well defined on the extended set $\hat{M}^{\mathsf{F}} \times \hat{E}_i$. Let $\hat{g}_i$ denote the extension of g_i to $\hat{M}^{\mathsf{F}} \times \hat{E}_i$. To write $\hat{g}_i$, one just repeats the formula for g_i. Each extended function is $\mathcal{C}^2$, as Property (2) requires.

Property (3). For every $(m, e) \in \hat{M}^{\mathsf{F}} \times \hat{E}$, the Jacobian associated with the equation system

$$\hat{g}(m, e) = 0$$

and, more generally, with the equation system

$$(*) \qquad \hat{g}_{i\ell}(m, e_i) = \delta_{i\ell}, \hat{g}_\ell^*(m, e_1) = \delta_\ell^*, i = 1, \ldots, n, \ell = 1, \ldots, L,$$

is

$$\hat{Q}(m; e) = \begin{pmatrix} \phi_{ww}^1 & 0 & \cdots & 0 & -I_L \\ 0 & \phi_{ww}^2 & \cdots & 0 & -I_L \\ \vdots & & \ddots & 0 & \vdots \\ 0 & 0 & & \phi_{ww}^n & -I_L \\ I_L & I_L & \cdots & I_L & 0 \end{pmatrix}.$$

Here I_L is an L-by-L identity matrix and the zero symbol denotes an L-by-L null matrix. The term ϕ_{ww}^i is a matrix and was defined as part of assumption (α 2). The entries in the bottom row are the partial derivatives of the components of Trader 1's "equilibrating" agreement function g^*. The last column contains partial derivatives with respect to prices. By Assumption ($\alpha 2$), each of the matrices ϕ_{ww}^i

has nonzero determinant. Hence the Jacobian $\hat{Q}(m; e)$ has nonzero determinant for all (m, e) in $\hat{M}^F \times \hat{E}$. That establishes Property (3).

Property (4). To establish the final property, we have to show that $\hat{Q}(m; e)$ satisfies a univalence condition. We shall use Corollary (ii), stated by Nikaido and found on page 379 of [15].[28]

That Corollary tells us that for any fixed e, and for any m in the open rectangular region $\hat{M}^F$, the Jacobian $\hat{Q}(m, e)$ has the univalence property if it meets the following two conditions (a), (b). Hence, if those conditions are met, any message in $\hat{M}^F$ which solves the equation system (*) is the only such message to be found in all of $\hat{M}^F$.

For brevity we denote the Jacobian $\hat{Q}(m; e)$ by $\hat{Q}$.

The two conditions are

(a) The determinant of $\hat{Q}$ is positive if it has an even number of rows (and columns) and negative if it has an odd number of rows (and columns).
(b) The symmetric matrix $\frac{1}{2}(\hat{Q} + \hat{Q}')$, where $\hat{Q}'$ is the transpose of $\hat{Q}$, is negative semi-definite.

We find that $\frac{1}{2}(\hat{Q} + \hat{Q}')$ is the following (bordered) diagonal matrix:

$$\begin{pmatrix} \phi^1_{ww} & & & & 0 \\ & \phi^2_{ww} & & & 0 \\ & & \ddots & & 0 \\ & & & \phi^n_{ww} & 0 \\ 0 & 0 & \cdots & 0 & 0 \end{pmatrix}.$$

This is negative semi-definite because we assumed (in $(\alpha 2)$) that each ϕ^i_{ww} is negative definite. So the above condition (b) is satisfied.

To verify condition (a), first define

$$\psi_i \equiv (\phi^i_{ww})^{-1}.$$

The matrix ψ_i is negative definite (if a matrix is negative definite then so is its inverse). Next premultiply $\hat{Q}$ by the following matrix product:

$$A \cdot A^{-1} = \begin{pmatrix} \phi^1_{ww} & & 0 \\ & \ddots & \\ & & \phi^n_{ww} \\ 0 & & I_L \end{pmatrix} \cdot \begin{pmatrix} \psi^1_{ww} & & 0 \\ & \ddots & \\ & & \psi^n_{ww} \\ 0 & & I_L \end{pmatrix}.$$

Since this product is the $(nL + L)$-by-$(nL + L)$ identity matrix, it follows that $A \cdot A^1 \cdot \hat{Q} = \hat{Q}$. Note that $A^{-1} \cdot \hat{Q} = T$, where

$$T = \begin{pmatrix} I_L & & 0 & -\psi_1 \\ & \ddots & & \vdots \\ 0 & & I_L & -\psi_n \\ \bullet\bullet & \bullet\bullet & \bullet\bullet & \bullet\bullet \\ I_L & \cdots & I_L & 0 \end{pmatrix}.$$

---

[28] The discussion in [15] makes use of results found in [2].

But we can modify T, by "elementary" operations, to obtain a matrix K whose determinant equals the determinant of T. To do so, successively subtract from the bottom row (the row below the line that is composed of bullets) each of the rows above the bottom row. We are left with

$$K = \begin{pmatrix} I_L & 0 & -\psi_1 \\ & \ddots & \vdots \\ 0 & I_L & -\psi_n \\ \bullet\bullet & \bullet\bullet\ \ \bullet\bullet & \bullet\bullet \\ 0 & \cdots\ \ 0 & \psi_1 + \cdots + \psi_n \end{pmatrix}.$$

The matrix K is block-triangular, so its determinant equals the product of the determinants of the diagonal blocks, i.e.,

$$\det K = \underbrace{(\det I_L) \cdot (\det I_L) \cdots (\det I_L)}_{n \text{ times}} \cdot \Big(\det(\psi_1 + \cdots + \psi_n) \Big)$$

$$= \det(\psi_1 + \cdots + \psi_n).$$

Note that $\psi_1 + \cdots + \psi_n$ is a sum of negative-definite L-by-L matrices, so it is negative-definite and L-by-L. Hence

$(+)$
$$\det K \begin{cases} > 0 \text{ if } L \text{ is even} \\ < 0 \text{ if } L \text{ is odd} \end{cases}.$$

Finally, we have

$$\det \hat{Q} = (\det A) \cdot (\det A^{-1}) \cdot (\det Q) =$$

$(++)$
$$(\det A) \cdot (\det T) = (\det A) \cdot (\det K).$$

But
$$\det A = (\det \phi^1_{ww}) \cdot (\det \phi^2_{ww}) \cdots (\det \phi^n_{ww}).$$

So, since each ϕ^i_{ww} is negative definite and L-by-L,

$$\text{sign of } \det A \begin{cases} \text{is positive if } L \text{ is even} \\ = \text{sign of } (-1)^n \text{ if } L \text{ is odd} \end{cases},$$

or more compactly

$$\text{sign of } \det A = \text{ sign of } (-1)^{Ln}.$$

That, together with $(+),(++)$ imply

$$\text{sign of } \det \hat{Q} = \text{ sign of } (-1)^{Ln+L}.$$

That means that the Jacobian $\hat{Q}$ satisfies the condition (a).

So both requirements of Nikaido's Corollary are met, Property (4) of Remark R1 is established, and the full mechanism indeed obeys Regularity Condition (c).

The satisfaction of Condition (c) by the full mechanism implies its satisfaction by the condensed mechanism. We shall use the abbreviations w, w', p to denote, respectively, the vectors $(w^1, \ldots, w^n), (w^1, \ldots, w^{n-1})$, and $(p_1, \ldots, p_L)$. We note first that

$$(\S) \qquad \text{for all } e \in E, \bar{g}_n(w', -\sum_{i=1}^{n-1} w^i, p) = g_n(w', p).$$

Next we note that if, for a given e and a pair (w, p), the agreement functions $\bar{g}_{i\ell}$ of the full mechanism take values $\delta_{i\ell}$ while the "equilibrating" functions g_ℓ^* equal zero, then (1) $(w, p) = (w', -\sum_{i=1}^{n-1} w^i, p)$, and (2) for the pair (w', p) the agreement functions of the condensed mechanism take, respectively, the same values $\delta_{i\ell}$. More precisely:

$$(\S) \quad \begin{cases} \text{if } \bar{g}_{i\ell}(w, p, e_i) = \delta_{i\ell}, i = 1, \ldots, n, \ell = 1, \ldots, L \text{ and } g_\ell^*(w, p, e_1) = \\ 0, \ell = 1, -, L, \text{ then } (w, p) = (w', -\sum_{i=1}^{n-1} w^i, p) \text{ and } g_{i\ell}(w', p, e_i) = \\ \delta_{i\ell}, i = 1, \ldots, n, \ell = 1, \ldots, L. \end{cases}$$

Now the satisfaction of Regularity Condition (c) by the full mechanism means that there exists a number $\bar{\delta} > 0$ and a C^2 approximate-solution function

$$\bar{t}^{\mathsf{F}} : E \times \underbrace{[-\bar{\delta}, \bar{\delta}] \times \cdots \times [-\bar{\delta}, \bar{\delta}]}_{n(L+1) \text{ times}} \to M^{\mathsf{F}},$$

with trade components $\bar{t}_{i\ell}^{\mathsf{F}}$ and price components $\bar{t}_\ell^{\mathsf{F}}$, $i = 1, \ldots, n, \ell = 1, \ldots, L$, such that

$$(\S\S) \quad \begin{cases} \text{for all } \delta_{i\ell}, \delta_\ell^* \in [-\bar{\delta}, \bar{\delta}], i = 1, \ldots, n, \ell = 1, \ldots, L \text{ and all } e \in E, \\[4pt] \qquad \bar{t}^{\mathsf{F}}(e, \delta_{11}, \ldots, \delta_{1L}, \ldots, \delta_{n1}, \ldots, \delta_{nL}, \delta_1^*, \ldots, \delta_L^*) \\[4pt] \text{is the unique message } m \text{ in } M^F \text{ satisfying the full-mechanism equation} \\ \text{system} \\[4pt] \qquad \bar{g}_{i\ell}(m, e_i) = \delta_{i\ell}, g_\ell^*(m, e_1) = \delta_\ell^*, i = 1, \ldots, n, \ell = 1, \ldots, L. \end{cases}$$

We obtain the condensed mechanism's approximate-solution function from the full mechanism's function $\bar{t}^F$ by setting the quantities $\delta_1^*, \ldots, \delta_L^*$ (which correspond to the equilibrating functions of the full mechanism) equal to zero. Thus we consider the function

$$\bar{t} : E \times \underbrace{[-\bar{\delta}, \bar{\delta}] \times \cdots \times [-\bar{\delta}, \bar{\delta}]}_{nL \text{ times}} \to M,$$

with trade components $\bar{t}_{i\ell}$ and price components $\bar{t}_\ell$, $i = 1, \ldots, n-1, \ell = 1, \ldots, L$, defined by:

$$\bar{t}_{i\ell}(e, \delta_{11}, \ldots, \delta_{1L}, \ldots, \delta_{n1}, \ldots, \delta_{nL})$$

$$= \bar{t}_{i\ell}^{F}(e, \delta_{11}, \dots, \delta_{1L}, \dots, \delta_{n1}, \dots, \delta_{nL}, \underbrace{0, \dots, 0}_{L \text{ times}})$$

and

$$\bar{t}_{\ell}(e, \delta_{11}, \dots, \delta_{1L}, \dots, \delta_{n1}, \dots, \delta_{nL})$$
$$= \bar{t}_{\ell}^{F}(e, \delta_{11}, \dots, \delta_{1L}, \dots, \delta_{n1}, \dots, \delta_{nL}, \underbrace{0, \dots, 0}_{L \text{ times}}).$$

It follows from (§) and (§§) that the C^2 function $\bar{t}$ so defined is indeed the approximate-solution function for which the condensed mechanism satisfies Regularity Condition (c). That is to say, for all $\delta_{i\ell} \in [-\bar{\delta}, \bar{\delta}], i = 1, \dots, n, \ell = 1, \dots, L$ and all $e \in E$,

$$\bar{t}(e, \delta_{11}, \dots, \delta_{1L}, \dots, \delta_{n1}, \dots, \delta_{nL})$$

is the unique message m in M satisfying the condensed-mechanism equation system

$$g_{i\ell}(m, e_i) = \delta_{i\ell}, i = 1, \dots, n, \ell = 1, \dots, L.$$

So the condensed mechanism indeed satisfies Regularity Condition (c).

Regularity Condition (d). Let t^F, t^c denote, respectively, the threads of the full mechanism and the condensed mechanism. (Recall that the thread is the exact-solution function and one obtains it from the approximate-solution function by setting all the "δ" terms equal to zero). The full mechanism has $n(L+1)$ message variables. It fails to satisfy Condition (d) since at every equilibrium message the n trade vectors sum to zero. Hence the image set $t^c(E) \subset \mathbb{R}^{n(L+1)}$ of equilibrium messages has dimension nL and does not contain an open set of $\mathbb{R}^{n(L+1)}$.

On the other hand, the condensed mechanism has nL message variables: L-component trade vectors for each of $n - 1$ traders and L prices. Assumption ($\alpha 4$) guarantees that the condensed mechanism's equilibrium-message set $t(E) \subset \mathbb{R}^{nL}$ contains an open set of $\mathbb{R}^{nL}$.

We conclude that the condensed mechanism Λ obeys all four of the regularity conditions.

Pareto optimality and individual rationality of the condensed mechanism's equilibrium outcomes. Suppose $\bar{m} = ((\bar{w}^1, \dots, \bar{w}^{n-1}), (\bar{p}_1, \dots, \bar{p}_L)) \in M$ is an equilibrium message of the condensed mechanism for the environment $\bar{e} \in E$, so that the trades which take place are given by $h(m) = (\bar{w}^1, \dots, \bar{w}^{n-1})$. It then follows from the definitions of M and of the condensed mechanism's agreement functions that $w^i \in S_{i\Delta}^{\#}, i = 1, \dots, n - 1$ and that
(4.13)

$$\phi_{\ell}^1(\bar{w}^1; \bar{e}_1) = \dots = \phi_{\ell}^{n-1}(\bar{w}^{n-1}; \bar{e}_{n-1}) = \phi_{\ell}^n\left(-\sum_{i-1}^{n-1} \bar{w}^i; \bar{e}_n\right), \ell = 1, \dots, L.$$

We first argue that there is a numeraire trade vector $(\bar{v}^1, \dots, \bar{v}^n)$ in the set

$$V^* \equiv \left\{(v^1, \dots, v^n) : \sum_{i=1}^n v^i = 0, v^i \geq V^i, \text{ all } i\right\}$$

such that the vector of holdings $(V^1 + \bar{v}^1, \dots, V^n + \bar{v}^n), (W^1 + \bar{w}^1, \dots, W^{n-1} + \bar{w}^{n-1}, W^n - \sum_{i-1}^{n-1} \bar{w}^i))$ is Pareto-optimal for $\bar{e}$ in the set of all vectors $(V^1 + v^1, \dots, V^n + v^n), (W^1 + w^1, \dots, W^{n-1} + w^{n-1}, W^n - \sum_{i-1}^{n-1} w^i))$ with $(v^1, \dots, v^n) \in V^*, (w^1, \dots, w^n) \in S^{\#}$, and $\sum_{i=1}^n w^i = 0$.

That follows from (1) the fact that every trader i has preferences represented by the function u^i satisfying $(\alpha 2)$; (2) the marginal-utility equalities in (4.13); (3) Assumption $(\alpha 3)$, which guarantees that every component of every one of the bundles $(W^1 + \bar{w}^1, \dots, W^{n-1} + \bar{w}^{n-1}, W^n - \sum_{i-1}^{n-1} \bar{w}^i)$ is positive; (4) the no-satiation property of those bundles (implied by our assumption, in $(\alpha 2)$, that every function ϕ_ℓ^i is strictly increasing); and (5) the first Welfare Theorem for pure exchange economies, as argued, for example in [10].

Next we turn to individual rationality. For every trader i, utility at the bundle held after the trades $\bar{w}^1, \dots, \bar{w}^{n-1}, -\sum_{i-1}^{n-1} \bar{w}^i, v^1, \dots, v^n$ take place is not less than utility at the endowment bundle. That is the case since (1) the conditions that we impose on ϕ^i imply that for traders $1, \dots, n-1$, the pair $(\bar{v}^i, \bar{w}^i)$, maximizes the utility $v^i + \phi^i(w^i; \bar{e}_i)$ subject to the constraint $w^i \in S_i^{\#}$ and the budget constraint

$$\sum_{\ell=1}^{L} \bar{p}_\ell \cdot (W_\ell^i + w_\ell^i) + (V^i + v^i) \leq \sum_{\ell=1}^{L} \bar{p}_\ell \cdot W_\ell^i + V^i,$$

while the analagous statement holds for trader n and the pair $(v^n, -\sum_{i-1}^{n-1} \bar{w}^i)$; and (2) as (4.8) stated, the no-trade vector belongs to the set $S_i^{\#}, i = 1, \dots, n$.

The full and condensed continuum Walrasian mechanisms in the illustrative two-trader linear-quadratic case. Detailed remarks on these mechanisms are provided in the second part of the Appendix. Here we present some of their main properties.

Recall that for $i = 1, 2, e_i = (\theta_i, \beta_i)$. The sets E_i are defined in the first part of the Appendix. Recall that trader i's utility function $u_i = v^i + \phi^i(w^i; e_i)$ is defined on the open set $S_i = \{(v^i, w^i) : -V^i < v^i < V^j + \gamma; -W^i < w^i < W^j + \gamma\}$, where $j \neq i$. Recall that

$$\frac{d\phi^i}{dw^i} = \theta_i - \beta_i w^i.$$

To define the message space of the condensed mechanism, we need the compact sets

$$S_{1\Delta}^{\#} = \{-W^1 + \Delta \leq w^1 \leq W^2 + \Delta\}, S_{2\Delta}^{\#} = \{-W^2 + \Delta \leq w^2 \leq W^1 + \Delta\},$$

where Δ satisfies

$$0 < \Delta < \min(W^1, W^2).$$

The message space of the condensed mechanism is the compact set

$$M = \left\{ m = (w^1, p) : w^1 \in S_{1\Delta}^{\#}, -w^1 \in S_{2\Delta}^{\#}; p^* - \Psi \leq p \leq p^{**} + \Psi \right\}.$$

(The terms p^*, p^{**} are defined in the Appendix). The agreement functions are

$$g_1((w^1, p), e_1) = \theta_1 - \beta_1 - p, \quad g_2((w^1, p), e_2) = \theta_2 + \beta_2 - p.$$

The outcome function h is a projection operator, i.e., $h(w^1, p) = w^1$. The solution to the equation system $g_1(w^1, p), e_1 = g_2(w^1, p), e_2) = 0$ is

$$\bar{w}^1 = \frac{\theta_1 - \theta_2}{\beta_1 + \beta_2}, \bar{p} = \frac{\theta_1 \beta_2 + \theta_2 \beta_1}{\beta_1 + \beta_2}.$$

It is readily checked that for any economy defined by an element e of the set E specified in the Appendix, the trade pair $(\bar{w}^1, -\bar{w}^1)$ is the only Pareto-otimal pair in $S_{1\Delta}^{\#} \times S_{2\Delta}^{\#}$.

Recall that when we demonstrate the full mechanism's satisfaction of Regularity Condition (c) (which implies that the above condensed mechanism also satisfies it), a major step requires the study of the Jacobian $\hat{Q}(m; e)$. In the present case that Jacobian is is as follows (with rows and columns appropriately indexed):

$$
\begin{array}{c}
\bar{g}_{11} \\
\bar{g}_{2} \\
g^{*}
\end{array}
\begin{array}{ccc}
w^1 & w^2 & p \\
\left(\begin{matrix}
-\beta_1 & 0 & -1 \\
0 & -\beta_2 & -1 \\
1 & 1 & 0
\end{matrix} \right)
\end{array}
$$

This has determinant $-\beta_1 - \beta_2$, which is nonzero, since $\beta_i > 0, i = 1, 2$. The matrix has an odd number of rows and the determinant is negative, so the first of Nikaido's conditions is met. Moreover, the symmetric matrix $\frac{1}{2}(Q + Q')$ is

$$
\begin{pmatrix}
-\beta_1 & 0 & 0 \\
0 & -\beta_2 & 0 \\
0 & 0 & 0
\end{pmatrix},
$$

which is negative semi-definite. So the second of Nikaido's conditions is met.

5 Applying propositions A–C: three exchange-economy propositions

We can now apply Propositions A-C to the condensed Walrasian mechanism and its approximations to obtain three exchange-economy propositions. Henceforth we generally omit the qualifier "condensed", which will be understood. Our exchange-economy propositions will use terms that were introduced in connection wirth Propositions A-C, in particular the terms "well-behaved goal function", "controlled round-off approximation", "superior", and "indifferent". We consider regular continuum mechanisms on our set E, each of whose elements define an n-trader exchange economy with L nonnumeraire commodities and strictly concave thrice differentiable quasi-linear utilities. We make assumptions $(\alpha 1) - (\alpha 4)$ for the exchange economies defined by E.

We can construct alternative continuum mechanisms on E which, like our continuum Walrasian mechanism, have the property that an equilibrium outcome for the economy described by any given e is a Pareto-optimal and individually rational trade vector in $\mathbb{R}^{L(n-1)}$. In particular the alternative mechanism may be a DR mechanism, where a message is a point e of E. We may let the DR mechanism's outcomes be the Walrasian trade vectors themselves. Then the DR mechanism's outcome function assigns a Walrasian trade vector to the message e and both mechanisms realize the same goal function γ, which assigns that Walrasian trade vector to e. Under our assumptions $(\alpha 1) - (\alpha 4)$, that goal function is well-behaved: it is C^2 and its image set includes an open set of $\mathbb{R}^{L(n-1)}$, the function's range.

Whatever outcome function we choose, the continuum DR mechanism has a larger message-space dimension than the Walrasian mechanism as long as J, the dimension of E, exceeds nL, the dimension of the continuum Walrasian mechanism's message space.

As long as the continuum Walrasian mechanism and the alternative continuum mechanism both realize well-behaved goal functions with ranges in $\mathbb{R}^{L(n-1)}$, we can apply Proposition A to the approximations of those two mechanisms without even requiring that the two goal functions be the same. If, moreover, the alternative mechanism is a DR mechanism, then we obtain the strongest of the superiority statements in Proposition A. Even if we are harsh to the Walrasian mechanism and choose rounded outcomes for its approximations, while we are kind to the DR mechanism by letting its outcomes be exact, the former still has lower error than the latter, if the former has a sufficiently fine mesh and the latter is not more costly than the former. That statement is Part I of our first exchange-economy proposition. Proposition A also assures us that if the alternative continuum mechanism does not have the DR property but has a message space of larger dimension than the continuum Walrasian mechanism, then the Walrasian mechanism's rounded-outcome approximations (and *a fortiori* its exact-outcome approximations) are superior (in the same sense) to the alternative mechanism's approximations, provided the latter are also of the rounded-outcome type. That statement is Part II of our first exchange-economy proposition.

Using techniques developed in earlier papers[29] one can show that that there is no regular continuum mechanism on our set E of n-trader $(L + 1)$-commodity exchange economies which realizes a Pareto-optimal individually rational allocation for each exchange economy e in E and has fewer than nL real message variables (i.e., has a message space of dimension less than nL, the dimension of the Walrasian mechanism's message space). That result, together with Part II of our first exchange-economy proposition, tells us that there is no regular continuum mechanism on E which realizes a well-behaved goal function that specifies Pareto optimality and individual rationality, has either more or fewer than nL real message variables, and has a complete class of controlled rounded-outcome approximations that is superior to a complete class of controlled rounded-outcome approximations of the Walrasian mechanism.

Exchange-economy Proposition 1. *Consider the continuum Walrasian mechanism on our J-dimensional class E of exchange economies. (There are L non-numeraire commodities, and n traders with fixed endowments whose preferences are given by strictly concave and thrice differentiable quasi-linear utility functions with parameters $(e_1, \ldots, e_n) \in E$). Let C be a complete class of controlled rounded-outcome approximations of the continuum Walrasian mechanism. Then*

---

[29] See [4],[5], [9], [14], [16].

(I) *if $J > nL$, the class $\mathcal{C}$ of controlled* **Walrasian rounded-outcome approximations** *is* **superior to** *a complete class of* **exact-outcome approximations of any continuum DR mechanism on** E *which realizes a well-behaved goal function whose range is* $\mathbb{R}^{L(n-1)}$

(II) *the class $\mathcal{C}$ of controlled* **Walrasian rounded-outcome approximations** *is* **superior to** *a complete class of controlled rounded-outcome approximations of* **any** *regular continuum mechanism on* E **which has a message space of dimension larger than** nL *and realizes a well-behaved goal function whose range is* $\mathbb{R}^{L(n-1)}$.

Next we turn to Proposition B, in order to study the increasing superiority of the Walrasian mechanism over a DR mechanism as the dimension of E grows. We consider a sequence $\{E^v\}$ of successively higher-dimensional exchange-economy environment sets $E^v = E_1^v \times \cdots \times E_n^v$, where

$$E_i^v = \{e_i^v = (e_{i1}^v, ..., e_{iK_i(v)}^v) \; : \; \bar{e}_{ij}^v \le e_{ij}^v \le \overset{=v}{e}_{ij}, j = 1, \ldots, K_i(v)\}$$

and $\bar{e}_{ij}^v < \overset{=v}{e}_{ij}$. For every v, the economies defined by the elements of E^v have the same properties as the economies defined by the elements of our set E. The ith component of the environment $e^v = (e_1^v, \ldots, e_n^v)$ in E^v is a vector $e_i^v = (e_{i1}^v, \ldots, e_{iK_i(v)}^v)$ of utility-function parameters for trader i, whose utility function is quasi-linear, with the nonlinear part denoted ϕ^{iv}. Thus in the economy defined by $e^v \in E^v$, trader i's utility for the trade (z^i, w^i) (where z^i denotes a quantity of the numeraire commodity and w^i denotes an L-tuple of nonnumeraire quantities) is $z^i + \phi^{iv}(w^i; e_i^v)$. Trader i's marginal utility for nonnumeraire commodity ℓ is denoted $\phi_\ell^{iv}(w^i; e_i^v)$. As the index v grows without limit, so does the dimension of E^v, i.e., $\lim_{v \to \infty} \sum_{i=1}^n K_i(v) = \infty$.

With the environment set E^v we associate a continuum DR mechanism $\Lambda^{*v} = \langle M^{*v}, (g_1^{*v}, \ldots, g_n^{*v}), h^{*v} \rangle$ and the (condensed) continuum Walrasian mechanism, denoted $\Lambda^v = \langle M^v, (g_1^v, \ldots, g_n^v), h^v \rangle$. We shall suppose that Λ^{*v} realizes the same goal function as Λ^v and that the endowments are fixed and stay the same for all v.

The message space M^{*v} equals E^v, while the message space M^v is the set

$$\{m = ((w^1, \ldots, w^{n-1}), (p_1, \ldots, p_L)) \; : \; w^i \in S_{i\Delta}^\#; i = 1, \ldots, n-1;$$

$$p_\ell^{*v} - \Psi^v \le p_\ell \le p_\ell^{**v} + \Psi^v, \ell \in \{1, \ldots, L\}\},$$

where

$$p_\ell^{*v} \equiv \min_{w^i \in S_i^\#, e_i^v \in E_i^v, i \in \{1,\ldots,n\}} \phi_\ell^{iv}(w^i; e_i^v), \; p_\ell^{**v} \equiv$$

$$\max_{w^i \in S_i^\#, e_i^v \in E_i^v, i \in \{1,\ldots,n\}} \phi_\ell^{iv}(w^i; e_i^v),$$

and $0 < \Psi^v < \min_{\ell \in \{1,\ldots,L\}} p_\ell^{*v}$.

To apply Proposition B, we have to restrict the sequence $\{E^v\}$ of exchange-economy environment sets. They have to obey conditions (i)-(vi) in the statement of that proposition. They also have to have properties guaranteeing that each member of the condensed-Walrasian-mechanism sequence obeys (vii)-(ix). In particular, condition (ix) requires that there exist two Ln-dimensional cubes in $\mathbb{R}^{Ln}$, called M and M', such that for every v, the message space M^v of Λ^v properly includes M' and is itself properly included in M.

The fact that endowments never change implies the existence of the "trade" part of the cube M. But we still need assumptions on the exchange-economy sequence which guarantee the existence of the "price" or "marginal utility" part of M as well as the entire inner cube M'. That is achieved by Part (b) of the following two-part assumption (#). Part (a), together with our definitions of E^v and Λ^v, insure satisfaction of (i)-(viii). To state Part (b) (which insures satisfaction of (ix)), we let Ω^v denote the set of all L-tuples $q = (q_1, \ldots, q_L)$ such that for some $e^v \in E^v$, some trader i, and some nonnumeraire trade vector $w^i \in S^{\#}_{i\Delta}$, the vector q equals the vector of marginal utilities $(\phi_1^{iv}(w^i; e_i^v), \ldots, \phi_L^{iv}(w^i; e_i^v))$. Our two-part condition is

(#)

    (a) There exist W, X, Y, Z, with $Y < W < X < Z$ such that for all v, all i in $\{1, \ldots, n\}$ and all j in $\{1, \ldots, K_i(v)\}$ we have
$$Y \leq \bar{e}_{ij}^v < W < X < \overset{=v}{e}_{ij} \leq Z$$

    (b) There exist an nL-dimensional cube $M' \subset \mathbb{R}^{nL}$ and an L-dimensional cube $Q \subset \mathbb{R}^L$ such that for all v and all Ψ^v with $0 < \Psi^v < \min_{\ell \in \{1, \ldots, L\}} p_\ell^{*v}$, we have $M' \subset M^v$, and we also have $\Omega^v \subset Q$.

Part (b) of (#) implies that for any choice of the positive numbers Ψ^v, which enter the definition of the Walrasian message spaces M^v, there is indeed a cube M which properly includes all the sets M^v as well as a cube M' which is properly included in all those sets.

Proposition B also imposes condition (xii) on the DR sequence $\{\Lambda^{*v}\}$. As we remarked in introducing that condition, it can be interpreted as a minimal-sensitivity bound for the goal function realized by Λ^{*v}, valid for all v. We shall consider the following stronger but simpler condition, which implies (xii):

(##)

    for some person i and some action component k, there exists $\tau > 0$ such that for every v and some j in $\{1, \ldots, K_i(v)\}$ we have
$$\left| \partial h_k^{*v} / \partial e_{ij}^v \right| \geq \tau.$$

Finally we shall need both of our continuum mechanism sequences to obey the "uniformly bounded derivatives" condition, defined in Part III and assumed in Proposition B.[30]

We are now ready to state the exchange-economy proposition which Proposition B implies.

Exchange-economy Proposition 2. *For n traders and L nonnumeraire commodities, consider a sequence $\{E^v\}$ of exchange-economy environment sets. Let the*

---

[30] Conditions (x), (xi), on the DR-mechanism sequence, are also assumed in Proposition B. But these are automatically implied by our assumption that Λ^v and Λ^{*v} realize the same goal function with range in $\mathbb{R}^{L(n-1)}$.

sequence $\{E^v\}$ meet the above condition (#). For each $v \geq 1$, let every member e^v of E^v define an exchange economy that has the same properties as the exchange economy defined by a member of our set E. (In particular, the economy satisfies assumptions $(\alpha 1) - (\alpha 4)$). Let the endowments $\{(V^i, W^i_\ell)\}_{i=1,\ldots,n,\ \ell=1,\ldots,L}$ be the same for every v and for all the exchange economies defined by the members of E^v. Let $\Lambda^v = \langle M^v, (g^v_1, \ldots, g^v_n), h^v \rangle$ be the **continuum (condensed) Walrasian mechanism** *on the exchange-economy set E^v. Let Λ^{*v} be a* **continuum DR mechanism** *on the exchange-economy set E^v with an action space the same as that of Λ^v. Let the sequences $\{\Lambda^v\}$, $\{\Lambda^{*v}\}$ have uniformly bounded derivatives. Let the DR sequence $\{\Lambda^{*v}\}$ obey condition (##). For all v, let the DR mechanism Λ^{*v} realize the same goal function as the Walrasian mechanism Λ^v.*

Then there exist a number $\hat{\epsilon} > 0$ and a function $\bar{r} : \mathbb{R} \times I^+ \to \mathbb{R}$ (where I^+ denotes the positive integers) such that for $0 < \epsilon < \hat{\epsilon}$, the exact-outcome **DR approximation** $\Lambda^{*v}_{\bar{r}(\epsilon,v)}$ *of Λ^{*v}* **has an overall error not higher than** *that of the exact-outcome* **Walrasian approximation** Λ^v_ϵ. *Moreover, for any such $(\hat{\epsilon}, \bar{r})$, there exists $\epsilon^* \leq \hat{\epsilon}$ such that:*

for all ϵ with $0 < \epsilon \leq \epsilon^*$, we have $\displaystyle \lim_{v \to \infty} \left[C\left(\Lambda^{*v}_{\bar{r}(\epsilon,v)} \right) - C(\Lambda^v_\epsilon) \right] = \infty,$

where, for an approximation $\mathcal{L}$, the symbol $C(\mathcal{L})$ denotes, as before, the finite number of equilibrium messages of $\mathcal{L}$.

Returning to the case of a fixed environment set E, it remains to apply Proposition C in order to consider the (possibly empty) class of regular continuum mechanisms on E that realize Pareto optimality and individual rationality, are exactly *tied* with the Walrasian mechanism with regard to number of real message variables, and are *distinct* from the Walrasian mechanism. Proposition C assures us that nothing is to be gained by considering that class. Finite approximations of members of that class are *indifferent*, in the sense we have defined, to finite approximations of the Walrasian mechanism.

Exchange-economy Proposition 3. *Consider our set E of n-trader $(L + 1)$-commodity exchange economies meeting assumptions $(\alpha 1) - (\alpha 4)$. A complete class of finite controlled rounded-outcome approximations of the* **continuum Walrasian mechanism** *on E is* **indifferent** *to a complete class of controlled rounded-outcome approximations of any other regular continuum mechanism on E, where the other mechanism,* **like the Walrasian mechanism, has nL message variables** *and realizes a well-behaved goal function whose range is $\mathbb{R}^{L(n-1)}$.*[31]

---

[31] In the continuum literature the typical technique for establishing message-space minimality of the Walrasian mechanism is as follows (See Hurwicz, [4], [5]). One considers a "test class" of exchange economies (for example, economies with utility functions that are Cobb-Douglas or are quasi-linear with a quadratic nonlinear part). One shows that a regular mechanism with lower message-space dimension than the Walrasian mechanism cannot realize Pareto optimality and individual rationality on the test class. Then, *a fortiori*, no such mechanism can realize that goal on any wider class that contains the test class. Unfortunately a parallel technique may not work in our problem. We cannot, for example, use the quasi-linear class as a "test class" in a parallel way. Given a wider class of exchange economies that includes the quasi-linear class, we cannot rule out the possibility that (i) for an approximation of the Walrasian mechanism, maximum error occurs *outside* the quasi-linear class, (ii) a not-more-costly

6 Concluding remarks

In the exchange economies we have studied, traders have strictly concave quasi-linear utility functions. That property permitted us to establish the uniqueness of the continuum Walrasian mechanism's solution function and its approximate-solution functions. Uniqueness is one of our regularity conditions and is not easy to establish.[32] It remains open whether the superiority of finite approximations of the Walrasian mechanism over approximations of alternative regular continuum mechanisms extends to a wider class of economies.

There are, moreover, other styles of approximation, in addition to the one we have studied.[33] It remains open whether we can obtain analogues of our results for those alternative styles while imposing regularity conditions on the continuum mechanisms which are weaker than the ones we have imposed. If so, it may be possible to obtain stronger results about the informational superiority of approximations of the Walrasian mechanism in a wide class of exchange economies.

While the subject is certainly not closed, our finite counterparts of the previous continuum statements substantially extend the claims one can make about the informational merit of mechanisms that use prices.

Appendix

1. Fulfilling assumptions $(\alpha 1) - (\alpha 4)$ in a two-person two-commodity economy with quadratic valuation functions

In the text we introduced the two traders' local environments $e_i = (\theta_i, \beta_i)$ where $\theta_i \equiv \alpha_i - \beta_i W^i$ and W^i is i's fixed endowment of the nonnumeraire commodity. We also introduced the utility functions,

$$v^i + \phi^i(w^i; e_i) = v^i + \left[(\theta_i + \beta_i W^i)(W^i + w^i) - \frac{1}{2}\beta_i (W^i + w^i)^2 \right].$$

Each trader's fixed numeraire endowment is V^i. We need the following sets:

$$\hat{S}_1 = \{(v^1, w^1) : -V^1 < v^1 < V^2 + \gamma; -W^1 < w^1 < W^2 + \gamma\}$$

$$S_1 = \{(v^1, w^1) : -V^1 + \sigma \leq v^1 \leq V^2; -W^1 + \sigma \leq w^1 \leq W^2\},$$

---

approximation of some non-Walrasian mechanism has a smaller maximum error outside the quasi-linear class even though its maximum error within the quasi-linear class exceeds that of the Walrasian approximation, (iii) as a consequence, the overall error of the non-Walrasian approximation on the entire wide class is less than the overall error of the Walrasian approximation.

[32] We note that most results about the uniqueness of Walrasian equilibrium, are achieved by imposing conditions on the excess demand functions, not on individual utility functions. (See [12], Section 5.7). But it is conditions of the latter sort that we need, in order to define the individual environment sets E_i on which our mechanism operates.

[33] In one alternative style of approximation, we assign to each e_i in E_i one of a finite collection of "surrogate environments". If $s(e_i)$ is the surrogate for e_i then in the finite approximation, with message set M_ϵ, person i agrees to a message $m \in M_\epsilon$ if and only if i would agree to m in the continuum mechanism at the environment $s(e_i)$.

$$\hat{S}_2 = \{(v^2, w^2) : -V^2 < v^2 < V^1 + \gamma; -W^2 < w^2 < W^1 + \gamma\},$$
$$S_2 = \{(v^2, w^2) : -V^2 + \sigma \le v^2 \le V^1; -W^2 + \sigma \le w^2 \le W^1\},$$
$$S_{1\Delta}^{\#} = \{w^1 : -W^1 + \Delta \le w^1 \le W^2 + \Delta\}, S_{2\Delta}^{\#}$$
$$= \{w^2 : -W^2 + \Delta \le w^2 \le W^1 + \Delta\},$$

where

(1) $\qquad \gamma > 0, 0 < \sigma < \min\{V_1, W_1, V_2, W_2\}; 0 < \Delta < \min(\sigma, \gamma).$

For each i, we choose the closed local-environment set, and the extended open one to be:

$$E_i = \{e_i = (\theta_i, \beta_i) : \bar{\theta}_i \le \theta_i \le \bar{\bar{\theta}}_i, \bar{\beta}_i \le \beta_i \le \bar{\bar{\beta}}_i\},$$
$$\hat{E}_i = \{e_i = (\theta_i, \beta_i) : \bar{\theta}_i - \zeta_i < \theta_i < \bar{\bar{\theta}}_i + \zeta_i, \bar{\beta}_i - \zeta_i < \beta_i < \bar{\bar{\beta}}_i + \zeta_i\},$$

where

(2) $\qquad 0 < \zeta_i < \bar{\theta}_i, 0 < \zeta_i < \bar{\beta}_i.$

We take the domain of the valuation function ϕ^i to be $\hat{S}^i \times \hat{E}^i$. On that domain ϕ^i is C^3 and so it satisfies Assumption $(\alpha 1)$. Assumption $(\alpha 2)$ requires that ϕ^i be strictly increasing at every w^i for which $-W^i < w^i \le W^j + \gamma, j \ne i$. That is the case if

(3) $\qquad W^2 + \gamma < \dfrac{\bar{\theta}_1 - \zeta_1}{\bar{\bar{\beta}}_1 + \zeta_1}, W^1 + \gamma < \dfrac{\bar{\theta}_2 - \zeta_2}{\bar{\bar{\beta}}_2 + \zeta_2}.$

(Even when trader i adds to his endowment of the non-numeraire commodity the upper bound in the set $\hat{S}_i$, trader i's utility is still increasing with respect to the holding of that commodity at the resulting bundle.) Assumption $(\alpha 2)$ also requires that the one-by-one matrix whose single entry is $-\beta_i$ be negative semi-definite. That is the case since $0 < \bar{\beta} \le \beta_i$.

We turn to Assumption $(\alpha 3)$. First notice that the equation which specifies equality of marginal rates of substitution, namely

(4) $\qquad \dfrac{d\phi^1(w^1; (\theta_1, \beta_1))}{dw^1} = \dfrac{d\phi^2(w^2; (\theta_2, \beta_2))}{dw^2},$

and the equation $w^1 + w^2 = 0$, are solved simultaneously by

$$\bar{w}^1 = \dfrac{\theta_1 - \theta_2}{\beta_1 + \beta_2}, \bar{w}^2 = -\bar{w}^1.$$

Assumption $(\alpha 3)$ requires that for every $e = ((\theta_1, \beta_1), (\theta_2, \beta_2))$ in $\hat{E}$, the pair $(\bar{w}^1, \bar{w}^2)$ satisfies:

$$\bar{w}^1 + W^1 > \sigma > 0, \bar{w}^2 + W^2 > \sigma > 0.$$

Using the definition of θ_i, it is readily verified that this is indeed the case if

(5) $\qquad (W^1 + W^2) \cdot \min(\bar{\beta}_1 - \zeta, \bar{\beta}_2 - \zeta) > \sigma + |\alpha_1 - \alpha_2|.$

Finally, we turn to Assumption $(\alpha4)$. Let $\phi^{i'}(w^i; e_i)$ denote $\frac{d\phi^i}{dw^i}$. We first obtain $\phi^{1'}(w^1; e_1) = \frac{\theta_1\beta_2 + \theta_2\beta_1}{\beta_1 + \beta_2}$. So for $e = (\theta_1, \beta_1, \theta_2, \beta_2)$, the set $\Phi(e)$ used in the statement of $(\alpha4)$ is a singleton, namely $\left\{ \left(\frac{\theta_1 - \theta_2}{\beta_1 + \beta_2}, \frac{\theta_1\beta_2 + \theta_2\beta_1}{\beta_1 + \beta_2} \right) \right\}$. Assumption $(\alpha4)$ requires that the set $\Phi(E)$ contain an open set of $\mathbb{R}^2$. We note that if

$$(6) \qquad \bar{\beta}_1 = \bar{\beta}_2 = \bar{\beta}, \bar{\bar{\beta}}_1 = \bar{\bar{\beta}}_2 = \bar{\bar{\beta}},$$

then for any β^* satisfying $\bar{\beta} \le \beta^* \le \bar{\bar{\beta}}$, the set $\Phi(E)$ includes the set

$$\Phi_{\beta^*}(E) \equiv \left\{ \left(\frac{\theta_1 - \theta_2}{2\beta^*}, \frac{\theta_1 + \theta_2}{2} \right) : \theta_1 \in [\bar{\theta}_1, \bar{\bar{\theta}}_1], \theta_2 \in [\bar{\theta}_2, \bar{\bar{\theta}}_2] \right\}.$$

Assumption $(\alpha4)$ is satisfied if

$$(7) \qquad \Phi_{\beta^*}(E) \text{ contains a nondegerate rectangle.}$$

To summarize: Assumptions $(\alpha1) - (\alpha4)$ are satisfied if the fixed endowments and the sets

$$\hat{S}_i, S_i, S_{i\Delta}^\#, \hat{E}_i, E_i$$

satisfy (1)-(7). An example of $\gamma, \sigma, \Delta, \beta^*$ and $(V_i, W_i, \bar{\beta}_i, \bar{\bar{\beta}}_i, \bar{\theta}_i, \bar{\bar{\theta}}_i, \zeta_i), i = 1, 2$, satisfying (1)-(7) is as follows:

$$\gamma = 1, \sigma = \tfrac{1}{100}, \Delta = \tfrac{1}{200}, V_1 = 3, W_1 = 4, \alpha_1 = \tfrac{54}{4}, \bar{\theta}_1 = 10, \bar{\bar{\theta}}_1 = \tfrac{52}{5}, V_2 = 4, W_2 = 3, \alpha_2 = \tfrac{51}{5}, \bar{\theta}_2 = \tfrac{48}{5}, \bar{\bar{\theta}}_2 = \tfrac{99}{10}, \bar{\beta}_1 =$$
$$\bar{\beta}_2 = \beta^* = \tfrac{1}{10}, \bar{\bar{\beta}}_1 = \bar{\bar{\beta}}_2 = \tfrac{1}{5}.$$

2. The full and condensed continuum Walrasian mechanisms for a class of two-person quadratic-valuation economies

We shall consider mechanisms on the set E just defined. As in the text, we let $S_i^\#, \hat{S}_i^\#$ denote the numeraire projections of the sets $S_i, \hat{S}_i$ just defined. Now we define

$$(8) \qquad p^* = \min_{w^i \in S_i^\#, e_i \in E_i, i \in \{1,2\}} \phi^{i'}(w^i; e_i) = \min(\bar{\bar{\theta}} - \bar{\bar{\beta}} W^1, \bar{\bar{\theta}} - \bar{\bar{\beta}} W^2).$$

In view of (1)–(3), $p^* > 0$. Define

$$p^{**} = \max_{w^i \in S_i^\#, e_i \in E_i, i \in \{1,2\}} \phi^{i'}(w^i; e_i)$$

$$(9) \qquad = \max(\bar{\bar{\theta}} + \bar{\beta}(W^1 - \sigma), \bar{\bar{\theta}} + \bar{\beta}(W^2 - \sigma)).$$

By (1)–(3), $p^{**} > p^*$. The message space of the full mechanism is the compact set

$$(10) \quad M^{\mathsf{F}} = \left\{ m = ((w^1, w^2), p) : w^i \in S_{i\Delta}^\#; i=1, 2; p^* - \bar{\Delta} \le p \le p^{**} + \bar{\Delta} \right\}.$$

Trader 1's agreement function $\bar{g}_1$ is a pair, namely $(\bar{g}_{11}, g^*)$, where

$$\bar{g}_{11}\left(((w^1, w^2), p); e_1\right) = \theta_1 - \beta_1 w^1 - p, \ g^*\left(((w^1, w^2), p); e_1\right) = w^1 - w^2.$$

For trader 2, we have $\bar{g}_2\left(((w^1, w^2), p); e_2\right) = \theta_2 - \beta_2 w^2 - p.$

The solution to the equation system

$$\bar{g}_1\left((w^1, w^2), p), e_1\right) = \bar{g}_2\left(((w^1, w^2), p), e_2\right) = g^*\left(((w^1, w^2), p); e_1\right) = 0 \text{ is}$$

(11) $$\bar{w}^1 = \frac{\theta_1 - \theta_2}{\beta_1 + \beta_2}, \bar{w}^2 = \frac{\theta_2 - \theta_1}{\beta_1 + \beta_2}, \bar{p} = \frac{\theta_1 \beta_2 + \theta_2 \beta_1}{\beta_1 + \beta_2}.$$

The argument given above, in connection with Assumption $(\alpha 3)$ shows that our assumed condition (5) implies $\bar{w}^1 \in S_{1\Delta}^{\#}, \bar{w}^2 \in S_{2\Delta}^{\#}$. By the definition of p^*, p^{**} in (8) and (9), we also have

$$\bar{p} \in [p^*, p^{**}],$$

since the satisfaction of the equations $\bar{g}_{11} = g^* = \bar{g}_2 = 0$ by $(\bar{w}^1, \bar{w}^2, \bar{p})$ implies that $\bar{p} = \phi^{i'}(\bar{w}^i; e_i)$, i=1,2, and since $\bar{w}^i \in S_{i\Delta}^{\#}, i = 1, 2$.

So the full mechanism covers E.

The condensed mechanism. The message space of the condensed mechanism is the compact set

$$M = \left\{ m = (w^1, p) \ : \ w^1 \in S_{1\Delta}^{\#}, -w^1 \in S_{2\Delta}^{\#}; p^* - \Psi \le p \le p^{**} + \Psi \right\}.$$

The agreement functions are

$$g_1(w^1, p), e_1) = \theta_1 - \beta_1 w^1 - p, \ g_2(w^1, p), e_2) = \theta_2 + \beta_2 w^1 - p.$$

For the outcome function h we have

$$h(w^1, p) = w^1.$$

The solution to the equation system $g_1(w^1, p), e_1 = g_2(w^1, p), e_2) = 0$ is

$$\bar{w}^1 = \frac{\theta_1 - \theta_2}{\beta_1 + \beta_2}, \ \bar{p} = \frac{\theta_1 \beta_2 + \theta_2 \beta_1}{\beta_1 + \beta_2}.$$

Note that $((\bar{w}^1, \bar{w}^2), \bar{p})$ is an equilibrium message for e in the full mechanism if and only if $(\bar{w}^1, \bar{p})$ is an equilibrium message for e in the condensed mechanism. Hence the fact that the full mechanism covers E (as just argued) implies that the condensed mechanism also covers E.

Regularity of the condensed mechanism. Regularity Conditions (a) and (b) are immediately verified. We turn to Regularity Condition (c). That is fulfilled for the condensed mechanism if it holds for the *full* mechanism. That is, in turn, the case if

Properties (1)–(4) of Remark R1 hold for the full mechanism. We now check those properties for the full mechanism.

Property (1). Define $\bar{M}^{\mathsf{F}} \equiv \{m = ((w^1, w^2), p) : w^i \in S_i, i = 1, 2; p^* \leq p \leq p^{**}\}$. As we remarked in the preceding discussion of coverage, $\bar{M}^{\mathsf{F}}$ contains an equilibrium message for every e in E. As Property **(1)** requires, the set $\bar{M}^{\mathsf{F}}$ is in the interior of M^{F}.

Property (2). We define the extended message space

$$\hat{M}^{\mathsf{F}} \equiv \left\{m = ((w^1, w^2), p) : w^i \in \hat{S}_i^{\#}; i = 1, 2; p^* - \tau < p < p^{**} + \tau\right\},$$

where $\Psi < \tau < p^*$. Defining each of the two extended functions $\hat{\bar{g}}_i$ by the same formula used to define $\bar{g}_i$, we have:

$$\hat{\bar{g}}_1 = (\hat{\bar{g}}_{11}, \hat{\bar{g}}^*), \hat{\bar{g}}_2 = \theta_1 - \beta_1 w^1 - p, \hat{\bar{g}}_2 = \theta_2 - \beta_2 w^2 - p, \hat{\bar{g}}^* = w^1 - w^2.$$

Each of the extended functions is $\mathcal{C}^2$ on $\hat{M}^{\mathsf{F}} \times \hat{E}$.

Properties (3) and (4). In the text we presented the Jacobian $\hat{Q}(m, e)$ for the condensed mechanism. Property **(3)** requires that its determinant be nonzero, which is the case, since $\beta_1 > 0, \beta_2 > 0$. For Property **(4)** (univalence) to hold it is sufficient that (a) the number of rows be odd and the determinant be negative (which is the case) and (b) that the symmetric matrix $\frac{1}{2}(\hat{Q} + \hat{Q}')$ be negative semi-definite (which, as noted in the text, is the case).

It remains to consider Regularity Condition (d) for the condensed mechanism. We showed in the preceding Section 1 of this Appendix that an exchange economy defined by an element of E satisfies Assumption $(\alpha 4)$. That implies that the condensed mechanism on E satisfies Regularity Condition (d). So all four regularity conditions are met by the condensed mechanism.

References

1. Calsamiglia, X.: Decentralized resource allocation and increasing returns. Journal of Economic Theory **14**, 263–283 (1977)
2. Gale, D., Nikaido, H.: The Jacobian matrix and global univalence of mappings. Mathematische Annalen **159**, 81–93 (1965)
3. Hayek, F.: The use of knowledge in society. American Economic Review **35**, 519–530 (1945)
4. Hurwicz, L.: On the dimensional requirements of informationally decentralized Pareto-satisfactory adjustment processes. In: Arrow, K.J., Hurwicz, L., (eds.) Studies in resource allocation processes. Cambridge: Cambridge University Press 1977
5. Hurwicz, L.: On Informational decentralization and efficiency in resource allocation mechanisms. In: Reiter, S. (ed.) Studies in mathematical economics. Washington, DC: Mathematical Association of America 1986
6. Hurwicz, L., Marschak, T.: Comparing finite mechanisms. Economic Theory (forthcoming)
7. Hurwicz, L., Marschak, T.: Finite allocation mechanisms: approximate Walrasian versus approximate Direct Revelation. Economic Theory **21**, 545–572 (2003)
8. Hurwicz, L.: Revisiting externalities. Journal of Public Economic Theory **1**, 225–245 (1999)
9. Jordan, J.: The competitive mechanism is dimensionally efficient uniquely. Journal of Economic Theory **28**, 1–18 (1982)
10. Koopmans, T.C.: Three essays on the state of economic science. New York: McGraw Hill 1957

11. Lange, O.: On the economic theory of socialism. In: Lange. O., Taylor, F.M., (eds.) On the economic theory of socialism. Minneapolis: University of Minnesota Press 1938
12. Mas-Colell, A.: The theory of general economic equilibrium: a differentiable approach. Cambridge: Cambridge University Press 1985
13. Mas-Colell, A., Whinston, M., Green, J.: Microeconomic theory. New York: Oxford University Press 1995
14. Mount, K., Reiter, S.: The informational size of message spaces. Journal of Economic Theory **8**, 161–192 (1974)
15. Nikaido, H.: Convex structures and economic theory. New York: Academic Press 1968
16. Osana, H.: On the informational size of messages for resource allocation processes. Journal of Economic Theory **17**, 66–78 (1978)
17. Walker, M.: On the informational size of message spaces. Journal of Economic Theory **15**, 366–375 (1977)

Information at equilibrium[*]

E. Minelli[1] and H. Polemarchakis[2]

[1] CORE, Université Catholique de Louvain, 1348 Lovain-la-Neuve, BELGIUM
(e-mail: minelli@core.ucl.ac.be)
[2] Department of Economics, Brown University, Providence, RI 02912, USA
(e-mail: herakles_polemarchakis@brown.edu)

Received: June 20, 2001; revised version: January 9, 2002

Summary. In a game with rational expectations, individuals simultaneously refine their information with the information revealed by the strategies of other individuals. At a Nash equilibrium of a game with rational expectations, the information of individuals is essentially symmetric: the same profile of strategies is also an equilibrium of a game with symmetric information; and strategies are common knowledge. If each player has a veto act, which yields a minimum payoff that no other profile of strategies attains, then the veto profile is the only Nash equilibrium, and it is is an equilibrium with rational expectations and essentially symmetric information; which accounts for the impossibility of speculation.

Keywords and Phrases: Nash equilibrium, Rational expectations, Common knowledge.

JEL Classification Numbers: D82.

1 Introduction

Private information may differ across individuals: it may be asymmetric.

If expectations are rational, individuals refine their information with the information revealed by the acts of others. If an event is common knowledge, individuals know it has occurred, they know that others know it, they know that others know that they know it, ...

---

[*] We wish to thank Pierpaolo Battigalli, Françoise Forges, Franco Donzelli, Leonidas Koutsougeras, Aldo Rustichini, Rajiv Vohra and Nicholas Yannelis for their comments.

Correspondence to: H. Polemarchakis

Thus, at a rational expectations equilibrium as well as for an event which is common knowledge, gains to knowledge from the exchange of information are exhausted.

Rational expectations were formalized by Radner (1979) in the context of walrasian equilibria. Common knowledge was formalized by Aumann (1976) without reference to the optimizing or strategic behavior of individuals.

Both rational expectations and common knowledge are powerful conceptual tools that lead to surprising and often similar conclusions. The impossibility of speculative exchange has been claimed by Milgrom and Stokey (1982) as a consequence of the common knowledge of individuals of their willingness to trade, and by Tirole (1982) as a property of equilibria with rational expectations.

Rational expectation is a property of the information of individuals at equilibrium. It applies across states of the world or of private information. Common knowledge is a property of events relative to the information of individuals. It applies at each state of the world or of private information.

The comparison of rational expectations and common knowledge should then be posed as follows: what events are common knowledge at a rational expectations equilibrium?

At a Nash equilibrium of a game with uncertainty and private information, according to the formalization of Harsanyi (1967 - 1968), individuals do not extract information from the acts of other individuals in the same round of play; this takes literally the simultaneity of moves. But it is naive.

At a Nash equilibrium of a game with rational expectations, individuals extract information from the simultaneous acts of other individuals.

Individuals know the strategies of others, and they observe their realized elementary acts; the no - regret condition that characterizes Nash equilibrium requires robustness to the new knowledge that individuals obtain at equilibrium. This is the motivation for the definition of a game with rational expectations.

A Nash equilibrium for a game with rational expectations abstracts from the strategic aspects of the revelation of information. This is in the spirit of Nash equilibrium, but may fall short of interpretation of rational expectations as a reduced form of a process of information revelation and learning; Dubey, Geanakoplos and Shubik (1987) and Forges and Minelli (1997a,b) explore the connection between dynamic or repeated games with asymmetric information and games with asymmetric information and rational expectations; Kalai (2000) discusses the issues involved.

At a Nash equilibrium of a game with rational expectations the information of individuals is essentially symmetric: any differences in information do not affect equilibrium acts; and the acts of individuals are common knowledge.

If the structure of payoffs in a games is such that at a Nash equilibrium, information is essentially symmetric, equilibria are à fortiori equilibria of the game with rational expectations, where the acts of individuals are common knowledge: this is the case for veto games, a powerful insight introduced and developed in Geanakoplos (1995).

Common knowledge of acts immediately implies consensus among individuals trying to guess the value of a random variable. Speculative behaviour, on the other hand, cannot be ruled out at a rational expectations equilibrium of a game. Indeed, an example shows that common knowledge of the equilibrium acts need not imply common knowledge of speculation. Speculation can thus occur at the Nash equilibrium of a veto game.

In a strong veto game, every individual plays its veto act at a Nash equilibrium, speculation cannot occur and, as a consequence, it cannot occur at a competitive equilibrium.

2 Games with rational expectations

A game with private information is a collection

$$G_{\mathcal{P}} = \{\mathcal{I}, \mathcal{S}, (\mathcal{A}^i, u^i, \mathcal{P}^i) : i \in \mathcal{I}\}.$$

Individuals are $i \in \mathcal{I}$, a finite set. States of the world are $s \in \mathcal{S}$, a finite set. A profile of private information is $\mathcal{P} = \{\cdots, \mathcal{P}^i, \cdots\}$, where $\mathcal{P}^i = \{P^i(s) : s \in \mathcal{S}\}$, is a partition of the set of states of the world that represents the private information of the individual. For an individual, an elementary act is $a^i \in \mathcal{A}^i$, and an act or a strategy is f^i, an element of the set of feasible strategies [1]

$$\mathcal{F}^i = \{f^i : \mathcal{S} \to \mathcal{A}^i, \text{ measurable with respect to } \mathcal{P}^i\}.$$

Across individuals, a profile of elementary acts is $a \in \mathcal{A}$, where $a = (\cdots, a^i, \cdots)$, and $\mathcal{A} = \times_{i \in \mathcal{I}} \mathcal{A}^i$, and a profile of strategies is $f \in \mathcal{F} = \times_{i \in \mathcal{I}} \mathcal{F}^i$.

The utility or payoff to the individual at $f \in \mathcal{F}$ is $u^i(f)$, and his utility function is $u^i : \mathcal{F} \to \mathcal{R}$.

For an individual, the complementary set of individuals is $\{-i\} = \{\mathcal{I} \setminus \{i\}\}$. At strategies $f^{-i} \in \mathcal{F}^{-i}$ by the complementary set of individuals, where $f^{-i} = (\cdots, f^{i-1}, f^{i+1}, \cdots)$, $\mathcal{F}^{-i} = \times_{i' \in \{-i\}} \mathcal{F}^{i'}$, and $f = (f^i, f^{-i})$, the optimization problem of the individual is to

$$\max u^i(f^i, f^{-i}),$$

$$\text{s.t. } f^i \in \mathcal{F}^i.$$

The solution to the optimization problem is $\varphi^i(f^{-i}) \subseteq \mathcal{F}^i$, which may be empty, when a solution does not exist, or not a singleton, when the solution is not unique. The choice or reaction correspondence is $\varphi^i : \mathcal{F}^{-i} \to \mathcal{F}^i$.

A Nash equilibrium is a profile of strategies, $f^{\mathcal{I}*}$, such that $f^{i*} \in \varphi^i(f^{-i*})$, for every individual.

Individuals optimize ex - ante, prior to the resolution of uncertainty, and the utility function evaluates profiles of strategies. Under conditions that are well understood, Debreu (1959), the utility function of an individual is additively separable across states of the worlds: $u^i(f) = \sum_{s \in \mathcal{S}} v^i(f(s), s)$. Under stronger conditions, Savage (1954), it has a state - independent expected utility representation:

---

[1] A function $f : \mathcal{S} \to \mathcal{A}$ is measurable with respect to a partition $\mathcal{P}$ if $P(s) = P(s') \Rightarrow f(s) = f(s')$.

$u^i(f) = E^i v^i(f(s))$, where the the probability measure under which expectations are computed is as much a characteristic that may vary across individuals as the cardinal utility index.

An ex - post formulation of games with uncertainty and private information is possible. For separable utility functions, the optimization problem of an individual at a state of the world is

$$\max \sum_{s' \in \mathcal{P}^i(s)} v^i((f^i(s), f^{-i}(s')), s'),$$

$$\text{s.t } f^i(s) \in \mathcal{A}^i.$$

For non - separable utility functions, an ex - post formulation is possible but contrived. With this ex - post formulation of games with private information, the solutions to the individual optimization problems coincide for s and s', with $\mathcal{P}^i(s) = \mathcal{P}^i(s')$. They yield unambiguously a solution to the ex - ante optimization problem, which, in particular, is measurable with respect to the information available to the individual [2].

Information is symmetric if , for some partition, $\mathcal{P}^0$, of the set of states of the world, $\mathcal{P}^i = \mathcal{P}^0$, for every individual: the information of individuals coincides. A game with uncertainty and symmetric information is

$$\mathcal{G}_{\mathcal{P}^0} = \{\mathcal{I}, \mathcal{S}, \mathcal{P}^0, (\mathcal{A}^i, u^i) : i \in \mathcal{I}\}.$$

If the utility functions of all individuals are separable across states of the world and information is symmetric, the game decomposes into a collection of games, indexed by the elements of the common partition.

At a Nash equilibrium of a game with private information, the information of individuals is essentially symmetric, with respect to a partition, $\mathcal{P}^0$, if the profile of acts is a Nash equilibrium of the game with symmetric information $\mathcal{P}^0$.

In a game with rational expectations, individuals refine their information with the information revealed by the elementary acts of other individuals at each state of the world.

For an individual, the feasible act correspondence is defined by [3]

$$\Phi^i(f^{-i}) = \{f^i : \mathcal{S} \to \mathcal{A}^i, \text{ measurable with respect to } \mathcal{P}^i \vee_{j \in \{-i\}} \mathcal{P}^{f^j}\}.$$

---

[2] An alternative ex-post formulation is for an individual to solve, at a state of the world, the optimization problem

$$\max \sum_{s' \in \mathcal{P}^i(s)} v^i((f^i(s), f^{-i}(s)), s'),$$

$$\text{s.t } f^i(s) \in \mathcal{A}^i;$$

solutions need not coincide even if $\mathcal{P}(s) = \mathcal{P}(s')$, and they need not yield a solution to the ex - ante optimization problem subject to the measurability constraint.

[3] A partition, $\mathcal{P}$, is at least as coarse as another, $\mathcal{P}'$ if and only if $P'(s) = P'(s') \Rightarrow P(s) = P(s')$; one writes $\mathcal{P}' \subseteq \mathcal{P}$. If $\{\mathcal{P}^k : k \in \mathcal{K}\}$ is a collection of Partitions, the join is defined as the partition $\overline{\mathcal{P}} = \vee_{k \in \mathcal{K}} \mathcal{P}^k$ such that $\overline{P}(s) = \overline{P}(s')$ if and only if $P^k(s) = P^k(s')$ for all $k \in \mathcal{K}$: it is the coarsest common refinement; the meet is the partition $\underline{\mathcal{P}} = \wedge_{k \in \mathcal{K}} \mathcal{P}^k$, the finest common coarsening. The partition, $\mathcal{P}_f$, induced by a function f is defined by $P^f(s) = P^f(s')$ if and only if $f(s) = f(s')$: it is the coarsest partition with respect to which the function is measurable.

A strategy is feasible for the individual if, whenever $f^i(s) \neq f^i(s')$, either $\mathcal{P}^i(s) \neq \mathcal{P}^i(s')$: the private information of the individual distinguishes states s and s' or, for some individual, $j \in \{-i\}$, $f^j(s) \neq f^j(s')$: the elementary acts of some other individual distinguishes states s and s'.

The optimization problem of the individual is[4]

$$\max u^i(f^i, f^{-i}),$$

$$\text{s.t. } f^i \in \Phi^i(f^{-i}).$$

The solution to the optimization problem is $\varphi^i(f^{-i}) \subseteq \Phi^i(f^{-i})$, and the choice correspondence is $\varphi^i : \mathcal{F}^{-i} \to \mathcal{F}^i$.

A profile of strategies, f, is feasible if $f^i \in \Phi^i(f^{-i})$, for all individuals.

A Nash equilibrium for the game with rational expectations is a feasible profile of strategies, $f^{\mathcal{I}*}$, such that $f^{i*} \in \varphi^i(f^{-i*})$, for every individual.

Proposition 1. *At a Nash equilibrium for a game with rational expectations, the information of individuals is essentially symmetric with respect to the information partition*

$$\mathcal{P}^* = \vee_{i \in \mathcal{I}} \mathcal{P}^{f^{i*}}.$$

Proof. If $f^* = (\ldots, f^{i*}, \ldots)$ is a profile of strategies which is a Nash equilibrium with rational expectations, then it is a Nash equilibrium for the game with symmetric information $\mathcal{G}_{\mathcal{P}^*}$, where $\mathcal{P}^* = \vee_{i \in \mathcal{I}} \mathcal{P}^{f^{i*}}$.

One argues in steps, for each individual:

1. By the definition, the function, f^{i*} is measurable with respect to the partition $\mathcal{P}^{f^{i*}}$, and, hence, also with respect to the finer partition $\mathcal{P}^*$.

2. By the definition of a Nash equilibrium with rational expectations, the function f^{i*} is measurable and optimal with respect to the partition $\tilde{\mathcal{P}}^i = \mathcal{P}^i \vee_{j \in \{-i\}} \mathcal{P}^{f^{j*}}$.

3. Since the partition $\mathcal{P}^{f^{i*}}$ is the coarsest partition with respect to which the function f^{i*} is measurable, the partition $\tilde{\mathcal{P}}^i$ is at least as fine as the partition $\mathcal{P}^{f^{i*}}$. It follows that the partition $\tilde{\mathcal{P}}^i \vee_{j \in \{-i\}} \mathcal{P}^{f^{j*}}$ is at least as fine as the partition $\mathcal{P}^{f^{i*}} \vee_{j \in \{-i\}} \mathcal{P}^{f^{j*}}$. Since $\tilde{\mathcal{P}}^i \vee_{j \in \{-i\}} \mathcal{P}^{f^{j*}} = \tilde{\mathcal{P}}^i$, while $\mathcal{P}^{f^{i*}} \vee_{j \in \{-i\}} \mathcal{P}^{f^{j*}} = \mathcal{P}^*$, the partition $\tilde{\mathcal{P}}^i$ is at least as fine as the partition $\mathcal{P}^*$.

4. Since the function f^{i*} is measurable with respect to the partition $\mathcal{P}^*$, it is measurable and optimal with respect to the partition $\tilde{\mathcal{P}}^i$, and the partition $\tilde{\mathcal{P}}^i$ is at least as fine as the partition $\mathcal{P}^*$, the function f^{i*} is measurable and optimal with respect to the partition $\mathcal{P}^*$.   □

At a Nash equilibrium for a game with rational expectations,

$$\mathcal{P}^i \vee_{j \in \{-i\}} \mathcal{P}^{f^{j*}} \subseteq \vee_{j \in \mathcal{I}} \mathcal{P}^{f^{j*}} \subseteq \mathcal{P}^{f^{j*}}.$$

---

[4] In a game with rational expectations, the two ex - post formulations of the game discussed above (see footnote 2) coincide, and they yield a solution to the ex - ante optimization problem subject to a measurability constraint which takes into account the information revealed at equilibrium.

Corollary 1. *A Nash equilibrium where the information of individuals is essentially symmetric with respect to a partition $\hat{\mathcal{P}}$ is an equilibrium for the game with rational expectations as long as the partition $\hat{\mathcal{P}}$ is, for every individual, (i) at least as fine as the partition $\mathcal{P}^i \vee_{j \in \mathcal{I}} \mathcal{P}^{f^{j*}}$ or (ii) at least as coarse as the partition $\mathcal{P}^i$.*

Proof. For every individual, the act f^{i*} is a solution to the optimization problem with respect to the partition $\mathcal{P}^i \vee_{j \in \{-i\}} \mathcal{P}^{f^{j*}}$, and thus f^* is a Nash equilibrium for the game with rational expectations.

For an individual, if $\hat{\mathcal{P}} \subseteq \mathcal{P}^i \vee_{j \in \{-i\}} \mathcal{P}^{f^{j*}}$, the relevant partitions are ordered as

$$\hat{\mathcal{P}} \subseteq \mathcal{P}^i \vee_{j \in \mathcal{I}} \mathcal{P}^{f^{j*}} \subseteq \mathcal{P}^i \vee_{j \in \{-i\}} \mathcal{P}^{f^{j*}} \subseteq \mathcal{P}^i,$$

while, if $\mathcal{P}^i \subseteq \hat{\mathcal{P}}$, since $\hat{\mathcal{P}} \subseteq \mathcal{P}^{f^{j*}}$,

$$\mathcal{P}^i \vee_{j \in \{-i\}} \mathcal{P}^{f^{j*}} = \mathcal{P}^i.$$

In either case, since f^{i*} is measurable and optimal with respect to the partition $\mathcal{P}^i$ as well as the partition $\hat{\mathcal{P}}$, it is measurable and optimal with respect to the partition $\mathcal{P}^i \vee_{j \in \{-i\}} \mathcal{P}^{f^j*}$.   □

A Nash equilibrium where the information of individuals is symmetric, but with respect to a partition $\hat{\mathcal{P}}$ which, for some individuals, fails to be either as fine as the partition $\mathcal{P}^i \vee_{j \in \mathcal{I}} \mathcal{P}^{f^{j*}}$ or as coarse as the partition $\mathcal{P}^i$, need not be an equilibrium with rational expectations.

Example 1. A game with private information is described by $\mathcal{I} = \{1,2\}$, $\mathcal{A}^1 = \{T, B\}$, $\mathcal{A}^2 = \{L, R\}$, $\mathcal{S} = \{1, 2, 3\}$, $\mathcal{P}^1 = \{\{1\}, \{2, 3\}\}$, $\mathcal{P}^2 = \{\{1\}, \{2\}, \{3\}\}$, $\pi = \{\frac{1}{3}, \frac{1}{3}, \frac{1}{3}\}$ and payoffs

$s=1$	L	R
T	5,1	5,2
B	1,1	1,2

$s=2$	L	R
T	2,1	2,2
B	3,1	3,2

$s=3$	L	R
T	3,2	3,1
B	1,2	1,1

The profile in which player 1 plays T in every state, while player 2 plays R in states 1 and 2, and L in state 3 is a Nash equilibrium, and information is essentially symmetric with respect to the partition $\hat{\mathcal{P}} = \{\{1, 2\}, \{3\}\}$. It is not a rational expectations equilibrium; player 1, with information $\mathcal{P}^1 \vee \mathcal{P}^{f^2} = \{\{1\}, \{2\}, \{3\}\}$, would play T in states 1 and 3, and B in state 2.   □

3 Common knowledge and rational expectations

If the information of an individual is described by the partition $\mathcal{Q}^i$, then, at a state of the world, $\bar{s}$, the individual knows an event, $\mathcal{E} \subseteq \mathcal{S}$ if

$$\mathcal{Q}^i(\bar{s}) \subseteq \mathcal{E}.$$

The states of the world at which the individual knows $\mathcal{E}$ is

$$\mathcal{K}^i(\mathcal{E}) = \{s \in \mathcal{S}, \text{ such that } \mathcal{Q}^i(s) \subseteq \mathcal{E}\}.$$

If $, \bar{s} \notin \mathcal{K}^i(\mathcal{E})$, the individual does not know $\mathcal{E}$ at $\bar{s}$: there exists a $s' \in \mathcal{Q}(\bar{s})$, such that $s' \notin \mathcal{E}$.

If there exist finite sequences of states of the world, $s_n, s_{n-1}, \ldots, s_1 = \bar{s}$, and individuals, $i_n, i_{n-1}, \ldots, i_1$, not necessarily distinct, such that $s_n \in \mathcal{Q}^{i_{n-1}}(s_{n-1})$, $\ldots, s_2 \in \mathcal{Q}^i(\bar{s})$, while $s' \in \mathcal{Q}^{i_n}(s_n) \setminus \mathcal{E}$, then, since $s' \in \mathcal{Q}^{i_n}(s_n) \setminus \mathcal{E}$, $s_n \notin \mathcal{K}^{i_n}(\mathcal{E})$. But then, since $s_n \in \mathcal{Q}^{i_{n-1}}(s_{n-1}) \setminus \mathcal{K}^{i_n}(\mathcal{E})$, $s_{n-1} \notin K^{i_{n-1}}(K^{i_n}(\mathcal{E}))$. Continuing in this manner, one obtains we that $\bar{s} \notin \mathcal{K}^{i_1}(\mathcal{K}^{i_2}(\ldots, \mathcal{K}^{i_n}(\mathcal{E})))$: individual i_1 does not know that individual i_2 knows that $\ldots$ individual i_{n-1} knows that individual i_n knows $\mathcal{E}$. Thus, for any finite sequence of individuals $i_n, i_{n-1}, \ldots, I_1$, individual i_1 knows that i_2 knows that ... i_{n-1} knows that i_n knows $\mathcal{E}$ at $\bar{s}$ if and only if, for any sequence of states of the world, $s_n, s_{n-1}, \ldots, s_2, s_1 = \bar{s}$, such that $s_n \in \mathcal{Q}^{i_{n-1}}(s_{n-1}), \ldots, s_2 \in \mathcal{Q}^{i_1}(\bar{s})$, $\mathcal{Q}^{i_n}(s_n) \subseteq \mathcal{E}$; equivalently, the event $\mathcal{E}$ contains $\underline{\mathcal{Q}}(\bar{s})$, the element of the meet or finest common coarsening of the individual partitions.

At a state of the world, $\bar{s}$, an event $\mathcal{E} \subseteq \mathcal{S}$ is common knowledge if

$$\underline{\mathcal{Q}}(\bar{s}) \subseteq \mathcal{E}, \quad \text{where } \mathcal{Q} = \wedge_{i \in \mathcal{I}} \mathcal{Q}^i.$$

A function, f, with domain the set of states of the world, is common knowledge at $\bar{s}$ if the event $f^{-1}(f(\bar{s})) = \{s \in \mathcal{S}, \text{ such that } f(s) = f(\bar{s})\}$ is common knowledge at $\bar{s}$. A function is common knowledge if it is common knowledge at all states of the world.

Corollary 2. *At a Nash equilibrium of a game with rational expectations, the strategies of all individuals are common knowledge.*

A Nash equilibrium where the strategies of all players are common knowledge is a Nash equilibrium for the game with rational expectations.

Proof. If $f^* = (\ldots, f^{i*}, \ldots)$ is a Nash equilibrium profile of strategies for the game with rational expectations, the information partition of an individual at equilibrium $\tilde{\mathcal{P}}^i = \mathcal{P}^i \vee_{j \in \{-i\}} \mathcal{P}^{f^{j*}}$, which is at least as fine as the partition $\vee_{j \in \{-i\}} \mathcal{P}^{f^{j*}}$. It follows that the meet of the individuals partitions at the equilibrium is at least as fine as the partition $\mathcal{P}^* = \vee_{i \in \mathcal{I}} \mathcal{P}^{f^{i*}}$. Since the act f^{i*} is measurable with respect to the partition $\mathcal{P}^*$, the result follows.

If $f^* = (\ldots, f^{i*}, \ldots)$ is a Nash equilibrium profile of strategies, and if the strategies of all individuals are common knowledge, then, for every individual, f^{i*} is measurable with respect to the meet of private individual partitions, $\underline{\mathcal{P}} = \wedge_{i \in \mathcal{I}} \mathcal{P}^i$. Since f^{i*} is optimal with respect to the partition $\mathcal{P}^i \subseteq \underline{\mathcal{P}}$, by corollary 1, the result follows. □

Corollary 2 and proposition 1 immediately imply that, at a Nash equilibrium in which acts happen to be common knowledge, information is essentially symmetric. This is an instance of the theorem that "common knowledge of actions negates asymmetric information about events" in Geanakoplos (1995).

Proposition 1 plays the same role with respect to this theorem that the result in Geanakoplos and Polemarchakis (1982) plays with respect to the theorem in Aumann (1976): even if, "to begin with"information is not symmetric and acts are not common knowledge, "eventually"acts are common knowledge and information is symmetric. The process of communication is not explicit. Rather, it is embedded in the definition of a game with rational expectations.

Example 2. The opinion game, (Geanakoplos and Polemarchakis (1982)), is a game with uncertainty and private information, $\mathcal{O} = \{\mathcal{I}, \mathcal{S}, (\mathcal{A}^i, u^i, \mathcal{P}^i) : i \in \mathcal{I}\}$, where $\mathcal{A}^i = (-\infty, \infty)$, and $u^i(f) = -\sum_{s \in \mathcal{S}} \pi(s)\, (f^i(s) - x(s))^2 = -\mathrm{E}_\pi (f^i - x)^2$, for a common prior probability measure, π, on the set of states of the world: individuals are guessing the value of x, a random variable.

At a Nash equilibrium, each player chooses the conditional expectation of the random variable, given his private information at each state of the world: $f^{i*}(s) = \mathrm{E}_\pi(x|\mathcal{P}^i(s))$.

Individuals may disagree, due to differences in their private information.

At a Nash equilibrium with rational expectations individuals choose a profile which, by proposition 1, is also a Nash equilibrium for the opinion game with symmetric information $\mathcal{O}_{\mathcal{P}^*}$), where $\mathcal{P}^* = \vee_{i \in \mathcal{I}} \mathcal{P}^{f^{i*}}$; at a Nash equilibrium with rational expectations $\varphi^{i*}(s) = \mathrm{E}_\pi(x|\mathcal{P}^*(s))$. Individuals "agree" in the opinion game with rational expectations.   □

4 Speculation

In a veto game, each individual has a veto strategy, e^i, that guarantees a level of utility $\overline{u}^i$:

$$u^i(e^i, f^{-i}) \geq \overline{u}^i, \quad f^{-i} \in \mathcal{F}^{-i}, i \in \mathcal{I},$$

and such that the profile e is ex - ante pareto optimal: at any feasible profile $f \in \mathcal{F}$ that gives to each individual at least $\overline{u}^i$, each individual obtains exactly $\overline{u}^i$

$$u^i(f) \geq \overline{u}^i, \quad \Rightarrow \quad u^i(f) = \overline{u}^i, \quad i \in \mathcal{I}.$$

At a Nash equilibrium of a veto game, all individuals attain the level of utility associated with their veto strategies.

For a veto game with private information and additively separable utilities, there is speculation at a state of the world, s, for a profile of strategies, f, if

$$\sum_{s' \in \mathcal{P}^i(s)} v^i((f^i(s), f^{-i}(s')), s') \geq$$

$$\sum_{s' \in \mathcal{P}^i(s)} v^i((e^i(s), e^{-i}(s'))s'),$$

$i \in \mathcal{I}$, with some strict inequality;

there is speculation, if there is speculation at some state of the world or, equivalently, the event of speculation,

$$\Sigma_f = \{s \in \mathcal{S} : \text{there is speculation at } s\},$$

is not empty.

Speculation can occur at a rational expectations equilibrium, as in the following example.

Example 3. A game with private information is described by $\mathcal{I} = \{1, 2\}$, $\mathcal{A}^1 = \{T, B\}$, $\mathcal{A}^2 = \{L, R\}$, $\mathcal{S} = \{1, 2\}$, $\mathcal{P}^1 = \{\{1\}, \{2\}\}$, $\mathcal{P}^2 = \{\{1, 2\}\}$, $\pi = \{\frac{1}{2}, \frac{1}{2}\}$ and payoffs

$s = 1$	L	R
T	$1, -1$	$-1, 0$
B	$1, -1$	$0, 0$

$s = 2$	L	R
T	$-1, 1$	$-1, 0$
B	$-1, -1$	$0, 0$

The choice of B in both states is a veto act for individual 1, and it guarantees $\overline{u}^1 = 0$; the choice of R in both states is a veto act for individual 2, and it guarantees $\overline{u}^2 = 0$. The profile in which individual 1 plays T in both states, while individual 2 plays L in both states is a Nash equilibrium and a rational expectations equilibrium; but, at state $s = 1$, there is speculation: individual 1 strictly prefers what he gets at equilibrium to what he obtains at the veto profile, while individual 2, given his information, is indifferent.   □

In the example, the acts of individuals are common knowledge, but not the event of speculation: at state $s = 2$, individual 1 prefers the veto profile.

Indeed, for a profile of acts, $f = (\cdots, f^i, \cdots)$, not necessarily a Nash equilibrium, and for a state of the world, s, the event of speculation cannot be common knowledge. Common knowledge of speculation at a state s implies that $\sum_{s' \in \underline{P}(s)} v^i(f(s'), s') \geq \sum_{s' \in \underline{P}(s)} v^i(e(s'), s')$, for all $i \in \mathcal{I}$, with some strict inequality. But then, by choosing strategies that coincide with f^i on $\underline{P}(s)$ and with e^i on the complement, all individuals would be at least as well off, and some strictly better off than at the veto profile, a contradiction. This is the argument of Milgrom and Stokey (1982) that speculation cannot be common knowledge. It is in the spirit, but stronger than the argument of Holmström and Myerson (1983) that ex-ante efficiency implies interim efficiency. As in the example, the occurence of speculation at a given state of the world need not contradict the interim efficiency of the veto profile.

Corollary 3. *In a veto game, the event of speculation cannot be common knowledge.*

In a strong veto game, each individual has a strong veto act, i.e. a veto act $e^i \in \mathcal{F}^i$ such that the veto profile is the only ex - ante pareto optimal profile:

$$u^i(f) \geq \overline{u}^i, \quad \Rightarrow \quad f^i = e^i, \quad i \in \mathcal{I}.$$

At a Nash equilibrium of a strong veto game, all individuals play their strong veto acts. In particular, speculation never realizes at a Nash equilibrium profile. This is the argument of Geanakoplos (1995) that Nash equilibrium suffices to prevent speculation in strong veto games:

Corollary 4. *The event of speculation cannot realize at a Nash equilibrium of a strong veto game.*

If the veto strategies of all individuals are measurable with respect to some partition, $\mathcal{P}^*$, at least as coarse as their private information, $\mathcal{P}^i \subseteq \mathcal{P}^*$, the veto game has a Nash equilibrium, e, at which information is essentially symmetric and which, as a consequence is a rational expectations equilibrium. In a strong veto game, this equilibrium is unique.

It remains an open question to characterize the class of games with private information, broader than the class of strong veto games, such that at a Nash equilibrium information is essentially symmetric. For such games, the distinction between a Nash equilibrium and a rational expectations equilibrium vanishes. Kalai (2000) investigates this question for a class of large anonymous games.

5 Competitive equilibria

A competitive economy with uncertainty and private information is

$$\mathcal{E}_{\mathcal{P}} = \{\mathcal{I}, \mathcal{S}, \mathcal{L}, (\mathcal{Z}^i, \mathcal{P}^i, u^i) : i \in \mathcal{I}\}.$$

Commodities are $l \in \mathcal{L}$, a finite set, of cardinality L. An elementary net trade or an elementary act for an individual is $z^i \in \mathcal{Z}^i$, a subset of the commodity space, and a net trade, across states of the world, is $f^i = (\ldots, z^i(s), \ldots)$.

An allocation or a profile of acts, $f = (\ldots, f^i, \ldots)$, is feasible if $\sum_{i \in \mathcal{I}} f^i = 0$.

A feasible allocation is pareto optimal if and only if there does not exist another, feasible, allocation, $\hat{f} = (\ldots, \hat{f}^i, \ldots)$, such that $u^i(\hat{f}^i) \geq u^i(f^i)$, for every individual, with some strict inequality.

If the no-trade allocation $e = 0$, is pareto optimal, there is speculation at an allocation of net trades, f, and, at some state of the world, s, if $\sum_{s' \in \mathcal{P}^i(s)} v^i(f^i(s), s') \geq \sum_{s' \in \mathcal{P}^i(s)} v^i(0, s')$, for all individuals, with some strict inequality.

Elementary or spot prices of commodities are $\pi \in \Delta^L$, the unit simplex, and commodity prices are $p : \mathcal{S} \to \Delta^L$.

In an economy with rational expectations, individuals refine their information with the information revealed by prices. The information revealed by prices is $\mathcal{P}^p$.

A competitive equilibrium for an economy with rational expectations, Radner (1979), is a pair, (p^*, f^*), of prices and a feasible allocation, such that, for an individual, f^{i*} is a solution to the maximization of utility over the set

$$\mathcal{B}^i(p) = \left\{ f : \mathcal{S} \to \mathcal{Z}^i : \begin{array}{l} f \text{ is measurable with respect to } \mathcal{P}^i \vee \mathcal{P}^p, \text{ and} \\ p(s)f^i(s) \leq 0 \text{ for all } s \in \mathcal{S} \end{array} \right\}.$$

Associated with an economy, Debreu (1952), there is a (generalized) Walrasian game with uncertainty and private information,

$$\mathcal{W}_{\tilde{\mathcal{P}}} = \{\tilde{\mathcal{I}}, \mathcal{S}, \mathcal{L}, (\mathcal{Z}^i, \mathcal{P}^i, u^i) : i \in \mathcal{I}, (\mathcal{A}^0, \mathcal{P}^0, u^0)\}.$$

Individuals are $i \in \tilde{\mathcal{I}} = \mathcal{I} \cup \{0\}$, and the profile of information partitions is $\tilde{\mathcal{P}} = \mathcal{P} \cup \{\mathcal{P}^0\}$; a profile of strategies is $\tilde{f} = (f, p)$. Individual $i = 0$ is the auctioneer; The set of elementary acts for the auctioneer is $\mathcal{A}^0 = \Delta^L$, the domain of elementary commodity prices, his utility function is $u^0(\tilde{f}) = \sum_{i \in \mathcal{I}} \sum_{s \in \mathcal{S}} p(s)f^i(s)$, his information is complete: $\mathcal{P}^0 = \{\{s\} : s \in \mathcal{S}\}$.

In the Walrasian game with rational expectations, individuals refine their information with the information revealed jointly by the acts of all other players, not only of the auctioneer.

An equilibrium, for the walrasian game with rational expectations is revealing if and only if prices, the act of the auctioneer, reveal the information revealed jointly by the net trades, the acts of all other individuals.

A revealing competitive equilibrium for the economy $\mathcal{E}_{\tilde{\mathcal{P}}}$ is a Nash equilibrium for the walrasian game, $\mathcal{W}_{\mathcal{P}}$, with rational expectations, and, at such an equilibrium, the information of individuals is essentially symmetric.

Corollary 5. *If the utility functions of individuals are separable across states of the world, and strictly quasi concave, speculation cannot occur at a competitive equilibrium of the economy with rational expectations.*

It suffices to observe that a competitive equilibrium allocation is a Nash equilibrium allocation of a well defined (generalized) game, while, under the stated conditions, if the no - trade allocation is pareto optimal, it is also a strong veto profile.

In the economy without rational expectations, the budget correspondence need not be measurable with respect to the individual information partition; solutions of the optimization problem of an individual at every state need not yield a solution to the ex-ante optimization problem[5]. A competitive equilibrium need not be a Nash equilibrium of any well defined generalized game, and speculation might occur[6].

The argument for no-speculation at a rational expectations equilibrium of an economy is thus the one in Geanakoplos (1995) and Tirole (1982), not the one in Milgrom and Stokey (1982): rational expectations are needed to guarantee that equilibrium allocations are Nash equilibria of a well defined (generalized) game; common knowledge, either of the acts, or of the event of speculation, is not the issue.

References

Aumann, R. J.: Agreeing to disagree. The Annals of Statistics **4**, 1236–1239 (1976)

Debreu, G.: A social equilibrium existence theorem. Proceedings of the National Academy of Sciences (USA) **38**, 886–893 (1952)

Debreu, G.: Topological methods in cardinal utility. In: Arrow, K. J., Karlin, S., Suppes, P. (eds.) Mathematical methods in the social sciences, pp. 16–26. Stanford: Stanford University Press 1959

Dubey, P., Geanakoplos, J., Shubik, M.: The revelation of information in strategic market games. Journal of Mathematical Economics **16**, 105–137 (1987)

Forges, F., Minelli, E.: A property of Nash equilibria in repeated games with incomplete information. Games and Economic Behavior **18**, 159–175 (1997a)

Forges, F., Minelli, E.: Self-fulfilling mechanisms and rational expectations. Journal of Economic Theory **75**, 338–406 (1979b)

Geanakoplos, J. D., Polemarchakis, H. M.: We can't disagree forever. Journal of Economic Theory **28**, 192–200 (1982)

---

[5] See footnote 2.

[6] Dubey, Geanakoplos and Shubik (1987) provide an example.

Geanakoplos, J. D.: Common knowledge. Handbook of game theory, Vol. 2, pp. 1427–1496. Amsterdam: North Holland 1995

Harsanyi, J.: Games with incomplete information played by Bayesian players, I–III. Management Science **14**, 159–182, 320–334, 486–502 (1967–68)

Holmström, B., Myerson, R.: Efficient and durable decision rules with incomplete information. Econometrica **51**, 1799–1819 (1983)

Kalai, E.: Private information in large games. Northwestern University, DP 1312 (2000)

Kreps, D.: A note on fulfilled expectations equilibria. Journal of Economic Theory **14**, 32–43 (1977)

Milgrom, P., Stokey N. L.: Information, trade and common knowledge. Journal of Economic Theory **26**, 17–27 (1982)

Radner, R.: Competitive equilibrium under uncertainty. Econometrica **36**, 31–58 (1968)

Radner, R.: Rational expectations equilibrium: generic existence and the information revealed by prices. Econometrica **47**, 655–678 (1979)

Savage, L. J.: The foundations of statistics. New York: Wiley 1954

Tirole, J.: On the possibility of speculation under rational expectations. Econometrica **50**, 1163–1181 (1982)

Nash and Walras equilibrium via Brouwer[*]

John Geanakoplos

Cowles Foundation for Research in Economics, Yale University, New Haven, CT 06520–8281, USA
(e-mail: john.geanakoplos@yale.edu)

Received: July 9, 2001; revised version: February 25, 2002

Summary. The existence of Nash and Walras equilibrium is proved via Brouwer's Fixed Point Theorem, without recourse to Kakutani's Fixed Point Theorem for correspondences. The domain of the Walras fixed point map is confined to the price simplex, even when there is production and weakly quasi-convex preferences. The key idea is to replace optimization with "satisficing improvement," i.e., to replace the Maximum Principle with the "Satisficing Principle."

Keywords and Phrases: Equilibrium, Nash, Walras, Brouwer, Kakutani.

JEL Classification Numbers: C6, C62.

Mordecai Kurz has been an inspiration for a whole generation of economists. I vividly remember many blissful summers at the IMSSS in Stanford, listening to the programs Mordecai masterfully put together. Those summer sessions defined economic theory for their time, and defined the standards of excellence we all tried to live up to. In retrospect, the late 70s and early 80s appear clearly as a golden era in the history of economic theory, and it is hard to believe things would have turned out so well if it weren't for IMSSS, and for Mordecai's energy, enthusiasm, and tenacity as its director.

---

[*] I wish to thank Ken Arrow, Don Brown, and Andreu Mas-Colell for helpful comments. I first thought about using Brouwer's theorem without Kakutani's extension when I heard Herb Scarf's lectures on mathematical economics as an undergraduate in 1974, and then again when I read Tim Kehoe's 1980 Ph.D dissertation under Herb Scarf, but I did not resolve my confusion until I had to discuss Kehoe's presentation at the celebration for Herb Scarf's 65th birthday in September, 1995.

Correspondence to: C. D. Aliprantis

1 Introduction

The standard proofs of the existence of Nash and Walras equilibrium (including the original proofs by Nash [19], Arrow and Debreu [2], and McKenzie [17]) rely on Kakutani's Fixed Point Theorem for correspondences. I show that a slight perturbation of the standard arguments enables one to work entirely with Brouwer's Fixed Point Theorem for continuous functions.[1]

Nash himself [20] gave a Brouwer fixed point proof of Nash equilibrium for the special case of matrix games. McKenzie [18] derived the existence of Walras equilibrium with production from Brouwer's Fixed Point Theorem. The only advantage of the maps I propose is that some readers may think they are simpler. For example, in my Walras existence proof the domain of the fixed point map is the price simplex. There is no need to enlarge the domain to include excess demands, as done by Gale [10] and Debreu [7], [8], or the demands of each consumer, as done in the generalized game proofs of Debreu [6] and Arrow and Debreu [2], or to add the auxiliary commodities introduced by McKenzie [18].[2]

In Section 2, the existence of Nash equilibrium in concave games is proved. Let a game $G = (u_n, \Sigma_n)_{n \in N}$ be described by its payoffs u_n and compact, convex strategy spaces Σ_n, for agents $n \in N$. The original proof by Nash relied on the best response correspondence $B_n(\bar{\sigma}_n, \bar{\sigma}_{-n}) = \mathrm{argmax}_{\sigma_n \in \Sigma_n} u_n(\sigma_n, \bar{\sigma}_{-n})$. My proof simply replaces B_n with a satisficing improvement *function*

$$\beta_n(\bar{\sigma}_n, \bar{\sigma}_{-n}) = \arg \max_{\sigma_n \in \Sigma_n}[u_n(\sigma_n, \bar{\sigma}_{-n}) - \|\sigma_n - \bar{\sigma}_n\|^2].$$

If u_n is concave in σ_n, it can easily be shown that β_n always moves agent n part of the way to his optimal response against $\bar{\sigma}_{-n}$. Moving all the way to a best response is irrelevant to demonstrating that a fixed point is an equilibrium. Section 1 also includes a discussion of earlier demonstrations of Nash equilibrium based on Brouwer's FPT for *matrix* games.

In Section 3 the existence of Walras equilibrium is proved for economies $E = ((u^h, e^h)_{h \in H}, (Y_f)_{f \in F}, (\theta_f^h)_{f \in F}^{h \in H})$ with quasi-concave utilities u^h and convex technologies Y_f. Let $M^h(p, \bar{p})$ be the minimum net expenditure household h must make at prices p beyond its Walrasian income $I^h(p)$ in order to achieve the same utility it would obtain if it faced prices $\bar{p}$ and income $I^h(\bar{p})$.[3] It is well-known that M^h is continuous in $(p, \bar{p})$ and concave in p for any fixed $\bar{p}$. Let $M(p, \bar{p})$ be the sum of the $M^h(p, \bar{p})$ over all households h. Let S be the price simplex. In Section 3

---

[1] Of course Kakutani's FPT can be derived from Brouwer's FPT, so in a sense all these standard proofs are derivable from Brouwer. But I mean there is a single continuous function, not involving any approximations and selection, whose fixed points are Walras equilibria.

[2] Thus in the proofs (10), (7), (8) the dimension of the domain of the fixed point map is $(L - 1) + (L - 1)$, where L is the number of commodities. In the proofs (6), (2), the dimension of the domain is $(L - 1) + (H + F)(L - 1)$, where H is the number of households and F the number of firms. In the proof (16) the dimension is $(L - 1) + F$. All of the proofs (10), (7), (8), (6), (2) are based on Kakutani's fixed point theorem. My proof uses Brouwer's fixed point theorem on a domain of dimension $(L - 1)$.

[3] Income is defined by $I^h(p) = p \cdot e^h + \Sigma_{f \in F} \theta_f^h \max_{y_f \in Y_f} p \cdot y_f$.

it is shown that the *function* $\varphi : S \to S$ defined for each $\bar{p}$ in S by

$$\varphi(\bar{p}) = \arg\max_{p \in S}[M(p,\bar{p}) - \|p - \bar{p}\|^2]$$

is continuous and has Walras equilibria as its fixed points.

The minimum expenditure function and its properties have been very closely studied since Hicks showed that the so-called Hicksian demand is more regular than the Marshallian demand. Intermediate textbooks often emphasize the duality between utility maximization and expenditure minimization. Precisely this duality guarantees (through the Maxmin theorem) that a fixed point of the function φ must be a Walras equilibrium. Nevertheless, though there are many closely related ideas to be found in the literature, to the best of my knowledge nobody has used the function M to demonstrate the existence of equilibrium.

To understand the genesis of the function M, let us temporarily suppose that the Walrasian demand correspondence $D^h(\bar{p})$, and the Walrasian supply correspondence $Y_f(\bar{p}) = \arg\max_{y_f \in Y_f} \bar{p} \cdot y_f$, and therefore also the Walrasian aggregate excess demand correspondence $Z(\bar{p}) = \sum_{h \in H}(D^h(\bar{p}) - e^h) - \sum_{f \in F} Y_f(\bar{p})$, are all single valued functions, which we denote by $d^h(\bar{p}), y_f(\bar{p}), z(\bar{p})$. (If utilities are strictly concave, and production sets strictly convex, this will be the case, assuming we enclose the economy in a compact space.) In that case we can define a continuous function $\psi : S \to S$

$$\psi(\bar{p}) = \arg\max_{p \in S}[p \cdot z(\bar{p}) - \|p - \bar{p}\|^2]$$

whose fixed points are Walrasian equilibrium prices, as we show in Section 4.

When $Z(\bar{p})$ is multivalued, there does not, at first glance, seem to be an analogue for ψ. However, define $D_+^h(\bar{p})$ as the set of all consumption bundles (budget feasible and not) that make agent h at least as well off as his Walrasian demands $D^h(\bar{p})$. Define the "better than excess demand correspondence" Z_+ by $Z_+(\bar{p}) = \sum_{h \in H}(D_+^h(\bar{p}) - e^h) - \sum_{f \in F} Y_f$, where firms choose anything feasible. A crucial advantage of Z_+ over Z is that it is lower semicontinuous as well as upper semicontinuous. We show in Section 3 that

$$\varphi(\bar{p}) = \arg\max_{p \in S}[\min_{z \in Z_+(\bar{p})} p \cdot z - \|p - \bar{p}\|^2]$$

defines a continuous *function* from the simplex to itself whose fixed points are Walrasian equilibria. In fact this is the same φ given earlier, since

$$M(p,\bar{p}) = \min_{z \in Z_+(\bar{p})} p \cdot z.$$

In the standard Kakutani existence proof pioneered by Debreu (see Arrow and Debreu [2]), the price player chooses p to maximize the value of a given excess demand z. The vector z is an independent argument in the fixed point map. In my proof the price player chooses p to maximize the cost of achieving a given social welfare $(v^h)_{h \in H}$, where v^h is a utility level for agent h. The $(v^h)_{h \in H}$ are in turn derived from prices $\bar{p}$, $v^h = v^h(\bar{p})$, the indirect utilities at Walrasian prices $\bar{p}$, so that prices are the lone independent variables.

The mapping φ naturally suggests a potential Lyapunov function $L : S \to \mathbb{R}$ defined by

$$L(\bar{p}) \equiv \max_{p \in S}[M(p, \bar{p}) - \|p - \bar{p}\|^2].$$

It might be interesting to establish conditions for the underlying economy guaranteeing that $L(\varphi(\bar{p})) < L(\bar{p})$ for all $\bar{p} \in S$, but this line of inquiry is not pursued here.

In Section 4 I examine several special economies with strictly quasi-concave utilities u^h, for which there are already standard proofs of Walras equilibrium based on Brouwer's FPT. In the first special case we also take the Y_f strictly convex, so excess demand $Z(\bar{p})$ is a *function* $z(\bar{p})$, as mentioned earlier. By replacing M with $N \equiv \min_{z \in Z(\bar{p})} p \cdot z$, obtaining

$$N(p, \bar{p}) \equiv \min_{z \in Z(\bar{p})} p \cdot z = p \cdot z(\bar{p})$$

we obtain the function ψ defined earlier.[4] The map ψ is quite different from the standard Brouwer map (deriving from Nash's matrix game map) that is exposited in most textbooks, but it turns out that $\psi(p)$ reduces to another one of the standard Brouwer maps, namely $h(p) = \text{Proj}_S(p + \frac{1}{2}z(p))$. But whereas it requires the Kuhn–Tucker theorem to verify that a fixed point of h is a Walras equilibrium, it is immediate that a fixed point of ψ is an equilibrium. Thus our perturbation $-\|p - \bar{p}\|^2$ still simplifies matters, even when dealing with excess demand *functions*. We apply similar maps in other special cases, e.g., with constant-returns-to-scale technologies (CRS).[5] In this case ψ turns out to be closely related to the maps used by Todd [25] and Kehoe [13] to compute equilibria of economies with fixed coefficient technologies.

The only technical point in this paper occurs in showing that the function $M(p, \bar{p})$ is continuous, which is tantamount to showing that the "better than" correspondence $Z_+(\bar{p})$ is upper semi-continuous (USC) and lower semi-continuous (LSC). This in fact is trivial, but I prove it after introducing a new lemma called the Satisficing Principle, which could perhaps stand just behind the Maximum Principle as a useful tool in the theory of choice, because it guarantees LSC and USC. The impression the student is sometimes left holding is that LSC is less central

---

[4] Note that for any pair $(p, \bar{p})$, $M(p, \bar{p}) \leq N(p, \bar{p})$; usually $M(p, \bar{p}) < N(p, \bar{p})$. Indeed when excess demand Z is a correspondence, as will typically be the case without further assumptions, $N(p, \bar{p})$ is not continuous. Even when $Z(\bar{p})$ is a function, and N is continuous, $M(p, \bar{p}) \neq N(p, \bar{p})$. The function N has nevertheless often been used to prove the existence of equilibrium. In one such approach the prices p are called "better" than the prices $\bar{p}$ if $N(p, \bar{p}) > 0$. Walras equilibrium then exists if it can be shown that this partial ordering on prices has a maximal element. The problem is thus reduced to one of maximizing a (nontransitive) binary relation, for which see Nikaido [22], Fan [9], Sonnenschein [24], and Aliprantis and Brown [1]. For a lucid exposition of these ideas, see Border [4]. Along these lines, see also the proof of the K–K–M–S theorem via Brouwer in Krasa and Yannelis [15]. For another proof of Walras equilibrium via Brouwer, that works even with infinitely many commodities, see Yannelis [26].

[5] An interesting feature of each successive Walras existence proof is that Brouwer's fixed point theorem must be augmented by Farkas' Lemma (when technology is given by a finite number of activities), the separating hyperplane theorem (when technology is given more generally by a cone), and the MinMax theorem (when technological possibilities are given by arbitrary convex sets).

than USC, but we should not forget that the Maximum Principle cannot be applied unless the budget correspondence of each agent is USC and LSC.

The Satisficing Principle supposes that an agent is maximizing a continuous utility $u_\alpha(x)$ subject to a constraint $x \in \beta(\alpha)$ over which he is locally nonsatiated. Suppose he is satisfied with a payoff $w(\alpha) < v(\alpha)$, where $v(\alpha)$ is the maximum achievable utility given the exogenous parameters α, and w is any continuous function. Then the correspondence $W(\alpha)$ of all choices achieving payoff at least $w(\alpha)$ is lower semi-continuous (LSC) as well as upper semi-continuous (USC) in α, provided that $\beta(\alpha)$ is. The Satisficing Principle complements the Maximum Principle, which guarantees that $v(\alpha)$ is continuous and that the set of choices achieving $v(\alpha)$ is USC but not necessarily LSC. One immediate application of the Satisficing Principle is that the Walrasian budget correspondence is LSC and USC when the endowment is strictly positive. More importantly, since the Walrasian indirect utility function $w^h(p)$ is continuous, and by nonsatiation, strictly less than the maximal utility $v^h(p) = v$ achievable without a budget constraint, the Satisficing Principle guarantees the LSC and USC of $D_+^h(p)$, and hence of $Z_+(p)$.

The Satisficing Principle is stated and proved in Section 5, where it is also used to give a Brouwer FPT proof that quasi-concave games have Nash equilibria. In some sense the whole idea of this paper comes down to replacing optimization with satisficing improvement; first for the game players and the auctioneer, by subtracting $||\sigma_n - \bar{\sigma}_n||^2$ or $||p - \bar{p}||^2$, and second for the households, in substituting $Z_+(\bar{p})$ for $Z(\bar{p})$.

2 Games and Nash equilibrium

2.1 Concave perturbation lemma

My proofs rely on the following concave perturbation lemma:

Concave perturbation lemma. Let $X \subset \mathbb{R}^n$ be convex, and let $\bar{x} \in X$. Let $u : X \to \mathbb{R}$ be concave. Then $\arg\max_{x \in X}[u(x) - ||x - \bar{x}||^2]$ is at most a single point, and if $\bar{x} = \arg\max_{x \in X}[u(x) - ||x - \bar{x}||^2]$, then $\bar{x} \in \arg\max_{x \in X} u(x)$.

Proof. Since u is concave in x, and $-||x - \bar{x}||^2$ is strictly concave in x, $[u(x) - ||x - \bar{x}||^2]$ is strictly concave, and $\arg\max_{x \in X}[u(x) - ||x - \bar{x}||^2]$ cannot contain two distinct points. Suppose $\bar{x} = \arg\max_{x \in X}[u(x) - ||x - \bar{x}||^2]$. Take any $x \in X$. By hypothesis, and by the convexity of X and the concavity of u, for any $0 < \varepsilon < 1$,

$$0 \geq \{u([(1-\varepsilon)\bar{x} + \varepsilon x]) - ||[(1-\varepsilon)\bar{x} + \varepsilon x] - \bar{x}||^2\} - \{u(\bar{x}) - ||\bar{x} - \bar{x}||^2\}$$
$$= u([(1-\varepsilon)\bar{x} + \varepsilon x]) - \varepsilon^2||x - \bar{x}||^2 - u(\bar{x})$$
$$\geq (1-\varepsilon)u(\bar{x}) + \varepsilon u(x) - u(\bar{x}) - \varepsilon^2||x - \bar{x}||^2$$
$$= \varepsilon(u(x) - u(\bar{x})) - \varepsilon^2||x - \bar{x}||^2$$

So

$$u(x) - u(\bar{x}) \leq \varepsilon||x - \bar{x}||^2 \text{ for all } \varepsilon > 0, \text{ so}$$
$$u(x) - u(\bar{x}) \leq 0 \qquad \qquad \square$$

2.2 Concave games

Let a game G among N players be defined by compact and convex strategy spaces $\Sigma_1, ..., \Sigma_N$ in finite-dimensional Euclidean spaces, and by continuous payoff functions $u_1, ..., u_N$, where for each $n \in N$, $u_n : \Sigma \equiv \Sigma_1 \times \cdots \times \Sigma_N \to \mathbb{R}$. We call G a concave game if for any fixed $\bar{\sigma}_{-n} \equiv (\bar{\sigma}_1, ..., \bar{\sigma}_{n-1}, \bar{\sigma}_{n+1}, ..., \bar{\sigma}_N) \in \Sigma_{-n} \equiv \Sigma_1 \times \cdots \times \Sigma_{n-1} \times \Sigma_{n+1} \times \cdots \times \Sigma_N$, $u_n(\sigma_n, \bar{\sigma}_{-n})$ is concave in σ_n.

Given a game $G = (\Sigma_1, ..., \Sigma_N; u_1, ..., u_N)$, a Nash equilibrium is a choice $\bar{\sigma} = (\bar{\sigma}_1, ..., \bar{\sigma}_N) \in \Sigma$ such that for all $n \in N$ and all $\sigma_n \in \Sigma_n$,

$$u_n(\bar{\sigma}) \geq u_n(\sigma_n, \bar{\sigma}_{-n}).$$

Theorem. *Every concave game has a Nash equilibrium.*

Proof. Define the function

$$\varphi_n : \Sigma \to \Sigma_n \text{ by}$$

$$\varphi_n(\bar{\sigma}_1, ..., \bar{\sigma}_n, ..., \bar{\sigma}_N) = \underset{\sigma_n \in \Sigma_n}{\arg\max}[u_n(\sigma_n, \bar{\sigma}_{-n}) - \|\sigma_n - \bar{\sigma}_n\|^2].$$

Observe that the maximand is the sum of a continuous, concave function in σ_n, and a negative quadratic function in σ_n, and hence is continuous and strictly concave. Since Σ_n is compact and convex, φ_n is a well-defined function. Furthermore, the maximand is continuous in the parameter $\bar{\sigma} = (\bar{\sigma}_1, ..., \bar{\sigma}_n)$, hence by the maximum principle, φ_n is a continuous function.

Now define $\varphi : \Sigma \to \Sigma$ by $\varphi = (\varphi_1, ..., \varphi_N)$. Clearly φ is continuous, and so by Brouwer's theorem it has a fixed point $\varphi(\bar{\sigma}) = \bar{\sigma}$.

By the concave perturbation lemma, for all $\sigma_n \in \Sigma_n$, $u_n(\sigma_n, \bar{\sigma}_{-n}) \leq u_n(\bar{\sigma})$. Hence $\bar{\sigma}$ is a Nash equilibrium. $\square$

Nash [19] suggested the correspondence $\psi_n : \Sigma \rightrightarrows \Sigma_n$ defined by $\psi_n(\bar{\sigma}) = \arg\max_{\sigma_n \in \Sigma_n} u_n(\sigma_n, \bar{\sigma}_{-n})$. Since u_n is not necessarily strictly concave, $\psi_n(\bar{\sigma})$ may contain multiple elements.

The maximand above is simply a perturbation of the Nash maximand. It guarantees that a player will always make some improvement when there is an opportunity to improve, but he will not necessarily move all the way to his best response. Another difference is that the Nash correspondence ψ_n throws away some information, since ψ_n actually is defined on Σ_{-n}. The map φ_n depends on all the coordinates, including Σ_n.

2.3 Matrix games

Two player matrix games are defined by $r \times s$ matrices A and B. Player α has strategy space $\Sigma_\alpha \equiv \{p \in \mathbb{R}^r_+ : \sum_{i=1}^r p_i = 1\}$ and player β has strategy space $\Sigma_\beta = \{q \in \mathbb{R}^s_+ : \sum_{j=1}^s q_j = 1\}$. The payoffs are defined by $u_\alpha(p, q) \equiv p'Aq$ and $u_\beta(p, q) \equiv p'Bq$. Since u_n is linear on Σ_n for $n = \alpha$ and β, these matrix games are indeed concave games.

Nash [20] showed that for matrix games, Brouwer's Fixed Point Theorem sufficed. He suggested using the excess return functions $z_\alpha(\bar{p}, \bar{q}) = A\bar{q} - (\bar{p}'A\bar{q})1$ and $z_\beta(\bar{p}, \bar{q}) = \bar{p}'B - (\bar{p}B\bar{q})1$, which specify the surplus each agent can get by playing each pure strategy instead of his designated mixed strategy. He then defined the map

$$f(\bar{p}, \bar{q}) = \left(\frac{\bar{p} + [A\bar{q} - (\bar{p}'A\bar{q})1]^+}{1 + [A\bar{q} - (\bar{p}'A\bar{q})1]^+ \cdot 1}, \frac{\bar{q} + [\bar{p}'B - (\bar{p}'B\bar{q})1]^+}{1 + [\bar{p}'B - (\bar{p}'B\bar{q})1]^+ \cdot 1} \right),$$

where for any vector y, $[y]^+$ is the vector with ith coordinate $\max(0, y_i)$, and 1 is the vector of all 1's, or just the scalar 1, depending on the context. A fixed point of the Nash map can be shown to be a Nash equilibrium by observing that $\bar{p}'[A\bar{q} - (\bar{p}'A\bar{q})1] = 0$. Indeed this same trick is copied in the now standard existence proof for Walrasian equilibrium, where it crops up as Walras law. The Nash map f exploits the special form of matrix games.

The map φ can be used for any concave game, not just matrix games. In the special case of matrix games, a short computation shows that it reduces to

$$\varphi(\bar{p}, \bar{q}) = h(\bar{p}, \bar{q}) \equiv \left(\Pi_{\Sigma_\alpha} \left(\bar{p} + \tfrac{1}{2}A\bar{q} \right), \Pi_{\Sigma_\beta} \left(\bar{q} + \tfrac{1}{2}\bar{p}'B \right) \right),$$

where $\Pi_K(x)$ is the closest point in K to x. The map h has already been used to prove the existence of Nash equilibrium in matrix games by Lemke and Howson [16], and to study the index of matrix game Nash equilibrium by Gul, Pearce and Stacchetti [12]. To see that φ reduces to h for matrix games, one needs to use the Kuhn–Tucker theorem. Indeed, one needs the Kuhn–Tucker theorem to verify that a fixed point of h is a Nash equilibrium.[6] But as we saw in the proof of our first theorem, using φ avoids the need for the Kuhn–Tucker theorem.

3 Walrasian economies

3.1 The Walrasian economy

Let us represent an economy by

$$E = \left\{ H, (X^h, e^h, u^h)_{h \in H}, F, (Y_f)_{f \in F}, (\theta_f^h)_{f \in F}^{h \in H} \right\},$$

where H is a finite set of households, $X^h \subset \mathbb{R}^L$ is the consumption set of household h, e^h is the endowment, and u^h is the utility function of agent $h \in H$, F is a finite set of firms, Y_f is the technology of firm $f \in F$, and $\theta_f^h \in \mathbb{R}_+$ is the ownership share of firm f by agent h, $\sum_{h \in H} \theta_f^h = 1$ for all $f \in F$. Following Arrow and Debreu [2], we assume in addition that $\forall h \in H$,

---

[6] By the Kuhn–Tucker theorem, $\varphi(\bar{p}, \bar{q}) = (\varphi_\alpha(\bar{p}, \bar{q}), \varphi_\beta(\bar{p}, \bar{q}))$ satisfies $A\bar{q} - 2(\varphi_\alpha(\bar{p}, \bar{q}) - \bar{p}) - \lambda e + \Lambda = 0$, where $\Lambda \geq 0$ is a diagonal matrix with $\Lambda_{jj} > 0$ only if $\varphi_{\alpha j}(\bar{p}, \bar{q}) = 0$. By the Kuhn–Tucker theorem, the map $h(\bar{p}, \bar{q}) = (h_\alpha(\bar{p}, \bar{q}), h_\beta(\bar{p}, \bar{q}))$ satisfies $-2(h_\alpha(\bar{p}, \bar{q}) - \tfrac{1}{2}A\bar{q} - \bar{p}) + \mu e + \Omega = 0$, where $\Omega \geq 0$ is a diagonal matrix with $\Omega_{jj} > 0$ only if $h_{\alpha j}(\bar{p}, \bar{q}) = 0$.

(1) X^h is closed, convex, and bounded from below: $\exists \underline{d}^h$ such that $\underline{d}^h \leq x$ for all $x \in X^h$

(2) $e^h \in X^h$ and $\exists d^h \in X^h$ with $d^h \ll e^h$

(3a) $u^h : X^h \to \mathbb{R}$ is continuous

(3b) u^h is quasi-concave, i.e., $[u^h(x) > u^h(y)$ and $0 < \lambda < 1] \Rightarrow [u^h(\lambda x + (1 - \lambda)y) > u^h(y)]$, for all $x, y \in X^h$

(3c) u^h is nonsatiated, i.e., $\forall y \in X^h$, $\exists x \in X^h$ with $u^h(x) > u^h(y)$ and for all $f \in F$,

(4) Y_f is a closed convex subset of $\mathbb{R}^L$, and $0 \in Y_f$ and furthermore,

(5) If $Y \equiv \sum_{f \in F} Y_f$, then $Y \cap \mathbb{R}_+^L = \{0\}$

(6) Irreversibility: $Y \cap -Y = \{0\}$.

3.2 Walras Equilibrium

A Walras equilibrium (WE) for the economy E is a tuple $(\bar{p}, (\bar{x}^h)_{h \in H}, (\bar{y}_f)_{f \in F}) \in \mathbb{R}_+^L \times \mathsf{X}_{h \in H} X^h \times \mathsf{X}_{f \in F} Y_f$ satisfying

(a) $\sum_{h \in H} \bar{x}^h \leq \sum_{h \in H} e^h + \sum_{f \in F} \bar{y}_f$

(b) $\bar{y}_f \in \arg\max_{f \in Y_f} \bar{p} \cdot y_f$, $\forall f \in F$

(c) $\bar{x}^h \in B^h(\bar{p}) = \{x \in X^h : \bar{p} \cdot x \leq \bar{p}e^h + \sum_{f \in F} \theta_f^h \max_{y_f \in Y_f} \bar{p} y_f \equiv I^h(\bar{p})\}, \forall h \in H$

(d) $\bar{x}^h \in \arg\max_{x \in B^h(\bar{p})} u^h(x)$.

By nonsatiation and quasi-concavity, we know that at a WE each agent spends all his income, so the budget inequality in (c) reduces to equality, and we therefore conclude that in a WE,

$$\sum_{h \in H} x_i^h < \sum_{h \in Y} e_i^h + \sum_{f \in F} \bar{y}_{fi} \Rightarrow \bar{p}_i = 0. \tag{1.1}$$

3.3 Easy consequences of the assumptions

It follows from (1.1) that we obtain an equivalent definition of equilibrium by strengthening the definition of equilibrium to require equality of supply and demand in condition (a), provided that we augment production by allowing free disposal, replacing Y with $\hat{Y} = Y - \mathbb{R}_+^L$. So without loss of generality we require equality in (a) but also assume

(7) Free disposal: $Y - \mathbb{R}_+^L = Y$.

As shown in Arrow and Debreu [2], assumptions (1)–(6) have the consequence that $\mathcal{A} \equiv \{(x^1, ..., x^H, y_1, ..., y_F) \in \mathsf{X}_{h \in H} X^h \times \mathsf{X}_{f \in F} Y_f : \sum_{h \in H}(x^h - e^h) - \sum_{f \in F} y_f \leq 0\}$ is compact. In view of the quasi-concavity of the utilities, restricting the consumption sets from X^h to $X^h \cap \hat{X}^h$ and restricting the technologies from

Y_f to $Y_f \cap \hat{Y}_f$ where $\hat{X}^h$ and $\hat{Y}_f$ are compact and convex and such that $\mathcal{A}$ is contained in the interior of $\mathsf{X}_{h \in H} \hat{X}^h \times \mathsf{X}_{f \in F} \hat{Y}_f$ gives rise to an economy $\hat{E}$ with exactly the same Walras equilibria as E. Thus without loss of generality, we may add assumption (8) and weaken assumption (3c):

(8) X^h and Y^f are compact for all $h \in H$ and $f \in F$,

which requires weakening (3c) to

(3c) $[(x^1, ..., x^H, y_1, ..., y_F) \in \mathcal{A}] \Rightarrow [\forall h \in H, \exists \hat{x}^h \in X^h, u^h(\hat{x}^h) > u^h(x)]$.

An implication of the convexity of X^h from (1), and the quasi-concavity of u^h from (3b), is that

(3d) u^h is locally nonsatiated in $X^h : \forall y \in X^h$, if $\exists x \in X^h$ with $u^h(x) > u^h(y)$, then $\exists \{x(n)\}_{n=1}^{\infty} \subset X^h$, $x(n) \to y$ with $u^h(x(n)) > u^h(y)$ for all n.

We list six more simple observations. All lemmas rely on assumptions (1)-(8). Lemmas 1 and 2 rely on the definitions of USC and LSC, and on the Satisficing Principle, all of which are deferred to Section 5.

Lemma 1. *The budget correspondence $B^h(p)$ is USC, LSC, nonempty valued, and compact-valued on $S = \{p \in \mathbb{R}_+^L : \sum p_\ell = 1\}$.*

Proof. This is a standard and trivial result. Instead of proving it directly, we note that it is a corollary of the satisficing principle proved in Section 5.

$$B^h(p) = \{x \in X^h : p \cdot x \le I^h(p)\} = \{x \in X^h : -p \cdot x \ge -I^h(p)\}.$$

Let $w(p) \equiv -I^h(p)$ be the satisficing threshold. Let $v(p) \equiv \max_{x \in X^h} -p \cdot x$ be the maximal threshold. Since $e^h \gg d^h$, for all $p \in S$, $w(x) = -I^h(p) \le -p \cdot e^h < -p \cdot d^h \le \max_{x \in X^h} -p \cdot x = v(p)$, so the lemma follows from the compactness of X^h, the continuity of $I^h(p)$, and the Satisficing Principle. $\square$

Let $v^h(p) \equiv \max_{x \in B^h(p)} u^h(x)$ be the so-called indirect utility function of agent h. Since $B^h(p)$ is USC and LSC, nonempty valued and compact-valued, by the Maximum Principle, $v^h(p)$ must be continuous on S. Furthermore, let

$$D^h(p) \equiv \arg\max_{x \in B^h(p)} u^h(x)$$

be the demand correspondence of agent h. Again by the Maximum Principle, $D^h(p)$ is USC. Unfortunately, $D^h(p)$ may not be LSC, as is well known.

A central element of the existence proof given in Section 3.4 is the replacement of the demand correspondence $D^h(p)$, which may fail to be LSC, with the "demand or better" correspondence $D_+^h(p)$, which is always LSC. McKenzie [18] used a similar correspondence.

Lemma 2. $D_+^h(p) = \{x \in X^h : u^h(x) \ge v^h(p)\}$ *is USC, LSC, and nonempty-valued for $p \in S$. Hence so is the better than excess demand $Z_+(p) = \sum_{h \in H} D_+^h(p) - \sum_{h \in H} e^h - \sum_{f \in F} Y_f$.*

Proof. The USC and nonemptiness of D_+^h follow immediately from the continuity of u^h. As for LSC, let $p(n) \to p$ and let $x \in D_+^h(p)$. Let $y(0) \in \arg\max\{u^h(y) : y \in X^h\}$. If $u^h(x) = u^h(y(0))$, then $u^h(x) \geq v^h(p(n)) \forall n$ and so letting $x(n) = x$ for $n \geq 1$ shows the LSC of D_+^h at p. If $u^h(x) < u^h(y(0))$, then by local nonsatiation (3d), $\exists y(m) \to x$ with $u^h(y(m)) > u^h(x)$ for all $m \geq 1$. Since the indirect utility v^h is continuous, $v^h(p(n)) \to v^h(p)$. Hence for $n \geq 1$ we can define $x(n) = y(m(n))$, where $m(n) \equiv \max_{0 \leq m \leq n}\{u^h(y(m)) \geq v^h(p(n))\}$. Then $x(n) \in D_+^h(p(n))$ and $x(n) \to x$, showing the LSC of D_+^h. The sum of USC (LSC) correspondences whose range is compact is also USC (LSC).

Lemma 2 can also be derived from the satisficing principle. Let $w(p) \equiv v^h(p)$ be the satisficing threshold, and let $v^*(p) = v^* = \max_{x \in X^h} u^h(x)$ be the maximum threshold. Apply the Satisficing Principle, noting that X^h and u^h are independent of p, and that $v^h(p)$ is continuous. $\qquad\square$

Lemma 3. *The minimum expenditure function*

$$M(p, \bar{p}) \equiv \min_{z \in Z_+(\bar{p})} p \cdot z$$

is continuous in $(p, \bar{p}) \in S \times S$, *and concave in* p *for any fixed* $\bar{p} \in S$.

Proof. Lemma 2 and the Maximum Principle guarantee the continuity of $M(p, \bar{p})$. For any fixed $\bar{p}$, $M(p, \bar{p})$ is the minimum of a family of linear functions in p, hence it must be concave. $\qquad\square$

Lemma 4. *For all* $\bar{p} \in S$, $Z(\bar{p}) \subset Z_+(\bar{p})$. *Hence* $M(\bar{p}, \bar{p}) \leq 0$.

Proof. Obvious. $\qquad\square$

The following Lemmas 5 and 6 show the role of the so-called "duality principle" that utility maximization and expenditure minimization are the same at points where nonsatiation holds. Lemma 5 also uses the linearity of unconstrained expenditure minimization.

Lemma 5. *If for some* $\bar{p} \in S$, *there is* $\bar{z} \in Z_+(\bar{p})$ *with* $\bar{z} \leq 0$, *then* $\exists \bar{x}^h \in X^h \, \forall h$ *and* $\bar{y}_f \in Y_f \, \forall f$ *such that* $(\bar{p}, (\bar{x}^h)_{h \in H}, (\bar{y}_f)_{f \in F})$ *is a Walrasian equilibrium.*

Proof. If $\bar{z} \in Z_+(\bar{p})$, then by definition there is $\bar{x}^h \in D_+^h(\bar{p}) \, \forall h \in H$, and $\bar{y}_f \in Y_f \, \forall f \in F$ with $\bar{z} = \sum_{h \in H} \bar{x}^h - \sum_{h \in H} e^h - \sum_{f \in F} \bar{y}_f$. Since $\bar{z} \leq 0$, nonsatiation obtains from (3c) and we deduce from local nonsatiation (3d) that $\bar{p} \cdot \bar{x}^h \geq I^h(\bar{p}) \, \forall h \in H$. But $\bar{z} \leq 0$ and $\bar{p} \in S$ implies that $0 \geq \bar{p} \cdot \bar{z} = \bar{p} \sum_{h \in H} \bar{x}^h - \bar{p}[\sum_{h \in H} e^h + \sum_{f \in F} \bar{y}_f] \geq \sum_{h \in H} I^h(\bar{p}) - \sum_{h \in H} I^h(\bar{p}) = 0$. Hence $\bar{p} \cdot \bar{x}^h = I^h(\bar{p}) \, \forall h \in H$ and $\bar{y}_f \in \arg\max_{y_f \in Y_f} \bar{p} \cdot y_f, \forall f \in F$. $\qquad\square$

It is worth noting that (assuming local nonsatiation), neither the quasi-concavity of the u^h nor the convexity of the Y_f played any role in proving Lemmas 1–5.

Lemma 6. *If for some* $\bar{p} \in S$, $\max_{p \in S} M(p, \bar{p}) = M(\bar{p}, \bar{p})$, *then* $\exists \bar{x}^h \in X^h \, \forall h$ *and* $\bar{y}_f \in Y_f \, \forall f$ *such that* $(\bar{p}, (\bar{x}^h)_{h \in H}, (\bar{y}_f)_{f \in F})$ *is a Walrasian equilibrium.*

Proof. We now invoke the convexity of the X^h and Y_f, and the quasi-concavity of u^h, to assert the convexity of $Z_+(\bar{p})$. The minmax theorem then guarantees that

$\exists \bar{z} \in Z_+(\bar{p})$ with $M(\bar{p}, \bar{p}) = \max_{p \in S} p \cdot \bar{z} = \bar{p} \cdot \bar{z} = \min_{z \in Z_+(\bar{p})} \bar{p} \cdot z$. Since by Lemma 4, $M(\bar{p}, \bar{p}) \leq 0$, we must have $\bar{z} \leq 0$ (if $\bar{z}_i > 0$, take $p_i = 1$). Hence by Lemma 5, $\bar{p}$ is a Walrasian equilibrium price vector. □

3.4 Existence of Walras equilibrium

We now construct an existence proof of Walras equilibrium for general quasi-concave preferences and convex production sets, that uses only the domain of prices S, and only Brouwer's fixed point theorem.

Theorem. Let $E = (H, (x^h, e^h, u^h)_{h \in H}, F, (Y_f)_{f \in F}, (\theta_f^h)_{f \in F}^{h \in H})$ be a Walras economy satisfying assumptions (1)–(6). Then E has a Walras Equilibrium $(\bar{p}, (\bar{x})_{h \in H}, (\bar{y}_f)_{f \in F})$.

Proof. Recalling that $Z_+(\bar{p}) = \sum_{h \in H} D_+^h(\overline{p}) - \sum_{h \in H} e^h - \sum_{f \in F} Y_f(\bar{p})$ is the at least as good as excess demand, and that $M(p, \bar{p}) = \min_{z \in Z_+(\bar{p})} p \cdot z$, define $\varphi : S \to S$ by

$$\varphi(\bar{p}) = \arg\max_{p \in S} [M(p, \bar{p}) - \|p - \bar{p}\|^2]$$

$$= arg\max_{p \in S} [\min_{z \in Z_+(\bar{p})} p \cdot z - \|p - \bar{p}\|^2].$$

Since (by Lemma 3) M is concave in p for any fixed $\bar{p}$, and $\|p - \bar{p}\|^2$ is quadratic, the maximand is strictly concave, so it has a unique maximum and $\varphi(\bar{p})$ is a function. Since by Lemma 3 M is continuous (equivalently, since $Z_+(\bar{p})$ is USC and LSC), φ is a continuous function. Therefore by Brouwer's fixed point theorem, φ has a fixed point $\bar{p}$.

At the fixed point $\bar{p}$,

$$M(\bar{p}, \bar{p}) = \max_{p \in S} [M(p, \bar{p}) - \|p - \bar{p}\|^2] = \max_{p \in S} M(p, \bar{p}).$$

where the last equality follows from the concavity of M in p and the concave perturbation lemma. By Lemma 6, $\bar{p}$ is a Walrasian equilibrium price vector. □

Again it is worth noting that the quasi-convexity of the X^h and the convexity of the Y_f played no role until the very last step where they guaranteed the convexity of $Z_+(p)$ at the single point $p = \bar{p}$. In traditional proofs of Walrasian existence, it is important to make sure that the excess demand correspondence is convex at every point p (otherwise there might not be a fixed point).

Aumann [3] gave a famous proof of Walras equilibrium without quasi-concavity (and without production) for an economy with a continuum of agents. He did it without using Kakutani's fixed point theorem, by adapting McKenzie's proof [18].

The existence proof just given can be extended to cover the case with convex Y_f, but without quasi-concave u^h, provided that we imagine that each agent h is now regarded as a continuum of identical replicas. I indicate the steps, without giving details. At the last step, when Lemma 6 is invoked, we must replace $Z_+(\bar{p})$ with its convex hull $co\{Z_+(\bar{p})\}$. Then apply the minmax theorem, obtaining $\bar{z} \in$

$co\{Z_+(\bar{p})\}$, as in Lemma 6. By Caratheodory's theorem, $\bar{z} = \sum_{i=1}^{L+1} \lambda_i \bar{z}^i$ where $\bar{z}^i = \sum_{h \in H} \bar{x}^{hi} - \sum_{h \in H} e^h - \sum_{f \in F} y_f^i \in Z_+(\bar{p})$, and the λ_i are nonnegative weights summing to 1. Regarding $\bar{x}^{hi}$ as the choice of a fraction λ_i of the agents of type h, and $y_f = \sum_{i=1}^{L+1} \lambda_i y_f^i \in Y_f$ as the choice of firm f, we get an equilibrium of the continuum compactified economy. However, without quasi-concavity, we can no longer be sure that an optimal consumption choice in the interior of the *compactified* consumption set is optimal in the original consumption set. So we must compute a different equilibrium for each compactification k. Then we let the size k of the compactifications go to infinity. Take convergent subsequences of the weights. For all those weights not converging to zero, take convergent subsequences of the $\bar{x}^{hi}(k)$, and of the $y_f(k)$. That limit is an equilibrium for the economy.

4 Comparisons to earlier proofs:
Walras equilibrium with strictly convex preferences

The main difference between the standard proofs of Walrasian existence and the proof just given in Section 3 is that the latter only requires Brouwer's fixed point theorem, applied to a domain of dimension $L - 1$. Another difference is that the latter proof has a natural "Lyapunov function" $L : S \to \mathbb{R}$ given by

$$L(\bar{p}) \equiv \max_{p \in S}[M(p, \bar{p}) - ||p - \bar{p}||^2].$$

I do not pursue the question of identifying conditions under which L declines under the dynamic $\bar{p} \mapsto \varphi(\bar{p})$.

Instead I turn to explaining the connection between my method of proof and the standard methods when excess demand is already a function. Taking advantage of the unicity of the excess demand, my proof can be modified to show its connection to earlier proofs.

In this section we specialize the general Walrasian economy given in Section 3 to cases where we can work with excess demand *functions*. For these cases it is already known that Brouwer's Theorem suffices to prove the existence of Walras equilibrium. But we show here that the perturbation $-||p - \bar{p}||^2$ can still simplify matters.

4.1 Pure exchange and strictly convex technologies

Let $S = \{p \in \mathbb{R}_+^L : \sum_{i=1}^L p_i = 1\}$ be the usual price simplex.

Let z be called an excess demand function whenever $z : S \to \mathbb{R}^L$ is a continuous function satisfying Walras Law: $p \cdot z(p) = 0 \ \forall p \in S$.[7]

---

[7] Suppose that, in addition to assumptions (1)–(7) from Section 2, for all $h \in H$,

$$[u^h(x) \geq u^h(y)] \Rightarrow [u^h(\lambda x + (1-\lambda)y) > u^h(y)]$$

if $0 < \lambda < 1$, $x \neq y$ and $x, y \in X^h$, and for all $f \in F$

$$[x \neq y \in Y_f, \ 0 < \lambda < 1] \Rightarrow [\exists z \in Y_f \text{ with } z \gg \lambda x + (1-\lambda)y].$$

We define a Walras equilibrium for the excess demand function z as a price vector $\bar{p} \in S$ satisfying

$$z(\bar{p}) \leq 0.$$

Note that by Walras Law, $z_i(\bar{p}) = 0$ unless $\bar{p}_i = 0$, in which case we may have $z_i(\bar{p}) < 0$.

Theorem. *Every excess demand function has a Walras equilibrium.*

Proof. Define the map $\psi : S \to S$ by

$$\psi(\bar{p}) \equiv \arg\max_{p \in S}[p \cdot z(\bar{p}) - \|p - \bar{p}\|^2].$$

Observe that the maximand is the sum of a linear function in p and a quadratic function in p, hence it is strictly concave and continuous in p. Since S is compact and convex, $\psi(\bar{p})$ is a single point, and so φ is a function. By the maximum principle, ψ is a continuous function (since the parameters $z(\bar{p})$ and $\bar{p}$ move continuously as $\bar{p}$ varies).

Hence by Brouwer's Fixed Point Theorem, ψ has a fixed point $\bar{p}$. By the concave perturbation lemma, $p \in S \Rightarrow p \cdot z(\bar{p}) \leq \bar{p} \cdot z(\bar{p})$. By Walras Law, $\bar{p} \cdot z(\bar{p}) = 0$, which implies $z(\bar{p}) \leq 0$. □

Debreu's [8] proof of Walras equilibrium uses the correspondence $\delta(z) = \arg\max_{p \in S} p \cdot z$. As Debreu said, δ is motivated by the principle that when there is excess demand in some commodity, $z_i > 0$, prices should go up, at least where excess demand is greatest. The only drawback to Debreu's construction is that $\delta(z)$ may be multivalued, thus forcing the use of Kakutani's Fixed Point Theorem. The function $\psi(p)$ is obtained by a slight perturbation of Debreu's construction.

The best known continuous function for proving Walras equilibrium is obtained by imitating the Nash [20] fixed point map for matrix games: $g_i(p) \equiv \{p_i + [z_i(p)]^+\}/\{1 + \sum_{j=1}^{L}[z_j(p)]^+\}$, where $[x]^+ = \max\{x, 0\}$, for $i = 1, ..., L$. A simple, but slightly awkward argument, using Walras law, shows that a fixed point of g is a Walras equilibrium.

The function $\psi(p)$ is (surprisingly) identical to the map $h(p) = \Pi_S(p + \frac{1}{2}z(p))$, where $\Pi_S(x)$ is the closest point in S to x.[8] By deriving ψ from the above maximization, one can see transparently that a fixed point is a Walrasian equilibrium. On the other hand, to show that a fixed point of h on the boundary of S is an equilibrium, the Kuhn–Tucker theorem must be invoked.

4.2 Production with constant returns-to-scale technologies

We now consider CRS production. A constant returns-to-scale (CRS) technology is a set $Y \subset \mathbb{R}^L$ such that Y is a closed, convex, cone ($y \in Y$ implies $ty \in Y$

Then $z(p) = \sum_{h \in H} D^h(p) - \sum_{h \in H} e^h - \sum_{f \in F} \arg\max_{y_f \in Y_f} p \cdot y_f$ is a continuous function satisfying Walras Law. In the special case $Y_f = \{0\} \; \forall f \in F$, we have a pure exchange economy.

[8] By the Kuhn–Tucker theorem, $\psi(\bar{p}) = \arg\max_{p \in S}[p \cdot z(\bar{p}) - \|p - \bar{p}\|^2]$ satisfies $(\psi(\bar{p}) - \bar{p}) = \frac{1}{2}z(\bar{p}) - \lambda e + \Lambda$ where $\Lambda \geq 0$ is a diagonal matrix with $\Lambda_{jj} > 0$ only if $\psi_j(\bar{p}) = 0$. Similarly by the Kuhn–Tucker theorem $h(\bar{p}) = \arg\min_{p \in S} \|p - [\bar{p} + \frac{1}{2}z(\bar{p})]\|^2$ satisfies the same equation.

for all $t \geq 0$; in particular, $0 \in Y$). Furthermore we suppose that Y allows for free disposal; $z \leq y$ and $y \in Y$ implies $z \in Y$. Finally, we suppose there is some $p^* \in S$ with $p^* \cdot Y \leq 0$, i.e., $p^* \cdot y \leq 0$ for all $y \in Y$.

A Walras equilibrium with production for an excess demand function, CRS-technology pair (z, Y) is a price $\bar{p} \in S$ such that $z(\bar{p}) \in Y$ and $\bar{p}Y \leq 0$. Note that by Walras Law the production plan $z(\bar{p})$ chosen makes zero profits, while alternatives either lose money or do no better.

The central example of a CRS-technology is an activity analysis production technology given by the matrix $B = [-I \ A]$ where I is the $L \times L$ identity matrix and A is an $L \times n$ vector of activities. Each column of the B matrix represents an "activity." Positive elements correspond to outputs, negative entries in B correspond to inputs. The first L columns of B represent pure disposal. The activity matrix B determines the CRS-technology

$$Y = \{Bx | x \in \mathbb{R}_+^{L+n}\}.$$

Clearly Y is a convex, closed cone allowing for free disposal. If for some vector $W \gg 0$, $\{x \in \mathbb{R}_+^{L+n} : Bx + W \geq 0\}$ is bounded, then there must be a $p^* \in S$ with $p^* \cdot Y \leq 0$.

Technology lemma. *If Y is a* CRS-*technology and for some vector $z \in \mathbb{R}^L$, $[p \in S$ and $pY \leq 0] \Rightarrow pz \leq 0$, then $z \in Y$.*

Proof. Suppose $z \notin Y$. Since Y is closed and convex, by the separating hyperplane theorem we can strictly separate Y and z, that is find some $\bar{p} \in \mathbb{R}^L$ such that $\bar{p} \cdot Y < \bar{p} \cdot z$. But Y is a cone, so $\bar{p} \cdot Y$ bounded above implies $\bar{p} \cdot Y \leq 0$; also $0 \in Y$, so we have $\bar{p} \cdot Y \leq 0 < \bar{p} \cdot z$. By free disposal, $\bar{p} \cdot Y \leq 0$ implies $\bar{p} \geq 0$. Scaling $\bar{p}$, we get $p \in S$ and $pY \leq 0 < p \cdot z$, contradicting the hypothesis. $\quad\square$

Theorem. *Every excess demand function,* CRS-*technology pair (z, Y) has a Walras equilibrium.*

Proof. We seek $\bar{p} \in S_Y \equiv \{p \in S : p \cdot Y \leq 0\}$ with $z(\bar{p}) \in Y$. By the technology lemma, it suffices to find $\bar{p} \in S_Y$ such that $p \in S_Y \Rightarrow p \cdot z(\bar{p}) \leq 0 = \bar{p} \cdot z(\bar{p})$.

By hypothesis, S_Y is nonempty. Furthermore, $S_Y \equiv \bigcap_{y \in Y} \{p \in S : p \cdot y \leq 0\}$ is the intersection of closed and convex sets, and so is closed and convex.

Define $\psi : S_Y \to S_Y$ by

$$\psi(\bar{p}) \equiv \underset{p \in S_Y}{arg\max}[p \cdot z(\bar{p}) - \|p - \bar{p}\|^2].$$

As we argued earlier, ψ is a continuous function. Since S_Y is compact and convex, Brouwer's Fixed Point Theorem guarantees ψ has a fixed point $\bar{p}$.

From the concave perturbation lemma, at the fixed point $\bar{p}$, $p \in S_Y \Rightarrow p \cdot z(\bar{p}) \leq \bar{p} \cdot z(\bar{p}) = 0$. $\quad\square$

The idea that Brouwer's theorem alone can be used to prove the existence of Walras equilibrium with production is due to McKenzie [18] who also used the set S_Y. His mapping is much more elaborate than ψ, but it allows for excess demand correspondences. McKenzie [18] showed that one could always reduce convex technologies to CRS-technologies by adding F auxiliary commodities, representing the contributions of the owners to each firm. The fixed point map must then be carried out in a simplex of dimension $L + F - 1$. In the above proof the domain is the original $L - 1$ dimensional simplex.

Todd [25] suggested the map $h(p) = \Pi_{S_Y} [p + z(p)]$. (A similar map is in Kehoe [13].) He showed by the Kuhn–Tucker theorem that a fixed point of h must be a Walras equilibrium, when Y is given by an activity analysis technology. The map ψ is identical, its only advantage being a perhaps more transparent proof that a fixed point is a Walras equilibrium (and the incorporation of general CRS Y).

4.3 Unbounded consumption sets, monotonic preferences and boundary behavior

In Sections 4.1 and 4.2 we assumed that the excess demand function z is continuous on all of S, including at $p \in S$ where some prices p_i may be zero. This will be true whenever utilities u^h are strictly concave, and consumption sets X^h are compact, as we indicated in Section 3.1. Some authors prefer to skip the step where we bound the consumption sets, preferring for aesthetic reasons not to invoke Assumption (8) (see Section 3.1). In its place they make the substantive assumption of strict monotonicity. I show now that the method of proof indicated in Section 4.2 still applies. To that end, let S^0 be the interior of S, and ∂S be its boundary. For every $\varepsilon > 0$, let $S^\varepsilon \equiv \{p \in S : p \geq \varepsilon 1\}$ be the trimmed simplex, and ∂S^ε its boundary, where $1 = (1, ..., 1)$.

We say that (z, Y) is an excess demand function, CRS-technology pair with proper boundary behavior whenever $z : S^0 \to \mathbb{R}^L$ is a continuous function satisfying Walras Law for all $p \in S^0$, and such that $\exists \varepsilon > 0$ and $\exists p^* \in S^\varepsilon$, satisfying

$$p^* \cdot Y \leq 0 . \tag{4.1}$$

$$p \in \partial S^\varepsilon \Rightarrow p^* \cdot z(p) > 0 , \tag{4.2}$$

When preferences are strictly monotonic, $p \to \partial S \Rightarrow$ some $z_i(p) \to \infty$. Since excess demand is bounded from below by the aggregate endowment of goods, strict monotonicity implies that for any $p^* \gg 0$, $p^* \cdot z(p) > 0$ if p is close enough to the boundary. Thus proper boundary behavior is automatically satisfied by excess demand functions derived from strictly monotonic preferences, provided we can find some strictly positive prices p^* at which $p^* \cdot Y \leq 0$. This latter condition is trivially verified if for example there is some indispensable input like labor that is never produced.[9]

Theorem. *Every monotonic excess demand function, CRS-technology pair with proper boundary behavior has a Walras equilibrium.*

Proof. S^ε is compact and convex. Hence $S_Y^\varepsilon \equiv S^\varepsilon \cap S_Y$ is also compact and convex. Define $\psi : S_Y^\varepsilon \to S_Y^\varepsilon$ by

$$\psi(\overline{p}) \equiv \arg\max_{p \in S_Y^\varepsilon} [p \cdot z(\overline{p}) - \|p - \overline{p}\|^2] .$$

---

[9] For a refinement of this boundary condition, see Neuefeind [19].

As before, ψ is a continuous function, hence it has a fixed point $\bar{p}$. Again by the familiar argument, $p \in S_Y^{\varepsilon} \Rightarrow p \cdot z(\bar{p}) \leq \bar{p} \cdot z(\bar{p}) = 0$.

If some $\bar{p}_i = \varepsilon$, then by proper boundary behavior, $p^* \cdot z(\bar{p}) > 0$, a contradiction, since $p^* \in S_Y^{\varepsilon}$. Hence $\bar{p} \gg \varepsilon 1$. But then by concavity of the maximand, $p \in S_Y \Rightarrow p \cdot z(\bar{p}) \leq 0$. By the technology lemma, $z(\bar{p}) \in Y$, so $\bar{p}$ is a Walras equilibrium. $\square$

5 The satisficing principle and quasi-concave games

5.1 The satisficing principle

Recall that the famous Maximum Principle asserts that the best response correspondence is upper semi-continuous (USC). The USC property is the crucial hypothesis in Kakutani's fixed point theorem for correspondences. Kakutani's theorem is used instead of Brouwer precisely because the best response correspondence may not be lower semi-continuous (LSC). What I show below is that if we replace maximization with almost maximization (satisficing), then the satisficing correspondence is LSC and USC.

Let $\mathcal{A} \subset \mathbb{R}^m$ and $X \subset \mathbb{R}^n$, and let $\psi : \mathcal{A} \rightrightarrows X$ be a correspondence associating with each $\alpha \in \mathcal{A}$ a subset $\psi(\alpha) \subset X$. We say that ψ is upper semi-continuous (USC) if

$$\left.\begin{array}{c} \alpha_n \to \alpha \\ x_n \to x \\ x_n \in \psi(\alpha_n) \end{array}\right\} \Rightarrow x \in \psi(\alpha)$$

for any $\{x_n, x\} \subset X$, $\{\alpha_n, \alpha\} \subset \mathcal{A}$. We say that ψ is lower semi-continuous (LSC) iff

$$\left.\begin{array}{c} \alpha_n \to \alpha \\ x \in \psi(\alpha) \end{array}\right\} \Rightarrow \begin{cases} \exists x_n \to x \\ x_n \in \psi(\alpha_n) \end{cases}$$

for any $\{\alpha_n, \alpha\} \subset \mathcal{A}$ and $x \in X$.

We say that ψ is USC or LSC at a point $\bar{\alpha} \in \mathcal{A}$ if the above conditions hold when $\alpha = \bar{\alpha}$. Clearly ψ is USC or LSC if it is USC or LSC at each point $\bar{\alpha} \in \mathcal{A}$.

Let $u : \beta \to \mathbb{R}$, where $\beta \subset \mathbb{R}^n$. We say that u is locally nonsatiated in β if for any pair $x, y \in \beta$ with $u(x) < u(y)$, there is a sequence $\{x(n)\}_{n=1}^{\infty} \subset \beta$ with $x(n) \to x$ and $u(x(n)) > u(x)$ for all n.

If β is convex and u is quasi-concave, then it follows immediately that u is locally nonsatiated in β.

Satisficing principle. Let $u : X \times \mathcal{A} \to \mathbb{R}$ be a continuous function, where $X \times \mathcal{A} \subset \mathbb{R}^n \times \mathbb{R}^m$. Let $\beta : \mathcal{A} \rightrightarrows X$ be a nonempty, USC and LSC correspondence. For each fixed $\alpha \in \mathcal{A}$, let $u(\cdot, \alpha)$ be locally nonsatiated in $\beta(\alpha)$. Let $v : \mathcal{A} \to \mathbb{R} \cup \{\infty\}$ be the maximum value function defined by $v(\alpha) \equiv \sup_{x \in \beta(\alpha)} u(x, \alpha)$. Finally, let $w : \mathcal{A} \to \mathbb{R}$ be continuous and satisfy $w(\alpha) < v(\alpha)$ for all $\alpha \in \mathcal{A}$. Then the correspondence $W : \mathcal{A} \rightrightarrows X$ defined by

$$W(\alpha) \equiv \{x \in \beta(\alpha) : u(x, \alpha) \geq w(\alpha)\}$$

is USC and LSC, and nonempty valued.

If in addition $\beta(\alpha) = \beta$ for all $\alpha \in \mathcal{A}$, and $u(x, \alpha) = u(x)$ for all $(x, \alpha) \in X \times \mathcal{A}$, then the same conclusion holds even with a weak inequality $w(\alpha) \leq v(\alpha) \equiv v$ for all $\alpha \in \mathcal{A}$.

Proof. The nonemptiness of W is evident. USC follows as in the maximum principle, and does not depend on the strict inequality $w(\alpha) < v(\alpha)$. Simply note that if $\{x_n \in W(\alpha_n)$ for all n, and $\alpha_n \to \alpha$ and $x_n \to x\}$, then by USC of β, $x \in \beta(\alpha)$. By hypothesis, $u(x_n, \alpha_n) \geq w(\alpha_n)$. Passing to the limit, and recalling the continuity of u and w, $u(x, \alpha) \geq w(\alpha)$, so $x \in W(\alpha)$.

To prove LSC of W, let $\alpha_n \to \alpha$. Suppose $\bar{x} \in W(\alpha)$ and $u(\bar{x}, \alpha) > w(\alpha)$. From the LSC of β, we can find $\bar{x}_n \in \beta(\alpha_n)$, $\bar{x}_n \to \bar{x}$. From the continuity of u and w, for large n, say, $n \geq N$, $u(\bar{x}_n, \alpha_n) > w(\alpha_n)$. Thus $\bar{x}$ can be approached by $\bar{x}_n$ in $W(\alpha_n)$ if $u(\bar{x}, \alpha) > w(\alpha)$. It remains to verify that any $\hat{x} \in W(\alpha)$ with $u(\hat{x}, \alpha) = w(\alpha)$ can be approached. If $v(\alpha) > w(\alpha)$, there is some $\bar{x} \in \beta(\alpha)$ with $u(\bar{x}, \alpha) > w(\alpha)$. By local nonsatiation, we can take a sequence of $\bar{x}(k) \in \beta(\alpha)$ converging to $\hat{x}$, with $u(\bar{x}(k), \alpha) > w(\alpha)$. Since each $\bar{x}(k)$ can be approached in $W(\alpha_n)$, so can $\hat{x}$.

If $u(\hat{x}, \alpha) = w(\alpha) = v(\alpha)$, then we must be in the additional case where β and u are independent of α. In that case, $\hat{x} \in W(\alpha_n)$ for all α_n, so $\hat{x}$ is trivially approachable. $\qquad \square$

The application of the satisficing principle to the Walrasian better than correspondence $D_+(p) = \{x \in X^h : u^h(x) \geq w(p)\}$, where $w(p) \equiv \max\{u^h(x) : x \in X^h, p \cdot x \leq I^h(p)\}$, is particularly simple, since then neither u^h nor $\beta(p) = X^h$ depends on p.

Corollary (Continuous correspondence lemma). *Let $X \subset \mathbb{R}^n$ be convex, and let $\mathcal{A} \subset \mathbb{R}^m$. Let $g_i : X \times \mathcal{A} \to \mathbb{R}$ be continuous, and convex on X for each fixed $\alpha \in \mathcal{A}$, for all $i = 1, ..., k$. Suppose that for each $\alpha \in \mathcal{A}$, there is $x(\alpha) \in X$ with $g_i(x(\alpha)) < 0$ for all $i = 1, ..., k$. Then the correspondence $B : \mathcal{A} \rightrightarrows X$ defined by*

$$B(\alpha) = \{x \in X : g_i(x, \alpha) \leq 0, \text{ for every } i = 1, ..., k\}$$

is USC and LSC.

Proof. Define $u : X \times \mathcal{A} \to \mathbb{R}$ by $u(x, \alpha) = \min_{1 \leq i \leq k}[-g_i(x, \alpha)]$. As the minimum of concave functions, u is concave on X for each fixed α, as well as continuous on $X \times \mathcal{A}$. Since $B(\alpha)$ is convex, u_α is nonsatiated on $B(\alpha)$. Furthermore, $v(\alpha) = \sup\{u(x, \alpha) : x \in X\} \geq u(x(\alpha)) > 0$, for all $\alpha \in \mathcal{A}$. Hence by the satisficing principle, $B(\alpha) = \{x \in X : u(x, \alpha) \geq 0\}$ is USC and LSC. $\qquad \square$

5.2 Quasi-concave games

In our definition of games given in Section 1, we can weaken the hypothesis that u_n is concave in σ_n to the hypothesis of quasi-concavity: $u_n(\sigma_n, \bar{\sigma}_{-n}) > u_n(\bar{\sigma}_n, \bar{\sigma}_{-n})$ implies $u_n(\lambda \sigma_n + (1-\lambda)\bar{\sigma}_n, \bar{\sigma}_{-n}) > u_n(\bar{\sigma}_n, \bar{\sigma}_{-n})$ for all $0 < \lambda < 1$. The result is called a quasi-concave game. We now use Brouwer's fixed point theorem to prove the existence of Nash equilibrium for all quasi-concave games.

Theorem. *Every quasi-concave game has a Nash equilibrium.*

Proof. Let $v_n(\bar{\sigma}_{-n}) \equiv \max_{\sigma_n \in \Sigma_n} u_n(\sigma_n, \bar{\sigma}_{-n})$ define a continuous function from Σ_{-n} to $\mathbb{R}$, called the "indirect utility function." Let $\delta_n(\bar{\sigma}) = v_n(\bar{\sigma}_{-n}) - u_n(\bar{\sigma})$, and let $\delta(\bar{\sigma}) = \max_{n \in N} \delta_n(\bar{\sigma})$. Clearly $\bar{\sigma}$ is a Nash equilibrium if and only if $\delta(\bar{\sigma}) = 0$. Let $\underline{\delta} = \min_{\sigma \in \Sigma} \delta(\sigma)$. Let $w_n(\bar{\sigma}_{-n}) = v_n(\bar{\sigma}_{-n}) - \frac{1}{2}\underline{\delta}$, and let

$$W_n(\bar{\sigma}_{-n}) = \{\sigma_n \in \Sigma_n : u_n(\sigma_n, \bar{\sigma}_{-n}) \geq w_n(\bar{\sigma}_{-n})\}.$$

Suppose G has no Nash equilibrium. Then $\underline{\delta} > 0$ and for all $\bar{\sigma} \in \Sigma$ and each n, $w_n(\bar{\sigma}_{-n}) < v_n(\bar{\sigma}_{-n})$. By the Satisficing Principle (which applies since u_n is quasi-concave, and thus locally non-satiated in any convex budget set), W_n is nonempty, USC, LSC, and convex-valued. Moreover, for all $\bar{\sigma}$ there is some player n with $u_n(\bar{\sigma}) < w_n(\bar{\sigma}_{-n})$, so $\bar{\sigma}_n \notin W_n(\bar{\sigma}_{-n})$. Define $\varphi_n : \Sigma_n \times \Sigma_{-n} \to \Sigma_n$ by

$$\varphi_n(\bar{\sigma}_n, \bar{\sigma}_{-n}) = \min_{\sigma_n \in W_n(\bar{\sigma}_{-n})} \|\sigma_n - \bar{\sigma}_n\|^2.$$

Clearly φ_n is a function, since W_n is convex-valued. Furthermore, if W_n is USC and LSC, then by the Maximum Principle, φ_n is a continuous function. Let $\varphi = (\varphi_1, ..., \varphi_N)$. If G has no Nash equilibrium, then φ is a continuous function with no fixed point, a contradiction. $\square$

References

1. Aliprantis, P., Brown, D.J.: Equilibria in markets with a Riesz space of commodities. Journal of Mathematical Economics **11**, 189–207 (1983)
2. Arrow, K., Debreu, G.: Existence of an equilibrium for a competitive economy. Econometrica **22**, 265–290 (1954)
3. Aumann, R.: Existence of competitive equilibria in markets with a continuum of traders. Econometrica **34**(1), 1–17 (1966)
4. Border, K.: Fixed point theorems with applications to economics and game theory. Cambridge: Cambridge University Press, 1985
5. Brown, D.J.: Acyclic choice. Cowles Foundation Discussion Paper No. 360, Yale University, mimeo (1973)
6. Debreu, G.: A social equilibrium existence theorem. Proceedings of the National Academy of Sciences **38**, 886–893 (1952)
7. Debreu, G.: Market equilibrium. Proceedings of the National Academy of Sciences **42**, 876–878 (1956)
8. Debreu, G.: The theory of value. New Haven: Yale University Press, 1959
9. Fan, K.: A generalization of Tychonoff's fixed point theorem. Mathematische Annalen **142**, 305–310 (1961)
10. Gale, D.: The law of supply and demand. Mathematica Scandinavica **3**, 155–169 (1955)
11. Gale, D., Mas-Colell, A.: An equilibrium existence theorem for a general model without ordered preferences. Journal of Mathematical Economics **2**, 9–15 (1975)
12. Gul, F., Pearce, D., Stacchetti, E.: A bound on the proportion of pure strategy equilibria in generic games. Mathematics of Operations Research **18**(3), 548–552 (1993)
13. Kehoe, T.: An index theorem for general equilibrium models with production. Econometrica **48**, 1211–1232 (1980)
14. Kehoe, T.: The existence, stability, and uniqueness of economic equilibria. Presented at the School Celebration at Yale, 1995
15. Krasa, S., Yannelis, N.C.: An elementary proof of the Knaster–Kuratowski–Mazurkiewicz–Shapley theorem. Economic Theory **4**, 467–471

16. Lemke, C.E., Howson, J.T.: Equilibrium points of bimatrix games. SIAM Journal of Applied Mathematics **12**, 413–423 (1964)
17. McKenzie, L.: On equilibrium in Graham's model of world trade and other competitive systems. Econometrica **22**, 147–161 (1954)
18. McKenzie, L.: On the existence of general equilibrium for a competitive market. Econometrica **27**, 54–71 (1959)
19. Nash, J.: Equilibrium points in N-person games. Proceedings of the National Academy of Sciences **36**, 48–49 (1950)
20. Nash, J.: Non-cooperative games. Annals of Mathematics **54**, 286–295 (1951)
21. Neuefeind, W.: Notes on existence of equilibrium proofs and the boundary behavior of supply. Econometrica **48**, 1831–1837 (1980)
22. Nikaido, H.: On the classical multilateral exchange problem. Metroeconomica **8**, 135–145 (1956)
23. Shafer, W.J., Sonnenschein, H.: Equilibrium in abstract economics without ordered preferences. Journal of Mathematical Economics **2**, 345–348 (1975)
24. Sonnenschein, H.: Demand theory without transitive preferences, with applications to the theory of competitive equilibrium. In: Chipman, J.S., Hurwicz, L., Richter, M.K., Sonnenschein, H.F. (eds.) Preferences, utility and demand. New York: Harcourt Brace Jovanovich 1971
25. Todd, M.: A note on computing equilibria in economies with activity analysis models of production. Journal of Mathematical Economics **6**, 135–144 (1979)
26. Yannelis, N. C.: On a market equilibrium theorem with an infinite number of commodities. Journal of Mathematical Analysis and Applications **108**, 595–599 (1985)

The application of CVM
for assessing the tele-health system:
an analysis of the discrepancy
between WTP and WTA based on survey data[*]

Masatsugu Tsuji[1] and Wataru Suzuki[1,2]

[1] Osaka School of International Public Policy, Osaka University, Osaka 560-0043, Japan
   (e-mail: tsuji@osipp.osaka-u.ac.jp)
[2] Japan Center for Economic Research, Tokyo, 100-0025, Japan

Received: February 2, 2001; revised version: November 2002

Summary. This paper analyzes the applicability of CVM (Contingent Valuation Method) for the economic assessment of the tele-health system. By focusing on the discrepancy between WTP (Willingness to pay) and WTA (Willingness to accept), we decompose the discrepancy into income and endowment effect according to the method used by Morrison [21, 22]. We also closely examine the endowment effect, and break it down into several components. It is noted that the portion, which was thought to be the endowment effect, contains imaginary as well as subsidy biases, and the endowment effect is not as large as previously estimated.

Keywords and Phrases: CVM, WTP, WTA, Tele-health, Imaginary bias, Strategic bias, Subsidy bias, Income effect, Endowment effect.

JEL Classification Numbers: I11, H43, C93.

1 Introduction

In the assessment of new medical services and technology, Cost Effective Analysis (CEA), Cost Utility Analysis (CUA), and Cost Benefit Analysis (CBA) are adopted

---

  * The authors are indebted to the officials of Katsurao Village, Fukushima Prefecture, especially Mr. Y. Ohara, and to its residents, who lent their support to our survey. Thanks are also due to Profs. H. Tsukahara of Meiji University, S. Nishimura of Kyoto University, and C. R. McKenzie of Osaka University for their comments on the earlier version of this paper. The authors are indebted to anonymous referees for their valuable comments and suggestions. The Grants of Scientific Research from the Ministry of Welfare and Labor and the Ministry of Education, Culture, Sports, Science and Technology are gratefully acknowledged.
Correspondence to: M. Tsuji

in the field of Health Economics. Although CEA is a simple method that compares costs and effects such as the rate of cure, effects should be compared with the same unit of measurement. On the other hand, since CUA evaluates its benefits in terms of Health-Related Quality of Life (HRQOL), it can compare the wide range of effects and has been widely adopted.[1] Some authors, however, have noted counterexamples against its theoretical basis and presumptions, and its appropriateness becomes problematic.[2]

The Contingent Valuation Method (CVM) and CBA based on this have clearer theoretical foundations and have recently been widely adopted in the field of Health Economics.[3] CVM is a method used to evaluate services and projects which are not traded on the market in pecuniary terms by asking the exact value they are willing to pay, and has been developed in the fields of Public Economics, Environmental Economics, and Experimental Economics. Although CVM has a strong theoretical basis, it tends to have a bias because it asks for concrete valuation and choice under fictitious circumstances. Regarding its methodology, therefore, more efforts have been made in the field of Environmental Economics, for example, so as to (i) clarify what kind of bias it has, and (ii) remove them.[4]

Health Economics, however, frequently adopts CVM, but little attention is paid to this methodological issue and research of other fields.[5] Thus, in this paper, we examine the discrepancy between WTP and WTA, which is critically important to the reliability of evaluating medical technology, and focus on the factors which gives rise to this discrepancy. Some research in Environmental Economics using CVM observe a discrepancy of 1.5 to 16 times between them, which should not be negligible as the theory suggests. This is one of the well-known unsolved issues in this field. The possible explanations are substitutability and the endowment effect, as have been argued, and in what follows we separate these two effects according to the method introduced by Morrison [21, 22]. We also analyze the endowment effect in detail using our own method, and show the different results from previous studies.

In this paper, we take the tele-health system as an example, which has been introduced to depopulated regions. This system is implemented as a pilot project so that inhabitants bear only low costs since it is subsidized by taxes. In this case,

---

[1] In HRQOL, the weight of preferences for health conditions is calculated using methods such as SG (Standard Gamble), TTO (Time Trade-Off), and VAS (Visual Analogue Scale), see Torrance [29].

[2] HRQOL, for instance, assumes that preferences for health and other conditions are independent of each other, but some empirical research show that this assumption is not satisfied in reality (Viscusi and Evans [33]). Presumptions of QOLY such as the additivity of the utility function, and independency between the utility of life span and that of quality of life are argued as not to being satisfied as well (Broome [4], Bleichrodt and Johannesson [2]).

[3] See Donaldson [5], Gafni [7], Johannesson et al. [10, 11], Golan and Shechter [8], Tolley et al. [28], O'Brien et al. [23], Kartman et al. [14, 15], Donaldson et al. [6], Phillips et al. [25], and Olsen and Smith [24].

[4] Regarding bias, strategic, imaginary, embedding, warm glow, information, interviewers, payment method, starting and ending point and response-ordering bias are well-known. Much effort has been made to design questionnaires so as to remove these biases.

[5] One of the few exceptions is Katman, Stalhammar and Johannesson [15]. They examine biases arising from the ordering of questionnaires, and reaction to the amount of goods.

since they already receive the service at low cost or without charge, it is difficult to receive an accurate valuation from them. This phenomenon is particular to the assessment of new medical technology.

In what follows, Section 2 provides a survey of previous studies in Environmental Economics and clarifies the issue. Section 3 presents data used in this paper, and an empirical analysis is given in Section 4. The conclusion is stated in the final section.

2 Discrepancy between WTP and WTA

In this section, a brief survey of WTP and WTA in Environmental Economics is presented, in addition to the framework of our analysis.

2.1 Survey of previous studies

WTP is defined as the maximum amount of money they are willing to pay for an object, and WTA the amount they are asking to give it up. In the context of Welfare Economics, the former is referred to as Compensated Valuation (CV), whereas the latter Equivalent Valuation (EV). Figure 1 shows these two concepts, and the horizontal axis denotes goods x that is evaluated, and the vertical axis composite commodity (income) m.[6] Let us suppose that initially a representative agent owns x_0 and income m_0, which is shown as $A=(x_0, m_0)$. U_0 denotes this indifference curve. When he buys one additional unit of x, $(x_0$ to $x_1)$, by keeping his utility the same as before, the maximum amount he is willing to pay is shown as $B = (x_1, m_1)$, namely, $m_1 - m_0$. Suppose the situation is that he already has one more unit, x_1. In this case, he owns m_0 as well as goods x_1, shown as $C = (x_1, m_0)$, which is on indifference curve U_1. When he loses one unit of x in keeping his utility level as before, he asks m_2, namely, to move from C to $D= (x_0, m_2)$. Thus, WTA is equal to $m_2 - m_0$.

It is apparent from Figure 1 that without the assumption that the marginal rate of substitution is constant, WTP differs from WTA because of the income effect. The amount of discrepancy, however, is negligible if the amount of expenditure on x is small enough as compared to his income, in addition to the usual assumptions on

---

[6] Bishop and Heberlein [1], for example, when estimating the WTP and WTA for the right of hunting geese in Wisconsin, obtained the result such that the average WTP is $21 and average WTA is $101 (WTA/WTP $\fallingdotseq$ 4.8). Row et al. [27] estimated WTP and WTA for a view. Their results are summarized as follows: average WTP for a declining view from 75 miles to 25 miles is $6.54, while WTA is $71.44 (WTA/WTP $\fallingdotseq$ 11); WTP from 50 miles to 25 miles is $3.53, while WTA is $46.63 (WTA/WTP/13). The largest discrepancy is found in the case of a declining view from 75 miles to 25 miles in sight of the power plant and WTP is $6.85, while WTA is $113.68 (WTA/WTP $\fallingdotseq$ 16). Brookshire et al. [3] also estimated WTP and WTA for hunters by asking them about the value of the hunting environment. WTP for the increase of possibility of meeting up with 0.1 to 1 deer during one hunting outing is $43.64, while WTA is $68.52 (WTA/WTP $\fallingdotseq$ 1.5). WTP from 1 to 5 deer is $54.60, while WTA is $142.60 (WTA/WTP $\fallingdotseq$ 2.5). These estimations are based on standard survey data, which is similar to ours. Their analysis, however, does not consider the income effect; therefore, they cannot determine how much of the discrepancy is due to the endowment effect or other effects.

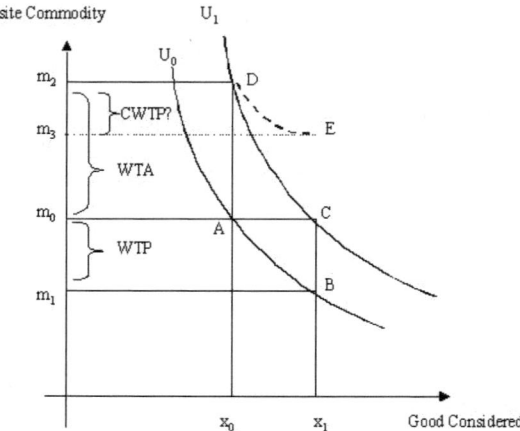

Figure 1. Concepts of WTP, WTA, and CWTP

the income elasticity of demand and the marginal rate of substitution (Randall and Stoll [26], for example). Survey researches conducted in Environmental Economics show that the discrepancy between the two is observed to be as large as 1.5 to 16 times – 7 to 8 times on average.[7] In order to explain this discrepancy, there are two reasons identified thus far: (i) substitutability (Hanemann [9]); and (ii) the endowment effect (Knetsh [16, 17] and Kahnemann et al. [12]).

Hanemann [9] shows that even if the income elasticity of demand is small, if the rate of substitution is smaller than this, WTP and WTA differ from each other due to the usual income effect. This can be easily understood if the indifference curve, U_0 in Figure 1, for example, is considered as having a shape where it is more vertical upwards from A, and more horizontal downwards from A. Thus, even under the usual assumption of the indifference curve, it is possible that the discrepancy between WTA and WTP becomes larger.

On the other hand, Knetsch [17] argues that the usual assumptions cannot explain this discrepancy so that concepts such as the 'non-reversible indifference curve' should be assumed, since the agent faces different indifference curves in the event he obtains and loses goods, respectively. The non-reversible indifference curve, for example, is shown as U_1 and the dotted line through D to E. In this case, WTA ($m_2 - m_0$), which is the movement from C to D, tends to be larger than WTP ($m_2 - m_3$) from D to E. This 'endowment effect' suggests that the initial assignment of property rights may serve as a kind of psychological illusion. That is, ownership itself adds value to the commodity, resulting in a higher selling price. Knetsch [17] thus adopts the 'Prospect Theory' suggested by Kahnemann and Tversky [13], which claims that people evaluate higher the goods they actually obtained. Kahnemann et al. [12] and Knetsch and Sinden [18, 19] try to verify this

---

[7] This bias arises from the format or design of questionnaires, and can be evaded by preparing appropriate questionnaires. In this paper, we pay attention to this problem by providing the clear explanations that our questionnaires are irrelevant to policy implementation.

hypothesis by experimental studies, but they test only this hypothesis and do not compare it with an alternative hypothesis. This procedure is not accurate.

Morrison [21, 22] examines the income effect (substitutability) and the endowment effect in one framework, and tries to extract which effect contributes to the discrepancy. Morrison conducts the following experiment with students: (i) provides them with mug cups without charge; (ii) asks WTA and pays its average amount to the students; and (iii) asks WTP again. Based on to this simple experiment, the income effect can be extracted, and WTP and WTA can be compared. The experiment is interpreted in Figure 1 as follows: Providing mugs and asking WTA imply the movement from C to D. In their reply, the amount $m_2 - m_0$ is given as WTA. Then, as this amount is paid to the students, they stay at D, and they are asked WTP. WTP at D is different from that at A, since the income level is different. WTP at D can be referred to as Compensated WTP (CWTP), since the income at D is compensated. If there is no endowment effect, CWTP is equal to $m_2.m_0$, since it is expressed by the movement along the same indifference curve U_1, namely, D to C. With the endowment effect, however, this movement is expressed from D to E, that is, WTP is equal to $m_2 - m_3$. Therefore, the existence of the endowment effect can be examined statistically by testing whether CWTP is the same as WTA. Morrison [21] obtains the result of CWTP being larger than WTA, and accordingly they verified the existence of the endowment effect.

2.2 Issues related to previous studies

It is rather easy to conclude that the discrepancy between WTA and CWTP is due to the endowment effect, because the discrepancy is affected not only by the characteristics of students such as income, gender, education, etc. which are not explicitly dealt with in the experiment, but also by a bias due to fictitious questionnaires. For example, if they understand that the policy depends on their answer, they will then reply with a lower WTP, and higher WTA. In this case, a strategic bias emerges (see Mitchell and Carson [20]).[8] In addition, as to whether a bias arises from either WTP or WTA is more difficult to imagine. If the respondents receiving compensation for giving up (WTA) is considered more unrealistic to imagine than for making a payment for something (WTP), WTA then tends to be less correct, and this gives rise to a discrepancy between the two. We refer this as being an imaginary bias in this paper, since there is difficulty in imagining it. In the case of tele-health, they adhere to current low charges because of subsidies, which can be a 'subsidy bias'. The amount measured as the endowment effect may contain these various biases, and care should be taken in these matters. In what follows, we will carefully examine the amount which was proved to be the endowment effect from these standpoints.

---

[8] As for the definition and characteristics of tele-health and tele-medicine, see Tsuji et al. [30, 31].

3 Data

The result of the experiment heavily depends upon the procedure and questionnaires. Here we explain the objectives and method of our survey conducted this
year regarding WTP and WTA.

3.1 Field research

The analysis here is based on our survey conducted in Katsurao Village in Fukushima
Prefecture, Japan in February 2001, as part of the research project of the Research
Group of Tele-medicine, Ministry of Welfare and Labor. Katsurao Village implemented the tele-health system as a part of the Ministry's project for the promotion
and implementation of medicine.[9] The system is manufactured by NEC and called
'*Sukoyakamate*'. Katsurao Village distributes the system not only to all households
with family members older than sixty-five, but also to those which desire to have
it. The tele-health system measures pulse, blood pressure, temperature, and elec-
trocardiogram, which are transmitted through the terminal to medical institutions.
Public nurses monitor these medical data, and provide health consultation. In depopulated regions such as Katsurao Village located far from any medical institution,
this system is highly valued.

With the cooperation of the village office, questionnaires were sent to all households which own the tele-health system on January 29, 2001, and it was requested
that they be submitted by the end of February. We sent it to 926 inhabitants of
325 households, and received replies from 492 persons (rate of reply was 53.1%).
Among those who sent replies, the number of people whose replies contained all
WTP, WTA, and CWTP totaled 230. Thus, the sample size of the analysis is 230.
Table 1 shows the descriptive statistics of sample properties.

3.2 Questionnaires

Questionnaires on WTP, WTA, and CWTP are designed in the following way. Questions 8, 9, and 10 are related to WTP, WTA, and CWTP, respectively.[10] Question
10 is related to WTP under the assumption that the amount given in the reply to
Question 9 is compensated. The method of reply is based on the Payment Card
Method. Although the dichotomous choice method, the two-stage dichotomous
choice method, and the multi-stage dichotomous choice method have become popular in CVM, in this survey the dichotomous method has not been adopted because
it includes complex questionnaires and is sent by mail.

---

[9] In Tsuji, Suzuki, Taoka, and Teshima [32], they adopt the multi-stage dichotomous choice questionnaires and estimate the probability of acceptance for tele-health service in Kamaishi City, Iwate
Prefecture, Japan. They obtain WTP for the project.

[10] Three indices also have peaks at ¥ 5,000 and ¥ 10,000 as Figure 2 shows, and this indicates another
'replying-exact-amounts' bias. Respondents' replies consisted of values closer to the abovementioned
amounts, so that they do not alter the importance of the average and median.

Table 1. Descriptive statistics for user profile

	Average	s.d.	Minimum	Maximum
Age	62.80323	15.14172	24	90
Sex	0.498525	0.5007369	0	1
Education	0.339286	0.4741731	0	1
Number of family members	4.780488	2.443362	0	22
Income	48.23248	60.08764	10	350
Asset	1279.67	1968.095	100	15000
Health condition	0.053672	0.2256889	0	1
Have chronic diseases or not	0.075209	0.2640962	0	1
Reason 1 of discrepancy between WTA and CWTP	0.412742	0.4930106	0	1
Reason 2 of discrepancy between WTA and CWTP	0.041551	0.1998383	0	1
Reason 3 of discrepancy between WTA and CWTP	0.030471	0.1721177	0	1
Reason 4 of discrepancy between WTA and CWTP	0.110803	0.3143241	0	1
Reason 5 of discrepancy between WTA and CWTP	0.207756	0.4062644	0	1

The sample size is the number of answers which included all of WTA, WTP, and CWTP in their replies. Education is a dummy variable, which takes 1 for graduates of junior high school, and 0 for others. Health condition is also a dummy variable, which takes 1 for sick or sickly, and 0 for otherwise.

[Questionnaire]

The following Questions 8, 9, and 10 are asked under fictitious assumptions such as "If the charges will be collected" or "If the system is taken away from you". They are purely academic so as to measure the pecuniary value of the system. The village office does not have such a plan.

Question 8

The costs of '*Sukoyakamate*' are financed by various government subsidies and there is no charge. Suppose that all subsidies are discontinued, if you were asked to pay the monthly charges, how much would you be willing to pay? Please choose the maximum amount from the numbers below.

¥ 0, ¥ 500, ¥ 1,000, ¥ 1,500, ¥ 2,000, ¥ 2,500, ¥ 3,000, ¥ 3,500, ¥ 4,000, ¥ 4,500, ¥ 5,000, ¥ 5,500, ¥ 6,000, ¥ 6,500, ¥ 7,000, ¥ 7,500, ¥ 8,000, ¥ 8,500, ¥ 9,000, ¥ 9,500, ¥ 10,000, ¥ 11,000, ¥ 12,000, ¥ 13,000, ¥ 14,000, ¥ 15,000, ¥ 16,000, ¥ 17,000, ¥ 18,000, ¥ 19,000, ¥ 20,000, ¥ 22,000, ¥ 24,000, ¥ 26,000, ¥ 28,000, ¥ 30,000, ¥ 35,000, ¥ 40,000, ¥ 45,000, ¥ 50,000, ¥ 55,000, ¥ 60,000, ¥ 70,000, ¥ 80,000, ¥ 90,000, more than ¥ 100,000

Question 9

The following is also fictitious. The government will change its policy and halt service of '*Sukoyakamate*,' and the machine will be taken away from you. Because of this change, the benefits you have been receiving will come to an end, and the government will compensate you for this. You will be paid a certain amount for the period you want to use '*Sukoyakamate*'. How much of this compensation per month will you receive for agreeing to have it taken away from you? Please choose the minimal amount from the list below.

(The amounts are the same as Question 8.)

Question 10

In continuation of the fictitious situation of Question 9, now that the retrieval '*Sukoyakamate*' is completed, you are to receive the same amount of compensation each month from the government, as indicated in your answer to Question 9. Katsurao Village wants to continue the same project as before, but the village will not be receiving any subsidies from the government. The village wants to charge you instead. How much would you be willing to pay per month? Please choose the maximum amount from the list below. Please remember that you are now richer than you were before you received the compensation from the government.

(The amounts are the same as in Question 8.)

4 Analysis

Here we examine our survey results and clarify the factors which causes the discrepancy between WTP and WTP. The discrepancy of our results from previous studies is also presented.

4.1 Discrepancy between WTP and WTA

Figure 2 shows the histograms of WTP, WTA, and CWTP. Three distributions have similar shapes, and three indices have peaks around ¥ 1,000.[11] The average and median of the above three variables are shown in the upper part of Table 2. As the average of WTP is ¥ 1,649 and that of WTA 5,932, the discrepancy is about 3.6, and this indicates that there is large discrepancy in our assessment of medical technology. As for median, on the other hand, they are all ¥ 1,000, so there is no discrepancy. In previous researches in Environment Economics and Experimental Economics, since the discrepancy was found in the median, our result is rather surprising. The possible interpretation is as follows: In Environment Economics

---

[11] It can be said that the endowment effect is included in WTA so it is not considered as being a bias, namely, it can be said that WTP and WTA are entirely different. Our standpoint is, however, based on policy issues relating to the introduction of a tele-health system in other depopulated regions in the future, so that WTP, rather than WTA, is a better concept. Therefore, we consider the endowment effect a bias, and make an attempt to identify it.

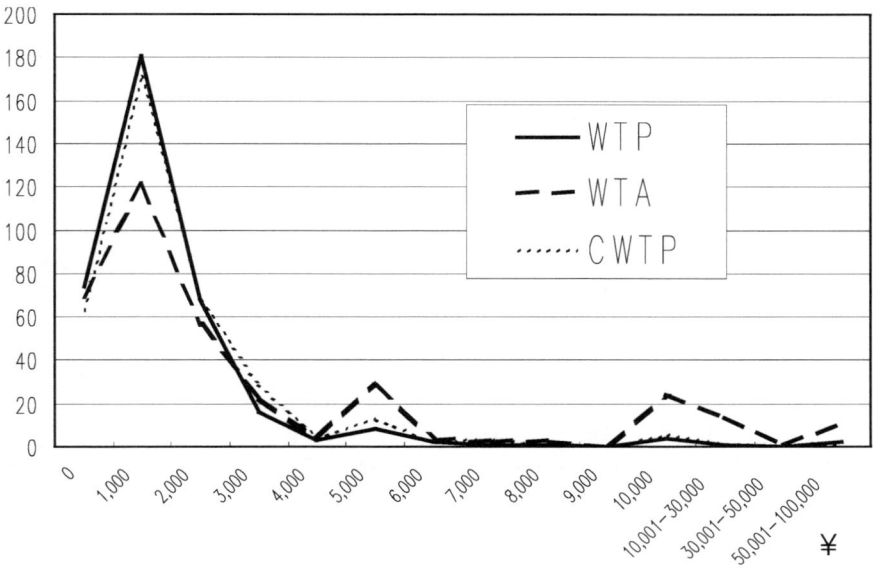

Figure 2. Histograms of WTP, WTA, and CWTP

and Experimental Economics, respondents are asked in general about WTP or WTA, but not both. In our survey, on the other hand, they are asked about WTP, WTA and CWTP together. It is possible for respondents to make their answers consistent with one another, so they may deviate from true values. In this sense, our analysis might have a methodological issue such as containing a 'least-chance-of-discrepancy' bias. Since our analysis is based on the questionnaire being asked to the same respondents, this is inevitable, otherwise it would not be possible to focus on factors of discrepancy as well as its relation with personal characteristics. By noting this point, we proceed with our analysis by paying attention to the average.

The discrepancy between WTP and WTA are decomposed into the following two parts:

$$\text{WTA} - \text{WTP} = (\text{WTA} - \text{CWTP}) + (\text{CWTP} - \text{WTP})$$

endowment effect   income effect (substitutability)

In the lower part of Table 2, the averages of (WTA-WTP), (WTA-CWTP), and (CWTP-WTP) are provided. The income effect indicated by (CWTP-WTP) is rather small such as ¥ 109, thus it is apparent that (WTA-WTP) is largely explained by the endowment effect of (WTA-CWTP).

4.2 Decomposition of discrepancy

Here we focus on the discrepancy between WTA and CWTP. As seen in the previous section, it is not certain whether the discrepancy between these two is due to the endowment effect, and further analysis is required. In our survey, we ask the following question when WTA and CWTP differs:

Table 2. Descriptive Statistics of WTA, WTP and CWTP

Statistics	
Average of WTA	5,932.1
Average of WTP	1,649.6
Average of CWTP	1,759.0
Median of WTA	1,000.0
Median of WTP	1,000.0
Median of CWT	1,000.0
Average of (WTA-WTP)	4,282.5***
Average of (WTA-CWTP)	4,173.1***
Average of (CWTP-WTP)	109.4n.s.

Note: *** indicates that the test of difference of the average is significant at the 1% level. n.s denotes it is not significant at the 10% level.

Question 11

We would like to ask a question to those who indicated smaller 'monthly charges' in their answer to Question 10 rather than 'monthly compensation' in Question 9. Please check the reasons for this discrepancy from the list below. You can select as many answers as possible.

1. Since this service was originally free, I feel hesitant to ask for 'larger monthly charges'.
2. It is trying to have the '*Sukoyakamate*' I once used taken away. Even if I know that I can use it by paying 'monthly charges,' I would rather ask for larger compensation than charges.
3. It says this is irrelevant from the actual plan, but if the government is going to determine the compensation based on this amount, it would be better to ask for a larger value.
4. It says this is irrelevant from the actual plan, but if the village is going to determine the charges based on this amount, it would be better to state a smaller value.
5. Since it is impossible for the government to pay compensation, the amount of compensation to be requested becomes a vague issue, while it is possible for the village to make charges; thus, I am prudent about making a decision on the monthly charges.

The above reasons correspond to the following biases, respectively: (1) subsidies bias; (2) endowment effect; (3) strategic bias (for WTA); (4) strategic bias (for WTP); and (5) imaginary bias. According to the survey, let us define five dummy variables corresponding to the above five questions in such a way that, if one checks the reasons, then it takes 1, and if one does not check, then it takes 0. We then follow the regression analysis with (WTA − CWTP) and these dummy variables so as to examine how dummy variables affect the discrepancy. In addition, we also take

Table 3. Estimation of (WTA-CWTP)

Dependent variable: FWTA-CWTP	Coefficient	t-value	p-value
Reason 1 of discrepancy	4649.933**	2.17	0.031
Reason 2 of discrepancy	17449.8***	3.58	0
Reason 3 of discrepancy	6832.154	1.02	0.308
Reason 4 of discrepancy	1172.133	0.34	0.731
Reason 5 of discrepancy	4907.724*	1.87	0.063
Age	−64.13085	−0.83	0.406
Sex	1154.33	0.56	0.575
Education	−2998.34	−1.22	0.224
Income	48.54432***	3.04	0.003
Asset	0.2078535	0.41	0.685
Number of family members	−436.3474	−1.06	0.291
Constant	3392.653	0.55	0.584
Adj.R^2	0.1024		
Number of sample	228		

Note: OLS estimation. ***, **, and * denote being significant at the 1%, 5%, and 10% levels, respectively.

into account the personal profile of people such as age, gender, income, education, assets, number of family members, etc. We estimate the following model with OLS:

$$[\text{WTA} - \text{CWTP}]_i = c + \sum_{j=1}^{5} \alpha_j DUM_{ji} + \beta Z_i + u_i \qquad (1)$$

where DUM_{ji} denotes the dummy variable corresponding to reason j, and Z_i a matrix of personal profile variables as explained.

The result of the estimation is summarized in Table 3. Among five dummy variables, it is clear that dummy 1 for reason 1 (the subsidy bias), dummy 2 for reason 2 (the endowment effect), and dummy 5 for reason 5 (the imaginary bias) are significant, and two strategic biases are not. Moreover, income provides a significant effect to the discrepancy.

4.3 Numerical representation of the discrepancy

Here, let us calculate exactly how much each reason affects (WTA − CWTP). According to Table 3, each coefficient shows that the subsidy bias is ¥ 4,650, the endowment effect ¥ 17,450, and the imaginary bias ¥ 4,908. By multiplying the average of each dummy variable, which are shown in Table 1, to the above value, we can observe the numerical value of the effect, namely, the subsidy bias is ¥ 1,919, the endowment effect ¥ 725, and the imaginary bias ¥ 1,016. The results show that the endowment effect is not very large. This does not necessarily deny the result

Table 4. WTA and WTP considering bias estimated

	Initial average	After considering bias	Amount of change
WTA	¥ 5,932	¥ 4,191	¥ 1,741
WTP	¥ 1,650	¥ 3,569	¥ −1,919
Discrepancy of (WTA-WTP)	¥ 4,283	¥ 623	

of the tests stated in Section 2, but requires the re-examination of the part which is supposed to be due to the endowment effect.

Next, we analyze how WTP and WTA undergo change when these three biases are taken into consideration. Since the subsidy bias is thought to affect WTP, this bias can be added to WTP. The endowment effect[12] and the imaginary bias, on the other hand, affect WTA, and we subtract these two from the original WTA. As a result, WTP becomes ¥ 3,569, and WTA ¥ 4,191; therefore, the discrepancy decreases to ¥ 623, which was originally ¥ 4,283. The discrepancy between the original average value and the average after bias is considered as being ¥ 1,741 for WTA, and -¥ 1,919 for WTP. This implies that both contain almost the same amount of bias (see Table 4). It is difficult to answer the question as to whether WTP or WTA is appropriate.

This analysis suggests that in evaluating medical technology which contains a subsidy bias, it is necessary to measure both WTP and WTA. The true value lies somewhere in between.

5 Conclusions

This paper applies the CVM method which is often used in Environmental Economics to evaluate medical technology, and we discuss its applicability. We focus on the discrepancy between WTP and WTA, which has been observed in previous studies, and decompose it to the endowment and income effect, according to the method utilized by Morrison [21, 22]. Then, the endowment effect is decomposed into various factors. As a result, the part which is thought to be the endowment effect, contains subsidy and imaginary biases, and the endowment effect itself is not very large.

We also examine whether WTP and WTA should be selected in evaluating medical technology. When various biases are considered, the discrepancy between WTP and WTA becomes so small that it is difficult to choose one from the other. Therefore, in evaluating medical technology which contains the subsidy bias, both WTP and WTA should be measured, since true value lies in between them.

---

[12] The imaginary bias itself affects both WTP and WTA, but the issue here is whether "WTA is more unrealistic than WTP". We can therefore conclude that it has a bias only to WTA.

References

1. Bishop, R.C., Heberlein, T.: Measuring values of extra-market goods: are indirect measures biased? American Journal of Agr. Economics **61**, 926–930 (1979)
2. Bleichrodt, H., Johannesson, M.: The validity of QALYs: an experimental test of constant proportional trade-off and utility independence. Medical Decision Making **17**, 21–32 (1996)
3. Brookshire, D.S., Randall, A., Stoll, J.R.: Valuing increments and decrements in natural resource service flows. American Journal of Agr. Economics **62**, 478–488 (1980)
4. Broome, J.: Qalys. Journal of Public Economics **50**, 149–167 (1993)
5. Donaldson, C.: Willingness to pay for publicly-provided goods: a possible measure of benefit. Journal of Health Economics **9**, 103–118 (1990)
6. Donaldson, C., Shackley, P., Abdalla, A.: Using willingness to pay to value close substitutes: carrier screening for cystic fibrosis revisited. Health Economics **6**, 145–159 (1997)
7. Gafni, A.: Willingness-to-pay as a measure of benefits: relevant questions in the context of public decision making about health care programs. Medical Care **29**, 1246–1252 (1991)
8. Golan, E.H., Shechter, M.: Contingent valuation of supplemental health care in Israel. Medical Decision Making **13**, 302–310 (1993)
9. Hanemann, W.M.: Willingness to pay and willingness to accept: how much can they differ? American Economics Review **81**, 635–647 (1991)
10. Johannesson, M., Jonsson, B., Borgquist, L.: Willingness to pay for antihypertensive therapy: results of a Swedish pilot study. Journal of Health Economics **10**, 461–474 (1991)
11. Johannesson, M., Johannesson, P.O., Kristrom, B., Gerdtham, U.G.: Willingness to pay for antihypertensive therapy: further results. Journal of Health Economics **12**, 95–108 (1993)
12. Kahneman, D., Knetsch, J., Thaler, R.: Experimental tests of the endowment effect and the coase theorem. Journal of Political Economy **98**, 1325–1348 (1990)
13. Kahneman, D., Tversky, A.: Prospect theory: an analysis of decision under risk. Econometrica **47**, 263–291 (1979)
14. Kartman, B., Andersson, F., Johannesson, M.: Willingness to pay for reductions in angina pectoris attacks. Medical Decision Making **16**, 248–253 (1996)
15. Kartman, B., Stalhammar, N., Johannesson, M.: Valuation of health changes with the contingent valuation method: a test of scope and question order effects. Health Economics **5**, 531–541 (1996)
16. Knetshe, J.L.: The endowment effect and evidence of non-reversible indiscrepancy curves. American Economics Review **79**, 1277–1284 (1989)
17. Knetshe, J.L.: Environmental policy implications of disparities between willingness to pay and compensation demanded measures of value. Journal of Environmental Economic Management **18**, 227–237 (1990)
18. Knetshe, J.L., Sinden, J.A.: Willingness to pay and compensation demanded: experimental evidence of an unexpected disparity in measures of value. Quarterly Journal of Economics **99**, 507–521 (1984)
19. Knetshe, J.L., Sinden, J.A.: The persistence of evaluation disparity. Quarterly Journal of Economics **102**, 691–695 (1987)
20. Mitchell, R.C., Carson, R.T.: Using surveys to value public goods: the contingent valuation method. Washington, DC: Resources for the Future/Johns Hopkins University Press 1989
21. Morrison, G.C.: Willingness to pay and willingness to accept: some evidence of an endowment effect. Applied Economics **29**, 411–417 (1997)
22. Morrison, G.C.: Understanding the disparity between WTP and WTA: endowment effect, substitutability, or imprecise preference? Economic Letters **59**, 189–194 (1998)
23. O'Brien, B.J., Novosel, S., Torrance, G., Streiner, D.: Assessing the economic value of a new antidepressant: a willingness-to-pay approach. PharmacoEconomics **8**, 34–45 (1995)
24. Olsen, J.A., Smith, R.: Who have been asked to value what? A review of 54 WTP-based surveys on health and health care. Paper presented to the Health Economists' Study Group Conference at University College Galway, July 1998
25. Phillips, K.A., Homan, R.K., Luft, H.S., Hiatt, P.H., Olson, K.R., Kearney, T.E., Heard, S.E.: Willingness to pay for poison control centers. Journal of Health Economics **16**, 343–357 (1997)
26. Randall, A., Stoll, J.R.: Consumer's surplus in commodity space. American Economic Review **70**, 449–455 (1980)

27. Rowe, R.E., d'Arge, R.C., Brookshire, D.S.: An experiment on the economic value of visibility. Journal of Environmental Economics Management **7**, 1–19 (1980)
28. Tolley, G.D., Kenkel, D., Fabian, R.: Valuing health policy: an economic approach. Chicago: University of Chicago Press 1994
29. Torrance, G.W.: Measurement of health state utilities for economic appraisal: a review. Journal of Health Economics **5**, 1–30 (1986)
30. Tsuji, M., Miyahara, S., Taoka, F., Teshima, M.: An estimation of economic effects of tele-home-care: hospital costs-saving of the elderly. Proceedings of the 10th World Congress of Medical Informatics, 825–833 (2001)
31. Tsuji, M., Teshima M., Mori, T.: Applications of telecommunications and multimedia technology in the fields of medicine and education: an international comparison based on field research of local governments. Osaka Economic Papers **49**, 1–21 (1999)
32. Tsuji, M., Suzuki, W., Taoka, F., Kamata, H., Ousaka, H., Teshima, M.: An empirical analysis of the assessment of the tele-home-care system and burden of costs based on a survey on Kamaishi City. Japan Journal of Medical Informatics (in Japanese) **22**, 79–86 (2002)
33. Viscusi, W.K., Evans, W.N.: Utility functions that depend on health status: estimates and economic implications. American Economic Review **80**, 353–374 (1990)

Similarity of endowments
and the factor price equalization condition[*]

Kwan Koo Yun

Economics Deptartment, SUNY at Albany, Albany, NY 12222, USA
(e-mail: yun@albany.edu)

Received: January 2, 2002; revised version: July 1, 2002

Summary. We give a geometric interpretation of the lens condition, proposed by Deardorff as a shortcut for checking the factor price equalization (FPE) condition. We identify the conditions under which the lens condition implies the FPE condition. If the FPE zone is not a neighborhood of the diagonal allocations, however, the lens condition is irrelevant despite the implication since the FPE condition (hence the lens condition) is unlikely to be satisfied in that case. We give precise conditions under which the lens condition is equivalent to the FPE condition and simultaneously, the FPE zone is a neighborhood of the diagonal allocations.

Keywords and Phrases: Factor price equalization, Lens condition, Similarity of endowments.

JEL Classification Numbers: F11.

1 Introduction

In a Heckscher-Ohlin model, the countries share constant returns to scale production technologies with no joint output. Goods are mobile across countries while factors are immobile. There are no transportation costs or tariffs. An integrated economy is a hypothetical economy where the factors as well as the goods are mobile. For simplicity, assume that the integrated economy has a unique competitive equilibrium. Then, the international economy has an equilibrium with equalized factor prices if and only if the integrated economy equilibrium outputs can be produced by the countries, using only their own factor endowments and using the integrated economy techniques (FPE condition).

---

    * The author would like to thank an anonymous referee, Bruce Dieffenbach and Michael Jerison for helpful comments.

In the two-country, two-good, two-factor case, the FPE condition is equivalent to the condition that the country endowments are in a parallelogram defined by the factor use vectors at the integrated economy equilibrium. The parallelogram contains the diagonal allocations where the relative compositions of country endowments are the same. The parallelogram has an interior (containing the diagonal allocations) if and only if the factor use vectors are linearly independent. In this case, factor prices equalize if the endowment compositions are sufficiently similar.

How do these statements generalize to a higher dimensional international economy? To answer this, we study the geometry of the factor price equalization zone (FPE zone) of endowment distributions that allows the international economy to reproduce the integrated economy equilibrium production. The FPE zone is a polytope (the convex hull of a finite number of elements in a Euclidean space) containing the diagonal allocations. The FPE zone is a neighborhood of the diagonal allocations if and only if there are at least as many goods as factors and the factor use matrix is of full rank. When the FPE zone is a neighborhood of the diagonal allocations, the factor prices are equalized if the relative compositions of endowments are sufficiently similar.

Deardorff [2] proposes the lens condition as a shortcut for checking the FPE condition: that the factor endowment lens is contained in the factor use lens. We give an interpretation of the factor use lens as a projection of the FPE zone. Similarly, the factor endowment lens is a projection of the polytope 'generated' by an endowment distribution. The factor use lens and the factor endowment lens are centrally symmetric. Using the lens condition, we give an algebraic condition under which factor prices do not equalize at an international equilibrium.

The lens condition is equivalent to the FPE condition if and only if either there are only two countries or the rank of the factor use matrix is equal to one, two or the number of goods. When the FPE zone is not a neighborhood of the diagonal allocations, however, the FPE zone is of smaller dimension than the endowment allocation space. Then, the lens condition is irrelevant despite the equivalence since the FPE condition (hence, the lens condition) is unlikely to be satisfied in that case. We identify the conditions under which the lens condition is equivalent to the FPE condition and, simultaneously, the FPE zone is a neighborhood of the diagonal allocations.

2 The geometry of the factor price equalization zone

Consider a Heckscher-Ohlin model with n goods, f factors and m countries. The world endowment of factors is ω, a strictly positive vector. Assume that there is a unique competitive equilibrium in the integrated economy and let v be the non-negative $n \times f$ factor use matrix in that equilibrium. The j_{th} row of v, v_j, is the factor use vector in producing good j. Let V be the non-negative $m \times f$ matrix of country endowments. The i_{th} row of V, V_i, is the factor endowment vector of country i. The full employment of endowments at the equilibrium means that $1_n v = 1_m V$, where $1_n \equiv (1, 1, \cdots, 1)$ is a row vector in R^n. A production share matrix λ is an $m \times n$ non-negative matrix satisfying $1_m \lambda = 1_n$. The matrix assigns country i to produce the fraction λ_{ij} of the integrated economy production of good

j. The i_{th} row of λ is λ_i. The space of allowable production share matrices is: $\Lambda \equiv \{\lambda \in R^{mn} | \lambda \geq 0, \sum_{i=1}^{m} \lambda_i = 1_n\}$.

Given a factor use matrix v, we ask what endowment distributions among countries would allow the international economy to duplicate it. Writing $\Lambda v \equiv \{\lambda v | \lambda \in \Lambda\}$, this is possible if and only if :

$$V \in \Lambda v \tag{1}$$

We call the relation (1) the *factor price equalization condition* and Λv the *factor price equalization zone*. If $V = \lambda v$ where $\lambda \in \Lambda$, then $1_m V = 1_m \lambda v = 1_n v$.

We study the factor requirement function that maps the production assignments (to countries) to factor requirements for countries to carry out the assignments using the integrated economy equilibrium production techniques. We can consider Λ as consisting of $\lambda = (\lambda^1, \cdots, \lambda^n)$, where λ^j, the j_{th} column of λ, belongs to the $m - 1$ dimensional standard simplex Δ_{m-1} in R^m. Thus, $\Lambda = \Delta_{m-1} \times \cdots \times \Delta_{m-1} \subset R^{mn}$. Since Λ is a product of polytopes, it is a polytope. The set of endowment allocations is: $M \equiv \{V \in R^{mf} | V_i \geq 0, \text{ all } i \text{ and } \sum_i V_i = \omega\}$. Since M is a bounded polyhedron, it is a polytope [8]. The spanned manifold of Λ, denoted $\mathcal{M}(\Lambda)$, is the set of all finite sums $\sum_{\iota=1}^{s} \alpha_\iota x(\iota)$ where $x(\iota) \in \Lambda$, for all ι and $\sum_{\iota=1}^{s} \alpha_\iota = 1$. The spanned manifold of M, $\mathcal{M}(M)$, is likewise defined. It is straightforward to check that $\mathcal{M}(\Lambda) = \{(x_1, \cdots, x_m) | x_i \in R^n, \text{ each } i \text{ and } \sum_i x_i = 1_n\}$ and $\mathcal{M}(M) = \{(y_1, \cdots, y_m) | y_i \in R^f, \text{ each } i \text{ and } \sum_i y_i = \omega\}$. For $x = (x_1, \cdots, x_m)$ in $\mathcal{M}(\Lambda)$, define: $L : x \longmapsto (x_1 v, \cdots, x_m v)$. It is easy to check that $L(\Lambda) \subset M$ and $L(\mathcal{M}(\Lambda)) \subset \mathcal{M}(M)$. The dimension of $\mathcal{M}(\Lambda)$ is $(m-1)n$ and that of $\mathcal{M}(M)$ is $(m-1)f$.

We define $D \equiv \{V \in R^{mf} | V_i = c_i \omega, \text{ for each } i \text{ where } c_i > 0 \text{ and } \sum_i c_i = 1\}$, the set of *diagonal allocations* of M. At a diagonal allocation, the relative compositions of endowments of countries are the same. If an allocation V is near D, the relative compositions of country endowments are similar.

Lemma 1. *The affine map $L : \mathcal{M}(\Lambda) \longrightarrow \mathcal{M}(M)$ is onto if and only if rank $v = f$.*

Proof. (Only if) If L is onto, then $n \geq f$. Choose any $c \in R^f$ and $(c, \omega - c, 0, \cdots, 0) \in \mathcal{M}(M)$. Then, there is $r_1 \in R^n$ such that $r_1 v = c$. Thus, v is onto R^f. (If) Consider $(y_1, \cdots, y_m) \in \mathcal{M}(M)$. Since $rank\ v = f$, there is r_i in R^n satisfying $r_i v = y_i$, for each i. We have $\sum_i y_i = (\sum_i r_i)v = \omega$. Let $r \equiv \sum_i r_i$ and $x_i \equiv r_i - \frac{1}{m}r + \frac{1}{m}1_n$, for each i. Then, $x_i v = y_i$, for each i and $\sum_i x_i = 1_n$. $\square$

The following theorem generalizes a result in the two country, two good world to higher dimensions. It describes when and in what sense, the factor prices equalize if the endowments of countries are sufficiently similar.

Theorem 2. *The factor price equalization zone $L(\Lambda)$ is a polytope in M containing the diagonal allocations. The FPE zone is a neighborhood of the diagonal in the endowment space if and only if $rank\ v$ is equal to f, the number of factors.*

Proof. The FPE zone $L(\Lambda)$ is a polytope since it is the affine image of a polytope. If V is in D, there are strictly positive numbers $\{c_i\}$ such that $\sum_{i=1}^{m} c_i = 1$ and $V_i = c_i\omega$, for all i. Choosing $\lambda_i = c_i 1_n$, for each i, we have $\lambda v = V$. Thus, $D \subset L(\Lambda)$. Note that λ is in the interior of Λ in $\mathcal{M}(\Lambda)$. If L is onto, L maps an open neighborhood of λ in Λ onto an open neighborhood of V in $\mathcal{M}(M)$ by the open mapping theorem. If L is not onto $\mathcal{M}(M)$, $L(\Lambda)$ is contained in a proper affine subspace of $\mathcal{M}(M)$ and is not full dimensional in $\mathcal{M}(M)$. The theorem now follows from Lemma 1. $\square$

We investigate the symmetry of $L(\Lambda)$. Let $\mathcal{A}$ be the class of non-negative, $m \times m$ matrices whose rows add up to 1_m. The class includes permutation matrices which contain only 0 or 1 in each entry and whose columns as well as rows add up to 1_m. We denote by e_i a column vector that has 1 in its i_{th} coordinate and 0 elsewhere.

Proposition 3. *An $m \times m$ matrix A defines a map $A : \Lambda \to \Lambda$ by $\lambda \to A\lambda$ if and only if A is in $\mathcal{A}$. Thus, if $V = (V_1, \cdots, V_m)$ is in $L(\Lambda)$, AV is also in $L(\Lambda)$ for any A in $\mathcal{A}$. In particular, $(V_{P(1)}, \cdots, V_{P(m)})$ is also in $L(\Lambda)$ for any permutation P of $\{1, \cdots, m\}$.*

Proof. (If) For any A in $\mathcal{A}$ and λ in Λ, $A\lambda$ is also in Λ since $A\lambda \geq 0$ and $1_m A\lambda = 1_m\lambda = 1_n$. (Only if) Suppose that $A\lambda$ is in Λ for each $\lambda \in \Lambda$. By choosing e_i as the first column of λ, Ae_i, the i_{th} column of A, is non-negative and $1_m Ae_i = 1$ for $i = 1, \cdots, m$. If V is in $L(\Lambda)$, $V = \lambda v$ for some $\lambda \in \Lambda$. Then, $AV = A\lambda v$ is also in $L(\Lambda)$ for any A in $\mathcal{A}$. $\square$

The symmetry of $L(\Lambda)$ implies that the projection of $L(\Lambda)$ to any i_{th} f coordinates has the same image. This image is the factor use lens discussed next.

3 The factor price equalization zone and the lens condition

Can we reduce the FPE condition to a simpler one, perhaps by exploiting the symmetry property of the FPE zone? When there are two countries and two goods (and any number of factors), for any $\lambda \in \Lambda$,

$$\lambda v = \begin{bmatrix} t_1 & t_2 \\ 1 - t_1 & 1 - t_2 \end{bmatrix} \begin{bmatrix} v_1 \\ v_2 \end{bmatrix} = \begin{bmatrix} t_1 v_1 + t_2 v_2 \\ \omega - t_1 v_1 - t_2 v_2 \end{bmatrix} \tag{2}$$

Here, $\omega = v_1 + v_2$. Thus, λv is completely determined by the parallelogram $P = \{t_1 v_1 + t_2 v_2 | 0 \leq t_1 \leq 1, 0 \leq t_2 \leq 1\}$. The condition $V \in \Lambda v$ is equivalent to $V_1 \in P$.

A straightforward approach to extending this idea to higher dimension is to consider the higher dimensional analog of an Edgeworth box. Given v, let $\mathcal{L}_m(v) \equiv \{(\lambda_1 v, \cdots, \lambda_{m-1} v) | \lambda_i \geq 0,$ all i and $\sum_{i=1}^{m-1} \lambda_i \leq 1_n\}$. Now, consider the projection map $\pi_{-m} : (x_1, \cdots, x_{m-1}, x_m) \in M \mapsto (x_1, \cdots, x_{m-1})$. The map π_{-m} on M is a homeomorphism onto its image since $(x_1, \cdots, x_{m-1}) \to (x_1, \cdots, x_{m-1}, \omega - \sum_{i=1}^{m-1} x_i)$ is a continuous inverse to π_{-m}. Since $(V_1, \cdots, V_{m-1}) = \pi_{-m}(V)$ and $\mathcal{L}_m(v) \equiv \pi_{-m}(L(\Lambda))$, $V \in L(\Lambda)$ if and only if $(V_1, \cdots, V_{m-1}) \in \mathcal{L}_m(v)$.

Proposition 4. *The FPE condition is satisfied if and only if $(V_1, \cdots, V_{m-1}) \in \mathcal{L}_m(v)$.*

In Equation (2), if V_1 is in P, $\{0, V_2, V_1 + V_2\}$ are also contained in P. Since the parallelogram P is convex and since the convex hull of $\{0, V_1, V_2, V_1 + V_2\}$ is $\{t_1V_1 + t_2V_2 | 0 \leq t_1 \leq 1, 0 \leq t_2 \leq 1\}$, $V_1 \in P$ if and only if $\{t_1V_1 + t_2V_2 | 0 \leq t_1 \leq 1, 0 \leq t_2 \leq 1\} \subset P$. For the case of m countries and n goods, Deardorff uses $\mathcal{L}(v) \equiv \{t_1v_1 + \cdots + t_nv_n | 0 \leq t_i \leq 1, \text{all } i\}$ and $\mathcal{L}(V) \equiv \{t_1V_1 + \cdots + t_mV_m | 0 \leq t_i \leq 1, \text{all } i\}$. Such a generalized parallelogram is called a *parallelotope* [1] and a *lens* by Deardorff. When $m = 2$, $\mathcal{L}_m(v)$ reduces to $\mathcal{L}(v)$. The relation $\mathcal{L}(V) \subset \mathcal{L}(v)$ is called the (Deardorff) *lens condition*.

A polytope P is *centrally symmetric* if there is a *center* c in P such that $c + x \in P$ implies $c - x \in P$. We describe the symmetric property of a lens. Let $I_n \equiv \{x \in R^n | 0 \leq x_i \leq 1, i = 1, \cdots, n\}$.

Proposition 5. *Suppose the aggregate endowment is ω. The factor use lens $\mathcal{L}(v)$ and the factor endowment lens $\mathcal{L}(V)$ are centrally symmetric polytopes with center at $\frac{1}{2}\omega$, containing the diagonal set $\{c\omega | 0 \leq c \leq 1\}$.*

Proof. Suppose that for some $a \in I_n$, $av = [(a^1, \cdots, a^n) - (\frac{1}{2}, \cdots, \frac{1}{2})]v + \frac{1}{2}\omega$. Then, $[-(a^1, \cdots, a^n) + (\frac{1}{2}, \cdots, \frac{1}{2})]v + \frac{1}{2}\omega = (1 - a^1, \cdots, 1 - a^n)v \in I_n v$. For each c, $0 \leq c \leq 1$, $(c, \cdots, c)v = c\omega$. A similar argument holds for the factor endowment lens. $\square$

If V is in $L(\Lambda)$, $\mathcal{A}V \equiv \{AV | A \in \mathcal{A}\}$ is contained in $L(\Lambda)$. The set $\mathcal{A}$ is a polytope since it is a product of simplices. Then, $\mathcal{A}V$ is a polytope since it is an affine image of a polytope. We say that the polytope $\mathcal{A}V$ is *generated* by V. Given $x = (x_1, \cdots, x_i, \cdots, x_m)$, where x_i is in R^f, consider the projection map $\pi_i : (x_1, \cdots, x_i, \cdots, x_m) \mapsto x_i$. For each country i, $\pi_i(L(\Lambda)) = \{\lambda_i v | \lambda \in \Lambda\} = \mathcal{L}(v)$ and $\pi_i(\mathcal{A}V) = \{a_i V | A \in \mathcal{A}\} = \mathcal{L}(V)$, where a_i is the i_{th} row of A. If $V \in L(\Lambda)$, then $\mathcal{A}V \subset L(\Lambda)$ and consequently, $\mathcal{L}(V) = \pi_i(\mathcal{A}V) \subset \pi_i(L(\Lambda)) = \mathcal{L}(v)$. Thus, the FPE condition implies the lens condition as Deardorff notes [2]. The converse, conjectured by Deardorff, is not true in general.

Lemma 6. *Suppose that the rank of v (resp. V) is r. Then, $\mathcal{L}(v)$ (resp. $\mathcal{L}(V)$) is contained in an r-dimensional subspace of R^f and is full dimensional there.*

Proof. By definition, $\mathcal{L}(v) = I_n v$. If $rank\ v = r$, then v^t, the transpose of v, maps R^n onto an r-dimensional subspace of R^f. Since I_n has an interior in R^n, v^t maps the interior of I_n onto an open set in the r-dimensional subspace of R^f. The same argument applies for V and $\mathcal{L}(V)$. $\square$

If $rank\ v < f$, $\mathcal{L}(v)$ is contained in a lower dimensional subspace of R^f, making it difficult to fit $\mathcal{L}(V)$ within $\mathcal{L}(v)$. We obtain a condition sufficient for the factor prices not to equalize at an equilibrium of the international economy.

Corollary 7. *Assume that there is a unique competitive equilibrium in the integrated economy. If $rank\ v < rank\ V$, the factor prices do not equalize at an equilibrium of the international economy.*

Proof. By Lemma 6, $rank\, v < rank\, V$ implies that the dimension of $\mathcal{L}(V)$ is larger than the dimension of $\mathcal{L}(v)$. Since the lens condition $\mathcal{L}(V) \subset \mathcal{L}(v)$ is necessary for factor price equalization, the factor prices cannot equalize in an equilibrium. $\square$

The lens condition implies the FPE condition if there are only two countries (Deardorff [2], Dixit-Norman [3]) or if there are at least as many factors as goods and the factor use matrix v is of full rank or if there are only two goods (Demiroglu and Yun [6]). However, there are cases where the lens condition is satisfied but the FPE condition is not (Demiroglu and Yun [6]). Qi [4], Xiang [7], Wong and Yun [5]) show that the lens condition implies the FPE condition when there are only two factors. We give an outline of Wong and Yun's constructive proof.

Lemma. Consider a Heckscher-Ohlin international economy with m (≥ 2) countries, n (≥ 2) goods and two factors. Assume that V_i, $i = 1, \cdots, m$ and v_j, $j = 1, \cdots, n$, are non-negative and non-zero, that $\sum_i V_i$ is strictly positive and that $\sum_i V_i = \sum_j v_j$. Then, the lens condition ($I_m V \subset I_n v$) implies the factor price equalization condition ($V \in \Lambda v$).

Outline of proof. Let $\{V_1, \cdots, V_m\}$ and $\{v_1, \cdots, v_n\}$ be groups of non-negative and non-zero vectors in R^2, each indexed in the order of increasing steepness such that $\sum_i V_i$ is strictly positive and is equal to $\sum_j v_j$. Let $s_i \equiv V_1 + \cdots + V_i$, $i = 1, \cdots, m$, and $s \equiv s_m$. Let Σ (resp. Γ) represent the arc, linearly joining $0, s_1, \cdots, s_m$ (resp. $0, v_1, \cdots, v_1 + \cdots + v_n$) in order. The curve Σ (resp. Γ) is the lower half of the boundary of the centrally symmetric lens formed by $\{V_1, \cdots, V_m\}$ (resp. $\{v_1, \cdots, v_n\}$). Thus, the lens condition is equivalent to the condition that Σ is above Γ; for each $(x_1, x_2) \in \Sigma$, there is $(x_1, x_2') \in \Gamma$ such that $x_2 \geq x_2'$. One can show that there exists x in Γ satisfying $x' \equiv x + V_m \in \Gamma$. For a, b in Γ with $a \leq b$, let $\Gamma[a, b] \equiv \{\gamma \in \Gamma | a \leq \gamma \leq b\}$ and define $\Sigma[a, b]$ similarly. We assign the production of $\Gamma[x, x']$ to country m. The resource requirement of the production assignments is V_m. Let $\Gamma[x', s] - V_m$ denote the translation of $\Gamma[x', s]$ by $-V_m$ and let $\Gamma_{m-1} \equiv \Gamma[0, x] \cup (\Gamma[x', s] - V_m)$. Since $\Gamma[x', s]$ forms the steepest part in Γ, line segments in Γ_{m-1} are ordered in increasing steepness. Also, $x + (s - x') = s_{m-1}$. Again, $\Sigma[0, s_{m-1}]$ is above Γ_{m-1}. Thus, the lens formed by $\{V_1, \cdots, V_{m-1}\}$ is contained in the lens whose lower boundary is Γ_{m-1}. We can now repeat the above procedure until country 1 is assigned Γ_1 which requires V_1. $\square$

Combining previous results, the lens condition implies the FPE condition either if there are two countries or two goods or two factors or if there are as many factors as goods and the rank of the factor use matrix v is full. Are there any other sufficient conditions? The following result states that, essentially, this is all we can ask of the lens condition.

Theorem. (Wong and Yun [5]) Consider a Heckscher-Ohlin international economy with m countries, n goods and f factors where m, n, f are two or more. Assume that V_i, $i = 1, \cdots, m$ and v_j, $j = 1, \cdots, n$, are non-negative and non-zero, $\sum_i V_i$ is strictly positive and that $\sum_i V_i = \sum_j v_j$. The lens condition is equivalent to the factor price equalization condition for the economy if and only if either $m = 2$ or $rank\, v \leq 2$ or $rank\, v = n$.

Outline of the Proof. We have seen that the FPE condition implies the lens condition. Therefore, for the equivalence of the lens condition and the FPE condition, we need to establish the sufficiency of the lens condition for the FPE condition.

(If) Suppose there are two countries. From the lens condition, $V_1 = \lambda_1 v$, for some row vector λ_1 in I_n. Then, $V_2 = (1_n - \lambda_1)v$. If $rank\ v = n$, $v^t : R^n \to R^k$ is one to one. From the lens condition, there is λ_i in I_n satisfying $V_i = \lambda_i v$, for each country i. Since v^t is one to one, $\sum_i V_i = 1_n v = \sum_i \lambda_i v$ imply $1_n = \sum_i \lambda_i$. If $rank\ v = 1$, all rows of v are proportional to $\sum_j v_j$. From the lens condition, each V_i is also proportional to $\sum_j v_j$. From this, the FPE condition obtains trivially. If $rank\ v = 2$, the image of $v^t : R^n \to R^f$, denoted as L, is two dimensional. It is straightforward to show that two linearly independent rows, say $\{v_{j*}, v_{j**}\}$, of v generate $v^t(R_+^n)$ by non-negative linear combinations. Let $h : L \to R^2$ be the bijective linear map defined by $h(v_{j*}) = (0,1)$ and $h(v_{j**}) = (1,0)$. Then, $h \circ v^t(I_n) \subset h \circ v^t(R_+^n) = R_+^2$. In particular, $\widetilde{v}_j \equiv h(v_j)$ and $\widetilde{V}_i \equiv h(V_i)$ are non-negative for all i, j. Since $\{v_j, V_i\}$ are non-zero vectors and h is an injective linear map, $\{\widetilde{v}_j, \widetilde{V}_i\}$ are non-zero. Using the linearity of h and $\sum_i V_i = \sum_j v_j$,
$$\sum_i \widetilde{V}_i = h\left(\sum_i V_i\right) = h\left(\sum_j v_j\right) = \sum_j \widetilde{v}_j \geq \widetilde{v}_{j*} + \widetilde{v}_{j**} = (1,1).$$ Applying the above Lemma to the lens condition $h \circ V^t(I_m) \subset h \circ v^t(I_n)$, there are $\{\lambda_{ij}\}, \lambda_{ij} \geq 0$, all i, j such that $\sum_i \lambda_{ij} = 1$, for each j and $\sum_j \lambda_{ij}\widetilde{v}_j = \widetilde{V}_i$, for each i. Since h is linear, $h(\sum_j \lambda_{ij}v_j) = h(V_i)$. Since h is one to one, $\sum_j \lambda_{ij}v_j = V_i$ for each i. So, the FPE condition is satisfied.

(Only if) We construct a counter-example for each case that does not meet the conditions of either $m = 2$ or $rank\ v \leq 2$ or $rank\ v = n$. Consider any natural numbers m, n, f, r satisfying $m > 2$ and $n > r > 2$ and $f \geq r$.

Choose $v_1 = (\overbrace{1, 0, \cdots, 0}^{r}, \overbrace{\frac{1}{r}, \cdots, \frac{1}{r}}^{f-r})$, $v_2 = (0, 1, 0, \cdots, 0, \frac{1}{r}, \cdots, \frac{1}{r}), \cdots, v_r = (0, 0, \cdots, 0, 1, \frac{1}{r}, \cdots, \frac{1}{r})$ and $v_j = \frac{1}{n-r}(1, 1, \cdots, 1)$, $j = r+1, \cdots, n$. Also,

choose $V_1 = (\overbrace{1, 1, 0, \cdots, 0}^{r}, \overbrace{\frac{2}{r}, \cdots, \frac{2}{r}}^{f-r})$, $V_2 = (1, 0, 1, \cdots, 1, \frac{r-1}{r}, \cdots, \frac{r-1}{r})$ and $V_i = \frac{1}{m-2}(0, 1, 1, \cdots, 1, \frac{r-1}{r}, \cdots, \frac{r-1}{r})$, $i = 3, \cdots, m$. Note that $\sum_i V_i = \sum_j v_j = (2, 2, \cdots, 2)$. One can show that the above $\{v_j\}$ and $\{V_i\}$ satisfy the lens condition. However, the FPE condition is not satisfied: Each $v_j, j > r$ requires all factors whereas no country has all factors in its endowment. Thus, for each $j > r$, no country can reproduce any part of v_j alone. $\square$

The above theorem gives exact conditions under which the lens condition implies (and thus is equivalent to) the FPE condition. When the FPE zone is not a neighborhood of the diagonal allocations, however, the FPE zone is contained in a lower dimensional affine subspace of $\mathcal{M}(M)$. In such a situation, the FPE condition (hence the lens condition under the same conditions) is not likely to be satisfied. The FPE zone has zero measure in $\mathcal{M}(M)$ while M has a positive Hausdorff measure since it has an interior in $\mathcal{M}(M)$. If the endowment distribution is chosen randomly in M according to a probability density function defined on M, it has zero probability of being in the FPE zone. Combining Theorem 2 with the theorem of Wong and Yun,

Theorem 8. *Consider a Heckscher-Ohlin international economy with m countries, n goods and f factors where m, n, f are two or more. Assume that $V_i, i = 1, \cdots, m$, $v_j, j = 1, \cdots, n$, are non-negative and non-zero, $\sum_i V_i$ is strictly positive and that $\sum_i V_i = \sum_j v_j$. The lens condition is equivalent to the factor price equalization condition and the factor price equalization zone is a neighborhood of the diagonal allocations if and only if EITHER $m = 2$ and $rank\, v = f$ OR $rank\, v = f = 2$ OR $rank\, v = n = f$.*

Proof. The lens condition is equivalent to the FPE condition if and only if $m = 2$ or $rank\, v \leq 2$ or $rank\, v = n$. On the other hand, the FPE is a neighborhood of the diagonal in the allocation space if and only if $rank\, v = f$. The theorem follows from combining the conditions. For example, combining $rank\, v \leq 2$ with $rank\, v = f$, we have $2 \leq f = rank\, v \leq 2$. $\square$

References

1. Brondsted, A.: An introduction to convex polytopes. Berlin Heidelberg New York: Springer 1983
2. Deardorff, A.: The possibility of factor price equalization revisited. Journal of International Economics **36**, 167–175 (1994)
3. Dixit, A.K., Norman, V.: Theory of international trade. Cambridge: Cambridge University Press 1980
4. Qi, L.: Conditions for factor price equalization in the integrated world economy. Review of International Economics (forthcoming)
5. Wong, S.-k., Yun, K.K.: The lens condition with two factors. Review of International Economics (forthcoming)
6. Demiroglu, U., Yun, K.K.: The lens condition for factor price equalization. Journal of International Economics **47**, 449–456 (1999)
7. Xiang, C.: The sufficiency of the lens condition for factor price equalization in the case of two factors. Journal of International Economics **53**, 463–474 (2001)
8. Ziegler, G.M.: Lectures on polytopes. Berlin Heidelberg New York: Springer 1995

Domestic and international strategic interactions in environment policy formation[*]

Kazuharu Kiyono[1] and Masahiro Okuno-Fujiwara[2]

[1] Faculty of Political Science and Economics, Waseda University, 1-6-1 Nishi-waseda, Shinjuku-ku, Tokyo, 169-8050, JAPAN
(e-mail: kazr@waseda.jp)
[2] Faculty of Economics, University of Tokyo, 7-3-1 Hongo, Bunkyo-ku, Tokyo 113-0033, JAPAN
(e-mail:fujiwara@e.u-tokyo.ac.jp)

Received: January 16, 2001; revised version: April 16, 2002

Summary. In this paper, we establish the most possilbe general formulation of the technology governing carbon-gas emission, giving rise to global external diseconomies, and ty to explore into the strategic interactions, both domestic and international, when an individual country decides on the environmental policies. Through the comparison among emission taxes, quotas, and standard in the perfectly competitive private economies, we find that the first two policies are equivalent but they are different in effects by virtue of what we may call the tax-exemption effect of emission standards. Such a difference in the policy effect further affects the other country's welfare through the global externalities, amplified through whether the government can precommit to either the emission tax or the emission standard.

Keywords and Phrases: Global warming, Emission tax, Emission quota, Emission standard, Strategic interaction, Tax exemption.

JEL Classification Numbers: D61, D78, H23, H77.

1 Introduction

The traditional literature on environment regulations as well as the standard policy evaluations assumes that the government can precommit to a certain policy variable before the private sector's decision. However the recent literature on the

---

[*] The authors thank the valuable comments by an anonymous referee. Ministry of Education and Science for its financial support is also greatly acknowledged.
Correspondence to: K. Kiyono

politicaly economy effects governing the policy formations emphasizes that there are no policies free from the strategic behavior by the private sector[1] . Surprisingly, there has been few literatures in environment economics to capture such theoretical perception.

Furthermore, the analysis in environment economics often employs too specified models concering the technology conditions concerning substitution between emissions of pollutants and other standads of inputs. The obtained results often hinge on the assumed specifications on abatement of pollutants, and the models employed fail to clarify the essential factors leading to those results.

In this paper, we formalize the economic role of pollutants yielding international external diseconomies, which we call the global warming gas emission[2] , as the unpaid factor of production à la Meade [4] in the most possibly general fashion, and elucidate the effects of tradtional emission control measures such as emission taxes, standards and quotas. This formulation, as we will see in this paper, enables us to clarify the differences among those three environment regulations in the easiest and clearest way.

The differences in the policy effects do no stem only from the government choice among the three emisson control measures but from the sequential-move structure in the government and private decisions. Those differences further yield differences in each country's policy decision through internationally strategic interdependece through global external diseconomies. In fact, a country's policy switch from emission taxes to emission standards controls directly affects the other country's choice of the effective emission tax rate, for such a policy by a country induces its private sector to produce more through the emission tax exemption effect under the emission standard policy and thus increases the marginal environment damage facing the other country.

We formalize the model and clarify the difference among the above emission control measures in Section 2. Comparison of the policy effects is done among them in Section 3, elucidating the tax exemption effect inherent to the emission standard policy in Section 3. In Section 4, we discuss the effect of strategic interactions between the government and the private sector on the formation of the domestic environment regulations. We will find that it critically affects the resulting equilibrium whether the government can precommit to a certain level of the emission control measure, particularly in the case of emission standards. Such differences in policy tools and the sequetial-move structure in the domestic policy formation lead to different equilibria in the presence of internationally strategic interactions as we will show in Section 5.

2 Basic model

We first establish a basic model for the discussion throughout the paper. Consider a world consisting of two countries, 1 and 2. Although the two countries are in

---

[1]   See Grossman and Helpman [1] for example.

[2]   See Barrett [5] for the literature and the issues involved in global warming.

autarky with regard to trade and investment[3] , they are involved in international interdependence through the globally wide-spread external diseconomies of global warming. Let us first construct each country's economic structure.

Each country produces and consumes two goods, X and Y. Sector Y produces good Y using the standard inputs, say labor, subject to constant returns to scale. By choosing the units of good Y, we may assume that a unit of good Y is produced with a unit of labor, so that when we choose good Y as the numeraire the wage rate is equal to unity, and the earned profit should be equal to zero under perfect competition as we formerly assume later.

Production of good X requires labor and emits global warming gas. We model this production structure à la Meade [4] so that emission of global warming gas is an unpaid factor of production. Take either of countries 1 and 2 and inquire into its cost structure of sector X. For the time being, we focus our attention to a closed economy until Section 5, so that we omit superscripts and subscripts i representing the name of the country.

2.1 Unpaid factors and emission taxes

As with sector X, let ℓ denote the input of labor, z the emission of global warming gas, and $f(\ell, z)$ the production function. We also let t denote the factor price of global warming gas emission, and x the output level. Then solution of the standard cost minimization problem yields the following total cost function:

$$C(x,t) = \min_{\ell, z} \{\ell + tz | f(\ell, z) \geq x\} .$$

In the absence of the environmental regulations, the private sector does not have to pay the factor costs on the global warming gas emissions, for they are unpaid factors in the sense of Meade [4]. But once the government imposes the emission taxes, the private sector needs to pay those taxes just as the prices of the standard factors of production. Thus theoretically speaking, the emission tax rate is nothing but the factor price of global warming gas emission.

2.2 Properties of cost functions

By Shepherd's lemma, $C_t(x, t)$ (where the subscript t indicates the partial derivative with respect to t) represents the emission of global warming gas by sector X when the economy produces x units of good X given the emission factor price t, which we often denote by $z(x, t)$. We assume:

Assumption 1. The total cost function $C(x, t)$ satisfies:

(A 1-1) $C(x, t)$ is strictly increasing and convex in x and strictly increasing and concave in t. In particular, $C_{tt}(x, t) < 0$, i.e., $z_t(x, t) < 0$.
(A 1-2) $C_{tx}(x, t) > 0$, i.e., $z_x(x, t) > 0$.

---

[3] Ishikawa and Kiyono [3] discusses the case in the presence of commodity trade.

(A 1-3) $C_{ttx}(x, t) < 0$, i.e., $z_{xt}(x, t) < 0$.
(A 1-4) $C_{txx}(x, t) \geq 0$, i.e., $z_{xx}(x, t) \geq 0$.

Assumption 1-1 is all standard. Especially, it says the demand for gas emission is decreasing in its own price given the output level. Assumption 1-2 implies that the global warming gas emission is a normal factor of production. That is, the global warming gas emission increases along with the output. Hereafter we call $C_{tx}(x, t) = z_x(x, t)$ the *marginal emission coefficient,* for it represents the increase in the global warming gas emission caused by a marginal increase in the output.

Assumption 1-3 then implies that the marginal emission coefficient is decreasing in the emission tax rate. This is because such a raise in the emission tax rate induces substitution between global warming gas and labor in production.

Lastly, Assumption 1-4 implies that the marginal emission coefficient is not decreasing in the output.

For the purpose of describing the production sector, we assume[4] :

Assumption 2. The markets are all competitive.

Then since sector Y is perfectly competitive subject to constant returns to scale, it earns zero profits. To describe sector X, we must specify what policies are employed against its global warming gas emission.

2.3 Emission controls

We are concerned with the following three types of emission controls on good X sector throughout this paper.

- Emission taxes
- Emission standards
- Emission quotas

Emission taxes.   The first policy is to impose a certain rate of emission tax, say t, on sector X. Then the profit earned in sector X is expressed by:

$$\tilde{\pi}^T(x, t, p) \overset{\text{def}}{=} px - C(x, t), \tag{1}$$

where p represents the market price of good X.

The associated government surplus, denoted by G_T, is given by:

$$G_T = tC_t(x, t). \tag{2}$$

Let us call the sum of the private profit (1) and the government surplus (2) the *social profit* earned in sector X. Then it is given by

$$\tilde{\pi}^T(x, t, p) + G_T = px - C(x, t) + tC_t(x.t). \tag{3}$$

---

[4] See Kiyono [2002] for the discussion on imperfect competition.

Emission standard. The second policy is to regulate the allowable volume of global warming gas emission for each possible output level. That is, it is expressed by a function $z = \phi(x)$ to assign an allowable volume of global warming gas emission to each output.

One of the extreme types of this policy is a control on the total volume of global warming gas emissions, which we call the *emission quota*. This is expressed by a constant-value function, i.e., $\phi(x) = \bar{z}$ where $\bar{z}$ is a positive constant.

In the succeeding discussion, however, our emission standard policies excludes those emission quotas, and focus our attention on the output-emission schedules which realize under emission tax policies. More specifically, we confine our attention to the output-emission schedule $(x, z) = (x, C_t(x, t))$. We call the emission standard expressed by such an output-emission schedule the *tax-equivalent emission standard*. Then the associated emission tax t represents the effective emission tax rate associated with the emission standard, which we call the *shadow emission tax rate*.

Given the shadow emission tax rate t, the profit earned in sector X is given by

$$\tilde{\pi}^S(x, t, p) \overset{\text{def}}{=} px - C(x, t) + tC_t(x, t). \tag{4}$$

The critical difference from the emission tax policy is that the firms are now exempt from the emission tax payment, which is shown by the third refund term.

The associated government surplus, denoted by G_S, is given by:

$$G_S = 0, \tag{5}$$

so that the social profit earned in sector X under the emission standard policy is expressed the same as in the emission tax policy (see (3)).

Emission quotas. The last policy is to impose a certain quota on the global warming gas emissions. Given this policy, the firm is constrained to produce the output subject to the emission quota. Using the production function, the profit maximization problem facing the representative firm is expressed as follows:

$$\max_{\{\ell, z, a\}} \{pf(\ell, z) - \ell | z \le \bar{z}\}, \tag{6}$$

where $\bar{z}$ denotes the emission quota.

As we have clarified, the firm can choose the combination of the unpaid factor of global waring gas emissions, z, and labor, ℓ. The choice depends on the relative factor price given by t. When the firm is not taxed on its emissions but it chooses the emission-labor pair different from the one given zero rate of emission tax, it takes into account what we may think as the shadow price of the emissions, the rate of which is equal to the marginal rate of technical substitution between emissions and labor. We express this shadow price with t as under the emission tax policy, which plays the same role as the shadow emission tax rate under the emission standard.

When the shadow emission tax rate is equal to t, the total production cost, denoted by $\bar{C}(x, t)$, is given by

$$\bar{C}(x, t) \overset{\text{def}}{=} C(x, t) - tC_t(x, t). \tag{7}$$

The first term represents the direct production cost given the emission tax rate t. But since the firm actually does not pay emission taxes, the associated tax payment is exempted as shown by the second term. Using the above cost function, the profit maximization problem under the quota, i.e., (7), is described in the following equivalent fashion.

$$\max_{\{x,t\}} \left\{ px - \bar{C}(x,t) | C_t(x,t) \leq \bar{z} \right\}. \tag{8}$$

In view of (4), the above problem can further be rewritten as below:

$$\max_{\{x,t\}} \left\{ \tilde{\pi}^S(x,t,p) | C_t(x,t) \leq \bar{z} \right\} \tag{9}$$

That is, the shadow emission tax rate plays the same role as the effective emission tax rate for the tax-equivalent emission standard. The government surplus under the emission quota policy, denoted by G_Q, is now given by

$$G_Q = 0, \tag{10}$$

so that the social profit earned in sector X is again the same as under the emission tax policy subject to the additional emission quota constraint.

2.4 Consumption and welfare

We consider the representative consumer, whose utility is assumed to measure the country's welfare, too. She consumes goods X and Y, and enjoys utility given by

$$U(x_c) - px_c + y_c - \theta D(Z), \tag{11}$$

where x_c (or y_c) denotes the consumption of good X (or Y), Z the world total emission of global warming gas, $D(Z)$ the world damage from global warming, and $\theta(> 0)$ the rate at which the individual (and thus the country) perceives as the own damage.

The utility function $U(x_c)$ is assumed to satisfy all the standard assumptions:

Assumption 3. The utility function $U(x_c)$ satisfies:

(A 3-1) $U(x_c)$ is strictly increasing.
(A 3-2) $U(x_c)$ is twice-continuously differentiable.
(A 3-3) $U(x_c)$ is strictly concave.

As with the world environment damage function $D(Z)$, we assume:

Assumption 4. The world damage from global warming satisfies:

(A 4-1) $D(Z)$ is strictly increasing.
(A 4-2) $D(Z)$ is twice-continuously differentiable.
(A 4-3) $D(Z)$ is strictly convex.

The individual maximizes the own utility (11) subject to

$$px_c + y_c = m, \tag{12}$$

where m denotes the own income. The first-order condition for utility maximization yields:

$$U'(x_c) = p, \tag{13}$$

which defines the inverse market demand function, $p = P(x_c) \, (= U'(x_c))$.

The total income m is the sum of (i) the profits earned in sectors X and Y, (ii) the government surplus which is transferred to the individual in a lump-sum fashion, and (iii) the total wage income. Since sector Y earns zero profits and the social profit earned in sector X is given by the righty-hand side of (3), the total income is in fact give by

$$m = px - C(x, t) + tC_t(x, t) + \bar{L}, \tag{14}$$

where L represents the constant initial labor endowment.

Let z^* denote the global warming gas emission by the foreign country. Then since $Z = z(x, t) + z^*$ holds, substitution of (12) and (14) into the utility function (11) gives rise to the following form of the country's welfare function:

$$\tilde{V}(x, t, z^*) = U(x) - C(x, t) + tC_t(x, t) - \theta D \left(C_t(x, t) + z^* \right), \tag{15}$$

where the wage income is left out from the above equation, for the labor endowment is assumed constant.

3 Comparison of three emission controls

Let us explore into the differences among the three policies
 – Emission taxes
 – Tax-equivalent emission standards
 – Emission quotas
using (1), (4) and (9).

3.1 Emission taxes

Let us first inquire into the properties of the market equilibrium under the emission tax policy. In view of the profit function (1) and the market demand function (13), the market equilibrium requires:

$$U'(x) - C_x(x, t) = 0. \tag{16}$$

The above condition defines the equilibrium output as a function of the emission tax rate, which we express by $x_T(t)$. In view of Assumptions 1 and 3, it satisfies:

$$x_T'(t) = \frac{C_{xt}(x, t)}{U''(x) - C_{xx}(x, t)} < 0, \tag{17}$$

where $x = x_T(t)$ on the RHS. That is, an increase in the emission tax rate decreases the equilibrium output through an increase in the marginal production costs. Such a relation is illustrated by the curve $x_T x_0$ in Figure 1.

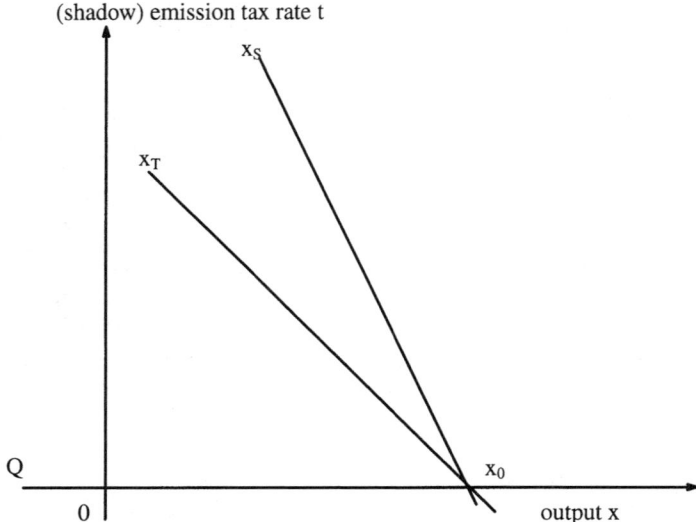

Figure 1. Emission taxes, standards and quotas

3.2 Emission standards

Second, take the case of the tax-equivalent emission standard with the shadow emission tax rate t. In view of the profit function (4), the equilibrium output should satisfy:

$$U'(x) - C_x(x,t) + tC_{tx}(x,t) = 0, \tag{18}$$

where we assume:

Assumption 5. The marginal production cost given any non-negative effective emission tax rate is increasing in the output,i.e.,

$$C_{xx}(x,t) - tC_{txx}(x,t) > 0.$$

This condition again defines the equilibrium output as a function of the shadow emission tax rate, which we express by $x_S(t)$. It should satisfy:

$$x'_S(t) = \frac{-tC_{ttx}(x,t)}{U''(x) - C_{xx}(x,t) + tC_{txx}(x,t)} < 0, \tag{19}$$

where use was made of Assumption 5.

Since the firms in sector X produces more than under the emission tax, the volume of global warming gas emission also becomes greater under the emission standard. Thus we have established[5] :

---

[5] This effect of emission standards are found by some authors, for example [6]. [2] also discusses the effects of pollution standards in various forms, but she does not explore into the differences in their effects on the outputs compared with the emission taxes, because she fixes the total output.

Proposition 1. *Given the same effective emission tax rate, the equilibrium output as well as the emission volume becomes greater under the tax-equivalent emission standard than under the emission tax.*

This result has been derived previously, but with restrictive technology conditions concerning emissions and other standard inputs[6] .

3.3 Emission quotas

Lastly consider the effect of imposing an emission quota given by, say $\bar{z}$. Given the market price p, the firm is constrained to choose the emission-labor combination subject to the imposed quota constraint. That is, its profit maximization problem is given by

$$\max_{\{\ell,z\}} \{pf(\ell,z) - \ell | z \leq \bar{z}\} \tag{20}$$

One should note that the choice of the emission-labor combination given the output level is equivalent to the choice of the marginal rate of technical substitution between emissions and labor insofar as the output is kept constant. When we express this marginal rate of technical substitution with t, the above constrained profit maximization is equivalent to:

$$\max_{\{x,t\}} \{px - C(x,t) + tC_t(x,t) | C_t(x,t) \leq \bar{z}\} \tag{21}$$

Let $\mathcal{L} = px - C(x,t) + tC_t(x,t) + \lambda(\bar{z} - C_t(x,t))$ where λ is the Lagrangian multiplier associated with the emission quota constraint. Then the first-order conditions for profit maximization are:

$$0 = p - C_x(x,t) + (t-\lambda)C_{tx}(x,t)$$
$$0 = (t-\lambda)C_{tt}(x,t)$$
$$0 = \bar{z} - C_t(x,t)$$

These conditions establish:

$$t = \lambda$$
$$p = C_x(x,t)$$
$$\bar{z} = C_t(x,t)$$

Note that the first condition in fact represents that the shadow emission tax rate t is, in fact, equal to the shadow factor price of the emission. And we find that when the market equilibrium condition is imposed, i.e., $p = U'(x)$, the second and third conditions above are equivalent to the equilibrium condition under the emission tax policy with the emission tax rate t. Thus we have ascertained in our framework the following tax-quota equivalence[7] .

---

[6] See Ulph [6] for example, which clearly points out the tax-exemption effect of emission standards.

[7] However this result critically depends on the assumption of perfectly competitive market. See Kiyono [3].

Proposition 2. *The emission tax and the emission quota policies are equivalent in the sense of achieving the same resource allocation insofar as the market is competitive and the resulting volume of the global warming gas is the same.*

Since the emission tax and the emission quota are equivalent, we confine our attention only to possible differences between the emission tax and emission standard policies in the succeeding discussion. We now make a further step in the next section to clarify their difference in the welfare effects by taking an explicit account of the strategic interactions between the government and the private sector in decision making.

4 Domestic strategic interactions in environment regulations

Insofar as we live in a market economy, the government cannot completely control the resource allocation. In the present case of environment regulations, the government can decide on the emission control measures but cannot directly enforce any other decision governing the market. Such an inherently private decision variable is the output in our framework. And the national welfare is affected by both decisions of the government and the private sector. Thus we must resort to the game theory for exploring into how the environment regulations are formed as well as the resulting resource allocation.

But before such an inquiry, it is of a great use to make clear what is the first-best outcome as the reference state of the world.

4.1 First-best outcome

Partial differentiation of the national welfare function (15) yields

$$\tilde{V}_x(x, t, z^*) = U'(x) - C_x(x, t) + (t - \theta D'(\cdot)) C_{tx}(x, t) \tag{22}$$

$$\tilde{V}_t(x, t, z^*) = (t - \theta D'(\cdot)) C_{tt}(x, t) \tag{23}$$

$$\tilde{V}_{z^*}(x, t, z^*) = -\theta D'(\cdot) \tag{24}$$

The first-best outcome, represented by the output-tax pair (x_B, t_B), should satisfy:

$$\tilde{V}_x(x_B, t_B, z^*) = 0,$$
$$\tilde{V}_t(x_B, t_B, z^*) = 0,$$

so that in view of (22) and (23) the following conditions must hold:

$$U'(x_B) - C_x(x_B, t_B) = 0 \tag{25}$$

$$t_B - \theta D' (C_t(x_B, t_B) + z^*) = 0 \tag{26}$$

The first-best outcome depends on the foreign emission of global warming gas z^*. However in view of (25) and (26), the first-best output depends only on the emission tax rate t. And the relation between the output and the emission tax

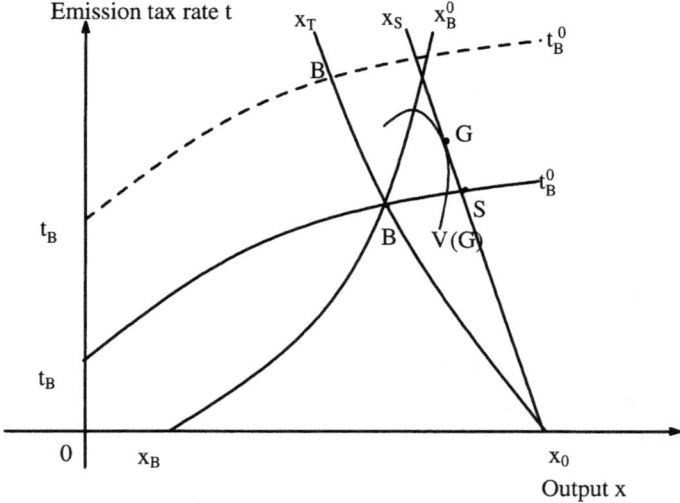

Figure 2. Emission controls in a closed economy

rate governed by (25) is the same as in the market equilibrium condition under the emission tax policy, (16), i.e, $x = x_T(t)$. Thus the effect of the change in the foreign emission is confined onto the first-best emission tax rate. We represent such a relation by $t_B = t_B(z^*)$ and $x_B = x_T(t_B)$.

The first-best outcome given the foreign emission is illustrated in Figure 2. The curve $t_B t'_B$ shows the emission tax rate satisfying $\tilde{V}_t(x, t, z^*) = 0$ for each output x given z^*. This schedule is upward sloping, for an increase in the output increases the country's perceived marginal environment damage through an increase in the associated emission. We call this schedule the *best emission-tax schedule*.

On the other hand, the curve $x_B x'_B$ shows the output maximizing the welfare given the emission tax rate and the foreign emission, i.e, satisfying $\tilde{V}_x(x, t, z^*) = 0$ for each emission tax rate given z^*. We call it the *best output schedule*. This schedule is, at least around its intersection with the best emission-tax schedule, upward-sloping. This is because an increase in the emission tax rate decreases the emission of global warming gas, leading to a decrease in the marginal damage due to the output increase[8].

The intersection of the two best schedules, point B, shows the first-best outcome given the foreign emission z^*.

In the following discussion, many of the equilibria are found along the best emission-tax schedule. But along the schedule, the volume of emission becomes greater as the emission tax rate gets higher. This is because the best emission tax rate is always equal to the country's marginal environment damage $\theta D'(\cdot)$, which

---

[8] More rigorously, let $\hat{x}(t, z^*)$ be the solution to $\tilde{V}_x(x, t, z^*) = -0$. Then application of the implicit function theorem yields:

$$\hat{x}_t(t, z^*) \propto (1 - \theta D'')C_{tt}C_{tx} + (t - \theta D')C_{ttx},$$

where the RHS is positive by virtue of Assumptions 1 and 4.

is strictly increasing in the output given the foreign emission under Assumption 4, $D''(\cdot) > 0$. We often resort to this result in the succeeding discussion, so that we sum it up in the following lemma.

Lemma 1. *Along the best emission-tax schedule, a rise in the emission tax rate leads to an increase in the global warming gas emission.*

4.2 Game-theoretic formulation

Given this feature of the first-best outcome, let us now inquire into the strategic interaction between the government and the private sector in forming the environment regulations. Its game theoretic exploration requires us to specify

  (i) Strategies of the players, the government and the private sector,
 (ii) Sequential-move structure specifying who moves first in decision.

Emission tax and standard games.   Since the private sector is confined to decide on the output, the first issue hinges on which emission control measure the government chooses, either the emission tax or the emission standard. When the government chooses the emission tax rate as its policy variable, the associated game is called the *emission tax game*. And when the tax-equivalent emission standard is chosen in the form of the shadow emission tax rate, the associated game is called the *emission standard game*.

Simultaneous-move and government-leader games.   As with the second issue, the problem is whether each player can precommit to the own strategic variable in advance. In the standard literature on the policy evaluation, the government is often assumed to move first as the so-called Stackelberg leader against the private sector. We call such a game the *government-leader game*.

  However as has been explored in the recent literature on the political economy of trade and industrial policies, the private sector may act as the leader and strategically distort the succeeding policy decision process. We may call it the *private-leader game*.

  In the present competitive market economy, this private-leader game is difficult to examine theoretically insofar as we keep the commodity market structure perfectly competitive. This is because the private sector's precommitment to a certain output level requires cartel formation by the firms in sector X, which changes the market essentially into a monopoly. For this reason, we do not take up the case of the private-leader game in the present paper[9].

  Thus the last possibility is that the two players move simultaneously, which we call the *simultaneous-move game*.

4.3 Emission tax game

Let us first take up the emission tax game in a simultaneous-move form. To obtain the equilibrium, we must define each player's reaction function. As with the gov-

---

[9] See the discussion of the private-leader game in Kiyono [3] for the case of imperfect competition.

ernment, it shows the best emission tax rate given the output. Clearly it is given by the best emission-tax schedule $t_B t'_B$ in Figure 2. As with the private sector, it represents the profit-maximizing output given the emission tax rate, which is given by $x_T(t)$ derived from the market equilibrium condition given the tax rate, (16). Thus by superimposing the curve $x_T x_0$ in Figure 1 on Figure 2, its intersection with the best-emission tax schedule gives the non-cooperative equilibrium for the simultaneous-move emission-tax game.

The equilibrium in fact coincides the first-best outcome B. This is because the market equilibrium condition (16) governing the private sector's reaction function requires that the tax-inclusive marginal production cost be equal to the market price, as is required for the first-best outcome (25).

As with the government-leader game, the government chooses the output-tax pair along the private sector's reaction curve so as to maximize the national welfare. Since the private sector's reaction curve passes through the first-best outcome point B, the associated equilibrium is again the first-best state.

We may summarize the above discussion in the following proposition.

Proposition 3. *Consider the emission-tax game. Then in either the simultaneous-move game or the government-leader one, the first-best outcome realizes as the equilibrium.*

Let us consider the effect of an increase in the foreign emission on the equilibrium. Since it has no effect either on the market demand or on the costs, the private sector's reaction function does not change even with a change in the foreign emission. It affects the government's reaction function only. An increase in the foreign emission increases the country's marginal environment damage, leading to a raise in the best emission tax rate. This implies an outward shift of the government's reaction curve to, say $t^*_B t^{*\prime}_B$ in Figure 2. This tax-increasing incentive of the government raises the equilibrium emission tax rate at the new equilibrium B^*, which decreases the output through an increase in the tax-inclusive marginal production costs. The country's emission of global warming gas emission falls, but the world total emission increases, which is reflected in the tax-rate increase.

The total effect on the welfare is negative. To demonstrate it, define the equilibrium national welfare function in the emission tax game as below:

$$V_T(z^*) \stackrel{\text{def}}{=} \tilde{V}\left(x_B(t_B(z^*)), t_B(z^*), z^*\right).$$

Then application of the envelope theorem yields:

$$V'_T(z^*) = \tilde{V}_{z^*}(\cdot) = -\theta D'(\cdot) < 0.$$

The results are summarized in the following proposition.

Proposition 4. *Consider the emission-tax game either in the simultaneous-move form or in the government-leader one. An increase in the foreign emission leads to (i) an increase in the emission tax rate, (ii) a decrease in the output, (iii) a decrease in the country's emission, (iv) an increase in the world total emission, and (v) the lower welfare.*

4.4 Emission standard game

Next consider the emission standard game first in the simultaneous-move form and then in the government-leader form.

Simultaneous-move game. In the emission standard game, the government chooses the shadow emission tax rate. Since a change in its rate alters the welfare as expressed by (23) and the government maximizes the welfare using only t, its reaction function is the same as in the emission tax game. That is, it is given by the best emission-tax schedule in Figure 2.

On the other hand, the private sector is now exempted from emission taxes, so that its output choice is governed by the associated market equilibrium condition (18). That is, its reaction function is given by $x_S(t)$, and is expressed by the curve $x_S x_0$ in Figure 1. Superimpose it over Figure 2. Then its intersection with $t_B t'_B$, denoted by S, shows the resulting non-cooperative equilibrium.

Since the private sector has an incentive to produce more under the tax-equivalent emission standard, the resulting equilibrium entails the higher (shadow) emission tax rate and the greater output than under the emission tax policy. Furthermore, the equilibrium S is located along the best emission-tax schedule, so that in view of Lemma 1 the country's emission also becomes greater. Thus we have established[10]:

Proposition 5. *Consider the simultaneous-move emission standard game. Then compared with the emission tax game, the equilibrium entails (i) the higher (shadow) emission tax rate, (ii) the greater output, (iii) the greater emission, and (iv) the lower welfare.*

The effect of an increase in the foreign emission is made clear as in the case of the emission tax policy. Its effects on the resource allocation are qualitatively the same as under the emission tax game, so that we now explore into its welfare effect. Let

$$V_S(z^*) \stackrel{\text{def}}{=} \tilde{V}\left(x_S(t_B(z^*)), t_B(z^*), z^*\right)$$

represent the equilibrium welfare in the present simultaneous-move emission standard game. Then in view of (18), application of the envelope theorem leads to:

$$
\begin{aligned}
V'_S(z^*) &= -tC_{tx}(x_S(t), y)x'_S(t)t'_B(z^*) - \theta D'(C_t(x_S(t), t) + z^*) \\
&= t\{1 + C_{tx}(\cdot)x'_S(t)t'_B(z^*)\} \\
&= -\frac{t}{1 - \theta D''(\cdot)}\{1 - \theta D''(\cdot)C_{tt}(\cdot)\} < 0,
\end{aligned}
$$

where $t = t_B(z^*)$ and use was made of $t_B = \theta D'(\cdot)$ from (26), $x'_S(t) < 0$ from (19), and Assumption 1. Thus, we have established:

---

[10] The third result has been noticed by [6] but only within a too specified technology conditions. In fact, his equilibrum is the one associated with the government-leader game in the succeeding section.

Proposition 6. *Consider the simultaneous-move emission standard game. Then an increase in the foreign emission leads to (i) the higher shadow emission tax rate, (ii) the smaller output, (iii) the smaller emission of the country, (iii) the greater world total emission, and (iv) the lower national welfare.*

Government-leader emission standard game. Now consider the case in which the government moves first. As we explained in the emission tax game, when the government moves as Stackelberg leader against the private sector, it choose the output-tax pair along the private sector's reaction curve $x_S x_0$ that maximize the national welfare. Such a point is given by point G where the iso-national welfare curve $V(G)$ touches the private sector's reaction curve.

The government understands in advance that the private sector's best-response output is decreasing in the shadow emission tax rate, so that a raise in the shadow emission tax rate leads to a decrease in the output. Since the output is socially excessive in production, the government attempts to precommit to the higher shadow emission tax rate than under the simultaneous-move game. This leads to the higher shadow emission tax rate and the smaller output than when it cannot make precommitment. The higher shadow emission tax rate reduces the equilibrium output and emission. But one should note that it is ambiguous in general whether the emission volume is smaller than under the emission tax policy.

The results are summarized in the following proposition.

Proposition 7. *Consider the government-leader emission standard game. Compared with the simultaneous-move game, the equilibrium entails (i) the higher shadow emission tax rate, (ii) the smaller output, and (iii) the smaller emission of global warming gas. But the country's emission volume may or may not be greater than under the emission tax game.*

Lastly consider the effect of an increase in the foreign emission. As under the emission tax game, it shifts the government's reaction curve outward to $t_B^* t_B^{*\prime}$. The graphical analysis is impotent to clarify the effect, so that we resort to the algebra. Given the private sector's reaction function $x_S(t)$ the national welfare is expressed in terms of the shadow emission tax rate t and the foreign emission z^* as below:

$$V_G(t, z^*) \overset{\text{def}}{=} \tilde{V}\left(x_S(t), t, z^*\right). \tag{27}$$

The first-order condition for welfare maximization with respect to the shadow emission tax rate t is given by:

$$\begin{aligned} 0 &= \frac{\partial V_G(\cdot)}{\partial t} \\ &= \tilde{V}_x(\cdot)x_S'(t) + \tilde{V}_t(\cdot) \\ &= -tC_{tx}(x, t)x_S'(t) + (t - \theta D'(\cdot))\, z_S'(t). \end{aligned} \tag{28}$$

This defines the equilibrium shadow emission tax rate as a function of the foreign emission, which we express by $t_G(z^*)$. (28) requires that this equilibrium shadow emission tax rate satisfy:

$$t_G(z^*) = \left\{ 1 + \frac{C_{tx}(\cdot)x_S'(t)}{C_{tt}(\cdot)} \right\} \theta D'(\cdot). \tag{29}$$

Application of the implicit function theorem to (28) implies:

$$t'_G(z^*) \propto -\theta D''(\cdot) z'_S(t) > 0,$$

which implies that an increase in the foreign emission leads to an increase in the shadow emission tax rate. Given this effect, the resulting effects on the resource allocation are the same under the emission tax game.

As with the welfare effect, the effect is determinate like in the simultaneous-move game. This is because the government decides on the shadow emission tax rate by taking the distortion of socially excessive production into account in advance. More rigorously speaking, application of the envelope theorem in view of (28) yields:

$$\frac{dV_G\left(t_G(z^*), z^*\right)}{dz^*} = \tilde{V}_{z^*}(\cdot) = -\theta D'(\cdot) < 0.$$

Thus we have established:

Proposition 8. *Consider the government-leader emission standard game. Then an increase in the foreign emission leads to (i) the higher shadow emission tax rate, (ii) the smaller output, (iii) the smaller emission of the country, and(iii) the lower national welfare.*

5 Non-cooperative emission controls

Let us extend the results in the previous closed economy case to an open economy one and explore into the strategic interdependence between two countries, 1 and 2. There are several possible cases for inquiry depending on the emission control measures employed by each country and the sequential-move structure in each country. However the discussion in the preceding sections implies that the following two common properties hold for every possible equilibrium in each country.

(Property i)   Given the foreign emission, the country's emission of global warming gas is decreasing in the emission tax rate as well as the shadow one.

(Property ii) An increase in the foreign emission leads to an increase in the country's emission tax rate as well as its shadow one.

Thus each country's best response (shadow) emission tax is decreasing in the other's. We represent this best response emission tax for country i by $R_i(t_j)$ $(i, j = 1, 2; j \neq i)$, and call it country i's reaction function. The associated reaction curve is depicted by the downward-sloping curves R_i $(i = 1, 2)$ in Figure 3.

As we have discussed in the previous sections, the equilibrium (shadow) emission tax rate given the foreign one, or alternatively the shape of the reaction curve depends on (i) the emission control measures employed by the governments and (ii) the sequential-move structure in each country's policy game. However, for making the discussion in the succeeding section sensible enough, we assume:

Assumption 6. In each possible case, each country's reaction curve has a slope greater than -1, so that the resulting equilibrium is stable.

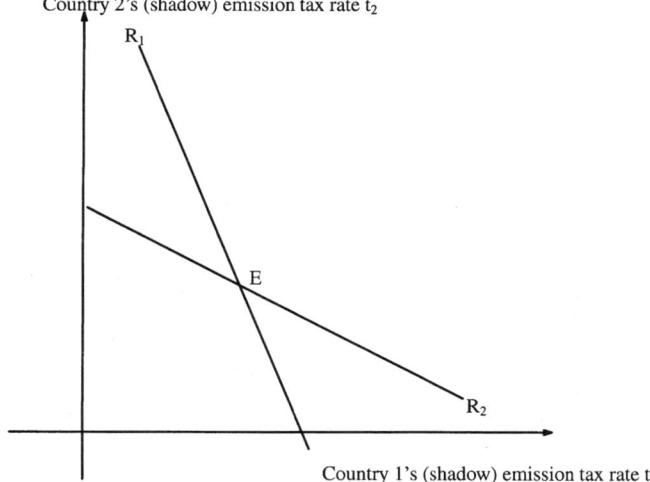

Country 2's (shadow) emission tax rate t_2

R_1

E

R_2

Country 1's (shadow) emission tax rate t_1

Figure 3. Non-cooperative equilibrium in international strategic interdependence

Let us first inquire into the case in which the private sector and the government move simultaneously.

5.1 Simultaneous-move games in each country

The first issue of interest is how a country's choice of emission control measures affects the equilibrium. As will be shown later, a country's policy choice affects the other country's best response, which has not been noticed in the previous literature.

We begin with the case in which both countries employ emission taxes. Country i's reaction curve for the emission tax is given by R_i^{TT} for $i = 1, 2$ and the associated equilibrium by E_{TT} in Figure 4.

Now what if country 1 switches to emission standards? The discussion in Proposition 5 implies that the (shadow) emission tax rate becomes higher. That is, country 1's reaction curve shifts outward, from R_1^{TT} to, say R_1^{ST} in the figure. But this is not the end of the story. For, the same proposition suggests that country 1's emission increases along with such a policy switch. This result, coupled with Property ii, implies that country 2 also has an incentive to raise the emission tax rate, which leads to the outward shift from R_2^{TT} to, say R_2^{ST} in the figure. The resulting equilibrium is now given by point E_{ST}.

When country 2 also switches to emission standards, the reaction curves shift further outward. Country 1's shifts to, say R_1^{SS}, and country 2's to, say R_2^{SS} with the associated equilibrium E_{SS}.

It is of importance to note that each government's best response tax should be set equal to the own marginal environment damage regardless of its emission control measures given the simultaneous-move structure in each country's policy game, i.e.,

$$t_i = \theta_i D'(Z) \ (i = 1, 2), \tag{30}$$

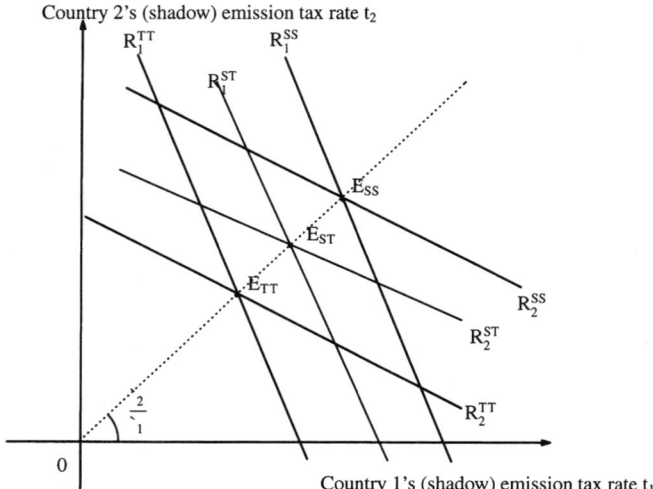

Figure 4. Simultaneous moves by the private sectors and the governments

so that the resulting equilibrium emission taxes should always satisfy:

$$\frac{t_1}{t_2} = \frac{\theta_1}{\theta_2}. \tag{31}$$

Outward shifts of the reaction curves under (31) leads to an increase in the (shadow) emission tax rate by both countries. Furthermore one should note that such an increase in the emission tax rate should be associated with an increase in the total world emission of global warming gas:

Proposition 9. *Suppose that in both countries the government chooses the emission tax rate as its emission control and the private sector decides on the output simultaneously. Then either country's policy switch from the emission taxes to the emission standards increases the (shadow) emission tax rates of both countries as well as the total world volume of global warming gas.*

How does each country's welfare change when either or both countries swtich from emission taxes to emission standards? Let us first consider the change from the initial emission-tax game equilibrium E_{TT} in Figure 4 to E_{ST} where only country 1 switches to the emission standard. By virtue of Proposition 5, such a policy switch increases country 1's emission, and lowers her own welfare. Proposition 4, then implies that such an increase in country 1's carbon-gas emission worsens country 2's welfare, leadint to the latter's emission-tax rate. Country 2's raise in the emission-tax rate, decreasing her carbon-gas emission, improves country 1's welfare as stated in Proposition 6. The total effect is that country 2 should be worse off at E_{ST} than at E_{TT} but that the change in country 1's welfare is ambiguous. Similar reasoning applies to the change from E_{ST} to E_{SS}. Thus we have established:

Proposition 10. *A country's switch from the emission tax to the emission standard worsens the other country's welfare but it is generally ambiguous whether the policy-switching country may be better off.*

However one can show that both countries are worse off at E_{SS} than at E_{TT}[11]. That is,

Proposition 11. *Compared with when the two countries choose the emission taxes, they both get worse off when they choose the emission standards.*

5.2 Government-leader games

Similar results hold also for the case in which the governments move before the private sector's decision. The difference comes from the equilibrium condition governing the emission tax rates for both countries when a government chooses emission standards and moves first.

As in the previous subsection, start with the case in which both government employ emission taxes. Then the reaction curves are the same as in Figure 4, which are reproduced in Figure 5. Suppose then that only the government of country 1 switches to emission standards. Then as stated in Propositions 6 and 7, country 1 finds an incentive to raise the shadow emission tax rate, so that its reaction curve shifts outward as under the emission tax policy, say to R_1^{ST} in Figure 5. Country 2's reaction curve is also affected for the same reason in the simultaneous-move game, so that it shifts outward to, say R_2^{ST}. The resulting equilibrium is shown by point E_{ST}. In view of (29) and (30), the resulting equilibrium (shadow) emission tax pair should satisfy:

$$\frac{t_1}{t_2} = \frac{\left\{ 1 + \frac{C_{tx}(\cdot)x'_S(t)}{C_{tt}(\cdot)} \right\} \theta_1}{\theta_2},$$

which is greater than $\frac{\theta_1}{\theta_2}$. Thus the policy switch to emission standards by only country 1 raises country 1's (shadow) emission tax rate relatively higher than under the emission tax policies as shown in Figure 5. This establishes:

Proposition 12. *Consider the government-leader game in each country where each government chooses emission taxes initially. Then either country's switch to emission standards raises its shadow emission tax rate relatively and absolutely than under the emission tax game.*

When both countries switch to emission standards, each country's reaction curve shifts further outward to, say R_i^{SS} for $i = 1, 2$. Thus there is an increase in the shadow emission tax rate by at least one country. And when the demand and cost conditions are sufficiently similar between the two countries, we find that there should be an increase in the shadow emission tax rate by each country. When it is the case, in view of Proposition 7 it is straightforward to establish that the shadow emission tax rate of each country should be higher than in the simultaneous-move game. Furthermore, Proposition 8 suggests that Pareto-improving international coordination requires the shadow emission tax rates to be raised further in both countries.

Proposition 13. *Consider the government-leader game in each country.*

---

[11] The proof is available unpon request.

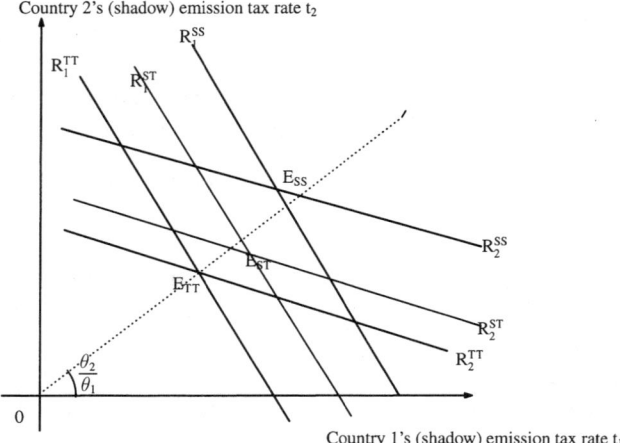

Figure 5. Government-leader game equilibria

1. *Compared with the emission tax game, the country switching to emission standards adopts the relatively and absolutely higher shadow emission tax rate.*
2. *When the demand and cost conditions are sufficiently similar, compared with the
   emission tax game, the shadow emission tax rates are higher in both countries.*
3. *In the above situation, Pareto-improving international coordination in environmental regulations requires each country to raise the shadow emission tax
   rate.*

The last remark is concerning the counter-parts to Propositions 10 and 11 in
the simultaneous-move games. In view of Propositions 7 and 8, it is straightforwar
to see that the counterpart for Proposition 10 holds. But as with the counterpart to
Proposition 11 the result is generally ambiguous because we have to know the sign
and value of the higher-order differentials of the final-demand and cost functions.

Proposition 14. *When the government commits to the environmental policies before the private sector's decision and both countries initially choose emisison taxes,
a country's switch from the emission tax to the emission standard worsens the other
country's welfare but its effect on the policy-switching country is ambiguous.*

6 Concluding remarks

Let us finally discuss possible possible implications of our analysis. The first is
the welfare effect of emission taxes and standards for an individual country. As
already discussed, the emission standards has the tax-exemption effect unlike the
emission taxes, leading to over-production of the final good and thus over-emisison
of carbon-gas. Thus even when the private-sector chooses to reduce the carbon-gas
emission through any voluntary efforts, unless it bears the marginal environment
damages, the country gets worse off because of the allocational inefficiency caused
by the tax-exemption effect.

Second, however, from the view-point of international strategic interactions, such a tax-exemption effect of emission standards may play an important role for a country. This is because in its presence there are two types of distortions facing each country, i.e., (i) the domestic distortion caused by environment damages and (ii) the strategic distortion induced through the global or transboudanry external diseconomies. Tinbergen's theorm of policy assignment tells us that there should be at least two policy variables to remedy those two distortions. A country may choose the emission standard to remedy the domestic distortion and the associated tax-exemption rate to make the best of the strategic position in the international interactions surrounding the enviroment policy formations.

References

1. Grossman, G.M., Helpman, E.: Special interest politics. Cambridge: The MIT Press 2001
2. Helfand, G.: Standards versus standards: The effects of different pollution restrictions. American Economic Review **81**(3), 622–634 (1991)
3. Kiyono, K.: Strategic effects and market structure in environment policy formations. Mimeo (2002)
4. Meade, J.E.: External econonomies and diseconomies in a competitive situation. Economic Journal **62**(245), 54–67 (1952)
5. Barrett, S.: The problem of global environmental protection. Oxford Review of Economic Policy **6**, 68–79 (1990)
6. Ulph, A.: Environmental policy and international trade. In: Carraro, C., Siniscalco D. (eds.) New directions in the economic theory of the environment, ch. 6, pp. 147–192. Cambrigde: Cambrigde University Press 1998

Firm reputation with hidden information[*]

Steven Tadelis

Department of Economics, Stanford University, Stanford, CA 94305-6072, USA
(e-mail stadelis@stanford.edu)

Received: July 31, 2001; revised version: December 20, 2001

Summary. An adverse selection model of firm reputation is developed in which short-lived clients purchase services from firms operated by overlapping generations of agents. A firm's only asset is its name, or reputation, and trade of names is not observed by clients. As a result, names are traded in all equilibria regardless of the economy's horizon The general equilibrium analysis links the value of a name to the market for services. This causes a non-monotonicity that precludes higher types from sorting themselves through the market for names, and leads to "sensible" dynamics: reputations, and name prices, increase after success and decrease after failure.

JEL Classification Numbers: D80, L14.

Keywords and Phrases: Reputation as an asset, Trade of names, Overlapping generations.

1 Introduction

A firm's name, which symbolizes its reputation, is considered to be one of its most valuable assets. A good reputation attracts clients, and often leads to higher prices and larger profit margins. However, by the time a firm's reputation is established, the individuals responsible may have left the firm and others may have replaced them. This is, after all, one interesting feature of firms: the separation of entity (the firm's name) from identity (the agents in the firm). This paper models and

---

[*] I thank Jon Levin, Eric Maskin and Drew Fudenberg for valuable discussions, and Heski Bar-Isaac for comments on an earlier draft. Financial support from the National Science Foundation (NSF grants SBR-9818981 and SES-0079876) is gratefully acknowledged. This paper replaces an older (and incomplete) working paper titled "Reputation with Hidden Information".

investigates the case in which a firm's name can outlive the agents who established it. Two particular questions are addressed: First, why would names, as reputation carriers, have value? And second, can the market for names sort out more able agents who would be willing to pay more for good reputations?

Most of the reputation literature in economic theory focuses on repeated games with complete or incomplete information (see Fudenberg and Tirole, 1991, Ch. 9). That is, a firm can "cooperate" (e.g., collude with other oligopolist in price setting, provide a high quality good to a consumer, or deter entry) or "deviate" (e.g., cut prices to capture market shares, provide a low quality good at a low cost, or accommodate entry). The focus usually is on whether reputational concerns lead firms to act cooperatively, and what conditions support such equilibria. However, the asset-like aspect of a firm's reputation was mostly ignored since it is commonly assumed that the firm is run by a long-lived agent. This in effect treats a firm's reputation as an individual's reputation.

A first attempt to develop a theory of the firm as a bearer of reputation was made by Kreps (1990). He offers a repeated game example that demonstrates how reputation (a firm's name) can become a tradeable asset. However, in his model equilibria exist in which the firm's name has no value and names are not traded. That is, the economic forces that would guarantee names to be valuable assets are not determined. Furthermore, there is no account of how a firm's reputation may increase (or decrease) in value following performance, as is commonly observed in reality.

Aoyagi (1996) develops a model similar in spirit to Kreps', but by introducing incomplete information, and imposing a reasonable belief refinement, he is able to establish a uniqueness result that selects the equilibrium in which a firm's reputation is traded and short lived agents carry the reputation from one generation to another. Interestingly, Aoyagi shows that the ex ante value of a firm with a sequence of short run owners is not lower, and may be higher, than the value when a single long run owner-operator is present.

In this paper I depart from the common game theoretic, partial equilibrium approach, and develop a model in which a large population of short-lived agents provide services to a large population of clients. Agents are characterized by the quality of their services. Hidden information about an agent's type generates asymmetric information, and clients use a firm's track record to form beliefs about the type of agent who is running the firm. The demographics of agents are captured by a heterogeneous overlapping generations model, so that when agents retire they can possibly sell their name (and track record) to other agents. A central assumption is that clients do not observe such trades, and, in a similar spirit, it is assumed that agents can change their names secretly to shed off poor past outcomes. In equilibrium clients will have correct beliefs about this behavior. This setup stresses the separation of a firm's entity from an agent's identity.

The paper offers two main results. First, the separation of entity from identity causes names to be valuable assets – this separation causes good track records to have value, so that they must be traded in *any* equilibrium. The second, more interesting result shows that names cannot act as sorting signals. More precisely, though it may seem that better agents would be able to gain more from acquiring a good reputation, it turns out that sorting more able types in this way is not possible in equilibrium.

Intuitively, when an agent is faced with the option of buying a name, this is weighed against the option of starting with a new name. The better the agent is, the better the future prospects from *both* alternatives. However, the degree to which good names deteriorate after failures, and new names increase in value after success, are determined by clients' beliefs about the composition of types behind these names. As the model demonstrates, when clients' beliefs about the composition of name buyers is too optimistic (successful names are bought primarily by more able agents), then Bayes' updating prescribes that successful names do not deteriorate fast enough after failed outcomes, while new names increase significantly in value after a success. This causes less able agents to have a higher willingness to pay for successful names, which is inconsistent with the optimistic beliefs of clients. Thus, by linking the two markets – the market for services and the market for names – in a general equilibrium framework, the present model is able to uncover novel features of the values of reputations, and the abilities of reputations to sort types. This failure to separate types leads to rather appealing dynamics that are mostly lacking in the standard game-theoretic models. Here, good reputations are created, and not assumed as part of an equilibrium, and these reputations are destroyed following bad performance. Furthermore, new firms establish reputations, and old successful firms eventually have their reputations destroyed.

A related paper is Mailath and Samuelson (2001), in which a firm's type changes over time without clients being aware of this change. This supports Markov Perfect equilibria in which "good" behavior is supported, and leads to sensible dynamics as described above. Their emphasis, however, is not on the forces that create an active market for names, and their model admits an equilibrium in which bad behavior prevails and reputations are not created. They also show that if type changes are endogenous then bad types are likely to value a very good reputation more than good types, demonstrating a no sorting result. Yet, to generate this result in a partial equilibrium framework they exogenously assume that good types have a better outside option, which arises endogenously in the model of this paper. In a recent working paper, Phelan (2001) develops a similar, yet simpler model to that of Mailath and Samuelson (2001) in which the type of agents who run a government changes exogenously, without the awareness of the agents in the economy (which are the clients in this case). A unique Markov Perfect equilibrium exists in which good government behavior increases its reputation, but eventually bad behavior causes the reputation to collapse, and this cycle begins again.

The model developed here extend the ideas presented in Tadelis (1999), in which a two-type finite horizon model is analyzed. However, the simplicity of that model leaves important robustness questions unanswered. This paper adds to the previous work in two important ways. First, a continuum of types is introduced which verifies that the results are not special to the two type case. Second, the analysis establishes the existence of equilibria in the infinite horizon economy, and shows that the qualitative results of the finite horizon model are robust.

The next section presents the model, and Section 3 establishes that names must be traded in all equilibria of a two-period model. Section 4 establishes the no sorting result and Section 5 analyzes the infinite horizon economy. Section 6 concludes and offers some thoughts for future research.

2 The model

In each period a client hires an agent to provide a service (or supply a good) for one period only. Agents differ in their probability of succeeding (e.g., delivering a "quality" good), which is the source of adverse selection. There is a continuum of clients and agents, and each agent i's type determines his probability of success (S), which is denoted by $\theta \in [0, 1]$ (and with probability $1 - \theta$ the agent fails (F)). Assume that the (Lebesgue) measure of agents is 1, i.e., $i \in [0, 1]$, and that for each agent i, θ_i is independently drawn from a uniform distribution over $[0, 1]$.

The price of supplying a service is determined competitively. To simplify, assume that the clients are on the long side of the market so that competition causes each client to pay her full surplus when transacting with an agent.[1] Following the classic adverse selection literature (see, e.g., Akerlof 1970) the following assumption is made,

Assumption A1. *Compensation cannot be based on the transaction's outcome.*

That is, problems of verifiability prevent the parties from writing outcome-contingent contracts. This implies that each client who employs an agent will pay up-front for the expected value of the service supplied. For simplicity assume that all clients are homogeneous; success generates a return of 1, and failure a return of 0.

Each agent in this economy runs a firm, which is represented by a *name*, and it is assumed that no two firms can share the same name. At the beginning of his lifetime an agent has two alternative choices: either *choose a new name* to represent his firm, implying that he will have no initial track record, or *buy an existing name* from an agent who is about to retire, inheriting the track record of that name. A name is, therefore, a vehicle over which clients will form beliefs about the type of agent behind the name, conditional on its track record. That is, clients will observe the track record of each firm, and use this information to update their beliefs. The following central assumption is made:

Assumption A2. *Shifts of name-ownership are not observable by clients.*

This extreme assumption can be weakened, as long as the impact of the current owner on the firm's past performance is uncertain. An alternative can be that clients can observe trade of names with some positive probability that is not too large. This, for example, can be generated from a model in which there are costs associate with observing ownership changes, and these costs are random and $i.i.d.$ across clients. In essence, A2 implies that when a client employs an agent who has a name with a history, she cannot determine whether the agent himself is responsible for that

---

[1] Any division of surplus in which the agents receive a positive fraction will not qualitatively alter the results. The analysis is simplified by assuming a uniform distribution. It will become clear from the proofs that the qualitative results do not depend on this simplifying assumption.

history or whether he has just bought it. This separation of a firm (entity) from an agent (identity) is further stressed by the following two assumptions:

Assumption A3. *At the beginning of each period every active agent can either choose to retain his past name or unobservably change it.*

Assumption A4. *An i.i.d set of agents of measure zero cannot change their name, and all other agents can costlessly change their name.*

Assumption A3 implies that once an agent chooses a new name, then the past record of this agent is erased and he can mingle with the new agents that enter the economy with a clean record. The symmetry between A2 and A3 is apparent: *any* shift of agents behind names, be it via trade or a name change, is hidden from clients. A4 refines the beliefs of clients in a sensible way: even if having a history is a zero probability event (e.g., all agents are expected to change their names) then beliefs are well defined by Bayes rule.

The demographics of the model follow an overlapping generations setup where agents are active in the economy for two periods, after which they leave the active economy for "retirement" (i.e., they value wealth when they retire). The total size of the population, and the distribution of types of agents is constant over time. Clients live for only one period and can observe only the firms' (names') track records. That is, by perfectly observing the sequence of successes and failures that were generated by the same name, clients will form beliefs about the type of agent who is currently running the firm with that name.

3 The market for names

A finite version of an OLG economy is analyzed first. This is important since it demonstrates that the value of reputations (names) is not due to a "bubble" that is supported by an infinite horizon model. Consider a two period model in which the size of the population of agents is constant and the features of an OLG model persist. The time line of this two period economy is described in Figure 1.

At date $t = 1$ the economy starts with agents from generations 0 and 1 choosing names for their firms, and then clients paying firms up-front for their services. Clearly, given that no prior information is available to the clients, they will pay the same wage to all firms. This equals the expected benefit from hiring a firm,

$$w_1 = \Pr\{\text{Success}\} = \int_0^1 \theta f(\theta) d\theta = \frac{1}{2},$$

```
                        t = 1      t = 2
        Generation 0:   |---------|
        Generation 1:   |---------|---------|
        Generation 2:             |---------|
```

Figure 1. A two period economy

which follows because θ is distributed uniformly (with density $f(\theta) = 1$), clients' value success at 1 and failure at 0.

At date $t = 2$ the analysis is less straightforward because some firms will have a past history while others will not. In turn, firms with a past history may fall into two categories: they can either be operated by a generation 1 agent who generated this history and continued with his name to the second period, or by an agent who just bought the name. Any equilibrium of this economy involves two markets at $t = 2$: the market for services in which clients pay (history dependent) wages for hiring firms, and the market for names in which agents pay prices for names with (possibly) different track records. Before proceeding with the two-period model, notice that since only past histories matter then two distinct names with the same history should generate the same expectations for future success at $t = 2$. Thus, let S denote any name at $t = 2$ with a past success, F with a past failure, and N a name with no past. The equilibrium wages of firms at $t = 2$ will be denoted by $w_2(h)$, $h \in \{S, F, N\}$, and the equilibrium prices of names at $t = 2$ will be denoted $v(S)$ and $v(F)$ respectively. Using rational expectations equilibria as the equilibrium concept, the following result is obtained:

Proposition 1. *S Names must be traded in all equilibria.*

Proof. Assume by contradiction that there exists an equilibrium in which no names are traded at $t = 2$. This implies that the value of a name with a past success must be zero since the supply of S names is positive and is equal (using a law of large numbers[2]) to the measure $\frac{1}{2}$ generated by half the agents from generation 0 who succeeded and then retired (without selling their name). Furthermore, half the agents from generation 1 continue with an S name while the other half continue with either a F name or a N name. (If $w_2(F) < w_2(N)$ then all Fs change their name, otherwise they may keep it.) The assumption of no trade in names implies that all agents from generation 2 start with a N name. Thus, $w_2(N) = \Pr\{\text{success}|N\} \leqslant \frac{1}{2}$ (It is equal only if no failure agents change their name. Otherwise, previous failures who are below average will cause a strict inequality). Furthermore,

$$w_2(S) = \Pr\{\text{success}|S\} = \int_0^1 \theta f(\theta|S) d\theta$$

where clearly $f(\theta|S) = \frac{f(\theta,S)}{\Pr\{S\}}$, and since $f(\theta, S) = \theta$ it follows that $w_2(S) = \frac{2}{3}$. Thus $w_2(S) > w_2(N)$, which in turn implies that any agent who has no history will be willing to pay (at least less than $\frac{1}{6}$) for an S name. This contradicts the no trade assumption.   Q.E.D.

The intuition goes as follows: since shifts of ownership are not observable (A2), if clients believe that no trade of names occurs then Bayes updating after a success is "good news", creating expectations of a higher probability of success. Thus, new agents will be willing to buy successful histories (names). With full observability

---

[2] Though standard in the macroeconomic literature, the use of a law of large numbers for a continuum of $i.i.d.$ random variables is known to be problematic using the Lesbegue integral (see Judd 1985). I too abuse this in the standard way.

of ownership shifts this result is no longer true – assigning beliefs to clients saying that only "bad" types (e.g., $\theta = 0$) buy names will support equilibria with no trade of names. This intuition illuminates an important point of this research program: non-observability of ownership shifts is a necessary condition for the market for names to be active in *all* equilibria. Notice that the result does not depend on the Uniform distribution assumption – any distribution would generate the contradiction of $w_2(S) > w_2(N)$ if no trade occurs. This generalizes Proposition 1 in Tadelis (1999) which was obtained for the two-type case.

Characterizing equilibria with a continuum of types is less straightforward than for a two-type setting. This follows because one cannot easily assign the supply (measure) of success names in different proportions to different types since all types in this setup are non-atomistic. Furthermore, Tadelis (1999) assumed that trade takes place *only* between retiring agents (of generation 0) and new agents (of generation 2). In a two-type setting this is without loss of generality, but this is a possibly restricting assumption for the more general model presented here. In particular, it may be that different types value names differently at any stage of their lifetime, and inter-generational, as well as intra-generational trade must be accommodated. For this one needs the following apparatus:

Definition 1. Let $\mu : [0,1] \to [0,1]$ be a *name mapping* such that $\mu(\theta)$ is the probability that an agent of type θ is assigned an S name at time $t = 2$, and the realized assignments add up to total supply,

$$2 \int_0^1 \mu(\theta)d\theta = 1 . \tag{1}$$

That is, a type θ agent who has not retired at $t = 2$ is assigned an S name (from the total measure of S names) with probability $\mu(\theta) \in [0, 1]$, and the total measure of S names is allocated. Namely, (1) is a market clearing condition for the market of S names. Also note that there is an inter-generational anonymity assumption embedded in the definition of a name mapping. Namely, agents of generations 1 and 2 are treated identically in period $t = 2$.

In the remainder of this section, we restrict attention to equilibria without trade in F names, and consider only equilibria in which no agent would want to buy an F name or continue with one. Proposition 1 established that S names must be traded, so this restriction is not unreasonable.[3]

Definition 2. An equilibrium is a quintuple $\langle \mu(\cdot), w_2(S), w_2(F), w_2(N), v(S) \rangle$ that satisfies:

1. $\mu : [0, 1] \to [0, 1]$ is a name mapping
2. Given $\mu(\cdot)$, $w_2(h)$ is consistent with Bayes' rule for $h \in \{S, F, N\}$
3. $v(S)$ is the market clearing price of S names.

Proposition 1 states that trade must occur in *all* equilibria, but no other characteristics of equilibria are implied, nor is existence shown. Furthermore, in equilibrium a name (or history) is a vehicle over which clients form beliefs of the

---

[3] Furthermore, since failures are generally "bad news", one can construct refinements on beliefs that would prohibit sale of F names. This is, however, beyond the scope of the paper.

population of agents behind the names. Thus, reputations are beliefs that must be Bayes-consistent in equilibrium. For the two period economy the following useful Lemma is established:

Lemma 1. $v(S) = w_2(S) - w_2(N)$ *in all equilibria.*

Proof. If $v(S) > w_2(S) - w_2(N)$ then any type θ agent who is active in the second period would strictly prefer not to buy an S name, contradicting Proposition 1. If $v(S) < w_2(S) - w_2(N)$ then any generation 2 agent (and any generation 1 agent who failed and changed his name) would prefer buying an S name to starting with an N name because wages are history dependent and do not depend on the agent's type. But this implies that the measure of demand is greater than 1 and the measure of supplied S names is $\frac{1}{2}$, violating market clearing. Since from Proposition 1 trade must occur, it must be at a price that causes indifference.   Q.E.D.

The intuition is rather simple. All agents get the same stream of benefits from buying an S name since there is only one period left, and the outcome of that period does not affect future payoffs. This in turn implies that S names are scarce, and for market clearing to obtain, the price of an S name is driven up to capture the benefit from having one. Lemma 1 is useful to characterize equilibria of the two period model. In particular, any name mapping $\mu(\cdot)$ will immediately determine the Bayes updating rules that pins down the wages $w_2(h)$, $h \in \{S, F, N\}$, and from Lemma 1 we know that $v(S)$ is pinned down by these wages. Thus, we can characterize an equilibrium by the a function $\mu(\cdot)$ as follows,

Proposition 2. *There is a continuum of equilibria. Furthermore, a name mapping* $\mu(\cdot)$ *is an equilibrium if and only if the resulting wages satisfy,*

(2.1) $w_2(S) - w_2(N) \geqslant 0$
(2.2) $w_2(F) - w_2(N) \leqslant 0$

Proof. Assume first that $\mu(\cdot)$ is an equilibrium. From Lemma 1, $v(S) = w_2(S) - w_2(N)$. Since names must be traded, it cannot be that $v(S) < 0$ otherwise trade would not occur (sellers of names would rather not sell them). Thus, (2.1) must hold. Since $\mu(\cdot)$ is assumed to be an equilibrium in which no F names are traded, this implies that (2.2) must hold, otherwise F names would be traded. The reverse direction follows immediately: if a name mapping $\mu(\cdot)$ results in wages that satisfy (2.1) and (2.2), then from Lemma 1 the resulting $v(S)$ is non-negative, and no F names would be traded. Now to verify that there is a continuum of equilibria, consider the following equilibrium: Let $\mu^*(\theta) = \theta$ for all $\theta \in [0, 1]$ be the "identity" name mapping that assigns an S name to every type of agent from generations 1 and 2 with the same probability as their type. This implies that

$$E[\theta|S] = \int_0^1 \theta f(\theta|S)d\theta = \int_0^1 2\theta^2 d\theta = \frac{2}{3}.$$

Similarly, by Bayes rule, $E[\theta|F] = E[\theta|N] = \frac{1}{3}$, and we can assume that all F realizations cause a name change (since A4 guarantees that beliefs are well defined for F histories we still have $E[\theta|F] = \frac{1}{3}$). This confirms that $\mu^*(\theta)$ is an equilibrium that satisfies $v(S) = w_2(S) - w_2(N) > 0$ and $w_2(F) - w_2(N) = 0$. By the strict inequality of (2.1), we can perturb $\mu^*(\cdot)$ to give slightly more weight to lower types resulting in lower values of $w(S)$ and higher values of $w(N)$, so that such a perturbation is an equilibrium. This implies that there are a continuum of equilibria.   Q.E.D.

The existence of a continuum of equilibria follows from the indifference result of Lemma 1, which in turn follows from the scarcity of names and from the fact that all agents have only on period left, in which there performance does not affect their wages. For performance to affect future wages, longer horizons are needed, which is the focus of the next two sections.

4 Reputational sorting

To analyze the ways in which current performance affects future wages, consider first a three period economy as shown in Figure 2.

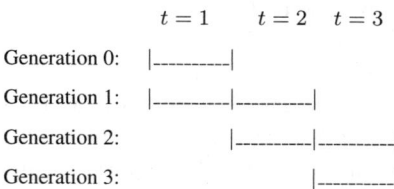

Figure 2. A three period economy

At dates $t = 1, w_1 = \frac{1}{2}$ as in the two period model. Also, the analysis of $t = 3$ in this three period model is identical to that of $t = 2$ in the two-period model, except that more names might be traded. Extending the notation of the previous section, let $w_t(h_t)$ denote the wage at time t to a firm with history h_t. Similarly define $E_t[\theta|h_t])$, where $h_1 \in \{S, F, N\}$ and $h_2 \in \{NN, NS, NF, SF, SS, FS, FF\}$, where NN is a new name that starts at $t = 3$. As before, we consider equilibria in which F names are not traded at $t = 2$ and similarly focus on equilibria in which FF, NF, and FS names are not traded at $t = 3$.[4] Equilibria will be defined as in the two period model with more name mappings, one for each name traded in equilibrium. For example, if NS, SF, and SS names are traded at $t = 3$, then an equilibrium is defined by four functions, $\mu_1^S(\cdot)$ and $\mu_2^h(\cdot)$, $h \in \{NS, SF, SS\}$ where $\mu_1^S(\cdot)$ is a name mapping at $t = 2$ and $\mu_2^h(\cdot)$ are the three mappings at $t = 3$.

Now consider an agent of type θ from generation 1 who chooses not to buy a name at $t = 2$. His expected lifetime flow of income is,

$$u^\theta(N) = w_2(N) + \theta w_3(NS) + (1 - \theta)w_3(NN)$$

<hr/>

[4] Assuming that F names are not trade at $t = 1$ does not imply that Fh names, $h \in \{S, F\}$, are not trade at $t = 2$. It does imply, however, that a client who observes a firm with a Fh name must believe that it is owned by a new agent from generation 3. Thus, by assuming "pessimistic" beliefs for names that should not be traded, it is guaranteed that Fh names will not be traded if F names are not traded at $t = 1$.

which follows if failures are "bad news" so that an agent who fails will change his name to get $w_3(NN)$ instead of $w_3(NF)$. Similarly, his net expected flow of income if he buys an S name at $t = 2$ is,

$$u^\theta(S) = w_2(S) + \theta w_3(SS) + (1 - \theta) \max\{w_3(SF), w_3(NN)\} - v(S) .$$

This follows because SF names may be valuable in equilibrium, in which case they will command a premium over new names and will be traded. Thus, the utility difference of a type θ agent from buying an S name versus not buying one at $t = 2$ is denoted by $\Delta u^\theta \equiv u^\theta(S) - u^\theta(N)$. Consider two types $\theta > \theta'$, and compare their utility differences as follows:

$$\Delta u^\theta - \Delta u^{\theta'} =$$
$$(\theta - \theta')(\underbrace{[w_3(SS) - w_3(NS)] - \max\{[w_3(SF) - w_3(NN)], 0\}}_{R}) . \quad (2)$$

Notice the bracketed term in (2) above which is denoted by R. If $R > 0$, then type θ gets a higher net utility from buying a good reputation (S name) than type θ'. In other words, if there is some θ' that is willing to pay $v(S)$ for an S name at $t = 2$ then so is every type $\theta > \theta'$. Similarly, if $R < 0$ then if there is some θ that is willing to pay $v(S)$ for an S name at $t = 2$ then so is every type $\theta' < \theta$. This observation implies that if successful reputations can be used as sorting signals, then when $R > 0$ this would enforce separating higher types through the market for names. Equilibrium analysis, however, implies that for higher-type sorting to occur, $R > 0$ must be consistent with such sorting. Similarly, $R < 0$ must be consistent with lower-type sorting for names to work as negative sorting signals.

Intuitively, it is tempting to consider the following argument: valuable reputations should be valued more by higher types, since they are less likely to fail and thus ruin this valuable asset. This would be analogous to the type of reputational equilibria that are analyzed in the repeated game framework of Kreps (1990). Namely, the reason opportunistic agents choose to perform well is precisely because they fear the loss of reputation, which leads to lower future payoffs. It turns out that the flavor of these results which were derived in a partial equilibrium (game theoretic) framework does not carry over to a model in which the value of a reputation is derived endogenously in a general equilibrium setting.

Before continuing, it is useful to determine an equilibrium selection mechanism for $t = 3$, since from Proposition 2 above it is clear that in the current setup there will be a continuum of equilibria for each valuable name at this trading stage. Furthermore, the following is a corollary of Lemma 1 for the three period model:

Corollary 1. *In all equilibria,*

$$v_3(SS) = w_3(SS) - w_3(NN) ,$$
$$v_3(NS) = w_3(NS) - w_3(NN) , \text{ and}$$
$$v_3(SF) = w_3(SF) - w_3(NN) .$$

This result follows immediately from the scarcity of names with respect to the full population of agents who are active in the last period of the economy. Note that in equilibrium $v_2(SF) < 0$ can be sustained if the composition of generators of these names without trade would imply that $w_2(SF) < w_2(NN)$ by Bayes rule. Of course, no seller would pay money to sell his name upon retirement, so such a price would never be observed in equilibrium.

The indifference of all types regarding whether to buy a name or not, as shown in corollary 1, suggests that there is no reason one type should be more likely to buy a name than another. If this is taken seriously to mean that in a frictionless market names are randomly assigned to the population of types that values these names at $t = 3$, then the composition of agents behind *each* of the names SS, NS, and SF should be the same.

It turns out that adding a second dimension of agent heterogeneity would naturally break the indifference demonstrated in Lemma 1 and Corollary 1, and pin down a unique equilibrium. For example, let agents vary with respect to their cost of maintaining a name versus starting a new one. This would capture the idea that when an agent wishes to transmit information about a name's history, he must incur some cost, whereas starting with a new name eliminates the need to do so.[5] Formally, let $\pi \in [0, \overline{\pi}]$ be the extra cost associated with maintaining an existing firm (name) that is *i.i.d.* distributed across all agents with the cumulative distribution function $\Phi(\cdot)$. An agent with cost π will buy an h name only if it is worthwhile given his costs, that is, only if

$$v_3(h) \leq w_3(h) - w_3(NN) - \pi,$$

implying that there will exist some $\pi^* \in [0, \overline{\pi}]$ such that all agents with cost $\pi < \pi^*$ will buy h names, and other agents will not, independent of their type. The equilibrium price of a h name will then be

$$v_3(h) = w_3(h) - w_3(h) - \pi^*$$

and the *i.i.d.* assumption guarantees that the buyers of names are a random selection of types. To continue with less notation, consider the limit case of $\pi^* \to 0$ so that $v_3(h) = w_3(h) - w_3(NN)$ is restored, and the focus of attention at terminal periods is the random matching equilibrium which is the limit of $\pi^* \to 0$. This follows easily, for example, from $\Phi(\cdot)$ being Uniform on $[0, \overline{\pi}]$ and letting $\overline{\pi} \to 0$. We can state the following result:

Proposition 3. *In any equilibrium of the 3-period model higher types cannot be strictly sorted at $t = 2$ through the market for S names. Furthermore, $R = 0$ in any equilibrium.*

Proof. First it is established that $R < 0$ cannot be part of any equilibrium. Assume in negation that $R < 0$. This implies that negative sorting should occur at $t = 2$ in the market for S names. However, negative sorting would immediately imply by Bayes rule that $w_2(S) < w_2(N)$, implying no trade of names. This contradicts the arguments of Proposition 1, so it must be that $R \geq 0$.
Second, it is established that SF names must be traded at $t = 3$ if positive sorting

---

[5] The "reverse" heterogeneity also works: an existing firm's name is known to some clients, which saves on advertisement costs of promoting a new business. This too would select a unique equilibrium.

occurs. Assume in negation that SF names are not traded at $t = 3$, and that $R > 0$ so that positive sorting occurs in equilibrium. A4 then implies that any SF name continuing at $t = 3$ must belong to a generation 1 agent who bought an S name and failed. Since positive matching implies that the highest types have purchased S names, it must be that,

$$E_{t=3}[\theta|SF] = \int_{\frac{1}{2}}^{1} \theta f(\theta|SF)d\theta = \int_{\frac{1}{2}}^{1} \theta 8(1-\theta)d\theta = \frac{2}{3}.$$

Now note that regardless of the composition of types who buy SS and NS names at $t = 3$, it must be that $E_{t=3}[\theta|NN] \leq E[\theta] = \frac{1}{2}$, which implies that $E_{t=3}[\theta|SF] > E_{t=3}[\theta|NN]$, contradicting that SF names are not traded when positive sorting occurs. Thus, SF names must be traded if positive sorting occurs. The same logic of Proposition 1 implies that NS and SS names must be traded at $t = 3$. To complete the proof, two cases need to be considered:
(i) Assume that all names $h \in \{SS, NS, SF\}$ are traded at zero prices. This implies that

$$w_3(SS) = w_3(NS) = w_3(SF) = w_3(NN), \tag{3}$$

which immediately implies that $R = 0$. Thus, if higher types are sorted in equilibrium at $t = 2$, it must be through the distribution of costs π which favors higher types, but this is inconsistent with the $i.i.d.$ assumption on the distribution of π.[6]
(ii) Assume that all names $h \in \{SS, NS, SF\}$ are traded at positive prices. The random matching mechanism implies that $w_3(SS) = w_3(NS) = w_3(SF) > w_3(NN)$, which in turn implies that $R < 0$, a contradiction.   Q.E.D.

Proposition 3 follows from the general equilibrium analysis of the model. Intuitively, when an agent is faced with the option of buying a name, this is weighed against the option of starting with a new name. The better the agent is, the better the future prospects from *both* alternatives. However, the degree to which good names deteriorate after failures, and new names increase in value after success, are determined by clients' beliefs about the composition of types behind these names. This is the key part of the argument: if clients believe that positive sorting occurs at $t = 2$, then the downward updating of SF names in period $t = 3$ is not severe. In the proof, this is computed using the Uniform distribution assumption. Note, however, that the method is rather general, and the driving force is that $E_{t=2}[\theta|SF] > E_{t=2}[\theta|NN]$ will be a result of positive sorting. This means that worse agents do not suffer big losses from buying an S name at $t = 2$, while they have poor prospects from starting their own new name. Thus, under these conditions of weak downward updating, lesser types would be willing to pay more than better types for S names, which contradicts positive sorting.

In a moral hazard, partial equilibrium setup such as Kreps, this link between the option of buying a name and the option of starting with a new name does not exist. In Kreps's model, an agent will buy the firm if and only if he intends to honor trust, but without buying a name he cannot build a trusting relationship. This equilibrium

---

[6] Without the iid assumption this would imply that higher types would buy names at $t = 3$, causing names to have positive value at $t = 3$, contradicting (3) above, and the proof would go through.

argument is based on the bootstrap nature of the so called "reputational" equilibrium that Kreps describes. Namely, the good reputation, or trusting equilibrium, is assumed at the beginning of time, and only agents who are willing to maintain it will be willing to pay a price for it. Furthermore, this price is set so as to deter abuse, but has no link to the service market that the agents provide. By linking the two markets – the market for services and the market for names – the present model is able to uncover novel features of the values of reputations, and the abilities of reputations to sort types.

5 Longer horizons

First, it is easy to see that the analysis of Proposition 1 extends as follows (the proof is a straightforward adaptation of the proof provided earlier),

Proposition 4. *For any finite or infinite horizon economy names must be traded in all periods of any dynamic equilibrium.*

The forces that drive Proposition 1 continue to work in any finite or infinite horizon economy – if no names are traded at some period t, then it must be that $E_t[\theta|NS] > E_t[\theta|N]$, which would create demand for these names. That is, if names that cause upward Bayesian updating (are "good news") are not traded, then they would have value, implying that they must be traded.

It turns out that if different reputations are identified with different histories then the formal analysis of equilibria in the infinite horizon model may seem potentially cumbersome. This is particularly true if there is no added heterogeneity as in Section 4 above since then different compositions of types can buy different names, where each of these would require a separate name mapping, market clearing condition, and non-negative price condition. However, by focusing on a simple form of Steady State Equilibria (SSE) and using the analysis of Section 4, it is possible to construct a simple equilibrium for the infinite horizon model, which is the purpose of this section.

Formally, let H be the set of all possible histories (including N, no history) and define a *reputation reduction* as a function that maps the set of all histories into equivalence classes. This can be done without putting restrictions on the actual histories that clients observe, but rather means that different histories will have the same reputation value supported by the correct beliefs of clients in the SSE constructed. Formally,

Definition 3. Let $r : H \rightarrow \{N, S, F, SF\}$ be a *reputation reduction* such that $r(N) = N$; $r(h) = S$ for all $h \in H_S \equiv \{NS, NSS, NSSS, NSSSS, ...\}, r(h) = SF$ for all $h \in H_{SF} \equiv \{NSF, NSSF, NSSSF, ...\}$, and $r(h) = F$ for all $h \in H_F \equiv H\backslash(\{N\} \cup H_{SS} \cup H_{SF})$.

In Definition 3 the infinite number of histories $h \in H$ are reduced into four possible equivalence classes: new names (N), names that had only successes (S), names that start with a sequence of only successes but their last realization was a failure (SF), and all other names (F).

Using this particular reputation reduction, a SSE is constructed as follows: Only names that never failed will be traded (S names), and names will be changed after the first failure occurs. The equilibrium will be supported by steady state wages $w(h)$, $h \in \{N, S, F, SF\}$, and all traded S names will trade at price $v(S)$. It will be confirmed that F and SF names will be worse than N names, so that these last two classes of names will not be traded and will be changed with probability one (recall A4). Following standard convention, agents in their first period will be called "young" while agents in their second period will be called "old".

In any such equilibrium, both old and young agents can potentially buy names, and both retiring agents and old agents (who succeeded when they were young) can sell their names. This was the case analyzed for the two and three period models. For analytical simplicity, assume in this section that old agents do not buy names or sell names; only retiring agents sell names, and only young agents buy names.

This restriction is without loss of generality. It will become apparent from the construction that one can construct equilibria where everyone sells names after success and then randomly buys the name back. This, however, seems less appealing. Indeed, if there is some transaction cost to buying and selling names then continuing agents who succeeded would rather hang on to their good name rather than sell it.

Formally, let $\mu_S(\theta)$ be the name mapping that determines the probability that a young type θ agent buys a S name. A SSE is then characterized by the tuple $\{w(h), v(S), \mu_S(\cdot)\}$, where $h \in \{N, S, SF, F\}$.

When considering such an equilibrium, a type θ agent who chooses to start with a new name faces the following (expected) stream of payoffs,

$$u^\theta(N) = w(N) + \theta w(S) + (1 - \theta)w(N) + \theta^2 v(S) + \theta(1 - \theta)v(S), \qquad (4)$$

which follows because his initial wage is $w(N)$, and his second period wage is either $w(S)$ or $w(N)$ depending on a first period success or failure (recall that any failure will cause a name change). When retiring, two consecutive successes, or a fresh success will create a name that sells at $v(S)$. Similarly, a θ type agent who buys an S name faces the following (expected) gross stream of payoffs (not including the price payed for a name),

$$u^\theta(S) = w(S) + \theta w(S) + (1 - \theta)w(N) + \theta^2 v(S) + \theta(1 - \theta)v(S). \qquad (5)$$

A type θ's willingness to pay for an S name is the difference between (5) and (4),

$$\Delta u^\theta \equiv u^\theta(S) - u^\theta(N) = w(S) - w(N),$$

and for $\theta' > \theta$,

$$\Delta u^{\theta'} - \Delta u^\theta = 0.$$

Notice that in the SSE under construction there is no sorting and all types are indeed indifferent between buying a name or not buying one, and the scarcity of names for sale implies that $v(S) = w(S) - w(N)$ as in Lemma 1 and Proposition 3. Indeed, if there were negative sorting then names would not sell since an N name would be more valuable than an S name. Similarly to Section 4, if there were positive sorting then SF names would sell at a premium since then we would have $w(SF) > w(N)$, and a similar equality to (2) would have a term like R that would be negative. This in turn would promote negative sorting, contradicting the positive sorting assumption. This clearly shows that Proposition 3 naturally extends to the infinite horizon period.

To calculate wages, recall that $w(h) = E[\theta|h]$. By Bayes' rule this expectation for S names is the weighted average of old agents who are continuing with the history S and young agents who bought it, and for N names it is the weighted average of old agents who changed their name and young agents who started with a new name. Letting $\lambda(\cdot)$ denote the (Lebesgue) measure of a set of agents, S_o denote the set of old agents who generated the history S and continue with it, and S_y the set of young agents who bought the history S, Bayes' rule implies that

$$E[\theta|S] = \frac{\lambda(S_o) \cdot E[\theta|S_o] + \lambda(S_y) \cdot E[\theta|S_y]}{\lambda(S_o) + \lambda(S_y)}. \tag{6}$$

Similarly, Let N_y denote the set of young agents who start with an N name, and N_o the set of old agents who changed their name, and Bayes' rule implies that

$$E[\theta|N] = \frac{\lambda(N_o) \cdot E[\theta|N_o] + \lambda(N_y) \cdot E[\theta|N_y]}{\lambda(S_o) + \lambda(N_y)}. \tag{7}$$

For the construction to work, it must be that a measure of $\frac{1}{2}$ of the young agents who fail every period (from the Uniform distribution of θ) change their name, so that $\lambda(N_o) = \frac{1}{2}$. Also, a measure $\frac{1}{2}$ of the old agents succeed, implying that there is a supply of S names of measure $\frac{1}{2}$. It follows immediately from market clearing that the measure of young agents who buy S names must be $\lambda(S_y) = \frac{1}{2}$. This simple steady state reasoning carries one step further: If half the young agents are buying S names, then the other half are not so that $\lambda(N_y) = \frac{1}{2}$. Similarly, if half of the young agents succeed regardless of their initial choice of buying a name or not, then from the uniform distribution and the equivalence class of h_S, it follows that $\lambda(S_y) = \frac{1}{2}$.

To continue, recall that the equilibrium selection of random matching (with the added heterogeneity) employed in the previous section implies that $\mu_S(\theta) = \frac{1}{2}$ for all $\theta \in [0, 1]$ of the young agents, which is consistent with a measure $\frac{1}{2}$ of S name buyers, and immediately implies that

$$E[\theta|S_y] = E[\theta|N_y] = \frac{1}{2}.$$

For old agents, however, Bayes updating is required. Using the uniform distribution it is easy to see that agents in the set of S_o are agents who previously succeeded, and agents in the set N_o are agents who previously failed, implying that

$$E[\theta|S_o] = \int_0^1 \theta f(\theta|S)d\theta = \frac{2}{3}, \text{ and } E[\theta|N_o] = \int_0^1 \theta f(\theta|F)d\theta = \frac{1}{3}. \tag{8}$$

Thus, all the values that are needed to compute $E[\theta|S]$ and $E[\theta|N]$ are given, and from (6) and (7),

$$E[\theta|S] = \frac{7}{12} > E[\theta|N] = \frac{5}{12}.$$

To complete the analysis it is necessary to show that all young agents who failed will indeed choose to change their names. By A4, any agent who appears with a history of SF, or of one failure (which is in the set H_F) must belong to the group N_o, and from (8) it follows that Bayes rule is well defined and,

$$E[\theta|SF] = E[\theta|F] = \int_0^1 \theta f(\theta|F)d\theta = \frac{1}{3}.$$

Finally, for all the other histories $h \in H_F$, they are not traded and therefore should not be observed (no agent would continue with one) and it is therefore possible to set $E[\theta|F] = \frac{1}{3}$ as out of equilibrium beliefs for these names, guarantying that they will not be traded.

The above is therefore a characterization of the unique equilibrium that arises from the selection of random matching as described in Section 4. The analysis also confirms that S names must be traded, an implication of Proposition 4, and that no sorting can occur in any equilibrium, even if we would abandon the equilibrium selection. Without the equilibrium selection, the SSE above will still be an equilibrium. However, as in Proposition 2, it is possible to slightly perturb the mapping $\mu_S(\theta)$ to support other equilibria, as long as $\mu_S(\theta)$ is bounded away from full sorting, enough to support the proposed equilibrium.

6 Concluding remarks

This paper stresses the point that it is necessary to tie the value of a firm's reputation to clients' perception of the quality of the firm's good or service. This departs from the partial equilibrium, game theoretic analysis that has been previously used to investigate the effects of a firm's reputation on the agents who own and run the firm. The general equilibrium analysis demonstrates that good reputations cannot serve as sorting devices that separate the better agents from the lesser agents. One way to think about this, in contrast to partial equilibrium game theoretic models, is that there is no exogenous *single crossing condition* with respect to purchasing names. Instead, the beliefs of clients determine in which direction the sorting will work: if clients have beliefs that are too "positive", then these beliefs would encourage the reverse type of sorting. Thus, a diverse mix of types must buy good names in equilibrium to generate consistent beliefs. Furthermore, these beliefs lead to sensible dynamics: good reputations are created following a firm's success, and destroyed following a firm's failure.

The basic idea of using a general equilibrium analysis to analyze the market for firm reputations was developed in Tadelis (1999). However, the simplicity of that model leaves important questions open as to the robustness of the results presented there. This paper extends the analysis in two important ways: a larger type space and infinite horizons. A different extension is considered in Tadelis (2002), where moral hazard is added into the general equilibrium setup with adverse selection. Aside from establishing a similar no sorting result, and the necessity of name trading, adding moral hazard endogenizes an interesting feature of the equilibrium demonstrated by Kreps (1990) in his model. Namely, short lived agents becomes "ageless" because the incentives provided by the market for names are equal to those created by future wage considerations of young agents who will not sell their names. The paper also addresses welfare analysis which is mute in the models without moral hazard. In particular, it is shown that providing clients with information about ownership shifts can reduce total surplus in the economy.

An interesting direction to take the modeling approach developed here would be the analysis of group reputations. If a group of agents all operate behind the same name (firm) and cannot be distinguished by the clients who purchase the firm's services, then the value of a group will depend not only on who is in the group, but on who is in other competing groups. Since updating of beliefs depends on both who is inside and outside of a group, the general analysis proposed here might be a fruitful way to address such issues. This is left for future research.

References

Akerlof, G.: The market for lemons: quality uncertainty and the market mechanism. Quarterly Journal of Economics **89**, 488–500 (1970)

Aoyagi, M.: Reputation and entry deterrence under short-run ownership of a firm. Journal of Economic Theory **69**, 411–430 (1996)

Fudenberg, D., Tirole, J.: Game theory. Cambridge: MIT Press 1991

Judd, K.: The law of large numbers with a continuum of iid random variables. Journal of Economic Theory **27**, 245–252 (1985)

Kreps, D.: Corporate culture and economic theory. In: Alt, J., Shepsle, K. (eds.) Perspectives on positive political economy, pp. 90–143 Cambridge: Cambridge University Press 1990

Mailath, G., Samuelson, L.: Who wants a good reputation? Review of Economic Studies **68**, 415–41 (2001)

Phelan, C.: Public trust and government betrayals. Federal Reserve Bank of Minneapolis Research Department Staff Report 283 (2001)

Tadelis, S.: What's in a name? Reputation as a tradeable asset. American Economic Review **89**, 548–63 (1999)

Tadelis, S.: The market for reputations as an incentive mechanism. Journal of Political Economy (forthcoming) (2002)

Structural breaks in the volatility of macroeconomic and financial data: The rule, not the exception[*]

Andrea Beltratti[1] and Claudio Morana[2]

[1] IEP, Bocconi University, Via Sarfatti 25, 20100 Milan, ITALY
(e-mail: andrea.beltratti@uni-bocconi.it)
[2] Facoltà di Economia, Dipartimento di Scienze Economiche e Metodi Quantitativi,
University of Piemonte Orientale, Via Perrone 18, 28100, Novara, ITALY
(e-mail: morana@eco.unipmn.it)

Received: August 29, 2002; revised version: September 25, 2002

Summary. In this paper we look at the empirical evidence in favor of structural breaks in the conditional volatility of some important macroeconomic and financial time series like currency returns, stock returns, output and inflation. We find strong evidence of both structural breaks and long memory in the break-free series. We use a variety of econometric methodologies, both parametric and non-parametric, in order to verify the robustness of our findings, which provide strong empirical evidence in favor of the Theory of Rational Beliefs.

Keywords and Phrases: Stock market volatility, Macroeconomic volatility, Long memory, Structural change, Rational beliefs.

JEL Classification Numbers: C32, F30, G10.

1 Introduction

The "rational expectations revolution" radically changed the models, ideas and economic policy prescriptions of most economists. Insisting on the need of formulating internally consistent expectations, derived from the same structure of the analytical model, eliminated one degree of freedom and forced economists to think coherently in terms of structural equations and expectations. This was certainly an achievement in the history of economic thought.

---

* We are grateful to Mordecai Kurz for comments on a previous version of this paper and to Olsen&Associates for making their high frequency data on currency returns available to us.
Correspondence to: A. Beltratti

In a standard rational expectations (from now on RE) model the representative agent is assumed to know the equilibrium mapping between the endogenous and the exogenous variables. This may be considered realistic only in the context of very stylized models. It is hard to believe a priori that agents living in real economic systems have accumulated all the knowledge and the information necessary to compute the equilibrium mapping, unless the economic system produces equilibria which are relatively stable and are observed for a long time. Incidentally Lucas [32] himself clarified that the hypothesis of RE was credible only in contexts of stable systems: "Economics has tended to focus on situations in which the agent can be expected to "know" or to have learned the consequences of different actions so that his observed choices reveal stable features of his underlying preferences....I think of economics as studying decision rules that are steady states of some adaptive process, decision rules that are found to work over a range of situations and hence are no longer revised appreciably as more experience accumulates.".

According to a large part of the economics profession a model should be evaluated on the basis of its predictions and not of its plausibility. In our opinion it is fair to say that a faithful and literal application of the rational expectations hypothesis has produced several models which are not consistent with observations. The main obstacle has been represented by the explanation of how financial markets work and determine asset prices. Application of RE in finance has gradually produced a long list of puzzles, among which the most famous is the equity premium puzzle. This is not surprising given that financial markets are determined by expectations in a fundamental way. The failure of the RE models to explain the dynamics of asset prices calls therefore into question the validity of the hypothesis made about the formation of expectations.

We have moreover observed another interesting phenomenon. In coincidence with the end of the bull market many academics and practitioners have started to question the size of the equity premium. Fama and French [12] for example suggest that maybe the large average stock return of the 1950-2000 was due to an unexpected shock to expected returns, associated with factors like the possibility to invest through mutual funds and diversify risk for ordinary investors and the increase in stock market participation. On the basis of data connected with dividends and earnings they suggest that a better estimate of the unconditional equity premium is a little larger than 4%, substantially lower than the historical estimate obtained from stock returns. This shows a remarkable fact: uncertainty about long run averages cannot be settled even with 50 years of data! In our view, if intelligent and equally informed people can come up with heterogeneous estimates of the mean of a variable observed for 50 years then there must be something wrong with the assumption of RE. Perhaps financial markets do not satisfy the conditions outlined by Lucas in the previous quote as those valid for the application of RE.

In the presence of model uncertainty then there must be heterogeneity of expectations. Financial theory however is largely based on the assumption of a representative investor, except for papers which have taken heterogeneity as a starting point in the context of partial equilibrium modeling of financial markets, see for example Harris and Raviv [18]. Recently there have been efforts to explain financial markets puzzles in the context of the behavioral finance literature. This literature, arising from the work of Shiller [45], draws from theoretical and empirical studies in psychology to model irrational behavior on the part of economic agents. It proposes therefore a dramatic departure from the standard methodology used in economic modeling, which assumes that agents are rational. The problem with this literature is that there are many types of irrational behavior which may be thought of and it may be difficult to choose the few which are relevant. It is like moving to nonlinear models and having to choose among infinite possibilities. Future research may limit the potentially interesting irrational behavior.

The theory of rational beliefs (from now on TRB) offers the possibility of introducing heterogeneity in expectations which is however constrained by consistency with the long run evidence provided by the data. Different agents may rationally carry different beliefs and expectations at each point in time and at the same time agree on the long run dynamics of the system. Rational expectations becomes a special (and not particularly plausible) case where all the agents hold the same dynamic model. The TRB is also compatible with the types of reasoning contemplated in the behavioral finance literature, except that it admits as scientifically interesting models only those which may produce forecasting rules compatible with the long run evidence of the economy.

How is this result achieved by the theory? The theoretical apparatus studied by Kurz assumes that the agents do not know the equilibrium mapping of the economy. More specifically agent k takes the true probability space (Ω, β, Π) as given. However he does not know the true probability space and acts on the basis of the belief that the probability space is (Ω, β, Q^k). The TRB "characterizes the set of all beliefs which are compatible with the available data". The assumption of rational expectations implies that $\Pi = Q^k$. Kurz shows that by doing empirical analyses agents can learn an empirical probability measure m. The empirical probability measure is consistent but not necessarily identical with the true probability measure. The existence of structural breaks may prevent the agents from learning the true probability measure. Agents may hold different theories of structural breaks and therefore may hold heterogenous probability measures, which are however all consistent with the empirical measure that can be learnt from the data. In simple words: agents learn all that is possible to learn but the objective truth may be impossible to learn so that agents act on the basis of beliefs which are not contradicted by the long run evidence but may be wrong at one or more points in time.

This brief description clarifies the importance of structural breaks, i.e. of nonstationarity of a stable probability measure. Even though non-stationarity of a probability measure might never be learnt, we believe that an empirical analysis of actual markets and economic data may be relevant to assess the practical importance of the TRB. If structural breaks can be found in actual data, then it is clear that the empirical underpinnings of this theory become even stronger. The possible existence of structural breaks cannot be dismissed a priori. There are several cases where popular econometric models have been shown to derive from incorrectly assuming away structural breaks. For example, the GARCH framework has been used for years without any economic justification for the persistence of volatility. Often researchers applying this model would find evidence in favor of permanent shocks to volatility, i.e. an integrated GARCH (IGARCH) model. However Diebold [10] and

Lamoreaux and Lastrapes [29] have pointed to neglected structural breaks as the explanation for the presence of IGARCH effects in the volatility process of financial assets. Similarly, Perron [38] had shown that the presence of a structural break may produce the impression that a time series includes a unit root. Structural breaks may therefore be more common than one would think; many successful econometric models may simply be the wrong way to account for the dynamic behavior of stochastic processes with structural breaks. This suspicion is confirmed by results like those obtained by Stock and Watson [47], which computes formal tests of instability and out-of-sample forecasts from sixteen different models using a sample of 76 US monthly postwar macroeconomic time series and finds "widespread instability in univariate and bivariate autoregressive models". In this paper we look at the evidence from currency and stock markets and test for the existence of structural breaks in the volatility of the series.

We draw on recent developments in the literature on high frequency data in finance which allows us to treat volatility as an observed variable and to use the large battery of models developed for time series analysis of observed variables. We test for the existence of structural breaks and long memory in the volatility of returns and of interesting macroeconomic variables. We conclude that there is widespread evidence in favor of the existence of structural breaks. Moreover the break-free series are often characterized by long memory and not by simple autoregressive models with few lags. We also find evidence that the structural breaks in macroeconomic variables do not appear to be related in a systematic way to structural breaks in stock volatility. The break-free stock return volatility is moreover largely generated by internal shocks, i.e. endogenous uncertainty is pervasive in the stock market. Our conclusion is that the data show most of the features on which the TRB is built upon.

2 Structural breaks and long memory

According to one popular definition, a stationary processes X_t is said to be long range dependent if the autocorrelation function is significantly different from zero at very long lags, that is if $\rho_X(\tau) = c_\rho \tau^{2d-1}$, where c_ρ is a positive constant, τ is the order of the autocorrelation and $d \in (0, 0.5)$ is the coefficient of fractional integration. The latter is related to the Hurst exponent by the relation $H = 1/2 + d$. Differently from I(0) processes, I(d) stationary processes do not show an exponentially fast decay of the autocorrelation function, but a slow hyperbolic decay. That is why these processes are often referred to as "long memory".

Diebold and Inoue [11] have recently proposed several models of structural change, which may lead to the detection of spurious long memory behavior when structural breaks are neglected in the modelling. The simplest model of structural change, which will be employed in the empirical analysis, is the break in mean model. Lets consider the process $X_t = \mu_i + \varepsilon_t$, where $\mu_i = \mu_1$ $0 \leq t \leq t_1$, ...$\mu_i = \mu_k$ $0 \leq t \leq T$, where ε_t is an I(0) processes and μ_i is an intercept term which can take several values according to the time period. The changing intercept term implies breaks in the unconditional mean of the processes X. Mikosch and Starica [33] have shown that the correlogram for a similar process, composed of two subsequent samples extracted from populations, each one strictly stationary, but with different means, converges in probability, for large lags, to a constant given by the squared difference of the two sample means. This points to an observational equivalence of certain type of structural change model and long memory processes in terms of the asymptotic behavior of the autocorrelation function. Granger and Hyung [14] have also found that such a process will show spurious evidence of long memory in terms of the value of the Hurst exponent, with the latter increasing with the number of neglected breaks occurring in the process. Since in our application we are interested in the conditional variance of financial returns and macroeconomic variables, the structural change model may be taken as referring to breaks occurring in the unconditional variance.

2.1 Methodologies for the estimation of the break process

Bai [3] has suggested an iterative procedure useful to detect structural breaks in the mean of a process. With reference to the model $y_t = \mu_i + x_t$ and $x_t = \sum_{j=0}^{\infty} a_j \varepsilon_{t-j}$, the suggested test statistic is

$$\sup F_T = \sup_{T\eta \leq \tau \leq T(1-\eta)} \frac{\bar{S}_T - S_T(k)}{\hat{\sigma}^2} \xrightarrow{d} \sup_{T\eta \leq \tau \leq T(1-\eta)} \frac{|B(\tau) - \tau B(1)|}{\tau(1-\tau)},$$

where $B(\cdot)$ is standard Brownian motion on $[0,1]$, $\eta \in \left(0, \frac{1}{2}\right)$, $\tau = \frac{k}{T}$ is the break fraction, $\bar{S}_T = \sum_{t=-n_1+1}^{n+n_2} (y_t - \bar{y})^2$, $S_T(k) = \sum_{t=-n_1+1}^{k} (y_t - \bar{y}_k)^2 + \sum_{t=k+1}^{n+n_2} (y_t - \bar{y}_k^*)^2$, $\bar{y}_k$ is the sample mean for the first $k+n_1$ observations and $\bar{y}_k^*$ is the sample mean for the last $n+n_2 - k$ observations, $\hat{\sigma}^2$ is a consistent estimator of $a(1)^2 \sigma_\varepsilon^2$.

The iterative procedure works as follows. In the first run the entire sample is explored and a first break point is determined. Then in successive runs the sub-samples determined by the break point are investigated and new break points are determined until the null of no structural change cannot be rejected.

Once the number and dating of the breaks has been determined, a break-free series can be computed as follows. Firstly, computes dummy variables (d_k) according to the dating of the breaks. Then, run the OLS regression $y_t = \alpha + \sum_{k=1}^{P} \theta_k d_k + \varepsilon_{t,y}$. The estimated break process is then simply the fitted value in the regression above.

An approach based on a Markov switching model has been suggested by Morana [35] and Timmerman [49]. Lets consider a k regime model for the unconditional mean, and let η be a vector consisting of the mean elements μ_i $i = 1, ..., k$ and the variance of the error process in the model $y_t = \mu_{s_t} + \varepsilon_t$, with $\varepsilon_t \sim N(0, \sigma^2)$. The transition between states is governed by a Markov chain whose realizations take on values in $\{1, ..., k\}$, $p(s_t = j | s_{t-1} = i) = p_{ij}$, with $\sum_{j=1}^{k} p_{ij} = 1$. Let $\mathbf{p} = (p_{11}, p_{12}, ..., p_{kk})'$ the $(k^2 \times 1)$ vector of transition probabilities. The econometrician is supposed to observe only the realizations of the variable y_t but not of the state s_t. The unknown parameters can be collected in the vector $\lambda = (\mathbf{p}', \eta')'$ and maximum likelihood estimates of the parameters of the model can be obtained via the Expectation-Maximization algorithm. See Hamilton [17] for further details.

The break process is then computed as $y_t = \hat{\mu}_t = \sum_{s=1}^{k} \hat{p}_{t,s} \hat{\mu}_s$ where $\hat{p}_{t,s,j}$ is the estimated probability that the observation t of process j belongs to state s and $\hat{\mu}_{s,j}$ is the estimated value of the mean in the sth state. By applying an argument presented in Ang and Bekaert [2], consistent estimation of the break process can be obtained by the Markov switching model if the omitted variables are not regime dependent. As found by Morana and Beltratti [37], the Markov switching approach seems to perform better than the Bai procedure, since the number of breaks may be underestimated by the latter procedure. In addition, the Markov switching model allows to estimate all the break points jointly, and to forecast the break process. Finally, tests and estimation of a common break process can be performed in the framework of the Markov switching model (Morana [36]).

Finally, Inclan and Tiao [24] have shown that under the null of homogeneous unconditional variance

$$D_k^* = \sqrt{T/2} \left(\frac{C_k}{C_T} - \frac{k}{T} \right) \xrightarrow{d} W^0 \quad k = 1, ..., T,$$

where W^0 is a Brownian bridge, $C_k = \sum_{t=1}^{k} r_t^2$, $r_t \sim NID(0, \sigma^2)$, and T is the sample size. From the limiting distribution of D_k the theoretical quantiles of the test statistic of interest, $\max_k |D_k^*|$, can be derived. Inclan and Tiao [24] find that the asymptotic 5% critical value (1.358) is a good approximation for sample size larger than 200 observations. As for the Bai [3] approach, the procedure suggests to determine the dating of the potential structural breaks in an iterative way and to estimate the break process by OLS, but differently from this latter approach, the method is suited to detect the presence of breaks in the variance of a series.

Therefore, while the Bai [3] and Markov switching approach are suited for modelling breaks occurring in realized variance processes, the Inclan and Tiao [24] approach is carried out using the returns series.

2.2 Estimation of the degree of persistence

Both parametric and semiparametric approaches have been proposed. The parametric approach can be traced back to the seminal work of Sowell [46] on ML estimation of ARFIMA models. One difficulty with this approach is the sensitivity of the estimates to the specification of the short dependent component of the model. In addition, even with modern computing facilities, in practice ML estimation may be unfeasible with very large dataset, as for instance the one available for financial time series. This is due to the required inversion of the variance-covariance matrix of the observations, which has dimension equal to the sample size.

On the other hand, semiparametric approaches do not require the specification of the short dependent component of the model for the estimation of the degree of persistence (the fractional differencing parameter) and are not affected by the dimension of the sample. Yet, semiparametric approaches are not free from drawbacks. The major difficulty with semiparametric methodologies is the selection of the bandwidth, i.e. the number of periodogram ordinates to be considered in the computation of the relevant statistics. As noted in the literature, there is a trade-off between bias and efficiency, with bias increasing with the bandwidth and efficiency decreasing with it. Recent advances in optimal bandwidth theory (Robinson [40], Henry and Robinson [20], Hurvich et al. [23], Delgado and Robinson [9], Henry [21]) have partially solved this problem, by allowing bandwidth selection in such a way that the MSE of the estimator is minimized. Yet, the empirical implementation of such procedure seems to suggest that also graphical procedures, as for instance the one suggested by Taqqu and Teverovsky [48], should be employed as a complementary tool. Below we briefly sketch the procedures more generally employed in the literature.

The local Whittle estimator proposed by Kunsch [27] and Robinson [42] requires the minimization of

$$Q\left(C, H_{LW}\right) = \frac{1}{m} \sum\nolimits_{j=1}^{m} \left(\log C \lambda_j^{1-2H} + \frac{\lambda_j^{2H-1}}{C} I\left(\lambda_j\right) \right)$$

where $I\left(\lambda_j\right)$ is the periodogram at frequency $\lambda_j = 2\pi j/T$, $j = 1, ..., m$, m is the bandwidth parameter, C is a positive constant. For $H < 0.5$ the process is antipersistent, for $H > 0.5$ it is long memory, and for $H = 0.5$ it is weakly dependent. It is shown that $\sqrt{m} \left(\hat{H}_{LW} - H \right) \xrightarrow{d} N\left(0, \frac{1}{4}\right)$.

Robinson [42] has shown that the local Whittle estimator dominates the averaged periodogram estimator and the log periodogram estimator under several respects. First it is asymptotically efficient, second it does not require the trimming of the lowest frequency estimates, third it does not assume Gaussianity, fourth it does not require the selection of a numerical value for any additional parameter.

The LM estimator (Robinson [43]) is an alternative estimator for H, with the same limiting distribution of the Local Whittle estimator. It is defined as

$$\hat{H}_{LM} = \frac{\sum_{j=1}^{m} \left(1 - 2v_j\right) I\left(\lambda_j\right)}{\sum_{j=1}^{m} \left(2 - 2v_j\right) I\left(\lambda_j\right)}$$

where $v_j = \log j - \frac{1}{m} \sum\nolimits_{j=1}^{m} \log j$. We denote this estimator as H_{LM} since it can be derived from the LM test of Lobato and Robinson [31].

Another estimator widely used in empirical work is obtained from the averaged periodogram, see (Robinson [39] and Lobato and Robinson [30])

$$\hat{H}_{AP,q} = 1 - \frac{1}{2 \ln q} \ln \left\{ \frac{\hat{F}\left(qm\right)}{\hat{F}\left(\lambda m\right)} \right\}$$

where $\hat{F}(\lambda) = \dfrac{2\pi}{n} \sum_{j=1}^{[\lambda n / 2\pi]} I(\lambda_j)$. The limiting distribution of $H_{AP,q}$ is

$$m^{1/2}\left(\hat{H}_{AP,q} - H\right) \xrightarrow{d} N\left(0, \frac{\left(1 + q^{-1} - 2q^{1-2H}\right)}{(\ln q)^2}\frac{(1-H)^2}{(3-4H)}\right)$$

for $\dfrac{1}{2} \leq H \leq \dfrac{3}{4}$, and

$$m^{2-2H}\left(\hat{H}_{AP,q} - H\right)$$
$$\xrightarrow{d} N\left(0, \frac{\left(1 - q^{2H-2}\right)}{(\ln q)}\frac{(1-H)\,\Gamma\left(2(1-H)\right)\cos\left((1-H)\,\pi\right)}{(2\pi)^{2-2H}}P\right)$$

as $T \to \infty$, where P is a random variable with unknown distribution, for $\dfrac{3}{4} < H < 1$.

A consistent but less efficient estimate of the fractional differencing parameter can be obtained by the log periodogram regression (see Geweke and Porter-Haudak [13] and Robinson [41])

$$\ln I(\lambda_j) = c + d\left(-2\log \lambda_j\right) + \mu_j \quad j = l, ..., m,$$

where l is a trimming parameter. It has been shown that $\sqrt{m}\left(\hat{d}_{LP} - d\right) \xrightarrow{d} N\left(0, \dfrac{\pi^2}{24}\right)$.

Finally, one can use the Lagrange Multiplier, Wald and Likelihood Ratio tests for the null $H_0 : H = 0.5$, against a two-sided alternative, which can be computed as follows

$$LM = m\left(\frac{\sum_{j=1}^{m} v_j I(\lambda_{jT})}{\sum_{j=1}^{m} I(\lambda_{jT})}\right)^2 \xrightarrow{d} \varkappa^2_{(1)},$$

$$W = 4m\left(\hat{H} - H\right)^2 \xrightarrow{d} \varkappa^2_{(1)},$$

$$LR = 2m\left\{\log \frac{\sum_{j=1}^{m} I(\lambda_{jT})}{\sum_{j=1}^{m} \lambda_j^{2H-1} I(\lambda_{jT})} + \left(2\hat{H} - 1\right)\frac{1}{m}\sum_{j=1}^{m}\log \lambda_j\right\} \xrightarrow{d} \varkappa^2_{(1)},$$

where $\hat{H}$ is the estimate of the Hurst exponent obtained from any of the above discussed methods.

2.3 Issues in implementation of testing for breaks and long memory

While the definitions are very clear, in practice it may be difficult to understand whether the data are generated by processes with breaks and long memory. One important difficulty in testing for structural breaks is that spurious structural change may be detected in a pure long memory process with the available tests for structural breaks, and, conversely, spurious long memory dynamics can be detected in weakly dependent processes subject to structural change using the available methodologies, both parametric and semiparametric.

For instance Kuan and Hsu [25] has shown that the Bai [3] test rejects the null of no structural change with probability one and is biased to select a break point in the middle of the sample when the process is actually characterized by long memory and no structural change. In addition, Granger and Hyung [14] have shown that the number of detected spurious breaks increases with the magnitude of the Hurst exponent and is zero only when the process is I(0). Two possible solutions indicated in the literature so far are to allow for long memory when testing for structural change and viceversa (Kuan and Hsu [26]), or, as suggested by Granger and Hyung [14], testing for a spurious break process by checking whether the break-free series is characterized by anti persistence, i.e. a negative fractional differencing parameter, while the actual series shows long memory (a positive fractional differencing parameter below 0.5). One problem associated with the former approach is that the available methodologies are suited to test for just one break point (see for instance Kuan and Hsu [26]). Differently, Taqqu and Teverovsky [48] have suggested an approach which can be employed when more than one structural break is present in the data. However, the methodology is not suited to estimate the location of break points, but rather to test whether a break process is present in the data.

Even resorting to the predicted hyperbolic decay of the autocorrelation function for long memory processes may lead in practice to unreliable results. In addition to the already quoted results of Mikosch and Starica [33], Granger and Terasvirta [16] and Diebold and Inoue [11] have provided some examples where a hyperbolic decay of the correlogram can be produced by a short memory non linear model. Moreover, Hosking [22] has shown that the limiting distribution of the autocorrelation function for a long memory process is not standard normal. This result has implications for testing the significance of the autocorrelation function using standard bounds. As suggested by Mikosch and Starica [34], the correlogram at long lags, although different from zero, may not be statistically significant. Using inappropriate critical values can, therefore, lead to invalid conclusions. Finally, Granger and Marmol [15] have shown that if the series is noisy, i.e. the series is composed of a long memory process plus a simple noise component, the autocorrelation function is underestimated at all lags, with the degree of underestimation depending on the variance of the noise component.

In addition, there is no reason to think of breaks and long memory necessarily as alternatives. Empirical evidence provided in a number of papers points in fact to the presence of both features in the volatility of financial assets and in macroeconomic time series, i.e. structural breaks, even if present in the data, may not fully account for shock persistence.

The most widely studied time series under this perspective is the inflation rate. Recent contributions in the literature (see for instance Hassler and Wolters [19], Baillie et al. [5], Morana [36]), suggest that the accurate modelling of inflation persistence requires a long memory model, possibly allowing for structural change.

Interestingly, Diebold and Inoue [11] have argued that even if the true data generating process shows weak dependence and breaks, a long memory model may be still useful for forecasting. In order to assess the empirical relevance of this conclusion, well specified competing models need to be contrasted over different forecasting horizons. Evidence from Morana and Beltratti [37] provide some empirical support to this conclusion, since modelling the break process does not seem to be important for very short term forecasting (1-step ahead), once the model considers a long memory component. However, at longer forecasting horizons (5–10 steps ahead) accounting for both long memory and structural change may lead to a superior forecasting model.

3 Structural breaks in volatility

There are two differences in our analyses of the currency and stock markets. The first is associated with our search for the potential association with macroeconomic variables: in the case of stock returns we try to evaluate the connection with the volatility of some macroeconomic variables, while in the case of currencies we do not try to relate volatility to the volatility of macroeconomic variables. There are some reasons for this choice. First, in the case of stock markets it is theoretically clear what are the economic fundamentals, i.e. dividends, interest rates, risk premia. The macroeconomic variables which presumably influence the time variation of these fundamentals may be selected among a relatively restricted set. In the case of currencies instead it is more difficult to isolate a little number of potential candidates, especially given that one should look at two countries rather than one. Given that here we are interested in finding evidence of structural breaks and not in testing for a specific structural model we do not pursue the study of the relations among breaks in currency volatility and breaks in macroeconomic variables volatility.

The second difference is due to the data sets. In both cases we focus on daily realized variance processes, even though we have different possibilities for estimating such realized variances. In the case of currencies, the raw data employed in this study are five minutes returns for the mark-dollar and the yen-dollar exchange rates over the period December 1, 1986 - December 1, 1996. The data have been provided by Olsen&Associates and are the same employed by Andersen et al. [1].[1] The daily return is defined as the logarithmic difference between exchange rates recorded at 21 p.m. G.M.T. A number of thin trading/non-trading days have been excluded from the sample, namely the weekend period (from Friday 21:05 GMT to Sunday 21:00 GMT), Christmas (December 24–26), New Year's (December 31–

---

[1] We are grateful to T. Andersen for kindly making us available the daily series employed in this study.

January 2), July Fourth (July 4 or July 3), Good Friday, Easter Monday, Memorial Day, Labor Day, Thanksgiving (and the day after). After data cleaning we are left with a sample size equal to 2445 complete days.

Daily realized variance processes have been obtained by summing five minutes squared returns over 280 successive observations, i.e. by computing $\sigma_t^2 = \sum_{n=1}^{280} r_{t,n}^2$. Theoretical results presented in Andersen et al. [1] and Barndorff-Nielsen and Shephard [6] show that the realized volatility process is a consistent (in the frequency of sampling) and asymptotically normal estimator of integrated variance. This implies that arbitrarily close approximation of integrated variance may be obtained at all the relevant horizons by summing high frequency squared returns. This approach has a clear advantage relative to the practice of squaring low frequency returns, since it yields virtually noise free measures of volatility, and allows therefore to model volatility as an observed variables.

In the case of the stock market and of the macroeconomic variables instead we do not have high frequency data and simply square the difference between the variable and the conditional mean. The realized volatility estimator has been previously employed by Schwert [44] to compute monthly realized variances for stocks by summing squared daily returns. More specifically, the estimated residuals from the conditional mean regressions are employed to construct realized variance processes for interest rates, stock market returns, and money growth, by summing the squared daily and weakly innovations over the corresponding months. For the rate of growth of industrial production and the inflation rate the squared monthly innovations have been employed. Although this yields noisy proxies of the volatility process, it allows to handle in the same statistical framework the volatility persistence analysis, i.e. the structural break and long memory analysis.

3.1 Currency returns volatility

We have computed summary statistics for daily returns, log realized variance and realized variance for the whole sample period 1986–1996, not reported for reasons of space. There is evidence of weak negative asymmetry and strong positive excess kurtosis and no sign of autocorrelation. Realized variance ranges in a much larger interval than returns. The kurtosis of log variance is lower than that of returns and much lower than that of simple variance. However normality is rejected for log variance as well. Finally, the Box-Ljung portmanteau test reveals a strong persistence in the variance and log variance processes. The autocorrelation function for the variance series (not reported for reasons of space) also reveals evidence of long memory. The autocorrelation function for both exchange rates decay slowly but is not statistically significant, according to the standard significance band, already after 60 lags. In addition it becomes negative for some long lags.

We next test for the existence of breaks in volatility by estimating a Markov switching model. Since the break process should reflect a low frequency volatility component, one may argue that neglecting high frequency dynamics, i.e. neglecting daily volatility extreme realizations, may lead to more reliable estimates. Indeed we have found that using monthly observations provided a reliable description of the break process for the two currencies.[2] After some experimentation a two-regime specification was selected for the monthly models. The results are reported in Table 1, Panel A. The estimated regimes are very persistent, i.e. the own transition probabilities are close to one for both variance processes. Secondly, the two processes show a similar duration for the low volatility state (231 days for the yen/dollar exchange and 199 days for the mark/dollar), while the mark/dollar show a higher duration for the high volatility state (203 days for the mark/dollar and 120 days for the yen/dollar).

The number of breaks detected for the mark is 9, while for the yen the number of breaks is 12. Finally, the yen shows higher volatility levels in both states than the mark.

The Inclan and Tiao [24] and Bai [3] approaches yield different results. The number of breaks estimated is 15 for the mark and 14 for the yen when the Inclan and Tiao [24] method is used, and 8 and 7, respectively, when the Bai [3] method is used. Figures 1 and 2 show that (a) the break processes obtained from the Inclan and Tiao and the Markov switching approaches are similar (apart from the longer high volatility period detected by the Markov switching model in the middle of the sample for the mark) and (b) little is lost by assuming recurrent breaks, i.e. by constraining the volatility switches between two levels, since the volatility levels detected by the two approaches are similar.

Are there common factors in the breaks to the two volatility series? In principle one could expect common breaks if the variance of the currency is associated with the variance of fundamentals and if breaks in the variance of the currency are due to breaks in the variance of the fundamentals. In this case the macroeconomic variables for the US economy would represent a common set of fundamental factors, which would be extended by the macroeconomic fundamentals for Germany and Japan. One would expect that breaks in the process of volatility of US macroeconomic series would affect both the mark/dollar and the yen/dollar volatilities, while breaks in, say, German fundamentals would not affect the yen/dollar volatility. Of course, lacking a structural model we cannot exclude the possibility that the volatilities of the two currencies are simultaneously determined by the volatilities of all the macroeconomic variables.

Some informal evidence can be obtained by looking at the previous results. The smoothed probabilities plotted in Figures 1 and 2 point to some overlapping of the same state between currencies. However the null of a single volatility break process seems implausible on the basis of the different behavior of the break processes in the early nineties, where the volatility processes for the two currencies appear to be in different regimes (the yen/dollar is in the low volatility regime, while the mark/dollar is in the high volatility regime). More formal evidence about the existence of a relationship between the two break processes comes from the prediction evaluation analysis, computed by means of a Probit model estimated using the smoothed probabilities for the mark/dollar and yen/dollar exchange rates (see Table 1, Panel B). The results show that the mark/dollar break process accurately predicts

---

[2] Monthly realized variance series have been computed by summing the five minutes returns over 5600 successive observations.

Table 1

Panel A: Markov-switching break processes, monthly data

	$\ln\sigma_t^2$ (mark/dollar)	$\ln\sigma_t^2$ (yen/dollar)
μ_1	−1.2651	−1.1420
μ_2	−0.5620	−0.4326
p_{11}	0.9950	0.9958
p_{22}	0.9950	0.9939
d_1	198.87	230.66
d_2	202.84	119.54

Panel B: Prediction evaluation

	Dep = 0	Dep = 1	Total	Dep = 0	Dep = 1	Total
Total	44	78	122	44	78	122
Correct	33	52	85	0	78	78
% Correct	75.00	66.67	69.67	0	100	69.93
% Incorrect	25.00	33.33	30.33	100	0.0	36.07
Total gain	75.00	−33.33	5.74			
% Gain	75.00	−	15.91			

Panel A reports the switching unconditional means (μ_i) with standard errors in brackets for the log realized variance processes ($\ln\sigma_{t_j}^2$) for the Deutsche mark-US dollar and Yen-US dollar exchange rates. p_{ii} are the own transition probabilities, while d_i indicates the duration of each regime.

Panel B reports statistics on the predictive ability of the Deutsche mark-US dollar break processes for the Yen-US dollar break processes. The last three columns report statistics for the case in which the Yen-US dollar exchange rate smoothed probabilities are regressed on a constant (constant probability model). The success cutoff (C) has been set to 0.5. Total is the total number of zero (Dep = 0) and one (Dep = 1) entries in the dependent variables; Correct is the number of accurate predictions for the zero (Dep = 0) and one (Dep = 1) entries; % Correct and % Incorrect denotes the percentages of accurate and inaccurate predictions, respectively; Total Gain denotes the difference between the percentages of observation correctly predicted by the two models, % Gain denotes Total Gain as a percentage of the numbers of predictive failures in the constant probability model.

the yen/dollar high volatility state 75% of the times and yen/dollar low volatility state 67% of the times. The percentage gain to move from a constant probability model, i.e. ignoring the yen/dollar volatility switching between regimes, to a switching model where the transition across states is explained by the mark/dollar volatility is only about 16%, supporting the lack of a close association between the break processes already noted above. A plausible conclusion is that both common and idiosyncratic elements explains the break process in the two currencies, with idiosyncratic factors dominating over the sample analyzed.

We next investigate the existence of long memory in addition to the presence of structural breaks. The frequency of observation is particularly important for this analysis. If we estimated the long memory properties from the monthly observations

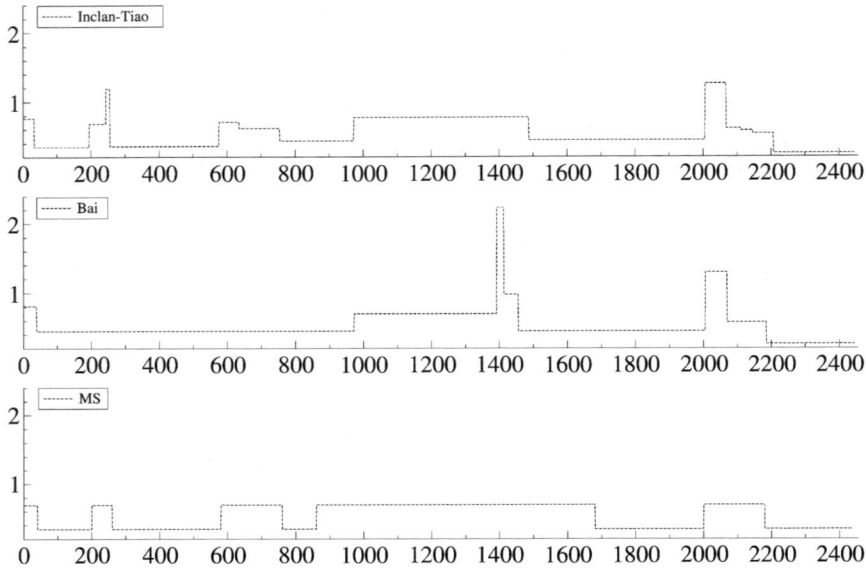

Figure 1. Break processes for mark/dollar realized variance

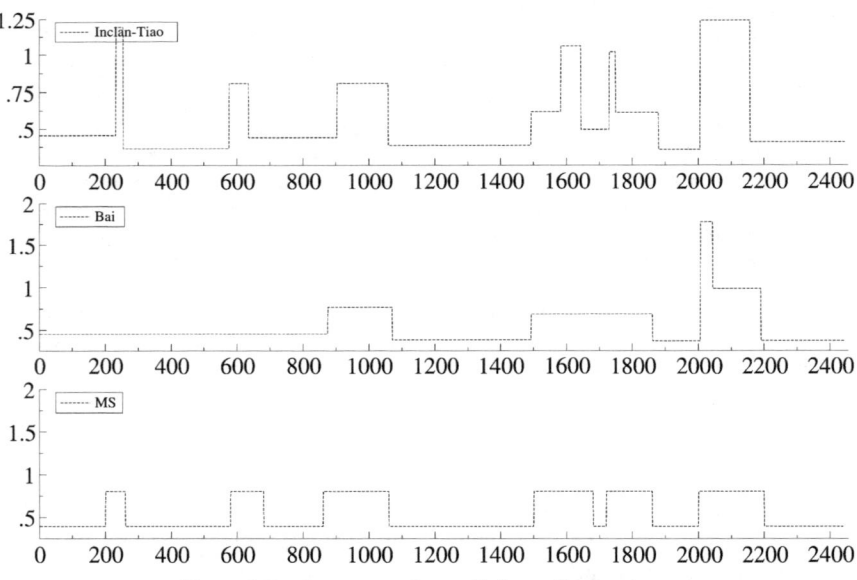

Figure 2. Break processes for yen/dollar realized variance

we would suffer from severe data limitations. Instead, we use the results of the monthly break analysis for a study of the long memory features of the daily series. From the monthly Markov switching model that we have just estimated a daily break process can be obtained by means of a step function, i.e. the daily smoothed probabilities are obtained from the monthly probabilities by keeping constant their

Table 2

Panel A: Stationarity and semiparametric tests: break-free series (mark/dollar)

	$\ln\sigma^2_{tIT}$ (mark/dollar)	$\ln\sigma^2_{tB}$ (mark/dollar)	$\ln\sigma^2_{tMS}$ (mark/dollar)
H_{LW}	0.7306 (0.0566)	0.7810 (0.0585)	0.8350 (0.0470)
H_{LP}	0.9180 (0.0096)	0.9351 (0.0118)	0.9395 (0.0072)
H_{AP}	0.7964 ($-$)	0.8318 ($-$)	0.8793 ($-$)
H_{LM}	0.7712 (0.0220)	0.7726 (0.0278)	0.7576 (0.0372)
H_A	0.8628 (0.0160)	0.8688 (0.0157)	0.8658 (0.0159)
W	23.3 (0.0000)	96.3 (0.0000)	48.1 (0.0000)
LM	54.7 (0.0000)	465.3 (0.0000)	204.6 (0.0000)
LR	40.6 (0.0000)	256.4 (0.0000)	118.7 (0.0000)

Panel B: Stationarity and semiparametric tests: break-free series (yen/dollar)

	$\ln\sigma^2_{tIT}$ (yen/dollar)	$\ln\sigma^2_{tB}$ (yen/dollar)	$\ln\sigma^2_{tMS}$ (yen/dollar)
H_{LW}	0.6370 (0.0581)	0.6790 (0.0598)	0.7370 (0.0395)
H_{LP}	0.6854 (0.0101)	0.7269 (0.0105)	0.8291 (0.0043)
H_{AP}	0.6120 (0.0650)	0.6590 (0.0586)	0.8175 (0.0094)
H_{LM}	0.6740 (0.0361)	0.7153 (0.0325)	0.7305 (0.0354)
H_A	0.8981 (0.0170)	0.9109 (0.0167)	0.9114 (0.0171)
W	23.3 (0.0000)	43.9 (0.0000)	42.5 (0.0000)
LM	54.7 (0.0000)	135.6 (0.0000)	146.2 (0.0000)
LR	40.6 (0.0000)	88.9 (0.0000)	74.3 (0.0000)

The table reports stationarity and semiparametric tests, with standard errors or p-values in brackets, for the daily break-free log realized variance ($\ln\sigma^2_{tj}$) of the Deutsche mark-US dollar and Yen-US dollar exchange rates, using the Markov switching ($j = MS$), Inclan-Tiao ($j = IT$) and Bai ($j = B$) approaches. H_i $i = LW, LP, AP, LM, A$, is the estimate of the Hurst exponent using the Gaussian semiparametric estimator, the log periodogram estimator, the averaged periodogram estimator, the LM estimator, and the $ARFIMA(0, d, 0)$ model. LM is the LM test; W is the Wald test; LR is the Likelihood Ratio test. In the implementation we used $q = 0.5$ for the averaged periodogram estimator and $l = 0$ for the semiparametric estimators.

value over each sequence of twenty observations. The transition matrix for the daily model is then obtained from the monthly one by means of the relation between the own transition probabilities and the duration of the regime (see Morana and Beltratti [37]).

In Table 2 we report the results of the stationarity analysis carried out on the break-free log variances. A first important result is that there is no evidence of spurious breaks for both currencies. In general the estimates of the Hurst exponent provided by the log periodogram regression are higher than those obtained from the other methods employed, and the LM, Wald and LR tests reject the null of weak dependence. On the other hand, the estimates obtained using the ARFIMA(0,d,0) model[3] are typically higher than those obtained using the semiparametric estimators. On the basis of the results of the specification analysis of the ARFIMA models it can be argued that the number of ordinates selected by the optimal bandwidth procedure is too low and that conclusions on the degree of long memory of the break free log variance series should be drawn preferably from the parametric estimator, rather than from the semiparametric estimators.[4] Second, for the Markov-switching break-free processes the ARFIMA estimator suggests that little persistence is explained by the structural breaks. The estimated fractional differencing parameters for the break-free volatility series are in fact smaller than those obtained from the raw series (not reported for reason of space), but still close to 0.40 (0.36 for the mark and 0.41 for the yen).

To gauge evidence on the joint presence of both long memory and structural change, we have carried out a robustness exercise aimed at assessing the impact of modelling the break process as a step function rather than as a smoothly evolving function. One possible advantage of this latter specification is that the conclusions on the long memory properties of the break-free processes should be less affected by the problem of accurately selecting the exact timing of the switches between regimes. To estimate a smoothly evolving break process, we have followed Beran [8] and employed a Kernel approach, obtaining results in line with those of the Markov switching model (see Morana and Beltratti [37]). As a further robustness analysis, we have computed the Hurst exponent for the break-free processes and plotted the estimates against the bandwidth (not reported for reasons of space). The plots show that in correspondence of the stable regions there is clear evidence of long memory according to all the estimators (estimates range between a minimum of 0.66 and a maximum of 0.91). Point estimates obtained by the ARFIMA model are 0.8450 (0.0169) and 0.8955 (0.0156), for the mark (30 and 10 parameters), 0.90 (0.0175) and 0.9329 (0.0163) for the yen. The estimates are in line with the results obtained using the Markov switching procedure, allowing to conclude against the evidence of spurious long memory, and validating the use of step functions to model break processes.

Overall the results suggest that accounting for structural breaks has an impact on the persistence properties of the variance series and that structural change is an important feature of the data which should be modelled. This result is also consistent with the findings of Beine and Laurent [7], using a Markov-switching FIGARCH model. It is interesting to note that our finding of long memory in

---

[3] A parsimonious ARFIMA(0,d,0) was found sufficient to model the raw and break-free processes according to the Ljung-Box test. Only for the break-free yen/dollar variance processes there is some evidence of serial correlation in the residuals. However, including up to five autoregressive terms did not improve the specification of the model. Moreover, the inclusion of the autoregressive terms caused the estimated fractional differencing parameter to assume negative values, results that is in contrast with the semiparametric analysis.

[4] Optimal bandwidth theory allows to select the largest possible bandwidth, avoiding the upper bias in the Hurst exponent due to the presence of short memory components in the process. Since the specification analysis suggests that short memory components are not a feature of the log variance processes, estimates of the Hurst exponent should be computed by using $T/2$ periodogram ordinates, which would ensure the highest efficiency. An alternative full band estimator is the ML estimator of Sowell [46], which is the most efficient across the various estimator employed in the paper.

the break-free series is in contrast with the finding of Granger and Hyung [14], although a possible explanation for this discrepancy can be found in the use of the log periodogram estimator and in the selection of a relatively low number of ordinates of the periodogram ($m = \sqrt{T}$ in their study). It can be concluded that both long memory and structural change are characteristics of the data generating process of the variance series considered.

3.2 Stock returns volatility

As noted above, the estimator for the conditional volatility of the macroeconomic variables is going to be based on the innovation process. Two steps were followed in the computation of the residuals. The first step is concerned with the detection of breaks in the conditional mean, while the second step is concerned with the modelling of the persistence of the break-free series by standard methodologies. In the second step the residuals employed in the computation of the volatility series were obtained by fitting parametric models to the break-free series. We have estimated ARFIMA models for all the series, apart from M1 growth, since the latter series was found to be well described by a weakly stationary process. The lag length of the ARMA component has been determined using the Schwarz information criterion, while the fractional differencing parameter has been fixed to the value determined by the semiparametric analysis. Two lags were selected for the short term and long term interest rates, while seven lags were selected for the Federal funds rate. On the other hand, a larger number of lags were selected for nominal money growth (thirteen), while a zero order ARMA component was selected for inflation and an AR(1) model was found to fit well the output growth series.

In Figure 3 we report the estimated conditional variance series for stock returns and various macroeconomic variables. To help the readability of the graphs, outlying observations were removed from the data and the variance processes for the Federal funds rate, nominal money growth, and industrial production were rescaled to match the range of variation of the stock volatility series.[5] In Figure 3 we plot only the Federal funds rate volatility, since the short term rate volatility process is similar to the one for the Federal Funds rate, while the long term rate volatility process shows a lower variability than the other two series. Coherent with Schwert [44], stock volatility tends to show a higher unconditional mean and standard deviation than the volatility process of the macroeconomic factors.

From Figure 3 it can be noticed that the different macroeconomic factors are associated in various ways with stock market volatility over different periods. For instance, the Federal funds rate and the stock market both go through a period of high volatility following the two oil price shocks. The same result holds for inflation and output growth. Interestingly, the relationship between M1 and stock volatility seems to be more stable than for the other macroeconomic factors, in particular for the last part of the sample, characterized by large volatility in both variables. Overall, these pictures seem to clarify why it may be hard to explain stock

---

[5] The outlying observations are October 1987 for the S&P500 returns volatility and January and February 1970, and September 2001 from M1 growth volatility.

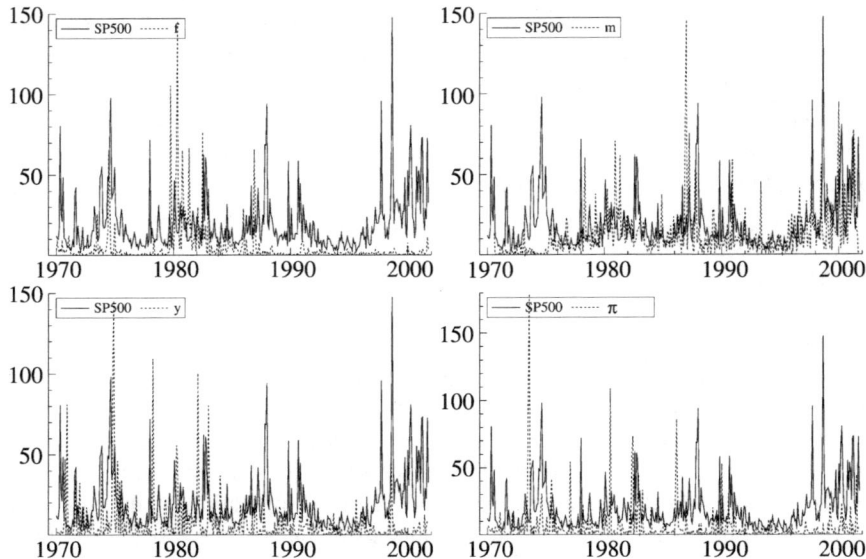

Figure 3. Realized variance processes (SP500: Standard and Poors 500 Index returns; m: M1 growth rate; y: indsutrial production growth rate; π: inflation rate). The volatility processes, apaprt from the inflation series, have been rescaled to match the range of the volatility of the S&P 500 retrurns

return volatility on the basis of a fixed-weight linear combination of the volatility of macroeconomic variables, as is done by the linear regression framework.

In Table 3 we have reported some summary statistics on the break processes in the log variance series estimated by means of the Markov switching approach. In order to model infrequent changes only, we decided not to estimate a break if the new regime lasted for less than a year. Interestingly, we were unable to find any significant break points in the inflation and industrial production volatility processes only. The Federal funds rate and the short term rate show similar break processes, while the break process for the long term rate is not closely related to those of the other interest rates. The break process for M1 growth also behaves differently from the other processes. No break points were detected in output growth by the Markov switching model, although imposing one break point in 1984:5 seems to be appropriate according to the graphical inspection of the actual data and the analysis of the long memory properties of the series.

In Figure 4 we plot the estimated break processes. The series are scaled in order to match means and ranges of the variance series. It is interesting to note that the dating of the break points in the variance of the factors and in the variance of returns is closely related. For instance, the increase in stocks volatility following the oil shocks is matched by an increase in the volatility of the Federal funds rate and the short term rate. In addition, the increase in volatility at the end of the 1980s and 1990s is matched by an increase in the volatility of M1 growth. On the other hand, the increase in volatility which took place in 1991 can also be related to the volatility of the factors, in particular to the Federal funds rate. Since the switch to the higher volatility regime lasted for less than a year, the break point was not estimated for

Table 3. Structural break analysis: log-variance processes

	s	l	f	m	$SP500$	y
μ_1	−4.0835	−4.3650	−2.5404	−1.4435	2.2041	−1.8132
	(0.1520)	(0.1239)	(0.0875)	(0.1119)	(0.0568)	(0.1761)
μ_2	−2.6181	−3.2159	−1.1434	−0.4152	3.2269	−2.5985
	(0.1146)	(0.0621)	(0.1534)	(0.1454)	(0.0758)	(0.2474)
μ_3	−0.2828	−1.6161	0.7457			
	(0.1388)	(0.1021)	(0.1690)			
p_{11}	0.9413	0.9572	0.9732	0.9824	0.9554	−
p_{22}	0.9450	0.9769	0.9175	0.9712	0.9463	−
p_{33}	0.9411	0.9652	0.9134			
$\Phi(\beta_1)$	0.6312	0.7080	0.6731	0.8144		
$\Phi(\beta_2)$	0.6398	0.5371	0.4579	0.1301		
$\Phi(\beta_3)$	0.1451	0.4767	0.1608			
break	1973/5 1975/3 1976/6 1977/8 1979/8 1982/12 1992/10 2000/4	1971/12 1974/11 1976/1 1978/8 1979/10 1986/7	1971/4 1973/1 1975/7 1979/10 1982/11 1986/5	1979/3 1983/5 1986/9 1987/12 1996/6	1973/3 1975/10 1979/12 1983/3 1986/4 1988/9 1990/8 1991/7 1997/2	1984/5

The table reports the estimated means (μ_i) in the different regimes ($i = 1, ..., 3$), the own transition probabilities (p_{ii}) and the estimated break points for the log variance processes for the short term interest rate (s), the long term interest rate (l), the Federal Funds rate (f), M1 growth (m), and the S&P500 returns ($SP500$). $\Phi(\beta_i)$ $i = 1, .., 3$ is the probability that the factor is in the volatility state i when the stock market is in the low volatility state. For interest rate series state $i = 1, .., 3$ denotes the low, average and high volatility state, respectively. For M1 growth $i = 1, .., 2$ denotes the low and high volatility state, respectively.

the Federal funds rate. Finally, stock market volatility and industrial production volatility regimes do not seem to be related, even though there is a clear visual association between the output gap and the regimes of stock return volatility. The volatility states are therefore in close accordance with the dynamics of the realized volatility and suggest a time-varying relation among stock returns volatility and macroeconomic volatility.

In Table 4 we report the results of the long memory analysis for the break-free log variance series. By comparing the estimates obtained using optimal bandwidth theory with those obtained in correspondence of the stable region, as suggested by Taqqu and Teverovski [48], it is possible to note that again optimal bandwidth theory leads to the selection of too conservative bandwidths. In general we do not find any evidence of spurious breaks. The point estimates of the Hurst exponent provided by the Taqqu-Teverovski method for the LM estimator are in the range 0.60-0.65 for inflation, the short term rate and the long term rate, 0.60-0.67 for the S&P500 returns, 0.57-0.60 for the Federal Funds rate, 0.49-0.53 for M1 growth, and 0.53-0.58 for the industrial production rate of growth. Hence, there is evidence of long memory in all the break-free log variance series, apart from M1 growth

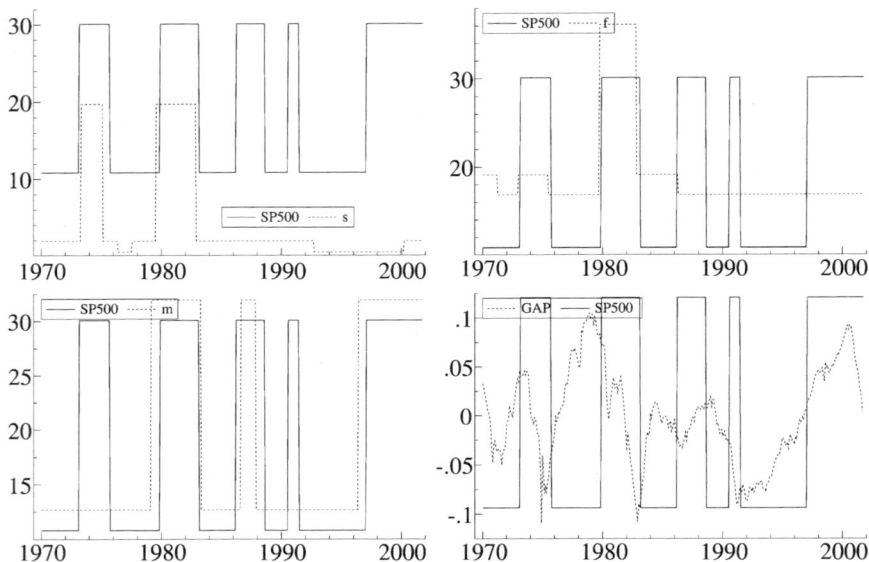

Figure 4. Estimated break processes (short term rate (s), lomg term rate (l), Federal Fund rate (f), M1 growth (m), output gap (GAP), S&P 500 returns (SP500))

and possibly industrial production growth, which, on the other hand, might be well described by weakly dependent processes.

4 Conclusions

The existence of economic regimes is in no way essential to the TRB, but many simulation models associated with the use of the TRB may be interpreted as giving rise to regimes of various types. It is therefore important to assess whether actual economic and financial time series may be characterized by the presence of regimes. Given that there is already evidence of systematic instability in macroeconomic time series, we have chosen here to explore the existence of regimes in the volatility process of macroeconomic and financial time series, including currencies. We have used where possible an estimate of time-varying volatility retrieved from the availability of intra-day high frequency data. When high frequency data are not available to us we have used the square of the residual of each series from a well-specified process for the conditional mean. At the methodological level we have compared various techniques in order to assess the robustness of the results. We have used the Markov switching model, methods for finding non recurrent breaks and methods which allow for a smooth evolution of the regime. We have also evaluated the results on the existence of breaks with semi-parametric techniques which also study the existence of long memory in the break-free volatility series.

Our results for currencies, US macroeconomic variables and the US stock market are coherent: there is robust evidence of structural breaks in the volatility processes. Moreover, the break-free series are not described by a simple autoregressive

Table 4. Semiparametric analysis and stationarity tests: volatility processes

	s_{bf}	l_{bf}	f_{bf}	$SP500_{bf}$	m_{bf}	π	y
	0.71	0.68	0.61	0.65	0.61	0.48	0.65
H_{LW}	(0.07)	(0.06)	(0.06)	(0.09)	(0.07)	(0.04)	(0.09)
	[52]	[60]	[60]	[29]	[58]	[176]	[51]
	0.68	0.68	0.58	0.55	0.50	0.61	0.62
H_{LP}	(0.01)	(0.02)	(0.01)	(0.01)	(0.02)	(0.01)	(0.02)
	[146]	[88]	[120]	[85]	[108]	[299]	[105]
	0.67	0.64	0.58	0.74	0.71	0.41	0.55
H_{AP}	(0.08)	(0.07)	(0.08)	(0.20)	(0.07)	(0.08)	(0.09)
	[54]	[67]	[62]	[24]	[74]	[91]	[47]
	0.64	0.65	0.58	0.55	0.55	0.54	0.55
H_{LM}	(0.07)	(0.04)	(0.06)	(0.07)	(0.07)	(0.03)	(0.07)
	[53]	[189]	[74]	[46]	[57]	[233]	[53]
W	4.04	17.43	1.83	0.45	0.60	1.20	0.57
	(0.04)	(0.00)	(0.18)	(0.51)	(0.44)	(0.27)	(0.44)
LM	7.70	35.94	2.57	0.55	0.75	1.39	0.72
	(0.01)	(0.00)	(0.11)	(0.46)	(0.39)	(0.24)	(0.40)
LR	7.09	28.64	2.76	0.65	1.02	1.59	0.40
	(0.01)	(0.00)	(0.10)	(0.42)	(0.31)	(0.21)	(0.77)

The table reports semiparametric estimates of the Hurst exponent and stationarity tests, with standard errors or p-values in brackets and selected bandwidth in square brackets. H_i $i = LW, LP, AP, LM, LP_m$ is the estimate of the Hurst exponent using the Gaussian semiparametric estimator, the log periodogram estimator, the averaged periodogram estimator, the LM estimator, and the multivariate log periodogram estimator, respectively. LM is the LM test; W is the Wald test; LR is the Likelihood Ratio test. In the implementation we used $q = 0.5$ for the averaged periodogram estimator and $l = 0$ for the log periodogram estimator. The variables are the break-free log-variance processes for the short term interest rate (s), the long term interest rate (l), the Federal Funds rate (f), and the S&P500 returns ($SP500$), the inflation rate (π), the rate of growth of industrial production (y) and the break-free log variance process for the rate of growth of M1. The data are monthly realized volatility processes for the period 1970:1-2001:9.

model. Rather, they are characterized by long memory, i.e. by a slow decay of the shocks. It is no wonder that many analysts and investors come up with many different models for analyzing data and with many different forecasts. The higher the complexity in the description of the relevant time series the higher the probability that different agents come up with different specifications of the same series. The interpretation of models arising from the application of the TRB in terms of economic and financial regimes is therefore a welcome step towards a realistic assessment of the empirical evidence.

Of course our analysis is only a first step towards a complete understanding of the dynamic evolution of the macroeconomic and financial time series. Other steps are necessary. First, we have only analyzed the dynamics of the second moment of the series. Given extensive evidence of non-normality of many time series, particularly in the financial sector, it is important to understand more about the dynamics of higher moments. Second, we have not studied here the connections between the volatility of the macroeconomic variables and the volatility of the financial variables. It is however important to understand whether there is a unidirectional link between the two groups or not. In principle one could expect that macroeconomic volatility forecasts stock market volatility, even though there may also be a link going from financial markets volatility to macroeconomic volatility. For example extended periods of high volatility in the currency and in the stock markets may affect the decisions of consumers and firms and cause periods of uncertainty in the real economy.

Third and most importantly, it would be useful to test for the implications of theoretical models, in order to see whether the restrictions imposed by the theories are rejected or not by the data. Measuring the existence of breaks and long memory is only an initial effort. However we need theory in order to understand the economic causes of breaks and persistency and the reactions of the economic agents to such events.

References

1. Andersen, T.G., Bollerslev, T., Diebold, F.X., Labys, P.: The distribution of exchange rate volatility. Journal of the American Statistical Association **96**, 42–55 (2001)
2. Ang, A., Bekaert, G.: Regime switches in interest rates. Journal of Business and Economic Statistics **20**, 173–182 (2002)
3. Bai, J.: Estimating multiple breaks one at a time. Econometric Theory **14**, 315–352 (1997)
4. Bai, J., Perron, P.: Estimating and testing linear model with multiple structural changes. Econometrica **66**, 47–78 (1998)
5. Baillie, R.T., Chung, C.F., Tieslau, M.A.: Analysing inflation by the fractionally integrated ARFIMA-GARCH model. Journal of Applied Econometrics **11**, 23–40 (1996)
6. Barndorff-Nielsen, O.E., Shephard, N.: Econometric analysis of realised covariation: high frequency covariance, regression and correlation in financial economics. Mimeo, University of Aahrus (2002)
7. Beine, M., Laurent, S.: Structural change and long memory in volatility: new evidence from daily exchange rates. Mimeo, University of Liege (2000)
8. Beran, J.: SEMIFAR models – a semiparametric framework for modelling trends, long range dependence and nonstationarity. Manuscript, University of Kostanz (1999)
9. Delgado, M.A., Robinson, P.M.: Optimal spectral bandwidth for long memory. Statistica Sinica **6**, 97–112 (1996)
10. Diebold, F.: Comment on "modeling the persistence of conditional variance" by R. Engle and T. Bollerslev. Econometric Reviews **5**, 51–56 (1986)
11. Diebold, F.X., Inoue, A.: Long memory and regime switching. Journal of Econometrics **105**, 131–159 (2001)
12. Fama, E.F., French, K.R.: The equity premium. Journal of Finance (forthcoming) (2002)
13. Geweke, J., Porter-Hudak, S.: The estimation and application of long memory time series. Journal of Time Series Analysis **4**, 221–238 (1983)
14. Granger, C.W.J., Hyung, N.: Occasional structural breaks and long memory. Manuscript, University of California, San Diego, Department of Economics (1999)
15. Granger, C.W.J., Marmol, F.: The correlogram of a long memory process plus a simple noise. Manuscript, University of California, San Diego, Department of Economics (1997)
16. Granger, C.W.J., Terasvirta, t.: A simple non linear time series model with misleading linear properties. Economics Letters **62**, 161–165 (1999)
17. Hamilton, J.D.: A new approach to the economic analysis of nonstationary time series and the business cycle. Econometrica **57**(2), 357–384 (1989)
18. Harris, M., Raviv, a.: Differences of opinion make a horse race. Review of Financial Studies **6**, 473–506 (1993)

19. Hassler, U., Wolters, J.: Long memory in inflation rates: international evidence. Journal of Business and Economic Statistics **13**, 1 (1995)
20. Henry, M., Robinson, P.M.: Bandwidth choice in Gaussian semiparametric estimation of long range dependence. In: Robinson, P.M., Rosemblatt, M. (eds.) Athens Conference in Applied Probability and Time Series Analysis, vol. II. Time series analysis. In memory of E.J. Hannan. pp. 220–232. Berlin Heidelberg New York: Springer 1996
21. Henry, M.: Robust automatic bandwidth for long memory. Journal of Time Series Analysis (forthcoming) (2001)
22. Hosking, J.R.M.: Asymptotic distributions of the sample mean, autocovariance and autocorrelations of long-memory processes. Journal of Econometrics **73**, 261–284 (1996)
23. Hurvich, C.M., Deo, R., Brodsky, J.: The mean squared error of geweke and Porter-Hudak's estimator of the memory parameter of a long memory time series. Journal of Time Series Analysis **19**, 19–46 (1998)
24. Inclan, C., Tiao, G.C.: Use of cumulative sum of squares for retrospective detection of change in variance. Journal of the American Statistical Association **89**, 913–923 (1994)
25. Kuan, C.-M., Hsu, C.-C.: Change point estimation of fractionally integrated processes. Journal of Time Series Analysis **19**(6), 693–708 (1998)
26. Kuan, C.-M., Hsu, C.-C.: Long memory and structural change: testing method and empirical estimation. Mimeo, National Central University of Taiwan (2000)
27. Kunsch, H.R.: Statistical aspects of self similar processes. In: Prohorov, Y., Sazanov, V.V. (eds.) Proceedings of the First World Congress of the Bernoulli Society, 1, pp. 67–74. Utrecht: VNU Science Press 1987
28. Kurz, M.: Endogenous economic fluctuations. Berlin Heidelberg New York: Springer 1997
29. Lamoreaux, C.G., Lastrapes, W.D.: Persistence in variance, structural change and the GARCH model. Journal of Business and Economic Statistics **8**, 225–234 (1990)
30. Lobato, I.N., Robinson, P.M.: Averaged periodogram estimation of long memory. Journal of Econometrics **73**, 303–324 (1996)
31. Lobato, I.N., Robinson, P.M.: A nonparametric test for I(0). Manuscript, Review of Economic Studies **65**, 475–496 (1997)
32. Lucas, R.E. Jr.: Adaptive behavior and economic theory. Journal of Business **59**, S401–S426 (1986)
33. Mikosch, T., Starica, C.: Change of structure in financial time series, long range dependence and the GARCH model. Manuscript, University of Groningen, Department of Mathematics (1998)
34. Mikosch, T., Starica, C.: Limit theory for the sample autocorrelations and extremes of a GARCH (1,1) process. Annals of Statistics (forthcoming) (2000)
35. Morana, C.: Measuring core inflation in the Euro area. ECB Working Paper Series, no. 36 (2000)
36. Morana, C.: Common persistent factors in inflation and excess nominal money growth and a new measure of core inflation. Università del Piemonte Orientale, SEMEQ Working Paper Series (2002)
37. Morana, C., Beltratti, A.: Structural change and long range dependence in volatility of exchange rates: either, neither or both? Università del Piemonte Orientale, SEMEQ Working Paper Series (2001)
38. Perron, P.: The great crash, the oil price shock, and the unit root hypothesis. Econometrica **57**, 1361–1401 (1989)
39. Robinson, P.M.: Semiparametric analysis of long memory time series. Annals of Statistics **22**, 515–539 (1994)
40. Robinson, P.M.: Rates of convergence and optimal spectral bandwidth for long range dependence. Probability Theory and Related Fields **99**, 443–473 (1994)
41. Robinson, P.M.: Log periodogram regression of time series with long range dependence. The Annals of Statistics **23**, 1048–1072 (1995)
42. Robinson, P.M.: Gaussian semiparametric estimation of long range dependence. The Annals of Statistics **23**, 1630–1661 (1995)
43. Robinson, P.M.: Comment. Journal of Business and Economic Statistics **16**(3), 276–279 (1998)
44. Schwert, G.W.: Why does stock market volatility change over time? The Journal of Finance **XLIV**(5), 1115–1153 (1989)
45. Shiller, R.J.: Stock prices and social dynamics. Brooking Papers on Economic Activity **2**, 457–498 (1984)

46. Sowell, F.: Maximum likelihood estimation of stationary univariate fractionally integrated time series models. Journal of Econometrics **53**, 165–188 (1992)
47. Stock, J.H., Watson, M.W.: Evidence of structural instability in macroeconomic time series relations. NBER Technical Working Paper 164 (1994)
48. Taqqu, M.S., Teverovsky, V.: Semi-parametric graphical estimation techniques for long memory data. In: Robinson, P.M., Rosemblatt, M. (eds.) Time series analysis in memory of E.J. Hannan, pp. 420–432, Berlin Heidelberg New York: Springer 1998
49. Timmerman, A.: Structural breaks, incomplete information and stock prices. University of California, San Diego, Discussion Paper, n. 2 (2001)

Effect of credible quality investment with Bertrand and Cournot competition[*]

Reiko Aoki

Department of Economics, University of Auckland, Auckland, NEW ZEALAND
(e-mail: r.aoki@auckland.ac.nz)

Received: February 8, 2000; revised version: February 14, 2002

Summary. We show how credible revelation and ability to commit to quality choice effect equilibrium qualities and welfare when product market is either Bertrand or Cournot competition. We show that results depend on the type of competition but not generally on the cost of quality function. We show that with Bertrand competition, the equilibrium qualities are lower with credible commitment. Competition is moderated and producer surplus is higher and consumer surplus lower. With Cournot competition, higher quality will be better but lower quality will be worse with credible commitment. Consumer surplus is always greater with credible commitment and if cost does not increase too quickly with quality, producer surplus will also increase. Thus credible commitment is a collusive device with Bertrand competition but it can improve social welfare with Cournot competition.

Keywords and Phrases: Vertical quality differentiation, Bertrand and Cournot competition, Commitment, Sequential vs. simultaneous choices.

JEL Classification Numbers: L1, D4, C7.

1 Introduction

The purpose of this paper is to analyze the effect of credible revelation and commitment of vertically quality investment on equilibrium qualities, profits and welfare.

---
* The idea of this paper originated in the weekly workshops of Mordecai Kurz at Stanford. I am forever in debted to Mordecai and fellow students – Luis Cabral, Peter DeMarzo, John Hillas, Michihiro Kandori, Steve Langois, Patrick McAllister, Steve Sharpe, Peter Streufert, Steve Turnbull and Gyu-Ho Wang – for their criticism and encouragement. I also benefited from comments from Yi-Heng Chen, Jin-Li Hu, Kala Krishna, Jinji Naoto, Thomas J. Prusa, and Shyh-Fang Ueng at various later stages of this work. Last but not least, I am grateful for the detailed comments of the referee.

We show how they depend on the nature of product market competition, Bertrand (price competition) or Cournot (quantity competition), when cost of quality is,

$$C(q) = kq^n,$$

where q is the quality level, k is a positive constant and n is any integer greater than 2. Product market competition matters because qualities are locally strategic complements with Bertrand competition but are locally strategic substitutes with Cournot competition in the relevant range. One expects that it may be more profitable to produce the inferior quality if rate at which cost increases with quality (parameter n) is large. Surprisingly, we find that our results are robust to size of n (with one exception). In equilibrium, since the cost function is smooth, any reduction of higher quality is accommodated by reduction of the lower quality.

We employ a duopoly game with two stages – the quality-setting first stage and the second stage in which sales are made. Products are vertically differentiated in quality. We compare situation where one firm can credibly reveal and commit to its quality choice (sequential move) and where this is not possible (simultaneous move) in the first stage.[1] We compare the subgame perfect equilibrium qualities under the two timing scenarios in the first stage and their welfare implications. The comparison is done with Bertrand second stage and Cournot second stage.

With Bertrand competition, profitability is determined by the relative distance of the two qualities i.e., extent of quality heterogeneity since firms price to compete for the marginal consumer. Both increase in high quality and decrease in low quality increase the distance of qualities or the heterogenity and this is profitable for both firms due to reduced competition. Thus depending on if the firm produces higher or lower quality, second stage equilibrium profit will be increasing or decreasing in own quality and decreasing or increasing in rival quality. With Cournot competition, increase in own quality will always increase the market clearing price while increase in rival quality always decreases the price. Thus second stage equilibrium profit is increasing in own quality and decreasing in rival quality, independent of if the firm produces higher or lower quality.

With both types of competition in the second stage, the first mover will choose to be the high quality firm in equilibrium. Recall that with both types of competition, high quality firm's profit is decreasing in low quality. Thus whether if the first mover increases or reduces own quality compared to simultaneous move is determined by if qualities are strategic complements or substitutes for the low quality (second mover). With Bertrand second stage, low quality firm's marginal profit with respect to own quality is negative because improvement of own quality reduces the difference in qualities. This negative effect is reduced when qualities are farther apart., i.e., cross derivative is *positive* (reduces absolute value of negative marginal profit). In order to reduce second mover quality, first mover *reduces* quality compared to the simultaneous choice.

---

[1] Previous analyses of endogenous quality choices have been confined to simultaneous choices (Gabszewicz and Thisse, 1980; Shaked and Sutton, 1982; Bonanno, 1986; Boyer and Moreaux, 1987; Motta, 1993). Motta has an upper bound $\bar{v}$ of qualities. This actually is a bound on revenue relative to cost. Thus $\bar{v}$ is the inverse of k in our paper. To be precise, $\bar{v} = 2/k$.

With Cournot second stage, low quality firm's marginal profit with respect to own quality is positive. This positive effect is greater when the qualities are closer: cross derivative is *negative* (increase of high quality increases distance which reduces positive marginal profit). In order to reduce second mover quality, first mover *increases* its quality compared to simultaneous move.

With both types of competition, it is more profitable to be producing the higher quality in simultaneous moves equilibrium for any n. Equilibrium qualities are lower with larger n, but as higher quality becomes less profitable, lower quality becomes even less profitable because the cost function is smooth. The first mover of sequential moves chooses to be the high quality firm for the same reason.

The effect of timing on consumer surplus also depends only on type of competition and not on n: sequential moves consumers surplus is smaller with Bertrand competition and it is greater is Cournot competition. The lower equilibrium qualities with sequential move translate to lower consumer surplus when second stage is Bertrand. sequential moves allows the first mover to internalize the competition, moderating the competition for the marginal consumer. This is bad for the consumer but implies producer surplus (aggregate profit) is greater with sequential moves for any n. There is no such competition for the marginal consumer with Cournot competition and the greater difference in qualities with sequential moves (higher higher quality, lower lower quality) does not effect consumers adversely. Consumers actually benefit from the better higher quality. The first mover, the high quality firm, will gain more than what second mover, the low quality firm, loses if cost of quality does not increase too much (n is small) and producer surplus is greater with sequential move. However if cost of quality does increases very quickly (n large), reverse is the case and producer surplus is smaller with sequential move. With Cournot competition, it is possible to increase both consumer and producer surplus by providing an opportunity to credible commit to quality if n is not too large.

This paper shows that the coordinating effect of sequential quality choice shown in Aoki and Prusa (1996) can be extended to any cost fucntion $C(q) = kq^n$ for any n and also to Cournot competition as long as n is not too large. However the effect on consumers will be reversed with Cournot competition. Aoki and Prusa argued that ability for a firm to credibly reveal and commit to quality was a coordinating device which hurts consumers. We have shown here that such an institution can also benefit consumers when market is Cournot competition and cost of quality does not increase too quickly.

The remainder of the paper proceeds as follows: In the next section we sketch a model of vertical quality differentiation with deterministic quality investment. In Section 3 we will characterize the equilibrium under the two timing scenarios when firms engage in Bertrand competition at the sales stage. In Section 4 we will do the same for Cournot competition. In both sections the method of analysis is characterization of the best-response correspondences. Only intuitive explanations are given. A complete analysis, including all proofs and a more detailed exposition can be found in Aoki (1995, 2001b). An examination of the robustness of the results is in Section 5.

2 The model

We employ the standard model of vertical quality differentiation with heterogeneous consumers (Gabszewicz and Thisse, 1980; Shaked and Sutton, 1982; Bonanno, 1986). There are two otehrwise identical firms, 1 and 2, that produce vertically differentiated products. In the first stage each firm chooses quality level, q_i, of its product at cost $C(q_i) = kq_i^n$, where $k > 0, i = 1, 2$ and $n \geq 2$ is an integer. We consider two cases:

   i) [simultaneous moves] q_i's are chosen simultaneously
   ii) [sequential moves] q_1 is chosen first and revealed, then q_2 is chosen.

In both cases, quality choices are revealed at the end of the first stage. Products are then produced and sold in the second stage. We consider two possible modes of competition in the second stage: Bertrand competition (firms choose prices) and Cournot competition (firms choose quantities).

We first describe the demand system. There is a continuum of consumers indexed by t which is uniformly distributed on $[0,1]$. A consumer with index $t \in [0, 1]$ consuming one unit of product with quality q_i at price p_i has surplus of $v(q_i, p_i; t) = q_i t - p_i$. A consumer will buy one unit of a product when surplus is positive and greater than the surplus from consuming the other product.

Specifically, a type t consumer will purchase the lower quality product, say q_j, if and only if $0 \leq v(q_j, p_j; t) > v(q_i, p_i; t)$ and will purchase the higher quality product, q_i, if and only if $0 \leq v(q_i, p_i; t) \geq v(q_j, p_j; t)$. The quantity of the high quality product sold is $x_i(p_i, p_j) = 1 - t_i(p_i, p_j)$ where $v(q_i, p_i; t_i(p_i, p_j)) = v(q_j, p_j; t_i(p_i, p_j))$. The quantity of the low quality product sold is $x_j = t_i(p_i, p_j) - t_j(p_j, p_i)$ where $0 = v(q_j, p_j; t_j(p_i, p_j))$.

For any pair of qualities, q_j and q_i, quantities (x_i, x_j) or prices (p_i, p_j) may be the choice variables. With Bertrand competition, firm i chooses p_i to maximize revenue $p_i x_i$. With Cournot competition, it will choose x_i. Since the focus of the paper is the timing of quality choices, we omit derivation of equilibrium of the second stage (Aoki, 2001a).

It was shown in Aoki (2001a) that unlike Bertrand competition, firms' profits is always increasing in firm's own quality and decreasing in that of rivals. This is true even when quality of lower quality firm increases. With Bertrand competition, firms' profitability is determined by extent of heterogenity of products, or the distance of qualities. This is because firms price to compete for the marginal consumer which become more fierce as products become more homogeneous. With Cournot competition, firms compete in quantities and prices adjust to clear the market. A firm's market clearing price will increase if its quality improves and will decrease if rival's quality improves. Thus profit is always increasing in own quality and decreasing in rival quality, even if the product become more homogeneous as result. What is also important for our result is that second order effect, i.e., how marginal profits change with quality, is the same with two types of competitions - magnitude of change in profit is greater when qualities are closer.

3 Quality choices under Bertrand competition

Subgame perfect Nash equilibrium requires that all actions constituting the equilibrium strategies are optimal in each subgame. In our context, this implies that we only need to examine the second stage revenue generated by the Nash equilibrium actions (prices or quantities) of the second stage subgames, one for each possible pair of qualities, (q_i, q_j). Let the equilibrium revenue function of the second stage subgame with quality pair (q_i, q_j) be $R^i(q_i, q_j), i = 1, 2$. We summarize the properties of the revenue function below.[2]

Lemma 1 *When there is Bertrand competition at the sales stage, the equilibrium revenue function is continuous for all (q_i, q_j) and twice continuously differentiable for all $q_i \neq q_j$.*

$$R^i(q_i, q_j) = \begin{cases} \frac{q_i q_j (q_j - q_i)}{(4q_j - q_i)^2}, & q_i \leq q_j \\ \frac{4(q_i)^2(q_i - q_j)}{(4q_i - q_j)^2}, & q_i \geq q_j \end{cases},$$

$$\begin{cases} R_i^i \gtreqless 0 \Leftrightarrow q_i \lesseqgtr \frac{4}{7}q_j, R_j^i > 0, R_{ii}^i < 0, R_{ij}^i = R_{ji}^i > 0, \ q_i < q_j \\ R_i^i > 0, R_j^i < 0, R_{ii}^i < 0, R_{ij}^i = R_{ji}^i > 0, \ q_i > q_j \end{cases}$$

A typical revenue function for firm 1 is depicted in Figure 1.[3] When firm 2's quality is q_2, firm 1's revenue is depicted by the thick revenue line; the thin revenue line depicts firm 1's revenue when firm 2's quality is slightly larger, say q_2'. Basically, with Bertrand competition, both firms prefer qualities to be farther apart. There is marginal gain when qualities become farther apart ($R_j^i > 0$ when $q_i < q_j$ and $R_j^i < 0$ when $q_i > q_j$). Similarly, (the absolute value of) marginal loss ($R_i^i < 0$ or $R_j^i < 0$) becomes smaller as qualities become farther apart ($R_{ij}^i = R_{ji}^i > 0$).

The payoff relevant for our analysis of the first stage is $\Pi^i(q_i, q_j) = R^i(q_i, q_j) - C(q_i)$. As shown in Figure 1 there are always two local maxima to the firm's payoff maximization problem: one below and one above the rival's quality level, denoted by $q^L(q_j) < q_j$ and $q^H(q_j) > q_j$, respectively. Both local maxima satisfy the first-order condition of maximization.[4] It follows from the signs of R_j^i's in (1) and the envelope theorem that $\Pi^i(q^H(q_j), q_j)$ is decreasing in q_j while $\Pi^i(q^L(q_j), q_j)$ is increasing in q_j. Thus, there is a unique $\hat{q}$ such that $\Pi^i(q^L(\hat{q}), \hat{q}) = \Pi^i(q^H(\hat{q}), \hat{q})$ and $\beta^i(q_j) = q^H(q_j)$ for $q_j \leq \hat{q}$ and $\beta^i(q_j) = q^L(q_j)$ for $q_j \geq \hat{q}$.

Since both local optima are interior solutions, total differentiation of the first-order condition of maximization and (1) yields the following.

Lemma 2 *When there is Bertrand competition in the second stage, the first stage best-response correspondence is,*

$$\beta^i(q_j) = \begin{cases} q^H(q_j) > q_j \text{ and increasing for } q_j < \hat{q}, \\ q^L(q_j) < q_j \text{ and increasing for } q_j \geq \hat{q}. \end{cases}$$

---

[2] Superscript on functions denote firm identity and subscripts denote partial derivatives.

[3] Related calculations and derivation of all figures are in Aoki (1995).

[4] Local concavity of $\Pi^i(q_i, q_j)$ in $\forall q_i \neq q_j$, $C'(0) = 0$, and $\lim_{q_i \to \infty} \Pi^i(q_i, q_j) = -\infty$ guarantees an interior solution to the maximization problem in q_i for $q_i \neq q_j$.

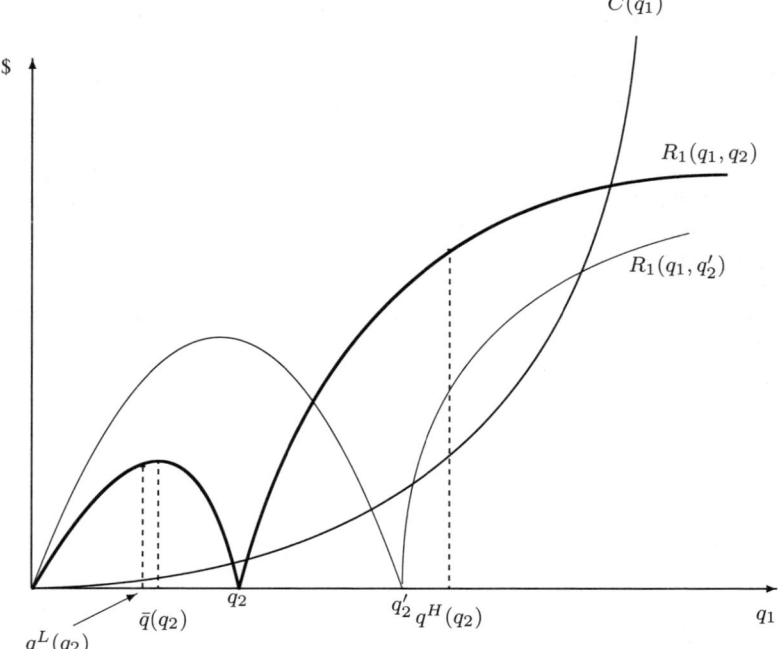

Figure 1. Revenue and cost functions with Bertrand competition

The best-response correspondence is shown in Figure 2. (Arrows indicate direction of increasing profit for firm 1 and iso-profit curves are those of firm 1.). In the terminology of Bulow, Geanakopolis, and Klemperer (1985), qualities are locally strategic complements when there is Bertrand competition in the sales stage (except at $q_j = \hat{q}$). Given our formulation, it is easy to show that the best-response correspondences actually intersect as in Figure 3, i.e., there are pure strategy subgame perfect Nash equilibria. In addition, it is possible to determine the profitability at the two equilibria.

Proposition 1 *There are two pure strategy Nash equilibria (E_{SIM}^1 and E_{SIM}^2) to the simultaneous quality choice game when there is Bertrand competition at the sales stage.*

(i) *The equilibrium qualities of the two equilibria are the same: $E_{SIM}^1 = (q_{SIM}^H, q_{SIM}^L)$ and $E_{SIM}^2 = (q_{SIM}^L, q_{SIM}^H)$ with $q_{SIM}^H > q_{SIM}^L$.*

(ii) *Both equilibrium qualities are decreasing in n.*

(iii) *It is more profitable to be producing the higher quality. That is, profit for firm i is higher at equilibrium E_{SIM}^i.*

The proof is in the Appendix. Revenue is higher for the higher quality firm for any pair of qualities from Lemma 1. Because the cost function is smooth, firms choose lower qualities when cost of quality is greater. But the difference in cost is never greater than difference in revenue.

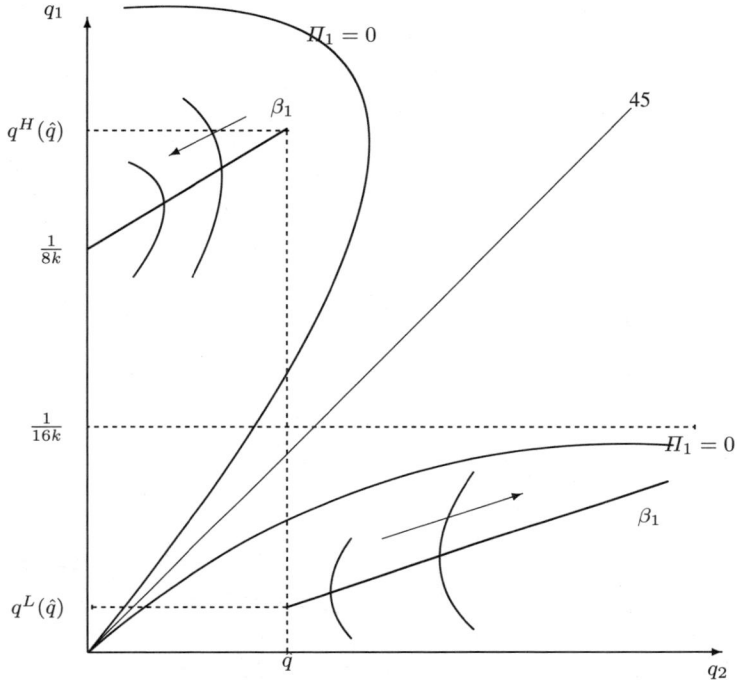

Figure 2. Best-response correspondences and iso-profit curves with Bertrand competition

We now characterize the equilibrium under sequential choice $E_{SEQ} = (q_{SEQ}^1, q_{SEQ}^2)$. Firm 1 will choose quality first, and then firm 2 will choose its quality after observing the choice of firm 1. There are two local optima to the constrained maximization problem of firm 1,

$$\max_{q_1} \Pi^1(q_1, q_2) \text{ subject to } q_2 \in \beta^2(q_1).$$

One of the maximum, F^H, involves firm 1 (first mover), choosing the higher quality, while firm 1 chooses the lower quality in the other local maximum, F^L. (Both points are depicted in Fig. 3.)

Proposition 2 *First mover will produce the higher quality in equilibrium of sequential quality choice. That is, F^H is the global optimum, i.e., $E_{SEQ} = F^H$.*

Proof is in the Appendix. Because the best-response correspondence is upward sloping (Lemma 2), E_{SEQ} lies to the southwest of E^1. By definition, the profit for firm 1 is greater under sequential choice than under simultaneous choice. From the iso-profit curves, it is easy to see that the profit for firm 2 is lower under sequential quality choice. Summarizing,

Proposition 3 *When there is Bertrand competition at the sales stage and firms choose qualities sequentially, the first mover will choose a quality higher than that*

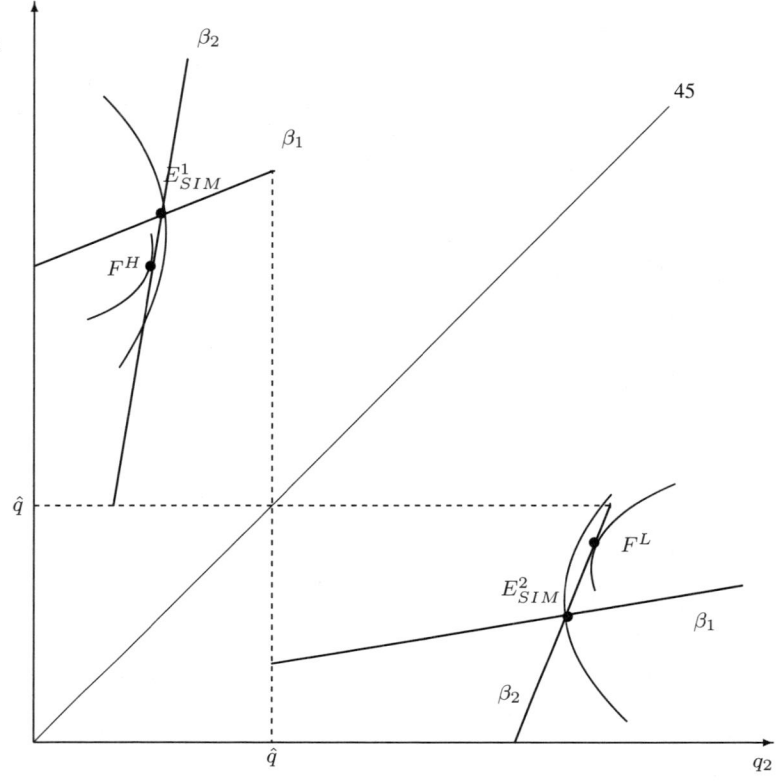

Figure 3. Best-response correspondences and equilibria with Bertrand competition

of the rival, i.e.,

$$q^1_{SEQ} > q^2_{SEQ}.$$

The qualities in the equilibrium under sequential choice are (pair wise) lower than the qualities of the simultaneous quality choice, i.e.,

$$q^H_{SIM} > q^1_{SEQ} \text{ and } q^L_{SIM} > q^2_{SEQ}.$$

The firm supplying the superior product earns greater profit and the firm supplying the inferior product earns less profit under sequential choice.

We can show that aggregate profit will be greater with sequential choice. When first mover reduces quality compared to simultaneous choice, the second mover responds by moving away, i.e., choosing a lower quality compared to simultaneous choice. With Bertrand competition, profitability depends on distance between the two qualities - competition for the marginal consumer. This competition is moderated with sequential choice and thus aggregate profit increases. By the same token, consumer surplus is smaller with sequential choice:

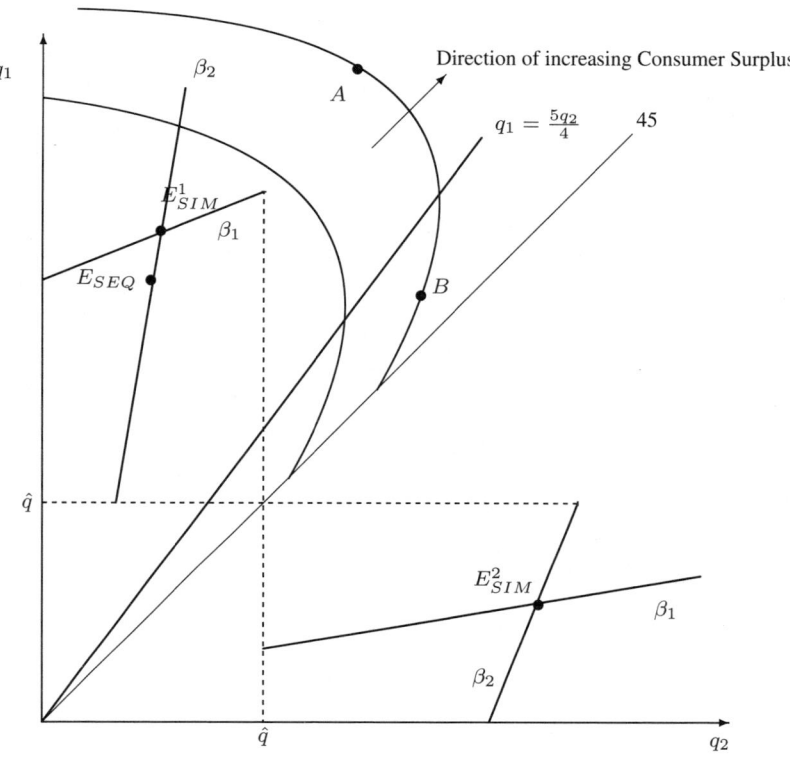

Figure 4. Iso-consumer surplus curves with Bertrand competition

Proposition 4 *Producer surplus is greater and consumer surplus is smaller with sequential choice than simultaneous choice when there is Bertrand competition in the sales stage. Summarising,*

Proof is in the Appendix. The iso-cosumer surplus curves are depicted in Figure 4. Even when both qualities become lower, if the relative distance become smaller, it is possible for consumer surplus to increase from greater competition for the marginal consumer. It is shown in the Appendix that this is never the case. Such qualities would be unprofitable for the first mover.

4 Quality choices under Cournot competition

The function $\tilde{R}^i(q_i, q_j)$, $i = 1, 2$ denotes equilibrium revenue from the second stage with Cournot competition. The payoff for firm i in the first stage is, $\tilde{\Pi}^i(q_i, q_j) = \tilde{R}^i(q_i, q_j) - C(q_i)$. The equilibrium revenue function has the following properties.

Lemma 3 *When there is Cournot competition at the sales stage, the equilibrium revenue function is continuous $\forall (q_i, q_j)$ and twice continuously differentiable $\forall q_i$*

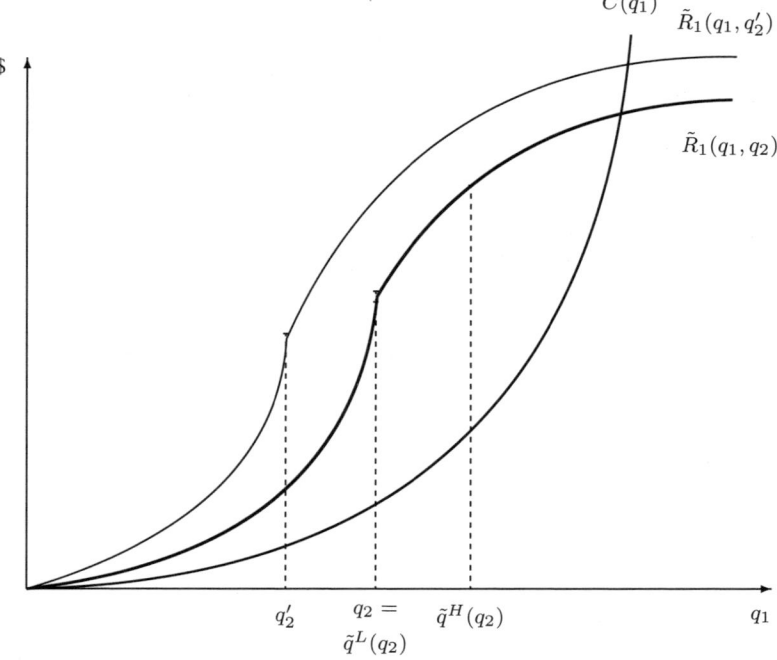

Figure 5. Revenue and cost functions with Cournot competition

$$\neq q_j.$$

$$\tilde{R}^i(q_i, q_j) = \begin{cases} \frac{q_i(q_j)^2}{(4q_j - q_i)^2}, & q_i \leq q_j \\ \frac{q_i(2q_i - q_j)^2}{(4q_i - q_j)^2}, & q_i \geq q_j \end{cases},$$

$$\begin{cases} \tilde{R}^i_i > 0, \tilde{R}^i_j < 0, \tilde{R}^i_{ii} > 0, \tilde{R}^i_{ij} = \tilde{R}^i_{ji} < 0, & q_i < q_j \\ \tilde{R}^i_i > 0, \tilde{R}^i_j < 0, \tilde{R}^i_{ii} < 0, \tilde{R}^i_{ij} = \tilde{R}^i_{ji} > 0, & q_i > q_j \end{cases}.$$

A typical revenue function for firm 1 is depicted in Figure 5. In Cournot competition, a firm's profit is increasing in its own quality and decreasing in rival's quality, independent of relative quality levels. As in the Bertrand case, magnitude of marginal change (in this case marginal *gain*, $\tilde{R}^i_i > 0$) becomes smaller when qualities become farther apart ($\tilde{R}^i_{ij} < 0$ for $q_j > q_i$ and $\tilde{R}^i_{ij} > 0$ when $q_i > q_j$).

As in the case with Bertrand competition, the best-response correspondence is given by one of the two local maxima of $\tilde{\Pi}^i$, $\tilde{q}^H(q_j) \geq q_j$ and $\tilde{q}^L(q_j) \leq q_j$. Note that the second order properties of the revenue function differs from the Bertrand case. Most notably, $R^i_{ii} > 0$ for $q_i < q_j$ and $R^i_{ij} > 0 \forall q_i$. We are able to make the following partial characterization of the best-response function.

Lemma 4 *When there is Cournot competition in the second stage, the first stage best-response correspondence satisfies,*

$$\tilde{\beta}^i(q_j) = \begin{cases} \tilde{q}^H(q_j) > q_j \text{ and increasing for } q_j \leq \frac{5}{27kn}, \\ \tilde{q}^L(q_j) < q_j \text{ and decreasing for } q_j \geq \frac{7}{27kn}. \end{cases}$$

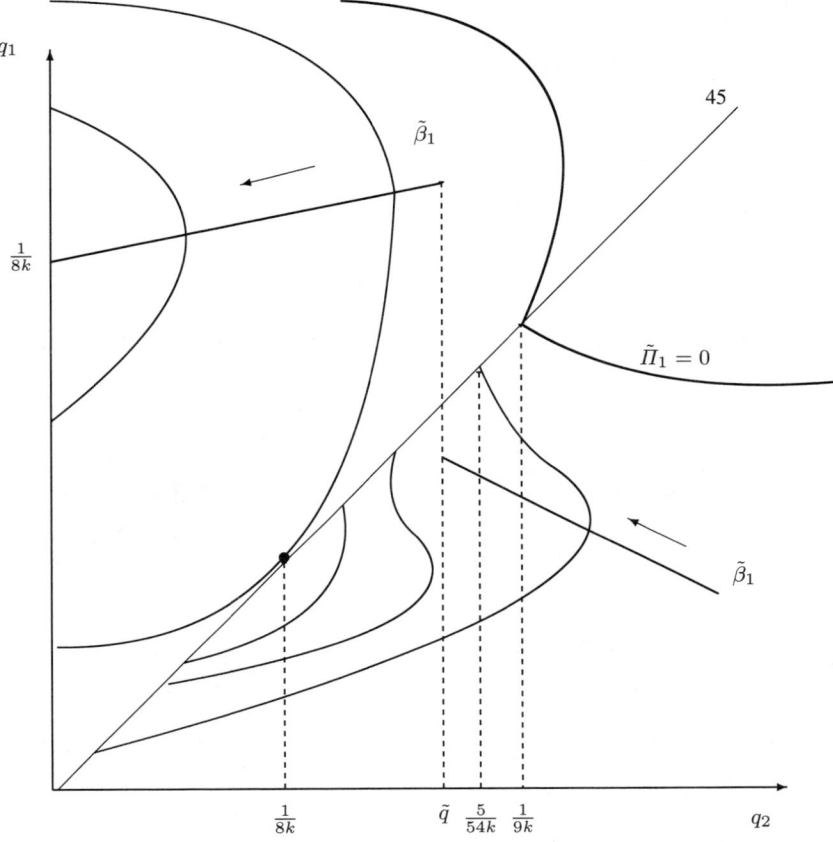

Figure 6. Best-response correspondence and iso-profit curves with Cournot competition

(See Fig. 6; arrow indicates direction of increasing profit for firm 1 and iso-profit curves are those of firm 1.)[5] Qualities are strategic complements for the higher quality firm and substitutes for the lower quality firm. The existence of the intersection of the two best-response correspondence (shown in Aoki (2001a)) implies existence of pure strategy equilibrium. In addition, the intersection lies in a region where the higher quality firm attains profit levels impossible for the lower quality firm to attain (see the iso-profit curve in Fig. 6). Thus we have,

Proposition 5 *For any n, there are two pure strategy Nash equilibria ($\tilde{E}^1_{SIM}$ and $\tilde{E}^2_{SIM}$) when there is Cournot competition at the sales stage.*

(i)   *The qualities of the two equilibria are the same: $\tilde{E}^S_{SIM} = (\tilde{q}^H_{SIM}, \tilde{q}^L_{SIM})$ and $\tilde{E}^2_{SIM} = (\tilde{q}^L_{SIM}, \tilde{q}^H_{SIM})$ with $\tilde{q}^H_{SIM} > \tilde{q}^L_{SIM}$.*

---

[5] Because $R^i_{ii} > 0$, local optimum may be a corner solution, $\tilde{q}^H(q_j) = q_j$. $R^i_{ij} > 0 \forall q_i$ means that $\tilde{\beta}^i$ may switch between the two local optima more than once. For $(\frac{5}{27nk})^{\frac{1}{n-1}} < q_j < (\frac{7}{27nk})^{\frac{1}{n-1}}$ both local maxima are interior solutions and $\beta^i(\cdot)$ must be determined by comparing $\Pi^H(q^H(q_j), q_j)$ and $\Pi^L(q_j, q^L(q_j))$.

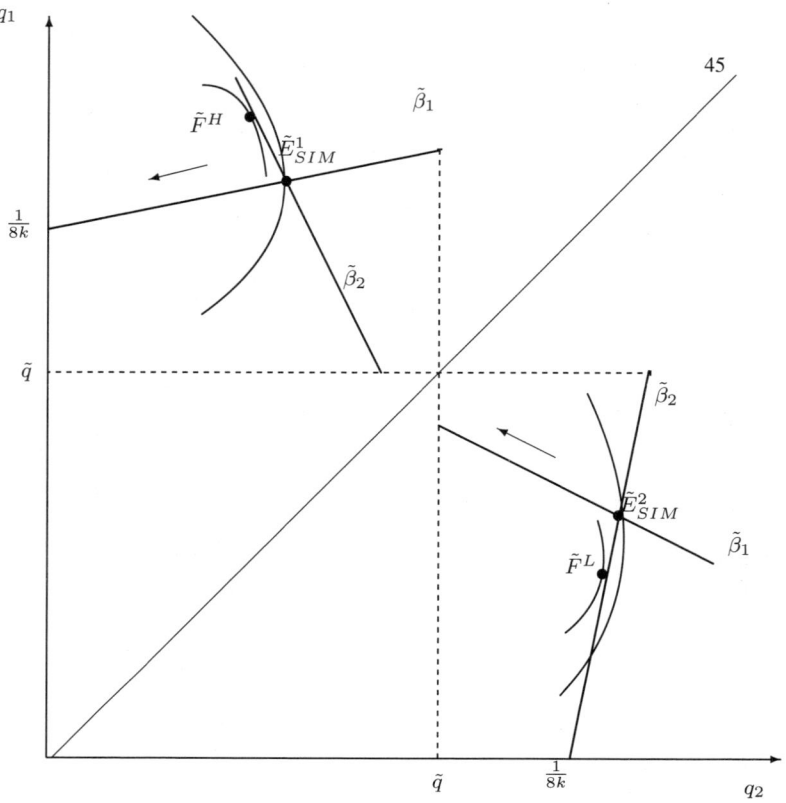

Figure 7. Best-response correspondences and equilibria with Cournot competition

(ii) The higher equilibrium quality ($\tilde{q}^H$) is decreasing in n. The lower equilibrium quality ($\tilde{q}^L$) may increase or decrease with n.

(iii) In equilibrium, it is more profitable to be producing the higher quality. That is, profit for firm i is higher at equilibrium $\tilde{E}^i_{SIM}$.

Proof is in the Appendix.

We now characterize the sequential quality choice equilibrium $\tilde{E}_{SEQ} = (\tilde{q}^1_{SEQ}, \tilde{q}^2_{SEQ})$ when firm 1 chooses quality before firm 2. Again there are two local optima to the constrained maximization problem of firm 1,

$$\max_{q_1} \tilde{\Pi}^1(q_1, q_2) \text{ subject to } q_2 \in \tilde{\beta}^2(q_1).$$

One of the optima, $\tilde{F}^H$, involves the first mover firm 1 choosing the higher quality and in the other, $\tilde{F}^L = (\tilde{q}^L_1, \tilde{q}^L_2)$, firm 1 chooses the lower quality (see Fig. 7).

The local optimum with higher quality, $\tilde{F}^H$, lies in a region of higher profit for high quality firm. Thus $\tilde{E}_{SEQ} = \tilde{F}^H$. Since firm 2's best-response function is downward sloping for $q_1 > q_2$, $\tilde{F}^H$ lies to the northwest of $\tilde{E}^1_{SIM}$. It is also clear from the direction of change along the best-response function that profit for firm 2 is lower at $\tilde{E}_{SEQ}$ than at $\tilde{E}^1_{SIM}$. So we have,

Proposition 6 *When there is Cournot competition in the sales stage and firms choose qualities sequentially, the first mover will choose a quality higher than that of the rival, i.e.,*

$$\tilde{q}^1_{SEQ} > \tilde{q}^2_{SEQ}.$$

The first mover's quality choice is also higher than the higher quality of the equilibrium under simultaneous choice. But the choice of the second mover will be lower than the lower quality of the equilibrium under simultaneous choice, i.e.,

$$\tilde{q}^H_{SIM} < \tilde{q}^1_{SEQ} \text{ and } \tilde{q}^L_{SIM} > \tilde{q}^2_{SEQ}.$$

The firm supplying the superior product earns greater profit under and the firm supplying the inferior product earns less profit under sequential choice.

Unlike with Bertrand competition, firms do not compete for the marginal consumer with prices. Thus although the qualities become fasrther apart with sequential choice, it does not hurt consumers. In fact, consumer surplus is greater with sequential choice since a higher quality is available. Producer surplus depends on cost quality. If n is small, then increase of first mover profit is greater than loss of second mover profit and producer surplus will be greater with sequential choice. If n is large, then the relationship reverses and producer surplus is lower with sequential choice:

Proposition 7 *Consumer surplus is greater with sequential choice than simultaneous choice when there is Cournot competition in the sales stage. Producer surplus will be greater if cost of quality is not too costly (n small) but it will be smaller otherwise (n large) with sequential choice.*

Proof is in the Appendix.

5 Concluding remarks

In both Cournot and Bertrand competition, the first mover chooses to be the higher quality. This is because it is more profitable (*profit* is greater) to be the higher quality firm for most pair of qualities. Under both types of competition, *revenue* is greater for the higher quality for any pair of qualities (Lemmata 1 and 3). The specified cost function guarantees that the cost of quality does not increase too quickly.

Our contrasting results with Berrand and Cournot competitions are driven by the differences in the slope of the best-response correspondence (Lemmata 2 and 4). Strategic complementary and substitutability are determined by the cross-derivative of the profit function which only depends on the revenue function. Thus results depending on this property are not only independent of the form of the cost function but also the specification of the demand system.

It is also not necessary that the values of the best-response correspondences be interior solutions. Since the envelope theorem does not hold at a corner solution, the direction in which the first mover would like to move from the equilibrium under simultaneous choice is not completely determined by the sign of R^i_{ij} or $\tilde{R}^i_{ij}$. But as long as this effect dominates, the results will follow.

6 Appendix

Proof of Proposition 1

Proof of (i). We need to show that,

Lemma 5 *The two best-response correspondences, $\beta^1(\cdot)$ and $\beta^2(\cdot)$, intersect.*

Proof of Lemma. Since $q_H(\cdot)$ is increasing and $q_H(0) = q^m$, $q^H(\cdot)$ and $q^L(\cdot)$ must intersect if

$$\beta^i(q^m) = q^L(q^m), \tag{A.1}$$

i.e., $\hat{q} < q^m$. Then from symmetry, the best-response correspondences must intersect (there will be two symmetric intersections). Equation (A.1) follows from,

$$\Pi^i(q^L(q^m), q^m) = \Pi^L(q^m, q^L(q^m)) > \Pi^i(q^H(q^m), q^m) = \Pi^H(q^H(q^m), q^m).$$

Since marginal profit is positive at $q_i = 0$ and $\Pi^i(0, q^m) = \Pi^L(q^m,) = 0$, and therefore $\Pi^i(q', q^m) > 0$ for some q' and it suffices to show that $\Pi^i(q^H(q^m), q^m) < 0$. We can write $q^H(q^m) = \alpha q^m$ for some $\alpha > 1$. Substituting this and $q^m = \left(\frac{1}{4nk}\right)^{\frac{1}{n-1}}$ into the profit function,

$$\begin{aligned}
\Pi^i(q^H(q^m), q^m) &= \frac{4\alpha^2(\alpha - 1)}{(4\alpha - 1)^2} q^m - k(\alpha q^m)^n \\
&= \left(\frac{4\alpha^2(\alpha - 1)}{(4\alpha - 1)^2} - k\alpha^n(q^m)^{n-1}\right) q^m \\
&= \left(\frac{4\alpha^2(\alpha - 1)}{(4\alpha - 1)^2} - \frac{\alpha^n}{4n}\right) q^m
\end{aligned}$$

This will be negative if $f_1(\alpha) < f_2(\alpha)$ where

$$f_1(a) = (a - 1)n, \quad f_2(a) = \frac{a^{n-1}(4a - 1)^2}{16}.$$

It is straightforward to show that $f_1(1) < f_2(1)$ and $0 < f_1'(a) < f_2'(a)$ for $a \geq 1$. Then it must be $f_1(\alpha) < f_2(\alpha)$ for all $\alpha \geq 1$.   □

Proof of (ii). Given concavity of the revenue functions, the best-response correspondences will move down as n increases.   □

Proof of (iii). We first prove the following lemma.

Lemma 6 *For (q_H, q_L), $q_H > q_L$, if $q_L = q^L(q_H)$ and $q_H \leq \left(\frac{7}{27nk}\right)^{\frac{1}{n-1}}$, then*

$$\Pi^H(q_H, q_L) > \Pi^L(q_H, q_L).$$

Proof of Lemma. If we let $q^L = q$, then $q^H = \alpha q$ for some $\alpha > 1$. We need to show

$$\Pi^H(\alpha q, q) = R^H(\alpha q, q) - C(\alpha q) > \Pi^L(\alpha q, q) = R^L(\alpha q, q) - C(q).$$

The differences in revenues and costs are,

$$\Delta R = R^H(\alpha q, q) - R^L(\alpha q, q) = \frac{(\alpha - 1)q\alpha}{4\alpha - 1},$$
$$\Delta C = C(\alpha q) - C(q) = (\alpha^n - 1)kq^n.$$

We need to show

$$\frac{\Delta C}{\Delta R} = \frac{(4\alpha - 1)(\alpha^n - 1)k}{\alpha(\alpha - 1)}q^{n-1} < 1, \tag{A.2}$$

Given the assumption on $q_H = \alpha^{n-1}q^{n-1}$,

$$\frac{\Delta C}{\Delta R} < \frac{(4\alpha - 1)(\alpha^n - 1)k}{\alpha(\alpha - 1)\alpha^{n-1}} \frac{7}{27nk}$$
$$= \left(4 - \frac{1}{\alpha}\right)\left(1 + \frac{1}{\alpha^2} + \cdots + \frac{1}{\alpha^{n-1}}\right)\frac{7}{27n}.$$

From (1) and $q = q^L(\alpha q)$, we know that $\alpha > \frac{7}{4}$. For any $\alpha > \frac{7}{4}$,

$$\frac{\Delta C}{\Delta R} < 4\frac{1 - (\frac{4}{7})^{n-1}}{\frac{3}{7}}\frac{7}{27n}.$$

It is straightforward to show that right-hand side is less than 1 for any $n \geq 2$. □

Now we apply this Lemma to $q_L = q^L_{SIM}$ and $q_H = q^H_{SIM}$. Equilibrium is on both $\beta^2(\cdot) = q^L(\cdot)$ and $\beta^1(\cdot) = q^H(\cdot)$. In particular, it satisfies the first condition of the Lemma, $q_L = q^L(q_H)$. Since $q^H(\cdot)$ is increasing, $\beta^1(\cdot) = q^L(\cdot)$ for $q \geq (\frac{7}{27nk})^{\frac{1}{n-1}}$, and $q^H((\frac{7}{27nk})^{\frac{1}{n-1}}) = (\frac{7}{27nk})^{\frac{1}{n-1}}$, and it follows that $q^H_{SIM} = q^H(q^L_{SIM}) = \alpha q \leq (\frac{7}{27nk})^{\frac{1}{n-1}}$. This is the second condition of the Lemma. Applying the Lemma, we get

$$\Pi^H(q^H_{SIM}, q^L_{SIM}) > \Pi^L(q^H_{SIM}, q^L_{SIM}).$$

□

Proof of Proposition 2

F^H solves:

$$\max_{q_1} \Pi^H(q_1, q^L(q_1)) \text{ subject to } q_1 \geq q^L(\hat{q}),$$

and F^L solves:

$$\max_{q_1} \Pi^L(q^H(q_1), q_1) \text{ subject to } q_1 \leq \hat{q}.$$

(Both points are depicted in Fig. 3). One or both local maximum can be corner solutions. Let $F^H = (q_H^H, q_L^H)$ and $F^L = (q_L^L, q_H^L)$. We need to show that $\Pi^1(F^H) = \Pi^H(F^H) > \Pi^1(F^L) = \Pi^L(F^L)$. F^L is on $\beta^2(\cdot) = q^H(\cdot)$ but $G^L = (q_H^L, q_L^L)$ will be on $\beta^1(\cdot) = q^H(\cdot)$. Since the two firms are identical, $\Pi^1(F^L) = \Pi^2(G^L)$. We can find a point $X = (x_H, x_L)$ on $\beta^2(\cdot) = q^L(\cdot)$ such that $\Pi^2(G^L) = \Pi^2(X) = \Pi^L(X)$.

We apply Lemma 6 to $q_H = x_H, q_L = x_L$. We chose X to be on $q^L(\cdot)$ and it satisfies the first condition, $x_L = q^L(x_H)$. We note regarding G^L and X that (i) both points are on the same iso-profit curve, (ii) the iso-profit curve is tangent to an upward sloping $\beta^1(\cdot)$ at G^L, (iii) the same iso-profit curve has slope 0 at X. This implies X must lie southwest of G^L. In particular, $x_H \leq (\frac{7}{27})^{\frac{1}{n-1}}$. The Lemma implies $\Pi^H(X) > \Pi^L(X)$. But $\Pi^1(X) = \Pi^H(X)$ and by definition of F^H, $\Pi^1(F^H) \geq \Pi^1(X)$. We have shown now,

$$\Pi^1(F^H) = \Pi^H(F^H) \geq \Pi^H(X) > \Pi^L(X) = \Pi^2(G^L)$$
$$= \Pi^L(G^L) = \Pi^1(F^L).$$

□

Proof of Proposition 4

Proof that producer surplus is greater with sequential choice. On $q_2 = \beta_2(q_1)$ (or $q_L = q^L(q_H)$) between E_{SIM}^1 and E_{SEQ},

$$\frac{\partial PS}{\partial dq_H} = \frac{\partial \pi_H}{q_H} + \frac{\partial R_L}{\partial q_H} > 0,$$

from (3) and $\frac{\partial \pi_H}{\partial q_H} > 0$. For the same segment,

$$\frac{\partial PS}{\partial dq_L} = \frac{\partial R_H}{dq_L} + \frac{\partial \pi_L}{\partial dq_L} < 0,$$

from (3) and $\frac{\partial \pi_L}{\partial q_L} = 0$. The iso-producer surplus curve is upward sloping on the segment. We compare the slopes of iso-producer surplus curve and $q_2 = \beta_2(q_1)$ at E_{SIM}^1,

$$d = |\frac{dq_H}{dq_L}|_{dPS=0}| - |\frac{dq_H}{dq_L}|_{q_L = q^L(q_H)}|$$
$$= 4\frac{q_H^2}{q_L^2} - \frac{q_H(48q_Lq_H + 7q_L^2 + 16nq_H^2 - 32nq_Hq_L + 7nq_L^2 - 16q_H^2)}{2(8q_H + 7q_L)^2}$$
$$= \frac{q_H(4q_H - q_L)(4(5-n)q_H + 7(n+1)q_L)}{2(8q_H + 7q_L)^2}$$

Since $q + L = q^L(q_H) < \frac{4}{7}q_H$,

$$d > (7(n+1) + 7(5-n))q_L > 0.$$

Best-response correspondence is flatter at E_{SIM}^1. This implies producer surplus is larger at E_{SEQ}.

□

Proof that consumer surplus is smaller. Consumer surplus when qualities (q_H, q_L) are available is,

$$\frac{dq_H}{dq_L}\Big|_{dCS=0} = \frac{q_H^2(28q_H + 5q_L)}{q_H(2q_H + q_L)(4q_H - 5q_L)}.$$

When the qualities are very close, $q_L > \frac{4}{5}q_H$, improvement of higher quality q_H reduces competition by making the products less homogeneous and will decrease consumer surplus. If the two qualities are sufficiently different $q_L < \frac{4}{5}q_H$, then benefit to consumers from better quality from better higher quality is greater than the loss reduction of competition. We observed previously that $q^L(q_H) < \frac{4}{7}q_H$. Firm 2's best-response correspondence, $\beta^2(q_1) = q^L(q_1)$ is in the region where the iso-consumer surplus curves are downward sloping. Thus E^1_{SIM} yields higher consumer surplus than E_{SEQ}.

Proof of Proposition 5

To show existence, we need to show that the two best-response correspondences intersect. This will follow from

$$\tilde{\beta}^2(q^m) = \tilde{q}^L(q^m) \tag{A.3}$$

$$\tilde{q}^L(q^m) < \frac{5}{27nk} \tag{A.4}$$

These two relationships imply $\tilde{\beta}^2(q^m) < \frac{5}{27nk}$. On the other hand, since $\tilde{\beta}^1(0) = \tilde{q}^H(0) = q^m$ and $\tilde{q}^H(\cdot)$ is increasing, $\tilde{\beta}^1(\frac{5}{27nk}) > q^m$. Then the two best-response correspondences must intersect.

Proof of (A.3). We need to show that

$$\tilde{\Pi}^H(\tilde{q}^H(q^m), q^m)) < \tilde{\Pi}^L(q^m, \tilde{q}^L(q^m)).$$

First we show that if α satisfies $q^H(q^m) = \alpha q^m > q^m$, then it must be that $\alpha \leq (\frac{28}{27})^{\frac{1}{n-1}}$. To show this, we first note that such α must satisfy first order condition of maximisation,

$$\tilde{R}^H_H(q_H, q_j) - C'(q_H) = \frac{(2q_H - q_j)(8q_H^2 - 2q_H q_j + q_j^2)}{(4q_H - q_j)^2} - nkq_H^{n-1} = 0,$$

which using $(q^m)^{n-1} = \frac{1}{4nk}$ becomes,

$$\frac{(2\alpha - 1)(8\alpha^2 - 2\alpha + 1)}{(4\alpha - 1)^3} = \frac{\alpha^{n-1}}{4}.$$

Left hand side is decreasing in α and equal to $\frac{7}{27}$ when $\alpha = 1$. Right hand side, which is increasing in α is equal to $\frac{1}{4}$ when $\alpha = 1$. Both sides will be equal to α such that $\frac{1}{4} \leq \frac{\alpha^{n-1}}{4} \leq \frac{7}{27}$. The second inequality implies what we need to show.

Now we show (A.3), by claiming that,

$$\tilde{\Pi}^H(\alpha q^m, q^m) < \tilde{\Pi}^L(q^m, \frac{q^m}{\alpha}), \tag{A.5}$$

for any $\alpha \le (\frac{28}{27})^{\frac{1}{n-1}}$. Then if we choose α as $\tilde{q}^H(q^m) = \alpha q^m$, (A.3) follows since $\tilde{\Pi}^L(q^m, \frac{q^m}{\alpha}) < \tilde{\Pi}^L(q^m, \tilde{q}^L(q^m))$.

In order to show (A.5), we define functions $f(\alpha)$ and $h(\alpha)$,

$$f(\alpha) = \frac{\alpha(2\alpha - 1)^2}{(4\alpha - 1)^2} - \frac{\alpha^n}{4n},$$

$$h(\alpha) = \frac{\alpha}{(4\alpha - 1)^2} - \frac{1}{\alpha^n 4n}.$$

Then

$$\tilde{\Pi}^L(q^m, \frac{q^m}{\alpha}) - \tilde{\Pi}^H(\alpha q^m, q^m) = q^m(h(\alpha) - f(\alpha)).$$

Note that $h(1) - f(1) = 0$. We are interested in $h(\alpha) - f(\alpha)$ for $1 < \alpha \ge \frac{28}{27}^{\frac{1}{n-1}}$. For this, we inspect the change in the difference,

$$h'(\alpha) - f'(\alpha) = -q^m \left\{ \tilde{\Pi}_L^L \frac{-1}{\alpha^2} + \tilde{\Pi}_H^H \right\}$$

$$= q^m \left\{ -\frac{1}{\alpha^2} \left(\frac{\alpha^2(4\alpha + 1)}{(4\alpha - 1)^3} - \frac{1}{\alpha^{n-1}} \right) \right.$$

$$\left. - \left(\frac{(2\alpha - 1)(8\alpha^2 - 2\alpha + 1)}{(4\alpha_1)^3} - \frac{\alpha^{n-1}}{4} \right) \right\}$$

Sign of this expression is the same as sign of

$$s = (4\alpha - 1)^3 (1 + \alpha^2 (\alpha^{n-1})^2)$$
$$- 4\alpha^{n-1} \alpha^2 \left((4\alpha + 1) + (2\alpha - 1)(8\alpha^2 - 2\alpha + 1) \right).$$

Using $1 < \alpha \le (\frac{28}{27})^{\frac{1}{n-1}}$, we have

$$s > (4\alpha - 1)^3 (1 + \alpha^2) - \frac{28}{27} 4\alpha^2 (8\alpha + 16\alpha^3 - 12\alpha^2).$$

This will be positive for any α such that $1 < \alpha \le (\frac{28}{27})^{\frac{1}{n-1}}$. This implies $h(\alpha) - f(\alpha)$ is positive for the same range of α, which includes α that satisfies $\tilde{q}^H(q^m) = \alpha q^m$. This proves (A.5). □

Proof of (A.4). $\tilde{q}^L(\frac{5}{27nk}) = \frac{5}{27nk}$ and $\tilde{q}^L(q_j)$ is decreasing in q_j for $q_j \ge \frac{5}{27nk}$. Since $q^m = \frac{1}{4nk} > \frac{5}{27nk}$, this implies $\tilde{q}^L(q^m) < \tilde{q}^L(\frac{5}{27nk}) = \frac{5}{27nk}$. □

Proof of (iii). Since $\tilde{R}_L^H < 0$ and $\tilde{R}_H^L > 0$, a firm's profit is monotonically increasing along the best-response correspondence $(\beta^i(\cdot))$ as rival quality decreases. (Direction of arrow in Fig. 4). Thus for firm 1, $\tilde{E}_{SIM}^1$ is more profitable than $\tilde{E}_{SIM}^2$. □

Proof of Proposition 7

Proof that consumer surplus is greater. When there is Cournot competition, consumer surplus for given pair of qualities is,

$$CS(q_H, q_L) = \frac{q_H(4q_H^2 - q_L^2 + q_H * q_L)}{2(4q_H - q_L)^2}.$$

The slope of the iso-consumer surplus curves is,

$$\frac{dq_H}{dq_L}\Big|_{dCS=0} = \frac{q_H^2(7q_L - 12q_H)}{2q_L^2 q_H + 16q_H^3 + q_L^3 - 12q_H^2 q_L}$$

Iso-consumer surplus curves are downward sloping since consumer surplus is increasing in both high and low qualities. In order to compare consumer surplus at E_{SIM}^1 and E_{SEQ} we need to compare the slopes of iso-consumer surplus and $\beta_1(q_2) = q^L(q_2)$. We can simplify the expression for the best-response correspondence using the first-order condition,

$$\frac{dq_H}{dq_L}\Big|_{q_L = q^L(q_H)} = \frac{q_H(2q_L(8q_H + q_L) - (n-1)(4q_H + q_L)(4q_H - q_L)))}{2q_L^2(8q_H + q_L)}$$

The difference between the two slopes is,

$$\frac{dq_H}{dq_L}\Big|_{dCS=0} - \frac{dq_H}{dq_L}\Big|_{q_L = q^L(q_H)} =$$
$$\frac{q_H(56q_L^3 q_H^2 + q_L^4 q_H - 48q_L^2 q_H^3 - 160q_H^4 q_L - q_L^5 + 128q_H^5)}{2q_L^2(8q_H + q_L)(2q_L^2 q_H + 16q_H^3 + q_L^3 - 12q_H^2 q_L)}.$$

This will be positive for

$$q_L < \left(\frac{5q_H}{27k(n-1)}\right)^{\frac{1}{n-1}},$$

which is condition derived from first-order condition. This means best-response correspondence is steeper, implying $\tilde{E}_{SEQ}$ is on a higher iso-consumer surplus curve than $\tilde{E}_{SIM}^1$.

Proof about producer surplus. On $q_2 = \beta_2(q_1)$ (or $q_L = q^L(q_H)$) between E_{SIM}^1 and E_{SEQ},

$$\frac{\partial PS}{\partial dq_H} = \frac{\partial \pi_H}{dq_H} + \frac{\partial R_L}{\partial dq_H} < 0,$$

from (3) and $\frac{\partial \pi_H}{\partial q_H} < 0$. For the same segment,

$$\frac{\partial PS}{\partial dq_L} = \frac{\partial R_H}{dq_L} + \frac{\partial \pi_L}{\partial dq_L} < 0,$$

from (3) and $\frac{\partial \pi_L}{\partial q_L} = 0$. Thus the iso-producer surplus curve is downward sloping on the segment. We compare the slopes of iso-producer surplus curve and $q_2 = \beta_2(q_1)$ at E_{SIM}^1,

$$
\begin{aligned}
d &= |\frac{dq_H}{dq_L}|_{dPS=0}| - |\frac{dq_H}{dq_L}|_{q_L=q^L(q_H)}| \\
&= \frac{2q_H(2q_H - q_L)}{(q_L)^2} - \frac{q_H(16(n-1)q_H^2 - 16q_Lq_H + -(n+1)q_L^2)}{q_L^2(8q_H - q_L)} \\
&= \frac{q_H((80+6n)q_H^2 - 8q_Hq_L + (n-3)q_L^2)}{2q_L^2(8q_H + q_L)}.
\end{aligned}
$$

This is decreasing in n for $q_H > q_L$. For any $q_H > q_L$, $n = 2$,

$$
d = \frac{q_H(48q_H^2 - 8q_Hq_L - q_L^2)}{2q_L^2(8q_H + q_L)} > 0,
$$

and when $n = 10$,

$$
d = \frac{-q_H(80q_H^2 + 8q_Hq_L - 7q_L^2)}{2q_L^2(8q_H + q_L)} < 0.
$$

The best-response correspondence is flatter when n is small and it is steeper for larger n. Producer surplus is larger at E_{SEQ} when n is small and it is smaller at E_{SEQ} when n is very large.

With Cournot competition, higher quality is higher but lower quality is lower with sequential choice: higher quality firm does better at cost to lower quality firm. When n is very large, the gain higher quality is not improved enough relative to lower quality so that aggregate profits decrease.   □

References

1. Aoki, R.: Sequential vs. simultaneous quality choice with Bertrand and Cournot competition. SUNY Stony Brook Working Paper Series (1995)
2. Aoki, R., Prusa, T. J.: Product development and the timing of information disclosure under U.S. and Japanese Patent Systems. Journal of the Japanese and International Economies **10**, 233–249 (1996)
3. Aoki, R.: Cournot and Bertrand competition with vertical quality differentiation. Department of Economics Working Paper No. 222, University of Auckland (2001a)
4. Aoki, R.: Equilibrium quality choices with generalized smooth cost function. Department of Economics Working Paper Series No. 222, University of Auckland (2001b)
5. Bonanno, G.: Vertical quality with Cournot competition. Economic Notes **15**, 68–91 (1986)
6. Boyer, M., Moreaux, M.: On Stackelberg equilibria with differentiated products: The critical role of the strategy space. The Journal of Industrial Economics **36**, 217–230 (1987)
7. Bulow, J., Geanakopolis, J., Klemperer, P.: Multimarket oligopoly: strategic substitutes and complements. Journal of Political Economy **93**, 488–511 (1985)
8. Gabszewicz, J., Thisse, J.-F.: Price competition, quality and income distribution. Journal of Economic Theory **20**, 340–359 (1979)
9. Motta, M.: Endogenous quality choice: price vs. quantity competition. The Journal of Industrial Economics **41**, 113–132 (1993)
10. Shaked, A., Sutton, J.: Relaxing price competition through product differentiation. Review of Economic Studies **49**, 3–13 (1982)

To each according to his needs: an axiomatic characterization*

Horace Wood Brock

Strategic Economic Decisions, Inc., 575 Park Avenue, Suite 705, New York City, NY 10021, USA
(e-mail: woody@sedinc.com)

Received: January 15, 2001; revised version: December 27, 2002

Summary. The age-old concept of allocating scarce resources according to "relative need" has proven notoriously difficult to formalize in a rigorous manner. This paper presents a new solution concept for this classical problem. After first providing a formal definition of what it means for one agent to be "everywhere needier than" another, we prove that the condition of being "needier than" is equivalent to being "more risk averse than". Next, we introduce and axiomatize a scalar measure of "relative neediness". This measure then becomes the basis for a needs-based allocation rule applicable to general n-person arbitration problems. The resulting allocation rule is closely related to the Nash bargaining solution, and does not entail explicit interpersonal comparisons of utility. In the final section, the new solution concept is embedded within a comprehensive theory of distributive justice.

Keywords and Phrases: Interpersonal comparisons of utility, Solution functions, Mathematical moral theory, Arbitration schemes.

JEL Classification Numbers: D6, D7.

1 Introduction

It was Aristotle in his *Politics* who first posed the question of "distributive justice", tautologically characterizing it as that state of affairs in which "each person receives his due". If he and most subsequent philosophers deftly dodged the issue of what it

---

* The author is indebted to Kenneth Arrow, Robert Aumann, Mordecai Kurz and Lloyd Shapley for helpful initial discussions of this theory. Carsten Nielsen was indispensable in commenting on the first written version of this paper and in tightening several of the arguments. Finally, Barry Feldman and John F. Nash, Jr. made helpful comments when a preliminary version of this paper was presented at the 13th International Game Theory Conference at Stony Brook, New York, July 2002.

means for each to receive his due, Karl Marx did not. In his *Communist Manifesto*, he famously thundered "From each according to his ability, to each according to his needs". But what exactly does this mean? Intuitively, it suggests that those citizens who are most capable should utilize their talents for the benefit of those who are needy – whether voluntarily or involuntarily.

Two questions arise. First, is Marx's dictum in fact a compelling characterization of distributive justice in the first place? Second, is his theory workable in practice? As to the second question, the verdict of twentieth century Marxist experiments is that it is not workable, due to well-known: incentive structure problems which derailed almost all real-world variants of communism. As to the first question, while most any compelling theory of justice will recognize the rival claims to the social product that are based on needs versus contribution, there has been surprisingly little analysis of the exact nature of these dual claims, and of how to resolve any tensions between them.

Indeed, what is the proper domain for allocation based upon relative need as opposed to relative contribution? Moreover, what exactly do we mean by allocation "according to" need/contribution? Can these two concepts be rendered analytically clear? Finally, can a theory of justice be constructed which integrates both of these distributive norms within their proper domains, thus resolving the zero-sum tension implicit in Marx's "from each/to each" imperative, and transforming the concept of distributive justice into a positive-sum game of sorts?

In a series of papers published in the later 1970s, the present author (Brock, 1978, 1979) set forth a new theory of justice that addressed all these issues, and that answered the last two questions in the affirmative. His theory attempted to resolve many of the paradoxes that arose in classical Utilitarianism, and in John Rawls' (1971) theory of justice. In his 1992 *Handbook of Game Theory* survey of mathematical moral theory, the late John Harsanyi (1992) compared and contrasted this new theory with his own version of Utilitarianism, and with Rawls' theory. Methodologically, whereas Harsanyi utilized Bayesian decision theory in his Utilitarian theory, and Rawls invoked the Min-Max principle from decision theory, the present author based his theory upon NTU–Value theory as developed by John F. Nash, Jr. and John Harsanyi in the context of bargaining, and by Lloyd Shapley in the context of the axiomatic valuation of n-person games.

In the present essay, the author wishes to respond to several of Harsanyi's comments, and in doing so to reformulate the portion of his theory that characterizes the concept of allocation according to relative needs. A new and axiomatic solution to this difficult problem is the central result of this paper, one that surprisingly does not entail interpersonal comparisons of utility.

In Section 2, we review and contrast the two principal types of cardinal utility functions that have been utilized and indeed axiomatized during the past century: deterministic utility functions constructed from a person's relative intensity of preference for the prizes at stake, and stochastic utility functions assessed from a person's preferences among lotteries. Our analysis of the deterministic utility function leads to a rigorous definition of what it means for a given person to be "more needy" than another. Theorem 1 establishes that the deterministic concept of a person's being more needy than another is formally equivalent to the stochastic concept of his being more risk averse. In Section 3, we revisit classical Nash-Harsanyi bargaining theory and, utilizing Theorem 1, establish in Theorem 2 that a party who is needier than another receives less from bargaining than does his counterparty. This formalizes the notion that a person who is needier gets "bargained down".

Building on this result in bargaining theory, we introduce in Section 4 a scalar measure of the relative neediness of each player in an n-person bargaining game, or equivalently, in an n-person arbitration scheme. It is shown in Theorem 3 that this measure of neediness uniquely satisfies several plausible axioms. Building on this measure, we establish our main result in Theorem 4 of Section 5: characterization of an allocation rule whereby each player receives a share of the payoff that is in strict proportion to his relative need. In the real world, this rule could be applied by New York City's Kenneth Feinstein who was charged in the winter of 2002 with distributing \$1.8 billion "according to need" to the families of fire-fighters and policemen lost in the 9/11 attack of 2001.

To anticipate, let an arbitrator confront the generic n-person arbitration problem A^n in which he must allocate some homogeneous medium (pie or money) y into n shares such that $0 \le y_i \le 1$ and $\sum y_i = 1$. He is assumed to know the cardinal utility functions $\{u_i\}$ of all the agents, and seeks that "equitable" allocation y^* such that, for every and any pair of agents (i, j), the allocation $\frac{y_i^*}{y_j^*}$ is proportionate to the needs of i relative to j, suitably characterized. Theorem 4 shows that the only solution to this problem satisfying numerous reasonable axioms is the allocation y^* whose i^{th} component is,

$$y_i^* = (x_i^*)^{-1} \cdot \left(\sum_j x_j^{*-1} \right)^{-1} \qquad \forall i = 1 \ldots n \qquad (1)$$

where x^* denotes the Nash allocation, i.e. the allocation that maximizes the generalized Nash product,

$$Max_x \prod_i u_i(x) \qquad (2)$$

of the pure (unanimity) bargaining game G^A naturally associated with the given arbitration problem A^n. [The prospect space U_A of achievable utility n-vectors in G^A is taken to be that of A^n so that the two problems are essentially identical.]

As will be seen later, y^* characterized in (1) is in fact the pairwise inverse of the Nash allocation x^* in the sense that, if i receives twice as much pie from bargaining as j, then j will receive twice as much as i according to (1). The intuition here is that, if bargaining by its nature penalizes a needier player by awarding him less, then an ethically normative theory will reward this player in inverse proportion to any such penalty.

In Section 6, we show how this new solution theory overcomes well-known counterexamples to the Utilitarian theory. Finally, Section 7 sketches how this new theory of relative needs forces a revision in the author's original theory of "full" distributive justice.

For simplicity of exposition, we shall assume throughout the paper that the arbitration problems (games) being analyzed possess homogeneous, divisible domains (e.g., pie or money) in which $0 \leq y_i \leq 1$ and $\sum y_i = 1$. Appendix A demonstrates how our solution concept can be extended to more general non-homogeneous domains.

2 Equivalence of relative neediness and risk aversion

To arrive at our main result in this section, we must first review the two differing routes to the concept of cardinally measurable utility that have been proposed during the twentieth century. First, there is the approach originally proposed by von Neumann and Morgenstern (1947), and significantly improved upon by Herstein and Milnor (1953). This approach is "stochastic" since uncertainty plays an essential role in constructing the desired utility function. In particular, numerical utilities for prizes are assessed via the determination of an agent's indifference probabilities for various lotteries.

The alternative and less well-known route to measurable utility stems from imposing an intensity-based preference ordering on interval differences for what is at stake. Following an exposition in Shubik (1984, pp. 421–422), suppose that prizes ranging continuously over [0,1] are at stake. Now let x, y, z, and w represent four points lying on this continuum (ordered in increasing value). In our notation, the statement $(x, y) \gtrsim (z, w)$ is then interpreted to mean that a change from outcome y to x is *weakly preferred* to a change from outcome w to z. In this set-up, a simple consistency axiom and a simple continuity axiom are joined by a more subtle "crossover axiom" to guarantee the existence of a cardinal utility function. The crossover axiom is introduced to formalize the important idea that it is interval differences that are being compared and ordered. It posits:

$$(x, y) \sim (z, w) \Leftrightarrow (x, z) \sim (y, w) \tag{3}$$

The existence theorem (originally proven by Shapley (1975)) states that, if these three axioms are satisfied, then there exists a cardinal utility function u from the domain X of prizes to the real numbers with the property that,

$$(x, y) \gtrsim (z, w) \Leftrightarrow u(x) - u(y) \geq u(z) - u(w) \tag{4}$$

Note that no probabilistic arguments arise in the construction of such a utility function. Deterministic utility functions of this second type have usually been thought of as "classical". That is, they capture a person's relative intensity of desire for the prizes at stake, and in particular his marginal utility for money, pie, or whatever. In contrast, utility functions assessed via a decision-maker's assessment of indifference probabilities are "modern", and capture a person's attitude towards risk taking.

2.1 "Neediness" and its Formal Equivalence to Risk Aversion

We now formalize the notion of relative neediness drawing on the intensity-based deterministic concept of utility of Shapley-Shubik. It is then demonstrated that relative neediness is formally equivalent to the Arrow-Pratt concept of absolute risk aversion. In doing so, we give support to the view of John Harsanyi (1953, 1975) that the two basic types of utility functions that have been introduced are very closely connected, and in certain cases are in fact equivalent in terms of the information they convey.[1] It will also be seen that, when agents are absolutely risk averse, the degree of concavity of an agent's utility function (relative to another's) is a proxy both for the degree of his relative neediness and of his risk averseness.

Definition. Consider two agents 1 and 2. Then agent 2 is said to be everywhere on [0,1] *more needy* than agent 1 if, $\forall x, y \in [0, 1]$ and $\forall \Delta_x, \Delta_y$, s.t. $(i)x \leq y$ and $(ii)x + \Delta_x \leq 1, y + \Delta_y \leq 1$ holds, then

$$(x + \Delta_x, x) \sim_1 (y + \Delta_y, y) \Rightarrow (x + \Delta_x, x) \succsim_2 (y + \Delta_y, y) \tag{5}$$

where the notation $\sim_i$ and $\succsim_i$ mean respectively "i is indifferent to" and "is weakly preferred by i to". When the agents satisfy the three intensity-based Shapley-Shubik axioms cited above, so that a cardinal utility function representation of these preferences on [0,1] exists, we can rewrite (5) as

$$u_1(x + \Delta_x) - u_1(x) = u_1(y + \Delta_y) - u_1(y)$$
$$\Rightarrow u_2(x + \Delta_x) - u_2(x) \geq u_2(y + \Delta_y) - u_2(y) \tag{6}$$

This definition says that whenever agent 1 is indifferent between a change Δ_x in consumption starting from a low level x of consumption and a change in consumption Δ_y starting at a higher level of consumption y, then agent 2 weakly prefers the change starting from the low level of consumption. This formalizes the classical idea dating back to Edgeworth and the early Utilitarians that a person who is (relatively) needier than another will possess a utility function that is (uniformly) more concave, i.e. will possess greater marginal utility for the first dollars of income than for subsequent dollars. As is well-known, welfare economists historically used this concept to justify a progressive income tax.

We can now establish our basic relationships between neediness and risk aversion and concavity.[2]

Theorem 1. *Suppose both agents have utility functions which have continuous 2nd order derivatives everywhere. Then agent 2 is everywhere on [0,1] more needy than agent 1 if and only if,*

$$-\frac{u_2''(x)}{u_2'(x)} \geq -\frac{u_1''(x)}{u_1'(x)}, \quad \forall x \in [0, 1] \tag{7}$$

---

[1] Specifically, our result reinforces the case for a "Strong Equivalence Principle", suggested to the author by John Harsanyi in a private discussion a decade ago: Assuming that an individual's preferences satisfy both the Shapley-Shubik axioms for decisions under certainty *and* the von Neumann-Morgenstern axioms for decisions under uncertainty – in particular the no love of gambling axiom – then the information contained in both utility functions is the same. Yet the results of Rabin (2000) may cast doubt on such equivalence.

[2] The author is indebted to Carsten Nielsen for helping to transform his original conjectures along these lines into the more rigorous result that follows.

This proposition is a consequence of Result A and Result B to be stated and proved now.

Result A.

Suppose $-\dfrac{u_2''(x)}{u'_2(x)} < -\dfrac{u_1''(x)}{u'_1(x)}$ for some $x < 1$. Then $\exists\, y > x$ and $\Delta_x, \Delta_y > 0$, s.t.

$$[u_1(x + \Delta_x) - u_1(x)] - [u_1(y + \Delta_y) - u_1(y)] = 0 \qquad \text{and}$$
$$[u_2(x + \Delta_x) - u_2(x)] - [u_2(y + \Delta_y) - u_2(y)] < 0 \qquad (8)$$

Proof.

$$\frac{\partial}{\partial y}\left[\frac{u'(x)}{u'(y)}\right] = -\frac{u'(x)u''(y)}{(u'(y))^2}$$

For $y = x$ this fraction equals $\dfrac{u''(x)}{u'(x)}$. Thus there is a $y > x$ (but close to x) s.t.

$$\frac{u_1'(x)}{u_1'(y)} > \frac{u_2'(x)}{u_2'(y)} \qquad (*)$$

Now $\dfrac{u'(x)}{u'(y)}$ is the derivative of the implicit function $\Delta_y(\Delta_x)$ defined by,

$$[u(x + \Delta_x) - u(x)] - [u(y + \Delta_y) - u(y)] = 0$$

when evaluated at $\Delta_x = \Delta_y = 0$. Thus, when we pick Δ_x and Δ_y sufficiently small but positive, s.t.,

$$[u_1(x + \Delta_x) - u_1(x)] - [u_1(y + \Delta_y) - u_1(y)] = 0$$

then because of $(*)$,

$$[u_2(x + \Delta_x) - u_2(x)] - [u_2(y + \Delta_y) - u_2(y)] < 0$$

Q.E.D.

As a consequence, when agent 2 is everywhere more needy than agent 1, agent 2's coefficient of absolute risk aversion is everywhere bigger than or equal to that of agent 1.

Next, in proving Result B just below, we use a theorem by Pratt (1964) which states that,

$$\frac{u_2''(x)}{u_2'(x)} \geq \frac{u_1''(x)}{u_1'(x)}, \quad \forall x \in [0, 1]$$
$$\Leftrightarrow \exists\, g : [0, 1] \to [0, 1] \quad \text{with} \quad g'(x) > 0, g''(x) \leq 0, \forall x \in [0, 1]$$
$$\text{s.t.} \quad u_2(x) = g(u_1(x)) \quad \forall x \in [0, 1]$$

The condition after the implication sign is one formal way of stating that u_2 is more concave than u_1.

Result B. Suppose $\dfrac{u_1''}{u_1'} \leq \dfrac{u_2''}{u_2'}$, $\forall x \in (0,1)$. Then agent 2 is everywhere more needy than agent 1.

Proof. Let for some $x \leq y$ and some $\Delta_x, \Delta_y \geq 0, u_1(x + \Delta_x) - u_1(x) = u_1(y + \Delta_y) - u_1(y)$. Then because $u_2 = g(u_1)$ for g increasing and concave and because $x \leq y$ which implies $u_1(x) \leq u_1(y)$ we have,

$$u_2(x + \Delta_x) - u_2(x) = g(u_1(x+\Delta_x)) - g(u_1(x))$$
$$\geq g(u_1(y+\Delta_y)) - g(u_1(y)) = u_2(y+\Delta_y) - u_2(y)$$

Q.E.D.

3 Relationship of "neediness" to bargaining theory

For some fifty years, it has been all but axiomatic that Nash-Harsanyi bargaining theory is irrelevant to moral theory. After all, bargaining theory is allegedly 'strategic' in nature, not 'ethical'. Moreover, bargaining theory reveals how rich people who are presumably less needy 'bargain down' poor people who are needier. Theorem 2 now establishes the precise sense in which this is true. It also opens the door to a recognition that bargaining theory is in fact central to ethical theory: the theorem will make it possible in Section 4 to characterize an intuitively appealing quantitative measure of relative neediness – one which does not entail problematic interpersonal comparisons of utility.

Theorem 2. *Consider an n-person pure bargaining problem where, for $i = 1, \ldots, n$, u_i is the $0 - 1$ normalized utility function of agent i with disagreement payoff $d_i = 0$. Let x denote the Nash bargaining allocation of the problem, that is, the allocation that maximizes the n-fold Nash product of the players' utilities. Then, (a) if player i is everywhere more needy than player j, $x_i \leq x_j$. That is, the needier agent receives less. Moreover, (b) the degree to which the needier player gets less is such that the resulting Nash allocation equalizes the proportional utility gains to the players.*

Proof (a). If we define $\bar{x} = x_i + x_j$, then x_i and x_j solve:

$$Max_x \ u_i(x)u_j(\bar{x} - x)$$

The First Order Condition for this problem is,

$$u_i'(x)u_j(\bar{x} - x) - u_i(x)u_j'(\bar{x} - x) = 0$$

If we differentiate the maximand twice we get,

$$u_i''(x)u_j(\bar{x} - x) - u_i'(x)u_j'(\bar{x} - x) - u_i'(x)u_j'(\bar{x} - x) + u_i u_j''(\bar{x} - x)$$

which is always less than 0, so that the Second Order Condition is always fulfilled. So all we need to show is that for $x = \frac{1}{2}\bar{x}$,

$$u_i'(x)u_j(\bar{x} - x) - u_i(x)u_j'(\bar{x} - x) \geq 0$$

so that we need to increase x beyond $\frac{1}{2}\bar{x}$ in order for the FOC to be fulfilled. To do this, exploit that $u_2 = g(u_1)$ for some g increasing and concave. Then the LHS of the FOC can be written,

$$u_i'(x)g(u_i(\bar{x} - x)) - u_i(x)g'(u_i(\bar{x} - x))u_i'(\bar{x} - x)$$

But if $x = \frac{1}{2}\bar{x}$ this expression reduces to,

$$u_i'(x)[g(u_i(x)) - u_i(x)g'(u_i(x))] \geq 0$$

since g is concave. Note that if g is strictly concave everywhere, then the above inequality is strict and agent 2 gets strictly less than agent 1. In other words, if agent 2 is everywhere strictly more needy than agent 1, then agent 2 gets strictly less than agent 1.

Proof (b). To understand exactly how much less the needier player will receive from bargaining, we must first introduce an invariant arising within cardinal value theory, and then establish its relationship to the Nash bargaining solution. For any given pair of cardinal utility functions u_i and u_j, for any $x \in [0, 1]$ (representing a particular slice of pie), and for any pie increment Δx, the entity $\left[\frac{\Delta u_j}{u_j} \div \frac{\Delta u_i}{u_i}\right]_x$ will be an *interpersonal invariant* in that the value of the bracketed entity evaluated at any given x will not change when the players' utility functions are multiplied be different constants. That is, the percentage gain in utility for of any one player divided by the percentage gain in utility of the other player remains unchanged by separate positive affine transformations, for any given value of x.

Given this fact, our result (b) follows from the fact that the Nash Bargaining solution will be the unique allocation that sets the value of this invariant $\left[\frac{\Delta u_j}{u_j} \div \frac{\Delta u_i}{u_i}\right]_x$ equal to unity. That is, the Nash allocation will be the unique allocation that *equalizes* the proportional utility gains of the players as defined just above. This is because the first-order conditions of the Nash product Lagrangean imply equality here.[3]

Note: We shall refer to this equal-proportional-gain property of the Nash theory as its *Bargaining Equity Property* since Shapley (1969), who first discussed a variant of this condition, deemed it an "equity" requirement lying at the heart of bargaining theory.[4] Q.E.D.

---

[3] Viewed geometrically, the first-order condition implies that the slope of the line passing from the origin to the solution $u(x^*)$ will have a slope equal in magnitude (but opposite in sign) to that of the line tangent to $u(x^*)$, a well known feature of the Nash solution.

[4] In his brief discussion, Shapley did not discuss the invariant we have introduced. This entity is arguably important in its own right. For it constitutes an *explicit* interpersonal comparison of players'

4 Axiomatic characterization of a measure of relative needs

Our ultimate goal (Sect. 5) is to produce an allocation rule whereby a homogeneous medium (e.g., pie) can be equitably divided into n shares on the basis of the relative needs of the parties involved.. This suggests that the desired needs measure take the form of a positive n-vector $\lambda \in E^{n+}$ with the property that the $n(n-1)$ ratios $\frac{\lambda_i}{\lambda_j}$ of its components constitute (pairwise) measures of the agents' relative needs for what is at stake. The requirement that fairness hold *pairwise* in an n-person problem follows from Harsanyi's (1963) demonstration that any n-person problem decomposes into $n(n-1)$ pairs of two-person subproblems, and that fairness must hold within each. What might be the nature of such a measure of relative needs, and what properties should it satisfy?

Consider an n-person arbitration problem A^n, or equivalently its associated n-person pure (unanimity) bargaining game G^A constructed by assigning to it the prospect space U^A of A^n. Assume that the allocation domain is homogeneous (e.g., pie or money). Assume also that the utility functions of all players possess continuous second derivatives, and that all n utility functions can be totally ordered by the "is everywhere needier on [0,1]" relation.

4.1 Proposed measure of relative needs

Let x^* denote the Nash bargaining allocation of the pure bargaining game G^A. Then the scalar measure λ_{ij} of the neediness of agent i relative to j for all pairs i, j that we propose is:

$$\lambda_{ij} \equiv \frac{\lambda_i}{\lambda_j} = \frac{x_j^*}{x_i^*} \qquad \forall \text{ pairs } (i, j) \tag{9}$$

That is, i's neediness relative to j's is the *inverse* of the ratio of their allocations from bargaining. We first give an intuitive justification for this measure. Thereafter, we give it a more formal justification.

Note: Further on, we suggest how this measure can be extended to more general cases where the players' utility functions are not totally ordered by the "is everywhere needier" relation.

Intuitive justification of λ: Theorem 2 tells us that bargaining by its very nature penalizes the needier player, or equivalently (by Theorem 1) the player who is more risk averse and thus possesses a lower Harsanyi-Zeuthen 'risk limit', or equivalently

---

proportional utility gains that plays an important role in bargaining theory. And this is true even though no "classical" interpersonal comparisons are entailed given its invariance under separate affine transformations, and under general linear transformations in the more general case where the disagreement payoff $d > 0$. Understanding the role of this invariant solves a riddle noted by Aumann and Kurz (1977) and by Harsanyi (1977): How can cardinal value theory solve problems that by their very nature seem to entail interpersonal comparisons (e.g., "How much more does my threat hurt you than yours hurts me?") whilst the underlying theory is invariant under separate linear transformations of the utilities? The answer is that a weak "proportional" interpersonal comparison *is* in fact needed, and existed within Value theory from the start in the form of the invariant cited above.

a greater Aumann-Kurz 'fear of ruin'. Now in the context of a 'pure' bargaining game, both explicit threats and coalition formation are ruled out by construction. The only difference between the players lies in their utility functions. Thus, the *only* reason for why one player can obtain more than another lies in his lesser neediness. This being true, the *magnitude* by which the needier player is penalized by the outcome of bargaining offers an intuitively appealing measure of *how much more needy* he is relative to the other. Thus, if the Nash allocation x^* awards only 1/4 the pie to i, and 3/4 the pie to j, and if the only reason why i gets one third as much as j is that i is more needy = more risk averse than j, then it seems reasonable to assert that player i *is three times needier* than player j in the context of the game at hand.

Formal justification: Intuitive support for the measure λ is reinforced by the fact that the proposed needs measure is the only measure that satisfies the following axioms.

Axiom A – *Independence of λ from ethically irrelevant variables.*
An equivalent but helpful restatement of this axiom would be:

Axiom A' – *Dependence of the measure λ on the preference structure of the players for what is at stake in the game G^A, and on nothing else.*

Comment: Asymmetries in payoffs to players in general n-person bargaining games can result from any of *four* possible asymmetries in the parameters of the underlying game, as Harsanyi (1977, p. 179) first pointed out: (i) Differences in the preferences of players for what is at stake; (ii) Differences in the threat power of the players in games in cases where there is no natural and agreed upon disagreement outcome; (iii) Differences in coalitional power; and (iv) Distortions in the rates of utility transfer due to "friction" or "taxation". Distortions of this kind alter the shape of the prospect space U and thus the outcome from bargaining.

From the standpoint of constructing a needs-based theory of equitable allocation, all but the first of these asymmetries are what philosophers call "morally arbitrary" (e.g., Rawls, 1971) and hence must be ruled out as inadmissible in the kind of theory we are constructing.[5]

Axiom B – *Concavity monotonicity.* If i is strictly needier than j in the precise sense of Theorem 1, and thus has a more concave utility function, then the desired needs measure should have the property that it deems i strictly needier than j. That is, $\lambda_{ij} > 1$.

---

[5] A fifth asymmetry might be thought to be differences in information, i.e., the structure of "incomplete" information if it exits. In a game proper, this would indeed matter. But in an arbitration problem it would not: the arbitrator will simply utilize best estimates of the agents' utilities and proceed as if complete information existed. Considerations of gaming would not arise as they would in bargaining proper.

Axiom C – *Symmetry.* If any two players i, j have identical cardinal utility functions, then their needs will be deemed equal, requiring that $\lambda_{ij} = 1$.

Axiom D – *Continuity.* Small changes in the value of the parameters of the arbitration problem of game (essentially, in the utility functions) should result in small changes in the values of the needs coefficients λ_{ij}.

Comment: In their discussions of "reasonable" arbitration schemes, Luce and Raiffa (1957, p. 123, Axiom iv) and Brock (1978) both postulate this axiom as essential in fair division problems. Yet both the Rawlsian lexicographic MINIMAX solution function and the Utilitarian function violate it.

Axiom E – *Operational significance (invariance).* The measure λ should be invariant under order-preserving linear transformations of the utilities.

Comment: This requirement is imposed for reasons of operational significance alone. An ethical theory is fundamentally different from an economic theory, and as Harsanyi (1955) and Hammond (1996) have convincingly argued, if classical interpersonal comparisons *must* be introduced, with all the problems of operationalizing them, then so be it. It turns out that they do not have to be.

Axiom F – *Uniqueness.* The measure λ should unique for any given problem.

Axiom G – *Proportionality.* Suppose i is needier than j. Then the degree to which i is deemed needier than j (i.e., the numerical value of λ_{ij}) must be proportional to a suitable measure of how much more i values what is at stake than does j.

Comment: While Axiom B above requires that $\lambda_{ij} > 1$ if i is needier than j, it offers no guidance as to the numerical value of λ_{ij}. We address this by requiring that its value be proportional to how much more i values what is at stake than does j. This of course gives specific meaning to our concept of "relative neediness" by equating it with "relative intensity of desire". The reciprocal of the Nash allocation will constitute such a measure of relative intensity of desire for reasons given in the proof below.

Theorem 3. *For the class of arbitration problems under consideration, the measure λ of relative needs characterized in (9) above satisfies these seven axioms, and it is the only measure that will do so.*

Proof. Axiom A (Independence of Ethically Irrelevant Variables) is satisfied because of the following two conditions. First, the vector λ is constructed from the bargaining allocation x^* generated in a game G^A that is assumed to be a *pure* (unanimity) n-person game. As already noted above, such games have been purified of asymmetries in coalitional strength and threat power. Second, the fact that we have assumed a homogenous domain (pie or money) ensures that there will be no "frictional" distortions to the shape of the prospect space due to transfer problems.

Asymmetries in the bargaining allocation of such games will thus reflect differences in the players' preferences for what is at stake, and nothing else, as required
by Axiom A.[6]

Theorem 2 above ensures that Axiom B is satisfied. It tells us that bargaining
will award *less* to the needier player. Since the scalar λ_{ij} is defined as the *inverse* of
the Nash bargaining allocation $\frac{x_i^*}{x_j^*}$, its value will clearly exceed unity whenever i is
needier than j. Satisfaction of the remaining four axioms C – F is directly inherited
from the Nash bargaining solution from which our measure λ is constructed. This
is obvious in the case of Axioms C, E, and F. Axiom D (Continuity) is satisfied
due to two facts: first, the Nash product maximand $\prod u_i$ is nonlinear (unlike the
Rawlsian and Utilitarian maximands); second, the boundary of the prospect space
U^A is regular given our assumptions about the players' utilities. Thus, there will
be continuity in the neighborhood of the point of tangency of the concave Nash
maximand with the convex space U^A at the solution point $u(x^*)$.

λ is most probably not the only measure satisfying the six Axioms A-F. For
example, a measure analogous to that of (9) based upon the Kalai-Smorodinsky
bargaining allocation should also satisfy these axioms. For uniqueness, a much
stronger axiom would seem to be needed, e.g., our Axiom G (Proportionality). We
now show that the reciprocal of the Nash allocation is a uniquely valid measure
of the players' relative desire for what is at stake, and that it thus satisfies this
final axiom. At the outset, note that what is needed is a *global* measure of relative
intensity of desire, a single summary measure suitable for use in an allocation
formula of the kind we are proposing below. Now the slope of the boundary of the
prospect space at any given point provides a *local* measure of relative intensity (i.e.,
of the local rate of utility transfer). Yet there will be an infinity of such measures,
one for each point on the boundary. Moreover, for any such local measure to be
a meaningful guide to allocation, the "correct" interpersonally calibrated utility
scales would have to be used from the outset. But how are these to be identified?
For these two reasons, these local measures are not satisfactory.

In contrast, the inverse of the Nash bargaining allocation offers a measure that
is global in the sense that it is an explicit function of the geometry of the entire
frontier. It can thus be thought of as synthesizing the infinity of local measures
into a single "sufficient statistic" capturing the relative intensity of desire of the
players. It is also invariant under linear transformations of the utilities. That this
measure is uniquely suitable for this role can be seen by analyzing how the Nash
allocation itself is arrived by the players. The process whereby this allocation is
arrived at via bargaining is one which tests how badly each player wants what
is at stake. It does so by testing how much risk each is willing to take to obtain
what is at stake, e.g. by testing the players' relative "risk limits" and/or "fear of
ruin". The equilibrium allocation arrived at is the unique allocation that equates
these entities, or equivalently by Theorem 2.B above, that equates the parties'
proportional utility gains. In this process, the more risk averse (more needy) person

---

[6] Brock (1978) originally introduced this concept of bargaining 'pureness' and argued that the role it
plays in his theory of distributional equity is analogous to that played by the concept of 'impersonality'
in the theories of Rawls and Harsanyi via their 'veil of ignorance' constructs.

gets *less* at equilibrium because he receives *proportionately more* utility from each increment of pie than does the less risk averse (less needy) player. It is for this reason of course, that our relative desire (needs) measure is the reciprocal the bargaining allocation. The Nash allocation is the only allocation possessing all these properties, and in this sense our measure of relative needs based upon it (9) uniquely satisfies Axiom G. Q.E.D.

Perhaps the most fundamental point we are making in proposing the measure λ is that the process of bargaining serves to *operationalize* the problem of assessing relative needs, provided importantly that bargaining is 'purified' of all morally arbitrary influences like threats and coalitions. In doing so, it offers a multi-person analogue to one of the principal achievements of modern decision theory in the case of single-person problems: the operationalization of the concept of cardinal utility via the process of assessing indifference probabilities over the domain of the utility function.

4.2 Extension to more general utility domains

The needs measure introduced above was defined in the special case where the entire set of players could be (weakly) ordered with respect to the relationship "is everywhere on [0,1] needier than". Geometrically, this can be shown to imply that none of the n concave utility functions of the players ever crosses any other function. Indeed, if i is strictly needier than j, then i's utility function will arc above that of j when both are plotted in the unit square. Can our measure of neediness (9) be extended to problems possessing more general utility functions, thereby extending the domain of Theorem 3? Yes.

Suppose that both i and j possess utility functions with positive first derivatives, but nothing more. In this case, i's utility function can intersect j's, and neither player can be deemed "everywhere needier" than the other. Suppose, also, that i's share of the pie from pure Nash bargaining is observed to be *less* than j's. We could then deem i to be *net more needy* on [0,1] than j. This being so, the reciprocal of the bargaining allocation once again becomes the relevant measure of relative need.

What might justify this weakening of our previous assumptions? Since the game played is a pure bargaining game, the *only* asymmetry between the players lies in their differing preferences for what is at stake. And if this difference causes i to settle for less than j, then i *must* in some sense be "net needier" than j since this will be the only explanation of why he gets bargained down. To sum up, Theorem 3 could be weakened to require an ordering of the player set by the "is net needier than" relation rather than by the stronger "is everywhere needier than" relation.[7]

---

[7] Mathematically, the following condition might be sufficient for player i to be 'net more needy' than j, as evidenced by his receiving less from bargaining than j does: $\int u_i(x)dx - \int u_j(x)dx > 0$, where integration takes place over the entire unit interval [0,1]. That is, if the total area under i's utility function is greater than j's, then he is 'net more concave', and he will get bargained down by j to less than half the pie. At the present, this is nothing more than a conjecture. Robert Aumann suggested pursuing this generalization of our theory.

5 'To each according to his needs' – the optimal arbitration scheme

In the main result of this essay, we now characterize an arbitration scheme (Raiffa, 1953) or equivalently an allocation rule which can be utilized by an arbitrator attempting to allocate a homogeneous medium among n people on the basis of their relative needs. More specifically, we seek an allocation rule that identifies an allocation y^* satisfying the needs-based fair-sharing rule,

$$\frac{y_i^*}{y_j^*} = \frac{x_j^*}{x_i^*} \equiv \lambda_{ij} \qquad \forall \text{ pairs } (i, j) \tag{10}$$

where λ_{ij} is the measure of the relative need characterized in Theorem 3, namely the inverse of the Nash bargaining allocation.

Theorem 4. *For the class of arbitration problems under consideration in this paper, there exists a unique allocation rule consistent with the fair-sharing conditions (10). This rule selects that unique allocation y^* whose i^{th} component satisfies*

$$y_i^* = x_i^{*-1} \cdot \left(\sum_j x_j^{*-1} \right)^{-1} \tag{11}$$

where the vector x^* is that allocation which maximizes the generalized Nash product

$$Max_x \prod_i [u_i(x) - u_i(d)] \tag{12}$$

of the pure bargaining game G^A associated with the given arbitration problem A, and where d is the 'no pie' disagreement outcome.

Proof. In the case of $n = 2$, the equitable allocation satisfying (10) is clearly the simple inverse of the Nash bargaining allocation. In the general case where $n > 2$, let us start off by utilizing the method of clearing fractions to rewrite (12) as

$$y_j^* x_j^* = y_i^* x_i^* \qquad \forall \, i \neq j \tag{13}$$

Now let κ be the common value of the $(y_i^* x_i^*)$ terms for all $i \neq j$. Then we have that

$$y_i^* = \kappa x_i^{*-1} \qquad \forall \, i \tag{14}$$

But there will be a unique κ such that the elements of y^* sum to unity. To determine this, note that

$$1 = \sum_j y_j = \kappa \left(\sum_j x_j^{*-1} \right) \tag{15}$$

which implies

$$\kappa = \left(\sum_j x_j^{*-1} \right)^{-1} \tag{16}$$

We can now substitute this into (14) to obtain

$$y_i^* = x_i^{*-1} \cdot \left(\sum_j x_j^{*-1} \right)^{-1} \tag{17}$$

as asserted in equation (11) of the above theorem.[8] Because the underlying mapping is bijective, y^* will be unique for any allocation x. Note also that (17) is symmetric: the reciprocal y^* of any Nash allocation x^* will have the original Nash allocation x^* as its reciprocal y^{**}. Q.E.D.

Comment: Due to its close relationship to the Nash bargaining solution, and due to Theorem 3 in particular, the arbitration scheme proposed above satisfies the five classical solution axioms of symmetry, continuity, invariance under linear transformations of the utilities, uniqueness, and efficiency.

6 Some Comparisons with the Utilitarian Theory

In attempting to test the "reasonableness" of the new solution concept, let us now contrast it with the Utilitarian solution. For both technical and conceptual reasons, it seems to yield a more satisfactory outcome in all the cases we have evaluated, and in doing so, the new allocation rule overcomes well-known objections to the Utilitarian function. To see why this is so, let us first consider the special case where all the agents possess isoelastic (concave) utilities of the form $u_i(x) = x^{\alpha_i}$. The first order conditions associated with the Nash product maximand $\prod_i x^{\alpha_i}$ yield

$$x_i^* = \frac{\alpha_i}{\alpha_i + \alpha_j}$$

$$\text{and} \quad x_j^* = \frac{\alpha_j}{\alpha_i + \alpha_j} \tag{18}$$

as the Nash allocation. Recall that our ethical solution y^* simply inverts this allocation so that j's "fair" allocation will be $x_j^* = \frac{\alpha_i}{\alpha_i + \alpha_j}$. Then it will be clear that, for virtually any realistic value of the parameters, an ethically appealing allocation results: the needier you are, the more you get. More specifically, the difference in the equitable allocations to the two players is strictly proportional to the difference in their concavity coefficients α_i. In the present isoelastic case, these coefficients are quantitative proxies for their relative neediness (recall Theorem 1).

Whereas our solution function makes extremely clear the subtle linkages between the concept of relative need, the nature of the utility functions, and allocations that seem "fair", the Utilitarian function fails to do so. Even in the present isoelastic case, there is no clear analytical relationship between the degree of concavity and the resulting allocations. This may be seen by noting the first order conditions associated with the Utilitarian maximand, conditions implying that the marginal

---

[8] The author is indebted to Professor Peter Veinott of Stanford for help with this argument. Professors Arrow and Aumann remarked to the author that this payoff to player i can be interpreted as the harmonic mean of the Nash allocations to the players.

utility of the players are equalized at the optimum. In the case of $n = 2$, where agent 1 receives pie share x and agent 2 receives share $(1 - x)$:

$$\alpha_1 x^{\alpha_1 - 1} = \alpha_2 (1 - x)^{\alpha_2 - 1} \tag{19}$$

where there is no clear relationship between the optimal allocation and the concavity parameters.

To be fair, one traditional source of appeal of the Utilitarian solution function is that it incorporates "urgency of desire" considerations as proxied by diminishing marginal utility, and as a result does allocate more pie (equivalently, a lower income tax rate) to an agent with a utility function that is more concave than that of another. Because of this, it is often implied that the Utilitarian solution allocates pie *in proportion* to relative need (Harsanyi, 1975, 1992). Our analysis suggests that this is not the case. Indeed, Utilitarian philosophers have never produced any quantitative measure of relative need, despite the fact that their models made use of classical interpersonal comparisons of utility which in principle should have facilitated the construction of such measures. Our solution does quantify need, yet it does not entail classical interpersonal comparisons.

Our solution concept also overcomes two other problems besetting the Utilitarian function, namely non-uniqueness and discontinuity. In the first instance, consider the extremely symmetric case where every agent has linear utility for pie and has identical interpersonally calibrated utility functions. Then the boundary of the prospect space will be a hyperplane with a negative $45°$ slope. Whereas our solution y^* will deem the equal-shares-for-all allocation as uniquely optimal in this case, the Utilitarian solution will deem *all* allocations to be equally optimal.

As for discontinuity, consider a modified hyperplane example where player i gets a utility payoff of $1 + \epsilon$ for consuming the entire pie, while each of the others gets a payoff of exactly 1 utile for the same outcome. In this case, *all* the pie will go to player i, and none to anyone else. Suppose now that an assessment error is discovered to have been made in interpersonally calibrating the utility functions: it turns out that i only gets $1 - \epsilon$ utile, whereas everyone else still gets 1 utile. Then i's share will shift *discontinuously* from 100% to 0%. Our Axiom 5 (Continuity) is thus violated. The nonlinear form of our solution concept (inherited from the nonlinear Nash product maximand)overcomes these problems.

To sum up, Utilitarianism emerges upon closer scrutiny to be what it was classically envisioned to be by the early Utilitarians themselves: a moral theory prescribing the production of the maximal amount of utility. It was not conceived as a theory of equitable allocation in proportion to needs, and does not comfortably fulfill this role. Nonetheless, under suitable regularity conditions, the allocations it prescribes are similar to those prescribed by our allocation rule (11) which *is* explicitly needs-oriented. This adds credibility to the claims of those who support Utilitarianism for distributional reasons. Finally, it should be noted that our new solution is itself Utilitarian in spirit in that it maximizes a weighted sum of the agents' utilities, and generates an efficient allocation. The utility weights are not all equal, however, as in the classical additive symmetric function $\sum u_i$. Rather, they depend *endogenously* upon the configuration of utility functions at hand. Moreover, no interpersonal comparisons of utility are required.

7 "Full" distributive justice revisited

We conclude this essay by revising the concept of Full Distributive Justice introduced by the author in his original theory of justice (Brock, 1978, 1979). The root
idea is that a comprehensive theory of justice must incorporate both allocation in
proportion to relative needs *and* allocation according to relative contribution. The
way this was achieved was to analyze an integrated two-stage decision procedure.
In Stage 1, construed as a Constitutional Convention, a constitution is selected from
a set of feasible constitutions. In Stage 2, the real-world game of life is played, with
all of its attendant bargaining, threats, coalitions, etc. But the play of this game is
subject to the provisos of the constitution selected in Stage 1.

We argued that allocation according to needs is appropriate in Stage 1, and now
modify our original characterization of relative needs by substituting the new allocation rule (11). We then proposed that allocation according to relative contribution
is appropriate in Stage 2, and we modeled this by requiring that the outcome of the
play of the Stage 2 game be a Shapley value allocation, or more specifically an NTU
Value allocation. This is because the Shapley Value of a game rigorously embodies
the norm of allocation according to relative contribution. The resulting outcome is
claimed to embody Full Distributive Justice in the sense that the "basic institutions"
chosen in Stage 1 incorporate needs-based equity whereas the outcome of playing
the induced Stage 2 game incorporates contribution-based equity.

The intuitive justification for this theory is that, during the constitutional deliberations of Stage 1, no one can claim to be contributing more than anyone else. For
all that transpires is a set of deliberations as to which of a set of mutually advantageous constitutions is "fairest". Absent differential contribution, fairness *reduces*
to considerations of relative need. Then in Stage 2, the criterion of relative contribution both should and does rise to the fore. Normatively, it dominates because it is
during this stage that differential contributions *are* made by the different agents, and
that claims of contribution-based equity legitimately arise. The incentive structure
problems that arise by ignoring these are celebrated both in theory and in history.
Descriptively, the criterion of relative contribution will be satisfied quite naturally
within Stage 2 because Shapley Value allocations result naturally from real-world
competitive market trading *and* from multilateral bargaining in politics *and* from
"mixed" political/economic systems.[9]

In integrating concepts of both needs and contribution, our theory incorporates
an important moral duality – one that highlights the primacy of needs. This duality
is that, whereas the needier person receives *more* than the less needy person in Stage
1 via our new theory, he will receive *less* in Stage 2. The latter is true because of the
formal equivalence of Shapley Value allocations and the allocations resulting from
general n-person bargaining games, and because the allocations from bargaining
are inversely proportional to relative need via Theorem 2 above. Since our Stage 1
arbitration problem occurs first, and its needs-based provisos bind what happens in

---

[9] Specifically, competitive market allocations are Shapley Values, via the Aumann (1975) correspondence theorems. So are the allocations resulting from "pluralistic" multilateral bargaining in the domain
of power politics, as Harsanyi (1963) first demonstrated. So, finally, are "mixed" political and economic
outcomes of the kind analyzed by Aumann and Kurz (1977).

the true game of life played in Stage 2, considerations of relative needs can be said to be primary in our theory. This is desirable in an avowedly normative moral theory such as the one we have proposed: It rids the theory of any might-makes-right odor.

Harsanyi (1992) rightly expressed concern that the outcome of our Stage 1 problem would unduly constrain the activities of agents during the Stage 2 game of life. How, he asked, might questions of individual liberty fit into the theory? The answer is that the utility functions of individuals incorporate their preferences for liberty, and this suitably *restricts* the feasible set of Pareto-improving constitutions from which an optimal (equitable) constitution will be selected in Stage 1. We failed to make this point clear in our original essays.

Appendix: Extensions to non-homogenous problems

We sketch here the manner in which our allocation rule (11) can be extended to allocation domains more general than the homogeneous (pie/money) domains considered above. The extension we propose is based upon a concept of utility payoff invariance. Two cases will be considered. First, there is the class of problems where the domain of arbitration is homogeneous *in principle*, but where practical restrictions make subdivision and/or transfer of pie impossible. Second, there is the class of problems where the domain of the utility functions is genuinely non-homogeneous. Thus, it might consist of alternative forms of representative government as opposed to alternative allocations of pie.

Restricted transfer domains: Let A^R denote a "restricted" arbitration problem in which (for simplicity) *all* of the pie ends up going to only *one* of n agents due to transfer restrictions. In this special case, efficiency will require the use of jointly randomized strategies, and this will generate a flat utility prospect space. The Nash solution awards each player an equal probability of winning the whole pie even when the players have very different utility functions. In short, since the Nash bargaining outcome x^* is *needs-insensitive*, no needs-respecting allocation y^* can be directly constructed from it.

In such a case, genuine needs-respecting equity can be achieved as follows. *First*, hypothesize that pie is in fact divisible and transferable. *Second*, determine the Nash allocation x^* of the associated homogeneous game $G(A^R)$. Third, utilizing (11), solve for the needs-equitable solution y^* that would be awarded by an arbitrator in this hypothetically homogeneous case. *Fourth*, take note of the pairwise utility ratios $\frac{u_i(y_i^*)}{u_j(y_j^*)}$ associated with y^*. *Fifth*, define R as the set of all jointly randomized pie-allocation strategies, with representative element $r \in R$. *Sixth*, define as the needs-equitable solution to the original problem the particular jointly randomized strategy $r^* \in R$ that generates a utility payoff $u(r^*)$ satisfying the following pairwise fair-sharing conditions:

$$\frac{u_i(r^*)}{u_j(r^*)} = \frac{u_i(y_i^*)}{u_j(y_j^*)} \qquad \forall \text{ pairs } i, j \tag{20}$$

In short, the randomized strategy r^* is the (unique) strategy which preserves the utility payoff ratios appearing on the right, the ratios that would be generated were the underlying domain truly homogeneous. Note, of course, that for any allocation x of pie, then

$$u_i(x) > u_i(r^*) \qquad \forall\, i, \text{ with } 0 < x < 100\,\% \tag{21}$$

since $u(r^*)$ will lie on the chord connecting the utility space coordinates (0,1 and 1,0). [This reflects the fact that the boundary of the prospect space U will be flat whenever randomized allocations alone are possible.] The invariant we are left with, therefore, is that of the *ratio* of the utility payoffs.

Non-homogenous Domains: In this fully general arbitration problem A^N, domains are non-homogeneous and complex. We sketch how to arrive at an approximation to needs-based equity. We end up with a solution that is utility-payoff-equivalent to that homogeneous problem which best approximates the given non-homogeneous problem A^N at hand. Here is the five-step procedure involved. *First,* construct the prospect space U^N of the non-homogeneous problem at hand. *Second,* consider the set $\{U^H\}$ of all possible prospect spaces corresponding to all possible homogeneous arbitration problems.

Third, find the element $U^{H^*} \in \{U^H\}$ which best approximates U^N according to some reasonable metric. For example, U^{H^*} could be chosen so as to minimize the area between the boundaries of the two prospect spaces. *Fourth,* utilize (11) to solve for the needs-equitable allocation y^* associated with U^{H^*}. Now note the utility payoff $u^*(y^*)$ associated with y^*, and create from it the utility payoff ratios $\left\{\frac{u_i^*}{u_j^*}\right\}$ for all pairs of agents. *Fifth,* determine that joint strategy ξ for the given non-homogeneous problem U^N that generates a payoff $u(\xi)$ with the property that

$$\frac{u_i(\xi)}{u_j(\xi)} = \frac{u_i^*}{u_j^*} \qquad \forall \text{ pairs } i,j \tag{22}$$

This is the strategy that best preserves the logic of our main theorem and is thus deemed the *quasi-needs-equitable* strategy. For the final utility payoff ratios are a "best approximation" to those that would have been generated were the domain in fact homogeneous.

Fortunately, in most real-world problems, a transferable "money" will be available so that, when choosing among tax regimes, the kinds of non-homogeneity problems that we have cited may well not arise, in which case the approximation (22) may not need to be constructed.

References

Arrow, K.J.: Essays in the theory of risk-bearing. Helsinki: Yrjo Jahnssonin Saatio 1965

Aumann, R.J.: Values of economies with a continuum of traders. Econometrica **43**, 611–646 (1975)

Aumann, R.J., Kurz, M.: Power and taxes. Econometrica **45**, 1137–1161 (1977)

Brock, H.W.: Social choice, distributive justice, and the theory of games with non-linearly transferable utility. Ph.D. Dissertation, Princeton University (1975)

Brock, H.W.: A new theory of social justice based upon the mathematical theory of games. In: Ordeshook, P. (ed.) Game theory and political science. New York: New York University Press 1978

Brock, H.W.: Game theoretic insights into ethics based on a new theory of justice. In: Brock, H.W. (ed.)
    Special Issue of: Theory and decision, entitled: Game theory, social choice, and ethics. Amsterdam:
    Reidel Press 1979

Hammond, P.J.: Consequentialist decision theory and utilitarian ethics. In: Farina, F., Hahn, F., Vannucci,
    S. (eds.) Ethics, rationality and economic behavior. Oxford: Clarendon Press 1996

Harsanyi, J.C.: Cardinal utility in welfare economics and the theory of risk-taking. Journal of Political
    Economy **61**, 434–435 (1953)

Harsanyi, J.C.: Cardinal welfare, individualistic ethics, and interpersonal comparisons of utility. Journal
    of Political Economy **63**, 309–321 (1955)

Harsanyi, J.C.: A simplified bargaining model for the n-person cooperative game. International Eco-
    nomic Review **4**, 194–220 (1963)

Harsanyi, J.C.: Can the MINIMAX principle serve as the basis for morality: a critique of John Rawls'
    theory. American Political Science Review **69**, 594–606 (1975)

Harsanyi, J.C.: Rational behavior and bargaining equilibrium in games and social situations. Cambridge:
    Cambridge University Press 1977

Harsanyi, J.C.: Rationality and morality. In: Aumann, R.J., Hart, S. (eds.) Handbook of game theory.
    Amsterdam: North-Holland 1992

Herstein, I.N., Milnor, J.: An axiomatic approach to measurable utility. Econometrica **21**, 29–297 (1953)

Kaneko, M., Nakamura, K.: The nash social welfare function. Econometrica **45**, 423–436 (1977)

Luce, R.D., Raiffa, H.: Games and decisions. New York: Wiley 1957

Pratt, J.W.: Risk aversion in the small and in the large. Econometrica **32**,122–126 (1964)

Rabin, M.: Risk aversion and the expected utility theorem: a calibration theorem. Econometrica **68**,
    1281–1292 (2000)

Raiffa, H.: Arbitration schemes for generalized two-person cooperative games. In: Kuhn, H.W., Tucker,
    A.W. (eds.) Contributions to the theory of games II, pp.361–387. Princeton: Princeton University
    Press 1953

Rawls, J.: A theory of justice. Cambridge: The Bellknap Press of Harvard University 1971

Shapley, L.: Utility comparisons and the theory of games. In: Guilbaud, G. (ed.) La decision. Paris:
    Centre National de la Recherche Scientifique 1969

Shapley, L.: Cardinal utility from intensity comparisons. Rand Corporation Memo #R – 1683 (1975)

Shubik, M.: Game theory in the social sciences: Cambridge: MIT Press 1984

Closed-loop equilibrium
in a multi-stage innovation race*

Kenneth L. Judd

Hoover Institution, Stanford, CA 94305, USA (e-mail: judd@hoover.stanford.edu)

Received: January 3, 2002; revised version: June 14, 2002

Summary. We examine a multistage model of an R&D race where players have multiple projects. We also develop perturbation methods for general dynamic games that can be expressed as analytic operators in a Banach space. We apply these perturbation methods to solve races with a small prize. We compute second-order asymptotically valid solutions for equilibrium and socially optimal decisions to determine qualitative properties of equilibrium. We find that innovators invest relatively too much on risky projects. Strategic reactions are ambiguous in general; in particular, a player may increase expenditures as his opponent moves ahead of him.

Keywords and Phrases: Multistage races, Perturbation methods, Dynamic games.

JEL Classification Numbers: C630, 0310.

1 Introduction

Innovation processes are important for economic growth and has been the subject of much study. The early work of Kamien and Schwartz, summarized in Kamien and Schwartz (1982), concentrated on the decision-theoretic problems associated with innovation and lead to the study of equilibrium of competition in innovation contained in Loury (1979) Lee and Wilde (1980), Reinganum (1981–1982), and Dasgupta and Stiglitz (1980a,b). These analyses examined one-shot innovation processes – as long as no competitor won, all competitors were equal. Also, they

---

* This is the final version of Judd (1985). The author gratefully acknowledges the comments of anonymous referees, Paul Milgrom, seminar participants at Northwestern University, the University of Chicago, the 1984 Summer Meetings of the Econometric Society, University of California at Berkeley, Stanford University, and Yale University, and the financial support of the National Science Foundation (SES-8409786, SES-8606581)

assumed that there was just one available innovation technology. More recently, Fudenberg et al. (1983), Lee (1982), Telser (1982), and Harris and Vickers (1985) examined multi-stage games. Bhattacharya and Mookherjee (1986) examine allocation of innovative effort across alternative projects in a one-shot, simultaneous move game. However, the specifications of R&D processes used in those models were limited for reasons of tractability.

This paper has two purposes. First, we examine the equilibrium of a race for a prize where each of two agents controls independent R&D projects. At each moment, each agent works to advance his own state of knowledge while knowing that of his opponent. The race ends when one of the firms has achieved a critical state of knowledge, here called "success." There is a social gain realized at that time and some of that gain is paid to the winner as a prize. This model is intended to be a stylized representation of a multi-stage R&D race, and we use it to address questions concerning firms' strategies and the allocation of resources across alternative investments.

Second, we use approximation techniques to precisely examine the nature of the subgame perfect equilibrium in some cases of our game. Global closed-form solutions to our general model are not known. The approximation techniques used below provide precise answers to interesting questions for an open set of games. While such an approach does not yield a global resolution of the issues, it does provide guidance as to what is true in some cases and points out the critical factors. The presentation of this analysis is itself a second independent purpose of this paper since it represents a general way to analyze subgame perfect equilibria of dynamic games without imposing economically unmotivated restrictions on critical model elements.

Even though this paper uses perturbation methods to analyze only multistage R&D races, the methods are quite general since they are based on a version of the implicit function theorem for analytic functions in Banach spaces. The perturbation methods described below can be used to solve many other dynamic games with state variables. We introduce the technique using R&D games since the application is clear and not excessively encumbered with complex notation. Budd et al. (1993), in independent work, also presented a perturbative analysis (using different asymptotic theorems) of a different dynamic game. These papers are just a couple of examples of the potential of perturbation and asymptotic methods for economic analysis.

Other numerical methods, such as those presented in Judd (1992), could be used to solve our R&D game. For example, Doraszelski (2002) uses projection methods from Judd (1992) to solve a generalization of Reinganum (1981). Perturbation methods and projection methods are complementary and both have a role. Perturbation methods need not make functional form assumptions and the results are theorems about an open set of cases, but that set may be small and not include some empirically interesting cases. Numerical procedures, such as projection methods, can examine a much more varied range of cases of a model but must make functional form assumptions, can examine only a finite number of instances, and suffers from numerical error. In general one would like to use both methods when studying a general model. See Judd (1997) for a more extensive discussion of the trade-offs between perturbation and alternative numerical methods. We focus on perturbation methods in this paper and the kind of qualitative results such an approach can produce.

Specifically, we find that if the prize to the innovator and the net social benefits are "small" (in a sense specified below) the model yields several results. First, if the prize equals the benefits, there is excessive innovation effort, a result common to innovation models of this nature. Second, since agents can be at differing levels of knowledge in our model, we would like to compare the relative efficiency of resource allocation across firms. We find that lagging firms are less efficient in that if there is to be a momentary subsidy of innovation effort, the first dollar of such a subsidy should go to the leading firm.

Since agents choose how to allocate resources across projects of varying riskiness, we examine the allocative efficiency of investment within firms. We find that there is relatively excessive investment in the riskier projects. We also want to know how each innovator reacts to his rival's advances. We find that if one firm advances, the other will surely increase its effort in risky projects, a movement contrary to the socially optimal reaction. However, we find that he may increase or decrease effort in less risky projects. This contrasts with the arguments in Fudenberg et al. (1983) and Harris and Vickers (1985) which find that each firm's effort is a decreasing function of his opponent's position. This finding shows that the special assumptions used in previous multistage models are critical for their results, and that some of their results do not hold up in general. This is a good example of how the perturbation method can find results missed by methods relying on closed-form solutions.

We also use this model to examine the nature of optimal R&D policy. First, we find that, in spite of the relative inefficiency of the lagging firm, it is optimal to let competition continue until some firm enjoys complete success. Second, we also find that the optimal prize asymptotically equals the social benefit when the social benefit is small.

Some of our results hold because the multi-stage nature of the game disappears if the prize is small. However, other features, particularly the nature of firms' reactions and the risk allocation decisions, are related critically to the multi-stage subgame perfect nature of our analysis. This indicates that we have successfully peeked into the nature of subgame perfect equilibrium in innovation races. Furthermore, we indicate how other approximations could be carried out, showing that the perturbation approach does not rely on the small prize assumption. The only requirement for the application of the approximation techniques used below is some example with a known closed-form solution. Since the key theorems and techniques are general, it is clear that our perturbation approach to closed-loop subgame perfect equilibrium analysis is of general applicability for game-theoretic analysis of dynamic strategic interaction.

Some of the features of our analysis will initially appear odd. In particular, we assume that the prize and social benefits are small, and we examine second-order terms of a Taylor series expansion instead of the more common linear terms. However, neither feature negates the usefulness of the perturbation approach. Perturbation analysis is often based on a degenerate case, particularly in the physical sciences where perturbation methods are commonly used. For example, the Einstein equations of general relativity theory are generally intractable. However, many of that theory's powerful implications, such as gravitational radiation, have come from computing the high-order terms of power series solutions to the general relativity field equations around the case of a universe with no matter and no gravity. If physicists find nearly empty universes to be informative then the case of zero prize in R&D races should be informative for economists. This kind of "nearly degenerate" approach combined with first-order approximations has often been useful in economics. In macroeconomics, we often implicitly assume that shocks are nearly zero and use linearizations of dynamic systems around their steady states to examine dynamic stochastic economies. In public finance, we often implicitly assume that taxes are nearly zero and say that the excess burden of a tax is proportional to the square of the tax rate. The usefulness of high-order approximations around the case of a zero prize will be apparent below.

Section 2 describes the general model and compares it with the multiperiod models of Fudenberg et al. (1983), Lee (1982), Telser (1982), and Harris and Vickers (1985). Section 3 gives an overview of the approximation technique which we utilize below and Section 4 demonstrates it in detail for a useful special case. We then examine the nature of our problem for the case of a small net social value, discussing in Section 5 the social optimum and in Section 6 the competitive outcome. Section 7 compares the optimum and equilibrium outcomes and Section 8 examines some implications for optimal social policy given rivalrous innovation. Section 9 discusses the relation of our analysis with other approaches, arguing that our approach gives a method to generalize solutions to problems which generate closed-form solutions. Section 10 concludes.

2 A multistage model of a race

We investigate a simple model of multi-state innovation with two firms. Competition takes the form of a race. The position of each firm is denoted by a scalar with firm one at $x < 0$ and two at $y < 0$. Success is defined by one firm crossing 0; therefore we assume x and y are initially both negative and that the current state of the race, (x, y), is represented by a point in the third quadrant of the plane. A firm can attempt to improve its position by investments which determine the probability of jumping to a better state of knowledge.

Jumps occur in two ways. There are *partial jumps* which will move a firm closer to the goal. If firm one (two) is at point $x < 0$ $(y < 0)$ and invests at rate u (v) on partial jump investment, then there is a partial jump with probability udt (vdt) during a dt time interval. If a partial jump occurs, there is a probability of $F(x)dt$ $(G(y)\,dt)$ of hitting 0 and otherwise there is a probability of $f(s, x)ds$ $(f(s, y)ds$) of landing in the interval $(s, s + ds)$, $s < 0$. We assume that the distributions of the partial jumps are ordered by first-order stochastic dominance, that is, if $x' > x$, then $f(s, x')$ first-order stochastically dominates $f(s, x)$. We assume that f is bounded above and there are no moves backwards; hence, if $s < x$, then $f(s, x) = 0$. Note that $F(x) = 1 - \int_{[x, 0)} f(x, a)ds$. $F(x)$ is increasing in x, by the stochastic ordering of f in x. We also assume that F is positive everywhere;

this says that there is always some chance of jumping to the finish from any state x. Our procedures could handle the more general case, but at substantial notational cost and little substantive gain.

There are also *leaps* from x (y) to 0. If firm one (two) is at point $x < 0$ $(y < 0)$ and invests at rate w (z) on leap investment, then firm one (two) leaps to $x = 0$ $(y = 0)$ with probability $wG(x)dt$, $(zG(y)dt)$ during a time interval dt. The leaps will be called more risky since whenever investment is such that leaps and partial jumps have the same expected jump, the expected gain in the value of any convex function of position is greater for leaps. For the sake of simplicity, we assume square cost functions. That is, firm one's costs and the social costs associated with its choice of u and w are $\alpha u^2/2 + \beta w^2/2$, where α, $\beta > 0$. The costs associated with firm two's choices equal $\alpha v^2/2 + \beta z^2/2$. The first firm to succeed receives a prize of P, with no prize for the loser. We assume that the social benefit from any success is $B > 0$ and that $\rho > 0$ is the social and private discount rate.

This model differs from earlier multi-stage models in substantial ways. In Lee (1982) and in Telser (1982), a firm may pull away in the sense that it may achieve an increasingly superior cost structure, but the leading firm has no advantage in achieving any other low level of costs. In this model, a firm may pull away from its competition and final success is easier to achieve the more advanced it is. The ability to pull away and attain some dynamic advantage is present in models analyzed in Fudenberg et al. (1983) and in Harris and Vickers (1985) but they both assume very special structures for innovation costs and limit the investment choices of innovators. In particular, innovation is a natural monopoly in Harris and Vickers' model in that society would only want one innovation project commanding resources, a feature which limits the ability to address issues in patent policy and the structuring of incentives for innovation. Under our assumptions, however, there is a social value to having resources allocated to each innovation project since the marginal cost of effort is zero when the effort level is zero for each project.

Both Fudenberg et al. (1983) and Harris and Vickers (1985) focus on conditions under which a firm will surely win the patent race once it has any small advantage over its competitor. The information lag model studied in Fudenberg et al. paper is closely related to our model. In both models no firm knows what the other firm is currently doing, but both know the position of its opponent at the beginning of each period. The models differ in that the state of each firm responds stochastically to his efforts whereas Fudenberg et al. assume a deterministic response. They also make an increasing cost assumption concerning the relationship between effort and progress, but must make restrictive assumptions to render the analysis tractable.

All previous dynamic models have assumed only one kind of research investment. By permitting alternatives of varying riskiness, we can compare the relative allocation of resources among projects of varying riskiness. Finally, we also determine how relative efficiency of the two firms is related to their relative position, finding that the lagging firm is less efficient. We address the issue of when a competition should be ended and a winner granted the monopoly right to the innovation, a question previously ignored.

This general model can be used to address several issues in the economics of races. Before analyzing our model we will first present our approximation approach.

3 Banach space approximations

The model described above is far too general to hope for a closed-form solution. Nor will the structure be sufficiently tractable so as to allow for comparative static analysis as in previous work. We will instead use basic approximation techniques to study our general model for cases near some tractable case. This section reviews the basic mathematics underlying our approach and discusses its usefulness.

The primary tool used below is the generalization of Taylor's theorem and the Implicit Function theorem in R to Banach spaces. Taylor's theorem for a real-valued function over R, says that if $f(x; z)$ is C^{n+1} in x on $[0, b]$ for all z, where we think of x as the variable in R and z as a parameter, then for any z and any $x \in (0, b)$ there is a $\xi \in (0, x)$ such that

$$f(x; z) = \sum_{k=0}^{n} f^{(k)}(0; z)\frac{x^k}{k!} + f^{(n+1)}(\xi; z)\frac{x^{n+1}}{(n+1)!}.$$

This states that the n-th degree polynomial in Taylor's Theorem is an $o(x^n)$ approximation of $f(x; z)$ for x near zero. In particular, properties such as positivity and convexity which hold for this approximating polynomial near zero also hold for $f(x; z)$ when x is near zero.

Since equilibria in our games will be expressed as a collection of functional equations of the equilibrium strategies, we will use the Implicit Function Theorem to compute equilibria for games close to games for which solutions are known. Generally, the Implicit Function Theorem states that f can be uniquely defined for x near zero by the equation $H(x, f(x); z) = 0$ if $H_1(0, f(0); z)$ exists and $H_2(0, f(0); z)$ is invertible. This allows us to implicitly compute the derivatives of f with respect to x as a functions of x and z, leading to a polynomial approximation for f.

However, our strategies are not going to be vectors of real numbers, but rather functions of the state variable, objects from infinite-dimensional spaces. It is necessary, therefore, to first introduce some terminology from nonlinear functional analysis and state the generalized Implicit Function Theorem for functions and power series over Banach spaces. Suppose that X and Y are Banach spaces, i.e., normed complete vector spaces. A map $M : X^k \to Y$ is k-linear if it is linear in each of its k arguments. It is a power map if it is symmetric and k-linear, in which case it is denoted by $Mx^k \equiv M(x, x, \ldots, x)$. The norm of M is constructed from the norms on X and Y, and is defined by

$$||M|| = \sup_{||x_i||=1,\ i=1,2,\ldots,k} ||M(x_1, x_2, \ldots, x_k)||$$

For any fixed x_0 in X, consider the infinite sum in Y:

$$Tx = \sum_{k=1}^{\infty} M_k(x - x_0)^k$$

where each of the M_k is a k-linear power map from X to Y. When the infinite series converges, T is a map from X to Y. It will be convenient to associate a real valued series, called its *majorant series*, with T

$$\sum_{k=0}^{\infty} ||M_k|| \, ||x - x_0||^k$$

The important connection between the power series for T and its majorant series is that T will converge whenever its majorant series does.

Definition 1. T *is analytic at* x_0 *if and only if it is defined for some neighborhood of* x_0 *and its majorant series converges for some neighborhood of* x_0.

With these definitions, we can now state the analytic operator version of the Implicit Function Theorem.

Theorem 2. *Implicit Function Theorem for Analytic Operators: Suppose that*

$$F(\varepsilon, x) = \sum_{n,k=0}^{\infty} \varepsilon^n M_{nk} \, x^k \tag{1}$$

defines an analytic operator, $F : U(0,0) \subset R \times X \to Y$, *where* $U(0,0)$ *is a neighborhood of* $(0,0)$ *in* $R \times X$. *Furthermore, assume that* $F(0,0) = 0$ *and that the operator* $M_{01} : X \to Y$, *representing the Frechet cross-partial with respect to* x *at* $(0,0)$, *is invertible. Consider the equation*

$$F(\varepsilon, x(\varepsilon)) = 0 \tag{2}$$

implicitly defining a function $x(\varepsilon) : R \to X$. *The following are true:*

1. *There is a neighborhood,* $\mathcal{V}$, *of* $0 \in R$, *and a number,* $r > 0$, *such that (2) has a unique solution of* $||x|| < r$ *for each* $\varepsilon \in \mathcal{V}$.
2. *The solution,* $x(\varepsilon)$, *of (2) is analytic at* $\varepsilon = 0$, *and, for some sequence of* x_n *in* X, *can be expressed as*

$$x(\varepsilon) = \sum_{n=1}^{\infty} x_n \, \varepsilon^n \tag{3}$$

   where the coefficients x_n *can be determined by substituting (3) into (1) and equating coefficients of like powers of* ε.
3. *The radius of convergence of the power series representation in (3) is no less than that of the analytic map,* $z(\varepsilon) : R \to R$, *defined implicitly for some neighborhood of* 0 *by*

$$0 = \sum_{n,k=0}^{\infty} \varepsilon^n \, ||M_{nk}|| \, z(\varepsilon)^k \tag{4}$$

Furthermore, for some sequence z_n *of real numbers,*

$$z(\varepsilon) = \sum_{n=0}^{\infty} \varepsilon^n \, z_n$$

represents the solution to (4) and $|z_n| > ||x_n||$.

Proof. See Zeidler (1986).   □

The actual execution of the mathematics in Theorem 2 turns out to be elementary since our task is reduced to recursive computation of x_n terms. The term-by-term approach alluded to in item 2 in Theorem 2 will be illustrated in the next section.

However, we should first discuss the value of such an approximation approach. Our objective below is to apply it to examine subgame-perfect equilibria in our model. In most of the analysis below, we will express equilibrium strategies and values as functions of the prize, P, social benefit, B, and the position, (x, y) and examine approximations for them around the case of a zero prize and no social value. At first blush, approximations based on such cases may appear useless since the case of a zero prize is degenerate. A number of considerations justify the effort and indicate the general value of this approach.

First, the approximations can provide counterexamples to conjectures. Suppose $g_1(P)$ and $g_2(P)$ are functions of interest, and it is initially conjectured that $g_1(P) > g_2(P)$. If we can show that $g_1(0) = g_2(0)$ and $g_1'(0) < g_2'(0)$, then there must be an interval of $P > 0$ where $g_1 < g_2$, contradicting the conjecture. This in fact will occur below when we discuss equilibrium reaction functions.

Furthermore, suppose g_1 depends on some function F, i.e., $g_1(P; F)$. More generally, one could identify conditions on F which lead to the "$g_1 > g_2$" conjecture failing. In models of dynamic competition, we often make special assumptions about the functional form of such F's. After deriving our results, however, we usually don't know exactlywhat general feature of the functional form was crucial. Our approach below will find exactly what features of all structural elements are critical for any results for the case of a small prize. Whenever the intuition gathered from such an analysis does not depend on P being nearly 0, then we have perhaps discovered a robust feature of the model. Generally, we study such approximations not because they are valid for nearly degenerate cases, but rather that they likely indicate patterns which continue to hold more generally.

Second, any analytical investigation of this model must focus on cases which are degenerate in some ways. Note that the models of Lee and Wilde, Reinganum, and Fudenberg et al. are all special cases of this general model (or some slightly different general model) which are degenerate in some dimension. For example, Lee and Wilde, and Reinganum implicitly assume that the success probability function $G(x)$ is independent of the position x, making position irrelevant. Also, $F(x)$ is essentially absent in their models, as if α were infinite. Each of these special cases are of interest despite their degeneracies. However, if we are interested, for example, in a precise look at how innovators react to each other's successes, it is valuable to look at cases in which there are as few unmotivated restrictions on the underlying stochastic structure as possible. It is unfortunate that we may have to assume a small prize, but that is the price we pay here to attain this particular goal. Finally, the technique that is exploited below can be used generally to develop a robustness analysis for all the special cases studied previously.

4 An example: the case of a single firm

In this section we analyze the case of a single firm. This will illustrate the analysis used below and will also be used later when we examine the optimal stage at which to end the race. Also, to cut down on inessential clutter, we will examine here only the simple case when β is infinite. The general solution will be displayed at a later point.

The case of a single innovator is a dynamic programming problem. If $M(x)$ is the value of position x to the firm, then the Bellman equation for M is

$$M(x) = \max_u \left\{ -\frac{\alpha u^2}{2} dt + M(x)(1 - \rho dt)\, (1 - u dt) \right. \tag{5}$$

$$\left. + (1 - \rho dt) u dt \left(\int_x^0 M(s) f(s, x) ds \right) + u dt P F(x) \right\}$$

where dt is the infinitesimal unit of time.[1] The individual terms of the maximand represent the expected value of innovative effort. If the rate of effort is u, R&D expenditures during dt equal $-(1/2)\alpha u^2 dt$. With probability $1 - u dt$ there will be no success, implying that the state of knowledge dt units of time in the future will remain x and the value will remain $M(x)$. The current unconditional expected value of that event is $(1 - \rho dt)(1 - u dt)M(x)$. With probability $u dt$ there will be a jump to some $s \in (x, 0]$. If x jumps to 0, an event with probability $F(x)$ conditional on a jump occurring, the immediate reward is P. Since the reward is immediate, no discounting occurs. If x jumps to a point $x' \in (s, s+ds)$, an event with a conditional probability of $f(s, x) ds$, the value becomes $M(s)$ in the next period. In the foregoing, $\int_x^0 \dots ds$ will represent $\int_{[x,0)} \dots ds$, thereby ignoring the atom at $x = 0$. We use this notation to distinguish reaching an intermediate stage from that of winning. Equation (5) therefore states that the value of a position equals the maximum expected current value of future positions net of current costs.

Solving the maximization problem in (5) shows that

$$\alpha u = \int_x^0 M(s) f(s, x) ds + P F(x) - M(x) \tag{6}$$

Substituting this first-order condition into the control equation yields

$$0 = \frac{1}{2\alpha} \left(\int_x^0 M(s) f(s, x) ds + P F(x) - M(x) \right)^2 - \rho M(x) \tag{7}$$

By standard dynamic optimization methods, there exists a unique such M.

We cannot generally find a closed-form solution for M in (7). We will instead use Theorem 2 to give us precise information about M for any F and an open set of values for P. Note that this fits our discussion above. If we assume that the value function M is in the Banach space of real-valued functions on the negative reals with the supremum norm, then the RHS of (7) is the sum of a linear and a bilinear

---

[1] We will employ the intuitive infinitesimal notation of equation (5). All the dynamic programming equations can be derived formally, as in Bryson and Ho.

operator acting on M and the real parameter P. To proceed in this fashion one should examine dimensionless versions of a problem since the concept of "small" should not depend on the choice of units. Define $m \equiv M/P$ to be the value of problem (5) relative to the prize. m is a dimensionless quantity representing the value of the game which will yield a substantive concept of small.

Rewritten in terms of m, (7) becomes the equation

$$m(x; p) = p \left(\int_x^0 m(s)f(s,x)ds + F(x) - m(x) \right)^2 \tag{8}$$

where $p \equiv P/2\alpha\rho$ is the size of the prize relative to the marginal cost of innovation and the cost of capital. Since the dimension of ρ is (time)$^{-1}$ and that of α is (dollars)$\times$(time), p is dimensionless and will be our measure of the prize. Since m, p, f, and F are all dimensionless, (8) is a dimensionless representation of (7). When p is zero, (8) yields the obvious solution, $m(x) = 0$. p may be zero either because P is zero or because $\alpha\rho$, the costs, are infinite. Focussing on p makes clear that we are not assuming that the prize itself is small but rather it is small compared to the rate of increase in marginal cost. This will imply that the prize is to the first order equal to the costs and that the net profits of an innovator are small relative to the prize. The interpretation that the prize just covers the opportunity costs of innovative activity makes our focus on small p more plausible.

Once we transform (7) into a dimensionless equation, we also must transform other variables of interest; in particular, the control variable, u. However, u is not dimensionless since it measures effort per unit of time and depends on the time unit. We can rewrite (6) into the dimensionless form:

$$\tilde{u} \equiv \frac{u}{\rho} = 2p \left(\int_x^0 m(s)f(s,x)ds + F(x) - m(x) \right) \tag{9}$$

where $\tilde{u}$ is the dimensionless rate of effort per normalized unit of time.

We now illustrate computing a local solution to (8). If $p = 0$, then $m = 0$. Applying the Implicit Function Theorem tells us that $m(x; p)$ is smooth in p for p near zero, and that we can approximate $m(x; p)$ for such p up to $O(p^n)$

$$m(x; p) \approx m(x; 0) + pk^1(x) + p^2k^2(x) + \ldots + p^nk^n(x) \tag{10}$$

where we define $k^n(x) \equiv \frac{1}{n!} \frac{\partial^n m}{\partial p^n}(x, 0)$. First note that $m(x; 0) = 0$ since a zero prize makes the optimal value of the problem zero.

Differentiating (8) with respect to p and evaluating at $p = 0$ shows that

$$k^1(x) = F(x)^2 \tag{11}$$

Taking a second derivative of (8) with respect to p, evaluating it at $p = 0$, and using the fact that $\partial m/\partial p(x; 0) = k^1(x) = F(x)^2$, we find that[2]

$$k^2(x) = 2F(x) \left(\int_x^0 F(s)^2 f(s,x)ds - F(x)^2 \right) \tag{12}$$

---

[2] Our notation will be burdened with many superscripts. Superscripts to functional names, as in $k^2(x)$, will represent distinct functions, and will *never* represent iteration as in $k(k(x))$. Superscripts to functional evaluations represent powers. Hence, $k^2(x)^3$ is the cube of the value of the function k^2 evaluated at x.

Continuing in this fashion, one can recursively compute $k^n(x)$ for any n justified by the known smoothness of m in terms of p. Note that *no* smoothness of m in x need be assumed.

It is usually quite tedious to do all the differentiation explicitly. A standard trick in perturbation analysis is to take the polynomial approximation for m in terms of p in (10), insert it into (8), and conduct the algebraic operations indicated in (8) to get an approximate polynomial representation of (8). Equation (8) then becomes

$$pk^1(x) + p^2 k^2(x) + \ldots =$$
$$pF(x)^2 + 2p^2 \left(\int_x^0 k^1(s)f(s,x)ds \right) - k^1(x)) + \ldots \right) \tag{13}$$

If we equate terms linear in p in (13), we find that $k^1(x) = F(x)^2$. Combining p^2 terms and using the computed solution for k^1 demonstrates (12). Continuing in this fashion will yield all k^n functions. Since this approach yields the terms of the Taylor series more efficiently, we will use it below.

From these expressions we may infer several obvious properties of the optimal control for small p. For example, if p is small, effort increases as one is closer to the finish. This follows from the observation that the $pF(x)$ term dominates in (9) since m is $O(p)$ implying that u rises as $F(x)$, and hence x, rises. Also, u falls and as α and ρ rise, an intuitive result since both represent costs. Using this approach, we next examine the total social optimum when we have two separate projects and two firms.

5 The social optimum

Let $W(x, y)$ be the social value function when current states are x and y. Then the Bellman equation becomes

$$W(x, y) = \max_{u,v,w,z} \frac{1}{2}(-\alpha u^2 - \alpha v^2 - \beta w^2 - \beta z^2)dt$$
$$+ udt \left(\int_x^0 W(s, y)f(s, x)ds + BF(x) \right)(1 - \rho dt)$$
$$+ vdt \left(\int_y^0 W(x, s)f(s, y)ds + BF(y) \right)(1 - \rho dt) \tag{14}$$
$$+ (wG(x) + zG(y))(1 - \rho dt)Bdt$$
$$+ (1 - \rho dt)(1 - (u + v + wG(x) + zG(y))dt)W(x, y)$$

The first-order conditions of (14) imply

$$\begin{array}{l} \alpha u = \int_x^0 W(s, y)f(s, x)ds + BF - W(x, y) \\ \beta w = G(x)(B - W(x, y)) \end{array} \tag{15}$$

αv and βz may be expressed similarly. Using the first-order conditions, (15), for u and w, and the corresponding conditions for v and z, (14) becomes

$$0 = (E_x\{W(s, y)\} - W(x, y))^2 / 2\alpha + (E_y\{W(x, s)\} - W(x, y))^2 / 2\alpha \tag{16}$$
$$+ (G(x)(B - W(x, y)))^2 / 2\beta + (G(y)(B - W(x, y)))^2 / 2\beta - \rho W(x, y)$$

where

$$E_x \{W(s,y)\} \equiv \int_x^0 W(s,y)f(s,x)ds + BF(x)$$

$$E_y \{W(x,s)\} \equiv \int_y^0 W(x,s)f(s,y)ds + BF(y)$$

Theorem 3. *There exists a unique solution, $W(x,y)$, to the social optimum problem, and $W(x,y)$ is analytic in B, α, β, and ρ.*

Proof. The RHS of (16) is an analytic operator on bounded functions over the nonpositive reals. When $P = 0$, the unique solution is $W = 0$. Furthermore, the cross Frechet derivative, first with respect to P then with respect to W, is $-\rho$, which is an invertible operator on bounded functions. Therefore, the conclusions follow from Theorem 2. □

We next compute an approximation for W. Suppose $W(x,y) = B(bh^1(x,y) + b^2 h^2(x,y) + \ldots)$ is the approximating series for W around $B = 0$, which exists by the Implicit Function Theorem. We let $b = B/2\alpha\rho$ be a dimensionless measure of the social value and $\gamma = \alpha/\beta$ be the dimensionless ratio of costs across projects, and use them to create a dimensionless representation of W/B. The linear term, h^1, is computed to be

$$h^1(x,y) = F(x)^2 + F(y)^2 + \gamma \left(G(x)^2 + G(y)^2\right) \tag{17}$$

and the investment rules are approximated to $O(b^2)$ by

$$\frac{u}{\rho} \approx 2bF(x) + 2b^2 \left(\int_x^0 h^1(s,y)f(s,x)ds - h^1(x,y)\right) \tag{18}$$

$$\frac{w}{\rho} \approx 2 \left(b - b^2 h^1(x,y)\right) \gamma G(x)$$

and similarly for v and z. The first-order approximations for u and w are as if the current hazard rate of immediate success was common to all stages since $\alpha u \approx BF(x)$ and $\beta w \approx BG(x)$ to $O(B)$. This indicates that the first-order behavior of this multi-stage game at any stage reduces to the behavior of a single-stage game. In particular, to a first order, the presence of other projects has no impact on investment rules. Intuitively, this is because for small B, effort levels are "small," the probability of success for any one project is "small," and by independence the probability of success by two projects is "small squared," hence negligible. Therefore, most of the interesting multi-stage questions we ask below will require examination of the h^2 function that appears in the $O(b^2)$ term.

Straightforward combinations of (17) and (18) prove Theorem 4.

Theorem 4. *For small B, the following hold for the optimal innovation policy:*

1. as $x(y)$ increase, $u(v)$ and $w(z)$ increase and $v(u)$ and $z(w)$ fall;
2. $w(z)$ is increasing and concave in B;
3. $u(v)$ is increasing in B but may be convex or concave in B;

4. W is increasing and convex in (x, y) if $F(x)$ and $G(y)$ are convex;
5. u and v (w and z) are decreasing in ρ and $\alpha(\beta)$;
6. and w and z are decreasing in α.

Particularly note that, if the two firms were managed in a socially optimal fashion, each firm would increase its efforts on both projects as it advances, and the other would decrease its effort. Also, the magnitude of these reactions are on the order of B^2. These features will be substantially different in the equilibrium of the R&D race.

6 Equilibrium of the innovation game

We next solve for the symmetric subgame-perfect equilibrium of the corresponding game. We are implicitly assuming that the current states of both firms are common knowledge since if we had assumed that no firm could observe the position of his competitor then the open-loop solution would be the correct equilibrium concept. While this common knowledge aspect is certainly valid in sports races, it may appear awkward here. It asserts that a firm may know how much its opponent knows without knowing exactly what its opponent knows. This is not an unrealistic description of matters in knowledge-intensive activities. Academics, for example, should not be uncomfortable with this assumption since they often judge colleagues' relative levels of knowledge about a subject without having an equivalent level of expertise in the area. In sum, we are assuming that firms may determine their relative positions without actually having access to each other's knowledge.

Let $V(x, y)$ represent the value to firm one of state (x, y). We will examine symmetric equilibria, implying that $V(y, x)$ will represent the value to firm two of state (x, y). We also limit our examination to equilibria which depend only on the current state of the game. The Bellman equation for firm one is

$$
\begin{aligned}
V(x, y) = \max_{u,w} \Big\{ &- \left(\alpha u^2 / 2 + \beta w^2 / 2 \right) \, dt + w G(x) P(1 - \rho \, dt) \, dt \\
&+ u \, dt \left(\int_x^0 V(s, y) f(s, x) ds + P F(x) \right) (1 - \rho \, dt) \\
&+ v \, dt \left(\int_y^0 V(x, s) f(s, y) ds \right) (1 - \rho \, dt) \\
&+ (1 - \rho \, dt) \left(1 - (u + v + w G(x) + z G(y)) \, dt \right) V(x, y) \Big\}
\end{aligned}
$$
(19)

The first-order conditions from (19) allow us to express firm one's strategy in terms of the value function:

$$
\alpha u(x, y) = \int_x^0 V(s, y) f(s, x) ds + P F(x) - V(x, y)
$$
(20)
$$
\beta w(x, y) = (P - V(x, y)) G(x)
$$

By symmetry, the strategies of firm two are

$$\alpha v(x, y) = \int_y^0 V(s, x) f(s, y) ds + PF(y) - V(y, x) \tag{21}$$

$$\beta z(x, y) = (P - V(y, x)) G(y)$$

Equilibrium is characterized by substituting the equations for strategies into the Bellman equation, which then reduces to

$$
\begin{aligned}
0 = {} & \frac{1}{2\alpha} \left(\int_x^0 V(s, y) f(s, x) ds + PF(x) - V(x, y) \right)^2 \\
& + \frac{1}{2\beta} (P - V(x, y))^2 G(x)^2 \\
& + \frac{1}{\alpha} \left(\int_y^0 V(s, x) f(s, y) ds + PF(y) - V(y, x) \right) \\
& \times \left(\int_y^0 V(x, s) f(s, y) ds - V(x, y) \right) \\
& - \left(\rho + \frac{(P - V(y, x)) G(y)^2}{\beta} \right) V(x, y)
\end{aligned}
\tag{22}
$$

Theorem 5. *There exists a $\bar{P} > 0$ such that for $P \in [0, \bar{P}]$, there is a unique symmetric subgame perfect equilibrium $V(x, y)$, which is analytic in P, α, β, and ρ, and represented as a solution to (22).*

Proof. Same as Theorem 3.   □

Theorem 5 is a strong result, but one which fits the focus on equilibria which depend on only the current state. Implicitly, we are ruling out equilibria where current actions depend on past history. This eliminates some reputation effects, trigger strategies, and other phenomena which can support implicit collusion in such infinite-horizon dynamic games. This is reasonable in the case of leap investment since such investments are unobserved and any cheating could be inferred only when a leap occurred, which would be too late. Some implicit collusion in partial jump investment may be possible since, as long as neither had won, each could infer cheating if the other seemed to be moving too quickly. The uniqueness result in Theorem 5 does not rule out the existence of reputational equilibria since it just says that there is a unique function $V(x, y)$ that is analytic in (x, y) and solves the equilibrium equations for small P.

Suppose $V(x, y) = P(pg^1(x, y) + p^2 g^2(x, y) + \ldots)$ is a Taylor series approximation of $V(x, y)$ for small p. By Theorem 5, such a representation exists and is unique for small p. Even though (22) is not expressed in p, it can be straightforwardly rewritten so that V/P, the dimensionless value of the game, depends on P, α, β, and ρ only through p and the dimensionless ratio $\gamma = \alpha/\beta$. By substituting this representation for V in (22) and equating coefficients of like powers, we find

$$g^1(x, y) = F(x)^2 + \gamma G(x)^2$$

$$g^2(x, y) = 2F(x) \left(\int_x^0 (F(s)^2 + \gamma G(s)^2) f(s, x) ds - F(x)^2 - \gamma G(x)^2 \right)$$

$$-2 \left(\gamma G(x)^2 + \gamma G(y)^2 + F(y)^2 \right) \left(F(x)^2 + \gamma G(x)^2 \right) \tag{23}$$

The equilibrium strategies are therefore approximated to $O(p^3)$ by

$$\frac{u(x, y)}{\rho} \approx 2pF(x) + 2p^2 \left(\int_x^0 g^1(s, y) f(s, x) ds - g^1(x, y) \right)$$

$$+ 2p^3 \left(\int_x^0 g^2(s, y) f(s, x) ds - g^2(x, y) \right) \tag{24}$$

$$\frac{w(x, y)}{\rho} \approx 2\gamma p \left(1 - pg^1(x, y) - p^2 g^2(x, y) \right) G(x)$$

and similarly for $v(x, y)$ and $z(x, y)$. This solution and its approximation now allow us to compare equilibrium with the social optimum and evaluate the competitive equilibrium allocation of resources.

7 Comparisons of the optimal and equilibrium outcomes

We next will compare the levels of innovative activity under social control with those levels in the game equilibrium. If $P = B$, the difference between innovative effort under competition, u^c, w^c, and the socially optimal levels, u^s, w^s, is expressed, up to $O(p^2)$, by

$$\rho^{-1}(u^s - u^c) \approx -2p^2 \left(F(y)^2 + \gamma G(y)^2 \right) F(x) \tag{25}$$

$$\rho^{-1}(w^s - w^c) \approx -2\gamma p^2 \left(F(y)^2 + \gamma G(y)^2 + \gamma G(y)^2 \right) G(x) \tag{26}$$

The difference between firm two's choices, v^c and z^c, and the optimal controls v^s and z^s, are similarly expressed. First note that there is excessive investment in all projects under competition, a conclusion common in these models. The excess is greater as either firm is closer to success. Also the excess investment relative to the socially optimal investment increases for each firm as the other firm is closer to success. These results are expected since each firm ignores the social value of the other's presence in the innovation process.

We also note that it is not clear which firm is more excessive in R&D investment. If E_{uv} is the difference, $(u^c - u^s) - (v^c - v^s)$, between the two competitor's excessive investment in their partial jump processes, then

$$E_{uv} / \left(2\rho p^2 \right) \approx F(y) F(x) \left(F(y) - F(x) \right) + \gamma \left(G(y)^2 F(x) - G(x)^2 F(y) \right)$$

to $O(p^2)$. If there are no leaps, then $G \equiv 0$ and $E_{uw} < 0$ if $x > y$, that is, the laggard's investment is more excessive than the leader's. This holds also if the leap and partial jump processes are sufficiently similar, in particular if $G = \lambda F$ for some

$\lambda > 0$. However, if $F(y)$ is small but $G(y)$ is not, then $E_{uw} > 0$, and the leader invests more excessively in partial jumps.

In relative terms, however, we can be more precise since

$$\frac{u^c - u^s}{u^s} \approx p\left(F(y)^2 + \gamma G(y)^2\right) \tag{27}$$

is increasing in y. $(w^c - w^s)/w^s$ is similarly found to be increasing in y. The dependence of $v^c - v^x$ and $z^c - z^s$ on x are symmetrically expressed. Therefore, the laggard's excess investment in both partial jumps and leaps expressed as a fraction of the socially optimal investment is greater. Theorem 6 summarizes these comparisons.

Theorem 6. *If B is small and $P = B$ then*

$$\frac{u^c - u^s}{u^s} > \frac{v^c - v^s}{v^s} \quad and \quad \frac{w^c - w^s}{w^s} > \frac{s^c - z^s}{z^s}$$

if and only if $x < y$.

These comparisons do not necessarily say anything about the efficiency of resource allocation given that there is competition. For example, in deciding whether to subsidize the current leader a social planner should consider its impact on the future nature of the distorted allocation of resources due to the competition. We next address this issue for the case $P = B$.

If $P = B$, the social value of the game is $V(y, x) + V(x, y)$ since all benefits of innovation are appropriated by the firms. At any position, the net social marginal values, $NSMV_u$ and $NSMV_v$, of u and w per dollar of expenditure equal the ratio of the net contribution to the social value and the marginal cost:

$$NMSV_u = \frac{\int_x^0 V(y, s) f(s, x) ds - V(y, x)}{\int_x^0 V(s, y) f(s, x) ds + PF(x) - V(x, y)} \tag{28}$$

$$NSMV_w = -\frac{V(y, x)}{P - V(x, y)}$$

where we use (20) to simplify expressions. Using our expansion for $V(x, y)$, (28) implies that, as p converges to 0,

$$p^{-1} NMSV_u \approx -g^1(y, x) = -F(y)^2 - \gamma G(y)^2 \tag{29}$$
$$p^{-1} NMSV_w \approx -F(y)^2$$

Symmetric expressions for $NMSV_v$ and $NMSV_z$ hold. If $x > y$ then $F(x) > F(y)$ and $G(x) > G(y)$, implying that $NMSV_z < NMSV_w$, and $NMSV_v < NMSV_u$. Therefore, the social value of more investment in either project is greater at the leading firm, even when we consider the distortions implicit in the competition.

Theorem 7. *If $P = B$ and P is small, social welfare at any stage would be increased by shifting innovation effort from the laggard to the leader. That is, if (x, y) is the current state and $x > y$, $V(x, y) + V(y, x)$ is increased if $u(x, y)$ is increased and $w(x, y)$ is decreased, and similarly for $z(x, y)$ and $w(x, y)$.*

Theorem 7 shows that any small temporary subsidy/tax scheme which reallocates effort towards the leader is socially desirable since combinations of subsidies and taxes can induce such a switch and the objective of $V(x, y) + V(y, x)$ ignore any redistributive component of such a policy. Therefore, in this limited sense, policy should favor the current leader over the laggard.

Another interesting issue which we can address in this model is that of the efficiency of the allocation of resources between the risky leaps and the less risky partial jumps. The social efficiency of the portfolio choice by firm one is determined by comparing the net social marginal values of u and v. $NSMV_u > NSMV_w$ iff $g^1(x, y) - \int_x^0 g^1(s, y) f(s, x) ds < F(x) g^1(x, y)$ which is true since $g^1(x, y)$ is increasing in x. Hence, there is an excessive share of resources allocated to the "risky" project. To get an intuitive grasp on this result, we should compare the social valuation of the intermediate stages with the equilibrium valuation by firm one. Since the difference between g^1 and h^1 is independent of x, we need to compare g^2 with h^2 to study differences relevant for one's portfolio choice between u and w. Straightforward manipulation of the expansions for V and W shows that, ignoring terms which are of $o(P^3)$,

$$V(x, y) - W(x, y) \approx 2p^2 \left(F(x)^2 + \gamma G(x)^2 \right) \left(F(y)^2 + \gamma G(y)^2 \right) P + Z(y) \quad (30)$$

where $Z(y)$ depends only on y. Therefore, $V - W$ is increasing in x for small p. First, this implies that investment is even more excessive than indicated by p^2 terms since the gap between social and private values of R&D is increasing at $O(P^3)$. Second, it indicates a bias towards risky R&D projects. Since this excess increases in x, those projects which are more likely to yield big jumps, holding the expected jump constant, will find their private value to be more excessive relative to their social value.

Theorem 8. *If $P = B$ and P is small, social welfare would be increased if resources were shifted from the risky R&D projects to the less risky projects.*

The last comparison we will make is between the optimal and equilibrium reactions of firms to each other's partial successes. Before using our approximations, note that our expression for firm one's equilibrium choice of w in equation (20), differs substantially from the expression for the social choice in equation (15), despite their formal similarity. In (15), it is clear that the optimal choice of w falls if the social value of the social position (x, y) increases but x, the position of firm one, remains unchanged. In particular, an advance in firm two's position will increase the social value, and hence lead to a reduction of expenditure at firm one on the leap investments. In (20), we find that expenditure on w will rise as the value of the game to firm one falls, which is the expected response to an advance by firm two. Hence, if the social and private value functions vary with position in the intuitive fashion, firm one will increase leap investments in response to an advance by firm two, even though the socially optimal response would be a reduction in effort.

Proving these conjectures globally would be quite difficult given the nonlinear nature of the expression for the equilibrium value functions. However, our approximations will immediately confirm them. Since $g^2(x, y)$ is independent of y, the dependence of u and w on y for small P, is determined by the dependence of g^3 on y, and is summarized in

$$\rho^{-1} u^c = \ldots + 2p^3 \left(F(y)^2 + \gamma G(y)^2 \right) \tag{31}$$

$$\times \left(\int_x^0 \left(F(s)^2 + \gamma G(s)^2 \right) F(s, x) ds - F(x)^2 - \gamma G(x)^2 \right)$$

$$\rho^{-1} w^c = \ldots + 2\gamma p^3 \left(F(y)^2 + \gamma G(y)^2 \right) \left(F(x)^2 + \gamma G(x)^2 \right) G(x)$$

where we have displayed all terms of $O(p^3)$ that depend only on y.

Theorem 9. *If $P = B$ and P is small,*

$$0 < \left| \frac{\partial u^c}{\partial y} \right| < -\frac{\partial u^s}{\partial y}$$

$$\frac{\partial w^c}{\partial y} > 0 > \frac{\partial w^s}{\partial y},$$

that is, firm one's equilibrium reactions are less than the optimal reactions in magnitude. Furthermore, $\partial u^c / \partial y$ is always positive and $\partial w^c / \partial y$ is of ambiguous sign. Symmetric results for firm two hold.

Proof. The comparisons of magnitude follow from the fact that $\partial u^c / \partial y$ is $O(p^3)$ by (31) but $\partial u^s / \partial y$ and $\partial w^s / \partial y$ are $O(p^2)$ by (18). The sign conditions for w^c and z^c follow from (31). If $F(s)$ and $G(s)$ are large relative to $F(x)$ and $G(x)$ for $s > x$, then the integral in (31) dominates and $\partial u^c / \partial y > 0$. However, if $F(s) \approx F(x)$ and $G(x) \approx G(x)$ for $s > x$, then $\int_x^0 (F(s)^2 + \gamma G(x)^2) f(s, x) ds \approx (F(x)^2 + \gamma G(x)^2) (1 - F(x))$ and $\partial u^c / \partial y < 0$ in (31). $\square$

In comparing the dependence of strategies on the positions of the firms, first note that there is no reaction of one firm to another's position to $O(p^2)$. Hence, the equilibrium reactions of the firms to each are smaller than the optimal reactions. Furthermore the direction may be wrong. In the case of leap investment, the reaction will always be in the wrong direction. This is intuitively seen from (20): we expect that as firm two advances, the value of the game to firm one, $V(x, y)$ decreases, thereby raising firm one's choice of w. In the social control case, the value increases as firm two advances, reducing the social choice for w.

However, the reaction of u is ambiguous. The reaction of a partial jump's control to the other firm's movement depends on just how different the stages are. If the stages are similar in that the probability of winning immediately per unit of effort with a leap, $G(x)$, or partial jump, $F(x)$, is nearly as large at x as at any later stage, then u will fall. On the other hand, if later stages have substantially greater likelihoods of getting one to success, then a firm's effort in partial jumps may increase as its opponent moves ahead. In the latter case, the improvement in the opponent's prospects prompts one to work harder, as if one must either work hard or concede the race. Also note that if a mean preserving spread in the probability weights $f(s, x)$ will increase the likelihood of a perverse reaction for u since the integral in (31) has a convex integrand.

This result about the reaction of u differs substantially from previous papers. Fudenberg et al. (1983) and Harris and Vickers (1985) make very specific assumptions about the R&D process and arrive at more definitive results. They made special assumptions since they were necessary to arrive at conclusions given their approach. The perturbation approach used here can handle models that are much more general in many dimensions. Of course, we assume that the prize is small, a loss of generality. The perturbation method does not uniformly dominate alternative analytical approaches, but the results here show that it can investigate new territory.

At this point we should expand on the appropriate interpretation of our juggling of these various orders of magnitude. For example, the fact that the reaction of u^c to y is zero at $O(p^2)$ and possibly nonzero only at $O(p^3)$ does not imply that reactions are generally unimportant and uninteresting when compared to the effects which show up at $O(p^2)$. In fact, in many games where reactions are generally important we would find that, as the payoffs go to zero, the reactions go to zero faster than other elements of equilibrium strategies. Only for nearly degenerate games does the order reflect the relative importance of various effects. Since our objective is to gain more general insight, we make no comparisons. On the other hand, one cannot infer that an $O(p^3)$ effect will eventually dominate any $O(p^2)$ effect since other, even higher, orders also contribute. Our goal in these calculations is to sign various effects and determine the critical structural elements for an open set of games, hoping to elicit general qualitative insights about the nature of the subgame equilibria. Arguments which mix various orders of magnitudes are either illegitimate or focus too tightly on the small p nature of the analysis.

8 Implications for social innovation policy

We next examine the optimal values of two parameters of social innovation policy, the portion of social benefits to be awarded to the winner and the stage at which a patent is to be granted. We will find that when B is small, the difference between the optimal P and B is negligible relative to B. This result validates our focus on the case $P = B$ in the previous section since it implies that all those results continue to hold for an optimally chosen P. In particular, this shows that the misallocation of resources between projects of varying riskiness will not change with an optimally chosen P. While these results are not surprising, it is instructive to show how to rigorously demonstrate them within our approach.

Let $P = \theta B$, i.e., θ is the portion of social benefits of innovation which the innovator is allowed to appropriate. We are making the simplifying assumption that this allocation of social benefits to the innovator can be made in a nondistortionary fashion. In the case of patents this is only valid if demand is inelastic. If a prize is awarded, this assumes that it is financed by nondistortionary revenue sources.

Presumably, θ is a parameter at least partially chosen by policy markers. Given that we found that there was excessive allocation of resources for innovation in the equilibrium of the innovation game, the optimal θ is never unity. Let W again represent the social value function except now we make explicit the dependence on

θ and B. Then

$$W(x,y,\theta,B) = - \left[\alpha(u^2 + v^2) + \beta(w^2 + z^2)\right] \frac{1}{2}dt \tag{32}$$

$$+(1 - \rho dt)\ (uF(x) + vF(y) + wG(x) + zG(y))\ Bdt$$

$$+(1 - \rho dt)\ (1 - (u + v + wG(x) + zG(y))\ dt)\ W(x,y)$$

$$+(1 - \rho dt)\left(u\int_x^0 W(z,y)f(z,x) + v\int_y^0 W(x,s)f(s,y)ds\right)dt$$

where u, v, and z are the equilibrium policy functions if the prize is θB.

We can use the characterization in (32) to generate some information about the optimal θ, $\theta^*(B)$, when B is small. $\theta^*(B)$ is defined by the equation $W_\theta(x,y,\theta^*(B),B) = 0$ This is not a completely trivial calculation since any θ is optimal when $B = 0$. Therefore we compute $\theta^*(0^+)$, the limit of θ^* as B falls to zero.

First, $\theta^*(B)$ is implicitly defined by $W_\theta(x,y,\theta^*(B),B) = 0$. For sufficiently small B, $\theta^*(B)$ is continuously differentiable by the Implicit Function Theorem applied to the equation $W_\theta(x,y,\theta^*(B),B) = 0$, since direct calculation shows that $W_{\theta\theta}$ is not zero and $W_{\theta B}$ exists for B close to zero. Since $\theta^*(B)$ is optimal for the initial position (x,y),

$$\lim_{B\to 0^+} \frac{W(x,y,\theta^*(0^+),B) - W(x,y,\theta,B)}{B^2} \geq 0 \tag{33}$$

for all θ. Since $W(x,y,\theta,B)$ and $W_B(x,y,\theta,B)$ both converge to 0 as B converges to 0, by l'Hospital's rule (33) equals

$$\lim_{B\to 0^+} \frac{W_{BB}(x,y,\theta^*(0^+),B) - W_{BB}(x,y,\theta,B)}{2}$$

for all θ. Therefore, $W_{BB}(x,y,\theta^*(0^+),0) - W_{BB}(x,y,\theta,0) \geq 0$ for all θ, implying that $\theta^*(0^+) \in \arg\max_\theta W_{BB}(x,y,\theta,0)$ and

$$W_{BB\theta}(x,y,\theta^*(0^+),0) = 0 \tag{34}$$

Direct substitution of the asymptotic equilibrium strategies into (34) shows that

$$\alpha\rho W_{BB}(x,y) = 4\left(\theta - \theta^2/2\right)\left(F(x)^2 + F(y)^2 + \gamma\left(G(x)^2 + G(y)^2\right)\right) \tag{35}$$

which is maximized at $\theta^*(0^+) = 1$. Therefore, when the prize is small, the social surplus maximizing policy gives nearly all of the social benefits to the innovator.

Note that this does not contradict our earlier result that innovation is excessive whenever the prize equals the benefit, just that the difference between the optimal prize and the social benefit goes to zero faster than the social benefit. This is not surprising since it just says that the externalities due to the competition over the rents fall more rapidly than B as B goes to zero. The primary point of this exercise is to illustrate how to determine the limit.

Second, further expansion of the social value function and application of l'Hospital's rule shows that the optimal θ falls more rapidly as B increases when

$F(x)$ and $G(x)$, the probability of an immediate success from the current position (assuming the firms begin at the same position), rises. This implies that the shorter the race, the smaller should be the winner's share under competition. Since the details entail only repeated applications of the foregoing calculations, they are omitted here.

Another crucial aspect of patent policy is the stage at which a patent is granted. A patent may be granted before final and complete success is achieved. In fact, in the existing patent system, a patent is granted when a description of an invention has been completed, before the development stages leading to a workable and commercial prototype have been achieved. This may be socially optimal if the effort of followers is so excessive and wasteful that it is better to force them out of the race, bearing the possible inefficiencies that may result when an innovator is given the monopoly early. In our model, this can be modeled by assuming that a patent is granted to the first firm which crosses $c \leq 0$. If $c = 0$, the firm must complete the project before acquiring a patent worth P. If $c < 0$, then a firm receives a patent at c and may finish development without any competition.

Proceeding as in the $c = 0$ case, we find that the equilibrium value function for the firms satisfies

$$0 = \frac{1}{2\alpha} \left(\int_x^c V(s,y)f(s,x)ds + \int_c^0 M(s)f(s,x)ds + PF(x) - V(x,y) \right)^2 \quad (36)$$

$$+ \left(\int_y^c V(s,x)f(s,y)ds + \int_c^0 M(s)f(s,y)ds + PF(y) - V(y,x) \right)$$

$$\times \left(\int_y^c V(x,s)f(s,y)ds - V(x,y) \right) \frac{1}{\alpha} - \rho V(x,y)$$

where $M(\cdot)$ is the monopoly value function computed in Section 4 with the extension to two instruments, u and v or w and z. We expand (36) as before for the case of a small social benefit and prize. We find that when B is small and $P = B$, the value of $V(x,y) + V(y,x)$, the social surplus value function, for $c < 0$ minus the value when $c = 0$ is approximated by

$$-F(y) \int_c^0 g^1(x,s)f(s,y)ds - F(x) \int_c^0 g^1(y,s)f(s,x)ds < 0 \quad (37)$$

Hence, the major factor is that if $c < 0$, the contest is ended early and the resulting loss in total effort is excessive relative to the cost savings. Note that this strict inequality depends critically on our standing assumption that $F(x) > 0$ for all x.

Theorem 10 summarizes our findings concerning optimal policy.

Theorem 10. *When B is small, the optimal policy is to award a prize only when the race is completely won and the prize should be nearly the entire social value of the innovation. Furthermore, the closer the innovators are to final success when competition begins, the less should be their share in the social benefit.*

While these conclusions are surely not globally true, we have shown their validity for an open set of problems and that contrary conjectures cannot be globally true. More important, we have shown how to address these questions for that set of problems. Other exercises, such as the impact of suboptimal innovation resource allocation on the optimal prize, can be conducted by straightforward examination of the higher-order terms of our expansion for W, the social planner's objective. In the interest of space, we leave such extensions to the reader.

9 Generalizations

There are many other exercises which could be used to demonstrate the applicability of perturbation methods. We examined one that most clearly illustrates the general approach advanced here. To indicate that the perturbation approach is not too specialized, we will now discuss some other possible applications.

All models with closed-form solutions are degenerate in some sense. When we use them we hope that the features that these tractable models ignore are not important. Perturbation methods can be used to test this presumption. Take, for example, the model used by Loury (1979) and Reinganum (1981). While it yields closed-form solutions for the quadratic cost specification, it abstracts from the possibility of intermediate stages, our focus here. Recall that our model with F and G equal to constant functions is exactly that model. To examine the importance of intermediate stages on the nature of equilibrium, we could have assumed that F and G deviated slightly from constant functions. This alternative would have allowed us to determine the nature of equilibrium for arbitrary prize but with only a small deviation from the implicit stage-independence assumption sometimes used.

Another possible generalization is allowing intermediate payments. The perturbation analysis conducted above could also allow intermediate payoffs since nothing we did used the absence of intermediate payoffs in an essential fashion; we focussed on the more simple payoff structure since our purpose was to present a robust analysis of the positional dynamics among competitors for one kind of race. A more general analysis with intermediate payoffs could generate insights, for example, into strategic implications of the learning curve; one approach would be to approximate the slow-learning case by knowing the solution to the no-learning case. However, we leave such an analysis to another study.

While this is certainly not an exhaustive list, it shows that perturbation methods are useful in examining the robustness of simple models generally, allowing us to add otherwise intractable elements to the analysis of a problem. While our analysis got started by examining the trivial case of no payoff, generally one can begin with any tractable case, making perturbation analysis a generally valuable tool for dynamic strategic analysis.

10 Conclusion

We have analyzed a simple closed-loop subgame perfect model of multi-stage innovation. We found the usual result of excessive innovative effort when the prize equals the social value. Under the assumption that the net social value of innovation is small, we have also found that there will be excessive risk-taking, that at any moment the following firm is a less efficient innovator relative to the leader, that the prize to the innovator should nearly equal social benefits, and that the competition should not be ended before one of the competitors has succeeded completely. While these results have obvious limitations on their generality, they do tell us that the contrary propositions cannot be generally true. While many of the results, e.g., the excessive investment when the prize equals the social benefit, follow naturally from the fact that these subgame perfect equilibria are close to some open-loop equilibria others, in particular the computation of the equilibrium reactions, are specific to the subgame-perfect solution. They have therefore given us a peek into the nature of subgame perfect equilibrium in such innovation models.

References

Bhattacharya, S., Mookherjee, D.: Portfolio choice in research and development. Rand Journal of Economics **17**, 594–605 (1986)

Budd, C., Harris, C., Vickers, J.: A model of the evolution of duopoly: Does the asymmetry between firms tend to increase or decrease? Review of Economic Studies **60**, 543–573 (1993)

Bryson, A. E., Ho, Y.: Applied optimal control. New York: Hemisphere Publishing 1975

Dasgupta, P., Stiglitz, J.: Uncertainty, market structure and the speed of research. Bell Journal of Economics **11**, 1–28 (1980a)

Dasgupta, P., Stiglitz, J.: Industrial structure and the nature of innovative activity. Economics Journal **90**, 266–293 (1980b)

Doraszelski, U.: An R&D race with knowledge accumulation. Hoover Institution, mimeo (2002)

Fudenberg, D., Gilbert, R., Stiglitz, J., Tirole, J.: Preemption, leapfrogging and competition in patent races. European Economic Review **22**, 3–31 (1983)

Harris, C., Vickers, J.: Perfect equilibrium in a model of a race. Review of Economic Studies **52**, 193–209 (1985)

Harris, C., Vickers, J.: Racing with uncertainty. Review of Economic Studies **54**, 1–21 (1987)

Judd, K. L.: Closed-loop equilibrium in a multi-stage innovation race. Discussion Paper No. 647, Kellogg Graduate School of Management, Northwestern University (1985)

Judd, K. L.: Projection methods for solving aggregate growth models. Journal of Economic Theory **58**, 410–452 (1992)

Judd, K. L.: Computational economics and economic theory: substitutes or complements? Journal of Economic Dynamics and Control **21**, 907–942 (1997)

Kamien, M., Schwartz, N.: Market structure and innovation. Cambridge: Cambridge University Press 1982

Lee, T.: On a fundamental property of research and development rivalry. University of California, San Diego, mimeo (1982)

Lee, T., Wilde, L.L.: Market structure and innovation: a reformulation. Quarterly Journal of Economics **94**, 429–436 (1980)

Loury, G. C.: Market structure and innovation. Quarterly Journal of Economics **93**, 395–410 (1979)

Reinganum, J.: Dynamic games of innovation. Journal of Economic Theory **25**, 21–41 (1981)

Reinganum, J. F. A dynamic game of R&D: patent protection and competitive behavior. Econometrica **50**, 671–688 (1982)

Selten, R.: Reexamination of the perfectness concept for equilibrium points in extensive games. International Journal of Game Theory **4**, 25–55 (1974)

Telser, L.: A theory of innovation and its effects. Bell Journal of Economics **13**(1), 69–92 (1982)

Zeidler, E.: Nonlinear functional analysis and its applications. Part I: Fixed point theorems. New York: Springer 1986

Modelling exchange of probabilistic opinions[*]

Hiroyuki Nakata

Mizuho-DL Financial Technology Co., Ltd., Otemachi First Square (East 16F),
5-1 Otemachi 1-chome, Chiyoda-ku, Tokyo 100-0004, JAPAN
(e-mail: hnakata@stanfordalumni.org)

Received: August 27, 2001; revised version: April 16, 2002

Summary. This paper studies how communication or exchange of opinions influences correlation of beliefs. The paper focuses on a situation in which agents communicate with each other infinitely many times without observing data. It is an extension to the 'Expert Problem' in Bayesian theory, where the informational flow is asymmetric. Moreover, this paper generalizes the existing literature of communication that employs the common prior assumption (CPA) by allowing for heterogeneous beliefs. Some basic convergence results are shown in contrast with the results obtained under the CPA. Furthermore, several economic implications of the basic results are provided.

Keywords and Phrases: Communication, Correlation of beliefs, Expert problem, Heterogeneous beliefs.

JEL Classification Numbers: D82, D83, D84.

1 Introduction

This paper studies how communication or exchange of probabilistic opinions influences correlation of beliefs. In so doing, we treat a probabilistic opinion as a datum (i.e. a realization of a random variable). With such a treatment, we explicitly model communication or exchange of probabilistic opinions among agents with heterogeneous beliefs.

To do so, we can consider two distinct situations. The first is a situation in which no additional data arrives while agents exchange opinions infinitely many times. In

---

* The results presented in this paper are taken from my Ph.D. thesis at Stanford University. I gratefully acknowledge the inspiration obtained from innumerable discussions with Mordecai Kurz about this subject. Also, I appreciate comments from Kenneth J. Arrow, Peter J. Hammond, Maurizio Motolese, Carsten K. Nielsen, Ho-Mou Wu and the anonymous referee.

this situation, agents update their beliefs solely by using the opinions of others. The second is a situation in which a set of data arrives as frequently as agents exchange opinions. In contrast with the first case, an agent in this case updates his belief by using *both* the newly arrived data *and* the opinions of others. This paper focuses on the first situation whilst [22] examines the second situation.

As far as the treatment of probabilistic opinions is concerned, this paper follows that of a literature in Bayesian theory known as the *Expert Problem*. This literature makes a rigid distinction between two groups of people. On the one hand, there are the *experts*, whose role is simply to provide probabilistic opinions (i.e. to 'speak'). On the other hand, there is the *decision maker* (agent, hereafter), whose role is to seek the opinions of experts (i.e. to 'listen'), and then to make decisions. The focus of the literature has so far been limited to the situations in which an agent acts only as a 'listener' whilst an expert acts only as a 'speaker'. In other words, the literature has confined its attention to asymmetric situations. (There are some works that study symmetric situations, albeit very few, e.g. [4].) Such asymmetry, however, is quite unusual. There are two reasons why this is so. First, agents do not know the relationships between their own information sets and the information sets of others. Thus, every agent may believe that some agents may have some information that he does not possess. Second, agents do not know the probability beliefs (measures) of others. Without such knowledge, every agent may believe that there are some agents who have 'better' probability beliefs.

In the economics literature, our paper is an extension to several existing papers on communication that employ the CPA, e.g. [2], [15], and [23]. They show that under the CPA, communication results in a merging of opinions even when asymmetric information is present *ex ante*. However, once we allow for heterogeneous beliefs, it has been shown that a merging of opinions does not occur generically (see [13]). With this in mind, we show that in fact the distribution of the forecasts will converge to a limit distribution, but not necessarily to a single mass point. Furthermore, we show that when agents exchange opinions without receiving data, each agent believes that in the limit, he will be able to forecast correctly what other agents' forecasts will be. Also, we prove that if an agent stops updating his belief then he is convinced that he can tell correctly what other agents' forecasts will be. This result applies to forecasts of practically all classes of random variables, both discrete and continuous.

Moreover, the result should be compared with the result of [7]. While [7] assume absolute continuity of probability measures over the infinite sequence, our paper merely assumes integrability of the limit random variables and a rationality axiom (Assumption 2; see below), which is much weaker than the absolute continuity assumption. The difference in the assumptions provides a sharp contrast in the results: while [7] yield a much stronger result of a merging of opinions similar to that of communication under the CPA, our result retains a diversity of opinions even in the limit.

Now let us explain some implications of our work for economic studies. In short, our work enables us to study in depth the mechanism that causes externalities in various contexts. Many economic studies have made it clear that externalities play a key role in a large class of economic phenomena that exhibit non-stationarity

– e.g. [26] and [12]. In their analyses, however, it is assumed that the economy experiences externalities in terms of actions and/or payoffs. Moreover, they do not capture correlation of beliefs as the mechanism that causes such externalities, since they do not allow for heterogeneous beliefs.

Unlike such previous studies on externalities, the current paper and [22] explicitly describe the mechanism that causes externalities rather than simply assuming that the actions and/or payoffs directly involve externalities. For instance, our work casts a severe criticism of its relevance, rather than its analytical soundness of the literature of *herd behaviour*,[1] which in general assumes an exogenous ordering of the timing of decision making – e.g. [3], [6], etc. Furthermore, while communication under the CPA is merely a process that removes asymmetric information, communication with heterogeneous beliefs is a mechanism that generates correlation of beliefs. And consequently, such correlation of beliefs is the very driving force that generates externalities. In [22] such an economic model is introduced, although it simultaneously aims to examine the second situation (i.e. exchange of opinions with data arrivals).

In the following section, first we reproduce the result of [16] in the literature of the *Expert Problem*. Then we describe our model and assumptions, and then provide the basic convergence results of the distribution of opinions. In Section 3, the dynamical behaviour of the distribution of opinions is studied. In addition, some implications for the literature of herd behaviour are provided, while a simple competitive economy is introduced to illustrate the effects of communication on the equilibria. Finally, Section 4 concludes the paper.

2 The model

In this section, we first reproduce the result of [16] followed by its interpretation. Then we prove the basic theorem about convergence of opinions/forecasts. We go on to prove that if an agent stops updating his belief then he is convinced that he can correctly tell what other agents' forecasts will be. To show the latter result, we utilize the results of [16]. In so doing, we begin with the simplest case of a probability of an event, and then extend the result to the cases of probabilities of discrete random variables and those of continuous random variables, in order.

2.1 The 'Expert Problem'

To begin with, let us reproduce Theorem 2.1 of [16]. Let Y^j denote the probability which the expert j will assign to the event $\{X = x\}$ (the opinion of the expert), let $Q^i\{X = x\}$ be the agent's probability of $\{X = x\}$ prior to learning Y^j, and let $Q^i\{X = x | Y^j = y^j\}$ represent the agent's probability of $\{X = x\}$ after learning $\{Y^j = y^j\}$. Also, let $dF^i(y^j)$ denote the agent's prior marginal probability distribution for Y^j. Before stating the theorem, let us introduce the

---

[1] Fashion may be a good example of herd behaviour.

following condition, which is essentially the same as the *rationalizability condition*
in economics (see for example [25]):

Consistency Condition (a modified version of [16]). *No matter what the unspecified marginal distribution dF^i for the expert's opinion Y^j is, there exists a joint
distribution for the random variable of interest X and Y^j which is compatible with
dF^i, satisfies the initial specifications and is such that the posterior probability
given the experts' opinions $Y^j = y^j$ is equal to $Q^i\{X = x|Y^j = y^j\}$.*

We maintain this condition throughout the paper. Now we are ready to state the
following theorem:

Theorem 1 (Theorem 2.1 of [16]). *Let the initial specifications consist of $Q^i\{X = x\}$ and $Z^{ij} = E_{Q^i}\{Y^j\}$, and assume that the support of dF^i is $[0, 1]$. Then $p_i^*(x; y^j) = Q^i\{X = x|Y^j = y^j\}$ satisfies the Consistency Condition if and only if*

$$p_i^*(x; y^j) = Q^i\{X = x\} + \lambda^i(x) \cdot (y^j - Z^{ij})$$

for some constant $\lambda^i(x)$ depending upon $Q^i\{X\}$ and Z^{ij}.

When an agent computes the full joint distribution of X and Y^j, the posterior
probability $p_i^*(x; y^j)$ has a unique formula depending upon the specification of the
joint distribution. However, when an agent does not compute the full joint distribution of X and Y^j, there are various possible formulae for the posterior probability
$g^i(x; y^j)$, i.e. there is some uncertainty about the functional form of $g^i(x; y^j)$. In
essence, Theorem 1 states that when an agent merely computes $Q^i\{X = x\}$ and
$Z^{ij} = E_{Q^i}\{Y^j\}$, instead of computing the full joint distribution of X and Y^j,
then $p_i^*(x; y^j)$ has the form of a linear function of y^j if and only if it satisfies the
Consistency Condition. Moreover, the full support assumption – i.e., the support
of dF^i is $[0, 1]$ – is not necessary for the proof (see [16]).

However, this surprisingly simple result must be interpreted with care. If we
follow the purely out-and-out Bayesian tradition, then we must assume that each
agent *knows* his full joint distribution of X and Y^j, although it may not be the
true distribution. However, [16] assume that the agent does not know or specify his
full joint distribution. Namely, [16] introduce uncertainty about each agent's own
belief. We note here that uncertainty is a rather vague concept. Along this line of
interpretation of uncertainty, [30] point out that [16] take account of uncertainty
about the functional form of $g^i(x; y^j)$ that results from the partial specification
of the joint distribution. Hence, we may interpret in such a way that Theorem 1
implicitly assumes that the agent is not convinced with his belief, and that he forms
a distribution over all possible distributions, although this is only one particular
interpretation of the uncertainty an agent is facing (i.e., there may not exist such
an explicit hierarchical structure of beliefs). With this interpretation, the linear
functional form above is in fact the expectation of $g^i(x; y^j)$ conditional on $\{Y^j = y^j\}$, i.e. $p_i^*(x; y^j) = E_{Q^i}\{g^i(x; y^j)|Y^j = y^j\}$, where the expectation is with
respect to the agent's posterior distribution of the function $g^i(x; y^j)$ given $\{Y^j = y^j\}$. Note that the construction of belief here must be compatible with the initial
specifications, and also it must satisfy the Consistency Condition. More precisely,
let $Q_q^i(\cdot)$ denote a lower-level distribution, and let U^i denote the class of all possible

lower-level distributions. Also, let $\mu^i(\cdot)$ denote the probability distribution over $Q_q^i(\cdot)$. Then, it must be that

$$\int_{U^i} Q_q^i\{X = x\}\mu^i(dQ_q) = Q^i\{X = x\}, \tag{1}$$

$$\int_{U^i} E_{Q_q^i}\{Y^j\}\mu^i(dQ_q^i) = E_{Q^i}\{Y^j\}, \tag{2}$$

and $Q^i\{\cdot\} := \int_{U^i} Q_q^i\{\cdot\}\mu^i(dQ_q^i)$ must satisfy the Consistency Condition. By construction, $p_i^*(x; y^j) = \int_{U^i} Q_q^i\{X = x | Y^j = y^j\}\mu^i(dQ_q^i | Y^j = y^j)$. Observe that Q_q^i does not necessarily specify the full joint distribution of X and Y^j either; Q_q^i is simply any probability distribution that is consistent with the initial specifications (for example, a distribution that specifies only up to the second moment). Thus, this construction of Q^i does not necessarily imply a specification of the full joint distribution of X and Y^j.

2.2 The basic theorem

In what follows, we prove the basic theorem on the convergence of the (distribution of) forecasts. We follow the literature of the *Expert Problem* with regard to the treatment of probabilistic opinions of others – i.e. we treat them as data (realizations of random variables). First, we explain the setup of the model, and then show the basic theorem. In doing so, we discuss the structures of the true and subjective probability spaces, too.

To begin with, we make the following two assumptions on behavioural rules. Namely,

Assumption 1.

(a) *Each agent i announces his forecasts truthfully.*
(b) *Each agent i believes that agent j's opinion is informative for all j initially.*

These assumptions will be maintained throughout the paper. By Assumption 1 (a), we put strategic concerns about the announcement of forecasts to one side, avoiding complications that would involve game theoretic considerations, which are not essential to our current focus. We are interested in situations in which any single agent believes that he cannot affect the whole system, and consequently other agents' decisions as in, for example, general equilibrium models.

Also, Assumption 1 (b) reflects the motivation for exchanging opinions explained in the introduction. Notice however that this assumption does not require such a belief to be common knowledge. In other words, an agent does not know if other agents know that he has such a motivation for exchanging opinions and so on. Instead, each agent *believes* that the opinions of others are informative because he *believes* that others know something that he himself does not know, or others have 'better' beliefs in terms of estimations than his own belief. These are exactly the motivations for the agents to listen to the opinions of others. Moreover, as it is unlikely for an agent to hold such structural knowledge, i.e. no agent knows what others know, such a motivation naturally implies exchange of opinions.

2.2.1 The structure of the model

In what follows, we show a basic result about convergence of forecasts/opinions. Let $(\Omega, \mathcal{F}, Q^i)$ be the (subjective) probability space of agent i, where Q^i denotes agent i's probability belief. Assume that there are finitely many I agents. Let X be some measurable/integrable random variable on $(\Omega, \mathcal{F}, Q^i)$ for all i, either discrete or continuous, in which each agent is interested. Assume that $(\Omega, \mathcal{F}, Q^i)$ is a standard space. Note that a measurable space $(\Omega, \mathcal{F})$ is said to be standard if $\mathcal{F}$ can be generated by a standard field, that is, if $\mathcal{F}$ possesses a basis. Then, by taking a generating standard field such that $\mathcal{G} := \{\mathcal{F}_l\}_{l=1}^{\infty}$, the space of all probability measures on $(\Omega, \mathcal{F})$ is Polish (see [17] for details; chapter 8 in particular). Also, let I_0^i be agent i's initial information, and let $Z_0^i := E_{Q^i}\{X|I_0^i\}$ denote the initial prior expectation of X with respect to Q^i. We treat the initial information I_0^i as a random variable, but it could be a σ-field more generically. All that matters is that it is part of the components that generate the σ-field at the beginning.

At the beginning of stage 0, each agent announces Z_0^i simultaneously, and this will be public information in the subsequent stages (the essence of the results does not change even if the announcements are about other statistics). Note in particular that although Z_0^j is a conditional expectation from agent j's point of view, it is a perceived announcement/forecast from agent i's point of view unless $i = j$. Thus, to avoid confusion, let Y_0^i denote a perceived announcement/forecast of agent i in stage 0, which is a random variable for all agents other than agent i. By definition, conditional expectation is a random variable. We emphasize, however, that Y_0^i is *not* a conditional expectation from the perspectives of agents other than agent i. In fact, Y_0^i is a family of random variables $(Y_0^i(1), Y_0^i(2), ..., Z_0^i, Y_0^i(i+1), ..., Y_0^i(I))$, where $Y_0^i(j)$ denotes the random variable Y_0^i from the perspective of agent j. However, we do not complicate the notation further, since whenever the notation Y_0^i is used, the identity of the agent who observes it is implicitly or explicitly understood. Using this notation, from the viewpoint of agent i, $\mathbf{Z}_0^{(i)}$ (vector of the forecasts of others), where superscript (i) denotes 'agents other than agent i', is really a random vector which we denote by $\mathbf{Y}_0^{(i)} := (Y_0^1, ..., Y_0^{i-1}, Y_0^{i+1}, ..., Y_0^I) \in \mathbf{R}^{I-1}$. After listening to the perceived announcements of others, each agent updates his belief by taking the public information into account – i.e. he computes the posterior expectation $Z_1^i := E_{Q^i}\{X|\sigma(I_0^i, \mathbf{Y}_0^{(i)})\}$, where $\sigma(I_0^i, \mathbf{Y}_0^{(i)})$ denotes the σ-field generated by the state space of $(I_0^i, \mathbf{Y}_0^{(i)})$. This posterior expectation becomes the prior expectation in the next stage $(s = 1)$. Then, agent i announces Z_1^i at the beginning of stage 1, which is Y_1^i for other agents, and it becomes public information in the subsequent stages. Agents iterate this procedure infinitely many times without observing (the realization of) the random variable of interest X. Thus, proceeding inductively, we can define the perceived announcement of agent i in stage s as a random variable Y_s^i (but not a conditional expectation) on any subjective probability space other than $(\Omega, \mathcal{F}, Q^i)$.

Now let $\mathcal{F}_s^i \subset \mathcal{F}$ be the σ-field generated by the state space of $(I_0^i, \mathbf{Y}_0^{(i)}, \ldots, \mathbf{Y}_s^{(i)})$, which is the information set of agent i at the beginning of stage $s+1$. Also, let $Y^j := \lim_{s \to \infty} Y_s^j$ denote agent j's limit perceived announcement, which is a random variable from other agents' points of view, although the proof for the existence of such a limit is provided later. Consequently, let $\mathcal{F}_\infty^i$ be the σ-field generated by the state space of $(I_0^i, \mathbf{Y}_0^{(i)}, \ldots, \mathbf{Y}^{(i)})$, which is a terminal σ-field in this updating process if the limit perceived announcement Y^j for all j on $(\Omega, \mathcal{F}, Q^i)$ exists. We note that a terminal σ-field here is not a tail σ-field, as opposed to the common usage of the term as another name of tail σ-field. Notice that $\sigma(Y^j) \subset \mathcal{F}_\infty^i$ by construction. To make our notation consistent, let $Z_s^i := E_{Q^i}\{X|\mathcal{F}_{s-1}^i\}$ and $Z_s^{ij} := E_{Q^i}\{Y_s^j|\mathcal{F}_{s-1}^i\}$ be agent i's conditional forecast of X and that of Y_s^j at the beginning of stage s, respectively. Note that although agent i's own past forecasts $(Z_0^i, \ldots, Z_{s-1}^i)$ could be explicitly included in the information set, we need not include them in the information set of the conditional forecast function, since by definition, Z_s^i is a function of I_0^i and $(\mathbf{Y}_0^{(i)}, \ldots, \mathbf{Y}_{s-1}^{(i)})$, and thus, it does not give any additional information. Note also that this is consistent with the subjective probability space defined above.

As we discussed in the previous section, we bring more structure to the beliefs of agents. Namely, we introduce distribution over distributions. This construction of beliefs is possible since we are assuming that the space $(\Omega, \mathcal{F})$ is standard (see [17], pp.264–265]). Let U^i be the collection of beliefs/distributions that assign positive probabilities to the event that the limit (of the) perceived announcement exists, i.e. $\lim_{s \to \infty} Y_s^j$ exists for all j. More precisely, let A^j denote the event that $\lim_{s \to \infty} Y_s^j$ exists, i.e. $A^j := \{\omega \in \Omega | \lim_{s \to \infty} Y_s^j \text{ exists}\}$, or in terms of realizations of $\{Y_s^j\}_s$, i.e. $\mathbf{y}^j \in \prod_{s=0}^\infty Y_s^j$, $A^j := \{\mathbf{y}^j \in \prod_{s=0}^\infty \mathcal{Y}_s^j | \lim_{s \to \infty} Y_s^j \text{ exists}\}$, where $\mathcal{Y}_s^j$ denotes the state space of Y_s^j. In particular, U^i is a collection of all probability distributions (measures) Q_q^i such that

$$Q_q^i\{A^j|\mathcal{F}_s^i\} > 0 \quad \forall j. \tag{3}$$

Furthermore, we assume that each agent forms a distribution (measure) over the collection of distributions (measures) Q_q^i, and let μ^i denote such a distribution, i.e.

$$Q^i\{\cdot\} = \int_{U^i} Q_q^i\{\cdot\}\mu^i(dQ_q^i),$$

and

$$Q^i\{A^j|\mathcal{F}_s^i\} = \int_{U^i} Q_q^i\{A^j|\mathcal{F}_s^i\}\mu^i(dQ_q^i|\mathcal{F}_s^i).$$

Therefore, U^i is a measurable set of probability measure μ^i, which is of course a finite measure. The above points are summarized by Assumption 2:

Assumption 2 (the Rationality Axiom). *Each agent i holds a distribution only over the collection of distributions Q_q^i that assign positive probabilities to the event that the limit of the perceived announcement Y_s^j exists for all j, i.e. A^j for all j.*

Although this assumption appears somewhat remote from rationality, and thus the name 'rationality axiom' may sound improper, we see below why it is appropriate. Moreover, we maintain this assumption throughout the paper.

Observe that this construction of beliefs, i.e. distributions over distributions, does not follow the pure Bayesian perspective, which assumes the lower-level distributions to be fixed because there should be no uncertainty about his own belief. Namely, the pure Bayesian view assumes that there is a probability mass such that $\mu^i\{Q_q^i = q\} = 1$ for some q, whilst we allow for some uncertainty about the lower-level distributions. Thus, Assumption 2 states that although each agent has some uncertainty about his lower-level distribution, he believes that any relevant lower-level distribution must not eliminate the possibility that the limit of the perceived announcement exists.

2.2.2 The structures of the probability spaces

Before proceeding, we discuss the relationship between the *true* probability space and the agents' subjective probability spaces. Because we are adopting the Bayesian view, albeit with more structure than usual, each agent's probability itself is subjective. We postulate that the truth does exist (as standard Bayesian theory does), and that the truth is objective in itself. Let $(\Omega, \mathcal{F}, \Pi)$ be the true probability space, where Π denotes the true probability measure. Then, the following assumption summarizes our view:

Assumption 3 (the Existence/Characterization of the Truth). *There exists the true (objective) probability space $(\Omega, \mathcal{F}, \Pi)$ such that any measurable random variable on a subjective probability space is measurable on it, including the limit of Y_s^i when it exists.*

The essential motivation for Assumption 3 is to make a logical connection between the subjective states of the world and the true states of the world. Namely, anything that is measurable subjectively must be measurable by the 'Truth', but the converse does not hold necessarily, and typically does not. This is in fact implicit in the construction of Bayesian theory, in which a subjective probability space must be a full description of the states of the world for the agent, while the true probability space must be a full description of the world objectively.

We note that there are two distinct possibilities concerning the structure of the true state space. That is, (a) we may draw a rigid distinction between the exogenous random variable X and the 'endogenous' random variables that are created by the agents' subjective beliefs. In this case, the 'Truth' (or any party who knows the true probability measure) knows the announcements fully. In other words, the Truth 'knows' the structure of the subjective beliefs of the agents, and thus, any extrinsic random variable does not act as an intrinsic random variable on $(\Omega, \mathcal{F}, \Pi)$. Hence, the Truth must be able to tell the announcements of the agents correctly at any time. This statement is captured by the following equation: $Z_s^i = Y_s^i = \tilde{\varphi}_s^i(I_0^1, ..., I_0^I, \mathbf{Y}_0, ..., \mathbf{Y}_{s-1})$ for all i, s, where $\tilde{\varphi}_s^i(\cdot)$ is a measurable function on $(\Omega, \mathcal{F}, \Pi)$. But, it is easy to check that this is equivalent to the statement that Y_s^i is a function of $(I_0^1, ..., I_0^I)$ on $(\Omega, \mathcal{F}, \Pi)$. On the other hand, (b) we can allow for an 'endogenous expansion of the state space'. In this case, the announcements of the agents Y_s^i will function as an intrinsic random variable on $(\Omega, \mathcal{F}, \Pi)$. Although the first view is valid yet, we claim that the second view is more natural as we explain from now on.

To clarify the discussion, we specify the probability spaces now. To do so, we assume that the probability spaces are standard Borel so that the sample space can be defined as a product space of the state spaces of the measurable random variables. Note that when a probability space is standard Borel, there is a one-to-one correspondence between the sample space and the state space of the random variables (see [17] for the definition of a standard Borel space). With this assumption, the subjective probability space of agent i is defined as $(\Omega^i, \mathcal{F}^i, Q^i)$, where

$$\Omega^i := X \times \left(\prod_{j=1}^{I} I_0^j \right) \times \left(\prod_{s=0}^{\infty} \mathbf{Y}_s^{(i)} \right),$$

and

$$\mathcal{F}^i := \sigma(X, I_0^1, I_0^2, ..., I_0^I, \mathbf{Y}_0^{(i)}, \mathbf{Y}_1^{(i)}, ...),$$

where X denotes the state space of random variable X and similarly for others. Furthermore, if we adopt the first view, the true probability space is defined as $(\Omega^{\Pi}, \mathcal{F}^{\Pi}, \Pi)$, where

$$\Omega^{\Pi} := X \times \left(\prod_{i=1}^{I} I_0^i \right),$$

and

$$\mathcal{F}^{\Pi} := \sigma(X, I_0^1, I_0^2, ..., I_0^I).$$

On the other hand, if we adopt the second view, the true probability space is defined as $(\Omega^{\Pi}, \mathcal{F}^{\Pi}, \Pi)$, where

$$\Omega^{\Pi} := X \times \left(\prod_{i=1}^{I} I_0^i \right) \times \left(\prod_{s=0}^{\infty} \mathbf{Y}_s \right),$$

and

$$\mathcal{F}^{\Pi} := \sigma(X, I_0^1, I_0^2, ..., I_0^I, \mathbf{Y}_0, \mathbf{Y}_1, ...),$$

with $\mathbf{Y}_s := (Y_s^1, Y_s^2, ..., Y_s^I)$. Hence, the sequence $\{\mathbf{Y}_s\}_{s=0}^{\infty}$ is a function of $(I_0^1, I_0^2, ..., I_0^I)$ when we adopt the first view, while it is a sequence of intrinsic random variables when we adopt the second view. One interpretation of the difference is that there is no uncertainty about the beliefs $(Q^1, Q^2, ..., Q^I)$ in the first case, while there is uncertainty about the beliefs in the second case.

To elucidate the reason why the second view is more natural, we relate our model to those with Harsanyi types. [8] construct the sample space as a product space of the common (underlying) space of uncertainty and the type spaces.[2] More

---

[2] Because their argument essentially relies on the Kolmogorov extension theorem (see for example [1]), countable additivity of measures is crucial. Also, to a larger extent, we need to care about the topological properties of the state space.

specifically, their construction is the following:

$$\Omega^{\Pi} := S \times \left(\prod_{i=1}^{I} T^i \right),$$

where S denotes the common state of uncertainty and T^i denotes agent i's type space T^i. In our model, S can be either $S := X \times I_0^1 \times \cdots \times I_0^I$ or $S := X$, depending on how we interpret the role of initial information $(I_0^1, I_0^2, ..., I_0^I)$ in the characterization of types. Namely, when we interpret that the informational asymmetry is about the objective random event, then the first interpretation follows, whilst the other interpretation comes into place when we interpret that the informational asymmetry is about types. However, this distinction is not essential, when we are to understand the role of announcements in the description of types. In fact, it is the overall structure of the true sample space that matters crucially. Namely, announcements do not play any role in describing types when the true sample space is $\Omega^{\Pi} := X \times (\prod_{i=1}^{I} I_0^i)$, whereas they do represent (at least part of) types when the true sample space is $\Omega^{\Pi} := X \times (\prod_{i=1}^{I} I_0^i) \times (\prod_{s=0}^{\infty} \mathbf{Y}_s)$. However, recall that we are interested in a situation where each agent does not know the beliefs of others. Therefore, each agent forms a belief about the beliefs of others. Note in particular that agents have a motivation to do so even when there is really no asymmetric information at all. Hence, the first view is not reasonable because there will be no type unless there is asymmetric information $(I_0^1, I_0^2, ..., I_0^I)$.

Moreover, under the CPA, the true probability space and all subjective probability spaces are identical, that is, $\Omega^{\Pi} = \Omega^i$, $\mathcal{F}^{\Pi} = \mathcal{F}^i$, and $\Pi = Q^i$ for all i. More specifically, the probability space should be defined as $(\Omega^{\Pi}, \mathcal{F}^{\Pi}, \Pi)$, where

$$\Omega^{\Pi} := X \times \left(\prod_{i=1}^{I} I_0^i \right),$$

and

$$\mathcal{F}^{\Pi} := \sigma(X, I_0^1, I_0^2, ..., I_0^I).$$

This is exactly the same as the first case above, where a rigid distinction between exogenous random variables and 'endogenous' random variables is made. Therefore, there is no type unless informational asymmetry is about the types themselves. In other words, if the common state of uncertainty is Ω^{Π} itself, there is no type in the model. It is now clear that in [15] and [23] where the common prior is assumed, communication is simply a revelation mechanism of the initial information $(I_0^1, I_0^2, ..., I_0^I)$ although $I = 2$ is assumed in their models. It is also obvious that their models can be interpreted as special cases of our model, although agents communicate back and forth, instead of announcing opinions simultaneously in their model. Moreover, we can see how restrictive the Harsanyi doctrine is in specifying the structure of probability spaces now. That is, the state space cannot be expanded endogenously. In fact, the doctrine eliminates all the interesting cases completely in which beliefs of the agents influence the structure of the sample space of the true probability space, i.e. endogenous expansion of the state space.

2.2.3 The basic theorem

Now we revert our attention to the limit behaviour of the forecasts in our model. In what follows, we relax the assumption that the probability spaces are standard Borel, and assume instead that the probability spaces are standard as we did at the beginning of this section. Now we have the following basic theorem on convergence of (the distribution of) the forecasts:

Theorem 2. *Suppose that for a random variable X, $E_{Q^i}|X| < \infty$ holds for all i. Then, under Assumptions 1,2 and 3, for every agent i,*

(a) *$Z_s^i \longrightarrow Z^i = E_{Q^i}\{X|\mathcal{F}_\infty^i\}$, almost surely and in L^1 with respect to Q^i;*
(b) *$Y_s^j \longrightarrow Y^j$ for all j, almost surely with respect to Q^i.*
   Assume also that for every agent i, $E_{Q^i}|Y^j| < \infty$ holds for all j. Then, for every agent i,
(c) *$Z_s^{ij} \longrightarrow Z^{ij} = E_{Q^i}\{Y^j|\mathcal{F}_\infty^i\}$ for all j, almost surely with respect to Q^i;*
(d) *$Z^{ij} = Y^j$ for all j, almost surely with respect to Q^i.*

Proof. Since $E_{Q^i}|X| < \infty$ for all i is assumed, it is clear that $\{Z_s^i\}_s$ is a uniformly integrable martingale (UI martingale) relative to the σ-fields $\mathcal{F}_s^i$ (see [9], for example). Thus, by the uniformly integrable (UI) martingale convergence theorem, Z_s^i converges to Z^i, almost surely and in L^1 with respect to Q^i. Hence, all of claim (a) is proved except the existence of the terminal σ-field $\mathcal{F}_\infty^i$ and the equality $Z^i = E_{Q^i}\{X|\mathcal{F}_\infty^i\}$.

Next, we prove that $Y_s^j \longrightarrow Y^j$ for all j, Q^i-a.s. holds. Note first that by Assumption 3, the claim $Z_s^j \longrightarrow Z^j$, Q^j-a.s. for all j implies that Y^j exists on $(\Omega, \mathcal{F}, \Pi)$ for all j. Consequently, we can take the complement of the event A^j in the following fashion: $\bar{A}^j := \{B_s^j(\epsilon), \text{ infinitely often for some } \epsilon > 0\}$, where $\{B_s^j(\epsilon)\}_{s=0}^\infty$ is a sequence of events $B_s^j(\epsilon) := \{\omega \in \Omega : |Y_s^j(\omega) - Y^j(\omega)| > \epsilon\}$. But, because we know that $Z_s^j \longrightarrow Z^j$ almost surely and in L^1 with respect to Q^j, it follows that Y^j exists Π-a.s. Hence, the intersection of $\sigma(I_0^i, \mathbf{Y}_0^{(i)}, \mathbf{Y}_1^{(i)}, \ldots)$ and $\bar{A}^j$ must be empty, i.e. $\sigma(I_0^i, \mathbf{Y}_0^{(i)}, \mathbf{Y}_1^{(i)}, \ldots) \cap \bar{A}^j = \emptyset$. It follows that $\lim_{s \to \infty} Q_q^i\{A^j|\mathcal{F}_s^i\} = 1$, since $Q_q^i\{A^j|\mathcal{F}_s^i\} > 0$ holds by Assumption 2.

Note that any conditional probability is a random variable and is integrable by definition. Moreover, we know that U^i is nonempty because Y^j exists on $(\Omega, \mathcal{F}, \Pi)$ for all j. Since U^i is a measurable set of finite measure, we can apply Lebesgue's bounded convergence theorem (or the dominated convergence theorem: see [18], for example). Therefore, we obtain for any (conditional) measure μ_t^i,

$$\lim_{s \to \infty} \int_{U^i} Q_q^i\{A^j|\mathcal{F}_s^i\}\mu_t^i(dQ_q^i) = \int_{U^i} \lim_{s \to \infty} Q_q^i\{A^j|\mathcal{F}_s^i\}\mu_t^i(dQ_q^i)$$
$$= 1.$$

Since this equation is true for any μ_t^i that satisfies the Rationality Axiom, it follows that

$$Q^i\{A^j|\sigma(I_0^i, \mathbf{Y}_0^{(i)}, \ldots, \mathbf{Y}_s^{(i)}, \ldots)\} = \lim_{s \to \infty} \int_{U^i} Q_q^i\{A^j|\mathcal{F}_s^i\}\mu^i(dQ_q^i|\mathcal{F}_s^i)$$
$$= 1. \tag{4}$$

Thus, claim (b) holds, and the existence of $\mathcal{F}_\infty^i$ is also proved. Thus, $Z^i = E_{Q^i}$ $\{X|\mathcal{F}_\infty^i\}$ Q^i -a.s. for all i. This establishes claim (a).

Since Y^j exists for all j on $(\Omega, \mathcal{F}, Q^i)$ and it is assumed that $E_{Q^i}|Y^j| < \infty$, it is straightforward that $E_{Q^i}\{Y^j|\mathcal{F}_s^i\}$ is a UI martingale relative to $\mathcal{F}_s^i$. Thus, again by UI martingale convergence theorem,

$$E_{Q^i}\{Y^j|\mathcal{F}_s^i\} \longrightarrow E_{Q^i}\{Y^j|\sigma(I_0^i, \mathbf{Y}_0^{(i)}, \ldots, \mathbf{Y}_s^{(i)}, \ldots)\}$$
$$= E_{Q^i}\{Y^j|\mathcal{F}_\infty^i\}, \forall j, Q^i - a.s., L^1,$$

where the last equation follows from the existence of $\mathcal{F}_\infty^i$.

Observe that $Y_s^j \longrightarrow Y^j$ for all j, Q^i-a.s. implies that $Z_s^{ij} \longrightarrow Z^{ij}$ for all j, Q^i-a.s. Note also that $Z^{ij} = E_{Q^i}\{Y^j|\mathcal{F}_\infty^i\}$ for all j, Q^i-a.s., since $\mathcal{F}_\infty^i = \sigma(I_0^i, \mathbf{Y}_0^{(i)}, \ldots, \mathbf{Y}^{(i)})$. Thus, claim (c) holds.

Moreover, observe that $\sigma(Y^j) \subset \mathcal{F}_\infty^i$ holds. It follows that $Z^{ij} = Y^j$ for all j, with probability $Q^i = 1$. Hence, claim (d) holds.   □

Assumption 2, which we call the rationality axiom, involves 'rationality', because an agent's belief will be incompatible with the fact that $Z_s^j \longrightarrow Z^j$ almost surely and in L^1 with respect to Q^j, when the assumption is not satisfied. To state, although agent i could hold a distribution such that $Q_q^i\{A^j|\mathcal{F}_s^i\} = 0$, such a distribution is contradictory to the fact that the realizations y_s^j remain unchanged for all s after reaching the last element Y^j. On the other hand, as long as $Q_q^i\{A^j|\mathcal{F}_s^i\} > 0$ is satisfied, $Q_q^i\{A^j|\mathcal{F}_\infty^i\} = 1$ holds. We therefore conclude that Assumption 2 is necessary and sufficient for a probability belief to retain compatibility with the data, which is perceived announcements here. Since compatibility with the data is frequently called the rationality condition, we call Assumption 2 'the rationality axiom'. Moreover, the hierarchical structure of beliefs allows for the agents to revise their beliefs to be compatible with the realizations. For example, if the limit random variable asserted by $Q_q^i = q$ is incompatible with the realizations, the agent will assign zero probability on such a belief in the limit, i.e. $\lim_{s\to\infty} \mu_s^i(Q_q^i = q) = 0$, although the limit itself is not necessarily directly conceivable to the agent (because an agent may not be able to tell that the process has reached the limit).

Theorem 2 (d) states that agents believe that in the limit they can forecast correctly what other agents' forecasts will be, when agents iterate exchanging their opinions infinitely many times. However, it does not state that agents can tell *a priori* what the limit forecasts of other agents will be. In other words, Theorem 2 (d) only states that agents believe that there will be no uncertainty about the functional form of the conditional forecast functions of others in the limit.

However, it is not necessarily the case that $Z^i = Z^j$ holds ($i \neq j$). In other words, agents do not necessarily hold the same forecast in the limit. Consider the following example. Suppose Humpty and Dumpty are trying to forecast the weather tomorrow, the probability of it being rainy, for instance. After reaching the limits, suppose Humpty always announces that his forecast of it being rainy tomorrow is 50% whilst Dumpty always announces 30%. Theorem 2 (d) states that Humpty believes that Dumpty's forecast will always be 30% with (Humpty's subjective) probability 1. There is, however, no reason for Humpty to make his opinion accord with Dumpty's opinion, since he does not know if Dumpty knows that Dumpty has all the knowledge/information that Humpty has. Moreover, Dumpty's forecast is still a random variable for Humpty, although Humpty correctly believes in the end that Dumpty's limit forecast will be a constant given the terminal information set. It is, however, impossible for Humpty to tell if Dumpty's forecast has reached a last element or not, although it is up to Humpty to *believe* that Dumpty's forecast has reached a last element. Of course, the same thing can be said for Dumpty's belief, too.

Another interpretation of Theorem 2 is that the result is in fact a merging of opinions on distinct sequences of random variables $(\{Z_s^1\}_s, ..., \{Z_s^I\}_s)$. Notice that Z_s^i is a random variable and is not known even from agent i's point of view in stage $s-1$ or earlier (because he is yet to listen to the opinions of others that are needed to compute Z_s^i), whilst $\{Z_s^i\}_s$ is an infinite sequence of (intrinsic) random variables Y_s^i from other agents' viewpoints. Therefore, the fact that $Z^i \neq Z^j$ (whilst $Z^{ij} = Y^j$) implies that the process reveals the difference between the agents' initial priors. The difference could be in either the unconditional probability distributions or the initial information sets. Nevertheless, since the initial information cannot be removed from the information set at any time in a decision theoretic sense when perfect recall is assumed, there is no place in which an unconditional probability has a substantial meaning with this respect unless the initial information set is empty.[3] In any event, each agent will, in the end, be convinced with respect to his subjective probability that he and the other agent have different priors, hence, different probability beliefs.

Observe also that Theorem 2 is analogous to the theorem by [7], which shows that absolute continuity of probability measures on an infinite sequence of random variables implies a merging of opinions, i.e. conditional probabilities converge to agree with each other. Theorem 2, however, exhibits a different result, since it does not assume absolute continuity, but assumes $E_{Q^i}|X| < \infty$ and $E_{Q^i}|Y^j| < \infty$ in conjunction with the rationality axiom and Assumption 3. It is therefore important to notice that this difference prevents agents' opinions from merging.

As is obvious from the proof of the theorem, the crucial step is to characterize the sequence of the conditional forecasts as a uniformly integrable martingale. In fact, as long as $E_{Q^i}|X| < \infty$ holds, we can claim that each agent is convinced that he will be able to pinpoint other agents' forecasts in the limit. It is however not necessarily the case that the forecasts of distinct agents agree with each other. As we discussed previously, it is not irrational at all for an agent to hold a different forecast from that of the forecasts of others in the limit, since each agent i typically believes that other agents believe that agent i himself has different information and/or belief than others have, and that no intrinsic data on X arrives, i.e. no empirical method to test the validity of the forecast functions exists, which uses the realization of X. Note that we distinguish the forecasts of agents from intrinsic data, although the agents treat the forecasts of agents as data.

Another point to note is that although we limited the contents of announcements to be the prior mean of each agent in every stage, essentially the same results hold even when the contents of announcements include other statistics, such as variances,

---

[3] See the discussion on the construction of prior probabilities by [14].

etc. Namely, as long as the statistics are measurable functions of Y_s^j and the function is in the L^1 space, the same conclusion holds.

2.3 The dynamics of the distribution of opinions

As we saw in the previous subsection, every agent is convinced that his forecast will converge to a random variable (a last element), and that he, in the limit, will be able to tell what other agents' forecasts will be. Nevertheless, the forecasts themselves do not merge in general. Then the question is how the distribution of the forecasts behaves dynamically as agents exchange their opinions.

To analyse such dynamics, we first prove that if an agent's belief that the opinions of others are informative vanishes within finitely many stages then he is convinced that he can tell exactly what other agents' forecasts will be. In so doing, we exploit the results of [16]. In particular, we examine four separate cases: (a) two agents with a binary random variable, (b) two agents with a generic discrete random variable, (c) more than two agents with a generic random variable and (d) two agents with a continuous random variable. In the end, we see that essentially the same conclusion holds in all cases.

Because of this result in conjunction with Assumption 1 (b), we need not consider the case of solitary agents, i.e. no real exchange of opinions from the beginning, since no forecasts will be changed when each agent believes that any other agent's opinion is not informative.

2.3.1 Case 1: Events/binary random variables

Consider a situation in which two separate agents are trying to assess the probability of an event. Let X be the indicator function of the event concerned, i.e. X is a binary random variable such that $\{X = 1\}$ indicates the occurrence of the event whilst $\{X = 0\}$ indicates its nonoccurrence. For the time being, we assume that both agents know that X is a binary random variable on $\{0, 1\}$. (Proposition 4 proves that this assumption is not crucial.) Let Q^i denote agent i's probability belief, and in particular let $Q^i\{X = 1|I_0^i\}$ denote agent i's initial prior belief concerning the occurrence of the event, where I_0^i denotes the initial information set of agent i. Let $Z_0^i := E_{Q^i}\{X|I_0^i\}$ denote the initial prior mean of X with respect to Q^i. Note that $Z_0^i = Q^i\{X = 1|I_0^i\}$ holds as X is a binary random variable on $\{0, 1\}$.

Now, instead of requiring each agent to compute the full joint distribution of X and $Y_s^{(i)}$, where $Y_s^{(i)}$ is the other agent's perceived forecast in stage s, we follow [16], [29] and [30] in requiring each agent to compute only his prior of X (or equivalently, Z_s^i for all s in our binary random variable case), and his prior mean of the other agent's forecast, $Z_s^{i(i)} := E_{Q^i}\{Y_s^{(i)}|\mathcal{F}_{s-1}^i\}$. By assuming such partial specifications, we implicitly assume that each agent forms a belief $\mu^i(\cdot)$ over beliefs Q_q^i, which belongs to U^i, which is a collection of beliefs that satisfy Assumptions 1 and 2, the Consistency Condition, and the initial specifications, as we saw earlier in this section. We now prove that if each agent stops updating his posterior of X then he is convinced that he can tell exactly what the other agent's forecast will be.

As is shown above, by Theorem 1 (Theorem 2.1 of [16]), the conditional forecast function in stage s has the form of a linear function of $y_s^{(i)}$,

$$Z_{s+1}^i = Z_s^i + \lambda_s^i \cdot (y_s^{(i)} - Z_s^{i(i)}), \qquad \forall i \in \{1,2\}, \tag{5}$$

for some constant λ_s^i depending upon Z_s^i and $Z_s^{i(i)}$, but not on $y_s^{(i)}$. From this result, we obtain the following proposition.

Proposition 1. *Suppose that every agent specifies his joint prior distribution over $\{X, Y_s^{(i)}\}$ only partially in every stage s given $\mathcal{F}_{s-1}^i$, providing the values Z_s^i and $Z_s^{i(i)}$, and assume that the marginal distribution for $Y_s^{(i)}$ has full support $[0,1]$ with respect to agent i's subjective probability in stage s. For every agent i, suppose that there exists a finite integer $\hat{s} \in \mathbf{N}$ such that $\lambda_{\hat{s}}^i = 0$. Then, $Z_s^{i(i)} = Y^{(i)}$ holds for all $s \geq \hat{s}$ with probability $Q^i = 1$.*

Proof. Observe first that if $\lambda_{\hat{s}}^i = 0$ for some finite integer $\hat{s}$, then by construction of Bayesian updating, $\lambda_s^i = 0$ for all $s \geq \hat{s}$. Thus, $Z_s^i = Z_{\hat{s}}^i$ for all $s \geq \hat{s}$, and it follows that $Z_s^i = Z^i$ for all $s \geq \hat{s}$.

Hence, if $\lambda_{\hat{s}}^i = 0$ for some finite integer $\hat{s}$, then it must be true that either (i) X is conditionally independent of $Y_s^{(i)}$ for all $s \geq \hat{s}$ given I_0^i, or (ii) $Y_s^{(i)}$ is a fixed random variable for all $s \geq \hat{s}$ given $\mathcal{F}_{\hat{s}}^i$ (i.e. $Y_s^{(i)}$ is $\mathcal{F}_{\hat{s}}^i$-measurable). Note that the first case does not involve conditional independence with $\mathcal{F}_{\hat{s}}^i$. It is clear that $Y_s^{(i)}$ is correlated with $Y_0^{(i)}$ for all $s < \hat{s}$ unless $Z_s^{i(i)} = Z_0^{i(i)}$ holds for all s, which implies $Z_s^{i(i)} = Y_s^{(i)}$, Q^i-a.s., for all s. This is because there is another updating process such that $Z_{s+1}^{i(i)} = Z_s^{i(i)} + \eta_s^i \cdot (y_s^{(i)} - Z_s^{i(i)})$, where η_s^i is some constant that is analogous to λ_s^i, and η_s^i is non-zero by Assumption 1 (b). Hence, if $Z_s^{i(i)} = Z_0^{i(i)}$ holds for all s, then case (ii) holds, and case (i) holds otherwise. Thus, case (i) is in fact a case in which X is conditionally independent of $Y_0^{(i)}$ given I_0^i.

Assumption 1 (b), however, rules out case (i), and thus, we claim that $Y_s^{(i)}$ is a fixed random variable given $\mathcal{F}_{\hat{s}}^i$ with respect to Q^i for all $s \geq \hat{s}$. It follows that $Z_s^{i(i)} = Z_{\hat{s}}^{i(i)}$ for all $s \geq \hat{s}$. Finally, from Theorem 2, we conclude that $Z_s^{i(i)} = Y^{(i)}$ for all $s \geq \hat{s}$ with probability $Q^i = 1$. Since i is arbitrary, the same claim holds for every i. $\square$

2.3.2 Case 2: Generic discrete random variables

Next we extend the result for the case of binary random variables to the case of more generic discrete random variables. Unless noted explicitly, we maintain the same assumptions as in the case of binary random variables. To save on notation, we employ the same notation as in the binary case above by modifying it to fit our current setting: X takes n distinct values, i.e. $X \in \{x_1, x_2, ..., x_n\} \subset \mathbf{R}$, where $x_1 < x_2 < \cdots < x_n$, and correspondingly, let $p_{0,k}^i := Q^i\{X = x_k | I_0^i\}$ denote agent i's initial prior for $\{X = x_k\}$, $k = 1, ..., n$. For the time being, we assume that both agents know that the random variable X is such a discrete random variable. Also, we assume that each agent announces only his forecast Z_s^i, not the full distribution of X. It follows that $p_{s,k}^i := Q^i\{X = x_k | \mathcal{F}_{s-1}^i\}$ denotes agent

i's prior for $\{X = x_k\}$ in stage s, where $Z_s^i := E_{Q^i}\{X|\mathcal{F}_{s-1}^i\}$ is the conditional expectation of agent i in stage s. Obviously,

$$Z_s^i = \sum_{k=1}^{n} p_{s,k}^i x_k \tag{6}$$

holds.

By Theorem 2.1 of [16] and Theorem 2 of [30], when each agent computes only the prior of the event $\{X = x_k\}$ and the prior mean of $Y_s^{(i)}$ in every stage, the only formula of posterior probability for an event $\{X = x_k\}$ in stage s that satisfies the Consistency Condition is the following:

$$p_{s+1,k}^i = p_{s,k}^i + \lambda_{s,k}^i \cdot (y_s^{(i)} - Z_s^{i(i)}), \quad \forall i \in \{1, 2\}, \tag{7}$$

where $\lambda_{s,k}^i$ is some constant that depends upon $p_{s,k}^i$ and $Z_s^{i(i)}$, but not upon $y_s^{(i)}$. Under this formulation, we have the following proposition:

Proposition 2. *Assume the same conditions as in Proposition 1, except that the marginal distribution for $Y_s^{(i)}$ has full support $[x_1, x_n]$ with respect to agent i's subjective probability in stage s. For every agent i, suppose that there exists a finite integer $\hat{s} \in \mathbf{N}$ such that $\lambda_{\hat{s},k}^i = 0$ for some k. Then, $Z_s^{i(i)} = Y^{(i)}$ for all $s \geq \hat{s}$ with probability $Q^i = 1$.*

Proof. The proof is essentially the same as the proof for Proposition 1. Just apply the same argument for Z_s^i to $p_{s,k}^i$ for some k by introducing $\lambda_{\hat{s},k}^i$ in place of $\lambda_{\hat{s}}^i$. □

2.3.3 Case 3: More than two agents

The previous propositions showed that the forecasts of forecasts of others should agree in the limit. We extend this result to the environment where more than two agents exchange opinions. Otherwise, we maintain the previous assumptions. Now, suppose that there are finitely many I agents who exchange opinions with each other. The result we propose below is in fact a straightforward extension of Theorem 3.2 by [16]. Again, we obtain a result such that if an agent stops updating his belief then he is convinced that he can tell other agents' forecasts correctly. By Theorem 3.2 of [16], when an agent computes only the prior of the event $\{X = x_k\}$ and the prior means of Y_s^j's, the only formula for the posterior of the event $\{X = x_k\}$ that satisfies the Consistency Condition is

$$Q^i\{X = x_k|\mathcal{F}_s^i\} = Q^i\{X = x_k|\mathcal{F}_{s-1}^i\} + \sum_{j=1,\neq i}^{I} \lambda_{s,k}^{ij} \cdot (y_s^j - Z_s^{ij}), \tag{8}$$

or

$$p_{s+1,k}^i = p_{s,k}^i + \sum_{j=1,\neq i}^{I} \lambda_{s,k}^{ij} \cdot (y_s^j - Z_s^{ij}), \tag{9}$$

where $p_{s,k}^i := Q^i\{X = x_k|\mathcal{F}_{s-1}^i\}$, and $(\lambda_{s,1}^{i1}, ..., \lambda_{s,n}^{iI}) \in \mathbf{R}^{n(I-1)}$ are constants that depend upon $Q^i\{X = x_k|\mathcal{F}_{s-1}^i\}$ and $(Z_s^{i1}, ..., Z_s^{iI}) \in \mathbf{R}^{I-1}$ (for $p_{s+1,k}^i$ to be a probability). Now, we have the following:

Proposition 3. *Assume the same conditions as in Proposition 1, except that the $I-1$ dimensional distribution for $\mathbf{Y}_s^{(i)}$ has full support $[x_1, x_n]^{I-1}$ with respect to agent i's subjective probability in stage s. For every agent i, suppose that there exists a finite integer $\hat{s} \in \mathbf{N}$ such that $\lambda_{\hat{s},k}^{ij} = 0$ for some k and for all j. Then, $Z_s^{ij} = Y^j$ for all j, for all $s \geq \hat{s}$ with probability $Q^i = 1$.*

Proof. The proof is essentially the same as that of Proposition 1 (and 2). Just apply the same argument for Z_s^i to $p_{s,k}^i$ for some k by introducing $\lambda_{\hat{s},k}^{ij}$ for all j (and for some k) in place of $\lambda_{\hat{s}}^i$. $\qquad\square$

This is the same conclusion as we had in the case of two agents, and the proof is essentially the same as that of Proposition 2. Hence, it should be understood that the case of two agents is rich enough to analyse the effect of general exchange of opinions.

2.3.4 Case 4: Continuous random variables

Now, we examine the case in which $X \in \mathbf{R}$ is a continuous random variable. The rest of the structure is the same as in the previous cases. For simplicity, we assume that only two agents exchange opinions, yet, essentially the same conclusion holds even if more than two agents exchange opinions as we saw in case 3.

Recall that a conditional forecast function $E\{X|W\}$ is a measurable function of W, and such a function can be approximated almost everywhere with respect to the law of W and arbitrarily well by some continuous function. Note that this is an application of Proposition 22 in [28, pp.69–70], which is a version of one of the Littlewood's three principles. Also, the collection of all polynomials (degree finite but otherwise arbitrary) is a dense subset of the set of continuous functions (see Example 4, section 6.3 in Chapter 2 of [19, p.48]). Thus, for any W there exists a sequence of polynomials which approximate $E\{X|W\}$ well. In our setting, the posterior Z_{s+1}^i is approximated well by a polynomial of $y_s^{(i)}$ in stage s. This fact plays a crucial part in the justification of the proposition below.

Before stating the proposition, let us introduce some notation: $(Z_s^i)^m := (E_{Q^i}\{X|\mathcal{F}_{s-1}^i\})^m$ and $(Z_s^{i(i)})^{<m>} := E_{Q^i}\{(Y_s^{(i)})^m \mid \mathcal{F}_{s-1}^i\}$ with $m \in \{1, 2, ..., M\}$, where $M \in \mathbf{N}$. Although in [16] it is assumed that X is a discrete random variable, we can consider an analogue to their result. First, partition the support of X by constructing intervals such as $a_k := [x_k, x_{k+1})$ where $-\infty < x_1 < \cdots x_k < x_{k+1} < \cdots < x_n < +\infty$. Next take limits by making these partitions 'finer'. This will end up with a formula with 'densities' since X is assumed to be continuous, i.e. $Q^i\{X \in a_k\} \longrightarrow Q^i\{X = x\}$ as $|x_k - x_{k+1}| \longrightarrow 0$ (by definition, we actually need absolute continuity of X to have a density; e.g. when the distribution function of X is the Cantor function, X is continuous but not absolutely continuous; however, the difference is not essential in our context). Note also that a distribution function is right-continuous by definition; thus the way we partitioned the support is consistent with this feature of distribution functions.

By Lemma 2.4 of [16], when agent i computes up to the Mth moment of the random variable $Y_s^{(i)}$, the only formula of the posterior that satisfies the Consistency

Condition is the following:

$$Q^i\{X = x|\mathcal{F}_s^i\} = Q^i\{X = x|\mathcal{F}_{s-1}^i\} \tag{10}$$

$$+ \sum_{m=1}^{M} \lambda_{s,m}^i(x) \cdot [(y_s^{(i)})^m - (Z_s^{i(i)})^{<m>}], \; \forall i \in \{1,2\},$$

or when X is absolutely continuous,

$$f_{s+1}^i(x) = f_s^i(x) + \sum_{m=1}^{M} \lambda_{s,m}^i(x) \cdot [(y_s^{(i)})^m - (Z_s^{i(i)})^{<m>}], \quad \forall i \in \{1,2\}, \tag{11}$$

where $f_s^i(x) := Q^i\{X = x|\mathcal{F}_{s-1}^i\}$ is the prior density in stage s with respect to Q^i, and $\lambda_{s,m}^i(x)$ are some constants that depend upon $f_s^i(x)$ and $((Z_s^{i(i)})^{<1>}, ..., (Z_s^{i(i)})^{<M>})$. Moreover, if we compute conditional expectations of X, i.e. Z_{s+1}^i and Z_s^i by using the densities $f_{s+1}^i(x)$ and $f_s^i(x)$, Z_{s+1}^i will be a finite polynomial of $y_s^{(i)}$ up to the degree of M. This construction is consistent with the above claim that any conditional expectation will be approximated well by a finite polynomial. Now, we have the following proposition, which is essentially the same as the previous propositions:

Proposition 4. *Assume the same conditions as in Proposition 1, except that the marginal distributions for* $(Y^{(i)})^m$ *have full support on the real line with respect to agent* i*'s subjective probability in stage* s *for all* $m \in \{1, ..., M\}$ *(i.e. agent* i *computes up to the Mth moment of the other agent's forecast* $Y_s^{(i)}$ *in every stage). For every agent* i*, suppose that there exists a finite integer* $\hat{s} \in \mathbf{N}$ *such that* $\lambda_{\hat{s},m}^i(x) = 0$ *for some* x *and for all* m*. Then,* $(Z_s^{i(i)})^{<m>} = (Y^{(i)})^m$ *for all* m*, for all* $s \geq \hat{s}$ *with probability* $Q^i = 1$*.*

Proof. Again, the proof is essentially the same as Proposition 1. Just apply the same argument for Z_s^i to $f_s^i(x)$ for some x by introducing $\lambda_{\hat{s},m}^i(x)$ for all $m \leq M$ (and for some x) in place of $\lambda_{\hat{s}}^i$. $\qquad\square$

Proposition 4 is important in the sense that we obtain the same conclusion for almost all practical conditional forecast functions that if each agent stops updating his belief then he is convinced that he can tell all moments of the forecasts of others correctly, although the degree of moment M is arbitrarily determined by the prior of each agent. As [16] argued briefly about this issue, each agent asserts his belief to set the upper bound M in the formula. In other words, it is the agent who determines the functional form of the conditional forecast function, and that his decision will be reduced to a decision on M, since the only form of conditional forecast function which satisfies the Consistency Condition must be constructed from the above formula, while a conditional forecast function should be approximated well by an arbitrary (finite) polynomial of $y_s^{(i)}$ in stage s. In fact, from equation (10), we obtain the following conditional forecast function:

$$Z_{s+1}^i = Z_s^i + \sum_{m=1}^{M} \bar{\lambda}^i{}_{s,m} \cdot [(y_s^{(i)})^m - (Z_s^{i(i)})^{<m>}] \quad \forall i, s,$$

where $Z_s^i = \int X dQ^i \{X | \mathcal{F}_{s-1}^i\} = \int x f_s^i(x) dx$ and $\bar{\lambda}^i{}_{s,m} = \int \lambda_{s,m}^i(x) x dx$.

Moreover, notice that Proposition 4 generalizes the previous propositions in the sense that we need not assume that each agent *knows* that a random variable is binary or discrete with some specific support. This is because it is a matter of subjective belief, *not* a matter of knowledge. Moreover, it is straightforward to extend the result to the case where there are more than two agents who exchange opinions with each other. Thus, we conclude that as long as an agent possesses a belief that satisfies the Consistency Condition, if he stops updating his belief then he is convinced that he can tell all the relevant moments of other agents' forecasts exactly, and that the relevance of the moments is determined by his own subjective belief.

3 Economic implications

In this section we examine the economic implications of the basic theorem (Theorem 2). First we study how the composition of population influences the dynamical behaviour of the distributions of opinions by introducing a specific updating rule that groups the behavioural rules into three classes: conformists, overshooters and contrarians. Then we examine whether it is reasonable to assume that agents believe that opinions of others are either unanimous or not by comparing the result to the implications of the CPA. Similarly, we examine whether it is reasonable to assume that agents believe that the opinions of others are either anonymous or not. Then, we evaluate the existing literature on herd behaviour in light of the results of the basic theorem. Finally, we introduce a simple competitive economy with financial assets to illustrate the effects of communication on the equilibria.

3.1 Population mix and the dynamics of opinions

We introduce several applications of the basic theorem to study the characteristics of exchange of opinions. Observe that the above results tell us that the essence of the characterizations of the updating processes is the same no matter what the random object of interest is believed to be, i.e. a discrete or a continuous random variable. We therefore assume for simplicity that each agent believes that the random object of interest is a discrete random variable, although the support of the random variable might be different across agents. Moreover, we assume that $\mathbf{Z}_s^{(i)} \neq \mathbf{Z}^{(i)}$ (or $\mathbf{Y}_s^{(i)} \neq \mathbf{Y}^{(i)}$ from agent i's perspective) for all s. With this assumption, we know from Propositions 1–4 that $\lambda_s^{ij} \neq 0$ for all j, and $Z_s^i \neq Z_{s-1}^i$ hold.

As pointed out previously, we can interpret exchange of opinions as a process that reveals the difference in the priors of the agents. To understand this point better, we consider the following simple updating process:

$$Z_s^i = \gamma_{s-1}^i \cdot \bar{y}_{s-1} + (1 - \gamma_{s-1}^i) Z_{s-1}^i \quad \forall i, s, \tag{12}$$

where $\bar{y}_{s-1} = \frac{1}{I} \sum_{j=1}^{I} y_{s-1}^j$. Observe that there are three possible categories of the behavioural rules of the agents;

(i) *conformists* if $\gamma_{s-1}^i \in (0,1]$,
(ii) *overshooters* if $\gamma_{s-1}^i \in (1,+\infty)$,
(iii) *contrarians* if $\gamma_{s-1}^i \in (-\infty,0)$.

Note that we exclude the case of $\gamma_{s-1}^i = 0$ since we assumed that $\mathbf{Z}_s^{(i)} \neq \mathbf{Z}^{(i)}$ for all s. Using this simple updating process, we examine several cases (assumptions) that are common in the literature of economics, and then discuss whether those cases are reasonable or not. Firstly, we examine a situation in which *unanimity of opinions* is assumed. In this case, an agent believes that the population of agents is unanimous; thus he believes that all other agents' opinions are random samples from a fixed population ($Z_s^{ij} = \alpha_s^i$ for all j, s in particular). Note also that the CPA or *rational expectations* is a special case of unanimity. In fact, we show below that the unanimity assumption yields the same implication as the CPA, although the assumption does not imply the CPA *per se*. After examining unanimity of opinions, we examine a situation in which *anonymity of opinions* is assumed.

Before proceeding, we investigate the characteristics of the dynamics of this updating process. For simplicity, we assume that $\gamma_s^i = \gamma^i$ for all s, and that $I = 3$. By construction, we have from equation (12)

$$\bar{y}_s - \bar{y}_{s-1} = \frac{1}{I} \sum_{i=1}^{I} \gamma^i \cdot (\bar{y}_{s-1} - Z_{s-1}^i) \quad \forall s. \tag{13}$$

This equation characterizes the population mean of the forecasts. Next, we derive the characterization of the dynamics of the population variance of the forecasts. Define

$$\Delta V_s := [\frac{1}{3} \sum_{i=1}^{3} (Z_{s+1}^i)^2 - (\bar{y}_{s+1})^2] - [\frac{1}{3} \sum_{i=1}^{3} (Z_s^i)^2 - (\bar{y}_s)^2], \tag{14}$$

while the difference between the sample variance in stage $s+1$ and that in stage s is $\frac{3}{2}\Delta V_s$. Then, by a straightforward calculation, we obtain

$$\Delta V_s = \frac{1}{3} \sum_{i=1}^{3} \gamma^i \cdot (\gamma^i - 2)(\bar{y}_s - Z_s^i)^2 - \frac{1}{9} [\sum_{i=1}^{3} \gamma^i \cdot (\bar{y}_s - Z_s^i)]^2. \tag{15}$$

It is obvious that if $\gamma^i \in (0,2)$ for all i, then $\Delta V_s < 0$ unless $\bar{y}_s = Z_s^i$ for all i. Thus, we obtain the following proposition;

Proposition 5. *When the population consists of conformists only, opinions will be narrowing unless opinions have merged already.*

Proof. Immediate from the claim that if $\gamma^i \in (0,2)$ for all i, then $\Delta V_s < 0$ unless $\bar{y}_s = Z_s^i$ for all i. Note, however, that $\bar{y}_s = Z_s^i$ for all i is a situation of merging of opinions. □

Next, we claim the following proposition, which is about the other extreme case, in which the entire population consists of contrarians.

Proposition 6. *When the population consists of contrarians only, opinions will be diverging.*

Proof. We prove the claim by deriving a lower bound of ΔV_s. By applying Chebyshev's inequality (see [5, p.27]), we obtain

$$\Delta V_s \geq \frac{1}{3}\sum_{i=1}^{3}\gamma^i \cdot (\gamma^i - 2)(\bar{y}_s - Z_s^i)^2 - \frac{3}{9}\sum_{i=1}^{3}(\gamma^i)^2 \cdot (\bar{y}_s - Z_s^i)^2$$

$$= \frac{1}{3}\sum_{i=1}^{3}(\gamma^i)^2 \cdot (\bar{y}_s - Z_s^i)^2$$

$$- \frac{2}{3}\sum_{i=1}^{3}\gamma^i \cdot (\bar{y}_s - Z_s^i)^2 - \frac{1}{3}\sum_{i=1}^{3}(\gamma^i)^2 \cdot (\bar{y}_s - Z_s^i)^2$$

$$= -\frac{2}{3}\sum_{i=1}^{3}\gamma^i \cdot (\bar{y}_s - Z_s^i)^2 = a.$$

Therefore, if $\gamma^i < 0$ for all i, i.e. the whole population consists of contrarians only, then $a > 0$, i.e. $\Delta V_s > 0$ for all s unless $\bar{y}_s = Z_s^i$ for all i. But, Assumption 1 (b) rules out the possibility that $\bar{y}_0 = Z_0^i$ for all i, and thus $\Delta V_s > 0$ for all s follows. $\square$

Note that Propositions 5 and 6 hold generically whatever the number of agents I is, although we assumed $I = 3$ for simplicity of calculations. Furthermore, although opinions will be diverging when all agents are contrarians, the degree of divergence is limited since Theorem 2 guarantees the existence of the limit distribution of opinions. Moreover, note that the lower bound of ΔV_s, which we derived above, is affected by γ^i as well as the distribution/location of Z_s^i. Thus, when the population is a mixture of different types of agents, the sign of the lower bound is not trivial.

Propositions 5 and 6 have important economic implications. As [20] show, when the (conditional) forecasts of agents in an asset market exhibit less diversity, they tend to fluctuate together inducing more volatile asset prices. On the other hand, they tend to cancel out each other's effect on market demand and resulting in lower price volatility when they are more diverse. These results in conjunction with Propositions 5 and 6 provide a somewhat counter-intuitive implication that the asset prices tend to be more volatile when the entire economy consists of conformists, whereas the prices are less volatile when the entire economy consists of contrarians. However, if we examine the implication more carefully, we can see that the result is supported by facts. For instance, it is frequently pointed out that when the majority of the market participants are using the same computer program to determine the portfolios, asset prices will be more volatile; 'Black Monday' is a good example. This situation clearly corresponds to the case of a conformists-oriented economy.

3.1.1 Is unanimity of opinions realistic?

First, let us define the term 'unanimity.' Unanimity is a family of beliefs that satisfies the following condition(s); (i) the agent treats other agents' opinions as random samples (from a fixed population), (ii) in particular, $Z_s^{ij} = \alpha_s^i$ holds for all j, where $\alpha_s^i \in \mathbf{R}$. To simplify the analysis, in addition to the two conditions, we assume that the λ's are the same across x_k's, when an agent assumes unanimity. Hence, when all agents assume unanimity, for every agent i, we have $\lambda_{s,k}^{ij} = \lambda_s^i$ and $Z_s^{ij} = \alpha_s^i$ for all j, s, k. Note that it must be the case that all agents assume unanimity so that an agent's belief be rational, otherwise he can detect that the other agents' opinions are not unanimous by simply looking at the past forecasts. For the time being, let us keep the number of agents generic — i.e. there are I agents exchanging opinions with each other. Hence, for every agent i, we obtain from equation (7)

$$p_{s,k}^i = p_{s-1,k}^i + (I-1)\lambda_{s-1}^i \cdot (\bar{y}_{s-1}^{(i)} - \alpha_{s-1}^i) \quad \forall s,k, \tag{16}$$

where $\bar{y}_{s-1}^{(i)} := \frac{1}{I-1}\sum_{j=1,\neq i}^{I} y_{s-1}^j$. It follows that

$$Z_s^i = Z_{s-1}^i + \bar{x}^i(I-1)\lambda_{s-1}^i \cdot (\bar{y}_{s-1}^{(i)} - \alpha_{s-1}^i) \quad \forall s, \tag{17}$$

where $\bar{x}^i := \sum_{k=1}^n x_k$ is a constant determined by the initial prior of agent i. Note that by construction of Bayesian updating, the support of X in every stage s is a subset of the support of the initial prior of X. We, therefore, assume that the support of the initial prior of X to be $\{x_1, ..., x_n\}$ with $x_1 < x_2 < ... < x_n$. To simplify notation, let $\tilde{\lambda}_s^i := \bar{x}^i(I-1)\lambda_s^i$. Then,

$$Z_s^i = Z_{s-1}^i + \tilde{\lambda}_{s-1}^i \cdot (\bar{y}_{s-1}^{(i)} - \alpha_{s-1}^i) \quad \forall s. \tag{18}$$

Condition (12) presumed a particular behavioural pattern, whilst condition (18) is a condition which follows from Proposition 3. The following proposition states the conditions that ensure compatibility between condition (12) and condition (18):

Proposition 7. *When agent i assumes unanimity of opinions, formula (12) satisfies the Consistency Condition iff $Z_{s-1}^i = Z_{s-1}^{ij} = \alpha_{s-1}^i$ for all j, and $\tilde{\lambda}_s^i = \frac{I-1}{I}\gamma_s^i$.*

Proof. The proof can be done by manipulating equation (12) directly. In particular,

$$\begin{aligned}
Z_s^i &= \gamma_{s-1}^i \cdot \left(\frac{I-1}{I}\bar{y}_{s-1}^{(i)} + \frac{1}{I}Z_{s-1}^i\right) + (1-\gamma_{s-1}^i)Z_{s-1}^i \\
&= \gamma_{s-1}^i \cdot \frac{I-1}{I}\bar{y}_{s-1}^{(i)} + (1-\gamma_{s-1}^i\frac{I-1}{I})Z_{s-1}^i \\
&= Z_{s-1}^i + \gamma_{s-1}^i \cdot \frac{I-1}{I} \cdot (\bar{y}_{s-1}^{(i)} - Z_{s-1}^i).
\end{aligned}$$

Thus, when unanimity is assumed, it is obvious that this formula coincides with equation (18), which is the only conditional forecast function that satisfies the Consistency Condition, if and only if

$$\gamma_{s-1}^i \cdot \frac{I-1}{I} \cdot (\bar{y}_{s-1}^{(i)} - Z_{s-1}^i) = \tilde{\lambda}_{s-1}^i \cdot (\bar{y}_{s-1}^{(i)} - \alpha_{s-1}^i). \tag{19}$$

It is straightforward that this equation holds if $Z_{s-1}^i = \alpha_{s-1}^i$ and $\tilde{\lambda}_{s-1}^i = \frac{I-1}{I}\gamma_{s-1}^i$. Thus, sufficiency is proved.

Now, we prove the necessity. Suppose that equation (19) holds. Note that the equation incorporates terms $(\gamma_{s-1}^i, Z_{s-1}^i, \tilde{\lambda}_{s-1}^i, \alpha_{s-1}^i)$, all of which are part of the decision functions of agent i, and thus depend upon $\mathcal{F}_{s-2}^i$. On the other hand, $\bar{y}_{s-1}^{(i)}$ is the realization of a random variable (from the perspective of agent i) in stage $s-1$, and hence, equation (19) must hold for all realizations $\bar{y}_{s-1}^{(i)}$ for a fixed set of variables, $(\gamma_{s-1}^i, Z_{s-1}^i, \tilde{\lambda}_{s-1}^i, \alpha_{s-1}^i)$. This is impossible unless the following argument holds. Equation (19) implies the following:

$$\bar{y}_{s-1}^{(i)}[\tilde{\lambda}_{s-1}^i - \gamma_{s-1}^i \frac{I-1}{I}] = \tilde{\lambda}_{s-1}^i \alpha_{s-1}^i - \gamma_{s-1}^i \frac{I-1}{I} Z_{s-1}^i.$$

Note that on the left hand side, the term $\bar{y}_{s-1}^{(i)}$ is out of agent i's control (i.e. out of agent i's Bayes decision) in stage $s-2$, whereas all terms other than $\bar{y}_{s-1}^{(i)}$ and I are decision functions of agent i depending upon $\mathcal{F}_{s-2}^i$. Hence, for this equation to hold for all realizations $\bar{y}_{s-1}^{(i)}$ in stage $s-1$, the condition such that $\tilde{\lambda}_{s-1}^i - \gamma_{s-1}^i \frac{I-1}{I} = \tilde{\lambda}_{s-1}^i \alpha_{s-1}^i - \gamma_{s-1}^i \frac{I-1}{I} Z_{s-1}^i = 0$ must hold. But, since $\tilde{\lambda}_{s-1}^i \neq 0$, $\tilde{\lambda}_{s-1}^i = \gamma_{s-1}^i \frac{I-1}{I}$ implies $\alpha_{s-1}^i = Z_{s-1}^i$. Hence, necessity is also established. $\square$

Proposition 7 states that when unanimity of opinions is assumed, equation (12) satisfies the Consistency Condition if and only if an agent does not distinguish himself from others and γ must satisfy a specific formula. Note that when I is large, $\tilde{\lambda}_s^i \approx \gamma_s^i$ must hold. Hence, for a large system, it does not matter whether an agent includes his own opinion in the population mean or not. Also, the first condition is a direct consequence of equation (12) when unanimity is assumed. By taking conditional expectations for both sides of equation (12) conditional upon $\mathcal{F}_{s-1}^i$, then we obtain $\gamma_{s-1}^i Z_{s-1}^i = \gamma_{s-1}^i \alpha_{s-1}^i$. Since we exclude the case that $\gamma_{s-1}^i = 0$, it follows that $Z_{s-1}^i = \alpha_{s-1}^i$.

Recall that there is another updating process such that $\alpha_s^i = \alpha_{s-1}^i + \eta_{s-1}^i \cdot (\bar{y}_{s-1}^{(i)} - \alpha_{s-1}^i)$ for all s, where η_{s-1}^i is some constant, which is analogous to λ_{s-1}^i. However, the condition that $Z_{s-1}^i = \alpha_{s-1}^i$ for all j requires the two updating processes to be identical. We conclude that the unanimity assumption is very restrictive, although Proposition 7 itself is valid only with equation (12). In fact, while equation (12) itself does allow for the agent to distinguish himself from others, the unanimity assumption rules out such a possibility. In this way, the unanimity assumption forces the agent to treat his own opinion as part of the unanimous opinions of the whole population. Moreover, when each agent can identify who announced which, for a single agent to assume unanimity of opinions, all other agents also must assume unanimity of opinions. Otherwise, the agent can check that such an assumption is not correct by simply looking at the realizations of the opinions of others. We further argue this feature by connecting to the CPA.

Under the CPA, [15] show that the posteriors will converge to a common posterior, even though their posteriors are based upon different finite information sets *a priori*. Furthermore, [23] generalizes their result for information sets described as σ-fields by using the martingale convergence theorem in the same way as we proved the basic theorem (Theorem 2) in the current paper. Under the CPA, therefore, any asymmetric information will be removed by exchanging opinions in the limit. Note that the opinions/information exchanged need not be the asymmetric information itself. Furthermore, this argument corresponds to the revelation principle when we regard the private information as the identification of 'types' of agents, since each agent will be able to map everything correctly so that the asymmetry is nullified. Nevertheless, the posterior with direct exchange of asymmetric information is not necessarily the same as the posterior with exchange of opinions other than the asymmetric information.

The unanimity assumption here implies that an agent must believe that his posterior coincides with his forecast of the posterior of others in every stage. Thus, there must be a common posterior in the limit, which is the same conclusion as the one under the CPA. Hence, the unanimity assumption is qualitatively the same as the CPA even with asymmetric information, at least in terms of the limit forecasts.

3.1.2 Is anonymity of opinions realistic?

Now, we assume that each agent believes that opinions of others are anonymous. In this case, each agent believes that there are several 'types' of agents; let $C^i \in [1, I]$ denote the number of types that agent i believes to exist, and let Z_s^{ic} denote the forecast of type c by agent i. In this case, $\lambda_s^{ij} = \lambda_s^{ic}$ for all j who belong to type c. Although we can introduce a generic distribution over C^i instead of assuming a probability mass $Q^h\{\# \text{ of types } = C^i\} = 1$, we do not employ this generalization, since it is not essential to our current analyses. Furthermore, we consider two separate cases; (i) each agent does not partition the samples/opinions of others, (ii) each agent partitions the samples into C categories by identifying other agents' types.

Case (i). In this case, each agent does not partition the samples, so the opinion to which he listens is the aggregate mean of the whole sample in every stage. Thus, for every agent i, we obtain

$$Z_s^i = Z_{s-1}^i + \sum_{c=1}^{C^i} \hat{\lambda}_{s-1}^{ic} \cdot (\bar{y}_{s-1}^{(i)} - Z_{s-1}^{ic}) \quad \forall s, \tag{20}$$

where $\hat{\lambda}_{s-1}^{ic} := (I-1)\bar{x}^i \lambda_{s-1}^{ic} \theta_{s-1}^{ic}$ with θ_{s-1}^{ic} denoting the fraction of type c in stage $s - 1$ that agent i believes to exist in stage $s - 1$. We provide the following proposition, which is a partial result of Proposition 7:

Proposition 8. *When agent i assumes anonymity of opinions, formula (12) satisfies the Consistency Condition if $Z_s^i = Z_s^{ic}$ for all c and $\sum_{c=1}^{C^i} \hat{\lambda}_s^{ic} = \frac{I-1}{I}\gamma_s^i$.*

Proof. It is obvious that the two formulae coincide if $Z_s^i = Z_s^{ic}$ for all c and $\sum_{c=1}^{C^i} \hat{\lambda}_s^{ic} = \frac{I-1}{I}\gamma_s^i$. $\qquad \square$

Observe that the first condition is the same as in the case of unanimity. The condition, however, is a sufficient condition, not a necessary condition. Hence, unlike unanimity we cannot claim that anonymity corresponds to the CPA. In fact, anonymity involves more complexities than unanimity does. Note that this case is valid only when the only available information is the aggregate mean of the whole sample. In other words, although an agent believes that there are several types, they are not capable of identifying the samples to sort them out in terms of types.

Case (ii). In this case, each agent partitions the samples into C^i categories by identifying other agents' types. Thus, each agent refers to the means that are different across types. Thus, for every agent i, we obtain

$$Z_s^i = Z_{s-1}^i + \sum_{c=1}^{C^i} \check{\lambda}_{s-1}^{ic} \cdot (\bar{y}_{s-1}^c - Z_{s-1}^{ic}) \quad \forall s, \tag{21}$$

where $\check{\lambda}_{s-1}^{ic} := (I-1)\bar{x}^i \lambda_{s-1}^{ic} \theta_{s-1}^{ic}$ and $\bar{y}_{s-1}^c := \frac{1}{\#c} \sum_{j \in c} y_{s-1}^j$, where $j \in c$ states that j belongs to type c and $\#c$ denotes the number of agents in category c. Notice that unlike case (i), the condition that $Z_s^i = Z_s^{ic}$ for all c, is not even a sufficient condition for equation (12) to satisfy the Consistency Condition in this case. Hence, we claim that there is no relationship with the CPA in this case. The result is striking, since existence of types does not really matter when the CPA is present, whilst it does matter with heterogeneous beliefs. In other words, asymmetric information cannot be nullified even in the limit when the CPA is absent. Note that an agent can identify the types in accordance with his beliefs as long as he can identify who submitted which forecast/opinion. Thus, case (ii) is more realistic than case (i), as it is unlikely for an agent to be entirely incapable of identifying the opinions.

As a conclusion, we claim that anonymity is a reasonable assumption, whilst unanimity is a polar, isolated case. In fact, when anonymity is assumed, agents hold beliefs over beliefs about types. This involves higher-order beliefs, and this is exactly the reason why anonymity does not correspond to the CPA. When the CPA is assumed, the type space is a single mass point for every agent, and thus, the updating process is identical across agents unless there is asymmetric information *a priori*. Moreover, even with asymmetric information *a priori*, such asymmetric information will nonetheless be nullified in the limit when the CPA is assumed. On the other hand, with the existence of higher-order beliefs, each agent's updating process is much more complex. More concretely, the updating processes of forecasts of forecasts of others are not identical across agents, and thus the dynamics will be more complex.

3.2 Herd behaviour

Now, we examine the results and/or the implications of the literature on herd behaviour in light of our results. Although the literature could explain in some ways how herding could occur, we claim that the standard analyses heavily depend upon implausible assumptions on the timing of decision making. We explain why the assumptions are implausible by referring to our basic theorem.

Although the literature has not described the phenomenon in a satisfactory manner, herd behaviour itself is a very interesting phenomenon, which can be observed on various occasions. Fashion can be considered as a typical example, since it may well not be the case that each agent's underlying preference changes drastically as

time passes (especially in a short period of time), although it apparently changes
to some extent. Simultaneously, it is not necessarily reasonable to say that people
simply love to conform with each other, i.e. their preferences are overwhelmingly
characterized by *conformism*. Note that conformism in terms of preferences is different from that in terms of beliefs. For example, consider a voter who has acquired
information about the predictions of the votes of others. Conformism in terms of
preferences means that the voter vote for the expected majority, whilst conformism
in terms of beliefs means that the voter updates his belief to be closer to the majority.
Yet he might vote for the minority if his preference is such that he is happier not
to vote for the majority with the updated belief. Thus, the economics profession
has tried to explain herd behaviour without introducing changes in preferences or
assuming conformism. The framework they used in common is the combination of
the CPA and asymmetric information, which is standard in economic theory today.
Namely, the literature defines herd behaviour as a behaviour with which an agent
ignores his private information but imitates other agents.

The most fundamental and crucial assumption shared widely in the literature is
that individuals make decisions sequentially, and the ordering is given exogenously.
For example, the second agent makes his decision after observing the first agent's
decision. Under such a structural assumption, it is shown that herding occurs generically, although the direction of herding is determined by early movers. In a sense,
herding is characterized as a knife-edge phenomenon in the literature.

However, this assumption is hard to justify. It is actually a shortcoming of the
literature, because it is better for each agent to move later. Moreover, the assumption
requires the agents to *know* that they make decisions one agent after another. A
question naturally comes into mind: How could agents ever obtain such knowledge
when agents are *a priori* symmetrical? If we follow the framework of the expert
problem (e.g. [16]), the first agent must be the most knowledgeable agent, the second
one is next to the first one in terms of knowledge and so on, i.e. $\mathcal{F}^1 \supset \mathcal{F}^2 \supset \cdots$,
or in words, there must exist a sequence of inclusion of sets. This must be the case,
since if agents believe that neither $\mathcal{F}^i \subset \mathcal{F}^j$ nor $\mathcal{F}^i \supset \mathcal{F}^j$ holds for some i, j, then
agents believe that agent i is willing to listen to j as well as agent j is willing to
listen to i (i.e. exchange of opinions). But, to retain consistency between the beliefs
of agents and such a sequence of inclusions, we need to require all agents to know
(or believe with probability one) that there exists such a sequence. This is a very
severe requirement, and is unreasonable as we pointed out in the introduction.

Once we relax (or remove) the assumption on the ordering of the timing of
decision making and the CPA, the natural situation will be the one described in
our model, i.e. exchange of opinions (or communication, simultaneous moves).
The basic theorem then goes on to show that opinions may not merge in general.
On the contrary, the distribution of opinions could be more diverse than the initial
distribution, although the distribution will converge in the limit. Moreover, even
though the entire population consists of conformists it is not necessarily the case
that a merging of opinions occurs. Hence, we conclude that herding is not a knife-
edge phenomenon as opposed to the characterization of the phenomenon in the
existing literature. In fact, this is much more realistic, since it is extremely unusual
to experience unanimity of opinions or actions in reality.

To complete the discussion on herd behaviour, we mention a couple of works that do not employ the standard assumption on the timing of decision making. First, [24] relaxes the assumption and assume symmetric communication instead. However, [24] introduces an arbitrary assumption on the class of agents – i.e. (a) 'independent' agents, and (b) imitators. In light of our results, the belief of an 'independent' agent corresponds to case (i) of anonymity above, whilst an imitator's belief corresponds to the case of unanimity. However, this is not a reasonable classification. Recall that all agents must believe unanimity for a rational agent to believe unanimity as we explained above. Hence, the assumption of [24] clearly violates this claim, and thus, is not reasonable as long as we are to sustain rationality of agents.

Another work that does not employ the standard assumption is [10]. They describe the communication structure as a random graph. However, what they are describing is really a description of coalition formation. All agents within a coalition hold the same belief; thus, in light of our results, $\lambda^{ij} = 0$ holds if i and j belong to the same coalition (i.e. they are connected in the random graph). [10] claim that the communication structure evolves slowly, but such an evolution is not modelled explicitly. Indeed, as long as such an evolution is random, the evolution of λ^{ij} is also random. This is a rather weird and/or arbitrary property in light of our results. A random λ^{ij} means that agent i sometimes believes that he does not need to listen to j, while sometimes he puts a lot of faith on j, and so on, but he changes his attitude randomly. We therefore conclude that the existing literature on herding is not reasonable in light of our results.

3.3 Endogenous expansion of the state space in a competitive economy

To illustrate how exchange of opinions/communication could impact the equilibria of an economy, we introduce a simple competitive equilibrium economy with financial assets. Consider a two-period exchange economy with a homogeneous consumption good and two financial assets. We assume that agents receive no endowment in the first period, but receive endowment of the consumption good in the second period. Furthermore, there are only two endowment states in the second period.

In the first period, agents exchange opinions about the endowment state infinitely many times, and determine their portfolios by observing the prices. We admit that the timing of the model is somewhat unnatural in the sense that agents observe the equilibrium prices and then exchange opinions infinitely many times based on the observed equilibrium prices, while the equilibrium prices will be determined by the portfolio choices that are affected by the exchange of opinions. However, the timing of price determination and determination of optimal choice by the agents is common in the general equilibrium models. Assume however that the initial announcements are made without referring to the announcements of others, while the subsequent announcements refer to the previous announcements of others. In the second period, each agent receives (or pays) returns on the financial assets (in terms of the consumption good) and endowment of the consumption good. As long as local non-satiation of preferences is satisfied, each agent then consumes all of them.

We assume that the two financial assets are linearly independent; thus, the economy has a complete market structure unless we allow for heterogeneous beliefs. Also, there is no asymmetric information other than about the beliefs. Hence, when exchange of opinions is absent, it is clear that the rational expectations equilibrium of this financial economy is equivalent with the equilibrium of an Arrow-Debreu economy. In this case, each agent i's portfolio is simply a function of the prices of the financial assets $\mathbf{P} := (P_1, P_2)$:

$$\theta^i = \theta^i(\mathbf{P}),$$

where the functional form of $\theta^i(\cdot)$ is affected by the probability belief $Q^i = \Pi$ for all i.

Let $W = 0, 1$ denote the endowment state. By construction, each agent i's forecast is defined as $Z_s^i := E_{Q^i}\{W \mid \mathbf{P}, \mathbf{Y}_0^{(i)}, \mathbf{Y}_1^{(i)}, ..., \mathbf{Y}_{s-1}^{(i)}\}$, while Z^i is indeed a perceived announcement for other agents – i.e. Y^i. It is clear that each agent's demand functions for the financial assets θ^i are functions of $(\mathbf{P}, \mathbf{Y}_0^{(i)}, \mathbf{Y}_1^{(i)}, ..., \mathbf{Y}_s^{(i)}, ...)$, that is, $\theta^i = \theta^i(\mathbf{P}, \mathbf{Y}^{(i)})$. Hence, the equilibrium prices of the financial assets are functions of the announcements $(\mathbf{Y}_0, \mathbf{Y}_1, ..., \mathbf{Y}_s, ...)$, where $\mathbf{Y}_s := (Y_s^1, Y_s^2, ..., Y_s^I)$:

$$\mathbf{P} = \phi(\mathbf{Y}_0, \mathbf{Y}_1, ...). \tag{22}$$

Before discussing the effects of exchange of opinions on the equilibrium prices, we prove that exchange of opinions has no impact on the rational expectations equilibrium.

Proposition 9. *Exchange of opinions has no impact on the rational expectations equilibrium.*

Proof. With rational expectations, it must be that $Q^i = \Pi$ for all i, where Π is the true probability measure. Therefore, it is straightforward that

$$Z_0^i := E_{Q^i}\{W \mid \mathbf{P}\} = E_{\Pi}\{W \mid \mathbf{P}\}, \quad \forall i. \tag{23}$$

It follows that

$$\begin{aligned}
Z_1^i := E_{Q^i}\{W \mid \mathbf{P}, \mathbf{Y}_0^{(i)}\} \\
= E_{\Pi}\{W \mid \mathbf{P}, \mathbf{Y}_0^{(i)}\} \\
= E_{\Pi}\{W \mid \mathbf{P}\} \\
= Z_0^i, \quad \forall i.
\end{aligned}$$

Hence, each agent i's portfolio becomes

$$\theta^i = \theta^i(\mathbf{P}, \mathbf{Y}_0^{(i)}, \mathbf{Y}_1^{(i)}, ...) = \tilde{\theta}^i(\mathbf{P}), \quad \forall i. \tag{24}$$

This is exactly the same as the portfolio in the rational expectations equilibrium in which exchange of opinions is absent.   □

To clarify the discussion, we describe the true and subjective probability spaces of this economy explicitly. As we discussed in Section 2, we need to sustain compatibility between the true and subjective probability spaces. In this particular economy, agent i's subjective probability space $(\Omega^i, \mathcal{F}^i, Q^i)$ should be defined as follows:

$$\Omega^i := \mathcal{P}_1 \times \mathcal{P}_2 \times \mathcal{W} \times [\prod_{s=0}^{\infty} (\mathcal{Y}_s^1 \times \cdots \times \mathcal{Y}_s^{i-1} \times \mathcal{Y}_s^{i+1} \times \cdots \times \mathcal{Y}_s^I)],$$

while $\mathcal{F}^i$ is the Borel σ-field of Ω^i thus defined. On the other hand, the true probability space should be defined as follows (we adopt the second view in Section 2):

$$\Omega^{\Pi} := \mathcal{W} \times \left(\prod_{s=0}^{\infty} \prod_{i=1}^{I} \mathcal{Y}_s^i \right),$$

and $\mathcal{F}^{\Pi}$ is the Borel σ-field of Ω^{Π} thus defined.

As shown above, on the true probability space, the equilibrium prices will be

$$\mathbf{P} = \phi(\mathbf{Y}_0, \mathbf{Y}_1, ...).$$

It is now clear that the equilibrium prices are influenced by the beliefs of the agents $(Q^1, Q^2, ..., Q^I)$. Moreover, we know from Theorem 2 that the forecasts/ announcements converge to $(Y^1, Y^2, ..., Y^I)$. Note however that the equilibrium prices are affected not only by the limit forecasts but also by the entire paths of the forecasts.

This is particularly important, because it indicates that the insight of the model is very different from that of a rational expectations model with sunspots. In a sunspot model, the equilibrium prices are functions of sunspots. Therefore, correlation of beliefs or communication has no impact on the equilibria of a sunspot model, while some exogenous random variables determine the equilibria. In this sense, sunspot models do not involve endogenous uncertainty, which is supposed to be generated by the beliefs of the agents of the economy. We stress the point that the entire paths of the forecasts in our model are generated by the agents themselves. In other words, they are not determined arbitrarily; thus, they are not sunspot variables. We therefore claim that the state space of the economy is expanded endogenously, and that, such endogenous uncertainty affects the equilibria.

Because the market opens virtually only once in this economy, the economy is essentially static; thus, dynamic consistency of the beliefs is not considered here. However, when we introduce a dynamical model, the issue arises. The literature of rational beliefs considers this issue while allowing for heterogeneous beliefs unlike the rational expectations literature. [22] studies the effects of communication on the rational belief equilibria by introducing an overlapping generations model with financial assets. The results are still partial, nevertheless indicate that communication does have a major impact on the equilibria. For the limited class of beliefs considered there, the model is capable of explaining large fluctuations of economic variables that are generated by correlation of beliefs through communication.

4 Conclusion

We have shown that the opinions of the agents converge to a limit distribution, but a merging of opinions does not necessarily occur. Moreover, each agent is convinced that in the limit he will be able to tell correctly what the opinions of others will be. The result is weaker than that of [7], reflecting the fact that the set of assumptions in our paper is much weaker than theirs (absolute continuity of measures). Also, by extending the result of [16], we showed that if an agent stops updating his posterior by referring to the opinions of others, then he is convinced that he can tell the opinions of others correctly.

While the results of the paper are fairly abstract, they have some interesting implications towards economic modelling.[4] We chose the literature of herd behaviour as an example of models whose assumptions are not very plausible in light of our theory. Namely, our result gives caveats to the models that employ the CPA and introduce asymmetric information without examining how plausible such assumptions are. We admit that asymmetric information is very important, but we at the same time emphasize that heterogeneity of beliefs could play a crucial role in explaining many economic phenomena that we have found somewhat puzzling. We briefly examined the effects of communication on the fluctuations of economic variables in a financial economy, while [22] exhibits more detailed analyses on this issue. Because there is virtually no way to estimate the subjective probability measures themselves, studies on heterogeneous beliefs need to hinge on abstract modelling. This is a challenging research area, but we believe that this is a very important area for the development of future economic studies.

References

1. Ash, R. B.: Real analysis and probability. San Diego, CA: Academic Press 1972
2. Aumann, R.: Agreeing to Disagree. Annals of Statistics **4**, 1236–1239 (1976)
3. Banerjee, A. V.: A simple model of herd behavior. Quarterly Journal of Economics **107**, 797–817 (1992)
4. Bayarri, M. J., DeGroot, M. H.: What Bayesians expect of each other. Journal of the American Statistical Association **86**, 924–932 (1991)
5. Berck, P., Sydsæter, K.: Economists' mathematical manual, 2nd edn. Berlin Heidelberg New York: Springer 1993
6. Bikhchandani, S., Hirshleifer, D., Welch, I.: A theory of fads, fashion, custom, and cultural change as informational cascades. Journal of Political Economy **100**, 992–1026 (1992)
7. Blackwell, D., Dubins, L.: A merging of opinions with increasing information. Annals of Mathematical Statistics **33**, 882–886 (1962)
8. Brandenburger, A., Dekel, E.: Hierarchies of beliefs and common knowledge. Journal of Economic Theory **59**, 189–198 (1993)
9. Breiman, L.: Probability. Philadelphia, PA: Society for Industrial and Applied Mathematics 1992
10. Cont, R., Bouchard, J. P.: Herd behavior and aggregate fluctuations in financial markets. Macroeconomic Dynamics **4**, 170–196 (2000)
11. Dalkey, N. C., Helmer, O.: An experimental application of the Delphi method to the use of experts. Management Science **9**, 458–467 (1963)

---

[4] There is wide spread technique known as the *Delphi method*. The empirical results of its applications give some support to the results of the paper. See [11], [21] and [27] for details.

12. Durlauf, S. N.: Nonergodic economic growth. Review of Economic Studies **60**, 349–366 (1993)
13. Freedman, D.: On the asymptotic behavior of Bayes estimates in the discrete case II. Annals of Mathematical Statistics **36**, 454–456 (1965)
14. French, S.: Updating of belief in the light of someone else's opinion. Journal of the Royal Statistical Society, Series A **143**, 43–48 (1980)
15. Geanakoplos, J. D., Polemarchakis, H. M.: We can't disagree forever. Journal of Economic Theory **28**, 192–200 (1982)
16. Genest, C., Schervish, M. J.: Modeling expert judgments for Bayesian updating. Annals of Statistics **13**, 1198–1212 (1985)
17. Gray, R. M.: Probability, random processes, and ergodic properties. New York, NY: Springer 1988
18. Halmos, P. R.: Measure theory. New York, NY: Springer 1974
19. Kolmogorov, A. N., Fomin, S. V.: Introductory real analysis (translated and edited by R. A. Silverman). Mineola, NY: Dover 1975 (c1970)
20. Kurz, M., Schneider, M.: Coordination and correlation in Markov rational belief equilibria. Economic Theory **8**, 489–520 (1996)
21. Martino, J. P.: Technological forecasting for decision making, 2nd edn. New York, NY: North-Holland 1983
22. Nakata, H.: A model of financial markets with endogenously correlated rational beliefs. Ph.D. dissertation, Ch. 3. On the dynamics of endogenous correlations of beliefs. Stanford, CA: Stanford University 2001
23. Nielsen. L. T.: Common knowledge, communication, and convergence of beliefs. Mathematical Social Sciences **8**, 1–14 (1984)
24. Orléan, A.: Bayesian interactions and collective dynamics of opinion: herd behavior and mimetic contagion. Journal of Economic Behavior and Organization **28**, 257–274 (1995)
25. Osborne, M. J., Rubinstein, A.: A course in game theory. Cambridge, MA: The MIT Press 1994
26. Romer, P. M.: Increasing returns and long run growth. Journal of Political Economy **94**, 1002–1037 (1986)
27. Rowe, G., Wright, G.: The Delphi technique as a forecasting tool: issues and analysis. International Journal of Forecasting **15**, 353–375 (1999)
28. Royden, H. L.: Real analysis, 3rd edn. Englewood Cliffs, NJ: Prentice-Hall 1988
29. West, M.: Modelling agent forecast distribution. Journal of the Royal Statistical Society, Series B **54**, 553–567 (1992)
30. West, M., Crosse, J.: Modelling probabilistic agent opinion. Journal of the Royal Statistical Society, Series B **54**, 285–299 (1992)

Effects of asset market structure on welfare and trading volume [*]

Kenneth L. Judd[1], Felix Kubler[2], and Karl Schmedders[3]

[1] Hoover Institution, Stanford, CA 94305, USA
   (e-mail: judd@hoover.stanford.edu)
[2] Department of Economics, Stanford University, Stanford, CA 94305, USA
   (e-mail: fkubler@stanford.edu)
[3] Kellogg School of Management, Northwestern University, Evanston, IL 60208-2006, USA
   (e-mail: k-schmedders@kellogg.northwestern.edu)

Received: April 2, 2002; revised version: October 8, 2002

Summary. We quantitatively explore how asset market structure affects risk-sharing, welfare, and trading volume in stylized rational expectations models. We examine five market structures: perfect asset trading, only bonds and equity, only equity, only bonds, and autarchy. We find a variety of results. First, trade in a single asset, either bond or stock, achieves most of the utility gain from going to perfect markets. Second, the value of adding a bond (stock) to the stock (bond) only economy is generally quite small, and has negligible price effects. Third, even in our simple model, the addition of a new asset may harm many traders. Fourth, adding a new asset may increase trading volume in old assets. Fifth, completing the market beyond a stock and bond may have substantial price and welfare effects, indicating that adding assets like options can have significant economic effects.

Keywords and Phrases: Incomplete asset markets, Welfare, Trading volume, Financial innovation.

JEL Classification Numbers: D52, G12.

1 Introduction

The function of asset markets is to implement risk-sharing and intertemporal trades. In this paper, we assess the gains from improvements in trading opportunities in asset markets for empirically plausible choices of tastes and asset returns. The

---

  [*] We are grateful to Mordecai Kurz for countless stimulating discussions on the subject of dynamic economic models. This paper is dedicated to him. We thank an anonymous referee for excellent comments on an earlier draft.
Correspondence to: K. L. Judd

theory of incomplete markets tells us that the welfare impact of a new asset is ambiguous in general. Hart (1975) presents an example where all agents lose from the introduction of a new asset. Elul (1995); Cass and Citanna (1998) extend this by showing that if asset markets are sufficiently incomplete and there is more than one consumption good then the distribution of utility may be arbitrarily perturbed by financial innovation. We examine these issues in an infinite horizon asset market model with an infinite number of consumption goods.

There has been recent interest in evaluating the incompleteness of asset markets and the welfare gains from improvements in capital markets. For example, several papers have tried to estimate the welfare gains from perfecting international asset markets; Wincoop (1999) surveys this literature. However, this literature assumes that all agents have the same preferences, differing only in their endowment processes. Also, some use unreliable numerical procedures to arrive at their estimates. For example, Tesar (1995) approximates stochastic equilibria with certainty equivalent, linearization methods which have, at best, only local validity for deterministic models. These methods are not reliable for stochastic models, as indicated by an example in Tesar (1995) where financial innovation resulting in perfect asset markets reduces expected utility for all investors, a violation of general equilibrium theory. In this paper we estimate the value of risk-sharing with heterogeneous tastes and use more flexible nonlinear computational methods. These solutions are also more reliable since they pass stringent Euler equation tests over the ergodic distribution of equilibrium asset holdings. We find that, generally, trade in a single asset, either a bond or stock, creates most of the potential utility gain from trade in risky assets. However, there are some cases where completing the market beyond a stock and a bond, i.e., adding derivatives like options, has a substantial marginal effect on utility. Adding new securities will affect the value of existing securities; we find that this effect is often small except when there is substantial negative correlation in idiosyncratic risks.

Volume is also a puzzle in dynamic general equilibrium. The Arrow-Debreu model of general equilibrium is unrealistic in that all transactions occur at the initial time. The Lucas (1978) asset pricing model is often used in financial economics, but is unsuitable for studying volume since its representative agent structure implies zero volume. Kreps (1982) examined the more realistic case where trading in a small number of assets can implement the Arrow-Debreu allocation. Even this model, however, does a poor job describing volume. Judd et al. (2001) show that volume is zero in the generic dynamically complete asset market because any asset structure rich enough to dynamically span the equilibrium consumption process is also able to span it statically. In particular, this result showed that portfolio rebalancing, i.e., individuals with common beliefs adjusting their portfolios in response to income shocks or new public information about future returns, cannot by itself be a source of trading. In this paper, we examine how changes in asset market structure will affect asset market volume. We find examples where adding a bond market to an existing stock market will lead to a higher volume of trade in stock. We also examine whether incompleteness in asset markets can generate economically substantial asset trading volume, and consider a couple of (nongeneric) examples of complete markets which do have trading in equilibrium. We never find economically significant levels of trading volume. Furthermore, we find that general equilibrium effects produce trade volumes and directions contrary to that predicted by partial equilibrium intuition.

This paper is a computational and quantitative exploration of several important issues in asset market theory. The results are limited in the coverage of the examples, but they display several patterns which merit further work. The paper is organized as follows: In Section 2 we describe the economic model under consideration. Section 3 provides an overview of our computational procedure. We report results on welfare in Section 4. Section 5 presents the results for trading volume. Section 6 concludes.

2 The economic model

We examine an infinite horizon pure exchange economy with heterogeneous agents and incomplete asset markets. Time is indexed by $t = 0, 1, 2, ...$ A time-homogeneous Markov process of exogenous states (y_t) takes values in a discrete set $Y = \{1, 2, \ldots, S\}$. The Markov transition matrix is denoted by Π. Let Σ denote the set of all possible histories σ of the exogenous states. A date-event σ_t is the history of states along a history σ up to time t, i.e. $\sigma_t = (y_0 y_1 \cdots y_t)$. There are S successors of any node σ_t, namely $\sigma_t s = (y_0 y_1 \cdots y_t s)$ for each $s \in Y$. Each node $\sigma_t, t \geq 1$, has a unique predecessor $\sigma_t^* = (y_0 y_1 \cdots y_{t-1})$. To simplify notation the event tree includes the root nodes' predecessor σ_0^*. In each date-event $\sigma \in \Sigma$ there is a single perishable consumption good.

We assume that there are finitely many types of infinitely-lived agents $h \in \mathcal{H} = \{1, 2, .., H\}$. Agent h's individual endowment at time event σ_t is a function $e^h : Y \to \mathbb{R}_{++}$ depending on the current state y_t alone. In addition, there is a single stock ("Lucas tree") in the economy paying dividends $d : Y \to \mathbb{R}_{++}$ depending on the current state y_t alone. The agents have initial holdings $\theta^h_{-1} \geq 0$ of the stock, which is in unit net supply. The aggregate endowment of the economy in state y_t is $e(y_t) = d(y_t) + \sum_{h=1}^{H} e^h(y_t)$. Occasionally it will be more convenient to write $e^h(\sigma)$. It will then always be understood that $e^h(\sigma) = e^h(y)$ where $\sigma = (\sigma^* y)$. Each agent h has a time-separable von-Neumann-Morgenstern utility function

$$U_h(c) = E \left\{ \sum_{t=0}^{\infty} \beta^t u_h(c_t) \right\}.$$

We assume that the Bernoulli functions $u_h(.) : \mathbb{R}_{++} \to \mathbb{R}$ are strictly monotone, C^2, strictly concave, and satisfy the Inada property, that is, $\lim_{x \to 0} u'(x) = \infty$. We also assume that the discount factor $\beta \in (0, 1)$ is the same for all agents and that expectations are taken under the true Markov probabilities.

Let the matrix

$$e = \begin{pmatrix} e^1(1) & \cdots & e^1(S) \\ \vdots & & \vdots \\ e^H(1) & \cdots & e^H(S) \end{pmatrix}$$

represent possible individual endowments. The vector of utility functions is $u = (u^1, \ldots, u^H)$. We collect the primitives of the economy as $\mathcal{E} = (e, d, u, \Pi, \beta)$.

2.1 Equilibrium concepts

In order to evaluate the welfare effects of incomplete markets we define an Arrow-Debreu equilibrium for complete markets.

Definition 1 *An Arrow-Debreu equilibrium for an economy $\mathcal{E}$ is a collection of prices $(p(\sigma))_{\sigma \in \Sigma}$ and a consumption allocation $(c^h(\sigma))_{\sigma \in \Sigma}^{h \in \mathcal{H}}$ such that markets clear and agents maximize, i.e.*

(1) $\sum_{h \in \mathcal{H}} c^h(\sigma) = e(\sigma), \forall \sigma \in \Sigma.$
(2) $(c^h(\sigma))_{\sigma \in \Sigma} \in \arg \max u^h(c)$
    s.t. $\sum_{\sigma \in \Sigma} p(\sigma)(c(\sigma) - (e^h(\sigma) + \theta^h_{-1} \cdot d(\sigma)) = 0$

In contrast to the Arrow-Debreu equilibrium we want to examine economies where the agents are restricted to transfer wealth across time periods and states of nature. In the first such model the agents can only trade their shares in the single stock but they are not allowed to perform short sales.

Definition 2 *A financial market equilibrium for an economy $\mathcal{E}$ with a single stock is a process of portfolio holdings and consumptions $(\theta^h(\sigma), c^h(\sigma))_{\sigma \in \Sigma}^{h \in \mathcal{H}}$ as well as asset prices $(q(\sigma))_{\sigma \in \Sigma}$ satisfying the following conditions:*

(1) $\sum_{h=1}^{H} \theta^h(\sigma) = 1$ *for all $\sigma \in \Sigma$.*
(2) For each agent h:

$$(\theta^h, c^h) \in \arg \max_{\theta, c} U_h(c) s.t.$$

$$c(\sigma) = e^h(\sigma) + \theta(\sigma^*) \cdot (d(\sigma) + q(\sigma)) - \theta(\sigma)q(\sigma)$$

$$\theta(\sigma) \geq 0.$$

We also examine economies where the agents cannot trade the stock but can only trade a bond. In these economies there is a single one-period bond at each node $\sigma \in \Sigma$ in zero net supply. The bond that is traded in period t pays a single unit of the consumption good in every state of nature in period $t + 1$. Agents face a borrowing constraint $\vartheta \geq -B$ for some positive number B for short positions in the bond. The definition of a financial market equilibrium with a single bond is analogous to Definition 2. (Agent h's endowment at node $\sigma \in \Sigma$ is given by $e^h(\sigma) + \theta^h_{-1} \cdot d(\sigma)$.)

Finally, we examine economies where the agents can trade two assets to achieve risk-sharing, namely the stock and a bond.

Definition 3 *A financial market equilibrium for an economy $\mathcal{E}$ with a stock and a bond is a process of portfolio holdings and consumptions $(\theta^h(\sigma), \vartheta^h(\sigma), c^h(\sigma))_{\sigma \in \Sigma}^{h \in \mathcal{H}}$ as well as asset prices $(q(\sigma), p(\sigma))_{\sigma \in \Sigma}$ satisfying the following conditions:*

(1) $\sum_{h=1}^{H} \theta^h(\sigma) = 1$ *for all $\sigma \in \Sigma$.*
(2) $\sum_{h=1}^{H} \vartheta^h(\sigma) = 0$ *for all $\sigma \in \Sigma$.*

(3) For each agent h :

$$(\theta^h, \vartheta^h, c^h) \in \arg \max_{\theta, c} U_h(c)$$

$$c(\sigma) = e^h(\sigma) + \theta(\sigma^*) \cdot (d(\sigma) + q(\sigma))$$
$$+\vartheta(\sigma^*) - \theta(\sigma)q(\sigma) - \vartheta(\sigma)p(\sigma)$$
$$\theta(\sigma) \geq 0, \vartheta(\sigma) \geq -B.$$

3 Computational procedure

In all examples below we assume that there exists a recursive financial market equilibrium for economies with one or two assets where asset prices and the agents' portfolio choices are functions of the last-period portfolio and the current state alone[1]. Throughout this paper we assume that there are two agents and we use the computational procedures developed in Judd et al. (1999, 2000) to approximate these equilibria numerically.

We briefly describe these procedures for the case of equity and borrowing. In order to examine the equilibrium behavior of our model we use a spline collocation algorithm to approximate portfolio policy functions $f = ((f_1^b, f_1^s), (f_2^b, f_2^s), \ldots, (f_S^b, f_S^s))$ and price functions $g = ((g_1^b, g_1^s), (g_2^b, g_2^s), \ldots, (g_S^b, g_S^s))$. We approximate $f(y, \theta^b, \theta^s) = (f^b(y, \theta^b, \theta^s), f^s(y, \theta^b, \theta^s))$ parametrically by functions

$$\hat{f}^b(y, \theta^b, \theta^s) = \sum_{i=1}^{n} \sum_{j=1}^{n} a_{ij}^{by} B_i(\theta^b) B_j(\theta^s)$$

and

$$\hat{f}^s(y, \theta^b, \theta^s) = \sum_{i=1}^{n} \sum_{j=1}^{n} a_{ij}^{sy} B_i(\theta^b) B_j(\theta^s)$$

and $g(y, \theta^b, \theta^s) = (g^b(y, \theta^b, \theta^s), g^s(y, \theta^b, \theta^s))$ by

$$\hat{g}^b(y, \theta^b, \theta^s) = \sum_{i=1}^{n} \sum_{j=1}^{n} b_{ij}^{by} B_i(\theta^b) B_j(\theta^s)$$

and

$$\hat{g}^s(y, \theta^b, \theta^s) = \sum_{i=1}^{n} \sum_{j=1}^{n} b_{ij}^{sy} B_i(\theta^b) B_j(\theta^s)$$

where n^2 is the number of terms used. The functions B_k are B-splines of order 4. These functions yield a linearly independent basis for one-dimensional cubic splines. For our approximations we use cubic splines instead of orthogonal polynomials since our price functions sometimes exhibit high curvature. For an overview on B-splines see Judd (1998) and de Boor (1978).

---

[1] Although with finitely many agents recursive equilibria of this type do not always exist (see Kubler and Schmedders, 2002) this is a standard assumption made in the literature.

Under our assumptions on short-sale constraints, the agents' Kuhn-Tucker conditions are necessary and sufficient for optimality. Using a simple change of variables (see Judd et al., 2000, for details) we can transform the Kuhn-Tucker conditions into a system of equations. A solution to these and the market clearing equations is an equilibrium. Under the assumption that a recursive equilibrium exists, the true policy functions f and g thus have to satisfy a functional equation $F(f, g, \kappa) = 0$, for a function κ which determines the Kuhn-Tucker multiplier as a function of the state. We hope to find approximating functions $\hat{f}, \hat{g}, \hat{\kappa}$ such that $\|F(\hat{f}, \hat{g}, \hat{\kappa})\|_\infty < \epsilon$ for some pre-specified $\epsilon > 0$.

Since we are not interested in the multipliers κ we do not attempt to approximate them with a function on our state space. To compute the coefficients $(a_{ij}^{by}), (a_{ij}^{sy}), (b_{ij}^{by})$ and (b_{ij}^{sy}) we select a finite collocation grid G of as many meshpoints $(\theta_i^b, \theta_j^s)_{i,j=1,\dots,n}$, as we have unknown coefficients for the approximating functions. We obtain 4 equations for each point and each income state. Notice that the problem has been transformed from finding functions f and g solving the Euler equations over the continuous state space to finding a zero of a system of $4 \cdot S \cdot n^2$ nonlinear equations that has the real coefficients $a = ((a_{ij}^{by}), (a_{ij}^{sy}))$ and $b = ((b_{ij}^{by}), (b_{ij}^{sy}))$ as unknowns.

The described system of equations can be very large - moreover if we leave $\hat{\kappa}$ unspecified we cannot solve it directly. To solve it, we follow a simple Gauss-Jacobi approach which can be motivated by the following economic intuition. It seems reasonable to hope that finite-horizon models with a very large number of time periods are a good approximation to the infinite-horizon model under consideration, although this fact has not been proven rigorously. A natural approach for computing portfolio holdings and prices in a finite-horizon model is backward induction. Therefore we compute the approximate policy functions $\hat{f}$ and $\hat{g}$ through an iterative process starting with some initial guess f_0 and g_0. In each iteration $k = 1, 2, \dots$, the algorithm solves the Euler equations for all collocation points $\theta = (\theta^b, \theta^s) \in G$ and income states $y \in Y$ by computing current portfolio decisions θ and corresponding asset prices q given functions $\hat{f}_{k-1}$ and $\hat{g}_{k-1}$ governing the policy process in the subsequent period. The coefficients a and b for the new functions $\hat{f}_k$ and $\hat{g}_k$, respectively, are then determined through interpolation. The algorithm terminates if

$$\max_{(\theta^b, \theta^s) \in G, y \in Y} \{|\hat{f}_k(y, \theta) - \hat{f}_{k-1}(y, \theta)|, |\hat{g}_k(y, \theta) - \hat{g}_{k-1}(y, \theta)|\} < \epsilon.$$

The key to the algorithm is then to solve the Euler equations at each collocation point effectively. The short-sale constraints often lead to an ill-conditioned system and frequently a homotopy method has to be used to find a solution (see Eaves and Schmedders, 1999, for an overview of homotopy methods).

Unfortunately there is no formal procedure that assures that the computed welfares are close to the actual equilibrium welfares. We choose the number of spline-nodes in such a way that the maximum relative error in the agents' Euler equations lies consistently below 10^{-8}. Since there are no formal techniques which can be used to evaluate how close the computed equilibrium price is to the true equilibrium price, we choose a small Euler equation error to minimize the influence of numerical error.

The complete market cases use the Negishi style method described in Judd et al. (2001). The Negishi method directly produces state contingent prices and consumption and asset holdings.

4 Welfare implications of market incompleteness

We assume specific values for the critical parameters of tastes and endowment processes and compute equilibrium for five economies: autarchy, a single asset (equity or bond), two assets (one stock and one bond), and complete asset markets.

4.1 Parameter specifications

We examine a few cases which are empirically plausible. In all cases we have in mind yearly frequency and assume that $\beta = 0.95$. We examine three kinds of examples.

Case 1: No idiosyncratic risk, heterogeneous risk aversions. One function of capital markets is to allocate aggregate risk. If asset markets are not complete, then first-best allocations equating state-contingent marginal rates of substitution cannot be implemented. In autarchy each agent holds a fixed, unchangeable, fraction of "equity."Adding a stock market allows for some reallocation of risk, but only to a limited extent. If the only asset is equity, then then equity is the only way to augment future consumption and equity trading must serve both risk-allocation and savings functions. Adding a bond would disentangle savings and risk allocation, allowing the equity market to allocate risk and the bond market to adjust the distribution of consumption. However, the bond and stock will generally not span the space of contingencies, and other securities, such as options, would allow agents to trade nonlinear functions of equity returns. We set

$$e^1 = e^2 = (1.5, 1.3, 1.5, 1.3, 1.5, 1.3)$$
$$d = (1, 1, 1.2, 1.2, 1.3, 1.3)$$

We assume that shocks to aggregate endowments are persistent (the probability of staying in the same income state is 0.8) while shocks to equity payoffs are i.i.d. This leads to the following transition probabilities

$$\pi(y|y') = \begin{cases} \frac{0.8}{3} & \text{if } y, y' \in \{2, 4, 6\} \text{ or } y, y' \in \{1, 3, 5\} \\ \frac{0.2}{3} & \text{otherwise.} \end{cases}$$

We consider the following values for (γ^1, γ^2): $(0.5, 2)$, $(1, 2)$, $(1, 3)$, and $(1, 4)$; and for initial portfolio holdings $\theta^1_{-1} = 0.2, 0.5, 0.8$. In these cases, we assume that no one can short the bond beyond $B = 1.5$.

Case 2: Common tastes, independent idiosyncratic risks. We next study an example where there is aggregate risk but all agents have the same tastes so all trade is due to idiosyncratic risks. This is the case often examined in the international finance literature on risk-sharing. When there is only equity the stock serves as a buffer stock of wealth to smooth income shocks as well as performing aggregate risk sharing. When there is both a stock and a bond, the bond can serve as buffer stock for individuals. Perfect markets go further and allow sharing of all risks. We take $\gamma^1 = \gamma^2 = 1.5$, assume endowments are

$$e^1 = \left(1, 1, 1.3, 1.3, 1, 1, 1.3, 1.3 \right)$$
$$e^2 = \left(1, 1.3, 1, 1.3, 1, 1.3, 1, 1.3 \right)$$

and assume that the state-contingent dividends are

$$d = \left(1, 1, 1, 1, 1.2, 1.2, 1.2, 1.2 \right).$$

As before, shocks to incomes are persistent, shocks to equity are i.i.d. Transition probabilities are given by

$$\mathbf{P} = \begin{pmatrix} 0.32 & 0.08 & 0.08 & 0.02 & 0.32 & 0.08 & 0.08 & 0.02 \\ 0.08 & 0.32 & 0.02 & 0.08 & 0.08 & 0.32 & 0.02 & 0.08 \\ 0.08 & 0.02 & 0.32 & 0.08 & 0.08 & 0.02 & 0.32 & 0.08 \\ 0.02 & 0.08 & 0.08 & 0.32 & 0.02 & 0.08 & 0.08 & 0.32 \\ 0.32 & 0.08 & 0.08 & 0.02 & 0.32 & 0.08 & 0.08 & 0.02 \\ 0.08 & 0.32 & 0.02 & 0.08 & 0.08 & 0.32 & 0.02 & 0.08 \\ 0.08 & 0.02 & 0.32 & 0.08 & 0.08 & 0.02 & 0.32 & 0.08 \\ 0.02 & 0.08 & 0.08 & 0.32 & 0.02 & 0.08 & 0.08 & 0.32 \end{pmatrix}.$$

Asset prices and consumption will depend on the initial wealth, particularly in the complete market case where the initial wealth determines the initial and permanent distribution of consumption. We consider three different initial distributions of equity; specifically, we consider $\theta^1_{-1} = 0.2, 0.5, 0.8$. Again, we assume that no one can short the bond beyond $B = 1.5$.

Case 3: Common tastes, negatively correlated idiosyncratic risks. This is the case considered in many macroeconomic models (e.g. Heaton and Lucas, 1996). We make the idiosyncratic risks perfectly negatively correlated so that they make no contribution to aggregate risk. There are 4 exogenous states. Aggregate endowments are given by $\bar{e} = (9.9, 10.5, 9.9, 10.5)$. We assume that the payoffs of the tree are $d(y) = 0.3 \cdot \bar{e}$, representing a thirty per cent share of capital in national income. Individual endowments are

$$e^1 = (1.386, 2.205, 5.544, 5.145)$$
$$e^2 = 0.7\bar{e} - e^1.$$

The transition probabilities are given by

$$\pi(y|y') = \begin{cases} 0.4 \text{ for } y, y' \in \{1, 2\} \text{ or } y, y' \in \{3, 4\} \\ 0.1 \text{ otherwise} \end{cases}$$

We assume that $\gamma^1 = \gamma^2 = 1.5$ and consider initial equity holdings for agent 1 equal to $\theta^1_{-1} = 0.2, 0.5, 0.8$. Again, we assume a shorting constraint of $B = 1.5$.

4.2 Welfare comparisons

For any initial distribution of wealth, we can compute the expected utility for each type under each asset market structure. We use these results to evaluate the welfare effects of changing asset markets by adding (or dropping) an asset given the distribution of wealth. For example, suppose that type 1 agents own one-third of all equity and are in state 1 of their idiosyncratic risk and that type 2 agents own two-thirds of equity and are in state 1 of their idiosyncratic risk, and that we want to know the value of adding bond market trading. We compute the expected utility of an individual investor under each market structure given this initial state, and take the difference across market structures as a measure of the impact to his welfare of a change in the asset market structure from this initial state. We argue that these results, conditional on the initial distribution of wealth and income state, are the most reasonable to examine in our dynamic context since if such a choice would ever present itself the decision would occur in some such distributional context. Another measure which may be considered is the expected long-run utility (defined as the expectation of utility with respect to the ergodic measure of wealth and income). This measure would ignore the transition process and implicitly assume that decision makers ignore the discounting in their utility functions.

We express all utility changes in terms of consumption equivalents. That is, if $U_i(y, \theta_-)$ is the expected utility to an agent in state (y, θ_-), the consumption equivalent is c^* such that $u(c^*) = (1 - \beta)U_i(y, \theta_-)$ since a constant consumption of c^* produces the same present value of utility. We then express the utility impact of adding an asset as the relative change in the consumption equivalent. More precisely, for an initial state (y, θ_-) we compute the consumption equivalent of autarchy, c_A^*, the consumption equivalent of the equity-only economy, c_S^*, the consumption equivalent of the bond-only economy, c_B^*, the consumption equivalent of the equity-bond economy, c_{SB}^*, and the consumption equivalent of the perfect asset market economy, c_P^*. The relative welfare gains of financial innovation in a state (y, θ_-) are represented by ratios

$$\frac{c_S^* - c_A^*}{c_A^*}, \frac{c_B^* - c_A^*}{c_A^*}, \frac{c_{SB}^* - c_A^*}{c_A^*}, \frac{c_P^* - c_A^*}{c_A^*}.$$

A natural hypothesis is that there are diminishing returns to new assets, that is, we expect to find, on average over the agents,

$$\frac{c_S^* - c_A^*}{c_A^*} > \frac{c_{SB}^* - c_S^*}{c_A^*} > \frac{c_P^* - c_{SB}^*}{c_A^*} > 0.$$

One expects this diminishing returns to hold also for individuals, but we also will look for cases where a new asset hurts someone. Hart (1975) showed that this is possible. Since we have an infinite number of commodities, it would not surprise us to find such utility reversals here also.

This comparison conveniently breaks up the set of possible assets into three groups. Trade in equity represents trade in actual physical assets, such as "Lucas trees" which produce dividends. Bonds are zero-net supply contracts to deliver a fixed amount of resources at a future date. The stock is a linear asset in the sense that its payoff equals the dividends, a key component of the aggregate state. The remaining assets are ones which are either not linearly related to dividends, such as options, or their payoffs depend on idiosyncratic risks, such as insurance. This description will help us understand the results below.

Tables 1–4 examine the economy in Case 1 above, that is, agents have different utility functions and different wealth, but have no idiosyncratic income risk. Table 1 examines the case where agent 1, the more risk averse type, has smaller initial wealth, Table 2 examines the case where initial wealth is equal, and Table 3 examines the case where agent 1 has greater initial wealth. The first panel of Tables 1, 2, and 3 displays the welfare gains from going from autarchy to various asset market structures. The row labeled "stock" presents the welfare changes for each agent type if trading in stock is added. The other rows report similar changes: "bond" refers to going from autarchy to trading a bond, "2 assets" refers to going from autarchy to trading in a stock and a bond, and "A-D" refers to going from autarchy to complete Arrow-Debreu markets. Each pair of columns in these tables refers to the distribution of welfare changes corresponding to the utility functions. For example, in the first panel under the heading "Welfare Gains Relative to Autarchy" the pair of columns under "(0.5,2)" in Table 1 tells us how each type fared under various market structures when $\gamma_1 = 0.05$, and $\gamma_2 = 2$; more specifically, agent 1 (2) enjoys a 0.056% (0.11%) gain in consumption equivalent with the introduction of equity and would enjoy a 0.062% (0.121%) gain in welfare by going from autarchy to complete Arrow-Debreu markets. The second panel in each table under the heading "Incremental Gains" decomposes the gain from going to complete markets into the incremental gains from going first from autarchy to one asset, then two assets, then complete markets. Here the pair of columns under (1,2) in Table 1 tells us that for agent 1 the introduction of equity produces a welfare increase equal to 93.8% of the welfare gain from going to complete markets. Adding a bond adds only 3.3% more of the potential gain, and completion of markets (which, in this case, is essentially the introduction of a set of options) produces the remaining 2.9% of the utility gain from complete markets.

Examination of Tables 1, 2, and 3 illustrate a number of points. First, neither type of agent gains much from asset trading in general. This result is undoubtedly tied to the fact that the equity premium in this example is small. Since we (like most other quantitative models of asset trading) do not model risk premia well in this model, we will focus on relative welfare changes.

Second, going from autarchy to one asset, two assets, or complete markets produces welfare benefits for both types of investors. This, however, is obvious since an investor can always keep his endowment and he can only gain from moving away from autarchy. Third, we do see diminishing welfare returns when we have either a stock or bond only and add the other asset. Both agents gain from the addition of the second asset in all of our examples. Fourth, the distribution of gains appears to depend on the distribution of the risk aversion parameter. If both agents have relative risk aversion in excess of 1 and there is substantial difference in risk aversion then the introduction of the second asset benefits the more risk tolerant agent more than the more risk averse agent, particularly when the former is more wealthy. In these cases it appears that the relatively risk tolerant agent benefits most from his

Table 1. Welfare effects for Case 1, $\theta^1_{-1} = 0.2$

(γ^1, γ^2)	(0.5,2)		(1,2)		(1,3)		(1,4)	
	Welfare gains relative to autarchy (%)							
stock	0.0559	0.110	0.913	0.0451	1.65	0.122	2.28	0.224
bond	0.0562	0.110	0.907	0.0450	1.64	0.122	2.28	0.225
2 assets	0.0675	0.113	0.945	0.0484	1.85	0.130	2.70	0.237
A-D	0.0615	0.121	0.973	0.0480	1.80	0.132	2.51	0.246
	Incremental gains (%)							
aut → stock	90.9	91.1	93.8	93.9	91.8	92.0	90.8	91.1
aut → bond	91.4	90.7	93.2	93.7	91.4	92.2	90.7	91.5
stock → 2 assets	18.9	2.2	3.3	7.0	11.0	6.3	16.9	5.3
bond → 2 assets	18.4	2.6	3.9	7.2	11.4	6.1	17.0	4.9
2 assets → A-D	−9.8	6.7	2.9	−0.9	−2.8	1.7	−7.7	3.6

Table 2. Welfare effects for Case 1, $\theta^1_{-1} = 0.5$

(γ^1, γ^2)	(0.5,2)		(1,2)		(1,3)		(1,4)	
	Welfare gains relative to autarchy (%)							
stock	0.0263	0.105	0.325	0.0324	0.738	0.110	1.07	0.212
bond	0.0276	0.104	0.329	0.0329	0.748	0.112	1.10	0.215
2 assets	0.0353	0.108	0.362	0.0370	0.912	0.120	1.42	0.227
A-D	0.0305	0.121	0.378	0.0376	0.855	0.128	1.24	0.246
	Incremental gains (%)							
aut → stock	86.2	86.2	86.2	86.2	86.2	86.3	86.3	86.3
aut → bond	90.3	85.6	87.2	87.5	87.4	87.5	88.8	87.2
stock → 2 assets	29.5	2.5	9.7	12.2	20.4	7.7	28.1	6.1
bond → 2 assets	25.4	3.1	8.7	10.9	19.2	6.5	25.6	5.2
2 assets → A-D	−15.7	11.3	4.1	1.6	−6.6	6.0	−14.4	7.6

willingness to absorb risk which the risk averse agents would have to absorb in the absence of a bond market. However, this pattern is reversed when one of the agents has risk aversion less than 1, as in the columns under $(\gamma^1, \gamma^2) = (0.5, 2)$. Here the agent absorbing the risk gains less from new assets. Fifth, one agent (usually the less risk averse type) may lose when we go from a stock and bond market to complete markets; we did not find any such perverse changes when we go from only a stock or only a bond to the bond and stock economy. More generally, these examples show that the welfare reversals noted in Hart (1975), Elul (1995), Cass and Citanna (1998) and elsewhere may arise in simple and natural contexts. Sixth, our examples show that the addition of derivative assets like options have nontrivial economic impact.

Table 4 displays the impact of financial innovation on stock prices for the three initial endowments examined in Tables 1, 2, and 3. The tables shows that adding a bond or completing the market will have negligible effect on asset prices; in fact,

Table 3. Welfare effects for Case 1, $\theta^1_{-1} = 0.8$

	Welfare gains relative to autarchy (%)							
(γ^1,γ^2)	(0.5,2)		(1,2)		(1,3)		(1,4)	
stock	0.0189	0.153	0.375	0.0758	0.597	0.181	0.766	0.308
bond	0.0220	0.149	0.386	0.0764	0.647	0.181	0.887	0.306
2 assets	0.0259	0.156	0.392	0.0829	0.737	0.193	1.038	0.327
A-D	0.0226	0.183	0.445	0.0901	0.713	0.216	0.913	0.369
	Incremental gains (%)							
aut $\rightarrow$ stock	83.6	83.5	84.3	84.2	83.8	83.6	83.9	83.6
aut $\rightarrow$ bond	97.3	81.4	86.7	84.8	90.8	83.6	97.2	82.9
stock $\rightarrow$ 2 assets	30.9	2.0	3.9	7.9	19.7	5.8	29.9	5.0
bond $\rightarrow$ 2 assets	17.2	4.1	1.5	7.3	12.7	5.8	16.6	5.7
2 assets $\rightarrow$ A-D	-14.5	14.5	11.8	7.9	-3.5	10.6	-13.8	11.4

Table 4. Stock price changes (%) for Case 1

	(γ^1,γ^2)	(0.5,2)	(1,2)	(1,3)	(1,4)
$\theta^1_{-1} = 0.2$	stock $\rightarrow$ 2 assets	-0.00434	-0.00118	-0.0194	-0.0169
	stock $\rightarrow$ A-D	0.00035	-0.00188	-0.00623	-0.0159
$\theta^1_{-1} = 0.5$	stock $\rightarrow$ 2 assets	-0.00386	-0.00156	-0.0179	-0.0199
	stock $\rightarrow$ A-D	-0.00174	-0.00183	-0.00923	-0.0212
$\theta^1_{-1} = 0.8$	stock $\rightarrow$ 2 assets	-0.00349	-0.00226	-0.0184	-0.0192
	stock $\rightarrow$ A-D	-0.00686	-0.00894	-0.0207	-0.0359

these changes are so small that they are probably within the numerical error of our algorithm; therefore, we conclude nothing other than that the effect is small.

Table 5 examines Case 2, where agents have the same tastes but individual risks which are partially correlated with the stock's dividend. Here the asset markets are used to facilitate sharing of idiosyncratic risk as well as allocation of aggregate risk. We report the utility gain to agent one when his initial endowment is $\theta^1_{-1} \in \{0.2, 0.5, 0.8\}$

This is adequate for our purposes since the agents have the same utility functions; therefore, the utility change to agent 2 when $\theta^1_{-1} = 0.2$ equals the utility change to agent 1 when $\theta^1_{-1} = 0.8$. As in Case 1, we find that a single asset, equity or bond, alone provides the majority of the total potential benefits. The addition of a second asset has a small impact on welfare, but the addition of the remaining assets will produce roughly 20% of the total gain from going to complete markets. These missing assets have an insurance quality, paying off to an agent in states where he has suffered a bad idiosyncratic shock. Here we see that insurance-type contracts are much more beneficial.

Table 6 examines Case 3 with common tastes but some aggregate risk together with perfectly negatively correlated idiosyncratic income shocks. Here we find much larger absolute gains to welfare from introducing assets. Introducing equity

Table 5. Welfare effects for Case 2

θ^1_{-1}	Welfare gains relative to autarchy (%)		
	0.2	0.5	0.8
stock	0.355	0.211	0.161
bond	0.343	0.211	0.168
2 assets	0.356	0.211	0.161
A-D	0.449	0.270	0.205
	Incremental gains (%)		
aut $\to$ stock	79.0	77.9	78.5
aut $\to$ bond	76.5	78.2	82.0
stock $\to$ 2 assets	0.3	+0.0	0.1
bond $\to$ 2 assets	2.8	−0.3	−3.4
2 assets $\to$ A-D	20.7	22.1	21.4

Table 6. Welfare effects for Case 3

θ^1_{-1}	Welfare gains relative to autarchy (%)					
	0.2		0.5		0.8	
stock	15.4	5.81	9.84	9.02	6.65	14.5
bond	7.1	3.64	5.26	4.17	4.66	5.7
2 assets	15.5	5.81	9.86	9.03	6.65	14.6
A-D	18.6	7.52	11.4	10.9	7.82	17.5
	Incremental gains (%)					
aut $\to$ stock	82.9	77.3	86.2	82.8	85.1	83.2
aut $\to$ bond	38.2	48.4	46.0	38.3	59.5	32.8
stock $\to$ 2 assets	0.6	0.1	0.1	0.1	0.0	0.3
bond $\to$ 2 assets	45.3	29.0	40.3	44.6	25.6	50.7
2 assets $\to$ A-D	16.5	22.6	13.7	17.1	14.9	16.5

alone has the greatest impact and is better than having just a bond market. Adding the bond market to an existing stock market has little impact, but final completion of risk markets produces a nontrivial gain. The new assets that complete the market are effectively insurance contracts allowing agents to neutralize the negatively correlated idiosyncratic risks. Since the idiosyncratic risks make no contribution to aggregate risk, the social cost of risksharing is small, and the gain from eliminating the idiosyncratic risks is large.

Table 7 displays the price effect on stocks from introducing a bond and from completing the market. We see that adding a bond to the stock market has no impact on prices, but completing the markets results in significant drop in price. Apparently the presence of the new options make the stock market less attractive as a way to allocate risks, driving down the price.

While we have examined only a few cases, they do make several points. First, welfare reversals are possible in standard contexts. Second, the addition of equity

Table 7. Stock price changes (%) for Case 3

θ^1_{-1}	0.2	0.5	0.8
stock $\rightarrow$ 2 assets	-0.20	-0.06	-0.12
stock $\rightarrow$ A-D	-10.04	-5.82	-7.22

alone produces the greatest social benefit, no matter what the motivating reason for asset trading, and sometimes a single bond market can also achieve substantial welfare gains over autarchy. Third, in the examples with idiosyncratic risk, the bond market had little additional value compared to the nonlinear assets which complete the span beyond stocks and bonds. This indicates, for example, that simple borrowing institutions have little social value without features such as forgiveness of debt in particularly bad states.

5 Volume in asset markets

We next compute the volume implications of alternative asset market structures. We examine trading volume in complete markets and incomplete asset markets.

5.1 Dynamically complete asset markets

Judd et al. (2001) examined asset trading volume in complete markets and found that asset trading volume was generically zero for long-lived assets such as equity. The intuition is that with complete asset markets, common beliefs, and intertemporally separable utility one can arrange a portfolio of assets which dividends produce the desired state-contingent consumption process. This result was surprising. Standard intuition says that investors should continuously trade securities in response to new information about expected future returns and their riskiness. A critical feature of their result is the large set of securities assumed. For generic dividend processes involving S states, we need S assets to span the security space. This genericity implies that the transition matrix among dividend states has no zero entries; that is, for each pair of states s and s', there is a positive probability of going from state s to state s'.

It is also common to assume that real markets contain fewer assets than the number of possible states but are still completed through dynamic trading. See Kreps, (1982) for an exposition of this approach. We now examine some examples where trading in a few assets can implement Pareto efficient dynamic consumption processes. Even though the examples will not be robust to changes in the transition matrix, they may correspond better with standard intuition than the generic analysis in Judd et al. (2001). Both examples assume $H = 2$ agents with CRRA Bernoulli-functions, where type one agents have relative risk aversion of $\gamma_1 = 0.5$ and type two agents have $\gamma_2 = 4$. The common discount factor is $\beta = 0.95$, implying a time period of about a year.

Our first example assumes $S = 5$ exogenous states. The first asset is a stock with dividends $d = (1, 1.5, 2, 2.5, 3)$. The Markov transition matrix is

$$\Pi = \begin{bmatrix} 0.5 & 0.5 & 0 & 0 & 0 \\ 0.5 & 0 & 0.5 & 0 & 0 \\ 0 & 0.5 & 0 & 0.5 & 0 \\ 0 & 0 & 0.5 & 0 & 0.5 \\ 0 & 0 & 0 & 0.5 & 0.5 \end{bmatrix}.$$

This example approximates a dividend following a random walk except for reflecting barriers when the dividend is at its highest and lowest possible values.

The stock is in unit net supply and initial endowments are equal, $\theta^1_{-1} = \theta^2_{-1} = 0.5$. The second asset is a riskless short-lived bond paying 1 unit of the consumption good in every state. Both agents have zero endowments of the bond, and they have no personal endowment of the consumption good.

A Negishi-style algorithm can be used to compute equilibrium (see Judd et al. 2001) since the assets are able to implement the equilibrium consumption process through dynamic trading. This will be true here with our stock and bond since there are only two possible future states at any time, and the stock and bond can never have identical returns. The Negishi algorithm tells us that the state-contingent consumption allocations are[2]

$$c^1 = (0.191, 0.572, 1.004, 1.457, 1.920)$$

$$c^2 = (0.809, 0.928, 0.996, 1.043, 1.080)$$

As expected, type one agents have more volatile consumption than the more risk averse type two agents. When dividends are low, type one agents consume much less than type two agents, but this is reversed in the high dividend states. This kind of reversal is expected since each type begins at time zero owning exactly half of all wealth. The state-contingent asset prices of the stock, q, and the bond, p, are

$$q = (17.49, 30.23, 39.93, 48.10, 55.27)$$
$$p = (0.749, 1.181, 1.024, 0.986, 1.020).$$

Let the end-of-period holding of stocks (bonds) by agent h in state s be denoted by θ^h_s (ϑ^h_s). The consumption patterns, the asset prices, and the state-contingent budget constraints imply that the state-contingent end-of-period portfolio of type one agents is

$$\theta^1 = (0.464, 0.502, 0.578, 0.604, 0.600)$$
$$\vartheta^1 = (-1.090, -1.797, -4.700, -6.085, -5.846)$$

The individual portfolios now vary across states. This does not contradict Judd et al. (2001) since their no-trade result was a generic result. The fact that the payoff

---

[2] None of the results are calibrated to match observed price processes. We use these examples since they strongly illustrate the key points.

matrix Π in this example has many zero entries is a nongeneric property and is obviously critical here. The fact that each state has only two possible successor states implies that it is possible for two assets to span the space of returns in the short-run, and that trading in those two assets may be able to span all possible consumption plans. In this example, the asset prices and returns all line up so that this is true.

The pattern of trading is interesting, illustrating a variety of factors. When dividends are high, the stock price is high, and the risk-tolerant type one agents hold the majority of the equity and are short in the bond market. Type one agents have high consumption and high net wealth in those states. This large holding of equity is necessary to finance type one consumption. As dividends fall type one agents unload their equity in order to finance consumption and pay off their debt. At the lowest dividend state, type two agents own the majority of the equity despite the fact that they are much more risk averse.

We should also note that this example is not a knife-edge case. In particular, a small change in any utility parameter or dividend parameter in this example will also produce a determinate equilibrium with trading. This holds since the critical matrices are nonsingular. The key, and nongeneric, assumption is the sparseness of Π.

The previous example had a simple random walk character with reflecting barriers. The mean and variance of returns changes over time, making it difficult to explain the asset movements. One simple conjecture is that changes in variance will induce trade with the more risk averse agent selling some of his holdings to the more risk tolerant agent. The next example examines this conjecture.

We assume that same utility functions for type 1 and 2 trades, but change the transition matrix and the stock dividends in the previous example to create an example where the mean future dividend is fixed but the variance changes. We assume the Markov transition matrix

$$\Pi = \begin{bmatrix} 0 & 0.5 & 0.5 & 0 \\ 0.5 & 0 & 0 & 0.5 \\ 0 & 0.5 & 0.5 & 0 \\ 0.5 & 0 & 0 & 0.5 \end{bmatrix}.$$

and assume that the new stock dividend process is $d = (1, 2, 4, 5)$. In this example the time t expectation of the time $t + 1$ dividend is always 3, but the variance of this next-period dividend can change across states. In states 1 and 3 the dividend in the next period is either 2 or 4, but is either 1 or 5 in states 2 and 4. If trading was induced by changes in variance, then we might expect the more risk tolerant type 1 agents to buy stock from type 2 agents after a transition from state 1 to 2 or from state 3 to 2.

The asset prices of the stock and the bond are

$$q = (17.77, 44.34, 75.87, 88.55)$$
$$p = (0.3030, 1.4146, 1.2935, 2.8248),$$

the portfolio policy of agent 1 is

$$\theta^1 = (0.5527, 0.5315, 0.5527, 0.5315)$$
$$\vartheta^1 = (-1.9462, -0.5912, -1.9462, -0.5912),$$

and the state-contingent consumption is

$$c^1 = (0.1527, 0.9370, 2.7821, 3.7363)$$
$$c^2 = (0.8473, 1.0630, 1.2179, 1.2637).$$

The volatility of asset prices changes substantially as states change. In state one, the asset price is low, but the price in the next period is either (approximately) 44 or 76, but in state two, the price is 44 with future price equally either 18 or 89. Similarly, expected price volatility is much higher in state four than in state two. Now the asset positions are identical in states 1 and 3 (2 and 4) since states 1 and 3 (2 and 4) have the same variance in returns.

The asset holding patterns are somewhat puzzling. The relatively risk-tolerant type one investors hold less equity in the riskier states 2 and 4. When the dividend process moves from state 1 to state 2, dividend and price variance increases and type one investors sell shares to type two agents who are more risk averse, and they reduce their leverage. Consider also the situation in state two. If the next period's state is (apparently less risky) state one, then type one agents increase their equity holdings. This seems odd from a portfolio rebalancing perspective. However, it is not surprising when we examine the consumption pattern. When the dividend process moves from state two to state one, type one consumption plummets from 0.9370 to 0.1527, leading type one agents to save and increase their asset holdings.

These two examples highlight some important points. First, portfolio trading patterns do not follow any simple rules, particularly those implied by portfolio demand theory. Equilibrium volume is strongly affected by equilibrium conditions. The equilibrium portfolios are determined by a variety of factors with riskiness of the assets being only one of them. As in Judd et al. (2001), equilibrium restrictions on prices substantially affect actual price and volume. Second, even in the cases where trading does occur and consumption volatility is large, volume is small in magnitude. Our examples encompass a wide variation in dividends and asset prices, and have agents with substantially different risk aversion. Still, the volume of trade is rather small. When one combines these examples with the generic no-trade result of Judd et al. (2001), it appears that trade in equity is small whenever markets are dynamically complete. We next examine volume with incomplete asset markets.

5.2 Incomplete asset markets

Trading volume will be nonzero when asset markets are incomplete. We examined the incomplete asset market models described in Section 2. We found that the ergodic measure of asset holding had large support; therefore, it is not necessary to consider the effects of initial wealth since they will diminish as equilibrium evolves into the ergodic distribution. This allows us to compute only the long-run expected volume, and compare it across asset market structures.

Table 8. Volume and returns for the case of two assets

	Stock volume (%)	Bond volume (%)	Bond market size (%)	Stock return (%)	Bond return (%)
Case 1 a	0.22	3.71	0.94	5.60	5.14
Case 1 b	0.31	0.0	1.0	5.78	5.10
Case 1 c	0.62	0.0	1.13	5.91	5.01
Case 2	10.6	0.4	0.41	5.7	5.1
Case 3	17.3	2.5	0.37	5.02	4.83

Table 9. Volume and returns for one asset only

	Stock volume (%)	Stock return (%)
Case 1 a	0.13	5.27
Case 1 b	0.19	5.72
Case 1 d	0.28	5.89
Case 2	3.5	5.83
Case 3	9.9	5.99

We create several indices to measure volume. First, volume in equity is defined to be the change in equity holding by type 1, where total stock is normalized to have size 1. Second, the size of the bond market is measured by the ratio between interest payments to aggregate consumption. This is a more proper measure of the size of debt market than, say, relative value since it measures the burden of debt on borrowers and the flow of debt service payments. Third, trade volume in bonds is defined by the percentage change in end-of-period bond holdings. Table 8 reports volume results for our various cases. Table 8 first examines three example from Case 1, where Case 1a, 1b, and 1c refers to the aversion combinations $(\gamma_1, \gamma_2) = (1, 2)$, $(1, 3)$,and $(1, 5)$. We see that there is little trade in the Case 1 examples where trade is motivated by heterogeneous risk aversion, much more stock trading in Case 2 with i.i.d. idiosyncratic risk where trade facilitates buffer stock savings, and even more trade in Case 3 where idiosyncratic risks are negatively correlated. Bond market size is small in all cases, and trading shuts down in some cases because traders tend to stay at their borrowing limits and trade only their equity.

Table 9 reports volume results for equity-only asset markets. Again we find small amounts of trade. We can compare Tables 8 and 9 and determine the relation between asset market completeness and volume. The interesting finding is that stock market volume is less in the equity-only case than in the two-asset market. Apparently the presence of bonds leads to more portfolio rebalancing by all agents. The total relation between trading volume and market structure must be non-monotonic since, as shown in Judd et al. (2001), there is no volume in both the autarchy and perfectly complete cases.

6 Conclusion

This paper examines the impact of financial innovation on asset trading volume and welfare in a simple dynamic general equilibrium model. We find that the benefits of financial innovation vary greatly depending on the nature of uncertainty. Asset trading has relatively small benefits in allocating aggregate shocks efficiently, but is more valuable when agents experience idiosyncratic shocks to non-asset income. We also find that trading volume is small even when asset markets are incomplete. The results of this study are tentative since it suffers, as do most asset market analyses, from an inability to calibrate the observed price of risk. While many of the results follow basic qualitative intuitions, they also show that the basic rational expectations models of asset markets have difficulty in quantitatively matching observations about prices and volumes in real markets.

References

1. Cass, D., Citanna, A: Pareto improving financial innovation in incomplete markets. Economic Theory **11**, 467–494 (1998)
2. deBoor, C.: A practical guide to splines. Berlin Heidelberg New York: Springer 1978
3. Eaves, B. C., Schmedders K.: General equilibrium models and homotopy methods. Journal of Economic Dynamics and Control **23**, 1249–1279 (1999)
4. Elul, R.: Welfare effects of financial innovation in incomplete markets economies with several consumption goods. Journal of Economic Theory **65**, 43–78 (1995)
5. Hart, O.: On the optimality of equilibrium when the market structure is incomplete. Journal of Economic Theory **11**, 418–443 (1975)
6. Judd, K. L.: Numerical methods in economics. Cambridge: MIT Press 1998
7. Judd, K. L., Kubler, F., Schmedders, K.: A solution method for incomplete asset markets with heterogeneous agents. Working paper (1999)
8. Judd, K. L., Kubler, F., Schmedders, K.: Computing equilibria in infinite horizon finance economies: the case of one asset. Journal of Economic Dynamics and Control **24**, 1047–1078 (2000)
9. Judd, K. L., Kubler, F., Schmedders, K.: Asset trading volume with dynamically complete markets and heterogeneous agents. Working paper (2002)
10. Kreps, D.: Multi-period securities and the efficient allocation of risk: a comment on the black-scholes option pricing model. In: McCall, J. (ed.) The economics of information and uncertainty. Chicago, IL: University of Chicago Press 1982
11. Kubler, F., Schmedders, K.: Recursive equilibria with incomplete markets. Macroeconomic Dynamics **6**, 284–306 (2002)
12. Tesar, L. L.: Evaluating the gains from international risk-sharing. Carnegie-Rochester Conference Series on Public Policy **42**, 95–143 (1995)
13. van Wincoop, E.: How big are potential welfare gains from international risk-sharing? Journal of International Economics **47**, 109–135 (1999)

Estimating the stationary distribution of a Markov chain[*]

Krishna B. Athreya[1] and Mukul Majumdar[2]

[1] Department of Operations Research and Industrial Engineering, Rhodes Hall, Cornell University, Ithaca, NY 14853, USA (e-mail: athreya@orie.cornell.edu)
[2] Department of Economics, Uris Hall, Cornell University, Ithaca, NY 14853, USA (e-mail: mkm5@cornell.edu)

Received: October 8, 2001; revised version: April 8, 2002

Summary. Let $\{X_j\}_0^\infty$ be a Markov chain with a unique stationary distribution π. Let h be a bounded measurable function. Write $\lambda_h = \int h \, d\pi$ and $\hat{\lambda}_{hn} = \frac{1}{(n+1)} \sum_0^n h(X_j)$. This paper explores conditions for the $\sqrt{n}$ consistency and asymptotic normality of the estimate of $\hat{\lambda}_{hn}$ of λ_h assuming the existence of a solution to the Poisson equation $h - \lambda_h = g - Pg$. Our framework covers the case of nonirreducible Markov chains arising in many growth models in economics.

Keywords and Phrases: Markov chains, Stationary distribution, Consistency, Asymptotic normality, Poisson equation, Martingale central limit theorem.

JEL Classification Numbers: C1, D9.

1 Introduction

Consider a Markov chain $\{X_n\}$ with a unique stationary distribution π which is not easy to compute analytically. An alternative is to estimate $\pi(A)$ for any subset A of the state space from observing the chain $\{X_j\}$ over a finite number of periods, say $0 \le j \le n$. A natural estimate of $\pi(A)$ is the sample proportion of visits to A (defined as $\hat{\pi}_n(A) \equiv \frac{1}{n+1} \sum_{j=0}^n I_A(X_j)$ where I_A is the indicator function of A). The asymptotic consistency property of such an estimate is based on laws of large numbers[1] that assert that, under certain conditions, *for π-almost*

---

[*] Thanks are due to Professors Rabi Bhattacharya, Nicholas Kiefer and Timothy Vogelsang on an earlier draft for helpful conversations, and a referee for insightful comments.
Correspondence to: M.Majumdar

[1] We have in mind several fundamental results on ergodicity and the strong law of large numbers that hold when the initial distribution of the Markov process is an invariant distribution (see, for example, Bhattacharya and Waymire, 1990, pp. 229–230).

all initial conditions x i.e., $X_0 \equiv x$, $\hat{\pi}_n(A)$ converges to $\pi(A)$ in a suitable sense. The usefulness of this claim may be open to question: as π is not known, and the support of π is often difficult to determine, or is a negligible set (see Sect. 4.4 for an example in which the support of the stationary distribution is a subset of $(0, 1)$, the state space, and has Lebesgue measure zero). Moreover, the context of dynamic economic models, the initial condition is *historically given*, and cannot be chosen by the observer[2] (to belong to the support of π).

This paper addresses the question of finding conditions under which $\hat{\pi}_n(A)$ is (i) a consistent estimator of $\pi_{(}A)$ for *any initial condition*, (ii) $\sqrt{n}$-consistent (i.e. $\sqrt{n}[\hat{\pi}_n(A) - \pi(A)]$ is stochastically bounded) and (iii) asymptotically normal. More generally, we consider a real-valued *reward function* h on the state space (which is assumed to be bounded and measurable), and would like to estimate $\lambda_h = \int h d\pi$, the expectation of h with respect to the stationary distribution. A natural candidate is the empirical average $\hat{\lambda}_{hn} \equiv [\sum_{j=0}^n h(x_j)]/n + 1$. One would like to assess the accuracy of $\hat{\lambda}_{hn}$, i.e., the order of magnitude of the error $|\hat{\lambda}_{hn} - \lambda_h|$.

When the Markov chain $\{X_n\}$ is *Harris irreducible* (see Orey, 1971), with respect to some nontrivial σ-finite measure, (this includes irreducible countable state space Markov chains) the techniques of regeneration due to Athreya and Ney (1978) and Nummelin (1978) can be exploited to find sufficient conditions (see the books by Nummelin, 1984, and Meyn and Tweedie, 1993) for $\sqrt{n}$ consistency and asymptotic normality of $\hat{\lambda}_{hn}$. However, there are many Markov chains that are generated by iterations of independent identically distributed random maps (also known as Iterated Function Systems (IFS)) that are in general *not irreducible* (see Bhattacharya and Lee, 1988; Athreya and Stenflo, 2000). This is especially true when the IFS consists of a finite or countable number of maps and the stationary distribution turns out to be a nonatomic one. Some of the best known stochastic dynamic models in economics – both descriptive and normative – fall into this category. We note that the literature on the "inverse optimal problem" identifies conditions under which a given IFS is "generated" by a stochastic dynamic programming model (see Mitra, Montrucchio and Privileggi, 2001, and the list of references). This line of research owes much to the pioneering efforts of Mordecai Kurz (1969).

The present paper is devoted to establishing $\sqrt{n}$ consistency and asymptotic normality of the estimate $\hat{\lambda}_{hn}$ under the key assumptions that h is bounded and that there is a bounded function g such that $h - \lambda_h = g - Pg$ (the so-called Poisson equation) holds. There are many nonirreducible chains that do satisfy these conditions. Also in some irreducible cases the present approach is an alternative to the regeneration approach referred to earlier. Kipnis and Varadlan (1986) treat the case when the chain is reversible with respect to the stationary distribution.

---

[2] Standard texts on development economics emphasize the *role of history* in understanding and explaining the evolution and institutions of economies (see, for example, Ray, 1998, Ch. 5) and, in a dynamic economic model it is natural to think of the initial condition as a product of history. A similar point was made in applications of ergodic theory to computer graphics: "we have no way of choosing the starting x" according to an invariant distribution; "in fact the idea is to start at some x and let a computer generated realization of the process draw a picture of the invariant distribution" (Elton, 1987, p. 482). The emphasis in Elton was to derive ergodic theorems starting from any state

After some preliminaries in Section 2, a variety of sufficient conditions for $\sqrt{n}$- consistency and asymptotic normality are presented in Section 3. In Section 4 we illustrate the applicability of the results in Section 3 by considering a class of Markov chains generated by iterations of independent identically distributed *monotone* maps on an interval $[c, d]$. When the chain satisfies the splitting condition introduced by Dubins and Friedman (1966), it has a unique invariant distribution π, to which the distributions of X_n converge in the Kolmogorov distance geometrically fast from any initial condition. This property is exploited in finding a bounded solution to the Poisson equation when $h(x) = x$ and $h(x) = I_A$. We also sketch in this section some interpretations of the results in examples of growth and cycles under uncertainty (see Stokey and Lucas, 1989; Ljungqvist and Sargent, 2000; Bhattacharya and Majumdar, 2001). All the proofs are in the last section.

2 Estimating the stationary distribution of a Markov chain

Let $\{X_n\}_0^\infty$ be a Markov chain with state space $(S, \mathcal{S})$, transition function $p(\cdot, \cdot)$ and a given initial distribution μ [*which may assign mass one to a single point*]. Recall that the transition function $p(\cdot, \cdot)$ satisfies: (a) for any $A \in \mathcal{S}, p(\cdot, A)$ is $\mathcal{S}$-measurable; and, (b) for any $x \in S, p(x, \cdot)$ is a probability distribution over $\mathcal{S}$. Assume that there is a *stationary probability distribution* (also called an invariant distribution) π on $(S, \mathcal{S})$, such that

$$\pi(A) = \int_S p(x, A)\pi(dx) \tag{2.1}$$

for every A in $\mathcal{S}$.

Suppose that we wish to estimate π from observing the Markov chain $\{X_j\}$ for $0 \leq j \leq n$ starting from a historically given initial condition. A "*natural*" estimate of $\pi(A)$ for any A in $\mathcal{S}$ is the *sample proportion of visits to* A, i.e.

$$\hat{\pi}_n(A) \equiv \frac{1}{n+1} \sum_0^n I_A(X_j) \tag{2.2}$$

where $I_A(x) = 1$ if $x \in A$ and 0 if $x \notin A$. We say that $\hat{\pi}_n(A)$ is a *consistent estimator* of $\pi(A)$ under $\mathcal{P}_\mu$ if $\hat{\pi}_n(A) \to \pi(A)$ as $n \to \infty$ in probability, i.e.,

$$\forall \varepsilon > 0, \quad \mathcal{P}_\mu(|\hat{\pi}_n(A) - \pi(A)| > \varepsilon) \to 0 \tag{2.3}$$

where $\mathcal{P}_\mu$ refers to the probability distribution of the sequence $\{X_n\}_0^\infty$ when X_0 has distribution μ. Assuming that $\hat{\pi}_n$ is such a consistent estimator, it will be useful to know the *accuracy* of the estimate, i.e., the order of the magnitude of the *error* $|\hat{\pi}_n(A) - \pi(A)|$. Under fairly general second moment conditions this turns out to be of the order $(\sqrt{n})^{-1}$ (i.e., the estimator is then called $\sqrt{n}$ *consistent*) and under some further conditions *asymptotic normality* asserting that

$$\sqrt{n}(\hat{\pi}_n(A) - \pi(A)) \overset{d}{\to} N(0, \sigma_A^2) \tag{2.4}$$

for $0 < \sigma_A^2 < \infty$ depending on A and possibly the initial distribution μ holds. This can then be used to provide confidence intervals for $\pi(A)$ based on the data $\{X_j\}_0^n$.

The above issues can be considered in a more general framework where the goal is to estimate

$$\lambda_h = \int h d\pi \tag{2.5}$$

the integral of a *reward function* h which we require to be a real valued bounded S-measurable function. A natural estimate for λ_h is the *empirical average*:

$$\hat{\lambda}_{h,n} \equiv \frac{1}{(n+1)} \sum_0^n h(X_j) , \tag{2.6}$$

As before it would be useful to find conditions to assess the accuracy of $\hat{\lambda}_{h,n}$, i.e. the order of the magnitude of $|\hat{\lambda}_{h,n} - \lambda_h|$. In particular, it is of interest to know whether this estimate is $\sqrt{n}$ consistent, i.e., whether $|\hat{\lambda}_{h,n} - \lambda|$ is of the order $(\sqrt{n})^{-1}$ and further whether *asymptotic normality* holds, i.e., $\sqrt{n}(\hat{\lambda}_{h,n} - \lambda_h)$ converges in distribution to $N(0, \sigma_h^2)$ for $0 < \sigma_h^2 < \infty$ depending on h. In the next section we state some results that provide precise conditions for the validity of the $\sqrt{n}$ consistency of and asymptotic normality of estimators $\hat{\pi}_n(A)$ and $\hat{\lambda}_{h,n}$.

3 Sufficient conditions for consistency and asymptotic normality

In this section we present some of sufficient conditions on h and p that ensure $\sqrt{n}$-consistency and asymptotic normality of $\hat{\lambda}_{h,n}$ defined in (2.6).

In what follows, if g is a S-measurable bounded real valued function on S, let

$$Pg(x) \equiv \int g(y)p(x, dy) = E[g(X_1)|X_0 = x] , \tag{3.1}$$

be the *conditional expectation* of $g(X_1)$ given $X_0 = x$. The *conditional variance* of $g(X_1)$ given $X_0 = x$ is defined by

$$\begin{aligned} Vg(x) &\equiv \text{ variance } [g(X_1)|X_0 = x] = E[(g(X_1) - Pg(x))^2|X_0 = x] \\ &\equiv P(g^2)(x) - (Pg(x))^2 \end{aligned} \tag{3.2}$$

3.1 $\sqrt{n}$ consistency

The following proposition provides a rather restrictive condition for the estimate $\hat{\lambda}_{h,n}$ of $\lambda_h[(2.5)-(2.6)]$ to be $\sqrt{n}$-consistent for *any* initial distribution μ and, in particular, for any historically given initial condition.

Proposition 1. Let π be a stationary distribution for $p(\cdot, \cdot)$, and h be a reward function. Let there exist a bounded S measurable function g such that the *Poisson equation*

$$h(x) - \lambda_h = g(x) - Pg(x) \quad \text{for all } x \text{ in } S \tag{3.3}$$

holds.

Then, for any initial distribution μ,

$$E_\mu(\hat{\lambda}_{h,n} - \lambda_h)^2 \equiv E_\mu \left(\frac{1}{n+1} \sum_{j=0}^n h(X_j) - \lambda_h \right)^2 = 0\left(\frac{1}{n}\right) \qquad (3.4)$$

Corollary 1. Under the hypothesis of Proposition 1, $\hat{\lambda}_{h,n}$ is a $\sqrt{n}$ *consistent* estimate of λ_h; that is, $\sqrt{n}(\hat{\lambda}_{h,n} - \lambda_h)$ is *stochastically bounded*, i.e., for $\forall \varepsilon > 0, \exists K_\varepsilon$ such that for all n

$$\mathcal{P}_\mu \left(\sqrt{n} \left| \frac{1}{n+1} \sum_0^n h(X_j) - \lambda_h \right| > K_\varepsilon \right) \le \varepsilon \qquad (3.5)$$

In addition, $\forall \varepsilon > 0$, as $n \to \infty$

$$\mathcal{P}_\mu \left(\left| \frac{1}{n+1} \sum_0^n h(X_j) - \lambda_h \right| > \varepsilon \right) \to 0 . \qquad (3.6)$$

A natural question is this: given a reward function h how does one find a bounded function g such that the Poisson Equation (3.3) holds? The following comments address this issue.

Rewriting (3.3) as

$$g = \tilde{h} + Pg$$

where $\tilde{h}(x) = h(x) - \lambda_h$ and iterating this we get

$$g = \tilde{h} + P(\tilde{h} + g) = \tilde{h} + P\tilde{h} + P^2 g .$$

This suggests that

$$g \equiv \sum_0^\infty P^n \tilde{h} \qquad (3.7)$$

is a candidate for a solution. This leads to the question of the convergence and boundedness of the infinite series $\sum_0^\infty P^n \tilde{h}$. This can often be ensured by the convergence of the distribution of X_n to π at an appropriate rate in an appropriate metric. For example, if the state space S is finite and the chain is irreducible then it is known (see, for example, Bhattacharya and Waymire, 1990) that for all i, j in S, $|p_{ij}^{(n)} - \pi_j| \le \alpha \lambda^n$ where α is a constant and

$$\lambda = \max\{|\lambda_i| = \lambda_i \quad \text{an eigenvalue of the transition probability matrix } P \text{ and}$$
$$|\lambda_i| < 1\} .$$

Similarly, if the Markov chain $\{X_n\}_{n \ge 1}$ is Harris recurrent and satisfies the Foster-Lyapunov type conditions then

$$\|P^n(x, \cdot) - \pi(\cdot)\| \le \alpha' \lambda^n$$

where α' is a constant, $\| \cdot \|$ is the total variation distance and $0 < \lambda < 1$ (see, for example, Meyn and Tweedie, 1993). For nonirreducible chains this point is elaborated in Remark 2 below and Section 4 (here the metric is not variation norm but weaker).

Corollary 2. Let $A \in \mathcal{S}$ be such that

$$\sup_{x \in S} |\mathcal{P}_x(X_m \in A) - \pi(A)| \le a_m, \sum_0^\infty a_m < \infty \tag{3.8}$$

Then g defined in (3.7) with $h(x) \equiv I_A(x)$ is bounded and satisfies (3.3) and hence (3.5)–(3.6) hold for *any initial distribution* μ.

Remark 2. In the case of Markov chains generated by iterations of *iid* monotone maps on a finite interval $[c, d]$ that satisfy the so-called splitting condition (See Sect. 4) it is known that

$$\sup_{y,x} |\mathcal{P}_x(X_m \le y) - \pi[c, y]| \le (1 - \delta)^m \tag{3.9}$$

for some $0 < \delta < 1$. It follows from (3.8) and (3.9) that the empirical distribution function $\frac{1}{n} \sum_0^{n-1} I_{[c,y]}(X_j)$ is a $\sqrt{n}$ *consistent* estimator of $\pi[c, y]$. Also, for $h(x) \equiv x$ it can be shown that g in (3.7) is bounded.

3.2 Asymptotic normality

The next proposition deals with asymptotic normality.

Proposition 2. Let h be a bounded reward function; let g satisfy (3.3) and be bounded. Let Pg, Vg be as defined in (3.1) and (3.2). Suppose that the initial distribution μ is such that under P_μ,

$$\frac{1}{n} \sum_0^n Vg(X_j) \text{ converges in probability to } \sigma^2 (0 < \sigma^2 < \infty, \text{ nonrandom})$$

$$\tag{3.10}$$

Then under such a $\mathcal{P}_\mu$

$$\sqrt{n} \left(\frac{1}{n+1} \sum_0^n h(X_j) - \lambda_h \right) \xrightarrow{d} N(0, \sigma^2) \tag{3.11}$$

Corollary 3. Let $A \in \mathcal{S}$ be such that (3.8) holds, and suppose the initial distribution μ is such that

$$\frac{1}{n} \sum_0 p(X_j, A)(1 - p(X_j, A) \text{ converge in probability to } \sigma^2,$$

$$0 < \sigma^2 < \infty \text{ nonrandom}$$

Then, under $\mathcal{P}_\mu$, $\sqrt{n} \left(\frac{1}{n+1} \sum_0^n I_A(X_j) - \pi(A) \right) \xrightarrow{d} N(0, \sigma^2)$.

4 Markov chains generated by iterations of I.I.D. Maps

We now consider Markov chains $\{X_n\}_0^\infty$ generated by iterations of an i.i.d. sequence of random maps. Let S be a closed bounded interval $[c,d](-\infty < c < d < \infty$, and $\underset{\sim}{S}$ the Borel σ-field of S. Let Γ be a family of maps from S into itself. Let $\{\alpha_n\}_1^\infty$ be a sequence of i.i.d. maps from Γ. For a given initial x, write

$$X_n(x) \equiv x \quad \text{for} \ n = 0$$

and

$$X_n(x) \equiv \alpha_n(\ldots \alpha_1(x)) \quad \text{for} \ n \geq 1 \tag{4.1}$$

Then for any x, $X_n(x)$ is a Markov chain with state space $S = [c,d]$ and the transition function $p(\cdot,\cdot)$ is given by

$$p(x,A) \equiv Prob\{\alpha_1(x) \in A\} \tag{4.2}$$

The initial state x can also be random (independently of the sequence $\{\alpha_n\}_1^\infty$ with some probability distribution μ. The distribution of X_n when the initial distribution in μ is denoted by μP^n.

4.1 Monotone maps and the splitting condition

Let Γ be a set of monotone maps from S into S; i.e., each element of Γ is either a nondecreasing function on S or a nonincreasing function.

Given two probability measures μ and ν on $\underset{\sim}{S}$, let $d_K(\mu, \nu)$ be the Kolmogorov distance, i.e.,

$$d_K(\mu, \nu) \equiv \sup_{x \in S} |\mu[c,x] - \nu[c,x])| \tag{4.3}$$

It should be noted that convergence in the Kolmogorov distance implies weak convergence.

Proposition 3 (Dubins and Feedman, 1966). Assume that the following splitting condition (**H**) holds:

(**H**) *There exist* $z_0 \in S, \delta > 0$ *and a positive integer* N *such that*
   (1) $Prob(\alpha_N \alpha_{N-1} \ldots \alpha_1 x \leq z_0, \forall x \in S) \geq \delta$
   and
   (2) $Prob(\alpha_N \alpha_{N-1} \ldots \alpha_1 x \geq z_0, \forall x \in S) \geq \delta$

Then:

(a) there is a unique invariant distribution π on $[c,d]$ of the Markov chain $\{X_n\}$; and

(b) for any initial distribution μ,

$$d_K(\mu P^n, \pi) \leq (1 - \delta)^{[n/N]} \tag{4.4}$$

where $[y]$ is the integer part of y.

The estimate (4.4) of the speed of convergence of μP^n to the invariant π plays a crucial role in applying Proposition 1 and its corollaries to the economic models that we describe now.

4.2 Models of growth and cycles

We should stress that Markov chains generated by *iid* maps arise "naturally" not only in descriptive dynamic economics, but also in the context of dynamic optimization under uncertainty, particularly when one wishes to study the evolution of states generated by an optimal policy function (see Majumdar, Mitra and Nyarko, 1989, for an extended list of references). We briefly outline two examples of applications of Proposition 3 (see Bhattacharya and Majumdar, 2001, Sect. III, for details, particularly for the verification of the splitting condition (H)).

Example 1. Let $\Gamma = \{F_1 \ldots F_N\}$ where each $F_i : R_+ \to R_+$ satisfies:

F.1. F_i is strictly increasing, continuous and there is some $r_i > 0$ such that $F_i(x) > x$ on $(0, r_i)$ and $F_i(x) < x$ for $x > r_i$.

F.2. for $i > i'$, $F_i(x) > F_{i'}(x)$ for all $x \geq 0$.

Let $Prob(\alpha_n = F_i) = p_i > 0 (1 \leq i \leq N)$.

As before, let $X_{n+1}(x) = \alpha_{n+1}(X_n(x))$. It is possible to show that this Markov chain $\{X_n(x)\}$ with state space $(0, \infty)$ has the following property: for each $x \in (0, \infty) X_n(x)$ enters the interval $[r_1, r_N]$ with probability one, and remains in $[r_1, r_N]$ forever. Hence from the perspective of long run analysis we can take $[r_1, r_N]$ as the effective state space. The splitting condition (**H**) is verified in Bhattacharya and Majumdar (2001).

Example 2. This example is motivated by the remarks of Solow in his celebrated paper (1956) and the subsequent work of Day (1982). Consider a Markov process with the state space $S = R_+$ and two possible laws of motion denoted by F and G (i.e., $\Gamma = \{F, G\}$) occurring with probabilities $\hat{p}$ and $1 - \hat{p}$ respectively $(0 < \hat{p} < 1)$. The law of motion F is monotone increasing and has an attracting positive fixed point (recall Fig. 1 of Solow, 1956); however, the other law G triggers cyclical forces and has a pair of locally attracting periodic points of period 2 (and a repelling fixed point: the precise assumptions are stated below). One may interpret F as the dominant long run growth law ($\hat{p}$ is "large"), and G as the law of short run cyclical interruptions. A numerical example is given in Section 5. But, first, we state the assumptions on F and G precisely and note their implications.

The law of motion that generates the growth process is represented by a continuous increasing function $F : [0, 1] \to [0, 1]$. We assume that

G.1. F has a fixed point $r > 1/2$ such that

$$F(x) > x \quad \text{for} \ 0 < x < r \,,$$
$$F(x) < x \quad \text{for} \ x > r \,.$$

Whether or not $F(0) = 0$ is not relevant for our subsequent analysis. Note that the trajectory from any initial x_0 converges to r; indeed, if $0 < x_0 < r$, the sequence $F^{(n)}(x_0)$ increases to r; whereas if $x_0 > r$, the sequence $F^{(n)}(x_0)$ decreases to r.

The law of motion that triggers cyclical forces is denoted by a continuous map $G : [0, 1] \to [0, 1]$. We assume

C.1. G is increasing on $[0, 1/2]$ and decreasing on $[1/2, 1]$.

C.2. $G(x) > x$ on $[0, 1/2]$

C.3. G has two periodic points of period 2 denoted by $\{\beta_1, \beta_2\}$, and a fixed point x^* which is a repelling fixed point of G, and no other fixed point or periodic point. Moreover, $\{\beta_1, \beta_2\}$ are locally stable fixed points of $G^{(2)}$.

Finally,

C.4. G has an invariant interval $[c, d](1/2 \le c < d < 1); c < \beta_1 < x^* < \beta_2 < d$. Also $r \in (c, d), r \notin \{\beta_1, x^*, \beta_2\}$.

Now, when we consider the evolution

$$X_{n+1} = \alpha_{n+1}(X_n)$$

where $\alpha_{n+1} = F$ with probability $\hat{p}$ and $\alpha_{n+1} = G$ with probability $1 - \hat{p}$. We can proceed as follows: for any initial $x \in (0, 1)$ the process $X_n(x)$ enters $[c, d]$ with probability one after a finite number of steps. Also, it is easy to see that $[c, d]$ is invariant under F. Hence, for the long run analysis of the evolution of X_n, we can take $[c, d]$ as the effective state space. The splitting condition (**H**) is verified by a careful consideration of the structure of the model (Bhattacharya and Majumdar, 2001). It should be stressed that while G is not a monotone function, on the (common) invariant interval $[c, d]$ both F and G are monotone (increasing and decreasing respectively).

Proposition 3 is, therefore, applicable directly to both examples. From (4.4) one has

$$\sup_{x \in [c,c]} |p^{(n)}(X, J) - \pi(J)| \le (1 - \delta)^{[n/N]}$$

where J is *any* subinterval of $[c, d]$. We can now apply Corollary 2 and conclude that the empirical distribution function $\frac{1}{n+1} \sum_0^n I_{[c,y]}(X_j)$ is a $\sqrt{n}$-consistent estimator of $\pi[c, y]$.

We now turn to the problem of estimating the "equilibrium mean" $\int y\pi(dy)$. Here the reward function is $h(x) = x$. Going back to (3.3) let us write $\tilde{h}(\cdot) \equiv h(\cdot) - \lambda_h$. Note that if we can ensure the convergence of the infinite series $\sum_{n=0}^{\infty} P^n \tilde{h}$ and its boundedness, then by defining

$$g(\cdot) \equiv \sum_{n=0}^{\infty} P^n \tilde{h}(\cdot)$$

we can satisfy (3.3).

With $\tilde{h}(z) \equiv z - \lambda_h \equiv z - \int_c^d y\pi(dy)$, assume, without loss of generality, that $c \geq 0$ [otherwise consider $\tilde{h}(z) = (z - c) - \int_c^d (y - c)\pi(dy)$]. Now,

$$
\begin{aligned}
(P^n\tilde{h})(x) &= E(\tilde{h}(X_n)|X_0 = x) = E(X_n|X_0 = x) - \int_c^d y\pi(dy) \\
&= \int_{[c,d]} Prob[X_n > u|X_0 = x)du - \int_{[c,d]} \pi((u,d))du \\
&= \int_{[c,d]} [1 - p^{(n)}(x, [c, u]) - 1 - \pi([c, u])]du \\
&= -\int_{[c,d]} p^{(n)}(x, [c, u]) - \pi([c, u])du
\end{aligned}
$$

Hence, $\sup_x |P^n\tilde{h}(x)| \leq \int_{[c,d]} (1 - \delta)^{[n/N]}du = (d - c)(1 - \delta)^{[n/N]}$. This ensures that g in (4.5) is well-defined and bounded over $S = [c, d]$. Corollary 1 now leads to the following conclusion: *the empirical mean $\hat{\lambda}_{h,n}$ is a $\sqrt{n}$-consistent estimate of $\lambda_h = \int y\pi(dy)$.*

4.3 Some additional results

We record first an important result applicable to the case where Γ consists of monotone nondecreasing functions on $[c, d]$.

Proposition 4. Let $S = [c, d]$ and Γ consist of monotone nondecreasing functions from S into S and assume that the splitting condition (**H**) holds. Then for any continuous function h with bounded variation, there is some bounded g satisfying (3.4). Moreover, independent of the initial distribution μ,

$$
\frac{1}{\sqrt{n}} \sum_{j=0}^{n-1} (X_j - \int y\pi(dy)) \to N(0, \sigma^2)
$$

and

$$
\sqrt{n}(\hat{\pi}_n([c, y]) - \pi([c, y])) \to N(0, \sigma^2)
$$

[where and $\hat{\pi}_n[c, y] = \frac{1}{n}\#\{j : 0 \leq j \leq n - 1, X_j \in [c, y]\}$ and $\sigma^2 = \int g^2 d\pi - \int (hg)^2 d\pi)$].

This proposition is a special case of Theorem 3.1 of Bhattacharya and Lee (who proved a functional central limit theorem given the assumptions listed above).

We turn to a pair of sufficient conditions for solving the equation $h - \lambda_h = g - Pg$ when the state space $S = [c, d]$ and $\{\alpha_i\}_1^\infty$ is sequence of i.i.d. maps from S to S (*not necessarily monotone*). Let $h = [c, d] \to R$ be absolutely continuous. Then $h(x) = h(c) + \int_c^x h'(u)du$ for $c \leq x \leq d$. Thus,

$$
E_x h(X_n) = h(c) + \int_c^d h'(u)P_x(X_n > u)du
$$

and

$$\lambda_h = h(c) + \int_c^d h'(u)\pi(u, d)du$$

so

$$|E_x h(X_n) - \lambda_h| \leq \int_c^d |h'(u)|\,|P_X(X_n > u) - \pi(u, d]|du$$

$$\leq \left(\int_c^d |h'(u)|du \right) d_k(P_x(X_n E \cdot), \pi)$$

yielding the following:

Proposition 5. Let $h = [c, d] \to R$ be absolutely continuous. Assume that

$$\sup_{c \leq x \leq d} \sum_0^\infty d_k(P_x(X_n(E \cdot), \pi(\cdot)) < \infty \ .$$

Then

$$g(x) = \sum_0^\infty (E_x h(X_n) - \lambda_h)$$

is well defined on $[c, d]$, and bounded and solves the equation

$$h - \lambda_h = g - Pg$$

In addition, if (3.10) holds,

$$\sqrt{n}(\lambda_{h,n} - \lambda_h) \xrightarrow{d} N(0, \sigma^2)$$

for some $0 < \sigma^2 < \infty$, nonrandom but depending on μ and h.

4.4 The support of an invariant distribution: an example

Example 3. Let $S = (0, 1)$ and Γ, the family of all admissible laws of motion consist of a pair of functions from the 'quadratic family', specified as follows:
$\Gamma = \{F_\mu, \hat{F}\}$ where

$$\hat{F}(x) = 2x(1 - x)$$

and

$$F_\mu(x) = \mu x(1 - x) \quad \text{where} \quad -\frac{1}{2} + \frac{1}{2}\sqrt{17} < \mu < 2 \tag{4.5}$$

Consider the Markov chain

$$X_{n+1} = \alpha_{n+1}(X_n) \tag{4.6}$$

where $\{\alpha_n\}_0^\infty$ is an i.i.d. sequence with the distribution $Prob(\alpha_{n+1} = \hat{F}) = p$ and $Prob(\alpha_{n+1} = F_\mu) = 1 - p$. For this Markov chain, it is verified by Bhattacharya and Rao (1993, p. 20, Example 2) that the splitting condition **(H)** holds and *the support of the invariant distribution is a Cantor set of Lebesgue measure zero.*

Example 4. Let $S = [0, 1]; F(x) = (3/4)x^{1/2}, G(x) = (3.1)x(1 - x)$. Then all the assumptions in Example 2 of Section 4 hold. The invariant distribution can be estimated through computer simulation.

5 Proofs

Proposition 1. Since h and g satisfy (3.3), i.e.

$$h(x) - \lambda_h = g(x) - (Pg)(x)$$

$$\frac{1}{n+1} \sum_0^n h(X_j) - \lambda_h = \frac{1}{n+1} \sum_0^n (g(X_j) - Pg(X_j))$$

$$= \frac{1}{(n+1)} \sum_1^n (g(X_j) - (Pg)(X_{j-1}))$$

$$+ \frac{1}{(n+1)} \sum_1^n (Pg(X_{j-1})) - Pg(X_j)) + g(X_0) - Pg(X_0)$$

Thus,

$$\frac{1}{n+1} \sum_0^n h(X_j) - \lambda_h = \frac{1}{(n+1)} \sum_1^n Y_j + \frac{1}{(n+1)} (g(X_0) - Pg(X_n)) \quad (5.1)$$

where $Y_j = g(X_j) - (Pg)(X_{j-1})$ for $1 \le j \le n$. By the Markov property of $\{X_n\}$ and the definition of Pg it follows that $\{Y_j\}_0^\infty$ is a *martingale difference sequence*, i.e.

$$E(Y_j | X_0, X_1, \dots X_{j-1}) = 0 .$$

Also, since h and g are bounded, $EY_j^2 < \infty$ for all j. Further $\{Y_j : 1 \le j \le n\}$ are uncorrelated. Thus, for any initial distribution μ

$$E_\mu \left(\sum_1^n Y_j \right) = 0 \quad \text{and}$$

$$E_\mu \left(\sum_1^n Y_j \right)^2 = V_\mu \left(\sum_1^n Y_j \right) = \sum_1^n V_\mu(Y_j) = \sum_1^n E_\mu V(Y_j) | X_0, \dots, X_{j-1})$$

where V_μ and E_μ stand for mean and variance under the initial distribution μ and $V(Y_j | X_0, \dots, X_{j-1})$ is the conditional variance of Y_j given $X_0, X_1, \dots, X_{j-1}$. Again by the Markov property and the definition of $(Vg)((x)$ in (3.3) of Proposition 1

$$E_\mu \left(\sum_1^n Y_j \right)^2 = \sum_1^n E_\mu(Vg)(X_{j-1}) = \sum_0^{n-1} E_\mu(Vg)(X_j) \quad (5.2)$$

Using $(a+b)^2 \le 2(a^2 + b^2)$ repeatedly, we get from (5.1) and (5.2) that

$$E_\mu \left(\frac{1}{n+1} \sum_0^n h(X_j) - \lambda_h \right)^2 \le \frac{2}{(n+1)^2} \sum_0^{n-1} E_\mu(Vg)(X_j)$$

$$+ \frac{4}{(n+1)^2} (E_\mu(Pg(X_n))^2 + E_\mu(g(X_0))^2) .$$

Now, by the boundedness of g the right side above is $0(\frac{1}{n})$ as was to be shown.

Corollary 1. By Chebychev's inequality,

$$P_\mu \left(\sqrt{n} \left| \frac{1}{n+1} \sum_0^n h(X_j) - \lambda_h \right| > K_\varepsilon \right) \leq \frac{n}{K_\varepsilon^2} E_\mu \left(\frac{1}{n+1} \sum_0^n h(X_j) - \lambda_h \right)^2$$

and by (3.4) the rightside is bounded by a constant multiple of K_ε^{-2}. Hence for $\forall \varepsilon > 0$, there exists a K_ε such that (3.6) holds. Similar calculation yields (3.6)

Proposition 2. From (5.1) and boundedness of $g(\cdot)$ it is enough to show

$$\frac{1}{\sqrt{n}} \sum_1^n Y_j \xrightarrow{d} N(0, \sigma^2)$$

As noted earlier, $\{Y_j : 1 \leq j < \infty\}$ is a martingale difference sequence. Thus it suffices to verify the hypothesis for Brown's Martingale central limit theorem (Bhattacharya and Waymire, 1990, p. 508). By the Markov property, $V(Y_j | X_0, \ldots, X_{j-1}) = (Vg)(X_{j-1})$. Since g is bounded, $\sum_{j=1}^n (EY_j^2 : |Y_j| > \varepsilon \sqrt{n} | X_0, \ldots, X_{j-1}) = 0$ for $\varepsilon > 0$ and n large. Thus Brown's conditions hold. So (3.11) follows. $\qquad \square$

Corollary 3. It is enough to note that

$$(Vg)(x) = \mathrm{Var}(I_A(X_1) | X_0 = x) = p(x, A)(1 - p(x, A)) .$$

References

Athreya, K.B., Ney, P.: A new approach to the limit theory of recurrent markov chains. Trans. Amer. Math. Soc. **245**, 493–501 (1978)

Athreya, K.B., Stenflo, O.: Perfect sampling for Doeblin chains. Technical Report, Cornell University (to appear in: Sankhya) (2000)

Bhattacharya, R.N., Lee, O.: Asymptotics of a class of Markov processes, that are not in general reducible. Annals of Probability **16**, 1333–1347 (1988)

Bhattacharya, R.N., Weymire, E.C.: Stochastic processes with applications. New York: Wiley 1990

Bhattacharya, R.N., Rao, B.V.: Random iterations of two quadratic maps. In: Cambanis, S., Ghosh, J.K., Karandikar, R.L., Sen, P.K. (eds.) Stochastic processes, pp.13–21. New York: Springer 1993

Bhattacharya, R.N., Majumdar, M.: On a class of stable random dynamical systems: theory and applications. Journal of Economic Theory **96**, 208–229 (2001)

Chung, K.L.: A course in probability theory, 2nd edn. New York: Academic Press 1974

Day, R.: Irregular growth cycles. American Economic Review **72**, 406–414 (1982)

Dubins, L.E., Freedman, D.: Invariant probabilities for certain Markov processes. Annals of Mathematical Statistics **37**, 837–858 (1966)

Elton, J.H.: An ergodic theorem for iterated maps. ErgodicTheory and Dynamical Systems **7**, 481–488 (1987):

Granger, C.W.J., Terasvirta, T.: Modelling nonlinear economic relationships. Oxord: Oxford University Press 1993

Kipnis, C., Varadhan, S.R.S.: Control limit theorem for additive functionals of reversible Markov processes and applications to simple exclusions. Communications of Mathematical Physics **104**, 1–19 (1986)

Kurz, M.: On the inverse optimal problem. In: Lecture notes in operations research and mathematical economics, vol. 11, pp. 189–201. Berlin Heidelberg New York: Springer 1969

Ljungqvist, L., Sargent, T.J.: Recursive macroeconomic theory. Cambridge, MA: MIT Press 2000

Majumdar, M., Mitra, T., Nyarko, Y.: Dynamic optimization under uncertainty; non-convex feasible set. In: Feiwel, G.R. (ed.) Joan Robinson and modern economic theory, pp. 545–590. London: MacMillan 1989

Meyn, S.P., Tweedie, R.L.: Markov chains and stochastic stability. Berlin Heidelberg New York: Springer 1993

Mitra, T., Montrucchio, L., Privileggi, F.: The nature of the steady state in models of optimal growth under uncertainty. CAE Working Paper, no. 01-04, Cornell University (2001)

Nummelin, E.: A splitting technique for Harris recurrent chains. Z. Wahrsch. **43**, 309–318 (1978)

Nummelin, E.: General irreducible Markov chains and nonnegative operators. Cambridge: Cambridge University Press 1984

Orey, S.: Limit theorems for Markov chain transition probabilities. London: Van Nostrand Reinhold 1971

Ray, D.: Development economics. Princeton, NJ: Princeton University Press 1998

Solow, R.M.: A contribution of the theory of economic growth. Quarterly Journal of Economics **70**, 65–94 (1956)

Stokey, N.L., Lucas, R.E.: Recursive methods in dynamic economics. Cambridge, MA: Harvard University Press 1989

Monte Carlo simulation of macroeconomic risk with a continuum of agents: the symmetric case[*]

Peter J. Hammond[1] **and Yeneng Sun**[2]

[1] Department of Economics, Stanford University, Stanford, CA 94305-6072, USA
(e-mail: peter.hammond@stanford.edu)
[2] Institute for Mathematical Sciences, National University of Singapore, 3 Prince George's Park,
Singapore 118402, REPUBLIC OF SINGAPORE, and
Department of Mathematics and the Centre for Financial Engineering, National University of
Singapore, Singapore, REPUBLIC OF SINGAPORE (e-mail: matsuny@nus.edu.sg)

Received: October 29, 2001; revised version: April 24, 2002

Summary. Suppose a large economy with individual risk is modeled by a continuum of pairwise exchangeable random variables (i.i.d., in particular). Then the relevant stochastic process is jointly measurable only in degenerate cases. Yet in Monte Carlo simulation, the average of a large finite draw of the random variables converges almost surely. Several necessary and sufficient conditions for such "Monte Carlo convergence" are given. Also, conditioned on the associated Monte Carlo σ-algebra, which represents macroeconomic risk, individual agents' random shocks are independent. Furthermore, a converse to one version of the classical law of large numbers is proved.

Key words: Large economy, Continuum of agents, Law of large numbers, Exchangeability, Joint measurability problem, de Finetti's theorem, Monte Carlo convergence, Monte Carlo σ-algebra.

JEL Classification Numbers: D80, C00, E00.

1 Introduction

Consider an economy with a continuum of agents. Suppose all agents face independent and identically distributed (i.i.d.) random shocks. In many economic

---

[*] Part of this work was done when Yeneng Sun was visiting SITE at Stanford University in July 2001. An early version of some results was included in a presentation to Tom Sargent's macro workshop at Stanford. We are grateful to him and Felix Kübler in particular for their comments. And also to Marcos Lisboa for several discussions with Peter Hammond, during which the basic idea of the paper began to take shape.

Correspondence to: P.J. Hammond

applications one would like to invoke an exact version of the law of large numbers, and claim that the fraction who experience each possible shock should almost surely equal the probability of that shock.[1] However, as pointed out by Doob in [10] and [11], with further elaborations in [14], [20] and [30], the usual construction of a process with a continuum of i.i.d. random variables creates fundamental measurability difficulties. The first concerns joint measurability – namely, except in some trivial cases, such a process can never be jointly measurable with respect to the completion of the usual product σ-algebra on the joint space of parameters and samples.[2] The second problem concerns sample measurability – as shown in Theorem 2.2 in [10], the collection of samples whose corresponding sample functions are not Lebesgue measurable has outer measure one, so Lebesgue measure offers no basis for a meaningful concept of the mean or the distribution of a sample function. That is, the sample function giving each agent's individual shock may not be typically Lebesgue measurable, and thus the fraction of agents associated with each shock may not be well-defined.

Ad hoc examples can be found in the literature to show that there is no contradiction between the independence condition and the essential constancy of sample distributions – i.e., between the condition and conclusion in a natural statement of the exact law of large numbers.[3] However, the conclusion of the exact law of large numbers can also fail badly in different versions of such ad hoc examples.[4]

In [28]–[31], some rich product probability structures on the joint space of parameters and of samples are used to make independence compatible with joint measurability. Such enriched product probability spaces extend the usual product probability spaces, retain the common Fubini property of product probability measures, and also accommodate an abundance of nontrivial independent processes. Both the sample and joint measurability problems are automatically resolved by the required Fubini property. Also, the desired exact law of large numbers holds if and only if the random variables are independent almost surely.

Consider a continuous parameter process with mutually independent random variables which, unlike [28]–[31], is not taken from a framework where the usual Fubini property is already satisfied. As pointed out in Remark 3 of [18], it may not be possible to ensure that such a process is measurable by extending the usual product σ-algebra while retaining the usual "two-way" Fubini property. However, it is shown in [18] that a natural "one-way Fubini" property does guarantee a unique meaningful solution to the joint measurability problem, even for processes with random variables that are independent in a very weak sense.

---

[1] See [14] and [20], as well as [1], pp. 2198–2199 and [29], pp. 502–503 for some well known references incorporating claims of this kind.

[2] See p. 57 in [11] and Proposition 1.1 in [30].

[3] See [1], [14], [17] and [20].

[4] See [31] for some detailed comments on such ad hoc examples, and on the difficulties of using a purely finitely additive measure-theoretic framework. In addition, the standard Birkhoff example shows that a continuum of mutually orthogonal random variables has Pettis integral equal to zero – see Example 5, p. 43 in [9] or Example 3.2.1, p. 33 in [32]. In this connection, [21] discusses some of the economics literature concerned with the Pettis integral.

The approach taken in this paper is inspired by the Monte Carlo method that is sometimes used to find numerical approximations to an ordinary multiple integral, especially when the integrand is of high dimension, or is complicated in some other way – see, for example, [16] for a recent survey. The basic Monte Carlo method computes the integral of a real-valued function by taking the average of the integrand evaluated at randomly selected points. Our purpose here is to extend this method in order to simulate macroeconomic uncertainty when many agents face individual risk which is modeled by a process with a continuum of random variables. Indeed, suppose that a sequence of the random variables is obtained by evaluating the process at randomly selected points of the parameter space. If the average of these random variables converges in the sense to be specified in Section 2 below, the process is said to be "Monte Carlo convergent".

Note that if a continuum of random variables forms an i.i.d. process, then any sequence taken from the continuum collection is obviously still i.i.d. When the common mean exists, the classical law of large numbers says that the arithmetic mean of the first n random variables in such a sequence converges to the common mean almost surely as n tends to infinity. This will trivially imply Monte Carlo convergence for the simple i.i.d. case.

The main purpose of this paper is to move beyond this obvious i.i.d. case and to consider Monte Carlo convergence for a large economy modeled by a continuum of (essentially) pairwise exchangeable random variables.[5] As shown by Proposition 2 in Section 5.3 below, such a process cannot be jointly measurable with respect to the usual product σ-algebra unless it is degenerate in the sense that almost all random variables are identical (and thus almost all agents' random shocks are perfectly correlated). However, based on some techniques developed in [18] to study a "one-way Fubini" extension of the relevant product probability space, several necessary and sufficient conditions for Monte Carlo convergence can be formulated and proved. We shall also define a corresponding Monte Carlo σ-algebra as that generated by the random variable whose value is the probability measure representing the appropriate almost sure limit of the empirical distribution of individual random shocks. This will be shown to simulate macroeconomic risk in the sense that individual agents' shocks, conditioned on this σ-algebra, are (essentially) independent. It follows as a corollary that the basic concept of stochastic independence will be characterized by triviality of the Monte Carlo limit – see Proposition 1 below.

In the rest of the paper, Section 2 describes the basic formulation and provides some essential definitions and assumptions. Then Section 3 describes some equivalence results in the basic i.i.d. case. Section 4 introduces and studies the basic properties of several concepts related to symmetry, pairwise exchangeability, and conditional independence. General results on the equivalence of these basic concepts with some special form of Monte Carlo convergence are presented in Section 5. Section 6 provides the proof of Theorem 1 through several lemmas. Finally, a concluding assessment appears in Section 7.

---

[5] Systematic applications of exchangeability in economics are proposed in [25]. Other economic applications can be found in [6], [19] and [22].

2 Basic formulation

2.1 Two probability spaces

Let $(T, \mathcal{T}, \lambda)$ denote a probability space which is to be regarded as the parameter space for a process. In the case of an economy with a continuum of agents, it is usual to regard T as the set of agents, and to take $(T, \mathcal{T}, \lambda)$ as the Lebesgue unit interval – i.e., $T = [0, 1]$, while λ is Lebesgue measure applied to the complete σ-field $\mathcal{T}$ obtained from the usual Borel σ-field by adding all sets which are sandwiched between two Borel sets of equal Lebesgue measure. However, none of our results rely on $(T, \mathcal{T}, \lambda)$ being this Lebesgue unit interval – instead, it can be an entirely general non-atomic probability space. Indeed, in statistical mechanics it might be natural to label each particle by its position at any particular time, in which case T should be a compact subset of $\mathbb{R}^3$ with its appropriate σ-algebra, and with a probability measure which is the appropriately normalized product Lebesgue measure. In economies with a continuum of agents, or games with a continuum of players, especially if there is incomplete information, one could take T to be a metric space of agents' or players' possible types, including some feature or label that uniquely identifies each agent or player. Or T could be a hyperfinite Loeb space which is used to index the economic agents, as discussed in [1].

Let $(\Omega, \mathcal{A}, P)$ be the sample probability space. For example, it can be the product of a continuum of copies of some other basic probability space, or some extension of this product, or some other space entirely. As usual in probability theory, it is not necessary to specify in detail what the sample probability space is provided some general existence issues are resolved. We assume that the σ-algebra $\mathcal{A}$ is *complete* in the sense that, if $A \in \mathcal{A}$ with $P(A) = 0$, then $A' \in \mathcal{A}$ for all $A' \subset A$. Let $(T \times \Omega, \mathcal{T} \otimes \mathcal{A}, \lambda \times P)$ denote the usual product probability space. For simplicity, the completion of this product space is denoted in the same way.

2.2 A Polish space

The variables of interest in a continuum economy usually describe agents' allocations, or else their characteristics such as endowments, preferences or utility functions, or discount rates. In a game with a continuum of players, the variables of interest usually describe strategies, or else characteristics such as payoff functions. We assume that all such variables are members x of some general metric space (X, d).

We shall assume that this metric space is *complete* in the sense that *Cauchy sequences* satisfying $d(x_m, x_n) \to 0$ as $m, n \to \infty$ must converge to some point in X, and also *separable* in the sense that X is the closure of some countable subset. A topological space is called a *Polish* space if it is homeomorphic to a complete separable metric space. Let $\mathcal{B}$ denote the Borel σ-algebra of the Polish space (X, d). Then $\mathcal{B}$ can be generated by a countable collection of open sets in X. From now on, we usually ignore the metric d and refer to $(X, \mathcal{B})$ as a Polish space.

It should be remarked that many rich spaces are Polish. Examples include finite sets, finite-dimensional Euclidean spaces with their Euclidean metric, and the space of real-valued continous functions on a bounded interval with the usual supremum norm. One other important example is the space of real-valued functions that are right continuous and have left-hand limits – when this space is given its (metrizable) Skorohod topology – see [4], for example. Thus, our theory will encompass most of the standard models encountered in macroeconomics, with a continuum of agents all facing individual stochastic processes whose values lie in a suitable function space. Such stochastic processes can be Markov chains, Brownian motion, general Ito processes, Ito processes with random jumps, etc.

2.3 π- and λ-systems

Given any nonempty set Y, a π-*system* $\mathcal{P}$ is a family of subsets of Y that is closed under finite intersections. Let $\mathcal{D}$ be a non-empty family of subsets of Y, and let $\mathcal{D}^{\pi}$ denote the family of all finite intersections of sets in $\mathcal{D}$. Then, it is obvious that $\mathcal{D}^{\pi}$ is a π-*system*, and that $\mathcal{D}$ and $\mathcal{D}^{\pi}$ both generate the same σ-algebra $\sigma(\mathcal{D})$.

A family $\mathcal{Q}$ of subsets of Y is said to be a λ-*system* if it satisfies: (i) $Y \in \mathcal{Q}$; (ii) for $A, B \in \mathcal{Q}$ with $A \subseteq B$, $B \setminus A \in \mathcal{Q}$; (iii) for any sequence $\{A_n\}_{n=1}^{\infty}$ of pairwise disjoint sets in $\mathcal{Q}$, $\cup_{n=1}^{\infty} A_n \in \mathcal{Q}$.[6]

A result we use many times in this paper is:

Dynkin's π–λ Theorem: If $\mathcal{P}$ is a π-*system* and $\mathcal{Q}$ is a λ-*system* that contains $\mathcal{P}$, then $\mathcal{Q}$ must contain the σ-algebra $\sigma(\mathcal{P})$ generated by $\mathcal{P}$.[7]

This theorem allows one to infer that if any two finite signed measures coincide on $\mathcal{P} \cup \{Y\}$ then, because it is easy to see that the family $\mathcal{Q}$ on which the two measures coincide must be a λ-system, in fact the two measures must coincide on the whole of $\sigma(\mathcal{P})$.

2.4 Countably generated and essentially countably generated σ-algebras

Let $\mathcal{D}$ be a non-empty family of subsets of a space Y. If $\mathcal{D}$ is countable, then the σ-algebra $\sigma(\mathcal{D})$ generated by $\mathcal{D}$ is said to be *countably generated*.[8] Note that $\mathcal{D}^{\pi}$ is still countable and also generates $\sigma(\mathcal{D})$. Thus, we can always assume that a countably generated σ-algebra is generated by a countable π-system. In particular, if Y is a Polish space as in Section 2.2, then its Borel σ-algebra $\mathcal{Y}$ is countably generated from a countable π-system $\mathcal{O}$ of open sets in Y.

Consider any mapping $f : \Omega \to Y$. Let $\sigma(f)$ be the smallest σ-algebra $\mathcal{C}$ such that f is measurable w.r.t. $\mathcal{C}$ on Ω and $\mathcal{Y}$ on Y. This σ-algebra $\sigma(f)$ is called the σ-algebra *generated* by f. Since $\sigma(f)$ is also generated by the countable family $\{f^{-1}(O) \mid O \in \mathcal{O}\}$, it is countably generated. On the other hand, if a σ-algebra $\mathcal{C}$ on Ω is countably generated, then there exists a Borel measurable mapping

---

[6] See, for example, [5], p. 41. This definition is easily seen to be equivalent to that of *Dynkin class*, in which (iii) is replaced by the requirement that for any sequence $\{A_n\}_{n=1}^{\infty}$ of sets in $\mathcal{Q}$ with $A_n \subseteq A_{n+1}$ for all n, $\cup_{n=1}^{\infty} A_n \in \mathcal{Q}$. See, for example, [8], p. 44.

[7] See, for example, [5], p. 42, or [8], p. 45.

[8] See the definition in [5], Ex. 2.11, p. 34.

$\theta : \Omega \to [0, 1]$ such that $C = \sigma(\theta)$ – see [5], Ex. 20.1, p. 270. In addition, if f is a random variable – i.e., f is measurable w.r.t. the σ-algebra $\mathcal{A}$ on Ω and the Borel σ-algebra $\mathcal{Y}$ on Y, then it is obvious that $\sigma(f)$ is a sub-σ-algebra of $\mathcal{A}$.

A sub-σ-algebra $C \subseteq \mathcal{A}$ is said to be *essentially countably generated* if it is the *strong completion* of some countably generated σ-algebra C', in the sense that $C = \{ A \in \mathcal{A} \mid \exists A' \in C' : P(A \triangle A') = 0 \}$. For simplicity, from now on we describe a σ-algebra as countably generated even when it is only essentially countably generated. Of course, the extra sets in the essentially countably generated σ-algebra only differ from sets in the original countably generated σ-algebra by some null sets.

Let $\mathcal{M}(X, \mathcal{B})$ be the space of Borel probability measures on a Polish space $(X, \mathcal{B})$. We assume throughout that this space is equipped with the topology of weak convergence of measures. Indeed, $\mathcal{M}(X, \mathcal{B})$ with this topology is itself a Polish space – see, for example, [5], pp. 72–73. The following result on the measurability of mappings taking values in $\mathcal{M}(X, \mathcal{B})$ is often implicitly used in the literature. Since we are not able to find a precise reference, we give a proof here for the sake of completeness.

Lemma 1 *Let μ be a mapping from Ω to the space $\mathcal{M}(X, \mathcal{B})$ endowed with the topology of weak convergence of measures. Let C be the smallest σ-algebra on Ω such that for each $B \in \mathcal{B}$ the mapping $\omega \mapsto \mu_\omega(B)$ is C-measurable from Ω to the real line with its Borel σ-algebra. Then $C = \sigma(\mu)$, the σ-algebra on Ω generated by μ.*

Proof. Let $\mathcal{F}$ be the family of closed sets in X. As shown in [4], p. 236, the specified topology of weak convergence on $\mathcal{M}(X, \mathcal{B})$ is generated by the family of subsets $\{\tau \in \mathcal{M}(X, \mathcal{B}) : \tau(F_i) < \rho(F_i) + \epsilon, i = 1, 2, \ldots, k\}$, where $\epsilon > 0, \rho \in \mathcal{M}(X, \mathcal{B})$, and the sets F_i are closed. It is also clear that $\omega \mapsto \mu_\omega$ is a measurable mapping into $\mathcal{M}(X, \mathcal{B})$ if and only if $\{\omega : \mu_\omega(F) < \rho(F) + \epsilon\}$ is measurable for each $\epsilon > 0$, each $\rho \in \mathcal{M}(X, \mathcal{B})$, and each closed set F in X. This is equivalent to the condition that $\omega \mapsto \mu_\omega(F)$ is measurable for each closed set F in X. This implies that $\sigma(\mu)$ is equal to C_0, defined as the smallest σ-algebra such that for each $F \in \mathcal{F}$ the mapping $\omega \mapsto \mu_\omega(F)$ is C_0-measurable. From this definition, it is easy to verify that $C_0 \subseteq C$, and hence $\sigma(\mu) \subseteq C$.

Consider the family $\mathcal{D} := \{B \in \mathcal{B} : \omega \mapsto \mu_\omega(B)$ is $\sigma(\mu)$-measurable$\}$. It is easy to verify that this family is a λ-system. We have just proved that $\mathcal{D}$ contains the family $\mathcal{F}$ of closed sets in X. Because $\mathcal{F}$ is a π-system, Dynkin's $\pi - \lambda$ Theorem implies that $\sigma(\mathcal{F}) \subset \mathcal{D}$. But $\sigma(\mathcal{F}) = \mathcal{B}$ by definition of the Borel σ-algebra. So $\mathcal{D} = \mathcal{B}$, implying that $\omega \mapsto \mu_\omega(B)$ is $\sigma(\mu)$-measurable for each $B \in \mathcal{B}$. Hence $C \subset \sigma(\mu)$.

Combining the results of these two paragraphs shows that $C = \sigma(\mu)$, as required.
□

2.5 Conditional expectations and regular conditional distributions

For the convenience of the reader, this subsection recapitulates the standard definitions of conditional expectation, conditional probability, and regular conditional distribution.[9]

Let C be a sub-σ-algebra of A, and f an integrable real-valued function on (Ω, A, P). An integrable real-valued function h on (Ω, A, P) is said to be the *conditional expectation* of f given C if h is C-measurable, and $\int_A f \, dP = \int_A h \, dP$ for all $A \in C$. This h is essentially unique and usually denoted by $\mathbb{E}(f|C)$.

For a π-system C^π that contains Ω and generates C, if φ is a C-measurable and P-integrable function on Ω that satisfies $\int_A f \, dP = \int_A \varphi \, dP$ for all $A \in C^\pi$, then Dynkin's π–λ Theorem implies that $\varphi = \mathbb{E}(f|C)$.

Let A be an event in A. The *indicator function* $1_A : \Omega \to \{0,1\}$ is defined by $1_A(\omega) = 1$ if $\omega \in A$, and $1_A(\omega) = 0$ otherwise. The *conditional probability* $P(A|C)$ of the event A given C is simply the conditional expectation $\mathbb{E}(1_A|C)$ of this indicator function.

Let f be a random variable from Ω to a Polish space X with Borel σ-algebra B. A mapping μ from Ω to $M(X, B)$ is said to be a *regular conditional distribution* (r.c.d.) for f given C if for each fixed $B \in B$, the mapping $\omega \mapsto \mu_\omega(B)$ is a version of $P(f^{-1}(B)|C)$ – i.e., $\mu_\omega(B) = \mathbb{E}(1_{f^{-1}(B)}|C)$. A classical result of Doob says that an r.c.d. exists if the mapping takes values in a "nice" space, including any Polish space with its Borel σ-algebra – see [13], pp. 33 and 230. In particular, an r.c.d. exists for f given C, which is denoted by $P(f^{-1}|C)$.

The following lemma is often used to compute conditional expectations – see [7], p. 223.

Lemma 2 *Suppose that f is a random variable from Ω to a Polish space X, and that $\mu = P(f^{-1}|C)$ is an r.c.d. for f given C. Let φ be any real-valued function on X such that $\varphi(f)$ is integrable on (Ω, A, P). Then $\mathbb{E}(\varphi(f)|C) = \int_X \varphi(x) \, d\mu_\omega(x)$.*

From Dynkin's π–λ Theorem, it is easy to obtain the following useful lemma.

Lemma 3 *Suppose that f is a random variable from Ω to a Polish space (X, B). Suppose that C is a sub-σ-algebra of A which is generated by a π-system C^π. Let B^π denote a countable π-system that generates B. Let μ' be an r.c.d. for f given C, and μ a mapping from Ω to $M(X, B)$ such that $\mu_\omega(B)$ is C-measurable for each $B \in B^\pi$. Suppose finally that $P(C \cap f^{-1}(B)) = \int_C \mu_\omega(B) dP$ for each $C \in C^\pi$ and $B \in B^\pi$. Then μ is also an r.c.d. for f given C, and $\mu_\omega = \mu'_\omega$ for P-a.e. $\omega \in \Omega$.*

Proof. For each $C \in C^\pi$ and $B \in B^\pi$, since μ' is an r.c.d. for f given C, one has

$$P(C \cap f^{-1}(B)) = \int_C 1_{f^{-1}(B)} dP = \int_C \mu'_\omega(B) dP.$$

Hence, for each fixed $B \in B^\pi$, one has $\int_C \mu_\omega(B) dP = \int_C \mu'_\omega(B) dP$ for all $C \in C^\pi$. Because C^π is a π-system generating C, for each $B \in B^\pi$ it follows that

---

[9] The reader can find more of their properties set out in Chapter 7 of [7] and Chapter 4 of [13].

$\mu_\omega(B) = \mu'_\omega(B)$ for P-a.e. $\omega \in \Omega$. But $\mathcal{B}^\pi$ is a countable π-system, so we can group countably many P-null sets together to show that, for P-a.e. $\omega \in \Omega$, one has $\mu_\omega(B) = \mu'_\omega(B)$ simultaneously for all $B \in \mathcal{B}^\pi$, and hence for all $B \in \mathcal{B}$ by Dynkin's π–λ Theorem. $\square$

2.6 A continuum of random variables

We assume throughout that the economic uncertainty of interest can be modeled as a process $g : T \times \Omega \to X$ with the property that, for each $t \in T$, the component mapping $\omega \mapsto g_t(\omega)$ is measurable, thus making every g_t a random variable defined on $(\Omega, \mathcal{A}, P)$. In this sense, provided that T has the cardinality of the continuum, we have a continuum of random variables g_t $(t \in T)$.[10]

2.7 Pairwise measurable probability

For simplicity, we also assume throughout that the process $g : T \times \Omega \to X$ has *pairwise measurable probabilities* in the sense that, for each $A \in \mathcal{A}$ and $B_1, B_2 \in \mathcal{B}$, the mapping $(t_1, t_2) \mapsto P(A \cap g_{t_1}^{-1}(B_1) \cap g_{t_2}^{-1}(B_2))$ is measurable w.r.t. the product σ-algebra $\mathcal{T} \otimes \mathcal{T}$ on the set of pairs $T \times T$.

To motivate this assumption, consider what happens when the pair of random variables g_{t_1} and g_{t_2} is sampled by drawing (t_1, t_2) at random from the product space $(T, \mathcal{T}, \lambda)^2 = (T \times T, \mathcal{T} \otimes \mathcal{T}, \lambda \times \lambda)$, before ω is drawn at random from $(\Omega, \mathcal{A}, P)$. The pairwise measurable probability condition implies that, for each $A \in \mathcal{A}$ and $B_1, B_2 \in \mathcal{B}$, there should be a well-defined joint probability that $\omega \in A$ and that $g_{t_1}(\omega) \in B_1$, $g_{t_2}(\omega) \in B_2$. Of course, this joint probability is given by

$$\int_{T \times T} P(A \cap g_{t_1}^{-1}(B_1) \cap g_{t_2}^{-1}(B_2)) \, d(\lambda \times \lambda)$$

It is important to realize that this measurability condition does *not* imply that, for P-a.e. fixed ω, the mapping $(t_1, t_2) \mapsto (g_{t_1}(\omega), g_{t_2}(\omega))$ is measurable with respect to the product σ-algebra $\mathcal{T} \otimes \mathcal{T}$ on $T \times T$, or even that almost all sample functions $t \mapsto g_\omega(t)$ are measurable with respect to the σ-algebra $\mathcal{T}$ on T.

2.8 Monte Carlo convergence and the Monte Carlo σ-algebra

We assume that the process g, and the associated continuum of random variables g_t $(t \in T)$, are intended to model an economy with many agents who face individual random shocks. It is natural to try to understand this process by considering what can be observed in a population formed by taking a random sequential draw from the agent space T. Such a general procedure may be called "Monte Carlo simulation" because of its similarity to the classical Monte Carlo method.

---

[10] Actually, most of the results presented in this paper seem to require only that g_t is a random variable for almost all t.

Take any set J in the product σ-algebra $\mathcal{T} \otimes \mathcal{B}$ on $T \times X$. Given a typical sequential draw $t^\infty \in T^\infty$, one can consider the finite sample $t_1, t_2, \ldots, t_n$ for each n. A relevant question then is whether the proportion of these n agents for whom the pair $(t, g(t, \omega))$ belongs to J will converge, as $n \to \infty$. This suggests the following:

Definition 1 *The process g is said to be* Monte Carlo convergent *if there is a function $\gamma : \Omega \to \mathcal{M}(T \times X, \mathcal{T} \otimes \mathcal{B})$ such that, for each $J \in \mathcal{T} \otimes \mathcal{B}$, the mapping $\omega \mapsto \gamma_\omega(J)$ is $\mathcal{A}$-measurable, and for λ^∞-a.e. sequence $t^\infty \in T^\infty$,*

$$\frac{1}{n} \sum_{i=1}^{n} 1_J(t_i, g(t_i, \omega)) \xrightarrow[P-a.s.]{} \gamma_\omega(J).$$

In this case, we say that the measure-valued random variable γ_ω is the Monte Carlo limit *of g.*

For the processes considered in this paper, one only need consider the convergence in Definition 1 for those sets J in the form of measurable rectangles $S \times B$ with $S \in \mathcal{T}, B \in \mathcal{B}$ (see Theorem 1 below). In this case, the convergence property simply holds for the fraction of the n randomly drawn random variables whose index t and value $x = g(t, \omega)$ lie in the sets S and B respectively. This convergence property seems to be a minimal requirement for Monte Carlo simulation of the process g to make any sense at all.

When g is Monte Carlo convergent, it is natural to focus attention on the uncertainty represented by the Monte Carlo limit probability measure.

Definition 2 *When the process g is Monte Carlo convergent, the Monte Carlo σ-algebra $\mathcal{C}^g$ generated by g is the smallest sub-σ-algebra of $\mathcal{A}$ such that the mapping $\omega \mapsto \gamma_\omega(J)$ is $\mathcal{C}^g$-measurable for each $J \in \mathcal{T} \otimes \mathcal{B}$.*

One important aim of this paper is to show that, conditioned on the information represented by $\mathcal{C}^g$, the randomness faced by individual economic agents in our symmetric setting is (essentially) i.i.d. In this sense, the Monte Carlo σ-algebra represents all the macroeconomic uncertainty.

3 The independent case

3.1 Monte Carlo simulation in the i.i.d. case

Let f be a mapping from $T \times \Omega$ to X with the property that, for each $t \in T$, the component mapping f_t is a random variable defined on $(\Omega, \mathcal{A}, P)$. Suppose that the random variables f_t $(t \in T)$ are i.i.d., with common distribution μ. That is, for each $n > 1$ and each $B_1, B_2, \ldots, B_n \in \mathcal{B}$, one has $P(\cap_{i=1}^n f_{t_i}^{-1}(B_i)) = \prod_{i=1}^n \mu(B_i)$ for any n points $t_1, t_2, \ldots, t_n \in T$.

Now let $\varphi : X \to \mathbb{R}$ be any μ-integrable function, with mean $m = \int_X \varphi(x) d\mu$. Then the functions defined by $h_t(\omega) := \varphi(f_t(\omega))$ (all $t \in T$) are also i.i.d. random variables, with common mean $m = \int_X \varphi(x) d\mu = \int_\Omega h_t(\omega) dP$.

Take any sequence $t^\infty = (t_1, t_2, \ldots,)$ from the countably infinite product space $(T, \mathcal{T}, \lambda)^\infty = (T^\infty, \mathcal{T}^\infty, \lambda^\infty)$ with $t_i \neq t_j$ for $i \neq j$. Then it is obvious that the sequence of random variables h_{t_i}, $i = 1, 2, \ldots$ is also i.i.d., with common mean m. Since λ is assumed to be non-atomic, the sequence h_{t_i} is i.i.d. for λ^∞-a.e. $t^\infty \in T^\infty$. When they are i.i.d., of course, the usual strong law of large numbers (see, for example, [13], Theorem 8.3 on p. 52) implies that the obvious sample average $\frac{1}{n} \sum_{i=1}^n h_{t_i}(\omega)$ converges P-a.s. to m as $n \to \infty$.

An important special case occurs when φ is the indicator function 1_B of some measurable set $B \in \mathcal{B}$. Then $\frac{1}{n} \sum_{i=1}^n 1_B(f_{t_i}(\omega))$ is the fraction of the n randomly drawn random variables whose value lies in B; this fraction must converge P-a.s. to the mean $\int_X 1_B(x) d\mu = \mu(B)$, which is the common probability that each $f_t(\omega)$ lies in B.

Furthermore, take any $S \in \mathcal{T}$. Then the usual strong law of large numbers implies that for λ^∞-a.e. $t^\infty \in T^\infty$, $\frac{1}{n} \sum_{i=1}^n 1_S(t_i)$ converges to $\lambda(S)$. Thus, for λ^∞-a.e. $t^\infty \in T^\infty$, the sequence $1_S(t_i)[1_B(f_{t_i}) - \mu(B)]$ $(i = 1, 2, \ldots)$ of uniformly bounded random variables is independent with mean zero. Another version of the law of large numbers (see [13], Theorem 8.2 on p. 52) therefore implies that the sequence $\frac{1}{n} \sum_{i=1}^n 1_S(t_i)[1_B(f_{t_i}(\omega)) - \mu(B)]$ converges P-a.s. to 0 as $n \to \infty$. Hence, for λ^∞-a.e. $t^\infty \in T^\infty$, $\frac{1}{n} \sum_{i=1}^n 1_S(t_i) 1_B(f_{t_i}(\omega))$ converges P-a.s. to $\lambda(S)\mu(B)$.[11] Of course, this result is just a version of the classical law of large numbers. Much more striking is the fact, shown below, that an "almost everywhere" or "essential" version of the i.i.d. condition is necessary for this convergence property to hold.

3.2 Necessary and sufficient conditions

The family of random variables g_t $(t \in T)$ is said to be *essentially pairwise independent* if the two random variables g_{t_1} and g_{t_2} are independent for $\lambda \times \lambda$-a.e. pair $(t_1, t_2) \in T \times T$.[12] If in addition there is a Borel probability measure μ on the Polish space $(X, \mathcal{B})$ such that g_t has distribution μ for λ-a.e. $t \in T$, then the process g is said to be *essentially i.i.d.*, and μ is the *essentially common distribution*.

Remark 1 Let f be a mapping from $T \times \Omega$ to X with the property that each component function f_t is a random variable defined on $(\Omega, \mathcal{A}, P)$. When the process f is i.i.d., it has pairwise measurable probabilities.

Proof. Let the mapping $(t_1, t_2, \omega) \mapsto F((t_1, t_2), \omega) := (f_{t_1}(\omega), f_{t_2}(\omega))$ define the process $F : T^2 \times \Omega \to X^2$ on the index space $(T, \mathcal{T}, \lambda)^2$ instead of on $(T, \mathcal{T}, \lambda)$. Because the random variables f_t $(t \in T)$ are mutually independent, it follows that the random variables $F_{(t_1, t_2)}(\omega)$ $((t_1, t_2) \in T^2)$ must be essentially pairwise independent. We treat this F as the process g in [18]. Take $E = T^2 \times A \times (B_1 \times B_2)$, where $A \in \mathcal{A}$ and $B_1, B_2 \in \mathcal{B}$. Part (1) of Theorem 1 in [18] implies that the

---

[11] As noted in the equivalence of conditions (2) and (3) in Proposition 1, this means that the process g is Monte Carlo convergent, with constant limit $\lambda \times \mu$.

[12] A condition of this type is called "almost sure pairwise independence" in [28] and [29].

mapping on T^2 defined by $(t_1, t_2) \mapsto P(H^{-1}_{(t_1,t_2)}(E_{(t_1,t_2)})) = P(A \cap f_{t_1}^{-1}(B_1) \cap f_{t_2}^{-1}(B_2))$ is λ^2-integrable, so measurable w.r.t. $\mathcal{T} \otimes \mathcal{T}$. $\qquad\square$

The following result is an obvious implication of Theorems 1 and 2 in Section 5 below. It states that an essentially i.i.d. process is characterized by degeneracy of the Monte Carlo limit.

Proposition 1 *The following three conditions are equivalent:*

1. *the process g is essentially i.i.d., with an essentially common distribution μ;*
2. *for each $S \in \mathcal{T}$ and $B \in \mathcal{B}$, one has, for λ^∞-a.e. sequence $t^\infty \in T^\infty$,*

$$\frac{1}{n}\sum_{i=1}^n 1_S(t_i)\, 1_B(g(t_i, \omega)) \xrightarrow[P-a.s.]{} \lambda(S)\mu(B);$$

3. *the process g is Monte Carlo convergent to the fixed product probability measure $\lambda \times \mu$ on $(T \times X, \mathcal{T} \otimes \mathcal{B})$.*

In an extended framework where the process g is jointly measurable and the usual Fubini property still holds, [28]–[31] show that essential pairwise independence is necessary as well as sufficient for an exact law of large numbers to hold. Proposition 1 is a counterpart of this result in the sequential or Monte Carlo setting considered in this paper. From another point of view, (1) $\Longrightarrow$ (2) is simply an obvious version of the "classical" law of large numbers restated in the continuum setting, while (2) $\Longrightarrow$ (1) is a converse of the classical law of large numbers in this setting.

4 Essential symmetry, pairwise exchangeability, and conditional independence

4.1 Essentially symmetric processes

The main focus of this paper is on general *essentially symmetric* processes $g : T \times \Omega \to X$ satisfying the condition that, for each $A \in \mathcal{A}$ and $B \in \mathcal{B}$, the probability $P(A \cap g_t^{-1}(B))$ is λ-a.e. independent of t. The following lemma characterizes such processes.

Lemma 4 *Let g be an essentially symmetric process. Then there exists a measurable mapping $\omega \mapsto \mu_\omega$ from $(\Omega, \mathcal{A})$ to $\mathcal{M}(X, \mathcal{B})$ such that, for each $A \in \mathcal{A}, B \in \mathcal{B}$, one has*

$$P(A \cap g_t^{-1}(B)) = \int_A \mu_\omega(B)dP$$

for λ-a.e. $t \in T$.

Proof. For each $A \in \mathcal{A}$ and $B \in \mathcal{B}$, define $c(A, B)$ as the common value of $P(A \cap g_t^{-1}(B))$, for λ-a.e. $t \in T$. For each fixed B, the mapping $A \mapsto c(A, B)$ is a measure on $(\Omega, \mathcal{A})$ which is absolutely continuous w.r.t. P. So there exists an essentially unique Radon–Nikodym derivative $\omega \mapsto \alpha^B(\omega)$ such that $c(A, B) = \int_A \alpha^B(\omega)\, dP$ for all $A \in \mathcal{A}$ and $B \in \mathcal{B}$.

Let $\mathcal{B}^\pi$ be a countable π-system that contains X and generates $\mathcal{B}$. Let $\mathcal{C}_0$ be the countably generated sub-σ-algebra of $\mathcal{A}$ generated by the family of mappings α^B ($B \in \mathcal{B}^\pi$). Let $\mathcal{C}_0^\pi$ be a countable π-system that contains Ω and generates $\mathcal{C}_0$. By grouping countably many λ-null sets together, we can find a set $T_0 \in \mathcal{T}$ with $\lambda(T_0) = 1$ such that, for each $t \in T_0$,

$$P(A \cap g_t^{-1}(B)) = c(A, B) = \int_A \alpha^B(\omega)\, dP \tag{1}$$

for all $A \in \mathcal{C}_0^\pi, B \in \mathcal{B}^\pi$. Fix any $t_0 \in T_0$. Let μ be the regular conditional distribution $P(g_{t_0}^{-1}|\mathcal{C}_0)$ of g_{t_0} given $\mathcal{C}_0$. This means that μ is a measurable mapping from $(\Omega, \mathcal{C}_0)$ to $\mathcal{M}(X, \mathcal{B})$ such that for each $A \in \mathcal{C}_0, B \in \mathcal{B}$,

$$\int_A 1_{g_{t_0}^{-1}(B)}\, dP = P(A \cap g_{t_0}^{-1}(B)) = \int_A \mu_\omega(B)\, dP. \tag{2}$$

Hence by Equations (1) and (2), we have

$$\int_A \alpha^B(\omega)\, dP = \int_A \mu_\omega(B)\, dP \tag{3}$$

for all $A \in \mathcal{C}_0^\pi, B \in \mathcal{B}^\pi$.

Since $\mathcal{C}_0$ is generated by the π-system $\mathcal{C}_0^\pi$, Dynkin's π–λ Theorem implies that Equation (3) is still valid for all $A \in \mathcal{C}_0, B \in \mathcal{B}^\pi$. For each $B \in \mathcal{B}^\pi$, since both $\alpha^B(\omega)$ and $\mu_\omega(B)$ are $\mathcal{C}_0$-measurable, essential uniqueness of the Radon–Nikodym derivative implies that $\alpha^B(\omega) = \mu_\omega(B)$ for P-almost all $\omega \in \Omega$. This means that Equation (3) still holds for all $A \in \mathcal{A}, B \in \mathcal{B}^\pi$, so

$$P(A \cap g_t^{-1}(B)) = c(A, B) = \int_A \alpha^B(\omega)\, dP = \int_A \mu_\omega(B)\, dP \tag{4}$$

for λ-a.e. $t \in T$. Since $\mathcal{B}$ is generated by the π-system $\mathcal{B}^\pi$, Dynkin's π–λ Theorem implies that for all $A \in \mathcal{A}$ and $B \in \mathcal{B}$, equation (4) still holds for λ-a.e. $t \in T$. □

4.2 Essential pairwise exchangeability

A collection of random variables is said to be *pairwise exchangeable* if there is a common distribution π on the product space $(X, \mathcal{B})^2$ such that all pairs of random variables from the collection have the same joint distribution π. A natural extension of this definition is to say that the process g is *essentially pairwise exchangeable* if there exists a common joint probability measure π on $(X, \mathcal{B})^2$ such that almost all pairs of random variables in $\{g_t : t \in T\}$ have the same joint distribution π – i.e., for $\lambda \times \lambda$-a.e. $(t_1, t_2) \in T \times T$, one has $P(g_{t_1}^{-1}(B_1) \cap g_{t_2}^{-1}(B_2)) = \pi(B_1 \times B_2) = \pi(B_2 \times B_1)$ for all $B_1, B_2 \in \mathcal{B}$.

The following lemma shows that essential pairwise exchangeability implies essential symmetry.

Lemma 5 *If the process g is essentially pairwise exchangeable, then it is essentially symmetric.*

Proof. Fix any $A \in \mathcal{A}$ and $B \in \mathcal{B}$. By definition of essential pairwise exchangeability, there exists a symmetric measure π on $(X, \mathcal{B})^2$ and a set T_1 with $\lambda(T_1) = 1$ such that, for each $t' \in T_1$, one has

$$P(g_{t'}^{-1}(B) \cap g_t^{-1}(B)) = \mathbb{E}(1_{g_{t'}^{-1}(B)} 1_{g_t^{-1}(B)}) = \pi(B \times B) \tag{5}$$

for λ-a.e. $t \in T$, and also

$$P(g_{t'}^{-1}(B)) = \mathbb{E}(1_{g_{t'}^{-1}(B)}) = \pi(B \times X) \tag{6}$$

Consider the Hilbert space $L_2(\Omega, \mathcal{A}, P)$, and let L be the smallest closed linear subspace which contains both the constant function $1 = 1_\Omega$ and also the family of indicator functions $\{ 1_{g_t^{-1}(B)} \mid t \in T_1 \}$. Let the function $h : \Omega \to \mathbb{R}$ be the orthogonal projection of the indicator function 1_A onto L, with $h^\perp$ as its orthogonal complement. By definition, $1_A = h + h^\perp$ where $h^\perp$ is orthogonal to each member of L. That is, $0 = \mathbb{E}(h^\perp 1) = \int_\Omega h^\perp dP$ and also $0 = \mathbb{E}(h^\perp 1_{g_t^{-1}(B)}) = \int_\Omega h^\perp 1_{g_t^{-1}}(B)dP$ for all $t \in T_1$. Because $1_A = h + h^\perp$, it follows that $\mathbb{E}(1_A 1_{g_t^{-1}(B)}) = \mathbb{E}(h 1_{g_t^{-1}(B)})$ for all $t \in T_1$, and also $P(A) = \mathbb{E}1_A = \mathbb{E}(1_A 1) = \mathbb{E}(h 1) = \mathbb{E}h$.

Next, because $h \in L$, there exists a sequence of functions

$$h_n = r_n 1 + \sum_{k=1}^{i_n} \alpha_n^k 1_{g_{t_n^k}^{-1}(B)} \ (n = 1, 2, \ldots)$$

with $t_n^k \in T_1$, as well as r_n and α_n^k $(k = 1, \ldots, i_n)$ all real constants, such that $h_n \to h$ in the norm of $L_2(\Omega, \mathcal{A}, P)$ – that is, $\int_\Omega (h_n - h)^2 dP \to 0$.

Let T_n^k be the set of t for which (5) holds when $t' = t_n^k$. By hypothesis, $\lambda(T_n^k) = 1$ because each $t_n^k \in T_1$. Define $T^* := T_1 \cap \left(\cap_{n=1}^\infty \cap_{k=1}^{i_n} T_n^k \right)$. Because T^* is the intersection of a countable family of sets all having measure 1 w.r.t. λ, it follows that $\lambda(T^*) = 1$. Also, for any $t \in T^*$, one has

$$P(A \cap g_t^{-1}(B)) = \mathbb{E}(1_A 1_{g_t^{-1}(B)}) = \mathbb{E}(h 1_{g_t^{-1}(B)}) = \lim_{n \to \infty} \mathbb{E}(h_n 1_{g_t^{-1}(B)}) \tag{7}$$

But (5) and (6) both hold whenever $t \in T^*$ and $t' = t_n^k$, so

$$\mathbb{E}(h_n 1_{g_t^{-1}(B)}) = r_n \mathbb{E}(1_{g_t^{-1}(B)}) + \sum_{k=1}^{i_n} \alpha_n^k \mathbb{E}(1_{g_{t_n^k}^{-1}(B)} 1_{g_t^{-1}(B)})$$

$$= r_n \pi(B \times X) + \sum_{k=1}^{i_n} \alpha_n^k \pi(B \times B).$$

It follows that $\mathbb{E}(h_n 1_{g_t^{-1}(B)})$ is independent of t, for all $t \in T^*$. But then $P(A \cap g_t^{-1}(B))$ must have the same property, by (7). Since $\lambda(T^*) = 1$, this completes the proof. $\square$

4.3 Essential conditional independence

For a given sub-σ-algebra $\mathcal{C} \subset \mathcal{A}$, two random variables $f_1, f_2 : \Omega \to X$ are said to be *conditionally independent* given $\mathcal{C}$ if, for every pair of Borel sets $B_1, B_2 \in \mathcal{B}$, the conditional probabilities satisfy

$$P(f_1^{-1}(B_1) \cap f_2^{-1}(B_2)|\mathcal{C}) = P(f_1^{-1}(B_1)|\mathcal{C}) \, P(f_2^{-1}(B_2)|\mathcal{C}) \qquad (8)$$

Lemma 6 *Suppose that $\mathcal{B}^\pi$ is a countable π-system that generates $\mathcal{B}$, and that $\mathcal{C}$ is a sub-σ-algebra of $\mathcal{A}$. For any two random variables $f_1, f_2 : \Omega \to X$, the following are equivalent:*

1. *the product r.c.d. $P(f_1^{-1}|\mathcal{C}) \times P(f_2^{-1}|\mathcal{C})$ is an r.c.d. for (f_1, f_2) given $\mathcal{C}$;*
2. *f_1 and f_2 are conditionally independent given $\mathcal{C}$;*
3. *equation (8) holds for all $B_1, B_2 \in \mathcal{B}^\pi$.*

Proof. (1) $\Longrightarrow$ (2) and (2) $\Longrightarrow$ (3) are obvious. (3) $\Longrightarrow$ (1) follows from Lemma 3. $\square$

The process g is said to be *essentially conditionally independent* given $\mathcal{C}$ if, for $(\lambda \times \lambda)$-a.e. $(t_1, t_2) \in T \times T$, the pair g_{t_1}, g_{t_2} is conditionally independent given $\mathcal{C}$. If, in addition, there is a mapping μ from Ω to $\mathcal{M}(X, \mathcal{B})$ such that μ is an r.c.d. for g_t given $\mathcal{C}$ for λ-a.e. $t \in T$, then g is said to be *essentially i.i.d. conditioned on $\mathcal{C}$.*

Lemma 7 *Suppose that there exists a countably generated σ-algebra $\mathcal{C}'$ such that the process g is essentially i.i.d. conditioned on $\mathcal{C}'$. Then g is essentially pairwise exchangeable.*

Proof. By hypothesis, there is a $\mathcal{C}'$-measurable mapping $\omega \mapsto \alpha_\omega$ from Ω to $\mathcal{M}(X, \mathcal{B})$ which is a version of the regular conditional distribution of g_t given $\mathcal{C}'$, for λ-a.e. $t \in T$. Since g is essentially i.i.d. conditioned on $\mathcal{C}'$, one has $P((g_t, g_{t'})^{-1}|\mathcal{C}')(\omega) = \alpha_\omega \times \alpha_\omega$ for $(\lambda \times \lambda)$-a.e. (t, t'). Thus, for each $V \in \mathcal{B} \otimes \mathcal{B}$,

$$P((g_t, g_{t'})^{-1}(V)) = \int_\Omega (\alpha_\omega \times \alpha_\omega)(V) \, dP.$$

Hence, $P((g_t, g_{t'})^{-1})$ is equal to the symmetric probability measure π defined by $\pi(V) := \int_\Omega (\alpha_\omega \times \alpha_\omega)(V) \, dP$ for each $V \in \mathcal{B} \otimes \mathcal{B}$. $\square$

5 Main results

5.1 First equivalence theorem

The first equivalence theorem is stated for a given measurable mapping μ from $(\Omega, \mathcal{A})$ to the space $\mathcal{M}(X, \mathcal{B})$ of measures on the Polish space $(X, \mathcal{B})$ equipped with the Borel σ-algebra corresponding to the topology of weak convergence of measures. The proof of this theorem is through Lemmas 8, 10 and 11 in the next section.

Theorem 1 *Suppose $\omega \mapsto \mu_\omega$ is a measurable mapping from $(\Omega, \mathcal{A})$ to $\mathcal{M}(X, \mathcal{B})$. Let $\mathcal{C}$ be the σ-algebra on Ω which is (countably) generated by this mapping. Then the following conditions are equivalent:*

1. *for each $A \in \mathcal{A}$ and $B \in \mathcal{B}$, one has $P(A \cap g_t^{-1}(B)) = \int_A \mu_\omega(B)\, dP$ for λ-a.e. $t \in T$;*
2. *the process g is essentially i.i.d. conditioned on $\mathcal{C}$, with $P(g_t^{-1}|\mathcal{C}) = \mu_\omega$ for λ-a.e. $t \in T$;*
3. *for each $S \in \mathcal{T}$, $B \in \mathcal{B}$, and for λ^∞-a.e. sequence $t^\infty \in T^\infty$, one has*

$$\frac{1}{n} \sum_{i=1}^{n} 1_S(t_i)\, 1_B(g(t_i, \omega)) \xrightarrow[P-a.s.]{} \lambda(S)\, \mu_\omega(B)$$

4. *the process g is Monte Carlo convergent, with Monte Carlo limit given by the product probability measure $\lambda \times \mu_\omega$ on $(T \times X, \mathcal{T} \otimes \mathcal{B})$.*

5.2 Second equivalence theorem

The second equivalence theorem uses the first to give necessary and sufficient conditions for the process to be essentially pairwise exchangeable. The equivalence of Conditions 1 and 4 below is a version of the classical De Finetti theorem which is appropriate in our setting, with a continuum of random variables.

Theorem 2 *The following four conditions are equivalent:*

1. *the process g is essentially pairwise exchangeable;*
2. *the process g is essentially symmetric;*
3. *there exists a measurable mapping μ from $(\Omega, \mathcal{A})$ to $\mathcal{M}(X, \mathcal{B})$, together with the corresponding countably generated σ-algebra $\mathcal{C} = \sigma(\mu)$, such that all four equivalent conditions of Theorem 1 are satisfied;*
4. *there exists a countably generated σ-algebra $\mathcal{C}'$ such that the process g is essentially i.i.d. conditioned on $\mathcal{C}'$.*

Proof. (1) $\implies$ (2) was shown in Lemma 5. By Lemma 4, (2) implies Condition 1 in Theorem 1, so (2) $\implies$ (3). Condition 2 in Theorem 1 trivially implies (4), so (3) $\implies$ (4). Finally, (4) $\implies$ (1) was shown in Lemma 7. $\qquad\square$

The following corollary follows easily from Theorems 1 and 2.

Corollary 1 *Assume that the process g is essentially pairwise exchangeable. Then*

1. *the Monte Carlo σ-algebra $\mathcal{C}^g$ equals $\sigma(\mu)$, where μ is the measurable mapping from $(\Omega, \mathcal{A})$ to $\mathcal{M}(X, \mathcal{B})$, as in the statement of Theorem 1;*
2. *the process g is essentially i.i.d. conditioned on $\mathcal{C}^g$.*

When a large economy with individual risk is modeled by an essentially pairwise exchangeable process g, this result shows that the corresponding Monte Carlo σ-algebra $\mathcal{C}^g$ does simulate macroeconomic uncertainty, in the sense that individual agents' random shocks are independent conditioned on $\mathcal{C}^g$.

5.3 Joint measurability implies perfect correlation

The following proposition shows that a process g satisfying the conditions stated above cannot be jointly measurable except in the completely trivial case when almost all the random variables g_t equal some fixed random variable, and so are perfectly correlated.

Proposition 2 *Suppose that the process g is $\mathcal{T} \otimes \mathcal{A}$-measurable, and satisfies any of the equivalent conditions of Theorem 2. Then there is a random variable α from Ω to X such that for λ-a.e. $t \in T$, $g_t(\omega) = \alpha(\omega)$ for P-a.e. $\omega \in \Omega$.*

Proof. By Theorem 2, condition 1 of Theorem 1 must be satisfied. So there exists a measurable mapping $\omega \mapsto \mu_\omega$ from $(\Omega, \mathcal{A})$ to $\mathcal{M}(X, \mathcal{B})$ such that, for each $A \in \mathcal{A}$ and $B \in \mathcal{B}$,

$$P(A \cap g_t^{-1}(B)) = \int_A \mu_\omega(B)dP$$

for λ-a.e. $t \in T$. Take any $S \in \mathcal{T}$. Because g is assumed to be $\mathcal{T} \otimes \mathcal{A}$-measurable, one can use the Fubini theorem and integrate the above equation over S to obtain

$$\int_{S \times A} 1_{g^{-1}(B)}d(\lambda \times P) = \int_S P(A \cap g_t^{-1}(B))d\lambda = \int_{S \times A} \mu_\omega(B)d(\lambda \times P).$$

Since all the measurable rectangles $S \times A$ form a π-system, Dynkin's π–λ Theorem implies that for each $B \in \mathcal{B}$ one has

$$\int_F 1_{g^{-1}(B)} \, d(\lambda \times P) = \int_F \mu_\omega(B) \, d(\lambda \times P)$$

for all $F \in \mathcal{T} \otimes \mathcal{A}$. So, by essential uniqueness of the Radon–Nikodym derivative, it follows that for each $B \in \mathcal{B}$ one has

$$1_{g^{-1}(B)}(t, \omega) = \mu_\omega(B) \tag{9}$$

for $\lambda \times P$-a.e. $(t, \omega) \in T \times \Omega$.

Let d be a metric on X and $\{x_n\}_{n=1}^\infty$ a dense sequence in X. Let $\mathcal{B}^\pi$ be the countable collection of all the open balls $B(x_n, 1/m)$ centered at x_n and with radius $1/m$, for $n, m \geq 1$. By grouping together countably many $(\lambda \times P)$-null sets, one can show that there exists a set $D \in \mathcal{T} \otimes \mathcal{A}$ with $(\lambda \times P)(D) = 1$ such that for each $(t, \omega) \in D$, equation (9) holds simultaneously for all $B \in \mathcal{B}^\pi$.

Consider any $(t, \omega) \in D$. Suppose $t' \in T$ is such that $g_\omega(t') \neq g_\omega(t)$. Then there exists a ball $B_0 \in \mathcal{B}^\pi$ such that $g_\omega(t) \in B_0$ but $g_\omega(t') \notin B_0$. Our hypotheses imply that

$$\mu_\omega(B_0) = 1_{g_\omega^{-1}(B_0)}(t) = 1 \neq 1_{g_\omega^{-1}(B_0)}(t') = 0$$

and so $(t', \omega) \notin D$. This proves that $g_\omega(t) = g_\omega(t')$ whenever $(t, \omega), (t', \omega) \in D$. Hence, for all ω such that the section $D_\omega \neq \emptyset$, there is a well-defined unique point $\alpha(\omega) \in X$ such that $g(t, \omega) = \alpha(\omega)$ for all $(t, \omega) \in D$, so for all $t \in D_\omega$ and also for all $\omega \in D_t$. Because $(\lambda \times P)(D) = 1$, the Fubini Theorem implies that the section D_ω satisfies $\lambda(D_\omega) = 1$ for P-a.e. $\omega \in \Omega$, and also $P(D_t) = 1$ for λ-a.e. $t \in T$. Hence, for λ-a.e. $t \in T$, one has $g_t(\omega) = \alpha(\omega)$ for P-a.e. $\omega \in \Omega$. Since g_t is $\mathcal{A}$-measurable for λ-a.e. $t \in T$, the function α must also be $\mathcal{A}$-measurable. □

6 Proof of Theorem 1

6.1 Proof that (1) $\Longrightarrow$ (2)

Lemma 8 *Suppose that for each $A \in \mathcal{A}$ and $B \in \mathcal{B}$, one has*

$$P(A \cap g_t^{-1}(B)) = \int_A \mu_\omega(B)\,dP \tag{10}$$

for λ-a.e. $t \in T$. Then the process g is essentially i.i.d. conditioned on $\mathcal{C}$, with $P(g_t^{-1}|\mathcal{C}) = \mu_\omega$ for λ-a.e. $t \in T$.

Proof. Let $\mathcal{C}^\pi = \{C_n\}_{n=1}^\infty$ and $\mathcal{B}^\pi = \{B_m\}_{m=1}^\infty$ be countable π-systems for $\mathcal{C}$ and $\mathcal{B}$ respectively. For each pair (m, n), there exists a set T_{mn} with $\lambda(T_{mn}) = 1$ such that for all $t \in T_{mn}$, equation (10) holds with $A = C_n$ and $B = B_m$. So for any $t \in T^* := \cap_{m=1}^\infty \cap_{n=1}^\infty T_{mn}$, equation (10) holds whenever $A = C_n$ and $B = B_m$, for all pairs (m, n) simultaneously. By Lemma 3 it follows that μ_ω must be a version of $P(g_t^{-1}|\mathcal{C})$, for all $t \in T^*$. It is also clear that $\lambda(T^*) = 1$.

Next, take any $t' \in T$, $B' \in \mathcal{B}$ and $C \in \mathcal{C}$. Because equation (10) holds when $A = C \cap g_{t'}^{-1}(B')$, it follows that

$$P(C \cap g_{t'}^{-1}(B') \cap g_t^{-1}(B)) = \int_C 1_{g_{t'}^{-1}(B')}\,\mu_\omega(B)\,dP$$

for λ-a.e. $t \in T$. Because μ_ω must be a version of $P(g_{t'}^{-1}|\mathcal{C})$ for all $t' \in T^*$, in particular $\mu_\omega(B') = \mathbb{E}(1_{g_{t'}^{-1}(B')}|\mathcal{C})$ for λ-a.e. $t' \in T$. But Lemma 1 implies that the mapping $\omega \mapsto \mu_\omega(B)$ is $\mathcal{C}$-measurable. It follows that $\int_C 1_{g_{t'}^{-1}(B')}\,\mu_\omega(B)\,dP = \int_C \mu_\omega(B')\,\mu_\omega(B)\,dP$ for λ-a.e. $t' \in T$. Thus, given any $C \in \mathcal{C}$ and any $B, B' \in \mathcal{B}$, for λ-a.e. $t \in T$, one has

$$P(C \cap (g_t, g_{t'})^{-1}(B \times B')) = \int_C (\mu_\omega \times \mu_\omega)(B \times B')\,dP \tag{11}$$

for λ-a.e. $t' \in T$. Because of the hypothesis in Section 2.7 that the mapping $(t, t') \mapsto P(A \cap g_t^{-1}(B) \cap g_{t'}^{-1}(B'))$ is $\mathcal{T} \otimes \mathcal{T}$-measurable for each $A \in \mathcal{A}$ and $B, B' \in \mathcal{B}$, it follows that (11) must hold on a $\mathcal{T} \otimes \mathcal{T}$-measurable set. But then the Fubini Theorem implies that for each $C \in \mathcal{C}$ and $B, B' \in \mathcal{B}$, equation (11) must hold $\lambda \times \lambda$-a.e. in $T \times T$.

In particular, for each triple (m, m', n) of positive integers, there exists $K_{mm'n} \in \mathcal{T} \otimes \mathcal{T}$ with $(\lambda \times \lambda)(K_{mm'n}) = 1$ such that for all $(t, t') \in K_{mm'n}$, equation (11) holds with $C = C_n$, $B = B_m$ and $B' = B_{m'}$. But then (11) holds for all pairs (t, t') in the intersection $K^* := \cap_{m=1}^\infty \cap_{m'=1}^\infty \cap_{n=1}^\infty K_{mm'n}$, which is a set whose measure w.r.t. $(\lambda \times \lambda)$ is 1.

Hence, for each $(t, t') \in K^*$, equation (11) holds for all $C \in \mathcal{C}^\pi$ and $B, B' \in \mathcal{B}^\pi$. Since $\{B \times B' : B, B' \in \mathcal{B}^\pi\}$ is a π-system that generates $\mathcal{B} \otimes \mathcal{B}$, Lemma 3 implies that $\mu_\omega \times \mu_\omega$ must be a version of $P((g_t, g_{t'})^{-1}|\mathcal{C})$. Putting $B' = X$ in (11) reduces it to $P(C \cap g_t^{-1}(B)) = \int_C \mu_\omega(B)\,dP$, so Lemma 3 implies similarly that μ_ω must be a version of $P(g_t^{-1}|\mathcal{C})$. The rest follows from Lemma 6. $\qquad\square$

6.2 Proof that (2) $\Longrightarrow$ (4)

Let $f : T \times X \to \mathbb{R}$ be a $\mathcal{T} \otimes \mathcal{B}$-measurable function with the property that the mapping $t \mapsto \mathbb{E}(f_t^2(g_t))$ defines an integrable function on T – i.e.,

$$\int_T \left[\int_\Omega f_t^2(g_t(\omega)), dP \right] d\lambda$$

exists. The following Lemma 10 proves a strengthened version of Condition 4, with the indicator function 1_J of any set $J \in \mathcal{T} \otimes \mathcal{B}$ replaced by f.

The result in Lemma 10 is somewhat similar to the classical law of large numbers for a sequence of i.i.d. random variables taking values in a Banach space – see [33], [15], and the detailed references in [21]. Lemma 10, however, involves a randomly selected sequence of real-valued square-integrable random variables $\omega \mapsto f(t_i, g(t_i, \omega))$ whose sample average converges almost surely on Ω, whereas the conclusion of the corresponding law of large numbers in a Banach space states only convergence in the Banach space norm (the L_2-norm in this case). Note that we need almost sure convergence here in order to prove Corollary 2 concerning weak convergence of the empirical distribution.[13]

The proof of Lemma 10 relies on the following elementary technical result.

Lemma 9 *Suppose that $\{a_i\}_{i=1}^\infty$ is a sequence of non-negative real numbers for which $\frac{1}{n} \sum_{i=1}^n a_i$ converges to the finite limit a as $n \to \infty$. Then the series $\sum_{n=1}^\infty a_n \frac{\log^2 n}{n^2}$ is convergent.*

Proof. We group terms and write $\sum_{n=1}^\infty a_n \frac{\log^2 n}{n^2}$ as $\sum_{m=1}^\infty b_m$, where each $b_m := \sum_{n=2^{m-1}}^{2^m-1} a_n \frac{\log^2 n}{n^2}$. Clearly, it is enough to show that $\sum_{m=1}^\infty b_m$ converges. But

$$b_m \leq \sum_{n=2^{m-1}}^{2^m-1} a_n \frac{\log^2 2^m}{(2^{m-1})^2} \leq 4^{1-m} (m \log 2)^2 \sum_{n=1}^{2^m} a_n = c_m \, m^2 \, 2^{-m}$$

where

$$c_m := 4 \log^2 2 \left(2^{-m} \sum_{n=1}^{2^m} a_n \right) \to 4a \log^2 2$$

as $m \to \infty$. This is enough to guarantee that $b_m \leq c \, 2^{-m/2}$ for a suitable value of the constant c, so the series $\sum_{m=1}^\infty b_m$ does converge. $\square$

Lemma 10 *Suppose that the process g is essentially i.i.d. conditioned on $\mathcal{C}$, with $P(g_t^{-1}|\mathcal{C}) = \mu_\omega$ for λ-a.e. $t \in T$. Let $f : T \times X \to \mathbb{R}$ be any $\mathcal{T} \otimes \mathcal{B}$-measurable*

---

[13] In fact, given any set $J \in \mathcal{T} \otimes \mathcal{B}$, the indicator function $1_J(t, g_t(\cdot))$ defines a mapping from T to the Banach space $L_\infty(\Omega)$ of essentially bounded random variables. Then, if Talagrand's condition of "proper measurability" were satisfied, we could apply the law of large numbers in $L_\infty(\Omega)$ – see [33], Theorem 8, p. 841. This would imply that the almost sure convergence property stated in Conditions 4 and 3 of Theorem 1 could be strengthened to convergence in the L_∞-norm. So far, however, we have not been able to verify Talagrand's condition in our context.

function with $\int_T [\int_\Omega f_t^2(g_t(\omega)) \, dP] \, d\lambda < \infty$. Then, for λ^∞-a.e. sequence $t^\infty \in T^\infty$, one has

$$\frac{1}{n} \sum_{i=1}^n f(t_i, g(t_i, \omega)) \xrightarrow[P-a.s.]{} \int_{T \times X} f(t, x) \, d(\lambda \times \mu_\omega). \tag{12}$$

Proof. Given the specified function f, for each $t \in T$ and $\omega \in \Omega$, define

$$\psi_t(\omega) := f_t(g_t(\omega)); \quad \varphi(t, \omega) := \int_X f_t(x) d\mu_\omega(x); \quad h_t(\omega) := \psi_t(\omega) - \varphi_t(\omega).$$

By hypothesis, ψ_t is square-integrable (and so P-integrable) on $(\Omega, \mathcal{A}, P)$ for λ-a.e. $t \in T$. Since $P(g_t^{-1}|\mathcal{C}) = \mu_\omega$, Lemma 2 implies that for λ-a.e. $t \in T$, one has $\mathbb{E}(\psi_t|\mathcal{C})(\omega) = \int_X f_t(x) d\mu_\omega(x) = \varphi(t, \omega)$ for P-a.e. $\omega \in \Omega$.

Now we apply the conditional version of Jensen's inequality for convex functions – see, for example, Theorem 10.2.7 of [12], or p. 225 of [13]. This inequality implies that $\mathbb{E}(\varphi_t^2) = \mathbb{E}[\mathbb{E}(\psi_t|\mathcal{C})]^2 \le \mathbb{E}(\psi_t^2)$ for λ-a.e. $t \in T$. Because ψ_t is square-integrable on $(\Omega, \mathcal{A}, P)$ for λ-a.e. $t \in T$, and $\mathbb{E}(\psi_t^2)$ is integrable on $(T, \mathcal{T}, \lambda)$, it is easy to see that φ_t and h_t have the same two properties. It follows from the joint measurability of f and the measurability of $\omega \mapsto \mu_\omega \in \mathcal{M}(X, \mathcal{B})$ that φ is $\mathcal{T} \otimes \mathcal{A}$-measurable. Then, because ψ_t is square-integrable on $(\Omega, \mathcal{A}, P)$ for λ-a.e. $t \in T$, and because $\mathbb{E}(\psi_t^2)$ is integrable on $(T, \mathcal{T}, \lambda)$, the Fubini Theorem implies that φ is square integrable w.r.t $\lambda \times P$.

Since g is essentially i.i.d. conditioned on $\mathcal{C}$, we have $P((g_t, g_{t'})^{-1}|\mathcal{C})(\omega) = \mu_\omega \times \mu_\omega$ for $\lambda \times \lambda$-a.e. $(t, t') \in T \times T$. Using Lemma 2 again,

$$\mathbb{E}(\psi_t \psi_{t'}|\mathcal{C})(\omega) = \int_{X \times X} f_t(x) f_{t'}(y) \, d(\mu_\omega(x) \times \mu_\omega(y)) = \int_X f_t \, d\mu_\omega \int_X f_{t'} \, d\mu_\omega.$$

From the definition of h_t, and because $\varphi_t = \mathbb{E}(\psi_t|\mathcal{C})$ for λ-a.e. $t \in T$, it is easy to see that

$$\mathbb{E}(h_t h_{t'}|\mathcal{C}) = \mathbb{E}(\psi_t \psi_{t'}|\mathcal{C}) - \mathbb{E}(\psi_t|\mathcal{C})\mathbb{E}(\psi_{t'}|\mathcal{C}) = 0$$

holds for $\lambda \times \lambda$-a.e. $(t, t') \in T \times T$. Hence there exists a $\mathcal{T} \otimes \mathcal{T}$-measurable set D such that $(\lambda \times \lambda)(D) = 1$ and $\mathbb{E}(h_t h_{t'}) = 0$ for all $(t, t') \in D$. Now define D^* as the set of all sequences $t^\infty = (t_i)_{i=1}^\infty \in T^\infty$ such that $(t_i, t_j) \in D$ for all $i, j \in \mathbb{N}$. An elementary argument shows that $\lambda^\infty(D^*) = 1$. Hence, for all $t^\infty \in D^*$, the random variables $(h_{t_i})_{i=1}^\infty$ are mutually orthogonal.

Now, since $\int_T \mathbb{E}(h_t^2) d\lambda < \infty$, the usual strong law of large numbers implies that for λ^∞-a.e. $t^\infty \in T^\infty$, $\frac{1}{n} \sum_{i=1}^n \mathbb{E}(h_{t_i}^2)$ converges to $\int_T \mathbb{E}(h_t^2) d\lambda$ as $n \to \infty$. Because $\frac{1}{n} \sum_{i=1}^n \mathbb{E}(h_{t_i}^2)$ converges, Lemma 9 implies that the moment condition $\sum_{n=1}^\infty \frac{1}{n^2} \log^2 n \, \mathbb{E}(h_{t_n}^2) < \infty$ of the strong law of large numbers in [11], Theorem 5.2, p. 158 is satisfied for λ^∞-a.e. $t^\infty \in T^\infty$. That result therefore applies to the random variables $(h_{t_i})_{i=1}^\infty$, because they are mutually orthogonal for λ^∞-a.e. $t^\infty \in T^\infty$. It implies that for λ^∞-a.e. $t^\infty \in T^\infty$, one has

$$\frac{1}{n} \sum_{i=1}^n h_{t_i}(\omega) \xrightarrow[P-a.s.]{} 0. \tag{13}$$

Because φ is square-integrable (and so integrable) on the product space $(T \times \Omega, \mathcal{T} \otimes \mathcal{A}, \lambda \times P)$, the Fubini Theorem implies that φ_ω is λ-integrable on T, for P-a.e. $\omega \in \Omega$; the usual strong law of large numbers then implies that for λ^∞-a.e. $t^\infty \in T^\infty$, $\frac{1}{n} \sum_{i=1}^n \varphi_\omega(t_i)$ converges to $\int_T \varphi_\omega(t) d\lambda(t)$. Using the Fubini theorem yet again, the relevant null sets can be interchanged, and so for λ^∞-a.e. $t^\infty \in T^\infty$,

$$\frac{1}{n} \sum_{i=1}^n \varphi_{t_i}(\omega) \xrightarrow[P-\text{a.s.}]{} \int_T \varphi_\omega(t) d\lambda(t). \tag{14}$$

By (13) and (14), for λ^∞-a.e. $t^\infty \in T^\infty$ we have

$$\frac{1}{n} \sum_{i=1}^n \psi_{t_i}(\omega) = \frac{1}{n} \sum_{i=1}^n h_{t_i}(\omega) + \frac{1}{n} \sum_{i=1}^n \varphi_{t_i}(\omega) \xrightarrow[P-\text{a.s.}]{} \int_T \varphi_\omega(t) d\lambda(t). \tag{15}$$

But $\int_T \varphi_\omega(t) d\lambda(t) = \int_T [\int_X f_t(x) d\mu_\omega(x)] d\lambda = \int_{T \times X} f(t,x) \, d(\lambda \times \mu_\omega)$, so the result follows from (15). $\square$

For each $x \in X$, let δ_x denote the degenerate probability measure attaching probability 1 to x. Then, for each single random draw $t^\infty \in T^\infty$ and each $\omega \in \Omega$, each measure $\mu_{t^\infty,\omega}^n := \frac{1}{n} \sum_{i=1}^n \delta_{g(t_i,\omega)}$ $(n = 1, 2, \ldots)$ is the empirical distribution of x given the n observations $g(t_i, \omega)$ $(i = 1, 2, \ldots, n)$. The following corollary says that μ_ω is identified as the (almost sure) weak limit of $\mu_{t^\infty,\omega}^n$.

Corollary 2 *For λ^∞-a.e. sequence $t^\infty \in T^\infty$, the empirical distribution $\mu_{t^\infty,\omega}^n$ converges weakly to μ_ω, for P-almost all $\omega \in \Omega$.*

Proof. We apply Theorem 6.6 on p. 47 of [27]. Because X is a Polish space, so in particular a separable metric space, this theorem implies that there exist a topologically equivalent metric on X and a sequence of bounded and uniformly continuous functions $\varphi_m : X \to \mathbb{R}$ $(m = 1, 2, \ldots)$ with the property that, for each $t^\infty \in T^\infty$ and each $\omega \in \Omega$, the distribution $\mu_{t^\infty,\omega}^n$ converges weakly to μ_ω if and only if the mean $\int_X \varphi_m(x) d\mu_{t^\infty,\omega}^n$ converges to $\int_X \varphi_m(x) d\mu_\omega$ as $n \to \infty$ for all $m = 1, 2, \ldots$ simultaneously.

For each fixed $m = 1, 2, \ldots$, because φ_m is measurable and bounded, the definition of $\mu_{t^\infty,\omega}^n$ and Lemma 10 together imply that for λ^∞-a.e. sequence $t^\infty \in T^\infty$,

$$\int_X \varphi_m(x) d\mu_{t^\infty,\omega}^n = \frac{1}{n} \sum_{i=1}^n \varphi_m(g(t_i,\omega)) \xrightarrow[P-\text{a.s.}]{} \int_X \varphi_m(x) d\mu_\omega. \tag{16}$$

Because one can group together countably many λ^∞-null sets, there exists a subset T_1^∞ of T^∞ with $\lambda^\infty(T_1^\infty) = 1$ such that for each sequence $t^\infty \in T_1^\infty$, Equation (16) holds for all m simultaneously. Consider any sequence $t^\infty \in T_1^\infty$. Again, because one can group together countably many P-null sets, for P-almost all $\omega \in \Omega$ one has $\int_X \varphi_m(x) d\mu_{t^\infty,\omega}^n \to \int_X \varphi_m(x) d\mu_\omega$ for all m simultaneously. This implies that for each sequence $t^\infty \in T_1^\infty$, the sufficient condition for $\mu_{t^\infty,\omega}^n$ to converge weakly to μ_ω is satisfied for P-almost all $\omega \in \Omega$. $\square$

6.3 Proof that (4) $\Longrightarrow$ (3) $\Longrightarrow$ (1)

Of course, (4) $\Longrightarrow$ (3) is obvious. To complete the proof of Theorem 1, therefore, we only need to prove the following lemma.

Lemma 11 *Suppose that for each $S \in \mathcal{T}$, $B \in \mathcal{B}$, and for λ^∞-a.e. sequence $t^\infty \in T^\infty$, one has*

$$\frac{1}{n}\sum_{i=1}^{n} 1_S(t_i)\, 1_B(g(t_i,\omega)) \xrightarrow[P-a.s.]{} \lambda(S)\, \mu_\omega(B) \tag{17}$$

Then for each $A \in \mathcal{A}$ and $B \in \mathcal{B}$, one has $P(A \cap g_t^{-1}(B)) = \int_A \mu_\omega(B)\, dP$ for λ-a.e. $t \in T$.

Proof. Integrating (17) w.r.t. ω over any measurable set $A \in \mathcal{A}$ yields the result that, for λ^∞-a.e. $t^\infty \in T^\infty$ one has

$$\frac{1}{n}\sum_{i=1}^{n} 1_S(t_i) \int_A 1_B(g(t_i,\omega))\, dP$$

$$= \frac{1}{n}\sum_{i=1}^{n} 1_S(t_i)\, P(A \cap g_{t_i}^{-1}(B)) \to \lambda(S) \int_A \mu_\omega(B)\, dP \tag{18}$$

Now, the hypothesis that probabilities are pairwise measurable clearly implies that $t \mapsto P(A \cap g_t^{-1}(B))$ is $\mathcal{T}$-measurable. It follows that for any $S \in \mathcal{T}$, the mapping $t \mapsto 1_S(t)\, P(A \cap g_t^{-1}(B))$ is also $\mathcal{T}$-measurable. By the usual strong law of large numbers, therefore,

$$\frac{1}{n}\sum_{i=1}^{n} 1_S(t_i) P(A \cap g_{t_i}^{-1}(B)) \to \int_T 1_S(t)\, P(A \cap g_t^{-1}(B))d\lambda$$

$$= \int_S P(A \cap g_t^{-1}(B))d\lambda \tag{19}$$

for λ^∞-a.e. $t^\infty \in T^\infty$. Because the two limits in (18) and (19) must be equal,

$$\int_S P(A \cap g_t^{-1}(B))\, d\lambda = \lambda(S) \int_A \mu_\omega(B)\, dP = \int_S \left[\int_A \mu_\omega(B)\, dP\right] d\lambda$$

for all $S \in \mathcal{T}$. By the essential uniqueness of the Radon–Nikodym derivative, it follows that

$$P(A \cap g_t^{-1}(B)) = \int_A \mu_\omega(B)\, dP$$

for λ-a.e. $t \in T$. $\square$

7 Concluding assessment

In mathematical economics, following the pioneering contributions of Vickrey [34] and Aumann [2], [3] respectively, it has become common to consider continuous density functions of relevant consumer characteristics, or more general economic models with a continuum of agents. These, of course, are mathematical abstractions which cannot hold exactly in any actual economy, with a finite set of agents. Nevertheless, they provide convenient approximations when used with appropriate care.

In models where agents face individual risk, the joint and sample measurability problems described in the introduction have made it difficult to provide rigorous foundations for the intuitively appealing idea that, with a continuum of agents, some version of the law of large numbers should hold exactly rather than approximately. The earlier work in [28], [29] and [31] shows that this obstacle can be overcome in an extended product measure-theoretic framework with the usual Fubini property. The only known examples of such a framework involve Loeb product spaces.

As mentioned in the introduction, if one adopts the asymptotic point of view by simply taking a randomly drawn sequence from a continuum of i.i.d. random variables, then the classical law of large numbers trivially applies to this sequence. This paper moves beyond this law to consider a large economy modeled by a continuum of essentially pairwise exchangeable random variables.[14] Our parameter space $(T, \mathcal{T}, \lambda)$ can be any atomless probability space, including the Lebesgue unit interval and hyperfinite Loeb spaces. Proposition 2 shows that the joint measurability problem cannot be avoided unless almost all agents' shocks are identical and thus perfectly correlated. Nevertheless, even without joint or sample measurability, it is shown that the "almost everywhere" or "essential" versions of the symmetry, pairwise exchangeability and conditional i.i.d. properties are all equivalent.

An important issue in macroeconomics is to devise a general mathematical framework allowing individual agents to face random idiosyncratic shocks which are independent when conditioned on suitably constructed random macroeconomic states. Ideally, the macroeconomic states should have a simple interpretation, and even be identifiable empirically. This paper shows how Monte Carlo simulation can achieve that purpose, with the macroeconomic states in the symmetric case considered here being just the weak limit of the empirical distributions obtained from a single random draw $t^{\infty} \in T^{\infty}$, as shown in Corollary 2.

The above results accord with a broad class of existing macroeconomic models. These include natural extensions to a continuum of agents of the models devised by Nielsen [26] involving simple independently distributed stationary (SIDS) processes. Such processes appear in a particularly simple and appealing class of rational belief equilibria of the kind considered by Mordecai Kurz and various collaborators – see especially [23] and [24]. Indeed, in such a continuum extension, a key part of the macroeconomic state would be the history of what proportions of agents have optimistic or pessimistic beliefs of various degrees at different times.

---

[14] We believe that even this type of symmetry condition can be greatly relaxed, as we plan to discuss in later work.

Finally, as a by-product of our work, we have shown that the fundamental probabilistic concept of (essential) independence constitutes a necessary condition for the classical sequential law of large numbers to hold. This converse result is entirely new in the extensive mathematical literature on the subject.

References

1. Anderson, R.M.: Non-standard analysis with applications to economics. In: Hildenbrand, W., Sonnenschein, H. (eds.) Handbook of mathematical economics, Vol. IV, ch. 39, pp. 2145–2208. Amsterdam: North-Holland 1991
2. Aumann, R.J.: Markets with a continuum of traders. Econometrica **32**, 39–50 (1964)
3. Aumann, R.J.: Existence of competitive equilibria in markets with a continuum of traders. Econometrica **34**, 1–17 (1966)
4. Billingsley, P.: Convergence of probability measures. New York: Wiley 1968
5. Billingsley, P.: Probability and measure, 3rd edn. New York: Wiley 1995
6. Chamberlain, G.: Econometrics and decision theory. Journal of Econometrics **95**, 255–283 (2000)
7. Chow, Y. S., Teicher, H.: Probability theory: Independence, interchangeability, martingales, 3rd edn. New York: Springer 1997
8. Cohn, D. L.: Measure theory. Boston: Birkhäuser 1980
9. Diestel, J., Uhl, Jr., J. J.: Vector measures. Providence, RI: American Mathematical Society 1977
10. Doob, J.L.: Stochastic processes depending on a continuous parameter. Transactions of the American Mathematical Society **42**, 107–140 (1937)
11. Doob, J.L.: Stochastic processes. New York: Wiley 1953
12. Dudley, R.M.: Real analysis and probability. New York: Chapman & Hall 1989
13. Durrett, R.: Probability: Theory and examples, 2nd edn. Belmont, CA: Wadsworth 1996
14. Feldman, M., Gilles, C.: An expository note on individual risk without aggregate uncertainty. Journal of Economic Theory **35**, 26–32 (1985)
15. Fremlin, D.H., Mendoza, J.: On the integration of vector-valued functions. Illinois Journal of Mathematics **38**, 127–147 (1994)
16. Geweke, J.: Monte Carlo simulation and numerical integration. In: Amman, H., Kendrick, D., Rust, J. (eds.) Handbook of computational economics, pp. 731–800. Amsterdam: North-Holland 1996
17. Green, E.J.: Individual level randomness in a nonatomic population. Economics Working Paper #ewp-ge/9402001 (1994)
18. Hammond, P.J., Sun, Y.N.: Joint measurability and the one-way Fubini property for a continuum of independent random variables. Stanford University, Department of Economics, Working Paper # 00-008 (2000)
19. Jackson, M.O., Kalai, E., Smorodinsky, R.: Bayesian representation of stochastic processes under learning: de Finetti revisited. Econometrica **67**, 875–893 (1999)
20. Judd, K.: The law of large numbers with a continuum of IID random variables. Journal of Economic Theory **35**, 19–25 (1985)
21. Khan, M.A., Sun, Y.N.: Weak measurability and characterizations of risk. Economic Theory **13**, 541–560 (1999)
22. Kohlberg, E., Reny, P.J.: Independence on relative probability spaces and consistent assessments in game trees. Journal of Economic Theory **75**, 280–313 (1997)
23. Kurz, M.: Rational beliefs and endogenous uncertainty. Economic Theory **8**, 383–397 (1996)
24. Kurz, M., Schneider, M.: Coordination and correlation in Markov rational belief equilibria. Economic Theory **8**, 489–520 (1996)
25. McCall, J.J.: Exchangeability and its economic applications. Journal of Economic Dynamics and Control **15**, 549–568 (1991)
26. Nielsen, C.K.: Rational belief structures and rational belief equilibria. Economic Theory **8**, 399–422 (1996)
27. Parthasarathy, K.R.: Probability measures on metric spaces. New York: Academic Press 1967
28. Sun, Y.N.: Hyperfinite law of large numbers. The Bulletin of Symbolic Logic **2**, 189–198 (1996)

29. Sun, Y.N.: A theory of hyperfinite processes: The complete removal of individual uncertainty via exact LLN. Journal of Mathematical Economics **29**, 419–503 (1998)
30. Sun, Y.N.: The almost equivalence of pairwise and mutual independence and the duality with exchangeability. Probability Theory and Related Fields **112**, 425–456 (1998)
31. Sun, Y.N.: On the sample measurability problem in modeling individual risks. Working Paper No. 99-25, Centre for Financial Engineering, National University of Singapore (1999)
32. Talagrand, M.: Pettis Integral and Measure Theory. Providence, Memoirs of the American Mathematical Society, No. 307 (1984)
33. Talagrand, M.: The Glivenko-Cantelli problem. Annals of Probability **15**, 837–870 (1987)
34. Vickrey, W.S.: Measuring marginal utility by reactions to risk. Econometrica **13**, 319–333 (1945)

A more reasonable model of insurance demand[*]

Donald J. Meyer[1] and Jack Meyer[2]

[1] Department of Economics, Western Michigan University, Kalamazoo, MI 49008, USA
(e-mail: donald.meyer@wmich.edu)
[2] Department of Economics, Michigan State University, East Lansing, MI 48824, USA
(e-mail: jmeyer@msu.edu)

Received: April 2, 2001; revised version: March 29, 2002

Summary. The analysis of the demand for insurance has been significantly affected by the failure to recognize the composite commodity theorem. This fact is demonstrated when the composite commodity theorem is used to modify the existing model so that insurance demand is more appropriately calculated. Comparative static properties of the modified model are derived and more reasonable comparative static results are obtained. Most importantly, in the modified model increases in wealth need not lead to a reduction in the quantity of insurance demanded even for decreasing absolute risk averse decision makers.

Keywords and Phrases: Insurance demand, Composite commodity theorem.

JEL Classification Numbers: G22, G11.

1 Introduction

The demand for insurance by an expected utility maximizing decision maker has been analyzed frequently during the past thirty-five years. This work includes [1, 2, 5–9]. In the majority of this analysis, one particular decision model has been used. In this model, the decision maker chooses the quantity of an insurance contract to purchase, and pays a specified price per unit for this insurance. This is the only decision that is explicitly modeled in the analysis. Within this context, many different questions concerning the optimal quantity of insurance have been addressed. These questions range from determining the effect of a change in wealth or the price of insurance on the quantity demanded, to evaluating how insurance demand is affected by the form of the indemnification function, the distribution of the loss, or the risk preferences of the decision maker.

---

[*] We thank the referee for helpful comments.
Correspondence to: J. Meyer

The analysis here introduces a small but significant modification to this standard insurance demand model. Recognizing that this is a demand model that includes three assets, the composite commodity theorem is used to specify the alternative to the purchase of insurance. That is, demand for insurance is calculated assuming that the quantities of each of the other assets included in the analysis can vary, but that their relative proportion in the portfolio is fixed. This is in contrast to the standard assumption where only the quantity of insurance and of the riskless asset can vary. The exact composite asset specified by the composite commodity theorem and its properties are described shortly.

There are two reasons for making this change to the standard model. One is theoretical, while the other is more practical. Theory, specifically the composite commodity theorem, suggests that if the demand for insurance or any other asset is to be analyzed under the assumption that only the quantity of insurance and of one other asset are variable, then that second asset should be a composite of all of the other risky and riskless assets in the decision maker's portfolio. Changing the standard model of insurance demand in this manner also has practical support in that the modified model yields more reasonable comparative static properties than those arising in the standard specification. Stating and demonstrating several of the more important comparative static theorems describing these properties is the main task of this paper.

The paper is organized as follows. First, in the next section, the standard insurance demand model is formulated as a portfolio decision to clearly show that there are three assets in this demand model. This is done mainly for exposition purposes, and the reformulation is carried out before any changes to the standard model are introduced. The portfolio formulation has the added advantage of allowing existing knowledge concerning optimal portfolios to be drawn upon. Furthermore, the portfolio setting is convenient for highlighting the fundamental difference between the model analyzed here and the standard one.

Section 3 examines the effect of a change in wealth and a change in the price of insurance on the reformulated demand for insurance. These two parameters, and their effect on the quantity demanded, are fundamental aspects of any demand analysis and have received considerable attention in the insurance literature. It is well known that in the standard model, decreasing absolute risk aversion implies that less insurance is demanded as wealth increases [8]. In fact, the quantity of insurance demanded can even increase with its own price in that model [1, 5]. On the other hand, with the modification introduced here, insurance is normal for a broad category of decreasing absolute risk averse decision makers. While the modified model does allow the possibility that insurance is inferior, this occurs only when the insured's risk aversion measure declines at a sufficiently rapid rate. Of course, the quantity of insurance demanded cannot increase as its price increases whenever insurance is normal.

Finally, in Section 4, a few additional comparative static theorems concerning the demand for insurance are presented. These include determining the effect of specific changes in the form of the indemnification function and the effect of a first-degree stochastic dominant change in the distribution of the loss on the demand for insurance. The analysis indicates that under usual risk preference assumptions, the comparative static findings in the modified model are somewhat more determinate than those derived in the standard model. Section 5 concludes and summarizes.

2 Models of insurance demand

With minor variation, most past studies of the demand for insurance have used the same decision model, one with a single decision variable. In this model, the insured maximizes expected utility from final outcome z by choosing the quantity of insurance, α. The final outcome is specified as $z = W_0 + M - x + \alpha[I(x) - P]$. In this expression, the parameter W_0 is the portion of initial wealth that is nonrandom, M is the value of the risky asset when no loss occurs, and x is the random loss whose support is in $[0, M]$. The insurance component of the expression, $\alpha[I(x) - P]$, consists of $\alpha \geq 0$ selected by the insured, an exogenously specified indemnification function, $I(x)$, which often is assumed to take a special form, and a price per unit of insurance, P, frequently assumed to be proportional to $E[I(x)]$.

As has been previously recognized, the linearity of this insurance demand model allows the model to be written instead as an equivalent model of portfolio choice. For the analysis carried out here, it is convenient to work in that context so the conversion to portfolio notation is made. This conversion is carried out in two steps. First, z is written as: $z = (W_0 - \alpha \cdot P) + V(M - x)/V + \alpha \cdot P(I(x)/P)$. In this step, terms are regrouped, and V is introduced to represent the price per unit for the insurable risky asset. V is some value between 0 and M. In the second step, the notation for return per dollar for each asset is introduced. Let $i(x) = I(x)/P$ denote the return from each dollar allocated to insurance, and $r_0(x) = (M - x)/V$ denote the return per dollar for the insurable risky asset. Each of these returns is risky and depends on the same random variable. The standard assumption is that the return per dollar from the riskless asset is one. Note that $r_0(x)$ is linearly decreasing in x. With this notation, the standard insurance demand model states that $\alpha \geq 0$ is chosen to maximize expected utility from $z = (W_0 - \alpha \cdot P)(1) + V \cdot r_0(x) + \alpha \cdot P \cdot i(x)$.

Writing the insurance demand model in this way makes it clear that the decision maker's portfolio consists of three assets: the riskless asset, the risky insurable asset and insurance. Thus, this demand model is one where total wealth or expenditure is allocated among three assets. Furthermore, it is clear that the way the standard model reduces the number of decision variables from two to one is by assuming that the quantity of the risky asset is fixed, and only the quantities of insurance and the riskless asset can vary. This is not the usual way to reduce the dimensionality of a demand model unless the asset whose quantity is held fixed is inconsequential to the analysis, which is clearly not the case. The assumption that the quantity of the risky asset is fixed, can and does significantly alter comparative static analysis. The response of the chosen quantity of insurance to changes in wealth, its own price, and other parameters is affected significantly. In a standard consumer setting, making this assumption is similar to determining the demand for hotdog buns, allowing the quantity of soda to vary, but holding the number of hotdogs fixed. While the analysis is feasible, the results would not typically be interpreted as demand analysis.

In demand analysis, the usual procedure for simplifying and reducing the number of decision variables to one is to assume that the prices of all assets other than insurance are held fixed, and to define a composite of the remaining assets as the second asset. This is precisely the change to the standard model that is made here. The composite commodity theorem is used to suggest how the three-asset insurance demand model can be reduced to a model involving only two assets, and to describe the characteristics of this second composite asset. While an even better procedure would be to analyze the demand for insurance allowing the quantities of all three assets to vary, this involves modeling a decision maker with two choice variables and leads to considerably more complicated analysis, and less sharp comparative static predictions. This has been analyzed by [2] for the case of coinsurance and by [6] for the deductible insurance case.

The composite commodity theorem indicates that when the quantities of insurance and only one other asset are assumed to be variable, then to more accurately preserve the comparative static properties of the general model, the second asset should be a particular composite of the several other assets that exist. This composite asset consists of the other assets in their optimal proportions. Therefore, for insurance demand, the composite asset is made up of both the riskless asset and the risky insurable asset. The return per dollar for this asset equals: $(W_0 - \alpha^* \cdot P)/(W_0 + V^* - \alpha^* \cdot P) + V^* \cdot r_0(x)/(W_0 + V^* - \alpha^* \cdot P)$, where V and α are at their optimal levels. Here V is both the number of dollars allocated to the risky asset and also its price since the number of units of the risky asset in the initial equilibrium portfolio is normalized to be one. The first term in this expression represents the return from the riskless portion of the composite asset, while the second is that from the risky portion. The composite commodity theorem then indicates that comparative static analysis should proceed assuming that the quantity of this composite asset can vary while its composition remains fixed.

To make the notation in the remaining sections of the paper as concise as possible, let $W = W_0 + V$ represent total initial wealth. This includes wealth held in the riskless and risky form, and let $r(x) = (W_0 - \alpha^* \cdot P)/(W_0 + V^* - \alpha^* \cdot P) + V^* \cdot r_0(x)/(W_0 + V^* - \alpha^* \cdot P)$ denote the return to the composite asset. Note that because $r_0(x)$ is linearly decreasing in x, $r(x)$ is also. With this notation, the modified model used throughout the analysis conducted here can be stated in a very compact way. The insured is assumed to choose $\alpha \geq 0$ to maximize expected utility from final outcome z, where $z = (W - \alpha \cdot P) \cdot r(x) + \alpha \cdot P \cdot i(x)$. Furthermore, an interior solution for this maximization is assumed to exist, and to be determined by the first order condition $Eu'(z) \cdot P \cdot (i(x) - r(x)) = 0$. $u'(z) \geq 0$ and $u''(z) < 0$ are assumed and thus the second order condition, $Eu''(z) \cdot P^2(i(x) - r(x))^2 < 0$, is satisfied.

This is a portfolio model with two risky assets. Two assumptions ensure that it is indeed a model of insurance demand, and these assumptions are maintained and used extensively throughout the comparative static theorems that follow. First, it is assumed that the return per dollar for the insurance asset is nonnegative and nondecreasing; that is, $i(x) \geq 0$ and $i'(x) \geq 0$ for all x. Since $i(x) = I(x)/P$, this is equivalent to requiring those same two properties for the indemnification function $I(x)$. This assumption specifies a defining characteristic of insurance. Insurance is an asset whose nonnegative return increases with the magnitude of the loss. No additional assumptions concerning $i(x)$ are used in the analysis conducted here.

The second assumption characterizes the selling side of the insurance market, and represents actions taken by the seller of insurance to prevent moral hazard. It is assumed that the final outcome, $z(x)$, is nonincreasing in x for all allowed choices of α and the given values for $W, P, i(x)$, and $r(x)$. This assumption indicates that the seller of insurance chooses an indemnification function, sets the price for insurance, and limits the choice of α so that the decision maker cannot prefer larger losses to smaller ones. An important difference between insurance, and an asset such as a put option, is precisely this restriction. While put options can be used to insure against losses, the analysis of the demand for put options is made more complex because one can choose to hold enough put options so that larger losses are preferred to smaller ones. Insurance demand models rule this out. By stating the assumption in this very general and very direct manner, many special sets of assumptions imposed in the insurance demand literature are incorporated into a single framework. Restrictions such as $I(x) = x, I'(x) \leq 1$ and $\alpha \leq 1$, are examples of assumptions that have been used to imply that $z(x)$ is decreasing in x for all choices of α.

No direct assumptions concerning $r(x)$ are made. The assumption that $z(x)$ is decreasing in x implies that $r(x)$ is also. As was noted earlier, the return to the composite commodity consisting of only the riskless and insurable risky asset, is in fact linearly decreasing in x. This linearity of $r(x)$ is not exploited in the analysis conducted here and is relatively unimportant unless $i(x)$ is also restricted to be linear or piecewise linear as is the case for coinsurance or deductible insurance. Consequently, as is argued more fully in the concluding section, the analysis conducted here holds in more general settings since the composite commodity could contain additional assets whose returns depend on x in a nonlinear manner.

3 Wealth and price effects

As indicated, the modified insurance demand model is a portfolio model with two risky assets. The proportion allocated to these two assets is chosen to maximize expected utility from final outcome $z = (W - \alpha \cdot P) \cdot r(x) + \alpha \cdot P \cdot i(x)$. In general, comparative static theorems cannot be demonstrated in such models unless significant restrictions are imposed. [3] and [4] verify this. The insurance assumptions, that $r(x)$ and $i(x)$ depend on the same random variable, and that the nonnegative $i(x)$ and $z(x)$ are nondecreasing and nonincreasing, respectively, provide the necessary restrictions.

Perhaps the most troublesome feature of the standard model of insurance demand is the implication that all decreasing absolute risk averse (DARA) decision makers purchase less insurance as their wealth increases. This is sometimes stated as "insurance is inferior under DARA", although this terminology is misleading or incorrect because the finding is derived under assumptions different from those typical when conducting demand analysis. The first theorem in this section indicates that this implication changes in a reasonable way when the standard insurance demand model is modified using the composite commodity theorem.

Theorem 1. $\partial\alpha/\partial W \geq 0$ when relative risk aversion is nondecreasing.

Proof. An interior solution is assumed and is characterized by $Eu'(z) \cdot P \cdot (i(x) - r(x)) = 0$. To determine the sign of the effect of increasing W, it is sufficient to determine the sign of $Eu''(z) \cdot (i(x) - r(x)) \cdot r(x) = -ER_A(z) \cdot r(x) \cdot u'(z) \cdot (i(x) - r(x))$, where $R_A(z)$ denotes the insured's absolute risk aversion measure. Moreover, when $[R_A(z) \cdot r(x)]$ is decreasing in x, this sign is positive. Let $R_R(z) = R_A(z) \cdot z$ denote the relative risk aversion measure. Then $R_A(z)r(x) = R_R(z)r(x)/z(x)$ and its derivative is: $R_R(z) \cdot [(z \cdot r' - r \cdot z')/z^2] + [r(x)/z(x)] \cdot R'_R(z) \cdot z'$. The entire second term in this expression is nonpositive under increasing or constant relative risk aversion since r and z are positive and z' is negative. The first term is also negative. To see this note that $z \cdot r' - r \cdot z' = [(W - \alpha \cdot P) \cdot r + \alpha \cdot P \cdot i] \cdot r' - r[(W - \alpha \cdot P) \cdot r' + \alpha \cdot P \cdot i'] = \alpha \cdot P(i \cdot r' - r \cdot i')$. With $r' < 0$ and $i' > 0$, this is negative and hence $R_A(z) \cdot r(x)$ decreases with x and $\partial\alpha/\partial W \geq 0$. $\square$

The risk preference assumption in this theorem, that relative risk aversion is nondecreasing, is a convenient way to limit the rate at which the absolute risk aversion measure decreases. The theorem indicates that among the set of DARA decision makers, insurance is normal for those whose absolute risk aversion measure does not decrease at too rapid a rate; that is, when $|R'_A|/R_A \leq (1/z)$ whenever $R'_A \leq 0$. This subset of DARA decision makers includes all those who are constant absolute and constant relative risk averse and those whose utility functions are convex combinations of these two important special cases. Furthermore, as in the standard model, when absolute risk aversion is increasing rather than decreasing, then insurance is also normal. Thus, in the modified model, insurance is normal for broad groups of decision makers, and the troublesome feature of inferiority is greatly alleviated by the change suggested by the composite commodity theorem.

An intuitive and verbal explanation for this result might be useful. In the standard model, where the quantity of the risky asset is held fixed and only the expenditures on insurance and the riskless asset can be varied, there are at least three distinct reasons why increases in wealth do not lead to a higher demand for insurance. First, with DARA more risk is desired and the only way to increase risk is to purchase less insurance. Second, the quantity of the insurable asset is kept fixed, so no additional insurance is needed to offset increased loss exposure. Finally, whether the additional wealth is used to purchase more insurance or more of the riskless asset, the insured's portfolio becomes less risky on average. To maintain the same average level of risk, purchasing less insurance is the only option.

In contrast, when expenditure on insurance and the composite risky asset can each be varied, the second and third of these reasons are reversed and the first is weakened. More wealth, if not allocated to insurance, implies that there is more of the risky, insurable asset to insure, and hence greater loss exposure. Furthermore, if the additional wealth is allocated only to the composite risky asset, then the insured's portfolio becomes riskier not only absolutely, but also on average. Finally, even though DARA implies that more risk is desired, now purchasing more of the composite asset can give this additional risk even when more insurance is purchased. Thus, one can interpret Theorem 1 as indicating that when the DARA effect is not too strong, the reasons for purchasing additional insurance dominate those for purchasing less.

Economists have also recognized that if insurance is inferior, that it is possible that the quantity demanded increases as its price increases. Papers on this topic have discussed the conditions on risk preferences that are consistent with this possibility. One necessary condition derived in a specialized setting is that relative risk aversion must be greater than one [5]. A more complex necessary and sufficient condition involves an expression including both the absolute risk aversion level and the level of the final outcome, and is related to relative risk aversion [1]. This latter condition is not simple to state or interpret and is not reproduced here. The only point drawn from this body of work is that the level of relative risk aversion is an important determinant of the relationship between the price of insurance and the quantity demanded in the standard model. The next theorem indicates that this is the case in the modified model as well.

Theorem 2. $\partial(\alpha \cdot P)/\partial P < 0$ *if relative risk aversion is less than or equal to one.*

Proof. Rewrite $z = (W - \alpha \cdot P) \cdot r(x) + \alpha \cdot P \cdot i(x)$ as $z = (W - A) \cdot r(x) + A \cdot i(x)$, where $A = \alpha \cdot P$ denotes the total expenditure on insurance. Treating A as the decision variable, an interior solution is characterized by $Eu'(z) \cdot (i(x) - r(x)) = 0$. Recall that $i(x) = I(x)/P$ and hence $\partial[i(x)]/\partial P = -i(x)/P$. Using this, the sign of the effect of increasing P on A is the same as the sign of $-[Eu'(z)i(x)/P + Eu''(z) \cdot (A/P) \cdot i(x) \cdot (i(x) - r(x))]$. Now $A(i(x) - r(x)) = z(x) - W \cdot r(x)$, which allows the expression to be written as $-(1/P)Eu'(z) \cdot i(x)[1 - R_R(z) + R_A(z) \cdot W \cdot r(x)]$ which is less than zero when $R_R(z) \leq 1$. Hence $\partial A/\partial P < 0$. $\square$

Corollary $\partial\alpha/\partial P < 0$ *if relative risk aversion is less than or equal to one.*

This corollary follows trivially since expenditure can only be reduced as price increases if there is a reduction in quantity purchased. This corollary can be demonstrated directly using the same steps as in the proof of Theorem 2 for the decision model where α is the choice variable.

Theorem 2 indicates that if relative risk aversion is low enough, not only does the quantity of insurance chosen decrease as its price increases, but the total expenditure on insurance decreases as well; that is, the elasticity of demand is greater than one. Thus, in the modified model, the price effect goes in the usual direction and strongly so for these decision makers. Combining the condition on risk preferences that implies normality in Theorem 1, with the condition in Theorem 2, it is clear that the modification to insurance demand analysis suggested by the composite commodity theorem implies that insurance is not Giffen under frequently assumed conditions on risk preferences. A few additional comparative static properties of the modified model are derived next.

4 More comparative static findings

The two primary comparative static questions of any demand analysis were discussed in the preceding section. Sometimes, however, the effects of other parameter changes are also considered. For insurance demand, a large number of additional questions have been considered, but most have been formulated in the context of a specific form for the indemnification function, usually the coinsurance or deductible form. A few questions, however, do not involve these specific forms, and two of these are examined here, including determining the effect of changing the indemnification function and of altering the distribution of the loss.

The first question addressed is related to Theorem 2 and its corollary. The specific question asked is when does an arbitrary increase in the return to insurance, holding price fixed, lead to an increased purchase of insurance? Such return increases occur when indemnification is increased. The method used to address this question has been used before in other contexts, and is particularly well suited for this analysis. The procedure used replaces the return to insurance, $i(x)$, by $[i(x) + \theta \cdot k(x)]$, and then asks how an increase in θ affects the demand for insurance. Doing this when $k(x)$ is an arbitrary nonnegative function is a convenient way to determine the effect of an increase in the indemnification function and hence the return to insurance.

Theorem 3. *If relative risk aversion is less than or equal to one, $\partial\alpha/\partial\theta \geq 0$ when* $k(x) \geq 0$.

Proof. Write z as $z = (W - \alpha \cdot P) \cdot r(x) + \alpha \cdot P \cdot [i(x) + \theta \cdot k(x)]$. When α is selected to maximize expected utility, the equation $Eu'(z)([i(x) + \theta \cdot k(x)] - r(x)) = 0$ determines α. The sign of the derivative of this expression with respect to θ determines the direction of the effect on α of increasing θ. This derivative is $[Eu'(z)k(x) + Eu''(z) \cdot \alpha \cdot P \cdot k(x) \cdot (i(x) + \theta \cdot k(x) - r(x))]$. Now $\alpha \cdot P \cdot (i(x) + \theta \cdot k(x) - r(x)) = z(x) - W \cdot r(x)$, which allows the expression to be written as $Eu'(z) \cdot k(x)[1 - R_R(z) + R_A(z) \cdot W \cdot r(x)]$. This is greater than zero when $R_R(z) \leq 1$.   □

Theorem 3 indicates that increasing the return to insurance by increasing indemnification leads to increased allocation to insurance under a standard assumption on risk preferences. The result is related to Theorem 2 where P is increased holding indemnification fixed, and hence reducing the return to insurance and the quantity demanded. The same condition on risk preferences and proof steps are used in the two theorems. The findings in Theorems 2 and 3 are not surprising since increased return might be expected to increase the demand for any asset. The next theorem, however, indicates the less intuitive result that increased riskiness for $i(x)$ is desired by all risk averse decision makers. Whether or not this leads to increases in the quantity of insurance selected, however, appears to be indeterminate. As in Theorem 3, the transformation methodology is employed to address this question.

Theorem 4. *If $u(z)$ is increasing and concave, $\partial Eu(z)/\partial\theta \geq 0$ whenever $k(x)$ satisfies*

$$\int_0^x k(x)dF(x) \leq 0 \quad \textit{for all} \quad x \text{ in } [0, M] \text{ and } \int_0^M k(x)dF(x) = 0 .$$

Proof. $\partial Eu(z)/\partial\theta = Eu'(z)(\alpha P)k(x)$. When $u(z)$ is increasing and concave, $u'(z)$ is nonnegative and increasing in x and hence this expectation is positive for the specified $k(x)$. □

The proof of Theorem 4 uses the insurance properties of the $z(x)$ function, and similar claims cannot be made for assets in general. The reason why this result holds for insurance and not other assets is that insurance is used to offset the risk associated with $r(x)$, and this offset is necessarily less than complete. Hence, increased variation in $i(x)$, which increases the risk offsetting ability of insurance, increases expected utility.

The mean preserving condition for indemnification functions imposed in Theorem 4 is a reasonable one since the price of insurance is fixed, and this price is often assumed to be proportional to expected indemnification. The result in Theorem 4 is consistent with and can be used to demonstrate the well-known result that the deductible form of insurance is optimal for all risk averse decision makers. Simply stated, the deductible form of indemnification is the riskiest form possible given the constraint that $i(x)$ and $r(x)$ satisfy the insurance assumptions. In the standard model, this is accomplished by assuming $I(x) \leq x, I'(x) \leq 1$ and that $E[I(x)]$ and P are fixed.

The changes to the return to insurance discussed in Theorems 3 and 4 do not alter the price and characteristics of the other assets in the demand model and hence are consistent with the use of the composite commodity theorem. For the final question addressed here this is no longer the case. Insurance demand studies have asked how the demand for insurance is affected by the nature of the risk involved. One version of this question asks how a first-degree stochastic dominant increase in x changes the quantity of insurance selected? Such a change in x affects not only the return on insurance, but also the return on the risky asset and hence the composite asset as well.

Theorem 5. *If relative risk aversion is less than or equal to one, first degree stochastic dominant increases in x lead to increases in α.*

Proof. It is sufficient to show that $u'(z)(i(x) - r(x))$ is increasing in x. Differentiating with respect to x one obtains $u'(i' - r') + u'' \cdot z'(i - r) = u'(i' - r')[1 - R_R \cdot z'(i - r)/((i' - r')z)]$. When $R_R \leq 1$, it is sufficient to show that $z'(i - r)/((i' - r')z) \leq 1$ or $z'(i - r) - (i' - r')z \leq 0$. Substituting for z and z' the expression is $[(W \cdot r' + \alpha P(i' - r')](i - r) - (i' - r')[(W \cdot r + \alpha P(i - r)] = W[r'(i - r) - r \cdot (i' - r')] = W[r' \cdot i - r \cdot i']$ which is always nonpositive. □

Since $i(x)$ is kept fixed even though the distribution of x has been altered, this theorem implicitly is holding indemnification and price of insurance fixed even though the loss distribution changes. The result can be interpreted as indicating that those decision makers with worse loss distributions in the FSD sense purchase more insurance when relative risk aversion is less than one. That is, adverse selection occurs. One can demonstrate a similar result with similar proof steps in the standard model. This FSD increase in x makes insurance, whose return is increasing in x, more desirable, and at the same time, makes the composite asset less desirable since $r(x)$ is decreasing. These two effects are reflected in the $u'(i' - r')$ term. The change in the loss distribution also reduces the real wealth level, however, and this effect is reflected in the $u'' \cdot z'(i - r)$ term. For normal goods, the size of the two effects must be compared, and the restriction on the magnitude of relative risk aversion ensures that the first effect dominates.

5 Conclusions

The analysis here has verified that the comparative static properties of demand models for insurance are sensitive to the assumption used to reduce the number of decision variables to one. The composite commodity theorem, and the limited set of results obtained in the demand model using it, indicates that more reasonable comparative static properties result when the second variable asset is specified as a composite asset rather than the usual assumption that it is the riskless asset. Making this point has been the main thrust of this paper. The analysis here in no way indicates that the results derived in the standard model are incorrect. Instead, the analysis indicates that the standard model is not the appropriate model to employ when discussing the *demand* for insurance.

Other comparative static properties remain to be examined. For instance, as in the standard model, it is very easy to show that more risk averse decision makers choose more insurance. Analysis of the special case of coinsurance, where $i(x)$ is linear, would likely make use of the linearity of $r(x)$, the return to the composite asset. As indicated, this linearity has not been exploited here, and as a result, the five theorems that have been demonstrated apply to insurance demand models with more than just these three assets. All that is required is that the return to the composite of assets other than insurance be decreasing in x. Finally, the special case of the demand for deductible insurance, where the quantity of insurance is selected by choosing a deductible level, does not fit the framework employed here and must be analyzed separately. It is likely that analysis of the properties of demand for this important special case would also be altered when addressed in the context of the composite commodity theorem.

References

1. Briys, E., Dionne, G., Eeckhoudt, L.: More on insurance as a giffen good. Journal of Risk and Uncertainty **2**, 415–420 (1989)
2. Eeckhoudt, L., Meyer, J., Ormiston, M.B.: The interaction between the demands for insurance and insurable assets. Journal of Risk and Uncertainty **14**, 25–39 (1997)
3. Hadar, J., Russell, W. R.: Diversification of interdependent prospects. Journal of Economic Theory **7**, 231–240 (1974)
4. Hadar, J., Seo, T. K.: The effects of shifts in a return distribution on optimal portfolios. International Economic Review **31**, 721–736 (1990)
5. Hoy, M., Robson, A.J.: Insurance as a giffen good. Economic Letters **8**, 47–51 (1981)
6. Meyer, D., Meyer, J.: The comparative statics of deductible insurance and insurable assets. The Journal of Risk and Insurance **66**, 1–15 (1999)
7. Meyer, J., Ormiston, M. B.: The demand for insurance in a portfolio setting. The Geneva Papers on Risk and Insurance Theory **20**, 203–211 (1995)
8. Mossin, J.: Aspects of rational insurance purchasing. Journal of Political Economy **76**, 553–568 (1968)
9. Smith, V. L.: Optimal insurance coverage. Journal of Political Economy **76**, 68–77 (1968)

Druck: betz-druck GmbH, D-64291 Darmstadt
Verarbeitung: Buchbinderei Schäffer, D-67269 Grünstadt